Proceedings of the International Conference on Industrial Engineering and Engineering Management

For further volumes:
http://www.atlantis-press.com

About this Series

Industrial engineering theories and applications are facing ongoing dramatic paradigm shifts. The proceedings of this series originate from the conference series "International Conference on Industrial Engineering and Engineering Management" reflecting this reality. The conferences aim at establishing a platform for experts, scholars and business people in the field of industrial engineering and engineering management allowing them to exchange their state-of-the-art research and by outlining new developments in fundamental, approaches, methodologies, software systems, and applications in this area and as well as to promote industrial engineering applications and developments of the future. The conferences are organized by CMES, which is the first and largest Chinese institution in the field of industrial engineering. CMES is also the sole national institution recognized by China Association of Science and Technology. Co-organiser of the conference series is the Tianjin University of Science and Technology.

Ershi Qi • Jiang Shen
Runliang Dou
Editors

Proceedings of the 21st International Conference on Industrial Engineering and Engineering Management 2014

ATLANTIS
PRESS

Editors

Ershi Qi
Tianjin University
Tianjin
People's Republic of China

Jiang Shen
Industrial Engineering Institution of CM
Tianjin University
Tianjin
People's Republic of China

Runliang Dou
Tianjin University
Tianjin
People's Republic of China

ISSN2157-3611   ISSN 2157-362X (electronic)
Proceedings of the International Conference on Industrial Engineering and Engineering Management
ISBN 978-94-6239-101-7   ISBN 978-94-6239-102-4   (eBook)
DOI 10.2991/978-94-6239-102-4

Library of Congress Control Number: 2014958285

Printed on acid-free paper

Preface

On behalf of the Chinese Industrial Engineering Institution, CMES, I am honored to welcome all the delegates of the 21st International Conference on Industrial Engineering and Engineering Management (IEEM 2014). It is your great efforts that brought out the proceedings of IEEM 2014, which records the new research findings and development in the domain of IEEM. What is more exciting, you are the experts or scholars with a significant achievement in the field. I believe that the proceedings will serve as a guidebook for potential development.

Being one of the most important international conferences in the field of industrial engineering and engineering management, it has been successfully held 20 times. Each conference was regarded as the greatest event in industrial engineering and engineering management. Today, we gather together in the beautiful city of Zhuhai to celebrate the 21st International Conference on Industrial Engineering and Engineering Management in the harvest season. The conference is sponsored by Chinese Industrial Engineering Institution, CMES and organized by Tianjin University of Science and Technology. The conference theme is "gathering the research results of the experts in the field, encouraging in-depth exchange of academic theories, exploring industrial innovation and development, promoting the world economic prosperity".

Now, industrial engineering theories and applications are facing ongoing dramatic paradigm shifts. The 21st IEEM responds to the reality by establishing an advantageous platform for experts, scholars, and business people in this area to exchange their state-of-the-art research in the field of industrial engineering and engineering management and by outlining new developments in fundamental, approaches, methodologies, software systems, and applications in this area to promote industrial engineering application and development in the future.

We extend our sincerest thanks to Atlantis Press for their generous support in the compilation of the proceedings. Next, we extend our sincerest thanks to Beijing Institute of Technology for holding such an excellent event. Finally, we thank all the delegates, keynote speakers, and the staff of the organization committee for their contribution to the success of the conference in various ways.

Chair of the conference Academic Committee   Ershi Qi
President of Chinese Industrial Engineering Institution, CMES

Organization

Honorary Chair

Yingluo Wang	Xi'an Jiaotong University, China
Zhongtuo Wang	Dalian University of Technology, China
Chongqing Guo	Tongji University, China
Shanlin Yang	Hefei University of Technology, China

Conference General Chairs

Dacheng Guo	Beijing Institute of Technology, China
Ershi Qi	Tianjin University, China

Program Chairs

Jiang Shen	Tianjin University, China
Yan Yan	Beijing Institute of Technology, China

Technical Committees

Anquan Zou	Changsha University, China
Bai-Sheng Chen	Takming University of Science and Technology, Taiwan
Bidyutkumar Bhattacharyya	Bengal Engineering & Science University, India
Binghui Liu	Xiamen University of Technology, China
Byoung-Kyu Choi	Korea KIIE Society, Korea
C.S. Lee	Taiwan Pingnan University, Taiwan
Changqing Li	Inner Mongolia University, China
Chen-Fu Chien	Taiwan Tsing Hua University, Taiwan
Cheng'en Wang	Chinese Academy of Sciences, China
Chen-guang Liu	Xi'an University of Technology, China

Chien-Hua Shen	Transworld Institute of Technology, Taiwan
Chih-Hsiung Hu	Taiwan Formosa University, Taiwan
Ching-Jong Liao	Taiwan University of Science and Technology, Taiwan
Chiu-tang Lin	Tungnan University, Taiwan
Chueh-Yung Tsao	Chang Gung University, Taiwan
Chuen-Sheng Cheng	Yuan Ze University, Taiwan
ChunChih Yeh	MingDao University, Taiwan
Chun-I Chen	I-Shou University, Taiwan
Chun-Mei Chou	Taiwan Yunlin University of Science and Technology, Taiwan
Chyuan Perng	Taiwan Quemoy University, Taiwan
Congdong Li	Guangzhou Jinan University, China
Dapeng Wei	Tianjin University of Science and Technology, China
Deng-maw Tsai	Taiwan Pingtung University of Science and Technology, Taiwan
Eric Min-yang Wang	Taiwan Tsing Hua University, Taiwan
ErryY.T. Adesta	International Islamic University, Malaysia
Ershi Qi	Tianjin University, China
Fajie Wei	Beihang University, China
Fansen Kong	Jilin University, China
Feifan Ye	Shaoxing University, China
Feng-Tsung Cheng	Taiwan Feng Chia University, Taiwan
Fen-Ru Shih	TaHwa University of Science and Technology, Taiwan
Fu Guo	Northeastern University, China
Fugee Tsung, Professor	Hong Kong University of Science & Technology, China
Fuh-Hwa Liu	Taiwan Chiao Tung University, Taiwan
Fuhou Zhao	Nankai University, China
Gang Liu	East China University of Science and Technology, China
Gongyu Chen	Sun Yat-sen University, China
Guang Cheng	Beijing Union University, China
Guofang Song	Shanghai University, China
Guoning Qi	Zhejiang University, China
Haicheng Yang	China Aerospace Science and Technology Corporation, Journal of Computer Integrated Manufacturing System—CIMS, China
Haoming Wu	University of Electronic Science and Technology, China
Hong-Da Lin	Chaoyang University of Technology, Taiwan
Hongming Zhou	Wenzhou University, China
Hsing-Pei Kao	Taiwan Central University, Taiwan

Hua Li	Xidian University, China
HuanChung Li	Asia-Pacific Institute of Creativity, Taiwan
Hui-Ming Kuo	Shu-Te University, Taiwan
Hung-Chin Lin	Vanung University, Taiwan
Hunszu Liu	Minghsin University of Science and Technology, Taiwan
Husheng Lu	Inner Mongolia University of Science and Technology, China
Hwan-Yann Su	Taiwan University of Kaohsiung, Taiwan
I-Ling Ling	Taiwan Chiayi University, Taiwan
Istvan Novak	SUN Microsystems, CEI-Europe AB, China
James T. Lin	TaiwanTsing-Hua University, Taiwan
Jau-Shin Hon	Tunghai University, Taiwan
Jen-Der Day	Taiwan Kaohsiung University of Applied Sciences, Taiwan
Jhy-Ping Jhang (Chin Ping, Chang)	Huafan University, Taiwan
Jian Li	Tianjin Polytechnic University, China
Jiang Shen	Tianjin University, China
Jian-long Chen	Fortune Institute of Technology, Taiwan
Jianqiao Liao	Huazhong University of Science and Technology, China
Jianxia Lu	Zhejiang University of Technology, China
Jianxin You	Tongji University, China
Jia-Yuarn Guo	Nan Kai University of Technology, Taiwan
Jie Lu	University of Technology, Sydney, Australia
Jinfeng Wang	Zhengzhou University, China
Jingzhu Zhang	China Machine Press, China
Jun Li	China Communications and Transportation Association, China
K. L. Mak	The University of Hong Kong, China
Kaichao Yu	Kunming University of Science, China
Kai-Ying Chen	Taipei University of Technology, Taiwan
Kun-chieh Wang	Overseas Chinese University, Taiwan
Kuo-En Fu	Taiwan Shoufu University, Taiwan
Leyuan Shi	University of Wisconsin-Madison, USA
Li Zheng	Tsinghua University, China
Li-Feng Xi	Shanghai Jiao tong University, China
Lihai Wang	Northeast Forestry University, China
Ling-Huey Su	Chung Yuan Christian University, Taiwan
Makoto Kawada	Meijo University, Japan
Min Zhu	Nanchang Aeronautical University, China

Mingxin Gao	Value Engineering Magazine, China
Min-Sheng Chen	Taiwan Yunlin University of Science and Technology, Taiwan
Mitchell M. Tseng	Hong Kong University of Science & Technology, China
Naiqi Wu	Guangdong University of Technology, China
Peiyu Ren	Sichuan University, China
Pin Zhuang	Nanjing University of Aeronautics and Astronautics, China
Qin Su	Xi'an Jiao tong University, China
Qing Tian	Harbin Institute of Technology, China
Qinghui Dai	North China Electric Power University, China
Qingquan Fang	Anhui University, China
Qixun Gao	Yanshan University, China
Quanqing Li	Zhengzhou Institute of Aeronautics, China
Quanxi Li	Jilin University, China
Qun Zhang	Beijing University of Science and Technology, China
Renzhong Tang	Zhejiang University, China
Ruigang Wang	Chinese Mechanical Engineering Society, China
Ruiyuan Xu	Hebei University of Science and Technology, China
Salvendy Gaviel	Purdue University, USA
Sasaki yuan	Central Japan Industry Federation, Japan
Sheau-Hwa Chen	Taiwan Tung Hwa University, Taiwan
Shengwei Hong	China Institute of Metrolog, Chinay
Sheng-Yuan Hsu	Chienkuo Technology University, Taiwan
Shihua Ma	Huazhong University of Science and Technology, China
Shiraya Masashi	Japanese Industrial Engineering Society, Japan
Shuhua Hu	Wuhan Polytechnic University, China
Shuping Yi	Sichuan Chongqing University, China
Sifeng Liu	Nanjing University of Aeronautics and Astronautics, China
Siqin Pang	Beijing Institute of Technology, Zhuhai, China
Soemon Takakuwa	Nagoya University, Japan
Soundar Kumara Pearce	The Pennsylvania State University, USA
Suicheng Li	Xi'an University of Technology, China
Ta-Chung Chu	Southern Taiwan University of Science and Technology, Taiwan
Tao Chen	Kainan University, Taiwan
Tien-Kuo Wang	University of Kang Ning, Taiwan
Tien-lun Liu	St. John's University, USA
Tongbing Ma	Shenyang Institute of Engineering, China
Tongshui Wu	Civil Aviation University of China, China
Tse-Chieh Lin	Lunghwa University of Science and Technology, Taiwan

Tsemeng Lin	Taiwan Tsing Hua University, Taiwan
Wei Liu	Nanjing Agricultural University, China
Wei Sun	Dalian University of Technology, China
Wei Xue	Wenzhou University, China
Weihua Gan	East China Jiao tong University, China
Weixuan Xu	Institute of Policy and Management, Chinese Academy of Sciences, China
Wen-Chin Chen	Chung Hua University, Taiwan
Wenying Ding	Beijing University of Science and Technology, China
Xiaodan Wu	Hebei University of Technology, China
Xiaowei Zuo	Chinese Mechanical Engineering Society, China
Xin Chen	Guangdong University of Technology, China
Xingsan Qian	Shanghai Polytechnic University, China
Xingyuan Wang	Shandong University, China
Xinmin Zhang	Shenyang University of Technology, China
Yeh-Chun Juan	Ming Chi University of Technology, Taiwan
Yenming J. Chen	Taiwan Kaohsiung First University of Science and Technology, Taiwan
Yeu-Shiang Huang	Taiwan Cheng Kung University, Taiwan
Yi-chuan Wang	TOKO University, Taiwan
Yi-Feng Chen	HsiuPing University of Science and Technology, Taiwan
Yijun Li	Department of management science of NSFC, China
Yi-Kuei Lin	Taiwan University of Science and Technology, Taiwan
Yiming Wei	Beijing Institute of Technology, China
Ying Chung Chang	ChienHsin University of Science and Technology, Taiwan
Yingqiu Xu	Southeast University, China
Yi-shuo Hung	Hsing-Kuo University, Taiwan
Yisong Zheng	Nankai University, China
Yongjiang Shi	Cambridge University, UK
Yoshimoto Mitsuhiro	WASEDA University, Japan
Yu Yang	Chongqing University, China
Yuchun Wang	FAW Car Co. Ltd., China
Yuejin Zhou	Nanjing University, China
Yuh-Jen Cho	ChungHua University, Taiwan
Yuh-Wen Chen	Dayeh University, Taiwan
Yunfeng Wang	Hebei University of Technology, China
Yuqiang Shi	Southwest University of Science and Technology, China
Zhen He	Tianjin University, China

Zhenguo Chen	Foxconn Technology Group, China
Zhibin Jiang	Shanghai Jiaotong University, China
Zhongyuan Wu	Tianjin Polytechnic University, China
Zhuangli Zheng	Guangdong University of Technology, China
Zuhua Jiang	Shanghai Jiao tong University, China

Organizing Committee Chairs

| Yonghe Jiang | Beijing Institute of Technology, China |
| Runliang Dou | Tianjin University, China |

Contents

Part III Engineering Management

Part I
Industrial Engineering Theory

Application on Product Information Storage of Expanded Linear List Based on Two-dimension Array

Neng-hao CAI

College of Management and Economics, Tianjin University, Tianjin, China
(banpai345@126.com)

Abstract - **Array can be regarded as a specific kind of extension of linear list. Based on two-dimension array, we extend the general linear list and present a new data structure which can be used to store and summarize product information. By implementing data storage system utilizing the new data structure, better logic association among data can be reflected, and it also facilitate the information retrieval process. This work gives a new way of thinking in product information storage.**

Keyword - **Data structure, linear list, product information storage, two-dimension array**

I. BACKGROUND

It is essential that a production-oriented enterprise collect detailed information of its products and utilize the information to optimize business strategies, to promote production efficiency or to orientate its market accurately [1]. With the expansion of the scale of production and market, more information of product are generated. So, in the enterprise informationization process, it is necessary to summarize product information through a more efficient data structure in order to achieve the efficient use of information [2].

Assume that a company launches n series of products based on market positioning, namely, N1, N2, N3 etc. The target consumers includes high-end consumers, medium-end consumers, low-end consumers who are divided by their consuming ability. In each series, several models are produced according to factors such as price or configuration (for example, N1 series may have several models, such as N1-a, N1-b, N1-c). According to real-life experience, under normal circumstances, if a series targets at higher end consumers, the number of its models tends to be limited. Because in the same enterprise, one model targets at higher end consumers has a higher price, and it has a more limited market, thus it may have a limited sales volume. In order to control cost, it will be unworthy to

launch more models.

In view of the above assumption, we intend to present a new data structure which can be used to store product information efficiently, when saving information to computers, the data structure should have several characteristics:

1 From a logical standpoint, different series of products should be sorted and stored according to the reference price of its target market, thus the market orientation of products can be reflected.

2 From a logical standpoint, to provide enterprise with information for reference, different models of products should be sorted and stored by price.

3 The comparison of prices between one particular model and the corresponding series which the model belongs should be facilitated.

4 Different models, different series of products can be compared in terms of price.

5 It should be easy to query and retrieval product information.

In fact, product series can be thought of as a base class, then the models of the series can be considered as its derivations. To simplify the complexity of the problem, we assume product series and product are of the same data type, the unique identity is product's name (or the name of product series), and only the price information are contained in the data type. The price of one series is the market reference price (or the average price in the industry) and the price of one product is its tag price.

II. PROBLEM ANALYSIS

The storage of product information can be implemented by varied data structures [3, 4].

For example, the implementation utilizing a tree structure is shown in Fig.1.

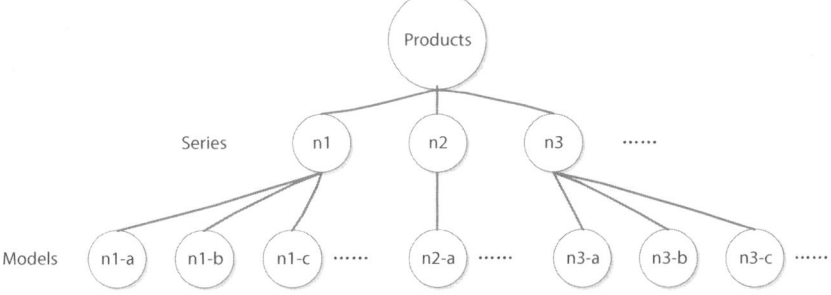

Fig.1. Utilizing tree structure to store information

By using the tree structure, the logic correlation among data can be well reflected as well as the relation

between series and models [5]. But considering that one model's tag price is within a certain range above or below the market reference price of its corresponding series, the tree structure cannot intuitively reflect the correlation between market reference price and product's tag price. Meanwhile, if we want to retrieval one specific model's information, we need to access down by the root node, thus retrieving through the tree structure is not flexible.

If the storage system utilize normal linear list based on one-dimension array, the approach is as shown in Fig.2:

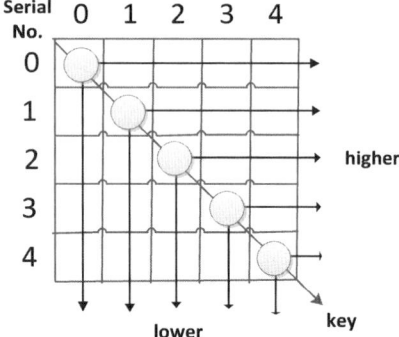

Fig.2. the linear list approach

Fig.3. the diagonal-major order storage structure based on two-dimension array

The linear list approach [6, 7] has a limitation of showing the correlation among series, by using the linear list approach we also need to create different tables corresponding to different series, thus it is not a flexible approach. But because the data structure is implemented by array, we can retrieval information very easily [8].

For the above analysis, we hold the view that a data structure based on two-dimension array would have certain advantages. Compared with one-dimensional array, the extra dimension can be utilized to reflect the correlation among series without losing the flexibility of retrieval and traversal via array index.

In general, two-dimension array has two ways to store information, respectively, row-major order storage and column-major order storage [9]. But in order to further enhance the data associativity, also considering the development of computer hardware level and that company has ability to estimate its scale of production, we can sacrifice some storage space and modify the two-dimension array by implementing a diagonal-major order storage.

As shown in Fig.3, the diagonal elements $a[i][i]$ $(i=0,1,2...)$ make the major storage order, product series are sorted (ASC) by price and stored in diagonal elements starting from $a[0][0]$. For models in one series $a[t][t]$, the information of those models whose price are higher than the series price are stored (ASC) in $a[t+1][t]$, $a[t+2][t]$, $a[t+3][t]...$; for those whose price are lower than the series price, their information are stored (DESC) in $a[t][t+1]$, $a[t][t+2]$, $a[t][t+3]...$

The diagonal elements sequence can be considered a normal one-dimensional list with all operations of one-dimensional list can be applied. While for a given series $a[t][t]$, both the horizontal sequence and the vertical sequences can also be seen as normal one-dimensional linear lists, thus the proposed approach has obvious advantages in retrieval and traversal process.

The proposed data storage structure make an intuitive and clear reflection of correlation among different models or different series also the correlation between models and series they belong to. Additionally, all the data operations such as insertion, deletion, updating are conducted via index, hence it will be more convenient and flexible [10].

It should be noted that when utilizing the proposed data structure to store product information, prior estimates of product series information and product models information need to be made so appropriate storage space can be allocated to the two-dimension array. If product information exceeds the pre-allocated storage space, the storage space for an array should be dynamically increased, which to some extent causes undesired operations.

It should also be noted that for element $a[i][i]$ $(i=0,1,2 ... ...)$ in major order, ie. the diagonal elements, with the increase of i, the corresponding row and column has less space for storage. Thus the diagonal elements should be inserted in ASC order by price starting from $a[0][0]$.

III. IMPLEMENTATION OF DATA STORAGE

TABLE I
DATA OF SERIES AND MODELS

(Unit: Price/Unit)

Product series	Model	Tag price
N1	N1-a	55
Average price: 50	N1-b	52
	N1-c	46
	N1-d	48
N2	N2-a	36
Average price: 30	N2-b	34
	N2-c	29
	N2-d	28
	N2-e	30
N3	N3-a	22
Average price: 20	N3-b	27
	N3-c	22
	N3-d	21
	N3-e	19
N4	N4-a	123
Average price: 100	N4-b	112

To illustrate how the data structure works, a set of product information is taken as an example. Assuming an enterprise's product information is as follows (Table I). The product series consists of N1, N2, N3 and N4. Among which N2 and N3 target at low-end consumers, N1 targets at medium-end consumers and N4 targets at high-end consumers. All models and their corresponding tag prices have been given in Table I. The long-term strategic plan of the enterprise is to launch a series N5 for the medium-end market, while phasing out the N2 series.

A. Initializing an empty data structure

This step is to estimate the produce information and allocate proper storage space, the initial empty data structure is shown as Fig.4

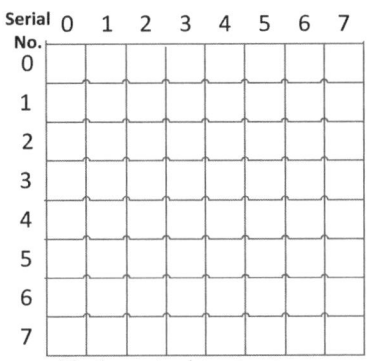

Fig.4.empty data structure

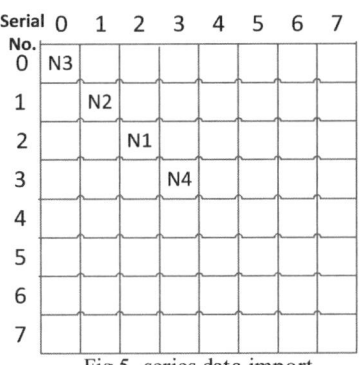

Fig.5. series data import

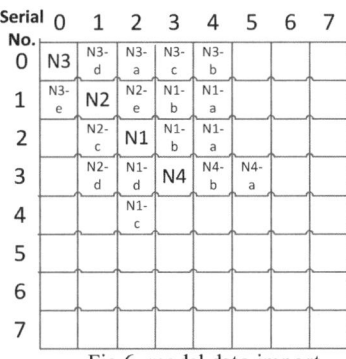

Fig.6. model data import

The series data should be inserted into the diagonal positions one by one, whenever a series information is inserted, comparison should be made in order to make sure that a[i][i] are sorted by price (see Fig.5). Then insert the information of models belong to each series, before insertion, comparison to the series price (average price in the industry) should be made. For those product whose tag price is higher than the series price, their information should be inserted into the corresponding row; for those whose tag price is lower than the series price, their information goes into the corresponding column. If the tag price is equal to the series price, we set a man-made rule that its information should be inserted into the corresponding row; then make comparison and sort the inserted data in the corresponding row or column (see Fig.6).

B. Data insertion

If we want to insert the information of N5 into the structure (assuming the average price of N5 series is 60), the insertion process is as follows:
(1) Compare N5's average price with prices of existing series, it can be seen that N5's price is higher than N1's price and lower than N4's price.
(2) Move N4's data as well as all data of models which belong to series N4 from the original position [p][q] to [p+1][q+1].
(3) Insert N5's data into the original position of N4 and import all N5 models' data sequentially.

C. Data deletion

If the enterprise plans to cease product N2-b of the N2 series, we can simply remove it from the data structure just like operating on a linear list. If it is the whole N2 series which is to be ceased, we just need to remove all N2 series data by deleting data of the corresponding row or column in a linear list operation way, then move all elements (including series' and models') which belong to series of higher price from the original position [p][q] to [p-1][q-1].

D. Data traversal and retrieval

To traverse data is simply to follow diagonal order and access data in the corresponding row or column via index. The retrieval process is also based on array indexing and there is no need to go into details here.

The implementation (C++ code) is as follows, comments are made to explain some detailed code and all the following code has been tested and validated on computer.

```cpp
class Product              // class of series or product model
{
public:
    char* name;                //series name or model name
    int price;                               //price
    Product(char*, int);                //constructor
};
class D_List               // two-dimension array storage
{
public:
    Product** elem;                // second rank pointer
    int num;                       //number of series
    int width[100];                // horizontal storage space
    int height[100];               // vertical storage space
    int listsize;               //initial side length of the table
};
//----implementation of constructor----
Product::Product(char* n, int p)
{
    name=n;
    price=p;
}
//----initialization of the two-dimension array----
int InitD_List(D_List &D, int n)
{
    //create pointer arrays
    D.elem = (Product**)malloc(n*sizeof(Product*));
    if(!D.elem)return 0;
    for(int i=0;i<n;i++){
        //for each row, allocate space for storage (n elements)
        D.elem[i] = (Product*)malloc(n*sizeof(Product));}
    D.listsize=n;               //side length of the table is n
    D.num=0;               //the initial number of series is 0
    for(int i=0;i<n;i++){
        //set the number of model to 0
        D.width[i]=0;
        D.height[i]=0;}
```

```
        return 1;
}
//----insertion of series data---
int InsertSeries(D_List &D, Product s)
{
        if(D.listsize<=D.num)return 0;
        for(int i=0;i<D.num;i++){
                //traverse each series
                if(s.price < D.elem[i][i].price){
                        MigrateSeries(D,i);
                        D.elem[i][i]=s;              //insert into position i
                        D.width[i]=D.height[i]=0;
                     return 1;}
        }
//if the price is higher than all existing series, insert into the far end
position
        D.elem[D.num][D.num]=s;
        D.width[D.num]=D.height[D.num]=0;
        D.num +=1;                //update the number of series
        return 1;
}
//----model data insertion---
int InsertProduct(D_List &D, Product s, Product p)
{
        int n;              //index to find the corresponding series
        for(int i=0;i<D.num;i++){
                if(s.name==D.elem[i][i].name)
                {n=i;break;}
                //find the series s belongs to
                if(i==D.num-1)return 0;}
        if(p.price >= D.elem[n][n].price) {
                //if price is no less than the series price, insert into the
corresponding row
                if(D.listsize<=D.width[n]-1)return 0;
                for(int i=0;i<=D.width[n];i++) {
                        if(p.price < D.elem[n][n+i].price){
                                MigrateProductW(D,n,i);
                                D.elem[n][n+i]=p;
                                return 1;}}
                D.width[n] +=1;        //update the number of models
                D.elem[n][n+D.width[n]]=p;
                // if the price is higher than all existing models, insert
                //into the far end position
                return 1;
        }
        else{
                //if the price is lower than the series price, insert
                //vertically
                if(D.listsize<=D.height[n]-1)return 0;
                for(int i=0;i<=D.height[n];i++){
                        //the following code is like the preceding code but
                        //in an opposite way
                        if(p.price > D.elem[n+i][n].price){
                                MigrateProductH(D,n,i);
                                D.elem[n+i][n]=p;
                                return 1;}}
                D.height[n] +=1;
                D.elem[n+D.height[n]][n]=p;
                return 1;}
}
//----move the series at position n horizontally to the right-----
int MigrateProductW(D_List &D, int n, int m)
{
        for(int i=D.width[n]+1;i>m;i--){
                D.elem[n][n+i]=D.elem[n][n+i-1]; }
        D.width[n] +=1;                //update the number of models
        return 1;
}
//---- move the series at position n vertically downwards ----
int MigrateProductH(D_List &D, int n, int m)
{
        for(int i=D.height[n]+1;i>m;i--){
                D.elem[n+i][n]=D.elem[n+i-1][n];}
        D.height[n] +=1;                // update the number of models
        return 1;
```

```
}
//---- move all the series behind position n rearward ----
int MigrateSeries(D_List &D, int n)
{
        if(D.listsize<=D.num)return 0;
        for(int i=D.num;i>n;i--){
                D.width[i]=D.width[i-1];
                D.height[i]=D.height[i-1];
                //update the numbers of models in both orientation for
                //the following loop
                for(int j=0;j<=D.width[i];j++){
                        //horizontally copy data
                        D.elem[i][i+j]=D.elem[i-1][i-1+j];}
                for(int j=0;j<=D.height[i];j++){
                        //vertically copy data
                        D.elem[i+j][i]=D.elem[i-1+j][i-1];}}
        D.num +=1;                //update the number of series
        return 1;
}
//----deletion of series at position n ---
int DeleteSeries(D_List &D, Product s)
// the following code is like the preceding code but in an opposite way
{
        int n;                                    //index n
        for(int i=0;i<D.num;i++){
                if(s.name==D.elem[i][i].name)
                {n=i;break;}                //find the corresponding series
                if(i==D.num-1)return 0; //if there is no such series, exit
        }
        //the deletion will not be conducted if n is less than 0 or no less
        //than the far end index
        if(n<0||n>D.num-1)return 0;
        for(int i=n;i<D.num;i++){
                D.width[i]=D.width[i+1];
                D.height[i]=D.height[i+1];
                //update the numbers of models in both orientation for
                //the following loop
                for(int j=0;j<=D.width[i];j++){
                        //horizontally copy data
                        D.elem[i][i+j]=D.elem[i+1][i+1+j];}
                for(int j=0;j<=D.height[i];j++){
                        //vertically copy data
                        D.elem[i+j][i]=D.elem[i+1+j][i+1];}
        }
        D.num -=1;                // update the number of series
        return 1;}
//note that this operation does not actually delete data of the series at
//the far end, but //the number of series (D.num) is updated and the
//data will not be accessible to users, //thus it is equivalent to deletion.
```

IV. CONCLUSION

This paper provides a new way of thinking to serve the storage of product information. The diagonal-major order storage method based one two-dimension array can better reflect the logical correlation. By utilizing array index, the traversal process of data is simplified. Although the proposed data structure is not so economical in storage space comparing to traditional tree structure or one-dimensional array structure, but given the current high levels of computer hardware, and that company often has ability to estimate its scale of production, the advantages of data structure are still visible.

REFERENCE

[1] Von Kohorn H. Product information storage, display, and coupon dispensing system: U.S. Patent 5, 249, 044[P].

1993-9-28.

[2] Strickler M T. Typically high-availability information storage product: U.S. Patent 6, 563, 706 [P]. 2003-5-13.

[3] YAN Wei-min, Wu Weimin. Data Structure(c-language version) [M]. Tsinghua University press, 1997.

[4] Kruse R L. Data Structures And Program Design In C+ [J]. 2001.

[5] Weiss M A. Data structures and algorithm analysis in C[J]. Pearson Education Asia, 2002.

[6] DA Linmei. The Implemention of Students' Score Query Based on Sequence List [J]. Computer Study, 2011 (2):

74-75.

[7] ZHANG Fu-xing, Sun Jia-xia. Importance of Generalized Lists in Data Structures [J]. Journal of Henan Institute of Science and Technology (Social Science Edition), 2006 (4): 103-104.

[8] Ford W, William F, Topp W. Data structures with C++ [M]. Simon & Schuster, 1995.

[9] Adam Drozdek. Data Structures and Algorithms in C++ (SECOND EDITION) [M]. Brooks/Cole, 2001.

[10]Reek K A. Pointers on C [M]. Addison-Wesley Longman Publishing Co., Inc., 1997.

Key Factors of K-nearest Neighbors Nonparametric Regression in Short-time Traffic Flow Forecasting

Jing-ting Zhong, Shuai Ling[*]

College of Management and Economics, Tianjin University, Tianjin, China

(ls5209@163.com)

Abstract - **Short-term traffic flow prediction plays an important role in route guidance and traffic management. K-NN is considered as one of the most important methods in short-term traffic forecasting, but some disadvantages limit the widespread application. In this paper, we use four tests to find the key factors of the K-NN method, which will give inspires to the further research to improve the method.**

Keywords - **K-nearest neighbors, key factors, short-term traffic flow forecasting**

I. INTRODUCTION

Shore-term traffic flow forecasting has played an important role in the intelligent transportation system (ITS). Traffic management departments can use it to design strategies of traffic control and traffic guidance, while travelers use it to make route choice. Under the above benefits, short-term traffic flow forecasting has become an attractive field both in traffic science and traffic engineering.

By definition, short-term traffic forecasting is the process to predict key traffic parameters such as speed, flow, occupancy, or travel time with a forecasting horizon typically ranging from 5 to 30 minutes at specific locations.

There are generally two kinds of approaches for short-term forecasting: parametric approach and non-parametric approach. The parametric approach assumes that there is an explicit forecasting mathematic by a set of parameters. Historical data is then used to find out a group of parameters which can minimum the forecasting error (gained by the historical data). Afterwards, the model can be used in the real-time forecasting.

Contrarily, instead of finding the explicit mathematic form the relationship between inputs and outputs, nonparametric approaches are data-driven and allow data to speak for itself [1] (Bosq 1996). The most popular nonparametric approaches includes nonparametric regression (NPR).

The NPR model is a data-driven approach. Instead of trying to compress all training data into a set of mathematical specifications (parametric approaches) or a certain network (ANN) through modeling process, it retains all historical observation and searches for the most similar case of the current state, based on which forecasting is then made. Oswald, Scherer et al. (2000) investigated the practical use of NPR model, and discussed some problem likely to encounter in the real world usage [2]. (Clark 2003) studies the multi-variant NPR forecasting, as well as the influence of neighbor size and the transferability of database, which are valuable for

the practical use of NPR model [3]. (Chang, Zhang et al. 2011) made three improvements, including the data organization and the search mechanism, for faster calculation and higher accuracy [4]. Other researches considered additional information in NRP forecasting, such as historical traffic state and traffic condition information, and stated these helps to reduce forecasting error [5, 6] (Abdulhai, Porwal et al. 2002; Gong and Wang 2002).

With the continuous deepening researches on K-NN, they become increasingly mature in short-time traffic flow forecasting and will be applied in intelligent transportation system [7, 8] (Friedman, Bentley et al. 1977; Van Der Voort, Dougherty et al. 1996). For example, with the help of nonparametric forecasting models, a real-time, on-line, self-learning forecast system can be implemented [9] (Zhu and Yeh 1998).

In practical applications, Although K-NN has many advantages, the application of nonparametric regression methods still have to pay attention to that the K-NN method involve massive data and calculations.

II. K-NEAREST NEIGHBORS NONPARAMETRIC REGRESSION

In fact, nonparametric regression is based on pattern matching and data mining. Suppose the short-time traffic flow forecasting has to predict the traffic flow of a section at next time epoch, the corresponding influencing factors $(f_1 \sim f_n)$ have to be explored firstly [10-12] (Bentley 1975; Bentley and Friedman 1979; Bentley 1990). These influencing factors may include traffic flows of the section and upstream section at previous time epoch, weather, road conditions, etc. In this paper, $f_1 \sim f_n$ are taken as the state components of the system, which compose the state vector of the system $(f_1, \cdots, f_n)$. The traffic flow at the forecasting time epoch (q) is called as decision attribute, which is determined by $(f_1, \cdots, f_n)$. In other words, the current state vector $(F = (f_1, \cdots, f_n))$ determines q at the forecasting time epoch. In this paper, F and q composed a pattern vector $(P = \{(f_1, f_2, \cdots, f_n) | q\})$.

Search k nearest neighbor (KNN) of the forecasting state (F_{pre}) in the historical pattern according to equation (1) and then predict by using q of KNN.

E. Qi et al. (eds.), *Proceedings of the 21st International Conference on Industrial Engineering and Engineering Management 2014*, Proceedings of the International Nonference on Industrial Engineering and Engineering Management, DOI 10.2991/978-94-6239-102-4_2, © Atlantis Press and the authors 2015

III. EXPERIMENTAL DATA

The Data used in this research is collected from website of University of Minnesota Duluth (http://www.d.umn.edu/tdrl/traffic/). Real-time traffic data have been collected since 1997 from over 4,000 double inductive loop detectors located around the Twin Cities Metro freeways. The road used for experiment is a section of I-35E South, intersecting with TH110 and Wagon Wheel trail. The network structure and detectors layout is shown in Fig.1.

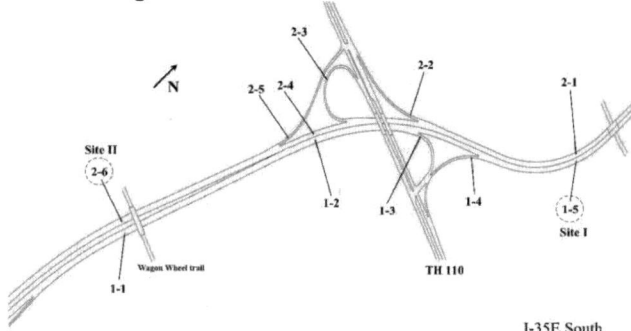

Fig.1. Twin Cities Metro freeways

Traffic data come from University of Minnesota Duluth (http://www.d.umn.edu/tdrl/index.htm). Based on the traffic management center of Minnesota, the Traffic Data Research Laboratory of University of Minnesota Duluth can provide all traffic data on expressways surrounding the University of Minnesota-Twin Cities. These data were collected continuously by more than 4,000 loop detectors around the whole year and compressed in exclusive format, enabling to output traffic flow and occupancy recorded by desired detector during the desired period.

A. Pattern composition

In this paper, data collected during 2012.03.01-2012.04.31 were used. We have recoded all the detectors in the map. Traffic flow at detector 1-5 and detector 2-6 are selected for experiments. For convenience, in the remaining of the paper we name them as site1 and site2 respectively.

With respect to the traffic flow forecasting at detector 1-5 and detector 2-6, the test adopted two compositions of state vector of modes: ① simple pattern: For data that is time series in nature, a state vector defines each record with a measurement at time [13,14] (Smith and Demetsky 1997; Smith, Williams et al. 2002); ② complicated pattern: both historical flow on time series of the forecasting section and flow of related upstream section are taken into account [6,15] (Abdulhai, Porwal et al. 2002; Li, Li et al. 2013).

Time-series pattern and temporal-spatial pattern under different state vectors were presented in the following.

For detector 1-5:

$$P^{15}_{Time-series} = \{f_{15}(t), f_{15}(t-1), f_{15}(t-2) \mid f_{15}(t+1)\} \quad (1)$$

$$P^{15}_{Temporal-spatial} = \{f_{15}(t), f_{15}(t-1), f_{15}(t-2), f_{12}(t), f_{12}(t-1), f_{14}(t), f_{14}(t-1) \mid f_{15}(t+1)\} \quad (2)$$

For detector 2-6:

$$P^{26}_{Time-series} = \{f_{26}(t), f_{26}(t-1), f_{26}(t-2) \mid f_{26}(t+1)\} \quad (3)$$

$$P^{26}_{Temporal-spatial} = \{f_{26}(t), f_{26}(t-1), f_{26}(t-2), f_{24}(t), f_{24}(t-1), f_{25}(t), f_{25}(t-1) \mid f_{26}(t+1)\} \quad (4)$$

where $f_{ij}(t)$ is the flow detected by detector $i\,j$ at t and $f_{ij}(t-1)$ is the flow detected by detector $i\,j$ at t-1.

In the KNN system design, nearest neighbors within a ring (centered at the forecasting pattern and $r = 50$) are searched firstly by combining fixed radius and fixed KNN search strategy. If k^* patterns were searched, then the number of nearest neighbors (k) can be determined by the following equation:

$$k = \begin{cases} k^* & k^* < 20 \\ 20 & k^* \geq 20 \end{cases} \quad (5)$$

B. Evaluation index of test results

RMSE is the common evaluation index of test results:

$$RMSE = \sqrt{\frac{1}{N}\sum_{i=1}^{N}(x_i - x_i)^2} \quad (6)$$

where x_i is true value, x_i is predicted value and N is number of patterns. RMSE is affected by the value range of experimental subject and couldn't reflect the absolute error. Therefore, this experiment used the mean absolute error (MAE) for evaluating the test results. MAE is dimensionless and insensitive to the value range of experimental subject.

$$MAE = \frac{1}{N}\sum_{i=1}^{N}\frac{|x_i - x_i|}{x_i} \quad (7)$$

IV. TEST DESIGN

A. Test1

The original traffic flow data collected during 2012.03.01-2012.04.30 were processed firstly to generate corresponding patterns. In the original state vector, states deviated significantly from the group are called as outliers. Although patterns including these outliers are real, they are against the test and deleted by using DBScan algorithm.

Density-Based Spatial Clustering of Applications with Noise (DBScan) is a representative clustering algorithm based on density. It doesn't need to know number of clusters in advance as it can identify all forms of clusters. More importantly, it can recognize outliers (red points in Fig.2).

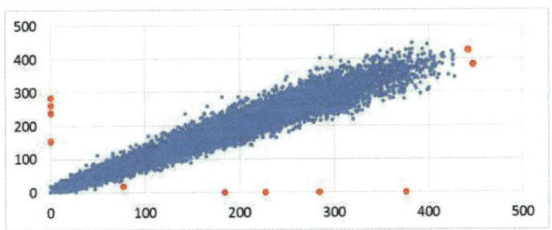

Fig.2. Two-dimensional projection of original state vector distribution (red points are outliers)

After the processing, the candidate data set (C) contains 15,500 data. Divide C in to 100 subsets randomly ($c_1, c_2, \cdots, c_{100}$ $c_1 \bigcup c_2 \bigcup \cdots \bigcup c_{100} = C, c_i \bigcap c_j = \phi \ i \neq j \ i, j \in 1, 2, \cdots, 100$).

Both KNN and ANN predicted the traffic flow on 2012.04.31 for 100 times and added a new data subset in the database after every prediction. Next, the data in the new database were used for KNN prediction and ANN training.

Fig.3 shows spatial distribution of state vectors with different data density.

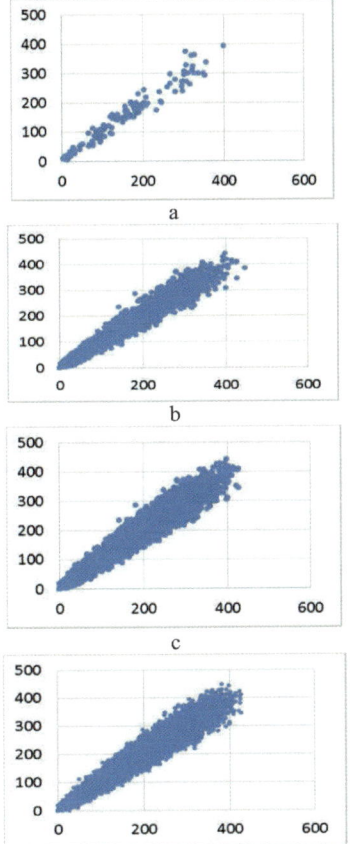

Fig.3. a 1% data density, b 25% data density, c 50% data density, d100% data density

Test results were listed as Fig.4:

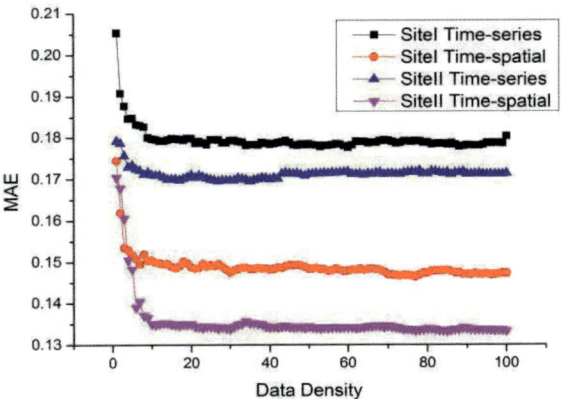

Fig.4. Mean prediction error against data density (MED)

B. Test2

Test design: Divide P into three independent subsets (P_A, P_B, P_C). Fig.5 is the two-dimensional projection of their state space.

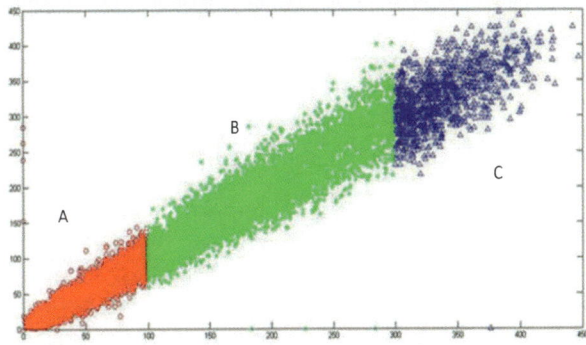

Fig.5. State space division

Decrease data in B at a rate of 1/15 and take the new dataset as the database of nonparametric regression forecasting system and training data of ANN system. Meanwhile, A and C are taken as the test set for prediction.

Test results were listed as Fig.6:

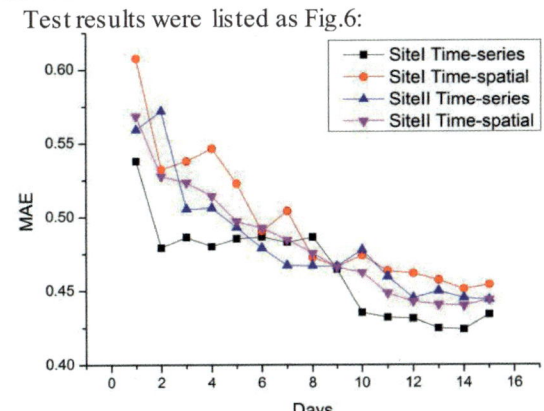

Fig.6. Mean prediction error against data density

C. Inspirations from tests

Viewed from the prediction nature, we can find that KNN system predicate traffic flow by averaging the neighboring patterns of the forecasting pattern. Longer running history of database will bring more similar neighboring patterns of the forecasting pattern.

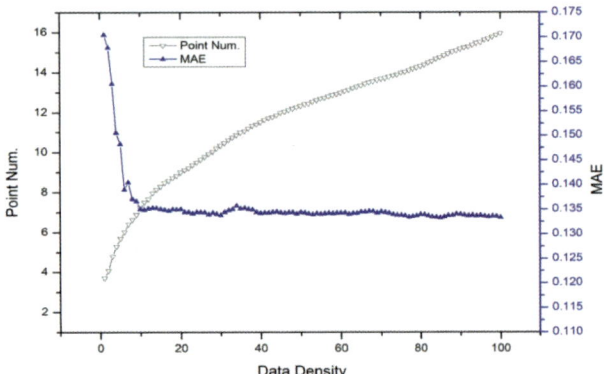

Fig.7. Comparison of prediction error and data density

Fig.7 shows the prediction results of $P_{Timporal-spatial}$ of KNN system at SiteII with random increasing. Point Num. represents the number of nearest neighbors searched by KNN system within the ring of $r = 45$ centered at the forecasting point and MAE represents the mean prediction error. When data density reached to the data density at turning point, MAE tends to be stable. At this moment, 7 nearest neighbors in average are searched within the ring. Subsequently, excessive nearest neighbors searched within the ring make no contributions to the prediction error reduction.

V. CONCLUSIONS

The short-time traffic flow forecasting performances of KNN and ANN methods are analyzed through tests, finding that:

1. The prediction accuracy of KNN system presents no linear improvement with the increasing of data size in the database. When data size increases to a certain level, the prediction accuracy of KNN will remain basically same.

2. The prediction accuracy of KNN is closely related with the state vector. More information contained in state vector will make it more representative and thereby brings higher prediction accuracy.

Additionally, this paper also explored some ways to further improve the prediction accuracy of ANN and KNN, such as optimizing composition of state vector, deleting data in high-data-density region, using representative data and using time-series data increasing.

REFERENCES

[1] Bosq, D. (1996). "Nonparametric statistics for stochastic processes." Lecture Notes in Statist.

[2] Oswald, R. K., W. T. Scherer, et al. (2000). "Traffic flow forecasting using approximate nearest neighbor nonparametric regression." Final project of ITS Center project: Traffic forecasting: non-parametric regressions.

[3] Clark, S. (2003). "Traffic prediction using multivariate nonparametric regression." Journal of transportation engineering 129(2): 161-168.

[4] Chang, G., Y. Zhang, et al. (2011). A summary of short-term traffic flow forecasting methods. 11th International Conference of Chinese Transportation Professionals (ICCTP 2011).

[5] Gong, X. and F. Wang (2002). Three improvements on knn-npr for traffic flow forecasting. Intelligent Transportation Systems, 2002. Proceedings. The IEEE 5th International Conference on, IEEE.

[6] Abdulhai, B., H. Porwal, et al. (2002). "Short-term traffic flow prediction using neuro-genetic algorithms." ITS Journal-Intelligent Transportation Systems Journal 7(1): 3-41.

[7] Friedman, J. H., J. L. Bentley, et al. (1977). "An algorithm for finding best matches in logarithmic expected time." ACM Transactions on Mathematical Software (TOMS) 3(3): 209-226.

[8] Van Der Voort, M., M. Dougherty, et al. (1996). "Combining Kohonen maps with ARIMA time series models to forecast traffic flow." Transportation Research Part C: Emerging Technologies 4(5): 307-318.

[9] Zhu, J. and A. G.-O. Yeh (1998). "A self-learning short-term traffic forecasting system." Horizons 72: 13.32.

[10] Bentley, J. L. (1975). "Multidimensional binary search trees used for associative searching." Communications of the ACM 18(9): 509-517.

[11] Bentley, J. L. (1990). K-d trees for semidynamic point sets. Proceedings of the sixth annual symposium on Computational geometry, ACM.

[12] Bentley, J. L. and J. H. Friedman (1979). "Data structures for range searching." ACM Computing Surveys (CSUR) 11(4): 397-409.

[13] Smith, B. L. and M. J. Demetsky (1997). "Traffic flow forecasting: comparison of modeling approaches." Journal of transportation engineering 123(4): 261-266.

[14] Smith, B. L., B. M. Williams, et al. (2002). "Comparison of parametric and nonparametric models for traffic flow forecasting." Transportation Research Part C: Emerging Technologies 10(4): 303-321.

[15] Li, L., Y. Li, et al. (2013). "Efficient missing data imputing for traffic flow by considering temporal and spatial dependence." Transportation Research Part C: Emerging Technologies 34: 108-120.

The Distribution of Residual Control Rights under Self-interest Behavior

Xing BI*, Jin DONG

Department of Management and Economy, Tianjin University, Tianjin, China
(bistar@126.com)

Abstract – **PPP (Public-Private-Partnerships) has been widely used because of its high investment efficiency, high degree of market and the advantages of optimizing the allocation of resources. In the pre-investment stage study of PPP projects, this paper considers the self-interested behavior of the private sector and the public sector and self-interest investment issues. In the study of the factors which influence residual control rights of PPP projects, we make self-interest investments as variables, then develop a mathematical model of the game to discuss how to allocate residual rights of control, and on this basis, we have proposed several measures to improve the cooperation efficiency in PPP projects.**

Keywords – **Efficiency, Public-Private-Partnerships, residual control rights, self-interest**

I. INTRODUCTION

Grossman, Hart and Moore have proposed incomplete contract theory (GHM), who think that residual control rights should be allocated to important or indispensable investment [1]-[2]. Hart supposes that the difference between private providers and the government package is the distribution of residual control rights. Concluded that Internal government services is preferred when the efforts of innovating cost bring large damage to quality, the efforts of innovating quality is very important and corruption is a serious problem for government procurement. When the efforts of innovating cost bring the quality fall within the scope of control, privatization is the preferred [3]. Besley and Ghatak think if the contract is not complete, then the ownership of public goods should be attributed to the high-valued side [4]. There are a lot of quasi-public goods in the real world, Francesconi and Muthoo think that optimal allocation depends on technological factors, the participant's evaluation of the goods, the degree of public goods [5].

II. OVERVIEW OF RESERCH

Zhang Zhe etc think that when firms have self-interest investments, different parameters corresponding to the share of bilateral cooperation which have different degrees of residual control rights, which is the key to improve the cooperation efficiency [6]. Sun Hui etc think that the income distribution program, the irreplaceable-degree of the partners' investment and the satisfaction of both sides on the projects' future earnings are the three factors that influence residual control rights provisions in PPP projects [7].

In the study of the relationship between government self-interest and non- government interest [8]-[10], foreign scholars have two extreme views. One view think that the government is representative of the public interest which deny self-interest of government; another view is to deny the public interest that the government is only seeking self-interest "economic man" [11]-[12]. Chinese scholars have done some researches on the self-interest of the government, there are three views. (1) Based on the nature of the government [13]. (2) The government should not have selfish behavior. Government self-interest and the public interest are in opposing state [14]. (3)The interests of government and social public are both coherence and oppositional [15].

III. MODELING OF ALLOCATION OF RESIDUAL CONTROL RIGHTS

Let the public sector G and the private sector N cooperate to provide certain public goods F, in the initial contract clearly states the allocation of initial control rights and the distribution of future income. In the investment stage, the two sides began to relationship-specific investments and which will be known by each other. Refer to Besley, Francesconi and Muthoo's model, includes the following three periods—time 0: the initial phase of the contract signed between G and N; time 1: specific investment stage; time 2: the allocation of co-income between G and N. let G and N are risk-neutral as BG shows.

A. *Time 0--initial phase of the contract signed*

Time 0, the initial contract will specify project future earnings distribution plan, and assign the initial control between G and N, we suppose that G has the ratio of residual control rights is π, then N has $i-\pi$, π is continuous, and $\pi \in [0,1]$.

B. *Time 1-specific investment stage*

Time 1, we suppose that the investment of G is t_g, e_g (self-interest investment), and the investment of G is t_n, e_n (self-interest investment), $C^i(t_i, e_i)$ is the cost function of t_i, e_i, and $C_1 > 0, C_{11} \geq 0$, the benefit of this project is B.

When $\pi = 1$, the benefit of the project is $B^G(t_g, e_g, t_n)$, and when $\pi = 0$, the benefit of the project is $B^N(t_g, e_g, t_n, e_n)$, t_i, e_i are function of π.

When the two sides dedicated public and private investment, they face with the following two Situations.

E. Qi et al. (eds.), *Proceedings of the 21st International Conference on Industrial Engineering and Engineering Management 2014*, Proceedings of the International Conference on Industrial Engineering and Engineering Management, DOI 10.2991/978-94-6239-102-4_3, © Atlantis Press and the authors 2015

One side, if only by the initial allocation of control to operate F, the benefit of the project is $B(t_g, e_g, t_n, e_n)$, at the other side, if both side have a positive communication during this process, the benefit of the project is $b(t_g, e_g, t_n, e_n)$, and we know $b(t_g, e_g, t_n, e_n) > B(t_g, e_g, t_n, e_n)$, we suppose that $B(t_g, e_g, t_n, e_n)$ is the linear combination of $B^G(t_g, e_g, t_n)$ and $B^N(t_g, e_g, t_n, e_n)$, and we let $B(t_g, e_g, t_n, e_n) = \pi B^G(t_g, e_g, t_n) + (1 - \pi) B^N(t_g, e_g, t_n, e_n)$.

At this moment, e_g will bring G the earnings $\Phi^G(e_g)$, and suppose that $\Phi_1^G(e_g) > 0, \Phi_{11}^G(e_g) \leq 0$, so as to that e_n will bring N the earnings $\varphi^N(e_n)$, and $\varphi_1^N(e_n) > 0, \varphi_{11}^N(e_n) \leq 0$; Then combined with China's actual situation, t_n will not only increase the project's revenue, but also to bring the private sector' image improvement and reputation of the community, this part of the benefit is $g^N(t_g)$, and $g_1^N(t_n) > 0, g_{11}^N(t_n) \leq 0$ .Self-interested investment will do harm to the benefit of PPP projects' income due to negative externalities, so we suppose that $B_1^N > 0, B_2^N < 0, B_3^N > 0, B_4^N < 0$; $B_{11}^N < 0, B_{22}^N > 0, B_{33}^N < 0, B_{44}^N > 0$.

When $\pi = 1$, marginal productivity of t_g is bigger than the situation when $\pi = 0$, so we suppose that $B_1^G - B_1^N > 0, B_{11}^G - B_{11}^N > 0$, and as to $B_2^G - B_2^N > 0, B_{22}^G - B_{22}^N > 0, B_3^G - B_3^N > 0, B_{33}^N - B_{33}^G < 0$

C. Time 2-the allocation of co-income between G and N

Time 2, cooperation decides the negotiations and consultations about whether transfer payments or not. If the negotiation is successful, consensus agreement, the benefit of G and N are as follow:

$$U_g(t, e) = \beta_g b(t_g, e_g, t_n, e_n) + \gamma \quad (1)$$
$$U_n(t, e) = \beta_n b(t_g, e_g, t_n, e_n) - \gamma \quad (2)$$

$\beta_i > 0, (i = g, n)$ is i's evaluation factor of F's future earnings, and we have $\beta_g + \beta_n = 1$, γ is transfer payments from the government public sector to the private sector, can be positive or negative. If the negotiation fails, then G and N operate F only by initial contract, then the certainly revenue of both sides are as follow:

$$\overline{U}_g = \beta_g B = \beta_g \left[\pi B^G + (1 - \pi) B^N\right] \quad (3)$$
$$\overline{U}_n = \beta_n B = \beta_n \left[\pi B^G + (1 - \pi) B^N\right] \quad (4)$$

By the analysis, negotiation of bilateral cooperation to bring the remainder of cooperation is $[b(t_g, e_g, t_n, e_n) - B(t_g, e_g, t_n, e_n)]$, for any π, t_g, e_g, t_n, e_n, at period one, we assume that the two sides cooperate in accordance with the remainder allocated to the Nash equilibrium, that is to be allocated in a 1:1 ratio, the two sides cooperate to get the remainder are $\dfrac{b-B}{2}$, Then G and N's revenue function are: $V^G = \overline{U}_g + \dfrac{b-B}{2}$, $V^N = \overline{U}_n + \dfrac{b-B}{2}$

So we get:

$$V^G = \beta_g \left[\pi B^G + (1 - \pi)B^N\right] + \frac{b-B}{2} \quad (5)$$
$$V^N = \beta_n \left[\pi B^G + (1 - \pi) B^N\right] + \frac{b-B}{2} \quad (6)$$

So, the ultimate benefit of both projects is:

$$U^G = V^G - C^G + \Phi^G$$
$$U^N = V^N - C^N + \varphi^N + g^N$$

Finishing get:

$$U^G = \beta_g \left[\pi B^G + (1 - \pi)B^N\right] + \Phi^G(e_g) - C^G(t_g, e_g) + \frac{b-B}{2} \quad (7)$$
$$U^N = \beta_n \left[\pi B^G + (1 - \pi)B^N\right] + \varphi^N(e_n) + g^N(t_n) - C^N(t_n, e_n) + \frac{b-B}{2} \quad (8)$$

In the case of cooperation when put in the best welfare maximizing public revenue should satisfy:

$$\frac{\partial U^G}{\partial t_g} = \beta_g[\pi B_1^G + (1 - \pi)B_1^N] + \frac{1}{2}[b_1 - \pi B_1^G - (1 - \pi)B_1^N] - C_1^G = 0 \quad (9)$$

Let the above formula be F_1, on the basis of F_1, for π and t_g partial derivative, since the model is assumed to set the four inputs of N and G are independently, then second order cross partial derivatives is 0,so we have $B_{mn}^G = 0, B_{mn}^N = 0, b_{mn}^G = 0, b_{mn}^N = 0, (m \neq n)$, for simplification, we get the following equation:

$$\frac{\partial F_1}{\partial \pi} = \beta_g\left[B_1^G + \pi B_{11}^G \frac{\partial t_g}{\partial \pi} - B_1^N + (1 - \pi)B_{11}^N \frac{\partial t_g}{\partial \pi}\right] + \frac{1}{2}\left[b_{11} \frac{\partial t_g}{\partial \pi} - B_1^G - \pi B_{11}^G \frac{\partial t_g}{\partial \pi} + B_1^N - (1 - \pi)B_{11}^N \frac{\partial t_g}{\partial \pi}\right] - C_{11}^G \frac{\partial t_g}{\partial \pi} \quad (10)$$

$$\frac{\partial F_1}{\partial t_g} = \beta_g[\pi B_{11}^G + (1 - \pi)B_{11}^N] + \frac{1}{2}[b_{11} - \pi B_{11}^G - (1 - \pi)B_{11}^N] - C_{11}^G \quad (11)$$

Similarly, let $\frac{\partial U^G}{\partial e_g} = F_2, \frac{\partial U^N}{\partial t_n} = F_3$,

$$\frac{\partial U^N}{\partial e_n} = V^G = \beta_g \left[\pi B^G + (1 - \pi)B^N\right] + \frac{b-B}{2} \quad (12)$$
$$V^N = \beta_n \left[\pi B^G + (1 - \pi) B^N\right] + \frac{b-B}{2} \quad (13)$$

$$U^G = V^G - C^G + \Phi^G$$
$$U^N = V^N - C^N + \varphi^N + g^N$$

So we get:

$$U^G = \beta_g \left[\pi B^G + (1 - \pi)B^N\right] + \Phi^G(e_g) - C^G(t_g, e_g) + \frac{b-B}{2} \quad (14)$$
$$U^N = \beta_n \left[\pi B^G + (1 - \pi)B^N\right] + \varphi^N(e_n) + g^N(t_n) - C^N(t_n, e_n) + \frac{b-B}{2} \quad (15)$$

$$\frac{\partial U^G}{\partial t_g} = \beta_g + \frac{1}{2}[b_1 - \pi B_1^G - (1 - \pi)B_1^N] - C_1^G = 0$$
$$b_{mn}^G = 0, b_{mn}^N = 0, (m \neq n) \quad (16)$$

So we get:

$$\frac{\partial F_1}{\partial \pi} = \beta_g\left[B_1^G + \pi B_{11}^G \frac{\partial t_g}{\partial \pi} - B_1^N + (1 - \pi)B_{11}^N \frac{\partial t_g}{\partial \pi}\right] + \frac{1}{2}\left[b_{11} \frac{\partial t_g}{\partial \pi} - B_1^G - \pi B_{11}^G \frac{\partial t_g}{\partial \pi} + B_1^N - (1 - \pi)B_{11}^N \frac{\partial t_g}{\partial \pi}\right] - C_{11}^G \frac{\partial t_g}{\partial \pi} \quad (17)$$

$$\frac{\partial F_1}{\partial t_g} = \beta_g[\pi B_{11}^G + (1 - \pi)B_{11}^N] + \frac{1}{2}[b_{11} - \pi B_{11}^G - (1 - \pi)B_{11}^N] - C_{11}^G \quad (18)$$

$$\frac{\partial U^G}{\partial e_g} = F_2, \frac{\partial U^N}{\partial t_n} = F_4$$

So we have $\frac{\partial F_2}{\partial \pi}, \frac{\partial F_2}{\partial e_g}, \frac{\partial F_3}{\partial \pi}, \frac{\partial F_3}{\partial t_n}, \frac{\partial F_4}{\partial \pi}, \frac{\partial F_4}{\partial e_n}$, finally got simplification:

$$\frac{\partial e_g}{\partial \pi} = \frac{-(\beta_g - \beta_n)(B_2^G - B_2^N)}{2[\pi(\beta_g - \beta_n)(B_{22}^G - B_{22}^N) + b_{22} - 2C_{22}^G + 2\Phi_{11}^G + (\beta_g - \beta_n)B_{22}^N]} \quad (19)$$

$$\frac{\partial t_n}{\partial \pi} = \frac{-(\beta_g - \beta_n)(B_2^G - B_2^N)}{2[\pi(\beta_g - \beta_n)(B_{33}^G - B_{33}^N) + (\beta_g - \beta_n)B_{33}^N + b_{33} - 2C_{11}^N + 2g_{11}^N]} \quad (20)$$

$$\frac{\partial e_n}{\partial \pi} = \frac{(\beta_g - \beta_n)B_4^N}{2(\beta_g - \beta_n)(1 - \pi)B_{44}^N - 4\varphi_{11}^N + 4C_{22}^N} \quad (21)$$

On the basis of the four equations above, we have four lemmas follow, discuss the relationship between t_g, e_g, t_n, e_n and π.

Lemma 1: the relationship between t_g and π

On the basis of $\frac{\partial t_g}{\partial \pi}$, when $\beta_g > \beta_n$, $\frac{\partial t_g}{\partial \pi} > 0$ is always true; when $\beta_g < \beta_n$ and $\pi > \frac{2C_{11}^G - b_{11} - (\beta_g - \beta_n)B_{11}^N}{(\beta_g - \beta_n)(B_1^G - B_1^N)}$, we'll have $\frac{\partial t_g}{\partial \pi} > 0$.

The lemma shows that when the project evaluation of G is lower than the evaluation of N and only in π satisfy certain conditions, then we can exert control motivation for investment in the public sector; otherwise, when the project evaluation of G is higher than the evaluation of N, just to give some control over the public sector, we will be able to endow the role of control over the public sector for investment incentives.

Lemma 2: the relationship between e_g and π

On the basis of $\frac{\partial e_g}{\partial \pi}$, regardless of the magnitude relationship between β_g and β_n. when $\pi > \frac{2C_{22}^G - 2\Phi_{11}^G - b_{22} - (\beta_g - \beta_n)B_{22}^N}{(\beta_g - \beta_n)(B_{22}^G - B_{22}^N)}$ then $\frac{\partial e_g}{\partial \pi} > 0$, when $\pi < \frac{2C_{22}^G - 2\Phi_{11}^G - b_{22} - (\beta_g - \beta_n)B_{22}^N}{(\beta_g - \beta_n)(B_{22}^G - B_{22}^N)}$, we'll have $\frac{\partial e_g}{\partial \pi} < 0$.

The lemma shows that when k larger and satisfy $\frac{\partial e_g}{\partial \pi} > 0$, then as π increases public sector self-interest investment incentive is to enhance, this is also in line with the social practice. Then when give the private sector over the public sector more residual control rights and satisfy $\frac{\partial e_g}{\partial \pi} < 0$, as π decreases, e_g increase.

Lemma 3: the relationship between t_n and π

On the basis of $\frac{\partial t_n}{\partial \pi}$, when $\beta_n > \beta_g$, $\frac{\partial t_n}{\partial \pi} > 0$ is always true; when $\beta_g > \beta_n$ and $\pi > \frac{2C_{11}^N - b_{33} - 2g_{11}^N - (\beta_g - \beta_n)B_{33}^N}{(\beta_g - \beta_n)(B_{33}^G - B_{33}^N)}$, we'll have $\frac{\partial t_n}{\partial \pi} > 0$, when $\pi < \frac{2C_{11}^N - b_{33} - 2g_{11}^N - (\beta_g - \beta_n)B_{33}^N}{(\beta_g - \beta_n)(B_{33}^G - B_{33}^N)}$, we'll have $\frac{\partial t_n}{\partial \pi} < 0$.

The lemma shows that when $\beta_n > \beta_g$, N has higher proportion of control, π can play a role in controlling the right incentives. When $\beta_g > \beta_n$, only π satisfy certain conditions, giving control over the private sector can play a role in control of incentive.

Lemma 4: the relationship between e_n and π

On the basis of $\frac{\partial e_n}{\partial \pi}$, regardless of the magnitude relationship between β_g and β_n, when $\pi > 1 - \frac{2(\varphi_{11}^N - C_{22}^N)}{(\beta_g - \beta_n)B_{44}^N}$, we'll have $\frac{\partial e_n}{\partial \pi} > 0$, when $\pi < 1 - \frac{2(\varphi_{11}^N - C_{22}^N)}{(\beta_g - \beta_n)B_{44}^N}$, we'll have $\frac{\partial e_n}{\partial \pi} < 0$.

The lemma shows that when give the private sector over the public sector more residual control rights and satisfy $\frac{\partial e_n}{\partial \pi} < 0$, as π decreases, e_n increases. In contrast, when the public sector over the private sector has more control and satisfies $\frac{\partial e_n}{\partial \pi} > 0$ as π increases, e_n increase.

From the above four Lemmas, we have drawn the following two propositions:

Proposition1

On the basis of Lemma 2 and Lemma 4, when the existence of self-interest investments in public and private sectors.

When $\beta_n > \beta_g$, π should be:

$$\pi \in \left[\frac{2C_{22}^G - 2\Phi_{11}^G - b_{22} - (\beta_g - \beta_n)B_{22}^N}{(\beta_g - \beta_n)(B_{22}^G - B_{22}^N)}, \; 1 - \frac{2(\varphi_{11}^N - C_{22}^N)}{(\beta_g - \beta_n)B_{44}^N} \right]$$

When $\beta_n < \beta_g$, π should be:

$$\pi \in \left[1 - \frac{2(\varphi_{11}^N - C_{22}^N)}{(\beta_g - \beta_n)B_{44}^N}, \; \frac{2C_{22}^G - 2\Phi_{11}^G - b_{22} - (\beta_g - \beta_n)B_{22}^N}{(\beta_g - \beta_n)(B_{22}^G - B_{22}^N)} \right]$$

The proposition has said that in the case of public service and self-interest investment are presented, the best way to distribute is the scope of the existence of a separate self-interest into the public sector and the private sector alone. If $\beta_n > \beta_g$, when π grows, self-interest investment can be reduced of the private sector, meanwhile, the self-interested behavior of public sector investment is also limited, and the project's negative externalities can be reduced also. When $\beta_n < \beta_g$, self-interested behavior of public sector investment can be limited, and the self-interest of private sector investment will be constrained to a certain extent.

Proposition2

When the existence of self-serving behavior of the public sector, the influence of government guidance for the cooperation efficiency of PPP Projects.

When $\beta_n > \beta_g$, π should be set in the range of proportions:

$$\pi \in \left[0, \; \frac{2C_{22}^G - 2\Phi_{11}^G - b_{22} - (\beta_g - \beta_n)B_{22}^N}{(\beta_g - \beta_n)(B_{22}^G - B_{22}^N)} \right]$$

When $\beta_n < \beta_g$, π should be set in the range of proportions:

$$\pi \in \left[\frac{2C_{22}^G - 2\Phi_{11}^G - b_{22} - (\beta_g - \beta_n)B_{22}^N}{(\beta_g - \beta_n)(B_{22}^G - B_{22}^N)}, \; 1 \right]$$

In both cases, the π is set over two ranges which can be achieved high cooperation efficiency in PPP projects. Excessive government intervention in any case would be inefficient for public-private partnership.

IV. CONCLUSION

To improve the cooperation efficiency of PPP projects, combined with the conclusions of the above model and the reality, we have proposed five measures. First, develop a distribution plan which is the best operational in advance; second, develop the supply of relevant policies and the implement them; third, clarify the responsibility co-border of both the public and private in PPP projects; fourth, improve the government regulatory system; fifth, standardize the use of the PPP model to provide public goods market environment.

REFERENCES

[1]   Grossman S,Hart 0.The costs and benefits of ownership: a theory of vertical and lateral integration [J]. Political Economy, 1986, 94: 691-719.

[2]   Hart O, Moore J. Property rights and nature of the firm [J]. Journal of Political Economy, 1990, 98(6):1119-1158.

[3]   Hart O, Shleifer A, Vishny R. The proper scope of government: theory and application to prisons [J]. Quarterly Journal of Economics, 1997, 112(4):1127-1161.

[4]   Besley T, Ghatak M. Government versus private ownership of public goods [J]. Quarterly Journal of Economics, 2001, 116(4): 1343-1372.

[5]   Francesconi M, Muthoo A. Control rights in complex partnerships [J]. Journal of the European Economic Association, 2011, 9(3): 551-589.

[6]   Zhang Zhe, Jia Ming, Wan Difang. The model research of allocation of residual control rights and cooperation efficiency in PPP [J]. Journal of Industrial Engineering and Engineering Management, 2009(3):23-29.

[7]   Sun Hui, Ye Xiuxian. Study of the allocation of residual control right sin the public-private partnership under in complete contracts [J]. Journal of Systems Engineering, 2013, 02.

[8]   Rousseau. Social contract theory [M]. Beijing: The Commercial Press, 1980.

[9]   Aristotle. Political science [M]. Beijing: The Commercial Press, 1965.

[10] James. E. Andersen. Public policy [M]. Beijing: HuaXia Press, 1990.

[11] Kenneth J. Arrow. Social choice and individual value [M]. Cui Zhiyuan. Chengdu: Sichuan People Press, 1987.

[12] Schumpeter. Capitalism, Socialism and democracy [M]. Beijing: The Commercial Press, 1979: 314.

[13] Ren Xiaolin, Xie Bin. Government selfishness logical paradox [J]. Journal of China National School of Administration, 2003(6): 32-36.

[14] Tu Xiaofang. The government interest theory - from the local government in the transition period from the perspective of [M]. Beijing: Peking University Press. 2008: 107-117.

[15] Klein B, Crawford R, Alchian A. Vertical integration, appropriable rents, and the competitive contracting process [J]. Journal of Law and Economics, 1978, 21(2): 297-326.

VPDs Operating System Modeling and Performance Analysis Based on Generalized Stochastic Petri Nets

Shuo Cheng[1,*], Bi-xi Zhang[1], Guang-hui Zhou[1], Sheng-qiang Hu[1,2]

[1]School of Management, Guangdong University of Technology, Guangzhou, 510520, China
[2]The College of Business Administration, Guangdong University of Finance and Economics, Guangzhou, 510320, China

(411120163@qq.com)

Abstract - **This paper put forward and analyzed the concept "virtual procedure database" (VPDs)operating system learning rate by using generalized stochastic Petri net, to analysis the performance of VPDs operating system .With the isomorphism relation between generalized stochastic Petri net and Markov chain, will meet the learning rate of the VPDs operating system of GSPN is transformed to the equivalent Markov chain model through using GSPN and Markov chain methods, to assess the performance of VPDs operating system, and verified by an example, provided the basis and reference for optimized VPDs operating system.**

Keywords - **Generalized stochastic Petri net, learning rate, Markov chain, virtual procedure databases**

I. INTRODUCTION

In the complicated and changeable demand environment, it became hot issues in business and academia that production operation system of enterprises quickly adapt to complex needs change of market and the reasonable distribution of the skills of the workers in the operating system. Jiang et al. put forward the concept- "virtual operating systems" (VPSs) [1, 2], and through the establishment of a self-adaptive on a single piece and small batch production VPSs integrated monitoring and dispatching system in virtual space, achieved real-time monitoring and dynamic adjustments scheduling [3]. Wang et al. (2006) established virtual process classification system by the processes characteristics, and the example showed the method of process quality controlled multi-varieties and small batch production [4]. This paper put forward the concept "virtual procedure databases" (VPDs) dynamic organization model of operating system, which is based on the workers skills and process equipment and other factors to construct VPDs, it composed production line according to the requirement of process flow, product quantity of work and the delivery date, selecting production formulated products required corresponding process elements from "virtual procedure" in the workers/equipment. This paper is mainly on the VPDs operating system for modeling, and its performance analysis.

For Petri-net has a powerful function of modeling and simulation, not only can describes the system state, and represents the system behaviors, is particularly suitable for distributed system modeling, In recent years, it has been widely applied in automation, mechanical manufacturing and other fields. Wang et al. used generalized stochastic Petri net through the re-manufacturing supply chain modeling, and analyzed the performance of the method on the supply chain [5]. This article uses VPDs operating system as the research object, firstly, described the specific process flow of the operating system, and then built the system Generalized Stochastic Petri net (GSPN) based on learning rate. Markov chain model which is isomorphic of the GSPN is proved, according to the theory in literature [6]. And analysis the specific process efficiency of operating system performance, based on the parameters of the steady state probability of Markov chain, find out the bottleneck, to adjustment and scheduling of the process and to optimize the process flow of employees in operating system.

II. VDPs MODEL BASED ON GSPN

Taking a notebook production enterprise' production operation as an example, the production line is acceptance of an order.

After receiving the order, the order will match the corresponding raw materials, then through the various steps operational requirements (6 steps) and different process steps, and each working procedure, are required to select and the corresponding process of workers from VPDs in the manual operation, and finally produced 2 kinds of products of this order.

With the repetition of workers to perform the same operation increases, each worker in the operating system improve the work efficiency will workers by the effect on learning effect, With reference to the literature [7], workers working time of single product and repeat the operation frequency is a negative exponential function distribution:

$$t_x = t_1 \cdot x^{\log b / \log a} \qquad (1)$$

Using VisObjNet tools to create corresponding GSPN workflow model (Fig.1). The meaning of places and transitions in GSPN model shown in Table I. The GSPN model shown in Fig.2 consists of 17 places and 16 transitions, transitions in link with *HUMAN* is instantaneous transitions, the other transitions are delay transitions, The transitions of the transfer rate is denoted by λ_i.

[1]National Natural Science Foundation of China (No.71271060), Guangdong Province Natural Science Foundation (No.S2012010009278)
[2]Guangdong Social Science Planning Project (No.GD13XGL17)

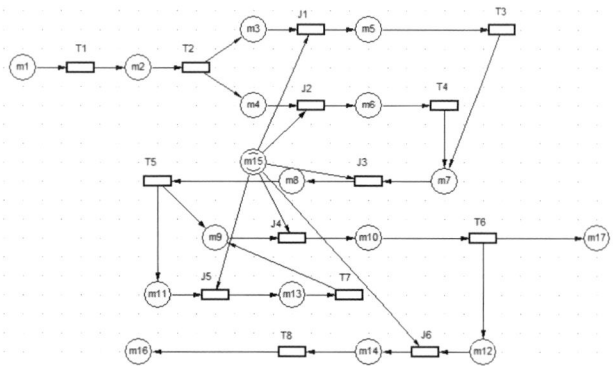

Fig.1. VPDs operating system GSPN model

TABLE I
THE MEANING OF PLACES AND TRANSITIONS

Place	Meaning
m1	Orders arrived
m2	Orders issued
m3/m4/m7/m9/m11/m12	Process first prepared
m5/m6/m8/m10/m13/m14	State of workers to complete second process
m16	State of products 1 completed
m17	State of products 2 completed
m15	State of workers assignment in VPDs
Transition	Meaning
Order	Accept orders and set to production workshop
T1/T2	Orders allocating
J i	Product process completed
T i	Process i transferring way to i+1

According to the principle of isomorphic between exponential distributed stochastic Petri nets and continuous time Markov chains, construct the effect of VPDs operating system of Markov chain models (Fig.2) in

learning rate. In this figure, rate of delay transition t_i is λ_i ($\lambda_i \geq 0$) in each process, the rate of instantaneous transition about the staff assigned to corresponding process is $\theta_i (\theta_i = 0)$.

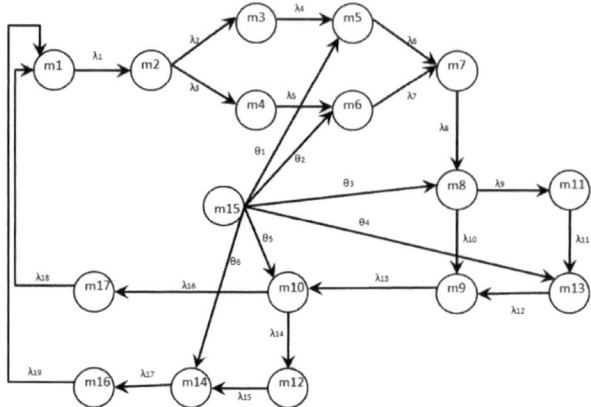

Fig.2. Markov chains model of VPDs operating system

According to every process completion time is exponential distribution, the transfer between different states caused by the delay transitions, according to Fig.2, building GSPN reachable marking, as shown in Table II. The number of tokens of P_i corresponding to M_i in Table II as shown in the matrix Q. According to literatures [5-6], into the formula (1), the solution of linear equations, can be stable probability of each place.

TABLE II
REACHABLE MARKINGS OF THE MARKOV CHAIN

	m1	m2	m3	m4	m5	m6	m7	m8	m9	m10	m11	m13	m12	m14	m17	m16
M1	1	0	0	0	0	0	0	0	0	0	0	0	0	0	0	0
M2	0	1	0	0	0	0	0	0	0	0	0	0	0	0	0	0
M3	0	0	1	1	0	0	0	0	0	0	0	0	0	0	0	0
M4	0	0	0	0	1	0	0	0	0	0	0	0	0	0	0	0
M5	0	0	0	0	0	1	0	0	0	0	0	0	0	0	0	0
M6	0	0	0	0	0	0	1	0	0	0	0	0	0	0	0	0
M7	0	0	0	0	0	0	0	1	0	0	0	0	0	0	0	0
M8	0	0	0	0	0	0	0	0	1	0	0	0	0	0	0	0
M9	0	0	0	0	0	0	0	0	0	1	0	0	0	0	0	0
M10	0	0	0	0	0	0	0	0	0	0	1	0	0	0	0	0
M11	0	0	0	0	0	0	0	0	0	0	0	1	0	0	0	0
M12	0	0	0	0	0	0	0	0	0	0	0	0	1	0	1	0
M13	0	0	0	0	0	0	0	0	0	0	0	0	0	1	0	0
M14	0	0	0	0	0	0	0	0	0	0	0	0	0	0	0	1
M15	1	0	0	0	0	0	0	0	0	0	0	0	0	0	0	0
M16	1	0	0	0	0	0	0	0	0	0	0	0	0	0	0	0

According to the GSPN Markov chain Reachable markings in Table II, get the corresponding transition transfer matrix:

$$Q = \begin{bmatrix}
-\lambda_1 & \lambda_1 & 0 & 0 & 0 & 0 & 0 & 0 & 0 & 0 & 0 & 0 & 0 & 0 & 0 & 0 \\
0 & -(\lambda_2+\lambda_3) & \lambda_2 & \lambda_3 & 0 & 0 & 0 & 0 & 0 & 0 & 0 & 0 & 0 & 0 & 0 & 0 \\
0 & 0 & -\lambda_4 & 0 & \lambda_4 & 0 & 0 & 0 & 0 & 0 & 0 & 0 & 0 & 0 & 0 & 0 \\
0 & 0 & 0 & -\lambda_5 & 0 & \lambda_5 & 0 & 0 & 0 & 0 & 0 & 0 & 0 & 0 & 0 & 0 \\
0 & 0 & 0 & 0 & -\lambda_6 & 0 & \lambda_6 & 0 & 0 & 0 & 0 & 0 & 0 & 0 & 0 & 0 \\
0 & 0 & 0 & 0 & 0 & -\lambda_7 & \lambda_7 & 0 & 0 & 0 & 0 & 0 & 0 & 0 & 0 & 0 \\
0 & 0 & 0 & 0 & 0 & 0 & -\lambda_8 & \lambda_8 & 0 & 0 & 0 & 0 & 0 & 0 & 0 & 0 \\
0 & 0 & 0 & 0 & 0 & 0 & 0 & -(\lambda_9+\lambda_{10}) & \lambda_9 & 0 & \lambda_{10} & 0 & 0 & 0 & 0 & 0 \\
0 & 0 & 0 & 0 & 0 & 0 & 0 & 0 & -\lambda_{11} & \lambda_{11} & 0 & 0 & 0 & 0 & 0 & 0 \\
0 & 0 & 0 & 0 & 0 & 0 & 0 & 0 & 0 & -\lambda_{12} & \lambda_{12} & 0 & 0 & 0 & 0 & 0 \\
0 & 0 & 0 & 0 & 0 & 0 & 0 & 0 & 0 & 0 & -\lambda_{13} & \lambda_{13} & 0 & 0 & 0 & 0 \\
0 & 0 & 0 & 0 & 0 & 0 & 0 & 0 & 0 & 0 & 0 & -(\lambda_{14}+\lambda_{16}) & \lambda_{14} & 0 & \lambda_{16} & 0 \\
0 & 0 & 0 & 0 & 0 & 0 & 0 & 0 & 0 & 0 & 0 & 0 & -\lambda_{15} & \lambda_{15} & 0 & 0 \\
0 & 0 & 0 & 0 & 0 & 0 & 0 & 0 & 0 & 0 & 0 & 0 & 0 & -\lambda_{17} & 0 & \lambda_{17} \\
\lambda_{18} & 0 & 0 & 0 & 0 & 0 & 0 & 0 & 0 & 0 & 0 & 0 & 0 & 0 & -\lambda_{18} & 0 \\
\lambda_{19} & 0 & 0 & 0 & 0 & 0 & 0 & 0 & 0 & 0 & 0 & 0 & 0 & 0 & 0 & -\lambda_{19}
\end{bmatrix}$$

Numerical Calculation

VPDs operating system is composed of process skills of workers and equipments matching with the skills, it is built to more complex systems based on requirements of product orders. In this paper, an order of production as example to analysis this paper used method, specific conditions such as shown in Fig.1, processing time of each step is determined by assigned to the workers on the process based on the effect of learning rate, and the processing time is a negative exponential distribution. The study select workers have the whole processes, and workers learning rate is set to a constant value k [8], each worker process completion time is generated by the random generator, then:

$\lambda_1 = \lambda_6 = \lambda_{10} = \lambda_{14} = 4$

$\lambda_2 = \lambda_5 = \lambda_9 = \lambda_{13} = \lambda_{15} = 4$

$\lambda_3 = \lambda_4 = \lambda_7 = \lambda_{11} = \lambda_{16} = 10$

$\lambda_8 = \lambda_{12} = \lambda_{17} = 5$

$\lambda_{18} = \lambda_{19} = 20$

There is building the steady state probability equation of GSPN based on each process completion time. then solute linear programming to get the steady state values of each process based on the steady state probability equations.

$$\begin{cases}
-\lambda_1 x_1 + \lambda_{18}x_{15} + \lambda_{19}x_{16} = 0 \\
\lambda_1 x_1 - (\lambda_2 + \lambda_3)x_2 = 0 \\
\lambda_2 x_2 - \lambda_4 x_3 = 0 \\
\lambda_3 x_2 - \lambda_5 x_4 = 0 \\
\lambda_4 x_3 - \lambda_6 x_5 = 0 \\
\lambda_5 x_4 - \lambda_7 x_6 = 0 \\
\lambda_6 x_5 + \lambda_7 x_6 - \lambda_8 x_7 = 0 \\
\lambda_8 x_7 - (\lambda_9 + \lambda_{10})x_8 = 0 \\
\lambda_9 x_8 - \lambda_{11}x_9 = 0 \\
\lambda_{11}x_9 - \lambda_{12}x_{10} = 0 \\
\lambda_{10}x_8 + \lambda_{12}x_{10} - \lambda_{13}x_{11} = 0 \\
\lambda_{13}x_{11} - (\lambda_{14} + \lambda_{16})x_{12} = 0 \\
\lambda_{14}x_{12} - \lambda_{15}x_{13} = 0 \\
\lambda_{15}x_{13} - \lambda_{17}x_{14} = 0 \\
\lambda_{16}x_{12} - \lambda_{18}x_{15} = 0 \\
\lambda_{17}x_{14} - \lambda_{19}x_{16} = 0 \\
\sum_{i=1}^{16} x_i = 1
\end{cases}
\quad
x_i = \begin{bmatrix}
0.086231 \\
0.049275 \\
0.019710 \\
0.123186 \\
0.024637 \\
0.049275 \\
0.137969 \\
0.057487 \\
0.022995 \\
0.045990 \\
0.172461 \\
0.038325 \\
0.076649 \\
0.061319 \\
0.019162 \\
0.015330
\end{bmatrix}$$

Selecting a subsystem PN' in the GSPN model of the operating system and subsystem P_{20} in module of operating system removal of P and t_1, and the average marker number of subsystem calculated in steady state condition.

$N = P(M(P_0 = 1)) + P(M(B_{11} = 1)) + P(M(B_{12} = 1)) + P(M(B_{21} = 1)) + P(M(B_{22} = 1))$
$\quad + P(M(B_{31} = 1)) + P(M(B_{32} = 1)) + P(M(B_{41} = 1)) + P(M(B_{42} = 1))$
$\quad + P(M(B_{51} = 1)) + P(M(B_{52} = 1)) + P(M(P_{20} = 1)) = 0.760471$

The number of markers (token number) into the subsystem in units of time is:

$\lambda = 4 \times P(M(B_{11} = 1)) + 4 \times P(M(B_{21} = 1)$

$= 4 \times 0.049275 + 4 \times 0.01971 = 0.275938$

The average delay time of subsystem is:

$T = N / \lambda = 0.760471 / 0.275938 = 2.755949$

The operation efficiency of all steps in VPDs can be reflected by its stability in the work of the state in the condition of probability, P_i has indicated that the system is in the working state, setting B_i that the efficiency of process i completed, then:

$$P(B_1) = P(m(B_{11}) = 1) + P(m(B_{12}) = 1) = 0.044347;$$
$$P(B_2) = P(m(B_{21}) = 1) + P(m(B_{22}) = 1) = 0.172461;$$
$$P(B_3) = P(m(B_{31}) = 1) + P(m(B_{32}) = 1) = 0.195456;$$
$$P(B_4) = P(m(B_{41}) = 1) + P(m(B_{42}) = 1) = 0.068984;$$
$$P(B_5) = P(m(B_{51}) = 1) + P(m(B_{52}) = 1) = 0.210786;$$
$$P(B_6) = P(m(B_{61}) = 1) + P(m(B_{62}) = 1) = 0.137969;$$
$$P(B_5) > P(B_3) > P(B_2) > P(B_6) > P(B_4) > P(B_1)$$

It can be obtained that the efficiency of each process ranking, it can be get workers is inefficient in the process 5, it should to reduce the number of workers in process 5, or arrange a part of workers in process 5 to have other extra processes; and workers in process 1 have high efficiency, it should increasing the number of workers or arrange workers assigned in other processes to process 1.

III. CONCLUSIONS

This paper calculated the stability of VPDs operating system by using the method of isomorphic GSPN and Markov Chain, using the Markov chain model, to transfer the GSPN model, simplify state space of the system, obtained the efficiency of workers in the learning rate and information of VPDs operating system, to conveniently find the bottleneck in production, and provide reference for VPDs operating system in staff scheduling and business restructuring.

REFERENCES

[1] Jiang, Z.B., Fung, R.Y.K., Tu, Y.L., etc. "A Framework for adaptive control of virtual production systems", In *Proceedings of the 3rd World Congress on Intelligent Control and Automation*. Hefei: 2000:138-142

[2] Jiang, Z.B., Fung, R.Y.K. "An adaptive agile manufacturing control infrastructure based on TOPNs-CS modeling", *The International Journal of Advanced Manufacturing Technology*, 2003, 18(10): 191-215.

[3] Lin Li, Zhibin Jiang. "Self-adaptive dynamic scheduling of Virtual Production Systems", *International Journal of Production Research*, 2007, 45(9): 1937-1951.

[4] Wang Li-ying, Sun Li, Wang Xiu-lun. "Multi-type & small batch procedure quality control method based on virtual procedure", *Computer Integrated Manufacturing Systems*, 2006, 12(8): 1263-1366, 1283 (In Chinese).

[5] Wang Wen-bin, Da Qing-li. "Remanufacturing Supply Chain Modeling and Analysis based on Generalized Stochastic Petri Nets", *Systems Engineering-Theory& Practice*, 2007, 12: 56-61 (In Chinese).

[6] Zuberek W M. "Performance evaluation using unbound timed Petri nets" //*Proceedings of the 3rd International Workshop on Petri Nets and Performance Models*, Kyoto, Japan, 1989.

[7] Lin Chuang. "Stochastic Petri Nets and System Performance Evaluation", Beijing: *Tsinghua University Press*, 2000:28- 35. (In Chinese)

[8] Zhang Bi-xi, Guang Ying-ying, "Song Jing. Optimization of batches processing modes for manual operating system considering leaning rate", *Systems Engineering-Theory & Pratice*, 2010, 30(4): 622-627 (In Chinese).

Research on Initiative Knowledge Spillovers of Supply Chain Entities Based on Service-oriented Manufacturing

Dan-li DU*, Qiu JIN, Hong-yan ZHAO

School of Economics and Management, Harbin Engineering University, P. R. China, 150001

(britrinjin@163.com)

A New Algorithm for the Risk of Project Time Based on Monte Carlo SimulationAbstract - **In the service-oriented manufacturing model, customer participation proposes higher requirements on the relationship between supply chain entities, and also provides more communication opportunities. The supply chain entities will have more opportunities and challenges to be considered as a whole, therefore, how to use the opportunities and advantages to enhance competitiveness is an important aspect to be considered. By putting forward repeated double auction theory and initiative knowledge spillovers, this paper points out that the process of cooperation game between supply chain entities should pay more attention to the long-term interests, and make full advantages of initiative knowledge spillovers which can promote trust and create new cooperation opportunities, so as to adopt some certain strategies to consolidate the supply chain partnership.**

Keywords - **Double auction theory, initiative knowledge spillovers, service-oriented manufacturing, supply chain**

I. INTRODUCTION

The fundamental question in the emerging field of strategic entrepreneurship is how firms combine entrepreneurial action that creates new opportunities with strategic action that generates competitive advantage [1]. Given the importance of knowledge for firms' competitive advantage, along with acquisitions and alliances, firms can access external knowledge via a less formal mechanism: knowledge spillovers [2]. Moreover, modern growth theory tells us that knowledge spillovers are crucial for the growth of high-income economies [3].

Knowledge spillovers can actively promote the development of different countries and industries macroscopically [4] [5], and also encourage innovative activities microscopically [6]. In this paper, we research on knowledge spillovers in the point of view of the supply chain in the service-oriented manufacturing model. Collaborations between the supply chain entities encourage innovative activities through leverage effect instead of delivering information alone, so as to enhance sustainable competitive advantage of the supply chain [7]. The knowledge can be mutually complementary between the supply chain entities, which is the key factor to influence knowledge innovation in the supply chain [8]. The Process of trust among supply chain entities' knowledge spillovers, help reduce the knowledge gap between supply chain entities, and also provide guarantee for knowledge innovation and collaborative innovation in the supply chain [9].

II. CORRELATION THEORY

A. Service-oriented manufacturing

As early as 1966, the United States economist Greenfield proposed the concept of Producer Services firstly. And then the concept of New Manufacturing was brought up by Duckers (1990) and perfected by Quinn (1996), which has been widely used [10]. The rising of the concept of service enhancement was proposed by Berger and Pappas in the late 1990s, who believed this kind of manufacturing is a foundation for knowledge creation [11] [12]. Yuan Qiong He and Kin Keung Lai (2012) constructed a conceptual model for service-oriented manufacturing and believed that the integration of business in the supply chain can affect the products' services. With the researching on service-oriented manufacturing and service enhancement, the concept of service-oriented manufacturing is gradually converging, which is based on manufacturing for services and also services for manufacturing.

Chinese scholars Sun Lin-yan and Li Gang pointed out the specific concept of service-oriented manufacturing which included three aspects: producer services, production of services and customer participation in the whole process, and the concept has been widely recognized in China.

1) Producer Services. Theoretically, it refers to the middle market investment services, which is not the final services for consumers but can be used for further production of goods and services [13].

2) Production of Services. It provides part of or all of their manufacturing services for other firms, and the cost of production is reduced by optimizing certain positions to improve production efficiency, thus strengthens the market strain capacity.

3) Customer Participation in the whole process. Productive services and service production of services only solve the problem of how to make the product to a certain extent, and how to rely on existing products services. But customer participation in the whole process can help the firms to find the customers' requirements and potential demands in easier way, and also change passive accepting feedback to offer products and services initiatively. Customers here includes the final consumers, the middle suppliers, distributors and the final users, etc., so customer participation in the whole process is an important feature of the service-oriented manufacturing.

The service-oriented manufacturing conceptual model [14] is shown in Fig.1 below:

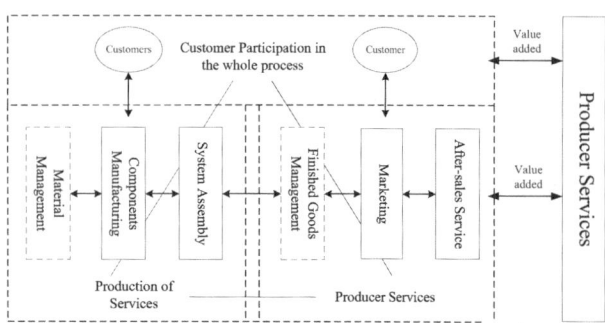

Fig.1. The service-oriented manufacturing conceptual model

B. Knowledge spillovers

In nowadays fast-growing information era, knowledge spillover phenomenon is widespread. During the process of communication and the application of knowledge, even if the individuals are not initiative to steal the owners' knowledge subjectively, they will be access to part of the knowledge to a certain way objectively [15].

According to the classification of knowledge spillovers by Verspagen [16], in this paper, knowledge spillovers are pure knowledge spillovers, which mainly refers to the individual knowledge itself which is imitated by other firms and embezzlement, and this kind of phenomenon is often caused by some objective reasons. This kind of knowledge spillovers is often inevitable to firms, even if they don't expect this to happen. Mansfield's investigation showed that in the sample, 60% of the patents and technological innovations have been copied in four years [17]. The characteristics of knowledge spillovers are inherent in knowledge. From the point of view of social development, the knowledge spillovers promote the social progress and development. Otherwise, from the point of view of knowledge owner, of course, the knowledge spillovers often represent the loss of interests, especially to that had been devoted much time and money to gain. Knowledge spillovers for a firm may cause the loss of its competitive advantage as it was.

C. Knowledge spillovers in the supply chain

In the service-oriented manufacturing mode, the supply chain entities tie together through a variety of contractual obligation or equities to be risk-sharing and have complementary advantages as a whole. The core aim of building service-oriented manufacturing supply chain is the customer demands that orient the knowledge spillovers and the creation of knowledge actively [18]. It not only considers the absorption capacity of the member firms, but also the knowledge spillover effect.

The supply chain entities need to choose the appropriate cooperation strategy from different ways, so as to seek innovative ways quickly [19]. Initiative knowledge spillovers in the supply chain entities need cooperation between each other, under the service-oriented manufacturing mode which includes

customer participation in the whole process. The supply chain entities need more close communication and cooperation to meet customer demands. However, for member enterprises of the supply chain themselves, the relationships among them establish under a certain contract, so there is often no sharing knowledge freely. It is hard to avoid knowledge transferring in the supply chain, including the parts not wanting others to get, because of the exist of the knowledge spillovers and knowledge exposes paradox phenomena. So there is a kind of game play process in the supply chain, which means the choice of strategies will be related to their collaborating objects.

III. METHODOLOGY

A. Double auction theory for strategy choice

Double auction model is used to discuss decisions in the situation of information asymmetry or incomplete information. In the supply chain, if M is the owner of the knowledge and N is a partner who is wanted, M hopes to make profit through the cooperation with N. N knows the value of the knowledge but M doesn't, which is evaluated to be V_b. M knows the acquisition cost of the knowledge but N doesn't, which is evaluated to be V_s. Defined the actual value ranges from 0 to 1, which means $V_b, V_s \in [0,1]$. The rule is: they bid at the same time, the bid price from N is P_b and is P_s from M. If $P_s \leq P_b$, they will have a deal at the price of $P = (P_s + P_b)/2$ (The utility is $V_b - P$ to N and $P - V_s$ to M); if $P_b \leq P_s$, they will not choose cooperation(The utility is Zero to both). N evaluates the price of the knowledge from the cost of knowledge acquisition and competitive advantage, while M evaluates from the effect of initiative knowledge spillovers [20].

Assuming that M and N are both rational, and pursue benefit maximization.

To N, for any $V_b \in [0,1]$, $P_b(V_b)$ must satisfy

$$\max_{P_b}[V_b - \frac{P_b + E[P_s(V_s)/P_b \geq P_s(V_s)]}{2}]P\{P_b \geq P_s(V_s)\} \quad (1)$$

$E[P_s(V_s)/P_b \geq P_s(V_s)]$ is the price N expects M to bid when N can offer more.

To M, for any $V_s \in [0,1]$, $P_s(V_s)$ must satisfy

$$\max_{P_s}[\frac{P_s + E[P_b(V_b)/P_b(V_b) \geq P_s]}{2} - V_s]P\{P_b(V_b) \geq P_s\} \quad (2)$$

$E[P_b(V_b)/P_b(V_b) \geq P_s]$ is the price M expects N to bid when N can offer more.

Assuming that the strategies form M and N are both linear. To N, the linear price strategy is $P_b(V_b) = a_b + c_b V_b$. To M, the linear price strategy is $P_s(V_s) = a_s + c_s V_s$.

According to formulas (1) and (2), we can get

formulas (3) and (4).

$$\max_{P_b}[V_b - \frac{1}{2}(P_b + \frac{a_s + P_b}{2})]\frac{P_b - a_s}{c_s} \quad (3)$$

$$\max_{P_s}[\frac{1}{2}(P_s + \frac{P_s + a_b + c_b}{2}) - V_s]\frac{a_b + c_b - P_s}{c_b} \quad (4)$$

So the equilibrium strategy of the Bayesian linear game for N is $P_b(V_b) = \frac{1}{12} + \frac{2}{3}V_b$.

The equilibrium strategy of the Bayesian linear game for M is $P_s(V_s) = \frac{1}{4} + \frac{2}{3}V_s$.

M and N play game around the value of the knowledge. The price will be accepted only if $P_b(V_b) \geq P_s(V_s)$, so we can get $V_b \geq V_s + \frac{1}{4}$.

According to Fig.2, we need $V_b \geq V_s$, the cooperation can be completed with benefit maximization only when $V_b \geq V_s + \frac{1}{4}$, so the scope of transactions will be less (without the shaded area), and the cooperation probability is only 56.25%, which means they may not reach benefit maximization through playing this game. So they need certain motivations to complete the cooperation.

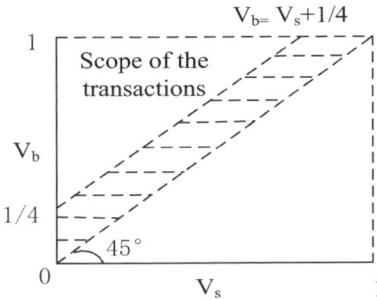

Fig.2. Scope of the transactions of the linear game for cooperation

B. Signaling effect of the motivations

We already know that the supply chain entities need certain motivations to maintain longtime corporations with each other. Assuming that N's input would be I, and the compositions of the input are knowledge, capital, human resources and other resources. M would be the "reputation one" with strong cooperation intention. N will response to the activity of M, which means N will cooperate if M chooses initiative knowledge spillovers. To M, the degree of initiative knowledge spillovers is random variable e(0<e<1). Because M is rational with cooperation intention, e can't be 0 or 1. The more e tends to be 1, the stronger cooperation intention M has.

When the degree of initiative knowledge spillovers tends to 1 infinitely, assumed the profit of the cooperation is G and the proportion is S to M. So the profit of M is SG, and the profit of N is (1-S)G to reach benefit maximization. We know M only spill a part of the knowledge, in fact the profit of the cooperation will be G' and the proportion will be the same. So the profit of M is SG', and the profit of N is (1- S)G' to reach

benefit maximization. And we can get the equation G'=eG. In the meantime, the game is still in the situation of incomplete information, and to M the probability of initiative knowledge spillovers is P. If N choose the cooperation need to meet the following condition: $P[(1-S)G] + (1-P)[(1-S)G'] > I$. So

$$G > I / [P(1-S) + (1-P)(1-S)e] \quad (5)$$

When $y = I / [P(1-S) + (1-P)(1-S)e]$, we can get the equation (6).

$$\frac{\partial y}{\partial e} = -\frac{I(1-P)(1-S)}{[P(1-S)+(1-P)(1-S)e]^2} = \frac{I(1-P)}{[P+(1-P)e]^2(1-S)} \quad (6)$$

We can know that $\partial y / \partial e < 0$, so y will decrease with the increase of e. In the meantime G>y, which means the requirement of N for cooperation profit value G gradually reduces along with the increase of the initiative. So N will choose cooperation with the awareness of the increase of the degree of initiative knowledge spillovers, and the requirement will also reduce. And the sincerity of cooperation is more important to N. In the service-oriented manufacturing mode, the supply chain entities need to cooperate as a whole in many activities, we can know from the analysis that in the case of lower expected returns, the cooperation also can be completed by showing the willingness of cooperation and the winning of trust initiatively.

C. Reputation effect of multiple batch initiative knowledge spillovers

We already know positive initiative spillover strategy will make a contribution to establishing a long-term relationship of cooperation. For the long-term relationship of cooperation of the supply chain entities in the service-oriented manufacturing mode, if it works well, the reputation will be higher. The firm and the staff on the basis of benefit maximization give up some interests to continue the cooperation and to achieve the effect of the long-term interest maximization. The supply chain entities' goal is to meet the customer demands maximally in a long-term partnership, the initiative knowledge spillovers must be multiple batches instead of only once.

Assuming that M is still the "reputation one" with strong cooperation intention, and N will response to the activity of M. In the paper, reputation is defined as biological neurons. According to the threshold of cell body, reputation receives a degree of ascension with high returns, and reputation measures through cell weight. We define that the reputation is g and $g = g + \Delta g$. Setting up the simulation function Δg:

$$\Delta g = \begin{cases} 0 & |r-r'| < a \\ \dfrac{1 - \exp[-(r-r')]}{1 + \exp[-(r-r')]} & |r-r'| \geq a \end{cases} \quad (7)$$

r is the rate of return of cooperation, r' is the average rate of return, a is the threshold. So the repeat bilateral auction game model is blow here.

$$P_{rb}(V_b) = \frac{1}{12} + \frac{2}{3}V_b + g_b \times \alpha \times P_{max} \tag{8}$$

$$P_{rs}(V_s) = \frac{1}{4} + \frac{2}{3}V_s - g_s \times \alpha \times P_{max} \tag{9}$$

P_{max} is the maximum profit of cooperation, α is constant and $\alpha \in (0,1]$.

M and N all choose a long-term cooperation and multiple - batch initiative knowledge spillovers, the profit of each batch of initiative knowledge spillover is a. Through the process of cooperation relationship, M and N are likely to cooperate the next time, and the probability is β ($0 < \beta < 1$). After this time, the probability will be β^2, and the times of cooperation are t. The total profit of M will be

$$a + a\beta + a\beta^2 + a\beta^3 + \cdots + a\beta^{t-1} = a(1 - \beta^t)/(1-\beta) \tag{10}$$

If M choose the initiative knowledge spillover only once, the profit is a'.

When $a(1-\beta^t)/(1-\beta) > a'$ ($1 + \beta + \beta^2 + \beta^3 + \cdots + \beta^{t-1} > a'/a$), M will not choose this strategy.

Assuming that the total amount of knowledge that M choose to spill is all the same and the spillover times are n. The total profit of M will be

$$\frac{a}{n} + \frac{a}{n}\beta + \frac{a}{n}\beta^2 + \frac{a}{n}\beta^3 + \cdots + \frac{a}{n}\beta^{nt-1} = \frac{a(1 - \beta^{nt})}{n(1-\beta)} \tag{11}$$

The profit of one time spillover is a'/n. So we can get:

$$1 + \beta + \beta^2 + \beta^3 + \cdots + \beta^{t-1} > a'/a \tag{12}$$

$$1 + \beta + \beta^2 + \beta^3 + \cdots + \beta^{nt-1} > a'/a \tag{13}$$

Because of $t > 1$ (nt-1 > t-1), we will find out that the value space of β is bigger.

So we know, when the total amount of knowledge that M choose to spill is all the same, the more times M spills the knowledge, the more profit M will get. So the multiple batch initiative knowledge spillovers is more competitive and good for the relationship of cooperation, and the probability of another time cooperation will be higher after several times of cooperation and the trust will also be more.

IV. RESULTS

From the above analysis, we can see that if the supply chain member companies only concerned with the immediate benefits during the process of the game, they'll miss the advantage of the formed supply chain to establish a long-term partnership opportunities. In the course of the game, when the members companies use the initiative knowledge spillovers in internal supply chain as a trusted investment, it often establishes a good relationship in the supply chain, so as to consolidate cooperation. When member companies take more attention to the long-term interests and the long-term

cooperative relationships during the game, the being sought collaborator N will give a slightly higher bid based on reputation, and be more likely to pay more. In the meanwhile, the one M who owns the knowledge will give a slightly lower bid just like accepting less. So we can improve the success rate and achieve win-win results.

For example, if M who owns the knowledge estimates the knowledge price as $V_b = 0.7$, bid for $P_b = 0.4$. Being-sought-collaborator N estimates the price as $V_s = 0.3$, bid for $P_s = 0.5$. Based on double auction game model, cooperation will be unable to be on its way. However, in repeated game model, as both sides focus on long-term interests and reputation, so they will give up some interests so as to make the cooperation successful. So the reputation $g_b = g_s = 0.3$, $\alpha = 0.1$, according to the formulas (8) and (9), it can be calculated that $P_{rb} = 0.55$ $P_{rs} = 0.35$, so the cooperation is successful and each side gains. However, because the two sides' revenues are greater than 0, so there will be a long-term repeated game to maximize the long-term interests. So it will let the other one get cooperation confidence if the member companies have a good reputation. In the meanwhile, how to build a good reputation is the key that the companies need to consider carefully. In the whole supply chain, member companies have more cooperation opportunities. It will establish a more solid relationship if the member companies use the cooperation knowledge spillovers better.

According to the formula (12) and (13), we can know the one that satisfies inequality (8)'s β interval necessarily satisfies the inequality (9). But on the other hand, it may not be established, because inequality (9)'s β interval values in the larger space. This situation indicates that, after a number of cooperation between M and N, because the strengthening of trust between the two sides, the range of β interval has been further extended. In the other words, in the past, it will promote bilateral cooperation only when the cooperation possibility number $\beta > 0.7$. But now, after the establishment of a long working relationship, the cooperation may be smooth only when the cooperation possibility number $\beta > 0.4$.

In other words, the knowledge owner M, through active, constantly, repeatedly showed willingness to cooperate, make the collaborators N to strengthen confidence in the success of cooperation. So they are very likely to choose to cooperate even so the probability of successful cooperation is not high. At the same time there will establish a good reputation and mutual trust relationship between M and N with the increasing frequency of cooperation. Thus, in the process of the game, two sides are willing to pay more attention to long-term benefits so as to establish a long-term cooperative relationship.

V. CONCLUSION

In summary, in the supply chain in service-oriented manufacturing model of customer participation, the partnership between member companies is long-term. Through the member companies change the passive knowledge spillovers into initiative knowledge spillovers, so to take full advantage of cooperation signal that initiative knowledge spillovers can promote supply chain entities involvement in the overall cooperation, gain trust, and expand cooperation to achieve long-term and maximized interests. Therefore, the supply chain entities can take the following measures to solid basis for cooperation.

1) Change the passive knowledge spillovers to initiative ones. Initiative spillovers emphasize the enthusiasm and initiation. Initiative side can show its willingness by additional investment in human resource, financial and other aspects. And it can also accelerate the investment time to show faith of cooperation, thus to contribute to cooperate and it is likely to get more investment opportunities for each other.

2) Choose multiple-batch initiative knowledge spillovers and spill the knowledge is bound to spill more. As a member of the supply chain, the internal member companies need to establish a more solid and trustful relationship, therefore they need to take initiative attitude for cooperation. It can provide some knowledge to member companies initiatively, letting the other one be aware of the cooperation enthusiasm and the company's strength characteristics, so as to establish a long-term relationship through a certain spillover strategy.

3) Use the full participation of customers to close the relationship between the members. The supply chain is a whole body, and its establishment often leads to solid cooperation, thereby reducing the cost to a certain extent and having rapid response. In the service-oriented manufacturing model, the full participation of customers not only provides customer knowledge to the whole supply chain, but also creates communication and coordination opportunities for the firms. Firms should make full use of the partnership and cooperate with others more actively, which will contribute to the establishment of a more reliable partnership.

ACKNOWLEDGEMENT

Fund projects: Supported by Heilongjiang Province Science Fund (G201109) and Three Personnel Training Projects of Harbin Engineering University in 2014

REFERENCES

[1] Hitt, M.A., D.L. Sexton, R.D. Ireland and S.M. Camp. Strategic Entrepreneurship: Integrating Entrepreneurial and Strategic Management Perspectives. Strategic Entrepreneurship: Creating a New Mindset [M]. M. Camp, Blackwell Publishers, Oxford. 2002. pp. 1-16.

[2] Juan Alcácer, Wilbur Chung. Location Strategies and Knowledge Spillovers [J]. Management Science. Vol. 53, No. 5, May 2007, pp. 760–776.

[3] Thomas Döring, Jan Schnellenbach. What Do We Know About Geographical Knowledge Spillovers and Regional Growth? A Survey of the Literature [C], Research Notes, Deutsche Bank Research. October 12, 2004 No. 14.

[4] Gerald A. carlino. Knowledge Spillovers: Cities' Role in the New Economy [J]. Business Review. 2001. 04:17-22.

[5] Ashish Arora, Andrea Fosfuri, Alfonso Gambardella. Specialized technology suppliers, international spillovers and investment: evidence from the chemical industry [J]. Journal of Development Economics. 2001, Vol.65:31-54.

[6] Cohen, W., Levinthal, D.A. Absorptive Capacity. A new perspective on learning and innovation [J]. Administrative Science Quarterly, 1990, Vol.35:128-152.

[7] Vachon, S., Klassen, R.. Environmental management and manufacturing performance: the role of collaboration in the supply chain [J]. International Journal of Production Economics. 2008,111 (2):299-315.

[8] Weck M.. Knowledge creation and exploitation in collaborative R&D projects: lessons learned on success factors [J]. Knowledge and Process Management, 2006, 13(4): 252.

[9] Stanley E. Fawcett, Stephen L. Jones, Amydee M. Fawcett. Supply chain trust: The catalyst for collaborative innovation [J]. Business Horizons 2012(55), 163-178.

[10] Ducker. P. E. The Emerging Theory of Manufacturing [J], Harvard Business Review, May-Junes. 1990.

[11] Pappas N, Sheehan P. The New Manufacturing: Linkages between Production and Service activities, Working for the Future: Technology and Employment in the Global knowledge Economy [M]. Melbourne: Victoria University Press, 1998.

[12] Nonaka I.. The concept of "Ba": building a foundation for knowledge creation [J]. California Management Review, 1998, 40(3): 15-54.

[13] Sun Linyan, Li Gang, Jiang Zhibin, Zheng Li, He Zhe. Service-Embedded Manufacturing: Advanced Manufacturing Paradigm in 21st Century [J]. China Mechanical Engineering, 2007, 18(19): 2307-2312

[14] Sun Linyan, Li Gang, Gao Jie. Service Manufacturing – Preliminary Theory & Business Practices [M]. Tsinghua University Press, 2009.3.

[15] Qi Hong-mei, Wang Hui-dong, Pang Shi-jun, A Strategy of Initiative Knowledge Spillovers Based on Cooperation [J]. Scientific Management Research.2004, 22(4):70-85.

[16] Verspagen B. Estimating international technology spillovers using technology flow matrices [J]. Weltwirtschaftliches Archiv, 1997, 133: 226-324.

[17] Kaz Miyagiwa, Yuka Ohno. Uncertainty, spillovers and cooperative R&D [J]. International Journal of Industrial Organization. 2002, 20: 855-876.

[18] Wu Bing, Liu Zhong-ying. Research on Knowledge Creation in the Supply Chain [J]. Journal of Information, 2007, (10):2-4.

[19] Kalyan Singhal, Jaya Singhal. Imperatives of the science of operations and supply-chain management [J]. Journal of Operations Management 2012(30) 237–24.

[20] Wang Hong, Chen Bao-guo, Ding Chang-chun. Evolutionary Game Analysis of Knowledge Sharing in Enterprise [J]. Science and Technology Management Research, 2007 (11): 217-220.

Study on the Relationship between the Government and Enterprises Based on Dynamic Game

Yicheng FENG

Department of Electric Machine and Electrical Engineering, University of Tsinghua, Beijing
(fengshuo19920618@126.com)

Abstract - **The relationship between the government and the enterprises determines the freedom and vitality of a country's economic development to a certain extent. Under the market conditions, the internal logic relationship exists between the government and enterprise, in view of the contradiction between the government want to maximize efficiency and the enterprises in pursuit of profit maximization, this article discuss the relationship between the government and enterprises from the perspective of game theory, build a three stage dynamic game model, and mainly analyzes the game result between the government and enterprises, get the best strategy choice of government and enterprises.**

Keywords – **Dynamic game, government and enterprises, strategy choice**

I. INTRODUCTION

Under the wave of economic globalization, our country's economy has made remarkable achievements, but there are still some problems in the enterprise development in our country, especially after China's accession to the WTO, the relationship between the government and enterprises will face greater challenges. Therefore, handle the relationship between the government and enterprises correctly is one of the important work in the future, it is related directly to whether our enterprise can become the main market players, whether the socialist market economy system is perfect, whether the economy can achieve sustainable development, etc. In the operation of the economy, the internal logic relationship exists between the government and enterprise, in view of the contradiction between the government want to maximize efficiency and the enterprises in pursuit of profit maximization, this article introduced the point of game theory, discuss the relationship between the government and enterprises, build a three stage dynamic game model, and mainly analyzes the game result between the government and enterprises, get the best strategy choice of government and enterprises.

II. GAME FEATURE ANALYSIS OF THE RELATIONSHIP BETWEEN THE GOVERNMENT AND ENTERPRISES

In the game of government and enterprises, as the game players, they respectively represent the policy makers and policy implementation object. To achieve its maximum benefits of the whole society, the government

has formulated corresponding macro and micro economic policies and measures that influence the development of enterprises. In the process of game, the enterprise can observe the behavior of the government through various means, to obtain relevant information, and then select strategy. The characteristics of game performance are as follows [1]:

1) The selection of payment function

The government's target of related policy is usually in pursuit of social benefits maximization, and the enterprise's behavior is want their own profit to maximization.

2) Select action space

As policy makers, government regulation of macro and micro economic use of fiscal, monetary, tax and other policies to encourage or inhibit the development of enterprise, and impact on enterprises to implement. The enterprises will choose the most favorable game strategies, through the positive or negative waiting to deal with the government's strategy, and finally achieve the balance of both government and business.

3) Analysis of the strategic effect

When government and enterprises select strategy, they need to advance game effect prediction and assessment, equivalent to the game players need to rational choice and decision [2]. The government in formulating relevant policy, select the corresponding policy tools, take the social and public reaction into account. And companies will adjusting the strategy in order to maximize their own efficiency.

4) Adjustment and coordination of strategy

The game process of the relationship between government and enterprise is essentially a dynamic game process, the process of strategic adjustment and coordination is equivalent to the strategy choice process of the government and the enterprises to carry out mutual echo and interaction.

III. ASSUMPTIONS AND ESTABLISHMENT OF DYNAMIC GAME MODEL OF THE RELATIONSHIP BETWEEN GOVERNMENT AND ENTERPRISES

A. Model assumption

1) Assume that model is complete information and dynamic game. Participants only enterprise E and

government G, both sides are rational: the enterprises in pursuit of profit maximization; The government want to maximize efficiency [3].

2) The type of enterprise development project space is {have prospects for development, no prospects for development}, and the probability of have prospects for development is P, the other probability is 1 - P.

3) The action space of enterprise is {positive, negative}, the probability of positive and negative is N and 1 - N. The action space of government is {encourage development, inhibit development}, the project which prospects for development, the probability of the government encourages the development is M1, the probability of the government inhibit the development is 1-M1; No prospects for development projects, the probability of the government encourages the development is M2, the probability of the government inhibit the development is 1 - M2.

4) There is no moral hazard between the two enterprises.

B. The establishment of dynamic game model of the relationship between government and enterprise

Assuming that enterprise and the government's game is three stage dynamic game model, the extended as shown in Fig.1. The data in parentheses are the payoff function of government and enterprises [4].

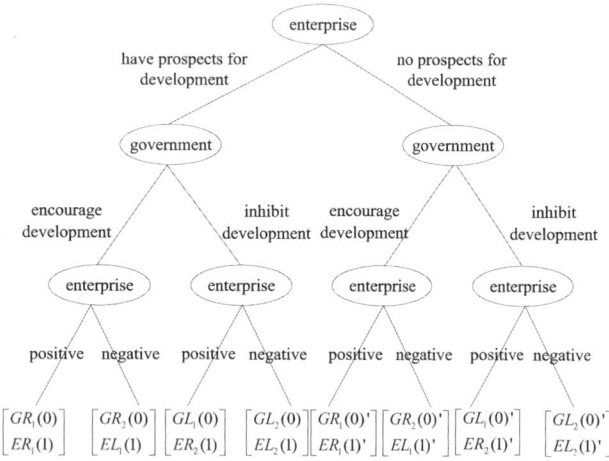

Fig.1. The dynamic game model of the relationship between government and enterprise

In the game, the enterprise first action, the enterprise has two selection strategies in the first stage of the game: develop promising projects or no prospects for development projects.

In the second stage of the game, the government has two strategies: encourage project development or inhibit the project development.

In the third stage of the game, the enterprise according to the government's strategy can choose the corresponding action. For promising projects, when enterprises actively strive for development the project, if the government to encourage attitude, the payoff function is $(GR_1(0), ER_1(1))$, if the government to restrain attitude, the payoff function is $(GL_1(0), ER_2(1))$; when enterprises negatively strive for development the project, if the government to encourage attitude, the payoff function is $(GR_2(0), EL_1(1))$, if the government to restrain attitude, the payoff function is $(GL_2(0), EL_2(1))$. For the projects which no prospects for development, when enterprises actively strive for development the project, if the government to encourage attitude, the payoff function is $(GR_1(0)', ER_1(1)')$, if the government to restrain attitude, the payoff function is $(GL_1(0)', ER_2(1)')$; when enterprises negatively strive for development the project, if the government to encourage attitude, the payoff function is $(GR_2(0)', EL_1(1)')$, if the government to restrain attitude, the payoff function is $(GL_2(0)', EL_2(1)')$ [5].

In the process of the game, there are two strategies for the government, four strategies for the enterprise, specific meaning as shown in Table I.

TABLE I
GAME STRATEGY SELECTION AND MEANING

The Strategy of the Enterprise	Meaning
(positive, positive)	The government choose the positive to encourage development
	The government choose the positive to restrain development
(positive, negative)	The government choose the positive to encourage development
	The government choose the negative to restrain development
(negative, positive)	The government choose the negative to encourage development
	The government choose the negative to restrain development
(negative, negative)	The government choose the negative to encourage development
	The government choose the negative to restrain development

Therefore, aiming at the prospects of the development project, we can set up corresponding strategic description of extended game respectively, as shown in Fig.2 and Fig.3.

enterprise

	(positive , positive)	(positive , negative)	(negative ,positive)	(negative, negative)
government encourage development	$GR_1(0), ER_1(1)$	$GR_1(0), ER_1(1)$	$GR_2(0), EL_1(1)$	$GR_2(0), EL_1(1)$
restrain development	$GL_1(0), ER_2(1)$	$GL_2(0), EL_2(1)$	$GL_1(0), ER_2(1)$	$GL_2(0), EL_2(1)$

Fig.2. Extended game payoff matrix of the promising projects

	(positive , positive)	(positive , negative)	(negative ,positive)	(negative, negative)
government encourage development	$GR_1(0)', ER_1(1)'$	$GR_1(0)', ER_1(1)'$	$GR_2(0)', EL_1(1)'$	$GR_2(0)', EL_1(1)'$
restrain development	$GL_1(0)', ER_2(1)'$	$GL_2(0)', EL_2(1)'$	$GL_1(0)', ER_2(1)'$	$GL_2(0)', EL_2(1)'$

Fig.3. Extended game payoff matrix of no prospects for development projects

IV. ANALYSIS OF THE RESULTS FOR DYNAMIC GAME RELATIONSHIP BETWEEN GOVERNMENT AND ENTERPRISES

A. The best strategy for the government

In general, the government mostly support for the projects with prospects for development, however, the government tend to inhibit those without prospects for development [6]. Therefore, in this game model, the best strategy of government is: M1 = 1, M2 = 0.

B. The best strategy for the enterprise

Based on the model analysis, we can know that when the N=0, the payoff function of the enterprise is [7-8]:

$$EL_2(1)' + [EL_1(1) - EL_2(1)'] \cdot P$$

When the N=1, the payoff function of the enterprise is:

$$ER_2(1)' + [ER_1(1) - ER_2(1)'] \cdot P$$

When $EL_2(1)' + [EL_1(1) - EL_2(1)'] \cdot P > ER_2(1)' + [ER_1(1) - ER_2(1)'] \cdot P$, that:

1) When $P < [EL_2(1)' - ER_2(1)'] / [ER_1(1) + EL_2(1)' - EL_1(1) - ER_2(1)']$ the N=0, so no matter the prospect of development project, the best strategy of the enterprise is negative, the payoff function is $EL_2(1)' + [EL_1(1) - EL_2(1)'] \cdot P$

2) When $P = [EL_2(1)' - ER_2(1)'] / [ER_1(1) + EL_2(1)' - EL_1(1) - ER_2(1)']$ the N=[0,1], so no matter the prospect of development project, the enterprise choose strategy is positive or negative, their payoff function are the same [9].

3) When $P > [EL_2(1)' - ER_2(1)'] / [ER_1(1) + EL_2(1)' - EL_1(1) - ER_2(1)']$ the N=1, so no matter the prospect of development project, the best strategy of the enterprise is positive, the payoff function is $ER_2(1)' + [ER_1(1) - ER_2(1)'] \cdot P$

C. Nash equilibrium

The game there are three groups of refined bayesian Nash equilibrium, as follows:

① N=0, M1=1, M2=0, so $P < P_0$

② N=1, M1=1, M2=0, so $P > P_0$

③ N=[0,1], M1=1, M2=0, so $P = P_0$

$$P_0 = [EL_2(1)' - ER_2(1)'] / [ER_1(1) + EL_2(1)' - EL_1(1) - ER_2(1)']$$

D. Equilibrium analysis of the general solution to the game model

1) As a rational economic man, when the probability of enterprise development project belongs to the development prospect is greater than, the best strategy of enterprise is positive, which make full use of the state's preferential policies and measures, promote the development of social economy and promote the enterprise benefit maximization, ensure that both sides of the payoff function for [10-11];

2) When the probability of enterprise developing project belongs to the no development prospect is less than, government will often choose inhibited attitude, the best strategy for the enterprise is to fight negatively against the policies and measures which inhibit the development of project formulated by government. In this case, government will obtain the minimum payment utility, while enterprises will obtain higher payment utility.

3) When the developing project belongs to neither the type with prospects for development nor the type with no prospects for development, the positive or negative policy measures that formulated by the government have little impact on it, so the best strategy for the enterprises is to choose between positive utilization and negative resistance.

V. CONCLUSION

In the game of the government and enterprises in our country, the government has always been in the active position, in addition, policies the government formulates often have an enormous effect on the development of enterprises. Therefore, in practice, when fomulate the measures that affect the economic development of the enterprise, the government has to be very careful. For the type with prospects for development, the government should provide preferential policy and technical support to encourage the development of the enterprise, for the type with no prospects for development, the government can limit, constrict, and raise taxes to inhibit the development of the enterprise. Also, enterprises should cope with the government's relative policies with awareness and coordination, stick to the flexibility of strategy, improve their countermeasures in time, and actively strive to make the government's relevant policies beneficial to their own development in a reasonable scope, so as to make harmonious development between the government and enterprises.

REFERENCES

[1] Zhifeng Qiu, Ning Gui, Geert Deconinck. Analysis of dynamic game played with inaccurate demand beliefs [J]. Applied Mathematics and Computation, 2014 (230): 530–541.

[2] Fengqun,Chenhong. Research on the effectiveness of coal mine safety management system based on Dynamic Game [J]. China Safety Science Journal, 2013, 02: 15-19.

[3] Guojia Liu. Study on dynamic game of complete information no emergency in supply chain [J]. Operations research and management science, 2012, 06: 105-111.

[4] Lizhong, Daming You, Weibai Liu. Dynamic game analysis of rural land transfer path [J]. Systems engineering, 2013, 04: 103-108.

[5] Cuihua Zhang, Lili Sun. Strategy Research on information disclosure quality distribution channel two retailers dynamic game [J]. Journal of management in Engineering, 2012, 04: 199-204.

[6] Yongyi Liu, Qiangwen Ma. Dynamic game analysis from the perspective of the ultimatum strategy [J]. Statistics and decision making, 2013, 17: 40-43.

[7] Pengbin ,Zhaozheng. Analysis of service outsourcing pricing based on two stage dynamic game [J]. Operations research and management science, 2012, 03: 154-158.

[8] Zhaodao Zhi, Baiyun Yuan, Liangjie Xia, Xinpeng Xie. Game in supply chain dynamic manufacturers compete to consider carbon emission constraints [J]. Industrial Engineering and management, 2014, 01: 65-71.

[9] Chengwen Ao. Study of grassroots governance rules and dynamic evolution perspective --Analysis Based on Dynamic Game Theory [J]. Modern management science, 2014, 02: 39-41.

[10] Shuqiang Wang,Yeli. Multi period dynamic game equilibrium analysis of asymmetric information on securities trading [J]. Mathematics in practice and theory, 2014, 04: 39-48.

[11] Huimei Zhang, Zigang Deng. Analysis of corporate social responsibility in the government, enterprises, social three party based on Dynamic Game [J]. Systems engineering, 2011, 06: 123-126.

The Method of Time Synchronization between Telecommunication Base Stations with GPS Common View

Xu Du[1,2,3,*], Yu Hua[1,2], Xiao-zhen Jin[1,2], Hai-ni Jia[4]

[1]National Time Service Center, Chinese Academy of Sciences, Xi'an, 710600, China
[2]Key Laboratory of Precision Navigation and Time Technology, Chinese Academy of Sciences, Xi'an, 710600, China
[3]University of Chinese Academy of Sciences, Beijing, 100039, China
[4]Xi'an Institute of Optics and Precision Mechanics of Chinese Academy of Sciences, Xi'an, 710119, China
(*xd050104104@ 126.com, hy@ntsc.ac.cn, jinxiaozhen@ntsc.ac.cn, laser_jiahaini@opt.ac.cn)

Abstract - **According to the demand of the navigation system of China telecom to realize high precision time synchronization between adjacent base stations, we come up with the method of long distance time transfer using GPS common view. The hardware design and software design of this system are introduced in detail. In this paper, we use the powerful multi-channel Resolution T receiver and the TDC-GP1 counter chip to form a set of GPS common view comparison system based on ARM platform. Preliminary results of the test show that the system has stable performance and the alignment accuracy is within 5 ns. Therefore, this system we design can realize high precision time synchronization between adjacent base stations.**

Keywords - **GPS Common view technology, high precision time, synchronization, telecommunication base station**

I. INTRODUCTION

In recent years, with the rapid development of mobile communication technology, the communication system is increasingly demanding for time synchronization especially the mature 3G technology and the widespread 4G technology [1], which requests higher precision of the time controlling in switching work between base stations. In terms of time transfer accuracy, satellite two-way method has the highest precision about nanosecond or sub-nanosecond level [2]. However, it takes up the satellite resources higher, and the equipment is complex and expensive, which restrict its use range to a great extent [3]. GPS common view time transfer technology has some advantages such as simple, cheap, timing of high precision, convenient use [4], etc. And it is the optimal method of long-distance clock comparison for its ratio of the transmission accuracy can reach a few nanoseconds [5].

Any deviation of time synchronization between telecommunication base stations will bring incalculable impact on users [6]. Therefore, all base stations must have a higher accuracy of time synchronization [7]. China telecom takes advantage of the existing communication base stations nationwide coverage, sets up high precision time synchronization between telecommunication base stations, to bring about indoor navigation, outdoor navigation and time service integration [8]. Thus it will have a great impact on the national economic construction and people's livelihood [9].

In this paper, we put forward a method of using the time interval measurement chip based on ARM platform instead of traditional time interval counter to achieve GPS common view technology. The alignment accuracy is within 5 ns. The GPS common view technology we studied can satisfy the high precision of time synchronization between base stations to a great extent.

II. THE PRINCIPLE AND DESIGN SCHEME OF THE GPS COMMON VIEW TECHNOLOGY

Common view technology is to point to that in the perspective of a GPS satellite, two atomic clocks of any two places on earth can use the time signal from the same satellite at the same to do time frequency comparison. The two GPS time receivers from the A ground and the B ground under the action of the same common view time schedule receive the same GPS satellite signal at the same time.

The output second pulses by the local atomic clock are sent to the receiver's built-in counter chip and in comparison with output second pulses by GPS receivers, to get the time difference Δt_{AGPS} and Δt_{BGPS} between the local time and the t_{GPS} respectively. The data of the B ground is sent by GPRS module in the form of text messages to the A ground. Subtracting the two types, time difference between the two atomic clocks is available.

The system structure of GPS common view time transfer receiver is shown in Fig.1. It is mainly consist of the GPS antenna, Resolution T receiver, TDC-GP1 counter chip, ARM platform (STM32) and rubidium clock.

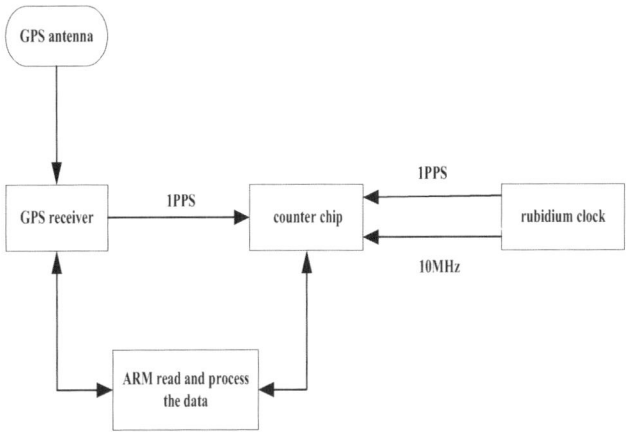

Fig.1. The system structure of GPS common view time transfer receiver

E. Qi et al. (eds.), *Proceedings of the 21st International Conference on Industrial Engineering and Engineering Management 2014*, Proceedings of the International Conference on Industrial Engineering and Engineering Management, DOI 10.2991/978-94-6239-102-4_7, © Atlantis Press and the authors 2015

When the system works, the first is to obtain accurate coordinates of the antenna and keep the coordinates of the receiver constant. The GPS receiver units receive the command sent by ARM chip and return the corresponding information to ARM chip. ARM chip is mainly used for demodulation, processing, measuring the GPS satellite signal, and to obtain ephemeris, the observation data of pseudo-range and clock correction. Rubidium atomic clock PRS10 provides 10 MHZ signal with high stability and 1 PPS signal.

In consideration of the cost of the device miniaturization and the technical costs, we use the time interval measurement chip TDC-GP1 (made by ACAM company of Germany) to realize the digital measurement of the time interval. TDC-GP1 counter chip is used to measure the difference between the output 1 PPS signal by the receiver and rubidium atomic clocks. The time difference between GPST and the specified star clock is obtained by reading the navigation message. And then the time difference between the local time and the specified star clock (REFGPS) is obtained. The above data is just the origin information measured by counter chip. It needs fitting and correction by upper software to calculate the parameters in the standard data file. The working principle of the system is shown as Fig.2.

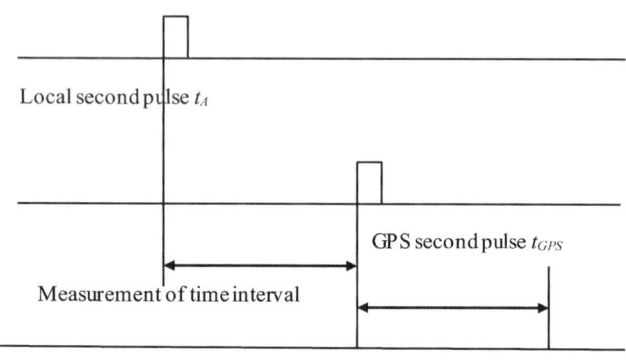

Satellite clock correction ΔT_{SV}

Fig.2. The figure of the system working principle

REFSV is the time difference between the local second pulse of the actual tracking midpoint and the tracking satellite. REFGPS is the time difference between the local clock of the actual tracking length midpoint and the GPS. REFSV and REFGPS are not only the most important parameters in GPS common view comparison, but also the reference data for common view comparison of a dual exchange file [10]. The calculation process of REFGPS is as following.

$$REFSV = t_A - t_{GPS} = (t_A - t_R) + (t_R - t_{GPS}) \qquad (1)$$
$$\Delta T_{SV} = af_0 + af_1(t - t_{OC}) + af_2(t - t_{OC})^2 + \Delta t_r \qquad (2)$$
$$\Delta t_r = F * e * (A)^{(1/2)} \sin E_k \qquad (3)$$
$$REFGPS = REFSV + \Delta T_{SV} \qquad (4)$$

t_A is the local reference time. t_{GPS} is the time of satellite. t_R is the time of receiver. ΔT_{SV} is the clock correction. $(t_A - t_R)$ is measurement of the counter. $(t_R - t_{GPS})$ can be converted by the information of the receiver. Δt_r is

the periodic relativistic correction. The parameters of af_0, af_1, af_2, t_{OC}, e, A, E_k can be obtained by satellite broadcasting data [11]. F is the constant parameter. The data of the B ground sends data to the A ground by GPRS module. Subtracting the two equations, the time difference is available. That is

$$\Delta t_{AB} = REFGPS(A) - REFGPS(B) \qquad (5)$$

III. MANAGEMENT AND THE DATA DESIGN OF THE PROCESSING SOFTWARE

The main job of developing time transfer receiver is the development of its data processing software. The function of the control part mainly includes the receivers' working mode, real-time data collection of satellite ephemeris and the time interval counter. The whole system running time keeps consistent with the international rules of common view observation time. Fig.3 is the processing flow of common view comparison system.

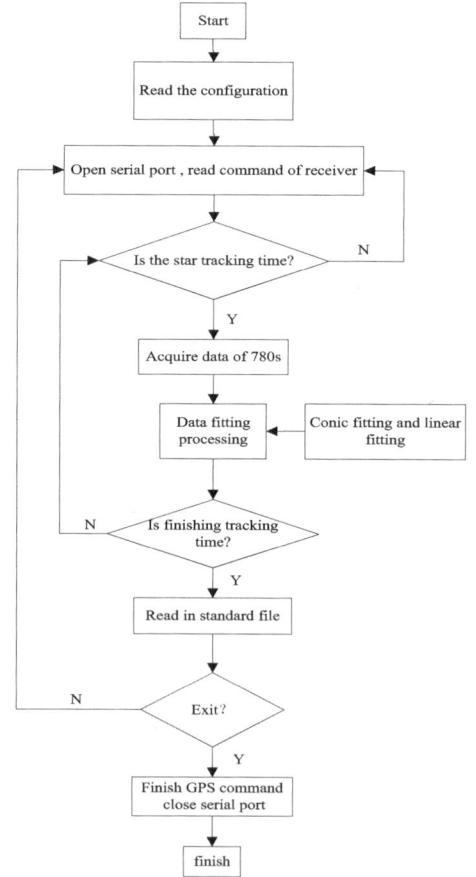

Fig.3. Software flow

Due to the influence of various measurement errors and measurement noises, the precision of common view system can't reflect the accuracy of atomic clock itself. Hence, to deal with the data can improve the alignment precision of GPS common view. The data processing software is to have the original data produced by GPS common view system necessarily processed and stored as a standard file formats. Its main function is to do fitting calculation of the measurement and correction of all kinds of time delay, for example ionosphere delay correction, troposphere delay correction, relativistic correction, satellite clock correction and the SAGNAC effect correction, etc [12]. The following is the main process of the software.

1) The data processing software sends commands to the receiver according to certain format, achieving the working mode of receiver, setting the output data and the signal type. The received data is encoded according to the fixed format, and the software can decode the data.

2) The data processing software can achieve real time data acquisition, including the measurement of counter chip, quantization error, ephemeris, and the observation data of pseudo-range and so on.

3) Because of the hardware limitation of the Resolution T receiver itself, there is a difference of about 15ns between the given physical second pulses and the actual second pulses. The receiver can provide a compensation value to modify this value, making the data smooth.

4) We put each time of 780s for tracking the satellite divided into 52 for 15 seconds, of the 15 number made quadratic least squares curve fitting. Then we take every midpoint numerical number of 15s and set numerical linear fitting of a total of 52 figures. The corresponding result of the fitting is the value of REFGPS for 389.5th seconds. The main purpose of the application of least square method is to make the variance of measurement results and the sum of squared residuals as a minimum.

5) The data processing software can complete correction of the GPS signal path delay. Mainly include the ionosphere, troposphere delay correction, periodic relativistic correction, geometric distance correction and clock correction [13], etc.

IV. THE RESULT OF ZERO-BASE LINE COMMON VIEW COMPARISON

Experiment of zero base common view comparison was carried in National time service center. Two receiver antennas were set on the roof of the research building where is without any shelter around but may be limited by some small interference radio frequency devices. Two sets of equipment are adopted 1 PPS and 10 MHZ which is offered by the same rubidium atomic clocks as a reference signal. The two sets of system are put in the same laboratory using the same software, also the same ionosphere and troposphere correction model. Equipment connection is shown in Fig.4.

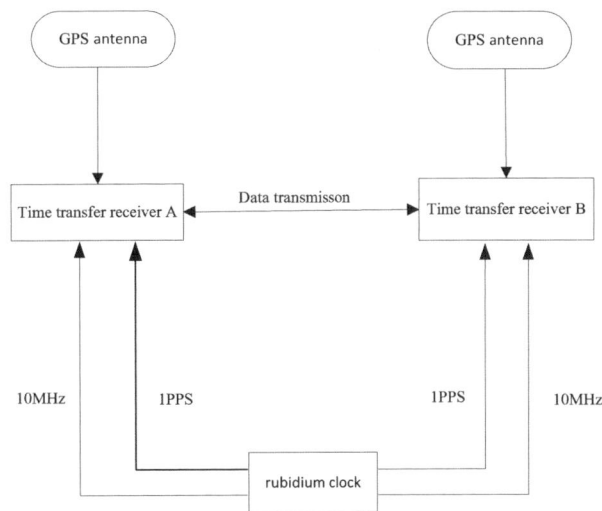

Fig.4. Zero-base line common view comparison method

Zero-base line comparison approach is to point to put the two time transfer receivers in the position close enough and to orientate the coordinates of the two antennas precisely [14]. The two receivers use the same time and the same frequency signal as reference signal.

Zero-base line common view comparison can eliminate the influence of ephemeris error, ionosphere delay error [15], troposphere delay error and the coordinate error of receiver antenna, primarily noises and the internal delay change of receiver system. Consequently, zero-base line comparison results can be used to detect the stability of receivers' noises and inner delay.

In the condition of having no accurate orientation for the antennas' coordinate, we can acquire the zero-base line alignment accuracy by subtracting the parameters *REFGPS* of standard files outputting by the two receivers and fitting the data by the least square method in the same satellite, the same MJD and at the same start-tracking time. As shown in Fig.5 that the comparison result of the two devices in the MJD (56819-56820) has no evident systemic change, with the data' RMS is 3.8ns. Thus it can be seen that our equipment is equally matched the international advanced common view receiver so far.

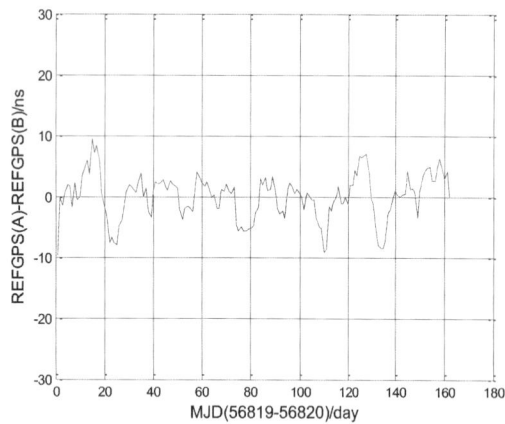

Fig.5. The comparison result of Zero-base common view

V. CONCLUSION

This paper introduces that we use the time interval measurement chip based on ARM platform instead of the traditional time interval counter in common view system, realizing device miniaturization. The preliminary Zero-base line results show that alignment accuracy of GPS common view receiver is about 3.8ns, which can realize the high precision time synchronization between base stations. Because of not having the receiver orientated precisely, it may affect the alignment accuracy to a certain extent. Also it may have the effect of random error for not having the original data smoothing treatment. We need to do further research of the data processing algorithms and to measure the coordinates of the antenna more precisely in order to further improve the alignment accuracy of the common view system.

REFERENCE

[1]   D W Allan, C Thomas, "Technical directives for standardization of GPS time receive software" [R]. Metrologia, 1994, 31:69-79.

[2]   David Allan, Marc Weiss, "Accurate time and frequency transfer during common-view of a GPS satellite" [A], Proc. 34th Frequency Control Symposium [C], 1980: 332-348.

[3]   ARIAS F, "BIPM comparison of time transfer technical" [C]. Frequency Control Symposium and Exposition 2005, 2005: 4.

[4]   ICD-GPS-200RC [S]. U.S. Naval Observatory, 2000, 4.

[5]   Qi-feng XU, "GPS satellite navigation and precise positioning" [M] (in Chinese), Beijing: People's Liberation Army publishing house, 1994.

[6]   Yue Zhang, Xiao-xun Gao, "Research on timing parameters for GPS common-view method" [J] (in Chinese), Measurement Journal 2004, 25(2), 167-170.

[7]   Lewandowski W, Azuobib J, "Time transfer and TAI" [C], In: Proc. IEEE International Frequency Control Symposium and Exhibition, 2000.

[8]   Brodford W. Parkinson, "Global positioning system: theory and application" [M], American Institute of Aeronautics, 1995.

[9]   Xu-hai Yang, "New GPS common-view time transfer system based on Motorola vp oncore receiver" [J] (in Chinese), Journal of Electronic Measurement and Instrument, 2005, 4, 15-19.

[10] Han Zhang, Yuan Gao, Jiang-miao Zhu, "Common-view comparison system based on the EURO-160 P3 code GPS receiver" [J] (in Chinese), Electronic Measurement Technology, 2007.11.

[11] Yu-ping Gao, Yi Qi, Zheng-ming Wang, "Multi-channel time transfer receiver NTSCGPS-1 development and testing" [J] (in Chinese), Global Position System, 2004.2.

[12] Gerard Petit, Claudine Thomas, Philippe Moussay John A, Davis, Mihran, Miranian, "Multi-channel GPS common view time transfer experiment: first result and uncertainty study" [J].

[13] Assistant Secretary of Defense, "Global positioning system standard positioning service performance standard", 2001, 4: 14-15.

[14] Xu-hai Yang, Yong-hui Hu, "Motorola vp oncore GPS receiver date receiving software" [J] (in Chinese), Time and Frequency, 2001, 6: 59-62.

[15] Jefferson D, Lichten S, Young L, "A test of precision GPS clock synchronization", In Proc.1996 IEEE Freq, Contr. Symp, 1996: 1206-1210.

Research on Random Mixed-model Two-sided Assembly Line Balancing Using Genetic Algorithm

Lei Wang[*], Kai-hu Hou, Wei-zhen Liao, Zheng-mei Jie, Cheng Chen, Ying-feng Zhang

College of Mechanical and Electronic Engineering, Kunming University of Science and Technology, Kunming, P.R. China

(498515366@qq.com)

Abstract -This paper presents a new mathematical for random mixed-model two-sided assembly line balancing. To minimize the cycle time of random mixed-model two-sided assembly line with the given number of workstation, the random mixed-model two-sided assembly line balancing problem of type II is studied. The influence of random factors on assembly line was convertedinto process time influence in the paper. Combined with random changes in product demand of different product, the united comprehensive process time was worked outby the method of weighted average. According to the comprehensive process time, the processeswere rearrangedto different workstations in the paper. To minimize the cycle time which as the objectives of the mathematical programming model, with constraints of process priorities, operational orientation and others, the genetic algorithms is used to work out the mathematical model. An instance of mixed-model two-sided automobile assembly line was given, which was optimized by the algorithm for optimization and compared the results of optimization before and after. The results verify the effectiveness of the algorithm for solving mixed-model sided assembly line balancing problem.

Keywords-Assembly line balancing,genetic algorithms,two-sided assembly line

I.INTRODUCTION

Assembly line is a sequence of several connected workstations convened in order to assemble different parts to final products[1-3]. The two-sided assembly line[4, 5] is a special type of assembly line in which tasks are assigned to workstations placed on both side of the assembly line. This type of assembly systems are used for manufacturing systems with large-sized products such as cars, buses, and trucks[6]. Mixed-model assembly lines [7, 8] allow for the simultaneous assembly of a set of similar models of a product, which may be launched in the assembly line in any order and mix. Random assembly line is that each process time is an uncertain value in the entire assembly procedure, and usually can only be used to express in a randomly distribution function. In other words, the influence of random factors on all the assembly line can be converted to impact on the process time. Assembly line balancing problem aims to assign tasks to workstations in order to balance the workload of the workstations. This paper mainly researches on mixed-model two-sided assembly line balancing problem under the influence of random factors.

II.THE ASSEMBLY LINE DESCRIPTION

In random mixed-model two-sided assemble line, processes are usually arranged on the right-side or left-side assembly lines and the longer operation time which adding all processes time in a work position on each side determines the cycle time of this work position. And the longest operation time position among all work positions determines the cycle time of the whole assembly line. The two-sided assembly line problem of this paper has the following assumptions[9]: ① allinputparametersoftheproblemareknown; ② operation idle time on a workstation cannot turn to another workstation;③one workstation can only complete one assembly task by the fixed operator.

Inordertofacilitatetheexpression,itwillbelistedtherele vantvariablesfollowing:tasksetPandP={1,2,…i}; $P_i = \{t_i$, d} meanstheiprocessandeachprocesscontainsthetimeanddir ectionofthesetwoproperties; P_E means the task canbeassignedtotheleft-sidedor theright-sidedof theassemblylineandassemblyline, $P_E \in P; P_L$ and P_R means task is assignedtotheleft-sidedandright-side of theassemblyline respectively, P_L , $P_R \in P; \mu_i$ and σ_i^2 arerespectivelythei-thaprocessofNproductsexp ectedvalueandvariancetotheaverageoperationtimeandsetlo cationjof P_j ontheassignedtask; T_j , $T_{j'}$ isthetotaloperationtimeofjandj'; P(i) is taski'alltheform ersequencesetoftask; B_j is aBooleanvariablewhichmeansst ationj'available,if P_j is not null $B_j = 1$, else $B_j = 0$;disoperatingposition, $d \in \{L, R, E\}$,Listheleft,Ristheright,Eis that whethertheleftandrightsidesarefeasible;Sisrandomvariable ,itsvaluebelongstotherange(0,1); x_{nik} is decisionvariable,w hichmeanstaskiofnproductsassignedworkstationserialnumb er; x_{wjd} meansthenumberofassignedtothepositionjandstatio nw,operatingbearingtaskd; q_n is the proportion of product n in the total demands.

III.MATHEMATICAL MODEL

To minimize the cycle time with the given workstation number, the mathematical model is built which is based on the satisfaction of two-sided assembly line's constraints [10].

(1)The process cannot be divided, and can only be assigned to a station.

E. Qi et al. (eds.), *Proceedings of the 21st International Conference on Industrial Engineering and Engineering Management 2014*, Proceedings of the International Conference on Industrial Engineering and Engineering Management, DOI 10.2991/978-94-6239-102-4_8, © Atlantis Press and the authors 2015

$$\begin{cases} P_J \cap P_I = \emptyset \\ \cup_k P_k = I, \quad k = 1,2,\dots,M \end{cases} \quad (1)$$

OR

$$\sum_{k=1}^{M} x_{nik} \ , \quad i = 1,2,\dots,I \quad (2)$$

(2)Process priority relationship constraints

$$\sum_{k=1}^{M} (k x_{njk} - k x_{nik}) \geq 0 \quad (3)$$

iisj'stightprocess,ortheorderprocess.

(3)Therelationshipbetweenthedemandsforeachtype

$$\sum_{n=1}^{N} q_n = 1 \quad (4)$$

(4)Anyproductofanylocationtimepluswaitingtimeislessthanthebeat

$$\max(T_{nk}) + \sum_{i=1}^{I} x_{nik} D_{nik}$$

$$= \max\left(\sum_{i=1}^{I} x_{nik} D_{nik}\right) + \sum_{i=1}^{I} x_{nik} D_{nik}$$

$$= \sum_{i=1}^{I} (\mu_{ni} x_{nik} + x_{nik} D_{nik})\delta \leq C \quad (5)$$

Hypothesis $t_{ni} \sim N(\mu_{ni},$
$\sigma_{ni}^2)$, D_{nik} isNkindsofproductprocessiassignedtoworkstationK,duetoitsassociatedwithlocationinsidetheworkprocedure,thepreambleoftheprocessiandisoperationiscompletedtimelaterthanprocessibegantoworktime,leadtoprocessbeforestartingtheprocessineedtowaitforthetimeinit,i=1,2,…,I; n=1,2,…,N; k=1,2,…,K.

(5)Averagecycletime

$$\overline{C} = \frac{T}{D} = \sum_{i=1}^{I} \sum_{n=1}^{N} q_n t_{ni} x_{nik} = \sum_{i=1}^{I} \sum_{n=1}^{N} \lambda_n \mu_{ni} x_{nik} \quad (6)$$

$\overline{C}$ isthefirstkstation'snkindsofproducts'averageproductiontime

(6)Theconstrainedbearing

$$x_{ni} \bmod 2 = \begin{cases} 0, & k \in I_R \\ 1, & k \in I_L \end{cases} \quad (7)$$

(7)Objectivefunction

$$\text{Min} C \quad (8)$$

IV. THE DESIGN OF THE GENETIC ALGORITHM

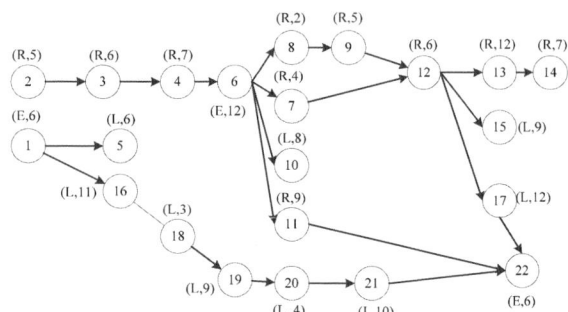

Fig.1. Precedence Relationship Chart of Example

In order to facilitate the expression of entire genetic algorithm design process, this paper will combine an example to genetic algorithm [11, 12] design. The priority process relation diagram show in Fig.1.

(1) Designcode

The traditional gene expression matrix for 1*n, but different encode is used in the paper. The traditional chromosome is represented by only a row. In this paper, chromosome is can be explained as a matrix with a 2*n matrix, the process that matrix of the first line represents the work elements; The second line represents the corresponding process orientation constraints, and its value is 0 (the process of the actual distribution to the right) and 1 (the process of the actual distribution to the left). An example is shown in Fig.1:

The assembly line process {1,5,16,6,8,1,0,18,19,20,21,17, 15} on the left side of the assembly line operation; The other assembly line process {2,3,4,11,9,7,12,13,14,22} on the right side of the assembly line operation, as shown in Fig.2.

1	5	2	3	4	16	6	8	11	9	10	18	19	20	21	7	12	13	17	14	22	15
1	1	0	0	0	1	1	1	0	0	0	1	1	1	1	0	0	0	1	0	0	1

Fig.2. AChromosome of the Example

(2)Generatetheinitialpopulation

The generation of initial population is generated randomly. Namely on the basis of the coding, coding of chromosome is not the same each time, so the combustion generated inside the body goes into a matrix, comprise the initial population, the size of the initial population can be set according to the needs of research[13].

(3)Decoding procedures

In the paper, genetic algorithm is used to solve the model with mixed-model flow two-sided assembly line balancing problem, as the name implies , decoding is after coding, below the premise that in the previous step and coding, decoding of the initial population.

Step1:According to the existing conditions calculate the beat.

$$C_{min} = \max\left\{ \left|\frac{T_{sum}}{k}\right|, \left|\frac{T_{suml}}{\frac{k}{2}}\right|, \left|\frac{T_{sumr}}{\frac{k}{2}}\right|, \max(t_{ni}) \right\} \quad (9)$$

$$T_{sum} = \sum_{n=1}^{N} \sum_{i=1}^{I} t_{ni} D_n \quad (10)$$

T_{sumr}, T_{suml}

$$\begin{cases} T_{sumr} = \sum_{n=1}^{N} \sum_{i=1}^{I} t_{ni} D_n \\ P(2,i) = 0 \end{cases} \quad (11)$$

$$\begin{cases} T_{suml} = \sum_{n=1}^{N} \sum_{i=1}^{I} t_{ni} D_n \\ P(2,i) = 1 \end{cases} \quad (12)$$

Among them, K is the number of workstations; T_sum is the sum of all process time; T_suml is on the left side of the sum of all process time; T_sumris on the right side of the sum of all process time; t_ni is N kinds of products of the ith a process time; D_n the nth demand of products; P(2 , i) is orientation constraint; P(2 , i)=0 is the right; P(2 , i)=1 is the left.

Step2:Conduct process distribution. Before distribution must determine its position, starting from the first position, one at a time distribution, working procedures are also follow the coding sequence. Distribution of each working procedure, we must calculate the station time, if the current detail process of time and the distribution of time, less than beat the lower limit, the distribution process can continue to this location. Otherwise, we need to open the next station. After opening location decide whether it is the last station, if so, the rest of the contractors that all processes are assigned to the station.

Step3:Calculate Cycle Time. After the completion of all process distribution, calculate the current beat, beat = max (location). Compared to the current beat and lower limit, if the beat is greater than the lower limit, the beat on the lower limit should be according to certain step length. Return to step 2.

Step4: Circulation. Loop Step2 and Step3 until beat is equal to the limit. Finish decoding.

(4)Fitnessfunction

Objective function is minimize the production rhythm for bilateral minimize beats the optimal chromosome as an assembly line, assigned to a workstation or so on both sides of the working hours is equal or close as far as possible, each step of process time if without a bigger difference, the best number is assigned to the two sides of the process as far as possible close.

So the fitness function, as shown below:

$$fitness = \frac{100}{C} + \frac{l}{\gamma} \qquad (13)$$

OR

$$fitness = \frac{100}{C} + \frac{\gamma}{l} \qquad (14)$$

Amongthem,Cisgetbeat;listhenumberofprocessesinth eassignedtotheleft; r isthenumberofprocessesinthedistributiontotheright.Weuse formula13when $l \le \gamma$;when $l \ge$ γ,weuseformula14tocalculate.

(5)Operatorselection

The classic roulette wheel selection method is used in the paper. By calculating the fitness of each individual fitness value of the sum of the proportion size to determine its descendants will remain. Individual the bigger the fitness of the selected the greater the probability that the more likely, it is to be preserved, On the other hand thrown away.

(6)Crossoveroperator

The way of simple random cross crossover operation is adopted, and the specific steps are as follows:

Step0:Read a pair of chromosomes from the initial population, P1 and P2, both matrix for 2 * i matrix, the first behavior process of real Numbers, the second behavior orientation constraint representation of a real number which i is the number of process.

Step1:Generate a random number k between 1 and i, k - 1 gene P1 and P2 reserved before the information remains the same.

Step2:k of P1 and P2 to the i a gene, the essence of the cross is after the P1' i - k + 1 gene in accordance with the order of the P2 to recode. In the same way, after the P2' I - k + 1 gene carried out in accordance with the order of P1 recode.

Step3:Cross generation and P1 and P2 there are both similarities and differences between the two chromosomes of C1 and C2.

Combining with the example, arbitrary generated P1 and P2, random number k = 6, intersecting C1 and C2, before and after the cross the following as shown in Fig.3 and Fig.4:

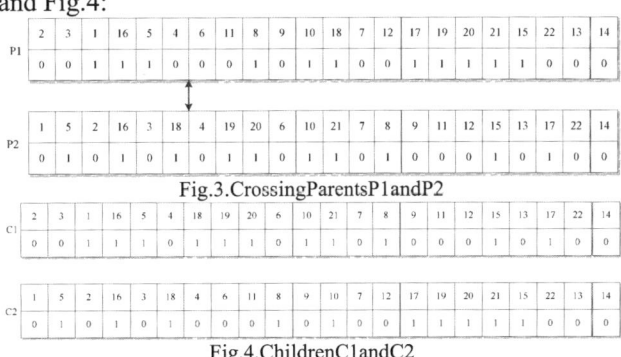

Fig.3.CrossingParentsP1andP2

Fig.4.ChildrenC1andC2

(7)Mutationoperator

The purpose of variation is to guarantee the diversity of population, makes each kind of solution can be produced, thus increasing the number of the feasible solution. In this paper, the variations of the specific steps are as follows:

Step0:Read from the initial population inside burning body P1.

Step1:Generate a random between 1 and i positive integer k.

Step2:P1 before k - 1 gene will retain the original information.

Step3:P1 of i - k + 1 after redistribution in the coding way get mutated chromosome C1.

The variations of before and after mutate results are shown in Fig.5:

Fig.5.ChromosomesThatBeforeandAfterMutain

(8)EndCondition

Programterminationconditionsforoperation,thebeatc onvergence,ortoreachacertainnumberofiterations,thedeter minationofthenumberofiterationsthroughmanytimestorunt heprogram,finallysettled.

(9) Validatearithmetic

The designed algorithm is feasible and effective, compared with other algorithms will be able to get more optimal solution, the algorithm validation.

Take the example verification[14], and comparing the calculation results of the original documents.

The original documents obtained by the beat and obtained by the author himself are 660 (with a working procedure of finish time for 660, according to the principle of working procedure is divided into the smallest rhythm is the process time), but the optimized smooth index fell from 81.07 to 76.99 is showed in Table

I, as you can see, the designed genetic algorithm is effective.

V. EXAMPLE

Former Y automobile manufacturing company C on an assembly line is equipped with a long period of 10, 20 workstations. This enterprise production program of its capacity for H = 100000 units/year; Working days D = 251 days a year; The number of hours per shift work T = 8 h; The current utilization of man-hours to Y = 0.87; N = 2 two shifts a day. Calculated in radix F =N*D*T*Y =2* 240*8*0.87 =3333 (h); Production rhythm t = 60 f/H = 60 * 3333/100000 = 2 (min/Taiwan). C assembly line assembly diagram show in Fig.6:

TABLE I
COMPARISON OF THE TWO SOLUTIONS

Workstation (side)	Solution of the original literature		Solution of Genetic Algorithm	
	Processes on each workstation	*Time of workstation(s)*	*Processes on each workstation*	*Time of workstation(s)*
1(1st left)	2, 4, 12	591	2, 8, 9, 21, 25	631
2(1st right)	1, 10, 24, 5, 21	657	1, 10, 12, 22, 24	611
3(2st left)	3, 15, 17, 18	650	4, 6, 13, 17, 19, 27	622
.	.	.	.	.
.	.	.	.	.
15(8st left)	56, 58	508	56, 58	508
16(8st right)	55, 57	508	55, 57	508

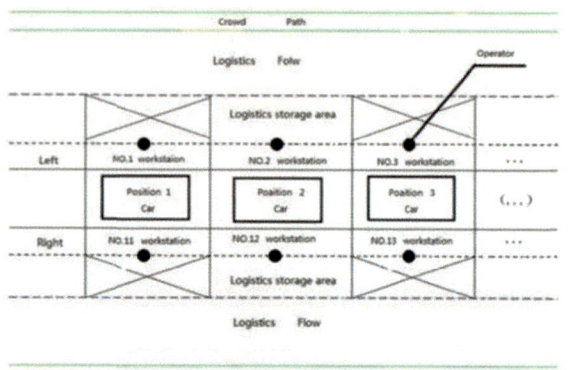

Fig.6.Assembly Schematic Diagram of C Line

As can be seen from the Fig.7, the existingscheme has many unreasonable place, the whole assembly line is not balanced, in the present set of beats C = 120 S fluctuating amplitude is bigger, small to 75.84 S to 75.84 S, which is beyond the beat of the workstation is the 11, so will often lead to stop line, productivity is low.

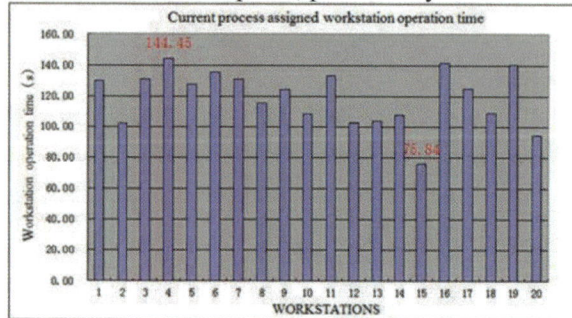

Fig.7.TheBarCharofWorkStationTimeaboutCurrentOperationScheme
Cmixed-modelflowbilateralassemblylineproductiono fA,B,Cthreeproductsneed110procedures,takeworksamplin

gsurveyonthisassemblyline20workstationrecord,tomeetthe requirementsoftheprecision,selectabsoluteaccuracyE=3%, calculatethenumberofobservationsfor464times,andthereco rdeddataaredatafittinganalysisworkstationhomeworktimeo beythenormaldistribution.Duetotheoperationbetweenthepr ocessareindependent,sotheprocessofoperationtimeis alsoobeythenormaldistribution,takingthemeanoperationti meofeveryworkasaproduct,processtime,asshowninTableII :

TABLE II
PROCESS TIME OF EVERY PRODUCTION ABOUT C ASSEMBLY LINE

Process number	Name of process	Process time of different products(s)		
		A	B	C
1	Fix front decorative plates	18.07	17.82	26.17
2	Assemble front-left plates	5.54	26.55	17.78
3	Assemble front-right plates	25.82	20.17	6.77
.	.	.	.	.
.	.	.	.	.
.	.	.	.	.
108	Paste engine warning signs	17.75	22.90	6.48
109	Paste brake warning signs	19.35	27.80	25.25
110	Fix front door armrests	11.84	21.76	15.48

Cbelongstothemixed-modelflowassemblyline,thepro ductionofA,B,Cthreevarieties,thedemandofeachvariety,ift heproductioncycleforAmonth,throughtheworksamplingan dthemethodofmathematicalstatisticsfoundthatthedemando fproductA,B,C,respectively,tooeylambdaA=300,lambda B=200,lambdaC=360Poissondistribution(specificdataproc essinginthispaper,donotdothis).UsingMATLABaccording tothedistributionofrandomlygeneratedA,B,Cthreevarieties ofdemand,productionMtimes.Astheresearchobjectofmixe d-modelflow,combinedwiththedemandoftheuncertaintyan dthevarietiesofprocesstimeisdifferent,sotocalculatetheinte gratedprocessoftime,toconverttherandomuncertainty,sotha tthefollow-upwork.Accordingtotheactualsituation,compre hensiveprocessoperationtimecalculationformulaofsuchast ype(15):

$$t_I = \frac{1}{M} \sum_{m=1}^{M} \frac{\sum_{n=1}^{N} t_{ni} D_{nm}}{\sum_{n=1}^{N} D_{nm}} \qquad (15)$$

Whichiisprocess,itsscopefori $\in$ [1,110].Nforthevarietiesofproducts,itsscopeisn $\in$ [1,3].mforthefirstmtimecalculation,itsscopeism $\in$ (1,m).Thisarticletaketherandomlygenerated100groupsofd ata,namely,M=100.Thispaperrandomlyproduce100setofda ta(duetotherandomnessofthedataofproductionandproducti ondataaregenerallytovaryeachtime,therefore,inthispaper,t hestochasticproductiondataisnottodoadetaileddescription), throughtheformula110proceduretheaverageoperationtime calculationintocomprehensiveprocessoperationtime.Assho wninTableIII:

TABLE III
COMPREHENSIVE PROCESS TIME

Process number	Name of process	Comprehensive process time(s)
1	Fix front decorative plates	21.40
2	Assemble front-left plates	15.49
3	Assemble front-right plates	16.54
.	.	.
.	.	.
.	.	.
108	Pasteengine warning signs	14.21
109	Pastebrake warning signs	23.77
110	Fix front door armrests	15.64

A.genetic algorithm
1)Algorithm process

Step1:Determine the size of the population of N, use the initial population is randomly generated way. Initialization is completed; compare all the individual species, according to its fitness value from big to small.

Step2:Use the method of roulette wheel selection. Select two individuals in the population as a parent, crossover and mutation operation to produce two individuals and deposit to the new populations. Repeat the process until a new population of individual number N.

Step3:The new offspring population set to replace the original population, according to the fitness value from big to small order.

Step4: Inspection algorithm termination conditions (set algebra) evolution, if satisfied, the output termination algorithm and the optimal solution. Otherwise, return to Step2.

This paper run genetic algorithm to get beat mean and minimum track as shown in Fig.8.

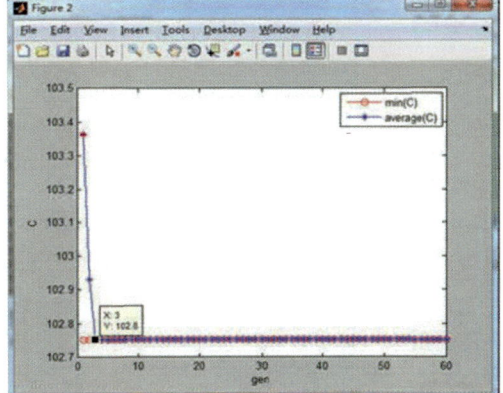

Fig.8. Tracing Figure for Average Cycle and Min Cycle in the Computation Process

2)After using genetic algorithm in the MATLAB software programming to solve the target:

```
function [obfv]=obfv(Chrom,T_average,C0)
[m,n]=size(Chrom);
w=1;
while w<=m/2
P0(1,:)=Chrom(2*w-1,:);
P0(2,:)=Chrom(2*w,:);
obfv(w,:)=decoding_main(T_average,C0,P0);
w=w+1;
end
```

The optimized solution was calculated as shown in Table IV.

TABLE IV
SOLUTION TO BALANCING PROBLEM OF TYPE II FOR LINE C

Position	Workstation	Process time(s)	Processes on workstation	Position	Workstation	Process time(s)	Processes on workstation
1	1	90.3504	1, 2, 4, 6, 8	6	6	100.2353	53, 57, 58, 59, 60, 65
	11	95.6772	3, 5, 7, 9, 11, 13		16	98.0637	72, 73, 74, 78, 79
2	2	101.0747	10, 12, 14, 16, 17, 18, 19, 20	7	7	95.853	66, 67, 68, 69, 71, 75
	12	101.8354	15, 21, 22, 23, 25		17	97.9428	84, 86, 89, 91
3	3	89.0099	24, 26, 28, 32, 36	8	8	86.7575	76, 77, 80, 81, 82, 83
	13	102.1121	27, 29, 31, 33, 35, 47		18	88.6864	93, 94, 95, 96, 97
4	4	100.8707	30, 34, 38, 40, 46, 48	9	9	88.2129	85, 87, 88, 90, 92, 99
	14	96.581	37, 39, 41, 43, 45, 49, 54		19	102.436	98, 100, 101, 102, 103
5	5	97.3211	42, 44, 50, 51, 52, 55	10	10	33.0291	106, 107
	15	94.8785	56, 61, 62, 63, 64, 70		20	93.7672	104, 105, 108, 109, 110

B. The analysis of C assembly line beats contrast before and after optimization

According to the Fig.7 and Fig.9, we can see each workstation is the production time before optimization fluctuation is bigger from the whole, the optimized production time is relatively balanced for each station, always stay at around 100 have smaller amplitude fluctuations, except individual station (station 10), and the optimized time of the work at various workstations are much lower than the original.

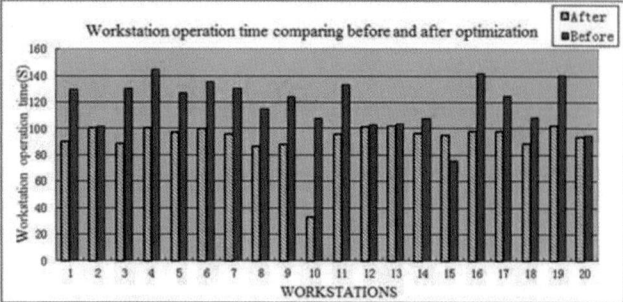

Fig.9.The Bar Char of WorkStation Time after Optimizing

The original time biggest workstation, namely the beat of 144.45 S; Optimized workstation biggest time that beat reduced to 102.44S, obtained the effective optimization; And the assembly line balance rate was improvedfrom 88.18%to 95.75%; At the same time it also shows that this algorithm the effectiveness of the mixed-model flow two-sided assembly line balancing problem.

Processes to different workstations, minimize assembly beat as objective function, and the corresponding number sequence model is established. Through reasonable design of genetic algorithm is used for its operation, it is concluded that the Y car model with mixed-model flow bilateral beats from 144 s on an assembly line, optimization in the 104 s, the beat was reduced by 27.78%, which proved the validity and reliability of the algorithm to solve the model with mixed-model flow bilateral assembly line balance problem provides a new way to provide the basis for the actual production operation

VI.DISCUSSION

Inthispaper,therandommixed-modeltwo-sidedautom otiveassemblylinebalancingproblemofYCompanyisthetar gettostudy.Random factors make the process time obey the normal distribution and convert the influence of random demands on assembly line balancing into process time influence. Considering those influences, this paper use the method of weighted average to revise the process time and the uncertain process time was converted into certain comprehensive process time.Under the condition of the workstation number, that process to satisfy the constraint conditions and reasonable distribution of different workstations, minimize assembly cycle time[15] as objective function, and use genetic algorithm to design and calculate the results, which provide a new way to solve the random mixed-modeltwo-sided assembly line balancing problem andprovide the basis for the actual production operation.

REFERENCES

[1] Ana Sofia Simaria, Pedro M. Vilarinho, "A genetic algorithm based approach to the mixed-model assembly line balancing problem of type II",Computers & Industrial Engineering 47(2004) 391-407.

[2] P.Th.Zacharia, Andreas C. Nearchou, "Multi-objective fuzzy assembly line balancing using genetic algorithms", Journal of Intelligent Manufacturing (2012) 23:615-627

[3] LaleOzbakir, Pinar Tapkan, "Balancing fuzzy multi-objective two-sided assembly lines via Bees Algorithm", Journal of Intelligent & Fuzzy System 21(2010) 317-329.

[4] UgurOzean, Bilal Toklu,"A tabu search algorithm for two-sided assembly line balancing", International Journal of Advanced Manufacturing Technology (2009) 43:822-829.

[5] Er-Fei Wu, Ye Jin, Jin-Song Bao, Xiao-Feng Hu, "A branch-and-bound algorithm for two-sided assembly line balancing", International Journal of Advanced Manufacturing Technology (2008) 39:1009-1015.

[6] Masoud Rabbani, Mohsen Moghaddam, and NedaManavizadeh, "Balancing of mixed-model-model two- sided assembly lines with multiple U-shaped layout", International Journal Advanced Manufacture Technology (2009) 59:1191-1210.

[7] S.AfshinMansouri, "A Multi-Objective Genetic Algorithm for mixed-model sequencing on JIT assembly lines", European Journal of Operational Research 167 (2005) 696-716.

[8] A. NoorulHaq, J. Jayaprakash, K. Rengarajan, "A hybrid genetic algorithm approach to mixed-model assembly line balancing", International Journal of Advanced Manufacturing Technology (2006) 28:337-341.

[9] AdilBaykasoglu, TurkayDereli, "Two-sided assembly line balancing using ant-colony-based heuristic", International Journal Advanced Manufacture Technology(2008) 36:582-588.

[10] Peng Hui, Xu Keli , Si Zhanhua, "Multi-objective optimization for mixed-model assembly line balancing problem based on GA", Modern Manufacturing Engineering 2011 No.11.

[11] Gregory Levitin, Jacob Rubinovitz, Boris Shnits, "A genetic algorithm for robotic assembly line balancing", European Journal of Operational Research 168 (2006) 811-825.

[12] Cao Zhenxin, Zhu Yunlong, "Application of Multiple Objective Genetic Algorithms in Sequencing Mixed Model Assembly lines" Computer Engineering (2005) Vol.31 No.22.

[13] Liang Hou, Yongming Wu, and Rongshen Lai, "Product family assembly line balancing based on an improved genetic algorithm", International Journal Advanced Manufacture Technology(2014) 70:1775-1786.

[14] Shi Peikuo, Fang Yexiang, and Huang Xiuling, "Assembly Line Balancing in a Toy Manufacturing Company", Industrial Engineering Journal 2009, 12(6): 127-130.

[15] P. Ji, M.T. Sze, and W.W. Lee, "A genetic algorithm of determining cycle time for printed circuit board assembly lines", European Journal of Operational Research 128 (2001) 175-184.

The Capacity Preparation in the Two-period Supply Chain with Remanufacturing Products and IOT

Fu-rong TAN, Han-jiang ZHANG*, Hong-xia ZHONG, Jia-yu ZHANG

School of Economy and Trade, Hunan University, Changsha, China

(zhanghj519@hnu.edu.cn)

Abstract - **In this paper, we analyze the production capacity and production decision issues in remanufacturing situations. A two-period model is introduced, in the first period, manufacturers produce and sell new products by using raw materials; in the second period, manufacturers produce new products and remanufactured products simultaneously by using used products recycled at the end of the first period. Under the circumstances that whether the demand for the second period products is greater than/less than the amount of recovery products which can be used for remanufacturing is uncertain, using manufacturers' capacity preparation in the two periods to get the optimal decision and the profit of manufacturers. By the way, we analyze the impact of each parameter on manufacturers' optimal decision variables. According to the optimal decision variables and profits in different conditions, we make parameter value simulation analysis.**

Keywords - **Capacity preparation, logistics and supply chain management, production decision, remanufacturing**

I. INTRODUCTION

Product recycling and remanufacturing issues have become hot issues of closed-loop supply chain researches in recent 10 years, among which some researchers don't take capacity preparation factors into consideration. Ferre and Swaminathan (2006, 2010) study two-period and multi-period models with one OEM and one independent remanufacturer, they investigate the effects of various parameters on equilibrium prices, profits and remanufacturing activities, analyze the process of manufacturers manufacture new products in the first period then not only manufacture new products but also produce remanufactured products simultaneously in the second period, to study remanufacturers' decision-making problems when there exist differences between new products and remanufactured products [1-2]. Cao et al. (2010) set up a two-stage dynamic game model, using a single cost function class to study the competition between the manufacturers and remanufacturers when they simultaneously enter a market with consumption transferring [3]. Majumde and Groenevelt (2001) [4], Ferguson and Toktay (2006) [5], Mitra and Webster (2007) [6] set up two-period models to study manufacturers' remanufacturing activities too. Based on the above research backgrounds and features of realistic products' life cycle, we divide it into a two-period model: in the first period, manufacturers just use new materials to manufacture and sell new products, products are in their initial sales and some products will retire for reaching their life cycle at the end of the first period; in the second period, the quantities of EOL (End Of Life) products available for recycling are the sales of the first period, manufacturers separately use new materials and recycled

EOL products to produce new products and remanufactured products and then get them onto the market, the upper limit of remanufacturing capacity is the amount of recyclable retired products in the first period, i.e. sales.

Among them, studying remanufacturing activities in consideration of capacity preparation has also been a hotspot in recent years. Franke et al. (2006) consider the remanufacturing capacity preparation issues in the mobile phone industry; they introduce a linear programming model to study issues of capacity preparation and production method planning [7]. Georgiadis et al. (2006) take the product life cycle and the recovery mode into account, analyzing remanufacturing activities in capacity shrink and expansion conditions; finally, they use system dynamics method to analyze the capacity planning problems [8]. Debo et al. (2006) take product life cycle and capacity preparation into consideration and study the impacts of product diffusion rate on the capacity requirements of new products and remanufactured products, whose capacity preparation demand are analyzed specifically [9]. Rubio S and Corominas (2008) find by modeling that the capacity preparation in manufacturing and remanufacturing activities can be adjusted in a given demand environment [10]. Yun Liang et al (2013) mainly study the optimal pricing strategy under the premise that closed-loop supply chain are faced with capacity constraints and limited recycling material and the new products can replace remanufactured products, then calculate and analyze the profit level of the supply chain participants and the whole supply chain [11]. Li et al. (2012) analyze a two-stage supply chain, in conditions that the capacity is limit and the manufacturer is in a dominated state, when the manufacturer's maximum production capacity reach the threshold, the entire supply chain are not affected by the capacity constraints; when the manufacturer's maximum production capacity is smaller than the threshold, the manufacturer raises the wholesale price, the retailer reduces the order quantity and the manufacturer's profit, the retailer's expected profit and the expected profit of entire supply chain all decrease [12]. Caner et al. (2013) set up a two-period model that manufacturers produce new products in the first and second period, products in the second period are remanufactured based on waste products recycled at the end of the first period, then further investigate the capacity and production decisions on remanufacturing plans under the assumption of product homogeneity [13].

We can see from related researches above that capacity preparation issues of manufacturers are valuable and are worthy of researching. Manufacturers need to produce new products and remanufactured products

simultaneously, while the two products will share labor resources, production facilities, equipment and warehouse space etc. How will manufacturers distribute the capacity among new products and remanufactured products when the demand and the supply are not matching? What's the impact of capacity preparation size on manufacturers' product quantity and profit? In view of these questions, a two period model is introduced to analyze both manufacturers' new product production in the first period and manufacturers' remanufactured products production and new products production simultaneously in the second period. Finally, we explore the optimal relevant decision variables of the two periods.

II. PROBLEM DESCRIPTION AND SYMBOLS

A. Problem Description

This paper analyzes a manufacturer produce new products (umbrella) and get them onto the market in the first period according to its capacity preparation, the length of this period corresponds to the working life of the product(umbrella); at the end of the first period, some parts (umbrella stand) of the product would be recycled; what's more, in the second period, manufacturers continue to produce new products and remanufactured products simultaneously (using recycled umbrella stand and cloths prepared by themselves to conduct remanufacturing activities) according to customers' demand.

To analyze and explain the model more clearly, some assumptions are introduced before presenting the model.

(1) In the first period, manufacturers produce and sell new products (umbrella) according to their capacity preparation, all capacity reserve are used for the production of new products; in the second period, manufacturers produce new products and remanufactured products(umbrella) simultaneously by using waste products (umbrella stand) recycled at the end of the first period.

(2) Manufacturers make decisions under the premise of completely knowing consumers' demand, manufacturers are completely rational and the goal of their decisions is to maximize their own utilities or profits.

(3) Assuming that consumers' perception of using new products and remanufactured products are homogeneous.

(4) In order to make the manufacturing process of new products and remanufacturing products have economic sense, this article assumes that $c_r < c_m$.

(5) In the first period, customers' demand function for new products is: $D_1 = M_1 - a_1 p_1$

(6) In the second period, customers' demand function for new products and remanufactured products is $D_2 = M_2 - a_2 p_2$

(7) The utility functions of the first period and the second period are:

$$Max\pi_1(q_{1n}) = (p_1 - c_n - c_Q)q_{1n} + \beta\pi_2^* \quad (1)$$

$$Max\pi_2(q_{2n}, q_{2r}, Q_n, Q_r) = (p_2 - c_n - c_Q)q_{2n} + (p_2 - c_r - c_{Qr})q_{2r} \quad (2)$$

B. Symbols

M_1 & M_2 refer to the market size of the first period and the second period respectively; a_1 and a_2 are price sensitive coefficients, p_1 and p_2 are prices.

c_Q is the cost for manufacturers to make capacity preparation for producing new products and remanufactured products. c_{Qr} is the cost for remanufacturers to make capacity preparation for producing remanufactured products (umbrella), such as the cost for cloths prepared by remanufacturers, $c_Q = c_{Qn} + c_{Qr}$.

$\tau\gamma q_{1n}$ is the quantity of recycled products (umbrella stand) which can be used for remanufacturing.

Q_1 is the amount of capacity preparation for the first period, Q_n is the amount of capacity preparation (the quantity of recycled products which is available for remanufacturing) for the second period that be similar with $\tau\gamma q_{1n}$. Q_r is the amount of capacity preparation (e.g. cloths prepared by manufacturers themselves) matching with $\tau\gamma q_{1n}$.

c_n is the unit cost of new products in the first and second period, c_r is the unit cost of remanufactured products.

Let $\tau \in (0,1]$, it shows the proportion of the recycled products suitable for remanufacturing i.e. τ is the effective recovery rate.

β is the discount factor of the second period.

γ is the ratio of the recycled part (i.e. umbrella stand) can be used for remanufacturing.

III. MODEL FORMULATION

A. The equilibrium solution on condition that consumers' demand for the second period products D_2 is less than the quantity of recycled products which can be used for remanufacturing.

At this time, $D_2 < \tau\gamma q_{1n}$. Due to an oversupply, as a rational actor, the manufacturer would not produce new products on account of remanufacturing cost advantage, they would only remanufacture products; at this time, customers' demand $q_{2r} = D_2$, $q_{2n} = 0$, the capacity preparation needed for remanufacturing is Q_r, i.e. $Q_r = D_2$. Then, the manufacturers' optimal decisions in the two periods are respectively obtained as follows:

$$Max\pi_1(q_{1n}) = (p_1 - c_n - c_Q)q_{1n} + \beta\pi_2^* \quad (3)$$

$$Max\pi_2^*(q_{2n}, q_{2r}, Q_n, Q_r) = (p_2 - c_n - c_Q)q_{2n} + (p_2 - c_r - c_{Qr})q_{2r}$$
$$= \left(\frac{M_2 - D_2}{a_2} - c_r - c_{Qr}\right)D_2 \quad (4)$$

The equilibrium solution is obtained as follows:

$$Q^* = q_{1n}^* = \frac{M_1 - a_1(c_n + c_Q)}{2} \tag{5}$$

$$Q_n^* = q_{2n}^* = 0 \tag{6}$$

$$Q_r^* = q_{2r}^* = D_2 \tag{7}$$

$$\pi_1^* = \frac{[M_1 - a_1(c_n + c_Q)]^2}{4a_1} + \beta\left(\frac{M_2 - D_2}{a_2} - c_r - c_{Qr}\right)D_2 \tag{8}$$

$$\pi_2^* = \left(\frac{M_2 - D_2}{a_2} - c_r - c_{Qr}\right)D_2 \tag{9}$$

This condition is: $D_2 < \tau\gamma q_{1n}$, i.e. $D_2 < \tau\gamma\frac{M_1 - a_1(c_n + c_Q)}{2}$.

B. The equilibrium solution on condition that consumers' demand for the second period products D_2 is equal to the quantity of recycled products which can be used for remanufacturing.

At this time, $D_2 = \tau\gamma q_{1n}$ On account that remanufactured products produced by manufacturers in the second period is exactly equal to the amount of consumer demand, in order to maximize profits, the manufacturer would rather remanufacture products than produce new products. At this time, customers' demand $D_2 = \tau\gamma q_{1n}$. i.e. $q_{2r} = \tau\gamma q_{1n}$, $q_{2n} = 0$. Q_r is the capacity preparation required for manufacturers' remanufacturing, $Q_r = \tau\gamma q_{1n}$. Then, manufacturers' optimal decisions in the two periods are respectively obtained as follows:

$$Max\pi_1(q_{1n}) = (p_1 - c_n - c_Q)q_{1n} + \beta\pi_2^* \tag{10}$$

$$Max\pi_2^*(q_{2n}, q_{2r}, Q_n, Q_r) = (p_2 - c_n - c_Q)q_{2n} + (p_2 - c_r - c_{Qr})q_{2r}$$
$$= \left(\frac{M_2 - \tau\gamma q_{1n}}{a_2} - c_r - c_{Qr}\right)\tau\gamma q_{1n} \tag{11}$$

The equilibrium solution is obtained as follows:

$$Q^* = q_{1n}^* = \frac{a_2[M_1 - a_1(c_n + c_Q)] + a_1\beta\tau\gamma[M_2 - a_2(c_r + c_{Qr})]}{2(a_2 + a_1\beta\tau^2\gamma^2)} \tag{12}$$

$$Q_n^* = q_{2n}^* = 0 \tag{13}$$

$$Q_r^* = q_{2r}^*$$
$$= \frac{a_2\tau\gamma[M_1 - a_1(c_n + c_Q)] + a_1\beta\tau^2\gamma^2[M_2 - a_2(c_r + c_{Qr})]}{2(a_2 + a_1\beta\tau^2\gamma^2)} \tag{14}$$

$$\pi_1^* = \frac{\left[a_2[M_1 - a_1(c_n + c_Q)] + a_1\beta\tau\gamma[M_2 - a_2(c_r + c_{Qr})]\right]^2}{4a_1a_2(a_2 + a_1\beta\tau^2\gamma^2)} \tag{15}$$

$$\pi_2^* = \frac{(2a_2 + a_1\beta\tau^2\gamma^2)[M_2 - a_2(c_r + c_{Qr})] - a_2\tau\gamma[M_1 - a_1(c_n + c_Q)]}{4a_2(a_2 + a_1\beta\tau^2\gamma^2)^2} \cdot$$
$$\left[a_1\beta\tau^2\gamma^2[M_2 - a_2(c_r + c_{Qr})] + a_2\tau\gamma[M_1 - a_1(c_n + c_Q)]\right] \tag{16}$$

The condition is: $D_2 = \tau\gamma q_{1n}$, i.e.

$$D_2 = \frac{a_2\tau\gamma[M_1 - a_1(c_n + c_Q)] + a_1\beta\tau^2\gamma^2[M_2 - a_2(c_r + c_{Qr})]}{2(a_2 + a_1\beta\tau^2\gamma^2)}$$

C. The equilibrium solution on condition that $D_2 > \tau\gamma q_{1n}$ consumers' demand for the second period products D_2 is greater than the quantity of recycled products which can be used for remanufacturing

At this time, $D_2 > \tau\gamma q_{1n}$. On account that remanufactured products produced by manufacturers in the second period is less than the amount of consumer demand, in order to maximize profits, the manufacturer will produce new products in the second period, at this time, customers' demand for remanufacturing products $q_{2r} = \tau\gamma q_{1n}$, customers' demand for new products is q_{2n}, $q_{2n} = D_2 - \tau\gamma q_{1n}$, capacity preparation required for manufacturers to produce new products (umbrella stand) is Q_n, $Q_n = q_{2n}$. The capacity preparation needed for new products and remanufactured products production by using cloths or other materials is $Q_r = \tau\gamma q_{1n} + q_{2n}$. So the optimal decision in the two periods is obtained as follows:

$$Max\pi_1(q_{1n}) = (p_1 - c_n - c_Q)q_{1n} + \beta\pi_2^* \tag{17}$$

$$Max\pi_2^*(q_{2n}, q_{2r}, Q_n, Q_r)$$
$$= (p_2 - c_n - c_Q)q_{2n} + (p_2 - c_r - c_{Qr})q_{2r} \tag{18}$$
$$= \left(\frac{M_2 - D_2}{a_2} - c_n - c_Q\right)(D_2 - \tau\gamma q_{1n}) + \left(\frac{M_2 - D_2}{a_2} - c_r - c_{Qr}\right)\tau\gamma q_{1n}$$

The equilibrium solution is obtained as follows:

$$Q^* = q_{1n}^* = \frac{M_1 - a_1(c_n + c_Q) + a_1\beta\tau\gamma(c_n + c_Q - c_r - c_{Qr})}{2} \tag{19}$$

$$Q_n^* = q_{2n}^* = D_2 - \tau\gamma q_{1n}^* = D_2 - \frac{\tau\gamma[M_1 - a_1(c_n + c_Q) + a_1\beta\tau\gamma(c_n + c_Q - c_r - c_{Qr})]}{2} \tag{20}$$

$$Q_r^* = \tau\gamma q_{1n}^* + q_{2n}^* = D_2 \tag{21}$$

$$\pi_1^* = \frac{[M_1 - a_1(c_n + c_Q) + a_1\beta\tau\gamma(c_n + c_Q - c_r - c_{Qr})]^2}{4a_1} + \beta D_2\left(\frac{M_2 - D_2}{a_2} - c_n - c_Q\right) \tag{22}$$

$$\pi_2^* = \frac{\tau\gamma(c_n + c_Q - c_r - c_{Qr})[M_1 - a_1(c_n + c_Q) + a_1\beta\tau\gamma(c_n + c_Q - c_r - c_{Qr})]}{2} + D_2\left(\frac{M_2 - D_2}{a_2} - c_n - c_Q\right) \tag{23}$$

The condition is $D_2 > \tau\gamma q_{1n}$ i.e.

$$D_2 > \frac{\tau\gamma[M_1 - a_1(c_n + c_Q) + a_1\beta\tau\gamma(c_n + c_Q - c_r - c_{Qr})]}{2}$$

IV. EQUILIBRIUM SOLUTION ANALYSIS

The decrease of remanufacturing cost C_r in the second period will lead to the increase of remanufacturing products quantities. When $D_2 < \tau\gamma q_{1n}$, owing to an oversupply, the remanufacturing cost C_r wouldn't affect the products quantity and capacity preparation of the first period. When $D_2 \geq \tau\gamma q_{1n}$, in the case of short supply, manufacturers will try to increase the capacity preparation and product quantity of the first period by reducing remanufacturing cost so that the amount of useful recycling products in the second period will increase.

The increase of new products cost in the first period c_n will cause a decrease of the new products production q_{1n}^* and the production capacity Q_1^* and show a reverse relationship with the quantity of the second period remanufacturing products q_{2r}^*. Owing to the increase of manufacture cost c_n, manufacturers choose to produce fewer products, thus the first period capacity preparation Q_1^* will decrease followed by q_{1n}^*, so the amount of remanufactured products q_{2r}^* decline.

There is an inverse relationship between the unit cost of capacity preparation c_Q in the first period and the amount of capacity preparation Q_1^*. Obviously, because the capacity preparation Q_1^* is quite few, the new products quantity of the first period q_{1n}^* will decrease with the decreasing of c_Q. The increase of remanufacturing capacity preparation of the second period c_{Qr} will cause the amount of remanufacturing products q_{2r}^* to decrease.

When $D_2 < \tau q_{1n}$, the recovery rate τ almost have no impact on the first period capacity preparation Q_1^* and the quantity of new products. When $D_2 \geq \tau q_{1n}$, owing to the cost for remanufacturing is lower than manufacturing, customers will prefer to buy remanufacturing products, thus the demand for remanufacturing products q_{2r}^* will increase progressively, so manufacturers will increase the recovery rate τ of products in the first period, the capacity preparation of the first period and the quantity of new products.

M_1, the market size of the first period, it is positively correlated with its products quantity q_{1n}^*. The larger the market capacity of the first period, the greater customer demand for new products q_{1n}^*, thus the amount of capacity preparation will increase progressively.

When $D_2 = \tau q_{1n}$, the market size of the second period M_2 can make the total production of the second period q_{2r}^* increase progressively.

TABLE I
THE OPTIMAL DECISIONS UNDER DIFFERENT
CIRCUMSTANCES

	$D_2 < \tau q_{1n}$	$D_2 = \tau q_{1n}$	$D_2 > \tau q_{1n}$
Q_1^*	23	26	24
q_{1n}^*	23	26	24
q_{2n}^*	0	0	$D_2 - 14$
q_{2r}^*	$D_2/3$	14	14
Q_n^*	0	0	$D_2 - 14$
Q_r^*	$D_2/3$	14	D_2
π_1^*	$700 + \frac{1}{3} D_2 (89 - 4D_2)$	2150	$984 + \frac{1}{3} D_2 (85 - 4D_2)$
π_2^*	$\frac{1}{3} D_2 (89 - 4D_2)$	759	$24 + \frac{1}{3} D_2 (85 - 4D_2)$

We can see from Table I that when $D_2 < \tau q_{1n}$, both manufacturers' decision variables in the two periods and

its optimal profits are the minimum. So in this case, there is an oversupply of remanufacturing products in the second period, owing to remanufacturing cost advantages, manufacturers would remanufacture products at this time rather than manufacture new products, and customers will choose to buy some remanufactured products. Accordingly, manufacturers' capacity preparation in the first period won't be too much and the profit won't be so high.

Detailed numerical calculation results are shown in Table I.

When $D_2 = \tau q_{1n}$, in order to maximize their profits, as rational actors, manufacturers won't produce new products in the second period, for reason that the supply of recycled products which can be used for remanufacturing exactly meet customers' need, customers' demand at this time will increase a bit in comparison with condition 1, so manufacturers' optimal capacity preparation as well as their new products production in the first period will increase a bit. In the two-period production process, there are no materials or products wasted. Manufacturers' profit increased in these two periods in comparison with condition 1.

When $D_2 > \tau q_{1n}$, the amount of recycled products which can be used for remanufacturing cannot meet the demand of consumers; manufacturers will choose to produce new products to meet the needs of the consumers in the second period. Since remanufacturing activity has a cost advantage and manufacturers have to make capacity preparation for new products, this will bring some costs, thus manufacturers' new products production at this time would not bring them too much profit.

V. CONCLUSION

This paper mainly studies in the remanufacturing supply chain that manufacturers manufacture new products and drive them to the market in the first period according to their capacity preparation; in the second period, manufacturers will make a rational choice between new products production and remanufactured products production on conditions that both customers' demand for products and the amount of recycled products which can be used for remanufacturing are uncertain, they will make capacity preparation and choose to conduct manufacturing or remanufacturing activities according to consumers' demand. A two-period model is developed to find manufacturers' optimal two-period decision variables and profits under different circumstances. Finally, numerical simulation calculations found that when the amount of recycled products which can be used for remanufacturing is greater than consumers' demand, manufacturers' two-period profits will be relatively low, manufacturers will cut down their capacity preparation in the first period so as to reduce their loss; when the amount of recycled products which can be used for remanufacturing is exactly equal to consumers' demand, manufacturers' profit will be relatively high, reasons are that there are no capacity preparation waste and resources waste in the two periods; when the amount of recycled products which can be used for remanufacturing is less than consumers' demand, in order to meet the market demand, manufacturers will choose to produce new products in the second period, while in comparison with condition 2, manufacturers' profit won't increase too much owing to the remanufacturing cost advantage, so, manufacturers can only increase their profits by increasing their capacity preparation in the first period.

We make analysis under the assumption that the market demand is certain, while how to analyze when the market demand is uncertain i.e. to do research with random variables description is the direction of our future research.

REFERENCES

[1] Ferrer G, Swaminathan J M. Managing new and remanufactured products [J]. Management Science, 2006, 52(1): 15-26

[2] Ferrer G, Swaminathan J M. Managing new and differentiated remanufactured products [J]. European Journal of Operational Research, 2010, 203(2): 370-379

[3] Jun C, Zhongkai X, Lisha L. Price and Quality Competition Between the New and Remanufactured Producer in the Closed-loop Supply Chain [J]. Chinese Journal of Management Science, 2010, 18(5): 82-90

[4] Majumder P, Grocnevelt H. Competition in remanufacturing [J]. Production and Operations Management, 2001, 10(2): 125-141

[5] Ferguson M E, Toktay L B. The effect of competition on recovery strategies [J]. Production and Operations Management, 2006, 15(3): 351-368

[6] Webster S, Mitra S. Competitive strategy in remanufacturing and the impact of take-back laws [J]. Journal of Operations Management, 2007, 25(6): 1123-1140

[7] Franke C, Basdere B, Ciupek M, Selinger S. Remanufacturing of mobile phones - capacity, program and facility adaptation planning [J]. International Journal of Management Science, 2006, 34(6): 562-570

[8] Georgiadis P, Vlachos D, Tagaras G. The impact of product lifecycle on capacity planning of closed-loop supply chains with remanufacturing [J]. Production and Operations Management, 2006, 15(4), 514-527

[9] Debo L G, Toktay L B, Van Wassenhove L N. Joint life-cycle dynamics of new and remanufactured products [J]. Production and Operations Management, 2006, 15(4), 489-513

[10] Rubio S, Corominas A. Optimal manufacturing - remanufacturing policies in a lean production environment [J]. Computers and Industrial Engineering, 2008(1), 234-242

[11] Yun L, Hong L, Xiaode Z. The Pricing Model of Closed-loop Supply Chain with Capacity Constraint under One-way Substitution Strategy [J]. Journal of Industrial Engineering and Engineering Management, 2013, 27(1): 114-120(in Chinese)

[12] Zheng Li, Xu Chen. Manufafacture's Optimal Pricing Policy and Retailer's Optimal Ordering Policy with Capacity Constraints [J]. Operations Research and Management Science, 2012, 21(5): 154-161(in Chinese)

[13] Caner S, Zhu S X, Teunter R. Capacity and production decisions under a remanufacturing strategy [J]. International Journal of Production Economics, 2013, 145(1): 359-370

Utility-Cost Model for Discrete Event Logistics Systems Simulation

Yu-kun Liu*, Meng-rui Shan

Logistics Engineering Center, School of Automation, Beijing University of Posts and Telecommunications, Beijing, China

(*lykbupt@163.com, shanmengrui@163.com)

Abstract – **A study on simulation fidelity, utility and cost is conducted to address simulation quality issues in Discrete Event Logistics System (DELS) context. First, a simulation lifecycle management (SLM) framework is constructed. Specially, the concept of fidelity is brought into the framework. Then, simulation fidelity, utility and cost are divided into multi-dimensions each according to DELS simulation features. Third, a dimension decomposition based relationship analysis between Fidelity-Utility and Fidelity-Cost is presented. A Utility-Fidelity mapping method is used to map simulation objective into fidelity requirement (FR). Fidelity design (FD) is strictly based on this FR, and simulation design is in turn based on FD. Finally, a Utility-Cost management model is presented to show how well a simulation implementation is done. The SLM framework is effective in guiding the simulation to achieve reasonable balance between utility and cost, which falls into the cost-effective equilibrium zone in the Utility-Cost management model.**

Keywords - **Cost, DELS, fidelity, simulation, utility**

I. INTRODUCTION

Despite the wide use of discrete event simulation (DES) in Discrete Event Logistics Systems (DELS) planning and design[1][2], the issues about the quality of simulation application are left unstudied. In a way, the quality of a simulation study can be evaluated by simulation utility (SU) it achieves and simulation cost (SC) it spends[3]. The lack of understanding about SU and SC leads to two kinds of consequences: insufficient SU and unnecessary high SC. Insufficient SU, in turn, usually results in large errors or even invalid simulations, and/or inefficiency in simulation modeling. This is why a study on Utility -Cost (U-C for short) equilibrium is valuable.

Fidelity (F) is the main drive of SU and SC. In general, the higher the fidelity, the higher the SU, and inevitably the higher the SC[4][5], thought this is not always the case due to the nonlinear relationships between F-SU (F-U for short) and F-SC (F-C for short). In this paper, the relationships between F-U and F-C are explored in DELS context, and a U-C model is built to address DELS simulation quality problems.

II. METHODOLOGY

A. Framework and perspective

For simulation quality, the lifecycle management of simulation is of great importance. A framework of simulation lifecycle management is given in Fig.1. It includes four stages: simulation preparation, simulation design, simulation implementation, and simulation quality management. How the simulation is prepared and designed largely decides the quality of a simulation study.

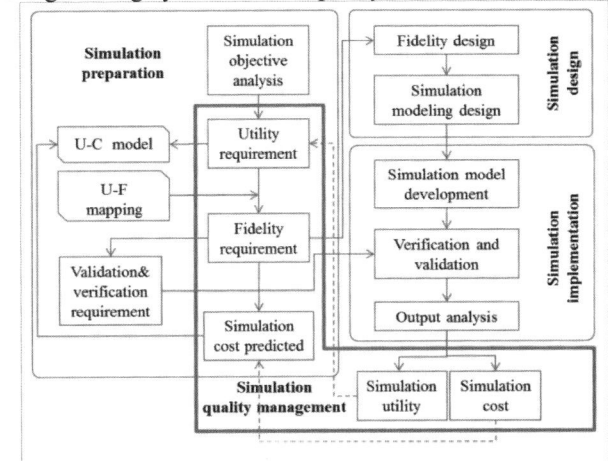

Fig.1. Framework of simulation lifecycle management

B. Concepts and dimensions
1) Fidelity and its dimensions

Generally, fidelity tells how well the simulation system represents the object system[6][7]. The concept of fidelity is recognized as early as 1960's. It was studied in the following half century, but still remains as a nebulous term used by simulation community. It could be the model itself, the behavior of the model or the simulation execution results[8][9]. In this sense, simulation fidelity is divided into experiment fidelity and model fidelity. Experiment fidelity tells how the simulation experiment scheme is designed to achieve the simulation objectives, while the model fidelity tells how well the model represents the object system. A simulation system with high model fidelity is capable to serve a simulation analysis with low experiment fidelity, though this is not economical, but a model with low model fidelity could not serve a simulation analysis with high experiment fidelity[10].

Theoretically, model fidelity is defined as the ratio between the simulation world and the object real world (existing or imagined reality)[11]. See formula (1).

$$Fidelity = \frac{M_S}{M_R} \qquad (1)$$

Where M_S is the simulation model and M_R is the reference model abstracted from the understood reality.

Simulation application in DELS domain is different from training field[12][13]. There is no hardware and/or man in the loop. The object system is usually a compound

¹ This research is supported by National Natural Science Foundation of China (Grant No. 61104055)

future system integrated by a set of queuing, inventory, transport network and other kind of systems. Usually, the purpose of simulation is to understand, analyze, evaluate and/or optimize the system. Based on the features of DELS simulation, fidelity is divided into the following dimensions as shown in Fig.2.

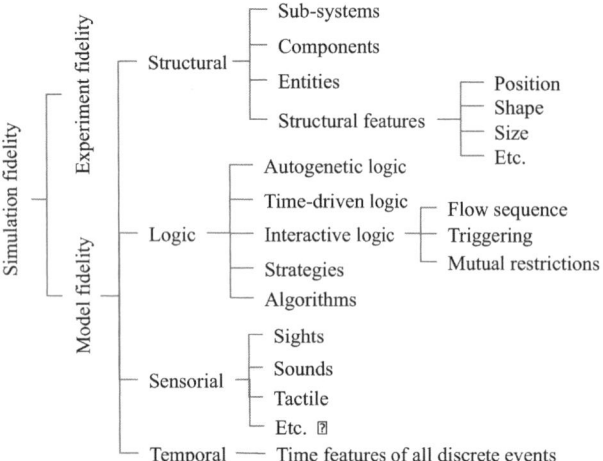

Fig.2. The dimensions of simulation fidelity

2) Simulation utility and its dimensions

Simulation utility is a blur and elusive concept, like fidelity itself. In a narrow sense, SU indicates how well a simulation implementation works for the given simulation objective pre-determined from a default viewpoint of the user. But a broader view of SU means more. In this study, the SU is divided into the follow dimensions shown in Fig.3. The relationships between different dimensions are complex. Some of the dimensions are correlated with others, while others could be regarded as independent elements. For example, the reusability is positively correlated with timeliness, while the interoperability is rather independent.

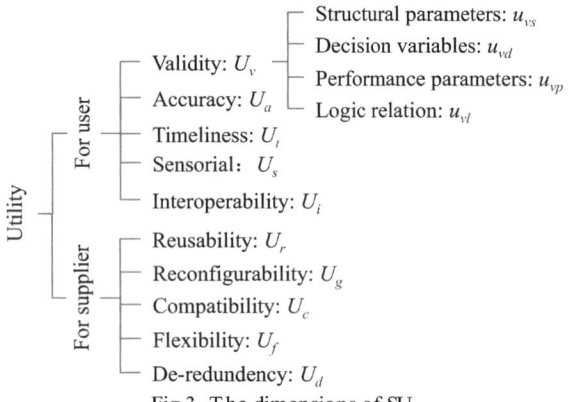

Fig.3. The dimensions of SU

3) Simulation cost and its dimensions

Simulation cost, as a concept, is much clear than SU. The resources used and the amount of time and manpower spent in simulation are the actual cost, which is usually measured in monetary value. A simulation is always expected to be cost-effective[14].

But when considering quantitatively cost evaluation, SC issue becomes much more complicated. It is difficult to measure SC in terms of monetary value based on time

and manpower spent because of the random and uncertain factors, such as intelligence, experience and skills. Objective measures are needed. In DELS context, investments for hardware, such as computers and software, are easy to be evaluated. The time and manpower cost in DELS simulation is spent in the simulation activities. In this sense, the essential activities in modeling and simulation (M&S) are the objective cost of the simulation.

Each of these M&S activities contains two parts: time to "know how" and the time to "do it". The two parts are usually disproportionate. The first part could be much greater than the second part. And the time for "know how" could be largely reduced by certain methods, skills and experiences.

Here, SC is divided in the dimensions shown in Fig.4. The dimensions are correlated. For example, if the computer has a higher configuration, the time to run the simulation could be reduced.

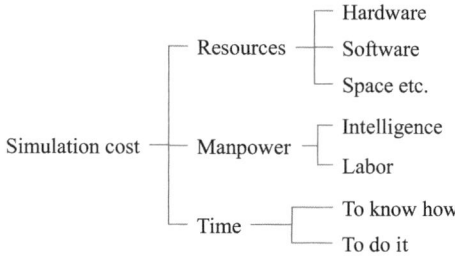

Fig.4. The dimensions of SC

C. Relationship analysis and mapping
1) Relationship between F-U

Fidelity	Simulation utility
Structural	Validity: U_v
	Accuracy: U_a
	Timeliness: U_t
Logic	Sensorial: U_s
	*Interoperability: U_i
	Reusability: U_r
Sensorial	Reconfigurability: U_g
	Compatibility: U_c
	Flexibility: U_f
Temporal	Deredundancy: U_d

———— positive correlation　　　 − − − − negative correlation
− · − · random correlation　　　 ·········· not correlated

*　For DELS context, interoperability is atypical and so omitted.

Fig.5. The relationship between F-U

Basically, the relationship between F-U is classified into three types: positive, negative and random correlation. See Fig.5. The complexity of the F-U relationship comes from two aspects. First, the model fidelity dimensions are not strictly independent. For example, the logic dimension is related to structural dimension. Second, the relationship between fidelity dimensions and utility dimensions is not clear.

An effective approach to understand the relationship is to split the correlation function into two parts based on field experience: the linear part and the nonlinear part. For example, the relationship between utility of validity (U_v) and the logic dimension of fidelity (f_l) is defined as formula (2).

$$U_v = F_1(f_l) + F_2(f_l) + G \qquad (2)$$

Where F_1 is the linear part, F_2 is the nonlinear part, and G is the part for other dimensions of fidelity.

Detailed analysis between each dimension of fidelity and dimension of simulation utility is carried out based on typical distribution center (DC) logistics system examples. There are random and uncertain factors for each DC examples. No common relation functions can be found for a general definition of F-U relationship. But it does have similar features for similar DC logistics systems. Therefor, the relationship analysis between fidelity and utility dimensions is a field knowledge based analysis. Details are omitted for sake of limited paper length.

2) Relationship between F-C

In general, an exponential relationship between F-C is accepted as a rough qualitative model in simulation community [15]. But the point of diminishing return is difficult to find. It is determined by the modeler or the simulation experts. This diminishing return point is very important for simulation experts to estimate the cost and decide the fidelity and expected cost. But in DELS context, the validity of cost estimation is largely depends on the experiences and skills of the simulation expert.

A further analysis of F-C relationship on individual dimension level is shown in Fig.6. Generally, the exponential relationship model also applies to the relationship between different dimensions of fidelity and dimensions of cost. Like the relationship of F-U, random and uncertain factors affect the relation between each dimension of fidelity and cost. Typical examples of DC logistics systems are used to help the exploration into the F-C relationship analysis, which helps to build up field knowledge.

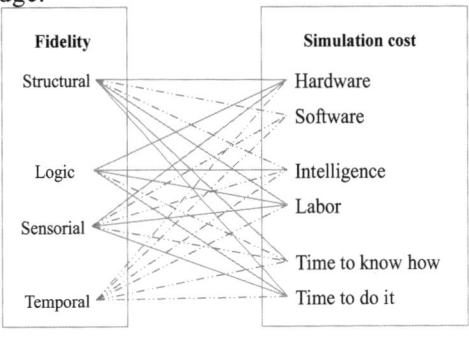

Fig.6. The relationship between F-C

3) U-F mapping

The first important process of simulation lifecycle management is to determine the fidelity requirement that matches the simulation objective. A U-F mapping method is presented here as Fig.7 to help determining the fidelity requirement. Dimension decomposition analysis is used to map simulation objective onto fidelity requirement. All dimensions are listed in detail, such as decision variables, performance parameters, accuracy, timeliness, sensorial and other requirements. The simulation objective determines all this elements. According to these requirements, the model structure, logic and quantitative relation are determined.

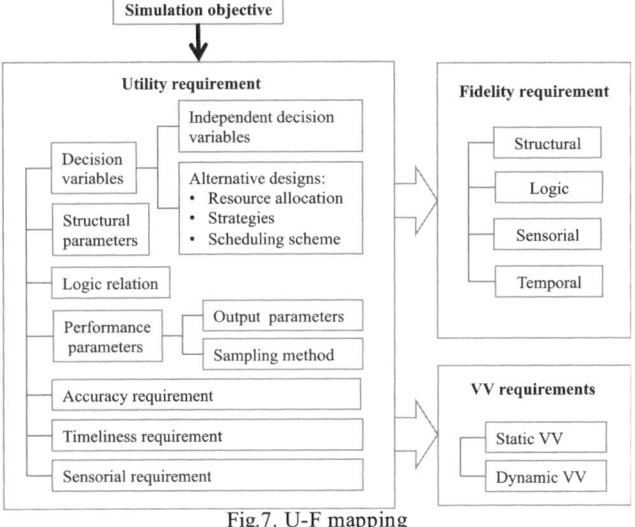

Fig.7. U-F mapping

III. RESULTS

By mapping the simulation objective onto fidelity requirement, a well-prepared fidelity design is possible. The simulation modeling and simulation experiments should be based on this fidelity design. The final possible result of the utility and cost of a simulation application would fall into different cases. A U-C model that summarizes the quality of simulation application in terms of utility and cost is shown in Fig.8.

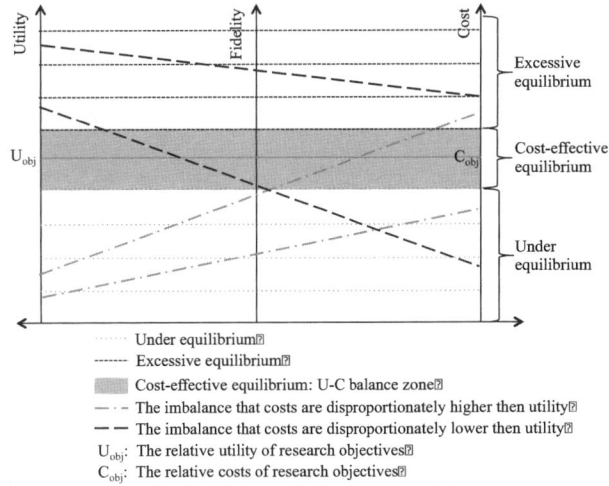

Fig.8. The simulation U-C model

Normally, the U-C relationship of a simulation application would be naturally balanced which is identified as a horizontal line. The position of the horizontal line could be higher or lower than the specific line determined by point U_{obj} and point C_{obj}, which is the theoretical optimal equilibrium of utility and cost of a given simulation study. If the simulation is not well designed and managed, inadequate fidelity and less cost might happen, and an under equilibrium state would be achieved; or excessive equilibrium state would be achieved when unnecessary high fidelity and cost happened.

Also, there might be imbalance cases. For example, if the straight line is positively sloped, the cost will be disproportionately higher than the corresponding utility of the simulation, and vice versa. A group of factors lead to such imbalance cases. For example, irrational pursue for 3D visual effect, lack of effective method to manage simulation modeling utility, wrong decision on simulation modeling design and lack of programing skills are the most recited factors and reason for unexpected high cost but not proportional utility.

At the same time, people are trying to find ways to reduce the simulation cost while keeping certain level of utility. This is often the case when modelers try to improve the simulation utility for themselves. For example, modelers are always seeking methods to improve reusability to effectively reduce modeling time in similar simulation applications. Theories and methods learned and experiences and/or skills gained of a simulation team would significantly reduce the cost when the team works on similar simulation projects. In this sense, balance is relative. The equilibrium is a balance under certain methodological restrictions. When new method improves the modeling efficiency, the cost will be reduced, and this is often called a new equilibrium instead of an "imbalance".

IV. CONCLUSION

In order to achieve high quality of simulation, i.e. cost-effective as well as sufficient utility, a deeper exploration into the relationship among three basic concepts about simulation, fidelity, utility and cost, is conducted. Relationship analysis based on dimensional structure decomposition indicates that the relationship between either two of the three is very complex and nonlinear. By relationship analysis between F-U and F-C, the bridge between utility and cost is built. A simulation lifecycle management framework is presented, and the concept of fidelity is brought into the framework. Under this framework, simulation objective is effectively mapped onto fidelity requirement, and then the fidelity requirement decides the model design. The framework is designed to guide the simulation practice to achieve a better balance between utility and cost. Finally, the U-C model provides a clear description and a better illustration of the quality of a simulation application.

REFERENCES

[1] C. N. Zhang, "Modeling and simulation of the warehouse system of logistics based on petri net" (in Chinese), Master dissertation, Nanjing Forestry University, Nanjing, China, 2007.

[2] Y. Fu, "Research on application mode and method of simulation in distribution center design" (in Chinese), Master Dissertation, Beijing University of Posts and Telecommunications, Beijing, China, 2011.

[3] J. Duncan, "Fidelity versus cost and its effect on modeling and simulation," presented at the *Evidence Based Research Inc*, Suffolk, VA, USA, 2006.

[4] D. Goncalves, "An approach to simulation effectiveness," in *Proceedings of 16th Annual International Symposium INCOSE 2006*, Orlando, FL, USA: INCOSE-International Council on Systems Engineering, 2006, vol. 1, pp. 167-178.

[5] J. Boles, R. Milligan, M. Hagenmaier, D. Eklund, "The advantages and costs of higher-fidelity turbulence modeling," in *Proceedings of the 2010 DoD High Performance Computing Modernization Program Users Group Conference*, Washington, DC, USA: IEEE, 2010, pp. 8-11.

[6] D. K. Pace, "Description and estimation/measurement of simulation fidelity," in *Proceedings of the Caltech V&V of Computational Mechanics Codes Symposium*, Pasadena, CA, USA: Army Research Office, 1998.

[7] H. Kim, L. F. McGinnis, Z. Chen, "On fidelity and model selection for discrete event simulation ," *Simulation: Transactions of the Society of Modeling and Simulation International*, vol. 88, no. 1, pp. 97-109, 2012.

[8] D. K. Pace, "Synopsis of fidelity ideas and issues," in *Proceedings of 1998 Spring Simulation Interoperability Workshop*, Orlando, FL, USA: SISO, 1998, 98S-SIW-071.

[9] F. Yu, Y. K. Liu, Z. Y. Su, X. G. Zhou, "Fidelity management and evaluation in logistics system," in *Automation and Logistics, 2009. ICAL'09. IEEE International Conference on*, Shenyang, China: IEEE, 2009, pp. 1662-1666.

[10] Z. C. Roza, "Simulation fidelity theory and practice," Ph.D. dissertation, TU Delft, Delft, Netherlands, 2004.

[11] Y. K. Liu, J. Chen, "Fidelity evaluation for DELS simulation models," in *Asia Sim 2013*, Berlin, Germany: Springer Berlin Heidelberg, 2013, pp. 391–396.

[12] D. Y. Zhang, Y. H. Yang, "Summary of substation's simulation and training" (in Chinese), *Automation of Electric Power Systems*, vol. 23, no. 23, 1999.

[13] Y. Gao, K. Tan, X. H. Pan, Z. L. Li, Y. J. Lin, et al, "Effect of arthroscopic knee joint surgery simulation training system on arthroscopic surgery training" (in Chinese), *Academic Journal of Chinese Pla Medical School*, vol. 34, no. 3, pp. 291-293, 2013.

[14] H. Pongracic, D. Marlow, T. Triggs, "Issues in cost-effectiveness and fidelity of simulation," in *Proceedings of the Second International SimTecT Conference*, S. Sabrina, Ed. Canberra, Australia: SimTecT 97 Organising and Technical Committee, 1997, pp. 221-226.

[15] J. Duncan, "Fidelity versus cost and its effect on modeling and simulation," presented at the *Evidence Based Research Inc*, Suffolk, VA, USA, 2006.

Improved Computer Methods for Sequencing Operations for U-shaped Assembly Lines

Li-yun XU[1,*], Loïc BÉLEC[1], Wei LIU[1], Qi-fang GU[2], Ai-ping LI[1]

[1]Institute of Advanced Manufacturing Technology, Tongji University, Shanghai, China
[2]Faw Jiefang Automotive CO.,Ltd. Wuxi Diesel Engine Works

(Lyxu@tongji.edu.cn)

Abstract - **U-shaped lines are considered to have a number of advantages over traditional lines. COMSOAL is a recurrent procedure for assembly lines balancing, and it stands for computer method for sequencing operations for assembly lines. In this paper, an improved U-COMSOAL is proposed to solve some problems of COMSOAL, the random picking part of the COMSOAL method is substituted by a ranked positional weight method, and takes both the predecessors and successors of the tasks into the account. Finally, a comparison of between the improved method and a classical method suggested by Ö. F. Baykoç was done, which illustrate the efficiency of U-COMSOAL.**

Keywords - **Assembly line balancing, heuristic, stochastic, U-shaped**

I. INTRODUCTION

In manufacturing, assembly line configurations can be seen as of the most ubiquitous and important issues in order to enhance production [1-9]. This problem is generally called ALBP, for Assembly Line Balancing Problem, and its consist on gathering tasks into workstations such that the sum of the processing times at each station does not exceed the station time, also named station cycle. A digraph as Fig.1 typically shows precedence relations between tasks. Digits, above the nodes, represent the processing times for theirs corresponding tasks. These tasks cannot be sub-divided and must be completed at their assigned stages.

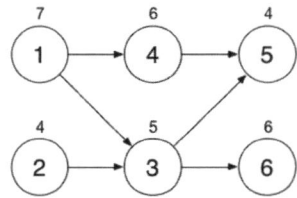

Fig.1. Example of a precedence network

There are two common versions of this problem: the first version is to minimize the number of workstations for a given cycle time, the second version is to minimize the idle time. As a reminder, the idle time is the non-productive time of employees or machines, or both, due to work stoppage from any cause. As an example, the Fig.2 shows an optimisation of the problem denounced by the Fig.1. In which, there is a cycle time of 10 time units with 4 workstations required.

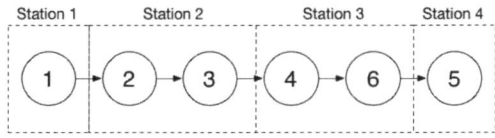

Fig.2. Straight assembly line solution to Fig.1

The advent of Just-In-Time has led industries to search for manufacturing technologies in order to satisfy lean concepts and high efficiency in both inventory and labour. So, because of JIT requirements, industrial companies are now designing theirs assembly lines as U-shaped assembly lines. There are many advantages to switch from straight assembly lines to U-shape. At first, they present a better potential for balancing. Also, they can improve visibility and communications: workers have a better point of view toward the assembly line. It does require fewer stages, there is more flexibility for adjustment, and eventual travels are shorter. Moreover the material handling is quite easier compared to straight lines. Fig.3 shows an optimal U-shaped assembly line for the example of network, Fig.1. One can see that there are only 3 stations required, compared to the 4 workstations required by the straight assembly line.

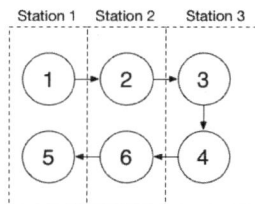

Fig.3. U-shaped assembly line solution to Fig.1

There are two types of classification for simple line balancing problems: type I and type II. In type I, the cycle time, seen as the pace of the production, tasks and theirs times, precedence relations are given. The objective of this is to minimize the number of stages. Indeed, an assembly line with fewer stations will result in a lower labour costs and it will reduce space requirements. Type I problems generally occurs when one has to design new assembly lines. To achieve the forecast demand, the number of workstations has to be lowered. For expansion, when demand is increased, type I problems also can be used to minimize the number of extra stations needed to install. In type II problems, the objective is to minimize the cycle time, whereas the number of stages (equivalently the number of employees) is fixed. The main purpose of it is to maximise the production rate. Type II balancing problem generally occur, when the organization wants to produce the optimum number of items, using a fixed number of stations without any expansion. In this latter type, it is necessary to consider subassembly lines. The most recurrent type of ALBP we can find in industry is the type I rather than the type II. But exact algorithms available for the same become intractable when the problem size increases.

E. Qi et al. (eds.), *Proceedings of the 21st International Conference on Industrial Engineering and Engineering Management 2014*, Proceedings of the International Conference on Industrial Engineering and Engineering Management, DOI 10.2991/978-94-6239-102-4_11, © Atlantis Press and the authors 2015

The type of SALBP (Single Assembly Line Balancing Problem) that is studied in this thesis is the type I. From a list of tasks submitted, with theirs times and precedence relations, we will analyse and implement solutions for U-shaped line balancing problem.

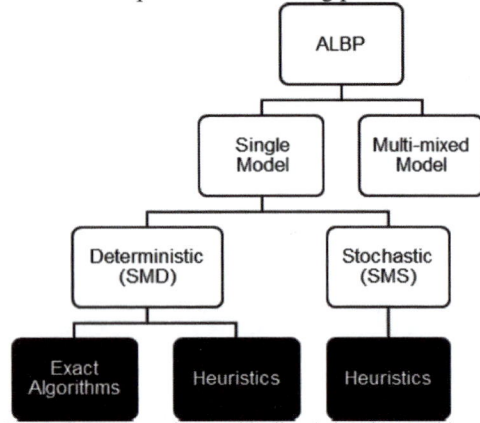

Fig.4. Summary of assembly line balancing problems

The Fig.4 is a summary of assembly line balancing problems. This paper deals with single model for ALBP that are deterministic. As of today, there is no exact algorithm, which permits to balance a U-shaped assembly line efficiently. Furthermore, there is also no way to solve this kind of problem exactly within a short time. Regarding multi-mixed model, it means assembly lines that combine many kinds of lines. In some factory, you can find straight lines combined with parallel lines, two-sided lines etcetera. This problem is even more elusive. However, we will try to give some elements of solution thereafter.

II. COMSOAL Method

COMSOAL stands for Computer Method of Sequencing Operations for Assembly Lines. It is a computer heuristic method, but also a stochastic method, originally reported as a solution approach to assembly line balancing problem. As a reminder, in probability theory, a purely stochastic system is one whose state is non-deterministic, so that the subsequent state of the system is determined probabilistically. Any system or process that must be analyzed using probability theory is stochastic at least in part. Stochastic systems and processes play a fundamental role in mathematical models [10-12] of phenomena in many fields of science, engineering, and economics.

In line balancing literature, recently, a few references [13-15] to COMSOAL are in the assembly line balancing area. As a solution method, COMSOAL quickly generates multiple feasible solutions and uses the best solution as its random picking as its final reported result. Picking randomly a task and constructing subsequent tasks generate sequences. New stations are opened when needed. Sequences that exceed the best solution are discarded. Better sequences become upper bounds. The main difference between COMSOAL and U-COMSOAL methods is the assignment of the tasks into the stations.

At the COMSOAL method, while the workstations are being constituted, only the tasks, the predecessors of which are assigned, are taken into account. However U-COMSOAL method takes both the predecessors and successors of the tasks into the account.

III. U-COMSOAL MODEL

The different steps for U-COMSOAL solution are as followed:

(1) Initializing the method

A index of the first sequence x is defined, an array A for tasks that still can be selected, the current cycle time C which has been calculated before running the method, the matrix WM we are working on, and the current station. We are using a duplicated matrix WM from the input M, because it allows us to keep a track on what was submitted.

```
% Start the method
function [] = starting()
    disp('starting')
    x=0; % first sequence
    UB=1000; % a big number
    c=C; % current cycle time
    A=TK; % task selected
    WM=M % working matrix
    station=1 % current station
    new_sequence();
end
```

(2) Starting a new sequence

A main method with no parameters was designed, input of Boolean values and output exemplifying the relevant criteria. It means that a sequential algorithm can run with adjacency matrix as input, and can deliver output as an idle time or a number of stations.

```
% New sequence
function [] = new_sequence()
    disp('new sequence')
    x=x+1; % x index of sequence
    A % display list A
    WM;
    precedence_feasibility();
end
```

The variable x is the index of the current sequence. At the first try, the U-COMSOAL code does not provide an optimal solution. After X sequences, we can consider that we have a good solution, but nothing says it is an optimal or even the best.

(3) Checking precedence feasibility

In order to establish which task can be assigned to the current station, we have to check for its precedencies and to determine if its time suits the current allowed time.

```
% Precedence feasibility
function [] = precedence_feasibility()
    disp('precedence feasibility')
    n=size(A);
    sizeA=n(2);
    B=[]; % reset B
    i=1;
    while i<=sizeA
    if (WM(1:end,A(i))==[0]) % look for predecessor
```

```
        B=[B A(i)]; % add task without predecessor from A to B
    elseif (WM(A(i),1:end)==[0]) % look for successor
        B=[B A(i)]; % add task without successor from A to B
    end
        i=i+1;
    end
        B % display list B
        time_feasibility();
end
```

```
            WM(r,1:end)=[0]; % update matrix WM column
            F(ii)=[]; % update list F
            c=c-t(r); % update cycle time
        if isempty(F)
                schedule_completion();
        else
                precedence_feasibility();
        end
    end
end
```

(4) Checking time feasibility

It means that the final array obtained can be longer than one element. So, we can call the list being constituted here a candidate list. If the candidate list, which has been constituted, is empty, the method has to start a new station. Indeed, the current cycle time is indexed on the station. It means that when you are assigning tasks to a station, the sum of operating times (from tasks assigned) cannot exceed the cycle time. Therefore, it is necessary to add a new station if no task is suitable for the current one. In the case where the candidate list is not empty, the method has to select a task.

```
% Time feasibility
function []=time_feasibility()
    disp('time feasibility')
    n=size(B);
    sizeB=n(2);
    F=[]; % reset F
    i=1;
while i<=size B
if (t(B(i))<=c)
        F=[F B(i)];
end
    i=i+1;
end
if isempty(F) % F empty means no successor or predecessor
    open_new_station();
else
    select_task();
end
end
```

(5) Open new station;

If there is not enough time remaining, the method has to open a new station. The current cycle time is indexed on the current station.

(6) Select the task suitable, and continue the loop

This number generated has just to be multiplied by the size of the array F, and to be ceiled, in order to pick up the task. There is no more random trick involved in this part. It constitutes the heuristic feature of this method.

```
% Select new task
function []=select_task()
    disp('select task');
    m=size(F);
    RN=rand(1,1); % select randomly an available task
    ii=ceil(m(2)*RN); %
    F(ii) % display selected task
    [p]=find(A==F(ii));
    [q]=find(B==F(ii));
    [r]=find(TK==F(ii));
    A(p)=[]; % update list A
    sizeA=sizeA-1;
    B(q)=[]; % update list B
    sizeB=sizeB-1;
    WM(1:end,r)=[0]; % update matrix WM row
```

(7) Schedule completion if no task remains.

It is used to measure the effectiveness of the current solution obtained. The lower the idle time is, the better the solution is.

This solution has been studied by Ö. F. Baykoç [13]. A main suggestion of enhancement is to implement a R.P.W. (rank positional weight) method to this first solution.

In fact, Talbot et al. [16] compiled a list of the numerical scoring functions that have been used by Hackman, Wee and Magazine [17], and other researchers in construction decision rules for assembly line balancing heuristics. Four of these are meaningful of the parallel station U-shaped line problem. Each of these rules assigns a score reflecting the importance of a task in terms of some measure such as time or number of tasks that they "control" with respect to assignment of tasks to workstations. The method suggested uses all following aspects with appropriate U-line adaptations: (1) Work element time, (2) Positional weight, (3) Number of followers, (4) Number of immediate followers.

In addition, it improves over earlier heuristic by immediately adjusting scores and re-ranking available tasks after each task assignment. The adjustment and the re-ranking is very important for U-lines since the assignment of tasks either end of the line can immediately affect ranking scores in the case of PW, NIF, and NF criteria [18].

Giving this method a R.P.W. implementation involves considering the Buxey constraint. It has been studied by Cheng et al. [19], Miltenburg [20], and Aese et al. [21]. It deals with the question of extending the current station to two parallel tasks. If the current stage has already been extended, this function returns immediately to the time feasibility – in order to check for other task suitable. Indeed, the Buxey constraint on minimizing duplicated equipment does not allow further expansion. In brief, the task cannot be assigned. If the time consumed by tasks already assigned to an unexpanded station requires the stage to be expanded beyond the minimum necessary to accommodate the current large task, expansion would also violate the Buxey constraint.

Buxey [22] was the first to study the practical aspects of parallel stations including costs of duplicated equipment and difficulties of layout and transportation. There is a constraint the bears his name. The Buxey constraint is a constraint that allows us to know if the current stage can be expanded. To check this, the simple following calculus has to be done.

$$q_{test} = \left\lceil \frac{t_i}{C} \right\rceil$$
$$t_{test} = slack + (q_{test} - 1) \times C$$

(1)

if $t_i \leq t_{test}$ *then the stage can be expanded*

t_i: time for task i ;

C: cycle time;

q_{test}: temporary value for number of stations n the current stage used to test if stage can be expanded;

t_{test}: temporary value used to test if stage can be expanded

If the task can be incorporated, the number of parallel stations in the stage is set equal to q, the slack time is adjusted accordingly, and the search to complete the load continues. The definition for the slack time is: term that refers to the time that an activity can be delayed. It also refers to the difference between the late and early start times of an activity.

The final algorithm, implemented by Matlab methods, can be found in appendix. It had been optimized by matrix operations. Indeed, Matlab, as its name indicates, is prone to manipulate matrix. The next part is then dedicated to compare existing solution to the one implemented in this study.

IV. RESULTS

One case study has been realized by Ö. F. Baykoç [13]. It deals with a Dishwasher Machine Facility, at Ankara. It gathers two assembly lines disposed as U-shaped lines. Daily outputs are as following.

Line 1: 200, Line 2: 900

The second line is described as composed of thirty-nine workstations in which thirty-nine workers are employed for a total number of tasks to complete the assembly is 183. In fact, working day is 8 hours with a 50 minute-lunch (a 20 minute break coffee is also provided). Thus, the production rate aimed at is 900 units per day.

Results obtained during this study show that the problem can be solved by 4 alternative solutions. However, one solution is better than the others as it can provide the targeted daily output, that is to say 900. In fact, other alternative solutions provide daily output from 894 to 897. When the assembly line is rebalanced with deterministic times in the U-line setting type, the number of stations to achieve 900 units of output decreased from 39 to 36. By applying the same input to our enhanced algorithm, we can demonstrate the importance of R.P.W. implementation (comparing these two methods). The number of output targeted remains the same as it constitutes an optimal number. By implementing the method given by Ö. F. Baykoç [13], we can compare the efficiency of our work with an existing solution's. This study is evaluated by the execution time. To do so, we evaluated the time complexity of the different methods. Both are equivalent to a quadratic. The frame of these methods can be summed-up in the following Table I.

TABLE I
TIME COMPLEXITIES

sequences	line browsing	loops
$O(1)$	$O(n)$	$O(n)$
$O(n^2)$		

Therefore, the time complexity can be written in a system for each method. These systems are as following. They are both quadratic, and n represents the number of tasks one can submit to them. The Fig.5 is a representation of the experimental running time for our solution.

$$\begin{cases} u(n) = 6.0314\,e - 05 \times n^2 + 0.00040978 \times n \\ + 0.0067479 \\ R^2 = 0.99586 \end{cases}$$

(2)

Equation (2): Running time for our solution

$$\begin{cases} u(n) = 8.2485\,e - 05 \times n^2 + 0.000245678 \times n \\ + 0.0024568 \\ R^2 = 0.98524 \end{cases}$$

(3)

Equation (3): Running time for Bayroç solution

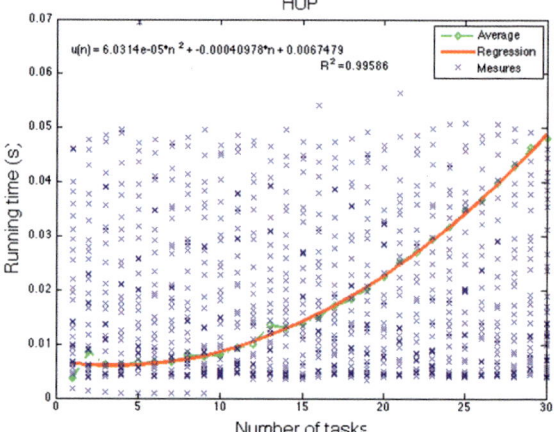

Fig.5. Experimental execution time

We can say that our solution is seen as more efficient. Not because of the R-square but because of the coefficient of the square term. This fact can be explained by method frameworks. The R.P.W. method initially implemented to a U-COMSOAL method (Baykoç's solution is based on it) gives an advantage. Browsing assembly line is shortened by the ranking processing. While a standard COMSOAL method takes into account every case - doing a loop, our enhanced method takes advantage of ranking scores to pick new solution. This method can be regarded as a pseudo ant colony algorithm, but it is none. Nevertheless, it does constitute an enhancement compared to existing solution. The improvement is based on the execution time, and not on the solution given. Therefore, this way of solving ALBP for U-shaped line is significant for bigger problems.

V. CONCLUSION

Nowadays, industrial companies are looking for the enhancement of their production at all levels. A solution often considered as one of the best, but also one of the most complex to manage because of its shape, is the U-shaped assembly line.

U-shaped lines can be interesting for a company because they can mainly help save workforce; companies are also trying to lower their workforce. To do it, they have many possibilities. One of them is to change their assembly lines - the major part of them is straight - to U-shaped lines. Thus, they respect the JIT philosophy. The main issue is to adapt theirs existing algorithms to a U-shaped problem. In this paper, we have adapted an already existing solution (the R.P.W. method) to a problem often neglected.

Numerous ways of improvement for the ALBP concerning U-shaped lines exist. Balancing assembly lines can be very complex, and therefore in order to provide a good solution to it, some have started to design algorithms that will reproduce the behavior of bacterial. Other persons use graph coloring to ALBP. There are many kinds of way to do it. The method we have dealt with in this paper represents the current best feasible solution. Furthermore, combining features provided by software as Tecnomatix Plant Simulation, edited by Siemens, could permit many improvements. Walking time, and more generally geometric dimensions associated with time are parameters ignored for their implementation complexity. Then taking them into account opens new ways of improvement. One could imagine a combination of two software like Matlab and Tecnomatix to solve more complex ALBP.

ACKNOWLEDGMENT

This work was supported by grants from the National Science and Technology Major Project of China [Grant No.2013ZX04012071].

REFERENCES

[1] D. Ajenblit, and R. Wainwright, "Applying genetic algorithms to the u-shaped assembly line balancing problem," In Proceedings of the IEEE Conference on Evolutionary Computation, pp. 96-101, 1998.

[2] P. Chiang, W. Kouvelis, and C. Chen, "An efficient heuristic for the u-shaped assembly line balancing problem in the just-in-time production environment," In Proceedings of National Annual Meeting to the Decision Sciences, pp. 1126, 1997.

[3] F. Guerriero, and J. Miltenburg, "The stochastic u-line balancing problem," J. Naval Research Logistics, Vol. 50, No. 1, pp. 31-57, 2003.

[4] H. Hwang, J. U. Sun, and T. Yoon., "U-line balancing with simulated annealing," In Proceedings of the First Asia-Pacific decision sciences institute conference, pp. 101-108, Hong Kong, 1996.

[5] G. J. Miltenburg, and J. Wijngaard, "The u-line balancing problem," Management Science, Vol. 40, No. 10, pp.1378-1388, 1994.

[6] K. Nakade, and K. Ohno, "Analysis and optimization of a u-shaped production line," Journal of Production Research Society of Japan, Vol. 40, No. 1, pp. 90-104, 1996.

[7] K. Nakade, and K. Ohno, "Stochastic analysis of a u-shaped production line with multiple workers," Computers and Industrial Engineering, Vol. 33, No. 3-4, pp. 809-812, 1997.

[8] K. Nakade, and K. Ohno, "An optimal worker allocation problem for a u-shaped production line," International Journal of Production Economics, Vol. 60-61, No. 2-4, pp. 353-358, 1999.

[9] A. Scholl, and R. Klein, "Ulino: Optimally balancing u-shaped jit assembly lines," International Journal of Production Research, Vol. 37, pp. 721-736, 1999.

[10] Ozcan, U., Toklu, B., "Balancing of mixed-model two-sided assembly lines," Computers & Industrial Engineering, Vol. 57, No. 1, pp. 217–227, 2009.

[11] Kim, Y.K., Song, W.S., Kim, J.H., "A mathematical model and a genetic algorithm for two-sided assembly line balancing," Computers and Operations Research, Vol. 36, No. 3, pp. 853–865, 2009.

[12] Wu, E.F., Jin, Y., Bao, J.S., Hu, X.F., "A branch-and-bound algorithm for two-sided assembly line balancing," International Journal of Advanced Manufacturing Technology, Vol. 39, No. 9–10, pp. 1009–1015, 2008.

[13] O. F. Baykoc, "Investigation the Behaviour of a Balanced Stochastic U-type Assembly Line Using Simulation," International Journal of Science & Technology, Vol. 3, No. 1, pp. 75-84, 2008.

[14] S. Chen, L. Plebani, "Heuristic for balancing U-shaped assembly lines with parallel stations," Journal of the Operations Research, Society of Japan, Vol. 51, No. 1, pp. 1-14. 2008.

[15] A. Bbaykasoglu, T. Dereli, "Simple and U-type assembly line balancing by using an ant colony based algorithm," Mathematical and Computational Applications, Vol. 14, No. 1, pp. 1-12. 2009.

[16] F. Talbot, "An integer programming algorithm with network cuts for solving assembly line balancing problem," Management Science, Vol. 30, pp. 85-99, 1984.

[17] S. Hackman, M. Magazine, and T. Wee, "Fast, effective algorithms for simple assembly line balancing problems," Operations research, Vol. 37, No. 6, pp. 916-924, 1989.

[18] Sparling, D., Miltenburg, J., "The mixed-model U-line balancing problem," International Journal of Production Research, Vol. 36, No. 2, pp. 485–501, 1998.

[19] Cheng, C.H., Miltenburg, J., Motwani, J., "The effect of straight- and u-shaped lines on quality," IEEE Transactions on Engineering Management, Vol. 47, No. 3, pp. 321–334, 2000.

[20] Miltenburg, J., "The effect of breakdowns on U-shaped production lines," International Journal of Production Research, Vol. 38, No. 2, pp. 353–364, 2000.

[21] Aase, G.R., Olson, J.R., Schniederjans, M.J., "U-shaped assembly line layouts and their impact on labor productivity: An experimental study," European Journal of Operational Research, Vol. 156, No. 3, pp. 698–711, 2004.

[22] G. Buxey, "Assembly line balancing with multiple stations," Management Science, Vol. 20, pp.1010–1021, 1974.

The Reuse Method of Design Knowledge Based on Knowledge Component

Ming-ming DONG*, Yan YAN, Guo-xin WANG, Jia HAO, Zhen-jun MING

Laboratory of Industrial Engineering, Beijing Institute of Technology, Beijing, China

(ming_ming_dong@163.com)

Abstract - **To speed up the product design efficiency, product designers would like to utilize the previous design knowledge. This requires a systematic and structured way for design knowledge representation and reuse. A new method which is named knowledge component to make design knowledge reused conveniently is presented in this paper. Knowledge component is a virtual reuse model which has specific function. The internal structure and working principle of knowledge component are discussed; meanwhile the involved design knowledge was analyzed in detail. The design process of barrel chamber was taken as an example to illustrate the executing process of knowledge component. Testing result shows that this method is feasible in terms of increasing design efficiency and helps the designers reuse the existing knowledge rapidly.**

Keywords – **Design automation, knowledge component, knowledge reuse, knowledge representation**

I. INTRODUCTION

Increasingly competitive and demanding markets are forcing enterprises to search for means to decrease time and costs for new product development, while satisfying customer requirements and maintaining design quality [1]. During the enterprises developing process, a large amount of design knowledge is accumulated, including product parameters, standards, specifications, templates, software using methods and so on [2]. Meanwhile, many new product development projects are based on the "variant design" where minor changes are made to the existing designs of the previous projects [3]. According to statistic, there are about 40% of product designs may reuse the existing components design directly, and 40% of product design may only make a minor changes, only 20% are entirely new design [4-6]. So we can see that it's greatly significant for enterprises to reuse the accumulated design knowledge if they want to design new products rapidly and win the market share. However, the existing knowledge managing and reusing methods, especially the complex modern products such as the planes, ships and so on, are far away to meet the practical requirement, which makes it extremely difficult, for even experienced design engineers, to trace the previous design routines. To achieve knowledge reuse during product design, much attention has been paid to integrate knowledge with product design process. Knowledge service and knowledge assistant are two important approaches for the integration [7-10]. In this paper, we present the method of knowledge component to implement the integrating.

The remainder of this paper is organized as follows. In Section II, the definition, structure, and invoking way of knowledge component will be introduced. In Section III, we will analyze the design knowledge involved in knowledge component in detail. In Section IV, the design process of barrel will be taken as an example to illustrate the feasible of knowledge component. In Section V, the conclusions are given.

II. KNOWLEDGE COMPONENT

A. The definition of knowledge component

Knowledge component is a virtual reuse model which has specific function. It can encapsulate different types of knowledge (design experience, templates, standards, etc.) in a structured way and accept parameters, execute actions and return corresponding results automatically under the control of drive program [2].

B. The structure of knowledge component

The structure of knowledge component is illustrated in Fig.1. It consists of 3 main parts, which can be represented as KC = {SC, E, S}. KC refers to the knowledge component.

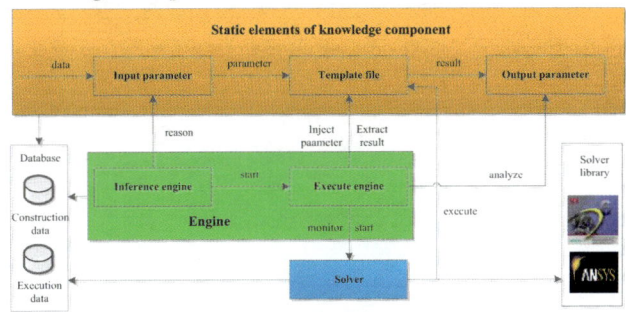

Fig.1. The structure of knowledge component

SC refers to the static elements of knowledge component, which includes input parameter, template file and output parameter. The function of input parameter is receiving data inputted by designer and representing design experience knowledge which can be used for reasoning design parameter. Template file can be parsed by particular solvers. For example, trail file can be used as the template of PRO/E; macro file can be used as the template of NX. Template is an important type of design knowledge which can record the previous design process. Output parameter is the container of the result.

E refers to component engine, which is the drive program of the knowledge component. It includes inference engine and execute engine. Inference engine can reason out design parameters by analyzing the designer's input data and previous design experience involved in the input parameter. Execute engine controls the executing

E. Qi et al. (eds.), *Proceedings of the 21st International Conference on Industrial Engineering and Engineering Management 2014*, Proceedings of the International Conference on Industrial Engineering and Engineering Management, DOI 10.2991/978-94-6239-102-4_12, © Atlantis Press and the authors 2015

process of knowledge component. It dynamically injects design parameters into the template file and calls for computing service from the solvers, when the computing is finished it will extract the results from the template file. After analyzing, the results are sent to the designers.

S refers to solver, which is used to complete a specific task. *A* solver is generally a software tool e.g. MATLAB, NX, ANSYS. Template file must be executed by the solvers.

C. The invoking way of knowledge component

Product design is a repeated iteration process which includes many stages, like modeling, simulation, optimize, and so on, so there are many repetitive works in the process [11]. We can encapsulate the repetitive work into knowledge component to improve the design efficiency. As shown in Fig.2, the invoking way of knowledge component is concise. In every stage of design process like validating parameter, generating model, analyzing model, and so on, the designers can invoke knowledge components which have been constructed from knowledge component management center. The invoke requests include two aspects of information, one aspect is the identifier of the knowledge component which we need to invoke, the other is the input data of the knowledge component. After sending the request, the designers need to do nothing to change or monitor the executing of the knowledge component except wait for the result returned by the management center.

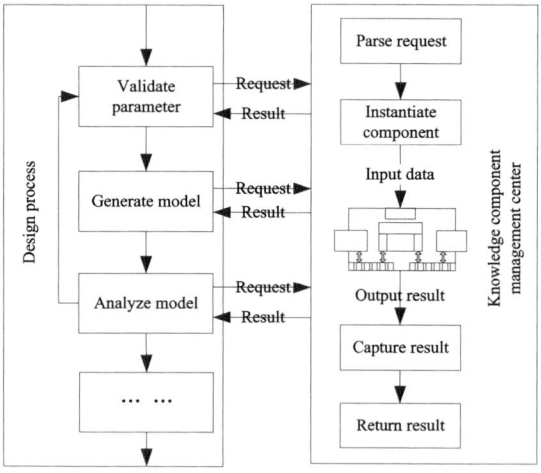

Fig.2. The invoking way of knowledge component

When the knowledge component management center receives the request, it parses the request and instantiates the corresponding knowledge component, and then injects the input data into the input parameters of the knowledge component, and then starts the driver program of the knowledge component. The knowledge component executes a series of actions automatically and outputs the results. At last, the knowledge component management center captures and returns the results to the designers. The entire life cycle of the knowledge component finishes.

III. DESIGN KNOWLEDGE INVOLVED IN KNOWLEDGE COMPONENT

Design knowledge can be acquired from various sources and generally requires an integrated representation for its effective and efficient reuse [12-13]. Team members' experience knowledge, together with the product design process knowledge, are very important intellectual properties of an enterprise and will tremendously improve the efficiency of future design projects if properly reused. In knowledge component, these two types of knowledge are respectively represented by input parameters and template files.

A. Experience knowledge

Compare with other types of knowledge, an obvious feature of experiences knowledge is the existence form. The majority of experience is stored in the designers' minds, so it's difficult to inherit and reuse. In order to solve this problem, we encapsulate this type of knowledge in form of production rules when modeling knowledge component input parameter. As in Fig.3, a production rule is a two-part structure using First Order Logic for reasoning over knowledge representation. "When" and "then" is the logical fields, "LHS" (Left-Hand Side) is the execution conditions of the rule, and "RHS" (Right-Hand Side) is the actions will be executed.

when
        LHS
then
        RHS

Fig.3. The structure of the rule

We take the barrel chamber design task as the example to illustrate the representation method of rules. In order to assure the sealing of chamber and the convenience of pumping bullet shell, the gap ($\Delta 1$, $\Delta 2$) between the barrel and the bullet shell must be kept in a reasonable scope, as showed in Fig.4. According to previous design experience, the gaps should be in accordance with formula (1)

$$0.05 \leq \Delta 1 \leq 0.30$$
$$0.05 \leq \Delta 2 \leq 0.35 \tag{1}$$

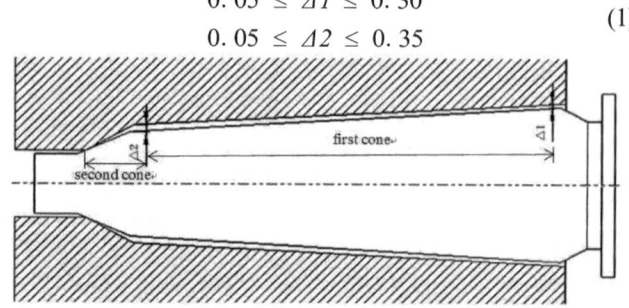

Fig.4. The assembly relation between barrel and bullet shell

The final rules are showed in Fig.5. In general, the parameters of the shell are known before designing the barrel. In "LHS", we can use some identifiers to indicate the diameters of shell, and then write the expression of target parameter using these identifiers and design experience in "RHS". When we design the chamber, the only thing we need to do is input the parameters of shell and design requirements of chamber into the knowledge component, then the engine will execute the rules automatically and the design experiences will be reused conveniently. The key advantage of this point is that using rules can make it easy to express solutions to difficult problems and consequently have those solutions verified. What's more, the experience knowledge can be much easier to maintain as there are changes in the future, as the knowledge is all laid out in rules.

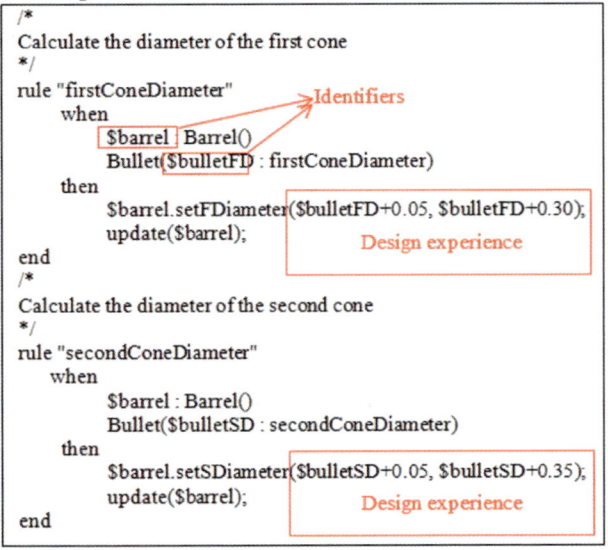

Fig.5. The content of rules

B. Design process knowledge

Design process knowledge is another important and abundant type of knowledge in enterprises. It records the whole solution and tools ever used to solve design problem, so we can rapidly complete design work by perfectly reproducing the previous design process [14-15]. Design process knowledge cannot represent by simple number or regular expressions, we represented it by template file in knowledge component. We take the macro file as the example to illustrate the modeling method of template. A streamlined template is shown in Fig.6. We need to pay attention to three aspects when modeling the template. First, we need to build the relationship between template and solver so that the template can be executed accurately. in this basic information we can find the information about solver. Second, we need to build the relationship between template and input parameter so that the design parameters can be injected into the template, in this example, we use ${example_T} to instead of variable T so that the template can receive parameters dynamically, which makes it more convenient to reuse the knowledge component. Third, we need to configure the path of the result file so that the engine can extract the result conveniently.

```
NX 7.5.0.32
Macro File: E:\chamber.macro
Macro Version 7.50                        Basic information
Macro List Language and Codeset: simpl_chinese 13
Created by DongMingMing on Mon Sep 02 17:27:01 2013
Part Name Display Style: $FILENAME
Selection Parameters 1 2 0.305441 1
Display Parameters 1.000000 16.880577 5.708333 -1.000000
****************
RESET
FOCUS CHANGE IN 1
MENU, 0, UG_FILE_OPEN UG_GATEWAY_MAIN_MENUBAR ! <MB/Toolbar>
FILE_DIALOG_BEGIN 0, ! filebox with tools_data
FILE_DIALOG_UPDATE 2
FOCUS CHANGE IN 1
FOCUS CHANGE OUT 1
FOCUS CHANGE IN 1|                        The path of result
FILE_DIALOG_END
FILE_BOX -2, example_prtFilePath   example_prtFilePath   0 ! Open
    SET_VALUE: 0 ! FSB item
    SET_VALUE: 1 ! FSB item
    SET_VALUE: 0 ! FSB item
FOCUS CHANGE OUT 1
...  ...
EVENT VALUE_CHANGED 0 0, 7340035, 0, 0, 0!
ASK_ITEM 7340035 (1 STRN 0) = "${example_T}"  !
EVENT FOCUS_OUT 0 0, 7340035, 0, 0, 0!       The input parameter
ASK_ITEM 7340035 (1 STRN 0) = "${example_T}"  !
... ...
MENU, 0, UG_FILE_QUIT UG_GATEWAY_MAIN_MENUBAR ! Pre actions<>
```

Fig.6. The content of template file

IV. CASE TESTING

In order to verify the feasibility of the proposed function of the knowledge component, we developed a prototype system and took the design process of barrel chamber as an example to test the executing process of a knowledge component, the process and result are shown in Fig.7: ① log in the system, select the "chamber design" knowledge component and double click to start; ② input the design requirements of the chamber and click the submit button, then the inference engine reasons the design parameters of chamber in the background; ③ examine all the parameters and change which does not suitable for requirements; ④ the selected knowledge component runs automatically; ⑤ the NX software starts

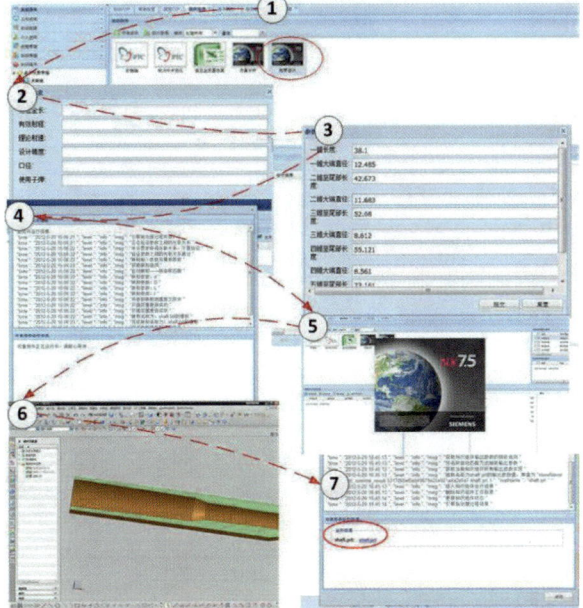

Fig.7. The executing process of knowledge component

itself; ⑥ the model of chamber is created; ⑦ a result file is returned to the designer. Of all the 7 steps, only 3 need to be operated by the designer, which indicates that the knowledge component is very easy to reuse. And the whole process only takes half a minute, shorter than operating in NX artificially.

V. CONCLUSION

It is a common sense that the previous design knowledge may play a crucial role in a product redesign process. Ensuring efficient knowledge reuse to support new product design, it is important to develop a systematic and structured way to represent the knowledge. In this paper, knowledge component is presented to implement this function. Knowledge component can represent design knowledge through its constituent elements and be executed automatically controlled by the engine. Designing product with knowledge component can rapidly reuse the design knowledge and greatly improve the efficiency of design, for it integrates design knowledge and design process and at the same time, making the CAD and CAE process run automatically.

It must be remarked that in the context of the present paper, only normalized design knowledge have been represented using knowledge component, whereas vague and unstructured design knowledge are not considered by the current approach. Actually the uncertainty and fuzziness often accompany the design knowledge. In our future work, we will study how to represent uncertainty and fuzziness knowledge so that knowledge component can have greater applicability.

ACKNOWLEDGMENT

This research work is supported by projects (A2220133001) funded by National Ministries and Commissions and projects (51375049) funded by the National Science Foundation of China. The authors are grateful to the anonymous reviewers for their kind comments on this paper.

REFERENCES

[1] M. Germani, M. Mengoni, and M. Peruzzini, An approach to assessing virtual environments for synchronous and remote collaborative design, Advanced Engineering Informatics, vol.26, pp. 793-813. 2012.

[2] HAO Jia, YANG Haicheng, YAN Yan, and WANG Guoxin, Study on product development process task oriented configurable knowledge component(in Chinese), Computer Integrated Manufacturing Systems, vol.18, no.4, pp.705-712, April, 2012.

[3] A. Al-Ashaab, M. Molyneaux, A. Doultsinou, B. Brunner, E. Martínez, F. Moliner, V. Santamaría, D. Tanjore, P. Ewers, and G.J. Knight, Knowledge-based environment to support product design validation, Knowledge-Based Systems, vol.87, pp.48-60, February, 2012

[4] M. Rezayat, Knowledge-based product development using XML and KCS, Computer-Aided Design, vol.32, pp.299-309, May, 2000.

[5] B.U. Haque, R.A. Belecheanu, R.J. Barson, and K.S. Pawar, Towards the application of case based reasoning to decision-making in concurrent product development (concurrent engineering), Knowledge-Based Systems, vol.13, pp.101-112, April, 2000.

[6] W.C. Regli, and V.A. Cicirello, Managing digital libraries for computer-aided design. Computer Aided Design, vol.32, pp.119-132, September, 1999.

[7] MENG Xianghui, and XIE Youbai, Embedded knowledge service supporting product development process (in Chinese), Computer Integrated Manufacturing Systems vol.15, pp.1049-1054, 2009.

[8] WANG Zhigang, LU Yiping, and ZHANG Xi'ai, Research on active knowledge assistant system oriented mechanical design process, Manufacturing Automation, vol.30, pp.12-15, 2008.

[9] SUN Chenyan, JING Shikai, and LIU Haibin, Complex product design process based on knowledge (in Chinese), Computer Engineering, vol.36, no.4, pp.283-288, 2010.

[10] ZOU Chunwen, ZHANG Shuyou, and LAI Jianliang, Research on process modeling of product structural design based on knowledge driven and its applications (in Chinese), Mechanical Engineering, vol.19, no.8, pp.929-944, 2008.

[11] G.S. Lynn, R.R. Reilly, and A.E. Akgun, Knowledge management in new product teams: practices and outcomes, IEEE Transactions on Engineering Management, vol.47, no.2, pp.221-230, 2000.

[12] WANG Hongwei, A. L. Johnson, and R.H. Bracewell, The retrieval of structured design rationale for the re-use of design knowledge with an integrated representation, Advanced Engineering Informatics, vol.26, pp.251-266, 2012.

[13] TANG Dunbing, ZHU Renmiao, TANG Jicheng, and XU Ronghua, Product design knowledge management based on design structure matrix, Advanced Engineering Informatics, vol.24, pp.159-166, 2010.

[14] LI Yuliang, ZHAO Wei, and SHAO Xinyu, A process simulation based method for scheduling product design change propagation, Advanced Engineering Informatics, vol.26, pp.529-538, 2012.

[15] TANG Guoxing, GUO Hun, and HU Jian, Study on a knowledge reuse based rapid product design process (in Chinese), Manufacture Information Engineering of China, vol.37, no.5, pp.38-42, and 2008.

Technology Satisfaction Measurement in Complex Product Development

Heng HE

Department of Management, Guangxi University of Science and Technology, Liuzhou, P. R. China, 545006
(heheng0772@163.com)

Abstract - **Complex product development is featured by technology innovativeness, large scale investment, long development cycle and uncertainty. Therefore, early technology assessment is essential for decision-making in the development project. This paper aims to provide a TSM (Technology Satisfaction Measurement) metric, which dynamically gauging the customer's satisfaction to the technology applied. The purpose of TSM is: 1) assessing market competitiveness of the product developed; 2) developing a synthesized metric to compose all of the technology measurement. Firstly, in this paper, a TRD (Technology Reliability Degree) distribution model has been presented. Secondly, a technology satisfaction distribution model has been built. Thirdly, TRD and TSM distribution model are synthesized to formulate the single technology's TSM. The fourth, weights were allocated to each of the technologies according to customer preference and thus a system technology satisfaction measurement was formulated. Finally, the method was exemplified by the application in the project of aircraft brake system development.**

Keywords - **Complex product development, complex product system, technology satisfaction measurement**

I. INTRODUCTION

Complex product System (CoPS) is defined as the product of complex customer requirement, complex composition, complex technology, complex manufacture ring process and complex project management, such as spacecraft, airplane, automobile, ship, complex Mechanical and electrical products, weapon system, etc. [1]. the complexity of CoPS has rendered the following features to the CoPS development project:

1) Technology innovativeness. There are more unknown knowledge area, more innovative point, and more exploration in the CoPS.

2) Large scale investment. The cost of a Cops development project may range from millions to billions.

3) Long development cycle. The process from project starting to prototyping and carrying on production may include many complex stages, and the cycle time may be several years, even a dozen years.

4) and uncertainty. These uncertainties include uncertainty of requirement, technology, cost, and market etc.[2].

These features have definitely complicated the decision-making for the CoPS development project.

Therefore, many researchers develop Key Performance Indicators (KPI) to monitor and control the CoPS development project [3]. Among all the KPIs, those measuring technology capability are especially important for CoPS development because of the technology complexity feature of the product.

Dynamic and effective measurement of system situation during the development process has become a key issue for the decision-making. The purpose of this paper is to provide a method which can dynamically and effectively measure the satisfaction of customer to the product being developed.

II. METRICS IN THE CoPS DEVELOPMENT PROCESS

According to literature, some technology metrics are static. it implies the performance of the technology can not be measured until the end of the development project, when the prototype or the product have been produced and put to validation or use in field [4]. Yet there some dynamic metrics is presented. In US military industry, TPMs (Technical Performance Measures) was used to manage technology development [5] [6]. In addition, TRL (Technology Readiness Level) was considered as an important indicator to decide whether the process should entry into next stage at milestone A, B, or C in the materiel acquisition process [7].

Literature [8] proposed that the technology developing of CoPS is a process of performance fluctuating, maturity growing, uncertainty reducing, and eventually the performance being stabilized in the design domain before the CoPS being put into operation. Evaluating technology performance in a development environment requires considering 3 factors: technology capability, technology maturity, and product requirement. The concept of TRD (Technology Reliability Degree) was present by combining TPM and TRL.

III. TECHNOLOGY SATISFACTION MEASUREMENT (TSM)

With the concept of TRD and TRD distribution model, metrics such as technology risk measurement, schedule measurement, and effort measurement, etc. can be developed. In this paper, however, they will be used to develop a Technology Satisfaction

E. Qi et al. (eds.), *Proceedings of the 21st International Conference on Industrial Engineering
and Engineering Management 2014*, Proceedings of the International Conference on Industrial Engineering
and Engineering Management, DOI 10.2991/978-94-6239-102-4_13, © Atlantis Press and the authors 2015

Measurement (TSM). TSM is a dynamic metrics. It can be define as the degree that customer will be satisfied with the technology performance of the product. Applying TSM has 2 purposes:

1) To evaluate market competitiveness of the product developed during the development process. This will help making decision about closing project, technology modification, or enhancing support.

2) To create a metric synthesizing all kinds of the technology measurement for the project. Technology metrics in CoPS development is not generally unique, therefore, how to composing them place difficulties for the project decision. PSM method synthesizes them by given the weights according to customer preference.

IV. TRD DISTRIBUTION MODEL

TRD can be defined as: according to current performance, the probability of the technology satisfying the requirement by the end of the development.

As Fig.1 shows, a requirement REQ related to a product PRD can be described as:

$$REQ:\ \mathrm{PRD}.ATT = \{a,b\} \qquad (1)$$

Where *ATT* is a measurable attribute of product; *a* and *b* are respectively lower limit and upper limit of the design domain. $\{a, b\}$ is the design domain.

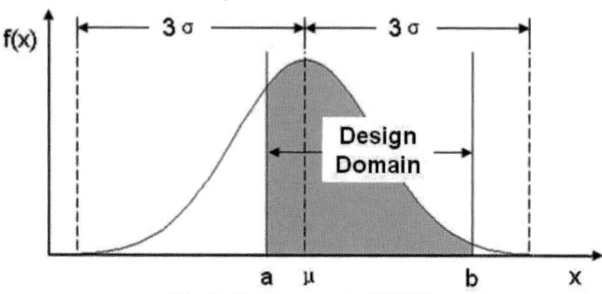

Fig.1. the concept of TRD

Technology TCH is the solution of requirement REQ, *x* is the technology capability. Considering technology uncertainty, the *x* value which the technology will finally gain is a random variable, Subjecting to the probability distribution function $f(x)$. for the purpose of simplification, let $f(x)$ is a continuous function and submits to the normal distribution. The TRD is calculated by the following formulas:

$$R = \int_a^b f(x)dx \qquad (2)$$

$$f(x) = \frac{1}{\sqrt{2\pi}\sigma} \exp\left(-\frac{(x-\mu)^2}{2\sigma^2}\right) \qquad (3)$$

Where:

R=TRD

$f(x)$ = technology capability distribution function;

μ = current measurement value of the technology capability. μ can be measured by calculation and analysis in the early stages(identified by TRL 1,2,3)

of the development project, or by testing and simulation in the middle stages (identified by TRL 4,5,6), or by validation or field operation in the later stage (identified by TRL 7,8,9).

σ is technology uncertainty. The larger the σ is, the more probable the technology will eventually deviate from current value. In engineering practice, σ can be determined by TRL. As TRL increasing, σ decreases. When TRL=9, σ=0. Development organization can specify the value of σ by formulating an σ-TRL relation curve according to related disciplines of the project.

V. TECHNOLOGY SATISFACTION DISTRIBUTION FUNCTION

When technology capability is *x*, the customer satisfaction to the technology can be calculated by $g(x)$. $g(x)$ is named as technology satisfaction distribution function. According to the shape of the function curve, TSM distribution function can be identified as 4 types: A) Horizontal, B) Trapezoid, C) Left Sloped, and D) right Sloped, As Fig.2 shows.

In which {a, b} is the design domain, s is customer satisfaction degree, $0 \le s \le 1$.

A Horizontal function curve means the customer satisfaction is make no deference to the technology capability (s=1) so long as the technology capability meet the design requirement ($a \le x \le b$).

A trapezoid function curve means the closer the x is to the region $\{c,d\}$, the higher the customer satisfaction. When $c \le x \le d$, the s gains its highest value (s=1).

A left sloped function curve means the greater the x is, the higher the customer satisfaction. When $x \ge c$, the s gains its highest value (s=1), c is therefore called satisfaction saturation point.

A right sloped function curve means the smaller the x is, the higher the customer satisfaction. When $x \le c$, the s gains its highest value(s=1), c is therefore called satisfaction saturation point.

VI. TSM FOR SINGLE TECHNOLOGY

According to the above statement, it can be known that when the value of technology capability get value x, the customer satisfaction will be $g(x)$. On the other hand, if current value of the technology capability is μ, and σ can be figured according to current TRL, and the possibility of technology capability of getting value x can be calculated by formula (3), so the formula of TSM for specific technology is:

$$s = \int_a^b f(x)g(x)dx \qquad (4)$$

TSM is the customer satisfaction degree to specific technology at the end of the development project, according to current technology capability value.

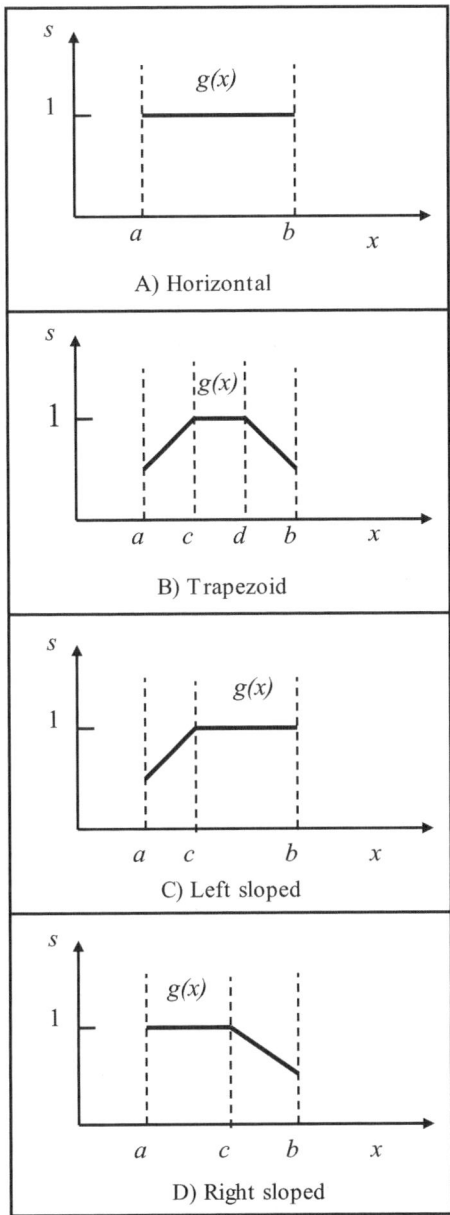

A) Horizontal

B) Trapezoid

C) Left sloped

D) Right sloped

Fig.2. the types of technology satisfaction distribution function

VII. COMPOSITION OF SYSTEM TSM (STSM)

Provided CoPS development system has several TPMs, any of the technology's TSM can be represented by s_i.

$$s_i = \int_{a_i}^{b_i} f_i(x) g_i(x) dx \qquad (5)$$

Where i=1, 2, ..., n

In order to synthesize each s_i into a STSM (marked by S), each s_i should be given a weight (marked by w_i) according to customer preference, w_i can be evaluated by Delphi method or analytic hierarchy process (AHP) method. The System TSM formula is:

$$S = \sum_{i=1}^{n} s_i w_i \qquad (6)$$

Where: S=STSM, $0 \le w_i \le 1$, and $\sum_{i=1}^{n} w_i = 1$

VIII. APPLICATION

An aircraft design company is going to develop a high quality aircraft anti-skid braking system. After investigation, they decided adopting the following technology, as listed in Table I. They hoped the new system will gain advantages on braking distance, weight and durability over the old system. To assess market competitiveness of the product being developed, STSM of both systems has been calculated for comparison.

Firstly the TPMs and TRLs of the two systems has been listed in Table I. Secondly, determine all parameters required by STSM calculation. Finally calculate STSMs of both systems, the result is shown as Table II.

Result shows the STSM of new system is much higher than that of the old system. This indicates new system will be more competitive than the old system, and suggests it is worthy to make strongly support for new system development.

TABLE I
TECHNOLOGY COMPARISON OF NEW SYSTEM AND OLD SYSTEM

Component	New system					old system		
	Technology	discipline	Design domain	Current TPM	Current TRL	Technology	Current TPM	Current TRL
Braking Device	Electrical Braking	mechanic-electronic	Weight (Kg) W={0,200}	W=160(Kg)	4	Hydraulic brake	W=300(Kg)	9
Anti-skid controller	Intelligent anti-skid control	Auto control	Braking distance(m) D={0,80}	D=70(m)	4	Differential braking control	D=80(m)	9
Brake disk	Powder metallurgic brake disk	Material	Brake disk life(h) L={1500, ∞}	L=1500(h)	5	C/C composite material	L=1200(h)	9

TABLE II
CALCULATING STSMS OF BOTH SYSTEMS

Component	Unit	New system			Old system			g(x)	a	b	w
		μ	TRL	σ	μ	TRL	σ				
Braking Device	W(kg)	160	4	20	300	9	0	Right Trapezoid	0	300	0.3
Anti-skid controller	D(m)	70	4	5	80	9	0	Right Trapezoid	0	80	0.4
Brake disk	L(h)	1500	5	50	1200	9	0	left Trapezoid	1200	99999	0.3
STSM		0.71			0.56						

Note: 1) The value of each σ is determined according to TRL and discipline. Because the TRL of old system is 9, so σ=0; 2) To make metrics of both systems suitable for comparison, design domain of new system should expand to include the TPM of old system; 3) The actual equations of TSM distribution function have not shown in this paper.

IX. CONCLUSION

TSM can dynamically evaluate the technology performances, so it often used in the early stage of the development to support decision-making. And STSM can be a very effective comprehensive metrics to understand the overall system technology situation.

REFERENCE

[1] LI Bohu, CHAI Xudong, ZHU Wenhai. Integrated Manufacturing System Technology of Complex Product [J]. Aeronautical Manufacturing Technology, 2002. (12): 17-40. [in Chinese]

[2] D. Hastings & D. Rhodes. Air Force/LAI Workshop on Systems Engineering for Robustness [EB/OL]. http://lean.mit.edu/index.php?option=com_docman&Itemid=776. 2004-6-8

[3] U. Dombrowski, K. Schmidtchen, and D. Ebentreich. Balanced Key Performance Indicators in Product Development [J]. International Journal of Materials, Mechanics and Manufacturing, Vol.1, No.1, PP.27-31, February 2013.

[4] Nadia Bhuiyan. A framework for successful new product development [J]. Jurnal of industrial engineering and management, 4(4), PP.746-770, 2011.

[5] Defense System Management College. Systems Engineering Fundamentals [M]. DSMC Press, 2001

[6] Garry J. Roedler, Cheryl Jones. Technical Measurement[R]. Technical Measurement, INCOSE, Lockhead Martin Corporation. 2005,12

[7] USA DoD Instruction 5000.2. Operation of the Defense Acquisition System [EB/OL]. http://www.deskbook.osd.mil, 2003.05.12.

[8] He Heng. Technology risk measurement in complex product development [C]. Proceedings of the 5th international conference on product innovation management. Wuhan: Hubei People Press, 2010, 286-291.

Multiscale Analysis of Reconfiguration for Reconfigurable Manufacturing Systems

Yan-di BAO*, Guo-xin WANG, Jing-jun DU, Yan YAN

School of Mechanical Engineering, Beijing Institute of Technology, Beijing, China
(charmanbyd@163.com)

Abstract - **Reconfigurable manufacturing systems (RMS) show multiscale characteristics on the granularity of reconfiguration. In order to assist manufacturing enterprises to appropriately select a reconstruction scale, the performance of manufacturing system was transformed into signal, which was disposed to be quantitatively expressed. On the basis of the characteristics and structure principles, the multiscale characteristics of RMS were proposed. Then a multiscale intrinsic model was established. The daily capacity was chosen as the production performance signal. Fourier transformation was used to reveal and quantitatively state the relationship between the reconfiguration scale and system performance. The model was then validated by means of a case-study.**

Keywords - **Fourier transform, multiscale, production performance, reconfigurable manufacturing system**

I. INTRODUCTION

In the 21st Century, in the circumstance of increasing global competition, manufacturing enterprises are confronting with rapidly changes in market demands [1]. Traditional paradigms, such as dedicated manufacturing lines and flexible manufacturing systems become inadequate in meeting the market demands on capacity and functionality, giving rise to redundancies and deficiencies in productive resources. To keep competitiveness in such a global competitive environment, manufacturing enterprises should utilize new manufacturing paradigms, which are costly effective and can rapidly respond to requirements. In 1999, the concept of reconfigurable manufacturing systems (RMS) was firstly proposed systematically by Koren [2, 3]. Recent years, reconfigurable machines [4-6], reconfigurable manufacturing cells [7-9], and reconfigurable system [10-14] were widely researched.

In this paper, the multiscale characteristics of RMS were described and a multiscale intrinsic mathematical model was established. Daily capacity, that is able to reflect production efficiency, was chosen as the production performance signal, so that the analysis process is simplified. The production performance signal is mathematically transformed to arrive at the quantitative relationship between reconfiguration scales and performance changes. The purpose intends to assist decision-makers in scales-selecting for reconfiguration.

Foundation item: Project supported by the National Natural Science Foundation, China (No. 51105039).

II. PRINCIPLES OF RECONFIGURATION

The construction of RMS includes equipment selection and distribution. Equipment selection is the procedure that matches the process design with the actual productive resources. Equipment distribution is the procedure that arranges the selected equipment, it's designed based on the product task and process requirements etc. In RMSs, products are grouped into families, each of which requires a system configuration [15]. When products are in the same family, the whole line reconfiguration is not necessary but some adjustments in the line will be needed as the products change.

Producers are primarily concerned about functionality and capacity. Generally, the reconfiguration of reconfigurable manufacturing tools (RMT) is oriented towards functionality, while the reconfiguration of production cells is oriented towards capacity. Machining precision is a type of functionality, which, in turn, will influence on the efficiency of a product line. When RMTs fail as regards precision, procedures such as updating or replacing the function modules are required. To improve capacity, adding or removing machines to match the new throughput requirements and concurrently rebalancing the system for each configuration, should accomplish the system reconfiguration [16].

A. Criteria selection

The high degree of flexibility of RMS is in terms of its capacity and functionality. With regard to the sophistication of RMS, productive function is no longer a primary issue, as there are various configurations, which could be selected for manufacturing configurable machines, which tend to cater for corresponding to requirements of processing techniques. The design of RMS is driven by external demand, since the aim of RMS excogitation is sustaining productive capacity with minimal redundancy and the lowest deficiencies.

Furthermore, productive functionality could be conveyed by productivity capacity. On the assumption that a current configuration of RMS does not possess the capacity for forging a particular output, then the capacity of the RMS for producing such an end product should be considered as being nil. On this basis, productive efficiency could evince the effect of the reconstitution and profitability of an enterprise, which should be considered as a dominant factor for a company, as a norm for measuring the performance of RMS productivity.

In dedicated manufacturing lines, the daily output of manufacturing systems is fixed. Based on RMSs, daily production changes, in response to external requirements. Regarding a RMS as an integrated item, without contemplating internal reconfiguration, the daily production could be treated as an index reflecting productive efficiency. Abiding by the criterion that, 'to reconstruct what is inferior', the effect of reconfiguration should be, 'visible' and demonstrated by daily productivity.

On the assumption that the operational use time of a manufacturing system per day is T and the number of workpieces produced during ΔT_i is n, then, the daily productivity of the whole system during ΔT_i could be calculated as Eq. (1),

$$N_i = \frac{nT}{\Delta T_i} \qquad (1)$$

According to economic theory, the external demand conforms approximately to the demand curve, which could be regarded as reflecting the lifecycle of the production. With external the needs augmenting, the productivity of RMS, cannot outweighs the outward demand, thus, the consignment could not be completed in a specific time and as a result, reconfiguration of the system should to be considered. On the assumption that the amount of orders in a day i is Q_{kji} and the requirement for delivering of the goods is before day j, the quantity of production in day m could be reckoned as formula (2),

$$R_m = \sum_{\substack{0<i<m, \\ m<j<n}} \frac{Q_{kij}}{j-i+1} \qquad (2)$$

As for RMS, the conceptual reconfiguration should achieve producing N_i at time i within the day m, fluctuating at, R_m according to the daily demand.

B. Production performance signal

On the basis of ignoring intermittent issues (**please check this carefully!) during the reconfiguration processing, the productive signal of manufacturing systems is incessantly changing, which could be regarded as a continuous-time signal. During the time domain, productivity could be considered as function $v(t)$ to time t, where the voltage of resistance is R, therefore, the average power is:

$$P = \lim_{a \to \infty} \frac{1}{2a} \int_{-a}^{a} \frac{v^2(t)}{R} dt \qquad (3)$$

When electrical resistance equals 1 ohm, the average power relates to the signal, as:

$$P = \lim_{a \to \infty} \frac{1}{2a} \int_{-a}^{a} v^2(t) dt \qquad (4)$$

And the power of the signal is:

$$E = \int_{-\infty}^{\infty} v^2(t) dt \qquad (5)$$

If a signal's energy has limits, it is called an energy signal; if its power is limited, it is called a power signal. A cycle of production, which is consistently limited, is a time-signal, therefore, in a life cycle of an RMS, the mechanical system's productivity signal is limited, which is a typical energy signal. If the ideal RMS is deemed to have ideal reusability and its life cycle is considered as being unlimited, then in ideal RMS life cycle, it has

unlimited consistent time-signal. Obviously, Eq. (4) has its limit and the RMS has unlimited energy, with limited power. The result is that, the productivity signal in whole life cycle is a power signal.

III. MULTISCALE CHARACTERISTIC OF RMS BASED ON PARTICLE SIZE

A. Intrinsic description

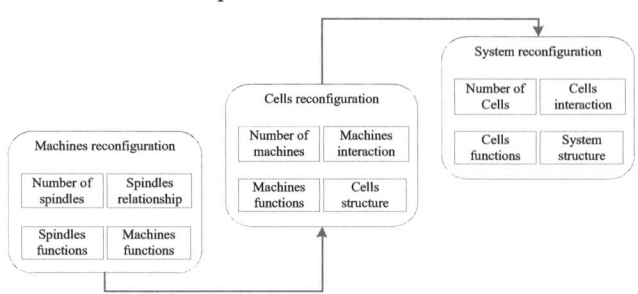

Fig.1. Characteristics of reconfiguration scale

The form of RMS presents multiscale features regarding time-scale and space-scale, therefore, the definition of multiscale should be made initially, when it is to be considered. When considering changes in different scales after a system has been divided into different scale group, in a space-scale, based on different granularity, it is possible to divide it into a system level, unit level and machine-tool level (Fig.1). Granularity is similar to the concept of granularity and scale in a landscape pattern, with the smaller the granularity, the more detailed is the research (Fig.2). The object of research changes from considering the layout of the plant, to the reconfiguration of reconfigurable manufacturing tools (RMT) according, to the operation cluster of a single workpiece.

Generally speaking, scale is the magnitude of time and space but to define scale in RMS in this manner is inappropriate. In a mechanical system, clearly the size of machines differs and different work pieces also have numerically level difference. This paper will analyze the relationship between RMS of systems, producing units for systems and machines for production units.

RMS is similar to a level system and therefore, the scale involved inclines towards the definition of organization scale and function scale, in Hierarchy theory. At different levels, different sub-systems have different functions and have a clear position in the complete system. Although they are different from typical time-scales, they can be considered as comparable in certain ways.

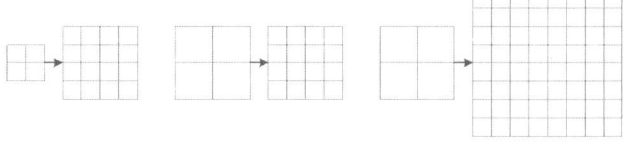

Change of amplitude   Change of granularity   Change of amplitude and granularity

Fig.2. Diagram of granularity and amplitude

In reconfiguration at the machine level, the object is the configuration of the machine tools. The process needs to consider the quantity of similar modules, types of module and the relationship of function-to-function and module-to-module in each machine. At the unit level, it is machines' quantity, function, structure and their inter-relationship that is important. At the system level, the object under consideration is a whole system's layout, which relates to the interaction of units' quantity, function, system structure and the like.

In practice, it is necessary to reconfigure to RMS by using the optimization rule, to choose that part which poses the need to reconfigure. The effect to the system is shown in the fluctuations of its performance. The reconfiguring of the machine tools of a unit should improve the performance of the unit, so long as the machine tools' performance is improved, which, in turn, Should improve the operation of the entire system. However, If the reconfiguring is relatively larger, (ex. The whole RMS's machines machinability becomes much better), this should bring obvious change to the system, by the accumulative effect.

The effect has, however, its own limitations. To reconfigure a unit at this time, should, however, significantly improve unit productivity.

B. Multiscale intrinsic model

In RMS, the operation relates to work pieces geometric features and forms mapping with them, which means that many operations relate to a single feature. An operation is similar to a process and if operations can be done by the same machine, this is called an operation cluster and can be considered as one-to-one relationship [7]. To ensure the structure of RMS, it is first necessary to design a craft for work pieces and then form an operation cluster. It is, then possible to bring order to the process, which will ensure the process route. Finally, the structure of the RMS is achieved. M, U, S is assumed as the structure of the machine, with unit and system, p_i as the operation cluster i. The process route for producing unit j is $R_j=[p_{j1}, p_{j2},...,p_{ji}]$, m_i as the structure of the machine for the operation cluster i, u_j as the structure of the producing unit j. $l(x)$ represents the information which is position or relative position of x. Assuming that the vector of the structure of the machine unit is $M_j=[m_{j1}, m_{j2},..., m_{ji}]$ and the vector of all producing units is $U=[u_1, u_2,..., u_j]$. For $X=[x_1, x_2,..., x_i]$, there is $L(X)=[l(x_1), l(x_2),..., l(x_i)]$ and thus, the function relationship is:

$f_M : p_{ji} \rightarrow m_{ji}$

$f_U : (M_j, L(M_j), [n_{j1}, n_{j2}...n_{ji}]) \rightarrow u_j$

$f_S : (U, L(U)) \rightarrow S$

n_{ji} represents number of machines in same position in unit j. Only if one exists, then $n_{ji} =1$. Since the RMS is changing with time, a function can be formed as:

$$m_{ji} = f_M(p_{ji}, t)$$

The function of a unit's structure is:

$$u_j = f_U(M_j, L(M_j, t), [n_{j1}(t), n_{j2}(t)...n_{ji}(t)])$$

For a system it is:

$$S = F_S(U, L(U, t))$$

Taking the system output function as f_c, then the C, the system output is:

$$C = F_c(S)$$

If the system is concerned with productivity, then C represents the system's productivity capacity. In the process of an RMS' design and configuration, the optimal goal is to allow the system meet external environmental changes. The deterministic variables would than become the RMS machine level, unit level and system structure. The constraint requirements are permitting the RMS' scale within certain limitations. As a result, an RMS optimization model is:

$$min \quad\quad F(S) = |F_c(S, t+t_0) - Q(S, t)|, \quad\quad (6)$$

$$s.t. \quad\quad M \in \{M_x \mid x \in N\}, \quad\quad (7)$$

$$U \in \{U_x \mid x \in N\}, \quad\quad (8)$$

$$S \in \{S_x \mid x \in N\}. \quad\quad (9)$$

In addition, $Q(S,t)$ represents the outer demand which S produced during the period of t. t_0 represents the response time to the external demand, which shows how rapidly RMS can react to external demand fluctuations. The result is that, the value is smaller than with a traditional mechanical system.

IV. SCALE DECOMPOSITION AND SELECTION

The implementation of unit reconfiguration signifies re-configuring a machine's layout in the workshop, which only operated as and when necessary, for instance, when significantly adjusting the performance of specific producing units or when the technicalities of a certain workpiece of some products has been substantially altered. The influence of reconfiguration of a unit scale for production performance is in terms of time-consumption, largeness of scale, which is reflected in the production performance signal, as demonstrated in the middle of the signal spectrum generated by unit-level reconstruction for production performance.

The frequency of system-level reconfiguration for a complete system is relatively low, with few operations during a lifecycle, this only occurring as regards producing cross-product processing or when the existing RMS configuration cannot meet the changing of external demands. The economic value of reconstruction is considerable, in that the cost for reconstruction is lower than the opportunity cost for the original production configuration. Nonetheless, reconfiguration for the whole system tends to incur the highest cost and have the lowest frequency. The reconfiguration should, however, be of benefit to production capacity and production function. With respect to the signal of production performance, the changing of system-level reconfiguration tends to be displayed in the forepart of the signal spectrum (Fig. 3).

With regard to any type of product, during a single lifecycle of the product, the longest interval between the adjacent effective unit-level reconstruction is T_c and the longest interval between the adjacent effective machine-level reconstruction is T_r. After every reconfiguration, the object will have its ramp up period and deterioration period one after another. Analyzing the graph of the production performance signal, it can be presumed reasonably that the fluctuation generated by a single reconfiguration of RMS corresponds to a half-cycle of the sinusoidal signal. The interval T reflecting the frequency domain being $\omega=\pi/T$. Each scale reconfiguration maps frequency domain, in which the different demarcation point is $\omega_c=\pi/T_c$ and $\omega_r=\pi/T_r$ respectively.

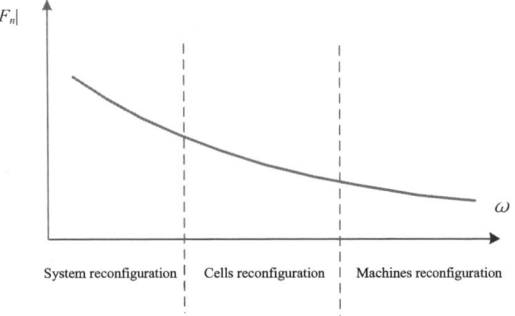

Fig.3. The tendency of Producing performance signal frequency spectrum and the distribution of each scale reconstruction

According to the regulated changing scope (the difference between the current daily output and daily demand) of an RMS system's performance, the corresponding amplitude of frequency spectrum will be $|F_n|=k\Delta$, (k is a constant). There is a value that corresponds to $|F_n|$ in the frequency spectrum, called ω and to be specific, if $\omega<\omega_c$, system-level reconfiguration were to be considered. On the other hand, if $\omega_c<\omega<\omega_r$, the unit-level reconfiguration should be taken into account, moreover, if $\omega>\omega_r$, the machine-level reconfiguration must be considered. Furthermore, if the frequency value ω, which corresponds to the amplitude $|F_n|$, spans multiple intervals, then the two forms of reconstruction should be considered. In this situation, issues such as cost, together with others factors should be took into consideration simultaneously, in terms of a reconstruction strategy.

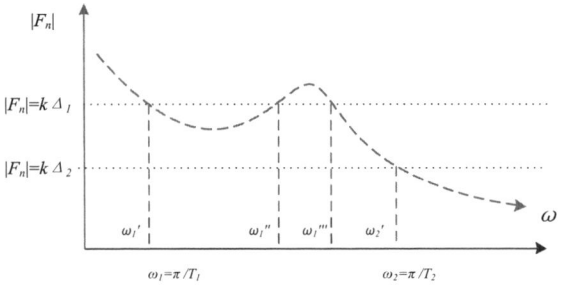

Fig.4. assistance decision-making of reconfiguration scales selection

Fig.4 illustrates the spectrum waveform for a particular RMS during a lifecycle. Such a wave pattern is representative (for instance, obtained from historical analysis) as a guiding value for reconstruction for a target system. One point to note is that if the periodic signal created by the prolongation indicates a single lifecycle, then the frequency spectrum tends to be discrete, which needs to connect each point as consecutive processing. On the contrary, if the analytic point is a full lifecycle, in which the wave shape of the frequency spectrum itself is continuous, it should be unnecessary for consecutive operation. With respect to two different external variations Δ_1 and Δ_2, the corresponding amplitudes are $|F_n|_1=k\Delta_1$ and $|F_n|_2=k\Delta_2$. From Fig.4, it can be seen that, with the variationΔ_1, there are relative frequencies designated as $\omega_1{}'$, $\omega_1{}''$, and $\omega_1{}'''$, and; with regard to variationΔ_2, the relative frequencies are designated as $\omega_2{}'$. Based on analyzing the specific interval for the respective frequency, as a result, as for variation Δ_1, the reconfiguration should be implemented for system-level or unit-level structure; as for variation Δ_2, the reconfiguration tends to aim at machine-level structure.

V. EXPERIMENT ANALYSIS

Because RMS still remains only at the theoretical level, there is, as yet no created RMS. Nonetheless, by using the method this paper recommends to analyse data from a mechanical system, it can, theoretical at least, give a possible indication the RMS' reconfiguration scale. Assuming production data for 100 days and taking every 4 hours as a unit to analyse daily production N_i in the i th 4 hours, which includes 600 samples (TABLE I).

TABLE I
PRODUCTION DATUM OF ONE ENTERPRISE

Number	t (hours)/4	Daily output N_i
1	1.00	3.12
2	2.00	4.57
3	3.00	5.67
4	4.00	5.65
...	...	...
...	...	...
...	...	...
599	596.02	11.35
600	597.01	10.57
601	600.99	7.25
602	602.01	4.30

The results of the example taken as sample are in accordance with the initial prediction. It is still possible; however, to establish if there is a significant scale effect from Low frequency reconfiguration on production capacity and that, then bigger range, the greater are the changes. With changes from 0 to 10, this means that the company has little unit RMS but machine-level RMS are not included in those of the system. In a traditional system, the adjustment of machines can be regarded as machine-level reconfiguration. The reconfiguration of the layout of manufacturing plant can be regarded as unit-level RMS because the structure of a production line is fixed and the company wills adjustment to combine machine tool and CNC, in an effort to improve productivity. This outcome totally different from a typical one because it has no reconfigurable and scalable features, which makes reconfiguration of the layout of mechanical plant more difficult. To sum up, Fig.6's prediction has its own logic and proves Fourier transform as being available for RMS.

The production ability signal in 100 days is shown as Fig.5:

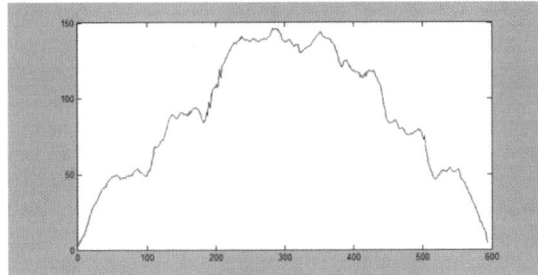

Fig.5. Production performance signal

After Fourier transformation by infrared, it is shown as Fig.6:

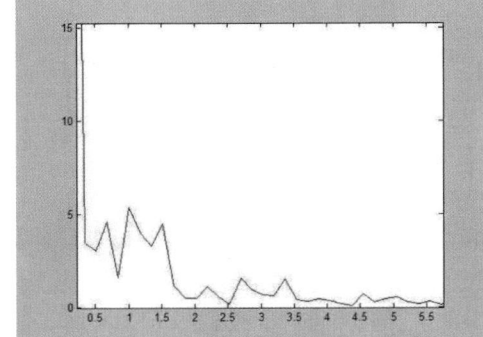

Fig.6. Spectrum curve of production performance signal

VI. CONCLUSION

This paper analyses the theory and process of RMS. It uses a mechanical system's daily production as an index of the production capacity of RMS and it collects production data regarding a productivity signal. By doing this, a spectrogram is created, which can reflect the relationship between reconfiguration's frequency and range. In addition, Fourier transform infrared has been proved to be useful in connection with the productivity signal of RMS and provide a simple way to assist a company to reconfigure range. It is also helps to make the collect and analyse of different ranges easier and it forms a sound basis for a multiscale model of RMS.

ACKNOWLEDGMENT

This research is supported by the National Natural Science Foundation, China (No. 51105039). The authors also express their thanks to the anonymous referees for their valuable time and constructive comments which improve the quality of the paper.

REFERENCES

[1]   Z. M. Bi, S. Y. T. Lang, W. Shen and L. Wang, "Reconfigurable manufacturing systems: the state of the art," International Journal of Production Research, vol. 46, no. 4, pp.967-992, 2008.

[2]   Y. Koren, U. Heisel, F. Jovane, T. Moriwaki, G. Pritschow, G. Ulsory and H. V. Brussel, "Reconfigurable Manufacturing Systems," *CIRP Annals - Manufacturing Technology*, vol. 48, no. 2, pp.527-540, 1999.

[3]   Y. Koren and M. Shpitalni, "Design of reconfigurable manufacturing systems," Journal of Manufacturing Systems, vol. 29, no. 4, pp. 130-141, 2010.

[4]   P. Spicer, Y. Koren, M. Shpitalni and D. Yip-Hol, "Design principles for machining system configurations," *CIRP Annals - Manufacturing Technology*, vol. 51, no. 1, pp.275-280, 2002.

[5]   K. Reuven, "Design principles of reconfigurable machines," *The International Journal of Advance Manufacturing Technology*, vol. 34, pp.430-439, 2007.

[6]   K. K. Goyal, P. K. Jain and M. Jain, "Optimal configuration selection for reconfigurable manufacturing system using NSGA II and TOPSIS," *International Journal of Production Research*, vol.50, no.15, pp.4175-4191, 2012.

[7]   L-U-R Ateekh-Ur-Rehman, "Manufacturing configuration selection using multicriteria decision tool," *International Jounal Manufacturing Technology*, vol. 65, no. 5-9, pp. 625-639, 2013.

[8]   J. Li, X. Z. Dai, Z. D. Meng, J. P. Dou and X. P. Guan. "Rapid design and reconfiguration of petri net models for reconfigurable manufacturing cells with improved net rewriting systems and activity diagrams," *Computer and Industrial Engineering*, vol.57, no.4, pp.1431–1451, 2009.

[9]   M. R. Abdi, "Layout configuration selection for reconfigurable manufacturing systems using the fuzzy AHP," *International Journal of Manufacturing Technology and Management*, vol.17, no.1/2, pp.149-165, 2009.

[10] R. Carin, "Reconfigurable production system design-theoretical and practical challenges," *International Jounal Manufacturing Technology*, vol.24, no.7, pp.998-1018, 2013.

[11] X. B. Zhao, J. C. Wang and Z. B. Luo, "A stochastic model of a reconfigurable manufacturing system: Part 1: A framework," *International Journal of Production Research*, vol. 38, no. 10, pp. 2273–2285, 2000.

[12] X. B. Zhao, J. C. Wang and Z. B. Luo, "A stochastic model of a reconfigurable manufacturing system: Part 2: Optimal configurations," *International Journal of Production Research*, vol. 38, no. 12, pp. 2829–2842, 2000.

[13] X. B. Zhao, J. C. Wang and Z. B. Luo, "A stochastic model of a reconfigurable manufacturing system: Part 3: Optimal selection policy," *International Journal of Production Research*, vol. 39, no. 4, pp. 747–758, 2001.

[14] X. B. Zhao, J. C. Wang and Z. B. Luo, "A stochastic model of a reconfigurable manufacturing system: Part 4: Performance measure," *International Journal of Production Research*, vol. 39, no. 6, pp. 1113–1126, 2001.

[15] R. Galan, J. Racero, I. Eguia and J. M. Garcia, "A systematic approach for product families formation in reconfigurable manufacturing systems," *Robotics and Computer-Integrated Manufacturing*, vol. 23, no. 5, pp. 489-502, 2007.

[16] W. C. Wang and Y. Koren, "Scalability planning for reconfigurable manufacturing systems," *Journal of Manufacturing Systems*, vol. 31, no. 2, pp. 83-91, 2012.

Benefiting from Supplier Supply Network Position: The Manufacturer Perspective

Sui-cheng LI, Na LI*

School of Economics and Management, Xi'an University of Technology, Xi'an Shaanxi, China
(lisc@xaut.edu.cn, *alice712@126.com)

Abstract - **The position of a supplier in the supply network is emerging as one of the important aspects of supplier management, but the role of managing supplier supply network position in improving the performance of manufacturers still remains unclear. This paper examines the impact of supplier supply network position on manufacturer performance and stresses the mediating role of supplier capability and supplier performance. Using structural equation modeling (SEM), this paper empirically tests a number of hypothesized relationship based on the data collected from 228 Chinese manufacturers. Results indicate that supplier supply network position does not have any significant direct effects on manufacturer performance, moreover, supplier capability and supplier performance plays a fully mediating role between supplier network position and manufacture purchasing performance.**

Keywords - **Manufacturer performance, supplier capability, supplier performance, supply network position**

I. INTRODUCTION

In today's business environment, reliance on supplier for manufacturing and innovation has become commonplace, the position of a supplier in supply network is emerging as one of important aspects of supplier management. In a supply network, supplier network position is defined as the outcome of the interaction relationships between a supplier and other actors, and then a supplier with the superior position in a supply network has more opportunities to learn from other actors and can access a large number of heterogeneous knowledge and information[1]. The network position also shapes the competitive priorities of a supplier and translates into resource advantage and capabilities. Thus a supplier with the superior position in a supply network can be expected to help its manufacturers obtain novel information and develop realistic marketing strategies[2,3].

Although existing researches have been realized that any supplier is embedded in a wide supply network, and suggested that the manufacturer should manage its suppliers in their structural network context, since the network surrounding of a supplier can affect the manufacturers' business decisions, behavioral choices, and economic outcomes[3,4]. While current researches on supplier network position emphasized individual itself as the research object and focused on the relationship between supply network position and innovation performance, the benefits of supplier network position and its role in improving the performance of manufacturers remain poorly understood.

Thus, this study addresses the research questions:

Does the supplier supply network position influence the performance of manufacturers? And if so, what's the mechanism? This remainder of this study is structured as follows. Section 2 reviews the theoretical background of this study. This is followed by the development of hypotheses. Next, we describe the research methodology and present the results. Finally, the last section draws some conclusions and suggestions for future research.

II. LITERATURE REVIEW AND HYPOTHESES

A. Supplier supply network position and manufacturer performance

Network position, a key variable of social network analysis, is the result of the interaction relationships between the actors in the network. With the deepening of the network research, scholars found that the various behavioral attributes of an actor, such as resource acquisition, strategy choice and innovation, can be interpreted as the function of its network position. The network position of an actor represents different opportunities to gain access to network resources and learn from other actors in the network, which plays an important role in operation and innovation of itself and partner firms.

Supplier supply network position refers to the structural position of a supplier in its supply network; it embodies the supplier's status and power that relates to other actors in network, and represents the opportunities and capabilities to obtain network resources. The effect of supplier supply network position on the performance of manufacturers is mainly manifested in the following two aspects: First, a superior position in a supply network promotes supplier knowledge base and absorptive capacity, then creating favourable conditions to satisfy the diversification demands of process improvement and new product development of manufacturers. Second, a supplier with the superior position could transfer network resources to manufacturers through the close manufacturer-supplier relationship, manufacturers can recognize the potential market opportunities or threats and review the competitiveness level of existing products and production technology so that they can promote product improvement and improve operational and innovation performance accordingly. Research hypotheses based on the above analysis:

H1. The supply network position of a supplier has a significant positive effect on the performance of manufacturers.

E. Qi et al. (eds.), *Proceedings of the 21st International Conference on Industrial Engineering and Engineering Management 2014*, Proceedings of the International Conference on Industrial Engineering and Engineering Management, DOI 10.2991/978-94-6239-102-4_15, © Atlantis Press and the authors 2015

B. Mediating role of supplier performance

A superior position can bring the abundant network resources to suppliers, but how to directly use suppliers' network resources is difficult for manufacturers. Studies have showed that the contribution that a supplier made to manufacturers depends on its deliverables—supplier performance, which is achieved by the integration of internal and external network resources of supplier. Supplier performance means whether a supplier is able to dispatch products with the agreed quality, cost, flexibility and innovation on time. As the initially external input, supplier product attributes and service performance will affect the final output performance of manufacturer[5,6], meanwhile, existing network research showed that a supplier with the superior position in a supply network could have opportunities to learn from other actors and gain access to new knowledge, which helped supplier provide the manufacturers with satisfying performance for a long time[7,8]. Research hypotheses based on the above analysis:

H2. Supplier performance plays a mediating role in the relationship between supplier supply network position and manufacturer performance.

C. Mediating role of supplier capability

Supplier capability refers to the supplier's ability to make use of its resources to meet the demands and business goals of manufacturers. From the perspective of the capability-based theory, supplier capability is a key factor to support its future business development and promote the improvement of manufacturer performance. There is the fact that different supplier capabilities can lead to different performance results[9], and directly or indirectly affect manufacturers' performance[10]. Only when having expertise, technical skills and knowledge resources can a supplier create unique competitive advantage in terms of R&D, production, network relationship and strategic capabilities, and launch more innovative, cheaper products than its competitors. In addition, manufacturers would arouse interest in collaborating with suppliers that do possess superior capabilities in order to capitalize on such supplier capabilities, avoid the inherent risks that are associated with partnering with incapable supplier, and create collaboration performance.

At the same time, supplier supply network position correlates significantly with supplier capabilities. A superior position is beneficial to strengthen the possibility of supplier learning and promote the effective integration between internal and external resources of supplier in order to realize supplier technical innovation and products upgrades, and improve the capabilities to meet the various demands of manufacturers. Research hypotheses based on the above analysis:

H3a. Supplier capability plays a mediating role in the relationship between supplier supply network position and supplier performance.

H3b. Supplier capability plays a mediating role in the relationship between supplier supply network position and supplier performance.

and manufacturer performance.

III. METHODOLOGY

A. Sample and data collection

To test the hypotheses on a broad empirical basis, we surveyed a cross-sectional sample of manufacturing firms located in China by means of face-to-face, onsite interviews and mail survey. A total of 276 questionnaires were filled out, and 48 were returned as nondeliverable, leading to an effective respond rate of 82.6%.

B. Variable definitions and measurement

All constructs for measuring independent and dependent variables were developed on the basis of previous literatures. The construct of supplier supply network position (SSNP) was designed to consider supplier extended supply network (which companies the supplier is connected to and how the connected companies fare in their own business) and how a supplier was embedded in its supply network. Based on the work by Choi and Kim[3], Gilsing et al.[11], Tsai et al.[12], Kim[13], the SSNP scale measuring used the five reflective items. Supplier capability (SC) scale was adapted from the prior research of Scheer et al.[14]. In this study, two items assessed supplier's product quality and delivery service, the other two items measured supplier capabilities in product improvement and new product design, the remaining two items evaluated the supplier capabilities to improve interaction quality and problem solving with the manufacturer through effective communication. The supplier performance (SP) scale was based on Ziggers and Henseler[15], Carter[16] and Shin et al. [17], using the five reflective items. The manufacturer performance (MP) scale measuring used five items from Azadegan[18]. All items were presented on 5-point Likert scales from 1= strongly disagree to 5= strongly agree.

C. Scale reliability and validity
TABLE I
DESCRIPTIVE STATITICS AND CORRELATIONMATRIX (N=228)

Variable	1	2	3	4
SSNP	(0.792)			
SC	0.234**	(0.784)		
SP	0.355**	0.311**	(0.721)	
MP	0.325**	0.360**	0.627**	(0.718)
Mean	4.0794	3.9250	4.0969	4.0676
S.D.	0.51175	0.58921	0.54743	0.43859

*significant at 0.05 level; **significant at 0.01 level.
( ) values on the diagonal are the square-root of AVE.

This paper used Cronbach's α and composite reliability (CR) to examine the reliability of each variable. Internal reliability was high, with Cronbach's α values between 0.804 and 0.888 and CR values between 0.8087 and 0.8932 for the subscales. We applied confirmatory factors analysis to evaluate the validity of the constructs. The overall fit of measured model was good (χ^2/df=1.3505; GFI=0.952; CFI=0.990; RMSEA=0.042), and the average variance extracted (AVE) for each variable value was above 0.5, indicating that all constructs have good convergent validity. Discriminant validity was supported as showed in TABLE I, all of the square root of AVE for each latent factor exceeded the respective correlation between factors. These results suggested that all measures exhibit satisfactory validity and can be used for hypothesis testing.

D. Statistical analysis

According to data in questionnaire survey, we tested our hypotheses based on structural equation modeling. The final model fitting was adequately supported. The χ^2/df index value of 1.355 and GFI index value of 0.948, both indices indicate an acceptable fit to the data. The RMSEA value of 0.043 at $p < 0.05$ indicates that the final model cannot be rejected at a high level of confidence. Furthermore other essential indices, such as CFI (0.979) and IFI (0.980), provide strong evidence that the fit between the structural model and the data is acceptable.

As depicted in TABLE II, the test results suggested that all hypotheses are supported. Supplier supply network position has not directly effect on performance of manufacturer ($\beta=0.285$, $p=0.736>0.05$), but the indirect effect generated by supplier capability and supplier performance is supporting, the value of indirect effects was 0.3353($0.328\times0.740+0.399\times0.232$), the results support H1, H2 and H3b. Furthermore, supplier capability also plays a mediating role in the relationship between supplier supply network position and supplier performance, the mediating effect intensity was 0.102 (0.399×0.256), the result supports H3a.

TABLE II
RESULTS OF HYPOTHESIS TESTING

	Standardized parameter estimate(β)	S.E.	C.R.	P
SSNP→SP	0.328	0.143	3.599	***
SSNP→SC	0.399	0.170	4.448	***
SSNP→MP	0.025	0.095	0.337	0.736
SC→SP	0.256	0.080	2.656	0.008
SC→MP	0.232	0.055	2.798	0.005
SP→MP	0.740	0.079	7.533	***

IV. RESULTS AND DISCUSSION

This study contributes to the existing literature on supplier management by investigating the impact of supplier supply network position on manufacturer performance and the mediating role of supplier capability and supplier performance. The research results show that supplier supply network position has a significant positive effect on manufacturer performance, meanwhile supplier capability and supplier performance play a fully mediating effect. Moreover, the supplier capability also plays a partial mediating role in the relationship between supplier supply network position and supplier performance.

There are two categories of managerial implications: First, this study tells managers that when evaluating suppliers, investigating supplier supply network position is generally beneficial, since a supplier's behaviors and performance depend on how it environs itself with other companies, especially its key suppliers and customers. Second, manufacturers should develop network awareness capability to identify their key suppliers' supply network position, to evaluate the informational and reputational values of the position, and to increase the level of integration between itself and supplier supply network.

The limitations should be acknowledged for future research. First, the questionnaire survey-based studies traditionally suffer from common method bias. Moreover, Sample data was collected from the manufacturers, it ignores suppliers' perspective. Finally, we empirically demonstrated that choosing a supplier who occupied a superior position in supply network is beneficial to improving manufacturer performance, but what specific managing practice that can promote this effect is still need for further exploration. Future research can probe into these issues.

ACKNOWLEDGMENT

This research was supported by National Science Foundation of China (Project Number: 71372172) and Humanity and Social Sciences Research of Ministry of Education (10YJA630085).

APPENDIX: KEY VARIABLES AND MEASURE

Supplier supply network position
(Cronbach's α=0.888; CR= 0.8932; AVE=0.6267)
SSNP1 suppliers who connected with our supplier fare in good condition.
SSNP2 customers who connected with our supplier fare in good condition.
SSNP3 our supplier interacts with its partner firms in supply network in high frequency.
SSNP4 the cooperation between our firm and other companies who are embedded in our supplier's supply network is brokered by our supplier.
SSNP5 Supply network ties generate significant influences on our supplier behaviors.
Supplier capability
(Cronbach's α=0.827; CR=0.8269; AVE=0.6142)
SC1 this supplier's product quality is excellent.
SC2 this supplier rarely delivers incorrect products.
SC3 this supplier could improve the features of its products our firm purchases each year.
SC4 this supplier could develop new technologies that enhance its products sourced by our firm.
SC5 this supplier could communicate with our firm effectively.
SC6 this supplier could timely provide our firm with information regarding problems it encounters.
Supplier performance
(Cronbach's α=0.838; CR=0.8420; AVE=0.5199)
SP1 compared with other suppliers, this supplier's product cost performance is competitive.
SP2 compared with other suppliers, this supplier's product quality is higher.

SP3 compared with other suppliers, this supplier delivers on time.

SP4 compared with other suppliers, this supplier's scheduling is flexible.

SP5 compared with other suppliers, this supplier's product is novel.

Manufacturer performance
(Cronbach's α=0.804; CR=0.8087; AVE=0.516)

MP1 using this supplier has enhanced our ability in reaching internal manufacturing cost reduction goal.

MP2 using this supplier has enhanced our ability in reaching defect rate reduction goals.

MP3 using this supplier has enhanced our ability in reaching delivery speed and reliability improvement goals.

MP4 using this supplier has enhanced our ability in responding to customization requests.

MP5 using this supplier has enhanced our ability in new product introduction time reduction goals.

REFERENCES

[1] Tsai, W. "Knowledge transfer in intraorganizational networks: Effects of network position and absorptive capacity on business unit innovation and performance". Academy of management journal, vol.44, no.5, pp.996-1004, 2001.

[2] Arya, B., Lin, Z. "Understanding collaboration outcomes from an extended resource-based view perspective: The roles of organizational characteristics, partner attributes, and network structures". Journal of management, vol.33, no.5, pp.697-723, 2007.

[3] Choi, T. Y., Kim, Y. "Structural embeddedness and supplier management: A network perspective". Journal of Supply Chain Management, vol.44, no.4, pp.5-13, 2008.

[4] Holmen, E., Aune, T. B., Pedersen, A.-C. "Network pictures for managing key supplier relationships". Industrial Marketing Management, vol.42, no.2, pp.139-151, 2013.

[5] Widener, S. K. "Associations between strategic resource importance and performance measure use: The impact on firm performance". Management Accounting Research, vol.17, no.4, pp.433-457, 2006.

[6] Garfamy, R. M. "Supplier selection and business process improvement: An exploratory multiple case study". International Journal of Operational Research, vol.10, no.2, pp.240-255, 2011.

[7] Uzzi, B. "Social structure and competition in interfirm networks: The paradox of embeddedness". Administrative science quarterly, 35-67, 1997.

[8] Zaheer, A., Bell, G. G. "Benefiting from network position: firm capabilities, structural holes, and performance". Strategic management journal, vol.26, no.9, pp.809-825, 2005.

[9] Hwang, D. W., Min, H. "Assessing the impact of ERP on supplier performance". Industrial Management & Data Systems, vol.113, no.7, pp. 1025-1047, 2013.

[10] Kannan, V. R., Tan, K. C. "Supplier selection and assessment: Their impact on business performance". Journal of Supply Chain Management, vol.38, no.4, pp.11-21, 2002.

[11] Gilsing, V. A., Duysters, G. "Understanding novelty creation in exploration networks—structural and relational embeddedness jointly considered". Technovation, vol.28, no.10, pp.693-708, 2008.

[12] Tsai, W., Su, K.-H., Chen, M.-J. "Seeing through the eyes of a rival: Competitor acumen based on rival-centric perceptions". Academy of Management Journal, vol.54, no.4, pp.761-778, 2011.

[13] Kim, D.-Y. "Understanding supplier structural embeddedness: A social network perspective". Journal of Operations Management, vol.32, no.5, pp.219-231, 2014.

[14] Scheer, L. K., Miao, C. F., Garrett, J. "The effects of supplier capabilities on industrial customers' loyalty: The role of dependence". Journal of the Academy of Marketing Science, vol.38, no.1, pp.90-104, 2010.

[15] Ziggers, G., Henseler, J. "Inter-firm network capability: How it affects buyer-supplier performance". British Food Journal, vol.111, no.8, pp.794-810, 2009.

[16] Carter, C. R. "Purchasing social responsibility and firm performance: The key mediating roles of organizational learning and supplier performance". International Journal of Physical Distribution & Logistics Management, vol.35, no.3, pp.177-194, 2005.

[17] Shin, H., Collier, D. A., Wilson, D. D. "Supply management orientation and supplier/buyer performance". Journal of operations management, vol.18, no.3, pp.317-333, 2000.

[18] Azadegan, A. "Benefiting from supplier operational innovativeness: The influence of supplier evaluations and absorptive capacity". Journal of Supply Chain Management, vol.47, no.2, pp.49-64, 2011.

The Analysis of the Industry-University-Research Cooperation of China Based on ISM

Peng Shen, Lu Feng, Ying Shen[*]

Economics and Management School, Changchun University of Technology, Changchun, China

(shenying@mail.ccut.edu.cn)

Abstract - **At present the most important of our national strategy is to deepen the reform of science and technology, to promote the closer integration of technology and economy and to accelerate the construction of the national innovation system, which is combined of enterprises, market and Industry-University-Research Cooperation. Our research on the Cooperation started late and there was only a little help to enhance China's comprehensive strength by the cooperation. Many factors are affecting the research results. This paper collects some mainly factors that affect the operation of the cooperation system by actual convey and analysis them by IMS, which provide the theoretical basis for the development of Industry-University-Research Cooperation. Cooperative system consists of subsystems of willingness to cooperate, cooperate resource subsystem and environment subsystems. The three subsystems work together to influence the development of the Industry-University-Research Cooperation system.**

Keywords - **Factors, Industry-University-Research Cooperation, SIM**

I. INTRODUCTION

Industry-University-Research Cooperation is an important component of the national innovation system. It is the key of the leading industry and strategic emerging industry in the country's overall development. But our country's cooperation has been in a low level over the years. There are many factors, interaction and mutual influence between them are complex. These factors form a more complex system. Chunhua Feng and others study from the cooperation and innovation and then to establish a Industry-University-Research Cooperation Theory which proposed by the government [1]; Xiaoli Guo think that it can improve our Industry-University-Research Cooperation by increasing corporate dominance and market leading through the combination study on the status and success of Japan. This makes a great contribution to our country [2]. But there is still no one summarize all of the factors together that may affect cooperation to analyze the relationship between each other systematic. This paper determines the impact factors of cooperation in accordance with the actual situation of China's Industry-University-Research Cooperation first, analyzes the influencing factors essentially by Interpretative Structural Model Method and then tries to find more effective measures to promote Industry-University-Research Cooperation

II. THE ANALYSIS OF COOPERATIVE INFLUENCING FACTORS

Xiaoyun Tang analyze the Industry-University-Research Cooperation on Macro perspective, he got five factors that risk investment mechanism is imperfect, the services of agency is imperfect, the necessary policies and regulations are not perfect, the conversion rate of research and development results need to be further improved, the lacks of governmental appropriate guidance and support and defined them as external factors of the Industry-University-Research Cooperation [3]. The lacks of research capacity of partners, cannot find a suitable partner, partners lack of integrity, the market value of R & D results are not high, lacks of the motivation and energy of cooperation. These five major factors are related to the participants of the Industry-University-Research Cooperation are named their own factors. The low level of cooperation, lack of technology transfer personnel, lack of equipment and funds, coordination and monitoring mechanisms, operating mechanism are imperfect, benefit distribution mechanisms are inadequate, unclear ownership of scientific and technological achievements, the lack of a platform for exchange, information of cooperation does not flow, these eight factors are common factors on both sides.

Azagra-Caro, Archontakis analyzed the influence of absorptive capacity of the region to the level of the Industry-University-Research Cooperation [4]. GDzisah and Etzkowitz analyzed the constraints of the cooperation by the "triple helix" theoretical, derived that the importance of industry and universities have become increasingly prominent, innovation policy is becoming results of the interaction, the role of the institutional environment is not obviously [5]. Hemphill and Vonotars, Judithsutz got that transaction savings and strategic motives are the main factors contributing to the Industry-University-Research Cooperation through opportunist theory [6]. Geisler also pointed out that the ideal willingness to cooperate, mutual trust and the psychological contract and previous history of cooperation, the degree of communication between partners, communication range, will influence the occurrence of cooperation [7].

Zhong Ling, Wang Zhanwu noted factors that restricting China research cooperation including research system is not perfect, the shortage of industrial capital markets combined with low college degree, intermediaries underdevelopment, poor flow of information, the promotion of science and technology talent shortage, imperfect scientific evaluation system [8]. Kong Yiping multiple regression model with SPSS software for analysis, and got that the research and development capabilities, size of the business, the technical characteristics of cooperation with the government policies have a direct impact on mode selection of the Industry-University-Research Cooperation [9]. Cui Xu,

Xing Li's survey showed the top three factors that affect the cooperation and got uneven distribution of rights and interests for 74.7%, technology is not mature enough for 36.8% decision management uncoordinated for 31.2% [10].

Fan Xia studied from the requirements of business cooperation threshold, by using of thresholds based on panel data regression model to make a conclusion that the company's own research capacity has a direct impact on the cooperation, which shows that companies with different sizes and different strategic intent have different influence on cooperation [11].

Scholars studied theoretical and empirical aspects of factors and draw some common conclusions that government support and relevant policies formulation, the main motivation for the cooperation parties and the organizational structure of cooperation have become important factors which would restrict the Industry-University-Research Cooperation. For information flow, capital investment, lack of talent, the impact of moral hazard, the distribution of benefits and other factors will have to be determined according to the actual situation [12-13].

TABLE I
THE RELATIONAL MODEL OF COOPERATION

Industry-University-Research Cooperation	The success rate of Industry-University-Research Cooperation	S1	
Willingness to cooperate	Dummy variables and intrinsic motivation of cooperation	S2	1
Cooperation resources	Basis of cooperation, including available resources come from companies and universities	S3	1
Cooperation Environment	Cooperative external environment, factors except for the two co-main, including government support, and environmental markets	S4	1
Venture Capital	A form of obtaining funds from the market. It can reduce the investment burden on the parties, and help to improve the cooperation willingness between the two sides.	S5	3,20
Transaction savings and strategic support policies	Cooperation in line with the concept of savings and compliance with national policy. This belongs to the cooperation environment	S6	4
Cooperative credit	It is dummy variable, namely the mutual cooperation relations. It has a direct impact on their willingness to cooperate	S7	12
Corporate research capacity	Ability of enterprises of their own research	S8	2,21
Research Capacity	Research capacity of universities that can make a contribution to the cooperation	S9	2,14,21
Development of cooperative	That is now a platform for cooperation. It can provide more convenience and security to the cooperation and increase willingness to cooperation between the two sides	S10	4,5
Government investment	It is also a part of the environment, which can increase sense of innovation and obtained money for universities, as well as improve the agents	S11	3,8,9,10
The level of cooperation	The level of cooperation between the two sides can use the advanced projects, complexity degree and time of cooperation to express	S12	2
History of cooperation	Due to historical cooperation partner of choice to produce subjective preferences	S13	7
Combination degree of universities and market	The knowledge of the needed content of research in the market.	S14	22
Corporate strategic intent	Enterprise's strategic intent is means that whether the cooperative enterprise views technology innovation as the main productive forces, which will affect the level of cooperation between the two sides as well as their willingness to cooperate enterprises	S15	2,12,17
Market demand for high-tech products	Belonging to the external environmental factors, and the potential for development of the market research results. It is the main driver of corporate willingness to cooperate	S16	2,4,14,22,21
Enterprises investment	Research funding invested by enterprises for cooperation	S17	3
Scientific concept of researchers	Whether researchers are trying to achievements in industry or completing the paper work and others	S18	2,12,21,22
The number of papers published and patent applications	The number of papers published and the high level of university patents published	S19	3,5,9,11
Interest distribution mechanism	Gain or loss sharing mechanism between universities and enterprises	S20	2
Characteristics of scientific content	High level of cooperation content between enterprises and universities	S21	12,22
High-tech industry index	The proportion of marke earnings that the scientific research transformed into practical products	S22	3,
Domestic GDP	China's GDP, reflecting the overall economic situation of China and the desire for technological innovation	S23	5,11,16
Direct benefits of cooperation to bring researchers	Through effective collaboration to increased revenue, which will increase the transformation idea of scientific research	S24	18

The current situation of our country is based on, from the perspective of cooperation to analyze and extract factors that influence the Industry-University-Research Cooperation. From the general sense of sociological sense, cooperation is an activity to achieve a common purpose between individuals, groups and communities, a mutually

supporting each other "joint action". Successful cooperation basic conditions are include: a common goal; unified understanding and norms; mutual trust and cooperative atmosphere; co-existence and development of certain material basis [14]. Firstly extracting factors from the goal of this basic condition: willingness to cooperate, cooperation model, cooperative goals, benefit distribution mechanisms, knowledge protection force, the flow of information flow and cooperation historical researchers. Second factor extracted from the material basis for cooperation: research capacity of enterprises, research capacity of colleges, universities and patent number of published papers, funding, benefit distribution mechanisms, scientific evaluation system, research ideas, technical cooperation project features, which funds invested sources include investment government, investment companies, venture capital on community. Lastly factor extracted from the environment for cooperation: government environmental policy, cooperation credit, development and research alliance, college degree combined with the market, the transaction savings and strategic motives, supervision and operation mechanism, history of cooperation, GDP. The interaction between them as shown in the Table I.

III. THE STRUCTURE MODEL OF RESEARCH SYSTEM IS AS FOLLOWS

For the relationship between the above factors, we calculated and processed by the model of SIM [15], finally got the hierarchical relationship between them as shown in Fig.1.

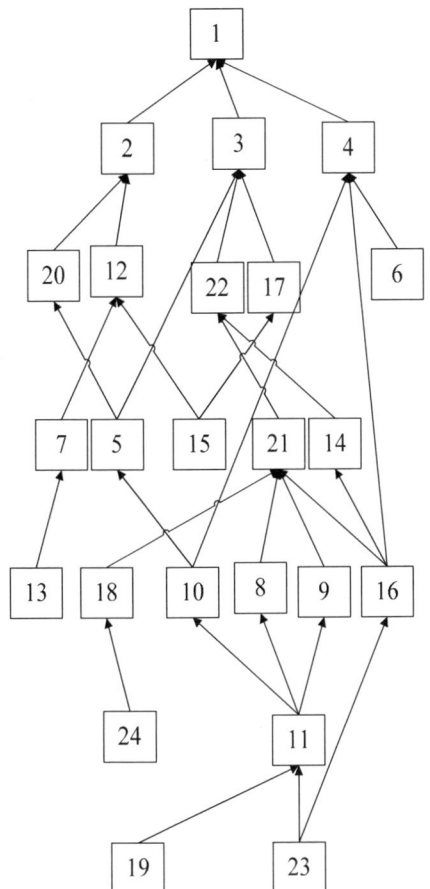

Fig.1. Progressive Struvture of Research Cooperation

IV. CONCLUSION

The analysis of the SIM result of the Industry-University-Research Cooperation:

Our goal of studying this system is to improve the efficiency of Industry-University-Research Cooperation. Based on the goal, there 22 influencing factors for this cooperative System, in the 22 factors, source conservation and policy of strategic support factors as external factors, and the rest 21 factors that constitute the three subsystems: subsystem of willingness to cooperate, collaborate resource subsystem and cooperate environmental subsystem, factors affect the system goals by the three subsystems.

1. Industry-University-Research Cooperation is affected directly by the willingness to cooperate, cooperation resources, cooperation environment. Other factors also affect the operation of the system by the three factors.

2. Assuming the state's policy is unchanged on T time, then the initial national policy on research cooperation would have a direct impact on the entire system in a long time, so the initial policy of the state is critical. So the policies made by relevant state departments must be combined with the actual situation, and the government must play a guiding role on the Industry-University-Research Cooperation

3. Cooperative alliances throughout the system have played a very important role in convergence. We can see the development of research alliance is a key part of the internal and external research cooperation will be linked together. So the emphasis on the development of research alliance is an important means to promote research cooperation. Especially in our country, because of the strength of the relatively large gap between research institutions, subject to geographical affect and influenced by the level of cooperation and co-habit with the impact of the previous form of cooperation obviously. Over the past few years, our study has improved the market conversion rate of our research, and brought great economic benefits. So it is necessary to increase government investment in research to improve the development platform and research alliances.

4. The main source of funding is government investment, venture capital, business investment. Venture capital funds mainly depends on the community, so the government's responsibility to play a good guide, business as the main role of research cooperation should increase investment in the cooperation.

ACKNOWLEDGEMENTS

Corresponding author is Ying Shen. The research work was supported by Academic Social Science and Humanities project of Ministry of Education of People's Republic of China (No.10YJA630048), the National Natural Science Foundation of China (No.71173025), Jilin province science and technology development plan project(No.20120646), Jilin province philosophy and social science planning projects (No.2012BS19) and Changchun city science and technology development plan project (No.2012199).

REFERENCES

[1] Feng Chunhua, "problems and countermeasures on the Industry-University-Research cooperation in Jilin Province"; in study of financial education; 25 Volume 6 November 2012

[2] Guo Xiaoli, "problems and analyze on the Industry-University-Research cooperation in Jilin Province" in Changchun University of Technology; 25 No. 1 January 2013

[3] Tang Xiaoyun, "Empirical Analysis and Countermeasures on Research Cooperative influencing factors" in Technology Management Research; 2009 5

[4] Joaquín M. Azagra-Caro, Fragiskos Archontakis, Antonio Gutiérrez-Gracia and Ignacio Fernández-de-Lucio," Faculty sup-port for the objectives of university–industry relations versus degree of R&D cooperation," in The importance of regional absorb-tive capacity [J]. Research

[5] Etzkowitz, H. & Leydesdorff, L. (1999), "The Future Location of Research and Technology Transfer," in Journal of Technology Transfer, 24: 111-123.

[6] Aokimasahi Harayama Yuko, "Industry-University Cooperation to Take on Here," from [J]. Research Institute of Economy, Trade and Industry, 2002(4): 42-49.

[7] Eliezer Geisler, "Industry-university technology cooperation- a theory of inter-organizational relationships," [J]. Technology Analysis & Strategic Management, 1995, 7(2): 217-229.

[8] Zhong Ling, Wang Zhanwu, "Analysis Collegiate constraints and Cooperative Mode," in Beijing University of Science and Technology Information; Social Science Edition; 2010-12-5

[9] Kong Yiping, "Factors of Cooperative Mode," in Chinese new technologies and products; 2010-4-10

[10] Cui Xu, Xing Li, "Our Industry-University-Research Cooperative Mode and constraints - Based on the government, enterprises, universities tripartite perspective," in Technology Management Research; 2010-3-23

[11] Fan Xia, "Complementary relationship between research cooperation and internal R&D - based in Guangdong Province," in the Ministry of Cooperative empirical; study science; 2011-5-15

[12] Gu Hai, "The analysis of restricting factor on research cooperation in the practice and countermeasures," Chinese technology industry; April 2001

[13] Liu Li, "Historical Review and discuss the nature of research cooperation," in Zhejiang University; Humanities and Social Sciences; February 2007

[14] Guo Xiaoli, "Stability of international energy cooperation," in Jilin University doctoral thesis

[15] Shen Ying, "The analysis of energy security cooperation in Northeast Asia on ISM," in Technical Economics and Management 2009 Section 6

Optimization of Supply Chain Network with Grey Uncertainty Demand by Improved Artificial Fish Swarm Algorithm

Jin-cheng FANG[1,*], Chang KE[1], Hua CHEN[2]

[1]Department of Transportation, Fujian University of Technology, Fuzhou, China
[2]Gulou District Party Committee Organization Department, Fuzhou, China
(fjcba@139.com)

Abstract - **This paper analyzed supply chain network optimization problem with the grey uncertainty of customer demand information, and established its mathematical models by grey system method. Then, an improved artificial fish swarm algorithm was put forward to solve the models based on binary encoding. Finally, the paper testified the effectiveness of the models and algorithm by using dynamic particle swarm optimization algorithm to solve the same calculation examples.**

Keywords - **Equal weight whitenization, grey uncertainty demand, improved artificial fish swarm algorithm, supply chain network optimization**

I. INTRODUCTION

With the economic globalization, the competition between enterprises turns to be more severe and the demand of customers varies from minute to minute. Under the volatile environment, how to reasonably plan supply chain network has become a hot topic in optimization management of supply chain system[1]. Grey system method is effective to tackle uncertain issues with inadequate sample data, poor information and shortage of cognitive experience [2, 3]. Therefore, this paper used the method of grey system in combination with artificial fish swarm algorithm, to study and solve the optimization of supply chain network with uncertainty of customer demand information as follows.

II. MATHEMATICAL MODEL ESTABLISHMENT OF SUPPLY CHAIN NETWORK OPTIMIZATION PROBLEM WITH GREY UNCERTAINTY DEMAND

A. Description of Problem

There is a four-stage supply chain network, which consists of M material vendors, N manufacturers, K distribution centers and L sales centers. The customer demand in each sales center is uncertain due to the variety of market competition. But historical data indicates that its demand information follows certain interval grey number distribution generally. Now, it requires to select some material vendors, manufacturers and distribution centers from the supply chain nodes, and meantime properly arrange their logistic relationship to accomplish the supply task at minimum cost.

B. Settings of Parameter

IDs for material vendor, manufacturer, distribution centers and sales centers are set as m, n, k and l

respectively. Set Raw material ID as i, and product ID as j. a_{imn} is unit price of raw material i purchased by manufacturer n from material vendor m; b_{jnk} is unit freight of product j from manufacturer n to distribution center k; c_{jkl} is unit freight of product j from distribution center k to sales center l; e_m^1 is maximum quantity of raw material provided by material vendor m; e_n^2 is maximum quantity of product produced by manufacturer n; e_k^3 is maximum amount of product distribution of distribution center k; f_n is fixed fees for the establishment and running of manufacturer n; r_k is fixed fees for the establishment and running of distribution center k; g_{jn}^1 is unit production cost of product j produced by manufacturer n; g_{jk}^2 is unit distribution cost of product j for distribution center k; h_{ij} is quantity of raw material i in need to manufacture product j per unit; P is maximum permissible quantity of manufacturers; Q is maximum permissible quantity of distribution centers; $d_{jl}(\otimes)$ is the grey quantity of product j demanded by sales center l, and its grey number interval is $[d_{jl}^-, d_{jl}^+]$.

C. Settings of Decision-making Variables

w_n: 0-1 variable, 1 denotes manufacturer n is selected, 0 is not; s_k: 0-1 variable, 1 denotes distribution centre k is selected, 0 is not; x_{imn}: quantity of raw material i purchased by manufacturer n from material vendor m; y_{jnk}: quantity of product j delivered from manufacturer n to distribution centre k; z_{jkl}: quantity of product j delivered from distribution centre k to sales centre l.

D. Establishment of Mathematical Model

Target function:

$$\min f = \sum_{i=1}^{I}\sum_{m=1}^{M}\sum_{n=1}^{N} a_{imn}x_{imn} + \sum_{j=1}^{J}\sum_{n=1}^{N} g_{jn}^{1}\sum_{k=1}^{K} y_{jnk} + \sum_{j=1}^{J}\sum_{k=1}^{K} g_{jk}^{2}\sum_{l=1}^{L} z_{jkl} +$$

$$\sum_{j=1}^{J}\sum_{n=1}^{N}\sum_{k=1}^{K} y_{jnk}b_{jnk} + \sum_{j=1}^{J}\sum_{k=1}^{K}\sum_{l=1}^{L} z_{jkl}c_{jkl} + \sum_{n=1}^{N} w_{n}f_{n} + \sum_{k=1}^{K} s_{k}r_{k} \qquad (1)$$

Constraint condition:

$$\sum_{i=1}^{I}\sum_{n=1}^{N} x_{imn} \le e_{m}^{1} \qquad \forall m \qquad (2)$$

$$h_{ij}\sum_{j=1}^{J}\sum_{k=1}^{K} y_{jnk} \le \sum_{m=1}^{M} x_{imn} \qquad \forall n \qquad (3)$$

$$\sum_{l=1}^{L} z_{jkl} \le \sum_{n=1}^{N} y_{jnk} \qquad \forall j,k \qquad (4)$$

$$\sum_{k=1}^{K}\sum_{j=1}^{J} y_{jnk} \le w_{n}e_{n}^{2} \qquad \forall n \qquad (5)$$

$$\sum_{l=1}^{L}\sum_{n=1}^{N} z_{nkl} \le s_{k}e_{k}^{3} \qquad \forall k \qquad (6)$$

$$\sum_{n=1}^{N} w_{n} \le P \qquad (7)$$

$$\sum_{k=1}^{K} s_{k} \le Q \qquad (8)$$

$$\sum_{k=1}^{K} z_{jkl} \ge d_{jl}(\otimes) \qquad \forall j,l \qquad (9)$$

$$x_{imn}, y_{jnk}, z_{jkl} \ge 0, \qquad \forall i,m,n,j,k,l \qquad (10)$$

$$w_{n}, s_{k} \in \{0,1\} \qquad \forall n,k \qquad (11)$$

In above models, formula (1) is target function, which ensures the minimum total cost of manufacturing and distribution in the supply chain network, and formulas (2)-(11) are the capacity constraints.

E. Whitening of Grey Parameter and Transformation of Mathematical Model

The grey variable $d_{jl}(\otimes)$ in the original mathematical formula (9) makes the constraint condition unclear and the original model unresolved. Generally, for a certain interval grey number $\otimes \in [a,b]$, its whitening value could be assigned as $\tilde{\otimes} = \alpha a + (1-\alpha)b$, where $\alpha \in [0,1]$ is position coefficient [4]. If the distribution information for the value of interval grey number is inadequate, it is usually whitened with equal weighted index, where the positioning coefficient $\alpha = 1/2$ [5]. Accordingly, $d_{jl}(\otimes)$ in the original model could be whitened with the function in formula (12). By the same token, the original formula (9) could be transformed into formula (13).

$$d_{jl}(\otimes) = (d_{jl}^{-} + d_{jl}^{+})/2 \qquad (12)$$

$$\sum_{k=1}^{K} z_{jkl} \ge (d_{jl}^{-} + d_{jl}^{+})/2 \qquad \forall j,l \qquad (13)$$

III. IMPROVED ARTIFICIAL FISH SWARM ALGORITHM AND ITS SOLVING PROCEDURE

A. Principle of Artificial Fish Swarm Algorithm

Artificial fish swarm algorithm (AFSA) is a kind of evolutionary computation technology based on swarm intelligence with the advantage of simple concept, easy implication, good robustness and high speed in finding the optimized solution [6, 7]. It has been widely applied in solving combinational optimization problem.

B. Encoding and Solving Procedure of Improved AFSA

AFSA, which is continuous intelligent algorithm, could not solve the discrete-type combinational optimization model. Therefore, this paper proposed an improved artificial fish swarm algorithm (IAFSA) based on binary encoding method [8], which encoded the individual artificial fish discretely. Hence, behaviors of artificial fish, which include following behavior, swarming behavior, preying behavior and random behavior, are selected in accordance with fitness evaluation of objective function. The problem will be solved after repeated iterative operation and replacement of bulletin with best state of artificial fish. The general solving procedure is as follows:

Step 1 Set parameters of AFSA, for details see [6];

Step 2 Define position vector $x = [x_1, x_2, \cdots, x_N, x_{N+1}, \cdots, x_{N+K}]$ for individual artificial fish. Function RAND is used to generate the artificial fish's initial position vector;

Step 3 Fuzzy function SIGMOID and random function RAND are employed for binary encoding of artificial fish, as shown in formula (14) and (15);

$$\text{sigmoid}(x) = \frac{1}{1 + \exp(-x)} \qquad (14)$$

$$y_i = \begin{cases} 0, if\ (\text{rand} \ge \text{sigmoid}(x_i)) \\ 1, otherwise \end{cases} \qquad (15)$$

Step 4 Let $w_i = y_i$, $s_j = y_{N+j}$. x_{imn}, y_{jnk} and z_{jkl} could be calculated by simplex algorithm. Having them substituted into objective function f(x), we could obtain the initial position of every artificial fish $(x, f(x))$. The best position state of every artificial fish $(x_{best}, f(x_{best}))$ is selected as the initial one in the bulletin.

Step 5 the operations on each artificial fish:

(1) Following behavior: Let $x^{(i)}$ denotes the artificial fish i current position. Its neighborhood area within perceptive scope is explored for the best neighboring position $x_{best}^{(j)}$. If $f(x^{(i)}) > f(x_{best}^{(j)})$ and the number of companions within $x_{best}^{(j)}$ area nof/N$< \delta$, let

$x^{(i)} = x^{(i)} + Random(Step) \times (x^{(j)}_{best} - x^{(i)}) / \left\| (x^{(j)}_{best} - x^{(i)}) \right\|$

and update state with $(x^{(i)}, f(x^{(i)}))$. Thus, following behavior succeeds, and turn to step 6. Otherwise, the behavior fails and turns to (2) for swarming behavior.

(2) Swarming Behavior: Set nof as the number of companions within $x^{(i)}$ area. If nof/N $< \delta$ and $f(x^{(i)}) >$ $f(x_c)$, let $x^{(i)} = x^{(i)} + Random(Step) \times (x_c - x^{(i)}) / \left\| (x_c - x^{(i)}) \right\|$. Thus, swarming behavior succeeds and the state will be updated. Otherwise, swarming behavior fails and turns to (3) for preying behavior.

(3) Preying Behavior: A position state $x^{(i)}_{new}$ is selected randomly within the perceptive scope of $x^{(i)}$. If $f(x^{(i)}) > f(x^{(i)}_{new})$ holds, let $x^{(i)} = x^{(i)} + Random(Step) \times (x^{(i)}_{new} - x^{(i)}) / \left\| (x^{(i)}_{new} - x^{(i)}) \right\|$. Thus, preying behavior succeeds and the state will be updated. Turn to step 6. Otherwise, select another position state randomly for preying. If artificial fish could not prey successfully after Try_num times of trial, turn to (4) for random behavior.

(4) Random Behavior: the artificial fish chooses a position randomly within the visual range and move to this new position x_{new}. If $f(x^{(i)}) > f(x_{new})$ holds, let

$x^{(i)} = x^{(i)} + Random(Step) \times (x_{new} - x^{(i)}) / \left\| (x_{new} - x^{(i)}) \right\|$

and the state is updated.

Step 6 Compare the current state of artificial fish with the one on the bulletin. If its current state is better, update the bulletin.

Step 7 If the iteration times has reached the maximum limit ITmax, end the iteration. Otherwise, turn to step 3.

IV. EXPERIMENT OF CALCULATION EXAMPLES

A. Example Description

The data in this example is cited from [9]. Suppose there are 3 material vendors, five candidate manufactures, five candidate distribution centers and four sale centers in the supply chain network, which mainly undertakes to manufacturer, distribute and sell two products A and B. Product A is made by raw material C and D in proportion of 2:1. Product B is made by raw material C and D in proportion of 1:1. Due to market fluctuation, the demands of products from sales centers are grey variables, whose grey demand interval is shown in Table I. Capacity constraint and fixed cost of each material vendor, manufacturer, distribution center and sales center is shown in Table II. Running cost between each node in the supply chain network is shown in Tables III-VI. The decision-maker is searching for the optimization design of supply and distribution network, to minimize the overall operation cost of the supply chain.

TABLE I
THE GREY INTERVAL OF PRODUCT A AND B DEMANDED BY EACH SALES CENTER

Sales Center i	Grey Demand Interval of Product A by Node i	Grey Demand Interval of Product B by Node i
1	$\otimes d_{11} \in [150, 200]$	$\otimes d_{21} \in [120, 150]$
2	$\otimes d_{12} \in [100, 160]$	$\otimes d_{22} \in [100, 260]$
3	$\otimes d_{13} \in [200, 250]$	$\otimes d_{23} \in [150, 170]$
4	$\otimes d_{14} \in [180, 200]$	$\otimes d_{24} \in [100, 150]$

TABLE II
CAPACITY CONSTRAINTS AND FIXED COSTS OF EACH NODE IN THE SUPPLY CHAIN NETWORK

Type of Node	S_1	S_2	S_3	P_1	P_2	P_3	P_4	P_5	D_1	D_2	D_3	D_4	D_5
Capacity Constraint	1500	1000	1500	400	550	490	300	500	530	590	400	370	580
Fixed Cost	0	0	0	1800	900	2100	1100	900	1000	900	1600	1500	1400

Note: In the table above, S_m, P_n and D_k represent the nodes of material vendor m, manufacturer n and distribution center k respectively in the network (the same below). Capacity constraint indicates the maximum capacity of supply, production and distribution. Fixed cost is the fees for establishment and running of the nodes.

TABLE III
UNIT PRICE OF RAW MATERIAL I PURCHASED BY MANUFACTURER N FROM MATERIAL VENDOR M

a_{1mn}	P_1	P_2	P_3	P_4	P_5	a_{2mn}	P_1	P_2	P_3	P_4	P_5
S_1	5	6	4	7	5	S_1	5	6	7	6	5
S_2	6	5	6	6	8	S_2	8	6	5	5	7
S_3	7	6	3	9	6	S_3	4	5	3	9	6

TABLE IV
UNIT FREIGHT OF PRODUCT J FROM MANUFACTURER N TO DISTRIBUTION CENTER K

b_{1nk}	D_1	D_2	D_3	D_4	D_5	b_{2nk}	D_1	D_2	D_3	D_4	D_5
P_1	5	8	5	8	5	P_1	3	5	3	5	5
P_2	8	7	8	6	8	P_2	5	6	6	8	7
P_3	4	7	4	5	4	P_3	3	5	4	4	3
P_4	3	5	3	5	3	P_4	3	4	3	5	3
P_5	5	6	6	8	3	P_5	4	5	5	6	3

TABLE V
UNIT FREIGHT OF PRODUCT J FROM DISTRIBUTION CENTER K TO SALES CENTER L

c_{1kl}	C_1	C_2	C_3	C_4	c_{2kl}	C_1	C_2	C_3	C_4
D_1	7	4	5	6	P_1	5	3	4	5
D_2	5	4	6	7	P_2	4	3	5	6
D_3	7	5	3	6	P_3	6	4	3	5
D_4	3	5	6	4	P_4	3	4	5	3
D_5	4	6	5	7	P_5	3	5	4	6

TABLE VI
UNIT PRODUCTION COST g^1_{jk} AND UNIT LOGISTICS DISTRIBUTION COST g^2_{jk} OF PRODUCT J

g^1_{jk}	P_1	P_2	P_3	P_4	P_5	g^2_{jk}	D_1	D_2	D_3	D_4	D_5
Product 1	1	1	2	2	2	Product 1	2	1	2	2	1
Product 2	1	1	2	2	1	Product 2	1	1	2	1	1

B. Experimental Results and Discussion

The problem is solved by the models and IAFSA mentioned above. The experimental parameters of IAFSA are set as: N =60, $ITmax$ =100, $Visual\_D$ =2, Try_num =30, $Step$=0.2, δ =0.8. The calculation results are shown as follows:

$$x_1 = \begin{bmatrix} 0 & 0 & 31 & 0 & 833 \\ 0 & 420 & 0 & 0 & 0 \\ 0 & 0 & 756 & 0 & 0 \end{bmatrix},$$

$$x_2 = \begin{bmatrix} 0 & 40 & 0 & 0 & 501 \\ 0 & 59 & 78 & 0 & 0 \\ 0 & 226 & 416 & 0 & 0 \end{bmatrix},$$

$$y_1 = \begin{bmatrix} 0 & 0 & 0 & 0 & 0 \\ 0 & 95 & 0 & 0 & 0 \\ 293 & 0 & 0 & 0 & 0 \\ 0 & 0 & 0 & 0 & 0 \\ 0 & 0 & 0 & 0 & 332 \end{bmatrix},$$

$$y_2 = \begin{bmatrix} 0 & 0 & 0 & 0 & 0 \\ 82 & 148 & 0 & 0 & 0 \\ 137 & 0 & 0 & 0 & 64 \\ 0 & 0 & 0 & 0 & 0 \\ 0 & 0 & 0 & 0 & 169 \end{bmatrix},$$

$$z_1 = \begin{bmatrix} 0 & 35 & 68 & 190 \\ 0 & 95 & 0 & 0 \\ 0 & 0 & 0 & 0 \\ 0 & 0 & 0 & 0 \\ 175 & 0 & 157 & 0 \end{bmatrix},$$

$$z_2 = \begin{bmatrix} 0 & 32 & 62 & 125 \\ 0 & 148 & 0 & 0 \\ 0 & 0 & 0 & 0 \\ 0 & 0 & 0 & 0 \\ 135 & 0 & 98 & 0 \end{bmatrix},$$

Optimum value $\min f = 36419$.

To verify the effectiveness of IAFSA algorithm, dynamic particle swarm optimization (DPSO) algorithm brought out by [10] is used for comparison and checking computation. After running algorithm DPSO and IAFSA each for 300 times of itineration calculation, their optimal results are both 36419. The calculation process is shown in Fig.1 and Table VII. The comparison of the iteration calculation of these two algorithms indicates that IAFSA has both advantages and disadvantages. Its solution has high precision and small relative error in average, and its algorithm has good robustness. However, it has some weakness as long response time in solution, low convergence speed and easy sinking into local optimum during iteration in later period.

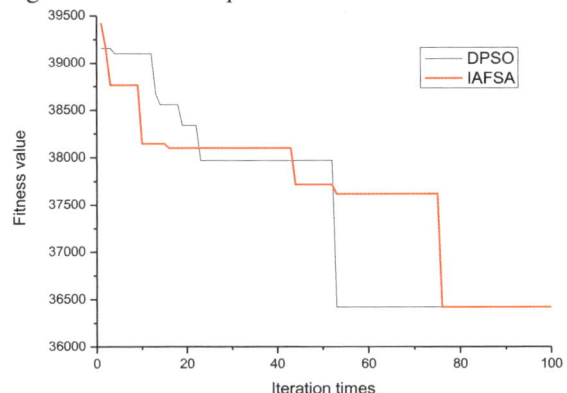

Fig. 1. The iteration process of optimum calculation by both algorithms

TABLE VII

COMPARISON OF THE CALCULATION RESULTS BY BOTH ALGORITHMS AFTER RUNNING 300 TIMES

Algorithm	OVF	AVF	T	R
IAFSA	36419	36822.41	16	5.33%
DPSO	36419	37259.05	23	7.67%

Note: OVF-the optimal value of fitness function, AVF-the average value of fitness function, T- the frequency of getting optimum result, R- the success ratio of getting optimum result.

V. CONCLUSION

The optimization of supply chain network with uncertain background is a research hotspot of supply chain management and decision. This paper analyzed and built the mathematical models for the optimization problem of supply chain network under the condition of grey demands by customers. To solve the models, an improved artificial fish swarm algorithm based on binary coding was presented. Through experiment of calculation examples and the comparison with DPSO algorithm, it indicated that the model and algorithm proposed by this paper were effective and feasible.

ACKNOWLEDGEMENT

This paper was supported by Scientific Research Fund of Fujian Provincial Education Department (Grant No. JA11192), and by Scientific Research Development Foundation Program for Youths, Fujian University of Technology (Grant No. GY-S10051).

REFERENCES

[1] J. Boissiere, Y. Frein, and C. Rapine, "Optimal stationary policies in a 3-stage serial production-distribution logistic chain facing constant and continuous demand," *European Journal of Operational Research*, vol. 18, no. 2, pp 608-619, Feb. 2008.

[2] J. L. Deng, *Grey Control System* (in Chinese). Wuhan: Huazhong University of Science and Technology Press, 1988, ch. 1, pp. 4-9.

[3] Q. S. Zhang, *Difference Information Theory of Grey Hazy Set* (in Chinese). Beijing: Petrolenum Industry Press, 2002, ch. 1, pp. 6-7.

[4] S. F. Liu, Y. G. Dang, Z. G. Fang, *Grey System Theory and Its* Application (in Chinese). Beijing: Science Press, 2004, ch. 5, pp. 113.

[5] X. X. Chen, "Hybrid grey multiple attribute group decision-making method based on evidential reasoning approach" (in Chinese), *Control and Decision*, vol. 26, no. 6, pp. 831-836, Jun. 2011.

[6] X. L. Li, F. Lu, G. H. Tian, and J. X. Qian, "Applications of artificial fish school algorithm in combinatorial optimization problems" (in Chinese), *Journal of Shandong University* (engineering science edition), vol. 34, no. 5, pp. 64-67, 2004.

[7] S. Farzi, "Efficient job scheduling in grid computing with modified artificial fish swarm algorithm," *International Journal of Computer Theory and Engineering*, vol. 1, no. 1, pp.13-18, 2009.

[8] Y. C. He, Y. Q. Wang, J. Q. Liu, "A new binary particle swarm optimization for solving discrete problems" (in Chinese), *Computer Applications and Software*, vol. 24, no. 1, pp. 157-159, 2007.

[9] A.Syarif, Y. S. Yun, and M.Gen, "Study on multi-stage logistic chain network: a spanning tree-based genetic algorithm approach," *Computers & Industrial Engineering*, vol. 43, no. 1, pp. 299-314, Jan. 2002.

[10] J.C. Fang, and Q.S. Zhang, "Vehicle routine problem under the condition of stock shortage and its improved PSO algorithm" (in Chinese), *Journal of Chongqing Technology and Business University* (natural sciences edition), vol. 26, no. 6, pp. 553-557, Jun. 2009.

Generalized Retarded Bihari-like Inequalities and Applications

Qing-ling Gao

College of Mathematics, Qilu Normal University, Shandong 250013, P. R .China

(wgb911@163.com)

Abstract - **Gronwall-Bellman-Bihari inequality is an important tool in the study of existence, uniqueness, boundedness, stability and other qualitative properties of solutions of differential equations and integral equations. We can found a lot of its generalizations in various cases from literature [1-3, 11-12]. More recently, many authors have made researches on retarded integral inequalities and obtained the plenteous results [4-10].**

Keywords - **Differential equations, integral equations, integral inequality**

I. MAIN RESULTS

Lemma ([8], Theorem 2.1) Let a, α be as in Theorem 1.1 of [8]. Assume $k, \omega \in C(R^+, R^+)$ are nondecreasing functions with $k(0) > 0, \omega(t) > 0$ for $t > 0$ and $\int_1^\infty \frac{dt}{\omega(t)} = \infty$. If $u \in C(R^+, R^+)$ satisfies

$$u(t) \le k(t) + \int_0^{\alpha(t)} a(t,s)\omega(u(s))ds, t \ge 0,$$

then

$$u(t) \le G^{-1}\left(G(k(t)) + \int_0^{\alpha(t)} a(t,s)ds\right), t \ge 0,$$

where $G(t) = \int_1^t \frac{ds}{\omega(s)}, t \ge 0.$

Throughout $R_0^+ = [0, +\infty], R^+ = (0, +\infty)$, and we use the notion $Dom(f)$ to denote the domain of function f.

Theorem 1.1. Let $a \in (R_0^+, R^+)$ and $\alpha \in C^1(R_0^+, R_0^+)$ be nondecreasing with $\alpha(t) \le t$ on R_0^+, $f, g, h \in C(R_0^+, R_0^+)$. Moreover, let $\omega \in C(R_0^+, R_0^+)$ be nondecreasing with $\omega(t) > 0$ for $t \ge 0$. If $u \in C(R_0^+, R_0^+)$ satisfies

$$u(t) \le a(t) + \int_0^t h(s)\omega(u(s))ds + \int_0^{\alpha(t)} f(s)\int_0^s g(\tau)\omega(u(\tau))d\tau ds, t \in R_0^+,$$

$$\tag{1}$$

then

$$u(t) \le G^{-1}\left(G(a(t)) + \int_0^t h(s)ds + \int_0^{\alpha(t)} f(s)\int_0^s g(\tau)d\tau ds\right), t \in [0, t_1],$$

$$\tag{2}$$

where $G(t) = \int_1^t \frac{ds}{\omega(s)}, t > 0$ and $t_1 \in R_0^+$ is chosen so that

$$G(a(t)) + \int_0^t h(s)ds + \int_0^{\alpha(t)} f(s)\int_0^s g(\tau)d\tau ds \in Dom(G^{-1}), \quad t \in [0, t_1],$$

where G^{-1} is the reverse of G.

Theorem 1.2. Let $f(t,s), g(t,s), h(t,s)$ be continuous on $(R_0^+ \times R_0^+, R_0^+)$ and nondecreasing in t for every s fixed, $a \in (R_0^+, R^+)$ and $\alpha \in C^1(R_0^+, R_0^+)$ be nondecreasing with $\alpha(t) \le t$. Moreover, ω as in Theorem 1.1, α is a diffeomorphism of R_0^+. If $u \in C(R_0^+, R_0^+)$ satisfies

$$u(t) \le a(t) + \int_0^t h(t,s)\omega(u(s))ds + \int_0^{\alpha(t)} f(t,s)\left(\omega(u(s)) + \int_0^s g(s,\tau)\omega(u(\tau))d\tau\right)ds, \ t \in R_0^+,$$

$$\tag{3}$$

then

$$u(t) \le G^{-1}\left(G(a(t)) + \int_0^t h(t,s)ds + \int_0^{\alpha(t)} f(t,s)\left(1 + \int_0^s g(s,\tau)d\tau\right)ds\right),$$

$$t \in [0, t_1],$$

$$\tag{4}$$

where $G(t) = \int_1^t \frac{ds}{\omega(s)}, t \ge 0$, G^{-1} is the reverse of G and $t_1 \in R^+$ is chosen so that

$$G(a(t)) + \int_0^t h(t,s)ds + \int_0^{\alpha(t)} f(t,s)\left(1 + \int_0^s g(s,\tau)d\tau\right)ds \in Dom(G^{-1}),$$

$$t \in [0, t_1].$$

Since the proofs of Theorem1.1 and Theorem1.2 are similar, we only give the proof of Theorem1.2.

Proof. Assume first that $t = 0$, then the conclusion is true. Fixing an arbitrary number $t_0 \in [0, t_1]$, we define a positive and nondecreasing function $z(t)$ on $[0, t_0]$ by

$$z(t) = a(t_0) + \int_0^t h(t_0, s)\omega(u(s))ds + \int_0^{\alpha(t)} f(t_0, s)\left(\omega(u(s)) + \int_0^s g(s,\tau)\omega(u(\tau))d\tau\right)ds$$

then

$$z(0) = a(t_0), u(t) \le z(t), t \in [0, t_0],$$

and

$$z'(t) = h(t_0, t)\omega(u(t)) + f(t_0, \alpha(t))\alpha'(t)$$
$$\left(\omega(u(\alpha(t))) + \int_0^{\alpha(t)} g(\alpha(t), s)\omega(u(s))ds\right)$$
$$\le h(t_0, t)\omega(z(t)) + f(t_0, \alpha(t))\alpha'(t)$$
$$\left(\omega(z(\alpha(t))) + \int_0^{\alpha(t)} g(\alpha(t), s)\omega(z(s))ds\right)$$
$$\le h(t_0, t)\omega(z(t)) + f(t_0, \alpha(t))$$
$$\alpha'(t)\omega(z(t))\left(1 + \int_0^{\alpha(t)} g(\alpha(t), s)ds\right)$$

E. Qi et al. (eds.), *Proceedings of the 21st International Conference on Industrial Engineering and Engineering Management 2014*, Proceedings of the International Conference on Industrial Engineering and Engineering Management, DOI 10.2991/978-94-6239-102-4_18, © Atlantis Press and the authors 2015

i.e.

$$\frac{z'(t)}{\omega(z(t))} \le h(t_0,t) + f(t_0,\alpha(t))\alpha'(t)\left(1 + \int_0^{\alpha(t)} g(\alpha(t),s)ds\right).$$

Integrating the above relation on $[0,t_0]$ yields

$$G(z(t_0)) \le G(z(0)) + \int_0^{t_0} h(t_0,s)ds$$
$$+ \int_0^{t_0} f(t_0,\alpha(s))\alpha'(s)\left(1 + \int_0^{\alpha(s)} g(\alpha(s),\tau)d\tau\right)ds.$$

Since G^{-1} is increasing on $Dom(G^{-1})$, the above inequality yields

$$z(t_0) \le G^{-1}\left(\begin{array}{l} G(a(t_0)) + \int_0^{t_0} h(t_0,s)ds \\ + \int_0^{t_0} f(t_0,\alpha(s))\alpha'(s)\left(1 + \int_0^{\alpha(s)} g(\alpha(s),\tau)d\tau\right)ds \end{array}\right),$$

$t \in [0,t_1]$. Since t_0 is arbitrary, taking $t = t_0$ in the above relation, using $u(t) \le z(t)$, we get

$$u(t) \le G^{-1}\left(\begin{array}{l} G(a(t)) + \int_0^t h(t,s)ds \\ + \int_0^t f(t,\alpha(s))\alpha'(s)\left(1 + \int_0^{\alpha(s)} g(\alpha(s),\tau)d\tau\right)ds \end{array}\right)$$

$$\le G^{-1}\left(\begin{array}{l} G(a(t)) + \int_0^t h(t,s)ds \\ + \int_0^{\alpha(t)} f(t,s)\left(1 + \int_0^s g(s,\tau)d\tau\right)ds \end{array}\right),$$

so inequality (4) is true.

Remark 1. Seting $\alpha(t) \equiv 0$ or $f \equiv 0$ or $g \equiv 0$ in Theorem1.1 and Theorem1.2, we obtain the Bihari inequality [11].

Remark 2. If $\int_0^\infty \frac{ds}{\omega(s)} = \infty$, then $G(\infty) = \infty$, the conclusions of Theorem1.1 and Theorem1.2 are valid on R_0^+.

Remark 3. If $h \equiv 0$ and $g \equiv 0$ in Theorem1.2, we obtain the Lemma.

Remark 4. If $h \equiv 0$ and $g \equiv 0$ in Theorem1.2 and $f(t,s) = b(t)m(s)$, we obtain Theorem1.2 in paper [8] immediately.

Theorem1.3. Let $a \in (R_0^+, R^+)$ and $\alpha_i \in C^1(R_0^+, R_0^+)$ be nondecreasing with $\alpha_i(t) \le t$ on R_0^+, $f_i, g_i, h \in C(R_0^+, R_0^+), i = 1,2,\cdots,n$. Moreover, let $\omega \in C(R_0^+, R_0^+)$ be nondecreasing with $\omega(t) > 0$ for $t > 0$. If $u \in C(R_0^+, R_0^+)$ satisfies

$$u(t) \le a(t) + \int_0^t h(s)\omega(u(s))ds$$
$$+ \sum_{i=1}^n \int_0^{\alpha_i(t)} f_i(s)\int_0^s g_i(\tau)\omega(u(\tau))d\tau ds, t \in R_0^+, (5)$$

then

$$u(t) \le G^{-1}\left(G(a(t)) + \int_0^t h(s)ds + \sum_{i=1}^n \int_0^{\alpha_i(t)} f_i(s)\int_0^s g_i(\tau)d\tau ds\right),$$

$t \in [0,t_1],$ \hfill (6)

where $G(t) = \int_1^t \frac{ds}{\omega(s)}, t > 0$ and $t_1 \in R_0^+$ is chosen so that

$$G(a(t)) + \int_0^t h(s)ds + \int_0^{\alpha_i(t)} f_i(s)\int_0^s g_i(\tau)d\tau ds \in Dom(G^{-1}),$$

$i = 1,2,\cdots,n, t \in [0,t_1]$, where G^{-1} is the reverse of G.

Theorem1.4. Let f_i, g_i, h, a, α_i be as in Theorem1.2. Moreover, ω as in Theorem1.1, α_i is a diffeomorphism of R_0^+, $i = 1,2,\cdots,n$. If $u \in C(R_0^+, R_0^+)$ satisfies

$$u(t) \le a(t) + \int_0^t h(t,s)\omega(u(s))ds$$
$$+ \sum_{i=1}^n \int_0^{\alpha_i(t)} f_i(t,s)\left(\omega(u(s)) + \int_0^s g_i(s,\tau)\omega(u(\tau))d\tau\right)ds, t \in R_0^+,$$

\hfill (7)

then

$$u(t) \le G^{-1}\left(G(a(t)) + \int_0^t h(t,s)ds + \sum_{i=1}^n \int_0^{\alpha_i(t)} f_i(t,s)\left(1 + \int_0^s g_i(s,\tau)d\tau\right)ds\right),$$

$t \in [0,t_1],$ \hfill (8)

where $G(t) = \int_1^t \frac{ds}{\omega(s)}, t \ge 0$, G^{-1} is the reverse of G and $t_1 \in R^+$ is chosen so that

$$G(a(t)) + \int_0^t h(t,s)ds + \int_0^{\alpha_i(t)} f_i(t,s)\left(1 + \int_0^s g_i(s,\tau)d\tau\right)ds \in Dom(G^{-1}),$$

$i = 1,2,\cdots,n, t \in [0,t_1]$.

Since the proofs of Theorem1.3 and Theorem1.4 are easy, we omit the details.

II. SOME APPLICATIONS

In this section, we give some applications of our results in the estimate of the solutions of differential equations with deviating argument, of which the earlier inequalities do not apply directly.

Example. We first discuss the following differential equation

$$\begin{cases} x'(t) = F(t, x(t), x(\alpha(t))) \\ x(0) = x_0. \end{cases}$$ \hfill (9)

with $x_0 \in R^n, F \in C(R_0^+ \times R^{2n}, R^n)$. Let $\alpha \in C^1(R_0^+, R_0^+)$ be a diffeomorphism of R_0^+ with $\alpha(t) \le t$ for $t \ge 0$. As in [12], let $[0,T]$ be the maximal interval of existence of of solution $x(t)$ to (5), which satisfies the initial condition $x(t) = x_0$, if $T < \infty$, then $\limsup_{t \to T} \|x(t)\| = \infty$. Using our conclusions we can give a sufficient condition of global existence of solutions of (9).

Proposition: Let

$$\left\|F(t,x,y)\right\| \le m(t) + h(t)\omega\left(\|X\|\right) + f\left(\alpha(t)\right)\int_0^t g(s)\omega\left(\|Y\|\right)ds,$$

$$(X,Y) \in R^{2n}, \quad h,f,g,\omega \in C\left(R_0^+, R_0^+\right)$$

and $m \in C\left(R_0^+, R^+\right)$ be nondecreasing, $\omega(t) > 0$,

$\int_{t_0}^{\infty} \dfrac{ds}{\omega(s)} = \infty.$ Then all solutions of (9) exist on R^+.

Proof. Proof by contradiction.

Let $[0,T]$ be the maximal interval of existence of a solution $x(t)$ for (9), which satisfies the initial condition $x(0) = x_0$. From the conditions, we have

$$\left\|x'(t)\right\| \le m(t) + h(t)\omega\left(\|x(t)\|\right) + f\left(\alpha(t)\right)\int_0^t g(s)\omega\left(\|x(\alpha(s))\|\right)ds.$$

Integrating the above relation on $[0,t], t \in [0,T)$, yields

$$\left\|x(t)\right\| \le \left\|x(0)\right\| + \int_0^t m(s)ds + \int_0^t h(s)\omega\left(\|x(s)\|\right)ds$$

$$+ \int_0^t f\left(\alpha(s)\right)\int_0^s g(\tau)\omega\left(\|x(\alpha(\tau))\|\right)d\tau ds.$$

Suppose $a(t) = \left\|x(0)\right\| + \int_0^t m(s)ds$, then

$$\left\|x(t)\right\| \le a(t) + \int_0^t h(s)\omega\left(\|x(s)\|\right)ds$$

$$+ \int_0^t f\left(\alpha(s)\right)\int_0^s g(\tau)\omega\left(\|x(\alpha(\tau))\|\right)d\tau ds,$$

$$= a(t) + \int_0^t h(s)\omega\left(\|x(s)\|\right)ds$$

$$+ \int_0^{\alpha(t)} f(s)\left(\alpha^{-1}(s)\right)'\int_0^s g\left(\alpha^{-1}(\tau)\right)\left(\alpha^{-1}(\tau)\right)'\omega\left(\|x(\tau)\|\right)d\tau ds,$$

$t \in [0,T)$. Applying Theorem 1.1, we get

$$\left\|x(t)\right\| \le G^{-1}\left(G(a(t)) + \int_0^t h(s)ds + \int_0^{\alpha(t)} f(s)\int_0^s g(\tau)d\tau ds\right), \quad t \in [0,t_1],$$

(10)

where t_1, G is chosen as in Theorem 1.1, (10) holds on $[0,t_1]$ for any $t_1 \in [0,T)$. So (10) also holds on $[0,T)$. Observe that the right hand side of (10), as a function in t, is bounded on the interval $[0,T)$. So is $x(t)$ on $[0,T)$, which contradicts the known conclusion $\limsup\limits_{t \to T}\left\|x(t)\right\| = \infty$. So all solutions of (10) exist on R^+.

III. CONCLUSION

In this paper, we have established a generalized retarded integral inequality of Bihari-like type. Our results generalize the existing results and give more convenient tools in the study of solutions to differential equations and integral equations.

We note that, in our opinion ,many new applications of Bihari-like inequalities are still to be found. We still have a lot of work to do.

REFERENCES

[1] B.G. Pachpatte, Explicit bounds on certain integral inequalities, J. Math. Anal. Appl.267 (2002) 48-61.
[2] B.G. Pachpatte, On some new inequalities retarded to a certain inequality arising in the theory of differential equations, J. Math. Anal. Appl.251 (2000) 736-751.
[3] B.G. Pachpatte, On some new inequalities retarded to certain inequalities in the theory of differential equations, J. Math. Anal. Appl.189 (1995) 128-144.
[4] R.P. Agarwal, S. Deng, W.Zhang, Generalization of a retarded Gronwall-like inequality and its applications, Appl. Math. Comput. 165(2005) 599-612.
[5] Rui A.C. Ferreira, Delfim F.M. Torres, Generalized retarded integral inequalities, Appl. Math. Lett. 22 (2009) 876-881.
[6] O. Lipovan, A retarded Gronwall-like inequalities and its applications, J. Math. Anal. Appl. 252(2000) 389-401.
[7] O. Lipovan, A retarded integral inequalities and its applicatins, J. Math. Anal. Appl. 285(2003) 436-443.
[8] O. Lipovan, Integral inequalities for retarded Volterra equations, J. Math. Anal. Appl. 322(2006) 349-358.
[9] Y.G. Sun, On retarded integral inequalities and their applicatins, J. Math. Anal. Appl. 301(2005) 265-275.
[10] W.S. Wang, A generalized retarded Gronwall-like inequality in two variables and applications to BVP, Appl. Math. Comput. 191(2007) 144-154.
[11] I. Bihari, A Generalization of a lemma of Bellman and its applicatoions to uniqueness problems of differential equations, Acta Math. Acad. Sci. Hungar. 7(1956) 71-94.
[12] R. Driver, Existence and continuous dependence of neutral functional differential equations, Arch. Rational Mech.Anal. 19(1965) 149-166.

Influence Factors Analysis of the 3D Printing Rapid Prototyping System Using Entropy Method and System Dynamics Model

Li-jie FENG[1,2], Ya-zhou WANG[1,*], Jin-feng WANG[1], Xue-qi ZHAI[1]

[1]Department of Management Engineering, Zhengzhou University, Zhengzhou, China
[2]Henan Provincial Coal Seam Gas Development and Utilization Co. Ltd., Zhengzhou, China
(wangyazhou1989@sina.com)

Abstract - **This paper mainly discusses the influence factors of 3D printing rapid prototyping system using entropy weight method and system dynamics model. Firstly, through survey we find the influence factors, next using entropy weight method to calculate index weight, then we use SD method to simulation and draw the causal diagram and loop diagram. Finally, we analyze the various influencing factors and find the key factors. The results shows that improving the advanced level of molding equipment and enlarged material R&D efforts are of important guiding significance for largely improving efficiency of forming and promoting the popularization of 3D printing rapid prototyping system.**

Keywords - **3D printing, entropy weight method, factor analysis, system dynamics model**

I. INTRODUCTION

With the development of science and technology, the 3D printing technology plays an increasingly important role. The 3D printing system that contains computers and 3D printing devices can be fast and accurate to print out new products in a short period of time, which will surpass the traditional process of the product development design and continue to work wonders. The 3D printing technology is widely used in medicine, machine manufacturing, the restoration and other industries. The accuracy and speed of print in all these areas have higher requirements, but the accuracy and speed of the 3D printing level is not high in reality ,for example, Some enterprises pursue speed so blindly that causes the printing precision and vice versa. The contradiction between speed and accuracy that reflects in the reality is print inefficiency, which not only hinders the development of 3D industry, but also wastes resources. Therefore, we need to find how to improve the efficiency of the 3D printing system to achieve the goal of 3D printing service in a good and fast way.

Many scholars at home and abroad have studied a lot on 3D printing rapid prototyping system, for example, Zhang Nan (2013) [1] explains the revolutionary change for the future design from the 3D printing on the concept of product design, design details, design process, etc; Wang ping (2013) [2] thinks the application of 3D printing in the field of education will help teachers to humanize model, and also can promote the student to study the level of ascension; Barry Berman (2012) [3] stresses the 3D printing has a great impact on the enterprise production by analyzing 3D printing which has an influence on the mass customization production and application in other areas;

Giovanni Cesaretti (2014) [4] etc. discusses the idea that building a lunar habitat by using of lunar soil in 3D printing; Suki (2012) [5] has made the analysis and suggestions of the future based on 3D printing technology by introducing the development situation at home and abroad with the survey data. Much more researches focus on the principle of 3D technology, application, development prospects, but to improve the efficiency of 3D printing system level research is little. So we need to strengthen this aspect of the research.

As 3D printing rapid prototyping is a complex and interlocking system, its efficiency is determined by internal and external factors which are interactional and inter-constraint, so these impacts on the efficiency of 3D printing system are inconsistent. The entropy weight method is objective and empowerment. According to the variation of each index, it can calculate the entropy weight of every index by using information entropy and through the entropy weight to revise the weight of each index and obtain more objective index weight. Meanwhile, the system dynamics based on system theory can reveal the change law of high order, nonlinear, multi feedback complex systems by a feedback loop to describe the structure of a system. Therefore, in this paper, we use [6] entropy method and system dynamic to set up the system dynamics model of 3D printing rapid prototyping system and use simulation software Vensim model for simulation in order to explore different factors on the impact of 3D printing rapid prototyping system and then direct the enterprise to improve the efficiency of 3 d printing rapid prototyping system.

II. 3D PRINTING RAPID PROTOTYPING SYSTEM INFLUENCE FACTOR ANALYSIS AND THE WEIGHT

A. 3D printing rapid prototyping system influence factor analysis

3D printing process can be divided into previous, medium and later period. In the previous period, we mainly pre design phase model for printing objects, including 3D structure model of 3D scanning and mapping software, construction which will have a great influence on the quality of printing products and printing efficiency level of the stage 3D model. The medium period is a key for 3D printing. We use 3D printing equipment translate the molding materials into material object according to the 3D mode. This process requires operating personnel, cooperation to complete printing equipments, so it will be affected by printing equipments, molding material, forming environment and personnel etc.

The later one is 3D printing products processing stage that puts forward a very high request to the stage of technical personnel skill level. Three interdependent stages, organic combination, finally achieve the successful operation of 3D printing rapid prototyping system, constantly through the print devices print materials into actual objects.

Combined with the analysis of the process, experimental operation experience and related enterprises [7-9], the factors affecting the 3D printing rapid prototyping efficiency into six parts: degree of specialization of the operator, advanced level of the equipments, the characteristics of the molding material, the complexity of the molding object, construction of 3D model, effects of environmental factors. These factors work together in the 3D printing rapid prototyping system, as is shown in Fig.1.

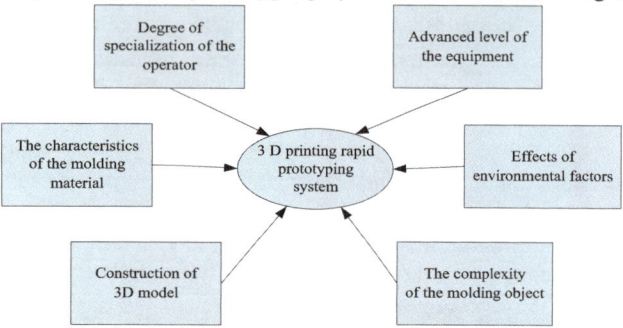

Fig.1. 3D printing rapid prototyping system model structure

(1) Specialization of the operators refers to 3D print related knowledge the operator has and operation proficiency of 3D print system, also includes scan or drawing of the print model in the early stage and the printing processing and the daily maintenance in the later stage;

(2) Advanced level of the equipment refers to 3D printing equipments in the rapid prototyping system includes the 3D printing rapid prototyping system characteristics of the equipments including 3D scanner and 3D printers, configuration, performance and the kinds of the resolution of the printer, printing process, such as advanced degree [10];

(3) The characteristics of the molding materials refer to the kinds of molding materials, thermal properties (heat capacity, thermal conductivity, heat of fusion and thermal expansion, the boiling point melting, etc.), mechanical properties (elastic modulus, tensile strength, impact strength, yield strength, fatigue strength resistance, etc.) and functional;

(4) The characteristics of the molding object refer to the target object's shape, size, complexity, the purpose of the forming object, color and the precision and quality requirements;

(5) Construction of 3D model refers to the detailed and complete degree of the 3D digital models which are designed or constructed by using 3D scanner in printed early or with the aid of design software;

(6) Effects of environmental factors refer to 3D printing rapid prototyping system environment requirements, including molding workshop of temperature, humidity, voltage, ventilation and lighting;

B. To determine weight of factors based on entropy weight method

With system dynamics simulation analysis, we first need to find the weight of index influence on the level of system efficiency. At present, the common methods to calculate the weights are Rough Sets, Principal Component Analysis (PCA), Analytic Hierarchy Process (AHP), the Variation Coefficient method and so on. These methods are flawed by subjective factors causing deviations, which are not conducive to accurately determine the weight of each index, moreover, they can't explain the results. The entropy weight method is a method to determine the index weight, which has strong objectivity, accuracy, and can better explain the results obtained. Such as Zhao Xiuli (2013) [11] etc. uses entropy method to determine the objective weight, then gets comprehensive weights, makes an analysis of supply chain logistics capability evaluation results, provides reference to promote the performance of supply chain logistics capability; Zhao Lei (2012) [12] carries on the research of sustainable land use in Huludao City, comprehensive evaluation and puts forward some suggestions of sustainable utilization of land resources in Huludao city in the future. Based on the advantages of entropy weight method, this paper uses the entropy weight method to determine the weight of each index, which provided support for the construction of the system dynamics model and the simulation. To calculate weight of entropy method comprises the following steps:

1) To gather data, $x_{ij}(i=1,2,...,m; j=1,2,...,n)$ means the index value of item i in the j year, set data x_{ij} into formula (1), calculate the proportion of each index a_{ij}

$$a_{ij} = X_{ij} \Big/ \sum_{i=1}^{m} X_{ij} \qquad (1)$$

2) Calculate the entropy of the i indicators h_i, set a_{ij} into formula (2), calculate the entropy of each index h_i,

$$h_i = - \sum_{j=1}^{n} a_{ij} \ln a_{ij} \Big/ \ln n \qquad (2)$$

3) Define weight CL_i, set h_i into formula (3), calculate the weight of each index CL_i

$$CL_i = 1-h_i \Big/ \sum_{i=1}^{m}(1-h_i) \quad 0 \le CL_i \le 1, \sum_{i=1}^{m} CL_i = 1 \qquad (3)$$

The magnitude of the weights CL_i reflect the relative importance of the factors on the system. Under certain conditions, the greater the weight CL_i, the greater the factors that influence the change on the system level. As in 3D printing rapid prototyping system, there are dynamic interaction relationships, so we need use system dynamics to build dynamic model which can reflect the influence each other between the index relationships.

III. SYSTEM DYNAMICS MODELING FOR 3D PRINTING RAPID PROTOTYPING INFLUENCE FACTORS

A. causal diagram and variable definition

The relationships in 3D printing rapid prototyping system are inter-constraint and interactional. And there are multiple relationships between positive and negative feedback, which 3D printing rapid prototyping system dynamics modeling can be founded [13].

1) Causal diagram of 3D printing rapid prototyping system

The level of printing rapid prototyping system is influenced by the workers' specialty, equipments, materials, objects, models and environment, and their causal relationship. Causal diagram is an effective way to analyze and understand system function and lay a good foundation for system flow. It is shown in Fig.2.

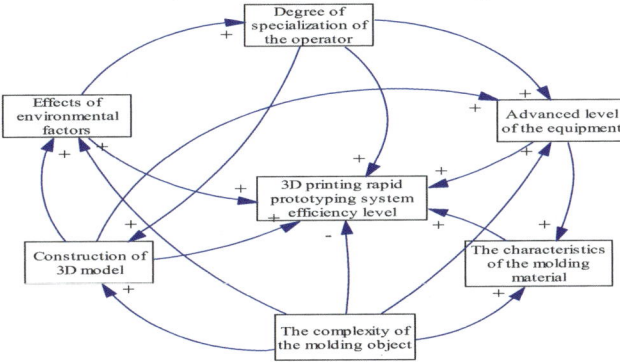

Fig.2. Causal diagram of 3D printing rapid prototyping system

Fig.2 shows that the relationships in 3D printing rapid prototyping system are not only inter-constraint and interactional, but also multiple positive and negative feedbacks which can make 3D printing rapid prototyping system dynamics modeling.

2) Variable-definition of 3D printing rapid prototyping system

The model variables can be defined by the causal diagram and loop diagram, so we can define the following mode variables, shown in Table I.

TABLE I
TABLE OF VARIABLES

Variables	definition	implication
auxiliary variable	ASP	3D printing rapid prototyping system efficiency level
level variable	L_1	Degree of specialization of the operator
	L_2	Advanced level of the equipment
	L_3	The characteristics of the molding material
	L_4	The complexity of the molding object
	L_5	Construction of 3D model
	L_6	Effects of environmental factors
rate variable	R_1	Increment of degree of specialization of the operator
	R_2	Increment of advanced level of the equipment
	R_3	Increment of the characteristics of the molding material
	R_4	Increment of the complexity of the molding object
	R_5	Increment of construction of 3D model
	R_6	Increment of effects of environmental factors
constant	Crm	The influence coefficient of degree of specialization of the operator on construction of 3D model
	Crs	The influence coefficient of degree of specialization of the operator on advanced level of the equipment
	Csc	The influence coefficient of advanced level of the equipment on the characteristics of the molding material
	Cdc	The influence coefficient of the complexity of the molding object on the characteristics of the molding material
	Cds	The influence coefficient of the complexity of the molding object on advanced level of the equipment
	Cdh	The influence coefficient of the complexity of the molding object on effects of environmental factors
	Cdm	The influence coefficient of the complexity of the molding object on construction of 3D model
	Cms	The influence coefficient of construction of 3D model on advanced level of the equipment
	Cmh	The influence coefficient of construction of 3D model on effects of environmental factors
	Chr	The influence coefficient of effects of environmental factors on degree of specialization of the operator

B. Flow and SD equation

3D printing rapid prototyping system flow can be made by the analysis of casual graph and variable definition, as is shown in Fig.3. We can establish model equations according to system flow diagram, as follows, auxiliary equations:

$$ASP.K = CL_1 \times L_1.K + CL_2 \times L_2.K + CL_3 \times L_3.K + CL_4 \times L_4.K + CL_5 \times L_5.K + CL_6 \times L_6.K \quad (4)$$

CL_i can be got by the above entropy method, CL_i means the weight of the index i   (5)

state equations:

$$L_1.K = (L_1.J + R_1 \times DT) \quad (6)$$

$$L_2.K = (L_2.J + R_2 \times DT) \times Crs \times Cms \times Cds \quad (7)$$

$$L_3.K = (L_3.J + R_3 \times DT) \times Csc \quad (8)$$

$$L_4.K = (L_4.J + R_4 \times DT) \quad (9)$$

$$L_5.K = (L_5.J + R_5 \times DT) \times Crm \times Chm \times Cdm \quad (10)$$

$$L_6.K = (L_6.J + R_6 \times DT) \times Cdh \quad (11)$$

K means the present tense and J is the past tense (Initial value). JK means the space of time from the past to the present. DT means simulation time step variables, which is the space of time it takes to answer from J to K.

The initial value equations:

$$L_1 = 78, L_2 = 80, L_3 = 81, L_4 = 82, L_5 = 84, L_6 = 86$$

$Crm = 1.28, Crs = 1.35, Csc = 1.41, Cdc = 1.25, Cds = 1.38, Cdh = 1.32, Cdm = 1.34, Cms = 1.25, Cmh = 1.30, Chr = 1.37$

Step DT=1(month), space of time is 36(months)

According to the parameter, the level increment quantity of all factors is $I_i \times \exp(-ASP/100)(i = 1, 2, ..., 6)$, and the increment rate in the system is 0.5. And according to dynamic development of the system level, the equation and value tends to be in accord with prototyping and finally will be obtained by the repeatedly- debugged historical statics. The increment rate can be expedited to 0.7 in simulation adjustment.

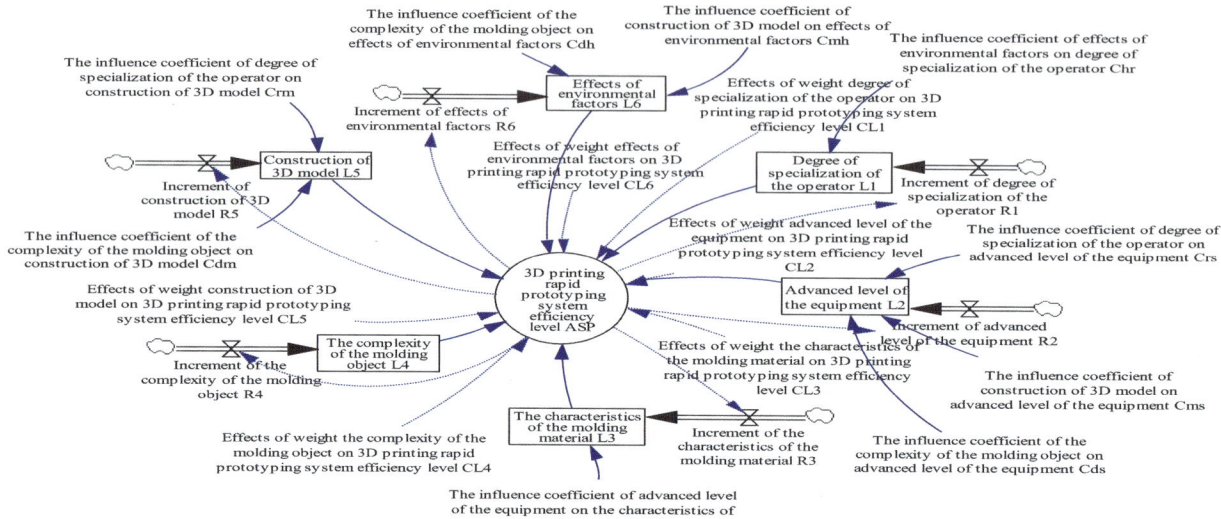

Fig.3. 3D printing rapid prototyping system dynamic flow diagram

TABLE II

THE IMPACT OF 3 D PRINTING RAPID PROTOTYPING SYSTEM

Index	Year 2011	2012	2013
Degree of specialization of the operator	3.7	4.0	4.2
Advanced level of the equipment	3.5	3.7	4.0
The characteristics of the molding material	3.4	3.6	3.8
The complexity of the molding object	3.8	4.0	4.1
Construction of 3D model	3.9	4.0	4.2
Effects of environmental factors	3.9	4.1	4.3

IV. THE ANALYSIS OF SIMULATION

A. To gather data

Before employing entropy method, the paper gathered data through expert evaluation method [14-15], which means that 50 experts in the field analyze and obtain relevant certain value through experience and 3 years of data from 2011 to 2013. $x_{ij}(i = 1, 2, ..., m; j = 1, 2, ..., n)$ is the targeted value of the target i in j year, the data is shown in Table II.

B. To determine weighing of all factors

The above targets are calculated and processed through entropy method which can get weighing of all factors CL_i (i=1, 2, 3,..., 6), which means the target i weighs the system. The results are as follows:

$CL_1 = 0.16074, CL_2 = 0.18433, CL_3 = 0.19302, CL_4 = 0.15708, CL_5 = 0.15508, CL_6 = 0.14975$

C. To analyze simulation and results

With system dynamics simulation software Vensim, we build the system dynamics flow chart of 3D printing rapid prototyping system (Fig.3). And with reference variables, we simulate factors of efficiency levels of 3D printing rapid prototyping system which takes 36 months.

In terms of initial values of variables and efficiency of enterprises, the initial value of every subsidiary factor increment rate is 0.5 (dimensionless value), which is based on the trend of system efficiency. In the same way, to be obviously contrast under different regulation ranges, the increment rate of regulation is 0.7 (dimensionless value), and then we analyze the level changes of regulation variables. The results are as follows:

1) The simulation analysis of increment rates of factors efficiency

Based on the variables and system dynamics flow chart simulation software, we conclude that the level trend of 3D printing rapid prototyping system efficiency is shown in Fig.4, the level trend of 3D printing rapid prototyping system efficiency factors is shown in Fig.5. In the initial conditions, the target value of the level of prototyping efficiency will be 90 in 24 months, so it takes a long and continuous term to improve the efficiency. Meanwhile, the level trend of all factors can be seen in Fig.5.

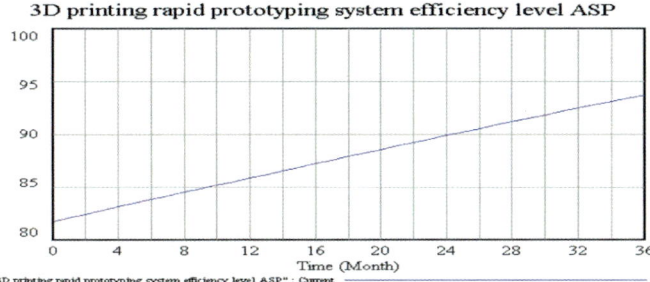

3D printing rapid prototyping system efficiency level ASP

Fig.4. 3D printing rapid prototyping system efficiency level trend chart

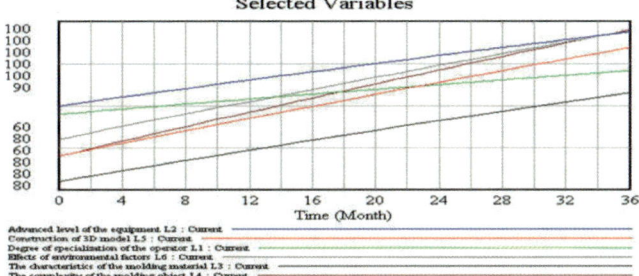

Selected Variables

Fig.5. The level trend of 3D printing rapid prototyping system efficiency factor

2) The simulation analysis of increment rates of all factors levels

In simulation, the increment rates of other factors remain the same, and the increment rates of selected factors levels are to be 0.7 one by one. With the software, we should know the influences of different factors on the levels of system efficiency. Current1, Current2, Current3, Current4, Current5 and Current6 represent the profession of operators, the advanced degree of equipment, materials, related subjects, objects and environment. The increment rate model construction is to be 0.7 one by one. The system efficiency levels are shown in Fig.6 when the increment rates of other factors remain the same.

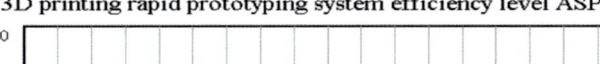

3D printing rapid prototyping system efficiency level ASP

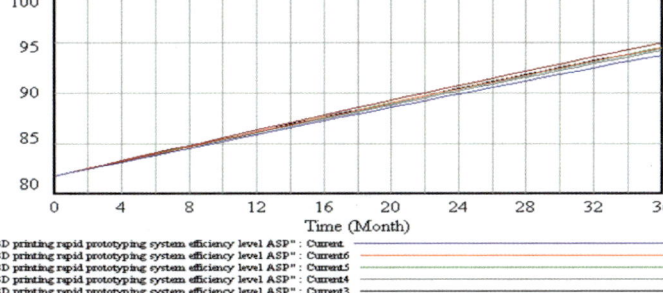

Fig.6. The addition of various factors on the impact of 3D printing rapid prototyping system efficiency level trend

3) The actual rates of all factors

The actual rate refers to the increment percentage of system efficiency levels every month in the same level

increment of other factors. On the basis of current in Fig.4, the average of system efficiency levels is 87.82452; the level values of every month in Fig.6 minus those of Current every month is the average, which is compared with the level average values of current, and the actual rate of factors will be got respectively. The actual rates of all factors are 0.003678, 0.007157, 0.004544, 0.002625, 0.004441, 0.004291, it can be seen that the equipment had the greatest influence on the efficiency of the molding system, followed by molding material, then forming model and forming environment, staff levels and forming objects. This is consistent with actual situation roughly of the 3D shape. The actual action rate contrast with the efficiency level about the different factors in complex system on the overall system accurately.

V. CONCLUSION

In this article, through the systematic analysis of the factors of efficiency level about 3D printing rapid prototyping system and comprehensive using entropy method and system dynamics model has carried on the simulation for the affecting factors and the relationship between influencing factors and system efficiency level, we find out the key factors influencing the system efficiency level at last. By improving the equipment of advanced level, increasing the intensity of materials research and development, the efficiency of 3D printing rapid prototyping system can be improved. It will be of a guiding significance for our 3D printing enterprises to improve production performance and promote the development of 3D printing industry.

REFERENCES

[1] ZHANG Nan, LI Fei. Influence of the development and application of 3D printing technology for the future product design [J]. Journal of Machine Design, 2013, (7): 97-99.

[2] WANG Ping. Study on 3D printing and its application in Education [J]. Distance Education in China, 2013, (8):83-87.

[3] Barry Berman. 3-D printing: The new industrial revolution [J]. Business Horizons, 2012, (55):155-162.

[4] Giovanni Cesaretti, EnricoDini, et al. Building components for an outpost on the Lunar soil by means of a novel 3D printing technology [J]. Acta Astronautica, 2013, 93(2014): 430-450.

[5] WANG Xue-ying. Analysis of the development and Prospect of 3D printing technology and Industry [J]. China High Technology Enterprises, 2012, (26): 3-5.

[6] WANG Zhong-hong, LI Yang-fan, ZHANG Man-yin. The status and development of China 3D printing industry [J], Economic Review, 2013, (01):90-93.

[7] GUO Zhen-hua, WANG Qing-jun, GUO Ying-huan. 3D printing technology and social manufacturing [J], Journal of Baoji University of Arts and Sciences (Natural Science Edition), 2013, (4): 64-70.

[8] DU Yu-lei, SUN Fei-fei, YUAN Guang, ZHAI Shi-xian, ZHAI Hai-ping. The development situation of 3D printing

material [J], Journal of Xuzhou Institute of Technology (Natural Sciences Edition), 2014, (1):20-24.

[9]   LU Bing-heng, LI Di-chen. Development of the Additive Manufacturing (3D printing) Technology [J], Machine Building & Automation, 2013, (4):1-4.

[10] CHEN Li-na. Study on "3Dprint" technology and industry development [J], China CIO News, 2013, (6): 105-106.

[11] ZHAO Xiu-li, GUO Mei. Performance Evaluation of Supply Chain Logistics Capability Based on Entropy [J], Science and Technology Management Research, 2013, (2): 200-202.

[12] ZHAO Lei, LIU Hong-bin, YU Guo-feng, LI Tie-zhu, SUN Wei, WANG Chuang. A Case Study on HuLuoDao City: Comprehensive Evaluation of Sustainable Utilization of Land Resources Via Entropy Method [J], Resources & Industries, 2012, 04: 63-69.

[13] ZHANG Shu-yan. Analysis of the factors affecting college students' online game addiction based on system dynamics analysis [D], Taiyuan University of Science and Technology, 2013.

[14] LI Jun. Weight Determining of Factors Influencing Grain Output Based on Entropy Weight Method [J], Journal of Anhui Agricultural Sciences, 2012, 11: 6851-6852+6854.

[15] WU Jian-ni. Research on Index System of Port Logistics Industry Cluster Competitiveness Based on Entropy Weight Method [J], Science and Technology Management Research, 2013, 06: 45-49+54.

Multi-objective Collaborative Optimization of Production Scheduling for Discrete Manufacturing

Xiao-ying YANG*, Xi WANG, Hua-yue SUN

School of Mechatronics Engineering, Henan University of Science and Technology, Luoyang, China
(lyyxy@haust.edu.cn)

Abstract - **Machinery industry enterprise as research background, this paper analyzed the feature of production scheduling for discrete manufacturing and proposed the collaborative multi-objective optimization problem. Based on the improved Taguchi loss function, nonlinear relationship of quality, delivery and cost was established. And the multi-objective collaborative optimization model of production scheduling was created as well as synthetically considering of discrete constraints. The effective solution was studied to solve the model integrating simulation modeling and genetic algorithm. Further, through the enterprise empirical study, the practicability and validity of the model and algorithm is verified. This study will improve the synergy degree among quality, delivery and cost three goals, and provide an effective theoretical method of production schedule for the discrete manufacturing.**

Keywords - **Collaborative optimization, discrete manufacturing, genetic algorithm, production scheduling, simulation modeling, Taguchi loss function**

I. INTRODUCING

Discrete manufacturing scheduling is optimization problem which has the characteristic of multi-part, multi-target, multi-constraint and randomness. In machinery industry, the typical enterprise such as heavy machinery enterprise, which designs and manufactures products according to customer order and demand, their products are multi-type, complex in structure, long production processes and long cycle time. Their machine is mainly universal [1].

Quality, delivery and cost become their schedule goals, process, time and resources as major constraint, there also comes with random uncertainty of insert urgent orders, withdrawals, and machine failure, the research of this issue becomes a worldwide NP-Hard problem [2] Currently most enterprises of this type schedule their production based on their experience and statically, it' difficult to guarantee duration, low efficiency, without science and lack of method of multi-objective optimization theory. Therefore, this research has both theoretical and practical value.

The NP-Hard problem of production scheduling optimization has been attracting scholars both home and aboard. But the production optimization problem of multi-objective scheduling is also the lack of research, especially for discrete manufacturing. The current research results focused on aspects related to the optimization model and algorithm [3-9], but most of the optimization goal which only consider a single goal, fewer with multi-target. Multi-target research mainly focus on weighted integrated optimization, such as

literature [10]; literature [11] used a hybrid intelligent algorithm to implement multi- resource scheduling constraints target the total duration of the project and WIP inventory; literature [12] used heuristic rules and genetic algorithm to optimize manufacturing cycle, resource utilization, manufacturing costs. Our research group is also continuing research on such issues as the literature [1, 13] used a simple multi-objective weighting method, but did not fully consider the relevancy between multi-objective. In actual production scheduling, every goal do not exist alone, they affect each other and are reciprocal relationship. Therefore, the research of this subject has certain challenges.

Taguchi quality loss function (TQLF) is to describe the quality loss or deviation squared deviation from the target value and quality characteristics [14]. It illustrates the nonlinear relationship between quality and the cost. Taguchi loss function has been successfully applied to model of the relationship between quality and cost [15]. This paper focused on collaborative optimization model of three goals by establishing the relationship of quality, time and cost based on improved TQLF. And it integrated simulation technique and genetic algorithms to solve global optimal problems, and verified the practicality and effectiveness of the model and algorithm through case study. This paper will better improve the synergy degree among quality, delivery and cost three goals, and provide an effective theoretical method of production schedule for the discrete manufacturing.

II. THREE GOALS RELATIONAL MODEL BASED ON IMPROVED TQLF

A. Model of quality-cost relation

Based on improved TQLF, the relationship model between quality and cost was built. Product costs include normal target cost and quality loss caused by process variation. We assumed that the normal target cost of parts j of the product i is named C_{ij}^{O} and quality loss is named C_{ij}^{q}, q_{ij} is quality, Δq_{ij} is variation of quality, k_{ij}^{q} is loss coefficient, μ_{ij}^{q} is target value of quality, σ_{q} is standard deviation of quality. So the model of quality -cost relation is as formula (1).

$$C_{ij}^q = \frac{1}{\sqrt{2\pi}\sigma_i^q}\int_{\mu_{ij}^q - \Delta q_{ij}}^{\mu_{ij}^q} k_{ij}^q \left(q_{ij} - \mu_{ij}^q\right)^2 \times \exp\left(-\frac{\left(q_{ij} - \mu_{ij}^q\right)^2}{2\left(\sigma_i^q\right)^2}\right) dq \qquad (1)$$

$$+ \frac{1}{\sqrt{2\pi}\sigma_i^q}\int_{\mu_{ij}^q}^{\mu_{ij}^q + \Delta q'_{ij}} k_{ij}^q \left(q_{ij} - \mu_{ij}^q\right)^2 \times \exp\left(-\frac{\left(q_{ij} - \mu_{ij}^q\right)^2}{2\left(\sigma_i^q\right)^2}\right) dq$$

B. Model of cost-time relation

Cost -time relation is similar with the quality-cost relation. It is the nonlinear relation. We assumed that the loss of the parts j of the product i named C_{ij}^t. t_{ij}, which is the product period, Δt_{ij} is variation of time, k_{ij}^t is loss coefficient, μ_{ij}^t is target value of time, σ_t is standard deviation of time. So the model of cost-time relation is as formula (2).

$$C_{ij}^t = \frac{1}{\sqrt{2\pi}\sigma_i^t}\int_{\mu_{ij}^t - \Delta t_{ij}}^{\mu_{ij}^t} k_{ij}^t \left(t_{ij} - \mu_{ij}^t\right)^2 \times \exp\left(-\frac{\left(t_{ij} - \mu_{ij}^t\right)^2}{2\left(\sigma_i^t\right)^2}\right) dt \qquad (2)$$

$$+ \frac{1}{\sqrt{2\pi}\sigma_i^t}\int_{\mu_{ij}^t}^{\mu_{ij}^t + \Delta t'_{ij}} k_{ij}^t \left(t_{ij} - \mu_{ij}^t\right)^2 \times \exp\left(-\frac{\left(t_{ij} - \mu_{ij}^t\right)^2}{2\left(\sigma_i^t\right)^2}\right) dt$$

III. MULTI-OBJECTIVE COLLABORATIVE OPTIMIZATION MODEL

In view of production scheduling optimization problem of machinery industry enterprise based on formula (1) and (2), the three goals (quality, time and cost) synergies optimization was made the total cost of a single goal. The collaborative multi-objective optimization model was established as formula (3).

$$C = \min\sum_{i=1}^{n}\sum_{j=1}^{n} \begin{bmatrix} C_{ij}^O + \frac{1}{\sqrt{2\pi}\sigma_i^q}\int_{\mu_{ij}^q - \Delta q_{ij}}^{\mu_{ij}^q} k_{ij}^q \left(q_{ij} - \mu_{ij}^q\right)^2 \times \exp\left(-\frac{\left(q_{ij} - \mu_{ij}^q\right)^2}{2\left(\sigma_i^q\right)^2}\right) dq \\ + \frac{1}{\sqrt{2\pi}\sigma_i^q}\int_{\mu_{ij}^q}^{\mu_{ij}^q + \Delta q'_{ij}} k_{ij}^q \left(q_{ij} - \mu_{ij}^q\right)^2 \times \exp\left(-\frac{\left(q_{ij} - \mu_{ij}^q\right)^2}{2\left(\sigma_i^q\right)^2}\right) dq \\ + \frac{1}{\sqrt{2\pi}\sigma_i^t}\int_{\mu_{ij}^t - \Delta t_{ij}}^{\mu_{ij}^t} k_{ij}^t \left(t_{ij} - \mu_{ij}^t\right)^2 \times \exp\left(-\frac{\left(t_{ij} - \mu_{ij}^t\right)^2}{2\left(\sigma_i^t\right)^2}\right) dt \\ + \frac{1}{\sqrt{2\pi}\sigma_i^t}\int_{\mu_{ij}^t}^{\mu_{ij}^t + \Delta t'_{ij}} k_{ij}^t \left(t_{ij} - \mu_{ij}^t\right)^2 \times \exp\left(-\frac{\left(t_{ij} - \mu_{ij}^t\right)^2}{2\left(\sigma_i^t\right)^2}\right) dt \end{bmatrix} \qquad (3)$$

s.t.
$$ts_{i,jrM_e}X_{i,jrM_e} \geq \left[ts_{i,j(r-1)M_e} + t_{i,j(r-1)M_e}\right]X_{i,j(r-1)M_e} \qquad (4)$$

$$1 \leq M_e \leq M \qquad (5)$$

$$t_{ij}(1+\zeta) \leq t'_{ij}, \forall t \in T, T \geq 0, 0 < \zeta < 1 \qquad (6)$$

$$X_{ijrM_e}\begin{cases} =1, Said\ machine M_e\ is\ selected \\ =0, Otherwise \end{cases} \qquad (7)$$

$$R_{ijrqM_e}\begin{cases} =1, process\ r\ is\ before\ q \\ =0, if not \end{cases} \qquad (8)$$

Formula (4) is process sequence, time constraints, components of the starting time of process r should before the starting time and the processing time of process $r-1$; formula (5) is equipment capacity constraints; formula (6) is uncertain constraints, considering random

factors ζ ; formula (7) and (8) are Moderator Variable.

IV. SIMULATION MODELING AND ALGORITHM DESIGN

A. Simulation Modeling(SM)

Based on Plant Simulation Software, the system physical model of discrete manufacturing was built, embedded mathematic model, solution algorithm and parameters of reality. The model was constantly debugging for calibration and validation.

B. Genetic Algorithm (GA)

1) Coding: Chromosome is made up of all parts of the code. Chromosome coding for genes are $n(n = 1, 2, ..., i \times j)$. Chromosome length is $i \times j$. The product parts p_{ij} process of encoding is r ($r = 1, 2, ..., q$), M_e is coding for machine equipment, its name $e(e = 1, 2, ..., N_e)$. The decoding $\left[ij, r, M_e, t_{ijrs}, t_{ijre}\right]$ is result space matrix of the work schedule. Machining process, equipment and its various parts of the starting time and the completion time is made by the matrix into a Gantt chart.

2) Fitness evaluation: Based on the adaptive function, the use of each individual in the population to adapt to the evolutionary search function value. Fitness function by the objective function formula (3) the scale transformation, Such as formula (9):

$$fitness(x) = \min C \qquad (9)$$

3) Genetic operations: Including population size, age level, mutation probability, generations, the crossover probability and termination of algebra, etc. When reach the preset maximum breeding, the number of algebraic algorithm automatically stop.

V. CASE STUDY

In this paper, the research of scheduling take the cement mill production of some machinery industry enterprises as an example. The grinding rotary part of cement mills includes three key components which are barrel, end cover, hollow shaft. Using P_i and $part_{ij}$ (i=1, 2, 3; j=1, 2, 3) represent products and product components. Main demand is using collaborative optimization model and integration algorithm to put forward the optimal production scheduling plan in a certain plan period.

There are 13 kinds of equipments, excepting two sets of M2, two sets of M4 and M12 and M13 are regardless of the constraint, others has only one set. Table I shows the processing craft and process time.

Depending on the product, equipment and technology and other basic information embedded in collaborative multi-objective optimization model, a simulation model system of Plant Simulation was created, shown in Fig.1.

TABLE I
PARTS PROCESSING CRAFT AND PROCESS TIME

N	Parts	The parts machining process/time (days)											
		1	2	3	4	5	6	7	8	9	10	11	12
1	Part11	M3/3	M4/3	M7/10	M8/13	M5/4	M11/5	M1/10	M10/5	M6/20	M9/8	M12/1	M13/4
2	Part12	M3/1.5	M4/1.5	M7/5	M5/3	M2/5	M13/3						
3	Part13	M1/8	M3/2	M2/6	M3/2.5	M2/10	M3/2.5	M2/2.5	M13/2.5				
4	Part21	M3/1	M4/1	M7/6.5	M8/8.5	M5/2	M11/2.5	M1/7.5	M10/3	M6/15	M9/6	M12/1	M13/2
5	Part22	M3/0.5	M4/0.5	M7/0.5	M5/1.5	M2/3	M13/1.5						
6	Part23	M1/6	M3/1	M2/4.5	M3/1	M2/2.5	M3/1	M2/1.5	M13/1.5				
7	Part31	M3/2.5	M4/2	M7/8.5	M8/11	M5/3	M11/4	M1/8.5	M10/4	M6/18	M9/7	M12/1	M13/3
8	Part32	M3/1.5	M4/1	M7/4	M5/2	M2/3.5	M13/2						
9	Part33	M1/7	M3/1.5	M2/5	M3/1.5	M2/9	M3/2	M2/2.5	M13/2				

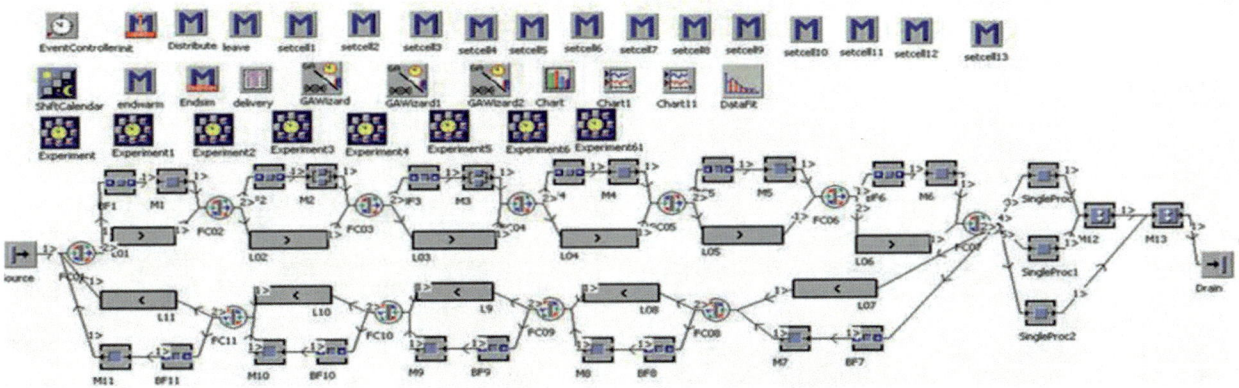

Fig.1. Simulation model of production system

The genetic algorithm embedded into the simulation model, a crossover probability is 0.9, mutation probability is 0.15, genetic race to 20, population size for 100. Optimization parameters according to the order of the parts to set up, the fitness functions for the collaborative optimization objective function. The fitness of convergence is obtained by simulation the optimized value as shown in Fig.2, the red line represents the optimal solution, the green line represents the average, the blue line represents the worst solutions, race number algebra in the diagram to the abscissa, ordinate is fitness values. The optimal fitness evaluation value is 27898.2 Yuan. The optimal solution Gantt chart is shown in Fig.3.

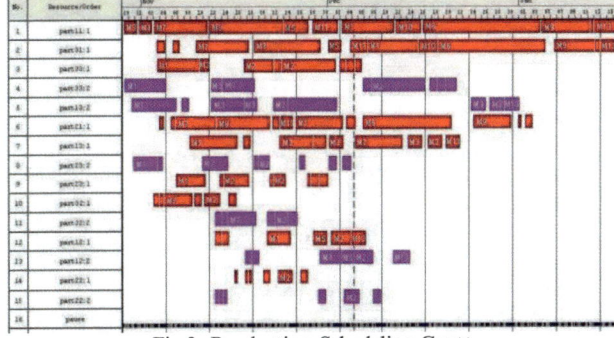

Fig.3. Production Scheduling Gantt

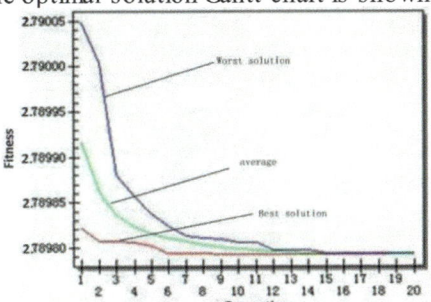

Fig.2. Fitness convergence values

This case applied the collaborative optimization model and its algorithm to simulation modeling and test, the optimal solution of the production schedule obtained. Will this method compared with enterprise artificial scheduling results, the results showed that the production schedule after the collaborative optimization, the final loss of 108340 Yuan for the enterprise to reduce the cost, improve the efficiency of production scheduling of nearly 55.8%. The test results showed that the collaborative optimization model and SM&GA is effective and practical.

VI. CONCLUSION

This paper focused on the difficult problem of production schedule for the discrete manufacturing. In view of the multi-objective optimization problem of the quality, time and cost is less studied, the non-linear relation model of quality, time and cost was established based on improved TQLF. So that the multi-objective collaborative optimization model of production scheduling was constructed, considering time constraints, equipment constraints, and uncertainty factors. And integrated SM&GA, the effective method was studied to solve the model. The validity of the model and algorithm had been verified through enterprise example. It proved that the optimization method has certain theoretical significance and practical value to cope with the production scheduling of discrete manufacturing.

ACKNOWLEDGEMENT

The authors gratefully acknowledge the support of the Henan Science and Technology Research Program, China (No.102102210487) and Luoyang of Henan province Science and Technology Program, China (No. 1101027A, No. 20130703).

REFERENCES

[1]   YANG Xiao-ying, SHI Guo-hong, WANG Xue. Combinatorial optimization of one-piece discrete production scheduling based on the lean logistics [J]. Industrial Engineering and Management, 2013, 18(3): 11-18. (Chinese)

[2]   Karl G Kempf, Pınar Keskinocak, Reha Uzsoy. Planning Production and Inventories in the Extended Enterprise: A State of the Art Handbook (Volume 1) [M]. Springer New York Dordrecht Heidelberg London, 2011.

[3]   LI Lin, HUO Jia-zhen. Multi-objective flexible Job-shop scheduling problem in steel tubes production [J]. Systems Engineering-Theory & Practice, 2009 (8): 117-126. (Chinese)

[4]   Susan K. Monkman, Douglas J. Morrice, Jonathan F. Bard. A production scheduling heuristic for an electronics manufacturer with sequence-dependent setup costs [J]. European Journal of Operational Research, 2008 (187): 1100–1114.

[5]   Jennifer Muñoz Blás. Development of a time-efficient heuristic method for production scheduling with resource constraints and changeover considerations [D].

INDUSTRIAL ENGINEERING UNIVERSITY OF PUERTO RICOM AYAGÜEZ CAMPUS December, 2007.

[6]   LI Chun, GE Mao-gen, ZHANG Ming-xin. Study on dynamic advanced planning and scheduling problem based on genetic and particle swarm optimization algorithm [J]. Journal of Hefei University of Technology (Natural Science), 2010, 33(1):5-9.

[7]   YI Jun, LI Taifu. Bacterial foraging optimization algorithm based on variable neighborhood for Job-shop scheduling problem [J]. Journal of Mechanical Engineering, 2012, 48 (12): 178-183. (Chinese)

[8]   Massimiliano Caramia, Stefano Giordani. A fast metaheuristic for scheduling independent tasks with multiple modes [J]. Computers & Industrial Engineering, 2010, 58: 64–69.

[9]   AntonioLova, PilarTormos, Mariamar Cervantes, FedericoBarber. An efficient hybrid genetic algorithm for scheduling projects with resource constraints and multiple execution modes [J]. Int. J. Production Economics 2009 (117): 302–316.

[10]   LEE Yan-fei,JIANG Zhi-bin. Production scheduling optimization with time-variable multiple-objectives for semiconductor wafer fabrication system [J]. Journal of Shanghai Jiao Tong University, 2008 (2): 209-213. (Chinese)

[11]   SHI Guo-hong, CHEN Jing-xian, MA Han-wu, CHEN Ling-qing. Optimization scheduling research of multi-resource-constrained project based on mixed-intelligence algorithm [J]. Journal of Engineering Design, 2008(4). (Chinese)

[12]   Lin Wei. Multi-objective constraint environment of production planning and scheduling method [D]. Donghua university master's degree thesis, 2008. (Chinese)

[13]   Zheng Boke, Xiaoying YANG. Optimization method of transfer batch adaptive job scheduling [J]. Journal of Henan University of Science and Technology: Natural Science, 2012, 33(2):17-21.

[14]   PAN Er-shun, LI Qing-guo. Improvement of Taguchi's loss function and its application in optimal economy production quantity [J]. Journal of Shanghai Jiaotong University, 2005, 39 (7): 1120-1122. (Chinese)

[15]   LI Jia-xiang, HAN Zhi-jun. Research on process control for multi-product and small-batch production based on Taguchi method [J]. Industrial Engineering and Management, 2009, 13(1): 31-35. (Chinese)

Job Shop Scheduling based on Improved Discrete Particle Swarm Optimization

Lvjiang Yin[1,2,*], Lijun Yang[2], Mingmao Hu [3]

[1]State Key Lab of Digital Manufacturing Equipment and Technology, Huazhong University of Science and Technology, Wuhan 430074, China
[2]Department of Economics and Management, Hubei University of Automotive Technology, Shiyan, China
[3]Department of Mechanical Engineering, Hubei University of Automotive Technology, Shiyan, China
(yinlvjiang@126.com)

Abstract - **Based on the basic particle swarm optimization algorithm and the analysis of the basic principles of job shop scheduling problem, an improved discrete particle swarm optimization (IDPSO) is proposed in the paper. In the IDPSO, the discrete particle location is updated by the operation of mutation and crossover based on the evolutionary mechanism of PSO, and a neighborhood algorithm is introduced to enhance the ability of local search. Finally a simulation results of an example proved the feasibility and validity of the algorithm.**

Keywords - **Job shop schedule, local search, particle swarm, simulation**

I. INTRODUCTION

The job shop schedule take the time allocation of resources into consideration to complete the task to meet certain performance index, under limited resource and task process constraints, it belongs to the most difficult combinatorial optimization issue. Specifically, in a dynamic production environment, some unexpected events may occur, it needs to change the old scheme. According to the current state of the system, assign the machine and change the order of process, then make the original plan and new reasonable cohesion, optimal operation target. Liu [1] has studied the particle swarm optimization and the application is discussed.

II. METHODOLOGY

A. Particle Swarm Optimization

Particle swarm optimization is an evolutionary algorithm based on population iteration , each individual is referred to as a particle in the particle swarm, and each particle is a point which moves to a certain rule in the target search space, by searching the speed and the location of the particle to find the optimal solution[2]. The basic particle swarm optimization algorithm process is as follows

Step1: The initialization of algorithm. It involves the location and the speed.

Step2: to the fitness function, evaluate the fitness value of each particle.

Step3: Compare the fitness value of each particle and the fitness value in the optimal position Opi, if the latter is better, choose it. Then compare the Opi of all particle and the global optimum gb, if the Opi is better, update the gb to Opi.

Step4: Update each particle's position and speed. Left and right-justify your columns. Use tables and figures to adjust column length. Use automatic hyphenation and check spelling. All figures, tables, and equations must be included *in-line* with the text. Do not use links to external files.

Step5: Judge it whether meet the termination conditions of the algorithm, if so, end the running and output the optimal solution. If not, turn to step2.

B. Job Shop Scheduling Problem [3,4]

Job shop scheduling is a problem of sorting job task on the basis of time. So, we can express the job work as a manufacturing process of a work piece, the mathematic model of job shop scheduling is as follows:

$$MinJ = Max(S_{ie_im} + T_{ie_im}) \quad i \in N \tag{1}$$

$$s.t. \ S_{ilq} - S_{ikq} \geq T_{ikq}, [O_{ikp}, O_{ikq}] \in P_i, l \in \{1, \cdots, m\} i \in N \tag{2}$$

$$S_{ilq} - S_{ikq} \geq T_{ikq} \ or \ S_{ikq} - S_{jlq} \geq T_{jlq}, [O_{ikq}, O_{jkp}] \in R_i \ i,j \in N, q \in M \tag{3}$$

$$r_i \leq S_{ijkl} \leq d_i - T_{ijq}, i \in N, q \in M, j \in \{1, \cdots, n_i\} \tag{4}$$

Formula one shows the minimize production cycle, formula two shows ordering constraint among all processes, the two different processes of the same work piece cannot be processed on two machines at the same time. Formula three shows resource constraint, each machine just process a procedure of an artifact, formula four means preparation time and delivery time constraints.

C. Calculation method

Problem of coding: Problem of coding is the primary key problem of designing particle swarm algorithm. Genm [5,6] proposed coding method based on process expression which using genetic algorithm to solve job-shop scheduling problem. Assumes that the workshop scheduling problem involves n work pieces and m machines, so each chromosome particle composed of genes that represent the integer value of the operation. And each artifact i appears m times, so we can know that the particles is a feasible solution. A problem of job shop schedule about three work pieces and three machines, as shown in Table I.

TABLE I
THE INSTANCE OF JOB SHOP SCHEDULE

Processing time			Machine sequencing			
Work piece	process			Work piece	process		
	1	2	3		1	2	3
j_1	3	1	3	j_1	m_1	m_1	m_2
j_2	3	5	2	j_2	m_2	m_3	m_1
j_3	2	3	3	j_3	m_3	m_2	m_3

From Table I, we can know that the first 2 correspond to the j2 of the first work piece is processed on the m1,the processing time is 1, the second 2 correspond to the j2 of the second work piece is processed on the m3,the processing time is 5,Then decode the chromosome particles, can get the maximum completion time is 12, as shown in Table II.

TABLE II
PARTICLE CODING

Particle dimension	1	2	3	4	5	6	7	8	9
Chromosome	3	2	2	1	1	2	3	1	3
Work number	1	1	2	1	2	3	2	3	3
Machine number	2	1	3	1	3	2	1	3	3

D. Particle position update strategy

The key factors of the position about the particle include influence part [7], cognitive part, society part, through the analysis can be obtained that the particle evolve to a new position is due to the current state, their best condition and the group best state mutual influence. So, based on the evolution optimization mechanism of the basic particle swarm algorithm, define the location update formula as follows:

$$X_i^{k+1} = c_2 \otimes f(c_1 \otimes g(\omega \otimes h(X_i^k), pB_i^k), gB^k) \quad (5)$$

ω: inertia weight factor, c_1: cognitive coefficient, c_2: social coefficient, $\omega, c_1, c_2 \in [0,1]$

E. Local search strategy

According to the particle coding and particle location update strategy, we find it suitable for the job shop scheduling of discrete particle swarm optimization algorithm. Meanwhile, it retains the advantages of fast convergence speed, easy to implement. Local search strategy is a key technology. So, to further improve and optimize the algorithm performance, we should do the local search strategy after the update. The procedure is as follows:

Step1: Randomly generate an integer $d_1 \in [1, len]$, len means the length of X_h. Set $m = d_1$, $X_h' = X_h$. Randomly generate an integer d between [0, 1], if $d = 1$, turn to Step2. If $d > 1$, turn to step3.

Step2: According to the following steps:

Step2.1: If $d_1 > 1$, randomly generate an natural number d2 between [0, d1-1]. It means the d1th gene start forward search to the d2th gene on the X_h, if $d_2 > 0$, go on.

Step2.2: Exchange the (m-1) th gene and the d2th gene on the X_h'.

Step2.3: Carry out m=m-1. if $m > (d_1 - d_2)$, return Step2.2.

Step3: Follow the step below.

Step3.1: If $d_1 < len$, randomly generate an integer d2 between [0, len-d1], it means the d1th gene start forward search to the d2th gene. If $d_2 > 0$, go to the step3.2.

Step3.2: Exchange the (m+1)th gene and the d1th gene on the X_h'. If $f(X_h') < f(X_h)$, set $X_h = X_h'$. Go to the step3.3.

Step3.3: Carry out $m = m+1$. If $m < (d_1 + d_2)$, go back Step3.2

III. RESULTS

A. The results of simulation and analysis

To test the performance of the improved discrete particle swarm algorithm, the numerical simulation tested and matched the eleven typical JSP test (FT06, FT10, FT20, LA01, LA06, LA11, LA16, LA21, LA26, LA31, LA36) respectively. The scale and theory optimal solution of the eleven test example as shown in Table III.

TABLE III
THE SCALE AND THEORY OPTIAL SOLUTION OF THE TEST

problem	n, m	C^*	problem	n, m	C^*
FT06	6,6	55	LA16	10,10	945
FT10	10,10	930	LA21	15,10	1046
FT20	20,5	1165	LA26	20,10	1218
LA01	10,5	666	LA31	30,10	1784
LA06	15,5	926	LA36	15,15	1268
LA11	20,5	1222			

Related algorithm parameters are as follows: the problem of population size is set to double the size, $G = 2 \times n \times m$, the maximum iterations $N_p = 1000$, inertia weight $\omega = 0.95$, cognitive coefficient $c_1 = 0.95$, social coefficient $c_2 = 0.9$. The stop criterion of the algorithm is the iterations become the maximum or the optimal 300 consecutive generation no longer updated.

The algorithm of this paper used are IDPSO, DPSO and BPSO, then random independently run 30 times for every test example. DPSO shows that it do not use local search algorithm on the basis of IDPSO, the algorithm parameters are consistent with IDPSO. BPSO is referring to the fundamental particle algorithm, The parameter is set to $\omega = 0.9, c_1 = c_2 = 2$.

B. Simulation Result

As a general rule, the optimization ability is the assessment criteria of the comparison test of the algorithm. By using relative error to represent the optimization ability of the algorithm, $(C_{max} - C^*)/C^* \times 100\%$, C_{max}: the optimal solution of multiple independent operation. C^*: the theory optimal solution of each test ,

$BRE = (BF - C^*) / C^* \times 100\%$: the optimal relative error percentage. $ARE = (AF - C^*) / C^* \times 100\%$: the average relative error percentage $WRE = (WF - C^*) / C^* \times 100\%$: the worst relative error percentage.

Among that, BF, AF, WF represent the optimal solution, the average optimal solution and the worst optimal solution when the algorithm performs thirty times respectively.

The paper also list the relevant results about HDE [8, 9] and HDPSO [10]. Then according to the calculation results, comparing the value of three kinds of algorithms:

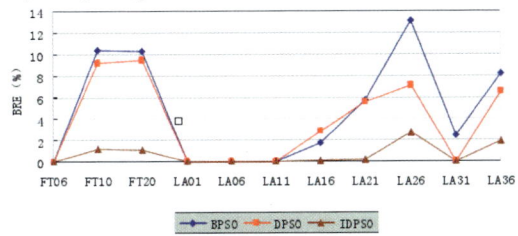

Fig.1. BRE comparison chart

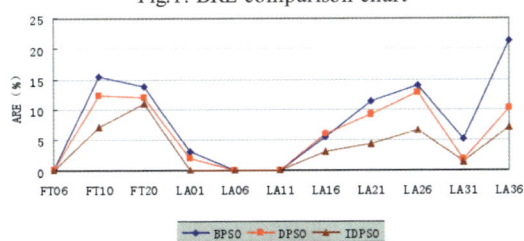

Fig.2. BRE comparison chart

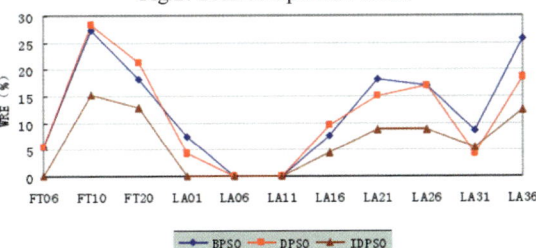

Fig.3. WRE comparison chart

From Fig.1, we can know that the BRE of IDPSO slightly greater than 4% except the FT20 example, the rest were greater than 4%. But some BRE of DPSO and BPSO all reach to about 10%. It shows that the global convergence ability of IDPSO better than DPSO and BPSO obviously.

From Fig.2 and 3, we can get that the IDPSO has a better optimization ability, and the ARE and WRE fluctuate less relatively, and the stability is perfect. In all, we can see that the IDPSO is competent on the optimizing.

IV. CONCLUSION

Based on the basic particle swarm optimization algorithm and the analysis of the basic principles of job shop scheduling problem, an improved discrete particle swarm optimization (IDPSO) is proposed in the paper. The future will use other algorithms for the exploration of the field

ACKNOWLEDGMENT

The authors thank the Academic Committee Chairs Prof. Ershi Qi and the anonymous reviewers for their constructive comments to improve the paper. This research is partially supported by the National Science Foundation of China (No. 51375004), and the ministry of education of humanities and social science project (No.14YJA630079), and the Doctoral Scientific Fund Project of Hubei University of Automotive Technology (No. BK201408, No. BK201301), and the scientific research project of Hubei province education office (No. Q20131801).

REFERENCES

[1] D. J. Beebe, "Signal conversion (Book style with paper title and editor)," in Biomedical Digital Signal Processing, W. J. Tompkins, Ed. Englewood Cliffs, NJ: Prentice-Hall, 1993, ch. 3, pp. 61–74.

[2] M. Akay, Time Frequency and Wavelets in Biomedical Signal Processing (Book style). Piscataway, NJ: IEEE Press, 1998, pp. 123–135.

[3] G. B. Gentili, V. Tesi, M. Linari, and M. Marsili, "A versatile microwave plethysmograph for the monitoring of physiological parameters (Periodical style)," IEEE Trans. Biomed. Eng., vol. 49, no. 10, pp. 1204–1210, Oct. 2002.

[4] V. Medina, R. Valdes, J. Azpiroz, and E. Sacristan, "Title of paper if known," unpublished.

[5] E. H. Miller, "A note on reflector arrays (Periodical style-Accepted for publication)," IEEE Trans. Antennas Propagat., in press.

[6] T. Menendez, S. Achenbach, W. Moshage, M. Flug, E. Beinder, A. Kollert, A. Bittel, and K. Bachmann, "Prenatal recording of fetal heart action with magnetocardiography" (in German), Zeitschrift für Kardiologie, vol. 87, no. 2, pp. 111–8, 1998.

[7] J. E. Monzon, "The cultural approach to telemedicine in Latin American homes (Published Conference Proceedings style)," in Proc. 3rd Conf. Information Technology Applications in Biomedicine, ITAB'00, Arlington, VA, pp. 50–53.

[8] F. A. Saunders, "Electrotactile sensory aids for the handicapped (Presented Conference Paper style)," presented at the 4th Annu. Meeting Biomedical Engineering Society, Los Angeles, CA, 1973.

[9] J. R. Boheki, "Adaptive AR model spectral parameters for monitoring neonatal EEG (Thesis or Dissertation style)," Ph.D. dissertation, Biomed. Eng. Program, Univ. Fed. Rio de Janeiro, Rio de Janeiro, Brazil, 2000.

[10] J. P. Wilkinson, "Nonlinear resonant circuit devices (Patent style)," U.S. Patent 3 624 12, July 16, 1990.

Study on Decomposition of Things and the Corresponding Matrix Equation

Kai-zhong Guo, Ran Li*, Jin-xin Li

Management Department, Guangzhou Vocational College of Science and Technology, Guangzhou, China

(ranli2006@126.com)

Abstract - **The concept of error matrix and the matrix representations of error decomposition and combination are proposed in the paper firstly, and then the paper points out that how the error occurs can be described as** $T(u) = u_1$**, and the paper focuses on studying how to get one element when the other two are known in this equation. In the last part of this paper how to get** T **is solved by the example about the things decomposition in traffic management of Guangzhou.**

Keywords - **Decomposition, error matrix, error decomposition and combination, matrix**

I. INTRODUCTION

1% of the error will lead to 100% of the failure, which means 100 minus 1 is not equal to 99 but equal to 0. Such as "broken window effect", "domino effect", "butterfly effect". These cruel facts make us realize that any small mistake in the management work will bring destruction [1-3].

There are many ways to solve the mistake, for example 2004 years ago too many bus-lines stopped in the same stand dock in Guangzhou and lead to traffic congest and traffic jam [4, 5]. Later, the traffic administrative department adopted the things decomposition method, the former one dock was divided into adjacent 2-5, and the buses could stand partly, so the road was not crowded. Another example, when people move a machine which is greater than the door into the house also use the things decomposition method, first to decompose the machine into subsystems which less than the door, and then to assemble them in the house[6-10].

How the error occurs can be described as $T(u) = u_1$, in which T means the decomposition, u means the object studied, and u_1 means the target object of the problem needs to be solved [11]. So to study the mechanism and the law of the error in the economy management is to get one element when the other two are known in this equation.

II. THE CONCEPT OF ERROR MATRIX

Define 1.1 suppose [12]

$$A = \begin{bmatrix} ((u_{111}, u_{112}, ..., u_{11k}), x_{11}) & ((u_{111}, u_{112}, ..., u_{11k}), x_{11}) & \cdots & ((u_{111}, u_{112}, ..., u_{11k}), x_{11}) \\ ((u_{111}, u_{112}, ..., u_{11k}), x_{11}) & ((u_{111}, u_{112}, ..., u_{11k}), x_{11}) & \cdots & ((u_{111}, u_{112}, ..., u_{11k}), x_{11}) \\ \cdots & \cdots & \cdots & \cdots \\ ((u_{111}, u_{112}, ..., u_{11k}), x_{11}) & ((u_{111}, u_{112}, ..., u_{11k}), x_{11}) & \cdots & ((u_{111}, u_{112}, ..., u_{11k}), x_{11}) \end{bmatrix}$$

be a $m \times n$ error matrix with k elements.

III. THE MATRIX REPRESENTATION OF ERROR LOGIC TRANSFORMATION

It is can be known that each column of the matrix can be defined as a decomposition transformation by the definition of matrix multiplication. And it can act on each element of the matrix on its right [13].

Define3.1

Suppose

$$(u, x) = (U_1, S_1(t), \bar{p}_1, T_1(t), L_1(t), x(t) = f((u(t), \bar{p}_1), G_u(t))) =$$

$$\begin{bmatrix} U_{10} & S_{10}(t) & \bar{p}_{10}(x_1, x_2, \cdots, x_n) & T_{10}(t) & L_{10}(t) & x_{10}(t) = f_{10}((u(t), \bar{p}_1), G_{U10}(t)) & G_{U10}(t) \\ U_{11} & S_{11}(t) & \bar{p}_{11}(x_1, x_2, \cdots, x_n) & T_{11}(t) & L_{11}(t) & x_{11}(t) = f_{11}((u(t), \bar{p}_1), G_{U11}(t)) & G_{U11}(t) \\ \cdots & \cdots & \cdots & \cdots & \cdots & \cdots & \cdots \\ U_{1r} & S_{1r}(t) & \bar{p}_{1r}(x_1, x_2, \cdots, x_n) & T_{1r}(t) & L_{1r}(t) & x_{1r}(t) = f_{1r}((u(t), \bar{p}_1), G_{U1r}(t)) & G_{U1r}(t) \end{bmatrix}$$

And

$$(V, y) = (V_2, S_2(t), \bar{p}_2, T_2(t), L_2(t), y(t) = f((v(t), \bar{p}_2), G_V(t))) =$$

$$\begin{bmatrix} (V_{20}, S_{20}(t)) & \bar{p}_{20}(x_1, x_2, \cdots, x_n) & T_2(t) & L_2(t) & y_{20}(t) = f_{20}((u(t), \bar{p}_2), G_V(t)) & G_{V10}(t) \\ (V_{21}, S_{21}(t)) & \bar{p}_{21}(x_1, x_2, \cdots, x_n) & T_{21}(t) & L_{21}(t) & y_{21}(t) = f_{21}((u(t), \bar{p}_2), G_V(t)) & G_{V11}(t) \\ \cdots & \cdots & \cdots & \cdots & \cdots & \cdots \\ (V_{2l}, S_{2l}(t)) & \bar{p}_{2l}(x_1, x_2, \cdots, x_n) & T_{2l}(t) & L_{2l}(t) & y_{2l}(t) = f_{2l}((u(t), \bar{p}_2), G_V(t)) & G_{V1r}(t) \end{bmatrix}$$

And

$$A = (u, x) = \begin{bmatrix} ((u_{11} & u_{12} & \cdots & u_{1k}), x_1) \\ ((u_{21} & u_{22} & \cdots & u_{2k}), x_2) \\ \cdots\cdots\cdots\cdots\cdots\cdots \\ ((u_{m1} & u_{m2} & \cdots & u_{mk}), x_m) \end{bmatrix}$$

$$B = (v, y) = ((v_{11} \quad v_{12} \quad \cdots \quad v_{1k}), y)$$

$$A \times B = \begin{bmatrix} ((w_{11} & w_{12} & \cdots & w_{1k}), x_1 \wedge y) \\ ((w_{21} & w_{22} & \cdots & w_{2k}), x_2 \wedge y) \\ \cdots\cdots\cdots\cdots\cdots\cdots \\ ((w_{m1} & w_{m2} & \cdots & w_{mk}), x_m \wedge y) \end{bmatrix}$$

Every element of the matrix above is null if $x_i \geq y(i = 1, 2, \cdots m)$, otherwise $((w_{i1} \quad w_{i2} \quad \cdots \quad w_{ik}), x_{i1} \wedge y) = ((u_{i1} \quad u_{i12} \quad \cdots \quad u_{ik}), x_i)$ and $v = (u_1 h u_2 h \cdots h u_m)$ are hold. Then A can be named as the decomposition transformation of B.

Define 3.2 Suppose [14]

$$A = \begin{bmatrix} ((u_{11} & u_{12} & \cdots & u_{1k}), x_1) \\ ((u_{21} & u_{22} & \cdots & u_{2k}), x_2) \\ \cdots\cdots\cdots\cdots\cdots\cdots \\ ((u_{m1} & u_{m2} & \cdots & u_{mk}), x_m) \end{bmatrix}$$

$$B = (((v_{11} \quad v_{12} \quad \cdots \quad v_{1k}), y_1) \quad ((v_{21} \quad v_{22} \quad \cdots \quad v_{2k}), y_2) \quad \cdots\cdots \quad ((v_{m1} \quad v_{m2} \quad \cdots \quad v_{mk}), y_m))$$

Then,

E. Qi et al. (eds.), *Proceedings of the 21st International Conference on Industrial Engineering and Engineering Management 2014*, Proceedings of the International Conference on Industrial Engineering and Engineering Management, DOI 10.2991/978-94-6239-102-4_22, © Atlantis Press and the authors 2015

$B \times A = (w,z) = ((w_1 \quad w_2 \quad \cdots \quad w_k), \ (x_1 \wedge y_1) \vee (x_2 \wedge y_2) \vee \cdots\cdots \vee (x_m \wedge y_m))$

and $\quad W_J = u_{1J} h u_{2J} h \cdots h u_{mJ} = v_{1J} h v_{2J} h \cdots h v_{mJ}$,

$z = (x_1 \wedge y_1) \vee (x_2 \wedge y_2) \vee \cdots\cdots \vee (x_m \wedge y_m)$, which is the target error value after the combination. For example if $y_i \geq x_i (i = 1,2,\cdots\cdots,m)$ and $z = \max(x_1, x_2, \cdots\cdots, x_m)$, then $z = (x_1 \wedge y_1) \vee (x_2 \wedge y_2) \vee \cdots\cdots \vee (x_m \wedge y_m)$ is met. Then A can be named as the combination transformation of B.

From here we can see that the relationship of decomposition transformation and combination transformation is just like the relationship of the multiplication which multiply things from right to left and the multiplication which multiply things from left to right in the matrix multiplication.

IV. THE EXAMPLE OF THINGS DECOMPOSITION

Suppose U = {Matters involved in the traffic of Guangzhou}, $S(t)$ = {The traffic of Guangzhou from 2007 to now}, $\vec{p}$ = {The location of Guangzhou traffic}, $T(t)$ = {Traffic congestion}, $L(t)$ = { a%(the road width), b% (the crossroads width, ..., ...)}, $G_U(t)$ = {The traffic rules and regulations, the goal of Guangzhou traffic system}, $x(t) = f((u(t),\vec{p},),G_U(t))$ [15, 16].

So $\quad A = (U, S(t), \vec{p}, T(t), L(t), x(t) = f(u(t),\vec{p}),G_U(t))$ is one error logic variable.

And

$$B = \begin{pmatrix} U & S_1(t) & \vec{p} & T(t) & L(t) & x(t) = f((u(t),\vec{p},),G_U(t)) & G_U(t) \\ U & S_2(t) & \vec{p} & T(t) & L(t) & x(t) = f((u(t),\vec{p},),G_U(t)) & G_U(t) \\ U & S_3(t) & \vec{p} & T(t) & L(t) & x(t) = f((u(t),\vec{p},),G_U(t)) & G_U(t) \\ U & S_4(t) & \vec{p} & T(t) & L(t) & x(t) = f((u(t),\vec{p},),G_U(t)) & G_U(t) \\ U & S_5(t) & \vec{p} & T(t) & L(t) & x(t) = f((u(t),\vec{p},),G_U(t)) & G_U(t) \\ U & S_6(t) & \vec{p} & T(t) & L(t) & x(t) = f((u(t),\vec{p},),G_U(t)) & G_U(t) \\ U & S_7(t) & \vec{p} & T(t) & L(t) & x(t) = f((u(t),\vec{p},),G_U(t)) & G_U(t) \\ U & S_8(t) & \vec{p} & T(t) & L(t) & x(t) = f((u(t),\vec{p},),G_U(t)) & G_U(t) \\ U & S_9(t) & \vec{p} & T(t) & L(t) & x(t) = f((u(t),\vec{p},),G_U(t)) & G_U(t) \end{pmatrix}$$

Therein, $S_1(t)$ = {Buses in Guangzhou traffic from 2007 to now}; $S_2(t)$ = {Private cars in Guangzhou traffic from 2007 to now}; $S_3(t)$ = {Lorries in Guangzhou traffic from 2007 to now}; $S_4(t)$ = {The road surface in Guangzhou traffic from 2007 to now}; $S_5(t)$ = {Road facilities in Guangzhou traffic from 2007 to now}; $S_6(t)$ = {Traffic management rules in Guangzhou traffic from 2007 to now}; $S_7(t)$ = {Pedestrian in Guangzhou traffic from 2007 to now}; $S_8(t)$ = {Traffic management software}; $S_9(t)$ = {Other relevant things in Guangzhou traffic from 2007 to now}.

$$T_{fsw} = \begin{pmatrix} U_{10} & S_{10}(t) & \vec{p}_{10(x_1,x_2,\cdots,x_n)} & T_{10}(t) & L_{10}(t) & x_{10}(t) = f_{10}((u(t),\vec{p}_1),G_{U10}(t)) & G_{U10}(t) \\ U_{11} & S_{11}(t) & \vec{p}_{11(x_1,x_2,\cdots,x_n)} & T_{11}(t) & L_{11}(t) & x_{11}(t) = f_{11}((u(t),\vec{p}_1),G_{U11}(t)) & G_{U11}(t) \\ \cdots & \cdots & \cdots & \cdots & \cdots & \cdots & \cdots \\ U_{18} & S_{18}(t) & \vec{p}_{18(x_1,x_2,\cdots,x_n)} & T_{18}(t) & L_{18}(t) & x_{18}(t) = f_{18}((u(t),\vec{p}_1),G_{U18}(t)) & G_{U18}(t) \end{pmatrix}$$

So we can get $T_{fsw} \times A' = B$, that is $T_{fsw} \times (U, S(t), \vec{p}, T(t), L(t), x(t) = f(u(t),\vec{p}),G_U(t)) = B$.

Because it is a things transformation, so elements of things column are different, but the rest are the same in the corresponding matrix. So we can get the thing transformation T_{fsw} by solving the following equation set:

$S_{10}(t) \cap S(t) = S_1(t)$;
$S_{20}(t) \cap S(t) = S_2(t)$;
$S_{30}(t) \cap S(t) = S_3(t)$;
$S_{40}(t) \cap S(t) = S_4(t)$;
$S_{50}(t) \cap S(t) = S_5(t)$;
$S_{60}(t) \cap S(t) = S_6(t)$;
$S_{70}(t) \cap S(t) = S_7(t)$;
$S_{80}(t) \cap S(t) = S_8(t)$;
$S_{90}(t) \cap S(t) = S_9(t)$.

It is can be known by the composition of this equation set that this equation set has solution only when
$S(t) \supseteq S_1(t) \cup S_2(t) \cup S_3(t) \cup S_4(t) \cup S_5(t) \cup S_6(t) \cup S_7(t) \cup S_8(t) \cup S_9(t)$
is hold, so let

$S_{10}(t) = S_1(t)$;
$S_{20}(t) = S_2(t)$;
$S_{30}(t) = S_3(t)$;
$S_{40}(t) = S_4(t)$;
$S_{50}(t) = S_5(t)$;
$S_{60}(t) = S_6(t)$;
$S_{70}(t) = S_7(t)$;
$S_{80}(t) = S_8(t)$;
$S_{90}(t) = S_9(t)$.

be hold, so the solution of this equation is:
$$T_{fsw} = B$$

We can known that this solution is coincide well with the practice, things decomposition is to divide $S(t)$ into $S_1(t) \cup S_2(t) \cup S_3(t) \cup S_4(t) \cup S_5(t) \cup S_6(t) \cup S_7(t) \cup S_8(t) \cup S_9(t)$. It is also further proved that to express the transformation by matrix is correct.

A can also be broken down further. For example:

$$\begin{pmatrix} U & S(t) & \vec{p} & T_1(t) & L(t) & x(t) = f((u(t),\vec{p}),G_U(t)) & G_U(t) \\ U & S(t) & \vec{p} & T_2(t) & L(t) & x(t) = f((u(t),\vec{p}),G_U(t)) & G_U(t) \\ U & S(t) & \vec{p} & T_3(t) & L(t) & x(t) = f((u(t),\vec{p}),G_U(t)) & G_U(t) \\ \cdots\cdots\cdots\cdots\cdots\cdots\cdots\cdots\cdots\cdots\cdots\cdots\cdots\cdots\cdots \\ U & S(t) & \vec{p} & T_n(t) & L(t) & x(t) = f((u(t),\vec{p}),G_U(t)) & G_U(t) \end{pmatrix}$$

Therein, $T_1(t)$ = {Traffic congestion caused by hardware}; $T_2(t)$ = {Traffic congestion caused by software}; $T_3(t)$ = {Traffic congestion caused by emergency}; $T_4(t)=$ {Traffic congestion caused by multi-factor} and so on.

V. CONCLUSIONS

Error- elimination transformation can be expressed by error matrix in the error eliminating theory, such as $T(u)=u_1$, and any unknown can be got by solving the matrix equation, which provides methods to avoid and to eliminate error in theory and practice.

ACKNOWLEDGMENT

This study was supported by the project items of science and technology of Guangzhou Vocational College of Science and Technology with the item number 2014ZR04 and the project items of vocation education of Guangdong Academy of Education with the item number GDJY-2014-B-b287.

REFERENCE

[1] Barwise,J, "Handbook of Mathematical Logic", Amsterdam: North- Holland Publishing Company, 1977.

[2] Cheng-yi Wang, "The introduction to fuzzy mathematics", Beijing: Beijing University of Technology Press, 1998.

[3] Kai-zhong Guo, Hai-ou Xiong, "Research on destruction connective of fuzzy logic error things", Fuzzy Systems and Mathematics Vol.20 (2): 34-39. 2006.

[4] Shi-yong Liu, Kai-zhong Guo, "Exploration and application of redundancy system in decision making-relation between fuzzy error logic increase transformation word and connotative model implication word", Advances Vol.39 (4): 17-29, 2002.

[5] Yong-qing Liu, Kai-zhong Guo, "The theory and method for researching conflict and error of the complex giant systems", Guangzhou: South China University of Technology Press. 2000.

[6] Kai-zhong Guo, "The error system", Beijing: Science Press, 2012.

[7] Hai-ou Xiong, Kai-zhong Guo, "Research on decomposition conversion connectors of fuzzy logic error domain", Fuzzy Systems and Mathematics Vol.20 (1): 24-29, 2006.

[8] Ka-izhong Guo, Shi-qiang-Zhang, "The theory of error set", Changsha: Central South University Press. 2001.

[9] Hong-bin Liu, Kai-zhong Guo, "One-element fuzzy error-matrix", Modeling Vol.27 (2):33-42, 2006.

[10]Kai-zhong Guo, Shi-qiang Zhang, "The theory and method for discriminating errors in decision-making of fixed assets investment", Guangzhou: South China University of Technology Press, 1995.

[11]Min Li, Kai-zhong Guo, "Research on decomposition of fuzzy error set", Advances, Vol.43 (26): 15-26, 2006.

[12]Kai-zhong Guo, Shi-qiang Zhang, "The introduction to Error-eliminating", Guangzhou: South China University of Technology Press, 1995.

[13]Kai-zhong Guo, Shi-yong Liu, "Research on laws of security risk- error logic system with critical point", Modeling Measurement & Control Vol.22 (1): 1-10, 2001.

[14]Shi-yong Liu, Kai-zhong Guo, "The substantial change of decision-making environment-mutation of fuzzy error system", Advances Vol.39 (4): 29-39, 2002.

[15]Yong Sheng, Wen Cheng, "Risk prevention and pitfalls avoidance", Beijing: Enterprise Management Publishing House, 1998.

[16]Zhong-kuan Zhao, "The introduction to mathematical dialectical logic", Beijing: Renmin University of China Press, 1995.

Study on Service Station Layout for Field Maintenance Service of Agricultural Machinery

Xin Li*, Jing-qian Wen, Yao-guang Hu, Xiao-feng Duan

The Institute of Industrial and System Engineering, School of Mechanical Engineering, Beijing Institute of
Technology, Beijing, China
(*2120130439@bit.edu.cn, wenjq@bit.edu.cn, hyg@bit.edu.cn, xfduan@bit.edu.cn)

Abstract - **This article is based on the characteristics of the field maintenance service for agricultural machinery, proposing a kind of emergency service mode in which agricultural machinery enterprise sets up emergency service station composed of maintenance vehicle in the rural to supple the existing fixed service ability at the busy farming season. By establishing the mathematical model and using an adaptive genetic algorithm, temporary service station location and service range are determined. The model is verified that enterprise can improve the service level and reduce the maintenance cost by an illustration.**

Keywords - **Adaptive genetic algorithm, agricultural fault, emergency service stations, location, site maintenance**

I. INTRODUCTION

With country supporting agriculture vigorously, agricultural mechanization has achieved an unprecedented development meanwhile after-sales maintenance service level of agricultural machinery also has got corresponding improvement. In order to improve the market competitiveness, agricultural machinery enterprises must improve the service level and increase farmers' customer satisfaction. Considering farm machinery fault's characteristics that include uncertainty of occurrence time, location and quantity, and farmers often have to spend a lot of time and cost to make agricultural machines resume their work. It not only hinders the normal of agricultural product operation, but also greatly reduces the customer satisfaction. In order to solve the problem, setting up emergency maintenance service system for agricultural machinery is the key, which emergency service station site selection is the primary task.

At home and abroad, the research about emergency facility location problem has had certain development. Reference [1] puts forward the emergency facility location problem for the first time. Since then, in view of the problem, the main research model has set covering model, maximum coverage model and value model, etc. At the same time, the optimization algorithm of the model proposed successively, including the accurate algorithm, analytic hierarchy process (AHP), genetic algorithm and ant colony algorithm, etc. Reference [2] puts forward a covering model to determine the service station location and the number of configured service personnel, in which minimize costs and ensure the response time. However, it didn't solve the model. Reference [3] puts forward facility location problem in the case of demand uncertainty, aiming at minimizes risk of rescue, but it didn't consider the influence of the existing rescue facilities. Reference [4]

constructs the loss minimization model, regarding the demand cannot meet the losses as the breakthrough point. Emergency system is mostly used in the rescue, medical, fire and other fields, but maintenance service research for agricultural machinery breakdown is very lacking [5].

This paper establishes a mathematical model based on agricultural machinery fault characteristics during the busy farming, considering the existing service station, and proposes an adaptive genetic algorithm to solve the multiple emergency service stations position in the rural. Thus, the emergency maintenance network for agricultural machinery is set up, which supplies service ability, reduces response time, improves service levels, and reduces the maintenance cost of the enterprise.

II. PROBLEM DESCRIPTION

At present, typical enterprise after-sales maintenance services of domestic agricultural machinery manufacturing uses a cooperation model that local dealers is given free agent to provide spare parts while agricultural machinery manufacturing enterprises pay the corresponding maintenance costs. In this case, dealers as service stations are directly responsible for after-sale maintenance of agricultural machinery.

Through the analysis of the characteristics of agricultural machinery fault, this article divides it into two categories: The farm machinery fault has influenced seriously the operation of agricultural machinery. So farmers have to wait to continue to work production after completion of maintenance; another type of failure does not affect the operation of agricultural machinery, in this case, the farmers are more willing to take time to rush in the harvest. For the first kind of failure mode, the paper establishes a kind of emergency maintenance system. Because of the uncertainty of the agricultural machinery fault's occurrence time, location and demand, and the limitless of dealer locations and service ability, servicer can't arrive timely, which leads to severely reduce the customer satisfaction. Especially in the growing season, agricultural machinery faults force farmers stop to wait for repair, and dealers are often located in the center of the city whose physical distance from the fault occurred hinders completing maintenance tasks in time. Thus, influence of agricultural work lowers the service level seriously.

Therefore, establishing emergency maintenance system by agricultural machinery enterprises during farming rush, namely establishing emergency service station in the rural, equipped with maintenance vehicle, is

E. Qi et al. (eds.), *Proceedings of the 21st International Conference on Industrial Engineering
and Engineering Management 2014*, Proceedings of the International Conference on Industrial Engineering
and Engineering Management, DOI 10.2991/978-94-6239-102-4_23, © Atlantis Press and the authors 2015

effective way to fill the inadequacy of ability of dealers and improve the ability of the enterprise service. In order to use this method, this paper solves the following problems: the location of the emergency service station, the service scope of emergency service station and existing service station.

III. MODEL

A. Assumption

– maintenance vehicle of dealers and farm machinery manufacturing enterprises can repair all types of faults;

– each farm machinery fault is repaired by only one service station;

– the speed of maintenance vehicle is constant;

– transportation cost is proportional to the driving range, due to the constant speed, so the transportation cost is proportional to the driving time;

– know agricultural machinery fault occurrence time and location;

– each service station service capacity is enough; the number of maintenance vehicle is infinite.

B. Mathematical model

Based on the above hypothesis, the mathematical optimization model was established as follows:

$$\min V_1 = \sum_{i \in I} \sum_{j \in J} a t_{ij} z_{ij} \tag{1}$$

$$\min V_2 = \sum_{j \in J} b_j x_j \tag{2}$$

$$\min V_3 = \sum_{j \in F} x_j + (1+\varepsilon) \sum_{j \in J-F} x_j \tag{3}$$

$$\sum_{j=J} y_{ij} = 1, \forall i \in I \tag{4}$$

$$t_{ij} y_{ij} \leq T \tag{5}$$

$$y_{ij} \leq x_j \tag{6}$$

$$x_j, y_i \in \{0,1\}, \forall i \in I, \forall j \in J \tag{7}$$

In formula: I is maintenance points set, i=1,2,3…; J is new emergency service station points set, j=1,2,3…; F is original service station points set; a is transportation cost per time; ε is a strength factor, the higher the cost of a new service station is, the greater the coefficient is; b_i is the fixed fee for establishing service station; T is the limited response time; t_{ij} is time of service station j to the maintenance point i; x_i indicates whether or not to set up the service station, if service station j is established, $x_i =1$, otherwise, $x_i =0$. y_{ij} indicates whether or not service station j services maintenance point i, if service station j services maintenance point i, $y_{ij} =1$, otherwise, $y_{ij} =0$.

In this mathematical model, (1) is the objective function which means that minimize the total time of response time; (2) is the objective function that minimize the total fixed charge; (3) is objective function that keep the existing service stations as much as possible, if they

must be retained for all, neglect (3). Equation (4) to (7) is restraint conditions, (4) makes sure each maintenance point has service station for its services; (5) represents the time of each emergency service station to its maintenance point must be no more than T;(6) ensures that service station can send its service to a maintenance point only under the condition that the candidate sets up service station.

C. The evaluation index

The layout of service point can be proved reasonable by service level basically. Economy mainly refers to the maintenance costs, due to the emergency service point is composed of car maintenance, so the fixed costs only include rental costs of staff recreation sites,so this article assumes that the maintenance cost combines transportation costs and the rental cost of emergency service station.

Service level mainly refers to the response time of service station, contains a single service point average response time T^j_{ave} which is calculated by (9) and the average response time of all service points T_{ave} which is calculated by (10).

$$T^j_{ave} = \sum_{i \in I} t_{ij} x_{ij} \Big/ n \tag{8}$$

$$T_{ave} = \sum_{j \in J} T^j_{ave} \Big/ p \tag{9}$$

Among them, n is the number of the maintenance station, p is the number of service stations.

IV. OPTIMIZATION ALGORITHM

A. Chromosome Coding

By analyzing the model, the binary coding method is a suitable method which is simple and feasible. The coded character set made by the binary symbols of one and zero is binary notation set {0, 1}. The individual genotype is a binary code strings. Chromosome coding length is determined by the number of candidate service stations; meanwhile, each code position corresponds to a candidate service station. Assuming n is the number of candidate service station, that is to say, each chromosome has n genes which are one or zero, and zero represents the service station is not to participate in the agricultural machinery service network while one represents it. For instance, chromosome coding was [010001] for six candidate service stations while represents the second station and the sixth candidate service stations are selected.

B. Fitness Function

Fitness function is a kind of criteria to distinguish individual species according to the objective function, so the fitness function is evaluated through the objective function. For more than one goal function, this paper uses the linear weighted sum method to make the multi-objective function into single objective function.

According to different target importance, this article gives different target weights intuitively to show the influence of different target tendency for service stations location, then integrats various aspects demanding to optimize service point location. Weighting function (10) as follows:

$$\sum_{k=1}^{3} \lambda_k = 1, \lambda_k \geq 0, k=1,2,3 \qquad (10)$$

λ_k is the weight of K objective function.

$$\min F = \sum_{k=1}^{3} \lambda_k V_k \qquad (11)$$

Objective function is (11), the fitness function (12) as follows:

$$Fit = C_{max} - F \qquad (12)$$

C_{max} is the biggest estimate of F.

C. Selection Operator

Genetic algorithm uses selection operator to weed out a group of individuals. This article uses the method combining the roulette wheel selection and optimal preservation strategy to select the next generation. First ,use the roulette wheel selection method which the probability of each individual in the next generation is equal to its fitness value's proportion in the total individuals fitness value.The higher the fitness value, the greater the likelihood of being chosen. And then use the optimal preservation strategy, that is, the highest fitness in current population individuals doesn't participate in crossover and mutation operations ,while replace the lowest fitness of the individual in the next generation populations directly.By this method can avoid destroying the fitness of the best individual in the group which due to the randomness of genetic operations,at the same time, improve the global search ability of the algorithm.

D. Crossover and mutation operator

Crossover operation is the main method to generate new individual, thus determines the genetic algorithm global searching ability; Mutation operation is auxiliary method to generate new individual, which determines the local search ability of genetic algorithm [6]. In order to ensure that "rapid convergence", genetic algorithm hope group as soon as possible get to the optimal state transition, this reduces the diversity of the individuals in the group and makes the algorithm "premature". Guarantee the global optimal, that is, maintain the diversity of the individuals in the group to avoid falling into local extremum, but that is at the expense of the convergence speed. Adaptive p_m, p_c are able to dynamicly provide the best value according to the centration of current fitness.Adaptive formula is (13) and (14):

$$p_c = \begin{cases} \dfrac{P_c}{1 - Fit_{min}/Fit_{max}}, & (\dfrac{Fit_{ave}}{Fit_{max}} > 0.8 \cap \dfrac{Fit_{min}}{Fit_{max}} > 0.8 \cap n<50) \\ p_c, & \text{,others} \end{cases} \qquad (13)$$

$$p_m = \begin{cases} \dfrac{P_m}{1 - Fit_{min}/Fit_{max}}, & (\dfrac{Fit_{ave}}{Fit_{max}} > 0.8 \cap \dfrac{Fit_{min}}{Fit_{max}} > 0.8 \cap n<50) \\ 0.9^{n-50} p_m, & \text{,others} \end{cases} \qquad (14)$$

The ratio of maximum and minimum fitness value can reflect the concentration of the whole group, that is, the more Fit_{min}/Fit_{max} closes to 1, the more group concentrates. The ratio of average and maximum fitness value can reflect the concentration of internal group, that is, the more Fit_{ave}/Fit_{max} close to 1, the more group concentrated.

At the beginning of the evolution, that is, the number of iterations is less than 50, for the sake of rapid convergence and finding global optimal solution, this article assumes that, if $Fit_{min}/Fit_{max} > 0.8$ and $Fit_{ave}/Fit_{max} > 0.8$, the group focused, and p_m, p_c carry out adaptive changes in accordance with the above formula(13), otherwise, keep the initial value. In late evolution, in order to guarantee convergence, use (14),among it,n is the number of iterations.

Cross intersection method ueses a single point, namely,in accordance with the crossover probability, pick two chromosomes in a population ,and then set an intersection within the scope of the chromosome length randomly.The parts of two individual carried on the exchange, so we generate two new individual [7, 8].

Mutation method is the basic variation which the value of certain genes in individuals changes, namely, according to the mutation probability, take randomly mutation point position, and then the value in the mutated gene changes opposite value. Finally, forms a new individual.

E. Termination conditions

Genetic algorithm has a variety of termination conditions, one is when the largest fitness value and the average fitness value tends to be stable; The second is when the genetic iterations or algebraic reaches preset value; The third is when a group of optimal fitness value reaches the a given value [9].

This paper termination method is: maximum fitness does not change in 30 generations, algorithm stops.

V. EXAMPLE VERIFICATION

Take the maintenance area responsibled by Laiyang Feng Ge Zhuang Chia agricultural repair shop as a example, take a set of maintenance data during busy season, such as Table I. Select eight candidates for emergency service points in the region based on the factors of actual geographical, social and economic,which location, rental costs are known [10-12].

Feng Ge Zhuang Chia agricultural repair shop is a identified service station, so the chromosome coding is a binary sequence of length 9 with a starting point 1. Parameters set, the initial value of crossover probability p_c is 0.5, and p_m is 0.05. Maximum estimated value of fitness C_{max} is 2500, the objective function weights are all set to 1.

Traveling speed of maintenance vehicles is 50 km / h, the cost per time is a = 2.5, the maximum response time T is 40min. Get results by Matlab.

TABLE I
MAINTENANCE POINTS DATA

Num.	Maintenance point	Latitude and longitude
1	Okara Village	121.373309,36.762793
2	Wali village	121.30197,36.752716
3	Young Village,	120.611628,36.742797
4	Chengyang Village	121.210728,36.801828
5	Matsuyama Village	120.875163,37.417165
6	Tingkou Village	121.022207,37.323749
7	Canggezhuang Village	121.014194,37.489735
8	Tangezhuang Village	120.603722,37.125921
9	Dayanjia Village	121.12938,36.692236
10	Xinda Village	121.352963,36.759533
11	Dongcun Village	121.235294,36.800354
12	Shangyijia Village	121.239111,36.949103
13	Suozhiqian Village	121.12994,36.911579
14	Songjia Village	121.341365,36.771907
15	Dayangjia Village	121.136732,36.695123
16	Wanliu Village	120.836442,37.00208
17	Daxiejia Village	120.836442,37.00208
18	Jinling Village	120.305229,37.400855
19	Taocun Village	121.142524,37.184617
20	Fangyuan Village	121.435988,36.781659
21	Chengyang Village	121.154707,36.827325
22	Liugezhuang Village	121.323358,36.811829
23	Daxinjia Village	121.374377,36.754733
	Total cost 2080	average time 3.2h

The allocation of Maintenace points is shown in Table II. From the Fig.1, the existing service point, namely Chia repair shop, is not allocated to any maintenance point, showing its location is unreasonable, but it is a fixed service station whose relocation needs a large costs, so in reality, we can regard it as a superior service station .When a more complex fault occurs which the underlying point of the emergency services can't afford to repaire, it can solve this kind of fault [13, 14].

TABLE II
TTE LOCATION OF MAINTANEANCE POINTS

Num.	Responsibled Maintenance points	Average response time (min)	Total average time (min)	Total cost(RMB)
1	no	no		
3	1,4,10,21,23	22		
5	3,8	28		
6	5,18	18.5	20.6	1427
7	6,7,17,19	19.5		
8	11,16,20	17.5		
9	29,12,13,14,15,22	18.3		

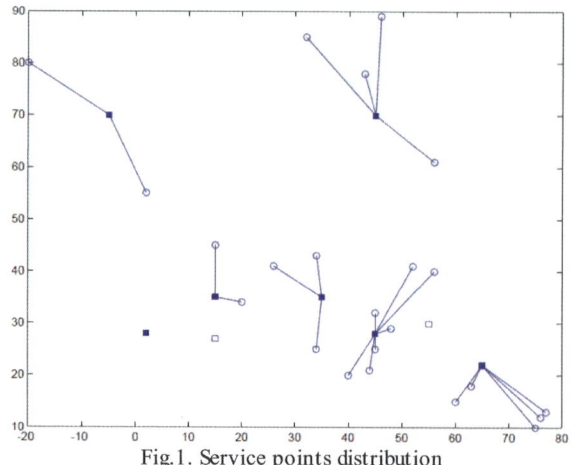

Fig.1. Service points distribution

Optimal iterative figure is Fig.2 which shows optimization process. Fig.1 shows service points distribution which is visual expression of Table II.

When using the only fixed service station,enterprise costs 2808 yuan, with an average response time of 3.2h, while as is shown in Table II, it can costs only 1427 yuan, with total average response time of 20.6 minutes by the new service station layout.Therefore, the proposed layout mode of service station is rational and effective [15].

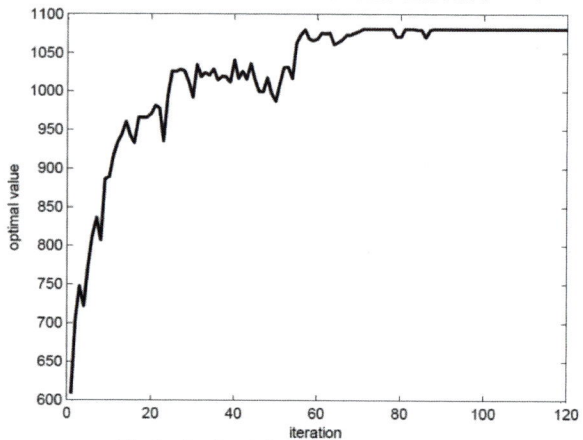

Fig.2. Optimal fitness iteration curve.

VI. CONCLUSION

This article proposes a kind of service model based on the characteristics of agricultural failure in the busy season, which the agricultural enterprises send a lot of maintenance vehicles to the country to build emergency service stations which can supply existing service capabilities. Then, establish a mathematical model, and use an adaptive genetic algorithm to obtain the location and scope of services emergency services station. That model is verified by an example of a certain maintenance that can reduce maintenance costs and improve service levels. But the article does not consider the number of maintenance vehicles configuration in each emergency repair service station, so the maintenance costs will not combine the cost of maintenance personnel. Future research will improve this aspect.

ACKNOWLEDGMENT

This research is funded by the Republic of China's National High-Technology Project (863) under contact number: 2013AA040402. Thank you for the cooperation of Foton Lovol International Heavy Industry Co., Ltd.

REFERENCE

[1] K. V. Ramani, "An interactive simulation model for the logistics planning of container operations in seaports," Simulation, vol. 66, pp. 291-300, 1996.

[2] Luo Fenglian. and Guo Qiang, "Emergency center location and with car problems,"(in China), Computer Engineering and Application pp. 241-244, 2011.

[3] M. S. Canbolat and M. von Massow, "Locating emergency facilities with random demand for risk minimization," Expert Systems with Applications, vol. 38, pp. 10099-106, 2011.

[4] K. V. Ramani, "An interactive simulation model for the logistics planning of container operations in seaports," Simulation, vol. 66, pp. 291-300, 1996.

[5] Zhang Yusong and Ma Ya Dong. "One kind of deal with unexpected demand for logistics activities center location model," (in China), Analysis and Decision, pp. 123-124, 2007.

[6] Liu Shuting. and Jin Taidong. "An improved adaptive genetic algorithm," (in China), Location Science, pp. 127-128, 2012.

[7] A. Apte, U. M. Apte, and N. Venugopal, "Focusing on customer time in field service: a normative approach," Production and Operations Management, vol. 16, pp. 189-202, 2007.

[8] T. Qunhong, G. R. Wilson, and E. Perevalov, "An approximation manpower planning model for after-sales field service support," Computers and Operations Research, vol. 35, pp. 3479-88, 2008.

[9] D. L. Haugen and A. V. Hill, "Scheduling to improve field service quality," Decision Sciences, vol. 30, pp. 783-804, 1999.

[10] Z. Yanxia and H. Jiazhen, "Customer service and distribution center location," in 2005 International Conference on Services Systems and Services Management, 13-15 June 2005, Piscataway, NJ, USA, 2005, pp. 315-17.

[11] C. Li-si and L. Zi-xian, "An emergency service center location model for vehicle repair with priority queuing rules and service level constrains," in 2011 IEEE 18th International Conference on Industrial Engineering and Engineering Management (IE& EM 2011), 3-5 Sept. 2011, Piscataway, NJ, USA, 2011, pp. 1333-6.

[12] L. Lin, L. Shixin, and T. Jiafu, "Basic models for solving distribution center location problems: a review," in 2007 International Conference on Service Systems and Service Management, 9-11 June 2007, Piscataway, NJ, USA, 2007, pp. 268-72.

[13] S. Sacone and S. Siri, "An integrated simulation-optimization framework for the operational planning of seaport container terminals," Mathematical and Computer Modelling of Dynamical Systems, vol. 15, pp. 275-93, 2009.

[14] J. M. Cabral, J. G. Rocha, J. E. Neves, and J. Ruela, "Scheduling algorithms to support QoS and service integration in sensor and actuator networks," in IEEE International Conference on Information Technology, 15-17 Dec. 2006, Piscataway, NJ, USA, 2007, pp. 529-34.

[15] M. R. Blumberg, "Technological developments and approaches to improving service quality," Biomedical Instrumentation and Technology, vol. 33, pp. 35-44, 1999.

On a Dual Sourcing Policy in Two-sided Disruptive Supply Chain

I-Chen Lin*, Yu-Hsiang Hung

Industrial Engineering and Management, Taipei University of Technology, Taipei, Taiwan
(*iclin@ntut.edu.tw, joaisn185@gmail.com)

Abstract - **In this study, we consider a single echelon supply chain in which a retailer replenishes inventory from two vulnerable suppliers whose supply could be affected by in/out-bound disruptions sporadically. Under a stochastic demand setting, a dual-sourcing policy with a supplementary supplier is introduced to mitigate the profit losses for the retailer. We built mathematical models to evaluate the performances of the retailer under single and dual sourcing policies respectively. The study shows that a risk-neutral retailer will prefer a dual sourcing policy that makes him carry more inventory from a more robust supplier when facing two-sided disruption risks. Furthermore, the retailer performances in terms of total profit are overwhelming under scenarios where the supplementary supplier is less, median and more robust to risks. Our mathematical models is examined through numerical examples and the results provide important guidelines for implementing sourcing policies when two-sided disruption risks is considered.**

Keywords - **Dual sourcing, inventory replenishment, supply chain disruption**

I. INTRODUCTION

Global competition features efficient operations in supply chain integration and coordination. To gain the competitive edge, more and more companies engage in improving the vendor-buyer relationship. Supply chain disruptions and hedging risks are considered the most pressing concerns facing business competing in today's global marketplace [1]. A disruption could result from an unplanned or unanticipated occurrence which brings huge and unintended losses for the supply chain in comparison with normal supply/demand complex interactions. Examples of these include the 9/11 terrorist attacks, the devastating earthquake and nuclear crisis in Japan 2011, severe flooding during monsoon season in Thailand 2011, the lightning strike at the Philips NV microchip plant in New Mexico, or the shutdown of all air traffic due to a volcanic eruption in Iceland. These events caused the variation of damages that decrease value-added activities in a chain. Therefore, building trustworthy relationships between suppliers, retailers and contractors becomes more and more important when dealing with disruption management and supply chain risks hedging [2].

A single entity in a supply chain can be vulnerable to supply disruptions, but if there is more than one supply source or if alternative supply resources are available, then the risk of disruption would be reduced [3]. Thus, our aim is to show the adaption of triangular relationships between one retailer and two suppliers who facing two-sided disruption risks are preferable than that in the one-to-one bilateral supply chain. We consider a dual-sourcing newsvendor model by building mathematical models in a one echelon supply chain facing stochastic end demand.

Our study concludes that the dual sourcing policy for the retailer can improve his performance when both suppliers are vulnerable to their in/out-bound environment risks.

Supply chain disruption management theories and practices provide alternative ways to hedge against supply side risks. One of the most common policies is flexible/multiple-sourcing that firms use it to a variety of strategic reasons, such as hedging against supply disruptions and safeguarding against predatory monopolistic practices [4]. While academics and practitioners argue that supply chains have become more vulnerable to disruption due to global competition in the last few years [5]. Building a multi-sourcing channel could significantly reduce the losses from supply chain disruptions. Cisco and Apple, leaders in the communication technology industry, provided good examples. Thanks to their sophisticated supply chain disruption management strategies, they were able to assess the impact of the disruption for their suppliers in 12 hours when the tsunami catastrophe struck Japan in March 2011 [6]. The sale for Apple's iPad2 went on just hours after the tsunami hit. With consolidated relationships with suppliers, Apple was capable of dealing with subsequent shutdowns caused by stock shortages and long delays in deliveries [7].

One of the vanguard and important studies concerning uncertainty framework is to characterize a product when seeking to devise the right supply chain strategies. His framework specifies the two main key uncertainties, i.e., demand and supply, and matches for supply chain strategies to the right level of demand uncertainties [8]. Another work as in [3] expanded this framework to include supply uncertainties. A dual sourcing policy is built to deal with such uncertainties. It implies that the buyer replenishes inventory from two suppliers in the same time (or may not). One of the supplies may dominate the other in terms of business share, quality, price, reliability, and others. Two significant research contributing in the field of inventory management for dual-sourcing policies are presented as in [9] and [10]. Both research considered a retailer who faces constant demand and replenishes from two identical-cost, capacitated suppliers subject to specific production failure rates. The assumptions in their models that demand is constant and the suppliers have identical cost structures are often not the case in reality. A recent comprehensive overview on supply disruption literature is provided by Snyder *et al.* [11]. There are research not only highlighted the long-term negative effects of supply chain disruptions but also contributed relevant insights into related issues such as supply chain disruption strategies [5] [12].

E. Qi et al. (eds.), *Proceedings of the 21st International Conference on Industrial Engineering and Engineering Management 2014*, Proceedings of the International Conference on Industrial Engineering and Engineering Management, DOI 10.2991/978-94-6239-102-4_24, © Atlantis Press and the authors 2015

In the past few years, there has been a substantial increase in supply chain disruption related fields applying different research methodologies to evaluate the impacts and its managerial implications. One stream of the existing publications are based their efforts on empirical qualitative studies, e.g., [13] and [14]. While some other studies utilize available secondary data bases and archival data to analyze the impact on supply chain disruptions [15].

While disruption management in supply chains has drawn extensive attention in academia in recent years, the modeling of multi-sourcing is still in an embryonic phase. Thus, we construct triangular retailer-supplier relationships by mathematical models to clarify the benefits of a dual sourcing policy under two-sided supply risks.

II. SINGLE AND DUAL SOURCING REPLENISHMENT

Our research will first present a single sourcing policy as the baseline under stochastic demand, and then we will construct a newsvendor model within a dual sourcing scenario by incorporating a secondary supplier under two-sided disruption risks. In this scenario, the retailer split his order between two non-identical suppliers simultaneously. Then, a numerical example will be performed to illustrate the performances of the retailer.

We consider in and out-bound disruption risks in a one-echelon supply chain. One type of risk is considered when the suppliers are vulnerable to its environment, i.e., natural disasters or unexpected accidents. The probability of such event occurrence is defined as α_i for each supplier i, where $i= 1$ or 2. When such an undesirable event happens, only a proportion of the initial ordered quantity could be delivered on time and thus the retailer will encounter a loss from goodwill cost due to the shortage. The other disruption risk, defined by r_i, is characterized by the features of supplier's recovery ability which presumes supplier's capacity to react and recover from a specific disruption event. Whenever a disruptive event happens in the environment of supply chain, at probability α_i, the supply of the product will endure a difficult situation. Therefore, the corresponding quantity delivered will diminish to a portion of retailer's initial order, i.e., $r_i Q_i$, where Q_i is the initial retailer's order quantity. The notations are as follow:

α_i the probability of environmental disruptions for supplier i ($i = 1, 2$),

r_i the recovery ability when supplier i faces disruptions ($i = 1, 2$) $0 < r_i < 1$,

T_i the on-time delivery rate for supplier i ($i = 1, 2$), $T_i = [1 - \alpha_i(1 - r_i)]$,

w_i the wholesale prices for supplier i ($i = 1, 2$),

g the goodwill cost due to product's shortage,

p the retail price,

s the unit salvage value,

x the customer demand with probability density function $f(x)$ and cumulative distribution function $F(x)$.

Without loss of generality, the retail price p is assumed greater than the wholesale price w_i ($p > w_i$) and the w_i will be greater than the salvage value s ($w_i > s$). We assume, for simplicity, there is zero order lead time for each supplier and zero setup cost for retailer. One single batch is allowed in a single selling period and no emergency order and backorder.

In a single sourcing scenario, the retailer needs to determine an order quantity Q_{SS} to the sole supplier in the beginning of sale season. Without any disruption, i.e., the probability $(1 - \alpha_1)$, the retailer's expected profit is obtained based on newsvendor analysis. When out-bound disruptions happen with probability α_1, the supplier can only deliver the quantity of $r_1 Q_{SS}$ to the retailer.

Taking both situations into account, we obtain the retailer's expected total profit in equation (1)

$$\pi_{SS}(Q_{SS}) = (1 - \alpha_1) \times [\int_0^{Q_{SS}}(px - w_1 Q_{SS} + s(Q_{SS} - x))f(x)dx + \int_{Q_{SS}}^{\infty}(pQ_{SS} - w_1 Q_{SS} - g(x - Q_{SS}))f(x)dx] + \alpha_1 \times [\int_0^{r_1 Q_{SS}}(px - w_1 r_1 Q_{SS} + s(r_1 Q_{SS} - x))f(x)dx + \int_{r_1 Q_{SS}}^{\infty}(pr_1 Q_{SS} - w_1 r_1 Q_{SS} - g(x - r_1 Q_{SS}))f(x)dx].$$
(1)

The retail is assumed to be a risk-neutral decision-maker, therefore he will order the quantity that maximize his total profit based on the necessary condition $\frac{\partial \pi_{SS}}{\partial Q_{SS}} = 0$. We have

$$(1 - \alpha_1)F(Q_{SS}^*) + \alpha_1 r_1 F(r_1 Q_{SS}^*) = \frac{(p - w_1 + g)}{(p - s + g)} \times T_1.$$
(2)

Proposition 1. In a single sourcing policy under stochastic demand scenario:

(i) The retailer's total expected profit function π_{SS} is concave in Q_{SS}.

(ii) The optimal order quantity Q_{SS}^*, that maximize π_{SS}, can be obtained by equation (2).

The proof of Proposition 1 can be easily obtained and is omitted. Based on the Proposition, the retailer thus will be guaranteed to achieve his global maximal profit based on a specific decision on Q_{SS}.

After considering the sole sourcing policy, we construct the dual sourcing scenario in a similar vein. When the retailer adopts a secondary supplier, his/her decision will be the order quantities i.e., Q_{S_1} and Q_{S_2}, simultaneously to supplier one and supplier two respectively. In the same time, supplier one and two face different disruptions probability α_1 and α_2 respectively. When disruptions happen, supplier one could only deliver quantity of $r_1 Q_{S_1}$ and supplier two can only deliver quantity of $r_2 Q_{S_2}$ to the retailer respectively.

Unlike the previous settings, there will be four possible circumstances in dual sourcing settings when the retailer place his orders to these two suppliers simultaneously:

(I) Both suppliers do not encounter any disruptions, with a probability $(1 - \alpha_1)(1 - \alpha_2)$, and the order quantity will be received on time without any shortage. The expect total profit for the retailer is π_{SD_a}, where

$$\pi_{SD_a} = [\int_0^{Q_{S_1} + Q_{S_2}}(px - w_1 Q_{S_1} - w_2 Q_{S_2} + s(Q_{S_1} + Q_{S_2} - x))f(x)dx + \int_{Q_{S_1} + Q_{S_2}}^{\infty}(p(Q_{S_1} + Q_{S_2}) - w_1 Q_{S_1} - w_2 Q_{S_2} - g(x - Q_{S_1} - Q_{S_2}))f(x)dx].$$
(3)

(II) Disruption could happen solely on supplier one and that makes him discount initial order quantity Q_{S_1} to $r_1 Q_{S_1}$ while the other supplier remains delivery quantity Q_{S_2}. The probability for this circumstance is $\alpha_1(1 - \alpha_2)$. Thus, the expect total profit for the retailer is π_{SD_b}, where

$$\pi_{SD_b} = [\int_0^{r_1 Q_{S_1} + Q_{S_2}} \left(px - w_1 r_1 Q_{S_1} - w_2 Q_{S_2} + s\left(r_1 Q_{S_1} + Q_{S_2} - x\right)\right) f(x) \, dx + \int_{r_1 Q_{S_1} + Q_{S_2}}^{\infty} \left(p\left(r_1 Q_{S_1} + Q_{S_2}\right) - w_1 r_1 Q_{S_1} - w_2 Q_{S_2} - g\left(x - r_1 Q_{S_1} - Q_{S_2}\right)\right) f(x) \, dx].$$

(4)

(III) Disruption could happen solely on supplier two and that makes him discount initial order quantity Q_{S_2} to $r_2 Q_{S_2}$ while the other supplier remains delivery quantity Q_{S_1}. The probability for this circumstance is $(1 - \alpha_1)\alpha_2$. Thus, the expect profit for the retailer is π_{SD_c}, where

$$\pi_{SD_c} = [\int_0^{Q_{S_1} + r_2 Q_{S_2}} \left(px - w_1 Q_{S_1} - w_2 r_2 Q_{S_2} + s\left(Q_{S_1} + r_2 Q_{S_2} - x\right)\right) f(x) \, dx + \int_{Q_{S_1} + r_2 Q_{S_2}}^{\infty} \left(p\left(Q_{S_1} + r_2 Q_{S_2}\right) - w_1 Q_{S_1} - w_2 r_2 Q_{S_2} - g\left(x - Q_{S_1} - r_2 Q_{S_2}\right)\right) f(x) \, dx].$$

(5)

(IV) Disruption could impact these two suppliers in the same time at a probability $\alpha_1 \alpha_2$. Thus suppliers will be forced to diminish their quantities delivered, i.e., $r_1 Q_{S_1}$ and $r_2 Q_{S_2}$ respectively. Thus the retailer's expect total profit is π_{SD_d}, where

$$\pi_{SD_d} = [\int_0^{r_1 Q_{S_1} + r_2 Q_{S_2}} \left(px - w_1 r_1 Q_{S_1} - w_2 r_2 Q_{S_2} + s\left(r_1 Q_{S_1} + r_2 Q_{S_2} - x\right)\right) f(x) \, dx + \int_{r_1 Q_{S_1} + r_2 Q_{S_2}}^{\infty} \left(p\left(r_1 Q_{S_1} + r_2 Q_{S_2}\right) - w_1 r_1 Q_{S_1} - w_2 r_2 Q_{S_2} - g\left(x - r_1 Q_{S_1} - r_2 Q_{S_2}\right)\right) f(x) \, dx].$$

(6)

Finally, taking four circumstances above into account, retailer's expected total profit now can be calculated by tallying up equations (3), (4), (5) and (6) as follows:

$$\pi_{SD}(Q_{S_1}, Q_{S_2}) = (1 - \alpha_1)(1 - \alpha_2)\pi_{SD_a} + \alpha_1(1 - \alpha_2)\pi_{SD_b} + (1 - \alpha_1)\alpha_2\pi_{SD_c} + \alpha_1\alpha_2\pi_{SD_d}.$$

(7)

In a dual sourcing setting, the decision for the retailer is to find optimal quantities $Q_{S_1}^*$ and $Q_{S_2}^*$ which ordered from supplier one and two to maximize his/her expected total profit. To maximize $\pi_{SD}(Q_{S_1}, Q_{S_2})$, we equate its partial derivatives to zero, i.e., $\frac{\partial \pi_{SD}}{\partial Q_{S_1}} = 0$ and $\frac{\partial \pi_{SD}}{\partial Q_{S_2}} = 0$. Since $\pi_{SD}(Q_{S_1}, Q_{S_2})$ is continuously differentiable and strictly concave on its domain, this approach works. We obtain equation (8) and (9) accordingly as follows.

$$(1 - \alpha_1)(1 - \alpha_2)F(Q_{S_1}^* + Q_{S_2}^*) + \alpha_1(1 - \alpha_2)r_1 F(r_1 Q_{S_1}^* + Q_{S_2}^*) + (1 - \alpha_1)\alpha_2 F(Q_{S_1}^* + r_2 Q_{S_2}^*) + \alpha_1\alpha_2 r_1 F(r_1 Q_{S_1}^* + r_2 Q_{S_2}^*) = \frac{p - w_1 + g}{p - s + g} \times T_1.$$

(8)

$$(1 - \alpha_1)(1 - \alpha_2)F(Q_{S_1}^* + Q_{S_2}^*) + \alpha_1(1 - \alpha_2)F(r_1 Q_{S_1}^* + Q_{S_2}^*) + (1 - \alpha_1)\alpha_2 r_2 F(Q_{S_1}^* + r_2 Q_{S_2}^*) + \alpha_1\alpha_2 r_2 F(r_1 Q_{S_1}^* + r_2 Q_{S_2}^*) = \frac{p - w_2 + g}{p - s + g} \times T_2.$$

(9)

Proposition 2. In a dual sourcing policy under stochastic demand scenario:

(i) The retailer's total expected profit function π_{SD} is concave in Q_{S_1} and Q_{S_2}.

(ii) The optimal order quantity, $Q_{S_1}^*$ and $Q_{S_2}^*$, that maximize π_{SD}, can be obtained by solving equations (8) and (9) together.

The proof of Proposition 2 is omitted. Based on the Proposition, the retailer thus will be guaranteed to achieve his global maximal profit based on specific decisions on $Q_{S_1}^*$ and $Q_{S_2}^*$ as it in a sole sourcing scenario.

III. NUMERICAL ANALYSIS

In this section, a numerical study was carried out to demonstrate the mathematical behavior of the proposed models and to gain some insights into the problem being studied. We adopted parameters from the example as in [16]. In order to analyze the mathematical models, we applied a free software, **R** (programming language) which used among statisticians and data miners for developing data and statistical analyses.

We assumed the customer demand is normally distributed with mean value 400 units and standard deviation 130 units. The "Environment Factor" parameters, i.e. retail price p, goodwill cost g and salvage value s are given $p = \$45$, $g = \$15$, $s = \$10$ respectively. To compare the performance of the parameters, we classify supplier's parameters (probability of disruptions risk α_i, recovery ability r_i and wholesale price w_i) into three different level of settings to measure the behavior of suppliers. The supplier one is assumed to be a less robust one with parameters setting featured highest disruption probability, lowest recovery ability from disruptions and hence a less expensive wholesale price accordingly. On the other hand, the second supplier is a more robust one with lowest probability of disruption, highest recovery ability from disruptions and hence a more expensive wholesale price accordingly. The setting of a median supplier will be in between. Table I shows these three level of parameters settings for the supplier and the retailer's optimal expected profits.

From Table I, we obtain the on-time delivery rate T_i by computing $T_i = [1 - \alpha_i(1 - r_i)]$ and the retailer's expected profit π_{SS}^* in single sourcing policy under three level of parameters settings. It is worth noting that the retailer's optimal profit in the median robust parameter setting is overwhelming in three circumstances.

Fig.1 presents the effects of supplier's disruption risk on retailer's decision and his optimal profit. It shows the higher probability of disruptions risk will make the retailer order more quantity for the sake of avoiding stock out. Meanwhile, the retailer's optimal total expected profit will declines when facing higher supply chain disruptions. Fig.2 shows a higher supplier's recovery ability will make the supply chain partnership more robust to disruptions risks by achieving a higher retailer expected profit. It is worth noting that the optimal order quantity is a concave function of the supplier's recovery ability.

In order to demonstrate the effectiveness of the implementation of the dual sourcing policy, we analysis the retailer's optimal total expected profit under three levels of supplier's parameters settings mentioned before and compare with that under a dual sourcing setting.

TABLE I
THE PARAMETERS SETTINGS AND RETAILER PROFITS

Supplier	α_i	r_i	$w_i(\$)$	T_i	$\pi_{SS}^*(\$)$
Less robust	0.5	0.1	21	55%	1,850.5
Median	0.3	0.6	23	88%	6,026.2
More robust	0.1	0.9	25	99%	5,734.6

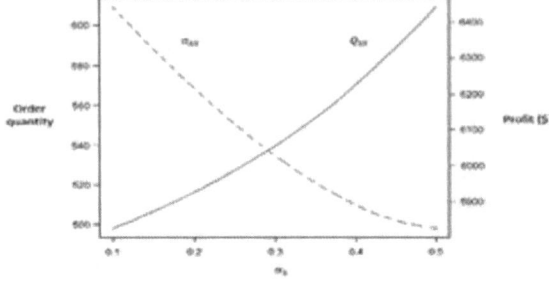

Fig.1. The effects of disruption risks

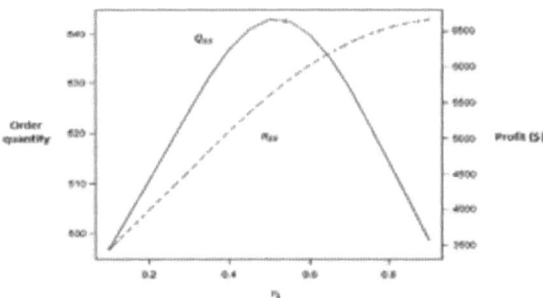

Fig.2. The effects of supplier's recovery capability

At first, we compute the optimal retailer's profits by choosing sole supplier under a less, median and more robust parameter settings. Then, in dual sourcing, we incorporate a secondary supplier and adopt his settings from less robust to more robust parameters in each case. There will be nine pairs of comparisons for retailer's optimal profits and the results are showing in the Table II. It can be seen obviously that the retailer's optimal total expected profit under dual sourcing settings is higher than that under settings at three levels for the secondary supplier. From the analysis, we conclude that the dual sourcing policy is always preferable compared to a single sourcing policy when suppliers are vulnerable to two-sided disruption risks under three levels.

To clarify the differences between the quantities ordered from the two suppliers result from the various disruption risks facing by the secondary supplier. The supplier one parameters are set at a median level and the supplier two is changing from the less robust one to more robust settings. The results show in Fig.3 the retailer will likely replenish more quantity from a supplier who is more robust to risks, and on the other hand reduce order quantity from the other supplier who is less robust. The retailer's profit increases when the secondary supplier faces lower disruption risks just as it in sole sourcing scenario.

TABLE II
COMPARISONS OF RETAILER PERFOR AMCE UNDER SINGLE AND DUAL SOURCING

Supplier one	Excepted profit ($)	Supplier two	Excepted profit ($)	Profit comparisons
Less robust	1,850.5	Less robust	4,308.8	Dual > Single
		Median	6,177.5	
		More robust	5,827.8	
Mediate	6,026.2	Less robust	6,177.5	Dual > Single
		Median	6,375.5	
		More robust	6,080.7	
More robust	5,734.6	Less robust	5,827.8	Dual > Single
		Median	6,080.7	
		More robust	5,741.8	

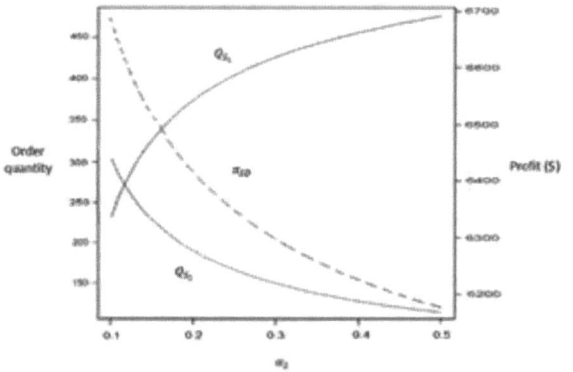

Fig.3. The effects of various disruption risks of the secondary supplier

IV. CONCLUSIONS

We propose a news vendor framework and illustrated the differences between sourcing strategies which will provide some insights in disruption management of supply chain and guidelines for companies. We build mathematical models under stochastic demand with two vulnerable suppliers and investigate the issues with sourcing strategies, e.g., single versus dual sourcing policies. To comply the reality, the in-bound and out-bound disruptions risks for suppliers are taken into account. It concludes, in a single sourcing policy, a higher disruptive risk for the supplier will prompt the retailer to increase retailer's order quantity while in the same time it undermines his total expected profit. Furthermore, a supplier with higher recovery capability will improve the retailer's expected profit by increasing replenishment quantity from such a supplier. It is worth nothing that a critical point exists in supplier's recovery ability that the retailer's order quantity declines inversely after the critical point.

In a dual sourcing policy, there are delicate trade-offs between the supplier's disruptive probability, recovery ability and wholesale price. A supplier with a higher disruptive probability makes the retailer order less quantity and that will lower the retailer total expected profit. Intuitively, a higher wholesale price will make such a supplier less attractive.

The retailer's optimal total expected profit is more sensitive to the supplier's recovery ability than the disruptive probabilities facing the supplier, while the retailer's optimal order quantities will be on the opposite of the story. It shows a supplier with higher recovery ability is more important than the levels of disruptive probabilities that the retailer should considered in a priority. In other word, the robustness of order fulfillment to risks for the suppliers or a consolidating partnerships between suppliers will be mostly important for the performance of the retailer.

The contributions of this research are as follows:

I. This research constructs the supply chain two-sided disruptions problems into mathematical models and generates propositions and insights that could help the retailer to make optimal decisions on order quantity.

II. We provide comparisons between single sourcing and dual sourcing policies under stochastic demand, and then facilitate parameter analyses to show the effects of various settings on the retailer's performances.

The possible extensions of the studies could consider more complicated supply chains with more tiers or more than one selling periods that will be more close to the generality of reality. Moreover, a multiple-sourcing policy or the adoption of partnership contracts under supply chain disruptions are still the prospective directions that can be investigated in the future.

REFERENCES

[1] C. Craighead, J. Blackhurst, M. J. Rungtsunatham and R. Handfield, "The severity of supply chain disruptions: Design characteristics and mitigation capabilities," *Decision Sci*, vol.38, no.1, pp. 131-156, 2007.

[2] J. W. Kamauff and R. E. Spekman, "The LCCS success factors," *Supply Chain Manag Rev*, vol.12, no.1, pp. 14-21, 2008.

[3] H.L. Lee. "Aligning supply chain strategies with product uncertainties," *Calif Manag Rev*, vol.44, no.3, pp.105-119, 2002.

[4] S. Veeraraghavan and A. Scheller-Wolf, "Now or later: A simple policy for effective dual sourcing in capacitated systems," *Oper Res*, vol.56, no.4, pp. 850-864, 2008.

[5] E. Simangungsong, L.C. Hendry and M. Stevenson, "Supply-chain uncertainty: A review and theoretical foundation for future research," *Intl J. of Prod Res*, vol.50, no.16, pp. 4493-4523, 2012.

[6] M. J. Sáenz, and E. Revilla, The supply chain management casebook: Comprehensive coverage and best practices in SCM. Cisco Systems, Inc.: Supply chain risk management, Financial Times Press, 2013, pp.80-96.

[7] E. Revilla, and M.J. Sáenz, "Supply chain disruption management: Global convergence vs national specificity," *J. Bus. Res*, vol.67, no.6, pp.1123-1135, 2014.

[8] M. L. Fisher, "What is the right supply chain for your product?" Harvard bus rev, vol.75, pp. 105-117. 1997.

[9] M. Parlar and D. Perry, "Inventory models of future supply uncertainty with single and multiple suppliers," *Nav Res Log*, vol.43, no.2, pp. 191-210, 1996.

[10] Ü. Gürler and M. Parlar, "An inventory problem with two randomly available suppliers," *Oper Res*, vol.45, no.6, pp. 904-918, 1997.

[11] L.V. Snyder, Z. Atan, P. Peng, Y. Rong, A.J. Schmitt and B. Sinsoysal, "OR/MS Models for Supply Chain Disruptions: A Review," Unpublished working Paper, 2012.

[12] B. K. Kaku, and B. Kamrad, "A framework for managing supply chain risk," *Supply Chain Manag Rev*, vol.15, no.4, pp. 24-31, 2011.

[13] S. M. Wagner, K. J. Mizgier, P. Arnez, "Disruptions in tightly coupled supply chain networks: the case of the US offshore oil industry," *Prod Plan Control*, vol.25, no.6, pp.494-508, 2014.

[14] A. Baghalian, S. Rezapour, and R. Z. Farahani, "Robust supply chain network design with service level against disruptions and demand uncertainties: A real-life case," *Eur J. of Oper Res*, vol.227, no.1, pp.199-215, 2013.

[15] A. Chaudhuri, B. K. Mohanty, and K. N. Singh, "Supply chain risk assessment during new product development: a group decision making approach using numeric and linguistic data," *Int J. Prod Res*, vol.51, no.10, pp. 2790-2804, 2013.

[16] A. Xanthopoulos, D. Vlachos, and E. Iakovou, "Optimal newsvendor policies for dual-sourcing supply chains: A disruption risk management framework," *Comput Oper Res*, vol.39, no.2, pp. 350-357, 2012.

The Analysis of Special Equipment Accident Factor Based on Rough Set

Jian Zhang[1], Hua-jie Li[2], Xue-dong Liang[2,*]

[1]Inspection Institute of Fujian Special Equipment, Fujian, 350008, China
[2]Business School, Sichuan University, Chengdu, 610065, China
(330596518@qq.com)

Abstract - **The exploration of special equipment accident factor is of great significance for preventing the happening of safety accident and ensuring the safety work of special equipment. In this paper, the special equipment accidents of calendar year are statistical analyzed, on this basis, special equipment safety factors are ascribed to management, environment, equipment and personnel. Through the processing of rough set, the incidence of special equipment safety accident levels influenced by the above four factors is concluded, and factor control measures are put forward accordingly.**

Keywords - **Factor, rough set, security, special equipment**

I. INTRODUCTION

Special equipment is the facilities referred for safety of life and greater danger. According to "Special equipment safety law of the People's Republic of China", special equipment includes boilers, pressure vessels (including gas cylinders), pressure pipes, elevators, lifting machinery, passenger ropeway, large-scale amusement facilities, etc, its character includes: 1, casualty which affecting public security is easily caused by the accident; 2, the risk is bigger, once an accident happens, it is likely to cause the group die or injury; 3, the using field is wide, involving all aspects of the national economy and people's life [1]. Therefore, the operation and safety in operation of special equipment is very important. However, special equipment safety accidents occur frequently, and bringing heavy losses to the people's life and property. The incidence of accidents will be effectively reduced by analyzing the reason of the accident and taking improvement measures.

At present, the cause analysis of the special equipment safety accident is mainly presents the two characteristics. (1) The cause analysis of some kind of special equipment safety accident or a specific accident is focused on. Tang Hongkai analyzes the accident reason of crane [2]. Xu Huoli analyzes the accident reason of boiler burst, and preventive measures are put forward [3]. Wang Peiyuan, Cheng Jun, Li Jianzhong, etc analyze the accident reason of tower crane happened in Tianjin [4]. (2) Qualitative analysis is mainly, lack of theoretical support and quantitative analysis model. In this paper, the special equipment accidents of calendar year are statistical analyzed, on this basis, special equipment safety factors are ascribed to management, environment, equipment and personnel. Reduction function of rough set is used to explore the relationship between

the cause of the accident and the accident quantity levels. Then the key factors leading to higher lever accident quantity are found out.

II. INTRODUCTION TO ROUGH SET THEORY

Rough set theory is put forward in 1982 by Polish scientist Pawlak [5], it is a new mathematical tool to study the fuzziness and uncertainty problem. The characteristics of this method is to use the information provided by the data itself, does not require any additional information or prior knowledge, through the indiscernibility relation and indiscernibility classes to determine the approximation of a given problem domain, so as to find out the inherent law of the problem.

Definition 1 $S = < U, A, V, f >$ is named as a knowledge expression system, U is an non-empty set, it is the set of all objects, known as the domain ontology; $A = C \cup D$ is all of the attributes of object in domain ontology, Subsct of C and D respectively refers to as the condition attributes and decision attribute sets, $V = \underset{a \in A}{U} V_a$ is the set constituted by attribute value, V_a means attribute value range of V_a, namely, the values range of a; $f: U \times A \rightarrow V$ is a information function, $f(a, x)$ determines the value of object x about attributes a.

Definition 2 $POS(X) = R_-(X)$ is called set X R–positive domain

$POS(X) = R_-(X)$ is called set X R –negative domain.

Definition 3 $K = (U, R)$ is a knowledge base, $P, Q \subset R$, knowledge Q depends on the knowledge of P by the degree of dependency $K(0 \leq K \leq 1)$, and denoted by $P => kQ$, If and only if

$$k = \gamma_p(Q) = card(POS_p(Q)/card(U))$$

If k = 1, it means knowledge Q depends entirely on the P;

If 0 < k < 1, it means Q part depends on the knowledge P;

If k = 0, it means knowledge Q completely independent in P.

Definition 4 P and Q is the equivalent relation collection in domain U, said $POS_p(Q) = \underset{X \in U/Q}{U} P_-(X)$

In rough set, the importance of attribute is measured by Changes in the classification of decision table after removing the attribute. If the classification of decision table change is big after the attribute is removed, it means the importance of this property is high, whereas the attribute importance is low [6]. By rough set attribute reduction, therefore, factors impacting on the quantity of accidents can be selected, and the influence size of the factors can also be distinguished. The reduction of decision table means making it has the same function between after the reduction of decision table and before reduction of the decision table, but decision table has fewer condition attributes after the reduction.

Specific processing steps are as follows:

(1) Eliminating duplicate rows;

(2) Simplifying the condition attributes of decision table (relative simplification), namely reducing some columns from the decision table;

(3) Eliminating redundancy value of each attribute in decision rules;

(4) Forming the decision rules, and revealing the relationship between condition attributes and decision attributes.

The study framework of this paper is as shown in Fig.1.

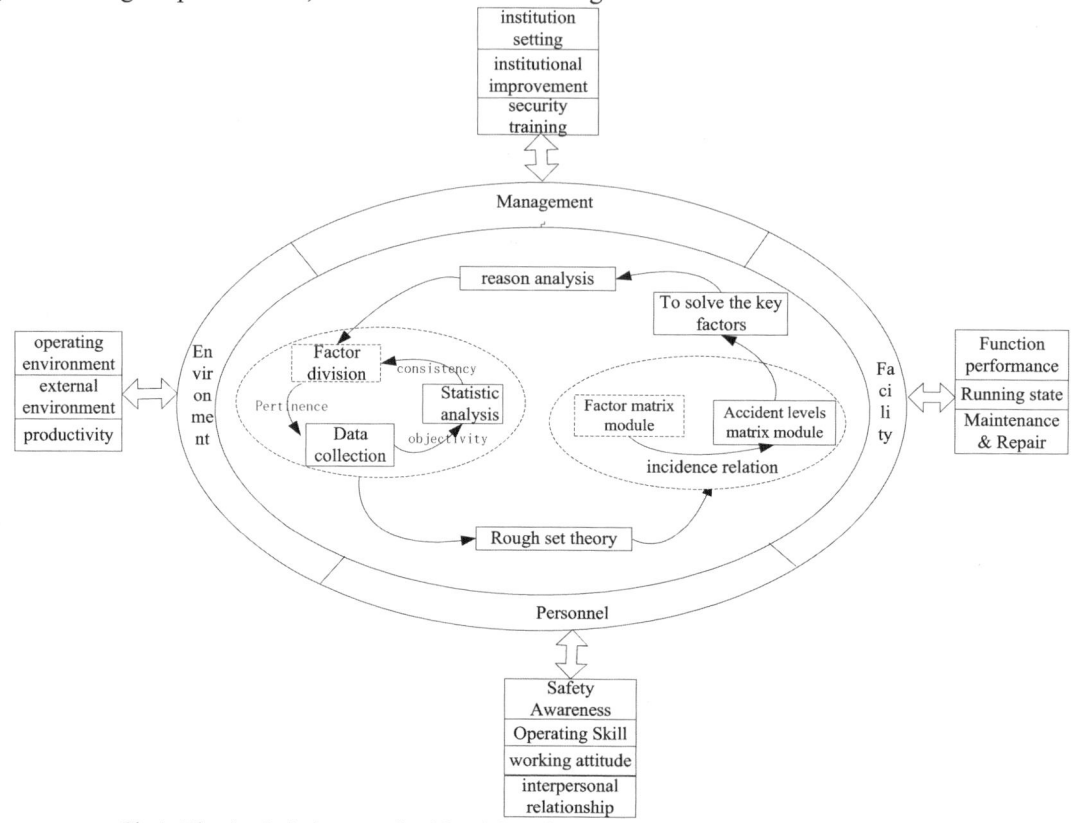

Fig.1. The Analysis framework of Special Equipment Accident Factor Based on Rough Set

III. DATA PROCESSING AND CALCULATION

A. Data processing

Every year in May, a national special equipment security situation report issued by AQSIQ (General Administration of Quality Supervision),the quantity of special equipment safety accident and the reason last year are revealed by the report. Through the analysis of these reasons, accident key factors can be attributed to the management, environment, equipment and personnel. Eight groups of statistical data from 2005 to 2012 are selected in the paper, Shown in Table I:

TABLE I
NUMBER OF ACCIDENTS CAUSED BY VARIOUS FACTORS FROM 2005-2012

Year	Management	Environment	Facility	Personnel	The total number of accidents
2005	219	8	22	81	274
2006	287	62	62	284	299
2007	249	0	56	149	256
2008	274	52	73	307	307
2009	308	45	47	271	380
2010	252	43	46	287	296
2011	238	0	0	245	275
2012	134	5	8	132	228

Case sets can be represented as $S = (U, A)$, it is a knowledge representation system, U is a finite set of objects, and represents a collection of vintage in this year. $A = C \cup D, C = \{C_1, C_2, C_3, C_4\}$ as the condition attribute set, represents the {management, environment, equipment, personnel}

$\{1, 2, 3\}$ is taken as condition attribute domain

Management factors C_1, $C_1 = \{1, 2, 3\} =$ {the small number, the reasonable number, the high number};

Environmental factors $C_2, C_2 = \{1, 2, 3\} =$ {the small number, the reasonable number, the high number};

Facility factors $C_3, C_3 = \{1, 2, 3\} =$ {the small number, the reasonable number, the high number};

Personnel factor $C_4, C_4 = \{1, 2, 3\} =$ {the small number, the reasonable number, the high number}.

$D = \{1,2,3\} =$ {the low level of accidents, the moderate level of accidents, the high level of accidents}

In order to use rough set for processing expediently, which level the accident number located need to be determined According to the overall situationthat factors caused the accident number, accident number level cased by various factors and accident number level in those year can be divided as Table II:

TABLE II
THE DIVISION ACCIDENT NUMBER LEVEL CASED BY VARIOUS FACTORS AND ACCIDENT NUMBER LEVEL IN THOSE YEAR

Attribute Number / Quantity level interval	1	2	3
Management C_1	$0 \leq C_1 < 200$	$200 \leq C_1 < 270$	270 and above
Environment C_2	$0 \leq C_2 < 20$	$20 \leq C_2 < 40$	40 and above
Facility C_3	$0 \leq C_3 < 30$	$50 \leq C_2 < 60$	60 and above
Personnel C_4	$0 \leq C_4 < 200$	$200 \leq C_4 < 300$	300 and above
Accident number D	$0 \leq D < 100$	$100 \leq D < 250$	250 and above

Therefore case set decision table of accident number cased by various factors can be shown as Table III:

TABLE III
CASE SET DECISION TABLE OF ACCIDENT NUMBER CASED BY VARIOUS FACTORS FROM 2005 TO 2012

U	C_1	C_2	C_3	C_4	D
1	2	1	1	1	2
2	3	3	3	3	3
3	2	1	2	1	1
4	3	3	3	3	3
5	3	3	2	3	3
6	2	3	2	3	3
7	2	1	1	2	2
8	1	1	1	2	1

B. Model calculation

For the same members in Table III, such as 2, 4, can be eliminated, and get the Table IV:

TABLE IV
THE DECISION TABLE AFTER ELIMINATING DUPLICATE MEMBERS

U	C_1	C_2	C_3	C_4	D
1	2	1	1	1	2
3	2	1	2	1	1
4	3	3	3	3	3
5	3	3	2	3	3
6	2	3	2	3	3
7	2	1	1	2	2
8	1	1	1	2	1

After eliminating duplicate members of the decision table, the degree of dependency decision attribute D depending on condition attribute C can be calculated and analyzed, whether condition attribute C can be omitted is examined, then the relative reduction between condition attributes C and decision attribute D can be determined, the repeat members in the decision table after simplified can be merged, the specific calculation is as follows:

$$U/\{C_1\} = \{\{1,3,6,7\}\{4,5\}\{8\}\}$$
$$U/\{C_2\} = \{\{1,3,7,8\}\{4,5,6\}\}$$
$$U/\{C_3\} = \{\{1,7,8\}\{4\}\{3,5,6\}\}$$

$$U/\{C_4\} = \{\{1,3\}\{4,5,6\}\{7,8\}\}$$
$$U/\{C_1, C_2, C_3\} = \{\{1,7\}\{3\}\{4\}\{5\}\{6\}\{8\}\}$$
$$U/\{C_1, C_2, C_4\} = \{\{1,3\}\{4,5\}\{6\}\{7\}\{8\}\}$$
$$U/\{C_1, C_3, C_4\} = \{\{1\}\{3\}\{4\}\{5\}\{6\}\{7\}\{8\}\}$$
$$U/\{C_2, C_3, C_4\} = \{\{1\}\{3\}\{4\}\{5,6\}\{7,8\}\}$$
$$U/\{C\} = \{\{1\}\{3\}\{4\}\{5\}\{6\}\{7\}\{8\}\}$$
$$U/\{D\} = \{\{1,7\}\{3,8\}\{4,5,6\}\}$$
$$posC\{D\} = \{1,3,4,5,6,7,8\}$$
$$k = \gamma C(D) = |posC\{D\}|/|U| = 7/7 = 1$$

D depending entirely on the C can be known By K = 1. The relative reduction between condition attribute C and decision attribute D can be determined as follows:

$$pos(C - \{C_1\})\{D\} = \{1,3,4,5,6\} \neq posC\{D\}$$
$$pos(C - \{C_2\})\{D\} = \{1,3,4,5,6,7,8\} = posC\{D\}$$
$$pos(C - \{C_3\})\{D\} = \{4,5,6,7,8\} \neq posC\{D\}$$
$$pos(C - \{C_4\})\{D\} = \{4,5,6,7,8\} \neq posC\{D\}$$

Because $pos(C - \{C_2\})\{D\} = \{1,3,4,5,6,7,8\} = posC\{D\}$, so C_2 is redundant variables, the classification ability of the decision table would not be altered after removing the attribute C_2, the decision table after removing C_2 is shown in Table V:

TABLE V
THE DECISION TABLE AFTER REMOVING REDUNDANT VARIABLES

U	C_1	C_3	C_4	D
1	2	1	1	2
3	2	2	1	1
4	3	3	3	3
5	3	2	3	3
6	2	2	3	3
7	2	1	2	2
8	1	1	2	1

The further calculation of rules reduction is as follows, the nuclear value of each rules can be calculated firstly
The first rule:
$$C_1[2]C_3[1]C_4[1] \rightarrow D[2];$$

$\{C_1[2], C_3[1], C_4[1]\} = \{\{1,3,6,7\}, \{1,7,8\}, \{1,3\}\}$;

$D[2] = \{1,7\}$,

$C_1[2] \cap C_3[1] = \{1,7\} \subset D$; $C_1[2] \cap C_4[1] = \{1,3\}$

$\not\subset D$; $C_3[1] \cap C_4[1] = \{1\} \subset D$.

It means that when removing C_1 and C_4 the rule $C_3[1]C_4[1] \to D[2]$ and $C_1[2]C_3[1] \to D[2]$ is compatible, while when remocing C_3, $C_1[2] \cap C_4[1] \to D[2]$ is not compatible, so the nuclear value of the first rule is C_3, Similarly, the nuclear value of each rule can be got as Table VI:

TABLE VI
THE DECISION TABLE AFTER BEING SIMPLIFIED

U	C_1	C_3	C_4	D
1		1		2
3				1
4				3
5				3
6			3	3
7	2			2
8	1			1

The rules as follows can be got from Table VI:

(1)$C_3[1] \to D[2]$

(2)$C_4[3] \to D[3]$

(3)$C_1[2] \to D[2]$

(4)$C_1[1] \to D[1]$

The rules that special equipment safety operation factors influence the accident quantity level can be concluded as follows:

(1) If the facility factor accident quantity level is low, the medium levels of accident quantity will be caused;

(2) If the personnel factor accident quantity level is higher, the higher levels of accident quantity will be caused;

(3) If the management factor accident quantity level is moderate, the moderate levels of accident quantity will be caused;

(4) If the management factors accident quantity level is low, the lower levels of the accident quantity will be caused;

(5) Environmental factors accident quantity level impact little on the quantity of accidents.

Therefore, in the process of special equipment safety management, personnel and management factors should be paid more attention, facility factors should be paid moderate attention to equipment, less attention can be paid to environmental factors.

IV.CONCLUSION

Based on the statistical analysis about the special equipment accidents, the accident key factors dividedinto management, environment, facility and personnel. Through the processing of rough set, management and personnel are the biggest influence factors, followed by facility, the smallest affecting factor is the environment. Therefore, userdepartment should strengthen the management of special equipment and personnel operating driving skills training, at the same time inspection and maintenance on special equipment should be down timely to ensure the safe operation of it. In addition, a safe environment of operating and using special equipment should be built. Through the above measures, the operation safety of special equipment, can be ensured, and the special equipment accidents quantity can be reduced.

REFERENCE

[1] Ding Shoubao, Liu Fujun. The current situation and prospect of special equipment inspection technology in our country [J]. Journal of China institute of metrology, 2008, 4 (19): 304-308.

[2] Tang Gongkai. The cause analysis of crane illegal operation death accident [J]. Journal of modern property management and new construction, 2013, 8 (12): 70-71.

[3] Xu Huoli. The reason analysis of the boiler burst accident and prevention measures [J]. Journal of energy and environment, 2011, 1:83-84.

[4] Wang Peiyuan Cheng Jun, li Jianzhong etc.The analysis of a tower crane overturning accident happened in Tianjin analysis [J]. Journal of failure and accidents, 2013, 5:92-94.

[5] Han mei, Deng Fangling, Han Yanhui etc. The research of the choice of regional logistics center city in Guangxi [J]. Journal of logistics technology, 2008, 10(27): 171-174.

[6] Wang Meng, He Yue. Supplier credit evaluation based on the theory of the rough sets and grey [J]. Journal of statistics and decision, 2011, 23 (347): 45-47.

Feature Recognition for Virtual Machining

Shixin XÚ[1,2,*], Nabil ANWER[1], Lihong QIAO[2]

[1]LURPA, ÉNS de Cachan, 94235 France
[2]SMEA, Beihang University, Beijing 100191, China
(*sxu@ens-cachan.fr, anwer@lurpa.ens-cachan.fr, lhqiao@buaa.edu.cn)

Abstract – **Virtual machining uses software tools to simulate machining processes in virtual environments ahead of actual production. This paper proposes that feature recognition techniques can be applied in the course of virtual machining, such as identifying some process problems, and presenting corresponding correcting advices. By comparing with the original CAD model, form errors of the machining features can be found. And then corrections are suggested to process designers. Two approaches, feature recognition from G-code and feature recognition from IPM, are proposed and elucidated. Feature recognition from IPM adopts a novel method of curvature based region segmentation and valuated adjacency graph. Recognized machining features are represented in conformance to STEP-NC. Feature recognition can help the virtual machining analysis in revealing potential defections in machining operations.**

Keywords – **Feature recognition, G-code, in-process model, STEP-NC, virtual machining**

I. INTRODUCTION

In a CAM environment, using a part's CAD model, after tool path strategies and cutting conditions are settled, the part's NC program can be generated. Then the NC program should be verified or proved in order to lower production risk. NC program proving by trial-and-error cycle on a physical machine is slow and costly. Sometimes, in manufacturing costly aerospace parts, trial-and-error based optimization might be unaffordable. The aim of virtual machining (VM) is to decrease the trials by simulating machining operations in digital environments prior to actual manufacture. VM is performed mainly on the establishment of the mathematical modeling of part-tool engagement geometry, machining process physics, structural and rigid body kinematic motion of the machine and work material properties. The simulation of machining operations in a virtual environment can predict dimensional surface errors left on a part, as well as maximum cutting forces, torque, power, vibration amplitudes in short time [1, 2].

We have been trying feature recognition techniques for virtual machining in two ways. One way is to carry out feature recognition on part programs (G-codes in conformance to ISO 6983) to rebuild the machining feature based model of the part. In some cases, the original CAM model could be neither available nor usable. For example, the part's archive is lost/damaged, or incompatible with upgraded systems; the part is designed and machined at shopfloors without aids of CAx tools. Therefore the rebuilt model can help understand and check the VM outcome. The other way is to perform feature recognition on the part's in-process model (IPM) resulting from the virtual machining. Then the extracted features can be used to evaluate form errors for machining and inspection. In both ways, the recognized machining features are expressed as STEP-NC features by the ISO 14649 standard for convenience of subsequent applications. The process diagram is illustrated in Fig.1.

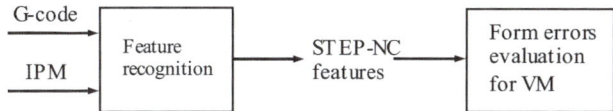

Fig.1. Process diagram of feature recognition for VM

The IPM, which is the geometric output data of the virtual machining, is usually a collection of unordered set of triangles, without topological information. Many NC simulation tools output the IPMs as STL files, like NCSIMUL, VeriCut. In fact, an IPM bears two types of features: surface micro features and machining features (such as holes, pockets, slots, steps). Surface micro features refer to textures, scallops, marks, veins, etc. on machined surfaces of the IPM. Fig.2 shows examples of surface micro features of a face milling. Surface micro features could be related to some cutting parameters, such as scallop height, stepover. From the patterns of the surface textures, we can determine whether the machining operation is stable, or is under chatter influence. If there are series of marks left on a machined surface, usually at its corners, that indicates jerks, revealing discontinuous feed rates along tool paths. Surface micro feature recognition involves image pattern recognition, etc., and in this paper we do not discuss it. Here we only use feature recognition approach to extract machining features from IPM. Thereafter we can identify potential machining operation problems by analyzing these features.

a)   b)

Fig.2. Examples of surface micro features: a) texture formed by stable milling of a plane; b) chatter marks left by unstable milling

Common feature recognition approaches take a solid model as their inputs, and search for pattern of faces and edges that obey certain topological or geometrical relationships [3]. But in this work we take a G-code program or the triangle mesh instead as the input of feature recognition. These need to design new methods to treat the G-code or the IPM data.

The remainder of the paper is organized as follows: Section II explains the feature recognition approach from

E. Qi et al. (eds.), *Proceedings of the 21st International Conference on Industrial Engineering*
and Engineering Management 2014, Proceedings of the International Conference on Industrial Engineering
and Engineering Management, DOI 10.2991/978-94-6239-102-4_26, © Atlantis Press and the authors 2015

G-code, Section III elaborates the feature recognition approach from IPM. Finally, we draw conclusions in Section IV and give some directions for further research.

II. FEATURE RECOGNITION FROM G-CODE

A. General strategies

G-code focuses on programming the cutting tool paths with respect to machine axes, rather than the machining process with respect to the part. Generally there is no information about cutting tools, the rawpiece' shape and its location, and the CNC controller type in the G-code, we should first supplement these for ensuring successful recognition. Then by analyzing hints (such as tool changes, speed changes, machining regions) in the G-code, workingsteps can be generated. In these workingsteps all operations are treated as freeform operations and features as toolpaths (which is converted into the data structure "toolpath" of freeform operations.), except those (like canned cycles) that can be easily attached to operations. Finally, we can extract machining features from the toolpath data by analyzing the machining regions, machining strategies, etc. By organizing the extracted feature data by ISO 14649 definitions, the STEP-NC features can be obtained.

B. Explanation of the approach

We devised an interpreter for the toolpath generation [4]. The interpreter emulates the execution of the given G-code one block by one block: if it meets a "G0" command, a "rapid movement" entity is created; if it meets one or several consecutive "G1"s or "G2|3"s, a "machining workingstep" which includes a freeform operation is created. The parameters of these commands are used as cutter location data stored in the toolpath list of the freeform operation. Sometimes computation is needed for obtaining the toolpaths.

Next, we deal with the extraction of manufacturing (machining) features from toolpaths and tool's geometry. In this phase, one freeform operation corresponds to one machining workingstep. Many freeform operations might correspond to one manufacturing feature because often there are several layers of a rough machining and a finish machining, which are needed to make a final feature. Hence one major procedure is to merge those freeform operations that machine the same feature, as well as the relevant workingsteps and rapid movements.

Some features can be easily extracted by the tool used. For example, if the tool type is for drilling, the feature is a hole; if the tool is a facemill, the feature is a planar face. If the tool is an endmill, which is a general and complex case, then analyse the x, y, z-values of the CL (cutter location) data in the toolpath list. If z-value varies and x, y-values keep constant, it mills a hole feature. If x, y, z-values all vary, it is a freeform milling operation for making a region of surface. If z-value keeps constant and x, y-values vary, it is a 2½D milling operation. The main grounds to find the remaining features (which can be planar faces, general outside profiles, closed pockets and

open pockets in STEP-NC.) in a 2½D milling operation are the cutting area (the tool's covering region for cutting movements) and the milling strategy, which are computed based on toolpath CL data.

C. An Example

In the example G-code of the study case part (Fig.3), 3 cutting tools (a drill, a reamer and an endmill) are used. By the tool types and the canned cycles in the code, we can extract a hole feature. By the cutting area and the milling strategy, we can identify a pocket feature.

```
G54 G90 G21 G40 G49 M5 M9                      Y100.
G0 Z100. (Move to the secure plane)           X85.
(To drill and ream a thru hole)               Y40.
T2 M6 (Use a spiral drill, diameter 20mm)     X55.
G43 H2 (Length compensation by 70mm)          Y100.
M8 M3 F900. S720                              X70.
G0 Z30.                                        Z0.
G90 G99 G81 X20. Y60. Z-18. R10.              G0 X69.532 Y47.815 (End of 1st layer)
G99 G81 X20. Y60. Z-36. R10. F1800.           ........(Code of next 4 layers omitted)
G99 G81 X20. Y60. Z-60. R10. F1350.
G1 Z10. F1800.                                (To finish pocket in 6 layers. 5mm/layer)
G80 G49 M5 M9 (end of drilling cycle)         (Bottom allowance 0.5, side allowance 1)
T3 M6 (Use a reamer, diameter 22mm)           G0 Z30.
G43 H3 (Length compensation by 50mm)          X74.890 Y60.285
M8 M3 S1080                                    Z15.
G90 G99 G85 X20. Y60. Z-60. R10.              (First 2 blocks: to run helical approach)
G80 G49 M5 M9 (End of reaming cycle)          G2 X77.2 Y55. Z-2. I-4.891 J-5.285
                                               G2 X70. Y55. I-3.60 J0.
(To cut a pocket, rough & finish)             G1 Y93.
T1 M6 (Use an endmill, diameter 18mm)         X78.
G43 H1 (Length compensation by 50mm)          Y47.
M8 S1200 M3 F2400.                            X62.
G0 Z30.                                        Y93.
X64.754 Y50.069                               X70.
Z15.                                           Y101.
(To rough pocket in 5 layers, 5.9/layer)      X85.
(First 2 blocks: to run helical approach)     G2 X86. Y100. I0. J-1.
G2 X77.2 Y55. Z-5.915.246 J4.932              G1 Y40.
G2 X70. Y55. I-3.60 J0.                       G2 X85. Y39. I-1. J0.
G1 Y90.                                        G1 X55.
X75.                                           G2 X54. Y40. I0. J1.
Y50.                                           G1 Y100.
X65.                                           G2 X55. Y101. I1. J0.
Y90.                                           G1 X70.
X70.                                           Z0. (End of 1st layer)
Y95.                                           .........(Code of rest of layers omitted)
X80.                                           G2 X55. Y101. I1. J0. (now Z-30.)
Y45.                                           G1 X70.
X60.                                           Z15. (End of finishing)
Y95.                                           M30
X70.
```

Fig.3. Case study and example G-code.

After the recognition from the G-code, we got two features: a hole and a pocket. Structuring the feature data by the STEP-NC standard, we can get the part21 file for the recognition result, like the following excerpt.

```
..........
DATA;
#0 = PROJECT ('EXECUTE EXAMPLE1',#1,(#2),$,$,$);
#1 = WORKPLAN ('MAIN WORKPLAN',(#11,#12,#13 #14),$ #3,$);
#2 = WORKPIECE ('BLOCK WORKPIECE',$,0.01 $,$,$,(#91 #92,#93,#94));
#3 = SETUP ('MAIN SETUP',#62,#60,(#4));
#4 = WORKPIECE_SETUP (#2,#63,$,$,());
#11 = MACHINING_WORKINGSTEP ('WS DRILL HOLE1',#60 #21,#32,$);
#12 = MACHINING_WORKINGSTEP ('WS REAM HOLE1',#60,#21 #33,$);
#13 = MACHINING_WORKINGSTEP ('WS ROUGH POCKET1',#60 #22,#34,$);
#14 = MACHINING_WORKINGSTEP ('WS FINISH POCKET1',#60 #22,#35,$);
#21 = ROUND_HOLE ('HOLE1 D=22MM',#2,#33 )#67,#70,#111,$,#25 );
#22 = CLOSED_POCKET ('POCKET1',#2,(#34,#35),#69 #71,(),$ #26,$,#112,#27);
#25 = THROUGH_BOTTOM_CONDITION ();
#26 = PLANAR_POCKET_BOTTOM_CONDITION ();
#27 = RECTANGULAR_CLOSED_PROFILE ($,#113,#114);
#32 = DRILLING ($,$,'DRILL HOLE1',15.,$ #44,#54,#51 $ $ $ $,$ #55);
#33 = REAMING ($,$,'REAM HOLE1',15.,$ #47 #54,#51,$,$,$,$,#56,.,T,.,$,$);
#34 = BOTTOM_AND_SIDE_ROUGH_MILLING ($,$,'ROUGH POCKET1',15.,$,
  #40,#57,#51,$,$,$,#58,6.5,$.,1.,0.5);
#35 = BOTTOM_AND_SIDE_FINISH_MILLING ($,$,'FINISH POCKET1',15.,$ #40,#57,#51 $,$,$ #59,2.,10.,$,$);
#40 = MILLING_CUTTING_TOOL ('ENDMILL_18MM',#41,(#43),80,$,$);
#41 = TAPERED_ENDMILL (#42,4,.RIGHT,.,F,.$,$);
#42 = MILLING_TOOL_DIMENSION (18.,$,$,29.,0,$,$);
#43 = CUTTING_COMPONENT (100,$,$,$,$);
#44 = MILLING_CUTTING_TOOL ('SPIRAL_DRILL_20MM',#45,(#43),90.,$,$);
```

```
#45= TWIST_DRILL (#46,2,.RIGHT.,.F.,0.84);
#46= MILLING_TOOL_DIMENSION (20.,31.,0.1,45.,2.,5.,8.);
#47= MILLING_CUTTING_TOOL (REAMER_22MM',#48,(#43),100.,$,$);
#48= TAPERED_REAMER (#49,6.,.RIGHT.,.F.,$,$ );
#49= MILLING_TOOL_DIMENSION (22.,$,$,40.,$,$,$);

#51= MILLING_MACHINE_FUNCTIONS (.T.,$,$,.F.,$,(),.T.,.$,$,());
#54= MILLING_TECHNOLOGY (0.04,.TCP.,$,-18.,$,.F.,.F.,.F.,$);
#55= DRILLING_TYPE_STRATEGY (0.75,0.5,2.,0.5,0.75,8.);
#56= DRILLING_TYPE_STRATEGY ($,$,$,$,$,$);
#57= MILLING_TECHNOLOGY (0.04,.TCP.,$,-20.,$,.F.,.F.,.F.,$);
#58= CONTOUR_PARALLEL ($,$,.CW.,.CONVENTIONAL.);
#59= CONTOUR_PARALLEL (0.05,.T.,.CW.,.CONVENTIONAL.);

#60= PLANE ('SECURITY PLANE',#61 );
#61= AXIS2_PLACEMENT_3D ('PLANE1',#90,#81,#82);
#62= AXIS2_PLACEMENT_3D ('SETUP1',#80,#81,#82);
#63= AXIS2_PLACEMENT_3D ('BLOCK WORKPIECE',#80,#81,#82);
#67= AXIS2_PLACEMENT_3D ('HOLE1',#97,#81,#82);
#68= AXIS2_PLACEMENT_3D ('DEPTH PLANE',#98,#81,#82);
#69= AXIS2_PLACEMENT_3D ('POCKET1',#99,#81,#83);
#70= PLANE('DEPTH SURFACE FOR ROUND HOLE1',#68);
#71= PLANE('DEPTH SURFACE FOR POCKET1',#68);
.............
```

III. FEATURE RECOGNITION FROM IPM

A. Introduction of the approach

The widely used mesh-based virtual machining simulation refers to that the IPM uses triangle mesh for processing and dynamic display, and hence the IPM can be output as collection of triangles. Most of commercial NC simulators provide an STL export interface for outputting the IPM data. During machining process simulation, one popular method is that the workpiece and the cutting tool are represented as discretized triangles for simplifying the computation and display. The geometric modeling of the cutter-workpiece engagement along the tool path are very important to the variation in chip thickness, and axial/radial depths of cut that are needed to evaluate force, torque, power, vibration and other process states along the tool path. The IPM mesh data has the following characteristics.

1) The IPM is a collection of triangles, among which have no sequence and topological relationship. The IPM sometimes contains offcuts, which should be identified and eliminated in the beginning of feature recognition. The mesh data usually contains degenerated triangles (i.e. isolated vertices and edges) and incorrectly oriented triangles, or lacks of triangles data resulting a gap on the surface.

2) Compared with the mesh data of a faceted CAD model, the IPM mesh data is more irregular and complex. For example, a rectangular planar face in a CAD model may be faceted as 4 triangles; but the counterpart in IPM might be like a waved surface composed by numerous triangles after machining. This is due to the scallops left on the face, forming small ridges and ruts, which are represented by tiny triangles in the IPM.

The overall method we designed is as follows: first, build a surface B-rep model by processing the IPM data; then, recognize machining features from the B-rep model.

Two approaches are considered to the feature recognition for virtual machining in our work. One is to attach attributes to triangles while carrying out the virtual machining. This approach is simple while practical when the number of the triangles is small. The other approach is to analyze the pure geometric mesh data of the simulation, on basis of discrete segmentation [5,6]. This is a general approach, not rely on a certain NC simulation tool. In this paper, we will focus on the latter.

B. Description of key procedures
1) B-rep triangular model building

We use the Open CASCADE [7] toolkits, an open source geometry development platform, to build the B-Rep triangular model from the IPM data. Each triangle in the STL file is transformed into a triangular face. When completed, a manifold, waterproof B-Rep triangular solid model is generated. This model has the following characteristics: In ruled surface regions and in planar regions at its curved boundaries, the model has many long and thin triangles resulting from the discretization; Vertices of triangles in planar and ruled regions are mostly on the part's boundary edges; Triangles in flat regions are sparse and relatively large, while dense and uniform in highly curved regions.

2) Curvature based region segmentation

Discrete curvature calculation is based on vertex vicinity. In triangle sparse areas, most of vertices are on boundary edges, belonging to two or more regions. Discrete curvature estimation in this case is not reliable, resulting in wrong segmentation. To avoid this, we detect the part's sharp edges. After the B-Rep triangular model construction, we compute the dihedral angle between two adjacent triangles. All the sharp boundary edges can be found by a threshold on dihedral angle. Using a propagating approach, preliminary regions demarcated by sharp edges can be obtained. Follow-up curvature calculation will be confined to use vertices only within such a region. Curvature calculation is improved and is more reliable since the neighborhood of each vertex is not disturbed by vertices in other regions.

On a surface the local shape around its point P is characterized by maximum and minimum principal curvatures (κ_{max}, κ_{min}) and by the two principal directions (d_{max}, d_{min}) corresponding to the tangent vectors for which the principal curvatures are obtained. The method used here adopts Cohen-Steiner's [8] and Zhao's [9] to compute the curvatures of a discrete shape. Shape index and curvedness [10] are two indicators derived from the principal curvatures. Shape index, ranged [-1, 1], is a quantitative measure of local surface type. Curvedness is a positive value that specifies the amount or intensity of the surface curvature.

Ten surface types (spherical cup, trough, rut, saddle rut, saddle, spherical cap, dome, ridge, saddle ridge, and plane) are defined based on shape index and curvedness [11] (Fig.4). Every vertex, except vertices of sharp edges, of the shape is assigned a surface type label during local surface type recognition. We use the curvedness value to detect smooth edges within a preliminary region. Vertices on smooth edges and transition regions have higher curvedness value, as thus a threshold can be given to detect them.

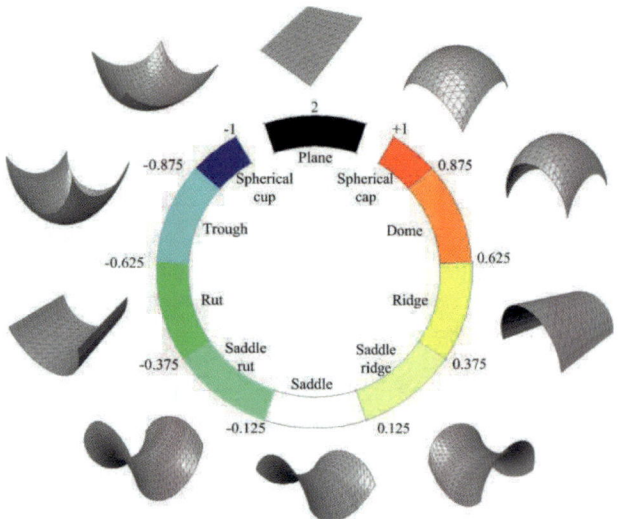

Fig.4. Surface types, shape index scales and color scales.

Then connected regions are generated from these clustered vertices. Two operations are performed: connected region growing to generate initial segmentation result, and region refining, which aims to reduce over-segmented regions and to improve the segmentation result. When a vertex yet unassigned with a region label is met, the vertex is marked with a region label by its neighbor condition. With this associated region label, the vertex is added to an existing region or creates a new region. We adopt the approaches reported in [5, 11] for the operations. Fig.5 shows a segmentation example.

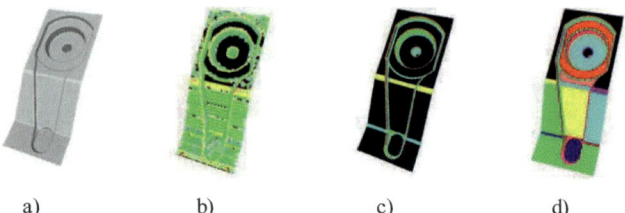

Fig. 5. A segmentation example: a) discrete shape; b) initial clustering; c) cluster refining; d) segmentation.

3) Machining Feature Recognition

To obtain the boundary representation model, we use the meshes' additional information, which help to know which meshes belong to the same surface and what is the type of the surface to be constructed. Edge loops that form the surface are identified and built. Surface reconstruction techniques are then used. Thus a continuous surface of the part is built. When all regions are exhaustively searched, the part's surface B-rep model is built. At present the surface types of the model can be planar faces, spheres, cones, cylinders and tori.

Machining feature recognition is performed on the surface B-rep model obtained above. Here we adopt the valued-adjacency graph approach [5], in which the B-rep model is converted into a planar graph where, e.g., its nodes represent faces and its arcs represent edges. Additional information, such as edge-xity (a measure of concavity/convexity), is incorporated into the graph.

A machining feature is regarded as a set of connected faces satisfying certain geometrical and topological conditions on the workpiece's B-Rep. These faces have topological links of concavity that form cavities or protrusions and are related to a machining volume. For each form feature, it is possible to define a set of convex edges that are the limits of the feature: boundary edges. Note that the definition of a feature is only based on its topology. Seven generic classes for form feature classification have been proposed. General and specific rules to define the classes are based on properties of planar graphs.

K-Protrusion class definition depicts formal rules (Fig.6c) that have been developed with strong emphasis in graph theory. The classification rule consists of the identification of a base face which is connected to all the wall faces by concave edges. The wall faces are linked to each other with either convex or concave edges that form inner protruding volumes (Fig.6 a and b). The rule can be easily described using cyclomatic and co-cyclomatic numbers of extracted planar graphs.

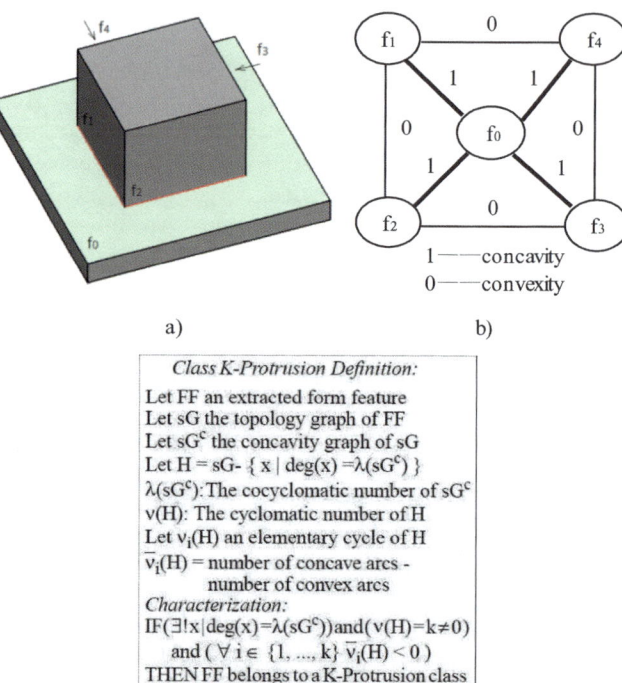

Fig.6. Adjacency graph: a) a part; b) the adjacency graph; c) K-Protrusion classification rule

The type and data structure of a machining feature adopt STEP-NC's definitions, so that the recognized feature data can be stored in conformance to the standard [12]. So far the types of machining features that can be recognized are closed pockets, open pockets, slots, steps, holes. A machining feature is then regarded as a set of connected faces satisfying certain geometrical and topological conditions. For each feature type, its feature template is defined. A graph-based and rule based manufacturability analysis approach is used for validating the results. For example, as for the manufacturability of a step, multiple interpretations due to 3 potential machining directions are available in the extracted feature data.

From the reconstructed surface B-rep model or the machining feature based model, we can easily obtain form parameters to be controlled, and compare them with the part's original CAD/CAM model. Hence we can evaluate the form errors under current machining operations.

C. Example

The part in Fig.7 has planar faces, holes, pockets, slots, etc. This figure shows some feature recognition results for the test part.

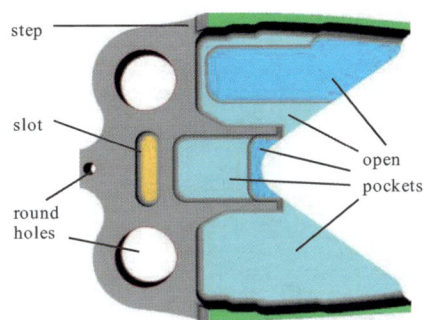

step

slot

round
holes

open
pockets

Fig.7. A test part

IV. CONCLUSION

This paper proposes that feature recognition techniques can be applied to virtual machining for identifying some machining operation problems. Two types of features exist in in-process model of a simulation: machining features and surface micro features. We focus on machining features, especially on feature recognition from IPM, which involves B-rep model construction, curvature based region segmentation, and valuated adjacency graph based feature recognition. By comparing with the original CAD/CAM model, form errors (such as over/undercut, out-of-tolerance dimensions) of the features can be found. So performing feature recognition on the VM output model can help the VM analysis in finding potential machining operation problems.

The work is in an initiating phase. Future works include: for the work of recognition from G-code, to increase the capacity of machining feature recognition, especially for region features; for the recognition from IPM, to develop specific approaches of region segmentation for other quadric/free-from surfaces and to test the approach on more complex parts, such as parts that are manufactured by five-axis and mill-turn machines. Future work will also verify that feature recognition can play a part for virtual machining.

ACKNOWLEDGMENT

This work is a part of the ANGEL FUI project funded by the French Inter-ministerial Fund and endorsed by top French competiveness clusters (SYSTEMATIC PARIS REGION "Systems & ICT", VIAMECA "Advanced Manufacturing" and ASTECH "Aeronautics & Space").

REFERENCES

[1] Y. Altintas, P. Kerting, D. Biermann, E. Budak, B. Denkena, and I. Lazoglu, "Virtual process systems for part machining operations," *CIRP Annals – Manuf. Tech.*, 2014

[2] Y. Zhang, X. Xu, Y. Liu, "Numerical control machining simulation: a comprehensive survey", *Int. J. of Computer Integrated Manufacturing*, vol.24, no.7, 2011, pp.593–609

[3] S. Joshi, "Graph-based heuristics for recognition of machined features from a 3D solid model". *Computer-Aided Design*, vol.20, no.2, 1988, pp.58–66

[4] S. Xu, N. Anwer, S. Lavernhe, "Conversion of G-code part programs for milling into STEP-NC", in *Proc. of Joint Conf. on Mech., Design Eng. & Adv. Manuf.*, Toulouse, France, June 2014

[5] S. Xu, N. Anwer, C. Mehdi-Souzani, "Machining feature recognition from in-process model of NC simulation" (Accepted for publication). *Computer-Aided Design & App.*

[6] H. Zhao, N. Anwer, P. Bourdet, "Curvature-based registration and segmentation for multisensor coordinate metrology", *Procedia CIRP*, 10, 2013, pp.112-118

[7] Open CASCADE, http://www.opencascade.org

[8] D. Cohen-Steiner, J.-M. Morvan, "Restricted Delaunay triangulation and normal cycle", In *19th Annual ACM Symposium on Computational Geometry'03*, 2003, 312-321

[9] H. Zhao, N. Anwer, P. Bourdet, "Curvature-based registration and segmentation for multisensor coordinate metrology", *Procedia CIRP*, 10, 2013, pp.112-118

[10] J.J. Koenderink, A.J. Doorn, "Surface shape and curvature scales", *Imaging and Vision Computing*, vol. 10, no. 8, 1992, pp. 557–565

[11] N. Anwer, Y. Yang, H. Zhao, O. Coma, J. Paul, "Reverse engineering for NC machining simulation," In *IDMME'2010—Visual Concept 2010*, Bordeaux, France

[12] ISO 14649 Part 10: General process data, 2002

Application of Cell Automaton in the Production Quality Inspection

Qing-hui DAI*, Lin-lin XING

School of Energy Power and Mechanical Engineering, North China Electric Power University, Baoding, Hebei, China, 071003

(*dcba6789@126.com, Xinglinlin163.2008@163.com)

Abstract - **On the basis of quality inspection, Cell Automaton theory was used to study quality management. This paper introduced the basic principles of cell automaton. The researcher established two-dimensional cell automaton model of product quality inspection, and defined cell morphology based on the state of the product, cell space based on product quality inspection, as well as cell neighborhood. The researcher constructed the transfer function by the combination of product quality factors of the production process, while the introduction of cell firmness, propensity score and other related concepts, and the design of appropriate parameters. Though the simulation, we verify the application of cell automaton in quality inspection. Research shows that, once the production factors change, it should be timely inspected in the initial stage. Otherwise, the substandard rate of product will increase with the passage of time.**

Keywords - **Cell automaton, computer simulation, quality inspection, transfer function**

I. INTRODUCTION

A. Cell Automaton Principles

Cell Automaton (CA) is a kind of Dynamics Systems which is defined in cell space composed by discrete and finite-state cells and evolves in discrete time dimension according to certain local rules [1]. The main feature of CA is that the time, space and states are discrete. The basic principles of CA system are: cell automaton system consists of independent cells; each cell is arranged in a regular grid; these cells all have a finite number of discrete states; each cell which follows the same state transfer rule updates; the cells only interact with their neighbors. CA which is the state discrete systems can reveal complex global properties. Thus it can be used for product quality inspection.

Distinguished by general Dynamics Systems, CA does not have determined mathematical functions, but operates systematically only by a series of identified models. The basic components comprise: cells, cell space, cell neighbors and cell rules.

B. Cells

The cell also called primitive, which is distributed in discrete one-dimensional, two-dimensional or multi-dimensional Euclidean space on the grid, and forms a discrete state; this is the most basic component of CA [2]. As shown in Fig.1. Each cell has its own cell discrete state at a fixed point in time. Discrete set is composed by assigning to discrete state of the cell, which can be {0, 1} binary form, or {-1, 0, 1} ternary form, also $\{x_0, x_1, x_2, \ldots, x_n\}$ in the form.

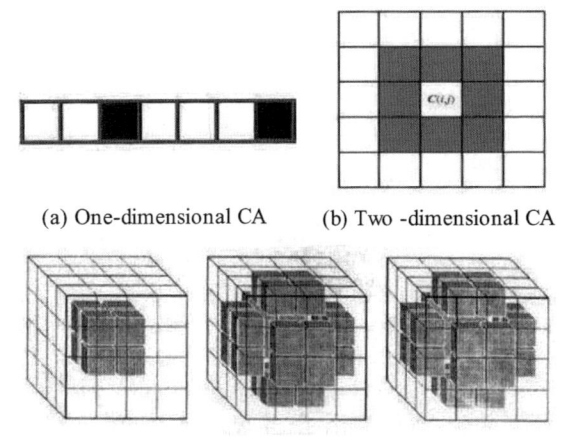

(a) One-dimensional CA   (b) Two -dimensional CA

(c) There -dimensional CA

Fig.1. Cell type of cell automaton

C. Cell Space

(1) *Geometric Division*. According to the theoretical point of view, the cell can occupy Euclidean Space of any set dimension; the research is generally conducted in one-dimensional or two-dimensional space at the present stage. The space dimension of one-dimensional CA is only one; two-dimensional CA presents multiple spatial structures whose common spatial configuration is triangular, square or hexagonal form. As shown in Fig.2. Because of the difference of various geometric shapes arranged in network, as well as advantages and disadvantages of their own, the characteristics and effect of the performance of Power System should be balanced. We should make a reasonable choice when selecting spatial structure.

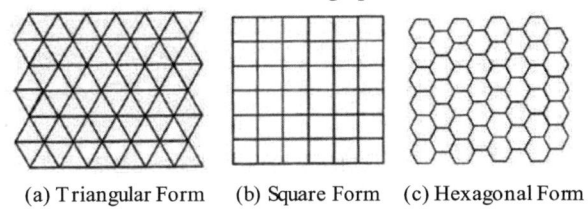

(a) Triangular Form   (b) Square Form   (c) Hexagonal Form

Fig.2. Two-dimensional Space Configuration

(2) *Boundary Conditions of Cells*. The common boundary conditions: periodic (or loop) boundary, fixed boundary, adiabatic boundary and mapping boundary [3]. As shown in Fig.3.

E. Qi et al. (eds.), *Proceedings of the 21st International Conference on Industrial Engineering and Engineering Management 2014*, Proceedings of the International Conference on Industrial Engineering and Engineering Management, DOI 10.2991/978-94-6239-102-4_27, © Atlantis Press and the authors 2015

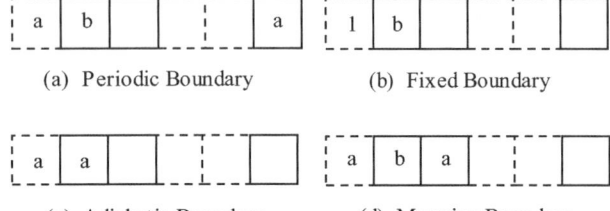

(a) Periodic Boundary   (b) Fixed Boundary

(c) Adiabatic Boundary   (d) Mapping Boundary

Fig.3. Boundary conditions of cell

(a) Periodic Boundary: refers to connecting the cell space of relative boundary. Such is often used for experiments in the associated theoretical analysis and simulation.

(b) Fixed Boundary: refers to the outer boundary cells are all taken to a fixed value.

(c) Adiabatic Boundary: refers to the states of the outer boundary neighbors always maintain consistent with the state of boundary cells, that is, with zero gradient condition.

(d) Mapping Boundary: refers to the states of the outer boundary neighbors are specular reflection with the axis of its boundary.

D. Cell Neighbors

Cells and cell space only show the static component of CA. In order to introduce "dynamic state" into the system, the evolution rules must be added to. In CA, the evolution rules are defined within the local area; that is, the state of a cell in the next time depends on the state of itself and its neighbors. Thus, the impact of a cell to other cells should be considered when the dynamic evolution of CA is in progress. So it is necessary to consider the local area of cell space which is named Cell Neighbors.

When considering influence from a cell to other cells, which each cell will affect what neighbors must be specified. In one-dimensional CA, usually using radius (abbreviated as r) to ascertain cell neighbors, and cells whose distance from a certain cell is less than r are regarded as cell neighbors of this certain cells. The definition of two-dimensional CA neighbor is more complex, and Von Neumann Type, Moore Type and Extended Moore Type are commonly used [4-6].

E. Cell Rules

CA is one of Kinetic motion model. The function is called the State Transition Function of CA, which defines the cell state of next time based on the current state of itself and the states of neighbors within specified range [7]. The function can be written as:

$$f : S_i^{t+1} = f\left(S_i^t, S_N^t\right)$$

From this function, f is the State Transition Function; s_i^t represents the state of the cell i at time t; and s_N^t represents the state of cell neighbor N of the cell i at time t.

CA is a dynamic system; with the change in time, every time although the physical structure system does not develop, the state changes. If you use a mathematical formula to represent CA, which can be summarized as a four-tuple, namely

$$A = (L_d, S, N, f)$$

Here, A represents a CA system; L_d represents cell space, d is the space dimension; S is the limited and discrete cell state set; N represents the combination of all cell neighborhood (including the center cell); f is a state transition function, which is the evolution of the rules [8].

II. ESTABLISHMENT OF CA BASED ON PRODUCT QUALITY INSPECTION

A. Cell Morphological Definition

Each cell represents a product. Each subject is defined three quality status. The cell state set is a finite set like A = {product status transition, lack of product status, product status suitable}, and the initial and final values of the unit cell can be expressed as A = {1, -1, 0}. In the simulation stage of evolution, according to the actual situation, the cell state value A allows to be given an absolute value less than or equal real number one.

B. Cell Space

The samples of 200 products were used as the test data of product quality inspection. Therefore, we used cell space of 200 for a group of one-dimensional lattice structure, and in the evolution of cells, we used the one-dimensional cell space traversal algorithms. In the entire product quality inspection, cell space which uses 200 samples for whole products is very limited, but that does not prevent the "holistic" test, with 200 samples reflect the overall product.

C. Cell Neighbors

The production line is regarded as a one-dimensional grid. In the pipeline, each product can accept influence from the quality level of every cell in the cell space, meanwhile, can affect other cells, realizing the impact of space and time. Due to the adoption of a one-dimensional grid, in order to simplify calculation, we choose every cell around the field, namely two neighbors are used as the conversion neighbors. Then using 200 one-dimensional cell networks and using cell traversal algorithm achieve the testing of all cells within cell space.

D. A Status Converting Method

(1) State Setting. Taking into the high stability requirements of the products, we define the cell transformation rules related to cell stability: $s_i(t)$ represents the state of cell (i) at time t, and represents the state of the products produced at this time. So $s_i(t)$ belongs to [-1, 1], and is defined as follows:

When $-1 \le s_i(t) < -0.5$, cell(i) is in the absent state;
When $-0.5 \le s_i(t) \le 0.5$, cell(i) is in the appropriate state;
When $0.5 < s_i(t) \le 1$, cell(i) is in a transitional state.

(2) Production Factors. In the actual production process, factors affecting the quality of products are shown as follows:

(a) Human factors. It includes people's quality consciousness, sense of responsibility, dedication, cultural quality, technical level, the operation proficiency and the ability of organization and management.

(b) Material factors. It refers to the quality of raw materials, blank, spare parts, standard parts and external components etc.

(c) Machine factors. It refers to the quality of equipment, process equipment, and other related production tools.

(d) Method factors. It refers to the quality of the production process, experimental analysis, measuring tools and testing instruments etc.

(e) Testing methods. It refers to the quality of the method for measuring and testing, measuring tools and testing instrument etc.

(f) Environmental factors. It refers to the temperature, humidity, dustiness degree, noise, vibration, radiation, and poison of air, the lever of cleanliness, beautification and civilization of labor environment.

(3) *Design of Transfer Function*. In fact, the value of $s_i(t+1)$ is not only have relation with neighbors, but also have relation with the value of tendency and firmness coefficient for cell itself [1]. Coefficient which influences the product quality factor is determined by the inherent properties of the model, it is represented by the following formula:

$$s_i(t+1) = w \times s_i(t) + q \times (f_1 + f_2 + f_3 + f_4 + f_5 + f_6) \times s_{i-1}(t) + s_{i+1}(t)$$

Among them, w is expressed as the product stability coefficient, reflecting the stability of the center cell for the next cell. Its value is [0, 1], and larger w indicates the stability of center cell stronger. When $w=0$, the stability of center cell is 0, that is very unstable and we can completely ignore the stability of center cell. In this case, $s_i(t+1)$ is completely obedient to the state of neighbors. When $w=1$, production is in a stable condition, which depends entirely on the central cell, and we can completely ignore the impact of neighbors. In this case, $s_i(t+1)$ is entirely submissive to the state of center cell; q is defined as the neighborhood coefficient, indicating that two neighbors are in the same position. Currently we use the two neighborhood, so $q=0.5$.

In actual production operations, f_i represents the product quality factors- the influence degree of human, machine, material, method, environment, test and other factors [9, 10]. However, in the actual production operations, the six kinds of factors will not operate simultaneously, usually no more than three changes. So in the process of simulating quality inspection of CA, we set f_1, f_2, and f_3 as the situation changes as up to three of six factors. The integration formula is as follows:

$$s_i(t+1) = w \times s_i(t) + q \times (f_1 + f_2 + f_3) \times s_{i-1}(t) + s_{i+1}(t)$$

(4) *Cell Traversal Algorithm*. CA can be used in the case of free moving or fixed neighborhood transformation [11, 12]. Just above conversion rules are applicable to the conversion of fixed cell neighborhood, but because of the liquidity of products, the conversion rules are more applicable at the free conversion neighborhood. Now cell traversal algorithm is as follows:

Initial: one dimensional array s (200);
For T=1, N;

Do i=1 to 200;
$s_i(t)$ exchange with $s_i(t+1)$;
Mapping new $s_i(t)$, $s_i(t+1)$ into $s_1(m)$, $s_1(m+1)$;
Update $s_1(m)$, $s_1(m+1)$ by Ep. (1);
End;
Output s (200);
End;

Among them, T is the number of cycles for this algorithm, and $N=2$. Derived by the algorithm, it ensures that each cell could make internal communication with others in the one-dimensional network on the process of moving in freedom neighborhood in every cycle, and using this algorithm will avoid double counting to achieve minimization. In cycles, however, the value will always be changing with the moving of cells after setting the basic distribution of specific values. So the stability of cell would not produce mutations with continually changing of its position.

III. SIMULATION

In the simulation, the stability coefficient of cell and the initial distribution of the weight value are random [13]. And once the formation of the initial distribution is confirmed, all experiments will be fixed. The experiment of moving cell to traverse space is as follows:

About the initial state setting: We initialize the amount of -1 to 35, 0 to 135 and 1 to 30, relevantly $w_1=0.65$, $w_2=0.60$ and $w_3=0.65$. The number of cycles T is 1 or 2, and the calculation of each circular convolution is 200 times.

Fig.4 shows the stability of the two cycles - the corresponding diagram of the cell numbers.

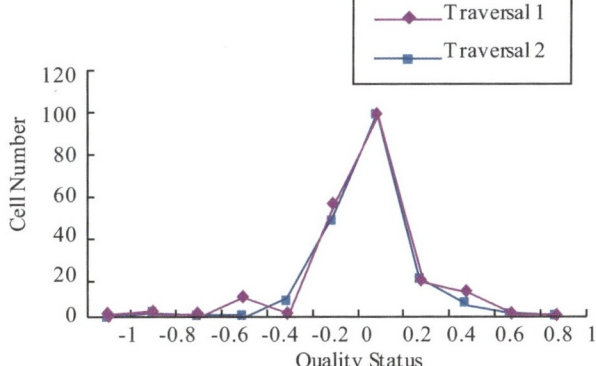

Fig.4. Rendering Quality Status

Fig.5 represents the cyclical trend graph at a certain time.

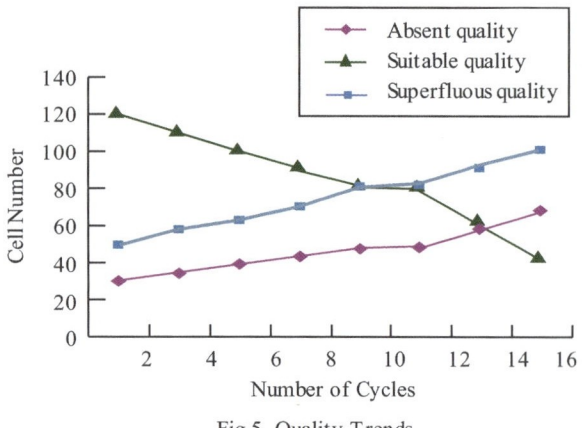

Fig.5. Quality Trends

IV. DISCUSSION OF SIMULATED EXPERIMENT

A. The Distribution of the Overall Tendency of Three Products

In the initial experiment design, "Suitable quality" products were significantly more than those of "superfluous quality" and "absent quality", then these three values were set 60:25:15. Under the above guidelines, observed from Fig.5, "suitable quality" products gradually had reduced and those of "superfluous quality" and "absent quality" had continually increased after experiencing a certain number of cycles. Among them, the "superfluous quality" products increased faster than the "absent quality" products.

B. The Influence of Changes in Quality Factors

Based on the above analysis, we can conclude: it is necessary to make the quality testing when one or more factors (human, machine, material method, environment and testing) of production have changed in the production process. If you neglected, the longer the production time lasted, the greater effect of the whole production line could appear in the production process. Therefore, once the production factors change, it should be timely inspected in the initial stage. Otherwise, the substandard rate of product will increase with the passage of time.

V. CONCLUSION

Based on the quality inspection, this article introduced CA, built quality-inspection CA models, and proposed cell transfer function and method of calculation designed by cell firmness. Through simulation analysis, it was concluded that quality inspection rules would change with production factors in the production activities; and when to inspect tightly, loose, generally, and even to do exemption are determined by the relevant factors and the stability with test result of the current production phase. Therefore, when an amount of production conditions changed in the production, product inspection should be carried out as soon as possible, and product quality status must be controlled from the source.

ACKNOWLEDGMENT

Thanks Jin-peng ZHAO for the important contribution to this article.

REFERENCES

[1] Wei FANG, Liu-jin HE, Kai SUN, Peng ZHAO. Internet public opinion propagation model study of cell automaton. Computer Applications, 2010, Vol. 30, No. 3, pp. 751-755. (In Chinese)

[2] Xiao-gang JIN. CA simulation of the design based on Matlab. Computer Simulation, 2002, Vol. 19, No. 4, pp. 27-30. (In Chinese)

[3] Jian-xun DING, Hai-jun HUANG. A cellular automaton model of public transport system considering control strategy. Journal of Transportation Systems Engineering and Information Technology, 2010, Vol. 10, No. 3, pp. 35-41.

[4] Werner Pries, Adonis Thanailakis, Howard C. Card. Group properties of cellular automata and VLSI applications. IEEE Trans. Computers, 1986, Vol. 35, No. 12, pp. 1013-1024.

[5] Jens U Wurthner, A-mal K Mukhopadhyay, Claus-Jürgen Peimann. A cellular automaton model of cellular signal transduction. Computers in Biology and Medicine, 2000, Vol. 30, No. 1, pp. 1-21.

[6] Zhong-jun WANG, Neng-chao WANG, Fei FENG, Wu-Feng TIAN. Evolutionary behavior of cellular automaton. Application Research of Computers, 2007, Vol. 24, No. 8, pp. 38-41. (In Chinese)

[7] Stahl W R. Algorithmically unsolvable problems for a cell automaton. Journal of Theoretical Biology, 1965, Vol. 8, No. 3, pp. 71-94.

[8] Pei-dong HUANG. The quality of the regulatory innovation model based on product quality factors. Quality and Standardization, 2013, Vol. 32, No. 2, pp. 40-43. (In Chinese)

[9] Xiao-jun WANG, Jie BAI. Factors affecting product quality and supervision model innovation of quality. Enterprise Reform and Management, 2014, Vol. 22, No. 7, pp. 14. (In Chinese)

[10] Lei YU, Hui-feng XUE, Xiao-Yan GAO, Gang LI. The research of infectious disease transmission model based on cell automaton. Computer Engineering and Applications, 2007, Vol. 44, No. 2, pp. 196-198. (In Chinese)

[11] Toshiaki Takayanagi, Hidenori Kawamura, Azuma Ohuchi. Cellular automaton model of a tumor tissue consisting of tumor cells, Cy-to-toxic T Lymphocytes (CTL), and cytokine Produced by CTL. IPSJ Digital Courier, 2006, Vol. 2, pp. 138-144.

[12] Yo-Sub Han, Sang-Ki Ko. Analysis of a cellular automaton model for car traffic with a junction. Theoretical Computer Science, 2012, Vol. 45, No. 4, pp. 54-67.

[13] Li-juan ZHANG, Na MENG, He-xiang ZHANG, Fu-Chang WANG. The simulation of epidemic model based on cell automaton. Computer Simulation, 2012, Vol. 29, No. 10, pp. 219-223. (In Chinese)

Study on Evaluation System of Product Quality Oriented Supply Chain

Xing-yu Jiang*, Dong-be Hu, Chao Gao, Xin-min Zhang, Li Li

School of Mechanical Engineering, Shenyang University of Technology, Shenyang, Liaoning, 110870

(xy_jiang9211@sut.edu.cn)

Abstract - **Seeing that the influence quality factors in operation process of supply chain are complex, multi-level, interconnection and interaction, which possess uncertainty and fuzziness obviously, product quality building process in supply chain is deeply analyzed, and evaluation system model of product quality oriented lifecycle in supply chain is established. On the basis, G1 is applied to determine weights of product quality indexes, which could solve the problem that it is not necessary to reorder the quality indexes in the case of their changes (increase or decrease); and intrinsic fuzziness of expert judgment is considered, fuzzy comprehensive evaluation (FCE) is applied to establish mathematical evaluation model of product quality oriented lifecycle in supply chain. Finally, the product of some Enterprise (lubricant station F335) is introduced as an example to demonstrate the rationality and validity of the method.**

Keywords - **Product quality, product lifecycle, supply chain**

I. INTRODUCTION

For globalization of world economy and individualization of customer's requirements, cooperation of enterprises has been closely, and market competition has changed into the competition among supply chains from the competition among enterprises [1]. Furthermore, integration of supply chain and quality management will be the very important factor of the future competition among supply chains [2]. Therefore, how to assure and constantly improve product quality, has been the primary task in operation process of supply chain, which is the key to gain market competitive advantage.

Since distribution of members in supply chain is dispersion and discontinuous, both the operation process and organizations of supply chain are dynamic changes. These lead to dynamic and several variable of product quality in supply chain. Therefore, how to constantly assure product quality is the key problem that enterprise in supply chain must face. Furthermore, evaluation of product quality is the most effective of continuous assurance and improvement of product quality. So, how, it is the important significance to realize the dynamic evaluation of product quality in supply chain for sustainable development of supply chain.

Study on product quality in supply chain has been the focus in the field of supply chain at present, scholars and enterprise technicians at home and abroad has carried out research to solve the problems of product quality in supply chain, and obtained some achievements. But, some disadvantages are existed, e.g. quality control in product collaborative design (Xiaoqing Tang, 2006) [3], research of collaborative quality evaluation based on supply chain (Jun Zhang, Aiping Song and Yimin Li, 2005) etc [4]. (1) These studies lack the deep analysis of product quality building process in supply chain; (2) The evaluation system of product quality mainly focused on the study of individual factor in supply chain, which lacks perfect evaluation system of product quality in the operation process of supply chain. (3) AHP and FCE are respectively applied to study the evaluation method of product quality in supply chain, which is difficult to solve the problems of complex, dynamic and multi-variable evaluation of product quality in supply chain.

Therefore, product quality building process in supply chain is further analyzed, the key factors influencing product quality in supply chain are found, evaluation system of product quality oriented lifecycle in supply chain is established. On the basis, for the problems of complex, dynamic and multi-variable quality index in supply chain, G1 is applied to determine the weights of dynamic and multi-variable quality index, which could solve the problem that it is not necessary to reorder the quality indexes in the case of their changes (increase or decrease); and intrinsic fuzziness of expert judgment is considered, fuzzy comprehensive evaluation (FCE) is applied to evaluate product quality oriented supply chain. Scientific, reasonable results of quality evaluation can be realized, and accord with actual.

II. PRODUCT QUALITY BUILDING PROCESS IN SUPPLY CHAIN

Design, production, sale and service after sale of product in supply chain could be accomplished in cooperation of all the members of supply chain [5, 6], objectively, product quality is assured and realized by all the members of supply chain [7]. Actually, building and realizing process of product quality is across the range of the whole supply chain. So, the factors influencing product quality in supply chain are throughout the whole process in product lifecycle included design, manufacturing, using.

Formation of product quality starts at market, customer requirements are collected and arranged by market investigation, that is the product quality customer expects; the customer requirements are transformed and mapped to engineering quality characteristics (engineering design requirements), product design is conducted, on the basis, process design is completed, so

E. Qi et al. (eds.), *Proceedings of the 21st International Conference on Industrial Engineering and Engineering Management 2014*, Proceedings of the International Conference on Industrial Engineering and Engineering Management, DOI 10.2991/978-94-6239-102-4_28, © Atlantis Press and the authors 2015

as to form product design quality; then entering production preparation, raw material and purchased parts are purchased, so as to form purchase quality; then entering product manufacturing, product is manufactured and assembled, so as to form manufacturing quality; finally, the product is sent to customer by sales, in this case, the product quality is the very final product quality that customer feels actually. It is shown as Fig.1.

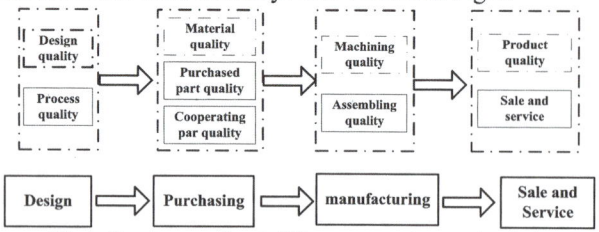

Fig.1. Product quality building process in supply chain

In the operation process of supply chain, the assurance of product design quality is provided by design department of core enterprise; product manufacturing quality is jointly assured by core enterprise, parts supplier and material supplier; quality of sale and service after sale is assured by distributors and retailers. Different quality is formed in various stages, the quality characteristics are transferred to next stage along quality flow, which is a part of product quality in next stage, in this way, final product quality is formed along the continuous quality transmission. Because of the

transmission order, product quality in former stage influence the one in next stage directly, downstream enterprise cannot interfere that product quality in former stage have effect on final product quality.

Therefore, the key factors that influence product quality in supply chain are determined, which includes product design quality, purchase quality (material and purchase parts), manufacturing quality, sale and service quality.

III. EVALUATION SYSTEM MODEL OF PRODUCT QUALITY ORIENTED SUPPLY CHAIN

Product quality evaluation oriented lifecycle in supply chain is the systematic evaluation activities of product quality in every stage of lifecycle in supply chain according to product integration information, by applying evaluation technology and method, with the support of the responding organization and persons, for the goal of quality improvement. There are lots of factors that influence product quality in every stage in operation process of supply chain, furthermore, these factors possess obvious dynamic, uncertainty and fuzziness. So, establishing the quality index system that is systematic, overall and hierarchical, is the foundation realizing scientific and effective product quality evaluation [8]. It is shown as Fig.2.

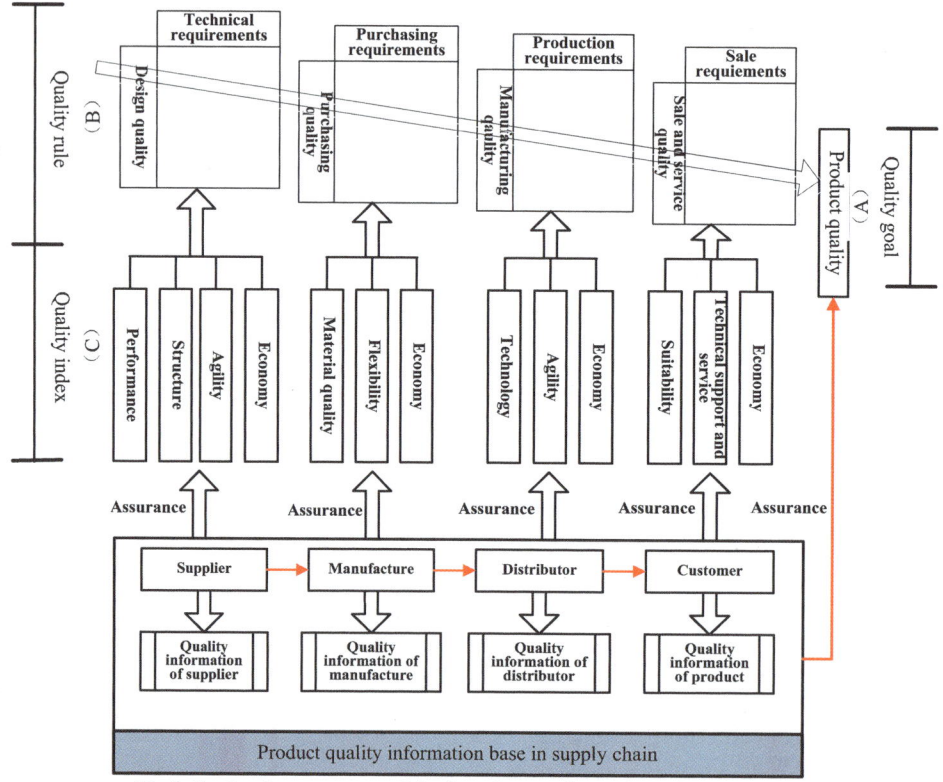

Fig.2. Evaluation system model of product quality oriented supply chain

Evaluation system model of product quality is compose of quality goal, quality rule and quality characteristic index, which is used to describe the hierarchical relationship of quality goal, quality rule and quality characteristic index. It could meet the requirement of product quality evaluation in every stage in operation process of supply chain, consideration to technicality, economy, customer demand and ecological environment etc., and realize multi-level, multi-view evaluation of product quality oriented lifecycle in supply chain.

IV. EVALUATION METHOD OF PRODUCT QUALITY ORIENTED SUPPLY CHAIN

AHP or FCE is mostly applied by existing evaluation method of product quality, but people's subjectivity is greatly in evaluation process of these method, which lead to deviation of objective fact. Furthermore, complex matrix and consistency analysis need to be structured, quality index order need to be reordered because of quality index change, quantity of quality index also need to be limited in using process of these method. These problems could greatly influence the efficiency and accuracy of quality evaluation.

G1 is a new method that determines index weight [9, 10]. Its basic steps are: quality indexes are ordered on their importance according to evaluation rule, ratios of importance degree of adjacent quality index are given, then quality index weight is obtained. So, it weakens the influence of people's subjectivity. Furthermore, it can solve the problem that it is not necessary to reorder the quality indexes in the case of their changes (increase or decrease), which is suitable for group judgment.

Since it need not structure complex matrix, test consistency, and limit quantity of quality index, G1 is more efficiency and easy to realize quality evaluation for complex, dynamic and multi-variable quality in supply chain.

Step 1: establishing quality evaluation model

Assume factor set of quality evaluation $U=\{U_1, U_2, U_3, U_4, U_5\}$, common set $V=\{V_1, V_2, V_3, V_4, V_5\}$.

Step 2: determining order relation

If important degree of quality index U_i is more than that of quality index U_j relative to some evaluation rule, then order relation is denoted by $U_i \succ U_j$; if relative to some quality evaluation rule, the relation equation of quality index $U_1, U_2 \ldots, U_m$ can be denoted as: $U_1^* \succ U_2^* \succ \ldots \succ U_k^* \succ \ldots \succ U_m^*$, $k=1,2,\ldots,m$, then order relation between quality index is determined according to "$\succ$". Where U_i^* is i quality index after $\{U_i\}$ is ordered according to order relation "$\succ$" (i=1, 2, 3, …, m).

For $\{U_1, U_2 \ldots, U_m\}$, order relation can be established by the follow steps:

1) Evaluation expert choose the most important quality index (only one) in $\{U_1, U_2 \ldots, U_m\}$, which is denoted as U_1^*;

…

k) Evaluation expert choose the most important quality index in remainder $[m-(k-1)]$ quality indexes, which is denoted as U_k^*;

…

m) After $(m-1)$th choose, the remainder quality index is denoted as U_m^*.

In this way, an order relation is only determined.

Step 3: Comparison judgment of relative important degree between adjacent quality index U_k and U_{k-1}

$$r_k = w_{k-1} / w_k, \qquad k = m, m-1, \ldots, 3, 2 \qquad (1)$$

Step 4: Calculating weight w_k

$$w_m = (1 + \sum_{k=2}^{m} \prod_{i=k}^{m} r_i)^{-1} \qquad (2)$$

$$w_{k-1} = r_k / w_k, \ k = m, m-1, \ldots, 3, 2 \qquad (3)$$

Where w_m is the weight of quality index k. Then weight set of quality index is denoted as $W=\{w_1, w_2, w_3, \ldots, w_k\}$.

Step 5: Establishing fuzzy comprehensive evaluation system.

Common set V and fuzzy relation R is quantization, and fuzzy comprehensive evaluation system is established.

Step 6: Fuzzy comprehensive evaluation

Final fuzzy evaluation is conducted, which is denoted as:

$$B = W \bullet R = W \bullet \begin{bmatrix} W_1 \bullet R_1 \\ W_2 \bullet R_2 \\ W_3 \bullet R_3 \\ W_4 \bullet R_4 \\ W_5 \bullet R_5 \end{bmatrix} \qquad (4)$$

Final scores of some quality index is denoted as A:
$$A = (B \bullet V) \qquad (5)$$

V. A CASE STUDY

The product of some Enterprise, Lubricant station F335 is introduced as an example to established hierarchy structure of evaluation index system of design quality. The evaluation index system of design quality is composed of total quality object, quality rule and quality index, which describes the hierarchy relationship and related attributes among quality object, quality rule and quality index. The system considers all influence quality factors of product life cycle, in order to make the system suit most products, that is to say, the system includes most indexes of general product basically.

In order to obtain objective evaluation result, expert group of Quality evaluation (20 experts) is retained to evaluate the design quality of the product, and both design quality evaluation set U={product performance product performance index (U_1), product structure index (U_2), product agility (U_3), product economy (U_4), product green properties (U_5)}, and common set V={Excellent, good, middle, qualified, unqualified} are established.

Suppose that quality expert i evaluate product performance product performance index (U_1), product structure index (U_2), product agility (U_3), product economy (U_4), product green properties (U_5) respectively.

Step 1: Importance ordering

$$U_1 \succ U_5 \succ U_2 \succ U_4 \succ U_3 \Rightarrow U_1^* \succ U_2^* \succ U_3^* \succ U_4^* \succ U_5^*$$

Step 2: Ratios of importance degree between adjacent quality indexes are given by quality expert:

$$r_1 = \frac{w_1^*}{w_2^*} = 1.3, \quad r_3 = \frac{w_2^*}{w_3^*} = 1.2, \quad r_4 = \frac{w_3^*}{w_4^*} = 1.2, \quad r_5 = \frac{w_4^*}{w_5^*} = 1.4$$

Step 3: Calculating weight w_j

Step 4: The weights of quality indexes (includes U_1, U_2, U_3, U_4, U_5) are obtained, which are as follows:

$$w_1 = w_1^* = 0.3007, \quad w_2 = w_3^* = 0.1927, \quad w_3 = w_5^* = 0.1147$$
$$w_4 = w_4^* = 0.1606, \quad w_5 = w_2^* = 0.2312$$

In this way, three grade indexes of the product design quality are ordered according to their importance by 20 quality experts, and Ratios of importance degree between adjacent quality indexes are given respectively, each quality expert determines weight of every quality index respectively. So, each quality expert obtain a table of quality indexes' weight, mean of each quality index's weight is obtained by each quality expert's opinion, then the comprehensive weight of each quality index is obtained. It is shown as Table I.

TABLE I
COMPREHENSIVE WEIGHTS OF QUALITY INDEXES

Second grad index	Third grad index	Comprehensive weight	Second grad index	Third grad index	Comprehensive weight
Product performance 0.2547	practicability	0.1237	Product agility 0.1887	Productivity	0.2798
	suitability	0.1207		Product handling	0.2076
	operability	0.1077		Reconfigurability	0.1912
	stability	0.1013		Parts generality	0.1887
	safety	0.0982		Serialization	0.1328
	adaptability	0.0773	Product economy 0.2365	profit	0.4501
	manufacturability	0.0735		cost	0.2500
	debugging	0.0674		price	0.2012
	fault testability	0.0652		Product lifecycle	0.1031
	assembling ability	0.0624	Product green property 0.1454	pollution	0.2698
	maintainability	0.0522		Harm to human health	0.2076
	technical support and service	0.0453		Optimum utilization of resource	0.1887
Product structure 0.1765	Reliability	0.2156		Sustainable utilization of energy	0.1887
	Product appearance	0.1856		Recycling and reusing	0.1452
	Structure similarity	0.1685			
	Machining ability	0.1524			
	Part assembling ability	0.1470			
	Structure compactness	0.1400			

The product design quality is graded by Common set V, which is quantified as V={0.9, 0.8, 0.7, 0.6, 0.3}. The final weights of the product quality indexes are obtained by establishing fuzzy comprehensive evaluation system. It is shown as Table II.

According to Equation (4), (5), the calculation result is equal to 0.73, then the evaluation score of the product design quality is 0.73, that is the grade of the product design quality is normal.

VI. CONCLUSION

(1) The key factors that influence product quality in operation process of supply chain are determined, which are design, purchasing, manufacturing, sale and service.

On the basis, evaluation system model of product quality oriented lifecycle in supply chain is established, which could reflect actual state of product quality in supply chain more completely and systematically.

(2) Aiming at complex, dynamic and multi-variable quality in supply chain, G1 is applied to determine weights of product quality indexes, which could solve the problem that it is not necessary to reorder the quality indexes in the case of their changes (increase or decrease); and intrinsic fuzziness of expert judgment is considered, fuzzy comprehensive evaluation (FCE) is applied to establish mathematical evaluation model of product quality oriented lifecycle in supply chain.

(3) The product of some Enterprise, Lubricant station F335 is introduced as an example to evaluate its design quality, the evaluation result is consistent with the actual condition of the enterprise. It is further verified that the method of G1 and FCE is reasonable and reliable. An effective way is provided for continuous assurance and improvement of product quality in supply chain.

ACKNOWLEDGEMENTS

Thank the anonymous referees for their careful reading and constructive comments on the paper. This work has been supported by Natural Science Foundation of Liaoning Province (2013020041).

TABLE II
FUZZY COMPREHENSIVE EVALUATION WEIGHTS OF F335

Second grad index	Third grad index	Comprehensive weight	Second grad index	Third grad index	Comprehensive weight
Product performance 0.2547	practicability	0.1237	Product agility 0.1887	Productivity	0.2798
	suitability	0.1207		Product handling	0.2076
	operability	0.1077		Reconfigurability	0.1912
	stability	0.1013		Parts generality	0.1887
	safety	0.0982		Serialization	0.1328
	adaptability	0.0773	Product economy 0.2365	profit	0.4501
	manufacturability	0.0735		cost	0.2500
	debugging	0.0674		price	0.2012
	fault testability	0.0652		Product lifecycle	0.1031
	assembling ability	0.0624	Product green property 0.1454	pollution	0.2698
	maintainability	0.0522		Harm to human health	0.2076
	technical support and service	0.0453		Optimum utilization of resource	0.1887
Product structure 0.1765	Reliability	0.2156		Sustainable utilization of energy	0.1887
	Product appearance	0.1856		Recycling and reusing	0.1452
	Structure similarity	0.1685			
	Machining ability	0.1524			
	Part assembling ability	0.1470			
	Structure compactness	0.1400			

REFERENCES

[1] Bin Dan, Lian-chun Ren, "Research on Quality Improvement of Manufacturer under Supply-chain Environment" (in Chinese), *Industrial Engineering Journal*, vol.2, pp. 1-5, 2010.

[2] Chun-chi Hsieh, Yu-te Liu, "Quality investment and inspection policy in a supplier-manufacturer supply chain", European J of Operational Research, vol.202, no.3, pp. 717-729, 2010.

[3] Mei-qing Wang, Xiao-qing Tang, "Research on Methodology of Quality Control in Product Design" (in Chinese), *Manufacturing Automation*, vol.9, pp. 12-14, 2003.

[4] Yuan Liu, "Construction of Collaborative Marketing Model Based on Supply Chain Management" (in Chinese), *Journal of Hubei Automotive Industries Institute*, vol.26, no.4, pp. 76-80, 2012.

[5] Zhen-jian Zhang, "Research on quality management information system oriented product lifecycle" (in Chinese), Aeronautical *Computing Technique*, vol.4, pp. 37-40, 2010.

[6] Wen-jie Yang, Yi-song Li, "Research on Life Circle- based Performance Evaluation System of the Core Enterprise in Agile Supply Chain" (in Chinese), Logistics *Technology*, vol.9, pp. 116-118, 2010.

[7] Tao Ding, Gong-chang Ren, Fang Wang, "Networked Manufacturing Based on Product Lifecycle Management" (in Chinese), *Modular Machine Tool & Automatic Manufacturing Technique*, vol.02, pp. 107-109, 2009.

[8] Xin-yu Jiang, Gui-he Wang, Hai-feng Zhao, Wei Ding, Wan-shan Wang, "Research on Comprehensive Quality Evaluation Method of oriented-Product Lifecycle" (in Chinese), *Journal of System Simulation*, vol.20, no.20, pp. 5581-5584, 2008.

[9] Xue-jun Wang, Ya-jun Guo, "Analyzing the consitstency of comparison matrix based on G1 method" (in Chinese), *Chinese Journal of Management Science*, vol.14, no.3, 65-70, 2006.

[10] Ya-jun Guo, *Method and theory of comprehensive evaluation* (Book style), Beijing, Science press, 2002. (in Chinese)

Robustness to Measurement Errors and Diversity in Resource Sharing in Time-driven Activity-based Costing

Xiao-dan Wu*, Xiang-jun Zuo, Dian-min Yue, Dan Liu

School of Economics and Management, Hebei University of Technology, Tianjin 300401, China

(*wxdan2000@sina.com, zuoxiangjunhaha@163.com)

Abstract - **The total cost error and the costing system's robustness were influenced by measurement errors under different resource sharing patterns. According to the analysis of Time-Driven Activity-based Costing calculation process, we propose the cost error models for variables of unit time and quantity to increase the costing system sensitivity to errors for the aspects of diversity studied, analyze the robustness to measurement errors under different resource sharing patterns. We also identify conditions where allocating costing system limited resources to cases characterized by high diversity in resource sharing patterns is detrimental to improve cost accuracy.**

Keywords - **Cost management, measurement errors, resource sharing, time-driven activity-based costing**

I.INTRODUCTION

The accuracy of cost analysis directly affects a firm to improve the cost and correct the process of production or service [1][2]. However, the indirect cost is so difficult to tracing its origin directly but easy to produce deviation, a wide deviation will directly affects the manager's decisions [3]. So in the different background of resource allocation, deviation will lead to the different cost errors eventually. The most common deviation is called measurement errors which are caused by the estimate [4]. Thus, it's crucial to analyze the total cost deviation caused by different measurement errors and the costing system's robustness under the different resource sharing patterns, use scientific cost accounting methods to build measurement errors analysis model.

With the production or service presenting personalized characteristics in the modern firm, time-driven activity-based costing (TDABC) arises at the historic moment. It can be more rapidly established than activity-based costing (ABC) and easier to be updated and maintained, at the same time, it can response the complex operation of actual situation more flexibly, eliminate tedious and subjectivity process in the allocation of resources to activities, fill the gaps that ABC produces more errors in computing [5], and improve the accuracy of cost accounting.

There were some valuable research results in the study of costing accuracy, including that there was setting or offsetting effect between measurement errors and other errors when changing variable values in ABC [4]; it proved the interaction between measurement errors and other errors by using simulation data, [6]; it proved that increasing the diversity of resource sharing would enhance the robustness to measurement errors in the cost drivers [7]; according to different production environments, choosing correlation method or cost rules can reduce cost

errors in ABC [8]. However, these studies are general research on ABC errors and there was not detailed research on resource sharing under some specific production environments. The conclusion that uses some specific experimental data or simulation method was vague and not general. In fact, it's clearer and has stronger commonality to choose a more accurate cost accounting method to build a theoretical model that analysis cost errors in the consideration of production or service environment. Such as the resource of doctors and nurses in hospital, compared with the nurses' cost, the doctors' is apparently much higher. If some division of activities is not so clear that the allocation of these two resources is confusing, it will not only cause a waste of resource and raise the cost, but also make the doctors' cost confused and lead to some greater measurement errors. At the same time, due to the difference of patients' condition or quantity every day, it's more reasonable to allocate cost based on TDABC.

Therefore, we study the measurement errors of cost in the firm based on TDABC and establish its model for cost errors. By setting two specific resource sharing ways (resource cost consumed evenly and resource cost consumed extremely unevenly) in the firm's production or service environment, we analyze the total cost error caused by different measurement errors and the robustness of costing system under different resource sharing patterns.

II. CONCEPT DEFINITION

A. Measurement Errors in Product Cost Estimates based on TDABC

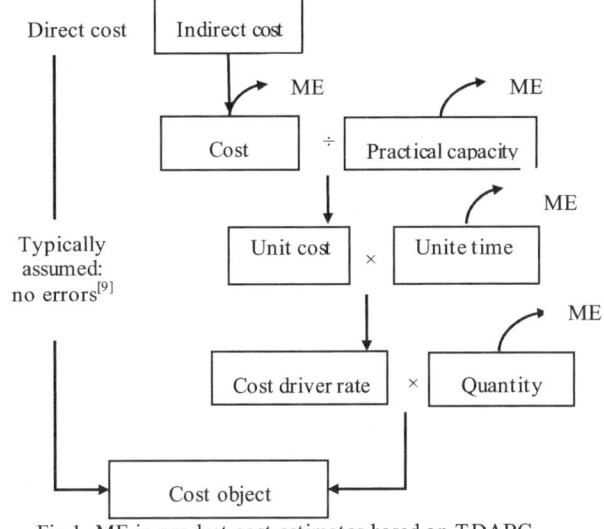

Fig.1. ME in product cost estimates based on TDABC

E. Qi et al. (eds.), *Proceedings of the 21st International Conference on Industrial Engineering and Engineering Management 2014*, Proceedings of the International Conference on Industrial Engineering and Engineering Management, DOI 10.2991/978-94-6239-102-4_29, © Atlantis Press and the authors 2015

Measurement errors (ME) occurs when variables to be measured are not supported by well-defined measurement guideline or measurement techniques, it is a kind of common errors in the indirect cost allocation [4]. In accordance with the accounting process of TDABC, measurement errors may exist in four variables as shown in Fig.1 [9]: (1) measurement errors in the cost (MEC); (2) measurement errors in the practical capacity (MET); (3) measurement errors in the unit time(MEU); (4) measurement errors in the quantity (MEQ).

B. Diversity in Resource Sharing

To illustrate the diversity in resource sharing, the parameter TCD (distribution of total cost) is introduced. We define that x_i represents the cost that ith resource pool consumed, $\overline{X}$ represents the mean value of the total indirect cost. α_{cw} is the sum of the absolute deviation of resource pools defined as $\sum_{i=1}^{n}\left|x_i - \overline{X}\right|$ and α_{cmax} is the maximum deviation defined as $\max_{i=1}^{n}\left|x_i - \overline{X}\right|$. There are no restricted conditions about each resource pool to cost consumption, then [6]:

$$TCD = \frac{\alpha_{cw}}{\alpha_{cmax}} = \frac{\sum_{i=1}^{n}\left|x_i - \overline{X}\right|}{\max_{i=1}^{n}\left|x_i - \overline{X}\right|}$$

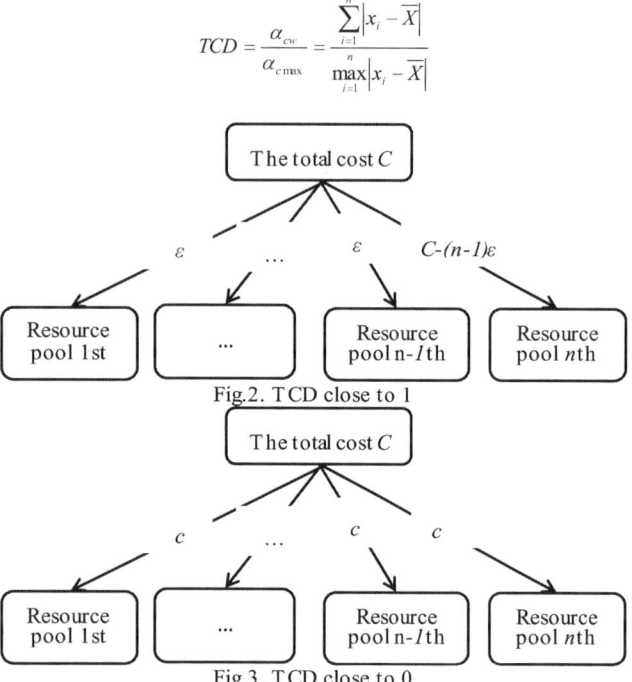

Fig.2. TCD close to 1

Fig.3. TCD close to 0

Accordingly, if TCD is close to 0, the cost is consumed more equal among resource pools, denoted by LTCD. If TCD is close to 1, then the cost is consumed much less equal among resource pools, denoted by HTCD. As shown in Fig.2 and Fig.3, it can be regarded as LTCD when the cost of each resource pool is $c(C=n \times c)$; it can be regarded as HTCD when most of the cost is consumed by nth resource pool but less is consumed by the first n-1 resource pools (ε is minimum rather than maximum. Because if ε is maximum, it illustrates that most of the cost is consumed evenly among the n-1 resource pools

and the remaining pool consumes little cost. It reflects the condition of LTCD and doesn't conform to the definition of HTCD.).

III. CONSTRUCTION OF ERRORS BENCHMARK MODEL

A. Construction of the Total Cost of Errors Formula

Suppose the total indirect cost is C, and there are m activities, n resource pools and s cost objects. So we can establish the errors benchmark model based on TD-ABC method as follows:

(1) Definition of the variables

The cost matrix of each resource pool is $C = \begin{bmatrix} c_1 & c_2 & \cdots & c_n \end{bmatrix}_{n\times1}$. The practical time of each resource pool is $T = \begin{bmatrix} t_1 & t_2 & \cdots & t_n \end{bmatrix}_{n\times1}$. The unit time matrix of activities is $U = \begin{bmatrix} u_{jk} \end{bmatrix}$ ($j=1,2,...,m$; $k=1,2,...,n$). The quantity matrix is $Q = \begin{bmatrix} q_{ij} \end{bmatrix}$ ($i=1,2,...,s$; $j=1,2,...,m$).

(2) Logical expression of computation

①Unit cost of each resource pool can be expressed as

$$C/T = \begin{bmatrix} c_k / t_k \end{bmatrix}^T \ (k=1,2,...,n)$$

Where $\dfrac{c_k}{t_k}$ is the unit cost of the kth activity, $k=1,2,...,n$.

②The distribution ratio of the cost driver can be expressed as

$$R = \begin{bmatrix} r_1, r_2, ..., r_m \end{bmatrix}^T = U_{m\times n} \times \frac{C}{T}_{n\times1}$$

Where $r_j = \sum_{k=1}^{n} u_{jk} \cdot \dfrac{c_k}{t_k}$ represents the unit activity cost of jth activity.

③The cost matrix of cost objects can be expressed as

$$CO = \begin{bmatrix} co_1, co_2, ..., co_s \end{bmatrix}^T = Q_{s\times m} \times U_{m\times n} \times \frac{C}{T}_{n\times1} \qquad (1)$$

Where $co_i = \sum_{j=1}^{m} q_{ij} \cdot r_j$ is the cost of ith cost object, q_{ij} is the quantity that the ith cost object consumes the jth activity. u_{jk} is the unit time of the jth activity in the ith resource pool. $i=1,2,...,s$. $j=1,2,...,m$. $k=1,2,...,n$.

④The total cost error of all the cost objects can be expressed as

We define the standard cost matrix of cost objects as $TC = \begin{bmatrix} tc_1, tc_2, \cdots, tc_s \end{bmatrix}^T$, simulated (false) cost matrix as $FC = \begin{bmatrix} fc_1, fc_2, \cdots, fc_s \end{bmatrix}^T$. Considering the offsetting relationship between the positive errors and negative errors of cost objects, the total error in the cost accounting can be expressed as [10]

$$\begin{aligned} EUCD &= \sqrt{(tc_1 - fc_1)^2 + (tc_2 - fc_2)^2 + \cdots + (tc_s - fc_s)^2} \\ &= \sqrt{\sum_{i=1}^{s} (tc_i - fc_i)^2} \end{aligned} \qquad (2)$$

Where tc_i gives the true cost accruing to cost object I in the true benchmark scenario, and fc_i is the (false) cost allocated to cost object I by the costing system approximation.

B. Construction of Measurement Errors Model

In this paper, we only construct and analyze measurement errors for two important variables: unit time and quantity.

(1) Measurement errors existing in the unit time

Due to the unit time is also obtained by estimated or interviewed, measurement errors may occur in some activities. Measurement errors in the unit time can come in several forms, but this article only choose one form so as to calculate and compare in the form of mathematical. Suppose that the unit time about each activity can produce errors, its scale is δ_j $(\delta_j \in (-1,1), j=1,2,\ldots,m)$ and its proportion of measurement errors is equal in all resource pools. So it can be expressed as

$$\begin{bmatrix} 1+\delta_1 & 0 & \cdots & 0 \\ 0 & 1+\delta_2 & \cdots & 0 \\ \vdots & \vdots & \ddots & \vdots \\ 0 & 0 & \cdots & 1+\delta_m \end{bmatrix}_{m \times m} \times \begin{bmatrix} u_{11} & u_{12} & \cdots & u_{1n} \\ u_{21} & u_{22} & \cdots & u_{2n} \\ \vdots & \vdots & \ddots & \vdots \\ u_{m1} & u_{m2} & \cdots & u_{mn} \end{bmatrix}_{m \times n}$$

When there are no measurement errors in the whole cost accounting, the standard cost matrix of cost objects can be expressed as

$$TC = Q_{s \times m} \times U_{m \times n} \times \frac{C}{T} = \begin{bmatrix} \sum_{j=1}^{m} q_{1j} \sum_{k=1}^{n} u_{jk} \frac{c_k}{t_k} \\ \sum_{j=1}^{m} q_{2j} \sum_{k=1}^{n} u_{jk} \frac{c_k}{t_k} \\ \vdots \\ \sum_{j=1}^{m} q_{sj} \sum_{k=1}^{n} u_{jk} \frac{c_k}{t_k} \end{bmatrix}_{s \times 1} \quad (3)$$

When there are measurement errors in the unit time, the false cost matrix of cost objects can be expressed as

$$FC = Q_{s \times m} \times (1+\delta)U_{m \times n} \times \frac{C}{T} = \begin{bmatrix} \sum_{j=1}^{m} q_{1j}(1+\delta_j) \cdot \sum_{k=1}^{n} u_{jk} \frac{c_k}{t_k} \\ \sum_{j=1}^{m} q_{2j}(1+\delta_j) \cdot \sum_{k=1}^{n} u_{jk} \frac{c_k}{t_k} \\ \vdots \\ \sum_{j=1}^{m} q_{sj}(1+\delta_j) \cdot \sum_{k=1}^{n} u_{jk} \frac{c_k}{t_k} \end{bmatrix}_{s \times 1} \quad (4)$$

So according to the formula (2), when the measurement errors exist in the unit time, the total error in the cost accounting can be expressed as

$$\mathrm{EUCD}_{\mathrm{MEU}} = \sqrt{\sum_{i=1}^{s} \left(\sum_{j=1}^{m} q_{ij} \cdot \delta_j \cdot \sum_{k=1}^{n} u_{jk} \frac{c_k}{t_k} \right)^2} \quad (5)$$

(2) Measurement errors exiting in the quantity

Due to the observation statistics of managers, the distribution of each activity on all cost objects may lead to measurement errors. Suppose that the quantity about each cost object can produce errors and its scale is δ_i $(\delta_i \in (-1,1), i=1,2,\ldots,s)$. So it can be expressed as

$$\begin{bmatrix} 1+\delta_1 & 0 & \cdots & 0 \\ 0 & 1+\delta_2 & \cdots & 0 \\ \vdots & \vdots & \ddots & \vdots \\ 0 & 0 & \cdots & 1+\delta_s \end{bmatrix}_{s \times s} \times \begin{bmatrix} q_{11} & q_{12} & \cdots & q_{1m} \\ q_{21} & q_{22} & \cdots & q_{2m} \\ \vdots & \vdots & \ddots & \vdots \\ q_{s1} & q_{s2} & \cdots & q_{sm} \end{bmatrix}_{s \times m}$$

When there are measurement errors in the quantity, the false cost matrix of cost objects can be expressed as

$$FC = (1+\delta)Q_{s \times m} \times U_{m \times n} \times \frac{C}{T} = \begin{bmatrix} (1+\delta_1) \sum_{j=1}^{m} q_{1j} \sum_{k=1}^{n} u_{jk} \frac{c_k}{t_k} \\ (1+\delta_2) \sum_{j=1}^{m} q_{2j} \sum_{k=1}^{n} u_{jk} \frac{c_k}{t_k} \\ \vdots \\ (1+\delta_s) \sum_{j=1}^{m} q_{sj} \sum_{k=1}^{n} u_{jk} \frac{c_k}{t_k} \end{bmatrix}_{s \times 1} \quad (6)$$

Of course, the standard cost matrix is the same with the former formula (3). So according to the formula (2), when the measurement errors exist in the quantity, the total error in the cost accounting can be expressed as

$$\mathrm{EUCD}_{\mathrm{MEQ}} = \sqrt{\sum_{i=1}^{s} \left(\delta_i \cdot \sum_{j=1}^{m} q_{ij} \cdot \sum_{k=1}^{n} u_{jk} \frac{c_k}{t_k} \right)^2} \quad (7)$$

IV. ANALYSIS FOR THE DIVERSITY OF RESOURCE SHARING

As mentioned in the first part, the consumption patterns of total cost is divided into two cases. Specifically, the contents are as follows:

a. (HTCD): Suppose that all the resource pools from 1 to $n-1$ consume ε units cost, but the remaining cost is consumed by the resource pool n. So $c_1=\ldots=c_{n-1}=\varepsilon$, $c_n=C-(n-1)\varepsilon$, (ε is minimal, $\varepsilon \to 0$);

b. (LTCD): Suppose that the total cost is evenly consumed by all the resource pools. So $c_1=\ldots=c_n=C/n=c$.

Next, we will discuss the effect of total cost consumption on measurement errors in each variable.

A. About the Unit Times of Activities

According to the formula (5), when measurement errors exist in the unit time, we calculate the total error in the cost respectively in *HTCD, LTCD*:

$$\mathrm{EUCD}_{\mathrm{MEU}-HTCD} = \sqrt{\sum_{i=1}^{s} \left(\sum_{j=1}^{m} q_{ij} \cdot \delta_j \cdot \sum_{k=1}^{n} u_{jk} \frac{\varepsilon}{t_k} + \sum_{j=1}^{m} q_{ij} \cdot \delta_j \cdot u_{jn} \cdot \frac{n(c-\varepsilon)}{t_n} \right)^2}$$

$$\mathrm{EUCD}_{\mathrm{MEU}-LTCD} = \sqrt{\sum_{i=1}^{s} \left(\sum_{j=1}^{m} q_{ij} \cdot \delta_j \cdot \sum_{k=1}^{n} u_{jk} \frac{\varepsilon}{t_k} + \sum_{j=1}^{m} q_{ij} \cdot \delta_j \cdot \sum_{k=1}^{n} u_{jk} \cdot \frac{c-\varepsilon}{t_k} \right)^2}$$

What does the two consumption patterns of resource effect on the measurement errors in the unit time? It needs to compare $\mathrm{EUCD}_{MEU\text{-}HTCD}$ and $\mathrm{EUCD}_{MEU\text{-}LTCD}$. We should discuss in two different conditions:

Hypothesis 1: The sum that the unit times of all activities in nth resource pool account for its practical time is less than the average value that the unit times of all activities account for practical time in all resource pools, it can be expressed as

$$\sum_{j=1}^{m} u_{jn} \cdot \frac{1}{t_n} \le \frac{1}{n} \sum_{j=1}^{m} \sum_{k=1}^{n} u_{jk} \cdot \frac{1}{t_k}$$

Based on the above hypothesis 1 and returning to the expressions $EUCD_{MEU\text{-}HTCD}$ and $EUCD_{MEU\text{-}LTCD}$, it must have the following expression established:

$$\min_k (c-\varepsilon) \cdot \frac{1}{n} \cdot \sum_{j=1}^{m} q_{ij} \cdot \delta_j \cdot u_{jk} \cdot \frac{1}{t_k} = (c-\varepsilon) \cdot \frac{1}{n} \cdot \sum_{j=1}^{m} q_{ij} \cdot \delta_j \cdot u_{jn} \cdot \frac{1}{t_n}$$

$$\therefore (c-\varepsilon)\cdot\frac{1}{n}\cdot\sum_{j=1}^{m}q_{ij}\cdot\delta_j\cdot\sum_{k=1}^{n}u_{jk}\cdot\frac{1}{t_k}\geq(c-\varepsilon)\cdot\frac{1}{n}\cdot n\cdot\sum_{j=1}^{m}q_{ij}\cdot\delta_j\cdot u_{jn}\cdot\frac{1}{t_n}$$

$$\therefore \sum_{j=1}^{m}q_{ij}\cdot\delta_j\cdot\sum_{k=1}^{n}u_{jk}\cdot\frac{c-\varepsilon}{t_k}\geq\left(c-\varepsilon\right)\cdot n\cdot\sum_{j=1}^{m}q_{ij}\cdot\delta_j\cdot u_{jn}\cdot\frac{1}{t_n}$$

$$\therefore \sqrt{\sum_{i=1}^{s}\left(\sum_{j=1}^{m}q_{ij}\cdot\delta_j\cdot\sum_{k=1}^{n}u_{jk}\cdot\frac{\varepsilon}{t_k}+\sum_{j=1}^{m}q_{ij}\cdot\delta_j\cdot\sum_{k=1}^{n}u_{jk}\cdot\frac{c-\varepsilon}{t_k}\right)^2}$$

$$\geq\sqrt{\sum_{i=1}^{s}\left(\sum_{j=1}^{m}q_{ij}\cdot\delta_j\cdot\sum_{k=1}^{n}u_{jk}\cdot\frac{\varepsilon}{t_k}+n(c-\varepsilon)\sum_{j=1}^{m}q_{ij}\cdot\delta_j\cdot u_{jn}\cdot\frac{1}{t_n}\right)^2}$$

$$\therefore EUCD_{MEU-LTCD}\geq EUCD_{MEU-HTCD}$$

Conclusion 1-1: When the sum that unit times of all activities in nth resource pool account for its practical time is less than the average value that the unit times of all activities account for practical time in all resource pools, the greater the diversity in resource sharing in resource pools, the smaller the MEU. It also suggests that as the diversity in resource sharing increases, the robustness of costing system will be strengthened.

Hypothesis 2: The sum that the unit times of all activities in nth resource pool account for its practical time is greater than the average value that the unit times of all activities account for practical time in all resource pools, it can be expressed as

$$\sum_{j=1}^{m}u_{jn}\cdot\frac{1}{t_n}\geq\frac{1}{n}\sum_{j=1}^{m}\sum_{k=1}^{n}u_{jk}\cdot\frac{1}{t_k}$$

Based on the above hypothesis 2 and returning to the expressions $EUCD_{MEU-HTCD}$ and $EUCD_{MEU-LTCD}$, it must have the following expression established:

$$\max_k(c-\varepsilon)\cdot\frac{1}{n}\cdot\sum_{j=1}^{m}q_{ij}\cdot\delta_j\cdot u_{jk}\cdot\frac{1}{t_k}=(c-\varepsilon)\cdot\frac{1}{n}\cdot\sum_{j=1}^{m}q_{ij}\cdot\delta_j\cdot u_{jn}\cdot\frac{1}{t_n}$$

$$\therefore (c-\varepsilon)\cdot\frac{1}{n}\cdot\sum_{j=1}^{m}q_{ij}\cdot\delta_j\cdot\sum_{k=1}^{n}u_{jk}\cdot\frac{1}{t_k}\leq(c-\varepsilon)\cdot\frac{1}{n}\cdot n\cdot\sum_{j=1}^{m}q_{ij}\cdot\delta_j\cdot u_{jn}\cdot\frac{1}{t_n}$$

$$\therefore \sum_{j=1}^{m}q_{ij}\cdot\delta_j\cdot\sum_{k=1}^{n}u_{jk}\cdot\frac{c-\varepsilon}{t_k}\leq(c-\varepsilon)\cdot n\cdot\sum_{j=1}^{m}q_{ij}\cdot\delta_j\cdot u_{jn}\cdot\frac{1}{t_n}$$

$$\therefore \sqrt{\sum_{i=1}^{s}\left(\sum_{j=1}^{m}q_{ij}\cdot\delta_j\cdot\sum_{k=1}^{n}u_{jk}\cdot\frac{\varepsilon}{t_k}+\sum_{j=1}^{m}q_{ij}\cdot\delta_j\cdot\sum_{k=1}^{n}u_{jk}\cdot\frac{c-\varepsilon}{t_k}\right)^2}$$

$$\leq\sqrt{\sum_{i=1}^{s}\left(\sum_{j=1}^{m}q_{ij}\cdot\delta_j\cdot\sum_{k=1}^{n}u_{jk}\cdot\frac{\varepsilon}{t_k}+n(c-\varepsilon)\sum_{j=1}^{m}q_{ij}\cdot\delta_j\cdot u_{jn}\cdot\frac{1}{t_n}\right)^2}$$

$$\therefore EUCD_{MEU-LTCD}\leq EUCD_{MEU-HTCD}$$

Conclusion 1-2: When the sum that the unit times of all activities in nth resource pool account for its practical time is greater than the average value that the unit times of all activities account for practical time in all resource pools, the greater the diversity in resource sharing in resource pools, the greater the MEU. It also suggests that as the diversity in resource sharing increases, the robustness of costing system will be weakened.

Obviously, the above two conclusions are also applicable to other resource pools and universal.

B. About the Activity Quantity

According to the formula (7), when measurement errors exist in the quantity, we calculate the errors of total cost respectively in HTCD, LTCD:

$$EUCD_{MEQ-HTCD}=$$
$$\sqrt{\sum_{i=1}^{s}\left(\delta_i\cdot\sum_{j=1}^{m}q_{ij}\cdot\sum_{k=1}^{n}u_{jk}\frac{\varepsilon}{t_k}+\delta_i\cdot\sum_{j=1}^{m}q_{ij}\cdot u_{jn}\cdot\frac{n(c-\varepsilon)}{t_n}\right)^2}$$

$$EUCD_{MEQ-LTCD}=$$
$$\sqrt{\sum_{i=1}^{s}\left(\delta_i\cdot\sum_{j=1}^{m}q_{ij}\cdot\sum_{k=1}^{n}u_{jk}\cdot\frac{\varepsilon}{t_k}+\delta_i\cdot\sum_{j=1}^{m}q_{ij}\cdot\sum_{k=1}^{n}u_{jk}\cdot\frac{c-\varepsilon}{t_k}\right)^2}$$

What does two different consumption patterns of resource effect on the measurement errors in the quantity? It needs to compare $EUCD_{MEQ-HTCD}$ and $EUCD_{MEQ-LTCD}$. We should discuss in two different conditions:

Hypothesis 3: the same as hypothesis 1, so it can be expressed as

$$\sum_{j=1}^{m}u_{jn}\cdot\frac{1}{t_n}\leq\frac{1}{n}\sum_{j=1}^{m}\sum_{k=1}^{n}u_{jk}\cdot\frac{1}{t_k}$$

Based on the above hypothesis 3 and returning to the expressions $EUCD_{MEQ-HTCD}$ and $EUCD_{MEQ-LTCD}$, it must have the following expression established:

$$\min_k(c-\varepsilon)\cdot\delta_i\cdot\frac{1}{n}\sum_{j=1}^{m}q_{ij}\cdot u_{jk}\cdot\frac{1}{t_k}=(c-\varepsilon)\delta_i\cdot\frac{1}{n}\cdot\sum_{j=1}^{m}q_{ij}\cdot u_{jn}\cdot\frac{1}{t_n}$$

$$\therefore (c-\varepsilon)\delta_i\cdot\frac{1}{n}\cdot\sum_{j=1}^{m}q_{ij}\cdot\sum_{k=1}^{n}u_{jk}\cdot\frac{1}{t_k}\geq(c-\varepsilon)\delta_i\cdot\frac{1}{n}\cdot n\cdot\sum_{j=1}^{m}q_{ij}\cdot u_{jn}\cdot\frac{1}{t_n}$$

$$\therefore \delta_i\cdot\sum_{j=1}^{m}q_{ij}\cdot\sum_{k=1}^{n}u_{jk}\cdot\frac{c-\varepsilon}{t_k}\geq(c-\varepsilon)\delta_i\cdot n\cdot\sum_{j=1}^{m}q_{ij}\cdot u_{jn}\cdot\frac{1}{t_n}$$

$$\therefore \sqrt{\sum_{i=1}^{s}\left(\delta_i\cdot\sum_{j=1}^{m}q_{ij}\cdot\sum_{k=1}^{n}u_{jk}\cdot\frac{\varepsilon}{t_k}+\delta_i\cdot\sum_{j=1}^{m}q_{ij}\cdot\sum_{k=1}^{n}u_{jk}\cdot\frac{c-\varepsilon}{t_k}\right)^2}$$

$$\geq\sqrt{\sum_{i=1}^{s}\left(\delta_i\cdot\sum_{j=1}^{m}q_{ij}\cdot\sum_{k=1}^{n}u_{jk}\cdot\frac{\varepsilon}{t_k}+n(c-\varepsilon)\delta_i\cdot\sum_{j=1}^{m}q_{ij}\cdot u_{jn}\cdot\frac{1}{t_n}\right)^2}$$

$$\therefore EUCD_{MEQ-LTCD}\geq EUCD_{MEQ-HTCD}$$

Conclusion 2-1: When the sum that the unit times of all activities in nth resource pool account for its practical time is less than the average value that the unit times of all activities account for practical time in all resource pools, the greater the diversity in resource sharing in resource pools, the smaller the MEQ. It also suggests that as the diversity in resource sharing increases, the robustness of costing system will be strengthened.

Hypothesis 4: the same as hypothesis 2, so it can be expressed as

$$\sum_{j=1}^{m}u_{jn}\cdot\frac{1}{t_n}\geq\frac{1}{n}\sum_{j=1}^{m}\sum_{k=1}^{n}u_{jk}\cdot\frac{1}{t_k}$$

Based on the above hypothesis 4 and returning to the expressions $EUCD_{MEQ-HTCD}$ and $EUCD_{MEQ-LTCD}$, it must have the following expression established:

$$\max_k(c-\varepsilon)\cdot\delta_i\cdot\frac{1}{n}\sum_{j=1}^{m}q_{ij}\cdot u_{jk}\cdot\frac{1}{t_k}=(c-\varepsilon)\cdot\delta_i\cdot\frac{1}{n}\cdot\sum_{j=1}^{m}q_{ij}\cdot u_{jn}\cdot\frac{1}{t_n}$$

$$\therefore (c-\varepsilon)\cdot\delta_i\cdot\frac{1}{n}\cdot\sum_{j=1}^{m}q_{ij}\cdot\sum_{k=1}^{n}u_{jk}\cdot\frac{1}{t_k}\leq(c-\varepsilon)\cdot\delta_i\cdot\frac{1}{n}\cdot n\cdot\sum_{j=1}^{m}q_{ij}\cdot u_{jn}\cdot\frac{1}{t_n}$$

$$\therefore \delta_i \cdot \sum_{j=1}^{m} q_{ij} \cdot \sum_{k=1}^{n} u_{jk} \cdot \frac{c-\varepsilon}{t_k} \leq (c-\varepsilon) \cdot \delta_i \cdot n \cdot \sum_{j=1}^{m} q_{ij} \cdot u_{jn} \cdot \frac{1}{t_n}$$

$$\therefore \sqrt{\sum_{i=1}^{s} \left(\delta_i \cdot \sum_{j=1}^{m} q_{ij} \cdot \sum_{k=1}^{n} u_{jk} \cdot \frac{\varepsilon}{t_k} + \delta_i \cdot \sum_{j=1}^{m} q_{ij} \cdot \sum_{k=1}^{n} u_{jk} \cdot \frac{c-\varepsilon}{t_k} \right)^2}$$

$$\leq \sqrt{\sum_{i=1}^{s} \left(\delta_i \cdot \sum_{j=1}^{m} q_{ij} \cdot \sum_{k=1}^{n} u_{jk} \cdot \frac{\varepsilon}{t_k} + n(c-\varepsilon) \cdot \delta_i \cdot \sum_{j=1}^{m} q_{ij} \cdot u_{jn} \cdot \frac{1}{t_n} \right)^2}$$

$$\therefore \mathrm{EUCD}_{MEQ-LTCD} \leq \mathrm{EUCD}_{MEQ-HTCD}$$

Conclusion 2-2: When the sum that the unit times of all activities in nth resource pool account for its practical time is greater than the average value that the unit times of all activities account for practical time in all resource pools, the greater the diversity in resource sharing in resource pools, the greater the MEQ. Meanwhile, it also suggests that as the diversity in resource sharing increases, the costing system's robustness will be weakened.

Obviously, the above two conclusions are also applicable to other resource pools and universal.

V. CONCLUSION

We use TDABC as an analysis tool and set two types of resource sharing situations to study the effect of various measurement errors on the cost accuracy. It's helpful for managers to analyze the effect of different resource consumption on the measurement errors and cost accuracy and understand the true cost of each product or service. Finally, it provides strong support for firm to establish scientific cost analysis and control systems, improve the level of cost management, and strengthen the competitive power.

Through modeling and analyzing, it finds that the total cost error caused by unit time is the same as by the variable of quantity. When the sum that unit times of all activities account for its practical time in the resource pool which consumed most of the cost is less than the average value that the unit times of all activities account for practical time in all resource pools, the greater the diversity in resource sharing in resource pools, the smaller the measurement errors existing in the unit time or quantity. It also suggests that as the diversity in resource sharing increases, the robustness of costing system will be strengthened. When the sum that the unit times of all activities account for its practical time in the resource pool which consumed most of the cost is greater than the average value that the unit times of all activities account for practical time in all resource pools, the greater the diversity in resource sharing in resource pools, the greater the measurement errors existing in the unit time or quantity. It also suggests that as the diversity in resource sharing increases, the robustness of costing system will be weakened.

We still use the medical resource as an example and define resource pools as a doctor pool and a nurse pool. Meanwhile, we use the time that the doctor spent on each patient as unit time. If some division of activities is confusing (some simple activities are finished by doctors instead of nurses), it may cause measurement errors in unit time. It not only makes the cost higher, but also makes the costing system's robustness weaker. In other words, the cost itself is not controlled reasonably and the cost accuracy is descended.

Meanwhile, our study has several limitations. Firstly, the cost consumption patterns are diverse in other different conditions. In addition, aggregation errors is essentially different from measurement errors [11] and they have different mathematical expressions. In the future, we can continue to study the effect of aggregation errors on the cost accuracy.

ACKNOWLEDGMENT

This work is supported by National Natural Science Foundation of China (71302169), Natural Science Foundation of Hebei (F2009000111) and Social Science Foundation of National Education Department (12YJC630235).

REFERENCES

[1] R. Cooper., R. Kaplan. How cost accounting distorts product costs [J]. Management Accounting, 1988, 69(4): 20-28.

[2] E. Shim, E. F. Sudit. How manufacturers price products. Management Accounting, 1995, 76(2): 37-39.

[3] R. A. Leitch, P. A. Philipoom, T. D .Fry. Opportunity costing decision heuristics for product acceptance decisions [J]. Journal of Management Accounting Research, 2005, 17: 95-117.

[4] S. Datar, M. Gupta. Aggregation, Specification and Measurement Errors in Product Costing [J]. The Accounting Review, 1994, 69(4): 567-591.

[5] R. Kaplan. Time-driven Activity-based Costing [J]. Harvard Business Review, 2004, 60(8): 27-38.

[6] E. Labro, M. Vanhoucke. A Simulation Analysis of Interactions among Error in Costing Systems [J]. The Accounting Review, 2007, 82(4): 939-962.

[7] E. Labro, M. Vanhoucke. Diversity in Resource Consumption Patterns and Robustness of Costing Systems to Errors [J]. Management Science, 2008, 54(10): 1715-1730.

[8] Balakrishnan R, Hansen S, Labro E. Evaluating Heuristics Used When Designing Product Costing Systems [J]. Management Science. 2011, 57(3): 520-541.

[9] Y. Hwang, J. H. Evans III, V. G. Hegde. Product cost bias and selection of an allocation base [J]. Journal of Management Accounting Research, 1993, 5(Fall): 213-242.

[10] C. Homburg. A note on optimal cost driver selection in ABC [J]. Management Accounting Res. 2001, 12:197-205.

[11] Y. M. Babad, B. V. Balachandran. Cost driver optimization in activity-based costing [J]. Accounting Review, 1993, 68(3): 563-575.

Applied Research on Virtual Ergonomics for Vehicle Based on SolidWorks

Liang ZHANG[1], Ding-dan WEN[2], Pei-zhou SUN[2], Ping ZHANG[2,*]

[1]School of Mechanical and Automotive Engineering, Hefei University of Technology, Hefei, China
[2]School of Architecture and Arts, Hefei University of Technology, Hefei, China
(Zhangp163@163.com)

Abstract - **This article focus on virtual ergonomics evaluation based on SolidWorks. The authors have created the parametric human models based on the software SolidWorks. These models have different percentiles; they can switch from one to another rapidly. Not only did the shape simulate from human beings, but also the motion of joints followed the people do. In vehicle design, virtual human model can be used to test if the product is better for people to use when it is just a digital model. So it is available and economic for the engineers and designers at work.**

Keywords – **Configuration, parametric human model, percentile, vehicle design, virtual ergonomics**

I. INTRODUCTION

Nowadays has got giant achievement which can create realistic artificial simulated environment by multi high technology. Virtual Reality In the field of ergonomics, Virtual Reality brought great improvement about human-computer interaction technique and provided a new convenient interaction means for the application. In the process, the human model plays more and more important role between human and products. It can give the virtual test and evaluation if it is comfortable or not, or how to fit the users in the process of product design. The designer can get the virtual information about the users to use the products before the products can be manufactured, then they can decide if to produce the product or to improve it [1]. This may reduce the cost and shorten the period of research and development. As you know, we have a lot of software to build human models, such as CATIA, POSER. But SolidWorks has great characteristics. It is a set of 3D software for mechanical design which is very popular to Industrial Designers, and more important thing is that has the more strong functions of modeling and virtual assembly, so you can test the product easily in SolidWorks [2][3][4].

II. VIRTUAL REALITY TECHNIQUE APPLYING IN ERGONOMICS

How does Virtual Reality technique combine with ergonomics? The main points lie in building prototype, virtual human model and virtual environment by Virtual Reality technique, and then give test and evaluation about ergonomics capability of design and biomechanics report about people at work. According to this, designer can keep enough health care and occupational safety to people. They display in the following aspects [5][6][7]:

1. Test and evaluation of workspace
2. Evaluation of environment effect
3. Analysis of kinematics and dynamics
4. Evaluation about ergonomics capability of comfort and operability
5. Design of human-computer interface
6. Virtual design, virtual manufacturing, virtual assembly, virtual repair

The frame model of Virtual Reality technique applying in ergonomics is shown in Fig.1.

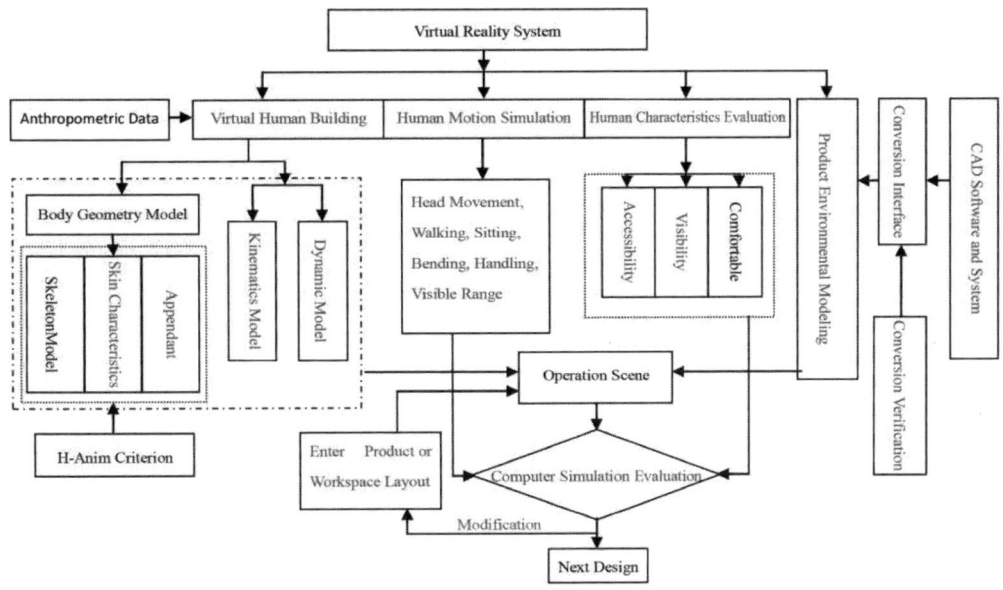

Fig.1. The frame of Virtual Reality technique applying in ergonomics

E. Qi et al. (eds.), *Proceedings of the 21st International Conference on Industrial Engineering and Engineering Management 2014*, Proceedings of the International Conference on Industrial Engineering and Engineering Management, DOI 10.2991/978-94-6239-102-4_30, © Atlantis Press and the authors 2015

III. SIMPLIFYING THE HUMAN MODEL AND CONFIRMING THE DIMENSIONS

A. Creating the parts of human models

As SolidWorks only can be used for mechanical design, in this study, we must simplify the models to some parts like a machine tool based on human body skeleton form. The model is divided into 15 parts: head, neck, trunk, left upper arm, left former arm, left hand, right upper arm, right former arm, right hand, left thigh, left shank, left foot, right thigh, right shank and right foot. These parts can be connected by some hinges, such as ball hinges, then the human models can be assembled together.

B. Confirming the dimensions

We confirm the control dimensions of parts according to the anthropometrics data. Some can be used directly, some must be deduced. Moreover, some dimensions cannot be deduced from the data directly, such as the neck length, which is not the key dimension and can be modified after assembly by checking the whole high of the model.

IV. PARAMETRIC DESIGN USING CONFIGURATION FUNCTION

Configuration function is a great feature of SolidWorks. The user can establish different dimension models by design tables; then modify the parameters according to the request to provide many designs, and switch one model to the other rapidly.

There are two important steps to process the models:

1) Choosing percentiles [8]

2) Creating parametric parts of the models (Table I shows the different dimension)

TABLE I
ANTHROPOMETRICS DATA (FOR EXAMPLE)   [mm]

	Items	5th %ile	50th %ile	95th %ile
1	Head length	210	230	250
2	Upper arm length	289	313	338
3	Former arm length	216	237	258
4	Palm length	163	173	183
5	Shoulder breadth	344	375	403
6	Thigh length	428	465	505
7	Crus length	338	369	403
8	Foot height	71	75	80

We create a model according to some percentile firstly, and then modify the dimensions in turn to gain the rest by the configuration function. For example, we choose the dimension of 5 percentile to create the upper arm model firstly, and add the configuration of 50 percentile and 95 percentile, then modify the dimensions of these two percentiles in different configurations, Fig.2 shows the models of upper arm for different percentiles [9].

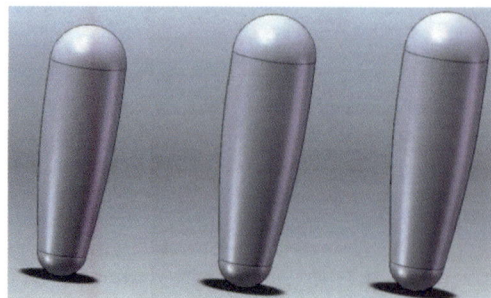

Fig.2. The models of upper arm for different percentiles

The others, such as forearms, thighs, etc, can be done like this. But the trunk is little bit complex, because it is the base parts of model, the others must be connected to it (As showing in Fig.3) [10].

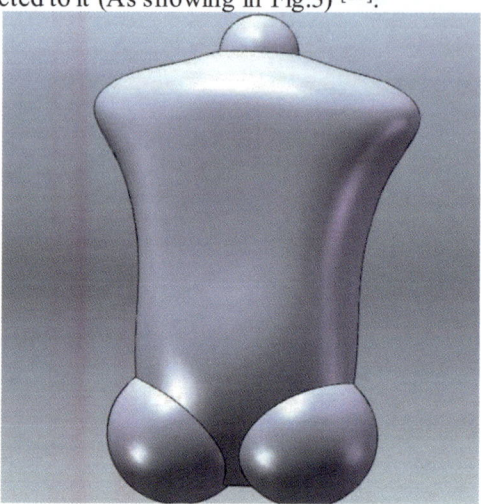

Fig.3. Half ball forms in assembly

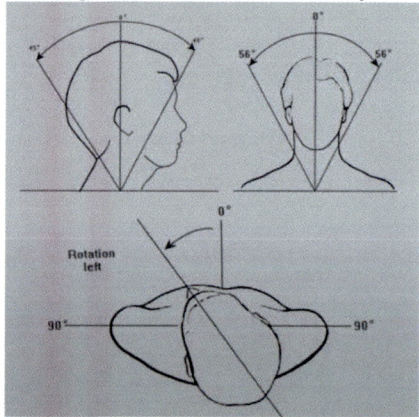

Fig.4. Three freedoms of neck joint

V. ANALYZING THE ASSEMBLY RELATIONSHIP BETWEEN PARTS AND THE ASSEMBLY

A. Analyzing the assembly relationship between parts

In the assembly, parts are connected by joints which are neck joints, shoulder joints, elbow joints, wrist joints, knee joints and ankle joints. In SolidWorks, a joint just can control movement in single plane. For example, the neck joint has three freedoms, which include flexion/extension, lateral bending and axial rotation. We use the head joint, neck joint and a ball to mate to implement, refer to Fig.4 and Fig.5, the ball control flexion/extension, neck joint control lateral bending and head joint control axial rotation [11][12].

According to this, we can deal with the mate at shoulder point, wrist point and the rest. For the one freedom joint, such as elbow point, it is easier to deal with, even without the ball parts.

B. Building the assembly

Loading the parts in turn to a new document, and make sure about the assembly relationship between the near two parts. For example, show you how to assemble the neck. Firstly, load the trunk, and make it fix, then load the ball restricting the movement range of the joint. Load the neck and control the mate relationship between them [13]. Finally, load the head restricting the movement range of the joint, as show in Fig.5.

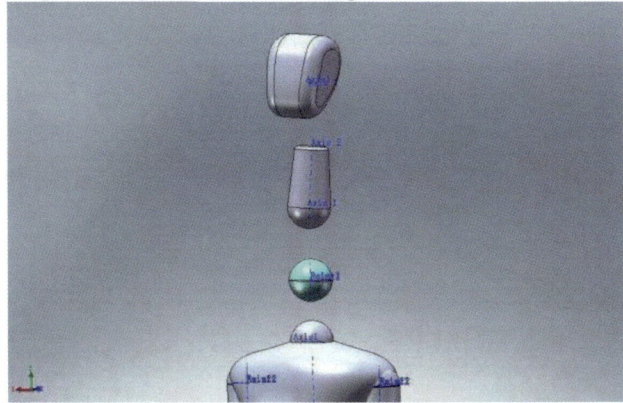

Fig.5. Parts mate in the neck

C. Building Configurations of the assembly

In the assembly, we need to build three configurations to fit three percentile models, and build the reference relations between the different configurations of the same part and the configuration of the assembly. Fig.6 shows the limited the movement range of the joints.

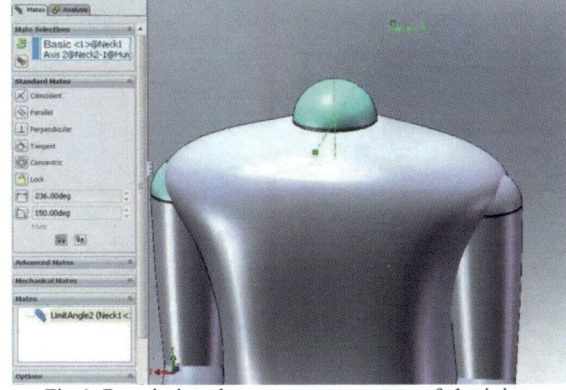

Fig.6. Restricting the movement ranges of the joints

VI. THE APPLICATION OF VIRTUAL HUMAN MODELS IN DIFFERENT PRODUCT DESIGN [14][15]

The application of virtual human models in different kinds of vehicle design is shown in Fig.7, Fig.8, and Fig.9. Among them, Fig.7 is shown the application of virtual human models for interference checking in forklift cab. Fig.8 is shown the application for evaluation of field of vision in fork lift truck. The shadows area showed the normal vision and mesh area showed blind area. Fig.9 is shown the application for evaluation of accessibility in forklift cab.

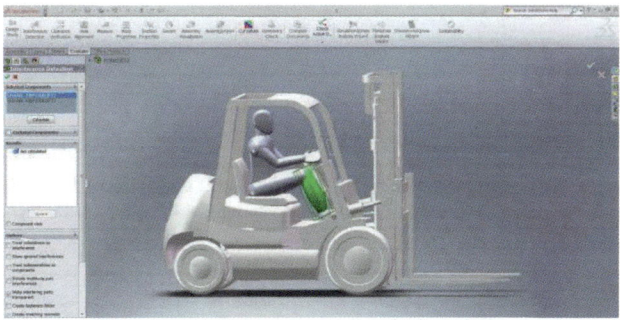

Fig.7. Application of virtual human models for interference checking in forklift cab

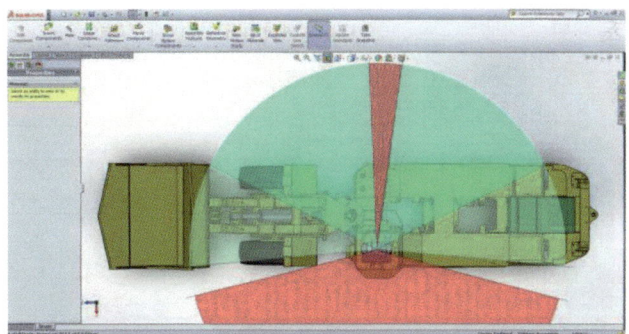

Fig.8. Application for evaluation of field of vision in fork lift truck

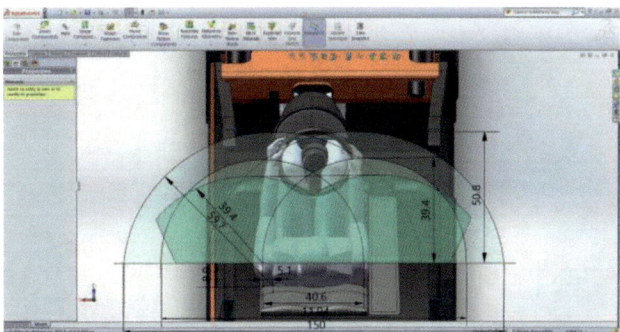

Fig.9. Application for evaluation of accessibility in forklift cab

VII. CONCLUSIONS

Nowadays, virtual ergonomics has got great achievement in industrial design. So the human model plays more and more important role in the design of interface between human and products. It can be used to test if the product is better for people to use. This article has created the parametric human models based on the software SolidWorks. These models have different percentiles; they can switch from one to another rapidly. Not only did the shape simulate from human beings, but also the motion of joints followed the people do. In the future, we have more and more work to do to develop the feature of virtual human models. More importantly, the virtual human models can be used to test the ergonomics characteristics for vehicle design effectively and economically.

REFERENCES

[1] Y.L. Ding: Ergonomics (Beijing Institute of Technology Press, 2000: 20-25) (In Chinese)

[2] Y. Cao: Proficient posts of SolidWorks 2007 (Chemical Industry Press, 2008: 338-339) (In Chinese)

[3] L.J. Wang and X.G. Yuan: Computer Applications and Researches, 2005:194-195 (In Chinese)

[4]   Y. Li.: Computer Applications and Software, vol. 25, 2008:78-79 (In Chinese)

[5]   M.Ji and CH.Q. Zeng: Mechanical Engineer, 2009:43-45 (In Chinese)

[6]   S.F. Gao and CH.L. Zhang: Machinery Design & Manufacture, 2006:134-135 (In Chinese)

[7]   Alvin R. Tilley: The measure of man & woman: human factors in design (Tianjin University Press, 2008: 11-17)

[8]   J.D. Ren and Z.J. Fan: Automotive Engineering, vol.28, 2006:647-651 (In Chinese)

[9]   J.CH. Wang: Ergonomics in products design (Chemical Industry Press, 2004:104-118) (In Chinese)

[10] CH.R. Liu: Applications of ergonomics (Shanghai: People's Art Press, 2004:115-125) (In Chinese)

[11] J.F. Huang, W. Tao, G. Zhao and P. Li: Standard for Applications of Virtual Reality in Man-machine Engineering (Chinese Journal of Ship Research, vol.3, no.6, Dec. 2008.) (In Chinese)

[12] S. Julier, J. Uhlmann and H. f. Durrant-Whyte: A new method for the nonlinear transformation of means and covariance in filters and estimators (IEEE Trans a C, 200, 45(3):477-482)

[13] M. Xu and S.Q. Sun: Progress of Research on Computer-Aided Ergonomics (Journal of Computer-Aided Design & Computer Graphics, vol.16, no.11, Nov. 2004: 1469-1474) (In Chinese)

[14] M. Grujicic, B. Pandurangan, X. Xie, A.K. Gramopadhye, D. Wagner, and M. Ozen: Musculoskeletal Computational Analysis of the Influence of Car-seat Design/Adjustments on Long-distance Driving Fatigue. (International Journal of Industrial Ergonomics vol.40 2010: 345–355)

[15] K.R. Wang and R.J. Ma: Virtual Reality Technology and Its Application in Agricultural Machinery Design. (Journal of System Simulation vol.18, Suppl. 2 Aug. 2006: 500–503) (In Chinese)

Intelligent Manufacturing Enterprises' Efficiency Evaluation of Technological Innovation: Research Based on DEA

Feng Liu, Jian Ning*

Institute of Industry Information, School of Economics, Hefei University of Technology, Hefei, Anhui, China
(ning2013@mail.hfut.edu.cn)

[1] *Abstract* - **In this paper, we utilize Chinese intelligent manufacturing enterprises' panel data during 2010-2013 to evaluate technology innovation efficiency. Studies have shown that the average technology innovation efficiency of intelligent manufacturing industry is in the rising stage, but the overall efficiency is still relatively low. Differences in technology innovation efficiency in various fields of intelligent manufacturing industry is relatively obvious. The development of CNC machine tool industry is relatively mature, while the technical development of intelligent manufacturing complete sets of equipment area is still in its infancy.**

Keywords - **CNC machine tools, intelligent manufacturing, innovation performance, robotics**

I. INTRODUCTION

According to the modern economic growth theory, technology progress and innovation are important factors in determining economic growth [1]. Since entering the 21st century, China has gradually become the world's manufacturing plant. In the guidance of the strategy of "secondary innovation", we not only have accumulated a wealth of manufacturing technology, but also accumulated abundant R&D knowledge and skills. With the increasingly fierce international competition and the development of economic and technological globalization, Chinese enterprises have to face the direct competition from global leader companies in foreign market even in the domestic market. If Chinese companies only take the learning style of imitation and reverse engineering to carry out a local search in order to catch up with technology and then achieve the middle of technical chain , they are usually trapped in a "chasing---backward---chasing again---fall behind again" cycle [2], and even fall into the trap of "local lock". You-lun Xiong believes that intelligent manufacturing representatives the dominant trend and inevitable result of manufacturing industry digitization, networking and intelligent, contains a wealth of scientific connotation (artificial intelligence, bio-intelligence, brain science, cognitive science, bionics and materials science), and it becomes high vantage point of high and new technology (internet of things, intelligent software, intelligent design, intelligent control, knowledge base, model libraries, etc.) and brought together a wide range of industrial chains and industrial clusters. It will be an important development direction of the new round of world scientific and technological revolution and the industrial revolution [3]. Xin Xie believes that intelligent manufacturing is in the basic of integration of modern sensor technology, network technology, automation technology, anthropomorphic intelligent technology and other advanced technology, through intelligent perception, human-computer interaction, decision-making and implementation of technology, achieve design process and manufacture process intelligent [4]. Chinese intelligent manufacturing industry is in a weak position in international competition and at the end of the industrial technology chain. Even some are in branched chain or supporting technology chain and assistive technology chain. On the one hand, enterprises are in the face of fierce competition from the global, they want to achieve technological catch-up by the introduction of high and new technology from other enterprises. On the other hand, in the reality background of a huge technology gap between developed countries, enterprises worry facing huge costs and risks, and then they may be caught in a dilemma while they carry out autonomous technological innovation. Technological innovation efficiency research will help us understand more clearly about its process of technological innovation inputs and outputs so that enterprises and the government sectors can make decisions.

Numerous empirical studies have proved that technology innovation is the main factors influencing enterprises' efficiency, so an army of enterprises are paying more attention to efficiency of technology innovation. From the angle of enterprise, innovation brings them competitive edge, prevent potential competitors from entering and seizing the market, thus high entry barriers are created and these enterprises take the lead in accumulating market knowledge, occupy a bigger market share and then get favorable development space. However, research on efficiency of technology innovation from microscopic perspective of enterprises is still scarce. Outstandingly, employing panel data on provincial level from 1985 to 1997, Li-an Zhou and Kai Luo (2005) conduct research on disparities between China's enterprise scale and innovation by adopting dynamic panel model [5]. They found that enterprise's scale bolsters innovation markedly, while non-state-owned enterprises rather than state-owned enterprises serve the positive function of enterprise scale's effects on innovation. Meanwhile, innovation bears significant cumulative effects and economic openness makes for level improvement of innovation activities. Using panel data of China's large and medium-sized manufacturing enterprises, Jefferson et al. (2006) study innovation performance's effects on enterprises. They confirm technology innovation indeed promotes development of enterprises, facilitates new product introduction, and improves enterprise

[1] Supported by "the Fundamental Research Funds for the Central Universities (J2014HGXJ0156)"

efficiency [6]. On the basis of a large number of questionaries, regarding manufacturing enterprises in Jiangsu province as samples, Jie Zhang, Zhi-biao Liu and Jiang-huai Zheng investigate key factors which affecting enterprises' innovation activities from the microscopic behavior level and found that there exists an obvious inverted U shape curve between enterprise scale and innovative input intensity, as well as "threshold effect". Rather than exerting stimulant effects on enterprises' innovation activities, industry build-up effect exerts certain negative effects at the present stage. Export factors forms complex influencing effects on enterprises' innovation activities. Adopting China's manufacturing panel data from 2000 to 2009, Zao Sun and Wei Song estimate effects of enterprises' R&D input on industrial innovative performance and found that in comparison with state-owned enterprise, private enterprises are provided with more strikingly positive correlation between R&D input and industrial innovative performance [7].

Above handful research contribute to our knowledge of influencing factors of Chinese enterprises' innovation activities and industrial innovation efficiency. Unlike above literature research, this paper analyzes the efficiency of technology innovation emphatically from the angle of intelligent manufacturing enterprises and key factors influencing the intelligent manufacturing enterprises' technological innovation efficiency, thus puts forward improvement measures on this basis.

In march of 2012, China issued "the 12th five-year" specialized planning of intelligent manufacturing's technology development, which pointed out that comprehensive research on intelligent manufacturing technology would be the core content of developing high-end equipment manufacturing industry and the necessity of promoting our country's shift from a big manufacturing nation to a great one. It means that technological innovation of intelligent manufacturing, which defines enterprises as the principals to implement technological innovation, has become a significant national strategy. In term of clear policy guidance, the follow-on question is what the current situation of Chinese intelligent manufacturing's overall technological innovation efficiency is. Whether Chinese intelligent manufacturing enterprises is in possession of a strong independent innovation strength or not? These issues will be studied and analyzed empirically.

Covering numerous fields, to have a foothold on high-end level, highlight the key points, focus the limited sight, seize the first chance of strategic development and boost the healthy and rapid development of intelligent manufacturing industry, the intelligent manufacturing lean its policy priorities to cultivation of high-end fields like high precision NC machine tools, 3D printing, and robots, which gain rapid development. According to latest data released by Gardner, the total output value of machine tools among world's 28 major producing countries or regions in 2012 is about $93.21 billion. China is still the world's biggest manufacturer. Japan is in the second place, with Germany on its heels. It indicates that NC machine tool industry possesses a comparatively mature development. Other development areas of intelligent manufacturing can learn from its development model.

II. RESEARCH METHOD

A. Selection of research method

Data Envelopment Analysis (Data Envelopment Analysis, DEA for short), put forward by famous American operational research experts A. Charnes and W. W. Cooper et al. [8], is A blend of mathematics, operational research, mathematical economics and management science knowledge. It's an efficiency evaluation method on the basis of relative efficiency concept. Depending on input and output data of analysis and decision making units, DEA employs mathematical programming (mainly contains: multi-objective programming, linear programming, generalized optimization with cone structure, class infinite programming, stochastic programming) to evaluate the relative efficiency of multiple input and multiple output decision making units [9]. DEA method has been widely used in various fields, its advantage lies in: (1) no need to estimate the production function of input and output, thus avoids error function; (2) no need to worry about dimensional normalization and the determination of index weight, thus ensures objectivity of evaluation; (3) DEA model has strong adaptability of multiple inputs and outputs' complex structure system; (4) it can not only evaluate efficiency value of each decision making unit, but also point out input and output adjustment direction and value of invalid decision making units, that is, how to get the same output with less investment or get more output with the same input [10].

Traditional DEA methods, such as the CCR model and the BBC model, when evaluating efficiency of decision making units, they do not consider the effect of slack variables, may cause deviation of efficiency measure. In this paper, we adopts efficiency evaluation model SBM of non-angular general scale reward to evaluate the innovative efficiency of each enterprise. Specific SBM model is as follows:

$$\min_{\lambda, s^-, s^+} \rho = \frac{1 - \frac{1}{m}\sum_{i=1}^{m} s_i^- / x_{io}}{1 + \frac{1}{s}\sum_{i=1}^{r} s_r^+ / y_{ro}}$$

Subject to
$$x_o = X\lambda + s^-$$
$$y_o = Y\lambda - s^+$$
$$\lambda \geq 0, s^- \geq 0, s^+ \geq 0$$
$$0 \leq \rho \leq 1, \rho \text{ is efficient value}$$

s_i^- reflects relaxation of the input variables, s_r^+ reflects relaxation of the output variable

B. The variables of input and output and the illustrations for data

1) The selection of input and output variables

Most of the literature using the output of patent, new product sales revenue, total revenue of enterprise and other business indicators to measure innovative performance of enterprise. Among many innovative input and output indicators, we choose four inputs and two outputs indicators based on the consistency of data availability and data in all decision-making units.

In the indicators of innovative input, the following indicators are widely used such as R&D investment, net assets invested (represented by NetAssets), R & D personnel investment, has patented inputs (represented by Patinvown). But because of data availability problem, R&D personnel are replaced by corporate employees (represented by Staffs), R & D investment, the net assets invested are in units of million. This article focuses on the relationship between the firm stale and innovative intensity. From the research method, the net assets or employees can be used to measure the scale of business. We try to comprehensively study the inherent relationship between the size of firm and the innovation of intelligent manufacturing efficiency. Existing literatures generally use R & D investment of cooperation or the number of engineers and scientists. Based on the availability of data, this article will use the R & D investment as a variable of input. As an innovation might inspire and spawned a number of new inventions, innovations in the past can permanently play a role in improving the new products or innovating the processes of production, so this article choose the number of patents as variables of input, making the results of experience are more reliable.

This paper choose the quantity of usual approval patents (represented by Patg) and operating incomes (represented by Incomes) to measure the indicators of output of Innovation, the operating income's units are million. As the capability of innovation and the potential signs of innovative economic value, the patents fully reflect the potential of long-term development in enterprise's several input and the potential performance of industrial innovation. It is more important that the objectivity of patent's standard makes the data of patents have good availability and strong comparability, and it can well reflect the dominant performance and the potential performance in enterprise's several input. Considering these above, we use the output of patents as the indicators to measure innovative performance of enterprise.

2) The source of sample data

In this paper, we use intelligent manufacturing enterprise's panel data in China in 2010-2013 to study. We get treated sample included 18 observations of 50 intelligent manufacturing enterprises, relevant statistical results are shown in Table I. These 50 companies are listed in the field of intelligent manufacturing enterprises, mainly engaged in CNC machine tools, robotics, 3D printing, industrial automation and control systems, intelligent control system software and other fields. The data used in this article such as corporation's R & D investment, net assets, number of employees, revenue are all from wind databases, the quantities of enterprise's patents are from the database offered by the Intellectual Property Press in State Intellectual Property Office of P. R. China which obtain by the service of searching for patents. From Table I we can see that during 2010 to 2013, the mean of R & D, Staffs, NetAssets, and Incomes of intelligent manufacturing enterprises increase year by year. Our intelligent manufacturing enterprises are in a state of rapid growth and put increasing emphasis on technological innovation of enterprises.

TABLE I
THE MAIN VARIABLES DESCRIPTIVE STATISTICS

Variables	Observations	Mean	Mean of 2011	Mean of 2012	Mean of 2013
R&D	18	4768.181	3060.287	5036.487	6207.769
Staffs	18	1694.946667	1563.88	1692.36	1828.6
NetAssets	18	105239.6883	96453.15	103089.4928	116176.422
Patinvown	18	79.04	83.92	34.64	118.56
Patg	18	30.89333333	27.84	34.64	30.2
Incomes	18	94521.2444	89891.9734	90423.3422	103248.4176

III. EMPIRICAL ANALYSIS

This section will use the SBM-NonOriented model to evaluate the efficiency of the input and the output in intelligent manufacturing enterprises. The results of technological innovation SBM efficiency values in each enterprise are shown in Table II. From an average score of each enterprise efficiency , the following enterprises get high points: SMTCL, Shanghai Automation Instrumentation Co., Ltd, Aritime, TRUKING, HUAHONGJT; several enterprises get low points: Suzhou Boamax Technologies Group Co., Ltd, SCIYON, Tatwah Smartech, Wisesoft Co., Ltd, CAC-CITC. China's "the 12th Five-Year Plan" determines the implementation of innovation-driven strategy to promote industrial upgrading and transformation through innovation, aiming at stimulating economic development. In this context, the Chinese government strongly supports the development of intelligent manufacturing enterprises, which enables the efficiency of enterprises' technological innovation to improve significantly. The efficiency of technological innovation SBM value increases from 0.347043 in 2011 to 0.553255 in 2012, and 0.671554 in 2013. It is worth noting that enterprises with higher SBM efficiency value of technological innovation belong primarily to fields like industrial automation, intelligent system control and CNC machine tools. This highlights a fact that various fields' development of intelligent manufacturing is unbalanced. In fields of CNC machine tools and industrial automation, there emerge a group of leading enterprises, such as SMTCL, Shanghai Automation Instrumentation Co., Ltd. and Aritime, which are provided with a high efficiency value of technology innovation. These outstanding enterprises basically build their state-level technology center or inherit the scientific payoff that the nation has obtained in its field.

They have strong technical power and fruitful achievements, and take on a number of national key technology research. On the other hand, enterprises with lower SBM efficiency value of technological innovation are new entrants and on a smaller scale. A short time-to-market disables them to get more money from the stock market and then a lack of research and development money leads to their low technical innovation value. As to high-tech enterprises, broader financing channels and more investment should be government's policy advantage to boost their entrance into the market.

TABLE II
SBM ENTERPRISES TECHNOLOGICAL INNOVATION EFFICIENCY VALUES OF 2011-2013

Enterprise	2011	2012	2013	Mean	Rank
SMTCL	1	1	1	1	1
Huazhong Cnc	0.226364	0.308381	0.654028	0.396258	31
HDCNC	0.302875	0.309029	0.276672	0.296192	40
KMTCL	0.140622	0.527022	0.202574	0.290072	41
Ght-china	0.01787	0.952993	1	0.656954	17
Himile Co., Ltd	0.02524	0.346781	0.413917	0.26198	44
DSBJ	0.028057	1	1	0.676019	15
Harbin Boshi Automation Co., Ltd	0.594319	1	1	0.864773	8
Mesnac Co., Ltd	0.23568	0.218829	0.646878	0.367129	32
Shanghai Automation Instrumentation Co., Ltd	1	1	1	1	1
Jiangsu Yawei Machine Tool Co., Ltd	0.165388	0.336647	0.575801	0.359278	33
Aritime	1	1	1	1	1
VMTDF	0.340398	0.833925	0.595317	0.58988	20
SIASUN	0.21915	0.343782	1	0.520977	22
Dalian Zhiyun Automation Co., Ltd	0.491131	0.413382	0.594237	0.499583	24
HANBELL	0.205344	0.232449	0.577536	0.338443	36
Guilin Guanglu Measuring Instrument Co., Ltd	0.363176	0.528416	0.590817	0.494136	25
Masterwork Machinery Co., Ltd	0.445647	0.547095	1	0.664247	16
RILAND	0.143223	0.359319	0.547403	0.349982	35
TONTEC	1	1	0.094141	0.698047	12
JMJJ	0.454579	0.841419	0.59906	0.631686	19
NTDY	0.351276	1	1	0.783759	9
SZSUNWIN	0.095341	1	1	0.698447	11
Suzhou Boamax Technologies Group Co., Ltd	0.151684	0.100129	0.065744	0.105852	50
FINCM	0.076726	0.316085	0.4478	0.280203	43
TRUKING	1	1	1	1	1
Dalian Rubber & Plastics Machinery Co., Ltd	0.728175	1	1	0.909392	7
Aerospace Hi-tech Holding Group Co., Ltd	0.220668	0.80904	1	0.676569	14
CAC-CITC	0.127377	0.11962	0.092376	0.113125	49
CHINARPM	0.115343	0.212444	0.59546	0.307749	39
Shanxi Qinchuan Machinery Development Co., Ltd	0.252344	0.468433	0.550956	0.423911	30
ZYS	1	0.440278	0.491641	0.643973	18
Fujian Haiyuan Automatic Equipments Co., Ltd	0.110338	0.248685	0.401228	0.253417	45
Suzhou Chunxing Precision Mechanical	0.251689	0.368433	0.355042	0.325055	37
WELLTECH	0.167813	0.317482	0.587379	0.357558	34
HUAHONGJT	1	1	1	1	1
Wisesoft Co., Ltd	0.108819	0.22787	0.171682	0.169457	47
IEFOREVER	0.118999	1	1	0.706333	10
Tatwah Smartech	0.086588	0.068758	0.253241	0.136196	48
CSG Smart Science&Technology Co., Ltd	0.043687	1	1	0.681229	13
Guodian Nanjing Automation Co., Ltd	0.928919	0.934546	1	0.954488	6
SF-AUTO	0.273858	0.394718	0.713043	0.46054	28
Iflytek Co., Ltd	0.065059	0.324485	0.977697	0.455747	29
TOPBAND	0.063886	0.522797	1	0.528894	21
Shenzhen Jieshun Science and Technology Industry Co., Ltd	0.10662	0.407459	1	0.504693	23
HODGEN	0.241011	0.033978	0.575593	0.283527	42
SZHITTECH	0.248124	0.191651	0.488662	0.309479	38
CHN-DAS	0.207088	0.337993	0.88701	0.477364	26
SCIYON	0.120593	0.382461	0.16872	0.223925	46
Dalian Sunlight Machinery Co., Ltd	0.691087	0.335938	0.38603	0.471018	27
Mean	0.347043	0.553255	0.671554	0.523951	

TABLE III
EACH VARIABLE'S MEAN VALUE OF RELATED ENTERPRISES IN 2011

Enterprises	R&D	Staffs	Net Assets	Patinvown	Patg	Incomes
mean value of relatively high efficiency enterprises	3698.24	3678.6	55388.08	63.4	62.6	249575.6
mean value of relatively low efficiency enterprises	2922.22	1435.72	33900.39	69.68	69.92	107273.9

TABLE IV
EACH VARIABLE'S MEAN VALUE OF RELATED ENTERPRISES IN 2012

Enterprises	R&D	Staffs	Net Assets	Patinvown	Patg	Incomes
Mean value of relatively high efficiency enterprises	8904.466	3645	64758.14	126	89.8	219081.7
Mean value of relatively low efficiency enterprises	4360.759	1498	45002.87	139.6	102.16	107160.4

TABLE V
EACH VARIABLE'S MEAN VALUE OF RELEVANT ENTERPRISES IN 2013

Enterprises	R&D	Staffs	Net Assets	Patinvown	Patg	Incomes
Mean value of relatively high efficiency enterprises	10140.73	3684.8	90678.29	215.8	53.6	210280.5
Mean value of relatively low efficiency enterprises	5723.595	1571.36	54348.08	241.76	59.52	104755.4

From Table III, IV and V, we can see that five companies, which have higher SBM efficiency in technology innovation than others, perform much better in three aspects, namely, R&D, number of employees and net assets, whose value are significantly higher than those of low efficiency enterprises. However, they are provided with smaller number of patents than the low efficiency enterprises. Enterprises which are higher efficiency in output are significantly more in operating income than lower efficiency enterprises, while approved patents are slightly less than lower efficiency of enterprises.

TABLE VI
REPRESENTATIVE ENTERPRISES' EFFICIENCY VALUE OF SBM TECHNOLOGICAL INNOVATION

Enterprise	2011	2012	2013	Mean
SMTCL	1	1	1	1
Mesnac Co., Ltd	0.235680028	0.218829	0.646878	0.367129232
SIASUN	0.219149602	0.343782	1	0.520977279
Masterwork Machinery Co., Ltd	0.445647324	0.547095	1	0.664247377
CAC-CITC	0.127376946	0.11962	0.092376	0.113124567
ZYS	1	0.440278	0.491641	0.643972885
SF-AUTO	0.27385848	0.394718	0.713043	0.46053958

We can estimate that the R&D investment and net assets have played a decisive role in the enterprises' technology innovation efficiency. We should note that SBM technology innovation efficiencies of Chinese Intelligent manufacturing enterprises are commonly lower. The average SBM technology innovation efficiency is only 0.6716 in 2013, which has a huge upside.

Table VI selects SBM technology innovation efficiency values of representative enterprises in various intelligent manufacturing fields. Thereinto, SMTCL, MESNAC, ROBOT, MKMCHINA, CAC-CITC, ZYS and BEIJING SIFANG are respectively representative enterprises in numerically-controlled machine tool, robot, 3D printing, intelligent manufacturing complete sets of equipment, and the key basic parts and automation fields. We can see from Table VI that the mean efficiency of Chinese representative enterprises of machine tool industry is highest, while that of Chinese representative enterprises of intelligent manufacturing complete sets of equipment is the lowest. There is more rapid development in the field of 3D printing and Robotics, concerned with Chinese policies to promote two industries. On the whole. The development of two fields, intelligent manufacturing complete sets of equipment and intelligent control system, are relatively weak, far behind with other intelligent manufacturing fields.

IV. CONCLUSION

In this paper, we employ DEA model to evaluate technological innovation efficiency of intelligent manufacturing enterprises. The results show that R&D investment, enterprise scale, and the number of existing patents (the cumulative effect of technology) are key factors affecting the efficiency of technological innovation of intelligent manufacturing enterprises in China. Meanwhile, unbalanced development appears in each field of intelligent manufacturing industry in China and the degree of technological development at early stage plays a vital role in its development. Irrational industrial R&D investment, financing difficulties, firm size, business management and other issues all restrict improvement of technological innovation efficiency. Accordingly, we propose constructive countermeasures and suggestions on technology innovation and development of intelligent manufacturing industry:

Firstly, the government should guide the cluster development of intelligent manufacturing industry and foster scale economies in various fields. The most maturely developed fields are CNC machine tools, intelligent control system, and industrial automation in intelligent manufacturing industry. A large number of outstanding enterprises spring up in these areas, which promotes their joint development and reduces the gap between international lead corporations and domestic ones.

Secondly, we should improve the industrial financing mechanisms so as to increase industrial capital investment. Fields of robotics, 3D printing and other intelligent manufacturing fields owned characteristics of high input, high-yield, and high-risk. Industry investment and financing mechanism is relatively simple. Its main channel is to mortgage from banks or financing through the SME board listed. Technology introduction and innovation of the industry can't do without a drive of large amount of capital, which should also presents a rising trend year by year. Shortage of funds has become one major factors restricting development of intelligent manufacturing SMEs. Therefore, we should broaden the financing channels for the industry, vigorously promote the development of securities markets, and promote potential companies to be listed.

Thirdly, we should strengthen human resource management of industrial technology, fostering high-level innovative talents of compounding type. The current social competition in the final analysis is the competition of human resources. Only enterprises with high-level innovative compounding talents can remain invincible. Personnel training and introduction mechanism should be improved. We can ameliorate industry environment from aspects of policy and funding; meanwhile, sound talent incentive mechanism should be established and equity incentive policies should be implemented.

Fourthly, government should strengthen support for companies and reduce unnecessary administrative intervention. Industry associations should cooperated with the government to formulate related industry supporting policies and carry out industrial strategy planning scientifically. Then reduce unnecessary administrative examination and approval so as to enhance productivity of enterprises.

It should be noted that DEA method is a measure of the relative efficiency rather than absolute efficiency. The specific size of the data does not indicate the absolute level of efficiency. In this paper, we selected a time period of data and the results will have a delayed effect. Due to data availability, only selected 50 representative companies are employed here. Therefore, this study can be further continued through increasing the number of foreign companies. In this way can you display the development status of intelligent manufacturing enterprises' technological innovation efficiency in a more detailed way, and the gap between foreign business leaders and domestic ones as well.

REFERENCES

[1] Romer P.M, "Endogenous Technological Chang (Periodical style)," *Journal of Political Economy*, 1990, 98(5).

[2] Shi-song Jiang, Li-min Gong, Jiang Wei, "Latecomer enterprises' capacity chase under the background of transition economy: a co-evolve model—taking Geely Group for example (Periodical style)" (in Chinese), *Management World*, 2011(4)

[3] You-lun Xiong, "Intelligent manufacturing (Periodical style)" (in Chinese), *Science & Technology Review*, 2013, 31(10).

[4] Xin Xie, "Intelligent manufacturing is in the ascendant (Periodical style)" (in Chinese), *Equipment Manufacturing*, 2013.

[5] Li-an Zhou, Kai Luo, "Enterprise size and innovation: from empirical data on a province level (Periodical style)" (in Chinese), *Quarterly Journal of Economics*, 2005, 4(3).

[6] Jefferson, Gary H., Hua-mao Bai, Xiao-jing Guan, and Xiao-yun Yu, "R&D Performance in Chinese Industry (Periodical style)" (in Chinese), *Economics of Innovation and New Technology*, 2006, 15(4).

[7] Zao Sun, Wei Song, "Impact of enterprises' R&D input on industry innovative performance (Periodical style)" (in Chinese), *Quantitative & Technical Economics*, 2012(4).

[8] Charnes A, Cooper W W, Rhodes E, "Measuring the efficiency of efficiency of decision making units (Periodical style)," *European Journal of Operational Research*, 1978, 2: 429-444.

[9] Quan-ling Wei, *Data envelopment analysis* (Book style) (in Chinese), Beijing: Science Press, 2004.

[10] Ying Feng, Jia-yi Teng, "Efficiency evaluation of technological innovation of high-tech industry in Jiangsu province (Periodical style)" (in Chinese), *Science of Science and Management of S&T*, 2010.

Evaluation Research on the Intensive Development of Regional Higher Education Based on Complex System Theory

Yun-ke Sun[1,2,*], Da-cheng Guo[3]

[1]School of Management and Economics, Beijing Institute of Technology, Beijing, China
[2]Development and Planning Office Beijing Municipal Commission of Education, Beijing, China
[3]Institute of Education, Beijing Institute of Technology, Beijing, China
(sunyk_qd@163.com)

Abstract - The intensive development of regional higher education was established based on complex system theory. Evaluate intensive development of regional higher educational based on its development level, coordination degree, sustaining ability by the basic theory of complex system, and establish the conceptual model and evaluation index system. Use the model to evaluate the intensive development of higher education in Beijing from 2000 to 2013. The result is able to reflect the characteristics of intensive development of higher education development in Beijing and provides support for next strategy in intensive development of higher education.

Keywords - Complex systems theory, evaluation, intensive development, regional higher education

I. INTRODUCTION

In recent years, along with the quality as the core strategy to promote intensive development of higher education, higher education has achieved a good and rapid development. Many of the intensive development of higher education research in academic field, but the main stay in the theoretical knowledge and qualitative analysis on the intensive development, the reports in the literature are mostly limited to the concept, content, influence factors, significance and macroeconomic policies, the lack of empirical research on promoting the intensive development of higher education [1-6]. The complex system is composed of many subsystems, the unit system connected closely, forming a network, has the characteristic of openness, dynamic, uncertainty [7-9]. From the view of system theory, higher education to achieve the intensive development needs of talent training, scientific research, social services and cultural inheritance and innovation in all aspects of system function. This paper starts from the theory of complex system and regional higher education, establish the system model of promoting the intensive development of regional higher education, exploring the theory of evaluate intensive development of regional higher education.

II. SYSTEM MODEL

Regional higher education is a complex system, composed of talent training subsystem, scientific research subsystem, social service subsystem, cultural inheritance and innovation subsystem, subsystem is composed of many factors [10]. Each subsystem and factor are mutual connection, mutual influence, promoting the higher education intensive development. The system is also affected by the external environment, the external environment input to the system teachers, students, funding, education facilities and other resources, and the higher education system output talent, scientific research, social services and cultural inheritance and innovation to the outside world. The system model of regional higher education is expressed as Fig.1.

Fig.1. The system model of regional higher education

To achieve the intensive development of regional higher education system, it's needed to ensure each subsystem of the complex system to realize the structure optimization, system stability and dynamic adjustment. The system is reflected by development level (DL), coordination degree (CD) and sustaining ability (SA).DL reflect the development level of the system, is a vector sum of each subsystem, and is used to Evaluate whether the regional higher education development in talents training, scientific research, social services and cultural inheritance and innovation. CD refers to efficiency between the system and subsystems, the factor and external factors, subsystem and subsystem, it can be used to judge whether the system and factors can be balance. SA is to grasp the DL and CD from the time dimension. It can be a judgment and the measure of region's effectiveness in a relatively long period of intensive development. The essence is how about the bearing capacity of environment when regional higher education keeping intensive development [11]. The concept model is expressed as follows [12]:

$$ID = \max f(D_1, D_2, D_3, T, L)$$

$$St : g_i(D_1, D_2, D_3) \le C_i \quad i = 1, 2, ..., 4$$

$$X_i \ge N_i, X_{ij} \ge N_{ij} \quad j = 1, 2, \cdots, n$$

ID as the system of regional higher education intensive development; S_1 as talent training subsystem; S_2 as scientific research subsystem; S_3 as social service subsystem; S_4 as cultural inheritance and innovation subsystem; D_1 as DL, $D_1 = f_1(S_1, S_2, S_3, S_4)$; D_2 as CD, $D_2 = f_2(S_1, S_2, S_3, S_4)$; D_3 as SA, $D_3 = f_3(S_1, S_2, S_3, S_4)$. T as time variable; L as spatial variables; C_i as the bearing capacity of each subsystem; N_i as the threshold value of each subsystem; N_{ij} as the threshold value of each factor.

III. THE SELECTION OF EVALUATION INDEX

According to the complex system theory, the system of regional higher education intensive development is not decided by a system (factors), but by multi factors, multi system results [13]. The collapse of any factor in the system, will affect the ability of regional higher education intensive development. According to the characteristics of intensive development and complex system, the intensive development of regional higher education can be evaluated from talent training, scientific research, social service, cultural inheritance and innovation, so the evaluation index system can be established.

Talent training is the core content of higher education, is the key factor of affecting the higher education intensive development. Scientific research, social service, cultural inheritance and innovation are the role of talent. Through talent training, higher education institutions provide talent support for the higher education function, conveying all kinds of talents for the society.

Scientific research is the driving force of the higher education intensive development, is the support of forming knowledge, updating talents training content, developing new subjects. Index system of scientific research intensive development should be structured from the results and the efficiency of knowledge creation, application and popularization.

TABLE I
EVALUATION INDEX OF HIGHER EDUCATION INTENSIVE DEVELOPMENT

Subsystem	Index
Talent training	Fund income; Proportion of new students; Teacher-student ratio; The number of full-time teachers; Proportion of graduate students; Proportion of senior professional; Total value of fixed assets
Scientific research	R&D staff; Number of patents; R&D efficiency; Grant funding; Published scientific papers; Total amount of patent sale; Award number
Social service	Entrusted research funds; Number of self-study students; scientific research; training; technology transfer contracts; Number of adult college students
Cultural inheritance and innovation	Social sciences research funding; Published papers in social sciences; Full-time teachers of Humanities and Social Sciences; Number of students in social sciences

Social service means higher education should meet the demand of the public and provide better services. It's the external power of higher education, and is the basic path of achieve higher education self-value. In order to reflect the status and changes of social service, the evaluation index should contain descriptive characteristics and changes.

Culture inheritance and innovation plays an important role to promote and enhance talent training, scientific research and social service in the human world outlook, outlook on life and values. It's contributed to the development of higher education in the higher levels.

The evaluation index of higher education intensive development is expressed as Table I.

IV. EVALUATION MODEL

According to the conceptual model, define D_1, D_2, D_3 as DL, CD and SA respectively. Define $0 \le D_1 \le 1$, $0 \le D_2 \le 1$, $0 \le D_3 \le 1$. The intensive development of regional higher education can be counted as follows formula.

$$ID = \sqrt[3]{D_1 \times D_2 \times D_3}$$

A. Model of development level

According to the synergetic, the change of system element has two effects: one is positive effect, another is the negative effect. Define order procedure variable as $u_i (i = 1, 2, \cdots, n)$, X_i as the value of each system element, α_i, β_i as influence boundary of subsystem index. The power function of system elements can be expressed as following [14]:

$$u_i = \begin{cases} 1 & X_i \ge \beta_i \\ X_i - \beta_i / \alpha_i - \beta_i & \alpha_i \le X_i \le \beta_i \\ 0 & X_i \le \alpha_i \end{cases} \quad (1)$$

$$u_i = \begin{cases} 1 & X_i \le \alpha_i \\ \beta_i - X_i / \beta_i - \alpha_i & \alpha_i \le X_i \le \beta_i \\ 0 & X_i \ge \beta_i \end{cases} \quad (2)$$

Type (1) is used to represent the parameters of X_i has a positive effect, type (2) is used to represent the parameters has a negative effect.

Due to the DL of regional higher education is used to evaluate the states of each subsystem. The utility function is defined as the distance from the initial stage 0, the farther the distance, and the greater the level of system development, so the calculation of regional higher education system DL by linear weighted model, such as formula:

$$D_1 = \sum_{i=1}^{n} w_i u_i$$

w_i as the weight of each element.

B. Model of coordination degree

CD can be obtained relative Haming distance:

$$C_{ij} = 1 - \sqrt{\alpha(u_{ij} - \bar{u}_{ij})^2 + \beta(u_{ij} - \bar{u}_{ij})^2}$$

Define C_{ij} as the CD between i and j. a_{ij} as the weight of each subsystem. The calculation of regional higher education system CD D_2 by linear weighted model, such as formula:

$$D_2 = \sum_{i=1}^{m-1} \sum_{j=i+1}^{m} (a_{ij} \times C_{ij})$$

$$\sum_{i=1}^{m-1} \sum_{j=i+1}^{m} a_{ij} = 1$$

C. Model of sustaining ability

Sustaining ability can be understood as the capacity requirements of system continue to future, system meet the minimum conditions to maintain sustained. So can describe any element x as follows: define N as number of range $[0, T]$ simulation, $n(T)$ indicate the number of times that lasts in time T, the number of $x(T) \geq x_0$ in N times' simulation. System duration in T time can be estimated by the relative frequency of continuous frequency, i.e.:

$$\hat{S}(T) = n(T) / N$$

x_0 as the threshold of the system sustaining. According to the Monte Carlo simulation [15], $S(T)$ is approximately obey Bernoulli distribution of random variables, so its estimate of standard deviation:

$$\delta_{\hat{S}} = \sqrt{\hat{S}(T)(1 - \hat{S}(T)) / N}$$
$$= \sqrt{n(T)(N - n(T)) / N^3}$$

w_i as the weight of x. The simulation of $\hat{S}(T)$ results can be obtained, therefore, SA:

$$D_3 = \sum_{i=1}^{n} w_i \hat{S}(T)_i$$

V. EMPIRICAL ANALYSIS

There is a specific application process in Beijing in order to illustrate the above theory analysis method.

Beijing is developed higher education areas in China. Since the enrollment expansion in 1999, the higher education scale has been enlarged. because the number of rapid expansion, the conditions for running the university is difficult to keep up with the pace, such as the shortage of teachers, the quality of employment competition etc.. After 2006, gradually higher education transferred to the road of intensive development, and taking the quality as the core.

The evaluation index data from 2000 to 2013 years are come from "Beijing Education Statistics", "higher school science and technology statistics compilation", "China Education Statistical Yearbook", "Beijing Social Sciences Yearbook".

First, evaluate the weight of each index by AHP and entropy information method, and determine the index threshold by experts grading method. Then evaluate DL of Beijing' higher education from 2000 to 2013 years according to the above model as shown in Table II:

TABLE II
2000-2013 DL OF BEIJING HIGHER EDUCATION

Year	2000	2001	2002	2003	2004	2005	2006
DL	0.25	0.24	0.23	0.31	0.34	0.34	0.45
Year	2007	2008	2009	2010	2011	2012	2013
DL	0.50	0.51	0.55	0.63	0.73	0.79	0.85

According to CD model, can evaluate CD of Beijing' higher education from 2000 to 2013 years as shown in Table III:

TABLE III
2000-2013 CD OF BEIJING HIGHER EDUCATION

Year	2000	2001	2002	2003	2004	2005	2006
CD	0.34	0.38	0.37	0.39	0.40	0.43	0.46
Year	2007	2008	2009	2010	2011	2012	2013
CD	0.53	0.56	0.58	0.64	0.66	0.69	0.72

Then the SA of Beijing' higher education from 2000 to 2013 years can be evaluated according to the above model as shown in Table IV:

TABLE IV
2000-2013 SA OF BEIJING HIGHER EDUCATION

Year	2000	2001	2002	2003	2004	2005	2006
SA	0.41	0.39	0.37	0.39	0.40	0.41	0.47
Year	2007	2008	2009	2010	2011	2012	2013
SA	0.52	0.55	0.57	0.63	0.64	0.71	0.76

Finally, according to the conclusion of DL, CD and the SA, the level of Beijing' higher education intensive development from 2000 to 2013 years can be can evaluated according to the above model as shown in Fig.2.

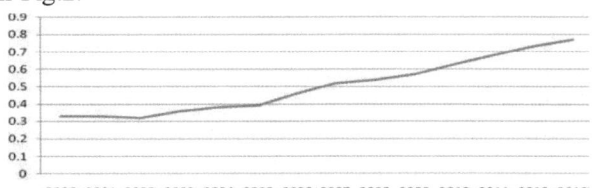

Fig.2. 2000-2013The Change of intensive development of Beijing Higher Education

Thus, higher education intensive development of Beijing in the wave style development state of low from 2000 to 2005. During this period, higher education is in the rapid epitaxial expansion stage of development, the school expanding and the number of students increasing, resulting in lack of education resources, restricts the improvement of the quality of personnel training, scientific research and social service functions are affected, cultural services and innovation are neglected, the intensive development of higher education is constrained. From 2006 to 2013, the state changed the policy of rapid expansion higher education, focusing on control scale of education, improving the quality of higher education, intensive development thought obtains the development and takes highly. Beijing has also introduced a series of policy to promote the intensive development of higher education. Intensive development has a good and rapidly increasing momentum.

VI. CONCLUSION

The research results of this paper have two mainly aspects: in theory, out of the limitations of pure theoretical research and policy of higher education intensive development, makes a systematic analysis of the regional higher education based on the complex system theory, put forward to research higher education intensive development from three aspects of development level, coordination degree and sustaining ability, establish evaluation system and the simulation model of the higher education intensive development. In practice, using historical data of Beijing from 2000 to 2013, calculated the situation of higher education intensive development, the result basically reflects the development situation of Beijing. On this basis, the intensive development of Beijing higher education is evaluated.

As we can see, the intensive development of higher education is a complex system with multi factor, multi level and multi target, including many indexes, variables and threshold, also need to design aspects of the scientific theory and method for quantitative analysis. This paper only made a beginning study on regional higher education intensive development in the perspective of the system. In the future, there needs more research in stochastic process, stochastic analysis, sensitivity of index analysis and random distribution of threshold, in order to make the model more precise.

REFERENCES

[1]   Yu Lan, Liu Li, "Three Ways to Achieve Connotative Development of Higher Education [J]," Research in Educational Development, Vol.7, No.7, pp.21-26, Jul.2013

[2]   Hui-bing Yang, "The problem of extensive development and the strategy of intensive development in higher education [J]," Journal of Hunan Agricultural University (Social Sciences), Vol. 14, No. 4, pp. 9-11, Apr. 2013

[3]   Xiu-ping Zhang, "Research in china's provincial higher education competitiveness [D]," Ph.D. dissertation, Management. Dalian University of Technology. Dalian, China, 2013

[4]   Mei Wu. "Research in evaluation index of college Intensive Development [J]," Journal of Tianjin Academy of Educational Science, Vol. 6, No. 6, pp. 22-25, Dec. 2011

[5]   Guo-dong Shi, "The Connotative Development of University from the Perspective of System Theory [J]."Journal of Changzhou university (social science edition), Vol. 14, No. 2, pp. 1-4, Mar. 2013

[6]   Li-ping Zhu, "On the Essence of the Sustainable Development of Higher Education [J]," Journal of Tianjin Normal University(Social Science), Vol. 195, No. 6, pp. 74-76, Dec. 2007

[7]   Xu-xiao Wu, "Research on the Intensive Development of Regional Central City Based on Complex System Theory [D]," Ph.D. dissertation, Management. Tianjin University. Tianjin, China, 2011.

[8]   Shan-jun Tian, "Construction and Optimization of the Regional Higher Education System: a Realistic Response to the National Education Plan [J]." Journal of Inner Mongolia University (Philosophy and Social Sciences), Vol. 46, No. 1, pp. 97-101, Jan. 2014

[9]   Shu-hua Liu, Xiao-li Yan, "Analyzing Disorder and Order of Higher Education system With Dissipative Structural Theory [J]," Heilongjiang Researches on Higher Education, Vol. 141, No. 1, pp. 5-7, Jan. 2006

[10] Hai-yan Hu, "China's University Development from the Perspective of System Science [J]," Journal of Higher Education, Vol. 33, No.5, pp. 1-7, May. 2012

[11] Luo Huil, Zhao Hai-feng, He Hao, Liu Lul, Gao Xiao-bin [J], Research on evaluation system and calculating methods of regional environmental sustainable degree--A case study of Yulin, Vol. 25, No.2, pp. 167-174, Mar. 2007

[12] Shi-jun Chen, "Research on the Sustainable Development of Regional Agriculture Based on Complex System Theory [D]," Ph.D. dissertation, Management. Tianjin University. Tianjin, China, 2007

[13] Liu Xiao-ping, Tang Yi-ming, Zheng Li-ping, "Survey of Complex System and Complex System Simulation [J]," Journal of System Simulation, Vol. 20, No.23, pp. 6303-6315, Dec. 2008

[14] Peng Fei, Yuan Wei, Hui Zheng-qin, "Research on Exponential Effective Function for Comprehensive Evaluation [J]," Statistical Research, Vol. 24, No.12, pp. 29-34, Dec. 2007

[15] LIU Fenqin, GU Peiliang, "tochastic Analysis Method on Defining Sustainability for Agricultural System [J]," Journal of Systems Science and Information Vol. 3, No.3, pp. 89-91, Mar. 2000

A Study of Coal Collection and Distribution System Based on Critical Path Method

Li-jing Zhang, Yan-rong Pang, Shuang Liu*

Department of Quality Technology Supervision, Hebei University, Baoding, China

(lianlianfushi@126.com)

Abstract - **This paper analyzed the electric energy coal collection and distribution system combined with the theory of CPM (Critical Path Method); model the electric energy coal multimodal transportation system. The objective of the paper is to study the transportation process on the energy coal under the collection and distribution networks of the port, and establish the energy coal port CPM networks, to solve the electric energy coal transportation optimal planning route to improve the efficiency of electric energy coal transportation.**

Keywords - **Collection and distribution system, CPM network, multimodal transport, optimal planning route, port, the electric energy coal**

I. INTRODUCTION

Due to China's energy structure and coal resources distribution, coal transportation becomes one of the most important determinants on healthy development of national economy, steady production behavior of enterprises and improvement of social stability. However, in order to accomplish coal transportation annual supply task the special pattern of transportation (such as "western coal shipping to the east" and "northern coal shipping to the south through port") has to be overcome [1-3]. To solve such problem under the social context of coal's deficiency supply chain which causes an increasing gap between suppliers and demanders (e.g. electric power corporations), how to increase the efficiency of coal transportation is becoming crucial. Take the electric energy industry as an example. In the field of electric energy, the corporations have to effectively utilize their resources to reduce the cost [4-6]. Further, they could also reduce the cost by planning an operational transportation scheme which can rationally increase the capacity and also establish a scientific coal allocation. This is a systematic and complex project. The purpose of this paper is to research on such questions. Using the theory of graph algorithm, combined with collection and distribution system, we model the electric energy coal transportation system. To create the electric energy industry coal transportation optimal planning model.

Coal transportation like freight commodity based models and trip based models as freight planning models can be classified (Holguin-Veras and Thorson, 2000). The commodity-based model estimates the freight tonnage production and attraction in each zone and estimates the tonnage flow between origin and destination pairs; coal transportation is of this type.

In literature, Hu (2005) finds a railway bottleneck when China uses coal from southern Mongolia's; Zhao and Yu (2007) analyze the road, railway and water

transportation of coal with respect to the distribution of China's coal resources, emphasizing that in the long run, the pipeline transportation of coal should be introduced on a large scale. However, Mou (2009) points out that the pipeline transportation of coal faces too many difficulties in China, most of which are policy related. Cheng et al. (2008) study China's inter- provincial coal flow and the driving force behind it, emphasizing that perfect transportation can facilitate coal regional flow. Li (2008) suggests that China construct strategic coal reserves in light of the long transportation distances and the possibility of natural disasters interrupting coal transportation [7-10].

This paper studies the China's coal flows and further considers future shifts in the coal supply zone and their influence on coal transportation arteries. The conclusion has practical value for the management of China's coal transportation.

II. METHODOLOGY

The critical path method (CPM) is an algorithm for scheduling a set of project activities. The related concept as followed.

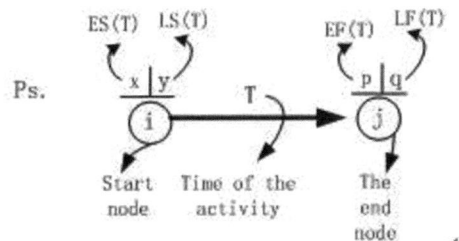

Fig.1. Concept of CPM representation

Professor Qi using the CPM network representation to said GPRs networks with minimum and maximum time restriction transform ideas. From the Fig.1, auxiliary condition with an arrow line said, if two working procedure i to j there have time constraints between. We can see the earliest process A representative start time x. The latest process A representative the start time y, the started node code is i, x and y according to downstream arrow calculation. x means estimated time, y means last start time, i means start node, T means time of the activity, p means estimated finished time, q means last finished time and j means the end node. The earliest finish time of the process A is p, q representative of the last finish time of process A, j is the end nodes code，p and q according to reverse the arrow calculation.

III. THE CASE STUDY

A. Problem statement

The coal resource distributes geographically unequally. China's coal resource locates mainly at the north and north-west areas. According to data from Ministry of Land and Resources (2008), the five provinces, Neimeng, Shanxi, Xinjiang, Shaanxi and Guizhou, have recoverable coal deposits 1014.8 billion tons, accounting 81.4% of total national reserve, and the total production of these five provinces is 1.57 billion tons, accounting 56% of national total production. Yet five key coastal provinces including Shandong, Jiangsu, Hebei, Guangdong and Zhejiang have a total supply gap of 0.8 billion tons. The gap must be met by coal mainly transported from the north and north- west areas. The electricity scarcity suffered by Yangtze River Triangle area during summer of 2011 is mainly caused by insufficient coal transportation capacity in the north China. Coal transportation has become the bottleneck of China's economy.

Because near sea transportation has a special status in China's coal transportation, we treat the ports handling the coal by the logic shown in Fig.2. Suppose i is the province that the coal is being transported out of and j is a province with a port that the coal is being transported to. We then have several options when transporting coal from i to j; transport it directly from i to j by railway or transport it from i to port A or port B by railway then by cargo ship to j. The price of sea transport is only about 1/3 of that by rail, thus the final choice of the route depends on the total transportation costs.

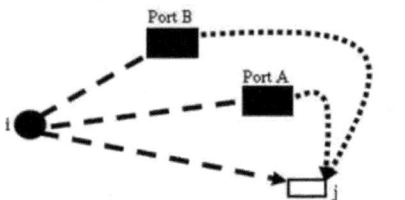

Fig.2. The path of coal transportation

B. Data

The data used in this paper study come from the following sources: data on coal production and consumption were obtained from the China Energy Statistica Year book 2009; and actual coal transportation data for comparison were obtained from Yearbook of China Transportation and Communication 2009.

C. Fundamental assumptions

We also use the notation given by Whitehouse G.E.as following:

T_A =single estimate of mean activity duration time.

ES_A =Earliest (activity) start time.

$EF_A = ES_A + T_A$ =Earliest (activity) finish time.

LS_A =Latest (activity) start time.

$LF_A = LS_A + T_A$ =Latest (activity) finish time.

$TF_A = LF_A - EF_A = LS_A - ES_A$ = Total activity float (slack).

The time that any given task may be delayed before it will affect the project completion time Lemma 1[6].

In the path, the length of the path:

$$a \rightarrow r \rightarrow s \rightarrow t \rightarrow ... \rightarrow u \rightarrow v \rightarrow e \rightarrow b$$

$$\mu(a,b) = ES_b - ES_a - FF_{\mu(a,b)}^{\Delta} \tag{1}$$

$$\mu(a,b) = LF_b - LF_a - FF_{\mu(a,b)}^{\Delta} \tag{2}$$

Theory 1[6]

$$\mu_i^* = ES_i \tag{3}$$

$$\mu_j^{\backslash\backslash} = LF_n - LF_j \tag{4}$$

Theory 2[6]

$$[AB] = \begin{cases} EF_A - LS_B, & EF_A \geqslant LS_B \\ 0 & EF_A < LS_B \end{cases}$$

Theory 3(Barycenter Theory) [6] the one which has the mini center of gravity of the two activities should be done first. The shorter the tardiness of total completion time is the better sequencing. If $C_A \leq C_B$, $A \rightarrow B$, otherwise, $C_B \leq C_A$, $B \rightarrow A$

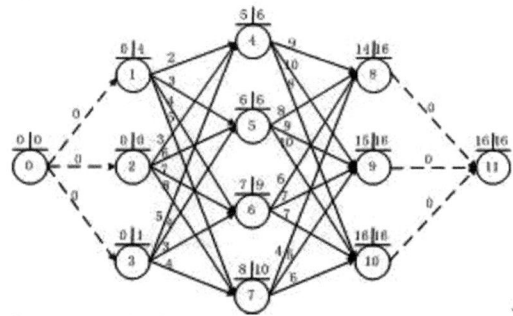

Fig.3. CPM network of coal Multimodal transport with Time value

Step 1: the electric energy coal was transported from Taiyuan to Fujian, from Data we can calculate the total transfer time to Qing Huang-dao port, Tianjin port, Huanghua port and Cao Fei-dian port, as shown in Fig.3.

$$T_4 = T(1-4-8) = 2+9 = 11$$
$$T_5 = T(1-5-8) = 3+8 = 11$$
$$T_6 = T(1-6-8) = 4+6 = 10 \tag{1}$$
$$T_7 = T(1-7-8) = 5+4 = 9$$

Choose T_7, $LS(T_i) = \min\{LS(T_i) \mid i \neq a, 1 \leq i \leq n\}$

From the Fig.4, the electric energy coal was transported from Inner Mongolia erodes to port by train, from Data we can calculate the total transfer time to Qing Huangdao port, Tianjin port, Huanghua port and Cao Fei-dian port.

$$T_{4'} = T(2-4-8) = 3+9 = 12$$
$$T_{5'} = T(2-5-8) = 6+8 = 14$$
$$T_{6'} = T(2-6-8) = 7+6 = 13 \tag{2}$$
$$T_{7'} = T(2-7-8) = 8+4 = 12$$

Choose $T_{7'\&4'}$, $LS(T_{i'}) = \min\{LS(T_{i'}) \mid i \neq a, 1 \leq i \leq n\}$

The electric energy coal was transported from Shenmu to Fujian, from Data we can calculate the total transfer time to Qing Huang-dao port, Tianjin port, Huanghua port and Cao Fei-dian port.

$$T_{4''} = T(3-4-8) = 5+9 = 14$$
$$T_{5''} = T(3-5-8) = 4+8 = 12$$
$$T_{6''} = T(3-6-8) = 3+6 = 9 \tag{3}$$
$$T_{7''} = T(3-7-8) = 4+4 = 8$$

$$\text{Choose } T_{7''}, LS(T_{i'}) = \min\left\{ LS(T_{i'}) \mid i \neq a, 1 \leq i \leq n \right\}$$

Comparing the total transfer time of the electric energy coal (formula 1,2and 3), to see the fattest way to destination Fujian, which is from Shenmu by Cao Fei-dian to Fujian used the shortest time.

Step2: the electric energy coal was transported from Taiyuan to Zhejiang, from Data we can calculate the total transfer time to Qinghuangdao port, Tianjin port, Huanghua port and Caofeidian port,

$$T_4 = T(1-4-9) = 2+10 = 12$$
$$T_5 = T(1-5-9) = 3+9 = 12$$
$$T_6 = T(1-6-9) = 4+7 = 11 \tag{4}$$
$$T_7 = T(1-7-9) = 5+5 = 10$$

$$\text{Choose } T_7, LS(T_i) = \min\left\{ LS(T_i) \mid i \neq a, 1 \leq i \leq n \right\}$$

The electric energy coal was transported from Inner Mongolia erodes to Zhejiang, from Data we can calculate the total transfer time to Qing Huang-dao port, Tianjin port, Huanghua port and Cao Fei-dian port,

$$T_{4'} = T_4 = T(1-4-9) = 2+10 = 12$$
$$T_{5'} = T_5 = T(1-5-9) = 3+9 = 12$$
$$T_{6'} = T_6 = T(1-6-9) = 4+7 = 11 \tag{5}$$
$$T_{7'} = T_7 = T(1-7-9) = 5+5 = 10$$

$$\text{Choose } T_7, LS(T_i) = \min\left\{ LS(T_i) \mid i \neq a, 1 \leq i \leq n \right\}$$

The electric energy coal was transported from Shenmu to Zhejiang, from Data we can calculate the total transfer time to Qing Huang-dao port, Tianjin port, Huanghua port and Cao Fei-dian port,

$$T_{4''} = T(3-4-9) = 5+10 = 15$$
$$T_{5''} = T(3-5-9) = 4+9 = 13$$
$$T_{6''} = T(3-6-9) = 3+7 = 10 \tag{6}$$
$$T_{7''} = T(3-7-9) = 4+5 = 9$$

$$\text{Choose } T_7, LS(T_i) = \min\left\{ LS(T_i) \mid i \neq a, 1 \leq i \leq n \right\}$$

Comparing the total transfer time of the electric energy coal (formula 4,5and 6), to see the fatest way to destination, which is from Shenmu by Cao Fei-dian to Zhejiang used the shortest time.

Stpe3: the electric energy coal was transported from Taiyuan to Guangzhou, from Data we can calculate the total transfer time to Qing Huang-dao port, Tianjin port, Huanghua port and Cao Fei-dian port,

$$T_4 = T(1-4-10) = 2+8 = 10$$
$$T_5 = T(1-5-10) = 3+10 = 13$$
$$T_6 = T(1-6-10) = 4+7 = 11 \tag{7}$$
$$T_7 = T(1-7-10) = 5+6 = 11$$

$$\text{Choose } T_{6\&7}, LS(T_i) = \min\left\{ LS(T_i) \mid i \neq a, 1 \leq i \leq n \right\}$$

The electric energy coal was transported from Inner Mongolia erodes to Guangzhou, from Data we can calculate the total transfer time to Qing Huang-dao port, Tianjin port, Huanghua port and Cao Fei-dian port,

$$T_{4'} = T(2-4-10) = 3+8 = 11$$
$$T_{5'} = T(2-5-10) = 6+10 = 16 \tag{8}$$
$$T_{6'} = T(2-6-10) = 7+7 = 14$$
$$T_{7'} = T(2-7-10) = 8+6 = 14$$

$$\text{Choose } T_{6' \& 7'}, LS(T_i) = \min\left\{ LS(T_i) \mid i \neq a, 1 \leq i \leq n \right\}$$

The electric energy coal was transported from Shenmu to Guangzhou, from Data we can calculate the total transfer time to Qing Huang-dao port, Tianjin port, Huanghua port and Cao Fei-dian port,

$$T_{4''} = T(3-4-10) = 5+8 = 13$$
$$T_{5''} = T(3-5-10) = 4+10 = 14 \tag{9}$$
$$T_{6''} = T(3-6-10) = 3+7 = 10$$
$$T_{7''} = T(3-7-10) = 4+6 = 10$$

$$\text{Choose } T_{6'' \& 7''}, LS(T_i) = \min\left\{ LS(T_i) \mid i \neq a, 1 \leq i \leq n \right\}$$

Comparing the total transfer time of the electric energy coal (formula 7, 8 and 9), to see the fattest way to destination, which is from Shenmu by Cao Fei-dian and Qiang Huang-dao to Guangzhou used the shortest time.

IV. CONCLUSION

Finally, based on the network optimization technology and related coal port research materials, the working operation processes have the maximum and minimum time constraint between each other under the GPR's network in the coal port. The purpose of this paper is to solve the electric energy coal transportation optimal planning route to improve the efficiency of electric energy coal transportation

ACKNOWLEDGMENT

This paper is supported by Hebei province department of soft science research project : "The strategic study on foreign trade enterprises to exploit the international market in Hebei province" (134576131D).

REFERENCES

[1] Mei Mudan, Yuan Shujie, Wei Wei. Prediction of Coal Mine Disastrous Accidents Based on Gray GM (1, 1) Model and its Algorithm, Coal Technology, 2010.

[2] Qi Jian-xun, LiXing Mei, CPM network in long road in order the theorem and the optimization of application, Beijing, 2008, P56-66

[3] Elmaghraby, S.E, The analysis of activity networks under generalized precedence relations, Management Science, 1992, vol.38, No.9, P1245-1263.

[4] Elmaghraby. S E (1964), An algebra for the analysis of generalized activity networks, Management Science, vol.10, P494-514.

[5] Roy, B. Graphes et ordonnancements. Rev. Francaise Recherche Operation, 1962, 25, P323-326.

[6] Park M. Capacity Modeling for Multimodal Freight Transportation Networks [D]. University of California Irvine，2005

[7] Rob Konings. Opportunities to Improve Container Barge Handling in the Port of Rotterdam from a Transport Network Perspective [J]. Transport Geography, 2007, 6(15): P43-54

[8] Shi-rong Zhang, Optimal control of operation efficiency of belt conveyor systems. Applied Energy, 2010, vol. 87, No.6, P1929-1937.

[9] S. Sakellaropoulos, A. P. Chassiakos, Project time-cost analysis under generalized precedence relations, Advances in Engineering Software, vol.35, 2004, P715-724.

[10] Munoz, Juan Carlos, Laval, et al. System Optimum Dynamic Traffic Assignment Graphical Solution Method for a Congested Freeway and one Destination. Transportation Research, Jan2006, vol.40 Issue 1, pp 1-15

Research on Service Mode of Auto Parts Technology Innovation Based on the Industrial Chain

You-yuan Wang[1,*], Xue-qin Xiao[2], Yu Zhou[2]

[1]Institute of Industrial Engineering, Nanchang Hangkong University, Nanchang, China
[2]School of Aeronautical Manufacturing Engineering, Nanchang Hangkong University, Nanchang, China
(yywnc@sina.com)

Abstract – **This paper introduces the basic concept of service pattern and content will be divided into service center mode Service center mode user mode, the introduction of the third-party service platform, research on auto parts technology innovation service mode based on the industrial chain is proposed in this paper. At the same time, construct the service mode of the system structure, expounds the service mode in detail the structure and function of architecture.**

Keywords – **Auto parts, industrial chain, service mode, service platform, technology innovation**

I. INTRODUCTION

The automobile industry is a highly driven, very long chain industry, involving more than 100 the development of related industries, is one of the mainstay of the national economy. And components is the foundation for the development of the auto industry, vehicle technology improvement, cost reduction depends on parts industry level of ascension [1].

However, the present situation of the auto parts industry in China is low melee, high-end, lack of the core components. This issue has seriously hampered the development of China's automobile manufacturing industry.

According to the four dimension contains proposed by Jinming Wu [2] auto parts industry chain, namely, analysis of supply chain, enterprise chain, spatial chain and value chain, it is concluded that lead to the problem is very important one reason is lack of technological innovation ability. And the innovation elements generation, transmission, distribution and conversion cannot do without good service mode [3].

Therefore, it need to study effective service model in order to improve the technical innovation ability of the auto parts industry chain, to make it out of the current industry situation. Scholars at home and abroad were studied

Analyzed such as Xiujie Wang [4] the difficulties encountered in the independent innovation of auto parts industry in China, puts forward the integration of administrative resources, the government-led macro-control service mode.

Irene, Gereffi and Humphrey [5-7] from the industrial perspective, such as suggested with the power of the enterprise itself into an alliance of industry technological innovation service mode to realize the whole industry chain to break.

Weber [8] is analyzed the role of technological innovation service platform for small and medium-sized enterprises in China, think platform service model can efficiently integrate various resources of small and medium-sized enterprises, improve enterprise competitiveness in an all-round way.

Wenxuan Yao [9] empirical analysis on the platform service mode of the meaning of auto parts industry, and using the product platform theory as the guidance, expounds the platform service mode of economy principle.

Above for the auto parts industry, technological innovation is different service models, but only solves the technical feasibility of innovation theory, other aspects are less mentioned. In this regard, the paper summarizes the basis of previous experience, the introduction of a third-party service platform, we propose a chain of auto part based on technical innovation and service model, to address shortage of auto parts industry chain technology innovation capacity issues provide a useful reference.

II. SERVICE MODEL

Service mode refers description [10] of users, service providers, service content, service strategy and other elements of the composition and the basic relationship. The service mode of the third-party service platform is to point to by the service platform will be the combination of the above four basic elements of the relationship between the, is a platform service system for the operation mode of the service activities.

Based on the essence of which is server processing value-added service content, forming the service products, and to provide for the use of user service strategy [11]. According to the four basic elements progression and function in a different way, could be divided into three service mode.

A. Services center mode

Services center model description is derived from services and services-centric service processes, as shown in Fig.1.

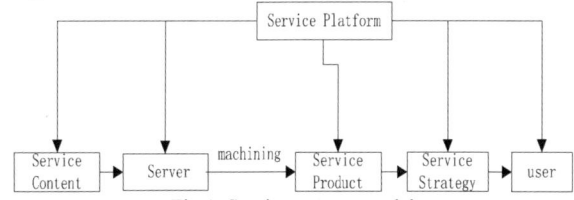

Fig.1. Services center model

Fig.1 can be seen in this service model, in this kind of service mode, service activities is the center of service content, focus on the service content of processing and service of production, occupies an important position in the service content. The service process, the outstanding is the service content itself, ignored the motility of the provider, not targeted improve the technological innovation ability of car parts industry chain.

B. Service hub mode

Service hub model is a departure from the service provider, and service provider-centric service model, as shown in Fig.2.

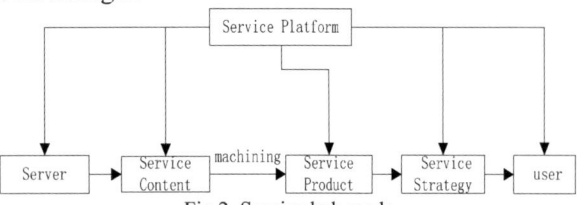

Fig.2. Service hub mode

This service model revolves around the elements of the service are started. Service providers based on their experiences, abilities and preferences process the contents for services, then a strategy of production products and services available to users. Service providers in the course of a dominant position in this service, all service providers in accordance with the wishes of the elements combined mutual relations, while the user is to be dominant.

In this process, notably the service themselves, ignoring the user's initiative, the participation of users is not enough to make their needs can not be fully met in the service process. This service mode is difficult to fundamentally change the difficult status quo of China's auto parts industry chain.

C. User center service model

Users for center service model is based on the user's needs and to meet user needs as the core objective of the service process, shown in Fig.3.

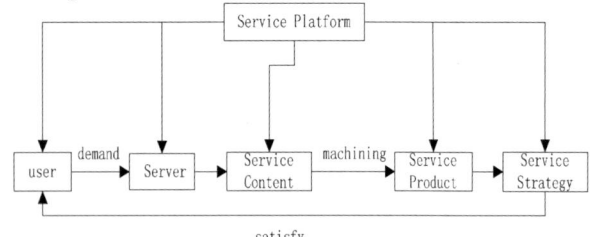

Fig 3. User center service model

Services are based on the user's needs, the production of goods and services provided to the user in some strategies to meet the needs of users. This is due to the need to meet the process and finally needed. In this process, the user needs and the use of services plays an important role, customer demand has become the starting point and destination service activities, the user's become an important guarantee to meet the need.

This mode is fully aware of the impact of personality factors in service activities in the user being and social environmental factors, attention to explore and meet user needs, attention to user choice of service offering, it is

China's auto parts industry chain technology innovation needed mode.

III. BASED ON THE ARCHITECTURE OF THE AUTO PARTS INDUSTRY CHAIN SERVICE MODEL INNOVATION

Based on the auto parts industry chain service model of technological innovation, mainly to address the following issues: ① the presence of a large number of users and service providers on the market, how to help the relationship between supply and demand sides to quickly build services; ② through the effective integration of service provider resources, to provide personalized services to users; ③ through the effective integration of user needs, to service users to provide enough resources; ④ services performed for the whole process to effectively monitor and ensure, improve service quality and value of services as possible.

A. Based auto parts chain service model innovation organization

From the perspective of the overall operation, based on the auto parts industry chain service model innovation respectively, by users, service providers, third-party service platform composed of three parts connected via the Internet, as shown in Fig.4.

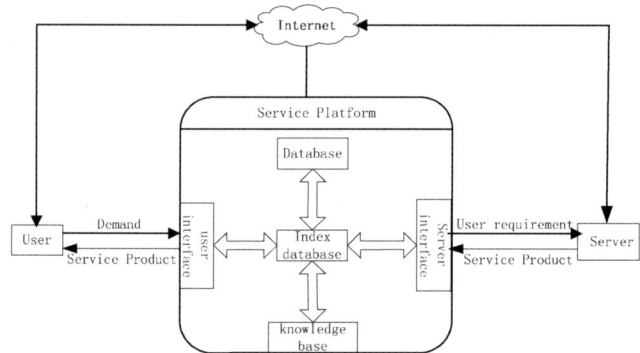

Fig.4. Based auto parts chain service model innovation organization

Third party service platform which is based on service-oriented architecture (SOA) system software architecture. All users and service providers through the service platform for the exchange, the service platform can accept a variety of user needs and recommend services and products, but also can receive services provided by various service offerings and recommend to user needs.

Users can be made to the service platform popular demands, its can also make individual needs, the service platform for processing according to the database, knowledge base, the index library service product needs of users. Services can opt to provide various services to the service platform products, in accordance with the needs of users can also be processed into a specific service content services and products, and provides strategies to use specific services through the service platform users.

Service platform to build a bridge between the user and the service provider service activities and value exchange, which the user needs as the core, to serve those who rely on the basic service content, service policies for the protection and efficient integration of the user and the service provider between service resources, while the QOS service process for effective supervision, so that the interaction between the two fully get added value.

B. Based auto parts chain service model innovation functional architecture

From the perspective of the reality-based auto parts chain technology innovation service modes are supported by technical expertise, industry resources sharing, "political research" collaboration of three parts, as shown in Fig.5.

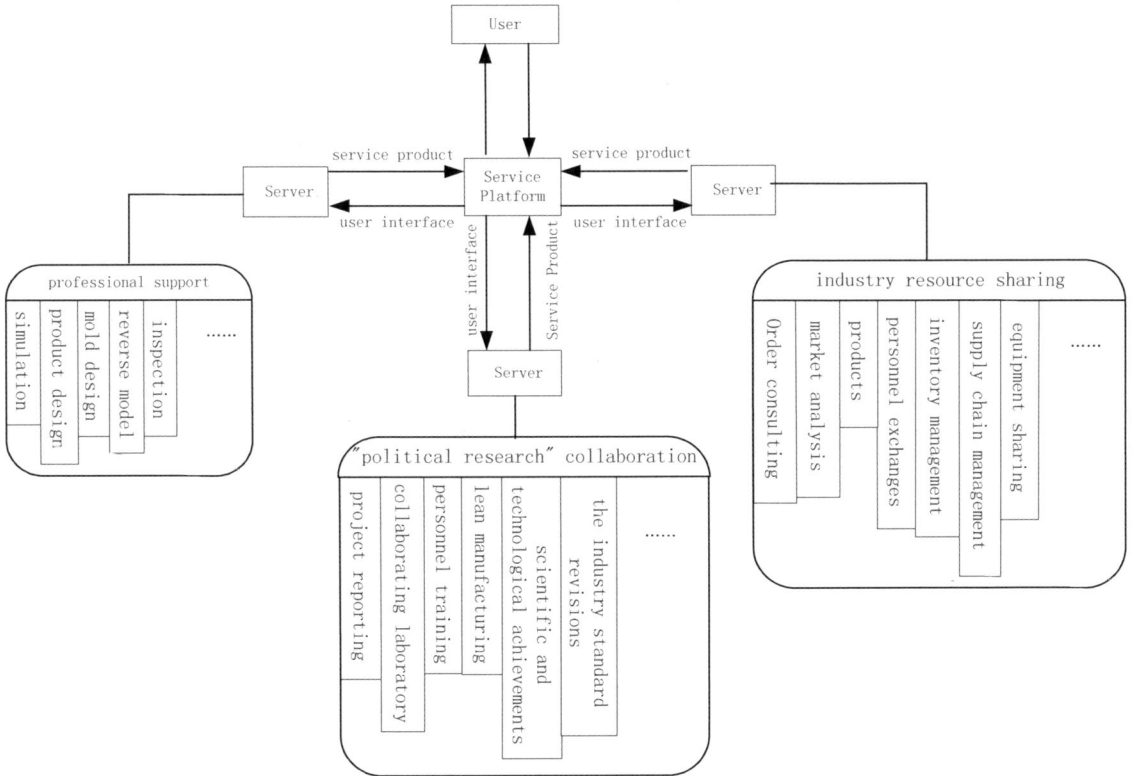

Fig.5. based auto parts chain service model innovation functional architecture

(1) Professional Technical Support: simulation, product design, mold design, reverse model, inspection and so on.

(2) Industry resource sharing: Order consulting, market analysis, products, personnel exchanges, inventory management, supply chain management, equipment sharing.

(3) "political research" Collaboration: project reporting, collaborating laboratory, personnel training, lean manufacturing, scientific and technological achievements, the industry standard revisions.

This service model leverages the various components of the automotive parts industry chain enterprises as the core, through the service platform effectively build up individuals, businesses, government, research institutes, universities and other value chain and achieve win-win cooperation, to enhance the car strategic objectives parts industry chain technology innovation capability.

IV. CONCLUSION

China's auto parts business for the current automobile companies far behind the situation, from the perspective of industry chain, this paper-based auto parts chain service model innovation. This model automobile parts enterprises oriented, by building a third-party service platform, design a reasonable organizational structure and functional architecture, integrating a variety of services resource industry chain, to enhance the technological innovation capability of auto parts to automobile companies to adapt to the trend of rapid development.

ACKNOWLEDGMENT

This work was financially supported by the National Science and Technology Support Program (No.2013BAF02B01), Scientific and Technological Support Projects of Jiangxi Province of China (No.20123BBE50095, No.20141BBE53005).

REFERENCES

[1] Yuncui Wang and Xiaohu Chang, "China auto parts industry development countermeasure research", *Science*

and technology and industry, vol. 12, no. 2, pp. 29–31, 2012.

[2] Jinming Wu and Shao Chang, "Industrial chain formation mechanism research – '4 + 4 + 4' model", *China industrial economy*, no. 4, pp. 36-43, 2006.

[3] Gordijn Jaap and Akkermans Hans, "Designing and evaluating e-business models", *IEEE intelligent Systems,* vol. 16, no. 4, pp. 11-17, 2001.

[4] Xiujie Wang, Xiaoyu Zhao and Derong Si Tu, "Promote the independent innovation capability of auto parts industry in China countermeasures", *Journal of economic enterprises*, no. 8, pp. 100–104, 2011.

[5] Gereffi Gary, Humphrey John and Sturgeon Timothy, "The governance of global value chains", *Review of international political economy,* vol. 12, no. 1, pp. 78-104, 2005.

[6] Humphrey John and Schmitz Hubert, "How does insertion in global value chains affect upgrading in industrial clusters", *Regional studies*, vol. 9, no. 36, pp. 1017-1027, 2002.

[7] Irene Ramos-Vielba, Manuel Fernández-Esquinas and Elena Espinosa-de-los-Monteros, "Measuring university-industry collaboration in a regional innovation system", *Scientometrics*, vol. 84, no. 3, pp. 649-667, 2010.

[8] Weber Lars Henning, "Knowledge transfer towards SEMs in China," Master thesis, Aalborg University & University of Twente, 2010.

[9] Wenxuan Yao, "Auto parts product platform research and case analysis," Master thesis, Shanghai Jiaotong University, 2012.

[10] Jianlong Chen, "Information service pattern research", *Journal of Beijing University: philosophy and social version*, vol. 40, no. 3, pp. 124-132, 2003.

[11] Wilson Tom, "Towards an information management curriculum", *Journal of Information Science*, vol. 12, pp. 203- 209, 1989.

Arterial Velocity Planning Algorithm based on Traffic Signal Information

SONG Wen[1], ZHANG Xin[1,*], TIAN Yi [2], ZHANG Xinn[1], SONG Jianfeng[3]

[1]School of Mechanical Electric and Control Engineering, Beijing Jiaotong University, Beijing, China;
[2]Department of Mechanical Engineering, Academy of Armored Forces Engineering, Beijing, China;
[3]School of Automation Engineering, Tianjin University of Technology and Education, Tianjin, China
(zhangxin@bjtu.edu.cn)

Abstract – **Fuel consumption is directly related to the vehicle velocity curve of arterial road, which includes many acceleration/deceleration patterns and idling times affected by traffic signals. With the development of modern wireless communication technology, the vehicle can get real-time information of the traffic signal information ahead. This paper creates an arterial velocity planning algorithm, which includes optimization calculation of the velocity curve part, calculation of acceleration-cruise-stop part and cruise-stop part, to suit the different conditions of the traffic signal phase and timing information. This algorithm sets running time and the acceleration as the optimization targets to ensure that the vehicle will avoid the sharp acceleration/decelerations and pass through intersections without coming to a stop. Thereby, the fuel consumption of vehicle with arterial velocity planning algorithm could be reduced by more than 14 %.**

Keywords - **Arterial road, fuel consumption, traffic signal information, velocity planning algorithm**

I. INTRODUCTION

Reducing vehicle fuel consumption can effectively reduce the oil consumption of urban transportation. Unlike free driving in the expressway, when a vehicle is driving on the arterial road, it is affected by traffic lights usually, and continuously makes sharp acceleration/deceleration, or idling at the traffic intersections [1-2]. The vehicle fuel consumption of idling at the traffic intersections is about $ 7.8 billion every year in U.S.A [3]. In order to reduce the vehicle fuel consumption on the arterial road, the following two points are needed:

1) Reduce sharp acceleration/deceleration as much as possible;

2) Avoid long idling time at the intersections;

The vehicle eco-driving system has been created to give dynamic speed advice to the driver including acceleration and deceleration, so as to reduce fuel consumption [4-7]. But, the traffic signals situation is not considered in the eco-driving systems, the vehicle eco-driving system could not help the driver to avoid the idling at the interaction. With the development of modern wireless communication technology, the vehicles can obtain real-time information of the traffic signals when approaching a traffic intersection. So the researchers plan velocity curve according to the obtained traffic lights information, and give the optimal velocity curve to driver. In 2009, Sindhura Mandava et al. proposed an arterial road velocity planning algorithm according to the traffic lights information which is 300m in front of the vehicle, and the acceleration is optimized. The sharp acceleration and deceleration can be reduced if the driver drives in

accordance with the velocity curve [4]. For the arterial road velocity planning algorithm that Matthew Barth et al. proposed in 2011, the idling is been reduced to a minimum value via the way of braking ahead of time, and the velocity is been controlled into a certain range, thus greatly improve the vehicle fuel economy [2].

These above papers set the distance of getting the traffic signal information to 300m, and take the short-distance velocity planning calculation. In fact, under the current communications situation, vehicle can get nearby real-time traffic signal information even it is thousand kilometers far away, so the velocity planning is not limited to the distance of signal interaction. This paper creates a long-distance velocity planning algorithm. When the vehicle enters into the current road section, it obtains the real-time traffic signal information of the closest traffic intersection at once, and makes the planning calculation to minimize the acceleration/ deceleration and idling.

II. ANALYSIS OF TRAFFIC SIGNAL AND VEHICLE OPERATION MODES

Assuming the distance between vehicle and the nearby traffic intersection is D_L, the current state of traffic light is green or red, as shown in Fig. 1. The time is $t_k^s, s \in \{g, r\}$, k is the number of light, s is the state of light, g demonstrates the green light, r demonstrates the red light [4].

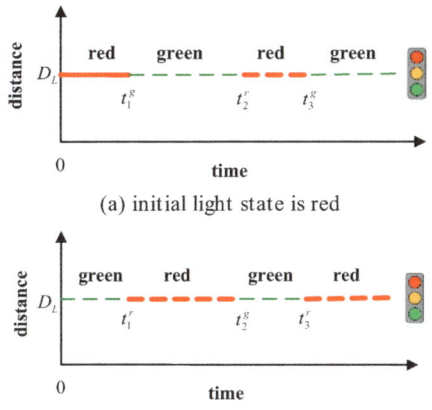

(a) initial light state is red

(b) initial light state is green

Fig.1. Schematic traffic signal information

If the initial velocity is v_0, the vehicle may cross the intersection by following four cases, as shown in Fig.2:

Case 1: The vehicle keeps acceleration and crosses the traffic intersection before the traffic signal turns to red.

Case 2: The vehicle cruises at velocity v_0, makes sharp deceleration when it arrives at the intersection and stops.

Case 3: The vehicle cruises at velocity v_0, coasts to the intersection and stops.

Case 4: The vehicle brakes ahead of time and drives at a relatively low velocity, which makes the vehicle cross the interaction without coming to a stop.

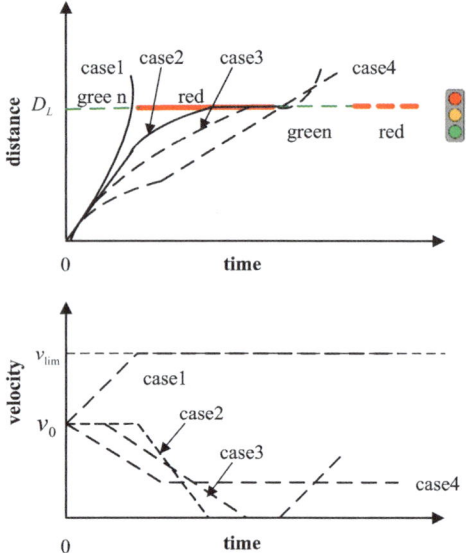

Fig.2. Schematic diagram of vehicle operation modes

The vehicle fuel consumptions of these four operation cases are different from each other. Case 1: Although the vehicle does not stop at the interaction, it needs accelerate sharply to cross the traffic intersection, results in a significant increase of fuel consumption. Case 2: The vehicle also makes sharp deceleration and idling at the traffic light for a long time. The fuel consumption is still high. Case 3: The coasting to intersection can make use of the kinetic energy of the vehicle to reduce the fuel consumption, but the vehicle still needs to idle. The fuel consumption of the restart vehicle is still high. Case 4: The vehicle can maximally avoid idling, and does not need to accelerate at a stop state, thus it has the minimum fuel consumption [2]. But in Case4, the deceleration value and deceleration time must be calculated accurately.

III. ARTERIAL VELOCITY PLANNING ALGORITHM

Firstly, the shortest time of vehicle arrives at the intersection is calculated. The vehicle firstly accelerates to the maximum velocity $v_{\lim}$, and drives at the max velocity to the intersection.

The acceleration time t_a is:

$$t_a = \frac{v_{\lim} - v_0}{a_h} \tag{1}$$

a_h is the maximum of acceleration.

The time of driving at the max velocity $v_{\lim}$ is:

$$t_c = \frac{1}{v_{\lim}} \left[D_i - v_0 \left(\frac{v_{\lim} - v_0}{a_h} \right) - \frac{1}{2} a \left(\frac{v_{\lim} - v_0}{a_h} \right)^2 \right] \tag{2}$$

The shortest time of vehicle gets to the intersection is t_p':

$$t_p' = t_a + t_c \tag{3}$$

The time area of vehicle crossing the intersection is $\begin{bmatrix} t_l & t_h \end{bmatrix}$. When the vehicle arrives at the intersection, the traffic signal is green:

$$\begin{cases} t_l = t_p' \\ t_h = t_k^r \end{cases}, \text{ if } t_{k-1}^g < t_p' < t_k^r \tag{4}$$

When the vehicle arrives at the intersection, the traffic signal is red:

$$\begin{cases} t_l = t_k^g \\ t_h = t_{k+1}^r \end{cases}, \text{ if } t_k^r \le t_p' < t_{k+1}^g \tag{5}$$

A. Optimization calculation of the velocity curve

In order to improve the fuel economy, the vehicle should cross the interaction without coming to a stop. This paper defines that the vehicle only accelerates or brakes once in each road section. So, the vehicle firstly runs at uniform velocity v_0, then accelerates or brakes, finally gets to the interaction at a uniform velocity v_1, $v_1 = v_0 + at_e$. t_e is the acceleration time. The velocity curve is as shown in Fig.3.

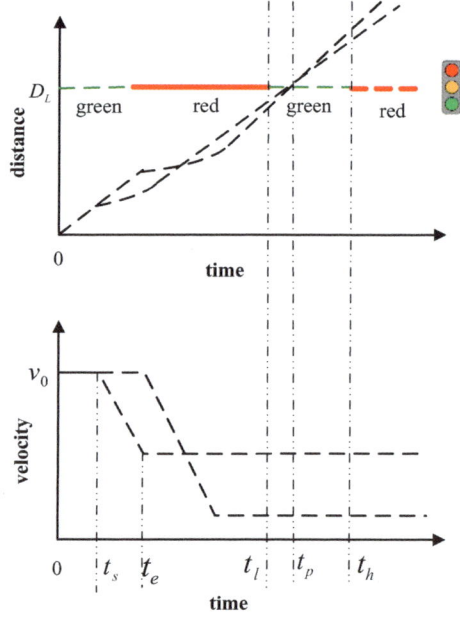

Fig.3. Schematic curves of velocity planning

It can be obtained from Fig.3 that the acceleration value and time are different, even though the time of vehicle crossing the traffic intersection is the same. t_s is the start time of accelerating, t_p is the time of crossing the intersection.

So this paper optimizes the vehicle's acceleration and crossing-time at the same time:

$$\text{Min} \quad \lambda_1 |a| + \lambda_2 t_p \tag{6}$$

Subject to:

1) $v_0 t_s + \left(v_0 t_e + \frac{1}{2} a t_e^2\right) + (v_0 + a t_e)(t_p - t_s - t_e) = D_L$

2) $0 \leq t_s \leq t_p$

3) $0 \leq t_e \leq t_p$

4) $0 \leq t_s + t_e \leq t_p$

5) $t_l \leq t_p \leq t_h$

6) $a_l \leq a \leq a_h$

7) $0 \leq v_0 + a t_e \leq v_{\lim}$

λ_1, λ_2 are the weight coefficients of a and t_p respectively, a_l is the minimum value of acceleration.

B. Calculation of acceleration-cruise-stop

In the optimization calculation of velocity curve, the cruising velocity is not considered. If the initial cruising velocity $v_0 < v_{low}$ and $t_s > t_{max}$, the vehicle should takes the acceleration-cruise-stop mode to avoid cruising at low velocity long time. Firstly, the vehicle accelerates in a' to the maximum velocity $v_{\lim}$, cruises at uniform velocity $v_{\lim}$, and brakes with deceleration a''. The vehicle starts to accelerate when the signal lights turn to green.

The acceleration time t_e' is:

$$t_e' = \frac{v_{\lim} - v_0}{a'} \tag{7}$$

The braking time t_e'' is:

$$t_e'' = \frac{v_{\lim}}{a''} \tag{8}$$

The cruising time is t_a':

$$t_a' = \frac{1}{v_{\lim}}\left[D_i - v_0\left(\frac{v_{\lim} - v_0}{a'}\right) - \frac{1}{2}a'\left(\frac{v_{\lim} - v_0}{a'}\right)^2 - \left(v_{\lim}\frac{v_{\lim}}{a''} - \frac{1}{2}a''\left(\frac{v_{\lim}}{a''}\right)^2\right) \right] \tag{9}$$

The waiting time t_w at the intersection is:

$$t_w = t_l - t_e' - t_e'' - t_a' \tag{10}$$

C. Calculation of cruise-stop

If $v_1 < v_{low}$ and $t_l - t_e - t_s > t_{max}$, the vehicle should takes the cruise-stop mode. The vehicle firstly cruising at uniform velocity v_0 and deceleration with a'' to arrive at the intersection, and start to accelerate when the signal lights turn to green.

The braking time of the vehicle t_e' is:

$$t_e' = \frac{v_0}{a''} \tag{11}$$

The cruising time t_c' of uniform velocity v_0 is:

$$t_c' = \frac{1}{v_0}\left[D_i - \left(v_0 \frac{v_0}{a''} - \frac{1}{2}a''\left(\frac{v_0}{a''}\right)^2\right) \right] \tag{12}$$

The waiting time t_w at the intersection is:

$$t_w = t_l - t_e' - t_c' \tag{13}$$

IV. RESULTS

This paper takes the route that from Zizhuyuan South Road to Huangchenggen Road of bejing in China, which includes 10 intersections, as shown in Fig.4.

Fig.4. Route

In this paper, the test time was April 5, 2014 7:00:00 AM, the vehicle located at the starting point (Zizhuyuan South Road). The initial velocity was 40km/h, the max velocity of the road was 50km/h, the max acceleration was 1.5 m/s^2, and the min deceleration was -2 m/s^2. In the traditional driver-controlled velocity curve calculation, the distance of the drivers get traffic signal information according to the visual and implement acceleration or deceleration operation is 75m [3]. The velocity curves are shown in Fig.5. The fuel consumption is shown in Table I:

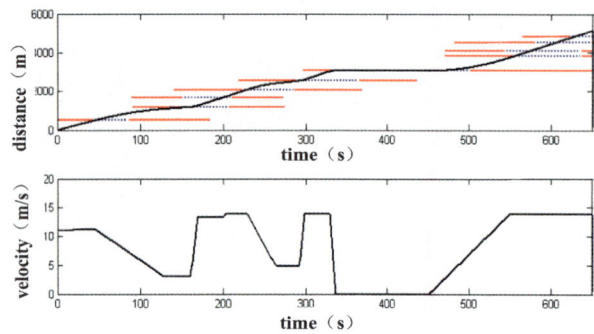

(a) With velocity planning algorithm

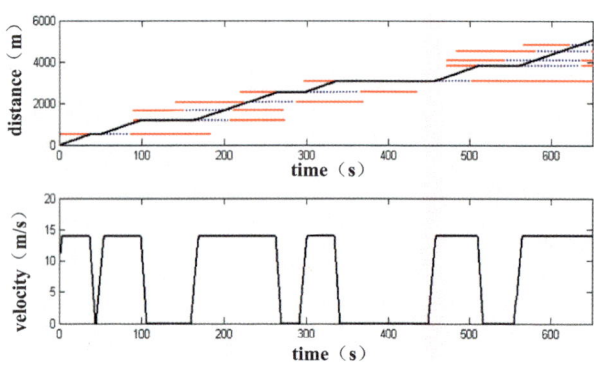

(b) Without velocity planning algorithm
Fig.5. Calculation results of the velocity curves

The vehicle takes 5 stop times when the vehicle takes the traditional driver-controlled mode without velocity planning algorithm, and the idle-running time is 215s, which is shown in Fig.5(b). But, it is shown in Fig.5(a) that in the curve established by the velocity planning algorithm, the vehicle just stops for once in the whole drive process, the idling time is 123s, and acceleration and deceleration rate are also eased.

TABLE I
RESULTS

	fuel consumption (L/100km)	reduce
Without velocity planning algorithm	6.3	/
With velocity planning algorithm	5.4	14.3%

Thereby, the fuel consumption of the vehicle with velocity planning algorithm is reduced by 14.3%.

V. CONCLUSION

The arterial road velocity planning algorithms presented in this paper have following advantages:

(1) By taking advantage of the recent developments in communication between vehicles and traffic lights, the velocity curve can be planned by the velocity planning algorithms. This algorithm ensures the vehicle to avoid sharp acceleration/deceleration and pass through intersections without coming to a stop.

(2) The velocity curve which is generated by the velocity planning algorithms presented in this paper can assist the driver to control the vehicle and greatly improve the fuel economy.

ACKNOWLEDGMENT

This work is supported by Beijing Natural Science Foundation (4122062) and National High Technology Research and Development Program of China (2012AA111106).

REFERENCES

[1] S. Fish, T.B.Savoie. Simulation-based Optimal Sizing of Hybrid Electric Vehicle Components for Specific Combat Missions [C]. IEEE Transaction on Magnetics, 2001 vol. 37: 485-488.
[2] Matthew Barth, Sindhura Mandava, et. al. Dynamic ECO-Driving for Arterial Corridors [C]// IEEE Forum on Integrated and Sustainable Transportation Systems Vienna, Austria, June 29 - July 1, 2011: 182-188
[3] Tae-Kyung Lee, Zevi Bareket, et al, Stochastic Modeling for Studies of Real-World PHEV Usage: Driving Schedule and Daily Temporal Distributions [C], Transactions On Vehicular Technology, 2012, 61(4), 1493-1502
[4] Sindhura Mandava, Kanok Boriboonsomsin, et. al. Arterial Velocity Planning based on Traffic Signal Information under Light Traffic Conditions[C]// Proceedings of the 12th International IEEE Conference on Intelligent Transportation Systems, St. Louis, MO USA, October, 3-7,2009:160-165
[5] Mohamad Abdul-Hak, Nizar Al-Holou, ITS Based Predictive Intelligent Battery Management System for Plug-In Hybrid and Electric Vehicles [C]// IEEE, 2009: 138-144
[6] Hesham Rakha, Raj Kishore Kamalanathsharma, Eco-driving at signalized intersections using V2I communication [C]// 2011 14th International IEEE Conference on Intelligent Transportation Systems, 2011
[7] Dening NIU, Jian SUN, Eco-Driving Versus Green Wave Speed Guidance for Signalized Highway Traffic: A Multi-Vehicle Driving Simulator Study [J], Social and Behavioral Sciences 2013: 1079 – 1090

A Centralized Scheduling Approach to Multi-Agent Coordination

Wei NIU[1], Ying-qiu XU[2,*], Jie WU[1], Ying-zi TAN[3]

[1] Department of Industrial Engineering, Southeast University, Nanjing, China
[2] RoboCup Research Group & School of Mechanical Engineering, Southeast University, Nanjing, China
[3] RoboCup Research Group & School of Automation, Southeast University, Nanjing, China
(101000232@seu.edu.cn)

Abstract - **The coordination in multi-agent system is the key to the global optimum and centralized coordination is regarded as the most natural and effective way to organized work among agents. In this paper we propose a centralized scheduling approach to manipulate centralized coordination among heterogeneous agents. The main contribution is that center agent, as information collector, processer and resource scheduler in this study, enacts centralized scheduling to run well. And clustering analysis based on artificial immune algorithm is applied to process information, moreover a series of schemes are suggested to ensure smooth scheduling. The effectiveness of the proposed method is shown through simulation results.**

Keywords - **Artificial immune algorithm, center agent, centralized scheduling, clustering analysis, multi-agent coordination**

I. INTRODUCTION

Multi-agent System is an important research aspect in the field of distributed artificial intelligence, in which a basic and important issue is to deal with coordination. Efficient coordination prevents agents form accidentally interfering with each other's sub goals while attempting to achieve a common goal. Nowadays two problems have been identified in executing multi-agent coordination, namely incomplete recognition of environment and lack of unified control.

In order to solve the problems mentioned above, this study investigated one typical multi-agent system - RoboCup Rescue Simulation System (RCRSS). Most current researches regarding multi-agent coordination are based upon RCRSS, and the numerous developed methods can be classified into three categories: Decentralized mutual adjustment, centralized direct supervision, and environment partitioning [1] [2] [3] [4]. Some outstanding research achievements include: Combinatorial auctions were used in [5] to achieve optimal task allocation, but the model also required large computational power and message bandwidth. MRL used BELBIC algorithm, which is a kind of motion learning method most effective in helping PF clearing blockades [6]. Q learning method is used by BonabRescue to train agents learn to act right in the right time and/or condition without communication with central control [7]. This method may make the agents more independent, but the Q-learning method requires a long time to start converging to good results. Inspired by the Partial Global Planning approach, LTI's hybrid task allocation approach encourages the agents to exchange information in order to reach a common conclusion [8]. RoboAKUT, however proposed Market-driven methods, focusing on the maximization of

the overall gain of a group of robots by coordination, collaboration and competition among them [9]. The decentralized approach is more flexible but not always preferable.

The main concern in the above mentioned research is that the active and critical role of center agent was demoted and sometimes ignored. Actually, ignoring the decision-making abilities of center agent, which is the same as platoon agents, equals to wasting resource. In the light of this, the study has come up with a method depending on center agent scheduling. Center agents collect information through communication between agents, with emphasis in civilian location information, because rescuing civilians is the main focus of simulation. Then clustering analysis is used to process the collected information. The cluster area is of heavy population, around which centralized scheduling will be carried out. There are two stages during the scheduling: one is to choose a cluster and the other is to choose a specific task. Therefore, center agent will play the role of information collector, processer and resource scheduler.

The rest of this paper is structured as follows. The next section introduces the RCRSS briefly. Section III talks about the implementation of clustering and scheduling. This is followed by the test results. And the last part is discussion and conclusion.

II. BRIEF INTRODUCTION TO RCRSS

RoboCup Rescue Agent Simulation provides a regional rescue under the environment of mass disaster [10]. Disaster may be defined as a crisis situation causing widespread damage which far exceeds our ability to recover. Its management is normally critical since it involves a very large number of heterogeneous entities in a complex and dynamic environment, where the information is incomplete and uncertain, and decision must be taken in a timely fashion. In those situations, coordination and cooperation play essential roles in assistance provisioning [11].

RCRSS provides a platform for disaster management where heterogeneous field (platoon) agents (police, fire brigades, and ambulances) cooperate with each other to deal with a simulated disaster scenario. Police force (PF) have to clear road blockades to provide access to the disaster sites, ambulance team (AT) have to rescue civilians, and fire brigade (FB) have to control the spread of fire and extinguish it. The simulator also provides center agents, a Police Office, a Fire Station, and an Ambulance Center. And center agents cannot interact directly with the world, but only communicates with platoon agents of its kind and with other centers [12]. Fig. 1 shows the mode of communication among agents.

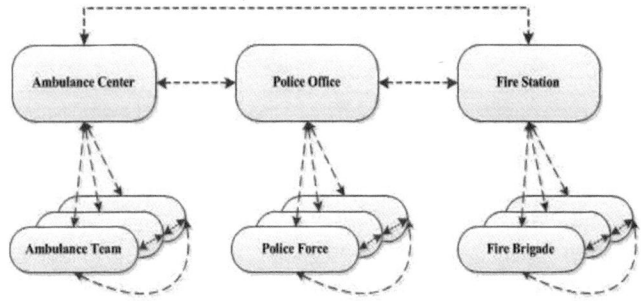

Fig.1. The mode of communication

So RCRSS is a partial global system. It is partial because the agents do not have all the environmental information available to make the best decision. On the other hand, it is global because the agents can exchange and obtain information through the centers which consolidates the environment information from all the agents [13].

III. METHODOLOGY

Based on the statements mentioned above, a variety of raw information floods to center agent. However, the disorganized information may distract decision making. So in order to find out densely populated area, clustering analysis is adopted to process the information of civilians' location for the key task-rescuing civilians. Then coordination will be made in these areas closely followed by purposefully. In this way, disorderly disintegration of all the agents can be avoided. The clustering analysis in this study is based on artificial immune algorithm, because compared to other clustering analysis methods, there are two advantages in this method: In this algorithm the search range is a wide and the convergence rate is moderate. On the other hand, memory cells can store excellent solutions, which can be used as initial antibodies in the next round of clustering. By this means, the whole process becomes dynamic.

A. Clustering analysis based on improved artificial immune algorithm

Cluster analysis or clustering is the task of grouping a set of objects in such a way that objects in the same group (called a cluster) are more similar (in some sense or another) to each other than to those in other groups (clusters). It is a main task of exploratory data mining.

In computer science, artificial immune systems (AIS) belong to computationally intelligent systems inspired by the principles and processes of the vertebrate immune system. The algorithms typically exploit the immune system's characteristics of learning and memory to solve a problem. The field of AIS is concerned with abstracting the structure and function of the immune system to computational systems, and investigating the application

of these systems towards solving computational problems from mathematics, engineering, and information technology[14][15].

The method this paper puts up with is described in the following parts.

1) Antigen recognition and antibody initialization

Antigens in the immune system correspond to the problems to be solved. In N data clustering analysis, first converts these data into N p dimensional vector sets $X=\{Ag_1, Ag_2, ..., Ag_n\}$ according to their attributes. In this paper these data correspond to the antigens Ag_i $(i \in [1, N])$ in the immune system. The purpose of this algorithm is to classify the N antigens.

Make a copy of antigens as the initial antibodies $Ab_k (k \in [1, M])$, here vectors in dimension p are seen as their coordinates, expressed in Loc, namely $Ag_i \Leftrightarrow Loc_i = \{l_i[1], l_i[2], ..., l_i[p]\}$. As a result, the structure of the antibody can be composed of antibody coordinate Loc_k and a set of its captured antigens AgS_k. C_k is the structure to describe cluster which is expressed by (1). k represents the number of clusters, Loc_k is the center of the class k.

$$C_k = < Loc_k, AgS_k >. \qquad (1)$$

2) Affinity calculation and antigen capture

Immune system produces a variety of antibodies, and the bonding strength between antigen and antibody is evaluated through the affinity. Generally the formula for computing the affinity is seen in (2).

$$(Ag)_k = 1/(1 + t_k). \qquad (2)$$

Among them, $(Ag)_k$ is the affinity between an antigen and antibody Ab_k whose value is between 0 and 1, and t_k is the combining strength. When $(Ag)_k = 0$, it suggests that the antibody and the antigen are ideal combined. t_k can be described as "distance", and the Euclidean distance $D_{ik}(Ab_k, Ag_i)$ is generally used to measure the distance between Ag_i and Ab_k:

$$D_{ik}(Ab_k, Ag_i) = \sqrt{\sum_{m=1}^{L}(Ab_{km} - Ag_{im})^2}. \qquad (3)$$

For each antigen Ag_i, evaluate its affinity with every antibody Ab_k, and $As_i = \{A_{ik} | k = 1, 2, ..., M\}$ is the set of affinities. Find the antibody Ab_{kmax} that corresponds to the largest value in As_i, and then add Ag_i into AgS_k, which contains antigens captured by Ab_{kmax}. Keep operating this process until every antigen is evaluated. Then calculate Loc_k', the centric of all the captured antigens according to AgS_k. At the end of this process, Ab_k can be expressed as $C_k = < Loc_k', AgS_k >$.

3) Cluster aggregation

The aim of cluster aggregation is to merge similar antibodies and recapture antigens.

If the captured antigens of an antibody are less than a certain value, then the antibody is relatively isolated and needs to be deleted.

4) Cloning and variation

Calculate the distance between all antibodies and choose the nearest $M/2$ antibodies for cloning and variation using (4), among which α is the variation rate.

$$Loc_{k1} = Loc_{k1} - \alpha(Loc_{k1} - Loc_{k2}). \qquad (4)$$

Then recalculate AgS_{k1} and AgS_{k2}. Add antibodies which are not selected but in variation into the memory cells. Make dynamic adjustment to α according to (5).

$$a = 0.2/\lceil gen/10 \rfloor. \qquad (5)$$

Among which $\lceil \rfloor$ means ceil. Obviously with the increase of generation, variation rate declines. So at the preliminary stage of clustering, the mutation rate is higher, the search range is bigger so that the antibodies can capture antigen in a wider range; when algorithm evolutes to a certain generation, the search range becomes narrow and the clusters are relatively stable.

Keep operating this process until comes to the stopping criterion.

During the simulation, civilians' locations are taken as antigens. The stable clusters are the outcome of the algorithm.

B. Center agent scheduling

Scheduling is a process of distributing tasks to agents, the main function of which is to maximize the efficiency of this task allocation and reduce its costs. Besides, task allocation plays a vital role in implementing coordination since the decomposition of objective into tasks is the most nature way to organize work among agents [14]. Thus, coordination can be brought out as long as effective assessment schemes for scheduling are made. Therefore scheduling, which depends on center, proposed in this paper aims to coordinate agents. There are two phases to implement scheduling: first, important clusters stand out after cluster assessed by a weighted model, and then specific tasks are clearly identified in the highlighted cluster for each related agent in the same manner.

The weighed model for cluster is mainly determined by the following factors: the size of area, population and distance to fire spots. The larger area, denser population and closer distance to fire spots all contribute to the importance of the cluster. In addition, there are some other influencing factors, including: the distance between clusters and agents, the distance between refuge and clusters etc. Hence, the importance of cluster is calculated as follows

$$Importance\ of\ Cluster =$$
$$\frac{10\left(\frac{density}{min\ density}\right)*area\ of\ cluster\ *x}{\frac{disToFire}{10}+distance+disToRefuge}. \qquad (6)$$

$$x = \begin{cases} 10 & if\ there\ is\ a\ fire\ spot\ in\ the\ cluster \\ 1 & otherwise \end{cases}$$

$$density = \frac{number\ of\ civilians\ in\ cluster}{area\ of\ cluster}$$

In which *min density* represents the minimum civilian density of all clusters, *distance* means the distance from cluster to agent, and *disToEntity* is the distance from cluster to the closest entity, for example *disToRefuge* is the distance from cluster to the closest refuge.

When important clusters are fixed, it is slightly more complicated to choose tasks for agents. Different kinds of agents deserve different weighed models.

With regard to fire brigade (FB), the importance of a burning building is calculated as follows:

$$Importance\ of\ Building =$$
$$\frac{area\ of\ building\ *f(building\ fieriness)}{distance+\frac{disToCluster}{10*Importance\ of\ Cluster}+disToGasStation/10}. \qquad (7)$$

$$f(building\ fieriness) = \begin{cases} 100 & if\ fieriness = 1 \\ 10 & if\ fieriness = 2 \\ 1 & otherwise \end{cases}$$

In which, *disToEntity* means the distance from building to the closest entity, for example *disToCluster* represents the distance from burning building to the closest cluster.

For ambulance team (AT), it is crucial to estimate whether it will be able to rescue the victims in time. If there is no fire spot in or near the cluster, ATs usually work independently; but if there is, AT must gather together to rescue the victim as fast as possible. At this time, only the following condition is met can the victim be rescued:

$$(RescueTime + 10)/TimeToBurn < n. \qquad (8)$$

RescueTime is the time it takes for AT to rescue the victim alone, *TimeToBurn* means the time till fire spreads to the location of the victim, n is the number of spare ATs.

As to the victims, the one who is at his last gasp needs to be given top priority. Equation (9) is adopted to rank the priority.

$$Importance\ of\ Civilian = 100/DyingTime. \qquad (9)$$

In which *DyingTime* is the victim's remaining time before death.

And the main task of police force (PF) is to assist other agents in order to quicken the pace of rescue by providing passive roads to disaster sites. In our tactics, the priority sequence is as follows:

a. the road with obstacles to cluster

b. the road to trapped agents

c. the road to burning building

d. the road to the buildings with trapped civilians

The two processes mentioned above make the important tasks in important clusters prominent from the partial and global view respectively. Then, center needs to match the tasks for each agent according to the importance of tasks so as to enhance holistic scheduling.

IV. RESULTS

We adopt the Eindhoven map in this study to run the simulation test and evaluate the proposed method. Initialized scenario is set as Fig.2 with few blockages: there are crowded victims and a covert fire spot in the lower-left corner of this map; while there are spare victims and an obvious fire spot in the top right corner. And the improved version introduces centralized scheduling to the original code of SEU_Jolly.

In order to weaken the side-effects, comparative tests are carried on the same map. Both versions of codes in this experiment repeat 10 times.

As is demonstrated in TABLE I, the test results of the proposed method shows that the improved version performs well and is of high stability. The following two comparative screenshots are provided to demonstrate the disparity of two codes in an intuitive manner.

TABLE I
TEST RESULTS

NO.	Original code	Improved code
1	46.03	90.79
2	47.12	93.04
3	46.35	91.66
4	45.97	89.30
5	45.66	91.13
6	46.14	91.21
7	44.88	92.47
8	47.54	91.82
9	48.01	91.46
10	46.25	90.53
Average Score	46.40	91.34

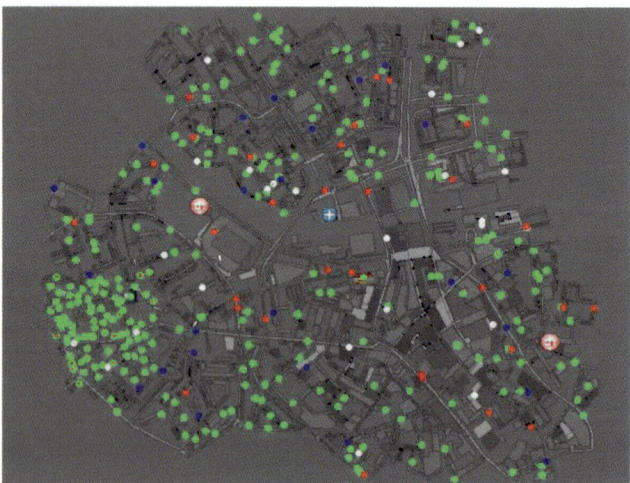

Fig.2. The initialized scenario of the Eindhoven map

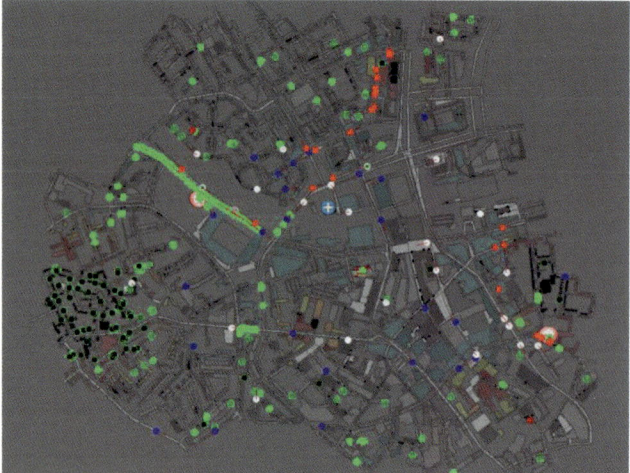

Fig.3. The screenshot at the 100th second of simulation with the original codes

Fig.3 shows the screenshot at the 100th second of simulation with the original codes. The blue, white, and red circles on the map represent respectively the PFs, ATs, and FBs. The bright green, dull green and black circles represent healthy, hurt or dead civilians in that order. The

yellow, orange, and maroon colors represent the increasing intensity if fire in buildings; black represents completely burned and destroyed buildings. We can see most FBs gathered at the top right corner of the map. However, the fire spot at the lower-left corner is ignored by FB due to the lack of global manipulation. What's more, ATs are scatted. As a result, all agents work independently with inefficiency.

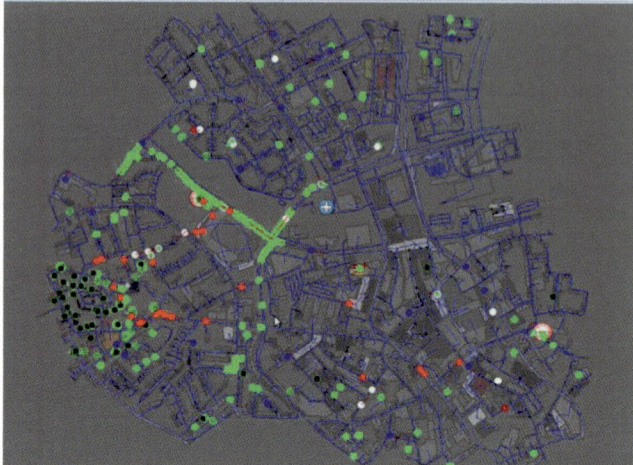

Fig.4. The screenshot at the 100th second of simulation with the improved codes

The performance of improved codes is really inspiring as is shown in Fig.4. From the screenshot at the 100th second of simulation with the improved, it is obvious that effective coordination between FB and AT is formed at the lower-left corner, the urgent area.

V. DISCUSSION AND CONCLUSION

This paper highlighted the importance of multi-agent coordination to MAS and the difficulties of its realization. The typical multi-agent system -- RCRSS was chosen as the research and test platform. We proposed an approach based on center scheduling to assist multi-agent coordination, in which center is treated as an information collector, processor and coordination scheduler. Simulation tests showed that the coordination method proposed in this study performed well in dealing with agent coordination in RCRSS.

Yet, there are still some issues existing in this approach which need to be addressed and improved in further research. For example, the process of aggregation and variation in the clustering analysis method can be further optimized; the weighed model about clusters and tasks could also be improved.

In addition, while center is default, more considerations should be taken into making agents act on behalf of centers in further work. And further research is needed to see whether it is practical for agents to group into teams to achieve partial coordination if communication is poor. Under such circumstance, whether partial coordination with limited information can contribute to the globally acceptable optimum.

REFERENCES

[1] Abbas Abdolmaleki, Mostafa Movahedi, Nuno Lau, and Luís Paulo Reis, "A distributed cooperative reinforcement learning method for decision making in Fire Brigade Teams," *RoboCup 2012: Robot Soccer World Cup XVI*, Springer Berlin Heidelberg, 2013, pp. 237–248.

[2] M Dias, A Stentz, "A market approach to Multi-robot Coordination," Technical Report CMU-RI-TR-01-26, Carnegie Mellon University, 2001.

[3] B Toledo,R Jennings, "Learning to select a coordination mechanism," in *Proceedings of the First International Joint Conference on Autonomous Agents and Multi-Agent Systems*, Bologna, Italy, 2002, pp. 1106-1113.

[4] Lix, L Soh, "The use of hybrid negotiation in resource coordination among agents," in *Proceedings of the 2003 IEEE/WIC International Conference on Intelligent Agent Technology ,IAT-03*, Halifax, Canada, 2003, pp. 133-139.

[5] Nair, R., Ito, T., Tambe, M., Marsella, S., "Task allocation in the RoboCup rescue simulation domain: A short note," *RoboCup 2001: Robot Soccer World Cup V*, Springer Berlin Heidelberg, 2002, pp.751-754.

[6] Maziar Ahmad Sharbafi , Caro Lucas , Abolfazel Toroghi Haghighat , Omid Amirghiasv , Omid Aghazade, "Using Emotional Learning in Rescue Simulation environment," in *Proceedings of World Academy of Science, Engineering and Technology ,S1307-6884*, 2006, 13, pp. 333-337.

[7] Farshid Faraji, Mohammad Reza Khojasteh, et al, "RoboCup Rescue 2010 Rescue Simulation League Team Description," RoboCup 2010 Symposium Proceeding CD [M/CD].from http://www.robocuprescue.org.

[8] Alan D. Barroso, Felipe de C. Santana, Victor Lassance, et al., "RoboCup Rescue 2013 LTI Agent Rescue Team Description," from http://www.robo.org./wiki/Rescue_Simulation_League.

[9] H. Levent Akın, Okan As,ık, et al, "RoboAKUT 2013 Rescue Simulation League Agent Team Description," from http://www.robo.org./wiki/Rescue_Simulation_League.

[10] Paulo Roberto Ferreira Jr., Fernando dos Santos, Ana L. C. Bazzan, Daniel Epstein, Samuel J. Waskow, "RoboCup Rescue as multiagent task allocation among teams: experiments with task interdependencies," *Auton Agent Multi-Agent System* , Springer US, 2010, pp.421–443.

[11] Hiroaki Kitano, "RoboCup Rescue: A grand challenge for multi-agent systems," in *MultiAgent Systems, 2000. Proceedings. Fourth International Conference on*, Boston: IEEE, 2000, pp. 5-12.

[12] Maitreyi Nanjanath, Alexander Erlandson, Sean Andrist, et al, "Decision and coordination strategies for RoboCup rescue agents," *Simulation, Modeling, and Programming for Autonomous Robots*, Springer Berlin Heidelberg, pp. 473–484.

[13] Andre H. Pereira, Luis Gustavo Nardin, Jaime Sim Simao Sichman, "Coordination of agents in the RoboCup Rescue: A partial global approach," in *Agent Systems, their Environment and Applications (WESAAC), 2011 Workshop and School of*, Curitiba, Brazil: IEEE, 2011, pp. 50-10.

[14] DeCastro LN, FJ, V., "Clonal selection algorithm with engineering applications," In *Workshop Proceedings of GECCO*, 2000, pp. 36-37.

[15] Guan-Chun Luh, Chun-Yin Wu, Wei-Wen Liu, "Artificial Immune System based Cooperative Strategies for Robot Soccer Competition," *Strategic Technology*, USA: IEEE, 2006, pp.76-79.

A Pilot Study on the Application of Cloud CRM in Industrial Automation

You-cheng SHAN[1,2,*], Chao LV[1], Wen-bo CUI[3]

[1] School of Management and Technology, Zhejiang Technical Institute of Economics, Hangzhou 310018, China
[2] School of Management and Economics, Tianjin University, Tianjin 300072, China
[3] China Zheshang Bank, Hangzhou 310006, China
(syc@tju.edu.cn)

Abstract - **This study mentions the new business mode and the business opportunity of the industrial enterprises which the cloud CRM creates. Also, this study introduces the concept of cloud CRM, analyzes the advantages of cloud CRM and local CRM and describes the designing principle and application mode of cloud CRM. In addition, the study discusses the effective way that the cloud CRM can be used in industrial automation and points out the development of industrial enterprises and improvement of techniques which are applied.**

Keywords - **Cloud computing, cloud CRM, cloud system, industrial automation, industrialization**

Industry and manufacturing requires the support of IT technology. The new web technology based on cloud computing provides a platform on which industry and manufacturing combines and organizes their resources effectively [1]. As a new generation of technology and one of the strategic industries set by government, cloud computing has successfully provides the industry with core competences on their business mode upgrading. In addition, cloud computing not only makes new way to manage enterprises IT foundation but also introduces new wisdom to run business. The appearance of cloud CRM provides CRM with techniques to make sure that clients need can be successfully fulfilled. This effectively limits business cost, improves business efficiency and creates new business mode and business opportunity.

I. CLOUD COMPUTING AND CLOUD CRM

A. Concept of cloud computing

As a rapidly developing technology, the concept of cloud computing has been modified and enlarged all the time since its appearance [2-4]. The University of California, Berkeley issued a report about the cloud computing, which believes that the cloud computing means not only the application in the form of services on internet but also the hardware and software in the data center which provides these services. These hardware and software are called "the Cloud". According to IBM, the cloud computing is a certain style of computing, whose foundation is the delivery of services, software and processing ability with public or private network. Google defines cloud computing as "the cloud of application". Initially, Google created the cloud computing in order to optimize its search engine which is Google's biggest business. But it wants to provide clients with the cloud computing as a type of service after it enlarges the scale of facilities.

B. Concept of cloud CRM

Cloud CRM, also known as cloud based CRM, web based CRM or online CRM, is the CRM system arranged on cloud system which provides different enterprises with CRM services via internet. Narrowly speaking, cloud CRM is the CRM services specifically designed for clients which is obtained via network (either Internet or local area network). Generally speaking, cloud CRM puts the CRM services specifically designed for clients on the server engine. Providers of the server manage the data and upgrading services. This mode makes clients input, obtain and handle data at any point [5].

C. Applications of Cloud CRM

Cloud computing will bring cloud CRM a series of developments and CRM provides will break the traditional limits to expands their business in the fields of SaaS, Online, SNS and so on. Without increasing the budget, cloud CRM makes the following come true: effectively expands business opportunity, cuts business burden of staff, optimizes business flow, decreases the mistakes on daily business running, shortens the service time, maximizes clients' satisfaction and loyalty and so on.

II. COMPARISON OF CLOUD CRM AND LOCAL CRM

A. Advantages of Cloud CRM

Compared with local CRM, cloud CRM has many advantages on providing services to medium and small-sized enterprises (SMEs).

(1) It has the database of high accessibility. By linking the data to cloud network, cloud CRM makes it easier for different departments to share and connect. Also, cloud CRM makes it possible for sales persons who are on business trip to access database to get information needed.

(2) Third party provides professional management services, which effectively cuts working burden. Also, service providers are able to configure system, manage the process of testing and data used in enterprise management application.

(3) It effectively limits business cost. It is highly possible that cost limiting is the primary reason that SMEs turns to cloud CRM. The way to pay is to use the service before paying according to the amount of clients monthly or yearly.

E. Qi et al. (eds.), *Proceedings of the 21st International Conference on Industrial Engineering and Engineering Management 2014*, Proceedings of the International Conference on Industrial Engineering and Engineering Management, DOI 10.2991/978-94-6239-102-4_37, © Atlantis Press and the authors 2015

(4) It consumes less resource. This saves more resource for the enterprises in data and information storage. It is obviously that the resource consumed when the data is saved on internet servers is much less than that consumed when the data is saved on local servers.

B. Advantages of Local CRM

Compared with cloud CRM, local CRM has many advantages on providing services to large industrial enterprises.

(1) Local CRM has the database of high controllability. This maximizes the ability of the enterprises on controlling clients' information. It is totally unnecessary for the enterprises to build database on public internet servers if they are able to construct database on private servers.

(2) Data is updated timely. There are many firms who prefer to handling the process of updating data timely, although cloud CRM is a practical tool for many applications of enterprises automation.

(3) It is quite safe. Local CRM plays a very good role when the information stored is used internal only.

(4) It is quite cost-effective. Sometimes it may not be able to save costs for enterprises if they pay according to the amount of users. Also, when the enterprise database is accessed by large amount of users every day, it will save more cost to set local system.

(5) It is much more immune to external interference. It is not exempt from external turbulence if the daily maintenance is outsourced. However, it cuts sharply the possibility that database is failed to access if the information system is constructed internally.

III. APPLICATION MODELS OF CLOUD CRM

A. Design principle of cloud CRM

Cloud CRM shows the principle of data-flow-centered when designing system. In order to show key steps of data flow in CRM system, it divides the system into different parts: data collection, data storage, data process and data delivery. Also, it connects all chunks with distributed data bus to exchange data effectively. As a result, the data flow and data exchange mode are designed clearly.

B. Design of cloud CRM

Cloud CRM is designed layer by layer. Logically, it divides the system into ability layer, platform pay and application lay. This way of design makes it possible to adjust the organization of every chunk in system according to clients needs.

(1) Ability layer. It consists of distributed data bus (ability to contact), distributed storage capacity (ability to store), distributed working flow and scheduling engine (ability to compute and arrange resource), etc.

(2) Platform layer. It contains consolidated data management and accessibility, consolidated service management framework, consolidated service delivery framework and automation management system, etc.

(3) Application layer. It structures the software chunks run on platform layer. And the application layer provides all these services.

C. Framework of cloud CRM application system

According to the above, cloud CRM application mode can be divided into 8 sub-systems shown on Fig.1.

(1) Maintenance management system. Built on cloud server, it contains configuration and control, priority management, alarming management, application management and fault management. It provides highly efficient maintenance service for system platform and business application.

(2) Distributed data bus system. Built on cloud server, it provides trustable data transformation, exchange and application collaborative platform. It supports the communicating and cooperating among all parts in the distributed data bus system.

(3) Data collection system. Built on cloud server, it contains configuration manager, data purification and data processing. It is the starting point of data driving framework. The data comes from the source like PC network, mobile network, community network and etc. Also, it purifies, organizes and converts the form of the input data according to configuration and stategy.

(4) Data storage system. Built on cloud server, it contains integrated data management system, cluster of relational database, distributed real-time database, distributed file system. It provides data storage service for data collecting, data analyzing and system platform.

(5) Data analysis system. Built on cloud server, it contains distributed computing model, data converting, data aggregating, data associating and data mining. It provides data analysis service for data storage, distributed working flow and strategy engine.

(6) Data service system. Built on cloud server, it contains service management system, data accessing management service, business logic service and Web service API. It provides data service management for service delivery, data storage and system platform.

(7) Service delivery system. Built on cloud server, it contains server, delivery management, load balancing. It provides data delivery service for system clients, system platform, data service and distributed working flow.

(8) Distributed scheduling system and working flow engine. Built on cloud server, it serves as the operator in the system. It provides the function of defining, scheduling, cooperating and operating for distributed working flow and strategy engine of large scale.

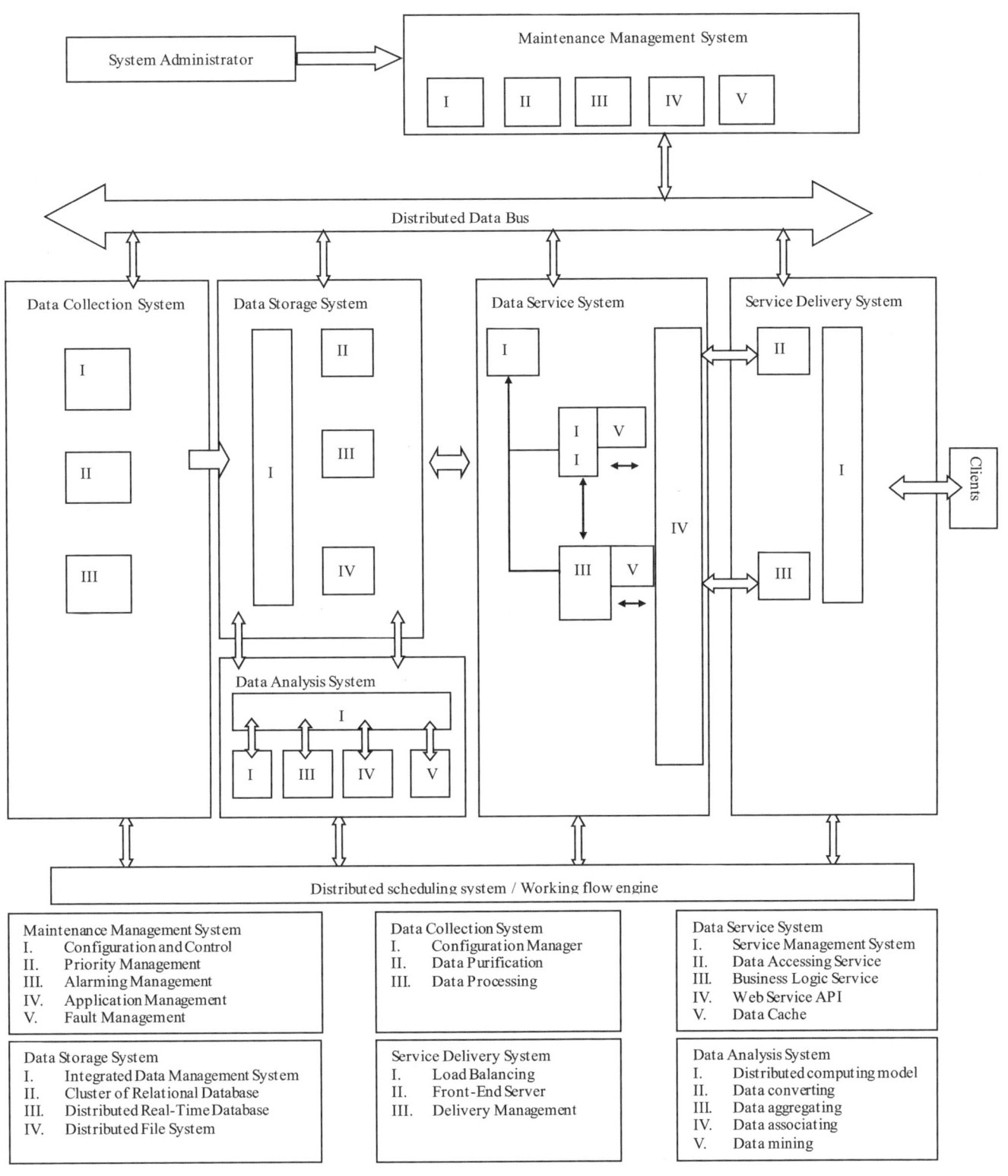

Fig.1. System Framework of Cloud CRM Application Model

IV. APPLICATION OF CLOUD CRM IN INDUSTRIAL AUTOMATION

A. Technology Developments and Business Startups based on Cloud Computing

Cloud computing platform and cloud computing service mode have become the primary mode in computing service in post-super computing era. The technology and service development created by it will have a great impact on the competition in industries and accelerate the upgrading of economic structure adjustment [6].

Industrial cloud computing platform and industrial cloud computing service provides services to SMEs, especially those who are lack the ability of production designing and developing. With the techniques and business mode of cloud computing, they build the advanced IT development tools on cloud platform so that industrial enterprises are able to rent network and reduce the barriers which prevent the firms from using the service. With industrial cloud platform, clients can learn advanced IT tools, obtain skill support from peers either with or without payment, get technical materials and design tools specifically for themselves. And all these help to run technology development and business startups.

B. Breaking point new technology system of industrial automation based new generation of cloud computing

Currently, cloud computing has been gradually becoming popular in the world. Although some people doubt the cloud computing, the solution provided by Amazon, Google has already tells us that it efficiently simplifies the computing process and integrates remote information and local information in just a few codes. Essentially the Andriod mobile phone has become client terminal equipment of the cloud computing. And the Apple is making effort to make ITune the cloud computing service.

It can be forecasted that the super business and super computer in scientific computation from all countries will soon be integrated into cell phones like those produced by Apple and Google. Therefore it is growing to the next breaking point that we design a new generation of the automation "cloud" system. Among many types of "cloud system", as a typical cloud CRM application, mobile CRM supports the accessibility via mobile phones, PC, tablet computer, which makes it possible to work on internet anywhere. Therefore, we can say that the era of mobile CRM has been there.

C. Development of industrialization based on Cloud CRM with IaaS, SaaS and PaaS modes

In terms of service type, the cloud computing can be classified as IaaS, Paas and SaaS [7]. Comparably, three types of CRM service systems can be developed for industrial enterprises.

(1) IaaS. IaaS providers give clients the cloud facilities which include multiple servers. This is a hardware collocation way in which clients buy the hardware from manufacturers.

(2) PaaS. PaaS provides development environment as a type of service. This is a distributed service platform where manufacturers provide development environment, server platform and hardware resource to clients. Clients make applications on the platform provided, transfer to other users via Internet and then run the applications on cloud computing platform.

(3) SaaS. SaaS providers put application software on their own servers. Industrial enterprises are able to not only obtain CRM services with low cost but also use these application software via all types of clients interface on client equipments like internet browser and email based on web page.

Based on IaaS, SaaS and PaaS, industrial enterprises are able to practice efficiently, cut their costs in designing and manufacturing, shortens the time span in the cycle of product upgrading and increases the resources efficiency so that they can develop their core competencies.

V. SUMMARY

The nature of the economic growth in China is industrialization, whose essential is the optimization of industries structure and the improvement of economic efficiency. The cloud computing makes it true that resource is integrated. Clients then use massive resource with low cost to meet their specific needs. Those resources are accessed through multiple channels. The "cloud computing" keeps high efficiency of computer resources used and saves the society's input. Therefore, in the current environment where the cloud computing is growing rapidly it is important and urgent to develop the application of the cloud computing on industrial automation. Meanwhile, CRM will merge with the new technologies like cloud computing, the internet of things and cloud computing to help the development of the industrial enterprises.

REFERENCES

[1] Deng Zhaohui, Liu Wei, Wu Xixing. Research and Appliation of Intelligent Grinding Cloud Platfrom Based on Cloud Computing [J]. China Mechanical Engineering, 2012, 23(1): 65-68

[2] Liu Peng. The cloud computing [M]. Beijing: Publishing House of Electronics Industry, 2010.

[3] Wang Peng, Huang Huafeng, Cao Ke. The cloud computing, the IT strategy of China in the future [M]. Beijing: People's Posts and Telecommunications Press, 2010.

[4] Yao Hongyu, Tian Suning, The cloud computing [M], Beijing: Publishing House of Electronics Industry, 2013.

[5] Wen Yiwen, Analysis of current status of CRM and its forecasting based on the cloud computing [D], Tianjin University, 2013.

[6] Zeng Yu, Wangjie, Wu Xixing, Research and Practice of Industrial Cloud Computing Platform [J]. China Mechanical Engineering, 2012, 23(1): 69-74

[7] Lei Baohua, Rao Shaoyang, Zhangjie etc. Deciphering cloud computing [M], Beijing: Publishing House of Electronics Industry, 2012.

Research on Quality and Efficiency of the Medical Decision-Making System Based on Robustness Theory

Jiang SHEN[1], Yanhong LIN[1], Man XU[2,*]

[1]Department of Management and Economy, Tianjin University, Tianjin, China
[2]Department of Industrial Engineering, Nankai University, Tianjin 300457, China
(td_xuman@nankai.edu.cn)

Abstract - **Based on analyzing the reasons for the vulnerability caused by medical decision-making system and the formation process and the factors leading to the vulnerability of the system and its impact methods and results, we can establish the interference model. Through analyzing the vulnerability of the health care system with robustness theory, we may ultimately find a way to improve the robustness of the system.**

Keywords - **Medical decision-making system, robustness, vulnerability**

I. INTRODUCTION

According to the study of medical institutions over the past 40 years, the American Institute of Medicine finds that 7% of medical patients had suffered serious medical errors. WHO reports that about 10 percent of patients in Canada, New Zealand and the United Kingdom and other countries suffer adverse medical events more than once every year. Therefore, how to improve health care quality and efficiency in the health care process has been an important research topic in the United States and other developed countries.

Due to poor economy in Chinese developed regions, the quality of medical services is difficult to be effectively guaranteed, which leads to medical errors and other adverse events by the diagnosis frequently. How to make our country's 2 percent of the world's medical resources to effectively cover 22 percent of the world population is an important issue facing the development of the medical profession.

II. LITERATURE REVIEW

Medical decisions system's quality and efficiency is a hot and difficult problems in the field of research, which is also a complex interdisciplinary problem. Many scholars have done a lot of research work on it and the robustness of complex systems is one of the important research content.

Robust control theory first proposed by Zames in 1981 [1], which is the most successful and relatively complete to solve robust problem currently. In the medical system, the robustness means the ability to adapt a range of different diagnostic or service requirements in a changing environment and external disturbances. In order to depth study the effect of the transfer efficiency to robustness, the following will review three aspects, including complex systems' anti-jamming capability, fault tolerance and redundancy of data structures.

A. Anti-interference ability of complex systems

In the knowledge uncertainty, a variety of random events on the system will produce different degrees of interference. Dismukes, et al. (1998) proposed several strategies to solve the system's vulnerability due to interference [2]. Morrow D, et al. (1994) analyzed the human expert blunders in medical decisions due to various aspects of external interference [3]. In order to solve the interference problem of complex systems, some scholars have established various graph model and goal programming, such as the space-time network graph model (Ahmad I Jarrah, et al., 2000) [4], interference resume game tree model, PERT chart model and use precise algorithms, heuristics (Xiangpei Wu et al, 2008) [5] and auction-type algorithm to solve the model. In addition, from the perspective of the theory of complex systems some scholars treat interference or signal as a function of the variable signal to describe interference problem, which is classifed into linear pulse interference function, pure step-type function, enhanced function and extended recession reduced impact type function (Guan ZH, et al., 2002) [6]. However, due to the complexity of the interference problem, the impact of interference metric and interference events and evaluation method is a problem, and modeling work is still a big challenge (Xiangpei Wu et al, 2007) [7].

B. Fault tolerance of complex systems

For medical decision system, Bogner (1994) considers emergency department is becoming important areas of fault tolerance research because of owning a large number of activities and special time constraints [8]. Information of medical decision systems is uncertain and complex, including patient vital signs parameters, diagnostic information and patient monitoring equipment to provide case information, etc., and the various emergency elements exhibit recessive, random or uncontrollable characteristics, such as the limitations of technology, data errors and inadequate communication between health care. These factors make communication in medical decisions system full of noise, which is likely to cause medical decision-making errors. But system fault tolerance can reduce or eliminate such errors in a certain degree in specific conditions. Medical decision system fault tolerance not only need to focus on the non-human faults but also artificial and unpredictable attacks (PENG Wen Ling, 2004) [9]. For instance, we can establish common methods include hierarchical fault-tolerant error detection and error handling framework to improve the system fault tolerance (Xuanhua Shi, 2006) [10].

However, research has yet to involve in solving fault-tolerant of system robustness, which is focus on the ability to maintain system stability when the environment changes within a certain range. So considering the impact of medical decision-making system robustness characteristics and designing and controlling medical decision robust system is important.

C. Information redundancy of complex systems

Uncertainty of knowledge in complex medical decision-making systems increases data redundancy, which leads to less efficient reasoning. How to efficiently eliminate data redundancy caused widespread concern (N. Hazon, 2008) [11]. For instance, though using the genetic algorithm in optimizing features to eliminate high-dimensional data attribute redundancy can improve search rate case, the case is easy to make into a local library in a state of disorder caused by reasoning under optimal accuracy decreased(GR Beddoe, 2006) [12]. Though using the neural network for the elimination of case-based reasoning similar matches redundancy can improve

matching efficiency, reasoning accuracy loss when the overall case neurons represented the maximum similarity is relatively low (R. Das, 2009) [13]. The method which dividing case into sub-case base library to overcome the case base retrieval redundancy improves retrieval efficiency, but the inference system sensitivity decreases (CA Tighe, 2008) [14]. These conventional methods to eliminate redundancy in the data to improve the efficiency of the system to some extent, but ignores the robustness of the system. For medical decision-making system, robustness means that reasoning is still able to maintain high accuracy and sensitivity when the system information is uncertain (such as becoming disordered) (K. Ziha, 2000) [15].

III. CONTENT OF THE STUDY

Questions of the Research can be further expanded, as shown in Fig.1. Several basic concepts should be explained:

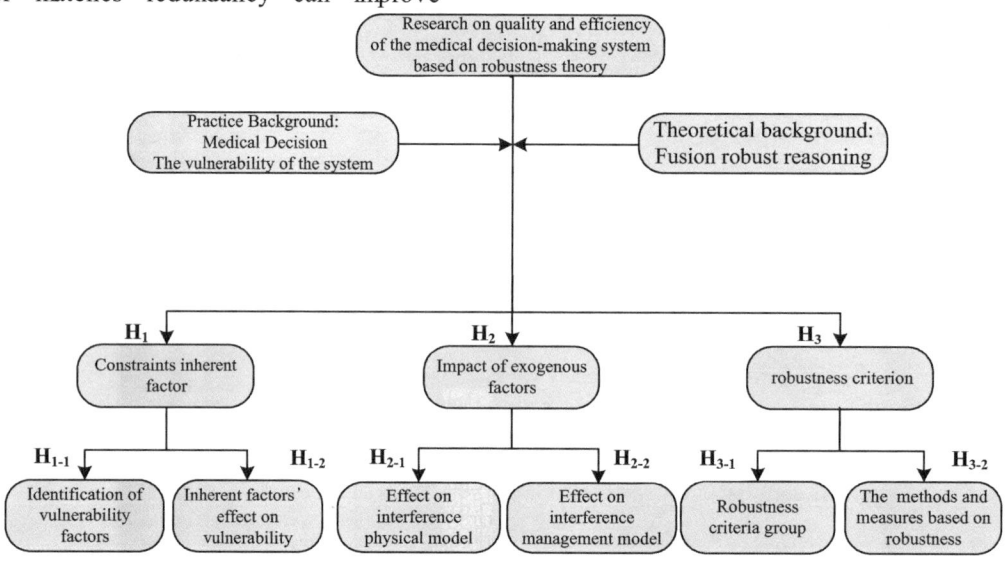

Fig.1. The expansion of the issues of the study

(1) Inherent factors of medical decision-making process (H1): refers to the internal behavior restrictions during medical decision-making process in the system, such as time constraints;

(2) Exogenous factors of medical decision-making process (H2): refers to the internal behavior restrictions during medical decision-making process out of the system, including hospital management factors and environmental factors;

Based on the above assumptions, this study intends to analyze complex medical decision-making system in the internal mechanism of the vulnerability of the phenomenon, using a variety of theories and methods of analysis. The main contents of the subject are divided into the following two aspects (C1, C2 for the contents of the code portion).

A. C1: Medical decision-making system vulnerability identification and interference transfer model

In this study, we determine the type of system interference by vulnerability identification method, and construct interference analysis model by using of system identification modeling. Specific research can be decomposed as follows:

C1-1: interference identification and classification of system's exogenous factors;

C1-2: qualitative study of vulnerability by the effect of the system exogenous factors, such as the manifestation of the key factors and influences (such as induction and inhibition) mechanism, etc.;

C1-3: quantitative study of vulnerability by the effect of the system exogenous factors. Study of the physical model studies of interference according to the time domain and frequency domain series expansion, and do some research on sensitivity of system interference (such as the source of interference and interference frequency, etc.);

C1-4: Study of interference delivery model of vulnerability management system, such as interfering factors' influence to the mechanism of fusion model and the system's critical value in the vulnerability critical state and so on.

B. C2: The method and measures of robustness to deal with the vulnerability of the system

To deal with the vulnerability of the systems and study measures of robustness, many factors need to be considered, such as spatial data redundancy system, tolerance and so on.

By adding Vulnerability constraints, we can establish transfer function model; Because interference transfer model can identify vulnerability in the robustness criteria group, this study intends to build target model in the robustness of the control, and use collaborative decision-making method to study system fault tolerance, anti-interference ability and robustness eliminating data redundancy. Specific research can be decomposed as follows:

C2-1: Study of fault-tolerant system's impact on robustness (such as improving the system tolerance);

C2-2: Study of target model's anti-interference ability under the system constraints;

C2-3: Study of spatial reasoning redundancy systems elimination's impact on vulnerability;

C2-4: Establish a system robustness criteria group and robustness measures.

IV. PROGRAM OF THE STUDY

Research programs by content broken down into the following two aspects:

A. Interference model-based vulnerability analysis

Based on the reasoning in the integration model, C1-1 modeling using system identification method to establish the model by adding random noise interference term, robust criteria group study and assess the impact of interference on the integration of the model error, the use of multiple criteria, such as the case similarity distinction , resolution and inference rules reliability study is based on the robustness of the decision-making and to study medical decision support system vulnerability assessment system. C1-2 simulated annealing genetic algorithm for medical decision-making system mistakenly classified, using FMEA method for medical decision-making system mistakenly retrospective, tracking and consequence assessment analysis.

B. The robustness of the system design to deal with the vulnerability

C2 using robust analytical methods to raise three robust threshold (in case similarity distinction vulnerability resolution and inference rules of harmonic maps in three dimensions on reliability) as constraints

added to the impact of the transfer of interference model to identify system fragility of the critical point of change, and thus eliminate the vulnerability of methods and measures from interference, such as fault tolerance and redundancy elimination of multi-angle study of the system. C2-4 use robust evaluation framework for complex medical decision-making system robustness criteria group analysis, analysis and evaluation index system robustness of complex medical decision-making systems, and improving the system's anti-interference ability and fault tolerance angle Research to improve the overall reliability of complex systems, to reduce the vulnerability of complex systems.

V. CONCLUSION

This study intends to study the quality and efficiency of medical decision-making system problems based on robustness theory, which is relatively lacking in domestic research. Combined with Chinese national conditions, we develop decision-making tools based on robustness that will help to improve the quality and efficiency of medical decision-making system, which is significant for the development of Chinese medical career.

ACKNOWLEDGMENT

This research was supported by the National Natural Science Foundation of China (Grant No. 71171143), National Natural Science Foundation of China Youth Project (Grant No. 71201087), Key Project of Science and Technology supporting program in Tianjin (Grant No. 13ZCZDSF01900) and the Fundamental Research Funds for the Central Universities (Grant No. NKZXB1458), China.

REFERENCES

[1] Kemin Zhou, Doyle J C. Robust and Optimal Control [M]. Defense Industry Press, 1999.

[2] Dismukes K, Young G, Sumwalt R. Cockpit interruptions and distractions: effective management requires a careful balancing act [J]. ASRS Direct line. 1998, 12:4–9

[3] Rasmussen J. Afterword. In: Bogner MS (ed). Human Error in Medicine. Hillsdale, NJ: L. Erlbaum Associates, 1994:385–93

[4] Ahmad I Jarrah, Jon Goodstein, Ram Narasimhan. An efficient airline re-fleeting model for the incremental modification of planned fleet assignments [J]. Transportation Science, 2000, 34(4): 349-363

[5] Xiangpei Wu, Qi Zhang, Qiulei Ding. Progress of interference management model and its algorithm [J]. System Engineering Theory and Practice, 2008, 28(010): 40-46.

[6] Guans Z H, Chen G, Yu X, et al. Robust decentralized stabilization for a class of large-scale time-delay uncertain impulsive dynamical systems [J]. Automatica, 2002, 38(12): 2075-2084.

[7] Xiangpei Wu, Qiulei Wu, Qi Zhang, Xuping Wang. Interference Management Research Review [J]. Management Science, 2007, 20(2): 2-7.

[8] Bogner M S E. Human error in medicine [M]. Lawrence Erlbaum Associates, Inc, 1994:1–11.

[9] Wenling Peng, Lina Wang, Huanguo Zhang. A Comparative Study of Network and Information robust tolerance applications [J]. Computer Engineering and Applications, 2004, 40(036): 151-153.

[10] Xuanhua Shi, Hai Jin, Weizhong Qiang. Universal Grid Fault Tolerance Framework [J]. Huazhong University of Science and Technology (Natural Science), 2006, 34(7):42-45

[11] N. Hazon, G.A. Kaminka. On redundancy, efficiency, and robustness in coverage for multiple robots [J]. Robotics and Autonomous Systems, 2008, 56(12): 1102-1114.

[12] G.R. Beddoe, S. Petrovic. Selecting and weighting features using a genetic algorithm in a case-based reasoning approach to personnel rostering [J]. European Journal of Operational Research, 2006, 175(2): 649-671.

[13] R. Das, I. Turkoglu, A. Sengur. Effective diagnosis of heart disease through neural networks ensembles [J]. Expert Systems with Applications, 2009, 36(4): 7675-7680.

[14] C.A. Tighe, A.Y. Tawfik. Using causal knowledge to guide retrieval and adaptation in case-based reasoning about dynamic processes [J]. International Journal of Knowledge-Based and Intelligent Engineering Systems, 2008, 12(4): 271-281.

[15] K. Ziha. Redundancy and robustness of systems of events [J]. Probabilistic Engineering Mechanics, 2000, 15(4): 347-357.

Comprehensive Evaluation Method for the Extended Indicators of Maintenance Management Based on the Standard Deviation of Weights

Hui ZHENG*, Xin-lu BAI

School of Mechanics, Tianjin University of Science & Technology, Tianjin, P. R. China
(tjzhenghui@tust.edu.cn)

Abstract - **The complex of machinery repair process restricts the development of machinery maintenance industry. While the machinery repair process management evaluation is the main factor in the maintenance of mechanical products. In this paper, based on the standard deviation of weights, the extended model is made to find out the importance of each maintenance indicators objectively and then determine the priority of planned maintenance mechanical products under comprehensive influence of various indicators reasonably.**

Keywords - **Correlation degree, comprehensive evaluation, deviation of weights, extension theory**

I. INTRODUCTION

With the speedy development of economy in our country, maintenance industry of mechanical products growing rapidly. In recent years, following the maintenance technology improving, maintenance mode optimization and maintenance strategy has attracted many scholars, however, there were few studies for maintenance assessment of mechanical products based on deviation of weights in different environment. Yang Dandan.et al [1-3] proposed an extension analytic hierarchy process, using format of extension sector number to get the judgment of experts subjective value, further to obtain the relevant value for each indicators through extension judgment matrix. Chen Bingfa.et al [4] proposed a new extended fault diagnosis method with gray correlation analysis and failure characters of mechanical products were analyzed by gray correlation analysis, obtaining an accurate correlation analysis through enlarging samples of fault diagnosis. Kong Fansen.et al [5] used extended analysis applying to engine cylinder body production line. Adopting up-to-down system analysis to define weight coefficients. Furthermore, maintenance strategies were developed according to the integration relationship degree. Zhang Xiong.et al [6] used extended method evaluating maintenance support capacity of helicopter, by establishing an extended matter element model under multiple indicators, gaining a scientific and reasonable evaluation process. Xiang Changcheng.et al [7] designed an extended immune algorithm, utilizing the theory of self and non-self recognition in immune system and immune clone selection, with the combination of extension matter-element analysis and extension set and applying in the fault diagnosis of

steam turbine. At the same time, the concept of extension distance has been put forward and on the basis of it the extension K nearest neighbor has been designed which can be used for data classification and fault diagnosis. Yang Yi et al [8] established an effective evaluation index system of maintenance support system by multi-level extension evaluation method. In the existing research results, most research has focused on using the extension method to specific analysis on the strategy of mechanical products maintenance and machine fault diagnosis, however, there are few research on maintenance evaluation of weights determination under different environment. The paper established a new model for maintenance management assessment, vertical determination for priority of planned maintenance products concerning about comprehensive influence of each indicators and made lateral correlation analysis based on the given standard indexes, until reached pre-determined criteria of turn around time (TAT), providing a reliable basis for ongoing analysis on maintenance industry of mechanical products.

II. EXTENDED ANALYSIS BASED ON THE STANDARD DEVIATION OF WEIGHTS

In the process of maintenance management of mechanical products, finding the delay time for repair and test is the key to solve TAT. By determining weights of each maintenance indicators and analyzing reasons of correlation analysis of each indexes for planned maintenance mechanical products, seeking the root of problem and putting forward improvement opinions for the quality of mechanical products maintenance.

In this paper, comprehensive assessment process of extended maintenance management divide into following two parts:

(1) Determine goal-index attribute matrix according to the given data and weights of each indicators were established through standard deviation, thus getting weight vector of evaluating matter element and value for each indicators of mechanical products, then can find goal-improving products based on the priority of planned maintenance mechanical products.

(2) Hierarchy structure was established according to maintenance management indicators of mechanical products, determine data scope of evaluation of matter element, put the improving

products as evaluation matter element, then putting each indicators of matter element into each subordinate set to evaluate under multiple indicators.

III. THE EXTENDED MODEL BASED ON STANDARD DEVIATION

A. Determination of weights coefficient

Adopting standard deviation method to determine the weights coefficients of evaluation indicators. Take mid-range of each interval determined by each classical domain subordinate set as attribute matrix H, which H_{ab} expresses the attribute value of products a under indicator b is r_{ab}:

$$\left(H_{ab}\right)_{n \times m} = \begin{bmatrix} r_{11} & r_{12} & \cdots & r_{1m} \\ r_{21} & r_{22} & \cdots & r_{2m} \\ \vdots & \vdots & & \vdots \\ r_{n1} & r_{n1} & \cdots & r_{nm} \end{bmatrix}$$

$$a=1, 2, \ldots, n; \quad b=1, 2, \ldots, m \qquad (1)$$

According to the charactor of maintenance products, all the evaluation indicators are cost type indicator. By cost type indicator dimensionless method to attribute matrix, then get the dimensionless attribute matrix $Z=(Z_{ab})_{n \times m}$. Let $W=(w_1, w_2, \ldots, w_m)^T$ be the weight vector of evaluation indicators and satisfied following the unit constraint conditions:

$$\sum_{b=1}^{m} w_b^2 = 1 \qquad (2)$$

Let S be the standardization of weight attribute matrix:

$$S = \begin{bmatrix} w_1 H_{11} & w_2 H_{12} & \cdots & w_m H_{1m} \\ w_1 H_{n1} & w_2 H_{22} & \cdots & w_m H_{2m} \\ \vdots & \vdots & & \vdots \\ w_1 H_{n1} & w_2 H_{n2} & \cdots & w_m H_{nm} \end{bmatrix} \qquad (3)$$

For the indicator b, standard deviation of each products attribute value S_b is defined according to matrix S:

$$S_b = \sqrt{\frac{1}{n} \sum_{a=1}^{n} \left(H_{ab} w_b - \frac{1}{n} \sum_{a=1}^{n} H_{ab} w_b \right)^2} = \sqrt{w_b \sigma_b} \qquad (4)$$

where,

$$\sigma_b = \sqrt{\frac{1}{n} \sum_{a=1}^{n} \left(H_{ab} - H_b \right)^2} \qquad (5)$$

$$H_b = \frac{1}{n} \sum_{a=1}^{n} H_{ab} \qquad (6)$$

If indicator H_m has few significant influence for all mechanical products, that is to say the indicator H_m has the same influence for all mechanical products. At the same time, the priority would has no effect based on this indicator. Obviously, this indicator may delete or make its weight coefficient zero. Conversely, if the indicator H_m has great difference for each goal of indicators, that is to say the indicator H_m has assignable influence for priority and this indicator should give a large value. Assume all the selected indicators have significant influence, then solve following optimization model can obtain the optimal weighted vector:

$$\max \phi(W) = \sum_{b=1}^{m} w_b \sigma_b \qquad (7)$$

$$\text{s.t.} \quad \sum_{b=1}^{m} w_b^2 = 1 \qquad (8)$$

By formula (8), the optimal weights vector can be obtained. Normalize above objective function value:

$$\varpi_b^* = \frac{\sigma_b}{\sum_{b=1}^{m} \sigma_b} , \quad b=1, 2, \ldots, m \qquad (9)$$

The results of normalization cannot be effected to priority.

B. Establish classical domain and joint domain of each evaluation matter element

According to the principle of analytic hierarchy process, hierarchy model for maintenance management optimization of mechanical products can be established. Suppose there are t mechanical products and s catalog quantity of products, then the catalog of evaluation of mechanical products can be denoted as N_{0s}, let C be evaluation indicators corresponding to s, and $V_{0st}=<x_{0st}, y_0 st>$ be the value range corresponding to C, where V_{0s} is the function of C, so the classical domain can be denoted as R_{0s}:

$$R_{0s} = \left(N_{0s}, C, V_{0s} \right) = \begin{bmatrix} N_{0s} & c_1 & <x_{0s1}, y_{0s1}> \\ & c_2 & <x_{0s2}, y_{0s2}> \\ & \vdots & \vdots \\ & c_t & <x_{0st}, y_{0st}> \end{bmatrix} \quad (10)$$

According to the standard matter N_0, joint domain can be denoted as R_p, where R_p is the matter element consisted by multiple standard matter. Owing to V_{0s} is the function corresponding to feature C, feature value range of joint domain is larger than each classical matter element.

$$R_p = \left(N_p, C, V_p \right) = \begin{bmatrix} N_p & c_1 & <x_{p1}, y_{p1}> \\ & c_2 & <x_{p2}, y_{p2}> \\ & \vdots & \vdots \\ & c_t & <x_{pt}, y_{pt}> \end{bmatrix} \quad (11)$$

Where, p represents summary of all the evaluation catalog, namely value range of joint domain contains all value range of evaluation catalogs s, let R_m be the evaluation matter element for mechanical product M:

$$R_m = \left(M, C, V \right) = \begin{bmatrix} M & c_1 & v_1 \\ & c_2 & v_2 \\ & \vdots & \vdots \\ & c_t & v_t \end{bmatrix} \quad (12)$$

C. Determination for correlation function and evaluation level

Define bounded interval $E_0 = (\alpha, \beta)$, let E_0 be the value range for each matter element of classical domain. Distance can be denoted as $\rho(e, E_0)$,

$$\rho(e, E_0) = \left| e - \frac{\alpha + \beta}{2} \right| - \frac{\beta - \alpha}{2} \quad (13)$$

According to formula (13) and characters and formulas of correlation function [8], correlation function can be denoted as $K(e)$:

$$K(e) = \frac{\rho(e, E_0)}{D(e, E_0, E)} = \begin{cases} \dfrac{\rho(e, E_0)}{\rho(e, E) - \rho(e, E_0) + \alpha - \beta}, & e \in E_0 \\[4mm] \dfrac{\rho(e, E_0)}{\rho(e, E) - \rho(e, E_0)}, & e \notin E_0 \end{cases} \quad (14)$$

Where, E_0 is the value rang of classical domain of each mechanical products corresponding to C, E is the value range of joint domain R_p corresponding to C.

According to formula (14), correlation degree of each indicators for evaluation matter N_i can be calculated. Afterward, make superiority evaluation based on detail actual requirements of maintenance mechanical products.

IV. CONCLUSIONS

By determining the standard deviation of weights, this paper can obtain the comprehensive priority of all maintenance products corresponding to each indicators, targeted to find maintenance products which have obvious influence on enterprises TAT, and providing effective data for ongoing lateral extended analysis. Each value range of hierarchy indicators were based on accumulated data and evaluation of indicators can be determined by extended analysis, this method can quantitatively and objectively reflect the degree of character for each indicator, providing the foundation for seeking the root influence of TAT.

REFERENCES

[1] YANG Dandan, Research on multi-attribute decision method based on extension mathematical theory [D]. Dalian Maritime University, 2010.

[2] ZHAO Lang, WANG Liang. Improvement for two faults of AHP [J]. Journal of XI'AN University of Science and Technology, 2007, 27(3): 507-510.

[3] YAN Dongfang, Extension Analytic Hierarchy Process and Its Applications [D]. Dalian Maritime University, 2012.

[4] YAO Yao, CHEN Bingfa, WANG Tichun. Study on the failure analysis of equipment based on extension theory [J]. Mechanical Science and Technology for Aerospace Engineering, 2014, 5: 012.

[5] KONG Fansen, CANG Anyu, ZOU Qing. Extension analysis method of maintenance strategy of engine cylinder body production line [J]. Computer Integrated Manufacturing Systems, 2007, 13(09): 0-0.

[6] ZHANG Xiong, ZHAI Jingchun, ZHANG Zongming, LI Wu. Application of extension method in evaluation of maintenance support capability of helicopter [J]. System Simulation Technology, 2011, 2: 018.

[7] XIANG Changcheng. Fault diagonsis and condition monitor based on extensis [D]. Chongqing University, 2008.

[8] YANG Yi, Wu Chang. Study of effectiveness evaluation of maintenance support system based on multi-level extension method [J]. Jouranl of air force engineering university. 2006, 7(6):76-80

[9] YANG Chunyan, CAI Wen. The extension engineering method [M]. Science Press, 1997.

Research on Key Technologies of 3D Geology Spatial DB Engine

Pei-gang Liu[1,2], Fei Hou[3], Mao Pan[1,*]

[1]The Key Laboratory of Orogenic Belts and Crustal Evolution, Ministry of Education, School of Earth and Space Sciences, Peking University, Beijing, China
[2]College of Computer & Communication Engineering, China University of Petroleum, Qingdao, China
[3]Shengli Oilfield Western Exploration Project Department, Dongying, China
(panmao@pku.edu.cn)

Abstract - **Analyzed existing spatial data models, and made sure the data model for the 3D geology spatial database engine; Referring to the existing relationship models, improved the topological relationship model and relationship calculations to 3D geology spatial data ; Discussed the existing space data management modes and their scopes, found the storage style and structure for 3D geology spatial data; Chose available spatial indexes and query /analysis strategy for 3D geology space database engine; Analyzed system structure and the characteristic of existing space database engine, put forward the structure for geology space database engine, and presented its work steps in the update and query/analysis processes.**

Keywords - **3D geology spatial data, spatial analysis, spatial database engine, spatial data model, spatial relationship model**

I. INTRODUCTION

For traditional database cannot support all requirements of spatial data because of the particularity of spatial data (such as timing, multidimensional), spatial database is called an occur. Most of existing spatial database are built on the basis of relational database management system(RDBMS), commonly known as the spatial database engine (SDE, first proposed by ESRI company), including ArcSDE, Oracle Spatial, SpatialWare, DB2 spatial extender, Informix Spatial DataBlade Module and SuperMapSDX+ [1].

Compared to the spatial data in geographic information system (GIS), underground geological spatial data is more complex. We can only get discrete data by special means (such as seismic, drilling) and then deduce all data in underground 3D space for the invisibility of underground solid. All analysis on the underground geological data is based on the deduced data, and the requirement of storing and processing these data is more special. This paper researched on the key technology in the design and development of 3D geological spatial database engine (GSDE), including data model, relational model, storage structure, spatial index, spatial analysis and system structure [2-5].

II. DATA MODEL OF 3D SPATIAL DATA

In the design of 3D geological database engine, the first problem is the expression of 3D geological spatial data, to establish a data model of 3D geological data. Spatial data model mainly includes three aspects as follows: 1) Geometric model, which describes the spatial morphology of geological bodies, such as shape, size, location, etc.; 2) Topology model, which expresses the relationship between geological bodies; 3) Property model, which reflects the property features of geological bodies, such as porosity, oil saturation [6].

Spatial data model is classified as wire frame model, surface model, entity model and hybrid model. Surface model and entity model are commonly used in geoscience. Surface model represents bodies with solid boundary, do not need to enumerate all internal point of bodies; entity model represents bodies by filling the body space with body element, which is easy to describe the property features.

At present the main data model includes CSG, TIN, Grid, TEN, HEX and Octree. Each model has its advantages and applicable scope. 3D geological model is commonly represented by mixed structure model such as TIN-Grid, TIN-CSG TEN-HEX, TIN-Octree, etc..
GSDE can adopt TIN-Octree mixed structure model. TIN is a kind of surface model, expressing the surface boundary of geological body and topological relation between bodies, it is a kind of vector data format; Octree is a kind of entity model, expressing internal structure and property features of geological bodies. In applications without internal properties such as visualization or spatial relation representation, the TIN model is enough; in applications involving internal properties of geological bodies, Octree model is adopt, which is a kind of raster data format.

In the TIN model, this paper subdivided four types of vector 3D geological data, point, line, plane, body into seven types as following: point (multipoint), line (multiline), closed line, polygon, nonclosed TIN plane, closed TIN plane, closed TIN body.

III. RELATION MODEL OF GSDE

ESRI defines 7 basic spatial relations in the interface of IRelationalOperator of ArcObjectS components: Equal, Contain, Within, Disjoint, Overlap, Touch and Cross.

Wu and Shi considered there are ten kinds of spatial topological relations in 3D space and there are twelve kinds of basic spatial relations which can represent these spatial topological relations [7]. The twelve kinds of basic spatial relations include Disjoint, Equal, Touch, Cross, In, Contain, Overlap, Cover, Cover by, Enter, Pass, Pass by. The ten kinds of spatial topological relations include point to point, point to line, point to plane, point to body, line to line, line to plane, line to body, plane to body and body to body [8].

In this paper, the spatial data are subdivided into 7 types with 21 types of spatial topological relation as following:
1) Point/point, point/line, point/polygon, point/nonclosed TIN plane, point/closed TIN plane, point / body (closed TIN);
2) Line/line, line/polygon, line/nonclosed TIN plane, line/closed TIN plane, line /body (closed TIN);
3) Polygon/polygon, polygon/nonclosed TIN plane, polygon/closed TIN plane, polygon / body (closed TIN);
4) nonclosed TIN plane/ nonclosed TIN plane, nonclosed TIN plane/closed TIN plane, nonclosed TIN plane / body (closed TIN);
5) Closed TIN plane/ closed TIN plane, closed TIN plane/ body (closed TIN);
6) Body (closed TIN) / body (closed TIN).

In view of the expression and processing requirements of geological entities, this paper defines nine kinds of basic spatial relations, including Disjoint, Equal, Contain, Inside, Overlap, OnBorder, Touch, Touch by and Cross, to express these 21 kinds of spatial topological relations. For convenience, the topological relations except Disjoint can be classified as a class. Then the spatial topological relation is summarized as two kinds: Disjoint and Intersect.

IV. STORAGE STRUCTURE OF GSDE

Spatial data is managed mainly by three modes [7], file mode, mixing file and database mode, database mode. And database included relational database, object-oriented database and object-relation database. The most commonly spatial data management mode is relational database mode or object-relation database mode. Representative of the relational database mode is ArcSDE, and representative of the object relation database mode is Oracle Spatial.

There are two methods to store spatial data in the relational database. One way is to split spatial entity into basic points, lines, planes and store them. Another way is to store it in a large field of binary data. The latter way is popular and it avoids connection search operation and retrieves property data retrieval fast. The drawback is that SQL does not support spatial retrieval, and access interface for spatial data should be developed.

Object-relational database itself provides support for objects, especially the Oracle Spatial supports three kinds of basic object types: points, lines and planes, and their aggregates. With the object-relational database, efficiency of storage and retrieval is very high.

Testing on 3D geological data shows that Oracle Spatial can't satisfy the 3D data model, as shown in Table I. In addition, Oracle Spatial does not support 3D index. Therefore GSDE only uses the relational database mode.

TABLE I
ORACLE SPATIAL'S SUPORT TO 3D GEOLOLY DATA MODEL

No.	Entity Type	Support or Not
1	MultiPoint	Yes
2	MultiLine	Yes
3	Polygon	Yes
4	Charactors	Yes
5	Tin	Yes
6	2D Raster	Yes
7	3D Raster	Not
8	BlockModel	Not
9	Volume	Not
10	Divided polyhedron	Not
11	Spatial property field	Not directly

V. SPATIAL INDEX AND ANALYSIS OF GSDE

Spatial index is a special data structure to store some (or a group of) property of geological entity, it can express some relation of spatial objects, such as location, shape etc.. Through the spatial index, spatial objects in accordance with query conditions can be quickly located. There are many kinds of spatial indexes, and spatial index technology typically include R-tree index, quad-tree index, grid index, which are established based on the minimum bounding rectangle of spatial entities. R-tree spatial index is famous and fast, but complex to maintain it, and it cannot guarantee the only path in the exact match query. Grid index is simple to implement and easy to maintain, and the lookup process is relatively simple. In order to improve the efficiency, multilevel grid index can be adopted [7-9]. Concluded that, R-tree index applies to the data with less variability, grid index applies to the data which changes frequently.

The characteristics of database management system and 3D geological spatial data model determines that the spatial index of 3D geological spatial database engine cannot be realized by the existing spatial index of relational database. So spatial index can only be realized by engine itself.

Spatial index itself is stored in the database. When the spatial database engine starts to work, the spatial index (such as R-tree index) is loaded into local backup; some index structures (such as the grid index) work by the way of SQL query, not being loaded to the local.

During the spatial analysis process, spatial retrieval is executed with spatial index according to conditions to find out the space object which conforms to requirements. Two-step algorithm is applied in the process of retrieval as following [8]: The first is filtering, which narrows the target area through the spatial index, eliminates spatial object which does not meet the conditions obviously, and gets a potential candidate set; the second is refining, which confirms the final result by detecting the accurate geometric information. The process is shown in Fig.1.

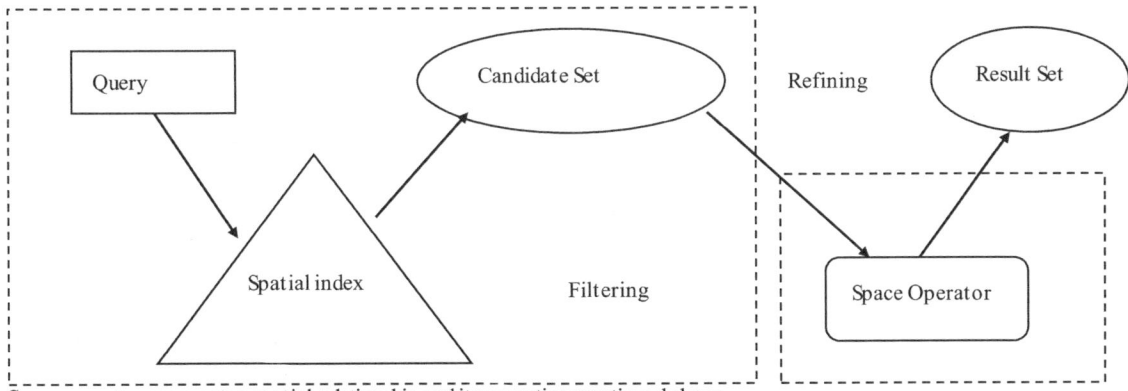

Space operator represents a spatial relationship and its operation mentioned above.

Fig.1. Two-step algorithm for Spatial Query

Considering some additional properties are stored in the factor / entity table, the method of union queries with property and spatial index can also be used to further improve the retrieval efficiency, as shown in Fig.2.

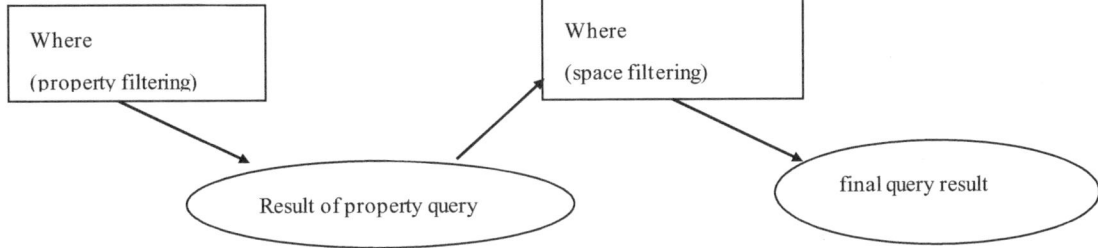

Fig.2. Spatial Query with Attributes Filter

VI. SYSTEMATIC STRUCTURE OF GSDE

A. Three Types of Spatial Database Engine

The development direction of SDE can be divided into three types, including built-in, middleware and module. Built-in SDE directly expands DBMS (RDBMS or ORDBMS), represented by Oracle Spatial; middleware adopts the three-layer structure mode, represented by ESRI company's ArcSDE; module SDE adopts the two-layer structure mode, represented by SuperMap SDX+. The three kinds of structures are shown in Fig.3 (Fig.3 is a simplified schematic, details see Fig.1, Fig.2 and Fig.3 in [2]).

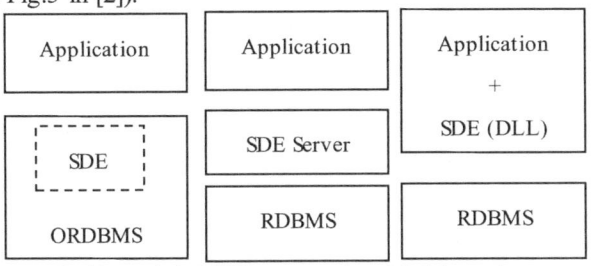

a. Inside mode b. Middleware mode c. Module mode
Fig.3. Three Structures for SDE

B. Structure Design of GSDE

The development of built-in SDE is too difficult and just the database vendor can do it. Middleware SDE adds a middle service layer, which makes it complex to use and requires developers to complete the security control. Module SDE uses the functions of database management system, such as data security policy, user management and concurrency control, and focuses on application. As a spatial database engine for professional applications, GSDE should be flexible, easy to use, efficient and available in different database management systems. Among the three kinds of development model, Module model fits the requirements best, and is the easiest to develop.

The working process of Module GSDE is shown in Fig.4. When data is updated, the engine is responsible for data processing (if necessary) and writes them to the database with SQL statements, and updates the local spatial index and ensures its synchronization with RDBMS. During the query/analysis process, the engine filters data according to the query command with spatial index, loads the filtered result set, and then refines and analyzes to get the final results. The process is shown in Fig.4.

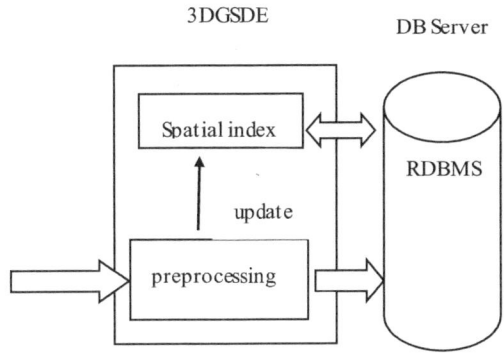

a. Insert

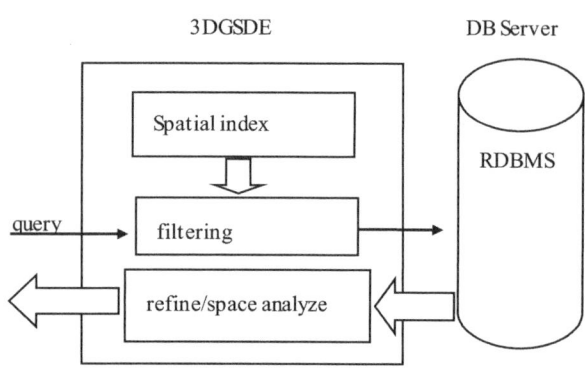

b. Query/Analyze

Fig.4. Work Flow of 3D Geology SDE a.Insert b. Query/Analyze

VII. CONCLUSION

Because of complexity and diversity, it is difficult to manage spatial data, especially 3D data. Because of the differences of Spatial data and processing requirements in different industries, spatial database engine must adapt to specific data types and data model to implement fast retrieval analysis in the vast amounts of data, it is a feasible way to develop professional spatial database engine according to the industry. The GSDE in this paper is designed for 3D geological spatial data. It adapts to the data model of 3D geological spatial data, satisfies the query, calculation, and processing requirements of mass data. It serves for virtual exhibition, spatial analysis, and other applications. The GSDE has been implemented, and applied in the Creatar platform which is developed by Information Geology Laboratory of Peking University, with good effect.

REFERENCES

[1] M.C. Wang, J. Zhao, Y.L. Li, "Analysis of spatial database engine and its solutions" (in Chinese), Geomatics World, vol. 8, no. 4, pp. 63–66, 2006.

[2] Q. Zhou, S.J. Li, Y.J. Li, L.Q. Ma, "Key technologies and development of spatial database engine" (in Chinese), presented at the *11th Annu. Meeting of China Association for geographic information system*, Peking, pp. 361-365, 2007.

[3] B. Li, Q.S. Wang, M. Feng, "The Anatomy of the Pivotal Technology in Spatial Database Engine" (in Chinese), Journal of Institute of Surveying and Mapping, vol. 20, no. 1, pp.35-38, 2003.

[4] H.Z. Ni, X. Z. Qiu, X.G. Cao, "Research of spatial database engine", Engineering of Surveying and Mapping, vol. 15, no. 1, pp.17-19, 2006.

[5] F. Fang, B. Wan, "The research and practice of spatial database engineering based on RDBMS", Software Guide, no.19, pp.17-19, 2005.

[6] Q. Wu,H. Xu,"3D geological modeling and virtual visualization"(in Chinese),Science Press, Peking,2011.

[7] X.C. Wu, W.Z. Shi, "Spatial Databases" (in Chinese), Science Press, Peking, 2009.

[8] S. Shekhar, S. Chawla, K.Q. Xie, X.J. Ma, D.Q. Yang, Spatial Databases (in Chinese), China Machine Press, Peking, 2004.

[9] X. He, "Key technology research on spatial database engine", Ph.D. Dissertatio, Institute of Computing Technology Chinese Academy of Sciences, Peking, 2006.

Dependency Evaluation of Disaster Prevention and Mitigation for Factors of Network Information Service Quality

Yang-xu Li[1,*], Zhuo-ning Zhao[2], Yuan He[1], Jing Liang[1]

[1]Management Faculty, Chengdu University of Information Technology, Chengdu, China

[2]Software Engineering College, Chengdu University of Information Technology, Chengdu, China

(wxmlyx@cuit.edu.cn)

[1]*Abstract* - The information service quality of internet disaster prevention and mitigation depends on the service of network platform. But there is the relation between processing usefulness and content easy use of service. So the hierarchical model of information service quality for internet disaster prevention and mitigation with useful index and easy use index is invalid. The paper applies ISM to analyze hierarchical and directed dependency of factors influencing information service quality for internet disaster prevention and mitigation, and use cross analysis to give set partition of factors based on their dependency, and on this basis, gives a quantitative measure and path evaluation of dependency. The result provides reference model for evaluating information service quality for internet disaster prevention and mitigation by the service process of network platform.

Keywords - Disaster prevention and mitigation, dependency, information service quality, ISM

I. INTRODUCTION

At the Internet age, network platform is an important channel for disaster prevention and mitigation information services. Government departments, meteorology, seismology, health and other disaster prevention and mitigation information producers, through the official website, the official microblogging platform, etc., provide information, knowledge and alerts about the prevention and mitigation for public consumers. In this process, maintaining the usefulness and ease of the information is the key to ensure public consumers disaster prevention and mitigation of network information service quality, and is an important link of service to implement the prevention and mitigation goals. In the service process of network platform, because usefulness and ease of use have association, the useful and easy to use index is difficult to maintain independence for evaluating the disaster prevention and mitigation of network information service quality. Based on network information service process and platform, discussing dependency of influencing factors about the information service quality has important implications for the proper evaluation of disaster prevention and mitigation of network information service quality.

[1]This research has been supported by Research Grant of Science & Technology Department of Sichuan Province (Program No. 2012ZR0120) and Development Fund of Philosophy and Social Sciences in the School (Program No. CCRF201004).

II. THE INFLUENCING FACTORS ABOUT THE INFORMATION SERVICE QUALITY OF INTERNET DISASTER PREVENTION AND MITIGATION

About researching on the evaluation of service quality of network information, Tao Lu [1] (2008), Yang Zhao [2] (2009), Meng-hua Liang [3] (2012), etc., When evaluating the information service quality, they fully considered the ease of use and usefulness. But Bo Gen [4] (2012) and Chun-ji Liu [5] (2013) used perceived usefulness and perceived ease of use in TAM model to construct the influence factors of information service. Thus, the ease of use and usefulness can be used as indexes to evaluate information service quality of internet disaster prevention and mitigation.

A. Usefulness

Usefulness means according to the possible harm of environmental disasters, the public consumers have the perception degree of information service utility of internet disaster prevention and mitigation. Regional matching degree, disaster harmfulness, update frequency, interpretability, user connectivity, amount of information, linkage, and so on, they are the main factors that influence the usefulness of information service quality of internet disaster prevention and mitigation.

From the disaster perspective, different regional climate and geographical environment is different, the local natural disasters have obvious regional characteristics. For the local economy and life, the destructive power of the similar disaster in different regions and different seasons is not same. Disaster harmfulness is bigger, public consumers will more pay attention on the information service quality of internet disaster prevention and mitigation. Visible, the usefulness of information service and the environment (that is regional matching degree) is positive correlation. From the information of disaster prevention and mitigation, its quality decides network information update frequency and the relevant information interpretability. From the public consumers, the ability of understanding information, the ability of obtaining information from the network platform, the quantity of obtaining information and the effect of guiding on actions (that is linkage) are terminal influence factors of information service of disaster prevention and mitigation.

B. Ease of use

Ease of use refers to the ease of access to the useful value of information service after the public consumers adopt the network information service of disaster prevention and reduction. So its influencing factors include in authority, accuracy, timeliness, platform connectivity, accessibility, intelligibility, etc.

From the information production, the information of natural disasters has a correlation with public consumers. The more authoritative the platform that publishes information, the easier the public consumer is convinced. The more accurate information, the easier the public consumer adopts the information. From the dissemination of information, the disasters are characteristic of sudden occurrence. In the disaster period, including pre disaster and post disaster, the information of disaster prevention and mitigation should be published by all platforms at the first time, so it can maintain the public consumer terminal platform connectivity. In order to cover a wider range of public consumers, the network channels that public consumers are most vulnerable to contact should be unblocked. From the content of information service, to ensure that the information service is ease of use, the information of disaster prevention and mitigation should be popular and easy to understand.

C. Dependent relationship between usefulness and ease of use

By reorganizing the influencing factors about the information service quality of internet disaster prevention and mitigation, we can acquire the hierarchical relationship of these factors, as shown in Fig.1.

In Fig.1, the hierarchical relationship can't reflect the dependent relationship between usefulness and ease of use. Dependent relationship refers to the input - output relationship of factors. The similarity of input and output is greater, and then the dependency of factors is stronger. The TAM model shows that perceived ease of use has a direct influence on the perceived usefulness [5]. In Fig.1, the factors of usefulness and ease of use are related with the information, and they have the interdependent relationship. This breaks the independence between factors. From the public consumer's point of view, unless the information service of internet disaster prevention and mitigation is easy to use, it can't display its value and usefulness. So there is the dependent relationship between usefulness and ease of use. In summary, the hierarchical relationship of factors is no longer suitable for analyzing the dependency between factors.

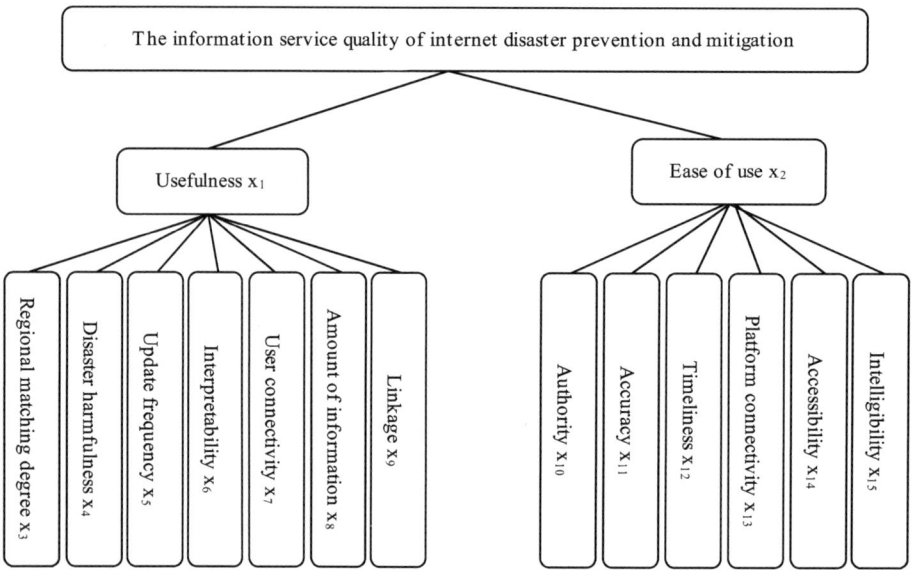

Fig.1. Hierarchical relationship of the influencing factors about the information service quality of internet disaster prevention and mitigation

III. ANALYSIS OF THE DEPENDENCY BETWEEN INFLUENCING FACTORS ABOUT THE INFORMATION SERVICE QUALITY OF INTERNET DISASTER PREVENTION AND MITIGATION

A. The analysis on directivity of factors' dependence

Factors' dependence has directivity. The directivity illustrates the relation of factors' dependence and the path of influencing the information service quality. So using interpretative structural modeling (ISM for short) [6] to analyze factors' directivity, we can get to the hierarchical structure of influence factors. The specific analysis steps are as follows:

(1) Through interviewing some experts in the field of disaster prevention and mitigation or network information services, the adjacency matrix for expressing the binary relation of factors is established, that is $A = (a_{ij})_{15 \times 15}$. According to this matrix, the reachability matrix "M" is established. Let x_i and x_j represent the influence factor i and j respectively, a_{ij} represents the influencing relationship of x_i and x_j, then

$$a_{ij} = \begin{cases} 1, & \text{The effect of } x_i \text{ to } x_j \\ 0, & \text{The effect of } x_j \text{ to } x_i \end{cases}$$

(2) Regional division. According to factors in beginning set B(x) and reachable set R(x) (or in ending set R(x) and ahead set A(x)), we can determine if the region could be divided.

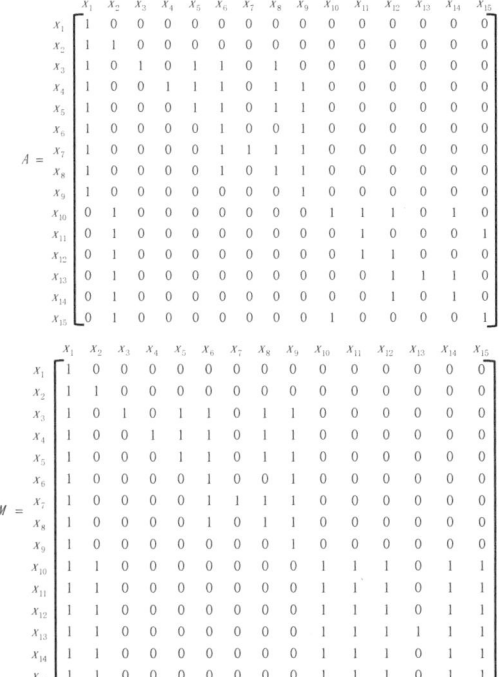

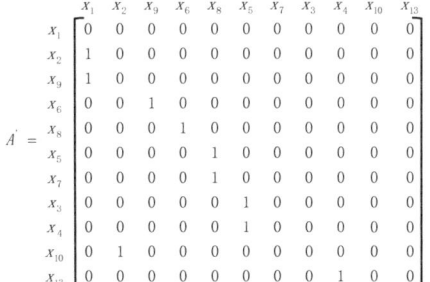

prevention and mitigation:

(P)=L1,L2,L3,L4,L5,L6,L7={x₁},{x₂,x₉},{x₆},{x₈},{x₅ ,x₇},{x₃,x₄,x₁₀,x₁₁,x₁₂,x₁₄,x₁₅},{x₁₃}

(4) According to the level division in the third step, we can adjust the row and column of reachable matrix, and then get a skeleton matrix "A'" by the contraction and detection of adjusted matrix.

(5) According to the skeleton matrix, we can establish the hierarchical structure of influence factors, as shown in Fig.2.

From the hierarchical structure model, we know influencing factors are divided in seven levels, from the bottom to the top, the arrow indicates the direct or indirect dependent relationship between the various factors. The usefulness in first layer directly depends on the ease of use and linkage in second layer, and indirectly depends on the factors in third and seventh layer. It is the essential factor to measure the information service quality of internet disaster prevention and mitigation. From the fifth layer to the seventh layer, user connectivity, regional matching degree, disaster harmfulness and platform connectivity don't depend on other factors, and they are depended on by other factors, so they are elementary factors to measure the information service quality of internet disaster prevention and mitigation.

(3) Level division, that is to determine the level of each element in the region. Let P be the set of a regional factors and $L_1, L_2, ..., L_i$, if using $L_1, L_2,...,L_i$ to represent the factor set of each level from high level to low level, then the level division can be expressed as $\prod(P)=L_1, L_2, ..., L_i$. Let $L_0=\Phi$, then $L_1=\{x_i \mid x_i \in P-L_0, C_0(x_i)=R_0(x_i), i=1,2,...,15\}$, $L_2=\{x_i \mid x_i \in P-L_0-L_1, C_1(x_i)=R_1(x_i), i<16\}$, ..., $L_k=\{x_i \mid x_i \in P-L_0-L_1-...-L_{k-1}, C_{k-1}(x_i)=R_{k-1}(x_i), i<15\}$. According to sub-matrix of set $P-L_0-L_1-...-L_{k-1}$, we can get common set $C_{k-1}(x_i)$ and the reachable set $R_{k-1}(x_i)$. Thus, we can get the result of level division for influencing factors about the information service quality of internet disaster

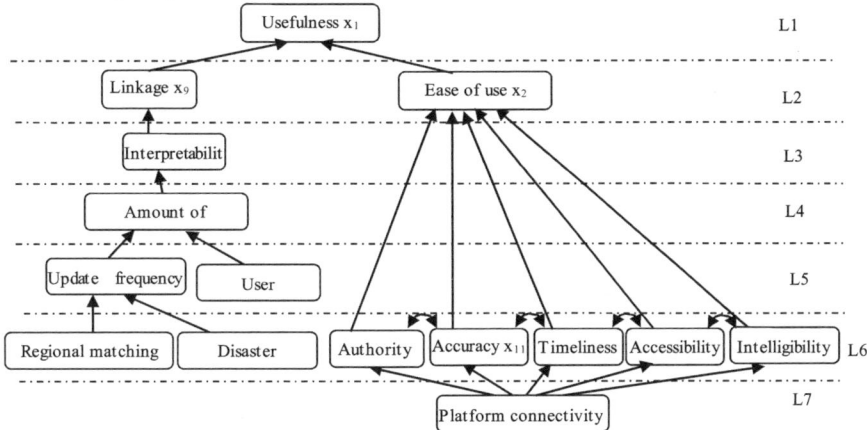

Fig.2. the hierarchical structure of influence factors about internet disaster prevention and mitigation

B. The analysis of dependent strength

The dependency of factors has some strong or weak distinction. If the dependency of factors is strong, the extent of influencing the information service quality of internet disaster prevention and mitigation is stronger and more direct. The input and output of factors can represent the dependent strength between factors. So we can use the number of input factors and output factors to estimate their strength and weakness. The number of input factors refers to the number of factors that influencing factor x_i directly or indirectly depends on; The number of output factors refers to the number of factors that directly or indirectly depend on influencing factor x_i. By cross analysis of two dimensions of input factors and output factors, we can get their partitioning. Through reachable matrix in part 3, we can obtained the number of input factors and output factors, as shown in Table I. Usefulness factor depends on all the other factors, but other factors don't depend on it. Its dependent strength is clear. So it is omitted in cross analysis. The result of cross analysis is as shown in Fig.3.

TABLE I
THE NUMBER OF INPUT/OUTPUT FACTORS

influencing factors	x2	x3	x4	x5	x6	x7	x8	x9	x10	x11	x12	x13	x14	x15	Number of output factors
x2	1	0	0	0	0	0	0	0	0	0	0	0	0	0	1
x3	0	1	0	1	1	0	1	1	0	0	0	0	0	0	5
x4	0	0	1	1	1	0	1	1	0	0	0	0	0	0	5
x5	0	0	0	1	1	0	1	1	0	0	0	0	0	0	4
x6	0	0	0	0	1	0	0	1	0	0	0	0	0	0	2
x7	0	0	0	0	1	1	1	1	0	0	0	0	0	0	4
x8	0	0	0	0	1	0	1	1	0	0	0	0	0	0	3
x9	0	0	0	0	0	0	0	1	0	0	0	0	0	0	1
x10	1	0	0	0	0	0	0	0	1	1	1	0	1	1	6
x11	1	0	0	0	0	0	0	0	1	1	1	0	1	1	6
x12	1	0	0	0	0	0	0	0	1	1	1	0	1	1	6
x13	1	0	0	0	0	0	0	0	1	1	1	1	1	1	7
x14	1	0	0	0	0	0	0	0	1	1	1	0	1	1	6
x15	1	0	0	0	0	0	0	0	1	1	1	0	1	1	6
Number of input factors	7	1	1	3	6	1	5	7	6	6	6	1	6	6	

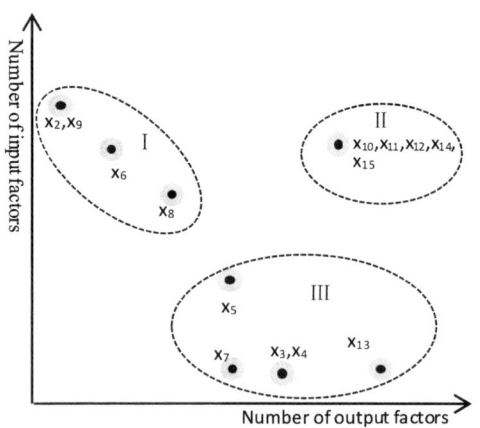

Fig.3. Result of cross analysis

By the cross analysis, the dependent strength of factors will be partitioned. The factors with strong dependence have a great effect on the information service quality of internet disaster prevention and mitigation; the factors with weak dependence are key factors of improving the information service quality of internet disaster prevention and mitigation. In part I, the input quantity of factors is many, the output quantity of factors is few, and their dependence is strong. In part II, the input quantity and output quantity of factors are more. They have strong dependence and are easy to interrelate. So the effect on the information service quality of internet disaster prevention and mitigation is not stable. In part III, the input quantity of factors is less, the output quantity of factors is more, and their dependence is weak. Other factors are dependent on them. So they play a key role in the information service of internet disaster prevention and mitigation.

The dependent strength of factors can be quantified. On the basis of similarity thought in literature [7], we can measure the similarity of input factor and output factor, the more the similarity, the dependence of factors is stronger. For factor i, let the number of input factors be I_i, let the number of output factors be O_i and let the dependent strength (similarity) be D_i, then

$$D_i = \frac{I_i}{I_i + O_i} \quad (1)$$

The dependent strength of influencing factors is as shown in Table II. The dependent strength of usefulness is maximal; it directly determines the quality of network information service. The dependent strength of factors, such as regional matching degree, disaster harmfulness, user connectivity, platform connectivity, they are the basic and important factor to measure the quality of network information service [8-10].

TABLE II
THE DEPENDENT STRENGTH OF INFLUENCING FACTORS

Influencing factor x_i	X_1	X_2	X_3	X_4	X_5	X_6	X_7	X_8
Dependent strength D_i	0.938	0.778	0.143	0.143	0.375	0.667	0.167	0.556
Influencing factor x_i	X_9	X_{10}	X_{11}	X_{12}	X_{13}	X_{14}	X_{15}	
Dependent strength D_i	0.778	0.462	0.462	0.462	0.111	0.462	0.462	

C. The evaluation method of dependence paths

According to dependence paths, the public consumers can evaluate the information service quality of internet disaster prevention and mitigation. The evaluation of single dependence path (such as x_3-x_5-x_8-x_6-x_9-x_1) reflects the information service quality of factors on this path. If the dependent strength of factors is strong, they are more likely to affect the service quality. The synthesis evaluation of all paths reflects the overall level of the information quality [11, 12].

If m represents the number of factors on the path, $D_i'(i = 1, 2, \cdots, m)$ represents their dependent strength, x_i represent the scaling value of factors (the values range of x_i from a to b, a represents the minimum value and b represents the maximum value), T represents the effect of these factors for the information service quality of internet disaster prevention and mitigation, then

$$T = \sum_{i=1}^{m} D_i' x_i \quad (2)$$

"D_i'" is the standardized dependent strength. The dependent strength of factors on different path has the different order of magnitude, so it is necessary to standardize for them. The processing method of Standardization is seen in [7].

T value reflects the information service quality of internet disaster prevention and mitigation. If T value is close to "b", it indicates the good service quality; conversely, if T value is close to "a", it indicates the poor service quality [13, 14].

D. applied analysis

Taking "http://www.eqsc.gov.cn/" for example, through observing this site and questionnaire, we can gather the scaling values of factors, such as Table III. These scaling values are from 1 to 5, and they respectively represent very poor, poor, good and excellent [15-17].

TABLE III
THE SCALING VALUES OF INFLUENCING FACTORS

influencing factor x_i	X_1	X_2	X_3	X_4	X_5	X_6	X_7	X_8	X_9	X_{10}	X_{11}	X_{12}	X_{13}	X_{14}	X_{15}
scaling value x_i	4	3	5	4	4	3	3	3	4	4	5	4	3	4	3

If a single path is used to judge the service quality of "Sichuan Earthquake Administration", such as "$X_3 \rightarrow X_5 \rightarrow X_8 \rightarrow X_6 \rightarrow X_9 \rightarrow X_1$". The dependent strength of factors on this path is expressed as $[X_3,X_5,X_8,X_6,X_9,X_1]$, then it is normalized as $[0.041,0.109,0.161,0.193,0.225,0.271]$. By equation (2), "T" is 3,688, so the service quality of this path lies between grade "good" and grade "excellent". If all paths are used to judge the service whole quality of this website, by the same method, "T" is 3.693. This shows that the service quality is above the average in general.

If T value of each path and whole website are compared, we can further judge the service quality and find out the way to improve the information service quality of internet disaster prevention and mitigation.

IV. CONCLUDING REMARKS

In this article, the influencing factors about the information service quality of internet disaster prevention and mitigation are constructed from two angle of the usefulness and ease of use. Using ISM, the hierarchical structure of influence factors is established. By this structure, the dependent relation and dependent strength between factors are analyzed. Through cross analysis, the partitioning of factors and the quantitative calculation methods of dependent strength are proposed, then the evaluation method of dependence paths are put forward. At last, through the application analyses, this evaluation method is shown to be exercisable and feasible.

Through the dependence path, the information service quality of internet disaster prevention and mitigation can be evaluated. If using some factors to evaluate, we can know the service level of part factors; if using all factors to evaluate, we can know whole service level. According to the service level, we can find the reasons of affecting the service quality and further improve the information service quality of internet disaster prevention and mitigation. In addition, according the evaluation results of dependence paths, we can compare all walks of life from the horizontal and vertical dimensions. This will provide the reference for improving the service quality of network information.

REFERENCES

[1]   Tao Lu, Xue Lei. The study of network information service quality evaluation and demonstration analysis. Documentation, Information & Knowledge, 2008(1)

[2]   Yang Zhao. The Evaluation of information service quality for industry information center website. Information and documentation services, 2009(6)

[3]   Meng-hua Liang. Research on the quality evaluation of information services about archives websites. Archives Science Bulletin, 2012(2)

[4]   Bo Gen. Analysis of influencing factors about consumers' online shopping intention based on TAM. Statistics & Decision, 2012(23)

[5]   Chun-ji Liu, Xue-gang Feng. A study on the behavioral intention of information search before Chinese tourists' outbound tours: based on the model of TAM, TPB and DTPB. Tourism Science, 2013(2)

[6]   Ying-luo Wang. System engineering. China Machine Press, 2011

[7]   Guo-hua Wang, Liang Liang. Decision theory and method. University Press of Science and Technology of China, 2006

[8]   Ze-jun Liao, Lin Bai. TAM-based empirical study on influencing factors of micro-blog usage of college students. Journal of Beijing University of Posts and Telecommunications (Social Sciences Edition), 2012(2)

[9]   Ke He, Jun-biao Zhang. Farmer demand for the low-carbon utilization of biomass. Resources Science, 2013(8)

[10] Xin Zhang, Wen Dong, Tian-he Chi, Hua-sheng Hong. Study on information service system for ocean disaster prevention and reduction in Taiwan strait area. Journal of Natural Disasters, 2009(6)

[11] Guang-xi Cao, He-lian Ding, Yu Guo. Meteorological disaster prevention and mitigation services benefit evaluation based on SEM. Application of Statistics and Management, 2013(6)

[12] Zhen-lian Zhang. Research on the index system of evaluating the quality of network information service. Journal of Information, 2005(2)

[13] Ze-mei Zhang. Evaluation system of information service qualities under the networked environment. Information Science, 2006(4)

[14] Yu-lin Dai, Qin Yao, Ruo-gu Kang, Zao Fan. An analysis of factors of social network information service quality. Journal of Information, 2008(1)

[15] Yu-ying Jiao, Xue Lei. Establishment and survey analysis of a consumer satisfaction-based evaluation model for network information service quality. Library and Information Service, 2008(2)

[16] Sheng-jin Shen. Research on the evaluation of network information service quality of university library. Information Research, 2012(12)

[17] Xiao-yan Yang, Jie Chen. ISM model of the knowledge flow element in the supply chain coordination. Soft Science, 2013(5)

An Ontology-based Semantic Retrieval Model for Fault Case

Qian-yun KE*, Qing LI, Jin-liang CHEN

School of Mechanical Engineering and Automation, Beijing University of Aeronautics and Astronautics, Beijing, China

(yjmymh2011@163.com)

Abstract - **The lack of a comprehensive retrieval method for aircraft fault case knowledge had caused some sharing and reusing problems. To deal with this, research about the method of knowledge representation and semantic retrieval method based on ontology was carried out. Ontology model of aircraft fault case knowledge was established according to the particularity of aircraft fault domain and the actual demand of knowledge sharing and reusing. With Chinese segmentation tools and Lucene retrieval tools, three main issues in semantic retrieval were solved, namely, fault case indexing, semantic extension and ranking mechanism. A new semantic similarity calculation method focus on semantic distance, hierarchical factor, coincidence of superior and subordinate concepts was proposed. The original ranking mechanism was modified according to the semantic similarity. The prototype system of semantic retrieval was developed based on this approach and was used successfully to raise the precision and recall values. Results show that user's requirements in semantic retrieval were satisfied.**

Keywords - **Ontology, semantic extension, semantic retrieval**

I. INTRODUCTION

The huge increase in the amount and complexity of fault case knowledge in the process of aircrafts' daily maintenance and support activities caused an excessive demand for tools and techniques that can obtain valuable information from huge amount of information accurately and efficiently. Current practice in information retrieval mostly rely on keyword-based search over full-text data, however, such a model misses the actual semantic information.

Ontology [1], proposed for knowledge representation, provides a common standard of concept understanding and supports the sharing and reusing of knowledge, and has become the backbone of semantic information retrieval. Currently, the research on ontology-based semantic retrieval has made certain achievements: Sonor Kara [2] applied the rule-based inference to the football field, which improved the performance of retrieval system, hui-ying gao [3] put forward the model of enterprise content retrieval, and made it possible to get the implied semantic information of user that the keyword-based method cannot get. However, when it comes to aircraft fault case, such a model loses its power. Combining the usability of keyword-based retrieval technologies is one of the most challenging areas in semantic searching, all the efforts towards increasing retrieval performance user-friendly will eventually come to the point of improving keyword-based semantic technologies.

To deal with this issue, ontology model of aircraft fault case was established according to the particularity of aircraft fault domain. The method of semantic similarity computation based on hierarchical structure and logic relations of ontology was carried out, and was used successfully to make semantic extension for query words. The similarity score calculation formula in Lucene was also improved on the base of semantic similarity, which made the query results become more accurate. The main concern in this study was achieving a high semantic performance in fault case domain. The method had been proved to have increased the precision and recall values.

II. FAULT CASE ONTOLOGY DESIGN

Fault case knowledge consists of concepts, logic relations and restrictions between concepts in the field of aircraft fault. It is an objective description of fault case features and their processes. Basic knowledge composition of fault case is classed in three categories:

1) *Basic information*: information source, fault location, fault date, plane type, user units, etc.

2) *Description information*: fault occur time, fault position, fault phenomenon, fault effect, etc.

3) *Process information*: fault discrimination, fault part, fault mode, fault cause, fault handing, etc.

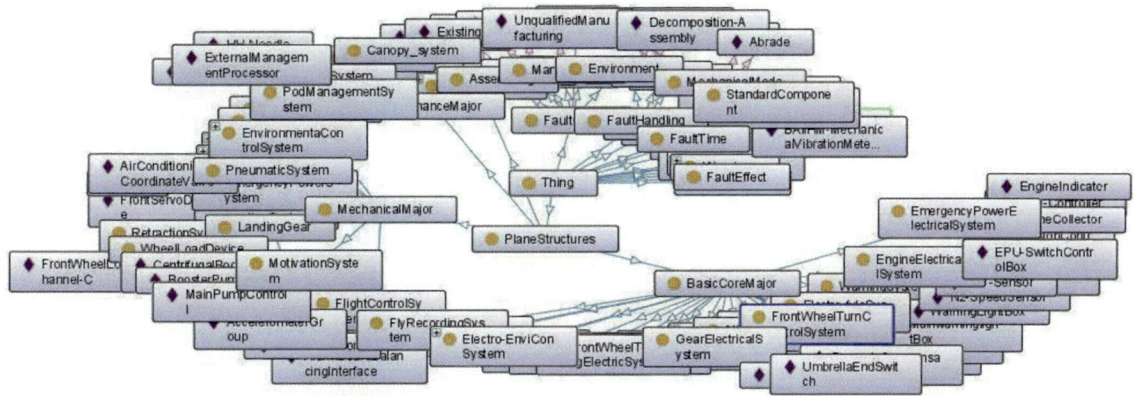

Fig.1. Fault case ontology model

All these knowledge were combined in strong hierarchical and logic relations. A central fault case ontology was designed according to this, which was utilized by nearly every aspect of the model, especially in semantic extension. Thus the overall performance of the model is highly dependent on its quality. According to the definition of ontology proposed by Gruber, ontology is a clear and formal specification of domain concepts [4], the ontology engineering phase was an iterative development process. Fault case ontology was finally ended up with containing 69 classes, 89 properties and hundreds of individuals [5]. Part of the structure can be seen in Fig.1.

III. SEMANTIC EXTENSION AND SEMANTIC SIMILARITY COMPUTATION

Semantic extension is one of the most important parts of ontology-based semantic applications. It is the process of expanding a keyword to two or more relevant words by synonym extension, ontology concept extension and ontology individual extension. Extending the query words semantically makes search results much more comprehensive. Obviously, as the extension words cannot replace the original words entirely, it is necessary to calculate the semantic similarity degree. We proposed semantic similarity calculation model on the basic of synonym extension, concepts extension and individual extension.

A. Synonym Extension

There are large numbers of colloquial expression for each terminology, such as abbreviations, nickname, idioms, etc. For example, "Fa Dong Ji" is commonly known as "Yin Qing". In order to get all the concepts, individuals and properties of ontology, Jena tools was used to parse the fault case ontology and generate a user custom dictionary [6]. Add synonyms and near-synonyms for the words in the dictionary, and define their similarity. Table I gives an idea about the similarity for different type of relations.

TABLE I
SYNONYM TYPES AND SIMILARITY

Relation Type	Similarity	Description
Synonyms	1.0	Has the same meaning, can replace each other
Near-synonyms strong	0.9	Can replace each other in most cases
Near-synonyms middle	0.8	Can replace each other in some cases
Near-synonyms weak	0.6	Can replace each other in few cases
Nickname	1.0	Can replace each other
Abbreviations	0.9	Can replace each other in most cases

B. Ontology Concepts Extension and Similarity

Ontology concepts extension is frequently used in semantic retrieval to express the user's query requirements. To obtain more comprehensive and effective query words, we need to calculate the similarity between concepts and take the ones that reach the threshold as new query words [7, 8].

Current studies on semantic similarity of ontology concepts are carried out in two main categories [9, 10], namely, structure-based method and feature-based method [11, 12]. Structure-based method is simple and high efficiency, unfortunately, this method relies on the integrity of the semantic links and semantic coverage, which restricts the improvement of accuracy. And the feature-based method is required to adjust the parameters to balance the proportion [13], resulting in poor generality. To make up the shortcomings of them, the model to calculate similarity with semantic distance, hierarchical factor and the coincidence of superior and subordinate concepts was carried out, which made it more comprehensive.

When two ontology concepts have common semantic features, we define them semantic similar, and use $sim(A,B)$ to represent the similarity between concept A and concept B. The similarity is defined as follows:

1) The value of $sim(A,B)$ belongs to [0,1], namely $sim(A,B) \in [0,1]$.

2) The similarity of two completely similar concepts is 1, namely $sim(A,B)=1$, if and only if $A=B$.

3) The similarity of two concepts that have no common features is 0, namely $sim(A,B)=0$.

4) Similarity is with symmetry, namely $sim(A,B)= sim(B,A)$.

1) Semantic Distance

Semantic distance refers to the length of the shortest path linking two concepts in the ontology diagram [14] when all the edge length of the diagram is 1. We use $dis(A,B)$ to describe the semantic distance of concept A and B. Generally speaking, $dis(A,B)$ represents a real number belongs to $[0,\infty]$. If $dis(A,B)=\infty$, $sim(A,B)=0$, and if $dis(A,B)=0$, $sim(A,B)=1$. Moreover, when the difference between two groups of semantic distance is fixed, the bigger the distance is, the smaller the difference between their similarities is, which means the function of similarity about semantic distance should be concave. To sum up, equation (1) is suitable for calculating similarity. For example, as is shown in Fig.2, $dis(J,G)=1$, $dis(J,F)=3$, then we can get $sim(J,G)=e^{-dis(J,G)}=e^{-1} \approx 0.36788$, and $sim(J,F)=e^{-dis(J,F)}=e^{-3} \approx 0.04979$

$$sim(A,B)=e^{-dis(A,B)} \qquad (1)$$

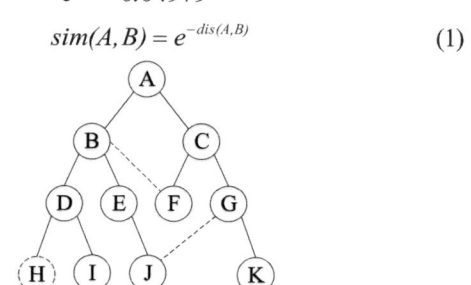

Fig.2. Ontology diagram.

2) Hierarchical Factor

In the hierarchy structure of ontology, the concepts were classified from coarse to fine, big to small, as a result, the similarity between concepts which are at the level far from the top level is greater than the ones nearer[15], meanwhile, the similarity increase with the decrease of level difference between the concepts. In other words, semantic similarity of any concepts increase along with the increase of the sum of their level, and decrease with the increase of their difference. Another thing to note is that, similarity should be between 0 and 1. Ultimately, equation (2) can explain the relation between similarity and concept level.

$$sim(A,B) = \frac{level(A) + level(B)}{2L(|level(A) - level(B)| + 1)} \quad (2)$$

$level(A)$ represents the level where concept A is. L represents the height of ontology. For example, the similarity of concept H and E in Fig.2 should be

$$sim(H,E) = \frac{level(H) + level(E)}{2L(|level(H) - level(E)| + 1)} = \frac{4+3}{2 \times 4 \times (|4-3|+1)} = 0.4375 \cdot$$

3) Coincidence of Superior and Subordinate Concepts

Coincidence of two ontology concepts, a percentage, is the ratio of their common superior and subordinate concepts numbers to all of their superior and subordinate concepts numbers. Obviously, similarity rises as coincidence increase. Considering the concepts with certain logic relations have greater similarity, the relative concept's direct superior and subordinate concepts should be taken into consideration when calculating their coincidence. Again, take Fig.2 as an example. Concept A, B, C, D, E, F, H, I, J are all the superior and subordinate concepts of B and F, among them, A, B, C, D, E, F are the concepts they share, thus, coincidence of B and F is 0.667. Introduce logarithmic function, and finally we get equation (3) as the similarity formula. Thus, the similarity of B and F is

$$sim(B,F) = \ln(1 + |\frac{con(B) \cap con(F)}{con(B) \cup con(F)}|) = ln(1 + 0.8) \approx 0.58779 \cdot$$

$$sim(A,B) = ln(1 + |\frac{con(A) \cap con(B)}{con(A) \cup con(B)}|) \quad (3)$$

$con(A)$ represents the set of all the superior and subordinate concepts of A.

Synthesize each kind of situation above, the comprehensive similarity of ontology concepts comes to be equation (4).

$$sim(A,B) = \alpha sim_1(A,B) + \beta sim_2(A,B) + \gamma sim_3(A,B) \quad (4)$$

$sim_1(A,B)$, $sim_2(A,B)$, $sim_3(A,B)$ represent the similarity of sematic distance, hierarchical factor and the coincidence of superior and subordinate concepts, respectively. α , β , γ are regulatory factors.

C. Ontology Individuals Extension

Ontology individuals extension is the third kind of sematic extension, if the query keyword matches any one of ontology individuals, then firstly obtain each property of the matched individual, and combine them with the

original keyword as a set of new query word, whose similarity with the original word is 1. After that, find the concept which the individual belongs to, and make semantic extension as the method mentioned in last section.

IV. SEMANTIC RETRIEVAL MODEL

For the detail process of semantic retrieval, a semantic retrieval model based on fault case ontology was established, as is shown in Fig.3. We adapt a semantic indexing approach based on Lucene indices. The idea is extending traditional full-text index with the extracted and inferred knowledge and modifying the ranking mechanism so that the fault case containing overall information gets higher rates. The details of the index structure and ranking mechanism are given in Section 4.1 and 4.2 respectively.

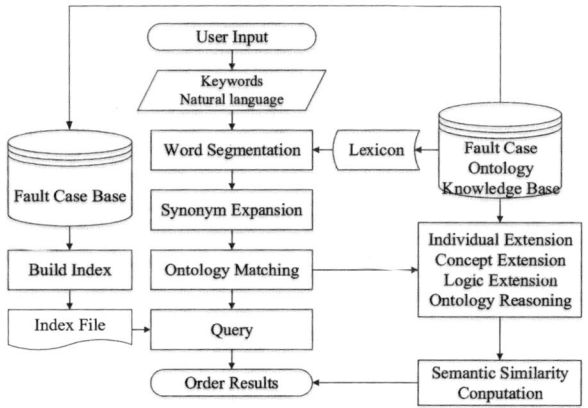

Fig.3. Semantic retrieval model for fault case

A. Index Structure

The structure of semantic index has utmost importance in the retrieval performance. We constructed a Lucene index such that aims at different kind of fault case knowledge. Generally, fault case were stored in database, retrieve attribute items can be divided into four types according to their different effect and different retrieval mode:

1) *Constraint condition attribute items*: This type do not need word segmentation to extract features, we mapped them with index files directly, fault part and plane type are typical examples.

2) *Content attribute items*: This type of knowledge need to go through word segmentation, and link the features to index files, for instance, fault phenomenon.

3) *Database location attribute items*: Location information of fault case knowledge, a parameter of database associate with index files.

4) *Primary key attribute items*: With global uniqueness. Index files are connected with database with real-time updates.

On the basic of different type of knowledge, we work out different mapping methods, they ultimately become feature-based indexing. This is especially important if the query is natural narrations. Segmenting the query into words, and linking them to features tolerate the incomplete query information, thus ensures at least the recall values.

B. Searching and Ranking

First of all, segment the query word [16]. To ensure the accuracy of the segmentation, Jena tools was used to parser ontology. At last, all the concepts, individuals and properties were parsed out, and added to segmentation thesaurus. Then words without real meaning, such as prepositions, quantifier, and adjectives, were filtered, then a new keywords group was born. Next, determine whether each keyword belongs to the ontology, and divide them into ontology words and non-ontology words.

In traditional keyword search, the query word usually contains nothing but the word itself, which severely restrict the recall ratio of search results. Since semantic extension is a key step of semantic retrieval, we extend ontology words and calculate the semantic similarity between them, once threshold similarity is reached, the words would be taken as new query words.

Indexed data usually contain just their raw text associated with that data. Lucene can easily handle such indices and its default ranking mechanism (5) gives usually good results.

q represents query word, $score(q,d)$ represents the ranking score of fault case d, $t.d$ represents a term of q, and $tf(t.d)$ is the term frequency in each fault case. $idf(t)$ represents the term frequency in inverted fault case. $boost(t.field.d)$ is the weight value of fault case field. $lengthNorm(t.field.$ represents the number of term in its field. $coord(q,d)$ is coordination factor. $queryNorm(q)$ is normalization constant of query words, it is the sum of squares of each item.

However, the extended words has complex indices, and should be handled carefully. Considering the extension word are associated with the original word in semantics, the ranking mechanism of Lucene may not be appropriate. For example, if an extension word has higher frequency, or an original word has too many extensions, their effect on the score may go over that of others', which would make the results far away from authenticity. In order to take advantages of our ontology-aided similarity, we slightly modified the default querying and ranking mechanism of Lucene. The improved equation (6) and (7) is far more suit for semantic retrieval model.

$$score(q,d) = \sum_{t.q}(tf(t.d) \times idf(t)^2 \times boost(t.field.d) \times lengthNorm(t.field.d)) \times coord(q,d) \times queryNorm(q) \tag{5}$$

$$score(q,d) =$$
$$\sum_{et.q}(tf(et.d) \times idf(et)^2 \times boost(et.field.d) \times lengthNorm(et.field.d) \times sim(et,t) \times w_{et.t}) \times coord(q,d) \times queryNorm(q) \tag{6}$$

$$w_{et.t} = \frac{sim(et,t)}{\sum_{et.t}sim(et,t)} \tag{7}$$

et represents extension word of t, $w_{et.t}$ is weight of et in all of t's extension words, which can be get by equation (7).

After analyzing large number of fault case, we concluded that different fault major have different fault rate. For example, the fault rate in avionics was much bigger than that of armament, which brought more fault case to avionic than armament. Therefore, simply calculating the frequency in all fault case is not enough, several fields were set on the basic of different major, each of which has different weight which depends on their fault case number. We called it $boost(t.field.d)$. It means that the field with less fault case would has bigger weight. Thus, the search results would rank as their score more reasonable.

V. CONCLUSION

In this paper, an ontology-based semantic retrieval model and its application to fault domain were constructed, which includes ontology construction, fault case indexing, semantic extension and semantic score ranking. During the application of the model in fault case domain, we observed that ontology-based semantic extension greatly boosts the precision and recall values. Moreover, the improved ranking mechanism further improve the performance and allow complex domain-specific queries to be handled successfully. Having observed the success in fault case domain, we presume that similar performance can be achieved in other domains as well by constructing relevant ontology and modifying the ranking mechanism for different domain. We also plan to further improve semantic retrieval by using information extraction technology to populate ontology.

REFERENCES

[1] Y. Yuehua. "Research on emergency intelligent information retrieval system based on domain knowledge model (Thesis or Dissertation style)" (in Chinese), Ph.D. dissertation, Beijing University of Posis and Telecommunicationns. Beijing, China, 2013.

[2] S. Kara, O. Alan, O. Sabuncu, S. Akpinar, N. K. Cicekli, F, N, Alpaslan, "An ontology based retrieval system using semantic indexing", *Inf. Syst,* vol. 37, no 4, pp.294-305, 2012.

[3] G. Huiying, Z. Jinghua, "Ontology-based enterprise content retrieval method (Periodical style)" (in Chinese), *Journal of Computers*, vol. 5, no. 12, pp. 314-321, Feb, 2010.

[4] Y. Liu. "Fuzzy ontology modeling methods and semantic information processing strategies (Thesis or Dissertation style)" (in Chinese), *Changsha: Central South University,* Changsha, China, 2011.

[5] Z. Mei, H. Jia, Y. Yan, L. Bo, "Ontology-based knowledge modeling" (in Chinese), *Transactions of Beijing Institute of Technology*, vol. 30, no. 12, pp. 1406-1408, Dec, 2010.

[6] G. R. Juan, P. Miguel, G. Jesus. "Ontology-based context representation and reasoning for object tracking and scene interpretation in video" (in Chinese), *Expert Systems with Applications*, vol. 6, no. 38, pp. 7494-7510, 2011.

[7] L. Wenqing, S. Xin, Z. Changyou, F. Ye. "A semantic similarity measure between ontological concepts" (in Chinese), *Acta Automation Sinica*, vol. 2, no, 38, pp. 229-235, Feb, 2012.

[8] D. Sanchez, M. Batet, A. Valls, "Ontology-driven web-based semantic similarity", *Journal of Intelligent Information Systems*, vol. 3, no. 35, pp. 383-413, 2010.

[9] G. Qian, C. Y. Im. "A multi-agent improved semantic similarity matching algorithm based on otology tree" (in Chinese), *Journal of Institute of Control, Robotics and Systems*, vol. 11, no 18, pp. 1027-1033, 2012.

[10] X. Jian, X. Zhuo, D. Zhaojun. "Research on term similarity computational methods based on network knowledge resources" (in Chinese), *Information Science*, vol. 30, no. 11, pp. 1745-1750, Nov, 2012.

[11] M. Batet, D. Sanchez, A. Valls. "An ontology-based measure to compute semantic similarity in biomedicine", *Journal of Biomedical Informatics*, vol. 1, no. 44, pp. 118-125, 2011.

[12] S. Haixia, Q. Qing, C. Ying. "Review of ontology-based semantic similarity measuring" (in Chinese), *New Technology of Library and Information Service*, vol. 1, no. 188, pp. 51-56, 2010.

[13] D. Sanchez, M. Batet, D. Isem. "Ontology-based semantic similarity: a new feature-based approach", *Expert Systems with Applications*, vol. 9, no. 39, pp. 7718-7728, 2012.

[14] A. Budanitsky, G. Hirst, "Evaluating WordNet-based measures of semantic distance", *Comput Linguistic*, vol. 1, no, 32, pp. 13-17, 2006.

[15] I. Choi, D. St-Onge. "Distillation using hierachy concept tree", In: *Proceedings of the 26th Annual International ACM SIGIR Conference on Research and Development in Information Retrieval*. Toronto, Canada, pp. 371-372, 2003.

[16] M. Yifu. ".Research on key technology and algorithms in unstructured information extraction based on cognition (Thesis or Dissertation style)" (in Chinese), *China University of Mining and Technology*, Beijing, China, 2013.

A Convergent Nonlinear Smooth Support Vector Regression Model

Li-ru TIAN*, Xiao-dan ZHANG

School of Mathematics and Physics, University of Science and Technology Beijing, Beijing, China
(tianliru19890321@163.com)

Abstract - **Research on the non-smooth problems in the nonlinear support vector regression. A nonlinear smooth support vector regression model is proposed. Using a generalized cubic spline function approach the non-smooth part in the support vector regression model. The model of the nonlinear smooth support vector regression is solved by BFGS-Armijo. Then, the approximation accuracy and the astringency of the generalized cubic spline function to the ε – insensitive loss function were analyzed. As a result, we found the four-order and six times spline function's approximation effect is better than other smooth functions, and the nonlinear smooth support vector regression model, which be proposed in this paper is convergent.**

Keywords - **Kernel, nonlinear, spline function, support vector regression**

I. INTRODUCTION

Support vector regression (SVR), a support vector machine (SVM) [1] for regression, has been widely applied to the fields of machinery fault diagnostic technique, dynamics environmental forecasting, and earthquake prediction. Based on VC dimension and structural risk minimization, it can solve some practical problems such as sparsity, nonlinearity, high dimension, etc. However, in practical applications, the training data sets in some important fields, such as telemetry data of rockets and missiles and data of human experimentation of vaccine, are sparse, of small size and contaminated by noise. This may decrease the generalization ability and the prediction accuracy of the algorithm. One way to solve this problem is to improve algorithm structure. Recently, several new SVM learning algorithms have been proposed for more powerful generalization ability. Such as proximal support vector machine(PSVM) [2], least squares support vector machine(LSSVM) [3], Primal twin support vector regression [4], Least squares twin support vector machines [5] and twin support vector regression(TSVR) [6] etc.

In this paper, we propose a new model of nonlinear smooth support vector regression. Generally, smooth support vector machine achieves smooth effect by approximating the square of ε – insensitive loss function. However, in this paper, we propose a new full smooth cubic spline function, it with 4-order differentiability, to approximate the ε – insensitive loss function directly.

II. THE MODEL OF NONLINEAR SMOOTH SUPPORT VECTOR REGRESSION

Given a training set $\{(x_i, y_i)\}_{i=1}^{m}$ with $(x_i, y_i) \in R^n \times R$ $(i = 1, 2, \cdots, m)$. Let $y = (y_1, y_2, \cdots, y_m)^T$. $x_i(i = 1, 2, \cdots, m)$ is denoted by $A_{m \times n}$, $y = (y_1, y_2, \cdots, y_m)^T$ is denoted by Y.

Recording $e \in R^m$ as a column vector components are all ones. Mark $\omega \in R^n$ as the weight vector, $b \in R$ as bias. For linear support vector regression machine [7] it, hoping to find the linear regression function $f(x) = \omega^T x + b$, makes small deviation from it with the training points. That is, to meet the constraint optimization problem

$$\min_{(\omega, b) \in R^{n+1}} \frac{1}{2}(\omega^T \omega + b^2) + Ce^T \xi \qquad (1)$$
$$s.t. \quad \left| Y - (A\omega + b) \right| - e\varepsilon \leq \xi, \ \xi \geq 0$$

Here $C > 0$ is a penalty parameter, ξ is a slack variable, which used to reflect the training points fall on the parallel insensitive interval.

For the nonlinear support vector regression machine [8], looking for the nonlinear regression function satisfies the constraint optimization problem

$$\min_{(\omega, b) \in R^{n+1}} \frac{1}{2}(\omega^T \omega + b^2) + Ce^T \xi \qquad (2)$$
$$s.t. \quad \left| Y - (K(A, A^T)\omega + b) \right| - e\varepsilon \leq \xi, \ \xi \geq 0$$

Here $K(x^T, x_i)$ is a nonlinear kernel [9], $K(A, A^T)$ is the kernel matrix. In (2) let the slack variable ξ be the following form:

$$\xi = \left| Y - (K(A, A^T)\omega + b) \right|_{\varepsilon} \qquad (3)$$

Here $|\cdot|_{\varepsilon}$ is the ε – insensitive loss function [7] $|x|_{\varepsilon} := (|x_1|_{\varepsilon}, |x_2|_{\varepsilon}, \ldots, |x_m|_{\varepsilon})^T$ and $|x_i|_{\varepsilon} = \max\{0, |x_i| - \varepsilon\}$. By substituting ξ in (2) by (3), then we can have an unconstrained optimization problem:

$$\min_{(\omega, b) \in R^{n+1}} \frac{1}{2}(\omega^T \omega + b^2) + Ce^T \left| Y - (K(A, A^T)\omega + b) \right|_{\varepsilon} \qquad (4)$$

However, $\left| Y - (K(A, A^T)\omega + b) \right|_{\varepsilon}$ in model (4) is not differentiable. Thus the model (4) is not smooth. In order to apply the fast optimization algorithm, we need to introduce a smooth function to approximate the ε – insensitive loss function to make the target model smooth. In this paper, to make (4) smooth, we achieved the following model:

$$\min_{(\omega, b) \in R^{n+1}} \frac{1}{2}(\omega^T \omega + b^2) + Ce^T f(Y - (K(A, A^T)\omega + b), k) \qquad (5)$$

$f(x)$ is a smooth function, which approximates the ε – insensitive loss function.

In literature [2, 10, 11], they solved the model as follows:

$$\min_{(\omega, b) \in R^{n+1}} \frac{1}{2}(\omega^T \omega + b^2) + \frac{C}{2} \sum_{i=1}^{m} \left| y_i - (K(A_i, A^T)\omega + b) \right|_{\varepsilon}^2 \qquad (6)$$

All of them, using smooth function approximate positive sign functions firstly, then transforming the function, in order to make $|\cdot|_{\varepsilon}^2$ smooth. The six

E. Qi et al. (eds.), *Proceedings of the 21st International Conference on Industrial Engineering and Engineering Management 2014*, Proceedings of the International Conference on Industrial Engineering and Engineering Management, DOI 10.2991/978-94-6239-102-4_43, © Atlantis Press and the authors 2015

polynomial function put forward in the literature [2] as follows:

$$P_0(x,k) = \begin{cases} \dfrac{1}{32k}(k^6 x^6 - 5k^4 x^4 + 15k^2 x^2 + 16kx + 5), & |x| < 1/k \\ (x)_+ & |x| \geq 1/k \end{cases} \quad (7)$$

The literature [11] put forward using 3-order spline function approximate the positive sign function. In literature [12], using the Sigmoid function to approximate the positive sign function. The above mentioned practices achieve the smooth model, but lead to the order of functions be too great, the computational be too complex. In this paper, to approach $\varepsilon-$ insensitive loss function, through the translation transformation of the 4-order six times spline function, we obtain the nonlinear smooth support vector regression based on the model (5).

III. PERFORMANCE ANALYSIS OF SPLINE SMOOTHING FUNCTION

A. Structure and accuracy analysis of smooth spline function

Definition 1 [13]: Let $(\cdot)_+$ be the positive sign function, $(x)_+ = \max\{0,x\}$, $k > 1$, $(0, y_1)$ is one point on the y axis. In the interval $[-1/k, 1/k]$, $-1/k$, 0, $1/k$ are nodes of $S(x,k)$. If $S(x,k)$ satisfies:

(1) $S^{(d)}(-1/k,k) = 0, d = 1, 2, \cdots, m$, $S(-1/k,k) = 0$, $S(0,k) = y_1$;

(2) $S^{(d)}(1/k,k) = 0, d = 2, \cdots, m$, $S'(1/k,k) = 1$, $S(1/k,k) = 1/k$;

(3) $S^{(d)}(0+0) = S^{(d)}(0-0), d = 1, 2, \cdots, m$.

Then $S(x,k)$ is called m-order full smooth spline function, which approximate $(x)_+$.

Lemma 1 [14]: Let $(\cdot)_+$ be the positive sign function, $k > 1$. $-1/k$, 0, $1/k$ are nodes in the interval $[-1/k, 1/k]$. Constructing 4-order six times smooth spline function $S_0(x,k)$ approximate $(x)_+$

$$S_0(x,k) = \begin{cases} -k^5 x^6/48 + 5k^3 x^4/16 - 5k^2 |x|^3/6 \\ \quad +15kx^2/16 + x/2 + 5/48k \end{cases} \quad |x| < 1/k \\ x_+ \quad\quad\quad\quad\quad\quad\quad |x| \geq 1/k \quad (8)$$

Theorem 1: Set $\varepsilon > 0, k > 1$, $x_0 = \varepsilon - 1/k, x_1 = \varepsilon, x_2 = \varepsilon + 1/k$ $x_3 = -\varepsilon - 1/k, x_4 = -\varepsilon, x_5 = -\varepsilon + 1/k$ are nodes in the interval $[-\varepsilon - 1/k, \varepsilon + 1/k]$, using the generalized Three-moment method, can construct 4-order six times spline function as follows

$$S(x,k) = S_0(-x - \varepsilon, k) + S_0(x - \varepsilon, k) \quad (9)$$

to approach $\varepsilon-$ insensitive loss function. Here $S_0(x,k)$ defined by (8).

Proof:

According to the definition of $\varepsilon-$ insensitive loss function $|x|_\varepsilon = \max\{0, |x| - \varepsilon\} = (x - \varepsilon)_+ + (-x - \varepsilon)_+$.

By lemma 1, it is easy to construct the 4-order spline function shown by (9) to approach $|x|_\varepsilon$.

Lemma 2 [14]: $x \in R$, $S_0(x,k)$ is defined by (8), then

(1) $S_0(x,k)$ meet the 4-order smoothness conditions;

(2) $S_0(x,k) \geq (x)_+$.

Theorem 2: $\varepsilon > 0, k > 1$, $S(x,k)$ is the smooth function shown by (9). $|x|_\varepsilon$ is the $\varepsilon-$ insensitive loss function, then

(1) $S(x,k)$ satisfies the 4-order smoothness condition in R;

(2) $S(x,k) \geq |x|_\varepsilon$, $x \in R$;

(3) $S(x,k) - |x|_\varepsilon \leq 5/48k$, $x \in R$.

Proof:

(1), (2) by Lemma 2 can be directly gain; (3) when $x \leq -\varepsilon - 1/k, x \geq \varepsilon + 1/k$ and $-\varepsilon + 1/k \leq x \leq \varepsilon - 1/k$, $S(x,k) - |x|_\varepsilon = 0$. Conclusions clearly established. When $x \in [\varepsilon - 1/k, \varepsilon]$, $S(x,k) - |x|_\varepsilon = S_0(x - \varepsilon, k) = -(\varepsilon k - kx + 5)(\varepsilon k - kx - 1)^5/48k$, let $a = k(x - \varepsilon)$, then $-1 \leq a \leq 0$, Thus $S_0(x - \varepsilon, k) = -(a+1)^5(a-5)/48k$ $S_0(x - \varepsilon, k)$ is monotone increasing, the maximum value is $S_0(0,k) = 5/48k$. when $x \in [\varepsilon, \varepsilon + 1/k]$, $S(x,k) - |x|_\varepsilon = S_0(x - \varepsilon, k) - x - \varepsilon = -(\varepsilon k - kx - 5)(\varepsilon k - kx + 1)^5/48k$ in the interval $[\varepsilon, \varepsilon + 1/k]$, $S(x,k) - |x|_\varepsilon$ is monotone decreasing. The maximum value is still the $S_0(0,k) = 5/48k$. As the same, in the interval $[-\varepsilon - 1/k, -\varepsilon + 1/k]$, the maximum value of $S(x,k) - |x|_\varepsilon$ is $5/48k$. To sum up, we know that, for arbitrary $x \in R$, $S(x,k) - |x|_\varepsilon \leq 5/48k$.

Comparison of different smooth function and approximate accuracy are shown by the figure and the table. Six times spline function are given from (9), six times polynomial functions is given in the style of Theorem 1, $P(x,k) = P_0(-x - \varepsilon, k) + P_0(x - \varepsilon, k)$, where $P_0(x,k)$ is provided by (7)

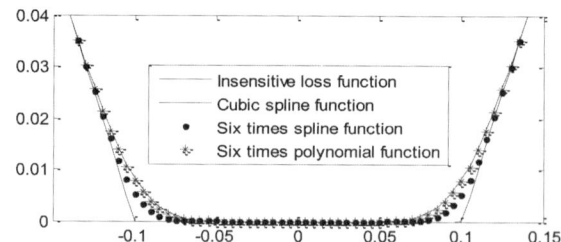

Fig.1. the overall approximation effect of the different smooth functions to the $\varepsilon-$ insensitive loss function ($\varepsilon = 0.1, k = 20$)

From Fig.1, we can see that six times spline's approximation effect is better than other smooth functions.

B. Convergence analysis of the model

In this section, to simplify the written certificate, we remember $x = [\omega, b]^T, \bar{A} = [A \ e], \bar{K} = [K(A, A^T) \ e]$

Theorem 3: Let $\overline{A} \in R^{m \times (n+1)}, Y \in R^{m \times 1}$, real functions $h(x) := R^n \to R$, $g(x,k) := R^n \times N \to R$ are defined as follows:

$$h(x) = \frac{1}{2}\|x\|_2^2 + Ce^T \left| Y - \overline{K}x \right|_\varepsilon \qquad (10)$$

$$g(x,k) = \frac{1}{2}\|x\|_2^2 + Ce^T S(Y - \overline{K}x, k) \qquad (11)$$

$S(x,k)$ is defined by (9),

(1) Optimization problem $\min h(x)$ exists optimum solution $\overline{x}$. Optimization problem $\min g(x,k)$ exists optimum solution $\overline{x}^k$ and $\lim_{k \to \infty} h(\overline{x}^k) = h(\overline{x})$.

(2) The optimal solution set of optimization problem $\min h(x)$ is D_h , then $\{\overline{x}^k\}$ exists a convergent subsequence $\{\overline{x}^{k_n}\}$ satisfies $\lim_{n \to \infty} \overline{x}^{k_n} = \overline{x}_h$, here $\overline{x}_h \in D_h$

Proof:

(1) Define the appropriate level set $L_\upsilon(g(x,k)) = \{x \mid x \in R^n, g(x,k) \leq \upsilon\}$ and $L_\upsilon(h(x)) = \{x \mid x \in R^n, h(x) \leq \upsilon\}$.

$S(x,k) \geq |x|_\varepsilon$, so that for any $\upsilon \geq 0$, they satisfy $L_\upsilon(g(x,k)) \subset L_\upsilon(h(x)) \qquad \subset \{x \mid \|x\|_2^2 \leq 2\upsilon\}$. Therefore, $L_\upsilon(g(x,k))$ and $L_\upsilon(h(x))$ are compact sets. Then optimization problems $\min h(x)$ and $\min g(x,k)$ exist optimal solutions . Let $\min h(x) = h(\overline{x})$, $\min g(x,k) = g(\overline{x}^k, k)$, for any $x \in R^n$, by theorem 2, we know

$$0 \leq g(x,k) - h(x) = Ce^T S(Y - \overline{K}x, k) - Ce^T \left| Y - \overline{K}x \right|_\varepsilon$$

$$= C\sum_{i=1}^{i=m} [S(y_i - \overline{K}x_i, k) - \left| y_i - \overline{K}x_i \right|_\varepsilon] \leq 5Cm/48k$$

So $\qquad 0 \leq g(\overline{x}^k, k) - h(\overline{x}^k) \leq 5Cm/48k$ and $0 \leq g(\overline{x}, k) - h(\overline{x}) \leq 5Cm/48k$ are established. Besides, $h(\overline{x}^k) \geq h(\overline{x})$, $g(\overline{x}, k) \geq g(\overline{x}^k, k)$, then $0 \leq h(\overline{x}^k) - h(\overline{x}) \leq h(\overline{x}^k) - h(\overline{x}) +$ $g(\overline{x}, k) - g(\overline{x}^k, k) = h(\overline{x}^k) - g(\overline{x}^k, k) + g(\overline{x}, k) - h(\overline{x}) \leq h(\overline{x}^k) - g(\overline{x}^k, k)$ $\leq 5Cm/48k$. So that we have $\lim_{k \to \infty} h(\overline{x}^k) = h(\overline{x})$.

(2) For any $k \in Z_+, k > 1, \|\overline{x}^k\|_2^2 / 2 \leq g(\overline{x}^k, k) \leq g(\overline{x}, k)$, $\{\overline{x}^k\}$ is bounded, then $\{\overline{x}^k\}$ has a convergent subsequence $\overline{x}^{k_n}$. Set $\lim_{n \to \infty} \overline{x}^{k_n} = \overline{x}_h$ then $\lim_{n \to \infty} h(\overline{x}^{k_n}) = h(\overline{x}_h) = \lim_{k_n \to \infty} h(\overline{x}^{k_n}) = h(\overline{x})$.

Therefore, $\overline{x}_h \in D_h$, that is to say, $\overline{x}_h$ is the optimal solution of optimization problems $\min h(x)$.

IV. CONCLUSION

In this paper, we proposed a new model of nonlinear smoothing support vector regression. Constructed a four-order and six times spline function. As a result, we found the four-order and six times spline function's approximation effect is better than other smooth functions, and the nonlinear smooth support vector regression model,

which be proposed in this paper is convergent.

REFERENCES

[1] V. Vapnik, "The nature of statistical learning theory", *New York: springer Verlag*, 1995.

[2] J. Z. Xiong, J. L. Hu , H.Q. Yuan, "Smoothing functions for support vector regressions (Chinese)", *PI&AI*. 2008, vol 21, no 3, pp.273-278.

[3] J. Suykens, V. Joos, "Least squares support vector machine classifiers", *Neural processing letters*, 1999, vol 9, no 3, pp 293-300.

[4] X. J. Peng,"Primal twin support vector regression and its sparse approximation", *Neurocomputing*, 2010, vol 73, no16, pp.2846-2858.

[5] K. Arun, M. Gopal, "Least squares twin support vector machines for pattern classification", *Expert Systems with Applications*, 2009, vol 36, no 4, pp.7535-7543.

[6] X. J. Peng, "TSVR: an efficient twin support vector machine for regression", *Neural Networks*, 2010, vol 23, no 3, pp.365-372.

[7] N. Y. Deng, Y. J. Tian, "New method of data mining-support vector machine (Chinese)", *Beijing: Science Press*, 2004.

[8] Y. J. Lee, W. F. Hsieh, C. M. Huang, " ε – SSVR: a smooth support vector machine for ε – insensitive regression", *Knowledge and Data Engineering, IEEE Transactions*, 2005, vol 17, no 5, pp.678-685.

[9] Z. J. Chen, Y. Cai, G. Jiang, "Study on SVM of complex Gaussian wavelet kernel function(Chinese)", *Application Research of Computers*, 2012, vol 29, no 9,pp.3263-3265.

[10] Y. Chen, J. Z. Xiong," A new method for solving the smooth functions of ε – insensitive support vector regression (Chinese)", *Computer engineering and science*, 2010, vol 32, no 8, pp.108-111.

[11] J. D. Shen," Research on a new function for smooth support vector regression (Chinese)", *Journal of China University of Metrology*, 2010, vol 21, no 2, pp.162-166.

[12] X. B. Chen, "Smooth twin support vector regression" *Neural Computing and Applications*, 2012, vol 21, no 3, pp.505-513.

[13] X. D. Zhang, S. Shao, Q. S. Liu, "Smooth support vecter machine model based on spline function (Chinese)", *Journal of University of Science and Technology Beijing*, 2012,pp.718-725.

[14] X. D. Zhang, H. L. Zhao, M. Wang, "A new smooth support vector machine and its application (Chinese)", *Mathematics in practice and theory*, 2014.

Game Analysis of Knowledge Resources Transactions

Jiawei Yan*, Yaoguang Hu, Jingqian Wen

School of Mechanical Engineering, Beijing Institute of Technology, Beijing, China

(*ys_keke@126.com, hyg@bit.edu.cn, wenjq@bit.edu.cn)

Abstract - **In order to realize the knowledge transaction in large research organizations, on the basis of the analysis of knowledge resources in large research organizations and research on related and abroad literatures, knowledge resource evaluation index system is established, calculation methods of index and weight are introduced, the model of knowledge seller and buyer is constructed. What's more, three different knowledge transaction modes are illustrated. Knowledge pricing process is explained by game theory model, the transaction price can be calculated using the transaction model. Overall, methods in this paper can increase the research capability and competitiveness of research organizations.**

Keyword - **Evaluation, game theory, knowledge resource, trade**

I. INTRODUCTION

People in large research institutions always use black box type way of negotiations in the process of knowledge resource deals, because there is not a unified standard and trading pattern. At the same time, Knowledge is created and innovated in the scientific research institutions. In order to promote the circulation of knowledge, mining the greatest value of knowledge, improve the promoting effect of knowledge in the organization's productivity and creativity, and promote formation of the organizational culture of knowledge management, I did some research.

Yaoguang Hu and Jingwen Li [1] have did some research in internal knowledge resources trading problem for scientific research institutions. They classified the knowledge resources and construct models to calculate the value range of knowledge resources. But in the actual transaction process, Buyers and sellers always need to bargain and bargain to determine the transaction price. Actually, this is a game process. So I analyzed the three kinds of knowledge resources transaction type (the contract transfer, the auction price transfer and the bidding price transfer) combined with the concept and principle of game theory.

II. GAME ANALYSIS OF THE CONTRACT TRANSFER

The contract transfer is that the deal is completed by the contract. In view of the game theory, Knowledge resources agreement transfer pricing process is a process of incomplete information dynamic game: Both sides of each other's information are incomplete. And then they test each other in the decision-making process, finally equilibrium result (if exist) and form the deal price.

A. Pricing factor analysis

1) Seller factor: The technical value of knowledge, the urgent degree of the seller's transfer of intellectual resources, license of knowledge resources;

2) Buyer factor: the use value of the knowledge resources, the buyer purchase urgent degree of knowledge resources, the buyer's patience;

3) Rationality factor: rationality assumption;

4) Competition factor: There is no cooperative game of both sides;

5) The supply and demand factor: Supply and demand affect the opportunity cost of both sides [2].

On the basis of buyers and sellers to estimate the value of knowledge resources, I focus on the factor of seller, and buyer.

B. Game process analysis

Before starting the analysis, we need to pay attention to an important hypothesis. That is the buyers and sellers of knowledge resources contract transfer is the specification in the sense of economic entities and they can make the right choice based on a rational man hypothesis. We used the Rubinstein's bargaining model [3] in the process of one-to-one pricing game analysis. Limited by the space, we just discussed the supply push transaction (the seller first bid, after the buyer bid) and demand-pull type transaction pricing process is similar with this. Our model is as follows:

(1) The basic elements of the game

a) Participant: Buyers and sellers of knowledge resources. We assume that when i = 1, the variable represents the seller and when i = 2, the variable represents the buyer.

b) Action: Because bargaining is a continuous process, so participant A and participant B will take turns to bid. At the beginning, the action of participant A is offer (Namely that make their own split share X_1 and participant B get $1-X_1$); the action of participant B have two options. One is accepting the price and the game is end; another is rejecting the price, and putting forward a new plan. Participant A also has a choice to choose to accept or reject. This will cycles until the price is accepted by both sides.

c) Information: Information is participant's quote.

d) Pay: Pay refers to the level of game participant's expected utility.

In this model, there is bargain cost. Conducting more than one round of the game, the income of each side will be one more discounted. We reflect this discount with the discount factor (participant A, δ_1; participant B, $\delta_2(0<\delta<1)$). Discount factor also reflects the buyers and sellers of urgency for knowledge transfer. The significance of the transaction is that the buyer and seller determine the distribution approach of the total surplus [4] (total surplus π = the buyer's highest expectations V_{Bmax} – the seller's lowest expectations V_{Smin}). When the game ends at t stage, the pay participants obtained is as follow:

Participant A: $\pi_1 = \delta_1^{t-1} X_i$ (1)

Participant B: $\pi_2 = \delta_1^{t-1}(1 - X_i)$ (2)

If t is an even number, i = 1; odd i = 2.

(2) Limited and indefinitely Game Analysis

According to the foregoing analysis, the expansion expression of the game is as shown in the Fig.1:

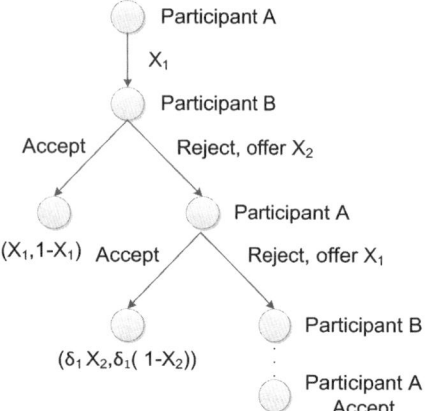

Fig.1. Game Tree of transfer agreement

For a limited game, we can use the reverse recursive method to solve the sub-game refining Nash equilibrium [5]. For example, If the game only have two stages, when t=2, Participant B offers the price and at the same time Participant A will accept it. Because Participant A have no choice to offer a new price. The Participant B obtained when t=2 is equal to δ_2 when t=1. So if Participant A bids $(1-x_1) \geq \delta_2$ at t=1, Participant B will choose to accept. Sub-game perfect Nash equilibrium outcome is Participant A get $x = x_1 = 1-\delta_2$, participant B get $1-x_1 = \delta_2$. The same can be obtained at any $T<\infty$ sub-game perfect Nash equilibrium. In general, when t is an odd number, a person's participation share is even larger than when t is even. That is to say: when the game has a deadline, the final bidding participants have the advantage.

According to proved Rubinstein, for the indefinite game, we can find the sub-game perfect Nash equilibrium based on the limited stage reverse recursion [6]. We make the following assumptions: when t=∞, participant A bids. For the participant A, he can get the smallest share is equivalent to the share when t=t-2. There are m = $1-\delta_2$ $(1-\delta_1 m)$

$$m = \frac{1-\delta_2}{1-\delta_1\delta_2}$$ (3)

When the largest share is the same with the smallest share, the result is a unique equilibrium.

$$X_1 = \frac{1-\delta_2}{1-\delta_1\delta_2} \qquad 1 - X_1 = \frac{\delta_2(1-\delta_2)}{1-\delta_1\delta_2}$$ (4)

When we give different values of δ (That means degrees of patience as well as the urgency of the transfer of knowledge resources is changing), we got the following conclusions:

a) Fixed δ_2 when δ_1 trend into an infinite, $X_1=1$;

b) Fixed δ_1 when δ_2 trend into an infinite, $X_2=1$;

Through the above analysis, we can draw that people with enough patience and knowledge transfer to the lower degree of urgency can always delay the time to obtain the highest residual value.

c) When $\delta_1 = \delta_2 < 1$, $X_1 = 1/(1-\delta) > 1/2$.

Through the above analysis, we can draw that when the game is indefinite and buyers and sellers have the same degree of patience and urgency of the knowledge transfer, the party taking the lead in negotiating pricing has first-mover advantage. This reminds us that in knowledge transactions distinguishing between demand pull and supply push type is meaningful [7].

In summary, the transaction price (P) of knowledge resources during transfer agreement process in large research institutions is the seller's minimum expectations plus seller's share or buyer's maximum expectations and buyer's share.

$$P = V_{Smin} + \pi X_1 = V_{Bmax} - \pi(1 - X_1)$$ (5)

Under the "one to one" knowledge resource transfer agreement transaction mode, we can draw the trading price of knowledge resources according valuation of knowledge resources, seller and buyer's patient extent and transfer of urgency, and finally promote knowledge exchange

III. GAME ANALYSIS OF AUCTION PRICING TRANSFER

Auction pricing transfer is the trading patterns that highest price obtained ownership of knowledge through public bidding. From the point of view of the game, auction pricing transfer is a three-stage incomplete information dynamic game: The seller gives the reserve price and sets the auction mechanism; buyer decides whether to participate in the auction; bidding.

A. Pricing factor analysis

a) Buyer: Buyer's intended use value brought by the knowledge resources;

b) Auction mechanism: Due to the asymmetry of information, Signal space which can reveal buyer's private information is infinite [8] and auction mechanism is infinite. So the seller can only make a choice in some representative auction mechanism;

c) The number of participants: The number of participants in the auction transaction will have some impact in price;

d) Buyer's bid: Buyers should consider other buyer's bid when bidding.

B. Auction mechanism design

Optimal design of the auction mechanism can be seen the process of signal game [9]. Auction mechanism design is that the seller must choose among an infinite number of signal space and define a configuration function whcih determines the transaction object and transaction price. It is almost impossible. Therefore, this paper only considers several common auction mechanisms. In general, the mechanism design problems include the following:

1) Participants: one seller with private Information and

several buyers without private Information;

2) Information: the type set of buyer I (Θ_i, i=1,2,...,n) and type space (Θ=X$_i\Theta_i$)'s spatial distribution (F(.));

3) Actions and sequence of actions: The vendor leads the action. They first determine the buyer's signal space (S=X$_i$S$_i$) and configuration function (y=(x(.),t(.)), x(.) is decision vector; t(.) is payment vector). And then buyers decide whether to participate in the auction;

4) Payment: We assume that the utility function is a Von neumann Morgenstein type [10] expected utility function.

By analyzing the game process of the auction, we can build the following model to describe the optimal auction mechanism.

$$\max_{x,t} E_\theta [\Sigma_{i=1}^n t_i (s^*_i, s^*_{-i})] \qquad (6)$$

s.t.

$$E_{-\theta_i}[\theta_i x_i(s^*_i, s^*_{-i}) - t_i (s^*_i, s^*_{-i})] \geq \bar{u}_i, \forall i \qquad (7)$$

$$E_{-\theta_i}[\theta_i x_i(s^*_i, s^*_{-i}) - t_i (s^*_i, s^*_{-i})] \geq$$

$$E_{-\theta} [\theta_i x_i(s^*_i, s^*_{-i}) - t_i (s^*_i, s^*_{-i})], \forall s_i \in S_i, \forall i \qquad (8)$$

Constraint one is rational participation constraint. Constraint two is buyer's incentive compatibility constraint.

According to "revenue equivalence theorem [11]", we choose the first-price sealed auction for analysis.

C. The first-price sealed auction's game analysis

The decision function is as follow:

$$(S_i, S_{-i}) = \begin{cases} 1, \text{when } S_i > S_j, \forall j \neq i \\ \frac{1}{m}, \text{when } m \text{ bid highest} \\ 0, \text{when } \exists S_j > S_i \end{cases} \qquad (9)$$

Transfer Function t$_i$:

$$t_i(S_i, S_{-i}) = \begin{cases} -S_i, \text{when } S_i > S_j, \forall j \neq i \\ 0, \text{when } \exists S_j > S_i \end{cases} \qquad (10)$$

From the game's point of view [12], the first-price sealed auction is an incomplete information static game between different types of buyer. Game model is as follows:

(1) The basic elements of the game

a) Participants: the number of buyer is n. The assessed value buyer proposed is V_{Bui} (defined in [0,1] uniform distribution function);

b) Action: The seller gives reserve price of knowledge resources. Buyer gives quotes (b$_i$, b$_i$=b$_i$(V$_{BUi}$), b$_i \in$ [V$_{Bmin}$,V$_{Bmax}$], i=1,2,...n).

c) Payment: Buyer's pay function is as follows:

$$u_i(b_i, b_j, V_{BUi}) =$$

$$\begin{cases} b_i - V_{BUi}, \text{when } b_i < b_j, \forall j \neq i \\ \frac{1}{m}(b_i - V_{BUi}), \text{when } m \text{ bid highest} \\ 0, \text{when } j \text{ makes } b_i > b_j \end{cases} \qquad (11)$$

(2) Analysis Process of Pricing Game

For the buyer, the first-price sealed auction is a symmetric static game [13]. And the buyer just needs to consider the symmetric equilibrium bidding strategy (b=b* (v)). For the buyer A, when the probability of the buyer B's offer less than buyer B's is $P\{b_j < b_i, \forall j \neq i\}$, the buyer's expected pay is:

$$u_i(b_i, b_{-i}) = (V_{BUi} - b_i)P\{b_j < b_i, \forall j \neq i\} \qquad (12)$$

According to the symmetry, b$_j$ = b*(v$_j$), so we can get the following result:

$$\begin{aligned} P\{b_j < b_i\} &= P\{b^*(v_j) < b_i\} \\ &= P\{v_j < b^{*-1}(b_i)\} \\ &= b^{*-1}(b_i) \\ &= \Phi(b_i) \end{aligned} \qquad (13)$$

So, we can get the optimal model of buyer's bid:

$$\max_{b_i} (v_i - b_i)[\Phi(b_i)]^{n-1} \qquad (14)$$

Differentiating:

$$(-\Phi^{n-1}) \cdot db_i + \{[\Phi(b_i) - b_i](n - 1)\Phi^{n-2}(b_i)\} \cdot d\Phi = 0 \qquad (15)$$

Initial conditions: Φ 0 =0. Bayesian equilibrium bidding strategy:

$$b^*(v) = \frac{n-1}{n} V_{BU} \qquad (16)$$

Clearly, as n increases, the buyer's bid approaching the buyer's given highest estimate.

IV. GAME ANALYSIS OF THE BIDDING PRICE TRANSFER

From the game's point of view, the bidding price transfer is a three-stage incomplete information dynamic game: the buyer give bidding information and floor price; the seller participate in bidding and give price; buyer select the successful bidder.

A. The basic elements of the game

(1) Seller: The value of knowledge resources;

(2) Buyer: Buyer give floor price and assess the seller;

(3) The number of bidding participants: The number of participants will have some impact in price;

(4) Seller's bid: Sellers should consider other buyer's bid when bidding

B. The bidding price transfer's game analysis

Game model is as follows:

(1) The basic elements of the game

a) Participants: the number of seller is n. The assessed value seller proposed is V$_{SWi}$ (defined in [0,1] uniform distribution function);

b) Action: the buyer gives floor price of knowledge resources. Seller gives quotes (P$_{Si}$ =P$_{Si}$(V$_{SWi}$), P$_{Si} \in$ [P$_{Smin}$,P$_{Smax}$], i=1,2,...n).

c) Payment: seller's pay function is as follows:

$$u_i(P_{Si}, P_{Sj} V_{SWi}) =$$

$$\begin{cases} P_{Si} - V_{SWi}, \text{when } P_{Si} < P_{Sj}, \forall j \neq i \\ \frac{1}{m}(P_{Si} - V_{SWi}), \text{when } m \text{ bid lowest} \\ 0, \text{when } j \text{ makes } P_{Si} > P_{Sj} \end{cases} \qquad (17)$$

(2) Analysis Process of Pricing Game

For the sellers, the bidding process is a symmetrical game static game [14]. What they should consider equilibrium bidding strategies: P$_{Si}$ =P$_{Si}$ (V$_{SWi}$). The probability of winning is as follows:

$$\prod_{j \neq i} P\{P_{Si} < P_{Sj}\} \qquad (18)$$

The bidder's expect pay is as follows:

$$u_i(P_{Si}, P_{Sj}) = (b_i - V_{SWi}) \prod_{j \neq i} P\{P_{Si}\} < P_{Sj}\} \qquad (19)$$

The optimal bid model of bidder is as follows:

$$max_{P_{Si}}(P_{Si} - V_{SWi}) \prod_{j \neq i} P\{P_{Si}\} < P_{Sj}\} \qquad (20)$$

When n=1, $P_{si}=P_{Smax}$. Solve differential:

$$P_{Si} = V_{SWi} + \frac{P_{Smax} - V_{SWi}}{n}$$

$$= V_{SWi} + \frac{V_{STi} + V_{SWi} - V_{SWi}}{n}$$

$$= V_{SWi} + \frac{v_{STi}}{n} \qquad (21)$$

Obviously, with the increase in the number of tender,

Bidder's offer must be reduced in order to win.

V. COMPARISON OF THREE TRADING PATTERNS

Finally, we compared the pricing process under three trading patterns of knowledge resources. The main indicators include [15]: way to trade, scope of application, subjectivity, transaction costs, trading rules complexity, and evaluation criteria. The result is as Table I:

TABLE I
COMPARISON OF THREE TRADING PATTERNS

content	the contract transfer	the auction price transfer	the bidding price transfer
way to trade	One to one	One to more	One to more
scope of application	Medium-value	Scarce knowledge resources	High-value. Especially apply to determine the research unit
subjectivity	medium	high	low
transaction costs	Low preparation costs; High negotiation costs	High preparation costs; medium negotiation costs	High preparation costs; low negotiation costs
trading rules complexity	Low	High	High
evaluation criteria	both parties agree	The highest bid	The lowest bid

VI. CONCLUSION

This paper presents a knowledge resource pricing methods for large research institution. We have built three kinds of knowledge resources transaction type (the contract transfer, the auction price transfer and the bidding price transfer). Combined with game theory, we finally established a knowledge resource pricing model and facilitate the trading of knowledge resources.

REFERENCE

[1]   Jingwen Li, Yaoguang Hu, Jialin Han. Knowledge Management Maturity Assessment in Research Institutions Using Analytic Hierarchy Process and Fuzzy Comprehensive Evaluation Method [J]. IEEM 2012, 2012.12, Hong Kong.

[2]   Dai Jun, Cheng Zhao-han. Studies and Constructs the Model of Trading Mechanism of Knowledge Market Inside Enterprise [J]. Forecasting, 2004, 4: 8

[3]   Xu xiang. The win-win of business negotiation [J]. Commerical Culture, 2009, 9: 104. (Chinese)

[4]   Grant Robert M. Toward a knowledge-based theory of the firm [J]. Strategic management journal, 1996, 17: 109-122.

[4]   Chen Bo. Research on the knowledge trade and management based on the theory of knowledge value [D]. Shanghai: Shanghai Jiao Tong University, 2007. (Chinese)

[5]   Sarvary Miklos. Knowledge management and competition in the consulting industry [J]. California Management Review, 1999, 41(2): 95-107.

[6]   Brooking Annie. On the Importance of Managing Intangible Assets as Part of Corporate Strategy [M]. London: Academic Conferences Limited, 2010: 137.

[7]   Lin Carol Yeh-Yun, Edvinsson Leif, National intellectual capital: a comparison of 40 countries [M]. New York: Springer, 2011:263-264

[8]   Ghandar Adam, Michalewicz Zbigniew, Using cellular evolution for diversification of the balance between accurate and interpretable fuzzy knowledge bases for classification [M]. Bankok: IEEE, 2011: 1481-1488.

[9]   Dekel E, Gul F. Rationality and knowledge in game theory [J]. Econometric Society Monographs, 1997, 26: 87-172.

[10] Fishburn P C, Kochenberger G A. Two-Piece Von Neumann-Morgenstern Utility Functions [J]. Decision Sciences, 1979, 10(4): 503-518

[11] Xiao Gang. Game Analysis and Pricing on Exchange of Enterprises' Property [D]. Tianjin: Tianjin University, 2008. (Chinese)

[12] Menezes F M, Monteiro P K. An introduction to auction theory [M]. Oxford University Press, 2005.

[13] Jofre-Bonet M, Pesendorfer M. Estimation of a dynamic auction game [J]. Econometrica, 2003, 71(5): 1443-1489.

[14] Zhang Xiangzhen, Cheng Linzhang. Knowledge value and compensation [J]. Wuhan University Journal, 2000, 53(3): 309-313. (Chinese)

[15] Zhang Fugui. Knowledge classification in the field of knowledge management [J]. Journal of Intelligence, 2001 (09): 5. (Chinese)

Three-Dimensional Decoration UCD Cross-Platform Interactive Software Design Methods Base on UCD in Consumption Patterns of Information

Zhendong Wu[1,2,*], Weiming Guo[1], Xiaoqun Ai[1]

[1]School of Design, JiangNan University, Wuxi, Jiangsu 21400
[2]College of Mechanical Engineering and Automation, HuaQiao University, Xiamen, Fujian 361021
(wzd888@gmail.com)

Abstract - **At present the information products and information services for the consumer object information consumption, has gradually become the mainstream of the times. In this paper, a UCD based furniture, home furnishing, display design, network marketing one-stop complete, means of realizing design research methods and technology to build information platform for user consumption of new consumption patterns and experience of home furnishing, furniture industry.**

Keywords - **Home-improvement intelligent, information consumption, user Experience, UCD**

2012 China's information consumer market has reached 1.7trillion Yuan, representing an increase in 29 percent in 2011. IT has been at all level of society to provide mankind with a convenient in the era of experience economy [1-3]. Virtual experience, online shopping has become a way of life, and based B2B, B2C e-commerce platforms, to create a more efficient experience environment, efficient, convenient, text (image) of shopping becoming consume trend [4, 5]. Therefore, the three-dimensional simulation of digital virtual technology and digital media technology which the representative of a new generation of information technology, were set up home design digital design and marketing platform based on UCD point of view and has an important practical significance.

I. AN INTERACTIVE DESIGN AND DEVELOPMENT OF THE PRINCIPLES OF COGNITIVE STRATEGIES

1) UCD Design Principles: UCD (User Centered Design), Refers to the user experience design process-centric design decisions, Emphasis on user preference of design patterns, from the users' needs and feelings of departure for software design and development. This program is for non-professional background of ordinary customers, Use of software development processes, information architecture, human-computer interaction, interaction with concise way, good visual experience with the user's habits [6].

2) The program is designed, firstly, to build an interior design and purchase platform; second, providing the user a good interactive experience in the process. Terry R. Schussle who an interaction designer noted: Interactive neither animation, nor the video, it is the user control and event experience. By ergonomics

with the application that enhance human cognitive rationalization and comfort in the use of the process [7-9]. The following specific development system for interactive design elements of analysis:

A. Psychological behavior

Interactive form of value system should be user during use could properly adapt to fully mobilize the user cognitive and emotional. Software system can be established from the emotional perspective of the man-machine relationship with the user in a more intelligent way feedback.

B. Consumer behavior

The interactive software system is to improve the process of buying furniture and home improvement customers design, the need for life behavior of users to do data analysis, such as: user base composition and differentiation, user information query behavior, user and social networks, users and information system, user satisfaction and user privacy and so on. Designers need to conclusion from the data, to determine the interaction of the functionality provided. Construction of organic information hierarchy, through rational consumption patterns, to find and explore information exchange and circulation order to optimize the audience consumer behavior.

C. Operating behavior

The software system ergonomics focuses on human-computer interface design. The operation of each step, directly facing the user is a GUI interface. GUI requires the user to make timely and accurate decisions feedback to improve the efficiency and speed of the man-machine dialogue; Reduce the cognitive load of people, simplify the steps; Formal request from the visual interface layout and beautiful, rational and efficient use of the display division, which is the basis of human-computer interaction.

II. INTERDISCIPLINARY APPLICATIONS

The software is interactive design category, technology modules including 3D design, interior design, interactive experience design, visual communication design, data processing and mining, as shown in Fig.1.

1) 3D design; 2) interactive experience design; 3) Computer underlying technology; 4) interior design; 5) visual design.

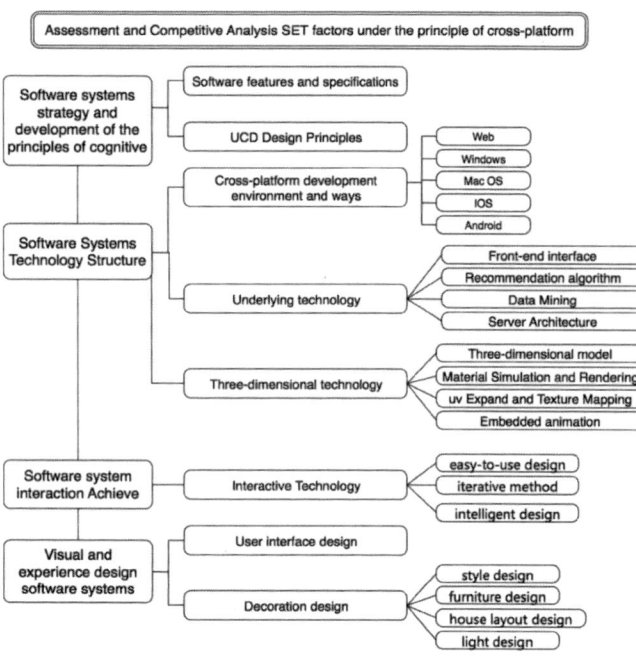

Fig.1. technology classification

III. DESIGN FOCUS AND PROCESS

Build the brand furniture networking and make it three-dimensional. The system will create massive furniture, home improvement database, each furniture and decorations have 3D models and display animation, you can enrich the print media and two-dimensional images, and these data with cross-platform data terminal operating system coordinated operation characteristics.

Explore operating a cross-platform terminal operating system. Display, design methods to buy, build furniture purchase and home design synchronization design. The system will purchase furniture and design processes unite through to build a digital marketing platform.

Conform to the domestic industry, digital lifestyle trend. The application of digital means to traditional physical industry, through human-computer interaction platform to enhance the flow of information users, merchants, agents, designers between. The software uses a paperless, non-materialized form of graphic image display and sales, Hosted by materialistic media to digital media, effectively reducing storefront Distribution flow, reduce logistics costs and the number of sales staff [10].

IV. KEY TECHNOLOGIES OF INTERACTION DESIGN

A. Systematic cross-platform software design.

With the development of computer technology, user-level operating system mainstream platforms Windows, Mac OS desktop systems as the main platform, In IOS, Android, Windows Phone and mobile terminals through different browser access. President of China Electronic Information Industry Development Institute Rowan in 2011 in the "direction of development of software and IT services, needs and tasks," the report pointed out that inter-terminal operating system platforms are becoming the commanding heights of the new industrial development. Information from the industry to the terminal fusion technology integration, network integration, service integration evolution, cross-platform terminal operating system market share will rapidly increase, becoming the new direction of development, and will determine the future ownership of industry dominance.

The software uses a unified back-end database server set up for data exchange and interoperability across the terminal operating system platform, Enhance the utilization efficiency of the software and user experience. Between the platform design based on their characteristics in response to the operating platform and hardware devices, Use as portable, powerful performance desktop computers, web-side rapid spread of mobile platforms and other platform features, Seamless interoperability across platforms, establishing wide adaptation, good user experience across terminal operating system.

B. 3D modeling of batch codes and pipeline design

"Content is king" is the purpose of the software is developed to build massive brand furniture corresponding 3D model library, determine its market application results. Therefore, the establishment of a scientific and rigorous three-dimensional model to build mass production and pipeline design specifications of the software development is the key issue to be addressed. The software modeling approach using direct modeling three-dimensional modeling software, the method is simple, accurate, polygon count models easy to control, After some of the more complex model uses a three-dimensional surface scanner, drawn by the second topology. 3D modeling of batch method specifications and pipeline design is primarily to establish a lightweight three-dimensional modeling methods and algorithms and norms, development of a use-scale production, with a common technology approach. LOD (Levels of Detail) technology and secondary manual optimization, a furniture models can be implemented on standard hardware platform for the lowest real-time 3D display.

C. Adaptive optimization of design data management.

Home design system can adapt to the customer to click and buy changes in the environment and automatically adjusts its structure and function in the process. To physical goods brand furniture manufacturers at home and abroad as the basis to establish one-dimensional simulation model for the commodity, establish real than on the database, and the establishment of relational database management systems, optimization and system management of massive data. The software chosen SQLite database development, it has an embedded, low resource occupancy characteristics, and can combine with many programming languages, such as C #, PHP, Java, etc., to support Windows / Linux / Unix, and so mainstream operating system to facilitate data optimization and system design structure.

D. Augmented reality effect of 3D Digital Design 3D home.

Realistic three-dimensional visual effects rendered interactive systems is a key factor in visual imaging software, the system complete home improvement home improvement program to restore the true maximum effect. Previous renderings exaggerated three-dimensional plane and landscaping, detached from reality, virtual reality products after visual effects monotonous, lacking a sense of space and a sense of texture and weight of furniture applications, acceptance of the audience is not high. Vision offers users the most direct reaction, augmented reality effect is directly related to the user's reliance on the product and life cycle, and the effect is more real, more able to help users achieve an ideal home improvement design.

V. CONCLUSION

Current design trends are the "creation of high-tech emotion, Excellent Experience." In the household sector across the terminal operating system platform home interactive software system development will promote the process of furniture, building materials, soft-mounted digital marketing industry, digital design, digital life. Make home design to adapt to new market demands under the information age, to meet user personalization, customization requirements and explore new business strategy. Meanwhile, the interactive software is also empirical research on user-centered design and good interactive experience model.

ACKNOWLEDGEMENT

Supported by Xiamen Colleges and universities scientific research institutes of scientific and technological innovation projects, A cross-platform interactive three-dimensional smart home improvement software system development (NO. 09SJD760008);

Supported by 2009 Project about the Philosophy Social Sciences of Higher education in Jiangsu Province (No. 09SJD760008);

Supported by 2010 Ministry of Education, Humanities and Social Sciences project (No. 10YJA760017);

Supported by 2012 Jiangsu Province Social Science Fund Project (No. 12LSB005);

Supported by the Key Project of Philosophy and Social Science Research in Colleges and Universities in Jiangsu Province (No. 2011ZDIXM046);

Supported by 2012 A Major projects of key research base of philosophy and Social Science in Colleges and universities in Jiangsu Province (No. 2012JDXM013).

REFERENCES

[1] Mark. Diani edited, Yao shou Teng translation, intangible society: the post-industrial world of design, culture and technology, Sichuan People's Publishing House, March 1998.

[2] Jonathan Cagan, Crige M. Vogel, Create breakthrough products - from product innovation strategy to project finalized, Machinery Industry Press, 2003.10.

[3] Alan Coope, Robert Reimann, David Cronin edited, songtao liu translated, About Face3 essence of interaction design, Electronic Industry Press, 2012.3.

[4] Lambert M Surhone, Mariam T Tennoe, Susan F Henssonow, User Experience Design, Beta script Publishing, 2011.7.

[5] Bill Buxton, Saul Greenberg, Sheelagh Carpendale, Nicolai Marquardt; Sketching User Experiences: the Workbook; Morgan Kaufmann Publishers In; Workbook; 2011. 12.

[6] Russ Unger, Carolyn Chandler, A Project Guide to UX Design: For User Experience Designers in the Field or in the Making; New Riders Publishing; 2nd Revised edition, 2012.3.

[7] Russ Unger, Carolyn Chandler; A Project Guide to UX Design: For User Experience Designers in the Field or in the Making; New Riders Publishing; 2nd Revised edition; 2012.3.

[8] Jessie James Garrett; The Elements of User Experience: User-Centered Design for the Web and Beyond; New Riders Publishing; 2nd Revised edition; 2010.12.

[9] William Albert, Thomas Tullis, Donna Tedesco, Beyond the Usability Lab: Conducting Large-scale Online User Experience Studies, Morgan Kaufmann, 1; 2010.1.

[10] Robert Schumacher, The Handbook of Global User Research, Morgan Kaufmann, 2009.10.

Part II
Industrial Engineering Practice

Evaluation of Wastewater Treatment Quality in the West Bank-Palestine Based on Fuzzy Comprehensive Evaluation Method

Xing Bi[*], Rabah A.M. Isaili, Qibin Zheng

School of Management, Tianjin University, Tianjin, China

(bistar@126.com)

Abstract - **Treated wastewater reuse has an important significance to solve the water shortage in the West Bank-Palestine. Based on fuzzy comprehensive evaluation method, the water quality of Al-Bireh WWTP is evaluated by referring to Chinese water quality evaluation criteria. It can help the Palestinians understand the current situation of sewage treatment and establish nationally appropriate water quality evaluation system.**

Keywords - **Fuzzy comprehensive evaluation, treated wastewater, west bank**

I. INTRODUCTION

A. Wastewater situation in the West Bank-Palestine

Water is a major concern in the West Bank (WB) since it's considered one of the poorest regions of the Middle East in terms of water resources. Water scarcity is mainly attributed to the Israeli control over the majority of the Palestinian water resources, as well as the arid and semi-arid climate conditions of the region [1]. The occupied Palestinian territories (Opt) are divided into sixteen Governorates. Eleven of these are located in the WB, the others in the Gaza Strip. The Oslo Accords divide the Opt into three types of areas, A,B and C. where areas A are under Palestinian control, areas B are under Palestinian administrative control and Israeli security control and areas C are under total Israeli control.

The wastewater sector in Opt has been neglected under Israeli occupation since 1967 with most attention focused on measures to solve water quantity and supply problems. A lack of wastewater treatment plants, of sewerage systems and wastewater collection for recycling, leads to the uncontrolled discharge of wastewater into the environment. There were insufficient financial resources within the Palestinian community to pay for new wastewater collection, disposal and treatment systems [2]. Israel collected taxes from the Palestinians through the Israeli Civil Administration, but they never spent the money on infrastructure for the Palestinian communities.

There are three old treatment plants in the West Bank, namely Ramallah, Jenin, Tulkarm[3,4]. All have operational difficulties and are not functioning effectively, and some are not functioning at all. Most of these plants are overloaded, under-designed or have experienced mechanical failures. In addition, the fourth plant, Al-Bireh, considerable new treatment plant, was constructed in the 2000 with support of German Government. It's over 22 dumdums of land. The system is designed to cover a population of 50,000 with enough capacity to serve future expansion. Daily wastewater flow rate 5000 m3/day.

B. Palestinian standards of Wastewater Quality

Wastewater treatment and reuse criteria differ from one country to another and even within a given country. Some of the main discrepancies in the criteria are, in part, due to differences in approaches to public health and environmental protection.

For a long time, Palestine did not have any specific wastewater regulations and references were usually made according to the WHO recommendations or to the standards of neighboring countries (as Egypt, Jordan) [5]. Recently, the Environmental Quality Authority, in coordination with Palestinian ministries and universities, has established specific wastewater reuse regulations. The draft of the Palestinian legislation for reuse of treated wastewater is still under study in the Palestinian Standard Institute [6].

Treated wastewater disposed by the sewage treatment plant is evaluated according to China's "Surface Water Quality Standards". The difference in standards is concluded between the two countries, contributing to the establishment of water quality evaluation system in the West Bank-Palestine.

II. METHODOLOGY

Fuzzy phenomena is everywhere in the nature, such as meteorological phenomena, land cover classification, spatial data quality, etc. Treated wastewater quality is also fuzzy. One cannot tell good or bad simply. Further, treated wastewater quality is affected by several factors and every type of factor has different effects on water quality [7]. As to this, a fuzzy comprehensive evaluation method is used in treated wastewater quality assessment. The method is a qualitative one and the following is the principle procedures of it:

1) *Establishing element set*: To find different factors in evaluating wasted water quality and put forward factor set:

$$U = \{u_1, u_2, \cdots, u_m\} \qquad (1)$$

In (1), u_i represents the i-th water quality indicator, m is the number of water quality indicators.

According to the Palestinian water environmental conditions and China's water quality evaluation, six indicators are determined as evaluation elements, which are TP, TN, DO, BOD_5, COD_{Mn} and NH_3-N.

$$U = \{TP, TN, DO, BOD_5, COD_{Mn}, NH_3-N\} \qquad (2)$$

2) *Establishing grade factor set*:

$$V = \{v_1, v_2, \cdots, v_n\} \qquad (3)$$

In (3), v_j is the assessment grade; n is the number of assessment grades.

E. Qi et al. (eds.), *Proceedings of the 21st International Conference on Industrial Engineering and Engineering Management 2014*, Proceedings of the International Conference on Industrial Engineering and Engineering Management, DOI 10.2991/978-94-6239-102-4_46, © Atlantis Press and the authors 2015

According to China's "Surface Water Quality Standards", five assessment grades can be determined[8]: I (v1), II (v2), III (v3) IV(v4) and V (v5):

$$V = \{I, II, III, IV, V\} \qquad (4)$$

3) *Establishing weight coefficient matrix*: Weight measures the size of the water pollution which a factor affects. The larger the weighting coefficient, the greater the impact on water quality .In fuzzy evaluation, every evaluation element has different contribution to image quality. Thus, the weight coefficient matrix of evaluation element is calculated according to "excessive multiples method" [9].

$$I_i = {c_i}/{s_i} \qquad (5)$$

In (5), I_i is a dimensionless number and represents the exceeding multiples of evaluation element compared with standard value. c_i is the monitoring value, and s_i is the mean of kinds of water quality standards limit.

Then to be normalized, the weight of each evaluation can be calculated:

$$w_i = {I_i}/{\sum I_i} \qquad (6)$$

The weight coefficient matrix is also determined:

$$W = \{w_1, w_2, \cdots, w_m\} \qquad (7)$$

4) *Establishing comprehensive evaluation matrix*: Due to the extent of water pollution and grading standards of water quality are vague, so it is reasonable to describe the boundaries with membership classification. It is a fuzzy mapping from U to V. r_{ij} represents the possibility of i-th water quality indicator can be evaluated as class j. The corresponding judgment matrix can be attained as follows.

$$R = \begin{bmatrix} r_{11} & r_{12} & \cdots\cdots & r_{1(n-1)} & r_{1n} \\ r_{21} & r_{22} & \cdots\cdots & r_{2(n-1)} & r_{2n} \\ \vdots & \vdots & \vdots & \vdots & \vdots \\ r_{m1} & r_{m2} & \cdots\cdots & r_{m(n-1)} & r_{mn} \end{bmatrix} \qquad (8)$$

Membership can be determined through the membership function and the membership functions are commonly described with a trapezoidal distribution. The membership of water quality with each category is calculated as follows [10,11].

$$\begin{cases} I & r_{ij} = \begin{cases} 1 & c_i \le s_j \\ \dfrac{s_{j+1}-c_i}{s_{j+1}-s_j} & s_j < c_i < s_{j+1} \\ 0 & c_i \ge s_{j+1} \end{cases} \\ \\ II \sim IV & r_{ij} = \begin{cases} 0 & c_i \le s_{j-1}, c_i \ge s_{j+1} \\ \dfrac{c_i-s_{j-1}}{s_j-s_{j-1}} & s_{j-1} < c_i < s_j \\ \dfrac{s_{j+1}-c_i}{s_{j+1}-s_j} & s_j \le c_i \le s_{j+1} \end{cases} \\ \\ V & r_{ij} = \begin{cases} 1 & c_i \ge s_j \\ \dfrac{c_i-s_{j-1}}{s_j-s_{j-1}} & s_{j-1} < c_i < s_j \\ 0 & c_i \ge s_{j-1} \end{cases} \end{cases} \qquad (9)$$

In (9), c_i is the monitoring value, and s_j is the standard value of j-th water quality indicator.

5) *Establishing Fuzzy Comprehensive Evaluation model*: After determining the fuzzy evaluation matrix R

and weight coefficient matrix W, fuzzy comprehensive evaluation model is also determined.

$$B = W \cdot R$$

$$= (w_1, w_2, \cdots, w_m) \begin{bmatrix} r_{11} & r_{12} & \cdots\cdots & r_{1(n-1)} & r_{1n} \\ r_{21} & r_{22} & \cdots\cdots & r_{2(n-1)} & r_{2n} \\ \vdots & \vdots & \vdots & \vdots & \vdots \\ r_{m1} & r_{m2} & \cdots\cdots & r_{m(n-1)} & r_{mn} \end{bmatrix} \qquad (10)$$

$$= (b_1, b_2, \cdots, b_n)$$

B is a fuzzy vector which not only represents all evaluation elements' contribution, but also reserves all degree of membership of every grade. Water levels should be evaluated for the class j, if $b_j = \max(b_1, b_2, \cdots, b_n)$.

III. APPLICATION&RESULTS

The Al-Bireh reuse demonstration project conducted the different aspects of reclaimed water use in irrigation by developing a set of different effluent polishing and irrigation techniques on crops. The primary goals of the project were to build the initial institutional relationships, raise the profile of wastewater reuse and compost use, and to develop the first stage of on-the-ground experience and capacity, in the field of wastewater reuse.

Based on the above fuzzy comprehensive evaluation model, the water quality of Al-Bireh WWTP is evaluated. The values of water quality indicators in Al-Bireh WWTP from June to November 2013 are shown in Table I.

TABLE I
ANNIVERSARY MONITORING VALUES OF WATER QUALITY FACTORS IN AL-BIREH WWTP

Index	TP	TN	DO	BOD$_5$	COD$_{Mn}$	NH$_3$-N
201306	0.037	1.124	4.16	2.76	1.04	0.087
201307	0.026	0.916	5.23	1.92	0.94	0.064
201308	0.024	1.295	8.53	1.65	0.77	0.169
201309	0.052	1.432	6.56	3.41	1.36	0.213
201310	0.050	1.355	6.02	4.72	1.41	0.146
201311	0.046	1.116	5.14	1.88	1.59	0.073

The values of the indicators used are the limit of qualities of the surface water environments in China. They are shown in Table II.

TABLE II
NATIONAL STANDARD OF QUALITIES OF THE SURFACE WATER ENVIRONMENTS

Index	TP	TN	DO	BOD$_5$	COD$_{Mn}$	NH$_3$-N
	$\le$	$\le$	$\ge$	$\le$	$\le$	$\le$
I	0.01	0.2	7.5	3	2	0.15
II	0.025	0.5	6.0	3	4	0.5
III	0.05	1.0	5.0	4	6	1.0
IV	0.1	1.5	3.0	6	10	1.5
V	0.2	2.0	2.0	10	15	2.0

According to the preceding formula, take the measurement data in June 2013 for example and weight coefficient for each evaluation factor is calculated.

W = {0.150,0.338,0.276,0.166,0.044,0.026}

The comprehensive evaluation matrix in June 2013 is as follows.

$$R = \begin{bmatrix} 0.000 & 0.520 & 0.480 & 0.000 & 0.000 \\ 0.000 & 0.000 & 0.752 & 0.248 & 0.000 \\ 0.000 & 0.000 & 0.580 & 0.420 & 0.000 \\ 1.000 & 0.000 & 0.000 & 0.000 & 0.000 \\ 1.000 & 0.000 & 0.000 & 0.000 & 0.000 \\ 1.000 & 0.000 & 0.000 & 0.000 & 0.000 \end{bmatrix}$$

Based on the evaluation of the weight distribution W , as well as the fuzzy evaluation matrix R, we can get the comprehensive evaluation result of the quality of the wasted water in June 2013.

$$B = \{0.166, 0.150, 0.338, 0.276, 0.000\}$$

The degree of membership of grade III is 0.338, which is the largest among five categories of water quality. Thus water levels in June 2013 should be classified as III.

Fuzzy comprehensive evaluation results of water quality of Al-Bireh WWTP from June to November 2013 are shown in Table III.

TABLE III
RESILT OF FUZZY COMPREHENSIVE EVALUATION

Time	I	II	III	IV	V	Result
201306	0.166	0.150	0.338	0.276	0.000	III
201307	0.128	0.230	0.305	0.000	0.000	III
201308	0.459	0.079	0.315	0.315	0.000	I
201309	0.311	0.311	0.306	0.306	0.000	III
201310	0.043	0.286	0.290	0.291	0.000	IV
201311	0.106	0.160	0.321	0.232	0.000	III

In summary, the results of water quality evaluation of Al- Bireh WWTP can be classified as grade III, just as shown in Table IV.

TABLE IV
RESULTS OF WATER QUALITIES EVALUATION OF AL-BIREH WWTP

Time	2013 06	2013 07	2013 08	2013 09	2013 10	2013 11	Mean
	III	III	I	III	IV	III	III

IV. DISCUSSION

In considering the weight index of evaluation element, the accuracy of model has been in restrictions to a certain extent, since the objective calculation of weight has been solve in the present.

V. CONCLUSION

Based on China's water quality assessment standards, take use of fuzzy comprehensive evaluation model to assess water environmental quality and describe the water quality classification. It can reflect the water quality under a variety of factors working together and solve the ambiguity of water environment evaluation, which may have great significance for West Bank-Palestine. Although the water quality of Al-Bireh WWTP reach Grade III, it still need to strive to improve the sewage disposal technology, improve the processing of water quality, and expand the scale of sewage treatment ,in response to the growing demand for water in Palestinian territories.

REFERENCES

[1] Zahra A, and Ahmad B A, "Water crisis in Palestine," *Desalination*, vol. 136, no. 1, pp. 93–99, 2001.
[2] McNeill, Laurie S., M. N. Almasri, and N. Mizyed, "A sustainable approach for reusing treated wastewater in agricultural irrigation in the West Bank–Palestine," *Desalination*, vol. 248, no. 1, pp. 315–321, 2009.
[3] Miller R, "Water use in Syria and Palestine from the Neolithic to the Bronze Age," *World Archaeology*, vol. 11, no. 3, pp. 331–341, 1980.
[4] McWhorter T J, del Rio C M, and Pinshow B, "Modulation of ingested water absorption by Palestine sunbirds: evidence for adaptive regulation," *Journal of Experimental Biology*, vol. 206, no. 4, pp. 659–666, 2003.
[5] Fatta D, Salem Z, and Mountadar M, "Urban wastewater treatment and reclamation for agricultural irrigation: the situation in Morocco and Palestine," *Environmentalist*, vol. 24, no. 4, pp. 227–236, 2004.
[6] Mogheir Y, and Abu Hujair T, "Treated Wastewater Reuse in Palestine," *International conference Water Value and Rights, Ramallah-Al-Bireh, Palestine May,* 2005.
[7] Zhai, Liang, and Xinming Tang, "Fuzzy Comprehensive Evaluation Method and Its Application in Subjective Quality Assessment for Compressed Remote Sensing Images," *FSKD*, 2007.
[8] Icaga Y, "Fuzzy evaluation of water quality classification," *Ecological Indicators*, vol. 7, no. 3, pp. 710–718, 2007.
[9] Gong L, and Jin C, "Fuzzy comprehensive evaluation for carrying capacity of regional water resources," *Water resources management*, vol. 23, no. 12, pp. 2505–2513, 2009.
[10] Mu Zheng, and Wang Fangyong, "Comprehensive Evaluation of River Water Quality based on Fuzzy Comprehensive Evaluation Model(in Chinese)", *Water Power*, vol. 35, no. 4, pp. 11–13, 2009.
[11] Lu Wenxi, Li Di, and Zhang Lei, " Application of fuzzy comprehensive evaluation based on AHP in water quality evaluation(in Chinese)", *Water Saving Irrigation*, no. 3, pp. 43–46, 2011.

Simulation-Based Environmental Analysis for Production Lot-Size Determination in Terms of Material Flow Cost Accounting

Run ZHAO[1,*], S. TAKAKUWA[2], H. ICHIMURA[2]

[1] Economics & Management School, Jiangsu University of Science and Technology, Zhen Jiang, China
[2] Graduate School of Economics and Business Administration, Nagoya University, Nagoya, Japan
(*zhaorun1982@hotmail.com, takakuwa@soec.nagoya-u.ac.jp, ichimura.hikaru@f.nagoya-u.jp)

Abstract - **In the modern manufacturing industry, environmental considerations are part of numerous phases of production. Inappropriate production lot-size determination can generate substantial scrapped overdue stocks and idle processing, which lead to serious environmental burdens. In this paper, Material flow cost accounting (MFCA), an environmental management accounting method, is adopted to reduce the amount of wastes that result from large overstocks and other wastes caused by current production lot-size determination are traced. For comparison with the conventional cost accounting used in the original simulation model, MFCA can identify negative products cost related to environmental impacts hidden in the production processes. Moreover, it is demonstrated that the proposed procedure of application of simulation with MFCA can also perform a dynamic analysis and a static analysis.**

Keywords - **Environmental management, material flow cost accounting, production lot-size, simulation**

I. INTRODUCTION

In the modern advanced manufacturing industry, the idea of green production has become increasingly important as part of sustainable development. It reflects a new production paradigm that employs various green strategies and techniques to achieve greater eco-efficiency. In green production systems, achieving zero emissions and reducing the environmental burden from production activities are thus important worldwide [1,2].

In a multi-variety and small-batch production system, it is recognized that an appropriate determination of production lot-size for different part types in different production stages is a complex problem [3, 4]. This complexity can easily lead to serious environmental problems with limited production resources. Because of inaccurate determinations, overstocks of unnecessary materials and intermediate products are often produced, causing huge material waste, idle energy consumption and stock scraps, which create substantial environmental burden [5]. Therefore, analyzing and determining an appropriate production lot-size to achieve both economic and environmental effectiveness are an important issue in the production research field that urgently needs to be solved.

In this paper, Material Flow Cost Accounting (MFCA) is adopted, among several environmental management accounting tools, and has received considerable attention for its effectiveness in improving both productivity and the harmony of environmental profitability [6]. Moreover, the MFCA standard has been granted ISO 14051 by the ISO secretariat to evaluate the environmental performance of the target production processes [7]. Consequently, in this paper, MFCA is introduced to study the environmental impacts of production lot-size determination through structuring simulation models in a multi-variety and small-batch production system. Within the MFCA framework, costs are calculated for not only good products but also non-product outputs or material losses. The former is referred to as "positive products," and the latter, as "negative products." MFCA visualizes the cost of producing non-product outputs or material losses and thus highlights areas of potential improvement [8, 9].

Although MFCA is powerful as an environmental management tool, it can perform only static analysis. Hence, this paper proposes that dynamic analysis as well as static analysis can be performed by constructing simulation models for the designated manufacturing systems and performing simulation analysis in terms of MFCA. A Real example is introduced for a forging manufacturing system which involves seven processes.

II. MFCA APPROACH REVIEW

Fig.1 shows the concept of MFCA. It is also a management information system that traces all input materials flowing through production processes and measures output in finished products and waste. In a processing-type production system, waste is generated in various steps of the production process. In particular, in the process of stocking and production, waste is substantially produced because materials and intermediate products that are overstocked as inventory may deteriorate in quality or be scrapped. Additionally, while materials or intermediate products are processed, residues or shavings may be generated. All of the wastes mentioned above are called "negative products" and lead to environmental burden [9, 10]. In MFCA, the idle processing, unnecessary energy and auxiliary material consumption caused during the waste generation are also called "negative products" and treated as environmental costs.

E. Qi et al. (eds.), *Proceedings of the 21st International Conference on Industrial Engineering and Engineering Management 2014*, Proceedings of the International Conference on Industrial Engineering and Engineering Management, DOI 10.2991/978-94-6239-102-4_47, © Atlantis Press and the authors 2015

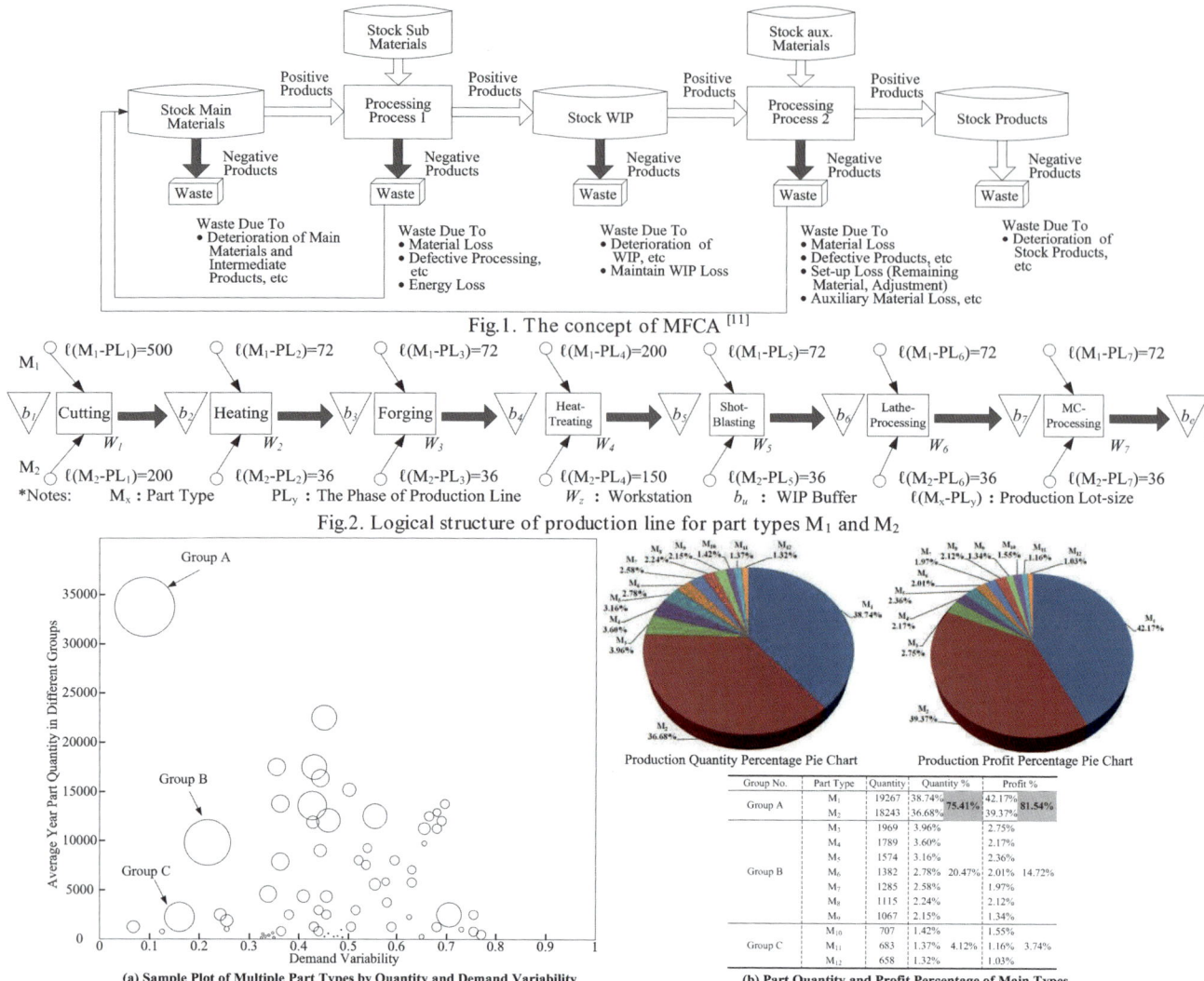

Fig.1. The concept of MFCA [11]

Fig.2. Logical structure of production line for part types M_1 and M_2

(a) Sample Plot of Multiple Part Types by Quantity and Demand Variability

(b) Part Quantity and Profit Percentage of Main Types

Fig.3. Relative statistical data on multiple part types from the current production

III. CASE STUDY

A. Multi-Variety and Small-Batch Production System

This paper considers a case of a certain multi-variety and small-batch production system, which is located in a precision component manufacturing workshop of a Japanese company. Fig.2 shows the logical structure of the current production line for part types M_1 and M_2. This production line mainly comprises seven workstations. To fulfill the requirements of part type diversification and rapid responses to market needs, different small production lot-sizes for M_1 and M_2 are used for each workstation and are denoted M_x-PL_y.

To satisfy diverse demands from different customers, hundreds of part types are produced, and corresponding production lines are designed. As Fig.3 (a) shows, the part types are divided into tens of groups because of changes in market needs. Parts in groups A, B and C have a large

production quantity and lower demand variability compared to the other groups. The economic benefit and productivity of these part types is crucial to the entire system. Fig.3 (b) shows that parts in group A account for over 75% of production quantity and 80% of profits. Consequently, in this paper, part types M_1 and M_2, composing group A, are selected as the research object.

B. Simulation Model Construction

Based on the characteristics and structure of the real production system, an original simulation model is constructed to analyze the current production problems, called the AS-IS model. The AS-IS model facilitates introducing MFCA to the production system to identify hidden environmental problems effectively over a long running time. This study uses the Arena simulation platform [12] to develop this AS-IS model comprising four parts, as shown in Fig.4.

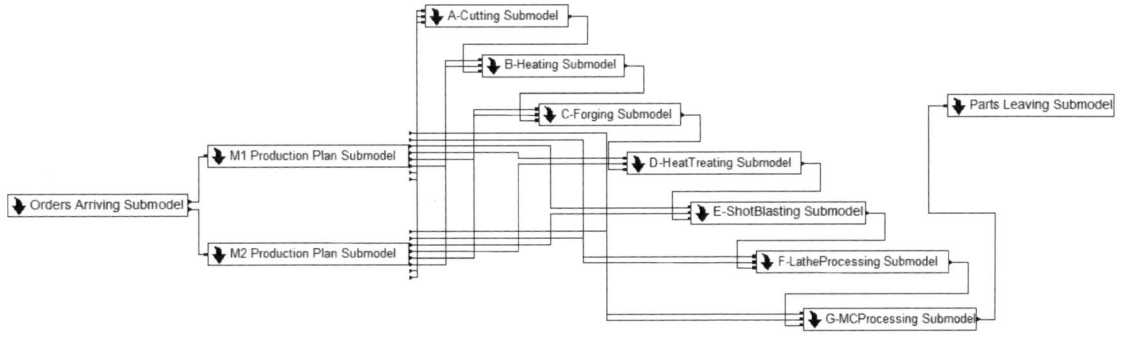

Fig.4. AS-IS simulation model

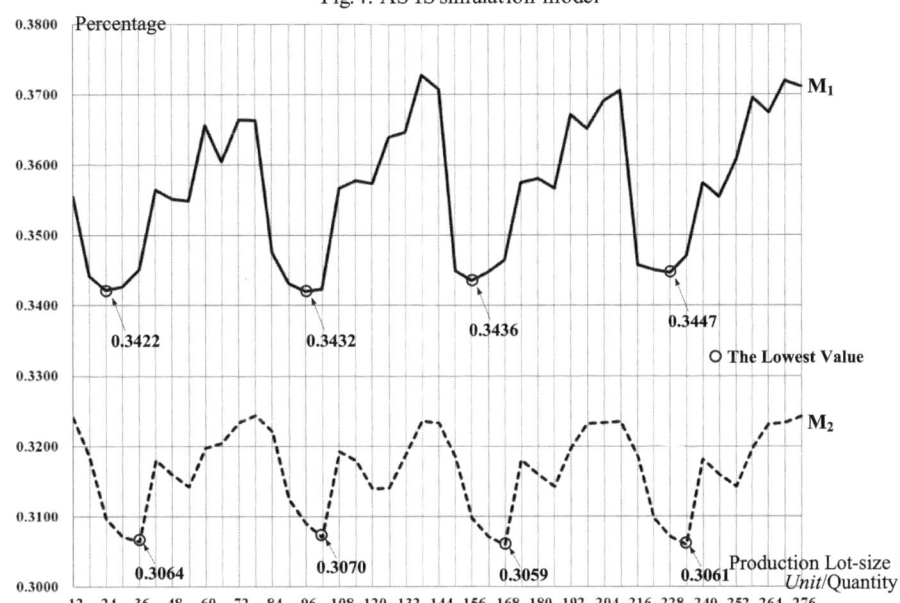

Fig.5. Negative products cost percentage of a unit part by regulating production lot-size

C. Simulation Analysis Using the Concept of MFCA

A reasonable production lot-size determination is crucial for production management. The study of production lot-size determination has thus received considerable attention from researchers recently. Azaron et al. developed a stochastic dynamic optimal programming algorithm for obtaining dynamic economic lot-size [13]. Nirmal and Tapan used multi-objective geometric programming to develop a multi-item finite production lot-size model [14]. Kämpf and Köchel used simulation optimization with a genetic algorithm as an optimizer to identify the optimal production lot-size [15]. However, these studies focused on obtaining an optimal algorithm for determining production lot-size, seldom considering the aspects of environmental performance.

Different production lot-size will produce different WIP inventory levels for different production stages. Hence, in this paper, a dynamic sensitivity analysis is used to analyze the changes in the negative products cost as a result of regulating the production lot-size. Additionally, in this case study, the production lot-size for the Cutting and Heat-Treating stations is set as a fixed value due to the current production schedule and

technological design. The production lot-size for the other stations can be regulated by running several different simulation scenarios. To reduce the reciprocal effects, the production lot-size of M_1 and M_2 in each workstation is regulated to the same value.

Negative products cost percentage of a unit part by regulating production lot-size is shown in Fig. 5. From Fig.5, for the total cost, the negative products cost percentage of the unit part is changed in a cyclical manner with changing production lot-size. For the unit part, the Negative products cost percentage changes along a declining curve. In one respect, this situation obeys the mass production mode that increasing the production lot-size generally reduces costs. The following are environmental considerations. First, through simulation monitoring and tracing with the MFCA method, the remaining overstock in inventory and the unusable idle processing corresponding to each production lot-size point are identified. Second, inapposite production lot-size generates substantial scrap and waste, thereby increasing the negative products and the environmental costs that are invisible during the production process and are easily overlooked by the conventional cost accounting

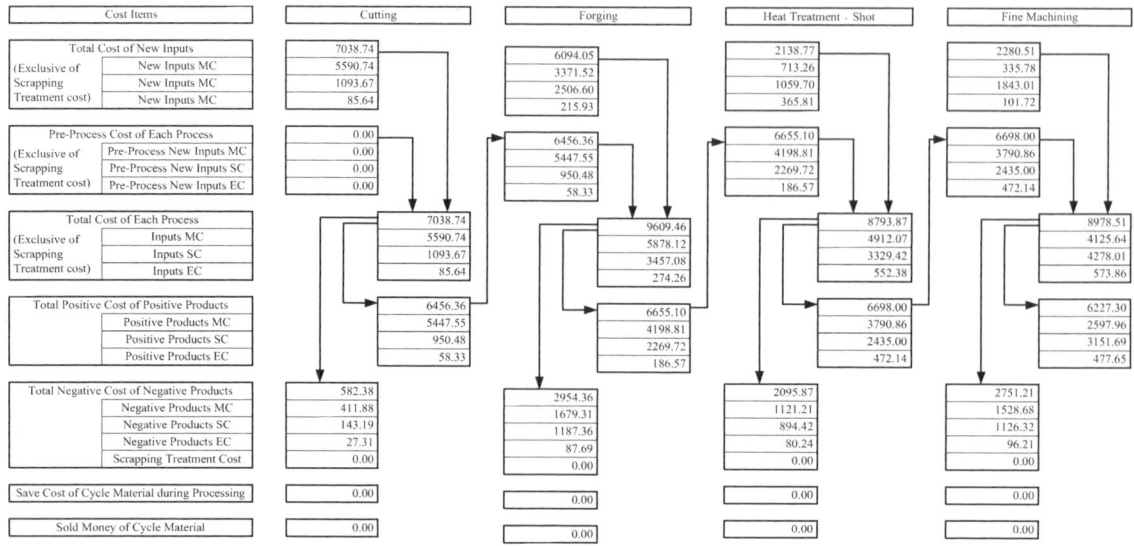

Fig.6. MFCA flowchart including calculation data (M₂-Lot Size=12) (Unit/JPY ￥)

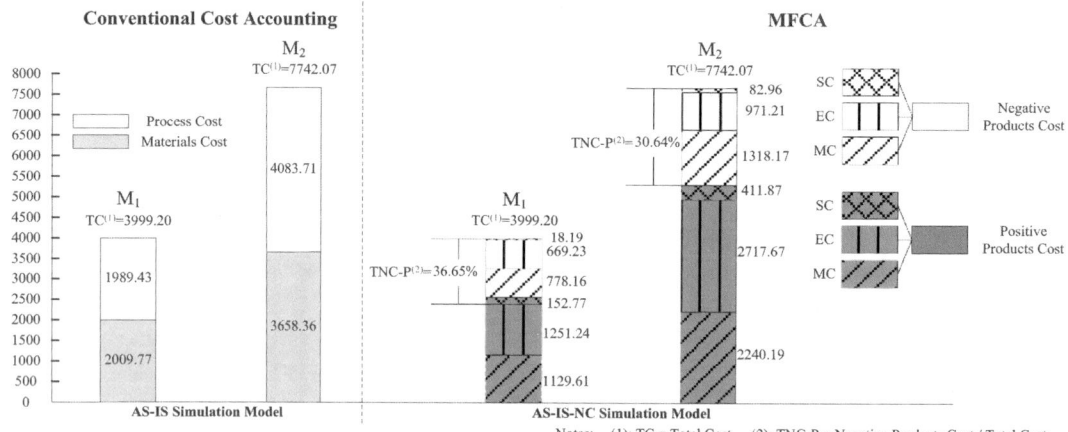

Fig.7. Cost results of unit part by comparing the AS-IS model and the AS-IS-NC model

method. Third, corresponding to the parts quantity distribution for the current order demand of each part type, there exists a relative appropriate production lot-size with the lowest negative products cost. Fourth, large increasing negative environmental costs will be reduced by increasing production lot size by, for example, using mass production mode. Final, in the real production system, the similar change curve for different part types in different processes will be changed because of various real random production factors, and the corresponding value will increase or decrease.

Furthermore, Fig.5 shows that the lowest or highest average negative cost value for different part types has a corresponding production lot size. To understand the concept of MFCA and apply calculation processes, an example is developed. Part M₂ is selected, and an extreme value situation (i.e., production lot sizes of 12) are considered, as shown in Fig.6. Based on the MFCA, the abandonment of the dead WIP stocks, unusable materials and idle processing are reflected as the generation of negative products cost in terms of monetary units, which are invisible during production. However, from these

MFCA flowcharts, the actual wastes and negative product costs generated during the production process are identified and clearly understood.

The AS-IS model is reconstructed to introduce the concept of MFCA by embedding a Monitor submodel, called the AS-IS-NC Model. All production operations are monitored, and all material flows are traced by this submodel. The final results of the AS-IS-NC simulation model are compared with those of the AS-IS model for two part types in Fig.7. Using MFCA can reveal invisible costs in the production processes; in particular, the negative products cost referring to environmental impacts becomes visible. For each unit part in the AS-IS-NC model, the negative products cost of M1 constitutes 36.65% of the total cost, and the negative products cost of M2 constitutes 30.64% of the total cost. Because the negative products cost is invalid for this production case, these high percentages indicate that the determination strategy for the production lot-size must be analyzed and improved to reduce the environmental burden by maintaining a low WIP inventory level.

After comparing the AS-IS-NC model and the AS-IS model, large negative products costs and environmental costs caused by the current production lot-size determination policy are identified. Running several different simulation scenarios yields a simulation analysis to analyze the changes in the negative products cost resulting from the regulation of the production lot-size. After observing the characteristics of change curves while gradually regulating the production lot-size, the declining change trend in negative products cost provides production managers with effective and strategic knowledge for determining appropriate production lot-size to maintain a low WIP inventory level and for considering both economic and environmental benefits.

IV. CONCLUSIONS

Using the concept of MFCA, a simulation analysis is performed to reduce the amount of wastes resulting from manufacturing activities. Within the framework of Material Flow Cost Analysis, costs are calculated for not only good products but also non-product outputs or material losses "negative products." A real case of a certain multi-variety and small-batch production system in a precision component manufacturing workshop of a Japanese company is presented. A simulation-based MFCA approach is adopted to study the environmental impacts of production lot-size determination for a multi-variety and small-batch production system in an actual forging factory. By applying the proposed approach, significant invisible wastes caused by inaccurate determinations of production lot-size are identified. It is demonstrated that the proposed procedure of application of simulation in terms of MFCA can also perform a dynamic simulation analysis along with a static analysis performed by using MFCA to increase production efficiency.

ACKNOWLEDGMENT

The authors wish to express their sincere gratitude to Mr. Y. Watanabe of Metal Worker Toa & Arai Company, Ltd. and Mr. Y. Murata of Fujita Health University for their cooperation in this research. Special thanks should be given to Prof. K. Kokubu of Kobe University for his suggestion on MFCA. This research was supported by the Grant-in-Aid of the Japan Society for the Promotion of Science (JSPS).

REFERENCES

[1] M. D. Ahmed, "A System Model for Green Manufacturing", *Journal of Cleaner Production*, Vol. 19, No. 14, pp. 1553-1559, 2011.

[2] "Simulation-based environmental cost analysis for Work-In-Process", *International Journal of Simulation Modelling*, Vol. 11, No. 4, pp. 211-224, 2012.

[3] A. Azaron, O. Tang, and R. M. Tavakkoli, "Dynamic Lot Sizing Problem with Continuous-time Markovian Production Cost", *International Journal of Production Economics*, Vol. 120, No. 2, pp. 607-612, 2009.

[4] R. Zhao, "Simulation-based impacts on environment costs caused by production lot size", In *Proceedings of International Conference Sustainable Manufacturing and Environmental Management*, pp.107-116, 2012.

[5] R. Zhao, H. Ichimura, and S. Takakuwa, "MFCA-based simulation analysis for production lot-size determination in a multi-variety and small-batch production system", In *Proceedings of the 2013 Winter Simulation Conference*, Edited by Pasupathy, R., Kim, S.-H., Tolk, A., Hill, R., and Kuhl, M. E., Piscataway, New Jersey: Institute of Electrical and Electronics Engineers, Inc, 2013.

[6] M. Nakajima, "Evolution of Material Flow Cost Accounting (MFCA): Characteristics on Development of MFCA Companies and Significance of Relevance of MFCA", *Kansai University of Business and Commerce*, pp. 27-46, 2009.

[7] Environmental Industries Office, Industrial Science and Technology Policy and Environment Bureau, Ministry of Economy, Trade and Industry, "Material Flow Cost Accounting MFCA Case Examples", Japan, 2010, http://www.jmac.co.jp/mfca/thinking/data/MFCA_Case_example_e.pdf.

[8] K. Kokubu, *Material Flow Cost Accounting* (In Japanese), Tokyo: Japan Environmental Management Association for Industry, 2008.

[9] K. Kokubu and H. Tachikawa, "Material Flow Cost Accounting" (Chapter 8), In *Manufacturing and Environmental Management*, Edited by S. Takakuwa, National Political Publishing House, 2012, Hanoi, pp. 265-283.

[10] X. Tang, and S. Takakuwa, "MFCA-based Simulation Analysis for Environment-oriented SCM Optimization Conducted by SMEs." In *Proceedings of the 2012 Winter Simulation Conference*, Edited by C. Laroque, J. Himmelspach, R. Pasupathy, O. Rose, and A.M. Uhrmacher. Piscataway, New Jersey: Institute of Electrical and Electronics Engineers, Inc, 2012.

[11] Environmental Industries Office, Industrial Science and Technology Policy and Environment Bureau, Ministry of Economy, Trade and Industry, Japan, "Guide for Material Flow Cost Accounting (Ver.1)", 2007 (a), http://www.meti.go.jp/policy/eco_business/pdf/mfca%20guide20070822.pdf.

[12] W. D. Kelton, R. P. Sadowski and N. B. Swets, *Simulation with ARENA*, 5th ed, McGraw-Hill, Inc., New York, 2010.

[13] A. Azaron, O. Tang and R. M. Tavakkoli, "Dynamic Lot Sizing Problem with Continuous-time Markovian Production Cost", *International Journal of Production Economics*, Vol. 120, No. 2, pp. 607-612, 2009.

[14] K. M. Nirmal and K. R. Tapan, "Multi-item Imperfect Production Lot Size Model with Hybrid Number Cost Parameters", *Applied Mathematics and Computation*, Vol. 182, No. 2, pp. 1219-1230, 2006.

[15] M. Kämpf and P. Köchel, "Simulation-based Sequencing and Lot Size Optimization for A Production-and-Inventory System with Multiple Items", *International Journal of Production Economics*, Vo. 104, No. 1, pp. 191-200, 2006.

Fault Tree Analysis on Boiler Water Shortage Accidents Based on Safety Ergonomic Theories

Yu-bin Ai, Xue-dong Liang*, Zhao-xia Guo

Business School, Sichuan University, P. R. China, 610064

(liangxuedong@scu.edu.cn)

Abstract - **The paper proposes the fault tree analysis method based on safety ergonomic theories in line with the actual conditions of serious consequences such as production interruption and personal injury caused by boiler water shortage accidents due to "human errors". In this method, firstly, the analysis is carried out on the accidents by means of safety ergonomic theories, with the focus on human hidden dangers in the accidents, so as to put forward the "human factors" causing the accidents. Secondly, analogy analysis is carried out on the accidents by means of historic information and data, so as to put forward the "non-human factors" such as equipment failure and design defects in the accidents. Thirdly, deductive reasoning is carried out for relevant factors by means of fault tree analysis method, so as to conduct qualitative analysis on them by constructing fault tree model, thus further confirming the cause logic among the factors and the rank of structure importance. Finally, the paper proposes the key measures for preventing and controlling such accidents, and also puts forward "3E countermeasures". Analysis is carried out on boiler water shortage accidents by this method. The results show that human factors, failing to install alarm and protection device or their failure are the key factors causing boiler water shortage accidents, which provides references and basis for the prevention work for such accidents in future.**

Keywords - **Boiler water shortage, fault tree analysis method, human factor, safety ergonomics**

I. INTRODUCTION

Boiler water shortage accidents is one of the common forms of major boiler accidents. It accounts for approximate 70% of boiler accidents according to statistical data [1]. There were 2710 boiler explosion accidents due to water shortage during the period from 1976 to 1987, which caused 1532 deaths and injuries to 5754 persons, bringing serious damage for social economy and personal safety [2].

Currently, the hidden dangers of humans are often ignored in the analysis on boiler water shortage accidents by frequently-used fault tree analysis methods, which lacks of effective analysis on human factors in boiler system. This will result in the deviation between the analysis results and real conditions in real application, which will reduce the credibility and reliability of the results, thus making it difficult to meet the requirement of accident analysis.

Therefore, this paper puts forward the fault tree analysis method based on safety ergonomic theories, and applies it to boiler water shortage accidents. Another factor - "human factor" which may cause the accidents is highlighted in this method. Firstly, the paper analyzes and

proposes the human factors causing boiler water shortage accidents by means of safety ergonomic theories on the basis of analogy analysis on the "non-human factors" such as the mechanical failure and design defect of the boiler system. Secondly, it analyzes and confirms the logical relationships among the factors and their influence degrees by means of fault tree analysis and based on the immediate causes and latent factors. Finally, it proposes the key measures for preventing and controlling such accidents, and also proposes corresponding "3E" improvement countermeasures for reducing human hidden dangers. This method not only increases the accuracy and credibility of analysis on accidents, but also provides theoretical basis for analysis and prevention for such accidents in future, which has important engineering significance for the safe and stable operation of boiler system.

II. ANALYSIS ON MAN-MACHINE HIDDEN DANGERS IN BOILER WATER SHORTAGE ACCIDENTS

Analysis on various accident cases of man-machine system shows that human factors are often the key link that causes the accidents; men's negative impact on the machines and environment may cause major accidents and result in severe consequences. Therefore, this paper emphasizes on analyzing and researching on the human factors in boiler water shortage accidents based on safety ergonomic theories and methods and in line with the operating features of boiler system, which aims at eliminating the human hidden dangers affecting the safe and stable operation of the system, and ensuring the safe and high-efficient operation of the system by improving human safety and reliability. This paper summarizes and confirms the main hidden dangers existing in the system by means of observation method, survey study method and analysis method etc. and in line with the habits and preferences of the operators in operation. The main hidden dangers include: ineffective safety management, imperfect systems and regulations, insufficient education and training, unreasonable design of man-machine interface [3].

(1) Ineffective safety management. Analysis on previous accident investigation reports indicates that the ineffective supervision and misprision of the supervisors constitute the latent factors which cause the accidents to a certain extent. Some supervisors on duty fail to perform their duties or fail to report the problems found during

E. Qi et al. (eds.), *Proceedings of the 21st International Conference on Industrial Engineering and Engineering Management 2014*, Proceedings of the International Conference on Industrial Engineering and Engineering Management, DOI 10.2991/978-94-6239-102-4_48, © Atlantis Press and the authors 2015

inspection in a timely manner, as a result, the dangers enlarge and finally evolve into major accidents, which brings serious damage to economy, society and personal safety. In boiler man-machine system, in case of failure of feed water system and leakage of blowdown when the firemen skive off work or fall sleep at work, but the supervisors with slack supervision fail to discover or fail to dispose in time after discovering, which will cause serious water shortage accidents.

(2) Imperfect systems and regulations. In production practice, the operators should strictly follow the uniform operation specifications. However, the flaws in the specifications are prone to causing misoperation or misjudgment of relevant operators, thus bringing bad consequences. During boiler operation, if the firemen carry out "boiler water testing" operation in line with the wrong or incomplete procedures, it will be prone to resulting in their inaccurate judgment on water shortage status; they may mistake severe water shortage for slightly water shortage and continue to add water to the overheated boiler, thus causing major boiler explosion accidents.

(3) Insufficient education and training. The self-qualities of the operators include professional skills level, psychological and physical qualities etc. The differences in the self-qualities will have an important influence on the accidents. Seen from the analysis on relevant accident cases, many major accidents have happened because of the operators' misoperation and misjudgment due to their nonstandard technological level as well as their improper disposal of the accidents due to their nervousness and panic. Meanwhile, the operators' poor safety awareness and insufficient conscientiousness are also one of the latent factors causing the accidents. In boiler man-machine system, the firemen who lack of safety awareness may forget to close blowdown valve or fail to inspect it, depart from work or fall sleep at work, give up "boiler water testing" operation, which will lead to the tragedies. Meanwhile, the operators may have insufficient understanding on the "boiler water testing" results and conduct wrong follow-up operations due to their limited individual knowledge and skills level, which will also cause severe boiler water shortage accidents.

(4) Unreasonable design of man-machine interface. Man-machine interface is the medium for information communication, execution and control as well as cooperative work between man and the equipment in man-machine system; it includes display device, control device and alarm. The operators can acquire the status information of the system via displayer or display screen to make decisions as well as control and operate. In case that the interface design is inconsistent with human's physical and physiological features, or there are design defects in the interface equipment, relevant personnel may fail to acquire the information, which may lead to wrong decision-making and operations. In the man-machine system of boiler, firstly, the inaccurate installation position of traditional bubble type water level gauge and the careless closing of steam and water cock will form false water level, which will affect the judgment of fireman supervisors. Secondly, the improper setting position of the water level gauge, unclear display of displayer, low sensitivity, tedious readings, complicated operation for instrumentation etc. will result in the wrong judgment and operation failure of operators. At last, if the alarm and protection devices are absent in the system, water shortage accidents will happen.

III. ANALYSIS ON MECHANICAL EQUIPMENT IN BOILER WATER SHORTAGE ACCIDENTS

Boiler system realizes safe, stable and high-efficient operation through man-machine coordination and interworking. Deviation in either party in the system will break the balance of the system, thus affecting the normal work and operation of the system. Normal operation of the feed water system of boiler can ensure fast water feed and timely water supplementing during the boiler's operation, it can also automatically regulate the water storage in steam pocket (boiler barrel) to maintain it above the safe water level. Analysis on relevant accident investigation reports shows that leakage of feed line, line clogging, failure of feed water pump and automatic feed water system etc. are the main causes for unstable and insufficient water feeding of the feed water system. If failing to troubleshoot in a timely manner, it will finally cause water shortage accidents. During the normal operation of the steam and water circulating system, the water will flow through such pipes as economizer, downcomer, water cooling wall, superheater and reheater successively, so as to generate high temperature and high pressure steam to provide power for steam turbine etc. Analysis on relevant accident investigation reports shows that the bursting of pipes due to the quality defect, being thinner by corrosion and clogging etc. will cause steam and water leakage of the boiler, which will affect the normal operation of the steam boiler, and even cause water shortage accidents.

IV. CONSTRUCTION AND ANALYSIS ON FAULT TREE OF BOILER WATER SHORTAGE ACCIDENTS

A. Confirming elementary events through analysis

Through the above-mentioned analysis, and based on water shortage accident investigation reports and related literatures, the set of elementary events for water shortage accidents is summarized and drawn as shown in Table I [4].

B. Constructing fault tree model

The logical relationships among the influencing factors are confirmed through analysis by means of deductive reasoning method [5]. Based on the principle of fault tree construction, the fault tree model of boiler water shortage accident is drawn as shown in Fig.1.

TABLE I
SET OF ELEMENTARY EVENTS

Set Of Elementary Events	Instruction Of Elementary Events
X0	The water level is lower than the minimum safe water level
X1	Failing to install the alarm or the alarm fails
X2	Failing to install water shortage protection device or the device fails
X3	The feed line is clogged
X4	Leakage in feed line
X5	Feed water pump fails
X6	Automatic feed water system fails
X7	Water failure
X8	The boiler tube is clogged
X9	The boiler tube wall is corroded
X10	The boiler tube has defects
X11	Failing to close blowdown valve
X12	The blowdown valve is not tightly closed
X13	Skive off work or fall sleep at work
X14	Depart from work
X15	The steam and water communicating pipe of the water level gauge is clogged
X16	The boiler water contains over salt and alkali
X17	Inaccurate installation position of the water level gauge
X18	The water level gauge is broken
X19	The steam and water cock of the water level gauge is closed
X20	Failing to conduct "boiler water testing"
X21	Misoperation in "boiler water testing"

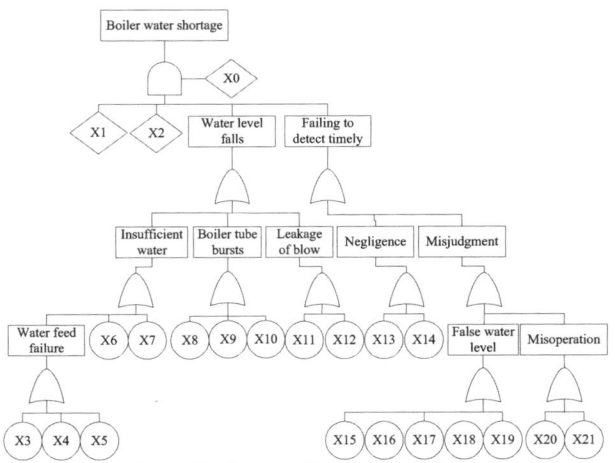

Fig.1. Fault tree of boiler water short

C. Qualitative analysis on fault tree of boiler water shortage accidents

1. Analysis on the minimum cutset of fault tree of boiler water shortage accidents

Step 1: List the structure function of the fault tree model of boiler water shortage accidents[6]:

$$Y = \phi(X) = X0X1X2(X3 + X4 + \cdots + X12)(X13 + X14 + \cdots + X21) \quad (1)$$

Step 2: Simplify the structure function by "Boolean algebra" rule:

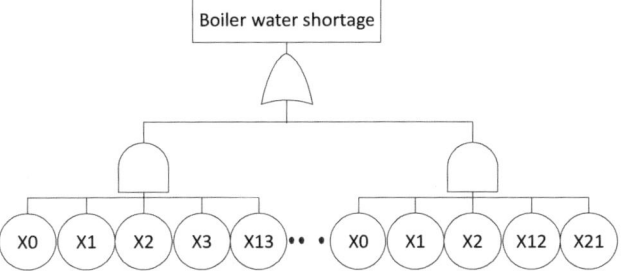

$$Y = \phi(X) = X0X1X2(X3 + X4 + \cdots + X12)(X13 + X14 + \cdots + X21)$$
$$= X0X1X2X3X13 + \cdots + X0X1X2X3X21$$
$$+ X0X1X2X4X13 + \cdots + X0X1X2X4X21$$
$$\vdots$$
$$+ X0X1X2X12X13 + \cdots + X0X1X2X12X21$$

Step 3: Get the minimum cutset of the fault tree:
$P1=\{X0, X1, X2, X3, X13\}$ $P2=\{X0, X1, X2, X3, X14\}$ … $P90=\{X0, X1, X2, X12, X21\}$, 90 sets in total.

Step 4: Get the fault tree model equivalent to fault tree of boiler water shortage accidents (minimum cutset indication), as shown in Fig.2.

Fig.2. Equivalent fault tree of boiler water shortage accidents (minimum cutset indication)

It can be known through the aforesaid analysis that there are in total 90 sets of all the possible modes which may cause boiler water shortage; for instance:
P1=\{X0, X1, X2, X3, X13\} indicates that: the alarm or water shortage protection device of the boiler system are uninstalled or fail, clogging of the feed line causes insufficient water feed, meanwhile, the firemen skive off work or fall asleep at work, which results in that the water level is lower than minimum safe water level, and finally causing water shortage accident.

2. Analysis on the minimum path set of the fault tree of boiler water shortage accidents

Construct success tree model according to the original fault tree model of boiler water shortage accidents, as shown in Fig.3.

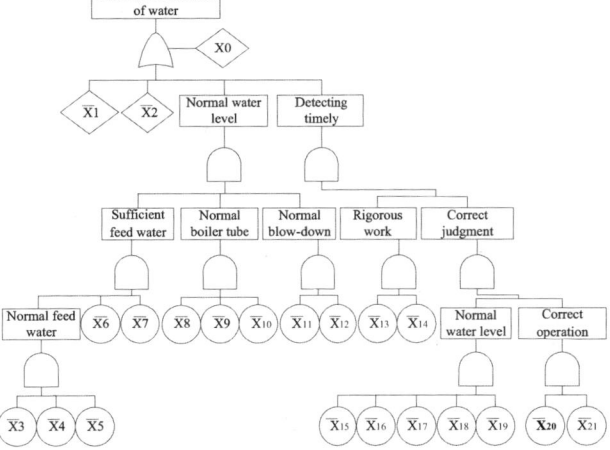

Fig.3 Success tree of boiler water shortage accidents

Step 1: List the structure function of the success tree:

$$Y = \Phi(\overline{X})$$

$$= \overline{X0} + \overline{X1} + \overline{X2} + \overline{X3 \cdot X4} \cdot \overline{X5} \cdot \overline{X6} \cdot \overline{X7} \cdot \overline{X8} \cdot \overline{X9} \cdot \overline{X10} \cdot \overline{X11} \cdot \overline{X12}$$

$$+ \overline{X13} \cdot \overline{X14} \cdot \overline{X15} \cdot \overline{X16} \cdot \overline{X17} \cdot \overline{X18} \cdot \overline{X19} \cdot \overline{X20} \cdot \overline{X21} \qquad (2)$$

Step 2: Get the minimum path set of the original fault tree (minimum cutset of the success tree):

Q1={X0} Q2={X1} Q3={X2} Q4={X3, X4, X5, X6, X7, X8, X9, X10, X11, X12} Q5={ X13, X14, X15, X16, X17, X18, X19, X20, X21} , 5 sets in total.

Step 3: Get the equivalent fault tree of the original fault tree (minimum path set indication), as shown in Fig.4.

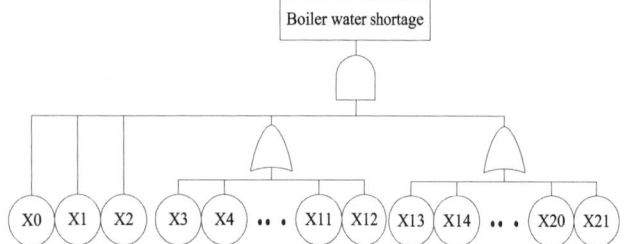

Fig.4 Equivalent fault tree of boiler water shortage accidents
(minimum path set indication)

It can be known through the aforesaid analysis that there are five sets of various effective measures for preventing and controlling boiler water shortage accidents; for instance:Q2={X1} indicates that: if the boiler alarm or alarm are installed in the boiler system and can work normally, then water shortage accidents can be effectively prevented.

$$I\phi(0) = I\phi(1) = I\phi(2) > I\phi(13) = I\phi(14) = I\phi(15) = I\phi(16) = I\phi(17) = I\phi(18) = I\phi(19) = I\phi(20) = I\phi(21)$$
$$> I\phi(3) = I\phi(4) = I\phi(5) = I\phi(6) = I\phi(7) = I\phi(8) = I\phi(9) = I\phi(10) = I\phi(11) = I\phi(12)$$

3. Analysis on the structure importance of the fault tree of boiler water shortage accidents

Since there are 90 sets of minimum cutsets of fault tree and 5 sets of minimum path sets in total, therefore, minimum path sets shall be used for analyzing the accidents.

Formula of structure importance [7]:

$$I\phi(i) = \sum_{X_i \varepsilon J} \frac{1}{2^{n_i - 1}} \qquad (3)$$

Where X_i——the *i*th elementary event

n_i——the number of the elementary events in the minimum path set where the *i*th elementary event is in

J——the minimum path set including the *i*th elementary event

The structure importance of the elementary events are got based on the analysis on boiler water shortage accidents in line with formula (3).

$$I\phi(0) = I\phi(1) = I\phi(2) = \frac{1}{2^{(1-1)}} = 1$$

$$I\phi(3) = \cdots = I\phi(12) = \frac{1}{2^{(10-1)}} \approx 0.00195$$

$$I\phi(13) = \cdots I\phi(21) = \frac{1}{2^{(9-1)}} \approx 0.00391$$

The weight ranking for the 21 elementary events are got according to the structure importance coefficients.

According to above analysis, it can be seen that the two factors of failing to install the alarm or the alarm fails (X1) and failing to install water shortage protection device or the device fails (X2) rank top in the weight ranking, and they are the main causes for boiler water shortage accidents. Meanwhile, they are also the key factors to be considered for effectively preventing such accidents. The influence degree of elementary events X13, X14, X15, X16, X17, X18, X19, X20 and X21 rank second, and the elementary events with the weakest influence degree are X3, X4, X5, X6, X7, X8, X9, X10, X11 and X12 [8].

V. "3E" COUNTERMEASURES FOR PREVENTING AND CONTROLLING BOILER WATER SHORTAGE ACCIDENTS

"3E" countermeasures (Education-educational countermeasures, Engineering-technical countermeasures, and Enforcement-management countermeasures) are countermeasures for preventing and controlling accidents based on safety ergonomic theories [9]. Starting from characteristics of man, "3E" countermeasures analyze man-machine-environment as a whole, and focus on improving human factors which may cause accidents to increase human safety and reliability, thus making man-machine system safer, more harmonious and high-efficient [10].

A. Technical Countermeasures

Technical countermeasures start from safety ergonomic theories to realize man-machine coordination and tactic cooperation by improving the mechanical equipment and related technologies in man-machine system, thus enabling the system to operate safely and high-efficiently.

(1)When designing the boiler system, it needs to take human factors into sufficient consideration, so as to make the man-machine system designed be suitable for human ability and operation requirement, and to endeavor to create a safe, comfortable, high-efficient and harmonious man-machine system. For instance, the main meters and instruments of the boiler system shall be installed at the place near horizontal sight line of the firemen to facilitate their reading and prevent misreading.

(2)When designing the boiler system, man-machine interface design shall be the key design contents, so that the man-machine interface designed can meet the operators' demand on reading as well as simple, accurate and fast operation. For instance, upgrade boiler display devices (steam temperature meter, bubble type water level gauge and pressure gauge) to digital meters, increase their reading sensitivity and accuracy; arrange the control apparatuses (emergency shutdown button or handle, valves) at the positions where are accessible for hands, meanwhile, their quantity shall be reduced in order to prevent misoperation.

(3) When designing the boiler system, key measures for preventing and controlling such accidents shall be taken [11]. Installing alarm and safe protection devices can effectively avoid the potential hidden dangers during the operation of the system, reduce equipment damage and personnel injury, and prevent the accidents from expanding. For instance, low water level alarm or low water level interlock protection device shall be installed at positions such as digital water level gauge and blowdown valve etc., high water level signaler, high pressure alarm and self-starting device for pump pit draining pump shall be installed at positions such as drain flash tank and continuous blowdown flash tank etc.

(4) When designing the boiler system, automatic supervision functions shall be explored. For instance, the system shall identify and accumulate the negative operations of operators. When the negative influences on the system are accumulated to some extent, the system will automatically lock out and refuse the negative operations, which will improve the collaborative capacity of man-machine system significantly, and it will also be conducive to operators' development towards a higher standard [12].

B. Educational countermeasures

All production activities are accomplished by man-machine system. Since man is the dominant factor, improving human quality is of great importance to safe production. Human qualities include professional quality of labor, cultural knowledge quality, safety quality, psychological and physical quality as well as physical health condition etc [13]. Enterprises should establish a long-term mechanism for safety education and skill training, and carry out safety education and trainings of specialized technologies in daily production and life on a regular basis, so as to increase the personnel's safety awareness, sense of responsibility as well as knowledge and operating skills. Enterprises should also assess the employees regularly, and only the employees who get the certificates subject to assessment can operate. Meanwhile, enterprises are also supposed to pay attention to the employees' psychological and physical status to enable them to develop towards the optimized comprehensive safety qualities.

C. Management countermeasures

Management work shall not only troubleshoot the hidden dangers of machinery equipment, but also strengthen supervision on personnel [14]. Firstly, the uniformed and perfected rules and regulations shall be made, including reward and punish system and operation specifications. Accountability system shall be implemented at all departments step by step; and the departments shall define their own working range and responsibility, in particularly the front-line production workers; in case of safety accident, relevant personnel shall be called to account. The personnel of all departments shall carry out corresponding work in strict accordance with the standard procedures and operation specifications, which can effectively avoid misjudgment and misoperation due to subjective factors, so as to avoid accidents. Taking the aforesaid measures can effectively reduce the probability of boiler water shortage accidents caused by operators' negligence and misoperation [15]. Secondly, carry out supervision and inspection work on a regular basis. During production process, the management department shall carry out regular inspection on the operators' implementation of safety regulations and operation specifications as well as the operation conditions of equipment, so as to eliminate hidden dangers in a timely manner. Taking the aforesaid measures can effectively avoid boiler water shortage accidents due to water feed failure and boiler tube bursting etc. Finally, the work shall be reasonably arranged. Enterprises are expected to scientifically arrange the work in line with the operators' personal conditions such as technical capacity, knowledge level and experience etc., so as to meet the requirement of the posts to the greatest extent [16]. The enterprises can give full play to men's subjective initiative to enable them work high-efficiently, safely and reliably.

VI. CONCLUSIONS

(1) Traditional fault tree analysis method often lacks of detailed analysis on human factors in man-machine system. Based on this problem, this paper proposes fault tree analysis method based on safety ergonomic theories, and applies it to boiler water shortage accidents. It confirms through analysis that the human factors causing such accidents include four aspects: ineffective safety management, imperfect systems and regulations, insufficient education and training, unreasonable design of man-machine interface.

(2) This paper constructs fault tree of boiler water shortage accidents and carries out qualitative analysis on it, and concludes the following conclusions: since there are many minimum cutsets of this accidents (90 sets in total), it can be known through analysis that boiler water shortage accident is prone to happening, and it is one of the common accidents of boiler system. Since the structure importance coefficients of elementary events of failing to install the alarm or the alarm fails (X1) and failing to install water shortage protection device or the device fails (X2) are the largest, it can be known through analysis that elementary events X1and X2 have the largest influence on boiler water shortage accidents; meanwhile, they are also the key measures for preventing such accidents.

(3) The safety measure formulation work for boiler system in the past fails to reinforce the safe operation ability of the system from the perspective of increasing human reliability in the system. Based on this problem,

this paper proposes "3E" countermeasures for preventing and controlling boiler water shortage accidents, aiming at ensuring the safe and stable operation of the entire system by increasing human safety and reliability.

ACKNOWLEDGMENT

This research is funded by the National Natural Science Foundation of China (71302143), China Postdoctoral Science Foundation (2012M521705), Postdoctoral Science Special Foundation of Sichuan Province and the Fundamental Research Funds for the Central Universities (skzx2013-dz07).

REFERENCES

[1]   Yang Lu, DaiYuebing. Analysis on and Prevention Countermeasures for Boiler Water-shortage [J]. Petrochemical Safety Technology, 2000, 29(1):30-32.

[2]   Zhao Xiaojiao, Zhang Nailu. Fault Tree Analysis of the Boiler's Absence of Water [C]. Conference Papers of 2007 Academic Annual Conference of China Occupational Safety and Health Association, 2007, 11(4): 343-346.

[3]   Su Yunqin. Analysis and Prevention for the Accident Boiler Water-shortage [J]. Railway Occupational Safety Health & Environmental Protection, 2003, 2(30): 93-95.

[4]   Zhou Guofen. Fault Tree Analysis of the Boiler's Water-shortage in Forest Region [J].Forestry Labor Safety, 1995, 25(3):35-37.

[5]   Wang Bin. Analysis on Boiler Explosion Accident and Its Safety Countermeasures [J]. Science and Technology Association Forum, 2010, 1(2): 34-36.

[6]   Han Jianguang. Fault Tree Analysis on Boiler Explosion [J]. Journal of Xinyang Normal University (Natural Science Edition), 1997, 4 (10):56-59.

[7]   Li Xinjian. Analysis on Steam Boiler Explosion due to Water-shortage Based on Fault Tree Analysis [J]. Digest of Management Science, 2009, 9 (2):214-245.

[8]   State Bureau of Machine — Building Industry, Quality Safety Supervision Division. Fault Tree Analysis and Application [M].Chinese Machinery Press, 1988.

[9]   Xie Rang. Application of Safe Ergonomics in Construction Safety Accident Prevention [J]. Guangzhou Construction, 2008, 2(36):39-42.

[10] Li Hailong, Hu Qinghua. Safety Ergonomics and the Application to Preventing Analyzing the Accident in Power [J]. Energy Conservation Technology, 2004, 3(22): 33-37.

[11] Shang Shaojun. Countermeasures for Prevention of Falling Accident of Object Lifted by Hoisting Machinery Based on Ergonomics [J].Safety, 2012, 2(15):12-15.

[12] Guo Xing, Chen Yiren. Analysis on the Reasons of and Countermeasures for Coal Mine Accidents by Ergonomic Theory [J]. Mining Safety & Environmental Protection, 2006, 4(33): 67-69.

[13] Chen Yiren, Ma Qiang. Analysis on the Reasons of and Countermeasures for the Coal Mine Accident by Accident Control Theory of Ergonomics [J]. Hebei Coal, 2006(38), 4:30-32.

[14] Li Farong. Application of Ergonomic Principles in Accident Analysis and Prevention [J]. Safety & Health, 2001, 9(4): 34-36.

[15] Zhao Gui. Improving the Safe Operation Level of Boilers by Ergonomics [J]. Safety, 2002, 6(8):11-16.

[16] Teng Hongfei, Wang Yishou. Key Supporting Technologies of Man-machine Cooperation [J]. Chinese Journal of Mechanical Engineering, 2006, 11(42): 1-9.

The Discussion on the Production Quality Management about Inner Tube Improvement of Plastic Company "S

Rui-yuan Xu, Hai-wei Gao*, Qing-na Meng

Department of Science and management, Hebei University of Science and Technology, Shijiazhuang, China

(1027736061@qq.com)

Abstract – **This thesis is mainly concentrated on analyzing the quality control department of plastic company S. And the author hope that Company S can get to a higher stage in the quality-control area by applying IE, quality management and other related knowledge. Then Company S can improve the unreasonable parts of it, and it will improve its working efficiency and the quality of its products, at the same time, the cost will also be reduced. Finally, Company S can enhance its market competitiveness in the industry. By all of the above, the author hopes that what the company finally gives the consumers will match its' company slogan, which say that "customer-oriented, quality first".**

Keywords – **Product quality, quality management, quality control, the inner tube production**

I. INTRODUCTION

After more than a month in S company internship, analyze the present situation of the production of the product and the enterprise product quality problem investigation and research. According to company's management target system, the sampling inspection, find out the unreasonable place [1], analysis of causes and improvement measures are put forward. According to company's quality management system and product sampling inspection method to detect the quality of your goods, the survey found that the enterprise did not adopt scientific and effective system of quality management method, to reduce costs [2], improve product quality, thus to achieve the ultimate goal of the enterprise.

II. THE INNER TUBE PRODUCTION PROCESS FLOW DIAGRAM

The inner tube production process flow is shown in Fig.1.

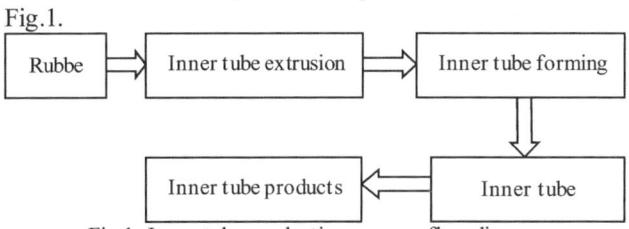

Fig.1. Inner tube production process flow diagram

III. THE INNER TUBE PRODUCTION PROCESS DEFECT ANALYSIS

TABLE I
THE SECOND WORKSHOP PROCESS DEFECT LIST IN MARCH

The production workshop	Production shift	sulfide	The molding	interface	Squeeze out	equipment	other	Waste tire in total	Grade A number	The total output of	Rate of grade A
The second workshop 1 area	A	3951	1177	140	668	35	4376	10347	771368	781715	98.68
	B	2932	1265	116	415	22	2957	7710	796731	804441	99.04
	C	3351	1248	81	339	0	3100	8123	740835	748958	98.92
The workshop 2 area	A	2790	1120	86	681	0	3383	8081	727523	735604	98.90
	B	2401	1195	54	468	0	2971	7072	802321	809393	99.13
	C	2931	1279	141	330	1	3768	8453	757415	765868	98.90

Through qualitative check member of the test statistics is unqualified emitting that sulfide in the process of inner tube production process of scrap rate is the highest, i.e. sulfide links exist a lot of quality problem [3]. Therefore, the use of quality management tools for analysis found that the problem of this link and effective improvements are put forward (Table I).

IV. INNER TUBE VULCANIZATION PROCESS STATUS

Inner tube sulfide is the last line of inner tube production process, this process caused by defective goods can't reuse raw materials [4]. So the quality control of the process is very important. The inner tube vulcanization about quality problem mainly has folds, uneven thickness, drying, expand, aging, spongy and sponge rubber edge phenomenon as well as the quality problems.

A. Using the diagram and cause and effect diagram analysis of the cause of quality control

Through the inner tire vulcanization process product inspection showed that the quality problem of the inner tube sulfide mainly include: skirts, uneven thickness, drying, expand, aging, spongy and sponge rubber edge, etc.

E. Qi et al. (eds.), *Proceedings of the 21st International Conference on Industrial Engineering and Engineering Management 2014*, Proceedings of the International Conference on Industrial Engineering and Engineering Management, DOI 10.2991/978-94-6239-102-4_49, © Atlantis Press and the authors 2015

B. Using the Pareto methods analysis main quality problems of tube vulcanization process

The first step: Collecting the number of unqualified products of Tire vulcanization process to produce in the first day (24 hours one day, three classes). Designing a data sheet, and fill the unqualified emitting in the sheet, then combined (Table II).

TABLE II
INNER TUBE VULCANIZATION PROCESS UNQUALIFED QUESTIONNAIRES

Unqualified type	Unqualified number	note
cleft	3	
Drying raw	5	
Uneven thickness	36	most
aging	8	
stigmas	30	more
Rubber edge	1	
spongy	4	
other	3	
combined	90	

The second step is to make a data table of Pareto chart, column in the table are the unqualified data, cumulative unqualified data. The unqualified percentage and cumulative percentage [5]. By quantity in order from largest to smallest. The data to fill in the table, the "other" item listed in the final (Table III).

TABLE III
UNQUALIFIED QUESTIONNAIRES

Unqualified type	Unqualified number	Cumulative number of unqualified	Ratio (%)	The cumulative ratio (%)
Uneven thickness	36	36	40	40
stigmas	30	66	33.3	73.3
aging	8	74	8.8	82.1
Drying raw	5	79	5.5	87.6
spongy	4	83	4.4	92
cleft	3	86	3.4	95.4
other	3	89	3.4	98.8
Rubber edge	1	90	1.2	100
combined	90		100	

The third step is make two longitudinal axis and a horizontal axis, labeled number (frequency) of the scale in the left vertical axis, the largest scale for the total number (frequency); Standard rate (frequency) scale in the right vertical axis, the largest scale was 100%.The total frequency of the scale on the left and The total frequency of the scale on the right have the same height [6]. On the horizontal axis coordinates according to the frequency size to draw a rectangle, the rectangle height represents the size of the frequency of the non-conformance (refer to Fig.2).

The last step is the judgment according to the figure of non-conformance. Can be seen from the diagram, the main quality problem is inner tube of type vulcanization process) and the thickness [7].

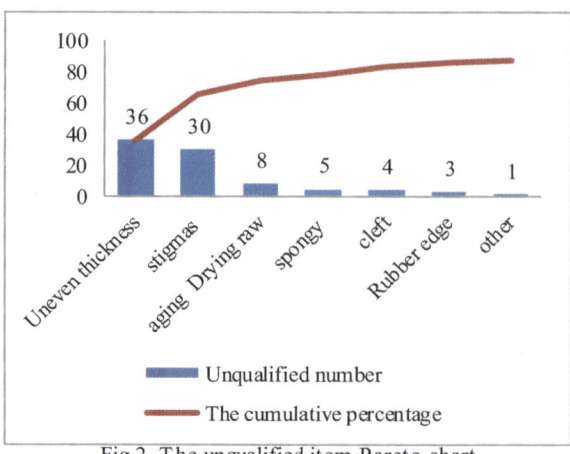
Fig.2. The unqualified item Pareto chart

C. Based on the causality diagram analysis of the cause of the inner tube thickness

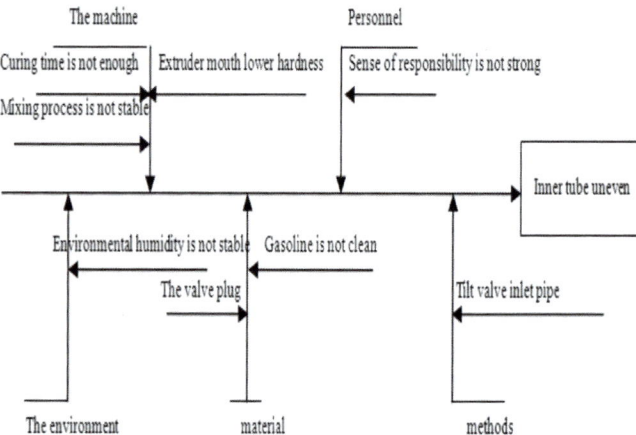
Fig.3. Inner tube thickness causal analysis diagram

Combined with the actual situation of tire vulcanization process analysis that caused the quality problem of the inner tube inner tube thickness vulcanization process is mainly(refer to Fig.3):

The extrusion process of extrusion machine equipment, and for discontinuous adhesive; the strength of the extruder mouth design is bad [8]. Mouth when extrusion deformation, which leads to the inner difference of correspondence and thickness, is not equal. Extrusion tire tube size is less than construction standards.

Aspects: (1) mixing operation method of instability [9]. Returns the glue mixed with a large proportion. Existing in production of rubber in the back glue dosage of instability, mixing a good film in the park on the conveyor belt is long wait for a reason, after the mixing of rubber and plastic value volatility leads to uneven thickness of tube semi-finished products (2) curing speed too fast, have to finalize the design before make the valve and thin wall parts of rubber to expand, the tire body local drum thick and thin value [10]. (3) The clamping, canning speed too slow. After making the first contact mould parts of the first heat contact mold parts of inflation. (4) After excessive storage of semi-finished products storage time is too long, thin fold bum.

V. THE QUALITY OF THE INNER TUBE VULCANIZATION PROCESS IMPROVEMENT MEASURES

Operation method: 1) the control returns when the dosage of the glue and regulate mixing proportion of time; 2) security extruder for the glue, prevent the glue supply disconnect or excess; 3) after mixing rubber in front of the supply of parking time should not be too long, lest affect the plasticity index of the rubber [11]; 4) to do a good job of equipment management, strengthen the maintenance extruder mouth; 5) control of sulfide before finalize the design speed; 6) to improve the speed of mold, canning; 7) reduce joint after the storage time of semi-finished products.

Machine and equipment: 1) improve the mouth strength design of extruder. Extruder molding workshop was directly by the motor start running process not only the demand is higher, difficulty is big, and easy to damage to equipment, electricity energy consumption is larger [12], it installed in 30 KW motor frequency converter, using variable frequency energy saving technology to reduce power consumption, and reduce the operation difficulty. 2) Extrusion tire tube size conform to the standards of construction. 3) is the main air volume of wind pressure on vulcanizing machine work, the original directly to the wind, the wind does not have any control measures and equipment when not working, air discharge [13] caused a large number of waste air and waste of electricity. Making its air volume will not discharge when not working, so fundamentally solved the waste air volume. In addition, on the basis of this and installed a pressure reducing valve in the pipe on the road [14], is composed of high voltage part into a low pressure wind, so that by the original using a high pressure air compressor and a low pressure air compressor into using a high pressure air compressor.

VI. CONCLUSION

Through application of quality management tools, Pareto diagram, cause and effect diagram and control chart to analyze the quality problem of the semi-finished products, is the quality of the product had the very big enhancement, greatly reduces the cost of materials, improve the production efficiency, increase the benefit of the company. Set up inner tube production line quality improvement system [15], unreasonable places in the inner tube production process for continuous improvement, constantly improve product quality. So through the inner tube production line for each process quality control and improvement of inner product percent of pass is greatly improved.

REFERENCES

[1] Shuping Yi. Fu Guo. Basic industrial engineering [M]. Beijing: mechanical industry publishing house, 2005

[2] Yiming Gong. Modern quality management [M]. Beijing: Tsinghai university press, 2007

[3] Xiansheng Qin. Quality management [M]. Beijing: science publishing, 2008

[4] ShuLin Kan. Basic industrial engineering [M]. Beijing: higher education press, 2005

[5] Ai Wu. The quality of management [M]. Guangzhou: Jinan University press, 2006

[6] JianXin You. Quality management [M]. Beijing: science press, 2008

[7] FengRong Zhang. Quality management and control [M]. Beijing: mechanical industry publishing house, 2006

[8] Qin Shu. Quality management and reliability [M]. Beijing: mechanical industry publishing house, 2006

[9] Weigong Li, Yang lian root. The quality of statistical technology [M]. Beijing: China metrology publishing house, 2006

[10] Changfeng Wang, Yinghui Li. Modern project quality management [M]. Beijing: mechanical industry publishing house, 2007

[11] ZhengMao Xiao. Quality control and improvement in field details [M]. Guangdong, Guangdong economic press, 2007

[12] Guangtai Li. Production field controls [M]. Shenzhen: the publishing house, 2006

[13] Harrington HJ. Business Process improvement [M]. New York: McGraw Hill, Inc, 1991

[14] Aikens C H. Facylity location models for distribution planning [J]. European of Operational Research, 1985, 22(3): 263-279

[15] Sanders MS, Mc Cormick E J. Human Factors in Engineering and Design [M]. McGraw Hill Publishing Co.1992

ADAMS Modeling and Dynamics Simulation of Mosaic Particle's Motion State

Jian-jun Bu[1,*], Xing-bo Wang[2]

[1] Department of Mechanical and Vehicle Engineering, South China University of Technology, Guangdong, Guangzhou 51, China

[2] Department of Mechanical Engineering, Foshan University, Foshan, China

(1057239154@qq.com)

Abstract - **According to dynamic movements and vibration of complex conveyer with Mosaic particles, simulation analysis of conveyer is carried out by ADAMS. First, three dimensional model of conveyer was established by the software of solidworks/proe, then it was lead into the software of ADAMS, the model was established by defining restriction, load, rigid-flexible coupling and contact among the parts of conveyer on the virtual machine of ADAMS .At last, the graphics of displacement, velocity of moving mosaic, acceleration of mass center of Mosaic particles were gotten by simulation ,which provides the certain foundation for the structure optimization of conveyer machine.**

Keywords - **Dynamics vibration, mosaic particles, simulation analysis, three-dimensional modeling**

I. INTRODUCTION

Mosaic is introduced from abroad where researches on design and production of Mosaic and automatic production line about mosaic have been accomplished roundly. However, the price of automatic paving machine imported from abroad is too expensive to purchase for small company except several large companies.

In recent years, domestic media also have some special reports about automatic mosaic paving machine [1]. For example, Chen Gangchang invented the patent technology "a Mosaic paving machine "in 1988. Han Qiwen invented the "bidirectional and fast glass Mosaic paving machine" whose application by placing the mosaic on the mold in1993.Guo Anhua invented " bricks paving machine" whose application by conveying mold with mosaic in 2007. Chen Yaozao designed a mosaic paving machine that worked like this: three conveyer belts was transporting at the same time three manipulators extracted mosaic from three conveyor belts and put them on the mold frame in order to an array in 2010. Zhang Liangfu, Zhang qiang in via-wisdom technology company in shenzhen invented the patent technology "automatic mosaic fixed-point paving equipment" that worked like this: equipment used the suction cup adsorb mosaic into corresponding position for mold. Mentioned equipments above have low efficiency and low success rate in 2013. In words, mosaic products were inexperienced from research and development and therefore cannot be promoted due to contemporary restrictions about technology. The lab devoted series of product research with associated with Mosaic particles, which Has authorized an invention patent "one kind of automatic sorting system" and patent number was ZL200810025703.4, also authorized a utility model

patent "multicolor and complex Mosaic paving device". Patent only describes device on the aspects of theory and can't effectively put all details and applied theory knowledge in detail on account of own properties. Article makes effective analysis of the theory on the transformation of mosaic's motion state under the action of vibration on the slope on the basis of paving mosaics inefficiently.

First, article establishes 3D model by Solidworks/Proe, then establishes Dynamics Simulation of multicolor and complex Mosaic paving device by ADAMS [2]. Through the simulation experiment and dynamic simulation, article obtained the relevant test data, the comprehensive detailed understanding of mosaic's motion state about the mosaic under the action of vibration on the slope, and provides an effective and efficient simulation animation which provides a certain theoretical basis for development and production on the future.

II. MODELING OFCOMPLEX PAVING MACHINE WITH MOSAIC PARTICLES UNDER THE ACTION OF VIBRATION

The modeling of simplified paving device under the action of vibration is shown in Fig.1.

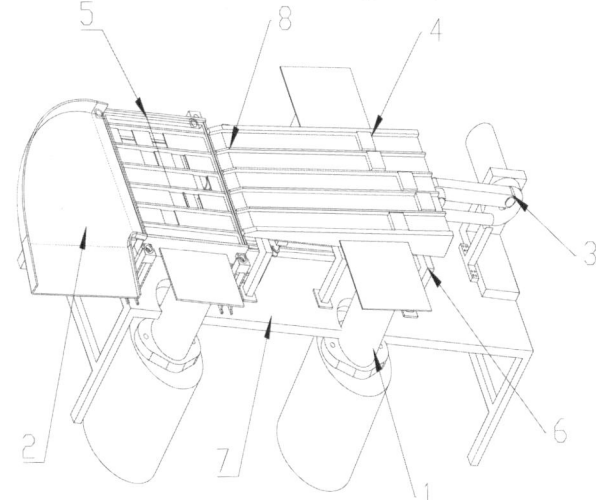

1 the electromagnetic vibrator; 2 the output ramp; 3 the crank slide block structure; 4 Mosaic particles; 5 mold; 6 guide rail; 7 support plate; 8 feeding track

Fig.1. the modeling of simplified paving device under the action of vibration

The mosaic paving device under the action of vibration overcomes the friction of feeding track by the function of mechanical vibration [3], [4], then falls into a mold with squares in a way. The purpose of this article is to study the moving behavior about the mosaic particles under different directions of vibration on the slope. The selected direction of vibration is shown in Fig.2, 3, 4. Mosaic particles having just entered into the feeding track entrance may appear the following situations:

(1) Side-boundary of the mosaic is parallel to that of the feeding track;

(2) Side-boundary of the mosaic isn't parallel to that of the feeding track

(3) A special status appears. Mosaic particles are turned 90 degrees.

The following is mechanics analysis in view of the three conditions above.

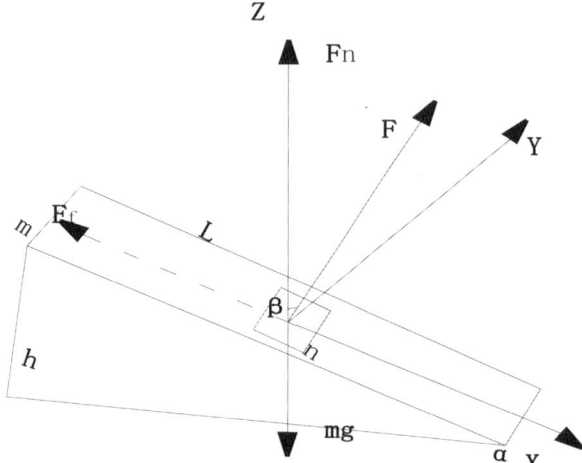

Fig.2. mechanics analysis when side-boundary of the mosaic is parallel to that of the feeding track

At this time, $F_n + F\cos\beta = mg$; $F\sin\beta > F_f$.

Mosaic particles slide along the track with accelerated state in Fig.2.

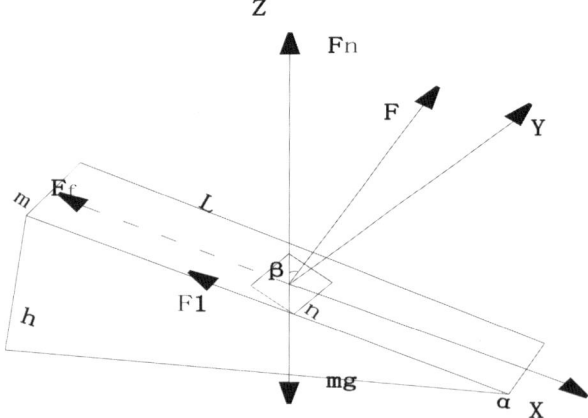

Fig.3. mechanics analysis when side-boundary of the mosaic isn't parallel to that of the feeding track

As shown in the Fig.3, $F\sin\beta$ and F_f are in a straight line, but they state a different direction with F_1, Mosaic particles slide along the track in accelerated motion. Force F_1 drives mosaic particles turned along the plane of slope until the degree=0. So, system is reaching the state of Fig.2 by this time; $F_n + F\cos\beta = mg$

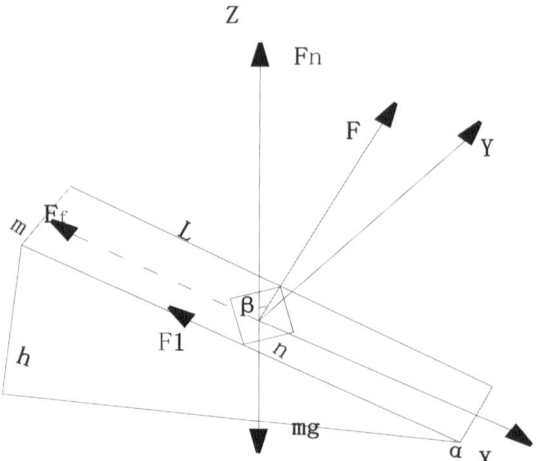

Fig.4. Mosaic particles just are turned 90degrees

As a special case, Mosaic rotates 90 degrees. In order to prevent mosaic stuck, we must ensure that side length of mosaic reaches the requirement: $m > \sqrt{2}n$.

III. THE DEFINITION OF FLEXIBLE BODY

In order to avoid mosaic particles, feeding orbits and mold crashing into each other under the action of vibration, the article defines the materials of feeding orbits and mold as flexible materials which have inherent damping and costs physical energy overcoming friction [5]. Based on theory of dynamics, simulation method and ADAMS software platform, a rigid-flexible coupling model is built for mosaic. In the process of simulation, rigid body is defined as loading mass [6], in contrast, flexible body is defined as object without mass. Considering that high-frequency vibration may cause Mosaic particles out of the way, this article regards the motion state under the action of low-frequency vibration as the aim of research.

IV.THE ESTABLISHMENT OF DYNAMIC MODEL

Based on the operational theory of mosaic paving device under the action of vibration, the Mosaic particles on the feeding track present a Spring - damper steady and forced vibration by FM vibrator [7]. Setting initial phase of FM vibrator presenting harmonic vibration is 0. The equation of motion:

$$s = A\sin\omega t \qquad (1)$$

Force produced by actuator is

$$F = m * s'' = -mA\omega^2 \sin\omega t \qquad (2)$$

Among them:

s — distance on the slope with vibration;

A — amplitude on the slope with vibration;

ω — angular velocity about FM vibrator;

t — time;

Note1: in formula (2), the type of minus sign shows that the direction of force is opposite with prescribed direction.

I need to make appropriate simplification by ignoring secondary factors with model before setting up dynamic model. As follows:

(1)The mass of spring in the case of dispersion is

small and negligible.

(2)All kinds of damping are equal to linear damping [8].

(3) Mosaic particles are equal to protons.

Based on aforementioned simplifying Rules of dynamic model, the article has accomplished establishment of dynamic model, as shown in Fig.5. Among them, Mosaic particles are defined as protons whose quality are M; Angle of slope is α; slope K where article establishes coordinate system X - Y - Z is seen as the initial planes; β is the angle between the Y direction and the direction of vibration; plane L is defined as a plane which is vertical with plane X; proton presents a spring - damper steady and forced vibration which can be equal to a combined action of spring - damper steady forced vibration along the direction of X and Y along the direction of the vector on the plane L.

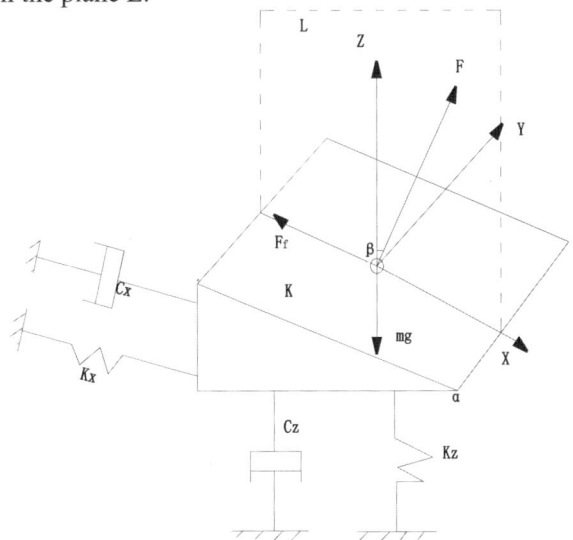

Fig.5. the establishment of dynamic model

The friction produces because of their interactions between the protons and slope K When protons can't jump from K. At this time, based on Lagrange equation, establishing differential equation is:

$$m\ddot{x}+k_x x+c_x \dot{x}+F_f = F\sin\beta+mg\sin\alpha \quad (3)$$

$$m\ddot{z}+mg\cos\alpha+k_z z+c_z \dot{z}=F\cos\beta+F_n \quad (4)$$

$$F_f = \mu F_n \quad (5)$$

Among them: m——the mass of mosaic;

$\ddot{x}/\ddot{z}$ —the acceleration in the direction of X/Z for mosaic;

$\dot{x}/\dot{z}$ —the velocity in the direction of X/Z for mosaic;

c_x/c_z —the damping coefficient in the direction of X/Z for the system;

k_x/k_z —the stiffness coefficient in the direction of X/Z for the spring;

F_n —support force slope imposing on the Mosaic;

F_f —friction slope imposing on the Mosaic;

μ —friction coefficient.

The displacement of proton not jumping out of slope is zero along the direction of Z, the displacement only in the direction of X exists, equation (4) reveals the system in the state of equilibrium, reforming equation (3), I get that:

$$m\ddot{x}+k_x x+c_x \dot{x}=F\sin\beta+mg\sin\alpha+\mu mg\cos\alpha+\mu F\cos\beta \quad (6)$$

Among them,

$\mu F\cos\beta = -\cot\beta.\mu\omega^2 mx$, $F\sin\beta = -m\omega^2 x$,

Reforming equation (6), I get that:

$$m\ddot{x}+c_x \dot{x}+(k_x +m\omega^2 +\mu m\omega^2 \cot\beta)x=mg\sin\alpha+\mu mg\cos\alpha \quad (7)$$

Equation (7) is typical 2-order linear inhomogeneous differential equation.

From equation (7) above, I get a conclusion: the frequency of vibration, different direction of vibration, the change of slope Angle and accuracy of system installation under the action of installation all affect motion state in the process of conveying mosaic particles.

Mosaic particles can jump out of the slope with the increase of vibration frequency or increasing amplitude.

Based on equation (4) $F\cos\beta = mg\cos\alpha$, meanwhile, F_n=0. The mosaic particles will jump out from the slope without the support and friction force which has reached the maximum static friction force [9].

Assuming the time quantum of mosaic particles jumping out of slope is from t_s to t_q,

$$A\omega^2 \sin\omega t_s \cos\beta = g\cos\alpha \quad (8)$$

Drawing a conclusion:

$$t_s = \frac{1}{\omega}\arcsin\frac{g\cos\alpha}{A\omega^2 \cos\beta} \quad (9)$$

When mosaic Jump out from slope, the movement behavior expressing in formula is:

$$F\sin\beta + mg\sin\alpha = m\ddot{x} \quad (10)$$

$$F\cos\beta - mg\cos\alpha = m\ddot{z} \quad (11)$$

After transformation,

$$\ddot{x} = -A\omega^2 \sin\omega t\sin\beta + g\sin\alpha \quad (12)$$

$$\ddot{z} = -A\omega^2 \sin wt\cos\beta - g\cos\alpha \quad (13)$$

Making the second integral equation for equation (11) and (12) in the time quantum that from t_s to t, equation of displacement for particle are obtained:

$$x = A\sin wt\sin\beta + \frac{1}{2}gt^2 \sin\alpha - A\sin\omega t_s \sin\beta - \frac{1}{2}gt_s^2 \sin\alpha \quad (14)$$

$$z = A\sin\omega t\cos\beta + \frac{1}{2}gt^2 \cos\alpha - A\sin\omega t_s \cos\beta - \frac{1}{2}gt^2 \cos\alpha \quad (15)$$

When the mosaic particles fall back on the slope, z

= 0. At this time,

$$A \sin \omega t_q \cos \beta + \frac{1}{2} g t_s{}^2 - A \sin \omega t_s \cos \beta - \frac{1}{2} g t^2 \cos \alpha = 0$$

(16)

The value of ωt is very small and close to zero so that $\sin \omega t \approx \omega t$. I get the approximate equation about time from jump out of slope to fall back on the slope by simplifying and sorting the equation of (15), (16):

$$\frac{1}{2} g t_q{}^2 \cos \alpha - A \omega t_q \cos \beta + A \sin \omega t_s \cos \beta - \frac{1}{2} g t_s{}^2 = 0$$

(17)

The equation (17) is quadratic equation. In the equation, only the variable of t_q is unknown, according to the actual situation, the solution of equation exists, so I can figure out the value of t_q.

V. SIMULATION AND ANALYSIS

Based on the analysis of the vibration/view in ADAMS, modeling and dynamics simulation of complex paving machine with Mosaic particles establishes. On the basis of the input channel, excitation function is defined. Similarly, on the basis of output channel, frequency response function is calculated. By the analysis of response function in the output position, I analyze the system of vibration. First, I import the Model for Mosaic in the ADAMS with simulation and then build the flexible body in ADAMS/Flex covering the original rigid body. I use rigid constraint between flexible and rigid bodies [10]. A coupled model of simulation with vibration is presented shown in Fig.6 as follows:

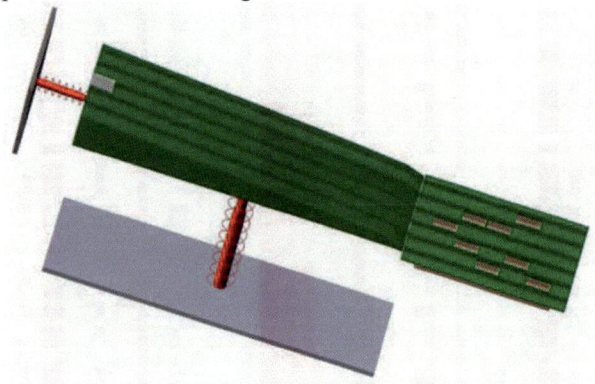

Fig.6. coupled model of simulation with vibration for mosaic particles

In the process of simulation with vibration, first of all, I need to define the material qualitative, as well as, the contact relationship between the various parts. Material selection is shown in Table I, the definition of contact for parts is shown in Table II, the definition of friction is shown in Table III.

TABLE I
THE DEFINITION OF MATERIALS

serial number	name of the material	material	Mass/g
1	Feeding track	flexible body	no
2	mold	flexible body	no
3	supporting plate 1/2	x6Cr13	1440/463
4	loose tooling	ceramics	198

TABLE II
THE DEFINITION OF CONTACT

contact	value
Contact Name	Contact 1/2
Contact Type	Solid to Solid
I Solid	Mosaic particle
J Solid	Feeding track/mold
Normal Force	Impact
stiffness	1.0E+0.08
Damping	1.0E+0.04
Exponent	2.2
Dmax	1.0E-0.04

TABLE III
THE DEFINITION OF FRICTION

item	value
Friction Force	Coulomb
Coulomb	on
Static Coefficient	0.2
Dynamic coefficient	0.15
Stiction transition	10
Friction transition	1000
Kx /Kz	5
Cx/Cz	0.2

In the ADAMS, the setting time of simulation is 1s, the steps of simulation is 150, the length of slope for feeding track is 800mm. Input channel function is defined as $s = 100 \sin 2t$, I define that $\alpha = 30^0$, $\beta = 10^0$. Based on the post processing in vibration/review, I get the displacement diagram in the direction of Z for mosaic particle, as shown in Fig.7. The velocity diagram in the direction of X, Y, Z respectively are shown in Fig. 8, 9 and 10. The acceleration diagram in the direction of Z is shown in Fig.11. The contact force diagram between mosaic particles and feeding track or loose tooling are shown in Fig.12 and 13.

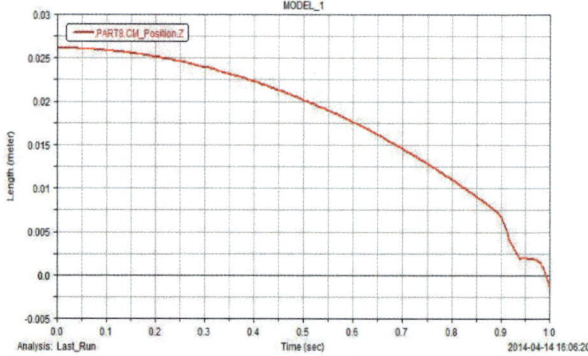

Fig.7. The displacement diagram in the direction of Z

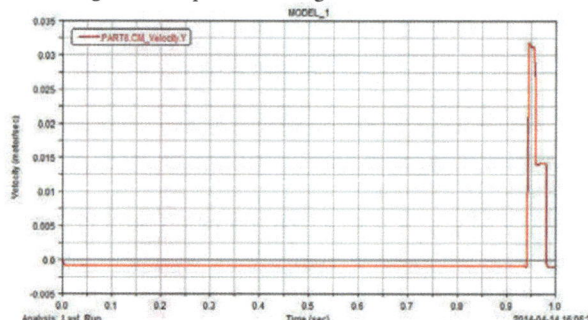

Fig.8.The velocity diagram in the direction of X

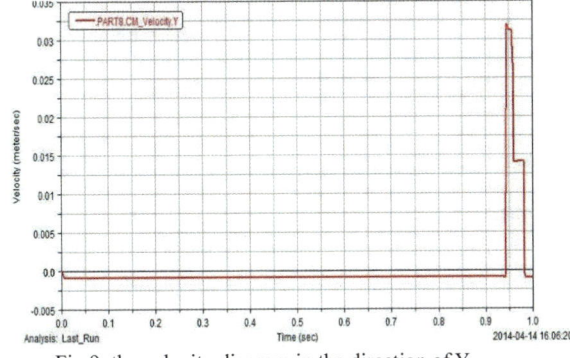
Fig.9. the velocity diagram in the direction of Y

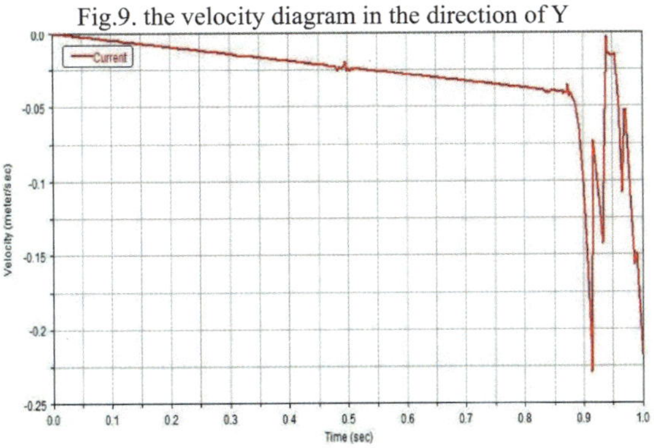

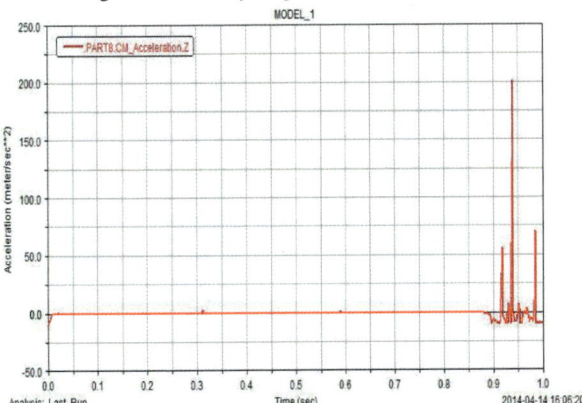
Fig.10. the velocity diagram in the direction of Z

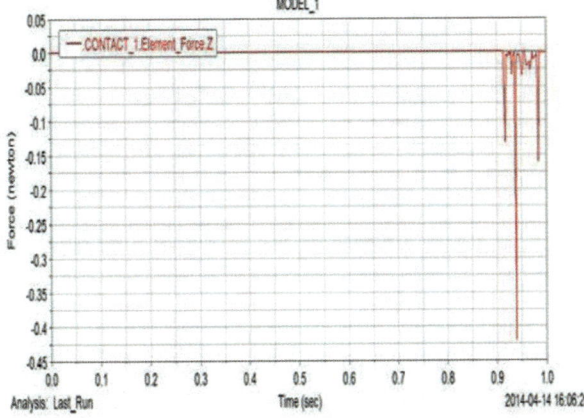
Fig.11. the acceleration diagram in the direction of Z

Fig.12. the contact force diagram between mosaic particles and feeding track

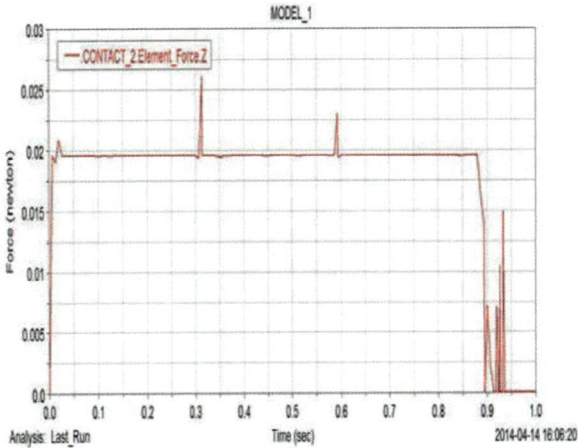
Fig.13. the contact force diagram between mosaic particles and loose tooling

By analyzing the diagrams, I draw several conclusions:

(1)   Setting the input channel function is s = 100 sin2t, as shown in Fig.7, 8, 9, 11 and 12, when the time pass from 0 to 0.9, the protons are not jumping up and down in the direction of Z, from the phenomenon described above, I draw a conclusion: the protons didn't jump out of the slope.

(2)   I can draw a conclusion from Fig.7 and 8: the velocity of mosaic should be constant for a constant ideally when the angle of slope is α, the velocity of mosaic in the direction of Z raises due to the effect of exciting force with vibration, and the component of the exciting force in the direction of X is greater than the friction force.

(3)   As shown in Fig.9, 10, 12 and 13, when the time pass from 0.9 to 0.92, mosaic particles jump   out from feeding track, the starting time t_s and ending time t_q are determined, which imported into equation (17) can verify the exactness of the equation.

(4)   when the time pass from 0.92 to 0.98 shown in Fig.11, 12 and 13, mosaic particles are falling into the mould and jumping up and down under the action of vibration until fell into the blank. As shown in Fig.10 and 11, the period of vibration is more than 0.1 s, and amplitude is about 0.05mm.

(5)   As shown in Fig.12, at the time of 0.91 s, the contact force is maximized value between mosaic and slope; as well as, at the time of 0.94 s, the contact force is maximized value between mosaic and mould, which occurs due to the simultaneous vibration of the feeding track and mould.

(6)   When the time pass from 0.98 to 1s, the contact force is zero between mosaic and others, and mosaic has maximal velocity, which show that mosaic has been out of the shop and laid-up completely well

According to the diagrams obtained from simulation and academic analysis, I understand that the time from the feeding track entrance to the shop for mosaic particles is a total of about 0.98-1seconds. In this article, the mosaic paving prototype use a whole row of pavement, therefore, paving mould whose standards is 5*5 takes about 5 seconds. The mould standards used in factory is 12*12.On that baseline, the cost of paving time is about 45seconds, considering the time of moving mould by A one-dimensional numerical control platform costing a certain amount of time, according to the standards that time of paving a mould completely is 50 seconds, the paving device can pave 1728 moulds in one day (including 24 hours) while normal workers (working10 hours a day) can pave about 700 moulds. The efficiency of paving device is more than doubled than manual manpower. Therefore, with the phasing-in of more automatic equipments, the work efficiency improves greatly, the cost of labor reduces, and the value of benefits increases.

Note 2: the conclusions above only are limited to the stage of feeding and pavement, not including the subsequent delivery.

VI. CONCLUSION

The article mainly Involves establishing a coupled model of simulation with vibration for mosaic particles, obtains the characteristics of kinematics and dynamics mechanism by dynamic simulation. Data obtained by analyzing diagrams are imported into equation (17) and verify the accuracy of the previously mentioned mechanical analysis. What's more, the article obtained the graphics of displacement, velocity and acceleration for mosaic by simulation under the action of vibration, so that it can be indirectly studied for the behavior of mosaic of particles, meanwhile provides a strong theoretical basis for the whole assembly system.

In this article, the results of simulation show that it is an effective method to design the structure of equipment by simulation with the virtual prototype, which also apply to the design of other products. The study of this article is limited to specific limitation, having some deficiencies, needing further optimization.

ACKNOWLEDGEMENTS

The research work is supported by National spark plan project 2013GA780052, Department of Guangdong Science and Technology under project 2012B010600018 and 2012B011300068, Foshan Bureau of Science and Technology under projects 2010C012, 2011AA100021 and 2012HC100131, and Chancheng government under projects 2011GY006, 2011B1023, 2011A1025, 2011A1030 and 2012B1011. The author sincerely present thanks to them all.

REFERENCE

[1] Du Yaoxue, Chen Qirui. The design and research on automatic production line of vitreous mosaic [J]. Mechanical design, Vol.5, pp.28-31, 1997.

[2] Li Zenggang. Introductory explanation and implementation for ADAMS [M]. Beijing: National defence of Industry Press, pp.125-150, 2007.

[3] Liu Jiao, Guo Shenghua, Zhang Wanlin. Mechanical Vibration [N]. Guide of science technology, Vol.23, pp: 43-47, 2013.

[4] Hu Shijie, Li Guirong. Discussion of vibration dynamics and wave mechanics [J]. Mining technology, pp: 115-116, 2010.

[5] Joyce Fang, Jen-San Chen. Deformation and vibration of a spatial elastica with fixed end slopes [J]. International Journal of solids and Structure, Vol.5, pp.824-831, 2013.

[6] Y.-S.Kim, K.Miura, S.Miura. Vibration, characteristics of rigid body placed on sand ground [J]. Soil Dynamics and Earthquake Engineering, pp.19-37. 2001.

[7] Ma Jisheng. The system modeling and dynamic equation iteration based on ADAMS [J]. Measuring and control technology, Vol.23, no.4, pp: 49-51, 2004.

[8] Wen Bangchun. The basic characteristics of the linear vibrator material movement and the selection and calculation of kinematics parameters. Mining machinery teaching and research section in Northeast institute of technology.

[9] Ren Lanzhu, Guo Bin. The kinematics parameters of vibration conveyor based on half analytical and numerical methods [J]. Coal mining machinery, Vol.25, no.7, pp.7-8, 2007.

[10] Cao Shuqian, Zhang Wende, Xiao Longxiang. The modal analysis， theory, experiment and application of the structure of the vibration [M]. Tianjin: Tianjin UP, 2001.

Design of White LED Lighting and Digital Communication System

Bo-hao Chen[1],[*], Kai-ru Lei[2]

[1]School of Information Science & Engineering, Shenyang University of Technology, Shenyang, China
[2]School of Information Science & Engineering, Northeastern University, Shenyang, China
([*]bhchenli@126.com, kellylei@126.com)

Abstract - **The essay proposed and introduced a visible light digital information communication system which uses white light of LED to be the medium, and is based on MCU. The essay introduced the working principle of the external circuit and the internal program of the communication system, and discussed the way to achieve the function of illuminate and signal interference problem between many illuminant. The research talked about the way to emission and received when the signal is transmitting. It also talked about relevant arithmetic and the method of modulate and demodulate, and disposed noise from signal. The communication system fulfills the task of illuminate, at the same time, transmits digital information at a short distance.**

Keywords - **Modulation and demodulation, photosensitive diode, visible light communication, white light LED**

I. INTRODUCTION

In recent years, the market quantity demanded for LED was increased than before. The advantages of LED, such as environmental protection, long operating life and high flux per LED make the LED to replace traditional lighting system, and to become a new trend of illuminate [1].

Visible light communication has two functions. It can complete not only a fundamental illumination but also an informational transmission [2]. Especially, the system never produces electromagnetic radiation, so that it can be used in some special places, to avoid electromagnetic interference. The system has an excellent security. As long as some objects obstruct the path of light, information can be protected and don't out leakage.

Using photosensitive diode which has light sensitive characteristic can pick out the frequency change of light. With a series of dispose to the signal, the information will be identified and restore.

II. THE WORKING PRINCIPLE OF THE SYSTEM

Visible light communication system is based on MCU. Two MCUs compute the information which comes from transmitter and receiver respectively. A high power LED is the light source of transmitter, and the photosensitive diode is the beginning of the receiver [3]. When it receives data, the analog signal will turn into digital signal by a series of external circuit. The data will be modulated into PPM code at transmitter. Constant current source driving circuit will send the signal to LED, so that LED lights in many different frequencies by PWM dimming, and carries information. Photosensitive diode receives visible light at receiver and changes its value of resistance, so that the special

frequencies which come from LED will be turn into digital signal and be analyzed by receiver's MCU and the system will accomplish the communication. The whole process of the system is shown in Fig.1.

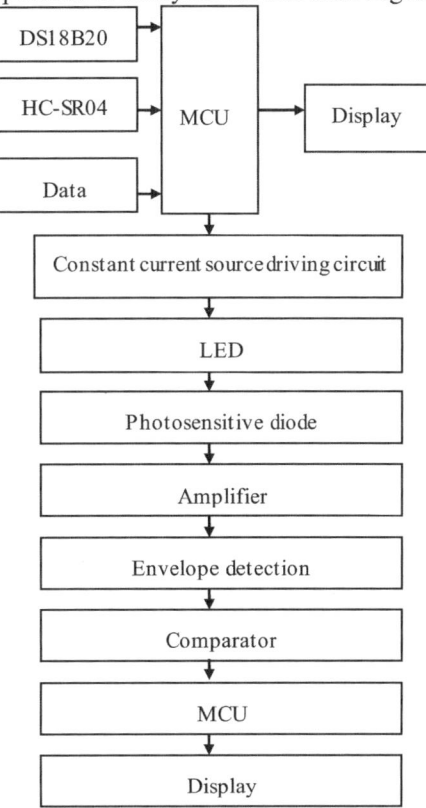

Fig.1. The composition of the system

III. THE WORKING PRINCIPLE OF TRANSMITTER

A. Illumination

Generally, people need 150-300lx illumination in offices and other indoor rooms to work and live. Especially, operating rooms and laboratories need 300-1500lx illumination to meet the normal work.

Traditional LED's power is low, which can only get the function of indication, it is inadaptable to illumination.

So the system should use a high power LED to be the light source [4]. In order to fulfill the task of illumination better and increase the area where is covered by light, make the light distribute equality as far as possible, multi-high power LED should constitute into a matrix. Because of the fast transmission rate, LED needs to change its switch on and off at a quite short time. According to the volt-ampere characteristics of LED, LED can not make an immediate response with the change of digital signal, there is a delay both on glowing and extinguishing [5]. The phenomenon makes the rising edge and falling edge of the signal exist some error. In order to avoid the case as well as ensure the veracity and the high-speed of the communication, it is necessary to choose an appropriate constant current source driving circuit. With LED's working time increased, its interior will produce hot, and affects the value of resistance.

If the LED is no longer working at the rated current value, it will reduce its service life and efficiency. Constant current source driving circuit can be a good solution to these problems [6]. In the current source driving circuit, with the MOS tube fast switching and the control of integrated chips, high-power LED's current value will not increased exponentially with the voltage signal which is in according to the volt-ampere characteristics, but a constant current value. So it can be used to show the voltage signal indirectly by current.

When determined the size of high power LED and driver circuit, the next step is to distribution the layout of light source, and the convergence of the scattered light. A 4 * 4 LED matrix can increase the light intensity effectively, and the interior lighting requirements can be achieved by a number of LED dot matrix. However the color of the walls are always be white, and white walls will reflect all colors of light, if there are too many indoor light sources, they will produce inter-symbol interference and multi-path effects, the visible light which comes from many light sources must affect each other [7]. In order to solve the problem, put a lens in front of LED, or set some mirrors around LED can gather the scatter light or the parallel light which is in many different directions to improve the lighting efficiency.

B. The test of light source's number affects on the information transmission success rate

In a closed interior space of 5m * 5m * 3m, the walls are white, 16 white LED which the rated power is 1W compose a light source. The receiver is placed in the center of the room, receiving visible signal 10 times. By increasing the number of light sources, analyze the impact of the number of light sources to information transmission success rate [8]. Besides, there is a uniform lighting in the room at the top. The information transmission success rate response of light source's number is shown in Fig.2 and the parameters of each component in the test are listed in TABLE I.

The experimental result shows that the increased number of light sources makes inter-symbol interference and multi-path effects become more severe, the communication success rate become lower.

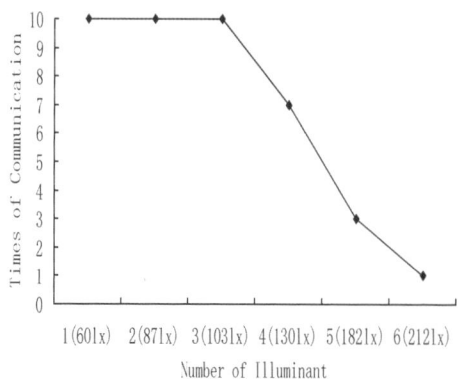

Fig.2. the information transmission success rate response of light source's number

TABLE I
RELATED PARAMETERS OF THE WHITE LED ARRAY AND VISIBLE LIGHT RECEIVER

Parameter	Value
LED array scale N*N	4*4
Rated power per LED	1.5W
Optical wavelength of the LED	450nm
Receiving area of photosensitive diode	1.5mm*1.5mm
Response time of photosensitive diode	5us
Divider resistance of photosensitive diode circuit	19K
Resistance of amplifier	4.23K
Comparison voltage of comparator	1.25V
Communication rate	500kb/s

C. Modulation

PPM coding is a process of modulate, the data can be converted into digital signals perfectly, and the way of modulate has a high fault tolerance. Even if there is some delay, the error can be reduced by setting some special instructions at receiver's MCU. By adding a high frequency carrier wave may reduce some noise of the signal during transmission [9]. Due to the LED and the photosensitive diode's character of fast response, it can transmit information after added the carrier wave. This process is a two-time modulation.

Multi-channel data can be transmitted at the same time by encoding in different ways, such as using frequency division multiplexing.

Initialize the timer of MCU through the internal procedures will set the width of the data codes, start codes and end codes. The digital data transmit to the drive circuit through one of the I/O port of MCU.

When the transmitter working, the MCU will be coded data into PPM coding, and data turn into a digital signal, complete the modulation [10]. The constant current source driving circuit makes the current which is through the LED carries the voltage signal. Finally, the LED luminance changes regularly and emits the modulation signal.

IV. THE WORKING PRINCIPLE OF RECEIVER

A. signal processing

First the light which is emitted by the transmitter is received by a photosensitive diode which has a photosensitive characteristic. According to the characteristics of resistance changes by the change of light intensity, the photosensitive diode can change the optical signal which is come from transmitter into an electric signal to the receiver. This is a process of photoelectric conversion [11]. Due to the special characteristics of the photosensitive diode, the frequency change of the light will show into a variation of the duration of the high electrical level. If the receiving area of the photosensitive diode is increased, more noise will be received, thereby causing the interference [12]. So the receiving area of the photosensitive diode should not be too large. Put the Fresnel lens in front of the photosensitive diode as an optical add concentrator may gather light better, and the focal point should fall on the surface of photosensitive diode [13]. The light around the interior space can be gathered by adding a plurality of Fresnel lens in each direction.

The receiver processing the signal is a significant process of filter noise and extract of useful information [14]. The bleeder circuit produces electrical signal, and send the signal into a low noise amplifier circuit for amplifying the signal. Because of the amplified signal of the amplifier will also be self-generated noise amplification at the same time, It should be possible to select a low noise amplifier. The amplified signal enters into an envelope detection circuit to filter out high-frequency carrier wave and restore the PPM code. This time the filtered wave still has some error and is disturbed by noise interference. So at the end of the receiving circuit, add the amplitude comparator will turn analog signal into digital signal, which is also a utility way to reduce the noise.

B. Demodulate

MCU calculate the time of high electrical level by timer initialization, using the counter function of MCU. If the crystal is 12M, the count unit is 1us, so as to distinguish the different types of PPM code. In the internal process, the data is received from the beginning of the start code is received, and stop receiving unless receipt the end code [14]. It is the process of demodulation and restores the information.

V. CONCLUSION

LED has plenty of advantages, such as low-cost, long life and more environmentally friendly. The design uses its advantages to transmit information while achieving the illumination function. Through the research and analysis of the LED voltage current characteristics of the delay, selected a suitable constant current source driving circuit to drive high power LED and transmit information. The paper talks about the influence of reflex walls and multi-path effects which is caused by multiple light sources affect on the system performance. Multi-path effects and inter-symbol interference were tested and analyzed. Through observation and signal processing at the receiver, use an external circuit to filter out noise, convert the analog

signal into a digital signal. LED white lighting digital communication system is incorporate with the lighting and communications. Furthermore, its environmentally friendly and efficient features will affect the field of communication in the future.

REFERENCES

[1] T.Mukai, S. Nakamura. White and UV LEDs. Oyo Buturi. 1999(68): 152-155,

[2] Elgala H. Indoor broadcasting via white LEDs and OFDM Consumer Electronics, IEEE Transactions on: 1127 - 1134.

[3] Mesleh, R LED nonlinearity mitigation techniques in optical wireless OFDM communication systems Optical Communications and Networking, IEEE/OSANov. 2012.

[4] Khoa Quang Huynh, Maximum likelihood detection upper bound for detect-and-forward relaying over AWGN channels. IEEE Transactions. 2011, 11: 705-709

[5] Quadeer A.A. Enhanced channel estimation using Cyclic Prefix in MIMO STBC OFDM sytems. Signal Processing and Information Technology (ISSPIT). 2010 IEEE International Symposium on 15-18 Dec. 2010: 277 - 282.

[6] Yang Jie. Optoelectronics and Image Pro Back to Results the Study on L-STBC-OFDM in HF Communication System. Optoelectronics and Image Processing (ICOIP). 2010 International Conference on 11-12 Nov. 2010: 641 - 644

[7] Kwon J K. Inverse source coding for dimming in visible light communication onreliable links [J]. IEEE Photon Technol,Lett, vol. Oct. 2010,22, pp. 1455-1457.

[8] Bai B, Xu Z, Fan Y. Joint LED dimming and high capacity visible light communi-cation by overlapping PPM [M]. Annu Wireless Opt. Commun. 2010 .5, pp. 71-75.

[9] Grubor J, Randel S, Langer K D. Broadband information broadcasting using LED-based interior lighting [J]. Light w Technol, vol.2008, 26, pp. 3883-3892.

[10] Cui K. Line-of-Sight visible light communication System design and demonstration. IEEE Int'l. Symp. Commun [C]. Sys. Networks and Digital Signal, 2010.7, pp. 621-25.

[11] Nanba S,Miyazaki N, Hirota Y. MIMO capacity estimation based on single and dual-polarization MIMO channel measurements [J]. APCC 2010, Oct, pp. 359-364.

[12] David A, Schmidt, Randall A. Minimum mean squared error interference alignment. IEEE Annual Asilomar Conference on Signals5Systems, Computers, Pacific Grove, California, USA, Nov 2009.

[13] Khoa Quang Huynh, Maximum likelihood detection upper bound for detect-and-forward relaying over AWGN channels. IEEE Transactions.2011, 6(9): 705 - 709

[14] McKay, M.R,Achievable Sum Rate of MIMO MMSE Receivers: A General Analytic Framework. Information Theory, IEEE Transactions. 2010, 1: 396-410.

Research on Data Analysis Model of Risk Assessment in Oil and Gas Drilling Operation

Jiadi Li[1,*], Tingke Li[1], Wei Li[2], Lin Jiang[3]

[1]School of Petroleum Engineering, Chongqing University of Science and Technology, Chongqing, China
[2]Safety, Environment & Technology Supervision Research Institute of Southwest Oil and Gas Field Company, PetroChina Co., Ltd. , Chengdu, China
[3]School of Computer Science, Southwest Petroleum University, Chengdu, China
(345914230@qq.com)

Abstract - **Through the establishment of multidimensional data model based on data warehouse, we can convert the original application-oriented data structures in the database for the analysis of multidimensional data structures. Multidimensional data model provide a good data analysis environment for risk assessment system in oil and gas drilling and overcome inadequate analytical capacity which can't satisfy complex analysis of risk assessment in oil and gas drilling operation compared with information system based on database technology. Because the model provides a good analysis of structures, risk assessment of oil and gas drilling operations can be analyzed from different perspectives and at different levels of the system related issues which provide an effective analytical tools and decision support for managers.**

Keywords - **Data analysis, data warehouse, drilling operation, multidimensional data model, risk assessment**

I. INTRODUCTION

From relevant research results in the present domestic, we have not yet seen the risk assessment software in drilling operation. There are two main reasons for this situation: First, the time on research of drilling operation risk is not long and start later; Second, the drilling risk is a complicated systematic project The key problems which need urgent solution in high-risk oil and gas drilling operations is to set up a suitable risk assessment model by using reasonable evaluation method. During the research, we also found that the risk assessment systems in existing other industry or areas of domestic are almost all data-based information system. Since the database system is only applied to undertake the daily manipulating application of raw data, and supports simple works of research and statistic, it is difficult to meet various complex analysis requirements; and for managers to provide effective and scientific decision basis for managers. In order to solve the two problems effectively, we bring data warehouse and online analytical processing (OLAP) technology into the implementation of risk assessment in drilling operation, put forward a solution to set up a multidimensional data model based on data warehouse. We can establish multidimensional data model in data warehouse through data extraction and conversion of oriented data in homology database according to professional model. And transform the data from the original application-oriented structure to the analysis-oriented multidimensional data structure, and provide the best data format [1, 2] for online analytical processing (OLAP). Only in this way can we provider a good analysis of the environment and analytical tools for

the risk analysis and assessment in drilling operation. Therefore, in view of the hazard, risk factors and possible characteristics of drilling accident in oil and gas drilling operation system, we should set up a suitable risk assessment model, as well as build a multidimensional data model for risk assessment based on data warehouse. It is the key problem needs to be solved in current oil and gas drilling operation. The study laid the foundation for risk assessment system and development platform in oil and gas drilling operation, and provider effective analytical tools and forecasting methods for scientific and systematic analysis, evaluation and forecast in drilling safety.

II. MULTIDIMENSIONAL DATA MODEL DESIGN AND IMPLEMENTATION OF RISK ASSESSMENT IN OIL AND GAS DRILLING OPERATION BASED ON DATA WAREHOUSE

A. Data warehouse based multidimensional data model design

1) Subject requirements analysis

The data of data warehouse is organized by subject-oriented. The subjects are criteria of data classification at higher level and usually reflect the key issues concerned by decision-makers. Following is the subject instance of "analysis of casualties" shows design procedure of multi-dimensional model.

2) Dimensions and granularity design

Dimension is an abstract of event-related factors in the relationship model. For example, analysis of casualties in oil and gas drilling operation relates to factors: time, geography, staff type, injury type, complex circs, well type and etc. Besides above factors, granularity should be taken into account as well since it concerns about the data layer issues. Granularity reflects the level of detail and data, for example, analysis of casualties concludes year, season, month, week and day analysis or statistical data. We can build dimension tables for those involved factors, which demonstrate its features via records attributes within the dimension table or hierarchy information. Granularity can be reflected through the division of each dimension level [3]. The following is commonly dimension and granularity design in analysis of casualties.

Time dimension (4 layers): year, quarter, month and day. See Fig.1 structure of time dimensional table.

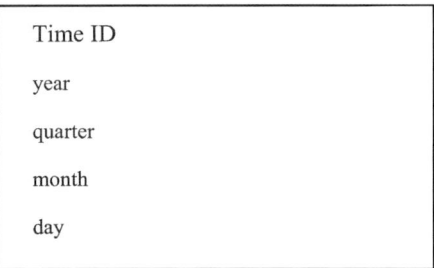

Fig.1. The structure of time dimension table

Geography dimension (4 layers): country, oil-field, block, and well. See Fig.2 structure of geography dimensional table:

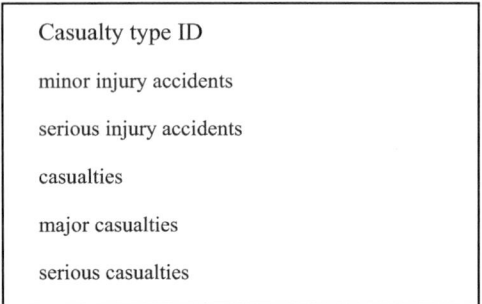

Fig.2. The structure of geography dimension table

Casualty type dimension: minor injury accidents, serious injury accidents, casualties, major casualties and serious casualties. See Fig.3 structure of casualty type dimensional table:

Fig.3. The structure of casualty type dimension table

Staff type dimension: operators, technicians and management personnel. See Fig.4 structure of staff type dimensional table:

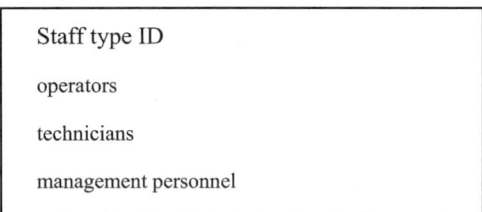

Fig.4. The structure of staff type dimension table

Complicated conditions' dimension: well blowout, well collapse, lost circulation, sticking and so on. Well type dimension: Vertical well, directional well and horizontal well.

3) Multidimensional data model design of casualty in oil and gas drilling operation

The fact tables and dimension tables are the basis for entire data model, so the division of fact tables and dimension tables must be reasonable. Dimension table attributes describe the properties of dimension itself, the fact tables includes value which indicators that which users want to know in the data warehouse, and numerical targets and additional characteristics.

The dimension table includes the time, geography, staff type, casualty type, complex circs and well type dimensional tables in accordance with requirements analysis. The fact table is casualty fact table, and each unit only stores a few points to each dimension table pointer: foreign key, these foreign keys corresponding to the primary keys which were stored in a different dimension tables. Measure value is the major value of multi-dimensional data sets, thus, the end-user browsing multidimensional data. Casualty fact table value uses the casualties.

At this point, the multidimensional data model of casualty in oil and gas drilling operation is designed. See Fig.5.

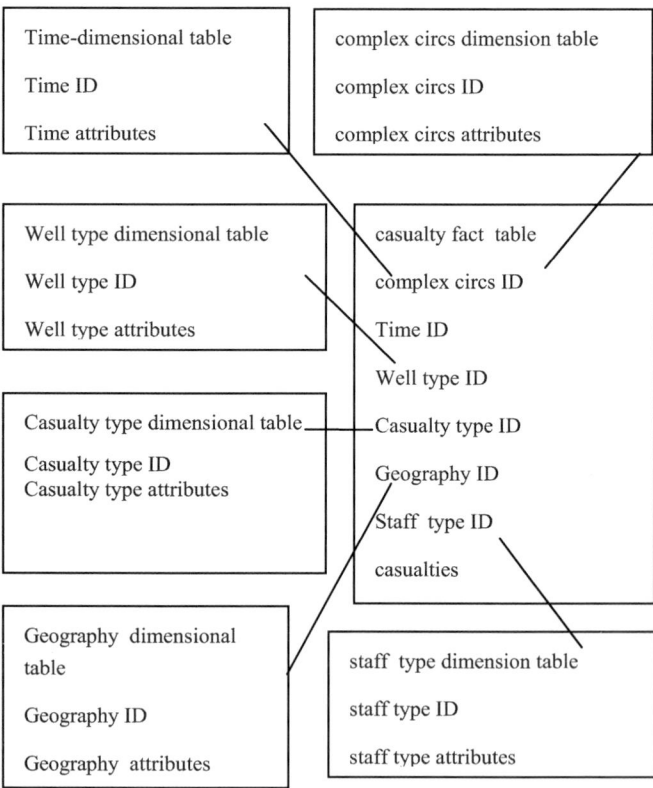

Fig.5. multidimensional data model of casualty in oil and gas drilling operation

B. The Implementation of OLAP System

1) OLAP multidimensional analysis engine

OLAP analysis engine is in the middle of data warehouse layer and user interaction layer, which manages the analyzing of user input, changes the user input into SQL statement and operates multi-dimensional in data warehouse, then the result set is returned to user interaction layer.

The workflow of OLAP engine is as follow [4, 5]. At first, the user input commands are changed into parameters, then parameters are transmitted input to multi-dimensional processing module, multi-dimensional processing module receives and analyzes the parameters, and generates the appropriate SQL statement. The appropriate SQL statement obtains the required data from the data warehouse and gets the multi-dimensional data sets. Tables and images are generated by processing this sets method according to user operation. Finally tables and graphs are returned to the user interaction layer. Workflow of OLAP multi-dimensional analysis engine is shown in Fig. 6:

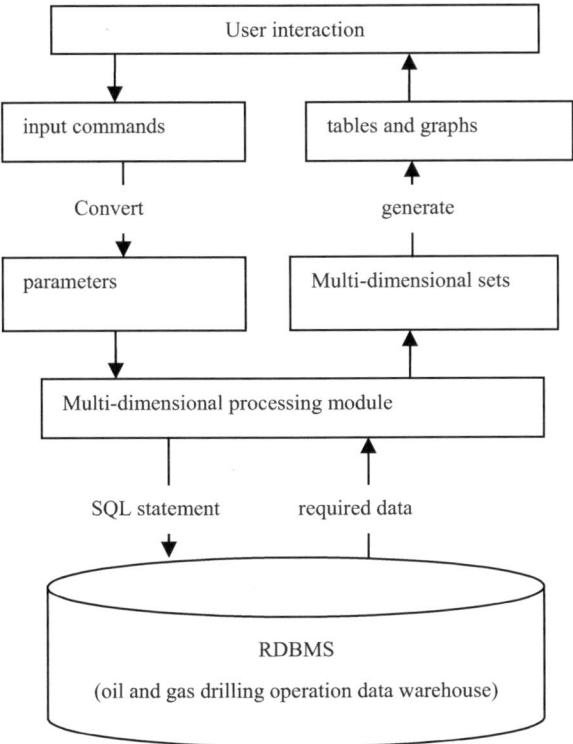

Fig.6. Workflow of OLAP multi-dimensional analysis engine

2) Way to access the database through the web

Method is that applying office web components (OWC, Office Web Components) to web pages, OWC is a set of components published along with Microsoft office; OWC offers a PivotTable control tool that can display the data from multi-dimensional data set by Analysis Services.

With Web-based OLAP analysis system the web page has been inserted two kinds of office web components [6]: PivotTable and Chartspace. PivotTable supports various OLAP operations including drilling, slicing and rotation which are easy to manipulate. Chartspace supports a variety of charts (including bar chart, pie chart, line chart, etc.), and it can be published blending with PivotTable.

III. DATA WAREHOUSE STRUCTURE OF RISK ASSESSMENT IN OIL AND GAS DRILLING OPERATION

Within the paper the data warehouse structure of risk assessment in oil and gas drilling operation applied B/S architecture, contrasted with C/S framework, the advantage of B/S architecture is obvious, it provides a method of Web-based channel for data exchanging, that make enterprise data and applications can be retrieved by any web-connected computer, which can exactly fit the requirements of the characteristics of world-wide-distributed of drilling professionals and administrators. This architecture of data warehouse structure of risk assessment in oil and gas drilling operation is shown following as Fig.7.

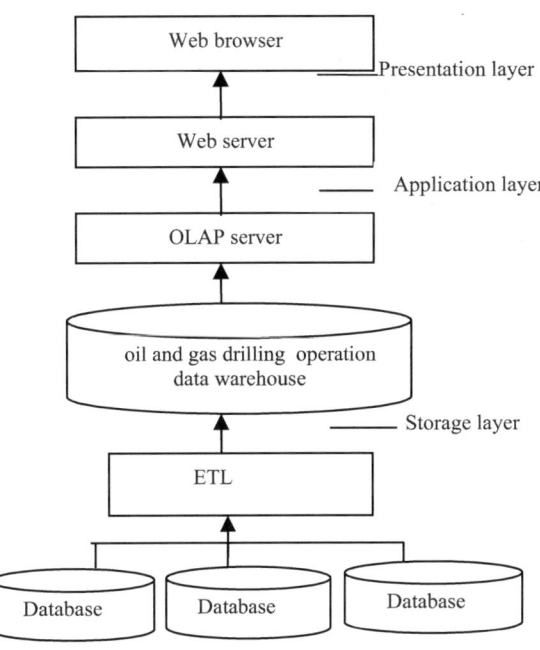

Fig.7. Architecture of data warehouse structure of risk assessment in oil and gas drilling operation

1) As illustrated in Fig.7, the fundamental layer is the storage layer (or data layer), which is generally divided into two parts: Online Transaction Processing (OLTP) databases, which is the source of analytical data. Data warehouse which is used by storing analysis data, the data is obtained from the OLTP database through the ETL tool. Thru this layer, OLAP data and OLTP data could be separated and would not impact on the efficiency of pristine system [7].

2) The second layer is the application layer, which consists of OLAP server and Web servers, OLAP server supports and manages the data processing engine of multi-dimensional data structures, and reads data from the relational database and mapping of a relational database to multi-dimensional logical model to display, so that the data is multi-dimensional formed when reporting.

3) The third layer is the browser. Applying web page to access multi-dimensional dataset and the query results

is displayed by table or graphics. The architecture of web-based OLAP not only allows end user to efficiently and easily access the data warehouse through the web browser, but also supports a unified interface and same style of web page, and the user can use data warehouse or OLTP databases via accessing web server by TCP/IP protocol that fits the demands of inter-regional and cross-platform.

IV. CONCLUSIONS

In order to meet the complex analysis requirements of risk assessment in oil and gas drilling operation, we put forward a multidimensional data model solution based on data warehouse. We bring the data warehouse and online analytical processing (OLAP) into the risk assessment system in drilling operation. The software architecture has applied relatively new technology of B/S mode, that various multi-dimensional data analysis through browsers by the method of inserting the PivotTable component and Chart components and linking to the analysis server to web pages that has been achieved, so it provides support and convenience that enterprise can effectively manage and make synthetic decisions. Furthermore, this system also realizes the spreading and the application of OLAP technology, which is a good tool for WWW environment. Users can efficiently access to data, and managers also get more useful information from large amounts of data, so the system has significant application value.

REFERENCES

[1] Jingmin Chen, Data warehouse and data mining technology, Beijing: Publishing House of Electronics Industry. 2002.

[2] Yu Lin. The principles and practice of data warehousing. Beijing: People's Posts Press 2003.

[3] Jian Li, Haode Liao, Bing Wang, The Design and Implementation of Web-based OLAP Drilling Analysis System [C]. 2010 Seventh International Conference on Fuzzy Systems and Knowledge Discovery Volume6, 10-12 August 2010, pp.2570-2573

[4] Jiang Xu. A J2EE-based OLAP architecture. Computer and Digital Engineering, 2006 (10), pp. 96-98.

[5] Rubo Hang, CUN Zuo, Yufang Sun. Research and Implementation of Web-based OLAP Technology. Computer Engineering, 2000 (10), pp. 7 - 8.

[6] Wei Zhao. Using Microsoft office Web components to generate complex diagrams in web. Computer Programming Skills& Intenance.2002 No. 4.

[7] Jian Li, Bihua Xu. ETL Tool Research and Implementation Based on Drilling Data Warehouse. 2010 Seventh International Conference on Fuzzy Systems and Knowledge Discovery Volume6, 10-12 August 2010, pp.2567-2569

Research on Improvement of Plant Layout Based on Process Analysis

Da-wei ZHOU[1,*], Li-tao WANG[1], Li-ying FENG[1], Dao-zhi ZHAO[2], Tian-song GUAN[3]

[1]School of Mechanical and Vehicular Engineering, Beijing Institute of Technology, Zhuhai, Zhuhai, China
[2]College of Management and Economics, Tianjin University, Tianjin, China
[3]Guangdong Yangxi Rural Credit Union, Yangjiang, China
(blzzdw@163.com)

[1]*Abstract* - **Analyzed the production processes and proposed a solution to improve plant layout, this paper applied process analysis to improve plant layout and explored another effective way to improve plant layout for small and medium manufacturing enterprises. A case explained the practicality and effectiveness of the method.**

Keywords – **Improvement, plant layout, process analysis, SMEs**

I. INTRODUCTION

In the plant facility layout design, SLP is applied to large-scale projects and SSLP middle and small projects. SLP and SSLP method can also be applied to improve the facility layout of the plant and shop. SLP method is tedious and complicated. Comparing with SLP, although SSLP was simplified, SSLP still has six steps and requires a lot of calculation and graphing [1-3].

Process program analysis method is commonly used in the program analysis of method study. It is a method which takes a product or a part of the manufacturing process in production system as a research object, based on the detailed observing, recording and analyzing for operations, inspections, transportations, storages and delays in processes, to research the improvement of operating processes [4-5]. With the dissemination and application of IE in Chinese enterprises, process program analysis is not only used to improve the flow processes, but also increasingly being used to optimize the layout of workshop facilities due to its simple and closely integrated with the production process [6]. But it is rarely used to improve the layout of plant facilities.

This paper takes an Guangdong YL Machinery Factory' actual improvement case as the basic material, based on process program analysis to study the improvement of the plant facility layout, in order to explore the simple, effective and practical way to improve.

II. ANALYSIS OF THE CURRENT STATUS AND IMPROVEMENT [7]

A. The basic situation of the enterprises

YL machinery factory mainly makes pressure vessels, pressure pipings, packing machinery' box-type products and so on. It has the casting, cutting, metal machining, welding, container, machine repair workshops

[1] Sponsored by Guangdong Experimental Comprehensive Reform of Mechanical Engineering and Automation

and the raw materials storage, covering an area of 141,000 square meters. Fig.1 shows the area of the factory workshop and warehouse layout.

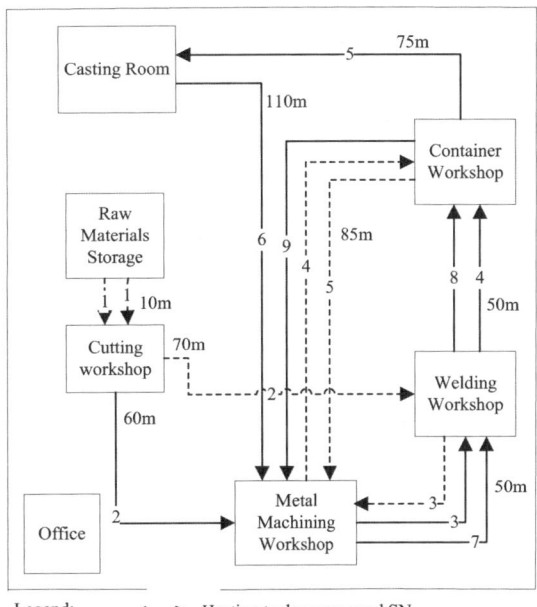

Fig.1. Plant layout and main production logistics chart (before improvement)

With the upgrading of the products, equipment had also been changed. In order to improve production efficiency, reduce production costs and further increase production capacity, this paper improved plant layout based on process analysis, under the conditions of the existing plant, equipment.

B. Production status

The main products of the YL are the heating cylinder and the heating tank. The following analyzes the heating tank as case.

The main production specification of the heating tank is $2328 \times 568 \times 672$. The parts are shown in Table I. The production process flow is shown in Fig.2.

Process charts of heating cylinder and heating tank are drawn. Due to limited space, this paper lists only the process chart of heating tank (as shown in Fig.3), some operations which do not affect layout improvement are simplified.

The primary main production logistics of heating cylinder and heating tank are shown in Fig.1.

E. Qi et al. (eds.), *Proceedings of the 21st International Conference on Industrial Engineering and Engineering Management 2014*, Proceedings of the International Conference on Industrial Engineering and Engineering Management, DOI 10.2991/978-94-6239-102-4_53, © Atlantis Press and the authors 2015

TABLE I
TABLE HEATING TANK AND ITS PARTS

Part No.	Product Name	Single piece mass / kg	Annual total number /piece	Annual total mass / kg
	Heating tank		108	75351.6
1	Side plate	71	108	7668.0
2	Base plate	178	108	19244.0
3	Faceplate	333	108	35964.0
4	Split-flow plate	14.1	432	6091.2
5	Others		3888	6404.4

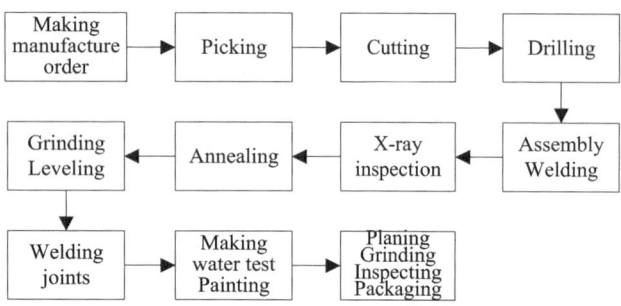

Fig.2. Heating tank production process flow chart

C. Analysis of the status and improvement

This section applied 5W1H, ESCRI (Eliminate, Simplify, Combine, Rearrange, Increase) principles to analyze the process diagram, including operations, inspections, transportations, storages and delays. And the following questions are found:

a. Longer transport distances, more transporting frequently, more waiting times.

b. Production cycle is longer, but effective operating time share very little in the entire production cycle, most of the time is moving and waiting.

The paper made the following improvements, based on the above processes and procedures and combined enterprise actual site, without changing the content of operating and inspecting:

a. Cutting machines in cutting workshop and drilling machines in metal machining workshop are moved to the raw materials storage, in order to reduce the transporting time and distance of raw materials and semi-finished products in operating.

b. The welding workshop and the container workshop swap places. In order to ensure the X-ray detection being between the tank welded and annealed, the X-ray detection equipment in container workshop is adjusted to the new welding workshop.

c. The processes of grinding and leveling are swapped with welding joints to reduce transport distance. Other production processes are unchanged.

d. The vacated cutting workshop is set up finished-parts storage. The problems that finished products were bumped into and WIP was stacked irregularly caused by the lack of space are solved in the metal machining workshop.

The improved process chart of heating tank is showed in Fig.4. The improved plant layout and main production logistics are showed in Fig.5.

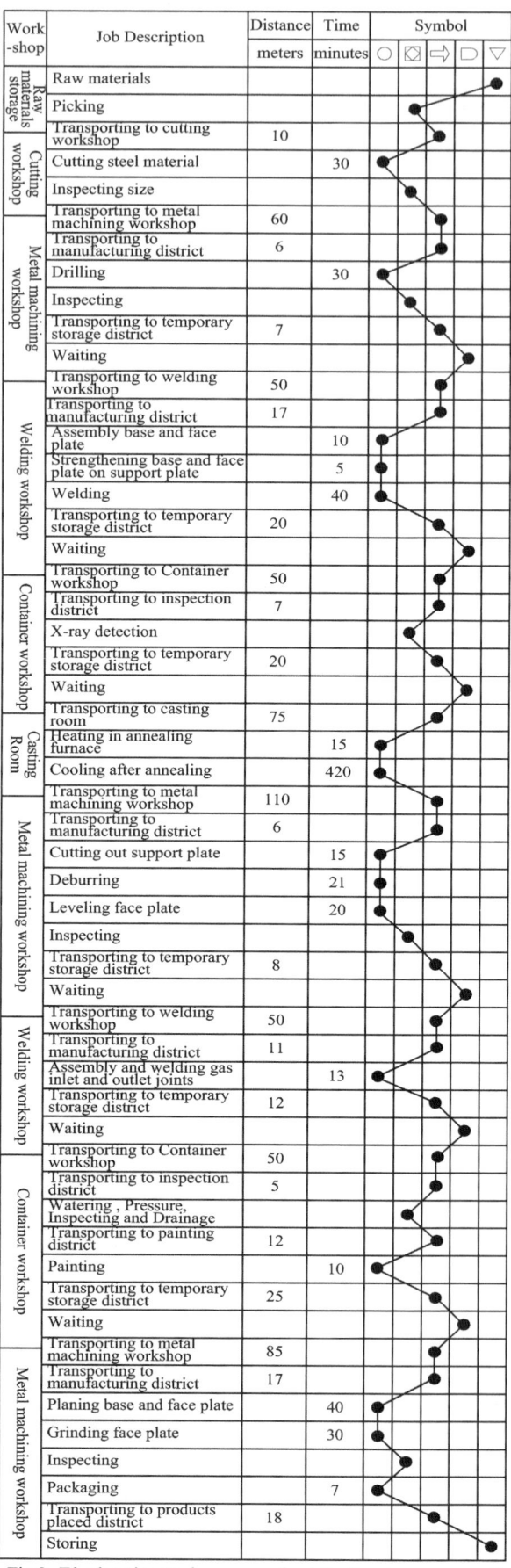

Workshop	Job Description	Distance (meters)	Time (minutes)
Raw materials storage	Raw materials		
Raw materials storage	Picking		
Cutting workshop	Transporting to cutting workshop	10	
Cutting workshop	Cutting steel material		30
Cutting workshop	Inspecting size		
Metal machining workshop	Transporting to metal machining workshop	60	
Metal machining workshop	Transporting to manufacturing district	6	
Metal machining workshop	Drilling		30
Metal machining workshop	Inspecting		
Metal machining workshop	Transporting to temporary storage district	7	
Metal machining workshop	Waiting		
Welding workshop	Transporting to welding workshop	50	
Welding workshop	Transporting to manufacturing district	17	
Welding workshop	Assembly base and face plate		10
Welding workshop	Strengthening base and face plate on support plate		5
Welding workshop	Welding		40
Welding workshop	Transporting to temporary storage district	20	
Welding workshop	Waiting		
Container workshop	Transporting to Container workshop	50	
Container workshop	Transporting to inspection district	7	
Container workshop	X-ray detection		
Container workshop	Transporting to temporary storage district	20	
Container workshop	Waiting		
Casting Room	Transporting to casting room	75	
Casting Room	Heating in annealing furnace		15
Casting Room	Cooling after annealing		420
Metal machining workshop	Transporting to metal machining workshop	110	
Metal machining workshop	Transporting to manufacturing district	6	
Metal machining workshop	Cutting out support plate		15
Metal machining workshop	Deburring		21
Metal machining workshop	Leveling face plate		20
Metal machining workshop	Inspecting		
Metal machining workshop	Transporting to temporary storage district	8	
Metal machining workshop	Waiting		
Welding workshop	Transporting to welding workshop	50	
Welding workshop	Transporting to manufacturing district	11	
Welding workshop	Assembly and welding gas inlet and outlet joints		13
Welding workshop	Transporting to temporary storage district	12	
Welding workshop	Waiting		
Container workshop	Transporting to Container workshop	50	
Container workshop	Transporting to inspection district	5	
Container workshop	Watering , Pressure, Inspecting and Drainage		
Container workshop	Transporting to painting district	12	
Container workshop	Painting		10
Container workshop	Transporting to temporary storage district	25	
Container workshop	Waiting		
Metal machining workshop	Transporting to metal machining workshop	85	
Metal machining workshop	Transporting to manufacturing district	17	
Metal machining workshop	Planing base and face plate		40
Metal machining workshop	Grinding face plate		30
Metal machining workshop	Inspecting		
Metal machining workshop	Packaging		7
Metal machining workshop	Transporting to products placed district	18	
Metal machining workshop	Storing		

Fig.3. The heating tank process chart (before improvement)

Work-shop	Job Description	Distance meters	Time minutes	Symbol ○ ◇ ⇨ □ ▽
Raw materials storage and Cutting workshop	Raw materials			
	Picking			
	Transporting to manufacturing district	5		
	Cutting steel material		30	
	Inspecting size			
	Drilling		30	
	Inspecting			
	Transporting to temporary storage district	5		
	Waiting			
Welding workshop	Transporting to welding workshop	75		
	Transporting to manufacturing district	17		
	Assembly base and face plate		10	
	Strengthening base and face plate on support plate		5	
	Welding		40	
	Transporting to inspection district	7		
	X-ray detection			
	Transporting to temporary storage district	13		
	Waiting			
Casting Room	Transporting to casting room	75		
	Heating in annealing furnace		15	
	Cooling after annealing		420	
Welding workshop	Transporting to welding workshop	75		
	Transporting to manufacturing district	11		
	Assembly and welding gas inlet and outlet joints		13	
	Transporting to temporary storage district	12		
	Waiting			
Metal machining workshop	Transporting to metal machining workshop	85		
	Transporting to manufacturing district	6		
	Cutting out support plate		15	
	Deburring		21	
	Leveling face plate		20	
	Inspecting			
	Transporting to temporary storage district	8		
	Waiting			
Container workshop	Transporting to Container workshop	50		
	Transporting to inspection district	5		
	Watering, Pressure, Inspecting and Drainage			
	Transporting to painting district	12		
	Painting		10	
	Transporting to temporary storage district	25		
	Waiting			
Metal machining workshop	Transporting to metal machining workshop	50		
	Transporting to manufacturing district	17		
	Planing base and face plate		40	
	Grinding face plate		30	
	Inspecting			
	Packaging		7	
	Transporting to finished-parts storage	60		
Finished parts storage	Storing			

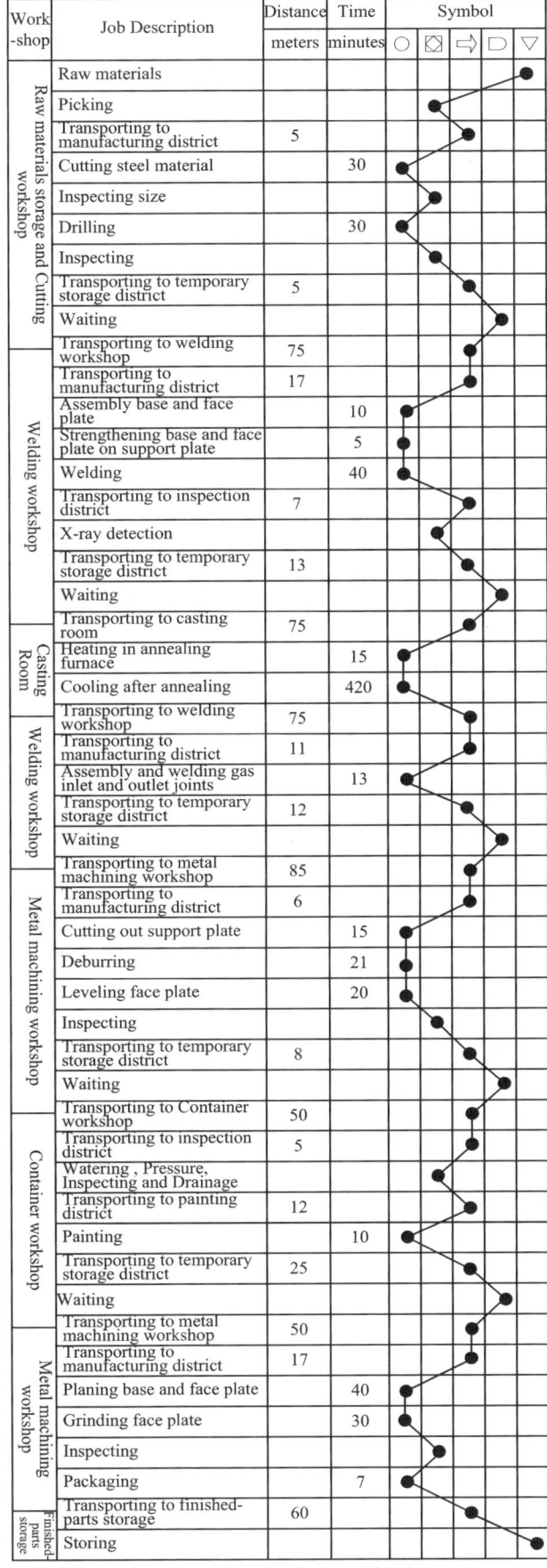

Fig.4. The heating tank process chart (after improvement)

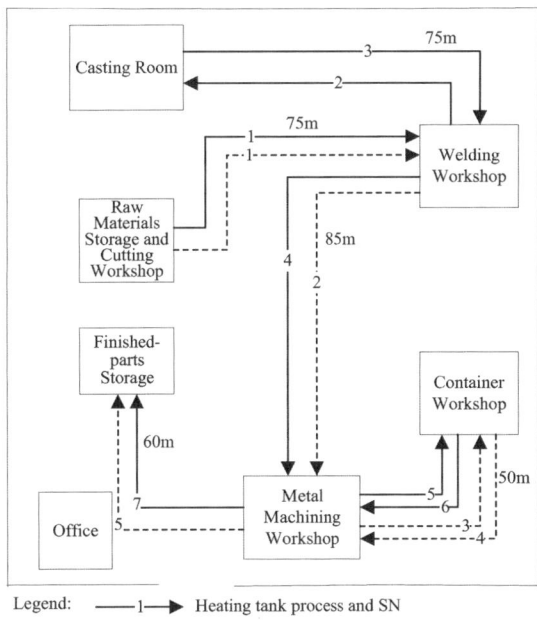

Legend: ——1——▶ Heating tank process and SN
----1---▶ Heating cylinder process and SN

Fig.5. Plant layout and main production logistics chart (after improvement)

III. ANALYSIS OF THE IMPROVED EFFECTIVENESS [7]

This section analyzes the effect of plant layout and facilities of the improved.

The operation statistics (as shown in Table II) before the improvement and after it for the heating tank can be generated by Fig.3 and Fig.4.

Comparison between Table II (a) and (b) shows that the improvement of the heating tank production, transporting time was reduced from 24 times to 20, transporting distance from 731 meters to 613, delay time from 6 times to 5. Similarly, transporting time, distance and delay time of heating cylinder production were also reduced.

TABLE II
OPERATING STATISTICS OF HEATING TANK
(a) before improvement

Job Content	Frequency/times	Time/minutes	Distance/meters
Operation	15	706	
Inspection	7		
Transportation	24		731
Storage	2		
Delay	6		

(b) after improvement

Job Content	Frequency/times	Time/minutes	Distance/meters
Operation	15	706	
Inspection	7		
Transportation	20		613
Storage	2		
Delay	5		

Comparison between Fig.1 and Fig.5 shows that the long-distance roundabout main production logistics of the heating tank is avoided and main production logistics of the heating cylinder is more smoothly after the improvement.

Labor costs, energy costs, loss and maintenance costs of handling equipment have reduced by the improvements. Single shift production capacity of the heating tank has been increased to 127 pieces per year from 108, the heating cylinder to 97 from 65. Production cycle is reduced, capital flow accelerated and product quality improved by the improvements. Annually, overall economic efficiency is increased of 1.1255 million RMB.

IV. DISCUSSION

Process analysis makes detailed records of the production site throughout the manufacturing process, and a detailed research and analysis of operations, inspections, transportations, storages and delays of the entire manufacturing process, focusing on the analysis of the hidden waste for transport distance and waiting etc. [8].

The case above extended the improvement object of process analysis from process improvements and workshop facilities improvements [9] to the factory layout improvement and optimization. It is mainly to optimize plant layout, to achieve the reduction in the hidden waste for transport distances and waiting times etc., in the case of the same production process.

The method of improvement is similar to the method of usual process analysis for the purpose of improving operation programs. It made adjustments on the basis of the flow chart of the program. The traditional analysis process made adjustment and optimization of operation processes or workshop facility layout under the premise of plant layout set up. In this paper, however, the research of process analysis for the purpose of improving plant layout optimizes plant layout through analysis of the program, in the case of the production process is essentially the same.

Being spread out above, during the analysis process procedures, according to the specific circumstances of the enterprise, considering the status of the plant facility layout, adjustment costs, and effectiveness adjusted, we timely determine whether for needing improvement and optimization of plant layout or not, namely whether for needing to achieve production operations improved, costs reduced and efficiency improved by optimizing the plant layout or not.

Both improvement of plant layout based on the process analysis and SLP (or SSLP) are made with established production process. The former is simple and intuitive, its job step accords with general process analysis, but the later requires a more complex calculation and drawing [10]. In SMEs been all about logistics relation, the implementation effects are the same of both.

Improvement of plant layout based on the process analysis has the following characteristics: (1) the method is similar to traditional process analysis, (2) it is more simple and convenient than the SLP(or SSLP), (3) the improvement effect is similar to other methods, production costs can be reduced effectively, production cycle shorten and production efficiency increased, and (4) to use other methods or cooperate with other methods will be still needful, in case of considering synthetic relationship is difficult due to production processes is different greatly, product variety more and batch similar.

V. CONCLUSION

The significance of this study is: (1) the application field of process analysis is extended, and (2) in addition to various existing methods of improving plant layout including the SLP, this new effective method can be provided.

In summary, the process analysis can be applied in improving plant layout easily and effectively. It can be used for small and medium enterprises that their production processes are similar or production lot is relatively concentrated in some kind of products. This method can effectively reduce the plan cycle of improvement formula and the computational workload. It can also reduce the cost and increase the efficiency of the improvement procedure.

REFERENCE

[1] Yao-xiang Zhu, and Li-qiang Zhu, *Facilities Planning and Enterprise Logistics*. Beijing: Mechanical Industry Press, 2004, pp.77–92

[2] Ding-zhong Feng, Neng Wu, Jia-jing Fan, and Mei-xian Jiang, "Layout design and simulation optimization of workshop facilities based on SLP and SHA", *Industrial Engineering and Management*, vol. 17, no.2, pp. 21–25, 2012.

[3] Da-wei Zhou, and Xiao-yang Zhang, "Application of production logistics-based SSLP in the facility layout of mechanical processing workshop", *Logistics Technology*, vol. 25, no. 6, pp. 68–70, 2006.

[4] Wei Xue, and Zu-hua Jiang, *Introduction to Industrial Engineering*. Beijing: Mechanical Industry Press, 2009, pp.56–60

[5] Qing-hua Kong, and Na Zhou, *Work Study Base and Case*. Beijing: Chemical Industry Press, 2009, pp.57–61

[6] Kang-qu Zhou, Xue-yun Zhai, Ji-an Liu, and Rui-juan Zhang, "Application of procedure analysis method to motorcycle crankshaft manufacturing", *Industrial Engineering Journal*, vol. 12, no.4, pp. 111–115, 2009.

[7] Tian-song Guan, *The Optimization Design of YL Machinery Factory Production System*, Zhuhai: Beijing Institute of Technology, Zhuhai, 2013, pp.1–2, 11–24, 33–40, 43

[8] Qi-ming Cai, Qing Zhang, and Pin Zhuang, *Fundamental Industrial Engineering*. Beijing: Science Press, 2009, pp.90–99

[9] Guo-zhang Jiang, *Fundament of Industrial Engineering*. Wuhan: Huazhong University of Science & Technology Press, 2010, pp.123–138

[10] Shou-feng Ji, *Modern Facilities Planning and Logistics Analysis*. Beijing: Mechanical Industry Press, 2013, PP.94–137

Comparison Study on Algorithms for Vehicle Routing Problem With Time Windows

Can YANG, Zhao-xia GUO*, Ling-yuan LIU

Business School, Sichuan University, Chengdu 610064, China

(zx.guo@alumni.polyu.edu.hk)

Abstract - **This paper investigates and compares heuristic algorithms for the vehicle routing problem with time windows. Three heuristic algorithms were proposed firstly through combining three classic heuristic algorithms with the cross exchange method (cross) and the 2-opt exchange heuristic (2-opt) respectively. These proposed algorithms are then compared with the three classical algorithms based on publicly available benchmark problems. The comparison results show that the effectiveness of the proposed algorithms and their superiority to the classical algorithms.**

Keywords - **Algorithm comparison, heuristic algorithms, Solomon benchmark problem, vehicle routing problem with time windows**

I. INTRODUCTION

The vehicle routing problem with time windows (VRPTW) is developed based on the basic vehicle routing problem (VRP) through adding a time window constraints. Recently, with the development of e-commerce distribution, the academic research and practical application of vehicle routing problem with time windows got much attention from industry and academia.

A large number of papers have been published in the VRPTW area. Some researchers have provided comprehensive surveys [1-4]. Early research of VRPTW mainly includes exact algorithms and classical heuristic algorithms. At present, global optimal solutions to VRPTW can be barely found by exact algorithms except for small-scale problems [5-6]. When the problem size is bigger, heuristic algorithms are usually used to find "nearly optimal solutions" or "satisfied solutions" since exact algorithms cannot find the optimal solutions within a reasonable time period. In the literature, three classical heuristics are commonly used, which are the savings algorithm [7], the sweep algorithm [8] and the insertion algorithm [9].

The core idea of savings algorithm is to combine two loop of the transportation problem into one circuit in turn, according to the principle of maximize the reduce of total distance. Optimize a new car until the current car arrived the load limit. Optimization process of savings algorithm can divide into parallel and serial manner.

Sweep algorithm was proposed by Gillett and Miller in 1974. Sweep algorithm has two phases of steps, which adopts polar coordinates to indicate the location of demand point, and then take a demand point as the starting point randomly, and make its Angle as zero. Then divide the service area according to the constraints of car capacity in clockwise or the reverse direction of the clock, and rank the demand points by Lin and Kernighan

exchange method, thus to construct the vehicle route schedule.

Insertion method is also called "farthest insertion method", which was proposed by Mole and Jameson in 1976 for solving vehicle routing problem. This method combined the idea of approaching method and savings method, insert customers into path one by one to build distribution route. This method firstly put the farthest point of terminal as the seed point of route, then take the minimum insert value point as the next point according to the concept of the adjacent point insertion method. Finally it decides the insert location according to the saving value generated from the general saving formula. Repeat the steps of selection and insert until vehicle capacity or time window is meet limit.

Some researchers also developed a large number of intelligent heuristic algorithms to solve the VRPTW, such as tabu search, genetic algorithm, simulated annealing algorithm and ant colony algorithm. Braysy and Gendreau [10-11] reviewed classic heuristic algorithms and intelligent heuristic algorithms for the VRPTW.

The classic heuristic algorithm seek satisfied solution by local search technology, it is simple and easy to realize, but easy to fall into local optimum. The intelligent heuristic algorithms are inspired by the idea of artificial intelligence, some of which have the potentials of finding globally optimal solutions through global search methods. However, the parameter settings of these algorithms are very complex and usually problem-dependent. Unfortunately, the parameter settings have a greater influence on the performance of algorithm. It is thus hard for these algorithms to be used in practice.

Comparing with intelligent heuristic algorithms, classical heuristic algorithms are usually easy-to-understand and easy-to-implement. However, the comparison study on classical heuristic algorithms has not been conducted before.

This paper thus presents a comparison study on heuristic algorithms for the VRPTW. Three novel heuristic algorithms are proposed firstly by combining three classical heuristic algorithms with the cross and the 2-opt respectively. The proposed algorithms are then evaluated and compared with the three classical algorithms based on publicly available benchmark problems.

The remainder of this paper is organized as follows. In section II, a simple introduction and mathematical model of the vehicle routing problem with time windows (VRPTW) is given. Three new heuristic algorithms are presented in section III. In section IV, we analyzed and

E. Qi et al. (eds.), *Proceedings of the 21st International Conference on Industrial Engineering and Engineering Management 2014*, Proceedings of the International Conference on Industrial Engineering and Engineering Management, DOI 10.2991/978-94-6239-102-4_54, © Atlantis Press and the authors 2015

compared the experiment results of the proposed algorithms and three classical algorithms. Concluding remarks, along with further research possibilities are given in the final section.

II. VEHICLE ROUTING PROBLEM WITH TIME WINDOWS

The VRPTW is an extension of the classical VRP. In the VRPTW, nodes are associated with more properties, and the solution has to satisfy more constraints. A service time s_i is considered; therefore, the vehicle has to stay at the location of customer c_i for a time interval at least s_i ($s_0 = 0$ is associated with the depot c_0) for service. A time window $\left[e_i, l_i\right]$ during which the service has to start is considered. Therefore, when a vehicle arrives at customer c_i earlier than e_i , it has to wait until the beginning of the time window to serve the customer. On the other hand, if a vehicle cannot arrive at c_i before l_i , the vehicle cannot serve c_i . At this time, customer c_i should be served by another vehicle. For depot c_0 , the time window is defined as that e_0 is the earliest start time, and l_0 is the latest return time of all the vehicles. The VRPTW has two objectives. The primary objective is to minimize the number of the vehicle routes V. The secondary objective is to minimize the total TD with the same number of routes.

The VRPTW can be stated mathematically as follows.

Define variables

$$x_{ij}^k = \begin{cases} 1, \text{ if vehicle k travels directly from i to j} \\ 0, \text{ otherwise} \end{cases}$$

$$y_j^k = \begin{cases} 1, \text{ if customer i is served by vehicle k} \\ 0, \text{ otherwise} \end{cases}$$

The goal of the VRPTW is to minimize

$$\min Z_1 = v \tag{1}$$

$$\min Z_2 = \sum_{i=0}^{n}\sum_{j=0}^{n}\sum_{k=1}^{v} t_{ij} * x_{ij}^k \tag{2}$$

s.t.

$$\sum_{i=0}^{n} x_{ij}^k = y_j^k \quad \forall k = 1, \cdots v, \forall j = 1, \cdots n \tag{3}$$

$$\sum_{j=0}^{n} x_{ij}^k = y_i^k \quad \forall k = 1, \cdots v, \forall i = 1, \cdots n \tag{4}$$

$$\sum_{i=0}^{n} x_j^k * q_i \quad \forall k = 1, \cdots v \tag{5}$$

$$\sum_{k=1}^{v} y_j^k = 1 \quad \forall i = 1, \cdots n \tag{6}$$

$$\sum_{k=1}^{v} y_0^k = v \tag{7}$$

$$t_i + w_i + s_i + t_{ij} = t_j \quad \forall i, j = 0, 1, \cdots n, i \neq j \tag{8}$$

$$e_i \leq t_i \leq l_i \quad \forall j = 0, 1, \cdots n \tag{9}$$

$$w_i = \max\left\{e_i - t_i, 0\right\} \quad \forall j = 0, 1, \cdots n \tag{10}$$

Objective (1) minimize the number of the vehicle routes v . Objective (2) minimize the total TD with the same number of routes. Constraint (3)-(4) denotes that every route only one vehicle pass. Constraint (5) denotes that the quantity of goods that each vehicle carries could not exceed the capacity. Constraint (6) stands for that each customer can only be served by one vehicle. Constraint (7) represents that all the routes start from the depot. Formulas (8)–(10) define the time window constraint, where t_i is the time when the vehicle arrives at node i ; w_i is the waiting time of a vehicle at the customer location until the time e_i ; s_i is the service time; and t_{ij} is the travel time between nodes i and j .

III. NEW HEURISTIC ALGORITHMS

This section describes how to develop the three new heuristic algorithms for the VRPTW by combining savings, sweep or insertion algorithms with 2 - opt and cross method respectively.

Fig. 1 shows the flowchart of the improved savings algorithm. The procedures involved are described below:

Step 1. Intialization: Generate initial solutions through savings algorithm;

Step 2. Crossover: Choose an initial path from the initial solution, and operate cross with another initial path;

Step 3. 2-opt Operation: Operate 2-opt to two new paths generated cross crossover operation respectively, find two new path with shortest total length;

Step 4. Judgment of total length: While the total length of the two new paths is shorter than the total length of step 2, replace the initial path with new path; while the total length of the two new paths isn't shorter than the total length of initial solution, retain the initial path;

Step 5. Judgment of crossover: Check whether the cross crossover operation has come to an end. If it's not true, return to step 3; if it's true, carry out step 6;

Step 6. Judgment of initialization: Check whether the entire initial path had carried out cross crossover operation between every two paths. If it's not true, return step 2; if it's true, output the final solution.

Comparing with the improved savings algorithm, the improved sweep algorithm and the improved insertion algorithm have the same procedures except for Step 1. The two improved algorithms generate initial solutions by using sweep and insertion algorithms respectively in Step 1.

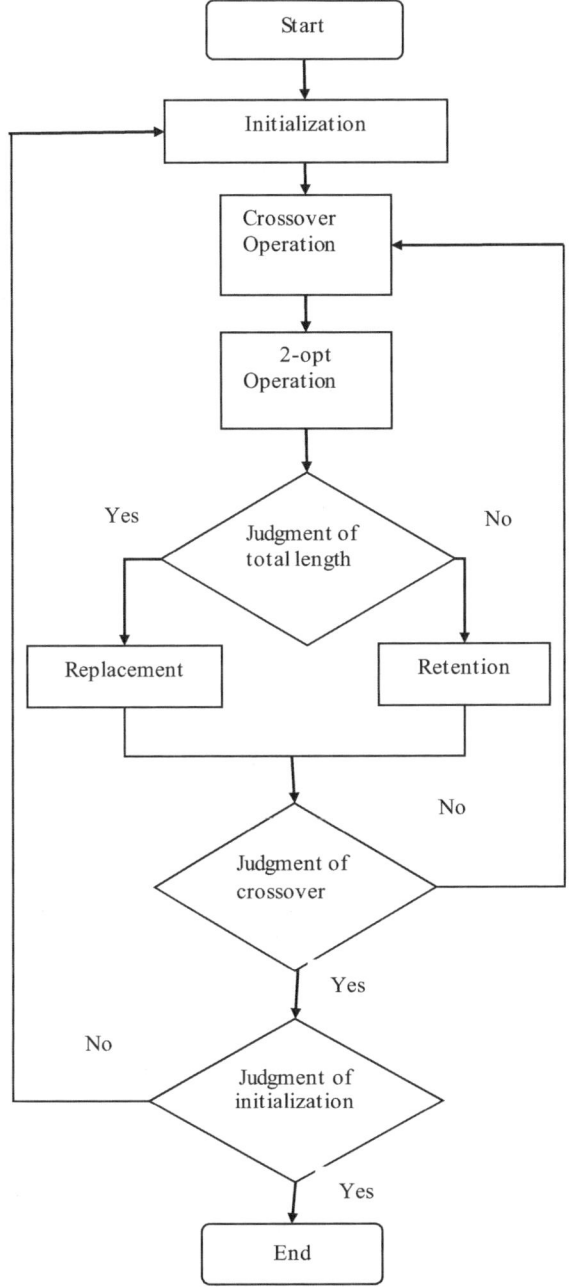

Fig.1. Flow chart of the improved algorithm

IV. COMPARISON EXPERIMENTS

In this part, we will analyze and compare the experiment results of three algorithms, and the data of experiment is from Solomon benchmark problem (http://web.cba.neu.edu/~msolomon/problems.htm). The Solomon benchmark problem includes six different types, totally conclude 56 data sets, and each data set consists of a parking yard, 100 customers. Data set C concludes a relatively concentrated customer base, data set R concludes some randomly distributed customers, and customers in the data RC concludes both relatively concentrated customer and randomly distributed customer. Data sets R1, C1 and RC1 has a relatively narrow time

window and the smaller car capacity, so a truck service only fewer customers; data sets R2, C2, and RC2 have relatively loose time window and larger vehicle capacity, so they can serve more customers.

A. Comparative analysis between the three algorithms

Here we only selected the R101 and R103 testing data set to test in account of the convenience, each data set test the top 25 customers, top 50 customers and 100 customers respectively. NV refers to the number of required vehicles, and TD refers to general travel distance.

The results generate from the three algorithms which haven't improved with cross crossover operation and 2-opt operation as shown in Table I. Although the difference of total travel length of three algorithms is not huge, but the number of required vehicle has great differences between three algorithms. In the problem of R101, insertion algorithm needs the least number of vehicles, the second one is savings algorithm, and sweep algorithm needs most vehicles; In R103 problem, insertion algorithm needs least vehicle, savings algorithm and sweep algorithm need the same vehicles.

TABLE I
RESULT COMPARING BETWEEN THREE UNIMPROVED ALGORITHMS

Problem	Savings		Insertion		Sweep	
	NV	TD	NV	TD	NV	TD
R101.25	9	637.9	9	790.3	14	850.6
R101.50	15	1152.6	12	1141.3	28	1708.4
R101.100	31	2002.4	21	1955.1	51	2953.8
R103.25	7	510.7	5	573.8	7	510.7
R103.50	12	920.8	8	932.8	12	920.8
R103.100	22	1511.3	16	1568.3	22	1511.3

The results generate from the three algorithms which improved with cross and 2-opt as shown in Table II. Compared with Table I, we can know that the results of three improved heuristic algorithm has significantly increased, and the vehicle number and travel distance of three algorithm is very close, only a little different.

TABLE II
RESULT COMPARING BETWEEW THREE IMPROVED ALGORITHMS

	Improved savings		Improved insertion		Improved sweep	
Problem	NV	TD	NV	TD	NV	TD
R101.25	9	673.7	8	646.1	8	655
R101.50	12	1153.2	12	1151.2	13	1209.4
R101.100	21	1842.5	21	1806.4	22	1886.9
R103.25	5	501.6	5	522	5	489.5
R103.50	9	856.7	8	883.7	10	917.9
R103.100	16	1373.8	16	1409.8	16	1377.4

Through comparison between Table I and Table II, we can know that, insertion algorithm performance better before improved with cross and 2-opt, and after improved with 2-opt and cross the results have promote a lot and there is no significant differences between the final results generated from three kinds of algorithm. That is to say, after generated an initial solution through insertion, savings and sweep algorithm and through improve with cross and 2-opt, we can get excellent and consistent solutions.

B. Comparison between the results generated from three improved algorithms and optimal solution in history

Since Solomon benchmark problem was published in 1987, many scholars test the algorithm using the baseline data, and the current optimal solution of 56 data sets the current optimal solution was published in the website of Solomon benchmark problem (http://web.cba.neu.edu/~msolomon/problems.htm).

Those current optimal solutions generated from 23 different algorithms, contains both accurate algorithm and heuristic algorithm.

Solomon benchmark problem includes six different types, totally 56 data sets. In consideration of our study is still ongoing, this paper randomly selected data sets R1, C1 and RC1 of Solomon benchmark problem, generated results through improved inserting algorithm, and compare the results with the current optimal solution, the comparison results as shown in Table III.

As shown in Table III, we can know that there almost unanimously between the solution generated from improved insertion algorithm (IS) with best known solution (BS). Compared with the best known solution, the solution generated from improved insertion algorithm only different in vehicle number V of question RC102.100, and the total travel distance only deviation between 0.19% and 12.34%.

TABLE III
COMPARISON BETWEEN IMPROVED INSERTION ALGORITHM AND CURRENT OPTIMAL SOLUTION OF SLOMON BENCHMARK PROBLEM

	BS		IS		Deviation	
Problem	NV	TD	NV	TD	V	D
R101.100	20	1637.7	20	1797.4	0	9.75%
R103.100	14	1208.7	14	1357.8	0	12.34%
R109.100	13	1146.9	13	1220.4	0	6.41%
C101.100	10	827.3	10	828.9	0	0.19%
C106.100	10	827.3	10	828.9	0	0.19%
C107.100	10	827.3	10	828.9	0	0.19%
RC102.10(	14	1457.4	15	1610.5	1	10.51%
RC107.10(	12	1207.8	12	1301.6	0	7.77%
RC108.10(	11	1114.2	11	1215.5	0	9.09%

V. CONCLUSION

Three heuristic algorithms were proposed firstly through combining three classic heuristic algorithms with the cross exchange method (cross) and the 2-opt exchange heuristic(2-opt) respectively, and those three algorithms solve the vehicle routing problem with time windows effectively. This is the first time to solve vehicle routing problem with time windows by combining classical heuristic algorithms with 2 - opt and cross. And we can

find that the proposed algorithms' solutions have significantly increased compared with the original algorithm by using Solomon benchmark problems as test problems. Furthermore, we found that the solutions generated by the 3 proposed algorithms were only slightly inferior to the best known solutions, which shows that the proposed algorithms are effective alternatives for the VRPTW.

In future studies, we will further optimize the three proposed algorithms, and compare them with intelligent heuristic algorithms as well.

ACKNOWLEDGMENT

The authors would like to thank the financial supports from the National Natural Science Foundation of China (Grant No.s 71302134, 71371130) and Sichuan University (Grant No. SKYB201301).

REFERENCES

[1] Golden B L, Assad A A. OR Forum-Perspectives on Vehicle Routing: Exciting New Developments [J]. Operations Research, 1986, 34(5): 803-810.
[2] Golden B L. Route Planning for Coast Guard Ships. Vehicle Routing: Methods and Studies. Studies in Management Science and Systems-Volume 16 [M]. 1988.
[3] Desrochers M, Lenstra J K, Savelsbergh M W P, et al. Vehicle routing with time windows: Optimization and approximation [J]. Vehicle routing: Methods and studies, 1988, 16: 65-84.
[4] Desrosiers J, Dumas Y, Solomon M M, et al. Time constrained routing and scheduling [J]. Handbooks in operations research and management science, 1995, 8: 35-139.
[5] Cordeau J F, Gendreau M, Laporte G, et al. A guide to vehicle routing heuristics [J]. Journal of the Operational Research society, 2002: 512-522.
[6] Golden B L, Wasil E A, Kelly J P, et al. The impact of metaheuristics on solving the vehicle routing problem: algorithms, problem sets, and computational results [M] // Fleet management and logistics. Springer US, 1998: 33-56.
[7] Clarke G, Wright J W. Scheduling of vehicles from a central depot to a number of delivery points [J]. Operations research, 1964, 12(4): 568-581.
[8] Gillett B E, Miller L R. A heuristic algorithm for the vehicle-dispatch problem [J]. Operations research, 1974, 22(2): 340-349.
[9] Solomon M M. Algorithms for the vehicle routing and scheduling problems with time window constraints [J]. Operations research, 1987, 35(2): 254-265.
[10] Bräysy O, Gendreau M. Vehicle routing problem with time windows, Part I: Route construction and local search algorithms [J]. Transportation science, 2005, 39(1): 104-118.
[11] Bräysy O, Gendreau M. Vehicle routing problem with time windows, Part II: Metaheuristics [J]. Transportation science, 2005, 39(1): 119-139.

Research on Order Allocation Problem with Multi-demand in Steel Processing and Distribution Centers

Fang-ping XU*, Wen-ying DING

Department of Mechanical Engineering, University of Science and Technology Beijing, Beijing, China
(xufangping123123@126.com)

Abstract - **With the continuous development of steel processing and distribution business, a reasonable order allocation among processing centers is an important means to save costs and improve the competitiveness for steel enterprises group. Aiming at this goal, a mathematical model was established to allocate orders by taking minimum costs of the enterprises group as the objective. In the model, the factors, such as production costs, transportation costs, orders delay costs etc. Genetic Algorithms was applied to deal with the model. A numerical example was used to verify the feasibility of the proposed approach. The results indicate that the proposed model and the algorithm can obtain satisfactory solutions. Orders allocation will provide effective production planning for the processing centers and help to optimize the steel enterprises group's overall resources and avoid vicious competition between the various processing centers.**

Keywords - **Genetic algorithms, order allocation, steel processing and distribution, steel enterprises group**

I. INTRODUCTION

With the continuous development of steel processing and distribution industries, processing centers increasing, customers' demand widely distributed, products also increased along with the diversified range of customers' demand. According to the diversity and frequency demands and the capacity of each processing center, steel enterprises group needs a reasonable allocation of orders that enable to spend a minimum cost.

Domestic and foreign scholars have done a large number of studies in order allocation methods. Duenyas [1] and Keskinocak [2] earlier to research this problem, proposed the capacity constraints queuing strategy. In Ji Xiao-Li's Literature [3], the factors, including production costs, transportation costs, inventory costs, orders delay costs etc, that affect the order allocation. She proposed a mixed integer linear programming model for orders dispatching in a supply chain with multi-product multi-order, and multi-period. XIANG Jinqian et al. studied the allocation of horizontal conglomerate's orders. A 0-1 programming model was established by taking maximum profit as the objective, in the model the factors, such as stock share, capital costs and operation fixed costs are taken into account [4]. LIU Xiaobing established a multi objective optimization model which maximize the total demands of assigned orders and the overall profits of enterprises group [5]. CHENG Fang-qi [6] and XIANG Wei [7] proposed an approach to proportionally allocate order based on the criterion of production load rate equilibrium. At the same time, XIANG Wei considered the uncertainty of demand and production capacity of enterprise actual operating process. The model which is built by JIANG Da-kui minimized a weight sum of the total lead time and the total cost. Meanwhile, he developed an integrated approach integrating tabu search and dynamic programming method to solve the problem [8]. Salman and others introduced the fuzzy theory, mathematics method, and intelligent algorithm in their literature [9-13].

Through the study of domestic and foreign literatures on this problem, considering the actual production capacity of enterprises and different product categories, an order allocation model is established in this paper.

II. PROBLEM DESCRIPTION

The multi-demand order allocation problem is a complex issue, which involves a number of centers, dozens of products and a lot of customers. Since the center location, the management level, as well as the differences in production lines, each processing center resulted in different types of products, even the same product, the production time, production costs and transportation costs are not the same in different centers. Assume that the internal and the external environment of centers are all in normal, production planning are arranged by orders without considering the lead time. In this context, assumes a planning cycle [1, T], order allocation shall be carried out with the cycle, the subsequent modeling process is also for the plan period. The orders to allocate in planning cycle are orders received from the previous cycle [14].

According to the characteristics of steel processing, we should as far as possible to reduce the processing waiting time and equipment downtime. Then assuming the production process must not be interrupted and the production planning are scheduled by EDD rules.

III. MODEL DEVELOPMENT

A. ASSUMPTIONS

Assuming there are a number of steel processing and distribution centers, located in different locations. The centers' production capacity and product types are known. Production costs of the same product in different centers are not the same.

The known data in the model include the distances from centers to customers and the transport prices at different distances. Centers' production capacities at each period and products' unit cost of production in different centers. The detailed order demands.

Several parameters and variables used in the order allocation model are expressed as follows:

$i = (1, 2, ..., I)$: Steel Processing and Distribution Center;
$j = (1, 2, ..., J)$: Customer Order;
$k = (1, 2, ..., K)$: Product Category;
$s = (1\text{-road}, 2\text{- railway})$: Mode of transportation;
C_{ik}: Unit production costs of product k in center i;
G_{jk}: Demand of product k in order j;
P_{ijs}: The base price from center i to customer j as the mode of transport is s;
F_{ijs}: The transport rates from center i to customer j as the mode of transport is s;
D_{ij}: The distance from center i to customer j;
T_j: The delivery date of order j;
t'_{ijk}: Production start in center i for product k order j;

t_{ijk}: the end of production for product k in order j;
β_{jk}: Delay penalty coefficient for product k in order j;
$W_i(t)$: Nominal capacity of center i at period t;
α_{ik}: Nominal capacity coefficient for product k in center i;
$g_{ijk}(t)$: Order quantity for product k allocated from order j to center i at period t.

B. MATHEMATICAL MODEL

1) *The objective function*: The main objective of this study is to spend a minimum total cost of the enterprises group, including the production and processing costs, transportation costs and orders delay costs, as in (1).

2) *Constraints*: Constraints limit the problem, including the machining centers' product limit and the capacity of each period. Subject to (2)-(5):

$$\min z = \sum_{k=1}^{K} \sum_{i=1}^{I} \sum_{j=1}^{J} x_{ijk} \left[C_{ik} G_{jk} + \sum_{s=1}^{2} \left(P_{ijs} + F_{ijs} D_{ij} G_{jk} \right) + \beta_{jk} C_{ik} G_{jk} Max\left(t_{ijk} - T_j, 0 \right) \right] \quad (1)$$

$$s.t. \begin{cases} x_{ijk} = 0\ or\ 1 & (2) \\ \sum_{i=1}^{I} x_{ijk} = 1 \quad (\alpha_{ik} > 0) & (3) \\ \sum_{t'_{ijk}}^{t'_{ijk}+N} g_{ijk}(t) = G_{jk} & (4) \\ 0 \le \sum_{k=1}^{K} \sum_{j=1}^{J} \alpha_{ik} g_{ijk}(t) \le W_i(t) & (5) \end{cases}$$

Where, $t_{ijk} = t'_{ijk} + N$ (6)

$$g_{ijk}(t) = \begin{cases} \dfrac{W_i(t) - \alpha_{ik'} g_{ij'k'}(t)}{\alpha_{ik}}, & t = t'_{ijk} \\[3mm] \dfrac{W_i(t)}{\alpha_{ik}}, & t'_{ijk} < t < t_{ijk} \\[3mm] G_{jk} - \sum_{t'_{ijk}}^{t_{ijk}-1} g_{ijk}(t), & t = t_{ijk} \end{cases} \quad (7)$$

Equation (2) is the product k in order j will be or not produced in center I;
Equation (3) shows an order of a product can only be done by a competent center;
Equation (4) represents an order for a product in the production time, the arranged production is equal to the quantity of the order.
Equation (5) represents the arranged production for a center is greater than or equal to zero and less than or equal to its nominal capacity in a period.

IV. SOLVING BY GENETIC ALGORITHMS

As the number of variables in the model is very large, it is difficult to solve by the traditional precise algorithm. Genetic algorithm (GA) is an effective method to solve such problems. Compared with other traditional optimization algorithm, genetic algorithm has strong global search capability. It can quickly search out the global optimal solution from the solution space [15].

Real number coding is adopted to solve this problem. The length of a chromosome is the amount of order demands which is not a fixed value. Genic value is the number of a center which is allocate to the demand. Therefore a chromosome represents an order allocation scheme.

The initial population is represented by a matrix: assuming the number of demands is J, the number of centers is I. Then the genetic value of each chromosome is a random positive integer between 1and I. When the population size is N, the initial population is represented by the matrix as shown in Fig.1. We must check the legitimacy of individuals in the population.

$$\begin{pmatrix} c_0(1) & c_0(2) & ...... & c_0(J) \\ c_1(1) & c_1(2) & ...... & c_1(J) \\ \vdots & \vdots & c_k(j) & \vdots \\ c_N(1) & c_N(2) & ...... & c_N(J) \end{pmatrix}$$

Fig.1. initial population

$$F(k) = 1 / \sum_{k=1}^{K} \sum_{i=1}^{I} \sum_{j=1}^{J} x_{ijk} \left[C_{ik} G_{jk} + \sum_{s=1}^{2} \left(P_{ijs} + F_{ijs} D_{ij} G_{jk} \right) + \beta_{jk} C_{ik} G_{jk} Max\left(t_{ijk} - T_j, 0 \right) \right] \quad (8)$$

The reciprocal of the objective function is set as a fitness function, as in (8).

Using the optimal preservation strategy and competition selection strategy as selection operator. Uniform crossover operator is used as shown in Fig.2.

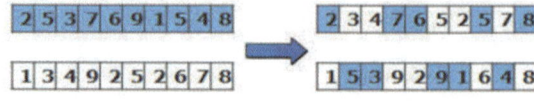

Fig.2. Uniform crossover

Adopting single point mutation, a chromosome gene to mutate in a range corresponding is randomly selected. When an illegal individual appeared, the original individual instead of the variant individual is retained to ensure the stability of population size.

When the genetic algorithm achieves the maximum number of iterations, the operation is stopped and the results are outputted.

Here is an example to illustrate the solution of the model. Assuming that a steel enterprises group has seven processing centers, five kinds of products, received eleven order demands in a planning cycle. Related information are shown in the Table I-VI. Set the population size: *psize* = 15, mutation probability: *Pm*=0.1, crossover probability: *Pc*=0.9, maximum iterations: *MaxGen*=300.

TABLE I
THE PROCESSING PRICES

Product	Center						
	1	2	3	4	5	6	7
1	120	190	140	∞	100	220	160
2	110	∞	240	180	∞	110	410
3	∞	450	130	150	160	120	∞
4	150	∞	150	230	100	∞	210
5	230	90	∞	340	170	130	290

TABLE II
THE NOMINAL CAPACITY COEFFICIENTS

Product	Center						
	1	2	3	4	5	6	7
1	0.8	0.9	0.6	0	1.3	1.3	1.2
2	0.9	0	0.9	0.8	0	1.2	1.1
3	0	0.7	0.9	1	1.1	1.1	0
4	1.1	0	1	0.9	1	0	0.9
5	0.7	1.1	0	0.7	1	0.9	0.8

TABLE III
THE NOMINAL CAPACITIES OF EACH PERIOD

Time	Center						
	1	2	3	4	5	6	7
1	20	16	30	12	25	17	30
2	25	20	31	21	23	20	29
3	50	20	20	32	18	36	12
4	33	10	26	12	40	11	26
5	17	14	33	15	17	40	19

TABLE IV
THE DISTANCE

Customer	Center						
	1	2	3	4	5	6	7
1	300	245	238	50	60	120	270
2	150	24	150	300	100	110	20
3	19	12	40	188	48	290	188
4	135	240	35	73	90	49	164
5	50	29	230	100	290	110	290

TABLE V
THE DEMAND INFORMATION

demand	order	Product	quantity	deadtime	penalty	transportation
1	1	1	70	1	2	1
2	1	2	60	4	5	2
3	1	5	50	4	6	2
4	2	3	90	5	2	2
5	2	1	60	2	4	1
6	3	2	50	4	3	1
7	3	4	20	1	5	2
8	4	2	100	1	6	2
9	4	5	50	2	2	1
10	4	3	70	5	3	1
11	5	4	50	5	6	2

TABLE VI
DELIVERY PRICES

transportation	Road		railway	
distance	Basic	rate	Basic	rate
0~50	1000	270	2500	250
50~100	1500	260	2600	240
100~150	2000	250	2700	230
150~200	2500	240	2800	220
200~250	3000	230	2900	210
250~300	3500	220	3000	200
300~350	4000	210	3100	190

Run the genetic algorithm it is concluded that the optimal result as shown in in Fig.3. The figures on the graph show the arrangement of production. The minimum total cost is 7398600. The order delay, including the demands of 1, 3, 5, 7, 8, 9. The changes of fitness value and least cost in the iterative process are shown respectively in Fig. 4 and 5.

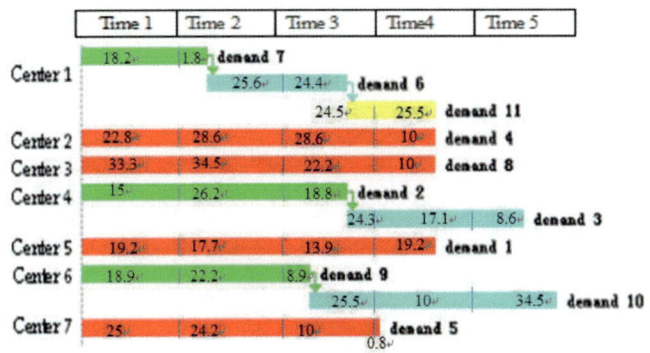

Fig.3. Results

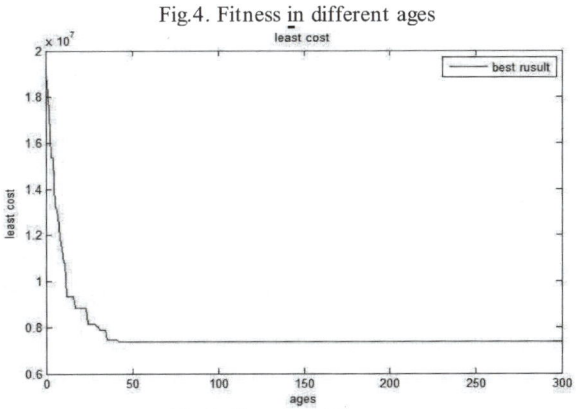

Fig.4. Fitness in different ages

Fig.5. Changes of least cost

V. CONCLUSION

The order allocation model takes the actual production capacity of centers and different demands of customers into account, which will solve the balance of group's production capacity and achieve a global optimization of resource. The enthusiasm to produce, as well as the ability to resist risks will be improved in the overall steel enterprises group.

REFERENCES

[1] Duenyas I. Single facility due date setting with multiple customer classes [J]. Management Science, 1995, 41(4): 608-619.

[2] Keskinocak P, Ravi R, Tayur S. Scheduling and reliable lead-time quotation for orders with availability intervals and lead-time sensitive revenues [J]. Management Science, 2001, 47(2): 264-279.

[3] JI Xiao li. Order Allocation Model in Supply Chain and Hybrid Genetic Algorithm [J]. Journal of Southwest Jiaotong University, 200, 40(6): 811-815.

[4] XIANG Jinqian, HUANG Peiqing, WANG Ziping. Order Allocation Model Based on Profit Maximization of Horizontal Conglomerate [J]. Journal of Southwest Jiaotong University, 2006, 41(2): 241-244.

[5] LIU Xiaobing, WANG Yuchun. Study on order allocation model in steel enterprises group [J]. Control and Decision, 2009. 11

[6] CHENG Fang-qi, WANG Hong-fei, YE Fei-fan. Research on order allocation model for horizontal virtual enterprise [J]. Mechanical& Electrical Engineering Magazine, 2009.4

[7] XIANG Wei, SONG Fa-shuai, YE Fei-fan. Simulation on Production Load Equilibrium Order Allocation within Multi-suppliers [J]. Journal of System Simulation, 2013.02

[8] JIANG Da-kui, LI Bo, TAN Jia-yin. Integrated optimization approach for order assignment and scheduling problem [J]. Control and Decision, 2013.02

[9] Salman Nazari-Shirkouhi, Hamed Shakouri, Babak Javadi, Abbas Keramati. Supplier selection and order allocation problem using a two-phase fuzzy multi-objective linear programming [J]. Applied Mathematical Modelling, 2013.05

[10] Wei Xiang, Faishuai Song, Feifan Ye. Order allocation for multiple supply-demand networks within a cluster [J]. Intell Manuf, 2013.01

[11] Feng-Cheng Yang, Kuentai Chen, Ming-Tzong Wang, Ping-Yu Chang, Kuo-Chih Sun. Mathematical modeling of multi-plant order allocation problem and solving by genetic algorithm with matrix representation [J]. Adv Manuf Technol, 2010.05

[12] Z.X. Guoa, W.K. Wongb, S.Y.S. Leung. A hybrid intelligent model for order allocation planning in make-to-order manufacturing [J]. Applied Soft Computing, 2012.07

[13] Weihua Liu, Haitao Xu, Xinyu Sun, Yi Yang, Yuming Mo. Order Allocation Research of Logistics Service Supply Chain with Mass Customization Logistics Service [J]. Mathematical Problems in Engineering, 2013.09

[14] Huanan zhou. Order assignment problem based on diversified demands [D]. Tianjin: Tianjin University. Master's thesis, 2010

[15] YOU Xuexiao, ZHONG Shounan. Schema Theory of the Decimal Coded Genetic Algorithm [J]. Wuhan Univ. (Nat. Sci. Ed.), 2005.

Study on the Performance of Chinese Manufacturing Low-carbon Technology Innovation in the Global Value Chain

Ke-xin BI [1, 2, *], Xiang-xiang WANG [1]

[1]School of Management, Harbin University of Science and Technology, Harbin, P. R. China
[2]School of Economics & Management, Harbin Engineering University, Harbin, P. R. China
(bikx@hrbust.edu.cn)

Abstract - **Based on global value chain, this paper builds manufacturing low-carbon technology innovation performance evaluation index system from the perspective of innovation input and output, then analyzes Chinese manufacturing low-carbon technology innovation performance in the global value chain with data reduction factor (DRF) and data envelopment analysis (DEA). The results show that there is a large difference among low-carbon technology innovation performance of manufacturing industries, and technical efficiency is not high. Resource allocation efficiency of Chinese manufacturing low-carbon technology innovation in the global value chain is low. And most manufacturing industries are in the increasing returns stage of scale, manufacturing industries should increase investment in low-carbon technology innovation, and improve the efficiency of resource allocation.**

Keywords - **Global value chain, low-carbon technology, low-carbon technology innovation performance**

I. INTRODUCTION

Global value chain is a global business network across the organization to realize the value of goods or services [1], and bring new impetus for the current world economy. The manufacturing countries have comparative advantages to participate in international economic production, to complete one or several aspects of commodity production process [2]. However, the value-added capabilities of different aspects are vary [3], the division of global value chain resulted in uneven economic development, environmental pollution transfer and other issues among countries. Since the signing of "United Nations Framework Convention on Climate Change" in June 1992, global climate change has become a world problem that must be considered when nations develop economy. Under the dual pressures of energy constraint and greenhouse gas emissions reduction, governments have introduced corresponding policies about energy consumption and greenhouse gas reduction, researched and developed the application of low-carbon technologies. Chinese manufacturing industries have rapidly developed along with a lot of energy consumption and greenhouse gas emissions, from the overall distribution of the global value chain perspective, Chinese manufacturing is at the low end of the value chain [4], and in response to climate change and environmental pollution, Chinese manufacturing lacks key green technology, especially low-carbon technology. In order to avoid being manipulated by others, it is necessary to choose the path of innovation and development of low-carbon technology. Therefore, the study of manufacturing low-carbon technology innovation performance in the global value chain provides a theoretical basis for the value upgrading of Chinese manufacturing in the global value chain and sustainable development, has important practical significance for Chinese economic and social development.

II. LITERATURE REVIEW

Carlo Pietrobelli (2011) studied the relationship between the global value chain and innovation system, thought that the relationship between the global value chain and innovation system is nonlinear, endogenous and mutual influence; the internal governance of global value chain is a dynamic phenomenon, constantly adjusts and changes, and the innovation system impacts this collaboration [5]. Gary Gereffi (2006) pointed out that although developing countries are at the low end part initially, it did not mean they will be forever locked in this link, the upgrading of global value chain depends on three variables: the complexity of transactions, the ability to codify transactions, and the capabilities of suppliers [6]. Sun (2010) further pointed out that companies in emerging economies should focus on product innovation, research and development, marketing and brand, rather than the OEM / ODM of global value chain, to increase their competitiveness, and then to shift the high-end position of global value chain [7]. NISHITANI K (2012) discussed the impact of greenhouse gas emissions reduction on firm value, thought that reducing more greenhouse gas emissions, companies are more likely to enhance corporate value [8]. Park (2009) researched that there was competition in manufacturing-related industry sectors, the development of energy-saving technology is also very important, and energy-saving and new technologies in the industrial sector have become the main driver for economic growth in the next generation [9]. Xiong-feng PAN (2011) indicated manufacturing should actively research and develop low-carbon technologies, while optimizing the industrial structure, and then the manufacturing industry structure upgrades to scale, low-carbon and high-end [10]. Bettencourt (2013) pointed out the growing of low-carbon energy technology market is not fast and economic opportunities promote patents and knowledge creation. It should be emphasized low-carbon energy technology market, market prices and policies affect carbon emission performance [11].

III. PROCESS ANALYSIS OF MANUFACTURING LOW-CARBON TECHNOLOGY INNOVATION IN THE GLOBAL VALUE CHAIN

The profit in the global value chain is not evenly distributed, and main profits flow to two ends of the value chain. The one end is R&D and design, and the other end is brand and marketing, while the middle part is manufacturing [12]. Also, some researchers found that the global value chain and carbon emissions had an inverse relationship, and the carbon emissions of manufacturing sector is more large [13], while the R&D and marketing aspects are relatively little [14]. Kline S.J and N. Rosenberg (1986) proposed chain model of innovative, thought that the main chain of the innovation process are potential market, the inventor/production probe analysis, detailed design and testing, re-design and production, and the market [15]. Technology innovation is the entire process from R&D to production, sales and realizing the market value, emphasizing the process and results [16]. It shows that the R&D, manufacturing, and marketing are important stages of the innovation process. Therefore, based on the above analysis, this paper divides manufacturing low-carbon technology innovation into three main stages by process in the global value chain, including global low-carbon R&D, global low-carbon manufacturing, global low-carbon marketing.

Griffith (2004) pointed out that R&D through innovation directly or technology transfer indirectly promotes growth [17]. Only through R&D of low-carbon technology innovation, it is possibility to take the science and technology knowledge into new products (processes, services) on the basis of absorbing advanced scientific achievements. The manufacturing of low-carbon technology innovation refers to transforming innovation achievements into products (processes, services) meeting the design requirements, and the majority components of manufacturing capacity play a key role for innovation performance [18]. The key criterion to judge the success of manufacturing low-carbon technology innovation is the extent of low-carbon products (processes, services) market realizing. The ability of marketing influences the low-carbon technology innovation transforming into commercial products [19], and is the key ability to link the products and consumers. According to global carbon three main stages which are global low-carbon R&D, manufacturing and marketing, the manufacturing low-carbon technology innovation performance in the global value chain under different stages of the innovation process are as follows: the global low-carbon R&D performance, global low-carbon manufacturing performance, global low-carbon marketing performance.

IV. THE BUILDING OF MANUFACTURING LOW-CARBON TECHNOLOGY INNOVATION PERFORMANCE EVALUATION SYSTEM IN THE GLOBAL VALUE CHAIN

A. The Principle of Constructing

Based on a global value chain perspective, this paper combines the characteristics of manufacturing low-carbon technology innovation performance, and indicators are selected considering the following four basic principles.

1) *Scientific principle*: Based on global value chain perspective, the manufacturing low-carbon technology innovation performance evaluation system scientifically summarizes low-carbon technology innovation process, and scientifically and standardizedly reveals low-carbon technology innovation performance.

2) *Objectivity principle*: Data obtained for manufacturing low-carbon technology innovation performance evaluation must be from reliable sources, really objective and minimize the subjective in the evaluation process to ensure the authenticity and accuracy of evaluation findings.

3) *Systematic principle*: Keep index system hierarchical, holistic and comprehensive to form a perfect indicator system. It is necessary to consider the different properties of different levels in the selection of indicators.

4) *Importance principle*: Ensuring index system systematic, comprehensive, it should seize the key factors that reflect the performance of low-carbon technology innovation, in order to evaluate Chinese manufacturing low-carbon technology innovation performance accurately and concisely.

B. Index System

According to the construction principles, and considering low-carbon technology innovation process from inputs to outputs throughout the whole event, establish appropriate evaluation system from the low-carbon technology innovation R&D, manufacturing, and marketing. As shown in Table I.

TABLE I
EVALUATION SYSTEM OF MANUFACTURING LOW-CARBON TECHNOLOGY INNOVATION PERFORMANCE IN THE GLOBAL VALUE CHAIN

Dimension	Index layer
Input	personnel input intensity of manufacturing global low-carbon technology innovation
	fund input intensity of manufacturing global low-carbon technology innovation
	the scale of foreign direct investment in manufacturing low-carbon technology innovation
	manufacturing worldwide carbon products (processes, services) production scale
	the degree of manufacturing worldwide carbon products (processes, services) using the existing international marketing channels
	he globalized degree of manufacturing low-carbon supply chain
Output	the effective international patent authorization number of manufacturing low-carbon technology
	manufacturing fossil energy consumption per unit of output in the global value chain
	manufacturing carbon intensity in international trade
	manufacturing international cooperation carbon capture and sequestration extent
	the proportion of low-carbon technology outputs account for the output value in manufacturing of international cooperation
	the growth rate of low-carbon transformation rate in manufacturing industry of international cooperation
	the share of manufacturing carbon products (processes, services) in the international market
	the total export of manufacturing global carbon products (processes, services)

V. THE EVALUATION OF CHINESE MANUFACTURING LOW-CARBON TECHNOLOGY INNOVATION PERFORMANCE IN THE GLOBAL VALUE CHAIN

A. EVALUATION MODEL

Firstly, based on the global value chain perspective, this paper uses data envelopment analysis to calculate Chinese manufacturing low-carbon technology innovation performance. The efficiency value calculated with C^2R model, BC^2 model can decompose it into pure technical efficiency (PTE) and scale efficiency (SE), in order to understand whether the reason causing the lack of efficiency is pure technical efficiency or scale efficiency [20]. Meanwhile DEA has two modes: input-oriented and output-oriented approach. Input-oriented model is the volume of outputs at a fixed premise to properly regulate and control inputs, and output-oriented model is the volume of inputs at fixed inputs to properly regulate and control outputs [21]. Considering the background of the global value chain, the characteristics of low-carbon technology innovation data, difficulty treatment, results analysis and other factors in manufacturing low-carbon technology innovation performance evulation, input control is easier than output control. Therefore, this paper selects BC^2 with input-oriented model. But when there is a strong linear relationship among the input or output indicators of manufacturing industry, it may reduce the credibility of evulation results. In order to avoid this situation, data reduction factor can be performed firstly [22]. Data reduction factor is a statistical technique that extracts the common factors from the variable group, with less several factors reflecting most information of the original variables and eliminating the correlation between variables.

B. DATA SYSTEM

Since innovation from input to the commercialization of new patents and new products (processes, services) usually requires a certain period, which means a lag from input to output. Some scholars believe innovation from input to output delays one year [23], some scholars believe that it delays two year [21]. Considering economic, ecological and social effects that low-carbon technology innovation brings, this paper chooses two years as the lag period, selects the 2007-2009 data as low-carbon technology innovation input, and the corresponding 2009-2011 data as low-carbon technology innovation output. In this paper, the source of data comes from Statistical Yearbooks, various development reports and authoritative press information and retrieval systems. This paper selects 29 manufacturing industries DMUs as empirical study except "Waste resources and materials recycling industry". The study objects and codes:1 Processing of food from agricultural products; 2 Manufacture of foods; 3 Manufacture of beverages; 4 Manufacture of tobacco; 5 Manufacture of textile; 6 Manufacture of textile wearing apparel, footware, and caps; 7 Manufacture of leather, fur, feather and related products ; 8 Processing of Timber,

Manufacture of wood, bamboo, rattan, palm, and straw products ; 9 Manufacture of furniture; 10 Manufacture of paper and paper products; 11 Printing, reproduction of recording media; 12 Manufacture of articles for culture, education and sport activity; 13 Processing of petroleum, coking, processing of nuclear fuel; 14 Manufacture of raw chemical materials and chemical products; 15 Manufacture of medicines; 16 Manufacture of chemical fibers; 17 Manufacture of rubber; 18 Manufacture of plastics; 19 Manufacture of non-metallic mineral products; 20 Smelting and pressing of ferrous metals; 21 Smelting and pressing of non-ferrous metals; 22 Manufacture of metal products; 23 Manufacture of general purpose machinery; 24 Manufacture of special purpose machinery; 25 Manufacture of transport equipment; 26 Manufacture of electrical machinery and equipment; 27 Manufacture of communication equipment, computers and other electronic equipment; 28 Manufacture of measuring instruments and machinery for cultural activity, office work; 29 Artwork and other manufacturing.

C. EVALUATION PROCESS

1) *Factor analysis*: Based on low-carbon technology innovation performance indicator system previously built and the original data, SPSS21.0 software is used to normalize all the original data, and then factor analysis. Manufacturing fossil energy consumption per unit of output in the global value chain, and manufacturing carbon intensity in international trade are negative indicators. Therefore, this paper uses linear conversion method to ensure the positive terms unity of output [24]. Specific formula is:

$$Y'_{ij} = -Y_{ij} + \max_{1 \le j \le m} Y_j \qquad (1)$$

Firstly, this paper tests KMO values, Bartlett's sphericity test and an explanation of the total variance are shown in Table II.

TABLE II
THE KMO AND BARTLETT'S TEST

	2007 input	2009 output	2008 input	2010 output	2009 input	2011 output
KMO measure of sampling sufficient	0.660	0.587	0.583	0.536	0.753	0.638
Bartlett's test of sphericity — Appro-ximate chi-square	456.075	536.948	453.024	536.948	520.734	583.34
df	15	28	15	28	15	28
Sig	0.000	0.000	0.000	0.000	0.000	0.000

KMO values are bigger than 0.5, the probability of Bartlett's sphericity test is 0.000, less than the significance level of 0.01, which means that there is a correlation among the original indexs, and is suitable for factor analysis. This paper selects eigenvalues greater than 1 as the standard of choosing factors. Principal component method is used to extract common factors of inputs and output indicators. The cumulative variance contribution rate of input indicators respectively reach 98.868%, 98.330%, 94.071%, and two input public factors extracted can illustrate all the variables. Using the same method to extract two output common factors, and the cumulative variance contribution rate reach 85.824%, 83.032%, 80.558%, covering almost all of the information, and the extraction effect is obvious.

2) *Data envelopment analysis*: Because BC^2 model requires data is greater than 0 and input, output indicators that have negative score value, it cannot directly use the DEA model for performance analysis. Firstly data is normalized by efficacy coefficient method [22], in order to avoid the impact of negative on the calculation. Specific formula is:

$$V'_{ij} = 0.1 + 0.9 \times \frac{V_{ij} - \min_{1 \le j \le m} V_j}{\max_{1 \le j \le m} V_j - \min_{1 \le j \le m} V_j} \quad (2)$$

29 manufacturing industries are decision making units (DMU_{j0}), x_j represents input of the manufacturing industries, y_j represents output of the manufacturing industries. Input and output meet:

$$x_j \in E^m, \ x_j > 0, \ j = 1, L, n ;$$
$$y_j \in E^s, \ y_j > 0, \ j = 1, L, n .$$

m and s denote the number of manufacturing industries input and output index, n is the number of decision making units, namely the number of manufacturing industries.

Corresponding linear programming:

$$(D_{BC^2}) \begin{cases} \min \theta \\ \sum_{j=1}^{n} x_j \lambda_j \le \theta x_0 \\ \sum_{j=1}^{n} y_j \lambda_j \ge y_0 \\ \sum_{j=1}^{n} \lambda_j = 1 \\ \lambda_j \ge 0, j = 1,2, L, n, \theta \text{ unlimited} \end{cases} \quad (3)$$

Its dual planning:

$$(P_{BC^2}) \begin{cases} \max \left(\mu^T y_0 - \mu_0 \right) \\ \omega^T x_j - \mu^T y_j + \mu \ge 0, j = 1,2, L, n \\ \omega^T x_0 = 1 \\ \omega \ge 0, \mu \ge 0, \mu_0 \text{ unlimited} \end{cases} \quad (4)$$

Introducing slack variables, s^-, s^+ represent redundant input value and insufficient output value, the dual linear programming as follows:

$$Min = \theta \quad (5)$$

$$s.t. \begin{cases} \sum_{j=1}^{n} \lambda_j x_j + s^- = \theta x_0 \\ \sum_{j=1}^{n} \lambda_j y_j - s^+ = y \\ \lambda_j \ge 0, j = 1, L, n \\ s^- \ge 0, \ s^+ \ge 0, \\ s^-, s^+ \text{ slack variables} \end{cases} \quad (6)$$

Using the normalized inputs, outputs data as basis, this paper uses DEAP2.1 software with input-oriented BC^2 model to calculate the efficiency value, and then analyzes low-carbon technology innovation performance of Chinese manufacturing in the global value chain. The results are shown in Table III.

TABLE III
THE EVALUATION RESULTS OF CHINESE MANUFACTURING
LOW-CARBON TECHNOLOGY INNOVATION PERFORMANCE
IN THE GLOBAL VALUE CHAIN

Dmu	2007-2009				2008-2010				2009-2011			
	Te	Pte	Se	Rs	Te	Pte	Se	Rs	Te	Pte	Se	Rs
1	0.665	0.990	0.672	irs	0.90	0.995	0.910	irs	0.81	0.998	0.816	irs
2	1.000	1.000	1.000	-	0.75	0.982	0.764	irs	0.57	0.992	0.574	irs
3	0.583	1.000	0.583	irs	1.00	1.000	1.000	-	0.58	0.982	0.600	irs
4	0.780	1.000	0.780	irs	0.91	1.000	0.919	irs	0.79	0.998	0.794	irs
5	0.712	0.997	0.714	irs	0.81	1.000	0.818	irs	0.83	1.000	0.836	irs
6	0.679	1.000	0.679	irs	0.73	0.987	0.743	irs	0.79	0.999	0.800	irs
7	0.722	1.000	0.722	irs	0.79	1.000	0.798	irs	0.80	0.999	0.802	irs
8	0.589	1.000	0.589	irs	0.74	0.981	0.759	irs	0.70	0.997	0.705	irs
9	0.741	0.991	0.748	irs	0.76	1.000	0.767	irs	0.67	0.995	0.677	irs
10	0.509	1.000	0.509	irs	0.75	0.950	0.798	irs	0.74	0.998	0.745	irs
11	0.748	1.000	0.748	irs	0.73	0.986	0.748	irs	0.69	1.000	0.694	irs
12	0.674	0.985	0.685	irs	0.72	0.973	0.742	irs	0.71	0.999	0.710	irs
13	1.000	1.000	1.000	-	0.75	1.000	0.753	drs	0.67	1.000	0.677	drs
14	0.735	0.741	0.992	drs	0.86	0.998	0.865	drs	1.00	1.000	1.000	-
15	0.750	0.980	0.766	irs	0.74	0.978	0.762	irs	0.54	1.000	0.546	irs
16	0.609	0.995	0.612	irs	0.74	0.998	0.750	irs	0.74	0.999	0.748	irs
17	0.548	0.944	0.581	irs	0.76	1.000	0.768	irs	0.59	1.000	0.590	irs
18	0.569	0.958	0.594	irs	0.74	0.964	0.770	irs	0.64	0.997	0.643	irs
19	1.000	1.000	1.000	-	0.89	1.000	0.893	irs	0.99	1.000	0.997	irs
20	1.000	1.000	1.000	-	1.00	1.000	1.000	-	1.00	1.000	1.000	-
21	0.695	1.000	0.695	irs	0.81	0.961	0.850	irs	0.64	0.950	0.677	irs
22	0.777	1.000	0.777	irs	0.80	0.982	0.820	irs	0.66	0.993	0.674	irs
23	0.885	0.955	0.926	drs	0.71	0.813	0.884	drs	1.00	1.000	1.000	-
24	0.830	1.000	0.830	irs	0.89	0.996	0.893	irs	0.76	1.000	0.766	irs
25	0.675	0.840	0.803	irs	0.76	0.792	0.971	irs	0.83	0.935	0.895	irs
26	1.000	1.000	1.000	-	0.94	1.000	0.947	drs	1.00	1.000	1.000	-
27	1.000	1.000	1.000	-	1.00	1.000	1.000	-	1.00	1.000	1.000	-
28	1.000	1.000	1.000	-	1.00	1.000	1.000	-	0.94	1.000	0.941	irs
29	0.824	0.998	0.827	irs	0.81	1.000	0.810	irs	0.79	0.999	0.791	irs

D. DISCUSSION

Based on the value changes and overall efficiency of 2007-2009, 2008-2010, 2009-2011, and according to the average of comprehensive technical efficiency, Chinese manufacturing low-carbon technology innovation performance in the global value chain is divided into four levels.

Comprehensive technical efficiency value is equal with 1, which belongs to the highest performance industries, including smelting and pressing of ferrous metals industry, manufacture of communication equipment, computers and other electronic equipment industry. Although smelting and pressing of ferrous metals industry is high energy consumption and pollution, but has a higher low-carbon technology innovation input and corresponding higher level of output. Communication equipment, computers and other electronic equipment manufacturing industry is China's earliest open and fastest-growing high-tech industry, representing the more important position in the national economy, with more patents, high technical efficiency.

Comprehensive technical efficiency value is 0.8 to 1, which belongs to the higher performance industries, including manufacture of non-metallic mineral products; manufacture of measuring instruments and machinery for cultural activity, office work; manufacture of electrical machinery and equipment; artwork and other manufacturing; processing of petroleum, coking, processing of nuclear fuel; manufacture of special purpose machinery; manufacture of tobacco; manufacture of raw chemical materials and chemical products; manufacture of general purpose machinery. Manufacture of non-metallic mineral products industry is one of the sunrise industries in modern society, high-tech and low environmental impact have become its development trend. Manufacture of measuring instruments and machinery for cultural activity, office work industry; manufacture of electrical machinery and equipment industry are high-tech industries, with higher levels of low-carbon technology innovation input and output.

Comprehensive technical efficiency value is 0.7 to 0.8, which belongs to the general performance industries, including manufacture of articles for culture, education and sport activity; manufacture of chemical fibers; smelting and pressing of non-ferrous metals; manufacture of beverages; manufacture of furniture; printing, reproduction of recording media; manufacture of textile wearing apparel, footware, and caps; manufacture of metal products; manufacture of transport equipment; manufacture of foods; manufacture of leather, fur, feather and related products ; manufacture of textile; processing of food from agricultural products. This paper mainly analyzes manufacturing industry of foods and manufacture of textile, both belonging to sunset industries. Low-carbon technology innovation input and output are in a weak position, the presence of labor, raw materials and other cost advantages are gradual loss, environmental pollution and ecological destruction and other issues. The use of existing international marketing channels and the degree of participation in the supply chain are still really weak.

Comprehensive technical efficiency value is less than 0.7, which belongs to the low performance industries, including manufacture of rubber; manufacture of plastics;

manufacture of paper and paper products; processing of timber; manufacture of wood, bamboo, rattan, palm, and straw products; manufacture of medicines. The innovation of manufacture of medicines needs to go through several phases with longer time and the higher cost.

VI. CONCLUSION AND SUGGESTION

Overall technical efficiency of Chinese manufacturing industries is quite different, and technical efficiency is not high. The average of 29 manufacturing industries's technical efficiency is 0.791, indicating that after the passage of a reasonable allocation of resources, and the premise of not reducing the current output, there are still 21.9% reduction space for low-carbon technology innovation inputs. On one hand, this reflects low efficiency of resource allocation of Chinese manufacturing low-carbon technology innovation in the global value chain. On the other hand, this shows there is much space for the improvement of competitiveness of Chinese manufacturing. And most of the industries are not in active size state, and for the whole, these industries should greatly improve the technical efficiency and scale efficiency, and should increase resources investment in low-carbon technology innovation to improve efficiency of resource allocation.

For communication equipment, computers and other electronic equipment manufacturing industry, Chinese high-tech industry policies create favorable environment and institutional safeguard for its development, and should continue to support the development of high-tech industries and encourage participation in high value-added sectors of the global value chain. For measuring instruments and machinery manufacturing industry for cultural activity, office work, electrical machinery and equipment manufacturing industry should improve technical efficiency and scale efficiency through the rational allocation of resources. Textile manufacturing industry should increase low-carbon technology innovation resources investment, and promote the sustainable development of the textile industry. Rubber and plastics manufacturing industry need to adjust the export structure of trade. Medicines manufacturing industry requires continuous R&D investment, further increases R&D intensity and improves low-carbon technology innovation ability.

REFERENCES

[1] UNIDO. "Industrial development report 2002/2003 competing through innovation and learning," *United Nations Industrial Development Organization*, pp. 3-112, 2003.
[2] Bei ZENG, Huan-jin CUI, "Why the evolution of Chinese industrial structure deviates from f international experience - based on the interpretation of the global value chains division" (in Chinese), *Finance and Trade Research*, no. 1, pp. 18-27, 2011.
[3] Ling-qing LIU, Li-wen TAN, Guan-qun SHI, "Rent, power and performance-Competitive advantage in the background

of global value chain" (in Chinese), *Chinese Industrial Economy*, no. 1, pp. 50-58, 2008.

[4] Sheng-qi ZHOU, Zhen-xian LAN, Hua FU, "The status of Chinese manufacturing industry in the international divisionof the global value chain - based Koopman, etc. GVC status index" (in Chinese), *International Trade Issues*, no. 2, pp. 3-12, 2014.

[5] C. Pietrobelli, R. Rabellotti, "Global value chains meet innovation systems: are there learning opportunities for developing countries," *World Development*, vol. 39, no. 7, pp. 1261-1269, 2011.

[6] G. Gereffi, J. Humphrey, T. Sturgeon, "The governance of global value chains," *Review of International Political Economy*, vol. 12, no. 1, pp. 78-104, 2005.

[7] S. L. Sun, H. Chen, E. G. Pleggenkuhle-miles, "Moving upward in global value chains: the innovations of mobile phone developers in china," *China Management Studies*, vol. 4, no. 4, pp. 305-321, 2010.

[8] K. Nishitani, K. Kokubu, "why does the reduction of greenhouse gas emissions enhance firm value," *Business Strategy and the Environment*, vol. 21, no. 8, pp. 517-529, 2012.

[9] C. W. Park, K. S. Kwon, W. B. Kim, "Energy consumption reduction technology in manufacturing-A selective review of policies, standards, and research," *International Journal of Precision Engineering and Manufacturing*, vol. 10, no. 5, pp. 151-173, 2009.

[10] Xiong-feng PAN, Tao SHU, Da-wei XU, "The carbon emission intensity change and factor decomposition of Chinese manufacturing" (in Chinese), *China Population, Resources and Environment*, vol. 21, no. 5, pp. 101-105, 2011.

[11] L. M. A. Bettencourt, J. E. Trancik, J. Kaur, "Determinants of the pace of global innovation in energy technologies," *PloS one*, vol. 8, no. 10, pp. e67864, 2013.

[12] Yong-ming HUANG, Wei HE, Ming NIE, "Chinese textile and garment enterprises' upgrade path option in the global value chain" (in Chinese), *China Industrial Economy*, no. 5, pp. 56-63, 2006.

[13] Xing PENG, Bin LI, "Study on carbon emissions effect with Chinese economy embedded manufacturing sector in the global value chain" (in Chinese), *Finance Research* , no. 6, pp. 18-26, 2013.

[14] Ai-ling GONG, "Study on the position in the global value chain division and export implied carbon of Chinese manufacturing" (in Chinese), *Economics and Management*, vol. 27, no. 8, pp. 72-76, 2013.

[15] S. J. Kline, N. Rosenberg, "An overview of innovation," *The positive sum strategy: Harnessing technology for economic growth*, pp. 275-305, 1986.

[16] Li-ru XHENG, Xuan ZHOU, "Value innovation: a new perspective on technology innovation strategy," *Science of Science Technology Management*, no. 8, 56-59, 2006.

[17] R. Griffith, S. Reding, J. Van Reene, "Mapping the two faces of R&D: productivity growth in a panel of OECD industries," *Review of Economics and Statistics*, vol. 86, no. 4, pp. 883-895, 2004.

[18] Jian-cheng ZHOU, "Research on the relationship between manufacturing capabilities and innovation performance of business: some empirical findings in China" (in Chinese), *Research Management*, vol. 25, no. 1, pp. 78-84, 2004.

[19] R.Yam, J. C. Guan, K. F.Pun, E. P.T.Tang, "An audit of technological innovation capabilities in Chinese firms: Some empirical findings in Beijing, China," *Research Policy*, vol. 33, no. 8, pp. 1123-1140, 2004.

[20] R. D. Banker, A. Chaenes, W. W. Cooper, "Some models for estimating technical and scale inefficiencies in data envelopment analysis," *Management Science*, no. 30, pp. 1078-1092, 1984.

[21] Jun-hong BAI, Ke-shen JIANG, Jing LI, Jia LI. "The analysis of environmental factors of regional innovation efficiency - Empirical study based on DEA-Tobit two-step method" (in Chinese), *Research and Development Management*, vol. 21, no. 2, pp. 96-102, 2009.

[22] Han-dong ZHANG, Jing ZHAO, "The study of transformation and upgrading of Hangzhou export competitive industries based on innovation performance" (in Chinese), *International Trade Issues*, no. 3, pp. 53-62, 2012.

[23] B. A. Kirchhoff, C. Armington, I. Hasan, S. Newbert, "The influence of R&D expenditures on new firm formation and economic growth," *Washington, DC: National Commission on Entrepreneurship*, no. 27, pp. 4-24, 2002.

[24] Zong-yu YE, "The select of positive indicators on and dimensionless method in multiple indicators comprehensive evaluation" (in Chinese), *Zhejiang Statistics*, no. 4, pp. 24-25, 2003.

Personnel Behavior and Modeling Simulation in the Emergency Evacuation

Xi-meng WU*, Jie WANG, Xiao-hong GUO

School of Safety and Environmental Engineering, Capital University of Economics and Business, Beijing, China
(wxmciir2009@sina.com)

Abstract - **The human behavior characteristics in the emergency evacuation, the features and the principles of Building EXODUS were introduced here through summarizing all kinds of evacuation models and methods so far developed both at home and abroad. Finally an example is given to show how to simulate emergency evacuation by Building EXODUS software. Results show that the study of evacuation behavior needs some knowledge of psychology and sociology, and the human behavior characteristics cannot be all included by evacuation models, thus the simulating evacuation models should be more comprehensive and more standard. The function of Building EXODUS software in simulating the behaviors of human is reliable and easy to find the bottleneck phenomenon at the exit, and the evacuation time can be decreased through changing the width of the door or giving some guidance to the evacuees.**

Keyword - **Building EXODUS, emergency evacuation, human behavior characteristics, simulating evacuation**

I. INTRODUCTION

With the rapid development of society, public security emergencies occur frequently due to the intensive urban population, unbalanced economic development and complex social interest relations. It is easy to cause the evacuees' following behavior and the panic psychology. Therefore how to make the emergency evacuation more effective and reduce the accident loss has become an important task.

The research on emergencies evacuation simulation is a complex problem referring to many subjects especially psychological and also is related to several subjects such as psychology and sociology [1]. The paper will introduce the human behavior characteristics, the evacuation models and the method of crowd evacuation simulation in the emergency evacuation, and simulate the evacuation through the Building EXODUS software to give the theoretical basis in designing the building evacuation and formulating the emergency evacuation plan.

II. THE HUMAN BEHAVIOR CHARACTERISTICS IN THE EMERGENCY EVACUATION

The study on human behavior characteristics in the emergency evacuation can be divided into three categories as is shown in Fig.1: crowd behavior characteristics, personnel behavior characteristics and personnel basic attributes. The study on the crowd behavior characteristics

is from a macro angle; personnel behavior characteristics is from a microscopic; personnel basic attributes refer to some characters in common, such as the physiological level, cultural level, personal qualities[2-3], et al.

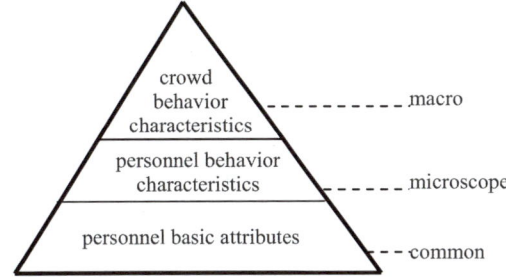

Fig.1. The human behavior characteristics in the emergency evacuation

The human behaviors can be affected by the surrounding environment and their own quality, which may have an effect on the whole evacuation. Therefore, the study on human behavior characteristics in the emergency evacuation is of great significance.

A. Following behavior

Following behavior which is irrational refers to the actions which are affected by others in an emergency evacuation [4]. Following behavior, to some extent, is benefit to the evacuation in the situation that the evacuees are not familiar with the environment; they can find the exits through following others. But it can also lead to serious consequences, such as overcrowded, reducing the speed, and even causing a serious crowded stampede accident.

B. Escaping together behavior

The escaping together behavior also occurred among people who have close social relationship with each other, such as family members ,friends and so on, and behaviors like waiting, wandering, helping each other, and communicating can also be found in the emergency evacuation [5]. The characteristic of the escaping together behavior is that the distance between people who are in the same group is short, while it's long between those who are in different groups (as is shown in Fig.2). It has affected the decision and action process of the evacuation on the macro, and walking speed and the necessary space of human body on the micro.

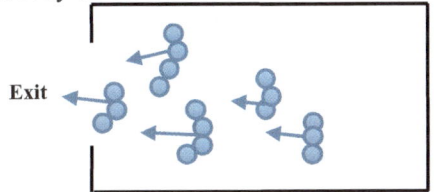

Fig.2. Escaping together behavior

The Science and technology innovation platform project (PXM2014_014205_000044); Teachers' Scientific Research Ability Promotion plan (2014); Academy of Metropolis Economic; Social Development project of Capital University of Economics.

E. Qi et al. (eds.), *Proceedings of the 21st International Conference on Industrial Engineering and Engineering Management 2014*, Proceedings of the International Conference on Industrial Engineering and Engineering Management, DOI 10.2991/978-94-6239-102-4_57, © Atlantis Press and the authors 2015

C. Pathfinding behavior

The pathfinding behavior is defined as a decision in choosing which way to go in an emergency [6]. In that situation each traffic facilities, such as exits, channel, stairs, and escalators, in the building can be regarded as an attractive medium and the pathfinding behavior obey the rule of the shortest path. But in fact the most sensible decision should depend on both the distance and the traffic condition.

D. Avoiding the obstacles behavior

The evacuation will moving to the destination in a straight line path, which is the shortest path, when the destination is determined, and if there is an obstacle on the path, the evacuation will change the primary path to avoid hitting the obstacle, at the same time, a new path will be generated [7]. The whole process is shown in the Fig.3.

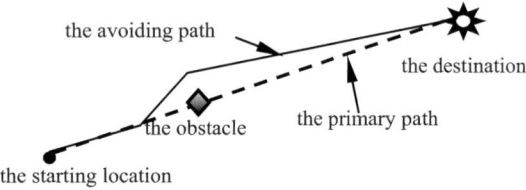

Fig.3. The behavior of avoiding the obstacles

E. Overcrowding at a" Bottleneck" behavior

The "Bottleneck" refers to the place where the wide channel suddenly switches to a narrow channel in a building [8]. When people walking through this place, it's easy to find that the speeds of the evacuees are slow down, and people gathered there in an arch shape. In the evacuation it's easy to find the bottleneck phenomenon at the places like the exits, corridors, channels, ramps, stairs, escalators, trails, ticket mouths, security checkpoints and so on.

F. Back for rescuing behavior

Back for searching behavior also occurred within the small group of people; the person who is outside will go back and find the companion, which is completely harmful to the whole emergency evacuation [9].

G. Panic behavior

According to the simulation experiment, the panic behavior can be defined as behaviors which are illogical, uncomfortable due to the fear of the emergency [10]. Excessive panic will reduce the ability to evacuate and the ability of the evacuees may be reduced or totally loss, even across the fence, chairs and so on. But moderate panic behavior contributes to the evacuation.

III. THE SIMULATION MODELS AND METHODS

A. Methods
1) The evacuation exercises

Amount of people are gathered in a building to evacuation pretending that an emergency has happened. In the whole evacuation researchers can do their study on the relationships among the evacuees' gender, age, walking habits through tracking the evacuees, what's more, the video of the evacuation can also contribute to the study on human evacuation characteristics [11].

The data which is got from the exercise, to a certain extent, is reliable and can provide basis for the simulation research. Actually, the behavior of the evacuees in the exercise is different from that in a real emergency for the reason that the evacuees will have panic psychological [12]. The randomness and variability of the behavior make the result unreliable in the only exercise. They need to make several exercises, but in fact it's not feasible.

2) The simulation models

On the computer a model can be established to simulating the evacuation by setting the data of the buildings and persons. The simulation evacuation models which are considered to be the most effectively and widely used methods for its operability and the consistency to the real situation [13]. Although the data and methods of different software are different, which lead to different result, after all, it's a reliable method which needs to be improved.

B. Simulation models introduction

There are more than twenty simulation models both home and abroad which can be divided into two categories [14]:

The models in which only the movement of human is considered, such as EVACNET+, TAKAHASHI'S MODEL, just take the evacuation capacity into consideration, and person in the model is regarded as a unconscious objection which automatically response to the external signal, the direction and the speed of the evacuee is only determined by physical factors, such as population density, the evacuation capacity of the exit, et al.

The models in which both the movement and the behavior of human are considered, like EGRESS, E-SCAPE, EXIT98, building EXODUS, in which not only the physical characteristics of the building, but also the characteristic of the human behavior (such as how to choose the exit and how to respond to the signals) are taken into consideration, the passenger in the model are seen as an active factor, which can respond to a variety of signals.

IV. EXAMPLE OF EMERGENCY EVACUATION

A. Introduction of the Building EXODUS software

Building EXODUS which is considered to be one of the most widely used simulation evacuation software in the world can be used to evaluate whether the architectural design meet the requirements, the analysis the evacuation performance of a structure and the evacuation efficiency of the building ,such as the supermarkets, hospitals, railway stations, schools, airports, et al.

Building EXODUS takes the interactions between people-people, people-fire and people-structure into consideration. Thus the behavior and movement of each passenger are determined by a set of heuristic rules, and these rules are divided into five kinds of interaction submodels, namely, passenger, movement, behavior, toxicity and hazard model, and they change with the geometric environment, as is shown in Fig.4.

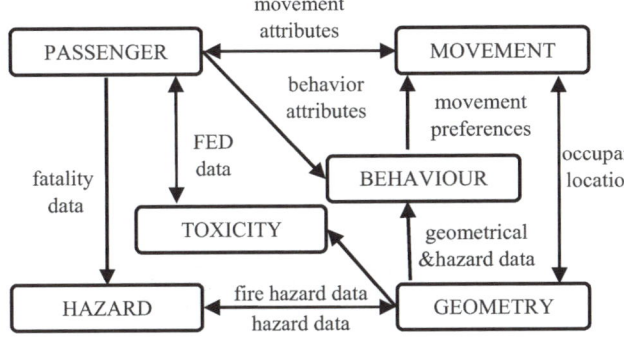

Fig.4. EXODUS submodel interation

Unlike other simulation evacuation software, Building EXODUS is a process simulation software in which the interactions between people-people, people-fire and people-structure are taken into consideration [15]. Therefore the attribute of evacuees and scenes can be truly simulated by building EXODUS, and it can get a comprehensive predicted result and plain it through tracing the details in the process of evacuation, what's more, the bottleneck location, the speed, the start time, the end time and the process can also be described by building EXODUS software [16].

B. Establish the structure

The research of the paper is a canteen of a university which is a two-story building structure. As is shown in Fig.5 and Fig.6, it has two stairs and five exits in the building. The author establish the structure model by setting the data of the building which are measured and modified according to the building EXODUS.

C. The number and parameters of the passenger

According to the survey results, the peak period are: 7:20 to 7:45, 11:50 to 12:50, and 17:30 to 18:30.The survey data are shown as follows:

1) The total number of emergency evacuation:

According to investigation statistics, the average numbers of people in the three periods are 477p, 1083p and 729p.

2) The age of evacuees:

The weight, height, reaction time, patience, and walking speed of different people are different for the different age and gender. According to statistics, the age of people in the canteen are shown in Table I.

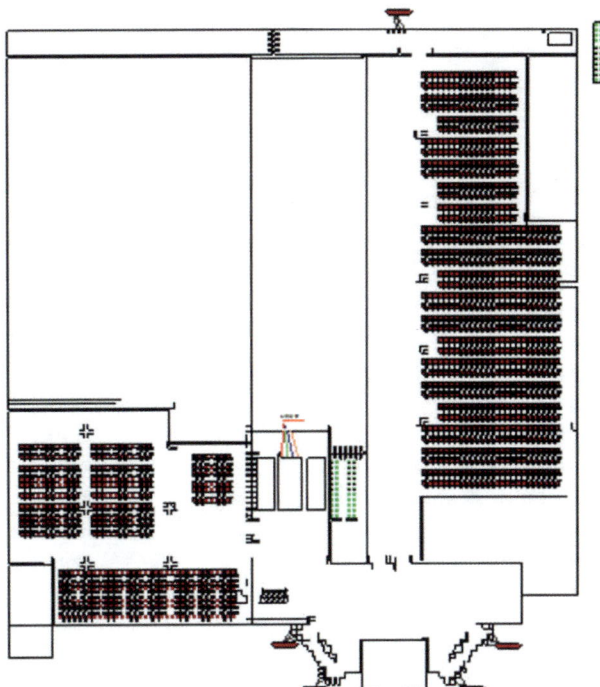

Fig.5. The structure diagram of the first floor

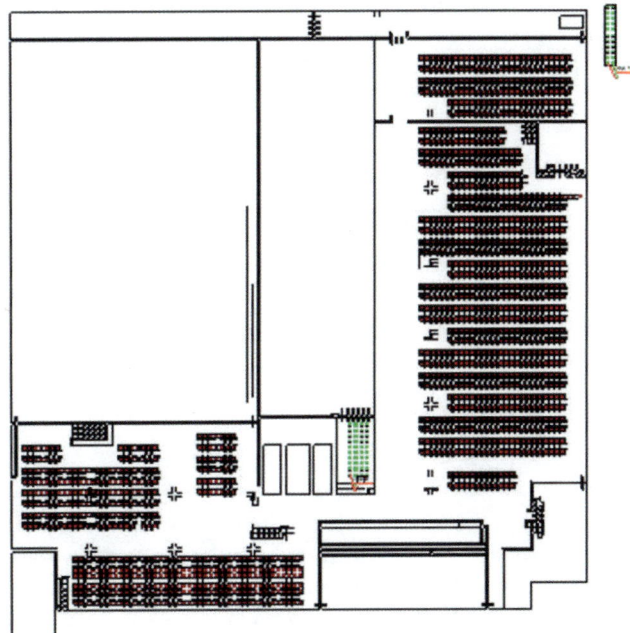

Fig.6. The structure diagram of the second floor

TABLE I
THE AGE OF EVACUEES

Male		Female	
Age	proportion	Age	Proportion
15-29	28%	15-29	55%
17~50	5%	17~50	9%
51~80	1%	51~80	2%

D. Simulation results and analysis

Assuming that in the period 11:50 to 12:50 an emergency happened, the author have an study on simulating the evacuation in an emergency in the building, The data of every escapee in every emergency exit (which is shown in Table II) can be outputted by building EXODUS.

TABLE II
THE USING INFORMATION OF EXITS

Exit	Total people of using	Passing times of the last people (s)
Exit1	427	332
Exit2	85	122
Exit3	22	41
Exit4	92	94
Exit5	457	435
In total	1083	435

According to the simulation results, the numbers of using Exit 1 and Exit 5 are the biggest and bottleneck phenomena happened there (as is shown in Fig.7), but there are no evacuees using the Exit 3 and Exit 4 after 94 seconds from the beginning, therefore, the width of Exit2, 3, 4 should be increased to attract more evacuees using these exits, which can also contribute to the evacuation efficiency of Exit 1 and Exit 5.

The low evacuation efficiency of the Exit 5 due to its out of its capacity of being used. Thus measures like leading some people using the other exits by radio or other methods.

After being improved, the result can be shown in Table III, according to which the evacuation efficiency is higher than before; the total evacuation time is reduced to 342 seconds from 435s seconds. That is to see increasing the width of the exits or guiding the evacuees is benefit to the evacuation.

TABLE III
THE USING INFORMATION OF EXITS AFTER IMPROVING THE MODEL

Exit	Total people of using	Passing times of the last people (s)
Exit1	320	332
Exit2	173	227
Exit3	159	206
Exit4	159	193
Exit5	270	342
In total	1083	342

V. CONCLUSION

(1) The paper suggests that the study of evacuation behavior need some knowledge of psychology and sociology, and the human behavior characteristics cannot be all included by evacuation models, thus the simulating evacuation models should be more comprehensive and more standard.

(2) The function of Building EXODUS, which is considered to be the most comprehensive in its function in the world, in simulating the behavior of human is reliable and it is indicated it's easy to appear a bottleneck phenomenon at the exit, and the evacuation time can be decreased through changing the width of the door or giving some guidance to the evacuees.

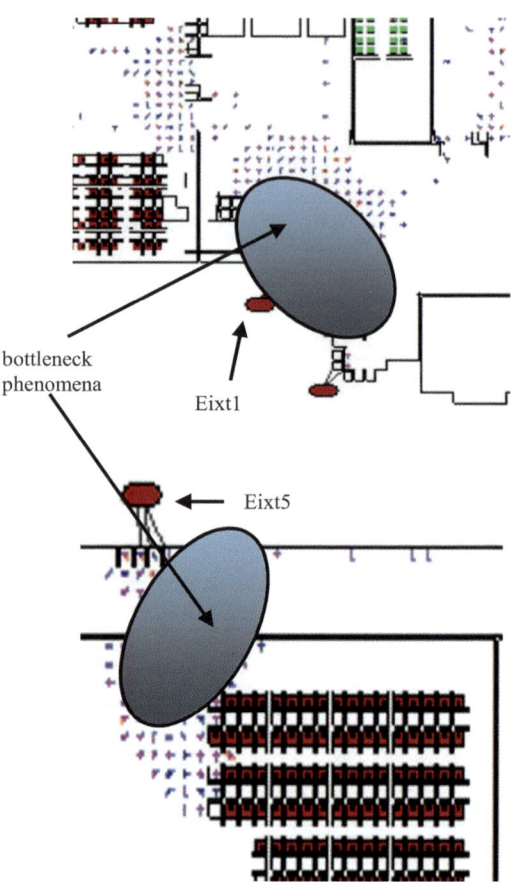

bottleneck phenomena

Eixt1

Eixt5

Fig.7. The bottleneck phenomena at Exit 1 and Exit 5

ACKNOWLEDGMENT

The authors would like to thank the Science and technology innovation platform project (PXM2014_014205_000044), Teachers' Scientific Research Ability Promotion plan (2014) and Academy of Metropolis Economic, Social Development project of Capital University of Economics and Business for providing financial support to this research.

REFERENCES

[1] Yu-min TIAN. Application of computer software in the practical teaching of crowd evacuation, Proceedings of Conference on Creative Education (CCE2011) [A].2011:55-56 (In Chinese)

[2] Xi-hong CUI, Jin CHEN, Qiang LI. Study on occupant evacuation modelin large public facilities [J]. Journal of basic science and engineering. 2006.12:94-100(In Chinese)

[3] Qing-mei HU, Wei-ning FANG, Guang-yan LI, Yu-quan JIA. Review on pedestrian behavior characteristics and crowding mechanism in public buildings [J]. China Safety Science Journal, 2008(08):68-73(In Chinese)

[4] Chi WANG. Study on safety evacuation during the fire of a subway station [D]. Beijing Jiaotong University, 2007 (In Chinese)

[5] Gui-fen WANG, Xian-li ZHANG, Wei-dong YAN. Building fire behavior modeling of EXODUS [J], Journal of Safety Science and Technology, 2011(8): 69-74 (in Chinese)

[6] Zhen WEI. Study of pedestrian traffic evacuation model based on behavioral characteristics [D].South China University of Technology, 2012 (In Chinese)

[7] Xiang XU, Min HUANG, An improve group intelligence pathfinding algorithm [J]. Computer Applications and Software, 2012(05): 139-142 (In Chinese)

[8] Xi-hong CUI, Qiang LI, Jin CHEN, Chun-xiao CHEN. Study on occupant evacuation model in large public place: to consider individual character and following behavior [J]. Journal of Natural Disasters, 2005(06):133-140 (In Chinese)

[9] Min XU. Analysis on characters of psychological behavior in dispersing disaster [J].Safety.2007 (06): 42-44. (In Chinese)

[10] Xia WANG, Zhi-min XIE, Xian-jun GUAN. Study on micro simulation of occupant evacuation in panic [A].2010 International Conference on Future Information Technology and Management Engineering, (FITME 2010). 2010: 40-43 (In Chinese)

[11] Fang XU, Dong WEI, Xing WEI. Analysis on the original data of crowd evacuation in public gathering places, China Safety Science Journal, 2008(4): 137-145 (In Chinese)

[12] Yu-min TIAN. On human behavior in fire accidents and computer simulation methods [J]. Journal of Safety and Environment, 2006(1): 26-30 (In Chinese)

[13] Li-li PAN, Tao CHEN. Discussion on research methodologies of evacuation [J]. Fire Science and Technology, 2010(03): 214-217 (In Chinese)

[14] Dao-liang ZHAO. Cellular automata simulation on special human behavior of occupant evaluation in emergency [D].University of Science and Technology of China, 2007 (In Chinese)

[15] Tao YU, Ying-feng ZHANG, Zun-ze HOU. Comparative study of evacuation time prediction methods [J].Fire Science and Technology, 2009(3): 35-40 (In Chinese)

[16] Guo-min ZHAO, Zhao-peng NI, Qing-song ZHANG. The study on evacuation discrete time computational model for subway station [J]. Journal of Disaster Prevention and Mitigation Engineering, 2010(2): 152-157 (In Chinese)

Research on Subway Traffic Dispatcher Team Error Issues in Accident Management

Feng LIN, Jie WANG*, Lian-feng XU

School of Safety and Environmental Engineering, Capital University of Economics and Business, Beijing, China
(grace_1984@163.com)

Abstract - **In order to prevent and reduce human errors of the subway dispatching system in the accident response process, a subway dispatcher team emergency response cycle framework was built and an application and analysis of the framework with a typical case was also provided. Results showed that different situations lead to different emergency response procedures and human errors. Through the proposed response cycle framework, several specific subway dispatcher team error modes were built. These error modes could be applied to find the weak links of the subway traffic dispatcher response cycle, and to help improve the human reliability of subway dispatching system.**

Keywords - **Accident management, emergency response cycle, subway traffic dispatching system, team error**

I. INTRODUCTION

Dispatching system is the central nervous of subway traffic system. In emergencies, the subway traffic dispatcher system forms a dynamic operation cycle by interaction with the outside world frequently. Once human errors happen, it will lead to the subway traffic dispatching system's delays or interruptions, or even worse accident consequences.

Human error researches mainly concentrated on the field of nuclear power plants, aerospace and other complex industrial areas. Such as Sasou [1] used the nuclear power plant simulator to qualitative research decision-making ability between of operators, inspectors, and managers; Y. H. J. Chang [2-6] established an IDAC model to realize probability forecast about the abnormal response of plant team. Ke-bing LIAO [7] created a crew behavior model to simulate and analyze the team's accident situation operation process; Ren-li LV [8] used the framework of threat and error management to analyzed the process and cause of air traffic human error; Li-cao DAI [9] analyzed and discussed the team error of the complex industrial systems. In the field of subway and rail, human error research is priority to pay attention to the individual error, such as Jie WANG [10-11] presented a dispatcher's human error behavior identification method based on emergencies scenario; Hai-tao WU [12] presented a human error identification method based on the cognitive process analysis which was applied in the high speed railway train dispatching system; Xin-ping WEI and Lei-zhen GUO [13] analyzed the reasons of human error from five aspects of biological rhythms, fatigue

---

[1] Graduate Science and Technology Innovation Plan of Capital University of Economics and Business; Science and technology innovation platform project (PXM2014_014205_000044); Teachers' Science and Technology Innovation Project

assignments, work sense of responsibility, professional quality, information error and the countermeasures.

Human reliability analysis (HRA) in complicated industrial system is an effective method for preventing human error. Team error is one form of human error. Subway traffic dispatchers need to complete their tasks independently and also need to make diagnosis and decision jointly for an accident. Therefore, researches on dispatcher team error issues in accident management make HRA more reasonable and perfect in the subway traffic system.

II. SUBWAY DISPATCHER TEAM EMERGENCY RESPONSE CYCLE FRAMEWORK

The process of subway traffic dispatchers' interaction with the outside world is a complicated and dynamical cycle, as is shown in Fig.1. Dispatcher team makes the subway traffic system to achieve the required state or solve emergencies through the response cycle.

(1) Outside information: The subway traffic dispatcher team needs to get information from the outside world before accident management, for example, the drivers tell the traffic fault information to dispatcher team through the communication equipment.

(2) Dispatcher team's response: Three stages are included, which are information processing (a), accident diagnosis and decision making (b) and the results performed (c).

Information processing (a): This is the foundation of team accident diagnosis and decision making. Firstly, accident information was collected; secondly, dispatchers had the crisis awareness, and the degree of crisis awareness is according to the collected information. The ability of information collection is depend on the sensitivity and agility of the dispatcher team's information cognitive, such as monitoring the subway running graph, dispatchers quickly found the fault signal; The ability of crisis awareness is depend on dispatcher team's judgment for the urgency of information, such as determine the accident's risk level.

Accident diagnosis and decision making (b): This step is the core of the dispatcher team's response cycle, and is also the guarantee and key of an effective response. The accident diagnosis and decision making of subway traffic dispatcher team in the emergency response belong to the "quick discussion make or real-time decision making". The decision goal was analyzed according to the information processing, and then the feasibility scheme was proposed to judge whether it was the best solution. If it is, the solution will be executed as a result; if it is not, more information will be got from the outside world and to perceive the degree of crisis awareness again.

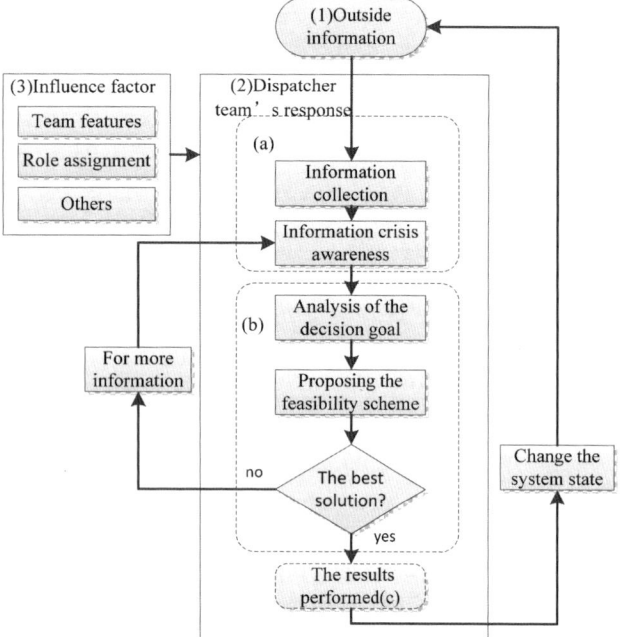

Fig.1. Subway dispatcher team emergency response cycle framework

The result performed (c): The performance of team response will be carried out in this step. After the dispatcher team's response is determined, the dispatchers interact with the outside world through telephone channels to deal with accidents, for example, a passenger fell into orbit suddenly, and traffic dispatchers ask the power to cut off the contact rail power by a specific telephone channel.

(3) Team characteristics and team role assignment also influence the emergency response quality of subway traffic dispatcher team. Team characteristics include team communication, support/support behavior (e.g., mutual supervision, mutual correction), team orientation (e.g., team target relative to the weight of personal goals), knowledge sharing ability, team cohesion (the comprehensive dimension of team unity, harmony and the get along way of team members [3]). Team role assignment means that a team is divided into three categories; these are decision maker, practitioner and consultant. Among them, decision maker is the leader of the team and has the final decision and the responsibility of the relevant authority; practitioner is the actual operation person to assist decision makers to solve the problem; the consultant provides technical support and advice to decision maker [3].

III. HUMAN ERROR MODE ANALYSIS OF THE SUBWAY TRAFFIC DISPATCHER TEAM

Dispatcher team error modes can be identified through the response cycle (as is shown in Table I), which are divided into three types of error modes: information processing error (includes information collection error and information crisis awareness error), accident diagnosis and decision-making error (includes analysis of

the decision goal error, proposing the feasibility scheme error and error of determining the best solution) and the results performed error.

TABLE I
TRAFFIC DISPATCHER TEAM ERROR MODE

Response process	Error mode	
Information processing (a)	Information collection error	1 inattention 2 mistake in comprehension 3 miscommunication 4 man-machine interface defects 5 time constraints of load 6 resource management defects
	Information crisis awareness error	1 business application error 2 lack of knowledge and experience 3 low level of situational awareness 4 high/low motivation
Accident diagnosis and decision making (b)	Analysis of the decision goal error	1 high/low self-confident degree 2 lack of knowledge and experience 3 knowledge base matching error
	Proposed the feasibility scheme error	1 scheme base matching error 2 situation forecast error 3 time constraints of load
	Determine the best solution error	1 high/low self-confident degree 2 situation forecast error 3 expert base matching error 4 excessive authority effect 5 time constraints of load
The result performed (c)	Performed error	1 lack of standardization in communication 2 adequate warning 3 without a comprehensive communication 4 resource management defects

IV. TYPICAL CASES STUDY

A. Summary

At a subway line, a subway train slipped from platform A to platform B. The traffic dispatchers taken measures to command all trains stop immediately, and notify the controller of station B to take measures to deal with it. Through the driver's processing, the problem train restored traction, and the traffic dispatcher commanded drivers to organize passengers to get off and drive the problem train into garage A. Through cooperating with other departments, the traffic dispatcher team was deal with sudden accident and let the whole route back into normal operation.

According to the case, emergency response of the subway traffic slip accident is shown in Fig.2.

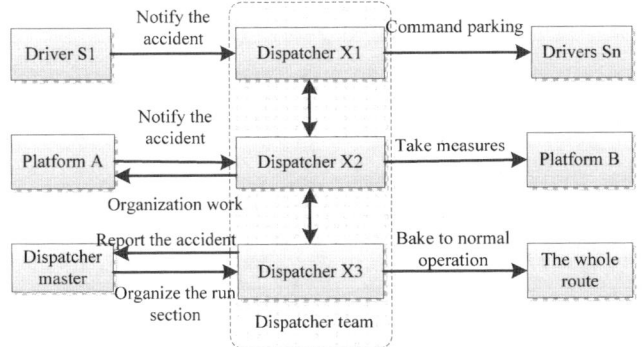

Fig.2. Emergency response of the subway traffic dispatcher team in the slip accident

B. Human error analysis

According to Table I, human errors in dispatcher response may occur, i.e., the bottleneck of the dispatcher emergency response reliability needs to be avoided by taking some measures. The conclusion is that different situations and inputting information lead to different emergency response procedures and human errors:

Different situations cause different human errors of emergency response. For example, dispatcher master notify dispatcher team to organize the run section, human errors may be occurred in information collection stage: inattention (e.g., did not notice the dispatcher masters' instructions, a-1), mistake in comprehension (e.g., the failure to correctly understand the instructions, a-2), human-machine interface defects (e.g., dispatcher team cannot instruction due to telephone malfunction, a-4). But when driver reports that fault has been ruled out, the most of commands has been completed, and the whole dispatcher team's attention is concentrated on the on-site troubleshooting report, therefore, team is without inattention. In addition, dispatcher team keep communication with drivers, so human-machine interface defects and mistake in comprehension cannot happen.

V. CONCLUSION

(1) The subway traffic dispatcher team emergency response cycle framework reflects the internal cognitive processes of the subway traffic system in accident management dynamically.

(2) The traffic dispatcher team error modes are benefit of discovering and preventing human error, and improving the reliability of subway traffic dispatching system.

(3) Researches on subway traffic dispatcher team error issues in accident management are still at the stage of qualitative analysis, and the improvement of the content and quantitative analysis will be proceed.

ACKNOWLEDGMENT

The authors would like to thank the Science and technology innovation platform project (PXM2014_014205_000044), Teachers' Science and Technology Innovation Project and Graduate Science and Technology Innovation Plan of Capital University of Economics and Business for providing financial support to this research.

REFERENCE

[1] SASOU K, NAGASAKE A, YUKIMACHI T. A study on the operating team activity of a nuclear power plant under abnormal operating conditions [J]. Safety science, 1993(81): 143-156

[2] Y. H. J. Chang, A. Mosleh. Cognitive modeling and dynamic probabilistic simulation of operating crew response to complex system accidents. Part 1: overview of the IDAC model [J]. Reliability Engineering and System Safety. 2007, 92(8): 997-1013

[3] Y. H. J. Chang, A. Mosleh. Cognitive modeling and dynamic probabilistic simulation of operating crew response to complex system accidents. Part 2: IDAC performance influencing factors model [J]. Reliability Engineering and System Safety. 2007, 92(8): 1014-1040

[4] Y. H. J. Chang, A. Mosleh. Cognitive modeling and dynamic probabilistic simulation of operating crew response to complex system accidents: Part 3: IDAC operator response model [J]. Reliability Engineering and System Safety, 2007, 92(8): 1041-1060

[5] Y. H. J. Chang, A. Mosleh. Cognitive modeling and dynamic probabilistic simulation of operating crew response to complex system accidents: Part 4: IDAC causal model of operator problem-solving response [J]. Reliability Engineering and System Safety. 2007, 92(8): 1061-1075

[6] Y. H. J. Chang, A. Mosleh. Cognitive modeling and dynamic probabilistic simulation of operating crew response to complex system accidents: Part 5: Dynamic probabilistic simulation of the IDAC model [J].Reliability Engineering and System Safety. 2007, 92(8): 1076-1101

[7] Ke-bing LIAO, Ai-qun LIU, Jie-juan TONG, Jun XIAO. Team error model in complex human-machine system and its quantitative analysis [J]. China Safety Science Journal.2007, 17(12): 42-48(in Chinese)

[8] Ren-li LV, Yan ZHOU, Mu ZHOU. Analysis of team errors in air traffic control [J]. China Safety Science Journal. 2009, 19(01): 64-70 (in Chinese)

[9] Dai-li CAO. Shu-dong HUANG. Li ZHANG. Si-rui GAO. Analysis on man-made error by work team in complicated industrial system [J]. China Safety Science Journal. 2004, 14(9): 20-23 (in Chinese)

[10] Jie WANG, Research on traffic dispatching system human error prediction for the subway emergency [D]. Beijing Jiaotong University [The dissertation]. 2012.12 (in Chinese)

[11] Jie WANG, Wei-ning FANG, Yan ZHANG. Identification and evaluation of human error influential factors in subway scheduling system [J]. China Safety Science Journal.2011, 21(8): 74-79(in Chinese)

[12] Hai-tao WU. He ZHUANG. Xia LUO. A human error identification method based on cognitive process analysis - application to dispatching system of high-speed railway train [J]. Journal of Safety Science and Technology.2014, 10(2): 99-105(in Chinese)

[13] Xin-ping WEI, Zhen-guo LEI. The reliability analysis of the railway dispatcher system [J]. Railway Operation Technology.2009, 15(4): 51-53(in Chinese)

Modeling and Analysis of Conceptual Design-Oriented Product Technology Network

Jian XIE*, Lin GONG, Zi-jian ZHANG

School of Mechanical Engineering, Beijing Institute of Technology, Beijing, China
(xiejian09ie@163.com)

Abstract - **Innovation in conceptual design period of products is an important indicator to show innovation capacity of one enterprise, and the improvement of relative technology is the base of it. By analyzing technical papers in a period, the relevant keywords can be summarized to make up a relevant technical keywords network, which may be very complex and include a large number of nodes and edges. This keywords network will show the key topics and hotspots in science and technology area currently or in the future. By building the keywords network which base on product conceptual design and analyzing the statistical result, some available techniques can be found to support product conceptual design. In this report an example of coal cutter is provided to discuss the feasibility and usability of this model.**

Keywords - **Conceptual design, complex network, modeling, statistical analysis, technical keywords network**

I. INTRODUCTION

Nowadays human society is facing to a new time knowledge economy. Due to globalization and competitions, the requirements of design and manufacturing are raised to a higher level than before. In order to improve the core competitive capacity of enterprises, the new knowledge of innovation should be used to design or manufacture products as soon as possible to make designs come true. The knowledge comes from many areas such as fluids, electronics and mechanics. How to understand and use it quickly is the key point of fast innovation methodology.

There are some knowledge models for this situation now [1,2]. However, it is difficult for product designers to understand them efficiently due to the complex relationship between different knowledge in the whole product system. At the same time, in conceptual design period the lack of key techniques and related knowledge makes some limitations for designers which make innovations of function or structure come true. Because complex network is the topology basis of a variety of complex systems, complex network theory plays an important role in complex system study. Therefore, complex network theory should be used to check the internal structure of complex system to realize the innovation of technology by methodology of topology and evolution.

In this research, complex network theory will be discussed to get some modeling approaches of technique network which are available for product conceptual design. By analyzing the complex features of them, typical innovative ideas or techniques can be summarized to show the hotspots and direction of study, which can help designers to complete the innovative product design in conceptual design period.

II. COMPLEX NETWORK THEORY

There is no uniform definition of complex network. Tao stated that complex network is a topology abstraction of large numbers of real complex systems. It should be more complex than regular network and random network. Because complex network is the topology basis of a variety of complex systems, the study of it is important and helpful for people to understand the complex nature of a complex system [3]. Jianmei thought that complex network is between the regular network and random network. It has some characteristics like large scale, rich data and emergence due to its complexity and relationships with many different areas [4]. Commonly, a complex network can be defined that a topology abstraction of real complex system, which is not regular or random network, with many nodes and edges.

Due to both the characteristics and evolution of complex systems can provide better analysis and research methods, it is found that many complex systems can be described by a complex network theory in reality, such as technical network [5, 6], biological networks [7, 8], social networks [9, 10]. Products can also be a complex network. For example, the constituent elements of the product can be considered as a node, and the relation of elements can be considered as edges connecting the nodes. At present, the domestic scholars have launched a related research. Mengjiang converted the element relation into the parameter relation of the design of product components. With the relation among quantitative analysis of complex networks, he proposed design navigation strategy based on parameter network [11]. Beibei applied complex network theory into real mechanical products systems to research, analyzing the topological characteristics and evolution of the product family structure. Moreover, she proposed the dimension of the universal product module components and how to build the main structure of the product family [12]. Xiang and Guanrong proposed local world evolving network model. By studying the World Trade Network they found that side priority connection is only valid in local-world network nodes instead of the global network [13]. Yue and others built a self-adapting biparticle graph evolution model to describe the research cooperation network between the research project leader and research assistant. This model considered every research project leader or research assistant to take a role in choice, decision making, competition and the evolution of itself and the entire network causing by it [14].

However, till now, there is not a simple way to be able to generate complex networks fully complied with the statistical characteristics. It is still far from automatically generating the innovation design of product system by varying parameters of complex networks. This article is also only carrying out the preliminary exploration from the perspective of network modeling and statistical indicators analysis.

III. TECHNOLOGY KEYWORDS NETWORK

Nowadays, with development and update of technology, more and more companies realize that simply satisfying the demand of consumers is not enough. Instead, developing products integrating multiple technologies or innovative products with differentiated technologies, in this way to expand the market. How to integrate related technology into product conceptual design and product innovation, which become the most important problem to enhance the competion power of companies and guiding the Industry development direction.

Researchers conduct experiments, analysis and research on the latest scientific or engineering problems in the research field. Then they summarize the conclusions by writing academic literature. To some degree, academic literature can figure out the hotspot in current or future research fields. By analyzing and summarizing academic literature, exploring viable techniques in order to provide reliable guidance for the conceptual design is possible. In an academic document, the key words are the expression of natural language and the core reflection of technical problem researches [15]. Technical words, as carriers of technical knowledge in the academic literature, contribute to a better research on topological structure and evolution of technical knowledge by building the keywords network technology in academic literature, providing innovative design solutions assists the product technology.

A. Model Construction

The core of complex network modeling is how will the relationship between objects in reality is abstracted as node and edge in the network. According to the relationship between literature and technology keywords, a literature-keywords bipartite network was established in order to achieve the technology keywords network model construction.

The relationship between academic literature and keywords can be expressed as $E_{K-P} = \{(K_i, P_j) \mid \theta(K_i, P_j) = KP_{i \times j}\}$. Among them, K is a set of keywords, K_i is a keyword and P is a set of literature. $KP_{i \times j}$ represents the strength of relationship:

$$KP_{i \times j} = \begin{cases} 1 & \textit{if the keyword } K_i \textit{ is in the literature } P_j \\ 0 & \textit{if the keyword } K_i \textit{ isn't in the literature } P_j \end{cases} \quad (1)$$

The relationship matrix of keywords and academic literature is $O = [K;P] = [KP_{i \times j}]_{n \times m}$, n and m are the number of elements in K and P .

The node in network is keyword and the edge is the co-ocurrence relation. The keywords network can be expressed by $G_K = (K, E_{K-K})$, wherein $K = (K_1, K_2, ..., K_n)$ represents the set of keywords and $E_{K-K} = \{(K_i, K_j) \mid \theta(K_i, K_j) = K_{i \times j}\}$ represents the set of edges. If $K_{i \times j}$ represent the extent of co-occurrence, that is, the number of keyword-pair that appear in the literature at the same time. The relationship matrix between the keywords is $O = [K;K] = [K_{i \times j}]_{n \times n}$, so:

$$\begin{aligned} O = [K;K] = [K_{i \times j}]_{n \times n} &= O[K;P] * O[K;P]^T \\ &= [KP_{i \times j}]_{n \times m} * [KP_{i \times j}]_{n \times m}^T \end{aligned} \quad (2)$$

$O[K;P]^T$ is the transposed matrix of $O[K;P]$.

B. Statistical index

Calculation and analysis of some statistical indicators of technical keywords network allowing us to get a comprehensive understanding of the network. We can analyze the network connectivity, average path length and central tendency with in the technical keywords network. And then, we can find the direction and hotspots of product innovation.

1) *Node degree and degree distribution:* In complex network, node degree is the number of nodes adjacent to the node, that is, the number of edges that is directly connected to the node. Complex network degree refers to the average level of all the node degrees in the network. The degree distribution $P(k)$ represents the ratio of the number of k degree nodes to total nodes number of network, which is the macro-characteristics of the network.

2) *Clustering coefficient:* Clustering coefficient measures the extent of network aggregation, referring to the edges connected probability between two nodes which connected by one node. A specific node clustering coefficient expressed as: in complex networks, the ratio of actual existence R edges of k neighbor nodes to most possible number of edges $E_i = k_i * (k_i - 1)/2$ among them. So,

$$C_i = R_i / E_i \quad (3)$$

And the clustering coefficient of network defined as:

$$C = \sum_{i=1}^{n} C_i / N \quad (4)$$

3) *Average path length:* If the network node can reach the node in turn at limited edges, this pathway which connected by several edges from to is called as one network path, and the number of edges to form this path is called the path length. As two single node may have many paths, so the shortest network path between any two nodes referring to the path which has least number of edges among all possible paths. The length was defined as the distance between node i and j , which represented by d_{ij} . The average path length means the average value

of all shortest path between two exist nodes. The formula is as follows:

$$\overline{L} = \frac{2}{N(N-1)}\sum_{i=2}^{N}\sum_{j=1}^{i-1}d_{ij} \qquad (5)$$

In addition, we also uses node centrality, betweeness centrality and closeness centrality to measure which nodes is the key indicators in important position, which nodes rated higher resources control ability and which nodes have strong capability against influence of other nodes. Their statistical formulas can be found in related literature [16].

IV. CASE STUDY

A. Problem Statement

In recent years, with the vigorous development of coal mining industry, shearer's function and quality are continuously improved, and its ability to adapt to the complex conditions is also rising. Faced with an increasingly complex coal mining industry competition, it is paid widespread attention by coal mining companies to design the shearer which has advantages of good braking performance, high rotation efficiency, speed control performance, energy-saving and low noise .Shearer's main functions include drop coal and deliver coal. In this section, we apply the above model to explore the related technologies which support delivering coal function. Through analyzing and studying related technologies, we can support conceptual design of shearer.

We choose "friction drive" as a keyword to establish technical keyword network, then specifically analyze the friction drive technology. Analysis process is as follows:

1). By means of the model, construct technical keyword network and visualize it.

2). Calculate the relevant statistical parameters of the technical words network.

3). Extract novel, effective and potential technical knowledge node, explain in-depth relationship between the network and its development trends, and explore the hot technology in the field of research to provide innovative support for the conceptual design of the product.

B. Data Source

In this paper, we choose WANFANG DATA as the data source of network samples, and select "friction drive" as retrieved keyword to search published journals and dissertations in the technical field of friction drive. The papers are published in the time period 1995-2014. Total number of journals and dissertations is 674, statistics shown below.

As can be seen from the Fig.1, from 1995 to 1999, study of the friction drive technology is still in its infancy. Since the 21st century, study of the friction drive technology has increased every year. Since 2006, number of journals and dissertations in the technical field of friction drive has confronted small fluctuations, but

overall stability of the number indicates friction drive technology is maturing nowadays.

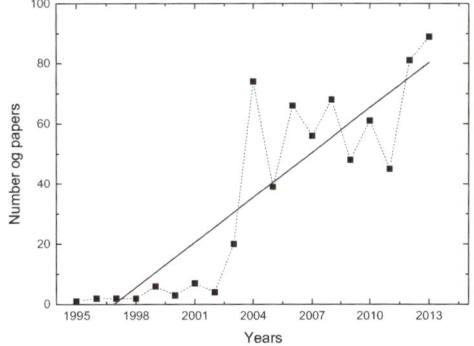

Fig.1. Published time and number of friction drive technology papers

C. Network modeling

Select above keywords of the academic literature as network nodes. After statistical analysis, all 674 technical papers include 3376 Keywords, average, each paper averagely contain 5 technical keywords. Then choose co-occurrence relationship between technical keywords as technical keyword network edges, after statistical analysis, the number of the edges is 6090.

As visualization can vividly demonstrate a variety of network features, we use Gephi to achieve visualization of friction drive technical keyword network. As Fig.2 shown below, the larger points and fonts represent the higher central degree of nodes. If two nodes are not connected, it indicates there is no co-occurrence relationship between them.

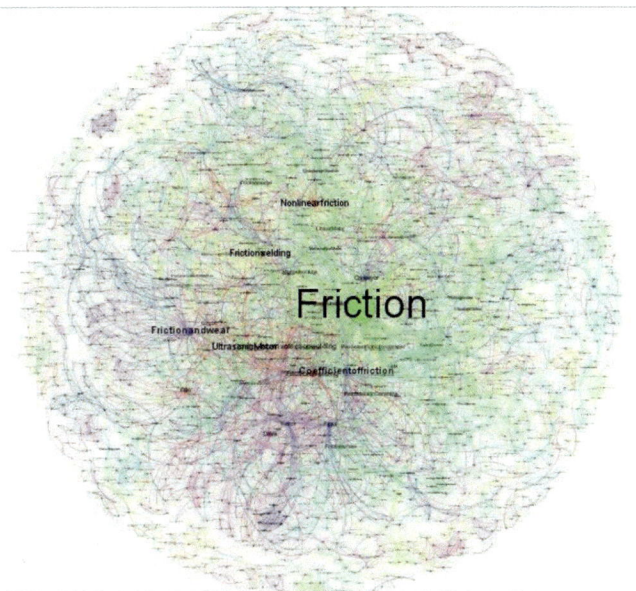

Fig.2. Friction drive technical keyword network

D. Network analysis

1) *Nodes Centrality Degree (NCD):* Node centrality degree refers to the times, rather than the numbers, of one

node occurrence with other nodes in the network. By calculating, it can be obtained that the average node centrality degree of the technological keywords network is 3.5 and the node centrality degree is 12142. Table I ranks the top 15 of node centrality degree that related to friction drive technology.

TABLE I
THE TOP 15 NODES OF NCD THAT RELATED TO FRICTION
DRIVE TECHNOLOGY

Ranking	Nodes	NCD	Relative NCD
1	Ultrasonic motor	32	0.928
2	Linear motor	31	0.899
3	Piezoelectric ceramics	26	0.754
3	Belt conveyor	26	0.754
5	Fuzzy control	20	0.58
6	Disturbance monitor	17	0.493
7	Servo system	12	0.348
7	Composite	12	0.348
7	Vibration control	12	0.348
10	Linear ultrasonic motor	11	0.319
10	Friction damper	11	0.319
12	Multi-Axis Controller	10	0.29
12	Compensation control	10	0.29
12	Self-adaption	10	0.29
15	Micro actuator	9	0.261

2) *Betweenness Centrality (BC):* Betweenness Centrality is used to analyze the node which locates on the line between two nodes. By calculating, it can be obtained that the average betweenness centrality of the technological keywords network is 5996.257 and the total betweenness centrality is 20681092. Also, we found out that the minimum betweenness centrality is 0, which means this kind of node does not have the ability to affect the co-occurrence of other two nodes. Table II ranks the top 15 of Betweenness Centrality that related to friction drive technology.

TABLE II
THE TOP 15 NODES OF BC THAT RELATED TO FRICTION
DRIVE TECHNOLOGY

Ranking	Nodes	BC
1	Piezoelectric ceramics	309973.719
2	Linear motor	235120.906
3	Belt conveyor	207567.531
4	Ultrasonic motor	156720.219
5	Fuzzy control	138307.688
6	Piezoelectric friction dampe	114585.750
7	Semi-active control	78230.414
8	Linear ultrasonic motor	76439.875
9	Disturbance monitor	72170.664
10	Magnetic suspension	71537.898
11	PID control	63730.746
12	Fuzzy control	56034.605
13	Monocrystalline silicon	55330.555
14	Motion control	51963.203
15	Composite	48838.309

3) *Closeness Centrality (CC):* Closeness Centrality represents the frequency of one node occurrences with other nodes. The smaller the value is, the greater the frequency of co-occurrence will be. The average Closeness Centrality of the technological keywords network is 3701505.5. Table III ranks the top 15 of Closeness Centrality that related to friction drive technology.

TABLE III
THE TOP 15 NODES OF BC THAT RELATED TO FRICTION
DRIVE TECHNOLOGY

Ranking	Nodes	CC
1	Piezoelectric ceramics	2029191
2	Fuzzy control	2029529
3	Ultrasonic motor	2029675
4	Linear motor	2029685
5	PID control	2030071
6	Linear ultrasonic motor	2030579
7	Traveling wave type ultrasonic motor	2030657
8	Servo system	2030671
9	Bang-Bang control	203072
10	Magnetic suspension	2030737
11	Electrohydraulic servo motor	2030803
12	Robust control	2030829
13	Minidriver	2030851
14	Belt conveyor	2030859
15	Magnetic force driving pump	2030863

4) *Integral analysis:* By calculating, the clustering coefficient of the network is 0.874, which indicates that the whole network has a very close relationship. The average path length is 6.041. This shows that six extra nodes will be needed in achieving co-occurrence form one node to another in friction drive technology network.

V. CONCLUSION

This paper analyzes the function of keywords network in charactering complex product system topology by summarizing the researches about the application of complex network in product design. Then, with the using of keywords network model of bipartite network building technology and the calculating of statistical Indicators, this paper analyzes the examples of coal mining with relevant technologies. In instance analysis, we found out some rules:

1) At present, hot research topics and trends of improving performance of the friction drive motor are mainly focus on ultrasonic technique, Linear technology and micro-electronic technique.

2) Hot research field and trends about improving performance of Motion control of the friction drive mechanism are mainly center on fuzzy control, motion control, Semi-active Control, PID control, compensation control and robust control.

3) Hot research topics and directions about reducing friction drive vibration performance are mainly concentrate on magnetic levitation technique and Piezoelectric friction damper technology.

4) According to the large clustering coefficient of Friction Drive Technology Network and small average path length, the Friction Drive Technology Network is proved to have small world-effect and shows a high efficiency of co-occurrence and responsiveness.

This study still has some limitations. In the future work, researches can make use of a complex network to go further study on evolution of technology Keywords network, clarify current stage of technology, use time series statistics of technology node to fit, study the rules of Evolution of Technology, Identify the evolution trends of technology, minimize the blindness in product conceptual design, eliminate unnecessary waste of time and resources and shorten the cycle of product design concept.

ACKNOWLEDGMENT

This research is supported by Basic Project of Technology of National Ministry of China (Grant No.2013208C001) and Fund of Basic Research of Beijing Institute of Technology. The authors also express their thanks to the anonymous referees for their valuable time and constructive comments which improve the quality of the paper.

REFERENCES

[1] T. Gao, S. Yang, Z. Xie, and H. Liu, "Research on product knowledge modeling method," *Journal of Computer Applications,* vol. 27, pp. 58-60, 2007.

[2] H. Chen, X. Chen and N. Mao, "Establishing a knowledge model of mechanical and electrical products based on KBE and WEB," *Mechanical & Electrical Engineering Technology,* vol. 40, pp. 29-32, 2011.

[3] T. Zhou, W. Bai, B. Wang, Z. Liu, and G. Yan, "Overview of complex network research," *Physics,* pp. 31-36, 2005-01-12 2005.

[4] J. Yang, "Comparison of research paradigms between complex network and social network," *Systems Engineering-Theory & Practice,* pp. 2046-2055, 2010-11-15 2010.

[5] S. Yook, H. Jeong and A. Barabasi, "Modeling the Internet's large-scale topology," *Proceedings of the National Academy of Sciences of the United States of America,* vol. 99, pp. 13382 - 13386, 2002.

[6] A. Barrat, M. Barthelemy, R. Pastor-Satorras, and A. Vespignani, "The architecture of complex weighted networks," *Proceedings of the National Academy of Sciences of the United States of America,* vol. 101, p. 3747, 2004.

[7] Z. N. Oltvai and A. Barabási, "Life's Complexity Pyramid," *Science,* vol. 298, pp. 763 - 764, 2002.

[8] R. Sharan and T. Ideker, "Modeling cellular machinery through biological network comparison," *Nat Biotechnol,* vol. 24, pp. 427-33, 2006-04-01 2006.

[9] A. L. Barabási, H. Jeong, Z. Néda, E. Ravasz, A. Schubert, and T. Vicsek, "Evolution of the social network of scientific collaborations," *Physica A: Statistical Mechanics and its Applications,* vol. 311, pp. 590 - 614, 2002.

[10] L. Adamic and E. Adar, "How to search a social network," *Social Networks,* vol. 27, pp. 187 - 203, 2005.

[11] M. Chen, "Parameter Network Modeling Technology and Its Application Research for Product Design," Master dissertation: Zhe Jiang University, 2010, p. 82.

[12] B. FAN, "Research on Key Technologies of Modular Product Platform Based on Network Analysis Method," PH.D: Zhe Jiang University, Hangzhou, China, 2011, p. 143.

[13] X. Li and C. Guanrong, "A local-world evolving network model," *Physica A: Statistical Mechanics and its Applications,* vol. 328, pp. 274 - 286, 2003.

[14] Y. He, P. Zhang, T. Xu, Y. Jiang and D. He, "A self-adaptive bi-particle graph model for scientific collaboration," *Acta Physica Sinica,* vol. 53, pp. 1710-1715, 2004.

[15] H. Jin, L. Zhou and Y. Wang, "Research on university library's information service for enterprise in China based on bibliometric analysis," *Sci-Tech Information Development & Economy,* pp. 105-108, 2014-02-10 2014.

[16] D. Zhu, D. Wang, H. Saeed-Ul, and H. Peter, "Small-world phenomenon of keywords network based on complex network," *Library and Information,* pp. 19-22+72, 2013-12-25 2013.

An Advanced Multi-objective Genetic Algorithm for Mixed-model Two-sided Assembly Line Balancing

Zheng-mei Jie[*], Kai-hu Hou, Yu-jie Zheng, Lei Wang, Cheng Chen, Ying-feng Zhang

Faculty of Mechanical and Electrical Engineering, Kunming University of Science and Technology, Kunming 650500, China

(jiezm.ok@163.com)

Abstract - **In this study, an advanced multi-objective genetic algorithm (AMGA) is proposed for solving type I mixed model two-sided assembly line balancing (MTALB-I) problem. Multi objective mathematical model is established and the AMGA is used to solve. The three minimal objectives are the number of stations, the balance index and the wait time. A real number combination based on task and operational direction, a dynamic mutation rate, the combination of the parallel selection method and arrangement selection are used in AMGA. The performance of AMGA is compared to the existing approaches. The results show that AMGA performs well. The algorithm improves the search efficiency of satisfactory solution.**

Keywords - **Genetic algorithm, multi-objective programming, two-sided assembly line balancing problem**

I. INTRODUCTION

Since Salveson first put forward the simple assembly line balancing problem in twentieth Century 50[1], the assembly line is derived from a variety of different types, including the two-sided assembly line and U-shaped assembly line. Compared with the single traditional assembly line balancing problem, the two-sided assembly line also belongs to the NP hard combinatorial optimization problems [2]. Moreover, mixed-model production let the TALB problem is more complicated. So the traditional assembly line balancing method is no longer applicable. We need new suitable methods for solving TALB problem. In 1993, Bartholdi put forward a kind of First Fit Riile (FFR) algorithm based on heuristic rules [3]. Since then, the researchers put forward lots of assembly line balancing problem algorithm. All the algorithms can be divided into two aspects [4], namely the exact solution method and the heuristic algorithm. Among them, genetic algorithm (Genetic Algorithm) is a stochastic global search method that imitating the selection mechanism of biology in the survival of the fittest. The source of this method is the simulation study of the biological system for computer. Professor Holland of the USA Michigan University first proposed this search mechanism in 1975 [5, 6]. At present, there are many methods based on genetic algorithm to solve Pareto optimal solution set of multi-objective optimization problem. However, from the view of solution performance, there is still room for improvement.

In order to solve the three minimal objectives, the number of stations, the balance index and the wait time, this article researches the TALB-I problem and establishes the multi-objective optimization model. Then an improved multi-objective genetic algorithm (AMGA) is proposed to solve the model.

II. CHARACTERISTICS OF MIXED-MODEL TWO-SIDED ASSEMBLY LINE

Two-sided assembly line (TAL) is an operation model that people can operate task in both sides of the assembly line. This is essentially the extension of traditional one-sided assembly line, through dividing the original one-sided assembly line operation area into left and right two relatively independent operation areas. The workers complete the assembly task parallel and independently in the respective operations area [7]. Every group of relative left and right position of TAL is called "corresponding station group". As shown in Fig.1, each "corresponding station group" is consists of left and right two corresponding station j and (j-1). The two-sided assembly line is composed of a series of "corresponding station group".

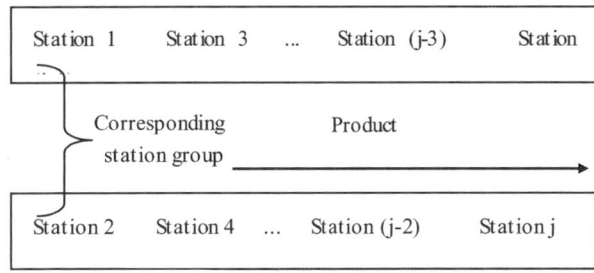

Fig.1. Two-sided assembly line

There are some advantages of two-sided assembly line compared to the one-sided assembly line: 1) It can shorten the length of assembly line and the product time; 2) It can raise the utilization rate of fixture and reduce equipment investment; 3) It can save the invalid labor time and improve the labor productivity of workers.

However, in addition to meet the "task priority relation constraint", " none separable constraint" and other traditional unilateral constraints, TALB problem also need to meet the constraints of its own, as follows: 1) According to the operating range constraint, the corresponding task need to be assigned to the left or right of the corresponding position on the assembly line; 2) Because of the precedence relationships, position task on both sides may cause a "waiting time"; 3) Whether the same side or different side concurrent operation constraint [8] may change average operation time.

III. MATHEMATICAL MODEL

The multi-objective optimization model of MTALB-I are shown in Eq.1 Eq. 2 and Eq. 3. The first optimization objective is to minimize the number of stations. The second optimization goal is to minimize the balance index. The third optimization objective is to minimize the wait time. Three objective functions are below:

$$min\ J_1 = j \tag{1}$$

$$min\ J_2 = \sum^{J_l} \sqrt{\frac{(tj-C)^2}{r}} \tag{2}$$

$$min\ J_3 = \sum_{i \in ws(j)} \sum_{j \in w} D_{ij} \tag{3}$$

The constraints of MTABL-I problem are below:

$$\cup WS(j) = T \tag{4}$$

$$\sum_{i \in WS(j)} t_{ij} + D_{ij} \le C \tag{5}$$

$$t_{ij} = \sum_{i \in ws(j)} \sum_{j \in w} max(t'_{im}) \cdot x_{ij} \tag{6}$$

$$t'_{im} = \frac{t_{ij}}{\substack{max(w_i) \\ i \in WS(j)}} \tag{7}$$

$$x_i mod2 = b \tag{8}$$

$$b = \begin{cases} 0, & k \in TR \\ 1, & K \in TL \end{cases} \tag{9}$$

$$TL = \{T_i(d) = L,\ i = 1,2,....,n\} \tag{10}$$

$$TR = \{T_i(d) = R,\ i = 1,2,....,n\} \tag{11}$$

$$x_k \le (x_i + a),\ \forall k \in P(i),\ \forall i \in T \tag{12}$$

$$a = \begin{cases} 0, & x_i \in 2n \\ 1, & x_i \in (2n+1) \end{cases} \tag{13}$$

In the above formula, C is the given takt time. x_i is the number of station which task i is assigned to. t_j is the operation time on J station. w_i is an integer greater than 0. W is the station set, wherein, $W=\{1,2,...,J\}$. $WS(j)$ is the task set assigned to the j station. ws_j is the Boolean variables. If $ws_j = 0$, we need not to open the station j, otherwise, $ws_j \ne 0$, we need to open the station j. T is task set, which one represents each task i has two attributes: operating time and range. m is the product model. t_{im} is the operation time of i task for M products. $d_i = \{L, R, E\}$, d_i represents the task can be assigned to the left, right or either side of the assembly line. ss_{ij} is the start time of the task i at station j. sp_{ip} is the starting time of the task i in the corresponding working group. D_{ij} is wait time caused by the preorder task of the its corresponding position, when the task i is assigned to station j. $D_{ij} = min(sp_{ip} - ss_{ij}, 0)$. When the task i was assigned to the j station, the value of x_{ij} is 1, otherwise x_{ij} is 0. t'_{im} is the average operation time, when i task of M product is assigned to the j station.

The constraints represent the following meaning: Eq.4 ensures that all tasks must be assigned to workstations. Eq.5 ensures that the sum of workstation task set total operating time and delay time are less than or equal to the takt time. Constraint Eq.6 represents the total operation time of all tasks assigned to station j. Eq.7, t'_{im} is the average operation time, when i task of m product is assigned to the j station. Eq.8 to Eq.11 ensures that task i must meet the requirements of assembly line operation range. Eq.12 and Eq.13 ensure that tasks meet the precedence constrains.

IV. ALGORITHM DESIGN

A. Initial population generation

To meet the priority relation constraint, using random method to generate the initial population, the steps are below:

Step1. In the operation element set I, find the operation elements that have no advance operation element or advance operation element have already finished work distribution. Then denote all these elements as A collection.

Step2. In the A collection, choose an operation elements randomly, followed by operation range.

Step3. Repeat steps1 and step2 until all the operation elements in the set A are distributed out.

B. Code

In the study of two-sided assembly line balancing problem, we have to consider not only which tasks assigned to the station, but also to consider the distribution range and the task of assembly sequence in the station. Therefore, a real number combination based on task and operational direction is used in this paper. In the encoding, the chromosome length is equal to two times the size of the problem. Odd chromosome gene value represents the assembly sequence of tasks and even bits represent task assigned range. As an example, the priority is shown in Fig.2. To satisfy the precedence constraints, a possible chromosome coding scheme is shown in Fig.3. The assembly orientation is d_i. $d_i = \{L, R, E\}$. E tasks randomly generate numbers 0 and 1. "0" represents the task on the right side of the assembly line. "1" represents the task on the left side of the assembly line.

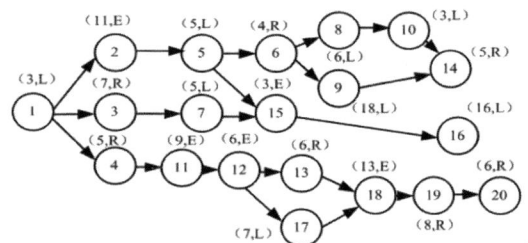

Fig.2. elemental task priority relation graph

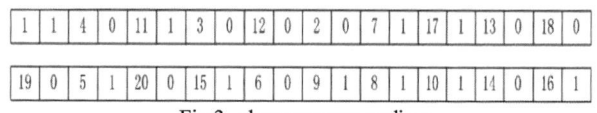

Fig.3. chromosome coding

C. Fitness function

The fitness function is to measure the quality of chromosome. The value of three functions is small, t the chromosome is better. Three fitness functions are below:

$$FitnF1 = \frac{1}{J_l} = \frac{1}{(k_0 + k_1)} \tag{14}$$

$$FitnF2 = 1/J_2 = 1 \left/ \left(\sqrt{\frac{\sum_{i=1}^{k_0}(C-T_0(i))^2}{k_0+k_1}} + \sqrt{\frac{\sum_{j=1}^{k_1}(C-T_1(j))^2}{k_0+k_1}} \right) \right. \quad (15)$$

$$FitnF3 = 1/J_3 = 1 \left/ \sum_{m=1}^{\min(k_0,k_1)}(T_0(m) - T_1(m))^2 \right. \quad (16)$$

D. Selection operator

Using classical Stochastic Tournament, each in accordance with the roulette wheel selection method for selecting a pair of individuals, and then let the two individual competitions, high fitness value is selected. This has been circulating, until the full number of selected population.

E. Crossover operator

This paper uses the most widely used chromosome two-point crossover method. First of all, setting a crossover probability, crossover probability is a numerical from 0 to 1. If this value is greater, then the frequency of crossover operation is greater. The crossover probability is P_c. We need according to the actual size of problems to set the reasonable probability of crossover. We often set the probability of crossover as a numeric value in the range of 0.8~1.0. The specific steps of the crossover operation is below:

(1) Read only odd bit sequence those represent sequence of operations. So the chromosomal has n gene.

(2) The odd gene is divided by k_1 and k_2 intersection randomly. So the paternal chromosome F_1 and F_2 that have n genes is divided into three parts as $[1, k_1]$, $[k_1+1, k_2]$, $[k_2+1, n]$.

(3) Find the same gene $[1, k_1]$ of F_2 in the F_1 and make up a new gene sequence according sorting way of F_1. Then replace the $[1, k_1]$ of F_2. Find the same gene $[k_1+1, k_2]$ of F_2 in the F_1 and make up a new gene sequence according sorting way of $F1$. Then replace the $[k_1+1, k_2]$ of F_2. Find the same gene $[k_1+1, n]$ of F_2 in the $F1$ and make up a new gene sequence according sorting way of F_1. Then replace the $[k_2+1, n]$ of F_2; So, it will be generate a new chromosome.

(4) It will be generate another new chromosome when we exchange the position of F_1 and F_2 in the step (3). We sign the two new chromosomes are C_1, C_2.

(5) End the crossover operation until selecting two chromosomes of maximum adaptive value though calculating the adaptive function of parent and offspring chromosome.

So, it's feasible that offspring chromosome though the cross transform. These offspring chromosomes are not only enough for precedence constraints, but also enough for operating range constraints.

F. Dynamic mutation operator

The Pareto optimal individuals of sub groups do not participate in individual mutation operation. The concrete steps for the other individual crossover operation is below:

(1) Reading the odd gene sequence that represents task sequence from chromosomes which have 2n gene.

(2) Choose the chromosome that will lead to gene mutation according to the chromosome mutation probability P_m.

(3) K is the position of gene mutation which is random integer of 1 to n.

(4) The gene before position K of chromosome remain unchanged, the gene after position K will be recombine according to the method of rebuilding the chromosome.

This paper uses the high dynamic mutation rate to get sufficiently large mutation and avoid the local optimal solution in the initial stage of the evolution [9]. With the increase of the evolution algebra, gene on chromosome become more and more better, which can reduce the population mutation rate and avoid disorderly evolutionary population. So, the mutation probability formula of chromosome generation i as follows:

$$p_{(m,i)} = \frac{Maxgen}{Maxgen+i} * p_m \quad (17)$$

Maxgen — Max generation of Genetic algorithm is set in the beginning

G. The termination condition

Let T be the largest iteration, t is the current algebra evolution, to determine whether the termination. The steps are below:

(1) If the t<T, then, with the new population as the initial population, to continue the selection, crossover and mutation operation.

(2) If t>T, output the optimal solution which is maximum fitness individuals in the iterative process. Then, the iteration is terminated.

H. Multi-objective optimization problem solving by AMGA

This paper combines the method of parallel selection and arrangement selection to solve the multi-objective optimization problem. The steps are below:

(1) Process of parallel selection. First, the whole population is divided into 3 parts of equal. And then calculate each sub objective functions of these 3 parts.

(2) Process of alignment selection. Firstly, the fitness value of individuals On 3 groups should be sorted respectively from big to small. Secondly, according to the ranking results, select M individual that are arranged in front of each population. Let 3 populations of M individuals with merger do the crossover and mutation. At last, the results of second step will be divided into three equal populations randomly.

Repeat the above steps and finally get the Pareto solutions of multi-objective optimization problems.

I. Decoding

In order to get solutions of problem, we need to decode the chromosome. According to the decoding process, first of all, judge operation range of i element in the two-sided assembly line. And judge whether it contains concurrent operation elements in the workstation sets.

(1). There are no concurrent operations, we calculate the work time and delay time, then to compare with the takt time. If the result is less than or equal to the takt time, the elements I can be assigned to this workstation. On the contrary, open the next workstation, the task will be assigned to the next workstation.

(2). If there are same side concurrent operations, the actual workstation time is the result of current workstation working time divided by the number of workers working simultaneously. Then follow the step (1).

(3). If there are different side concurrent operation element, so the work elements are assigned to the other corresponding workstation. And the operation element in the corresponding station group must begin at the same time. Then follow the step (1).

Repeating the decoding process can be sequentially decodes the chromosome.

V. ALGORITHM VALIDATIONS

(1) Verify the first objective function: open the least station number. This paper analyzes and compares AMGA with the first Bartholdi's FFR algorithm and Qin Xing min's TRPW algorithm [10]. By solving the P_{24} problem currently published, verify the validity of AMGA for solving the objective function of minimum open position number. P_{24} problem is from reference [11]. The task precedence diagram of P_{24} is shown in Fig.4.

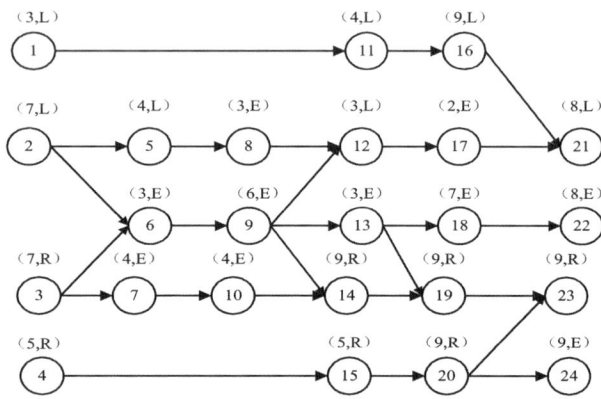

Fig.4. The task precedence diagram of P24

TABLE I
THE RESULTS OF THREE KINDS OF ALGORITHMS

Problem	takt time C [12]	Satisfactory solution of FFR [12]	Satisfactory solution of TRPW [12]	Satisfactory solution of AMGA
P_{24}	30	6	6	5
	35	5	6	5
	40	4	4	4

The population size (number), iterative algebraic (Maxgen), the crossover (PC) probability and mutation probability (PM) required for the verification algorithm were determined by orthogonal experiment design. Result of the population number is 20. Maxgen algebra is 200 generations. Crossover probability is 0.8 and mutation probability is 0.2.

The calculation results by using three different rhythms are shown in Table I.

So, as the Table I shown, compared with FFR algorithm and TRPW algorithm, the performance of the AMGA algorithm is better. By setting appropriate parameters, we could get a satisfactory solution compared with others. This algorithm can provide better design scheme.

(2) Verify the second objective function: minimal balance index. This paper solves the problems of the currently published P_{19}, P_{19} problem from reference [13]. By comparing the standard PSO algorithm and PSO-SA algorithm with the AMGA in this paper, verify AMGA method for solving the objective function index is effective. The task precedence diagram of P_{19} is shown in Fig.5.

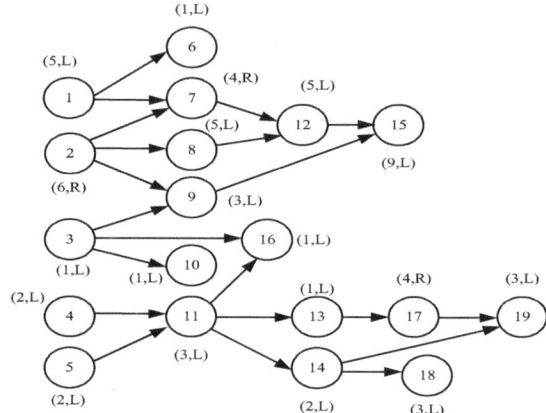

Fig.5. Precedence graph of P19

The calculation results by using three 3 rhythms are shown in Table II.

TABLE II
VERIFICATION OF ALGORITHM

Workstation number	1	2	3
AMGA task set	{1, 3, 4, 5, 11, 13, 14, 16}	{2, 7, 17}	{6, 8, 9, 10, 12, 15, 18, 19}
AMGA workstation time	1.60	1.53	1.86
takt time		2	
AMGA balance index		0.3653	
standard PSO balance index [14]		53.336	
PSO-SA standard balance index [14]		1.5	

From the Table II, compared with the balance index of standard PSO algorithm and PSO-SA, AMGA balance index get big improvement. The balance index improved 52, 9087 than standard PSO algorithm, and improved 1.0272 than PSO-SA. So AMAG algorithm in balance index is verified the validity.

(3) Verify the third target function: minimal wait time. Same as above, this paper solves the problems of the currently published P_{20}, P_{20} problem is from reference [15]. We can get that AMGA is Effective

algorithm to solve the third objective function. The task precedence diagram of P20 is shown in Fig.6.

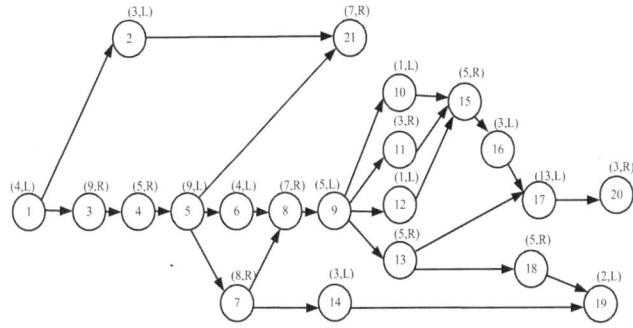

Fig.6. Precedence graph of P20

TABLE III
THE COMPARATIVE RESULT

Problem	The number of work elements	takt time C (s)	Result of PSO (s)[12]	Result of AMGA (s)	Improvement (s)
P_{20}	21	15	89	85.273	3.727

The results are solved by the AMGA and PSO algorithm is shown in Table III. The Table III shows that the calculation results of waiting time of AMGA improve 3.727s than PSO, and verify the effectiveness of AMGA.

VI. CONCLUSIONS

This paper analysis the characteristics and complexity of type I mixed model two-sided assembly line balancing problem. And multi-objective mathematical model is established to minimize the number of stations, the balance index and the wait time. An advanced multi-objective genetic algorithm (AMGA) is proposed to solve the problem. It improves the satisfaction solution search efficiency. With Compared to other published mixed two-sided assembly line methods, verify the validity of the AMGA, and the method has innovation and theoretical value, which can provide reference for solving the assembly line balancing problem.

REFERENCE

[1]   Salveson M E. The assembly line balancing problem [J]. Journal of Industrial Engineering, 2012, 6(16): 18-25

[2]   Ghosh S, Gagnon R J. A comprehensive literature review and analysis of the design, balancing and scheduling of assembly systems [J]. International Journal of Production search, 1989, 27(4): 630-670

[3]   Bartholdi, J. Balancing two-sided assembly lines: A case study [J]. International Journal of Production Research, 1993, 31(10): 2447-2461

[4]   Ghosh S, Gagnon R J. A comprehensive literature review and analysis of the design, balancing and scheduling of assembly systems [J]. International Journal of Production Research, 1989, 27(4): 630-670

[5]   Xuan guangnan, Cheng runwei. Genetic algorithm optimization and engineering [M]. Beijing: Tsinghua University press, 2004: 89-226

[6]   Li Minqiang, Kou Jisong, Lin Dan. Theory and application of genetic algorithm [M]. Beijing: Science Press, 2002:24-98

[7]   Kim Y K, Kim Y, Kim Y J. Two-sided assembly line balancing: a genetic algorithm approach [J]. Production Planning & Control, 2000, 11(1):44-53

[8]   Hu bin. Research on large-scale complex assembly line balancing problem [D]. Shanghai: Shanghai Jiao Tong University, 2006

[9]   Leung Y, Gao Y, Xu ZB. Degree of population diversity-a perspective on premature convergence in genetic algorithms and its Markov-chain analysis [J]. IEEE Trans Neural Network, 1997, 8(5), 1165-1176

[10]   Qin X M, Jin Y. A heuristic method for two-sided assembly line balancing problem [J] Journal of Shanghai Jiaotong University (Science), 2005, 10(1), 61-65.

[11]   Lee TO, Kim Y, Kim YK. Two-sided assembly line balancing to maximize work relatedness and slackness [J]. Computers & Industrial Engineering, 2008, 40(3): 273-292

[12]   Wu EF, Jin H, Xu AM, Hu XF. based on Two-sided assembly line balancing of modified genetic algorithm [J]. Computer integrated manufacturing system, 2007, 12(2): 268-274

[13]   Thomopoulos N T. Mixed Model Assembly Line Balancing with smoothed station assignments [J]. Management Science, 1970, 16(9): 593-603.

[14]   Yu Q L, Research on the problem of stochastic mixed model assembly line balancing based on hybrid particle swarm algorithm [D]. Chengdu, Xi'an Jiaotong University. 2012

[15]   Ponnambalam S G, Aravindan P, Naidu G, Mogileeswar G. Multi-objective genetic algorithm for solving two-sided assembly line balancing problem [J]. International Journal of Advanced Manufacturing Technology, 2000, 16(5): 341-352

Study on the Application of the Operator CWGA for Selecting a Supplier in Supply Chain Management

Shen Hang[1,*], Liu An-Min[2]

[1]The School of Economy and Management, Hunan Institute of Technology, Hunan Hengyang, 421001, China
[2]The School of Mechanical Engineering, Hunan Institute of Technology, Hunan Hengyang, 421001, China
(shenhang0115@hotmail.com)

Abstract - **The quality for the selection of a supplier is related to the producing continuity and sector's performance of an enterprise. This paper uses the operator CWGA for selecting the supplier in the supply chain management. This method emphasizes the data itself and the location of data in the all. It can make full use of the assessment information and effectively make the gap among various suppliers to choose the best suppliers. This method is characterized by strong practicability & maneuverability in the supplier selection process.**

Keywords - **Group decision, operator CWGA, supply chain management, supplier**

I. PREFACE

By the speeding up of the globalization of economy and the deepening up of the marketing process, the core of the market competition has evolved to the competition in supply chains from the one in enterprises [1-3]. Under the circumstances, the higher profit would be given to the one who is able to take the best means for the costing control and has more customers. As a result, it makes many enterprises to emphasize the supply chain management and tries to enhance their core competitive power. The key issue of supply chain management now is the one of supplier selection that is the most important. The supplier is referred to those who directly supply raw materials, equipment, and other resources and services to the enterprise. They can be an enterprise and its branch unit, a privately owned company including a manufacturer, a seller and the other mediate agent. To guarantee normal production and continuous scientific research, it is necessary for an enterprise to have a list of reliable suppliers, whose role of importance is obvious [2-4].

It is a very important and complicated process to select a supplier in the supply chain management. It relates to the success or failure of an enterprise whether or an agile, competitive and tolerant supplier may be found out. After the decision making information is given by experts, the most thing is to concentrate the information. The basic characteristic of the operator WGA (Weigh of Geometric Average) is that first the weigh on the importance of data is made, and then followed by the concentration. The basic characteristic of the operator OWGA (Order

Weigh of Geometric Average) is that first the given data are reordered from large to small, weighed on the data position, and then followed by the concentration. The former focuses on the data itself, and the latter focuses on the data position. Both methods are not complete. However the operator CWGA (Combined Weight of Geometric Average) takes the advantage of both operators to make more reasonable tackling of the information. In this paper, the operator CWGA is applied in the supplier selection of the supply chain management. The practicability and feasibility are analyzed for this method based on their given evaluation [5].

II. BASIC KNOWLEDGE

The following related basic knowledge are first introduced before the supplier selection method is given:

Definition 1 [6] WGA:
$$WGA_w(a_1,\cdots,a_n) = \sum_{j=1}^{n} a_j^{w_j},$$
here $w = (w_1, w_2, \cdots w_n)$ is exponential weight vector, $w_j \in [0,1]$, $\sum_{j=1}^{n} w_j = 1$, the function WGA is called as the operator of weigh of geometric average.

Definition 2 [6] OWGA:
$$OWGA_w(a_1,\cdots,a_n) = \sum_{j=1}^{n} b_j^{w_j},$$
here: $w = (w_1, w_2, \cdots w_n)$ is exponential weigh vector related to OWGA, $w_j \in [0,1]$, $\sum_{j=1}^{n} w_j = 1$, and b_j is the jth maximal element in set of $a_1,\cdots,a_n$, the function OWGA is called as the operator of order weight of geometric average.

Definition 3 [6] CWGA:
$$CWGA_{w,l}(a_1,\cdots,a_n) = \sum_{j=1}^{n} b_j^{l_j},$$
here: $l = (l_1, l_2, \cdots, l_n)$ is exponential weight vector related to OWGA, and b_j is the jth maximal element in set of $a_i^{n w_i}$ $(i = 1,2,\cdots,n)$, here: $w = (w_1, w_2, \cdots w_n)$ is exponential weight vector in set of $a_1,\cdots,a_n$,

$w_j \in [0,1]$, $\sum_{j=1}^{n} w_j = 1$, n is the balance factor, and the function CWGA is called as the operator of combination weight of geometric average in n dimension space.

III. SUPPLIER SELECTION METHOD IN THE SUPPLY CHAIN MANAGEMENT

Suppliers play an important role in the supply chain management. The enterprises' running environment is diversified due to the differences in running orientation, culture, and atmosphere. Naturally the selection of a supplier must take the internal and external situations of the enterprise into account. Although the methods taken may differ among enterprises, the basic steps are that first the supplier selection evaluation index system is established, and then weight of the index is determined and evaluation result is summarized, and finally suppliers are ordered for the selection. The analysis for supplier selection is given as follows:

G.W. Dickso [1], the pioneer in supplier selection research, gave 23 evaluation criteria in supplier performance by analyzing questionnaires on the purchasing agents and managers. Based on the continuous studies of researchers and practical experiences in this field, thus the supplier selection evaluation indexes in the supply chain management can generally be classified as follows [1-4]:

(u1) Good Quality Factor: The good quality has an influence on marketing rates and competitive advantages. It is a necessary condition to guarantee the normal production and running whether or not the qualities of purchased goods meet the requirements. So, good quality factor is a precondition in supplier selection.

(u2) Cost Factor: It can effectively reduce the running costs to select a supplier who provides a low selling price. You can acquire more room for the profits and good prices. Therefore the cost factor is one of the main factors. However it doesn't mean that the supplier is the best with a low selling price. You should have a consideration on costs, qualities, and so on.

(u3) Delivery Timing Factor: Delivery in time from suppliers can ensure the continuity of production and running, and also influence the stock lever of a company. So the ability of delivery in time for a supplier is an important factor in supplier selection as well.

(u4) Service Level Factor: The selection of a supplier means not only the selection of its goods but also its services. Therefore it should be taken into account that technical support, agile order services, operation training, maintenances and upgrading.

(u5) Other Factors: the other factors are those such as producing capacity, risk taking capacity, financial situations, special processing ability, and project management ability.

Determination of Evaluation Index Weight:

There are some research achievements in the method of determination of evaluation index weight. The widely used method is presented by Satty T.L., i.e., by comparing the importance of the evaluation indexes with each other the judgment matrix is obtained and the importance of each index is found out, in other words, Analytic Hierarchy Process (AHP) [5]. By the development of AHP and the presentation of fuzzy mathematics, people have gradually realized that a person's consciousness has certain uncertainty and fuzzy nature in comparing the index importance. Thus the Fuzzy Hierarchy Analytic Process is presented by combining the fuzzy nature and hierarchy analytic process, i.e., a method of applying belonging lever to describe the degree of importance between two indexes—FHAP. FHAP method means that the fuzzy inter-compensation judgment matrix $R = (r_{ij})_{n \times n}$ is obtained by the selection of scaling 0.1-0.9 to describe the importance lever. Further this matrix leads to evaluation index weight, namely, $w_i = \frac{1}{n}\left(2\sum_{j=1}^{n} r_{ij} - n + 1\right)$. From this formula the supplier evaluation index weight vector in the supply chain management can be given as $w = (w_1, w_2, \cdots, w_n)$.

IV. SUPPLIER SELECTION METHOD IN THE SUPPLY CHAIN MANAGEMENT BASED ON THE OPERATOR CWGA

With the study of supply chain management, the technologies and methods about the supplier selection have acquired certain achievements. There are mainly three methods, i.e., quantitative analysis, qualitative analysis, and the combination of both. But these methods are all based on individual evaluation. With the development of science and the complexity of the production process itself, it is easy to lead to wrong judgment due to individual lack of knowledge or bias for a person's decision-making. Therefore a group of experts from related fields in different majors and subjects should be employed to make the choice of suppliers in the process. This way will effectively avoid the mistaken evaluation results due to the lack of experiences or documents. In the supplier evaluation process with many participators, former models only take the evaluation information of experts into account by a simple weight. This method is not reasonable. When synthesizing the evaluation results of experts, we not only need to consider the degree of effectiveness of the information given by experts, but also should consider the order of decisive information given by each expert. Meanwhile, we should try to make the gaps among the assessing values of the suppliers as big as possible in objectively and reasonably selecting evaluation method. By doing this, the excellent suppliers may appear on. In the process of concentrating decision information given by the group of experts, we use the operator Combined Weight Geometric Average. The procedures of selecting suppliers based on the operator CWGA are given in details as follows [7-9]:

Step1 Index Weight Determination: Comparing the important lever among (u1) Good Quality Factor, (u2) Cost Factor, (u3) Delivery Timing Factor, (u4) Service Level Factor, and (u5) Other Factors, and constructing fuzzy compensation judgment matrix, i.e., $R = (r_{ij})_{n \times n}$, and obtaining these five evaluation indexes' weight vector, i.e., $w = (w_1, w_2, \cdots, w_5)$.

Step2 Acquisition of Evaluation Information: employing m experts to form a group (the weight of experts is $e = (e_1, e_2, \cdots, e_m)$), and to evaluate the n candidate suppliers and obtain m evaluation matrix, i.e., $P^k = (p_{ij}^k)_{n \times 5}, k = 1, 2, \cdots, m$.

Step3 Individual Expert Evaluation Information Concentration: for the kth evaluation matrix $P^k = (p_{ij}^k)_{n \times 5}$, using operator WGA concentration, and obtaining the k experts' synthetic evaluation vector for every suppliers as $h^k = (h_1^k, h_2^k, \cdots, h_n^k)$, here, $h_i^k = \sum_{j=1}^{5} (p_{ij}^k)^{w_j}$.

Step4 Synthesizing All Evaluation Information: Selecting exponential weight vector $\lambda = (\lambda_1, \cdots, \lambda_m)$, and selecting m experts to form vector for the evaluation of the ith supplier as $h_i^1, h_i^2, \cdots, h_i^m$, letting $g_i^j = (h_i^j)^{me_j}$, for γ_i^j ($j = 1, 2, \cdots, m$) we order decreasingly and get vector $\rho_i^1, \rho_i^2, \rho_i^3, \cdots, \rho_i^m$. Therefore, the ith supplier's synthetic evaluation value of the operator CWGA is $z_i = \prod_{j=1}^{m} (\rho_i^j)^{\lambda_j}$, $i = 1, 2, \cdots, n$.

Step5 Optimal Selection: According to the order in magnitude of the synthetic evaluation values $z_1, z_2, \cdots z_m$ based on the operator CWGA for the n suppliers, we can finally select the supplier whose synthetic evaluation value is the maximum by the optimization rule.

V. CONCLUSIONS

In the process of selecting a supplier, all kinds of factors should be taken into account such as quality, costs, delivery, and services including (QCDS rules). According to the company's needs, we also should make an investigation into registered capitals, equipment, staff quality, production capability, and know-how of advanced technology. The best way for these is to go to with the supplier on site and pick out the suitable supplier and refuse those undesirable. In the last, experts from different fields are employed and the suitable evaluation method is used to find out the one you need of those candidates. Meanwhile, the selection of a supplier is a very complicated process. In this paper, the combined weight geometric average operator CWGA, which takes both the data itself and its position in order into account, is applied to the selection of supplier. It makes the full use of the information of experts, and can also pull the gaps of the candidate suppliers with each other, and make the excellent ones out, and provide the decisive opinions for the final decision maker.

ACKNOWLEDGMENT

This research is partially supported by Heng Yang Science and Technology Plan (2013KJ47).

REFERENCES

[1] Xu Jie, Tian Yuan, Editted, Purchase and stock [M]. Qin Hua University Publisher, 2004.9

[2] Ma Shi Hua, Lin Yong, Editted , Supply Chain Management [M], High Eduation Publisher, 2006.8.

[3] Huang Fu Hua, Deng Sheng Qian, Editted, Modern Enterprise Material Circulation Management [M], Science Publisher, 2010.2

[4] Tanlada, J., Paosuosks, J., et al., Material Curculation Management - Supply Chain Process uniformation [M], Bei Jing Mechanical Industry Publisher, 1999.

[5] Ying Chuan Wu, Fuzzy Multi-Rules Group Decision Making Mehtod Study [D], Xi An Science and Technology University, 2007.

[6] Xu Zhe Shui, Da Li Qing, A Combination Weight Geometric Average Operator and Application [J], SouthEast University Journal (Nature Science Eition), 2002, 32(3).

[7] Yao Yua, Study on e-commerce evaluation system based on AHP [J]. scientific progress and strategies, 2009,26(10):129-133.

[8] Fu Li Fang, Feng Yu Qan, and Wu Qiu Fen , DEC-based e-commerce system evaluation method and quality diagnosing method [J], Harbin Normal University Natural Science Journal, 2007(3).

[9] Chen wen lin, E-commerce evaluation and e-commerce achievement industry comparison [J], Shen Zhen University Journal (Human Social Science Version), 2009(2).

Modeling and Machining End Mill via Object ARX

Qi-rong LUO[*], Xing-bo WANG

Department of Mechanical Engineering, Foshan University, Foshan City, Guangdong province, China 528000
(1183992513@qq.com)

Abstract - **The article presents progress of 3-dimensional modeling and 5-axle machining an end mill based on the mathematical model of the end mills via Object ARX programming; the C++ programming scheme and the key technical issues are introduced in detail which can be a reference for modeling and simulating other kinds of mills.**

Keywords - **5-Axle machining, end mill, modeling, Object ARX**

I. INTRODUCTION

It is known that cutters determine the quality and efficiency of machining and they are the fundamental tools in machining, especially in numerical control machining [1]. With the development of technology in numerical control machining, parts that are machined are more and more complicated and the demands for high-performance machining-cutters have been continuously increased. In the past, cutters of high precision and high performance were mainly made from the abroad and it bared the development of the nation's industries. Nowadays, a lot of researches and developments on the mills have been made and great achievements are made. For example, paper [2] put forward three CAD methods forming cutter profile with processing spiral surface which are Boolean operation, slice method and radius method and two methods of data extraction which are extraction-point and extraction-curve. Papers [3]-[11] introduce mathematical model of end mill. In terms of method of modeling, papers [5]-[7] use mixed programming method of Matlab and VC++; Paper [7] and [9] use VBA; Paper [8] uses VB to carry on the second-development to CATIA; Paper [10] and [11] use VC++ to carry on the second development to UG.

As it knows that, AutoCAD is a widely used designing tool in industries. It can be very good in cooperating with the other manufacturing software such as Master CAM, UG and so on. Hence it is a valuable task to develop software in AutoCAD and make it cooperate with the machining center. AutoCAD enables its users to develop their own software with a development kit called ARX, which is the short name of ObjectARX that is provided by AutoDesk. Actually, there are many samples of software developed with ARX, as reference [12-15]. In terms of such a thought, we develop such software to model and machine the end mill in the AutoCAD. This paper introduces the technical details.

II. MATHEMATICAL MODEL

A. Spiral mathematical model

To model and simulate an end mill, it is necessary to build its mathematical model first. The cutting edges of a mill are spiral curves and the surface of the mill is a spiral surface that is scanned by spiral and groove. By theory of computer aided geometric design, it is better to use parametric curves and surfaces for a CAD entity, as paper [10] did. Therefore, we build the mathematical model by a parametric curves and surface. By correcting some minor misstates in the mathematical models introduced in paper [10], we rebuild the mathematical models for flat end-mill, conic end-mill and bulb end-mill.

A point on surface of revolution revolves around its axis to take rotation motion is cutting edge spiral. The following Fig.1 is cutting edge spiral.

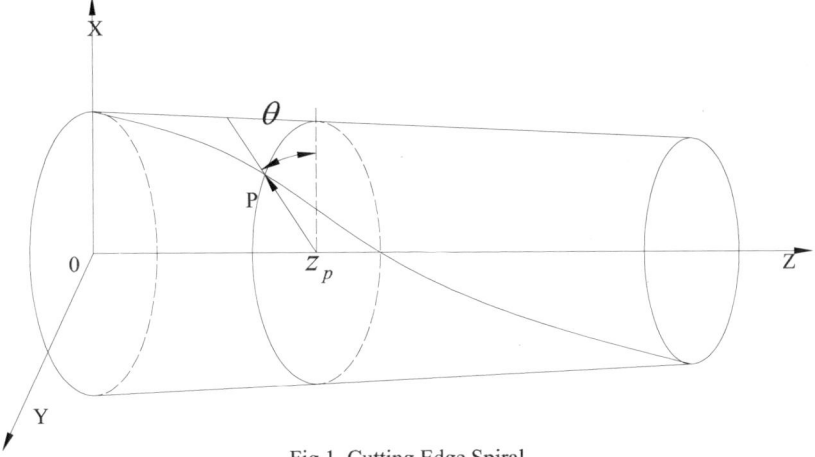

Fig.1. Cutting Edge Spiral

Let $P(x, y, z)$ be a point on cutting edge spiral, then equation of cutting edge spiral is

$$\begin{cases} x = f(z)\cos\theta \\ y = f(z)\sin\theta \\ z = z \end{cases} \quad (1)$$

In the equation (1), $f(z)$ is the radius of random point on cutting edge spiral and θ is the angle between radius and X-axle.

B. End surface mathematical model

  Taking a four teeth end mill as an example, Fig.2 is its end surface. The end surface is consist of five parts: a rake face AB, a bottom land arc BC, a first-arc CD, the edge of second back facet DE and the edge of first back facet EF.

  Let R be cutter radius, N be number of teeth , h be height of teeth, γ_0 be tool orthogonal rake, α_1 be first tool orthogonal clearance, α_2 be second tool orthogonal clearance, L_1 be first back facet, L_2 be second back facet, R_c be radius of chip flute and R_f be radius of tooth surface arc. According to above value, the value of A, B, C, D, E and F are calculated so that integrated end surface is made.

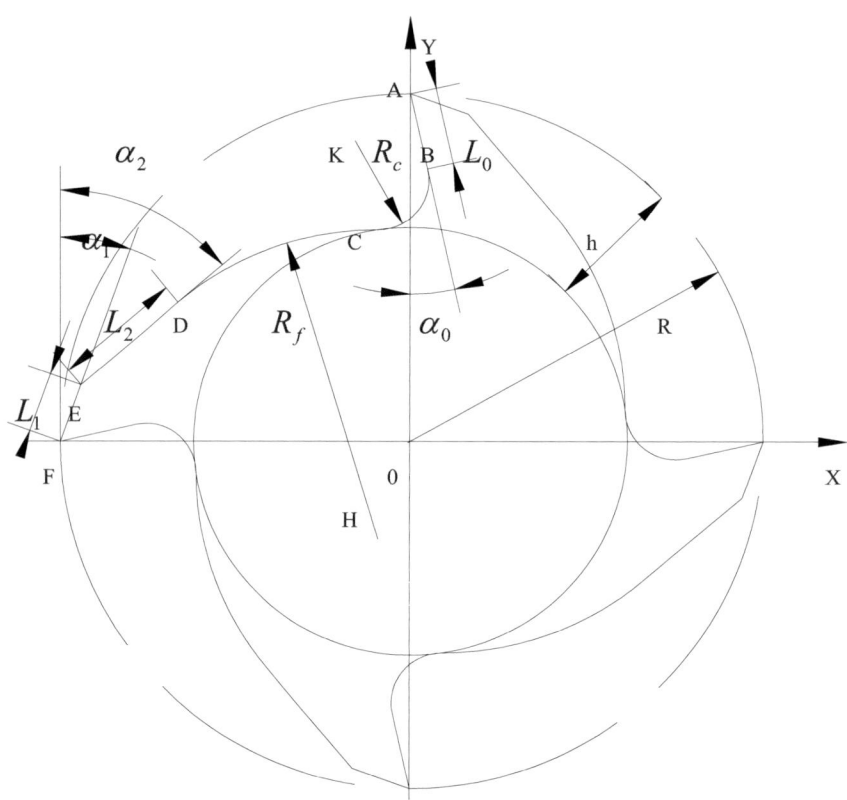

Fig.2. End surface

At $z = 0$ end surface, we can get $A(x_A, y_A) = (0, R)$. To get the value of B, rake face L_{AB} must be calculated. Hence, line AB is extended firstly, and then line OS is made vertically with extension. At last, line OS is made paralleled with line BJ that is as same as line OS. The following Fig.3 illustrates it.

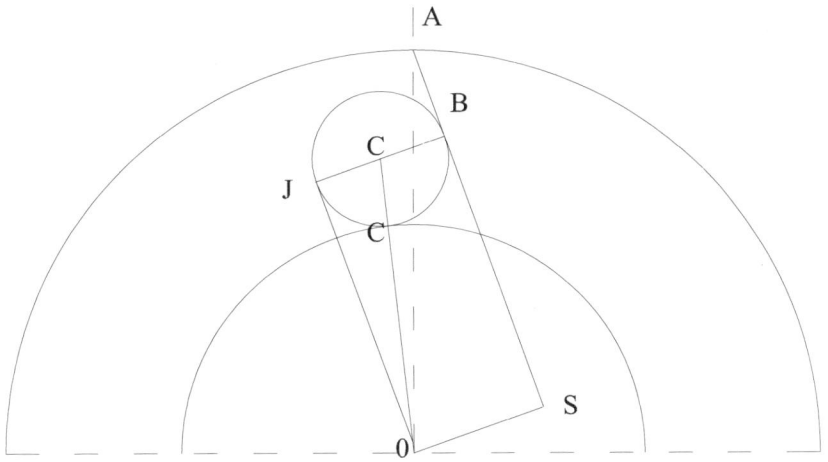

Fig.3. Solution to rake face

$$L_{BS} = \sqrt{(L_{OC} + R_C)^2 - (L_{OC} - R_C)^2} \qquad (2)$$

$$L_{AB} = L_{AS} - L_{BS} = R\cos\gamma_0 - \sqrt{((R-h)+R_C)^2 - (R\sin\gamma_0 - R_C)^2} \quad (3)$$

According to expression (2) and (3), point B obviously yields

$$\begin{cases} x_B = R - L_{AB}\cos\gamma_0 \\ y_B = -L_{AB}\sin\gamma_0 \end{cases} \qquad (4)$$

Basing on A, B and R_c, the C of coordinate value yields

$$\begin{cases} x_C = x_B - \dfrac{R_c}{\sqrt{1+\left(\dfrac{x_B - x_A}{y_B - y_A}\right)^2}} \\[4mm] y_C = y_B + \dfrac{R_c}{\sqrt{1+\left(\dfrac{x_B - x_A}{y_B - y_A}\right)^2}}\dfrac{x_B - x_A}{y_B - y_A} \end{cases} \qquad (5)$$

Point F will be gotten when point A rotates $\dfrac{360^o}{N}$. The value is such that

$$\begin{cases} x_F = x_A\cos\left(\dfrac{360^o}{N}\right) - y_A\sin\left(\dfrac{360^o}{N}\right) \\[4mm] y_F = x_A\sin\left(\dfrac{360^o}{N}\right) + y_A\cos\left(\dfrac{360^o}{N}\right) \end{cases} \qquad (6)$$

According to the length of first back facet and its angle, point E is calculated. Its value is

$$\begin{cases} x_E = x_F + L_1\cos\alpha_1 \\ y_E = y_F - L_1\sin\alpha_1 \end{cases} \qquad (7)$$

According to the length of second back facet, its angle and the value of point E, point D yields

$$\begin{cases} x_D = x_E + L_2\cos\alpha_2 \\ y_D = y_E - L_2\sin\alpha_2 \end{cases} \qquad (8)$$

Because arc CD is tangency with line DE and arc BC, the relation is that

$$\begin{cases} \dfrac{y_E - y_D}{x_E - y_D}\dfrac{y_H - y_D}{x_H - x_D} = -1 \\[3mm] (x_H - x_D)^2 + (y_H - x_D)^2 = R_f \\[3mm] \sqrt{(x_C - x_H)^2 + (y_C - y_H)^2} = R_f + R_c \end{cases} \qquad (9)$$

The point H will be calculated. The value is that

$$\begin{cases} x_H = x_D + \dfrac{R_f}{\sqrt{1+\left(\dfrac{y_D - y_E}{x_D - x_E}\right)^2}}\dfrac{y_D - y_E}{x_D - x_E} \\[6mm] y_H = y_D - \dfrac{R_f}{\sqrt{1+\left(\dfrac{y_D - y_E}{x_D - x_E}\right)^2}} \end{cases} \qquad (10)$$

At last, point C will be calculated by the value of point H, point C and R_c. Its value is

$$\begin{cases} x_c = x_K + \dfrac{x_H - x_K}{y_H - y_K}\sqrt{\dfrac{R_c^2(x_H - x_K)^2}{(x_H - x_K)^2 + (y_H - y_K)^2}} \\[6mm] y_c = y_K + \sqrt{\dfrac{R_c^2(x_H - x_K)^2}{(x_H - x_K)^2 + (y_H - y_K)^2}} \end{cases} \quad (11)$$

III. STRUCTURE OF SYSTEM

According to process of modeling and simulating grinding of a mill, the system structure is designed as Fig.4, which is composed of 7 modules. The first is a data-input module that receives the original designed from the designer. The second is a data preprocessing and restricted module that judges whether the data is right. The third is a 3D model module that gets 3D model. The fourth is a machining path module that is zigzag path. The fifth is a 3D mesh module that gets 3D mesh. The sixth is simulation that is process of grinding end mill. The last is NC code module.

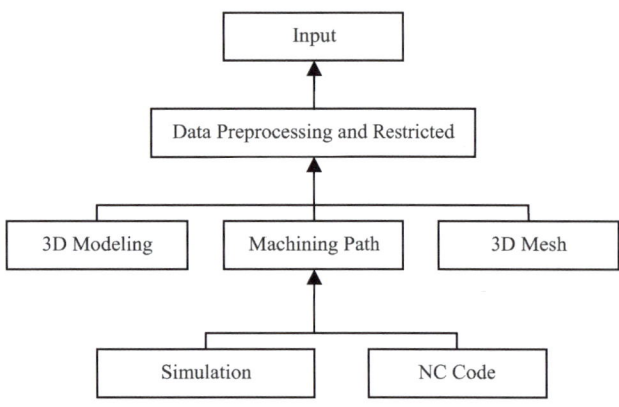

Fig.4. Structure of five-axis grinding system

IV. SOME KEY TECHNOLOGIES

The system choose VC++ as developing tool, taking the need of international standards into consideration, because most commercial CAD or CAM software choose Visual C++ as development platform. Therefore, several key technologies such as 3D modeling, stroke milling machining and mechanism simulation achieve function of system.

A. 3D Modeling

3D modeling technology is to describe the shape of objects in computer. It is practical in AutoCAD to describe an object by 3D model, hence we use 3D mesh model too. For the 3D model, we have to get an end surface as Fig.2 in AutoCAD. And then the end surface is stretched the length of a little piece. After that, the solid we get is moved the length and rotated an angle. At last, all solids are united. That is the process of 3D model. For 3D mesh, A PolygonMesh is an $M \times N$ mesh, where M represents the number of vertices in a row of the mesh and N represents the number of vertices in a column of the mesh. So it is important to figure out the vertices. The following Fig.5 is 3D model and Fig.6 is 3D mesh model.

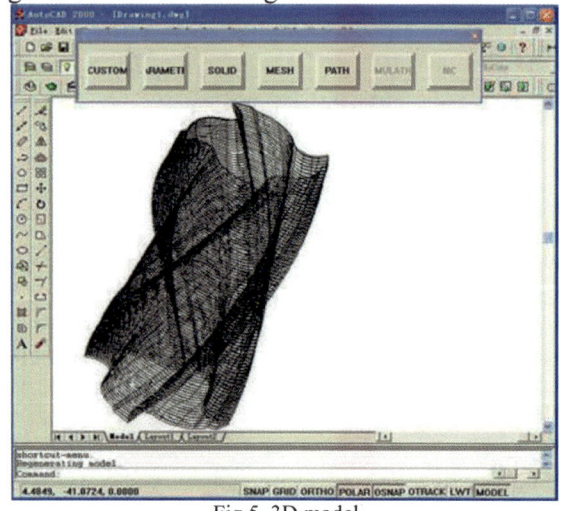

Fig.5. 3D model

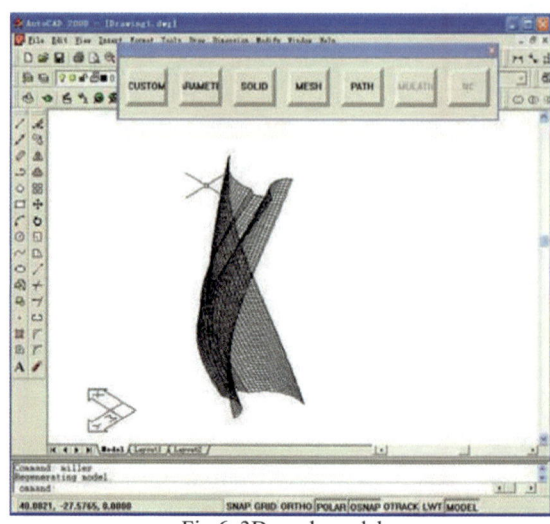

Fig.6. 3D mesh model

B. Stroke Milling Machining

It is necessary to choose tool path reasonably in numerical control machine because it influences the precision of part and machining efficiency directly. There are two main methods, one is zigzagging and the other is contouring. In this system, zigzag is used because it fit the shape of the mill. To generate the path, it is important to figure out the data of cutter location points. The following Fig.7 is the machining path of zigzag.

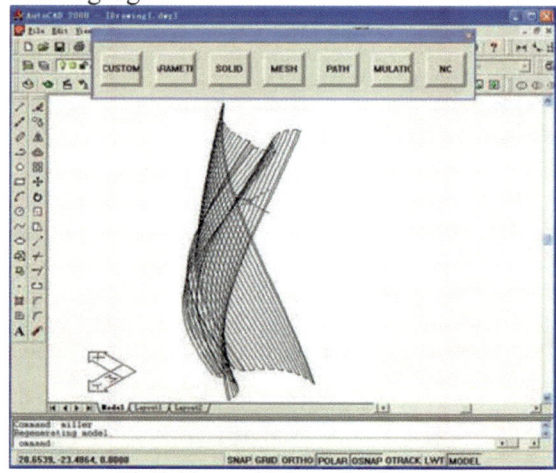

Fig.7. The machining path of zigzag

C. Machining Simulation

Simulation of machining a mill is relatively a complex task. Since grinding of a mill is more complicated than the others, it is necessary to simulate the grinding process. There are some points that are taken care of. The first point is interference between the grinding header and the path. The second point is that the normal vector of every point is paralleled with cutters'. When these problems are solved, NC code is got. The following Fig.8 and Fig.9 describe how to carry out simulation module and generate NC code.

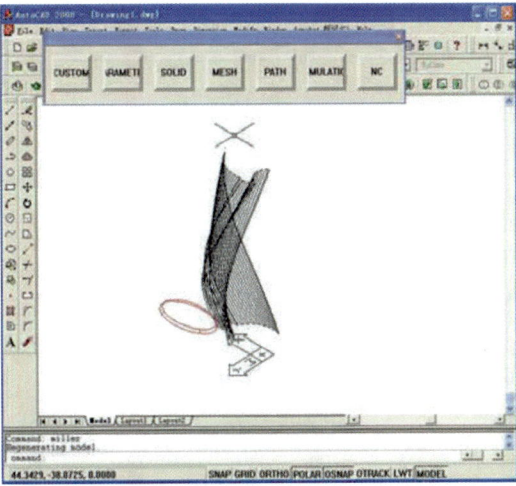

Fig.8. Simulation

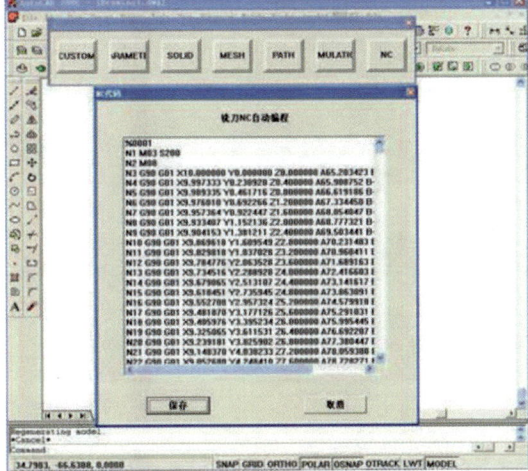

Fig.9. NC code

V. CONCLUSION AND PROSPECTION

It takes long time to search and develop CAD/CAM software systems that have absolute self-owned intellectual property. Because devices and software systems form foreign are mature in grinding and techniques and the introduction of devices and software from abroad into domestic are anywhere. The five-axis grinding system is coming. The system will show 3D model, motion simulation and generate NC codes through inputting some key parameters, but it still have some problems. We hope it will make a contribution to the field of CAD/CAM.

REFERENCE

[1] Lei Yuanzhong, Recent research advances and expectation of mechanical engineering science in China, Journal of Mechanical Engineering, Vol 45, No 5, pp.1-11, 2009

[2] Ma Tianjun, Research on CAD method to generate manufacturing tool of spiral surface, ZhenZhou University, 2011

[3] Liu Jianjun, Li Rong, Cheng Xuefeng, Jin Xiaobo, Ding Guofu, Research on CNC grinding simulation technology of torus end milling cutter's end edges, Modern manufacturing engineering, 2012

[4] Han Zhengfeng, Research on mathematical model of ball-nose end milling cutter based on four-axis grinding, Guizhou University, 2008

[5] Sun ChangFu, Research on grinding trajectory modeling and programming of integral flat-end mill, HaRBin Institute of Technology, 2013

[6] Chen Fang, Research on CNC grinding complex shaped cutter using spherical grinding wheel, HuaZhong University of Technology, 2009

[7] Chen FengJun, The research on simulation and mathematical model of ball-end cutters based on five-axis numerical control grinder, HuNan University, 2006

[8] Mi Rong, Research on 3D parametric design system of the end mill, Southwest Jiaotong University, 2011

[9] Qiu YingBin, 3D solid modeling for complex shaped cutters, HuaZhong University of Technology, 2007

[10] Hei Daquan, A study on parametric design and geometry parameters optimization of ball-end cutter, Hunan Univerisity, 2012.5

[11] Gao Changcai, Study on 3D parameterized design system for helix-bladed milling cutter base on the UG, Dongbei University, 2006.1

[12] Chai Zhongkui, Yang Yi, Liao Yougui, CAD gallery management system of ObjectARX based heat exchange equipment, Value Engineering, 2012

[13] Wang Yao, Researching data generating and extracting method of 3D pipe ends models based on ObjectARX, HarBin Engineering University, 2013

[14] Zeng Hongbing, A CAD study of shield tunnel in oil and gas pipeline based on ObjectARX, Southwest JiaoTong University, 2013

[15] Long Juan, Liao Guoming, Zhou Jianxin, Parametric design of riser system based on ObjectARX custom entity, Foundry Technology, Vol.33, No.12, pp 1412-1414, 2012

Optimization of Urban Traffic Lights System Based on Human Factors Engineering

Ya-fang ZHENG, Ling SONG, Yan XU, Yu-ping ZHANG, Li-zhen ZHANG*

Department of Industrial Engineering, Shanghai Ocean University, Shanghai, China

(lzzhang@shou.edu.cn)

Abstract - **In order to construct convenient, safe, high-efficiency, low-carbon, environmental friendly and diversified urban transport systems, based on the knowledge of human factors engineering, aiming at the deficiencies of the existing traffic lights, this paper optimized the waiting time, designed a convenient button and linkage induction door to prevent pedestrian from running the red light, used the wind/solar hybrid generation system instead of electricity to save energy, and took other measures to realize the improvement of the existing traffic lights system.**

Keywords - **Human factors engineering, improvement, transportation system, traffic lights**

I. INTRODUCTION

So far, Japan, Europe and the United States have competed to invest a lot of money and manpower, established corresponding organizations engaging in development and application of relevant aspects, and have achieved some achievements, developed a series of relatively stable traffic control systems [1]. These abroad traffic control systems, after long-term application, technology has been relatively mature and the reliability of the system is relatively high, in the specific application in our country, also show some inadaptability. Chinese scholars are dedicated to studying urban traffic control systems [2].

As the traffic flows in different directions intersects at road intersections, causing traffic behaviors such as traffic conflicts, interflow and split-flow, etc, road network intersections become the "defiles" of road capacity and the "black dots" of traffic accidents. In urban traffic management and control, intersection management is an important and indispensable component [3]. The stand or fall of intersection management will be the key to solve traffic congestion problems. Capacity of signal intersection and block level of vehicles through the intersection are directly affected by the traffic lights control methods [4]. In line with the principle of the pertinence, feasibility, timeliness and necessity, how to not only maintain both low carbon and environmental protection, but also ensure the convenience for the participants and managers has become a study direction of development and reform of road traffic signal lights [5].

II. DEFICIENCIES OF THE EXISTING TRAFFIC LIGHTS

According to the results of data analysis and questionnaires, domestic traffic lights system has the following disadvantages:

(1) Does not consider the tolerance limit of pedestrian for waiting time that may cause bad trip emotion and serious consequences in traffic disputes.

(2) The phenomena of vehicles and pedestrians running red lights are common occurrences.

(3) Does not design the special indication methods for special groups (such as color blindness and blind, etc.).

(4) Design of traffic lights pole is unreasonable.

(5) Most illuminant of traffic lights are the electric incandescent lamps and low voltage tungsten-halogen lamps, whose low luminous efficiency, high power consumption and low brightness lead to waste of resources and huge financial investment [6].

Aiming at the above problems, some design measures have been taken to improve from the perspective of human factors engineering in this paper.

TABLE I
REASONABLE WAITING TIME

Period	Direction	Steering	Actual traffic time (units / min)	Saturated traffic (vehicles /min)	Total loss time	Best cycle	Best green time (sec)
7:00~9:00 17:00~19:00	North and South	Straight ahead	19	70			50
		Turn Left	14	65	4s	178s	48
	East and West	Straight ahead	27	70			60
		Turn left	5	65			20
9:00~17:00	North and South	Straight ahead	12	70			11
		Turn left	15	65	7s	51s	12
	East and West	Straight ahead	20	70			18
		Turn left	5	65			10

E. Qi et al. (eds.), *Proceedings of the 21st International Conference on Industrial Engineering and Engineering Management 2014*, Proceedings of the International Conference on Industrial Engineering and Engineering Management, DOI 10.2991/978-94-6239-102-4_63, © Atlantis Press and the authors 2015

III. IMPROVEMENT OF TRAFFIC LIGHTS SYSTEM

A. Waiting time optimization of traffic lights

According to human factors engineering, consider of the emotional aspects of psychological factors, to reduce the bad feelings of vehicles and pedestrians caused by unsuitable waiting time, which may result in a negative impact on people's work, study and life [7], this paper conducted a questionnaire, observed and recorded relevant data at the traffic lights, treated these data with statistics to come to a more reasonable duration of traffic lights. From the junction of Huinan town Xuanhuang highway and Nantuan highway [8], the experimental data are shown in Table I.

In addition, some intersections have fewer pedestrians to across, setting a button on traffic lights column [9], as shown in Fig.1. When pedestrian need to cross the road, they can press this button, then the light will turn green automatically, which will let drivers know someone across the street. Not only will it reduce pedestrian waiting to cross the road in time, but also can enhance the crossing security level, according to human factors engineering GB10000-1988 body dimensions of each value, loaded this button at 1.2m~1.4m of the traffic signal pole [10-12].

For the problem that blind people cannot know the time length of the traffic light, we install a voice broadcast device in traffic lights. When the blind people press this button, the player immediately offer low voice sound signal stimulus for the blind to help them cross the road. It can reduce the traffic accident occurred because of the invisible traffic lights.

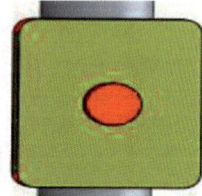

Fig.1. the botton for pedestrian crossing road

B. the design of linkage induction door

In order to control of pedestrian not to jaywalk at the crossroads, set the linkage induction door which is similar to the subway induction door at the two endpoints of crossing, and set up three meters long of the fence on the two sides of the sensor doors as shown in Fig.2. When the lights turn red, the induction door will close to prevent the pedestrians; and when the lights turn from red to green, door responds to open and pedestrians can smoothly pass. Referred to the book of ergonomics in which has the human body measurement data, the height that we design of the induction door is 1.2m.

C. Set of beep and lights of different shapes

Contraposing the question that color-blind people can't distinguish colors right, we design the different traffic lights into different shapes. The red light is designed into triangle which means to need caution,

yellow is designed into circle warning and the green light is designed into square which means peaceful as shown in Fig.3. This improvement can take care of the color-blind people.

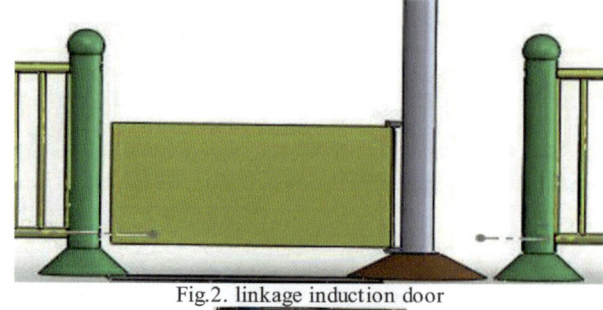

Fig.2. linkage induction door

Fig.3. traffic lights with different shapes

D. Adjustment installation height of traffic lights

In order to protect road safety, the mounting height of lights should ensure that vehicle, non-motor vehicle drivers and pedestrians on the road can clearly see the road traffic lights. According to the field of vision and horizon knowledge and specifications in setting up and installing road traffic lights, vehicle traffic lights take cantilevered installation, the installed height is 5.5m, and the installed location is 50 meters away from the stop line to import, non-motor vehicle lights installed height of 2.5m to 3m. If road don't have vehicle trails and non-motorized vehicles isolation belt, non-motor vehicle lights should be attached to the lights poles guiding motor vehicles to pass through, and crosswalk installation height will be 2m to 2.5m [13].

E. Use of wind/solar hybrid generation system

Due to LED traffic lights have strong light penetration, no glare, high reliability, energy saving, easy instructions and other salient advantages, we recommend using LED traffic lights [14-16].

To take advantage of solar and wind energy, traffic lights designed for wind and solar systems to use solar and wind power for traffic lights in order to achieve the purpose of energy saving and emission reduction, Fig.4 shows the design of wind and solar system.

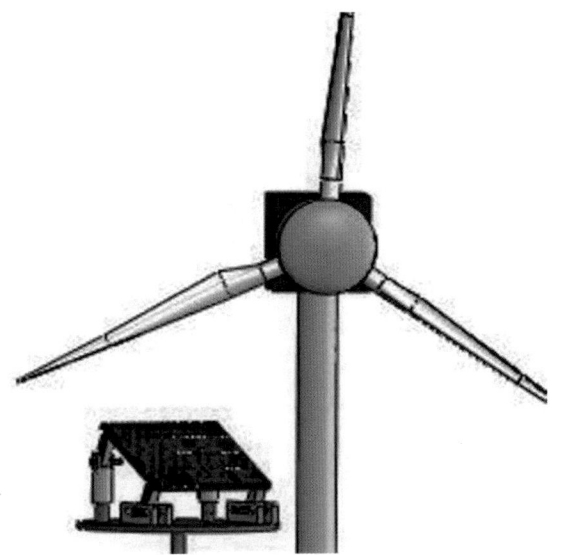

Fig.4. wind/solar hybrid system

IV. CONCLUSIONS

Starting from the human point of view, combined with China's national conditions, mainly according to the knowledge of human factors engineering, some designs and improvements are done for traffic lights system on the crossroads These adopted methods are reducing pedestrian waiting time, preventing pedestrian from running red light, helping blind and colorblind people cross the road safely, more comfortably to observe traffic lights and energy saving. These five aspects have improved the existing traffic lights one by one. The optimization of the existing traffic lights will achieve the purpose of improving people- vehicles pass flow, reducing congestion, improving efficiency and strengthening humanization, intelligence and low-carbon.

ACKNOWLEDGEMENTS

This work was financially supported by National Training Programs of Innovation and Entrepreneurship for Undergraduates. Corresponding author Lizhen ZHANG guided the total work.

REFERENCES

[1] Lin-xi Lee, Hai-jun Gao, Fei-yue Wang, "Two adjacent intersection traffic signal coordination control", *AAS*, vol.29, no.6, pp.947-952, 2003. (Chinese)

[2] Jin-Song Xue, Ji-Wan Xu, Da-hai Chen, "Modeling techniques of urban traffic network and optimal control", *Information and Control*, vol.10, no.3, pp.12-18, 1981. (Chinese)

[3] Xi Liu, Zuo-jun Zou, "Urban traffic coordination control", *AAS*, vol.12, no.4, pp.430-437,1986.(Chinese)

[4] Zhi-yong Liu, *Intelligent transportation theory and its application*, Science Press, 2003, pp.125-186. (Chinese)

[5] Ji-kai Yi,Yuan-bin Hou. *Intelligent control technology*, Beijing University Press, 1999, pp.192-200. (Chinese)

[6] Wen-xing Zhu, Lei Jia, Xiao-qing Wu. "Urban trunk road traffic control multi-objective optimization" *Shandong University (Engineering Science)*, vol. 34, no.3, pp.72-76, 2004. (Chinese)

[7] Zhi-yong Liu, *Intelligent Traffic Control Theory and Applications*, Science Press, 2003, pp.78-80. (Chinese)

[8] Pei-kun Yang, Shu-sheng Zhang, *Traffic management and control*, People's Communications Press, 1995, pp.102-107. (Chinese)

[9] Ming-yuan Bian, Si-zhong Chen, Han-jun Luo, "Intelligent transportation systems and the development" *Wuhan Journal of Management*, vol.23, no.1, pp.67-70, 2003 (Chinese).

[10] Xin-hong Shi, Bergen Cai, Jian-cheng Mu. "Intelligent Transportation Systems" *Northern Jiaotong University*, vol.26, no.l, pp. 29-34, 2004. (Chinese)

[11] Pei-jian,Yang, Bing Wu, *Traffic management and control*. People's Communications Press, 2003, pp.89-92. (Chinese)

[12] Yan Bai, Qing-chao Wei, Qing-yun Qiu. "Discuss urban transport development based on Green Transportation" *Beijing Jiaotong University (Social Science Edition)*, vol.6, no.2, pp.10-14, 2006. (Chinese)

[13] Tong-mei Zhou, Wen-bin Yan. "The development trend of road traffic control research" *Journal of public transportation university (natural science edition)*, vol.12, no.4, pp.49- 53. (Chinese)

[14] De-wang Chen, Shu-ming Tang, Xiao-yan Gong and so on. "Urban development and prospect of intersection traffic signal control study" *white Automation Expo*, vol.4, no.2, pp.48-50, 2004. (Chinese)

[15] Bian-shun Jing. *Road Traffic Control Project*, People's Communications Press, 1995, pp.89-98. (Chinese)

[16] Jing Liu, Wei Guan. "Reviewed traffic flow forecasting methods", *Highway and Transportation Research*, vol.5, no.1, pp.82-85. (Chinese)

Bus Scheduling Optimization Based on Improved Strength Pareto

Xiao-yue YANG [1], Xin-yu LI[1,*], Jun LIANG[1], Cheng PENG[2]

[1]Department of Industrial and Manufacturing System Engineering, Huazhong University of Science and Technology,
Wuhan, China

[2]Department of Industrial Engineering, Tsinghua University, Beijing, China

(lixinyu@mail.hust.edu.cn)

Abstract - **Bus Scheduling Problem involves conflictive objectives for bus agencies and passengers. To obtain solutions from this multi-objective problem, an optimizer SPEA2, was employed to provide a Pareto set of reliable bus schedules. The proposed approach was applied to a specific bus route using historical passenger flow data. The result obtained in this work was compared to the results provided by different optimization strategy in a quoted research, which was to solve a similar model with Immune Artificial Algorithm. Furthermore, the approach to execute real-time control was studied and an online platform was established to publish the bus transit information to improve service quality.**

Keywords - **Bus scheduling, bus transit information system, optimization, SPEA2**

I. INTRODUCTION

The Single Depot Bus Scheduling Problem in this paper was formulated as a nonlinear constrained multi-objective optimization problem where the objective was to maximize the profit of bus operators while minimize passenger dissatisfaction and waiting cost simultaneously. The constraints include lower bound of loaded passenger capacity, limitation of waiting time and passenger backlog.

A number of researches were done on the bus scheduling problem. Some [1] treated it as single objective optimization problem while others [2-5] treated it as multiple objectives problem. Some made research on single depot scheduling [6] and others focused on multi-depot problem [7]. Zhiwei Yang, Shengchuan Zhao and Qian Zhao [2] considered both profit of both passengers and bus carriers and converted the problem into a single objective one by linear combination of different objectives as a weighted sum. Ming Wei, Wenzhou Jin, Weiwei FU and Xiaoni Hao [3] proposed the multi-depot bus scheduling problem with route time constraints, which goal is to minimize the number of vehicles, deadhead and waiting time. Improved Ant Colony Algorithm was used to obtain the solutions. Derong Tan, Jing Wang, Hanbo Liu and Xingwei Wang [4] studied a bi-objective problem, to minimize the number of dispatched buses and average waiting time of passengers, and used Genetic Algorithm on a bus route to optimize the bus departure schedule of Zibo city.

In this study, Improved Strength Pareto Evolutionary Algorithm was developed to provide a Pareto set of explicit bus schedule. Concerning its implementation, it has few parameters to regulate and performs well on convergence and diversity. Another important advantage of employing SPEA2 is that the result of the multi-objective design is a set of feasible solutions, which allows the schedule maker to select the most adequate according to interest and actual conditions of the bus agency. The feature illustrates the practicality and flexibility of the proposed method.

Moreover, to establish a complete system for the bus agency, the approach to execute real-time control based on expert system was explored and an online platform was established to publish the bus transit information for the sake of passenger convenience.

II. METHODOLOGY

A. The Bus Scheduling Problem

1) Notations

In order to describe the problem more clearly, the notations to be used are defined as:

l : number of period of time in a day

l_i : time period i

T_i : interval time in time period i

n_i : times of departure in time period i , the total departure times of a bus in transit in one day

t_i : departure interval in time period i

$t_{i\max}$: max departure interval time in time period i

l_i : upper limit of waiting time that passengers can accept during period i

Q : rated passenger capacity of a bus

$q_{\min}$: minimum of carrying passenger capacity

$q_{\max}$: maximum of carrying passenger capacity

C_m : upper limit of carrying capacity that passengers can accept

m : number of stations on the bus route

m_j : stop j

$\partial_{i,j}$: average time of passengers' getting on or getting off the bus at stop j in time period i

θ : ratio of passenger getting off the bus to total passenger capacity

η : cost of each single trip for a bus

ε : price of a single ticket

$\beta_{i,j}$: passenger arrival number at stop j in time period i

$\tau_{i,j}$: number of passengers arrive at stop j in time period i

P : total passenger capacity on a bus route the whole day

P' : number of passengers that are dissatisfied with the waiting time

$x_{i,j}$: in-vehicle passenger capacity between stop $j-1$ and stop j in time period i

$y_{i,j}$: number of passengers that get off the bus in time period i

$Q^\wedge$: number of passengers that are dissatisfied with crowding of the bus

q_i : average ratio of carrying passenger capacity to rated capacity in time period i

$\xi_{i,j}$: buses stopping time at stop j in time period i

S : total length of the bus route

κ_1 : cost of waiting the vehicle in every minute

κ_2 : cost of staying in vehicle in every minute

ζ_1 : passenger flow in time period i

2) Assumptions
- The maximum capacity of a bus is constant.
- One day is partitioned into several periods of time according to the passenger flow distribution and the departure interval in a certain period of time is the same.
- The passengers arrive at or get off bus stop j in time period i by *Poisson Distribution* and the time of passengers getting on and off the bus is exponentially distributed.
- The bus runs at a constant speed on the bus route.

3) Objectives
A reliable schedule designed for a given bus route has to balance between the utility of transit agencies and society, so two kind of component is considered in this paper, including the profit of bus agencies, satisfaction of passengers and waiting cost of both in-vehicle and boarding passengers. To solve this problem, a multi-objective model was built.

- *Objective 1: profit of bus operators*

$$w_1 = \frac{\varepsilon * P - \eta * n}{\eta * n} * 100\%$$

The ratio of bus operator's gross profit and operational cost is used to measure the input-output ratio of a bus agency.

- *Objective 2: passenger satisfaction*

$$w_2 = 1 - (\frac{P' + Q^\wedge}{2 * P} * 100\%)$$

Passenger satisfaction is related to dissatisfaction, since the sum of these two opposite sides is 1. Dissatisfaction is calculated here to formulate passenger satisfaction indirectly for the sake of convenience. Passenger dissatisfaction is divided into two parts in the same weight, which are endurance on waiting time and crowding.

- *Objective 3: passengers' waiting cost [8]*

$$w_3 = \frac{\kappa_1}{2} * \sum_{i}^{N} \sum_{j}^{m} \beta_{i,j} * t_i^2 + \sum_{i=1}^{N} \left(\frac{S}{v_i} * \kappa_2 * \zeta_i \right)$$

The waiting cost considers both boarding and in-vehicle passengers, which are expressed as two parts in the expression. One is the waiting time cost and the other represents the cost of loss in terms of in-vehicle passengers.

4) Constraints
To maintain the normal operation of the bus agency and ensure passenger satisfaction, three constraints were defined as followed:

- *Ratio of carrying capacity to rated capacity*

The ratio of carrying passenger capacity to rated capacity should not be lower than the minimum allowed.

$$q_i \geq q_{\min}$$

This can be also expressed as:

$$t_i \geq \frac{T_i * Q * q_{\min}}{\sum_{j=1}^{m} (\tau_{i,j} - y_{i,j} + x_{i,j})}$$

- *Limitation of passengers' waiting time*

Departure interval should be no more than a specified value in each period of time.

$$t_i \leq t_{i\max}$$

- *No passenger backlog*

To ensure that no passengers will be held at the bus stop, the carrying capacity should be larger than number of passengers arrived at the station.

$$\max\{\xi_{i,j}\} \leq t_i \leq \frac{T_i * Q * q_{\max}}{\sum_{j=1}^{m} (\tau_{i,j} - y_{i,j} + x_{i,j})}$$

Based on the above analysis, the *Bus Scheduling Problem* can be described as the following multi-objective problem:

$$\max w_1 = \frac{\varepsilon * P - \eta * n}{\eta * n} * 100\%$$

$$\max w_2 = 1 - (\frac{P' + Q^\wedge}{2 * P} * 100\%)$$

$$\min w_3 = \frac{\kappa_1}{2} * \sum_{i}^{N} \sum_{j}^{m} \beta_{i,j} * t_i^2 + \sum_{i=1}^{N} \left(\frac{S}{v_i} * \kappa_2 * \zeta_i \right)$$

$$s.t. \begin{cases} t_i \geq \dfrac{T_i * Q * q_{\min}}{\sum_{j=1}^{m} (\tau_{i,j} - y_{i,j} + x_{i,j})} \\ t_i \leq t_{i\max} \\ \max\{\xi_{i,j}\} \leq t_i \leq \dfrac{T_i * Q * q_{\max}}{\sum_{j=1}^{m} (\tau_{i,j} - y_{i,j} + x_{i,j})} \end{cases}$$

B. Solution Methodology
SPEA is one of multi-objective genetic algorithms, which are widely applied to solve multi-objective optimization problem for the advantage of simplicity, robustness, global optimization and etc.

SPEA2 is an improved version of SPEA presented by Zitzler and Thiele [9] in 2001. Improved fitness assignment, archive truncation method and a nearest neighbor density estimation technique was proposed to overcome the shortage of previous method and develop a powerful algorithm. The optimization was performed by SPEA2 in this paper.

1) Multi-objective Optimization

Each scheduling scheme, obtained from our approach to solve the bus scheduling problem, is a solution which represents an objective vector composed of bus carrier's profit, passenger satisfaction and waiting cost. The goal is to find the Pareto-optimal set of scheduling schemes. The Pareto-optimal set is constituted by the nondominated solutions, or to say the scheduling schemes. The nondominated solution, which is referred to as Pareto-optimal, will be found if the following conditions are satisfied:

- The feasible solution has a higher value of bus carrier's profit and passenger satisfaction than other feasible solution
- The feasible solution has a lower value for passenger waiting cost than other feasible solution

2) SPEA2

The algorithm schema of SPEA2 in this work can be described as below [10, 11].

Step1: *Initialization*. Generate an initial population P, and create the empty archive $\overline{P}$. Here the population size is 100, the archive size is 80 and the maximum number of generations is 100.

Step2: *Copy*. Copy nondominated members of P to $\overline{P}$.

Step3: *Fitness assignment*. Calculate the fitness values of individuals in the external Pareto set $\overline{P_t}$ and the population P_t.

Step4: *Selection*. Select two individuals at random from the updated external set $\overline{P_{t+1}}$ and compare their fitness. Select the better one and copy it to the mating pool.

Step5: *Crossover and Mutation*. Apply problem-specific crossover and mutation operators.

Step6: *Termination*. Stop if the maximum number of generations is reached, else go to Step2.

3) Best Compromise Solution

Upon obtaining the Pareto-optimal set of bus scheduling scheme, the presented approach provides scheduling operator with one solution from the set as the best compromise solution before implementation.

III. CASE STUDY & RESULTS

After collecting passenger flow data from No.518 bus route in Wuhan and analyzing them, we input the statistical data into the example. The techniques in this paper were developed based on MATLAB and C++ language.

TABLE I
SOME OF THE ACQUIRED PASSENGER INFORMATION

	Stop1	Stop2	Stop3	Stop4	Stop5	Stop6	Stop7
Distance/km	0	1.6	0.5	1	0.73	2.04	1.26
5:00-6:00 up	371	60	52	43	76	90	48
down	0	8	9	13	20	48	45
6:00-7:00 up	1996	376	333	256	589	594	315
down	0	99	105	164	239	588	542
7:00-8:00 up	3626	634	528	447	948	868	523
down	0	205	227	272	461	1058	1097
8:00-9:00 up	2064	322	305	235	477	549	271
down	0	160	123	169	300	634	621
9:00-10:00 up	1186	205	166	147	281	304	172
down	0	81	75	120	181	407	411

In this case study, there are 14 stops along the NO. 518 bus route, and the bus runs from 5:00 to 23:00. According to the acquired statistics, we divide the whole day into five time period. The interval time of every period is $T_i = (60, 180, 420, 120, 300)$, $i = 1, 2, 3, 4, 5$. The max departure interval time is $t_{i\max} = [15, 5, 10, 5, 15]$. The upper limit of waiting time that passengers can accept is $l_i = [10, 4, 8, 4, 10]$. Rated passenger number is 75. The bus ticket is 1.0, and the cost of each bus trip is 70.

Some of the acquired passenger information is shown in TABLE .

Based on SPEA2 method, we solved the multi-objective problem of this case study, and the result is shown in TABLE .

The solution in TABLE shows that intervals and times of departure have been modified slightly to guarantee times of departure is integers.

According to our survey, the interval time of our solution is satisfactory, and this method is much more convenient than traditional manually made timetable method.

TABLE II
OPTIMAL TIMETABLE OF NO.518 BUS ROUTE

Period of Time	Interval Time (min)	Times of Departure
05:00-06:00	9	7
06:00-09:00	3	60
09:00-16:00	7	60
16:00-18:00	4	30
18:00-23:00	17	18

IV. DISCUSSION

In order to evaluate the effect and efficient of the algorithm and method applied in this research, we compare our method with Classical GA and a method put forward by Zhiwei Yang et al. [1]. The comparison results are shown in TABLE .

In the research of reference [1]错误!未找到引用源。, the author researched on Dalian bus transit and proposed optimization on bus scheduling based on artificial immune algorithm. On-trip cost of passengers expected satisfaction of passengers and bus carriers are listed to make a contrast of previous schedule and new proposed one.

Since the weight of profits of passengers and bus carriers in the quoted research was 0.3 and 0.7 separately, the selection of the Pareto solution follows this criterion.

From TABLE we may conclude that there is little difference in time table between our method and GA method. However, our method concentrates more on the satisfaction of passengers. Though the expected satisfaction of bus carriers in our method (4.05998) is smaller than GA (5.05938), the satisfaction of bus carriers is much better.

TABLE III
CONTRAST BETWEEN EFFECTS OF DIFFERENT OPTIMAL BUS SCHEDULES

Item	SPEA2	Classical GA	Method comparison	
			This paper	Quoted Research
Bus Departure Interval	[9,3,7,4,17]	[13,3,10,4,14]	The same for specific time period	Different for every trip
Times of Departure	[7,60,60,30,18]	[5,60,42,30,22]	Determined by the interval times	Determined by all the departure time
Expected Satisfaction of Bus Carriers	4.05998	5.05938	Lower priority	Same priority
Expected Satisfaction of Passengers	1	-0.32762	Higher priority	Same priority
On-trip cost of passengers	34774.567	116523.970	Consider the on-trip cost of customer	/

V. IMPLEMENTATION

A bus scheduling system is usually consisted of hardware and software parts. Passenger data are collected from hardware part and transmitted to the software part to make further analysis, decision and control. The developed system integrates scheduling and information publishing. The scheduling part is based on the profit of customers and bus carriers, the function of which is to generate regular bus timetable for a bus route and make real-time control through information feedback. In order to make convenience for passengers, information like specific bus arrival time is published on an online platform for passengers to view.

A. Bus Scheduling Generation

Bus scheduling is a routine departure timetable made for a specific bur route, according to historical passenger flow data. The fleet follows this timetable every day except when real-time scheduling is needed.

• *Departure Timetable*

To meet transit agencies' own demand and principle, parameters in the objective function can be set through the interface before generating the departure timetable.

• *Block*

Meanwhile, the sequence of buses to pull-out is the other important part of departure scheme. In order to make arrangement on the fleet, timetable, fleet size and rest time are several necessary factors to formulate the block. The Gantt chart of block shows specific departure time for every bus in the fleet, which is the visualization of arrangement of bus sequence. The model discussed in this paper is Single Depot Vehicle Scheduling model, so scheduling approaches like cross-route scheduling and empty driving are not allowed. The formation of block follows the principle of First-In-First-Out. Fig.1 gives an example of Gantt chart of a block.

B. Real-time Control

Real-time scheduling is a kind of supervision and control on buses in transit when abnormal circumstances happen and lead to deviation on actual interval time due to various causes. Knowledge-based expert system is introduced to help operators make real-time control.

Through information feedback like real-time position, the expert system is enabled to analyze and find out abnormal circumstances. Temporary scheduling schemes including temporary timetable, dispatching special buses like express buses and etc. can be provided to operators as reference [12].

1)   Real-time Geographical Position

In order to get real-time geographical position of a bus in transit, the method of coordinates establishing and transformation is introduced. The real-time coordinates of a bus in transit are got from GPS devices on the bus and send to bus scheduling center. By transforming the GPS coordinates into one-dimensional coordinates, they can be used to diagnose abnormal circumstances and forecast arrival time at next several stations.

2)   Knowledge-based Expert System

The knowledge-based expert system is constructed with three key components: a knowledge base, an inference engine, and rule sets. The expert system aggregates the knowledge of common knowledge of vehicle scheduling and expertise on scheduling. The scheduling rule focuses on critical stations [13]. Real-time control model is applied in the rules to make scheduling on the critical stations.

Fig.2 illustrates how inference engine works in this system, or to say - the procedure of reasoning about an abnormal circumstance from recognition to formation of temporary scheduling timetable.

One of the abnormal circumstances is deadhead. X. J. Eberlein [14] proposed a strategy to solve real-time deadheading problem to determine at dispatching time which vehicles to deadhead and how many stations to skip.

C. Online Information Publishing

Routine bus scheduling is a kind of reference for passengers. However, in order to achieve expected effect of scheduling, passengers needed be informed with more exact information like arrival time about the bus in transit. So temporary timetable made through real-time scheduling and arrival time forecasting is also to be published on an online platform for passengers to check by themselves [15]. This information can be published on electronic board or other kind of termination

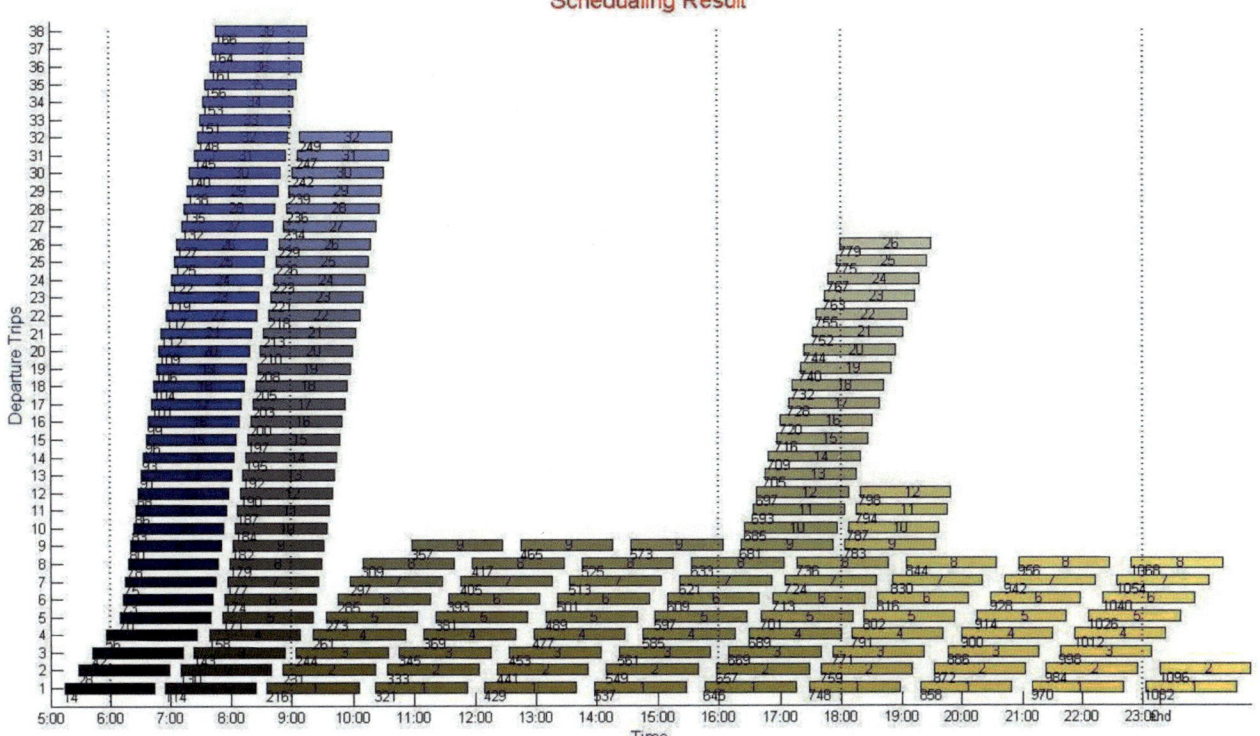

Fig.1. Gantt chart of Block

VI. CONCLUSION

This paper proposed a new methodology, based on SPEA2, to solve the bus scheduling problem, which simultaneously targets on profit of passengers and bus agency. The multi-objective optimization is designed to form the departure timetable, which is the tabular display of the scheduling scheme.

A case study was constructed using the data from a bus route in Wuhan city and the results were compared to the one in another research that uses Immune Artificial Algorithm to solve the problem. The comparison result shows slight superiority in the aspect of customer satisfactory.

What's more, a bus transit information system developed in this paper is able to be applied to make guidance on the single line scheduling work. It contains three parts, which are scheduling generation, real-time control and online information publishing. Block and departure timetable are the routine work of scheduling. In order to make temporary scheduling, knowledge-based expert system is introduced. The inference engine is designed to promptly diagnose the abnormal circumstances of buses in transit and real-time control model is employed to execute real-time control on critical bus stops. The information publishing part is mainly designed for passenger convenience.

Future research will be focused on integrating more factors to enhance the model and developing more efficient algorithm for solving the problem.

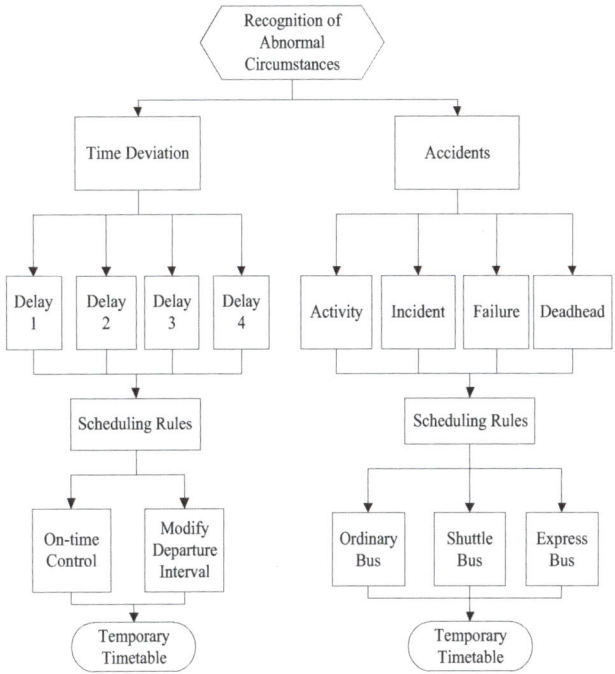

Fig.2. Inference Engine

ACKNOWLEDGMENT

This research was sponsored by National University Students' Innovative Training Project Fund, Huazhong University of Science and Technology. The Operations Research and Optimization Team of Huazhong University of Science and Technology is also appreciated.

REFERENCE

[1] Xu, J., Liu, H., & Teng, J. (2008, August). A vehicle scheduling model and efficient algorithm for single bus line. In Power Electronics and Intelligent Transportation System, 2008. PEITS'08. Workshop on (pp. 458-461). IEEE.

[2] Yang, Z., Zhao, S., & Zhao, Q. (2008, October). Research on bus scheduling based on artificial immune algorithm. In Wireless Communications, Networking and Mobile Computing, 2008. WiCOM'08. 4th International Conference on (pp. 1-4). IEEE.

[3] Wei, M., Jin, W., Fu, W., & Hao, X. N. (2010, July). Improved ant colony algorithm for multi-depot bus scheduling problem with route time constraints. In Intelligent Control and Automation (WCICA), 2010 8th World Congress on (pp. 4050-4053). IEEE.

[4] De-rong, T., Jing, W., Han-bo, L., & Xing-Wei, W. (2011, December). The optimization of bus scheduling based on genetic algorithm. In Transportation, Mechanical, and Electrical Engineering (TMEE), 2011 International Conference on (pp. 1530-1533). IEEE.

[5] Yue Song, Jihui Ma, Wei Guan, Tao Liu and Shi Chen. A Multi-objective Model for Regional Bus Timetable Based on NSGA2. Computer Science and Automation Engineering. 2012

[6] Haghani, A., Banihashemi, M., & Chiang, K. H. (2003). A comparative analysis of bus transit vehicle scheduling models. Transportation Research Part B: Methodological, 37(4), 301-322.

[7] Haghani, A., & Banihashemi, M. (2002). Heuristic approaches for solving large-scale bus transit vehicle scheduling problem with route time constraints. Transportation Research Part A: Policy and Practice, 36(4), 309-333.

[8] Gang, T. O. N. G. (2005). Application Study of Genetic Algorithm on Bus Scheduling [J]. Computer Engineering, 13, 29-31.

[9] Eckart Zitzler, Marco Laumanns, and Lothar Thiele. SPEA2: Improving the Strength Pareto Evolutionary Algorithm. Swiss Federal Institute of Technology (ETH), Zurich, Switzerland. Technical report TIK-Report. No.103. May 2001.

[10] Eckart Zitzler, Lothar, "Multiobjective Evolutionary Algorithms: A Comparative Case Study and the Strength Pareto Approach", IEEE Transactions on Evolutionary Computation, Vol. 3, No. 4, November 1999.

[11] Muhammad Tami Al-Hajri, M, "A. Abido. Multiobjective Optimal Power Flow Using Improved Strength Pareto Evolutionary Algorithm (SPEA2)", *11th International Conference on Intelligent Systems Design and Applications*, 1097-1103, 2011.

[12] Malachy Carey, "Optimizing Scheduled Times Allowing Response", Transportation Research Part B 32: 329~342, 1998.

[13] Yan, Shangyao, and Hao-Lei Chen. "A scheduling model and a solution algorithm for inter-city bus carriers." Transportation Research Part A: Policy and Practice 36.9 (2002): 805-82

[14] .X.J.Eberlein, N.H.M.Wilson, C.Barnhart and D.Bernstein. The Real-Time Deadheading Problem in Transit Operations control. Transportation Research Part B: Methodology. 1998

[15] Sutanto Soesodho, Nahry, "Optimal Scheduling of Public Transport Fleet at Network Level", Journal of Advanced Transtation, 34(2): 297-323, 1999.

Simulation of an Astronaut's Connecting Bolt Action in Orbit

Ai-ping YANG[1,*], Xin ZHANG[2], Guang CHENG[1], Hui-min HU[2], Chau-Kuang CHEN[3]

[1]Department of Industrial Engineering, Beijing Union University, Beijing, China
[2]Ergonomics Laboratory, China National Institute of Standardization, Beijing, China
[3]Department of Institutional Research, Meharry Medical College, Nashville, Tennessee, USA
(jdtaiping@buu.edu.cn)

[1] *Abstract* - **This study primarily focuses on the simulation of an astronaut connecting bolt action in orbit. The upper limb model was established, which included three segments of upper arm, forearm, and hand, as well as three revolute joints of shoulder, elbow, and wrist. Joint kinematical parameters during one period were obtained by using the numerical solution of reverse kinematical equation. The corresponding virtual upper limb model was established and simulated using LifeMOD software. Thus, the simulation results can be validated by comparing the simulation trajectories of the mass center of hands with that of hands' known motion. Additionally, the joints' strengths can be compared when hands move in different directions of the human body coordinate system. The results are useful to astronaut's orbital mission planning and training in the ground.**

Keywords - **Joint strength, LifeMOD, simulation, virtual upper limb model**

I. INTRODUCTION

Tasks such as performing orbital scientific experiments or engagFing spacecraft maintenance activities need astronauts' participation. A statistical analysis on astronauts' extravehicular activities shows the largest number of astronauts' task is connecting actions, which include the following three types: (1) string, soft belt and harness connection or bundled; (2) plugs, hooks and pins connecting the two parts; and (3) bolts connecting the two parts in spacecraft. A tie or wiring harness connection can be simplified to finger movements while plug or pin connections can be simplified as the motion of hand center of mass along a linear or circular path, A bolt operation can be simplified as the hand motion along an arc or circular helix trajectory. These connecting operations usually serve as one of astronauts' basic activities in orbit [1-5]. Due to being in a microgravity environment along with expensive costs in space exploration, it is impossible to conduct these types of experiments in orbit. The approach of computer simulation to astronaut movements has been widely used [6-9]. The advantages of using computer simulations include capabilities in viewing all sights, shortening design time, and ease of operation. This study primarily focuses on the simulation of an astronaut connecting bolt based on the multi-rigid-body system theory by using the biomechanical software LifeMOD. Simulation results have a significant value for on-ground training and planning of astronaut tasks in orbit.

---

[1] This work is sponsored by National Science and Technology Support Program No.2014BAK01B04

II. SIMULATION METHODS AND STEPS

A. Task descriptions and terminal motion model along cylindrical helical trajectory

This task involves an astronaut tightening or loosening bolts operation by using his hand under his feet fixed in orbit, which terminal motions can be seen as hand movements along a circular line or a circular helix. Hand centroid motion along the circular line trajectory can be achieved by joint plane motion or three-dimensional motion. The movement along the circular helix trajectory itself is a space motion, which requires the joint participation of three-dimensional motion. Based on bolts' connecting operation of light loads, the task is simplified as an upper limb three-dimensional motion under the trunk fixed. Therefore, the astronaut movement can be treated as an upper limb motion with the mass center of hands move along circular helix trajectories.

Fig.1 shows cylindrical spiral motion model of the upper limb with hand movement along the coronal plane normal direction in human body coordinate planes. Hand centroid does uniform circular motion in the sagittal plane of the body coordinate system, while doing uniform linear motion along the coronal plane normal direction. The terminal trajectory curve is shown in Fig.1. The circle is the projected trajectory in the sagittal plane. Cylindrical helix motions of the other two coordinate axes in the human body coordinate system are similar to the above.

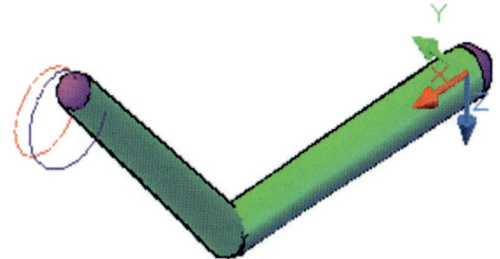

Fig.1. Cylindrical spiral motion model of mass center of hands along the coronal plane normal direction

Cylindrical spiral motion equations for mass center of hands in the human body coordinate axis direction are described in the equations (1), (2) and (3).

The initial point coordinates of hand centroid is set (x_0, y_0, z_0), with $v_i (i = x, y, z)$ being uniform linear motion speed, $\theta = \omega t$ being the circular motion angle, and ω being Circular motion angular velocity. Equation (1) represents the mass center of hand rotating counterclockwise along circular path in the sagittal plane of the body coordinates system while doing uniform motion along the coronal axis,

$$\begin{cases} x = x_0 + r(1 + \cos\theta) \\ z = z_0 + r\sin\theta \\ y = y_0 + v_y t \end{cases} \quad (1)$$

Equation (2) represents the mass center of hand rotating counterclockwise along a circular path in the coronal plane of the body coordinates system, while doing uniform motion along the sagittal axis,

$$\begin{cases} y = y_0 + r(1 + \cos\theta) \\ z = z_0 + r\sin\theta \\ x = x_0 + v_x t \end{cases} \quad (2)$$

Equation (3) represents the mass center of hand rotating counterclockwise along a circular path in the horizontal plane of the body coordinates system, while doing uniform motion along the vertical axis,

$$\begin{cases} x = x_0 + r(1 + \cos\theta) \\ y = y_0 + r\sin\theta \\ z = z_0 + v_z t \end{cases} \quad (3)$$

Given the geometry, the hand centroid motion is known, along with quality and inertia parameters for each segment of the human upper limb. Using multi-body system kinematics and dynamics theory, the motion parameters of shoulder, elbow and wrist joints of the human upper limb can be solved.

B. Coordinate systems and degrees of freedom of the upper limb model

Each joint (shoulder joint, elbow joint and wrist joint) in the upper limb of human body has the following three revolute degrees of freedom: rotations around the transverse axis, sagittal axis, and frontal axis in the human body coordinate system, respectively. Since the shoulder of the space suit has a rotary joint structure and the rotation rate of human body joints is generally small, each joint rotation around the transverse axis can be fixed, so the upper limb model with six-degree of freedom can be established. In this model, each joint of the upper limb can rotate around the sagittal axis and the frontal axis in human body coordinate system.

The inertia frame of the upper limb model is set in shoulder joint while the local joint frames are set in all three joints of shoulder, elbow, and wrist. The coordinate systems and degrees of freedom of the upper limb motion model are shown in Fig.2, where s_i (i=1, 2, 3) denotes three joints of the upper limb model. q_i (i=1, 2, ..., 6) denotes joint angles of the upper limb model, x_i, y_i, and z_i (i=0, 1, 2, 3) are joint inertial and local coordinates of the model.

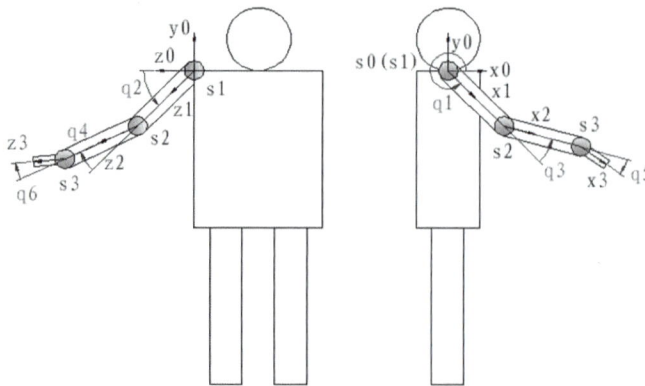

Fig.2. Frames and degrees of freedom of the upper limb model

C. Reverse kinematical model

The object of the reverse kinematical model is to solve numerically the joints' movement by the motion of the mass center of hands; the joints angles drive the follow-up corresponding virtual model established by the LifeMOD platform. Based on the multi-rigid-body system theory, joint kinematical parameters of the upper limb model can be calculated by the reverse kinematical equation [6, 7]. The joints' angular velocities of the upper limb model can be calculated by the methods in [7]. Corresponding joint angles and angular accelerations can be obtained by the integration and derivation to the joints angular velocities, respectively.

Fig.3, Fig.4 and Fig.5 are joint angle curves of upper limb while hand centroid moving along cylindrical helix trajectories in body coordinates axes, among which a circle radius is assumed 0.075m, the period is 20s.

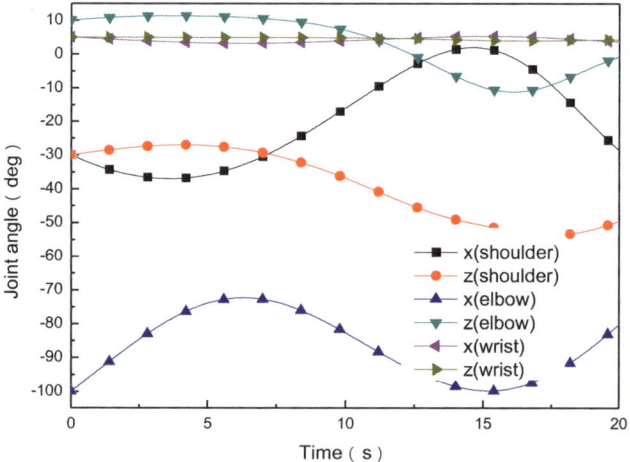

Fig.3. Joint angle curves of upper limb while hand centroid moving along cylindrical helix trajectory in sagittal axis direction

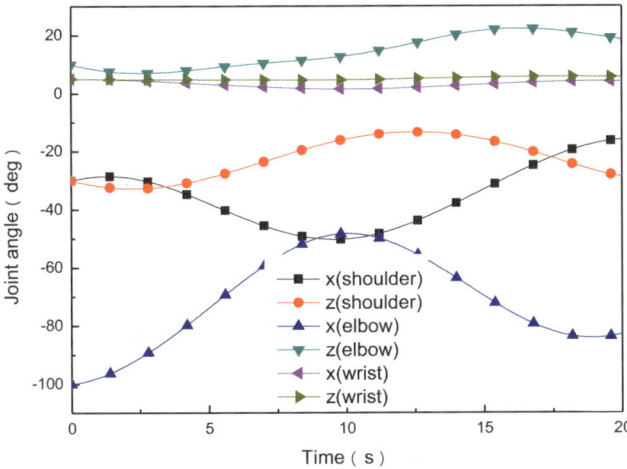

Fig.4. Joint angle curves of upper limb while hand centroid moving along cylindrical helix trajectory in vertical axis direction

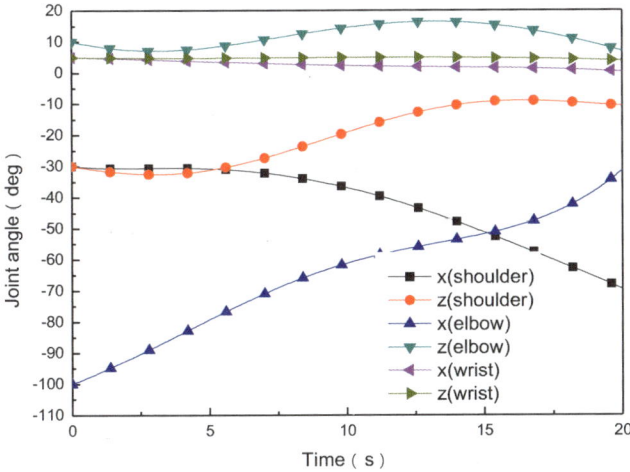

Fig.5. Joint angle curves of upper limb while hand centroid moving along cylindrical helix trajectory in horizontal axis direction

D. Simulation of an astronaut connecting bolt action

1) Simulation flow of astronaut movements: Fig.6 illustrates the simulation flow chart of astronaut movements. The first step is to calculate joint kinematical parameters of joint angles, joint angular velocities and joint angular accelerations by solving the equation of the motion of mass center of hands based on the inverse kinematics model. The second step is to establish the corresponding virtual model by are obtained via the biomechanical software LifeMOD. Next, the dynamic simulation is performed by driving the virtual model with joint angles calculated during one period. Finally, simulation results are verified by comparing the simulation trajectory of the mass center of hands with that of hands known motion, and joint strengths while hands moving along the different motions of the human body coordinate plane are obtained via the biomechanical software LifeMOD [10-12].

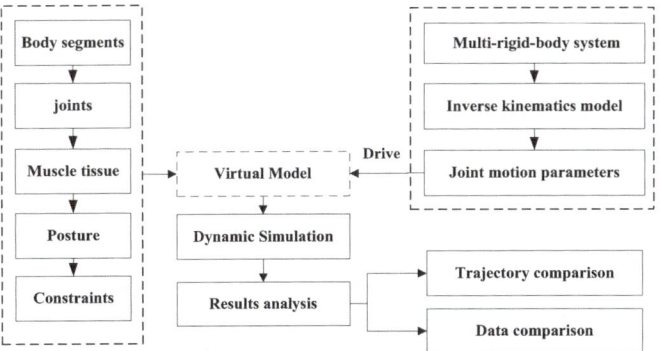

Fig.6. Simulation flow of astronaut movements

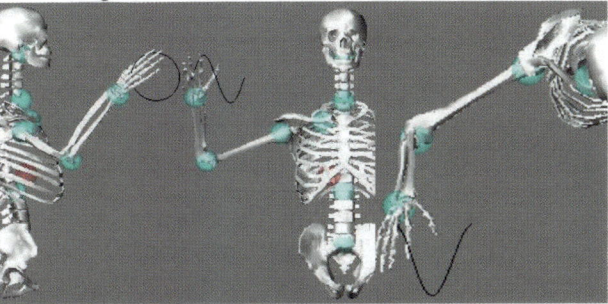

Fig.7.Simulation trajectory of the mass center of hands along x axis

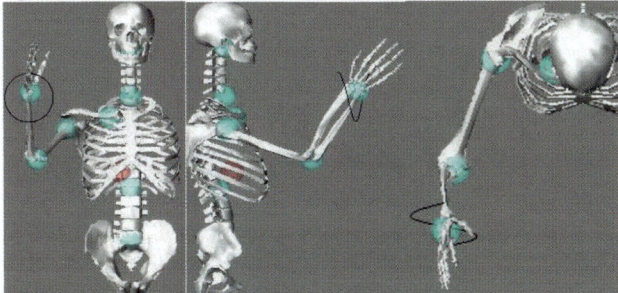

Fig.8.Simulation trajectory of the mass center of hands along z axis

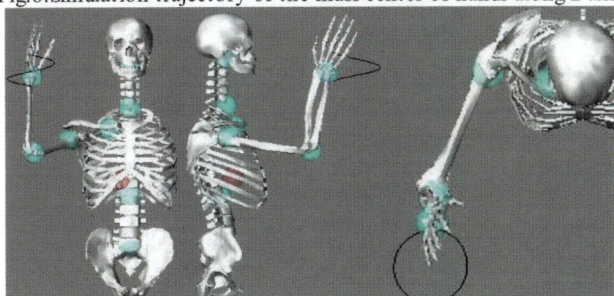

Fig.9. Simulation trajectory of the mass center of hands along y axis

2) Simulation of an astronaut connecting bolt: This study primarily focuses on the simulation of connecting bolt by an astronaut. The simulation process can be accomplished within 20 seconds of the time interval. The simulation terminal model can be seen as 3 cylindrical helix motions for the mass center of hands moving along three coordinate axes in the human body coordinate system. Using the aforementioned methods and LifeMOD software, dynamical simulations of the 3 cylindrical helix actions can be performed. The simulation trajectories of hands motions are shown in Fig.7 is simulation trajectory of hands moving along the horizontal axis in the human body coordinate system. Fig.8 is simulation trajectory of hands moving along sagittal axis in the human body coordinate system. Fig.9 displays the simulation trajectory of hands moving along vertical axis in the human body coordinate system. It is shown that connecting bolt by an astronaut using his arm under microgravity environment can be achieved according to these animations and simulation trajectories of the mass center of hands.

E. Results analysis and discussion

The joints' strengths of the upper limb model during one period can be obtained via the biomechanical software LifeMOD.

1) Impact on joints' strengths along different coordinate axis: Fig.6 shows projections of shoulder joint forces when hands move along the normal directions in three coordinate planes of human body coordinate system respectively. There are nine curves in Fig.10, they include x, y and z curves signed with blackbody symbols which denote shoulder joint force projections when hand is moving along the horizontal axis of human body coordinate system; x, y and z curves signed with hollow symbols which denote shoulder joint force projections when hand is moving along the vertical axis of human body coordinate system; and x, y and z curves signed with cross and hollow symbols which denote shoulder joint force projections when hand moving along the sagittal axis of human body coordinate system. The values of the peak projections between the curve signed with blackbody symbols along horizontal axis and the curve signed with cross symbols and hollow symbols along vertical axis are basically the same, while there is smaller peak joint force projection when hand is moving along sagittal axis which curves signed with hollow symbols. It shows less joint forces in the normal direction of transverse plane than that in both the sagittal and frontal planes.

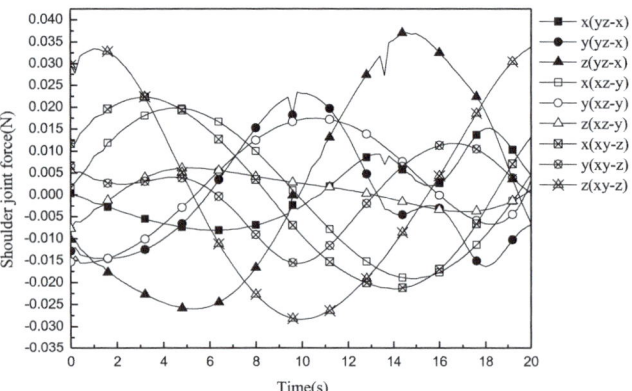

Fig.10. Projections of shoulder joint forces along different axes

2) Impact on joints' strengths with different motion velocities: Fig.11 and 12 show the projections of shoulder and elbow joint forces while hands are moving with different velocities along cylindrical helix trajectory in transverse plane of human body coordinate system. The motion period for these projections was 10s and 20s, respectively; the corresponding move length was 20cm and 10cm. There are six curves in each graph denoting three coordinate's projection of joint forces of shoulder

and elbow respectively. From the y-axis component of two graphs, larger joint forces are required when hands move with higher velocities.

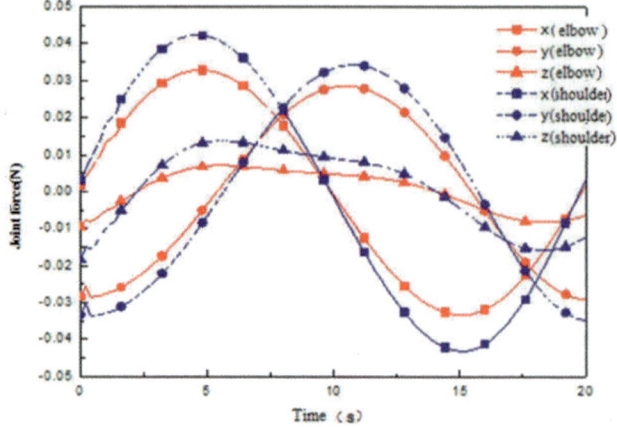

Fig.11. Projections of shoulder and elbow joint forces when the period is 20s, and the run distance is 10cm

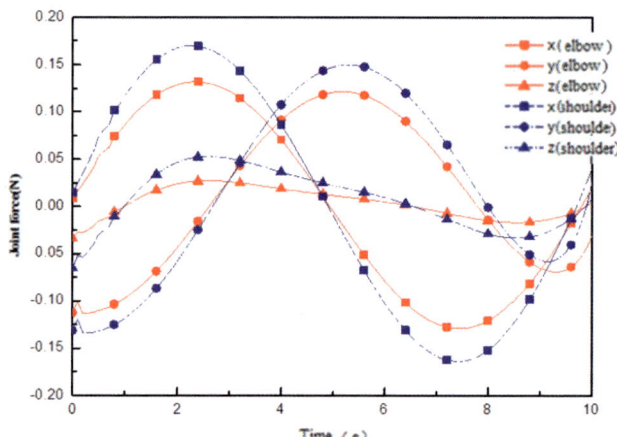

Fig.12. Projections of shoulder and elbow joint forces when the period is 10s, and the run distance is 20cm

III. CONCLUSION

1) Simulations of cylindrical helix motions of an astronaut upper limb outside the spacecraft under a microgravity environment were achieved in this study. Firstly, the joints' kinematical parameters were calculated by the reverse kinematical equation; secondly, simulations of the corresponding virtual model were performed based on LifeMOD platform. Finally, simulation results were verified by comparing the simulation trajectory of the mass center of hands with that of hands' known motion. Also, the joint strength comparisons were performed when hands move along different motion directions in the human body coordinate system.

2) Three cylindrical helix motions were simulated, which include hands moving along three axes in the human body coordinate system. The joint forces under different motion directions were compared such as hands moving along different axes and different motion velocities. The results show that less joint forces are

needed when an astronaut's hands move along cylindrical helix path in the sagittal axis. More joint forces are needed when hands move with higher velocities. It has a great value for on-ground training and planning of astronaut tasks in orbit.

3) The model used in this study has some limitations, that is, the fixed lower limb and torso which lead to inconsistency with human motions where most body joints are involved. Therefore, more research needs to be developed towards the simulation of the astronaut tasks that whole body joints participate under feet fixed.

REFERENCES

[1]   "Hubble Space Telescope Servicing Mission 1 Media Reference Guide" by NASA in Lockheed Martin Missiles & Space, 1993.

[2]   "Hubble Space Telescope Servicing Mission 2 Media Reference Guide" by NASA in Lockheed Martin Missiles & Space, 1997.

[3]   "Hubble Space Telescope Servicing Mission 3A Media Reference Guide" by NASA in Lockheed Martin Missiles & Space, 1999.

[4]   B. Nelson, M. Higashi, P. Sharp. "Hubble Space Telescope Servicing Mission 3B Media Reference Guide" in Lockheed Martin, Lockheed Martin Missiles & Space, 2002.

[5]   B. Nelson, M. Higashi, P. Sharp. "Hubble Space Telescope Servicing Mission 4 Media Reference Guide" in Lockheed Martin Missiles & Space, 2009.

[6]   A.P.Yang, C.X.Yang, P.Ke, "Modeling and Verification of an Astronaut Handling Large-mass Payload" in Chinese Journal of Mechanical engineering, vol. 23, no. 4, pp.517-523, Jun. 2010.

[7]   A.P.Yang, C.X.Yang. "Numerical Simulation of an Astronauts' Task" in Proc. 9rd Conf. Engineering and Technological Innovation. Orlando, FL, 2008, pp. 323-327.

[8]   A.P.Yang, Chau-Kuang Chen, Guang Cheng, Simulation of Astronaut's Movement for Orbital Replaceable Units, proceedings of 2011 IEEE 18th international conference on industrial engineering and engineering management, Changchun. 1749 – 1753, 2011.

[9]   LI Jing-wen, DING Li, YANG Ai-ping, Biomechanical simulation and verification of astronaut extravehicular activities, Journal of Medical Biomechanics, Vol.27 No. 4, Aug. 2012.

[10] Z.H. Ji, E. Zhang, Z.H. Liu. "Simulation and Experiment Research on Human Riding Comfort for Vehicle Vibration in Dynamic Environment", in Proc. 9rd Conf. Computer-Aided Industrial Design and Conceptual Design. KunMing, YN, 2008, pp. 175-181.

[11] Y.H. Ho, L.C. Hsiehl, H.W. Wu. "Study of the Simulation in Kinematics and Kinetics in Normal Gait using various parameters" in Proc. IEEE 35rd Conf. Bioengineering. Boston, MA. 2009, pp. 31-32.

[12] Y. Su, J.G.Qian, Y.W.Song. "The Application of LifeMOD" in Chinese Journal of Nanjing Institute of Physical Education, vol. 6, no. 4, 2007, pp. 1-3.

Research on the Coal Mine Production Logistics Security Status Based on Key Resources Recognition

Jin-feng WANG[1], Yun-fei AN[1,*], Li-jie FENG[1,2], Xue-qi ZHAI[1]

[1]Management Engineering Institute, Zhengzhou University, Zhengzhou, China
[2]Henan Provincial Coal Seam Gas Development and Utilization Co. Ltd., Zhengzhou, China
(ayf1990@163.com)

Abstract - **To identify the safety of coal mine production logistics system and point out the focus of the safety work, this paper establishes index system of resources based on analysis of coal mine production logistics system. The evaluation system can be improved by entropy weight method through identification of key resources. Finally, according to the improved evaluation system, coal mine production logistics system is evaluated through the support vector machine (SVM) classification algorithm, to distinguish system safety level and provide a reference for coal mine production logistics safety management.**

Keyword - **Coal mine production logistics, entropy weight method, key resources recognition, support vector machine (SVM)**

I. INTRODUCTION

There is a big demand of coal resources in our country, thus the coal mining has become particularly important. But frequent disasters with high mortality have been seriously influenced on the coal mine production in the coal industry. In the coal mine accident, nearly a quarter is the result of logistics. Therefore, the logistics is an important link in safety production of coal mine, has had a huge impact on coal mine safety production [1]. Coal mine production logistics refers to the whole process of coal from working face to outward transport [2], which can realize all materials of whole coal mine production process space transfer function. Therefore, in order to promote the improvement of safety production and eliminate safety loophole in coal mine, it's necessary to distinguish between the secure state of coal mine production logistics system.

At present, researches on evaluation of coal mine have many achievements. F. D. Wu [3] adopted Fuzzy - AHP method to determine the weights of evaluation factors of coal mine, in establishing the evaluation index system constructed evaluation model on the basis of safety. For providing a reference for coal mine provide a reference for coal mine, Li Xinchun [4] used system dynamics to build coal mine safety investment driving mechanism model and evaluated the safety investment driving mechanism of coal mine. Through technology archives information management in the importance of the safety evaluation, Liu Nan [5] introduced the technology of archives information in the application of the safety evaluation. Li Bin [6] and so on made the SVM model

applied to comprehensive evaluation of the essence of coal mine safety management which can be qualitative question to carry on the quantitative evaluation, the result to reasonably reflected the essence of coal mine safety management of the status quo. According to the statistics and analysis on the Chinese coal mine accidents in recent years, Chen Kun [7] constructed the fuzzy comprehensive evaluation model of coal mine enterprise safety culture. It concluded that coal mine enterprise safety culture index system of education training and rewards and punishment system were the key to the construction of safety culture. Jian-ning Gao [8], ensured the safety of the mine and formulated prevention and cured countermeasures by using the grey comprehensive evaluation model of entropy queuing analysis to carry out the possibility of coal mine accidents. Yang Wei [9] got coal mine safety evaluation value and determined the coal rank through artificial intelligence neural networks method.

In conclusion, the current domestic and foreign scholars did many researches of the overall safety factor of coal mine had made certain achievements, but the study of evaluation in coal mine production logistics system is relatively rare. Zhao Mingzhong [10] has made the evaluation in coal mine production logistics system by the BP neural network. However, because of the BP neural network's learning efficiency low, easy to fall into local minimization and so on shortcomings, this method has too long training time and failure occurs in actual use. Therefore, this article employs other evaluation methods, in order to solve the shortcomings of BP neural network method to evaluate the coal mine.

This paper, firstly, uses the entropy weight method to identify the key safety resources for coal mine production logistics system, and then builds a model of support vector machine (SVM) classification and uses the identified key resources on the system to evaluate system safety state. Finally, verify the effectiveness of key resources in the security rating and evaluation accuracy through the instance. Results can reflect the coal mine production system safety situation objectively and accurately, and provide guidance for coal mine safety production.

II. COAL MINE PRODUCTION LOGISTICS SYSTEM SECURITY RESOURCES

In this paper, based on analyzing the characteristics of each subsystem, it preliminary screens subsystem security state security resources from equipment, technology, personnel and so on: Mining subsystem includes 4

---

[1] Acknowledgments: This research is sponsored by National Nature Science Fund (No. 71271194) and Science & Technology Program of Zhengzhou City (141PPTGG343).

influence factors such as mining machinery, mining technology management, the relationship between the mining, mine ventilation; Ventilation subsystem includes 3 influence factors such as ventilation safety monitoring, the ventilation technology management, equipment configuration; Geological measure in the prevention and control of water subsystem includes 3 factors such as technical personnel, drainage institutions, emergency rescue capabilities. Auxiliary subsystem includes 3 influence factors such as security capacity of transport, electrical and mechanical safety management, ability to predict environment.

Finally, the evaluation index system of coal mine safety production logistics resources are determined.

III. ESTABLISHING OF COAL MINE PRODUCTION LOGISTICS SAFETY STATE EVALUATION MODEL BASED ON IDENTIFICATION OF KEY RESOURCES

A. The Fusion principle of entropy weight method and SVM

First, this paper selects the key security resources of system. Next, according to the selected security resources, the paper establishes new evaluation index system. Finally, it takes advantage of new evaluation index system to evaluate the system.

At present, there are many domestic about index evaluation methods: One kind is subjective value method. This method usually uses qualitative ways of comprehensive consulting score. However, it is difficult to reflect the objective relationship among the objects because of strong subjectivity. Another kind is objective method. This method is effective because it avoids the deviation caused by the subjective factor and is more objective [11]. Therefore, this paper uses the entropy weight method to calculate the history data of coal mine production logistics system security resources and come about the weight of each resource, so as to realize the recognition of critical resources. Filtrate indexes of evaluation system according to the crucial degree of system resources.

The coal mine production logistics system security evaluation is a typical nonlinear, high dimension classification problem. There are particularities of coal mining, high cost of collecting data, and the sample size is small. SVM is recognized as the best tool to solve this problem.

B. Evaluation model

This paper combines entropy weight method and support vector machine (SVM) effectively. Entropy weight method screens the key safety indexes, and SVM is adopted to training samples and build coal mine production logistics system security status recognition model. As shown in Fig.1.

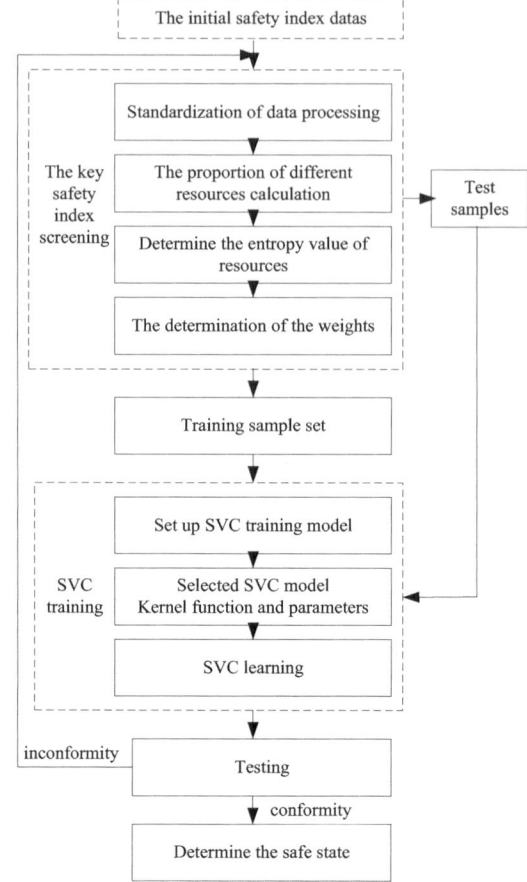

Fig.1. Framework of coal mine production logistics system security state identification model

1) The identification of coal mine production logistics system key resources based on the entropy weight method:

Summarize each system security resources associated the number of accidents. $x_{ij}(i=1,2,\ldots,n; j=1,2,\ldots,m)$ is the number of accidents associated the resource i in the j year.

Data normalization process

$$\begin{cases} x'_{ij} = x_{ij} / Maxx_{ij} \\ x'_{ij} = Minx_{ij} / x_{ij} \end{cases} \quad (1)$$

Calculate each year the proportion of various system resources a_{ij}

$$a_{ij=} \frac{x'_{ij}}{\displaystyle\sum_{i=1}^{m} x'_{ij}} \quad (2)$$

Calculate the h_i.

$$h_i = - \frac{\displaystyle\sum_{j=1}^{m} a_{ij} \ln a_{ij}}{\ln m} \quad (3)$$

Define weights CL_i.

$$CL_i = \frac{1-h_i}{\displaystyle\sum_{i=1}^{n}(1-h_i)}, \quad 0 \le CL_i \le 1 \quad (4)$$

2) *Establish the evaluation model based on SVM:*

Support vector machine (SVM) is proposed initially to solve the problem of two-dimensional linear separable. It evolved from the linearly separable case optimal classification plane, and the basic idea of this kind of classification method can be expressed in Fig.2. Rectangles and triangles represent two classes of samples. In the figure, rectangles and triangles represent two classes of samples, H is the line of classification. H₁ and H₂ across every kind of sample points which are the nearest to the line of classification and parallel to classification line. The distance between them is called classification interval. The effect of the optimal classification line is not only required to separate two kinds of samples correctly, but also make the largest classification interval [12].

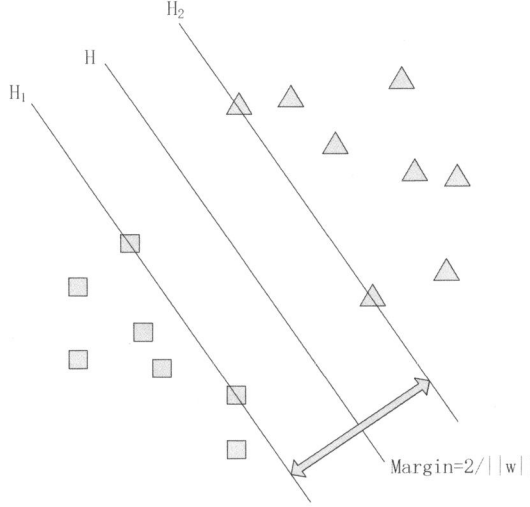

Fig.2. The optimal classification plane sketch

Classification for line equation is $xw + b = 0$. After normalization, it makes sample set of linearly separable $(x_i, y_i), i = 1, 2, \ldots, n, x \in R_d, y \in \{+1, -1\}$ satisfy:

$$y_i \left[(wx_i + b) \right] - 1 \geq 0 \qquad (5)$$

At this time, classification interval is equal to $2/\|w\|$. To obtain largest interval, it is equivalent to get the minimum of $\|w\|^2$. The optimal classification plane satisfies (5) and make $1/(2\|w\|^2)$ achieve the minimum classification plane. The training sample points in H₁ and H₂ have become support vectors. According to Lagrange optimized method, the optimal classification plane problem can transform into dual problem. The constraint conditions:

$$\begin{cases} \sum_{i=1}^{n} y_i a_i = 0 \\ a_i \geq 0, (i = 1, 2, \ldots, n) \end{cases} \qquad (6)$$

To solve the maximum function:

$$Q(a) = \sum_{i=1}^{n} a_i - \frac{1}{2} \sum_{i,j=1}^{n} a_i a_j y_i y_j (x_i x_j) \qquad (7)$$

a_i is Lagrange multiplier corresponding to each sample.

Through find inequality constrained quadratic function optimization problem only solution, it can get the optimal classification function, the optimal classification function is get:

$$f(x) = \text{sgn}(wx + b) = \text{sgn}\{\sum a_i^* y_i (x_i x) + b^*\} \qquad (8)$$

SVM can be gradually extended to solve the multiple classification and nonlinear problems. The basic idea is to use nonlinear transform defined by inner product function to conform input space into a high dimensional space. Find nonlinear relationship between the input variable and output variable in the high dimensional space. By the theory of fonctionelle, as long as satisfy the Mercer conditions, it can use kernel function $K(x_i, x_j)$, which is know the specific transformation, to realize the linear classification without increase of computational complexity after the nonlinear transformation [13]. At this time, the objective function (9) turns into:

$$Q(a) = \sum_{i=1}^{n} a_i - \frac{1}{2} \sum_{i,j=1}^{n} a_i a_j y_i y_j K(x_i x_j) \qquad (9)$$

The corresponding function classification becomes:

$$f(x) = \text{sgn}(wx + b) = \text{sgn}\{\sum a_i^* y_i K(x_i x) + b^*\} \qquad (10)$$

At present, there are three kinds of SVM commonly used kernel functions [14]:

Polynomial kernel function: $K(x_i, x) = (x_i x + 1)^d$

Radial basic kernel function: $K(x_{i,} x) = \exp(-\frac{\|x - x_i\|^2}{\sigma^2})$

Two-layer NN kernel functions: $K(x_i, x) = \tanh(kx_i x + \theta)$

This paper will use radial basic kernel function as a kernel function because of its characteristics of strong local learning ability [15].

According to the standard in coal mine production logistics safety evaluation framework, it uses the technology simulation to generate sufficient number of evaluation index sequences.

In the number k values evaluation grades, it sets the lower limit and upper limit of evaluation index values are a_j^k and b_j^k. y_j^k are their levels of the corresponding evaluation. There is the evaluation index random simulation formula:

$$x_{ij}^k = \text{rand}(n_k)(a_j^k - b_j^k) + b_j^k \qquad (11)$$

i is capacity index sequence generated by a certain evaluation level. $i = 1, 2, \ldots, n_k$. k is number of evaluation grade. j is evaluation index number. According to formula (11), it can put the number k evaluation class into n_k group (x_{ij}^k, y_j^k). And then rearrange its subscript to get a new sequence (x_{ij}, y_i). The new sequence serves as the training sample.

In the SVM model, it regards x_{ij} evaluation index in stay evaluation system as the input of the model. Evaluation rating severs as the output of the model.

IV. EMPIRICAL ANALYSIS

According to the analysis of all coal mine accidents in a certain area in the recent 3 years, this paper summarizes the number of accidents associated to each system security resources. Show as TABLE I.

TABLE I
STATISTICAL TABLE OF COAL MINE PRODUCTION LOGISTICS SYSTEM RESOURCES ACCIDENT ASSOCIATED NUMBER

Resource \ Year	2011	2012	2013
mining machinery C_1	6	7	6
Mining technology management C_2	2	1	1
the relationship between the mining C_3	9	6	6
ventilation C_4	10	9	8
ventilation safety monitoring C_5	8	8	7
ventilation technology management C_6	7	6	7
equipment configuration C_7	7	8	7
technical personnel C_8	7	7	6
drainage institutions C_9	7	6	6
emergency rescue capabilities C_{10}	1	2	1
security capacity of transport C_{11}	7	5	6
electrical and mechanical safety management C_{12}	7	8	6
ability to predict environment C_{13}	1	1	1

Through the formula of entropy weight method, the data in TABLE I change into TABLE II.

TABLE II
THE WEIGHT OF COAL MINE PRODUCTION LOGISTICS SYSTEM SECURITY RESOURCES

CL_1	CL_2	CL_3	CL_4	CL_5	CL_6	CL_7
0.092	0.019	0.094	0.102	0.098	0.093	0.097

CL_8	CL_9	CL_{10}	CL_{11}	CL_{12}	CL_{13}
0.094	0.092	0.019	0.090	0.095	0.008

It gets rid of non-critical ($CL_i \leq 0.05$) system resources C_2, C_{10}, C_{13} to form a new system security evaluation system. Evaluate the system using the system security evaluation system.

Coal mine production logistics system security state is the embodiment of the various resources effective utilization value. And resources utility value (UTV) can be shown as the product of total investment (TI) of resource and resource utilization rate (RUR):

$$UTV = TI * RUR \quad (0 < RUR \leq 1)$$

RUR is evaluation by experts.

Calculate the region's UTV of every coal mine production logistics system. For the convenience of statistics and establishing of evaluation standard, it normalizes UTV and gets the relative resource utilization rate ($RRUR$):

$$RRUR = (TCMI * TMUR) / TREUVIN * 100\%$$

$TCMI$ is the coal mine investment. TMUR is the mine utilization rate. $TREUVIN$ is top resources effective utilization value in China.

Coal mine production logistics system security state is divided into three grades according to the relative resource utilization. According to the related history data

and expert assessment, it builds the evaluation framework. Shown in TABLE III.

TABLE III
GRADES AND STANDARDS OF EVALUATION

No.	C_1	C_3	C_4	C_5	C_6	Corresponding evaluation
1	≥75%	≥80%	≥85%	≥85%	≥80%	Safe
2	≥55%	≥60%	≥70%	≥65%	≥60%	General
3	<55%	<60%	<70%	<65%	<60%	Dangerous

No.	C_7	C_8	C_9	C_{11}	C_{12}	Corresponding evaluation
1	≥80%	≥80%	≥75%	≥80%	≥85%	Safe
2	≥65%	≥60%	≥55%	≥60%	≥70%	General
3	<65%	<60%	<55%	<60%	<70%	Dangerous

"Safe" state represents excellent security situation in coal mine production logistics system, which is far better than the national standard level. The accident probability is much lower than others. "General" state represents domestic standard security situation level in coal mine production logistics system and keeps the average risk of accident. "Dangerous" state represents Coal mine production logistics system security is having a serious shortage of resources allocation. Security level is far lower than the average and prone to accidents.

Select 45 coal mines in the same place and the 30 as the training sample, 15 (K_1, K_2, ..., K_{15}) as testing samples. Relevant test sample data are shown in TABLE IV.

TABLE IV
PRODUCTION LOGISTICS SAFETY RESOURCE ASSESSMENT FOR EACH MINE (%)

	C_1	C_3	C_4	C_5	C_6	C_7	C_8	C_9	C_{11}	C_{12}
K_1	82	92	94	93	94	87	85	98	90	93
K_2	79	85	86	91	93	87	84	95	89	88
K_3	78	82	87	90	91	86	85	89	86	86
K_4	78	84	85	88	89	86	84	83	81	90
K_5	77	69	80	88	85	85	65	83	79	77
K_6	77	65	78	87	84	84	64	83	78	72
K_7	77	65	76	87	84	83	64	81	75	80
K_8	76	64	76	79	79	75	63	80	74	78
K_9	73	64	75	74	74	73	62	78	74	75
K_{10}	69	64	73	69	74	63	62	77	73	76
K_{11}	69	63	57	59	68	59	61	75	72	65
K_{12}	67	52	55	55	65	59	59	74	46	64
K_{13}	67	45	53	53	64	49	57	73	34	63
K_{14}	65	44	48	44	55	39	46	55	28	56
K_{15}	63	34	28	44	35	23	23	43	26	53

LibSVM [16], which is designed By Taiwan University, Lin Chih-Jen, has a few adjustable parameters and characteristics of fast calculation. It provides the interactive test (including many kinds of pattern recognition based on regression and one-to-one algorithm) function.

Through the matlab 8.0 software, this paper uses the Grid search method and PSO method to solve the data respectively.

The initial parameters are shown in TABLE V.

TABLE V
THE SVM INITIAL PARAMERER SETTINGS

Method	Range of initial parameters
Grid search method	$cmax=8$, $cmin=-8$, $gmax=8$, $gmin=-8$
PSO method	ga maxgeneration=100, ga popsize=20, $0<c<100$, $0<g<100$

Through the calculation, the results are shown in TABLE VI.

TABLE VI
THE OUTPUT OF LIBSVM TABLE

	K_1	K_2	K_3	K_4	K_5	K_6	K_7	K_8
Grid search method	1	1	1	1	2	2	2	2
PSO method	1	1	1	1	2	2	2	2

	K_9	K_{10}	K_{11}	K_{12}	K_{13}	K_{14}	K_{15}
Grid search method	2	2	3	3	3	3	3
PSO method	2	2	3	3	3	3	3

The evaluation results of two methods are basically identical. The evaluation results and the actual situations are much the same. Therefore, the results show that evaluation through the key resources of the coal mine production logistics system is effective, so that the system resources selected have a huge impact on the system. At the same time, this evaluation method can be applied to find the coal mine production logistics system security level, to provide guidance for the work of coal mine production logistics security.

REFERENCES

[1] Zhang Taifa, Mu Lihua, Zhang Hongyan. Analysis and study on coal mine accidents and prevention measures [J]. China Mining Magazine2012, (03):28-31

[2] Yang Zhihong, Shao Bin. Research on Coal Production Logistics System [J], Coal Technology, 2012, 01:271-272.

[3] F. D. Wu, N. L. Hu. Study on the model of safety evaluation in coal mine based on Fuzzy-AHP comprehensive evaluation method [J]. Proceedings -2011 International Conference on Mechatronic Science, Electric Engineering and Computer, 2011:1671-1674.

[4] Li Xinchun, Liu Quanlong. Research on the Analysis and Evaluation of Safety Input Dynamic System in Coal Mine Enterprises [J]. Science & Technology and Economy, 2014, 02:91-95.

[5] Liu Nan. Theory of technical archives information management work of coal mine safety evaluation [J]. Shaanxi Coal, 2013, 02:135-136+132.

[6] Li Bin, Wang Zhijun. SVM model for comprehensive evaluation of coal mine inherent safety management and its application [J]. Mining Safety & Environmental Protection, 2013, 05:117-120.

[7] Chen Kun, Xu Longjun, Yi Jun. The evaluation of coal mine enterprise safety culture based on the principle of SMART [J]. Journal of Safety and Environment, 2010, 06: 226-230.

[8] Gao Jianning, Li Chengwu. Grey entropy model applied in the evaluation of coal mine [J]. Safety in Coal Mines, 2007, 09:87-90.

[9] Yang Wei, An Mingyan, Wang Qiuju. Quantitative analysis of artificial intelligence neural networks in risk assessment of gas accidents in coal mine [J]. Opencast Mining Technology, 2007, 05:57-59.

[10] Zhao Zhongming. Evaluation of coal mine production logistics system based on evidence theory and neural network [D]. Zhengzhou University, 2010.

[11] Jiang Huiyuan, Wang Hao. Evaluation of Supply System of Inland Water Transport Based on Entropy Proportion Means [J]. Waterway Engineering, 2008, 06:1-6.

[12] Lu Min, Zhang Zhanyu. Evaluation of sustainable utilization of water resources based on SVM [J]. Hydroelectric Energy, 2005, 05:18-21+4.

[13] Zhang Chaoyang. Study on evaluation of product innovation ability of private enterprises and improvement measures [D]. Tianjin University, 2009.

[14] Sun Huali, Xie Jianying, Xue Yaofeng. A Customer Satisfaction Degree Evaluation Model Based on SVM in Logistics [J]. Journal of Shanghai Jiaotong University, 2006, 04: 684-688.

[15] Liang Liming, Xia Yuchen. Liver Disease Identification Based on Hybrid Kernel SVM [J]. Industrial Control Computer, 2013, 09: 97-99.

[16] Zhu Peigen, Mei Weijiang, Shi Xiufeng, Bian Jinying. Research on the method of effective power increase of the alternative fuel forecast based on LibSVM [J]. Journal of Shihezi University (Natural Science), 2012, 05:657-660.

Universal Simulation Model of Autonomous Vehicle Storage & Retrieval System

Sai-peng ZHANG, Ning ZHAO*, Yin-fan ZHAO

School of Mechanical Engineering, University of Science and Technology Beijing, Beijing, China

(zhning@sina.com)

Abstract - **Autonomous vehicle storage and retrieval system (AVS/RS) is a relative new solution for automatic storage and retrieval, which is based on autonomous vehicle technique. Based on conventional simulation technique, warehouse designers have to build different simulation models according to different design scenarios. New simulation technology is needed to improve the efficiency. A universal simulation model based on modularization is proposed, which can quickly study different design scenarios and estimate the performances (the transaction cycle time and the utilization). Finally, the model is validated through a case study.**

Keywords - **Autonomous vehicle storage and retrieval system, modularization approach, simulation**

I. INTRODUCTION

Autonomous vehicle storage and retrieval systems (AVS/RS) is a material-handling facility which composed of vehicles, lifts and storage racks [1]. The key distinction of AVS/RS relative to traditional automated storage and retrieval systems (AS/RS) is the movement patterns of S/R devices [2]. In AVS/RS, the unit loads are handled by vehicles moving horizontally, moving vertically by lifts. Differently in AS/RS, the unit loads are handled by crane and simultaneously move in horizontal and vertical dimensions. Compare with AS/RS, several transactions can be arranged simultaneously in AVS/RS and therefore better performances can be reached.

With the difference of movement patterns, the design of AVS/RS is difficult to traditional AS/RS. In order to satisfy the performance requirements (transaction cycle time, utilization of vehicle and lift, etc), the designers need to decide the number of aisles, tiers, columns and the number of vehicles and lifts. Furthermore, the speed, acceleration/deceleration, storage/retrieval strategies of vehicles and lifts are also need to consider. There have been some studies to this design problem, such as:

"Tier to tier" and "tier captive" are the two main configurations of AVS/RS [3]. In the tier to tier configuration, the number of vehicle is always less than the number of tiers. Conversely, in the tier captive configuration, the number of vehicle is equal to the number of tiers.

The first study on the performance of an AVS/RS was presented by Malmborg (2002) [2]. With reference to a "tier to tier" configuration, a state equation model was built to estimate the performance indicators: vehicle utilization and transaction cycle time. Furthermore, Malmborg (2003) extended the state equation model by considering the proportion of dual command cycles in

autonomous vehicle storage and retrieval systems [4]. Cycles that include both storage and retrieval transaction are referred to as dual command (DC) cycles, conversely, cycles that include single transaction are defined as single command (SC) cycles.

Due to the complexity of calculation, more and more studies use queuing theory to build the models. Kuo et al. (2007) presents a "tier to tier" configuration system based on the imbedded queuing theory to estimate the transaction cycle time [5], and the imbedded queuing theory is the combination of M/G/V and G/G/L. Besides, Fukunari and Malmborg (2009) apply the network queuing to the "tier to tier" configuration, and estimate the performances of system[6].

As mentioned above, the analytical model can be built quickly and estimate system performances efficiently. And simulation is an accurate way to estimate system performances, but time consuming and inefficiency are the shortcoming. Therefore, in most papers, the combination of simulation model and analytical model is applied to the comparison and analysis. Namely, researchers build the analytical models to estimate the system performances, and results are validated via simulation. Take Kuo et al. (2007) and Zhang (2009) as an example, both of them combined the analytical model and simulation model to estimate the performance [5, 7]. However, there are also many papers just rely on simulation models to estimate the system performances. Such as, Ekren (2010) built the model based on simulation to estimate the effects of rack configuration [8]. Besides, Ekren and Heragu (2010) also provided model based on simulation to estimate transaction cycle time in function with dwell point policy and I/O points location [9].

There are many factors affect the system performance. Such as type of command cycles [4-7], rack configuration and storage policy [9-13]. As for the system performance indicators, researches mainly focus on the following aspects: storage and retrieval efficiency, utilization, transaction cycle time and waiting time. Roy (2012) presents "tier to tier" configuration model based on half-open network queuing theory, and provides the calculation method of lift and vehicle waiting time [12]. In addition, some papers also take the length of queues into consideration [13].

However, how to build a universal simulation model to validate different design scenarios is still lacking. The warehouse designers have to build one simulation model to one design scenario, which is time consuming. Furthermore, design analysis is always neglected because of the inefficiency of simulation.

E. Qi et al. (eds.), *Proceedings of the 21st International Conference on Industrial Engineering and Engineering Management 2014*, Proceedings of the International Conference on Industrial Engineering and Engineering Management, DOI 10.2991/978-94-6239-102-4_67, © Atlantis Press and the authors 2015

The main purpose of this paper is to offer a universal modularization simulation model to research the AVS/RS [14]. In addition, we also estimate the following transaction performance indicators: the transaction cycle time, the vehicle and the lift utilization, assuming single aisle tier captive configuration [15-16].

The paper is organized as follows. Section II presents a modularization approach based on simulation to research the system. The cases studied by modularization simulation are described in Section III and Section IV presents the conclusions.

II. SIMULATION MODELING

As mentioned in Section I, simulation is an accurate way to estimate performance, but time consuming and inefficiency is the shortcoming. In this section, a modularize approach to AVS/RS is reported, which could quickly transform the simulation model to different design scenarios [17-18], and the simulation software used in this paper is Em-plant simulation (version 9.0).

Fig. 1 presents an AVS/RS system. As the figure illustrates, Lifts are mounted at fixed positions on the periphery of the storage, and the input/output point is located at the first tier beside each lift. In a single aisle, a single lift is installed, and the number of vehicles is equal to the number of tiers. The first position at each tier of storage serves as a buffer: one buffer handles the unit loads which have been retrieved called buffer out, the other one, located in the other sides of the lift, handles the unit loads to be stored called buffer in.

Fig.1. An AVS/RS system

In order to develop a simulation model that can be transformed via different values assignment, modularization approach is used in this paper. The aim of this approach is dividing simulation into several basic modules and each module can assemble with others by interface [19]. The simulation model for different design scenarios can be assembled by modules via different input values. The simulation model can be determined by four kinds of basic modules, namely: orders module, warehousing information module, management module and rack module. Details about these modules are as follows:

A. Orders Module

Before operating the AVS/RS system, we need read the storage orders and the retrieval orders. Due to the both orders have the same format and the similar functions, so two kinds of orders are set in same module. The order module dialog box is shown in Fig.2, order types are

divided into storage orders and retrieval orders, and the order data come from four areas, namely: local data, external data, user-defined and automatic generation.

In order to add the reusability of modules, two kinds of data interface are used: the Excel data interface and the Socket data interface.

The local data can be written into system through the Excel interface by input the file name. As for the external data, the system can achieve telecommunications with external data by Socket interface. When User-defined button is selected, the order information is directly written into the system. Another way is Automatic generation, when you click this button, the system will generate an order with 50-line automatically. The name of goods is composed of alphabet and numbers, and the number of goods is generated randomly.

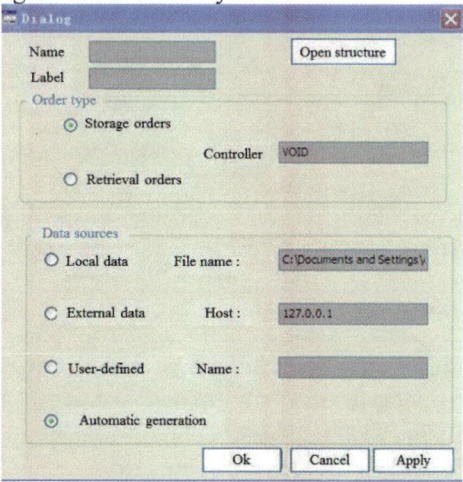

Fig.2. Order module dialog box.

TABLE I
THE RANDOM PARAMETERS OF ORDERS

Order type	Goods types (Evenly distributed)		Number (Evenly distributed)	
	The lower limit	The upper limit	The lower limit	The upper limit
Multi- varieties and large batches	1	13	7	11
Multi- varieties and small batches	1	13	1	5
Fewer- varieties and large batches	1	5	7	11
Fewer- varieties and small batches	1	5	1	5

Fig.3. The icon of order module

In order to better simulate the actual system, we provide four kinds of order types, the details are shown in Table I.

To facilitate the users, the order module is added to the software user-defined toolbar, and the icon of module is shown in Fig.3.

B. Warehousing Information Module

The warehousing information module also provides three kinds of data generation types: local data, external data and automatic generation. As shown in Fig.4, this module has the same data transmission modes with order module.

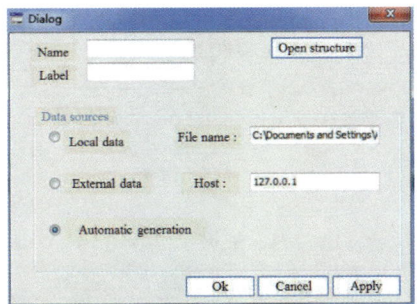

Fig.4. Warehousing information module dialog box

In the warehousing information table, the following information is designed: the fourth column is defined as tier number, and the seventh column is column number per aisle, and the fifth column is the bays I/O direction relative to vehicle. There are only two kinds of values in the fifth column, 0 notates the left side of vehicle and 1 notates the right side. Obviously, any bays in the system can be positioned by the fourth, seventh, fifth columns. In addition, the third column is goods name stored in the corresponding bays, and the second column is the corresponding number of this type goods.

Similarly, to facilitate the user, the warehousing information module is added to the software user-defined toolbar, and the icon of module is shown in Fig.5.

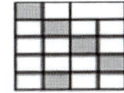

Fig.5. The icon of warehousing information module

C. Management Module

All kinds of order information and warehousing information are connected by management module. Besides, this module can also provide storage and retrieval strategies for the system, and re-integration of orders information and warehousing information is needed before each transaction which can be operated through clicking the register button. Obviously, the management module is the system core component. The management module dialog box is shown in Fig.6, and Fig.7 is the icon of module.

In the actual situation, the order number is uncertain. In order to simulate situations with a steady stream of orders, this module also provides a dynamic remote order interface for the system.

As for the storage and retrieval strategies, this module provides five kinds of transaction strategies for storage and retrieval transaction, respectively. In addition, the storage and retrieval strategies are designed according to the heuristic. Therefore they have openness, and are integrated with simulation system. Details about storage and retrieval strategies are shown in Table II.

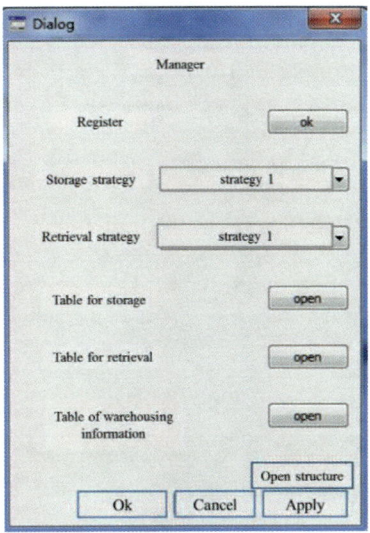

Fig.6. Management module dialog box

Fig.7. The icon of management module

TABLE II
FIVE KINDS OF STORAGE AND RETRIEVAL STRATEGIES

Number	Storage strategies	Retrieval strategies
1	The bay containing the same goods has priority from bottom to top tiers.	The bays located in the lower tiers take priority over higher tiers
2	The half-full bays have priority from bottom to top tiers	The half-full bays have priority over empty bays from bottom to top tiers.
3	The empty bays have priority from bottom to top tiers.	The full bays have priority from bottom to top tiers.
4	The bays with the shortest length from I/O point have priority	The bays with the shortest length from I/O point have priority.
5	The bays involving retrieval transaction have priority.	The bays involving storage transaction have priority.

D. Rack Module

In order to develop a simulation model that can be transformed via different value assignment, the rack module is designed, and the system model for different design scenarios can be assembled automatically by basic components via different input values.

The rack module builds the foundation of simulation model. As can be seen from Fig.8, through this module, the user can choose rack configuration: "tier captive" or "tier to tier". Besides, three type of parameter interfaces are presented, the first is rack-related parameter interface, the second is vehicle-related parameter interface, and the last is lift-related parameter interface, specific values are as follows.

- Rack-related parameter: number of tiers, number of columns, storage unit capacity, length of system and height of system.
- Vehicle-related parameter: number of vehicles, the maximum vehicle velocity, the acceleration and the deceleration of vehicle.
- Lift-related parameter: number of lifts, the position of lifts, the maximum lift velocity, the acceleration and the deceleration of lift.

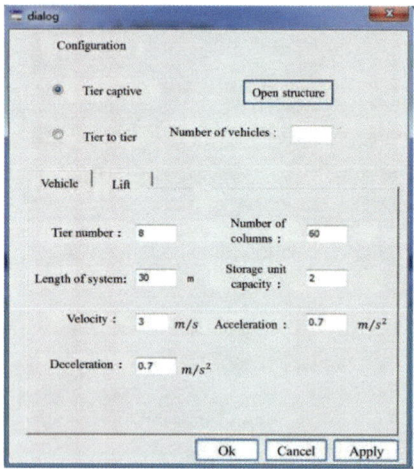

Fig.8. Rack module dialog box

Fig.9. The icon of rack module

Fig.10. The icon of rack basic modules

Obviously, the efficiency of simulation is improved significantly by this module. The icon of rack module is shown in Fig.9, and the Fig.10 shows the icon of rack basic modules.

According to the above module division, we will present a transaction flow chart in Fig.11. First, the system will generate storage orders, retrieval orders and warehousing information tables, then all these data will be integrated and sent to management module. Second, the system will assign tasks to vehicles and lift, and ready to performance transactions, furthermore, the corresponding information will update after each transaction operating. Finally, the updated information will provide feedback and integrate in management module again.

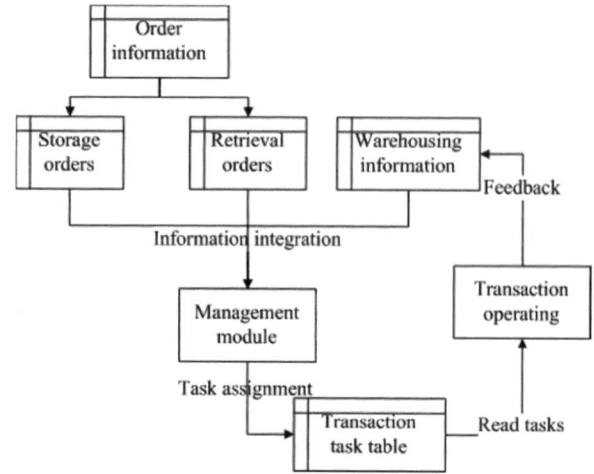

Fig.11. The transaction flow chart

III. CASE STUDY

In this section, we will apply the modularization simulation approach to a single aisle AVS/RS with tier captive configuration. In this paper, only single retrieval transaction will be taken into account. The motivation of this choice is twofold. First, retrieval transactions represent the most critical activities compared with storage transactions, and they cannot be postponed. Second, once a single retrieval transaction has been modelled, the process may be easily extended to the case of single storage transaction. The flow chart of retrieval transaction can be seen in Fig.12.

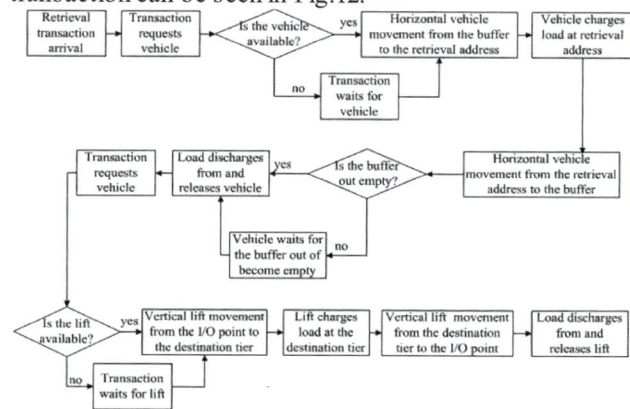

Fig.12. The flow chart of retrieval transaction

The notation used in the studied single aisle "tier captive" AVS/RS is as follows.

V_{tran} time allowance for charging and discharging load from vehicle

L_{tran} time allowance for charging and discharging load from lift

T number of tiers

C number of columns per aisle

μ_w unit width clearance per bay including allowances

μ_h unit height clearance per bay including allowances

v_h maximum horizontal velocity of vehicle

a_h acceleration deceleration of vehicle

d_h deceleration of vehicle

v_v maximum vertical velocity of lift

a_v acceleration of lift

d_v deceleration of lift

The main assumptions of the "tier captive" AVS/RS are:

- The number of unit loads handled per cycle by lift and vehicle is 1;
- Each bay and each buffer can hold two unit loads;
- The dwell point policy of lifts and vehicles is the point-of-service-completion (POSC).
- The transfer time of unit loads to and from vehicles is assumed to be 2s each.
- The transfer time of unit loads to and from lifts is assumed to be 3s each.

TABLE III
MAIN DATA FOR THE SCENARIO ANALYSIS

Variable	Unit of measure	Data
μ_w	m	0.5
μ_h	m	0.8
v_h	m/s	2
a_h	m/s^2	0.5
d_h	m/s^2	0.5
v_v	m/s	2
a_v	m/s^2	0.5
d_v	m/s^2	0.5

The number of tiers and number of columns have been taken as constant, and equal to 8 and 30, respectively. Besides, four kinds of order types have been examined, which can be seen from Table II. Retrieval strategy can be defined as RS_i (i is equal to the corresponding number in Table II), and there are five options according to Table II. Therefore, twenty scenarios have been examined, and data reported in Table III have been considered.

The models are running for 10 independent replications, and the confidence is 90%. The simulation results with respect to the transaction cycle time, as well as lift and vehicle utilization are reported in Table IV.

TABLE IV
THE SIMULATION RESULTS WITH RESPECT TO THE TRANSACTION CYCLE TIME, AS WELL AS LIFT AND VEHICLE UTILIZATION

Order types	RS_i	Transaction cycle time [s/transactions]	Vehicle utilization [%]	Lift utilization [%]
Multi-varieties and large batches	1	24.78	7.71%	38.33%
	2	24.66	7.68%	38.54%
	3	25.92	7.42%	40.66%
	4	23.75	6.63%	46.94%
	5	24.79	7.71%	38.33%
Multi-varieties and small batches	1	20.76	9.74%	22.12%
	2	20.36	9.69%	22.53%
	3	24.77	7.71%	38.35%
	4	18.24	5.64%	54.90%
	5	20.73	9.74%	22.12%
Fewer-varieties and large batches	1	26.55	6.93%	44.58%
	2	26.62	6.94%	44.48%
	3	26.68	6.98%	44.20%
	4	26.62	6.91%	44.77%
	5	26.56	6.93%	44.58%
Fewer-varieties and small batches	1	23.63	7.91%	36.75%
	2	23.77	7.94%	36.52%
	3	24.77	7.49%	40.07%
	4	22.71	6.22%	50.26%
	5	23.61	7.91%	36.75%

IV. CONCLUSION

In this paper, a universal simulation model is used to research the AVS/RS. The aim of the model is estimate AVS/RS performance efficient and convenient. Through the modularization AVS/RS, users just change input values according to different scenarios, and the corresponding models will be generated automatically. The case study showed that the simulation models could quickly estimate the performance of different AVS/RS design scenarios. Therefore, this study proposed a useful way to help warehouse designers selecting the best AVS/RS design solution.

ACKNOWLEDGMENT

This paper is supported by the natural science foundation of China (71301008), Beijing natural science foundation (9144030) and Beijing higher education young elite teacher project.

REFERENCE

[1] Kees Jan Roodbergen, Iris F.A.Vis, "A survey of literature on automated storage and retrieval systems (Periodical style)," European Journal of Operational Research., vol. 194, pp. 343-362, 2009.

[2] Malmborg, C.J, "Conceptualizing tools for autonomous vehicle storage and retrieval systems (Periodical style)," International Journal of Production Research, vol. 40, no. 8, pp. 1807-1822, 2002.

[3] G. Marchet, M. Melacini, S. Perotti, and E. Tappia, "Analytical model to estimate performances of autonomous vehicle storage and retrieval systems for

product totes (Periodical style)," International Journal of Production Research., vol.50, no. 24, pp. 7134-7148, 2012.

[4] Malmborg, C. J, "Interleaving dynamics in autonomous vehicle storage and retrieval systems (Periodical style)," International Journal of Production Research., vol. 41, no. 5, pp.1057-1069, 2003.

[5] Po-Hsun Kuo, Ananth Krishnamurthy, Charles J. Malmborg Charles, "Design models for unit load storage and retrieval systems using autonomous vehicle technology and resource conserving storage and dwell point policies (Periodical style)," Applied Mathematical Modelling., vol.31, pp.2332-2346, 2007.

[6] MiKi Fukunari, Charles J. Malmborg, "A network queuing approach for evaluation of performance measures in autonomous vehicle storage and retrieval systems (Periodical style)," European Journal of Operation Research., vol. 193, pp. 152-169, 2009.

[7] Li Zhang, Ananth Krishnamurthy, Charles J. Malmborg, Sunderesh S. Heragu, "Variance-based approximations of transaction waiting times in autonomous vehicle storage and retrieval systems (Periodical style)," European J. Industrial Engineering., vol.3, no. 2, pp.146-169, 2009.

[8] Banu Yetkin Ekren, Sunderesh S. Heragu Charles. "Simulation-based regression analysis for the rack configuration of an autonomous vehicle storage and retrieval system (Periodical style)," International Journal of Production Research, vol. 48, no, 21, pp. 6257-6274, 2010.

[9] Banu Y. Ekren, Sunderesh S. Heragu, Ananth Krishnamurthy, Charles J. Malmborg, "Simulation based experimental design to identify factors affecting performance of AVS/RS (Periodical style)," Computers & Industrial Engineering., vol. 58, no.1, pp. 175–185, Feb. 2010.

[10] Debjit Roy, Ananth Krishnamurthy, Sunderesh Heragu, Charles Malmborg, "Impact of zone on throughput and cycle times in warehouses with autonomous vehicles (Presented Conference Paper style)," *5th Annual IEEE Conference on Automation Science and Engineering.* Bangalore, pp.449-454, Aug.2009.

[11] M. Fukunari, C. J. Malmborg, "An efficient cycle time model for autonomous vehicle storage and retrieval systems (Periodical style)," International Journal of Production Research., vol.46, no.12, pp.3167-3184, 2008.

[12] Debjit Roy, Ananth Krishnamurthy, Charles J. Malmborg, "Performance analysis and design trade-offs in warehouses with autonomous vehicle technology (Periodical style)," IIE Transaction, vol.44, no.12, pp.1045-1060, 2012.

[13] Banu Y. Ekren, Sunderesh S. Heragu, "Performance comparison of two material handling systems: AVS/RS and CBAS/RS (Periodical style)," International Journal of Production Research., vol.50, no.15, pp.4061-4074, 20112.

[14] Pratt, D. B. Farrington, P. A. Basnet, C. B. Bhuskute, H. C. Kamath, M. Mize, J.H, "A framework for highly reusable simulation modeling: separating physical, information, and control elements (Presented Conference Paper style)," *Proceedings of the 24th Annual Simulation Symposium.* IEEE, New Orleans, pp.254-261, 1991.

[15] Carliss Y. Baldwin, Kim B. Clark, *Design Rules: the Power of Modularity* (Book style). Beijing, China Citic Press, 2006

[16] Hecker, Mary-Ellen, "Modular Simulators: How to make it work (Presented Conference Paper style)," *IEEE*

Proceedings of the National Aerospace and Electronics Conference. Dayton, IEEE, pp. 1080-1085, 1985.

[17] Arturo I. Concepcion, "A hierarchical computer architecture for distributed simulation (Periodical style)," IEEE Computer Society., vol. 38, no. 2, pp.311-319, 1989.

[18] ZHAO Ning, DONG Shaohua, WANG Guohua, " AS/RS planning based on three-phase simulation (In Chinese)", Mechanical Engineering School. University of Science and Technology Beijing., vol.29, no.10, pp.1054-1059, 2007.

[19] Ulgen, Onur M, Thomasma. T, Otto. N, "Reusable models: making your models more user-friendly (Presented Conference Paper style)," *1991 Winter Simulation Conference Proceedings*, Phoenix, AZ, PP.148-151, 1991.

Research on Cylinder Buffer Control Based on AMESim

Chao Song *, Ming He

Department of Mechanical Engineering, Beijing Institute of Technology, Beijing, China
(songchao918@126.com)

Abstract – **The cylinder buffer at the end of the working stroke of a manual spot welding was studied based on AMESim in this paper. Due to the special working environment, the buffer control needs to be simple and reliable, this paper used the way of exhaust throttle at the end of the working stroke to achieve buffer control. The improved buffer system was analyzed and the mathematic model was established. And the model diagram was simulated by AMESim. Through the analysis of simulation model, the simulation parameters were determined on the buffering effect of the improved system. And the cylinder buffer capacity was improved effectively after the system was made better. The pneumatic system was made to adapt to different conditions when the throttle control value opening time and the throttle value area were adjusted.**

Keywords - **AMESim, buffer, cylinder motion, modeling and simulation**

I. INTRODUCTION

The cylinder is widely used in the pneumatic system, one of its advantages is effective driving mechanism of high speed motion. High speed is one of the inevitable development trend of modern cylinder [1-2]. Improving the cylinder speed is not difficult, as long as increasing the effective area of exhaust channel can realize high speed driving [3]. The key question is how to make the high speed movement of the cylinder to be stopped in the terminal without developing severe impact. When the operation condition of the large inertia load or high speed, the buffer is particularly important. So, in order to effectively avoid the impact of institutional components deformation and ensure the safety of the system, it is necessary to deal with buffer control in the high speed movement mechanism and make the striking velocity at the end of movement within a certain range. So, we need to study the process of high speed movement at the end of the buffer and determine the appropriate buffer control parameters. To make the pneumatic mechanism achieve high speed movement, and control the impact velocity [4-6].

II. THE IMPROVED DESIGN OF CYLINDER BUFFER STRUCTURE

The actual spot welding cylinder structure diagram is shown in Fig.1, the cylinder is mainly composed of piston, steel cylinder, the front and rear end cover and the piston rod [7]. This structure of cylinder has a collision at the end of piston movement.

To establish the simulation model in the AMESim software after making changes to the existing cylinder system. As shown in Fig.2:

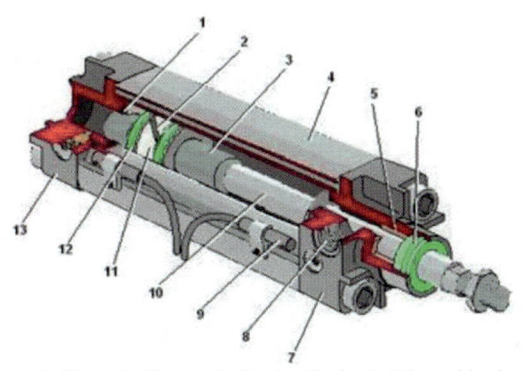

1. Plunger 2. Piston 3. Plunger 4. Steel cylinder 5. The guide sleeve 6. The dustproof plug 7. The front cover 8. Port 9. Sensor 10. Piston rod 11. Wear ring 12. Sealing ring 13. The rear end cover
Fig.1. Cylinder structure

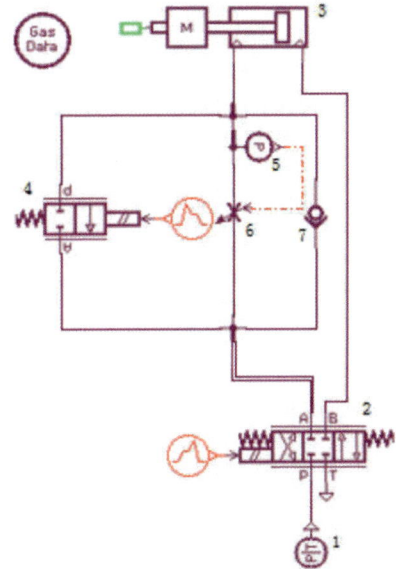

1. Air source 2.Three-position four-way directional control valve 3.Cylinder 4.The throttle control valve 5.Pressure relay 6.Throttle valve
Fig.2. Pneumatic system diagram

The cushion process of the system is: when the cylinder is in high-speed motion, the gas of the rod chamber exhausts through the throttle control valve; throttle control valve closes when it is closed to the end of piston movement. Due to gas pressure of the rod chamber did not reach the setting opening pressure of the pressure relay, so the rod chamber pressure increases rapidly, absorbs the impact energy and the cylinder speed decreases. When the rod chamber pressure reaches the setting pressure of the pressure relay, the gas drain away from the throttle valve, at the same time the gas energy of the rod chamber is released. When the gas pressure of the rod chamber go down to the setting pressure of the

E. Qi et al. (eds.), *Proceedings of the 21st International Conference on Industrial Engineering and Engineering Management 2014*, Proceedings of the International Conference on Industrial Engineering and Engineering Management, DOI 10.2991/978-94-6239-102-4_68, © Atlantis Press and the authors 2015

pressure relay, the throttle valve no longer works, a small amount of residual gas is consumed by the pipeline of the pneumatic system. The buffer of the system not only ensures movement speed of the cylinder, but also provides throttling exhaust at the end of the movement.

III. THE MATHEMATICAL MODEL OF THE SYSTEM

Due to the compressibility of gas, so the model should be set according to the basic characteristics. And it should be analyzed according to the basic theory of gas dynamics and thermodynamics characteristics [8]. Due to the compressibility of gas, the change of gas pressure directly affects the gas density, and gas in the energy transmission and throttling process will cause the changes of the gas flow [9]. Therefore, we need to do the necessary simplification in the establishment model [10]. (1) Charging and discharging process of the cylinder follows the ideal gas law; (2) Thermodynamic process of gas follows the adiabatic process. The pipeline is converted to the two chamber volume of the cylinder in the derivation of mathematical model and making the value components to be equivalent synthesis [11].

A. The Energy Equation

Charging and discharging process of cylinder chamber is a kind of thermal process of a variable system. According to the equation $KRT_sdM_s=Vdp+kpdV$ and $Q_{m1}=dM_s/dt$, it is available of the rodless chamber pressure equation:

$$\dot{p}_1=\frac{kRT_sQ_{m1}}{V_1}-\frac{kp_1}{V_1}\dot{V}_1 \qquad (1)$$

$$V_1=A_1\left(x_{10}+x\right) \qquad (2)$$

Where V_1 is an air intake chamber volume, m³; A_1 is the rodless cavity area; p_1 is the absolute pressure of the cylinder cavity, Pa; x is piston displacement, m; x_{10} is the equivalent length of the cylinder clearance volume, m; Q_{m1} is the mass flow rate of the intake pipe, kg/s; T_s is the air source temperature, K;

Similarly, the pressure equation of the rod chamber is shown as follows:

$$\dot{p}_2=-\frac{kRT_2Q_{m2}}{V_2}-\frac{kp_2}{V_2}\dot{V}_2 \qquad (3)$$

$$V_2=A_2\left(x_{20}+L-x\right) \qquad (4)$$

Where V_2 is the exhaust chamber volume, m³; A_2 is the exhaust cavity area, m²; p_2 is the absolute pressure of the exhaust cavity, Pa; x_{20} is the equivalent length of the exhaust cavity clearance volume, m; Q_{m2} is the mass air flow of the exhaust pipe, kg/s; T_2 is the temperature of the exhaust cavity, K; L is the cylinder stroke, m.

For the drive system has an initial pressure difference, the initial pressure of the exhaust cavity is p_s, and the initial temperature is T_s. Based on the relationship of isentropic-process state parameters, they must satisfy the equation: $T_2=T_s\left(p_2/p_s\right)^{\frac{k-1}{k}}$.

B. The Dynamics Equation

According to Newton's second laws, the equations of motion for cylinder piston are shown as follows:

$$\begin{cases} \ddot{x}=\dfrac{\left[p_1A_1+p_0\left(A_2-A_1\right)-p_2A_2-F\right]}{M} \\ \left(x=0\cap p_1A_1+p_0\left(A_2-A_1\right)>p_2A_2+F\right) \\ \cup\left(0<x<L\right) \\ \ddot{x}=\dfrac{\left[p_2A_2+p_0\left(A_2-A_1\right)-p_2A_2+F\right]}{M} \\ \left(x=L\cap p_1A_1+p_0\left(A_2-A_1\right)+F<p_2A_2\right) \\ \ddot{x}=0 \\ \left(x=0\cap p_1A_1+p_0\left(A_2-A_1\right)\leq p_2A_2+F\right) \\ \cup\left(x=L\cap p_1A_1+p_0\left(A_2-A_1\right)+F\geq p_2A_2\right) \end{cases} \qquad (5)$$

Where M is the mass of cylinder piston and drive components, kg; p_0 is the atmospheric pressure, Pa; F is the force acting on the piston, N.

C. The Mass Flow Equation

The flow of pneumatic components can be expressed as follow:

$$Q_m=\frac{A_ep_u}{\sqrt{RT_u}}\psi(\sigma) \qquad (6)$$

$$\psi(\sigma)=\begin{cases} \sqrt{2\sigma(1-\sigma)} & b<\sigma=\dfrac{p_d}{p_u}\leq 1 \\ \dfrac{\sqrt{2}}{2} & \sigma=\dfrac{p_d}{p_u}\leq b \end{cases} \qquad (7)$$

Where P_d, P_u is the pressure of upstream and downstream, MPa; T_u is the upstream temperature of the pipe system, K.

D. The System Simulation Model and the Discussion about the Simulation Results

This paper simulates the pneumatic system by using the AMESim, the AMESim is a high-level modeling and simulation software launched by the French IMAGINE company and it provides a complete platform that involves the system and engineering, at the same time it can provide the system model in various fields for the user including pneumatic, mechanical control, etc. This version of the paper is LMS Imagine Lab AMESim REV13. In the process of simulation, the cylinder diameter is 65mm, the piston stroke is 50mm, and the air pressure is 0.7 MPa.

Fig.3 and Fig.4 are the simulation curves when the mass of load is 30kg. The solid line shows the velocity and displacement of cylinder rod before the improvement of the pneumatic system, and the dotted line represents the velocity and displacement of cylinder rod after the improvement of the pneumatic system. There is a collision and rebound before the improvement of the pneumatic system. It can be seen that the speed reduced to 0 m/s and then the piston has a reverse movement which is shown in the solid line of Fig.3 and the piston has a rebound that is shown in the solid line of Fig.4. After the improvement of the pneumatic system, the speed curve is moderate, as shown in the dotted line of Fig.3. Finally, the displacement curve that is shown in Fig.4 gradually reaches the maximum stroke.

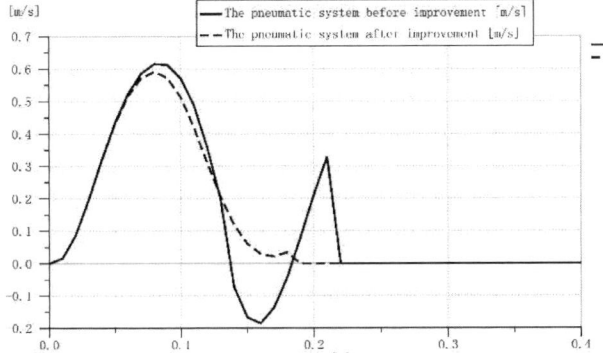

Fig.3. Cylinder speed curves.

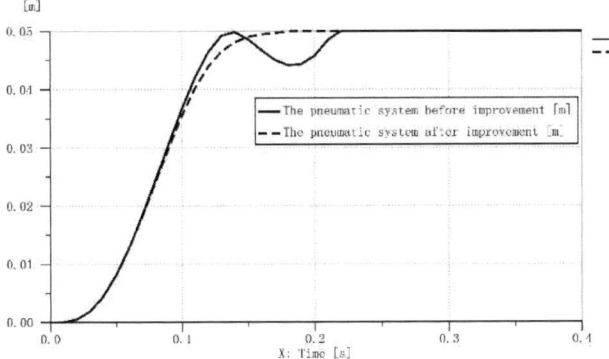

Fig.4. Cylinder displacement curves.

In the process of simulation, the parameter settings of the improved pneumatic system are the same as the original system. When the rodless cavity is inflated, and the throttle control value is opened, the piston gets an accelerated motion. Closing the throttle control value after it is been kept open for 0.1 second. At this time, the gas pressure of the rod cavity rises, when the pressure reaches the system setting pressure, the gas in the rod cavity exhausts from the throttle value. When the opening area of the throttle value is 6 mm², it can provide an effective buffer. In the process of simulation, the change of throttle value area has a great impact on the velocity curve and displacement curve. When the piston area and the cylinder stroke change, the throttle value opening area should be considered firstly to realize the pneumatic system buffer.

IV. CONCLUSION

In this paper, making a study of the exhaust throttling pneumatic system modeling and simulation, and the simulation model is verified right.

(1) Establishing the motion process mathematical model of the single rod cylinder. The buffer impact of exhaust process is researched. The simulation parameters are used the real parameters of the actual cylinder to verify the correctness of the mathematical model.

(2) Proposing a scheme about the improvement of the existing manual spot welding pneumatic system. The scheme can solve effectively the cylinder buffer at the end of the piston movement. This way of exhaust throttle can also absorb the impact energy.

(3) Adjust the system through the throttle control value opening time and the value flow of the throttle value, in order to adapt the system to different conditions. Therefore, adjusting the appropriate buffer system in this paper is easier than the ordinary cylinder.

REFERENCES

[1] Y. S. Wang, "Study on the Pneumatic System of Rail Grinding Platform Based on the Co-simulation of AMESim and ADAMS" (in Chinese) [D]. Beijing: Beijing Jiaotong University, 2012.

[2] X. Y. Liu, "Study on Modeling and Simulation of Pneumatic Cylinder under Low Speed Stick-slip" (in Chinese) [D]. Harbin: Harbin University of Science and Technology, 2012

[3] N. Zhao, "A Research on Proportional value-based Pneumatic Positioning Control of Linear Actuator" (in Chinese) [D]. Lanzhou: Lanzhou University of Technology, 2009.

[4] C. B. Jing, "Simulation Study of Hydraulic Free-Piston Engine Based on AMESim/MATLAB" (in Chinese) [J].Journal of System Simulation, 2009, 21(23):7681-7685.

[5] SINGH. R, "Modeling of an Impulse-absorbing Pneumatic Cylinder" [J]. Journal of Sound and Vibration, 1982(4):598-600.

[6] Whitehead. J. C, "Hydrogen Peroxide Gas Generator Cycle with a Reciprocating Pump" [R]. AIAA2012-3702.

[7] L. Liao, "The Research and Implementation for Performance Test Platform of Cylinder under the Shock Load" (in Chinese) [D]. Chengdu: School of Mechatronics Engineering, 2013.

[8] J. C. Ream, C. M. Liao, "A Study on the Speed Control Performance of Servo-pneumatic Motor and the Application to Pneumatic Tools" [J]. The International of Advanced Manufacturing Technology, 2010, (29):7-8.

[9] Q. L. Huang, J. X. Li, Y. Zheng, M. X. Yang, "Numerical Simulation of Unsteady Inner Flow Field in Reciprocating Piston Pump" (in Chinese) [J]. Journal of Chongqing University of Technology (Natural Science), 2011, 25(3):6-10.

[10] LUOXH, CAOSP, ZHUYQ. "Simulation on the Impact Pneumatic Cylinder with a Reservoir" [J]. Chongqing Uneven Ed, 2009, 7(1):47-51.

[11] W. L. Quan, X. X. Yao, F. Lin, "Modeling and Linearity Analysis of Ram-air Servo System" (in Chinese) [J]. ACTA ARMAMENTARL II, 2010, 31(8):1125-1129.

Design of the Integrity Inspection System for the Resistor Color Code

Lin-wei MAO*, Yao-guang HU, Jing-qian WEN, Jia-wei KE

School of Mechanical Engineering, Beijing Institute of Technology, Beijing, China

(maolinwei@139.com)

Abstract - **To reduce the inspection cost and make sure the electronic components have high quality and performance, it is important for the research and design of more advanced automation quality inspection systems. In this paper, an integrity inspection system for the resistor color code is proposed and the detailed structure and performance of the system is studied and analyzed. The resistor rotation strategy is adopted in the integrity inspection system. The integrity inspection system could be popularized to make the resistor quality management more easy and stable.**

Keywords - **Equipment design, integrity detector, resistor color code, rotation**

I. INTRODUCTION

There is a growing demand for the electronic components, with the rapid development of electronic, new material and intelligent technologies. The manufacturing factories of electronic components are dispersed factories, producing different kinds of electronic components with different sequence of operations. All the manufacturing processes are arranged according to the purchase orders from clients. The producing process of electronic components is very fast based on the popularized automatic production line. However, the duplicate production of electronic components is very common for the semi-manufactured products and finished products. The raw materials and production technologies have a huge influence to the quality of finished products [1~2].

The increase of electronic component varieties and outputs leads to the competition of different manufacturing factories. Therefore, the requirement of electronic component quality is being stricter. The quality and reliability of electronic components have a great influence on the performance of electronic products, especially for the aerospace and military electronic products. The manufacturing factories of electronic components have to attach more importance to the quality and reliability managements of electronic components. The quality management runs all through the selecting, testing, purchasing, inspecting, assemblage, debugging and failure analysis processes of electronic components. To ensure the high quality of electronic components, the controlling system must be a closed system, including the selection of components, internal quality evaluation, secondary screening, destructive physical analysis, failure analysis, quality tracking and quality databases sections. All the sections are all connected together. One section could affect all the other sections. Therefore, anyone of the sections is indispensable and important. The finished product quality of electronic components is determined by the closed system and all the sections.

To make sure the electronic components have high qualities and performances, the quality testing and analyzing processes of all sections should be accurate, reliable and efficient. With the increasing demand for electronic components, the automation equipments are becoming widely used in the manufacturing process. In the manufacturing factory with a large-lot production, the automation quality inspection system has a high accuracy and efficiency than manual inspection, without the problem of human weariness resulted by manual inspection. In the long run, the advanced automation quality inspection system could also reduce the inspection cost. Therefore, the research and design of more advanced automation quality inspection systems become a tendency in the study of quality control process of electronic components.

At present, the research all over the world on the online visual inspection of metal film resistor are mainly focused on the identification of numerical value of resistances based on the resistor color code [3-7]. The visual inspection system for the imperfection of resistor color code has seldom been studied. Therefore, a new integrity inspection system of resistor color code for the metal film resistor is developed in this paper. The integrity inspection system is composed mainly by two parts, which are image acquisition module and mechanical execution module [8].

II. PROBLEM STATEMENT

Objective analysis of the integrity inspection system

This study aims at developing a system for integrity inspection of resistor color code. The detailed equipment of the inspection system will be studied and proposed.

To inspect the integrity of resistor color code, the completed image of the resistor color code is the most important part [9]. However, the image from all around the resistor color code of metal film resistor could not be obtained by only one shoot of the optical camera. Therefore, this study develops a new strategy by capture three images from different positions and different angles to get a completed image of the resistor color code [10]. If only one optical camera is available, the metal film resistor must be rolling forward in the image capture area to make sure all the surrounding color code is completely captured by the optical camera. Therefore, while the optical camera is capturing the images, the metal film resistor should circumferentially rotate for one circle [11], as Fig.1 shown.

The circumferential rotation is generally driven by a mechanical clamping system. The mechanical clamping system has many merits, such as easy controlling, precise rotating, mature driving and other technologies. However, the metal film resistor is too light in weight for the mechanical clamping system to get control. In the production lines, the metal film resistor will be produced, checked and recorded with a large amount [12~14]. Therefore, it is hard to make the metal film resistor get rotating.

Fig.1. Image capture system

III. METHODOLOGY AND RESULTS

A. Strategy for resistor rotation

It is necessary to simplify the mechanical clamping system to cut down the cost of equipment. In this study, a new mechanical clamping system is proposed.

The basic function of the new system is fixing both ends of the metal film resistor in a groove and then driving the groove to make the resistor rolling and moving forward. The rotating speed of the resistor is controlled by the speed difference between the groove and conveyor. Therefore, the metal film resistor is rotating while moving forward on the conveyor. Then it is easy to capture the completed image of the resistor color code by simply placing the optical camera over the conveyor [15]. The new mechanical clamping system with a groove is shown in Fig.2.

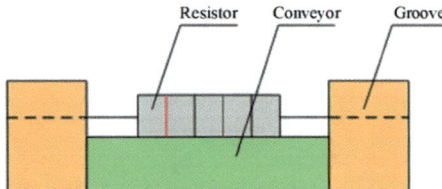

Fig.2. Mechanical clamping system

The detailed rotating situation of the resistor is mainly decided by two factors. The first one is the speed difference between the groove and conveyor, and the second one is the frictional force between the resistor and conveyor. Therefore, the mechanical clamping system with a high performance could be realized by controlling the speed and surface roughness of the conveyor.

B. Mechanical actuator

Based on the proposed mechanical clamping system, the integrity of the resistor color code will be examined by a fixed optical camera while the resistor is rolling through on the conveyor. The function of the mechanical actuators is to make sure the resistor could rotate through under the optical camera smoothly and steadily.

Fig.3 shows the schematic structure of the mechanical actuator system. The mechanical actuator is composed by an electrical motor, powered shaft system, driven shaft system, supporting system, transmission system, detecting system and rolling system.

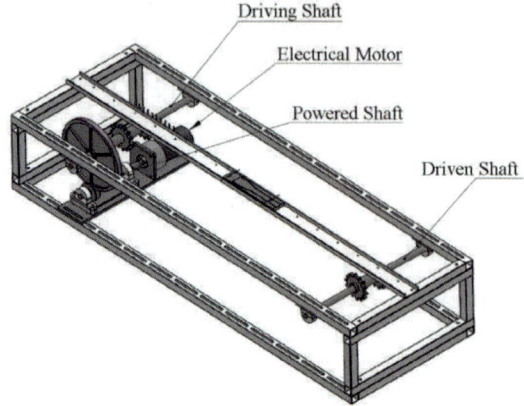

Fig.3. Mechanical actuator system

In the industrial process, the electrical motor is a direct current electromotor with the rotational speed of 30 rpm. The power shaft is connected with the electrical motor directly. The power shaft exports the shaft power to the driving shaft by a gear system with the drive ratio of 1:4. The rotational speed of driving shaft is 7.5 rpm. Two chain gears with the same specification are located on the driving shaft. The number of the chain gear teeth is 14. Another two chain gears with the same specification are located on the driven shaft. The shaft power is delivered from the driving shaft to the driven shaft by two chains between the chain gears. The operation speed of chains is 22.2 mm/s.

The resistor rolling system is the most important part in the mechanical actuator system. The structure of resistor rolling system is shown in Fig.4.

Fig.4. Resistor rolling system

The movement of resistor is driven by the iron bar fixed on the chain. When the resistor reaches the friction block, the resistor starts rolling and moving forward. It is a pure rolling process when the resistor is rotating on the friction block. The integrity inspection of the resistor color code is accomplished after the resistor rolling for one circle.

IV. CONCLUSION

The integrity inspection of the resistor color code is very important for the quality management in the electronic component producing process. It is necessary to perform automation production instead of artificial production to ensure the high quality of electronic components. In this study, an integrity inspection system for the resistor color code is proposed. The detailed structure and performance of the system is studied and analyzed. The resistor rotation strategy is adopted in the integrity inspection system. The optical camera could receive a completed image of all surround the metal film resistor. It is very easy and reliable to analyze the received images and manage the quality of metal film resistors. The integrity inspection system could be popularized to make the resistor quality management more easy and stable.

ACKNOWLEDGMENT

This work was financially supported by the Industrial Engineering Laboratory of Beijing Institute of Technology.

REFERENCE

[1] Du GL, Zhang P. Hot-rolling strip steel end online vision measurement device used in on-site production environment, has visual information processing system that is connected with coil detection device monitoring system by universal serial bus wire [J]. Robotics and computer-integrated manufacturing, 2013(29): 484-492.

[2] Zhu Y, Hu C, Hu J, et al. Accuracy and simplicity oriented self-calibration approach for two-dimensional precision stages. IEEE Trans Ind Electron, 2012(60): 2264-2272.

[3] Meng Yan, Zhuang Han qi. Autonomous robot calibration using vision technology. Robotics Comput-Integr Manufacturing, 2007(23): 436–46.

[4] Rui bo, He Ying jun Zhao, et al. Kinematic-parameter identification forserial-robot calibration based on POE formula. IEEE Trans Robotics, 2009(3): 411–423.

[5] Meng Yan, Zhuang Han qi. Autonomous robot calibration using vision technology. Robotics Comput-Integr Manufacturing, 2007(23): 436–46.

[6] Chen Pei-Yin. VLSI Implementation of an edge-oriented images caling processor. IEEE Trans Very Large Scale Integr (VLSI) Syst, 2009(17): 1275–84.

[7] Andreff N, Martinet P. Unifying kinematic modeling, identification, and control of a Gough–Stewart parallel robot into a vision-based frame work. IEEE Trans Robotics, 2006(22): 1077–86.

[8] Ma DL, Liu CS, Zhao Z, et al. Rolling friction and energy dissipation in a spinning disc. Proceedings of the Royal Society A-Mathematical Physical and Engineering Sciences, 2014(470):1364-1471.

[9] Chen JS, Li CS. Control of Surface Thermal Scratch of Strip in Tandem Cold Rolling. Chinese Journal of Mechanical Engineering, 2014(27): 738-744.

[10] Ni DR, Chen DL, Wang D. Tensile properties and strain-hardening behaviour of friction stir welded SiCp/AA2009 composite joints. Materials Science and Engineering A-Structural Materials Properties Microstructure and Processing, 2014(608): 1-10.

[11] Zhang SH, Song BN, Wang XN. Analysis of plate rolling by MY criterion and global weighted velocity field. Applied Mathematical Modelling, 2014(38):3485-3494.

[12] Zhang DH, Cao JZ, Xu JJ. Simplified Weighted Velocity Field for Prediction of Hot Strip Rolling Force by Taking into Account Flattening of Rolls. Journal of Iron and Steel Research International, 2014(21): 637-643.

[13] Ernesto A, Mazuyer D, Cayer-Barrioz J. The Combined Role of Soot Aggregation and Surface Effect on the Friction of a Lubricated Contact. Tribology Letters, 2014(55): 329-341.

[14] Li FX, Liu YZ, Yi JH. Modeling the Thermal Fields of Deposited Materials during the Spray Rolling Process. Metallurgical and Materials Transactions A-Physical Metallurgy and Materials Science, 2014(45A): 4012-4021.

[15] Branscomb and D. G. Beale, "Fault detection in braiding utilizing low-cost USB machine vision," Journal of the Textile Institute, vol.102, pp.568-581, 2011.

Differences in Episodic Memory between Patients with Alzheimer's Disease and Normally Aging Individuals

Min-Sheng CHEN*, Wei-Ru CHEN

Department of Industrial Engineering & Management, Yunlin University of Science and Technology, Douliou, Taiwan
(chens@yuntech.edu.tw)

Abstract - **There are numerous problems associated with a greater prevalence of old age, one of which is dementia. The prevalence of dementia increases year by year, and patients with Alzheimer's disease account for the largest proportion of patients with dementia. In early stages of Alzheimer's disease patients have significant episodic memory impairment. An experiment was conducted to explore the differences in episodic memory performance between patients with Alzheimer's disease and normal aging. In the experiment, participants carried out a place-object word pair recall task. Different consistency of word pairs and delays before recall was allowed were investigated. Overall, Alzheimer's disease patients showed impairments in episodic memory performance. The results showed better performance for consistent word pairs, and for immediate recall. Based on these results, we offer some suggestions for auxiliary devices and cognitive therapies that might be used in the future for patients with Alzheimer's disease.**

Keywords - **Alzheimer's disease, consistency, episodic memory, recall delay**

I. INTRODUCTION

The prevalence of dementia increases year by year and Alzheimer's disease patients account for the largest proportion of those with dementia. Because of brain damage associated with Alzheimer's disease, patients have deficits in episodic memory [1], which is evident from early in the disease process [2]. Episodic memory refers to memory regarding specific events, places, and times [3]. Baddeley and Hitch (1974) [4] proposed that there are three parts of short-term memory, a visuo-spatial sketch pad, a phonological loop, and a central executive unit. The visuo-spatial sketch pad is responsible for visual and spatial memory; the phonological loop is responsible for auditory and semantic memory; and the central executive unit is required to coordinate and control information delivery between the visuo-spatial sketch pad and the phonological loop. Baddeley (2000) [5] proposed the concept of an episodic buffer; he suggested that there are multidimensional memory buffers, involving both phonological and visual memory units. Other scholars [1, 6, 7] have used various stimulus bindings (e.g., different times, sequences, and modalities) in visual memory tasks to explore the episodic buffer model. They found that people can bind the different stimulus dimensions, and difficulty levels vary, depending on the nature of the stimuli.

To explore episodic memory, studies have used variations on word pair recall and recognition tasks. Stories, word pairs, and individual words are manipulated in recall and recognition tasks to assess episodic memory, as well as geometric pictures. The Wechsler Scale of Adult Intelligence [8] also uses a word pair task to assess episodic memory. Bäckman (2001) [9] used a concrete word recall and recognition task to assess the episodic memory performance of patients. Participants should free recall the word that they saw after the presentation of stimuli. Then, there would present some words as stimuli, and participants were asked to determine whether these words ever appeared in the free recall task or not. In contrast, Nordahl et al. (2005) [10] used pictures whose content was arbitrarily colored red or green to assess the episodic memory performance of patients with mild cognitive impairment. To increase the connection between color and picture, participants were asked to suggest reasons for the coloration based on their own personal experiences. In a subsequent recall task, participants we required to state the color each picture had during its initial presentation. It was found that patients had significant impairment in their memory performance. Small and Sandhu (2008) [11] explored the differences in episodic and semantic memory performance between Alzheimer's disease patients and normally aging controls. Two factors, familiarity and time, were manipulated in their study. They used common, unique, dated, and contemporary pictures instead of words as the stimuli. They found that patients and controls performed best on common and dated words in a naming task. This indicates that the long-term memory of patients may be intact.

However, the semantic recall tasks which only word or unrelated word pairs used in prior research may have been too simple to accurately assess episodic memory. To improve upon prior tasks, the present study took connection concept into account; we manipulated word pairs that consisted of a place noun and an object noun. The specific place and the specific object constituted an episodic memory. Participants were tasked to recall the specific object. To assess the ability of episodic memory to accurately discriminate between Alzheimer's disease patients and individuals aging normally, these stimuli were designed to be relevant to the real life of participants.

II. METHODOLOGY

To explore differences in episodic memory between patients with Alzheimer's disease and individuals aging normally, a place-object word pairs recall task was used. The consistency of word pairs and recall delay were manipulated in the experiment. We investigated the recall of given words under different conditions.

A. Participants

Twenty-four individuals participated in the experiment. They were divided into two groups: Alzheimer's disease patients (n = 12; mean age =76.42

E. Qi et al. (eds.), *Proceedings of the 21st International Conference on Industrial Engineering and Engineering Management 2014*, Proceedings of the International Conference on Industrial Engineering and Engineering Management, DOI 10.2991/978-94-6239-102-4_70, © Atlantis Press and the authors 2015

years, SD = 5.35; 8 women), and a normal aging group (n = 12; mean age = 73.67 years, SD =5.789; 5 women). The Alzheimer's disease patients were recruited from the National Taiwan University Hospital, Yun-Lin Branch. All patients were in early stage Alzheimer's disease; their score on the Clinical Dementia Rating scale (CDR) was 1. Normally aging controls were chosen from the Association for the Older in Yunlin. All participants in the normal aging group had no known organic brain disorder.

B. Materials

Forty-eight word pairs, which consisted of one place noun matched with one object noun, were manipulated in the experiment. The word pairs were divided into two categories: consistent word pairs and inconsistent word pairs. Consistent pairs were those where the place and object would normally appear together (e.g., class room and black board). Conversely, inconsistent pairs were places and objects that would not normally appear together (e.g., toilet pillow). And there would broadcast an auditory episodic context simultaneously when every word pair appeared. This background sounds were played to establish an episodic context for each pair. The auditory episodic context was a sentence which consisted of the place and object of the word pairs. (e.g., the word pair is class room and black board; the episodic context is that teacher writes his name on the blackboard.)

C. Experimental Design

The experiment used a 2×2×2 factorial design. The between-participants factor was health status (i.e., Alzheimer's disease versus normal aging). The within-participants factors were consistency of word pairs and recall delay. Consistency of the word pairs had two levels, namely consistent or inconsistent. Recall delay also had two levels, namely immediate or delayed recall, denoting whether participants recalled the corresponding word of a pair immediately after the stimulus disappeared, or whether a 15 s pause was interposed before recall. Dependent variables consisted of recall accuracy and response times. Participants received twelve trials of each experimental condition. To minimize learning effects and fatigue, the order of presentation of conditions was counterbalanced.

D. Procedure

First, the purpose and procedure of the experiment was explained to the participants, and their written consent sought. The participants were given two practice trials. The experimenter confirmed the recall accuracy of these trials before beginning the formal experiment. A notebook PC was placed 35 cm from the participants. The stimuli, which were produced using E-Prime 2.0 software, were presented on the center of the screen. A trial schematic is shown in Fig.1. First, instructions were presented on the screen, to inform the participants what they should do in the experiment. Then, a fixation cross was shown. Next, a word pair was presented on the screen for 6 s, accompanied by an auditory episodic context.

Participants were to remember the object noun corresponding to the place noun in every trial. In the immediate recall condition, participants stated the object of the word pair immediately after the stimulus disappeared. However, in the delayed recall condition there was a 15 s pause after the stimulus before verbal recall was allowed.

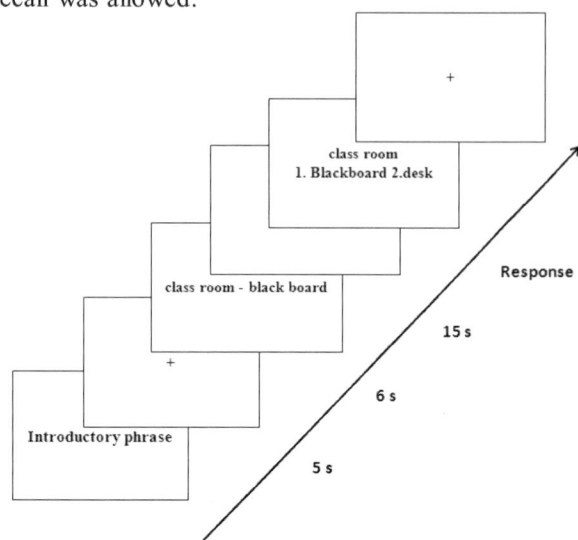

Fig.1. Procedure for the word pair recall task with delay

III. RESULTS

A. Data Analysis

Response time and accuracy data were collected. We conducted analysis of variance (ANOVA) on the data using SPSS 2.0. When significant interactions between factors were identified, a least significance difference test (LSD) was used to compare the different levels of each factor.

B. Response time

We conducted an ANOVA on the response times of participants (Table I). The analysis revealed a significant interaction between health status and recall delay (F (1, 22) = 5.1, p = 0.034). An LSD post-hoc test revealed that the performance of Alzheimer's disease patients was better for immediate recall (mean = 7427 ms) than delayed recall (9486 ms; F (1, 11) = 6.884, p = 0.024). However, performance of the normal aging group was not significantly difference for the immediate versus delayed recall (F (1, 11) = 0.908, p = 0.361). For both levels of delay, the Alzheimer's disease patients were slower to respond than the normal aging group (immediate: F (1, 22) = 11.226, p = 0.003; delayed: F (1, 22) =14.176, p = 0.001).

The analysis revealed a significant main effect of health status (F (1, 22) = 13.375, p = 0.001). Participants experiencing normal aging were faster (mean = 2022 ms) than Alzheimer's disease patients (mean = 8456 ms). That is, Alzheimer's disease patients had relatively poorer episodic memory performance (Fig.2).

Additionally, the main effect of recall delay was also significant (F (1, 22) =7.755, p = 0.011). Overall, participants were faster for immediate recall than delayed recall. Nevertheless, there were no differences in performance contingent on the consistency of the word pairs (F (1, 22) = 0.083, p = 0.776).

TABLE I
ANOVA RESULTS FOR RESPONSE TIMES OF ALZHEIMER'S DISEASE PATIENTS AND NORMALLY AGING PARTICIPANTS ON A WORD PAIR RECALL TASK

Source	Type III SS	df	MS	F	p
Between					
HS	993636293	1	993636293	13.375	0.001*
Error	1634368223	22	74289465		
Within					
Consistency	322005	1	322005	0.083	0.776
Consistency × HS	615	1	615	0.000	0.990
Error	85585795	22	3890263		
RD	31042352	1	31042352	7.755	0.011*
RD × HS	20412400	1	20412400	5.100	0.034*
Error	88058089	22	4002640		
Consistency × RD	4024668	1	4024668	0.307	0.585
Consistency × RD × HS	3826659	1	3826659	0.291	0.595
Error	288844864	22	13129312		
Total	3150121963	95			

Notes: HS=health status; RD=recall delay; *p<.05

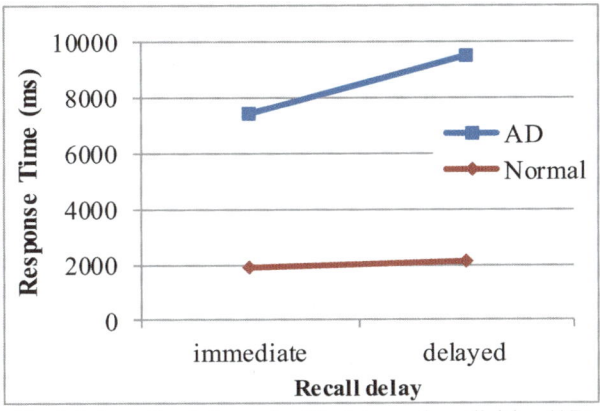

Fig. 2. Interaction between health status and recall delay (AD: Alzheimer's disease patients; Normal: normal aging)

C. Accuracy

Similarly to the response time data, we conducted an ANOVA on the accuracy data of participants (Table II). The analysis revealed that the interaction between health status and word pair consistency was significant (F (1, 22) = 4.475, p = 0.046). The normally aging participants performed perfectly on consistent word pairs (accuracy = 1.000). This group also performed well on inconsistent pairs (accuracy = 0.977). Thus, for normal aging, there was no difference between consistent and inconsistent word pair recall. However, Alzheimer's disease patients had poorer accuracy for the inconsistent condition (accuracy = 0.924) than the consistent condition (accuracy = 0.958). A LSD post-hoc test revealed that the performance of Alzheimer's disease patients was

statistically significant between the two consistency conditions (F (1, 11) = 5.901, p = 0.033). Furthermore, Alzheimer's disease patients and the normal aging group had different performance on inconsistent word pairs (F (1, 22) = 7.141, p = 0.014). However, there was no difference between the two groups for consistent word pairs (F (1, 22) = 3.828, p = 0.063).

TABLE II
ANOVA RESULTS FOR RECALL ACCURACY OF ALZHEIMER'S DISEASE PATIENTS AND NORMALLY AGING PARTICIPANTS ON A WORD PAIR RECALL TASK

Source	Type III SS	df	MS	F	p
Between					
HS	0.079	1	0.079	6.544	0.018*
Error	0.265	22	0.012		
Within					
Consistency	0.009	1	0.009	6.673	0.017*
Consistency × HS	0.006	1	0.006	4.475	0.046*
Error	0.029	22	0.001		
RD	0.002	1	0.002	0.605	0.445
RD × HS	0.004	1	0.004	1.182	0.289
Error	0.066	22	0.003		
Consistency × RD	0.000	1	0.000	0.025	0.877
Consistency × RD × HS	0.001	1	0.001	0.207	0.653
Error	0.071	22	0.003		
Total	0.532	95			

Notes: HS=health status; RD=recall delay; *p<.05

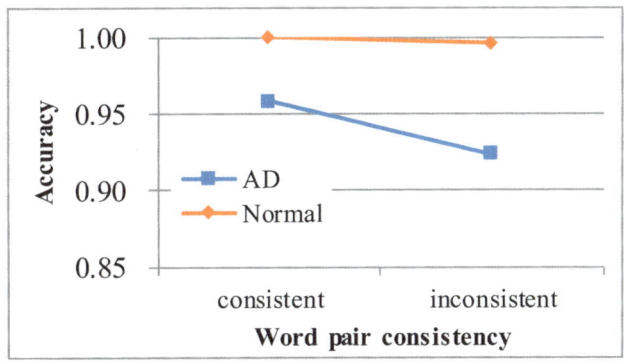

Fig.3. Interaction between health status and word pair consistency (AD: Alzheimer's disease patients; Normal: normal aging)

The main effect of health status was significant (F (1, 22) = 6.544, p = 0.018), indicating that Alzheimer's disease patients and people undergoing normal aging differed in episodic memory ability. We found the mean recall accuracy for normal aging was greater than that of Alzheimer's disease patients (normal aging = 0.998; Alzheimer's disease patients = 0.941) (Fig.3).

There was also a significant main effect of consistency (F (1, 22) = 6.673, p = 0.017). Accuracy was higher when word pairs were consistent. This implies that the familiarity with consistent word pairs improved performance, regardless of health status (consistent = 0.979; inconsistent = 0.960).

IV. CONCLUSIONS

The purpose of the present study was to explore the differences in episodic memory ability between Alzheimer's disease patients and people experiencing normal aging. In all conditions of the word pair recall task, the normal aging group performed better than the Alzheimer's disease patients. Alzheimer's disease patients needed more time to recall the corresponding word of a pair, and they made more mistakes in recall. These results are similar to those of Dhanjal et al. (2013) [12], whose Alzheimer's disease patients also exhibited worse performance in an episodic sentence retrieval task than individuals aging normally. Dhanjal proposed that episodic memory has a close relationship with the auditory attention and semantic message handling systems. The auditory cortex of Alzheimer's disease patients is damaged so that these systems consequently decline. Many researchers have found that the episodic memory impairment of Alzheimer's disease patients is related to cortical changes, which are accompanied by a decline in some cognitive abilities. However, the results of present study showed that recall accuracy of Alzheimer's disease patients was high (accuracy= 0.941), indicating that they could recall episodic memories correctly but they could not do the episodic memory task endurably Some studies have shown that early stage Alzheimer's disease patients have reduced attentional abilities, including deficits in sustained attention, selective attention, and divided attention [13]. Thus, Alzheimer's disease patients need to spend more attentional resource in episodic memory tasks compared with those aging normally.

The results of the present study also showed that Alzheimer's disease patients had poor performance when recall was delayed. Karlsen et al. (2010) [1] also manipulated different recall delays in a geometric shape and color connection task. The findings were similar to the present study, in that participants performed better when there was no delay imposed before recall. John et al. (1996) [14] proposed two resources, primary and secondary memory, used to store and retrieve memories. The primary memories recede in time in the delayed recall condition. There only the secondary memory remains in the brain after a period of time. Thus, participants are expected to take longer to retrieve such memories. John et al. named this phenomenon a recency effect.

In the present study, it was found that participants, specifically the Alzheimer's group, performed worse in an episodic memory recall task when the word pair was inconsistent. That is, Alzheimer's disease patients made more mistakes when the word pair was inconsistent. Sarah et al. (2013) [15] also found that Alzheimer's disease patients had poor performance when they performed an associated episodic memory retrieval task. The patients' cortical damage would lead to their declining ability to make episodic memory connections. As such, they struggled to recall intact episodic memories and made numerous incorrect memory retrievals. Furthermore,

Wang et al. (2013) [16] proposed that people usually use their past experiences to form memory recall strategies. In the present study, participants could not use their past experiences for inconsistent word pairs which is different from the normal condition appeared in daily life to recall the memory. Furthermore, these inconsistent words might interfere participants in memory process.

Overall, when information is provided to Alzheimer's disease patients, the information should be as closely related to their personal experiences as possible. Familiar experiences help patients to connect and retrieve memories quickly and correctly. Additionally, one should take account of the importance of recency effects when patients recall memories.

REFERENCES

[1]   P. J. Karlsen, R. J. Allen, A. D. Baddeley & G. J. Hitch, (2010). Binding across space and time in visual working memory. *Memory and Cognition*, 38, pp. 292–303.

[2]   N. S. Clayton, A. Dickinson (1998). Episodic-like memory during cache recovery by scrub jays. *Nature*, 395, pp. 272-74.

[3]   A. D. Baddeley, R. J. Allen & G. J. Hitch (2011). Binding in visual working memory: The role of the episodic buffer. *Neuropsychologia*, 49(6), pp. 1393-1400.

[4]   A. D. Baddeley & G. J. Hitch (1974). Working memory. In G.H. Bower (Ed.). The psychology of learning and motivation: Advances in research and theory. *New York: Academic Press*. vol. 8, pp. 47–89.

[5]   A. D. Baddeley (2000). The episodic buffer: A new component of working memory? *Trends in Cognitive Sciences*, 4(11), pp. 417-423.

[6]   R. J. Allen, A. D. Baddeley & G. J. Hitch (2006). Is the binding of visual features in working memory resource-demanding? *Journal of Experimental Psychology: General*, 135(2), 298.

[7]   R. J. Allen, A. D. Baddeley & G. J. Hitch (2009). Cross-modal binding and working memory. *Visual Cognition*, 17(1-2), pp.83-102.

[8]   D. Wechsler (1987). Wechsler Memory Scale-Revised. New York: Psychological Corporation.

[9]   L. Bäckman, B. J Small & L. Fratiglioni (2001). Stability of the preclinical episodic memory deficit in Alzheimer's disease. *Brain*, 124, pp.96-102.

[10] C. W. Nordahl, C. Ranganath, A. P. Yonelinas, C. DeCarli, B. R. Reed, & W. J. Jagust, (2005). Different mechanisms of episodic memory failure in mild cognitive impairment. *Neuropsychologia*, 43(11), pp.1688-1697.

[11] J.A. Small & N. Sandhu (2008). Episodic and semantic memory influences on picture naming in Alzheimer's disease. Brain and Language, 104, pp 1-9.

[12] N. S. Dhanjal, J.E. Warren, M. C. Patel & J. S. Richard (2013). Auditory cortical function during verbal episodic memory encoding in Alzheimer's disease. *Annals of Neurology*, 73(2), pp.294-302.

[13] J. P. Richard (2000). The nature and staging of attention dysfunction in early (minimal and mild) Alzheimer's disease: relationship to episodic and semantic memory impairment, *Neuropsychologia*, 38, pp.252-271.

[14] D. W. John, D. B. Alan & R. H. John (1996). Analysis of the episodic memory deficit in early Alzheimer's disease:

Evidence from the doors and people test, *Neuropsychologia*, vol. 34, No. 6, pp. 537 551.

[15] G. Sarah, C. Fabienne, F. Dorothe´e, P. Christophe, S. Eric & B. Christine (2013). Item familiarity and controlled associative retrieval in Alzheimer's disease: An fMRI study, *cortex*, 49, 1566 -1584.

[16] H. M. Wang, C. M. Yang, W.C. Huang, , C. C. Kuo& C. Hung, (2013) .Use of a Modified Spatial-Context Memory Test to Detect Amnestic Mild Cognitive Impairment. *PLoS ONE*, 8(2).

Optimal Capacity Allocation in an Assembly Job-shop using a Gradient-based Heuristic

Liang Huang[1,*], Xin Shi[2]

[1] School of Control Engineering, Northeastern University at Qinhuangdao, Qinhuangdao, P. R. China
[2] Liaoning Branch, China Huanqiu Contracting & Engineering Corp., Fushun, P. R. China

(n-xyz@163.com)

Abstract - **This paper presents a new capacity allocation approach to design or redesign an assembly job-shop with stochastic orders and processing times. The solutions for capacity allocation can be adding or removing workers and machines at every work stations. A bi-criteria objective function comprising tardiness penalty and fixed costs is used to evaluate each solution. A modified simulated annealing procedure coupled with a simulation model is used, and convergence is accelerated through a gradient-based heuristic based on bottleneck analysis. A case study in an assembly job-shop for refrigeration equipments is provided to reveal the feasibility and validity of the new approach.**

Keywords - **Assembly job-shop, bottleneck analysis, capacity allocation, gradient-based heuristic, simulated annealing**

I. INTRODUCTION

In recent years, market demand for manufacturing products is increasingly diversified and customized. The production scheduling and control is complex when a large fraction of products are manufactured in an assembly job-shop under a make-to-order production policy, as it processes a high variety of jobs at low volumes with different routings and different due dates. In many studies, it is generally assumed that the capacity at each work station is determined. However, in practice, it is often needs to be changed dynamically [1, 2]. This paper will address optimal planning for capacity allocation in an assembly job-shop to support long term (several months to years) decisions under a given dispatching rule with stochastic orders and processing times.

For capacity allocation, most problems need to allocate multiple capacities of different work stations simultaneously. These are complex optimization problems. Some studies use simulation models as well as meta-heuristics algorithms in the design of the manufacturing systems. Arakawa and Chen [3, 4] presented a simulation model for job-shop scheduling incorporating capacity adjustment. In their study, pattern search method and genetic algorithm (GA) are used to restrict capacity. Yang et al. [5] used the particle swarm optimization (PSO) algorithm for integration of production scheduling and capacity planning in a job-shop. Shahabudeen et al. [6] set the parameters of a multi-product Kanban system using simulated annealing (SA); the parameters include the number of machines at each work station. In another study of Shahabudeen et al. [7], they set similar parameters of an assembly line using GA. In all these

studies, the optimization algorithms usually use local search to reach the optimum solution from an initial solution. Coupled with simulation models, many alternatives were examined by simulation in the search procedure. This generally caused the algorithms to be time consuming in solving large-scale problems.

An effective neighborhood-generation method is helpful in accelerating convergence and controlling the run time of the local search procedure. In this paper, bottleneck analysis is used as approximate discrete gradients of the objective function of the weighted tardiness. A modified simulated annealing is also presented, in which the neighborhood-generation is guided by the approximate discrete gradients in order to reduce the run time of the optimization procedure. Based on the proposed method, in this paper a capacity allocation tool is implemented by using Microsoft SQL Sever 2008, which consists of an optimization model, a simulation model, a bottleneck analysis method and a modified simulated annealing named gradient-based simulated annealing (GBSA). This capacity allocation tool is applied to an actual mechanical assembly workshop and the results reveal the feasibility and validity of the new approach.

II. OPTIMIZATION MODEL

In this section, we present the conceptual optimization model for capacity allocation in an assembly job-shop. A central issue is how much capacity should be allocated at each workstation not only to satisfy the customer demand but also with a lowest cost.

In this study, the alternatives for capacity allocation are assumed to be adding or removing workers and machines. It is assumed that in an assembly job-shop that consists of m workstations, a linear array $s = [c_1, c_2, \ldots, c_m]$ is the solution vector of the capacity allocation problem, where c_j is the alternative number of the capacity level at workstation j, for $j = 1, 2, \ldots, m$. The solution space is the set of discrete vectors s, denoted as S.

For make-to-order production, the weighted tardiness is commonly used as the performance measure of a job-shop. In this study, a purpose of capacity allocation is to fulfill the due dates of all jobs as much as possible through reducing the weighted tardiness. Assuming that n jobs will be manufactured in an m-workstation assembly job-shop in a q-month period, we can formulate the first objective-function component to represent the weighted tardiness of the assembly job-shop as follow:

$$z^{\mathrm{T}}(s) = \sum_{l=1}^{p} w_l^{\mathrm{TP}} \sum_{i \in I_l} n_i^{\mathrm{LS}} \max(x_i^{\mathrm{C}}(s) - x_i^{\mathrm{D}}, 0), \qquad (1)$$

[1] This paper is supported by "the Fundamental Research Funds for the Central Universities (N130323018)".

where w_i^{TP} is the tardiness penalty weight for job i, $x_i^{\mathrm{C}}(s)$ is the completion time of job i in solution s, and x_i^{D} is the due date of job i. In the capacity allocation tool, each w_i^{TP} is assumed to be given in the q-month planning period.

Another objective of capacity allocation is to reduce the cost of the allocated capacity, which mainly consists of the depreciation of machines and the salaries of operators. Based on the monetary values of the depreciation per month per machine w_j^{M} and the salaries per month per operator w_j^{O} at each workstation j in the q-month period, for a solutions s with the number of machines $n_j^{\mathrm{M}}(s)$ and the number of operators $n_j^{\mathrm{O}}(s)$, the second objective-function component can be expressed as

$$z^{\mathrm{C}}(s) = q\sum_{j=1}^{m} w_j^{\mathrm{M}} n_j^{\mathrm{M}}(s). \tag{2}$$

The two objective terms are considered together in this study. Hence, the optimization model with a bi-criteria objective function is

$$\min z^{\mathrm{T}}(s) + z^{\mathrm{C}}(s) \tag{3}$$

$$\text{subject to: } s \in S. \tag{4}$$

III. GRADIENT-BASED SIMULATED ANNEALING

Kirkpatrick et al. [8] firstly presented SA in 1983. In its neighborhood search, SA accepts inferior solutions according to a probability in order to bypass local optimums. But it is time consuming in solving large-scale optimization problems coupled with simulation models.

The result of the bottleneck analysis can aid in pointing toward the direction of the maximal decrease of the objective function. Taking this result as an approximate gradient of the objective function, the neighborhood generation in the meta-heuristic procedure for capacity allocation can be guided to accelerate convergence and thus, reduce the computing time. A similar approach can be found in a few other studies on the hybrid algorithms, each of which consists of a gradient-based method and a meta-heuristics algorithm [9]. But these hybrid methods do not deal with the condition of an absence of an effective method to calculate the gradient of the objective function except by simulation. In this paper, using the results of the bottleneck analysis as approximate gradients, we couple the approximate gradients with simulated annealing (SA) and present a hybrid method named gradient-based simulation annealing (GBSA) as follows:

Step 1: Input the control parameters of GBSA: Initial temperature T_{i}, termination temperature T_{f}, cooling rate α, freeze limit Φ and accept limit β. Take T_{i} as current temperature T. Generate an initial solution s_0. In the experiments of this study, the s_0 is generated according to the practical current solution in the actual case and generated according to the mean capacity requirement added 20% protective capacity in each randomly generated case. Perform a simulation to calculate the objective-function value z_0 of the solution s_0.

Step 2: Detect the bottlenecks in the assembly job-

shop. To detect and measure the shifting bottlenecks in an assembly job-shop, a state equations method has been presented by Huang et al. [10]. Although this method is not an exact one, it is very robust, easy to apply and has the ability to detect the bottlenecks in steady state systems or non-steady state systems.

Step 3: Suppose there are M solutions s_{0h} (h=1, 2, ..., M) neighboring to solution s_0 (only the capacity at a workstation is modified by plus 1 or minus 1 in the ordinal number of the solution space). Based on the results from the bottleneck analysis, an estimated objective-function value $z(s_{0h})$ for each neighbor s_{0h} is computed by state equations. Then, a new solution s_1 is chosen from the M solutions according to a probability shown as follows:

$$\mathrm{P}(s_1 = s_{0h}) = \frac{\left(z_{\max} - z(s_{0h})\right)^{\gamma}}{\sum_{h=1}^{M}\left(z_{\max} - z(s_{0h})\right)^{\gamma}}, \tag{5}$$

where $z_{\max}$ is the maximum in $z(s_{0h})$, h=1, 2, ..., M. Therefore, the neighbor of a better estimated objective-function value has a higher probability to be chosen in order to accelerate convergence. Parameter γ in Eq. (5) is a new control parameter used to adjust the impact of the estimated gradient on the neighborhood generation. Based on pilot experiments, we observe that when the objective-function value has a large improvement in the previous iteration indicating that the guidance of the gradient works well at this stage of the search procedure, γ should be set to a larger value to make full use of the guidance of the gradient, or else γ should be set to a smaller value to have a better chance to move from one local minimum area to another one. For this consideration, in this study γ is set to 1 at the beginning of the search procedure and will be adjusted at each iteration as stated in Step 4.

Step 4: Perform a simulation to calculate the objective function value z_1 in the new solution s_1. Let $\Delta z = z_1 - z_0$. If $\Delta z < 0$, the new solution s_1 will replace the current solution s_0; otherwise, apply a probability $\mathrm{P(A)} = e^{-\Delta z/KT}$, where K is a constant, to determine whether the new solution will replace the current one. Set $\gamma = |\Delta z|/(|\Delta z|)_{\max}$, where $(|\Delta z|)_{\max}$ is the maximum among all the $|\Delta z|$ values in the past iterations.

Step 5: The current temperature T is adjusted after every Φ iterations according to α. If it's below T_{f} or the solution has not been improved for too many consecutive iterations to overstep β, stop the neighborhood search; otherwise, go to Step 2.

Step 6: Report s_0 and z_0 as the final solution and its objective function value, respectively.

In GBSA, the neighborhood-generation is according to a probability obtained by the bottleneck analysis, which is different from the random method in the classical SA. This method speeds up the search for a better solution in the area with the most potential while still allows the search to move away from a local area to another. Thus, the neighborhood search may stop earlier as controlled by β and the computing time is reduced.

IV. COMPUTATIONAL EXPERIMENTS

In this paper, a case study is tested using the proposed GBSA. The case consists of 3 types of orders and 5 work stations. The parts of the products are fabricated at work stations 1 to 4 and the products are assembled at work station 5.

There are 3 to 10 machines at each of the 5 work stations. The scheduling method used in this workshop is a dispatching rule, earliest due date with the tie broken by first come first service (EDD/FCFS), for it is very easy to be applied in a dynamic assembly job-shop with stochastic demand and processing times. Within a work station, the scheduling is complex in this workshop. For we have not enough detailed records about it, according to the production manager's suggestion, we make an assumption that a task can always make full use of the capacity within a work station and the processing time of the tasks processed at the work station will decrease/increase linearly with adding/removing capacity to the work station.

In the simulation model, inter arrival times of the orders and processing times of the tasks are generated in exponential distributions; constraints of lead times, tardiness penalties per hour and depreciation of machines are set to be fixed values. These data is shown in Table I and Table II.

The simulation software was developed in Microsoft SQL2008. The simulation for any given solution was performed in the duration of 24000 hours. The simulations were all performed in a personal Pentium IV computer with 2.4G CPU and 2G memory. The mean simulation time of each simulation (including the time for bottleneck analysis) is 32.5 seconds in this case.

According to the pilot runs, two groups of control parameters are used to both traditional SA and GBSA. Therefore, there are 4 kinds of algorithm with different control parameter values or different neighborhood-generation methods applied to the case study, which is denoted as A1, A2, A3, and A4. Their control parameter values are shown in Table III. The results are shown in Table IV.

TABLE III
CONTROL PARAMETERS

Type		Control Parameter Values				
		T_i	T_f	α	Φ	β
A1	GBSA	1	0.1	0.9	10	20
A2	GBSA	1	0.3	0.7	5	10
A3	traditional SA	1	0.1	0.9	10	20
A4	traditional SA	1	0.3	0.7	5	10

TABLE IV
RESULTS OF THE COMPUTATIONAL EXPERIMENTS

	Objective Function Value (1000 RMB)	Run Time (minute)
A1	4250	42.37
A2	4287	37.08
A3	4420	225.54
A4	4587	89.40

V. CONCLUSIONS

In this paper, a modified SA, named GBSA, is used as an optimization tool to optimize capacity allocation in an assembly job-shop. With less computing time, GBSA found better solutions compared to the traditional SA. These results show that the proposed method can often finds better solutions with a shorter computation time compared to the traditional method. These optimal solutions for capacity allocation can be very useful to support decisions in performing tradeoffs between the tardiness penalty and the cost of capacity allocation.

TABLE I
DEMAND REQUIREMENTS AND TARDINESS PENALTIES

Product Type	Mean inter arrival time of orders (hour)	Constraints of lead time (hour)	Tardiness penalty (RMB/hour)
1	30	60	25
2	50	70	20
3	70	80	15

TABLE II
PROCESSING TIMES AND DEPRECIATION OF MACHINES

Product Type	Work Station	Mean processing time (hour)	Depreciation of machines (1000 RMB)
1	1	2.25	20
	2	2.00	20
	3	2.50	10
	4	1.75	10
	5	5.00	30
2	1	1.25	7.5
	2	1.25	7.5
	3	2.00	15
	4	1.50	10
	5	4.50	35
3	1	1.75	10
	2	1.25	7.5
	3	2.25	10
	4	1.25	10
	5	7	40

REFERENCES

[1] C. H. Yeh, "Schedule based production," *International Journal of Production Economics*, vol. 51, no. 3, pp. 235–42, 1997.

[2] M. Huang, M. Song, N. N. Zhou, and X. W. Wang, "Capacity allocation strategy of a single facility with private information in random environment", *Control Theory and Applications*, vol.31, no.4, pp.444-50, 2014.

[3] M. Arakawa, M. Fuyuki, and H. Nakanishi, "An optimization-oriented method for simulation-based job shop scheduling incorporating capacity adjustment function," *International Journal of Production Economics*, vol. 85, no. 3, pp. 359–69, 2003.

[4] H. W. Cheng and M. Arakawa, "Genetic algorithm with a changing grid for facility location problems in a two-stage logistics system with restricted inventory capacity," *Journal of Japan Industrial Management Association*, vol. 62, no. 3, pp. 135–44, 2011.

[5] Y. H. Yang, F. Q. Zhao, Y. Hong, and D. M. Yu, "Integration of process planning and production scheduling with particle swarm optimization (PSO) algorithm and fuzzy inference systems," in *ICMIT 2005: Control Systems and Robotics*, Chongqing, China, pp. 2292–97.

348

L. Huang and X. Shi

[6] P. Shahabudeen, K. Krishnaiah, and M. Thulasi Narayanan, "Design of a Two-Card Dynamic Kanban System Using a Simulated Annealing Algorithm," *International Journal of Management system*, vol. 21, no. 10-11, pp. 754-59, 2003.

[7] P. Sivasankaran and P. Shahabudeen, "Study and analysis of GA-based heuristic applied to assembly line balancing problem," *Journal of Advanced Manufacturing Systems*, vol. 13, no. 2, pp. 113-31, 2014.

[8] S. Kirkpatrick, C. D. Gelatt, Jr., and M. P. Vecchi, "Optimization by simulated annealing," *Science*, vol. 13, pp. 671–80, 1983.

[9] K. F. C.Yiu, Y.Liu, and K. L.Teo. "A hybrid descent method for global optimization." *Journal of Global Optimization*, vol. 28, no. 2, pp. 229-38, 2004.

[10] L. Huang, Y. Gao, F. Qian, S. Z. Tang, and D. L. Wang, "Configuration selection for reconfigurable manufacturing systems by means of characteristic state space," *Chinese Journal of Mechanical Engineering*, vol. 24, no. 1, pp. 23-32, 2011.

Process Optimization for Wiring Technology of Aircraft Harness Based on ECRS Principle

Pei Gao[*], Jing-qian Wen

The Institute of Industrial and System Engineering, School of Mechanical Engineering, Beijing Institute of Technology, Beijing, China

(peggy_xiang@163.com)

Abstract - **In this paper, wiring technology during aircraft harness manufacturing is regarded as the research object, shortcomings of the existing technology have been analyzed, such as confusing procedures and low efficiency. To improve the situation, an optimization scheme was proposed from process flow and human-machine interface based on ECRS Principle. Methods used consist of merging operations with similar contents, adjusting processing sequence and optimizing the human-machine interface, in order to simplify the operation procedures and reduce the working procedure time. Finally, the feasibility of the scheme was verified through developing simulation model in eM-Plant. The study offers effective thinking for the improvement of aircraft harness wiring technology.**

Keywords - **Aircraft harness, ECRS Principle, eM-Plant, wiring technology**

I. INTRODUCTION

Harness, which is known as "nervous system" of aircraft, is a group of electric cables used in power transmission between aircraft electrical systems. With the rapid development of electronic technology, harness structures become more and more complex, while processing cycle still need to be reduced [1], which has set a higher demand on harness processing technology. However, aircraft development is in the mode of more variety and minor batch, most of the operations are performed manually since the production line is in low level of mechanization and automation, leading the problems of low efficiency and resource waste. How to optimize the production process and improving efficiency have been paid much more attention by enterprises.

Wiring procedure is the process to connect related terminals with wires according to design requirements. Generally, there are various types of wires to be dealt with, such as room temperature line, higher temperature line, shielded wire, and power line and so on, which are of different number of cores and wire diameter. During the process, workers need to understand the operation instruction for each wire firstly, selecting, cutting and marking the corresponding wire after that. Then perform wiring following the designated path on a special worktable. Because of the involved mass information processing, the wiring has become a bottleneck restricting efficiency.

This paper taking the wiring technology of aircraft harness as research object, in which redundant and repetitive operations have been analyzed based on ECRS Principle, and the processing sequence of related objects has been redesigned as well. Besides, in accordance with

the problem that workers need to process much information during wiring, existing information display mode was optimized to seek more harmonious human-machine interface. Finally, with statistical data as time parameter, build wiring process models before and after optimization in the simulation platform eM-Plant to verify the scheme.

II. CHARACTERISTIC ANALYSIS OF THE TRADITIONAL WIRING TECHNOLOGY

At present, the wiring is still mainly done by hand work. According to the topological structure of harness, wires of different types and diameter will be connected to the appointed terminal on worktable. Therefore, instruction for wiring technology include the information of wires, terminals and the topology, which can be represented as a triple "$O_{harness} = \{Wir, Con, Top\}$", where Wir is an information set of wires while Con is an information set of components, and Top represents the topology. The first two kinds of information are displayed as process reports on computer screen. Partial contents are shown as Table I. And the topology is drawn on a form board by handcraft showing as Fig.1.

TABLE I
AN EXAMPLE OF WIRING INSTRUCTION TABLE

Terminal 1	Pin1	Wire number	Color	Specification	Terminal 2	Pin2
1FXA	1	C210A	WH	20	10DXA	1
1FXA	2	C211A	WH	20	10DXA	2
1FXA	3	C212A	WH	22	10DXA	3
1FXA	4	C213A	WH	22	10DXA	4
2FXA	A	C320A	WH	18	10DXA	5
2FXA	B	C320B	WH	18	10DXA	6
…	…	…	…	…	…	…

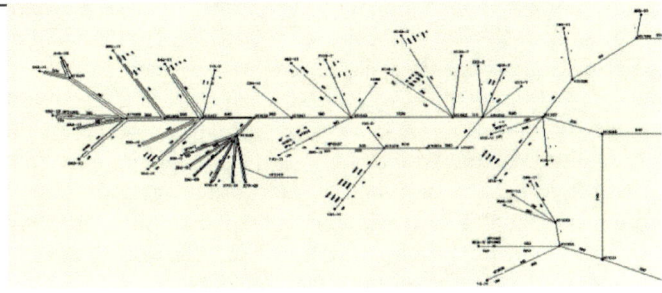

Fig.1. Form board for showing topology

Wires for aircraft harness are hard to distinguish by shape or color. Each wire is uniquely identified by wire number given from designers, for example, multi-core wires need to be manually tagged with marker sleeves while single-core wires print the number on insulating layer besides wearing marker sleeves. In actual operation, workers will select wire one by one and locate the position of its conjunctive two terminals on topological structure diagram following the wiring instruction. After determining the accurate length, cut and fix the wire on dowels by knotting. Then binding and other procedures will be carried out followed. The concrete process is given in Fig.2.

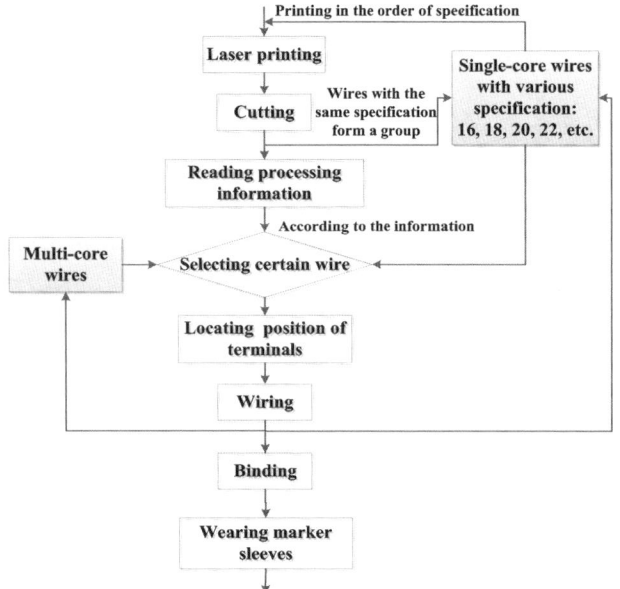

Fig.2. Flowchart for wiring and related procedures

Traditional wiring technology is design for manual operations, wire length need to be determined on worktable manually. Therefore, wiring procedure attends to reducing wire type change, and sort wiring sequence by the name of terminal in order to operate easily. Actually, harness length is measured on prototype in design phrase, and with the measured results to draw the form board according to 1:1 scale. Usually, there is big error in harness length and the manufacturing process is random. Analyze the existing wiring technology and can find the following main problems:

1) *Repetitive operations*: As the connection bridge, harness transmits power as well as signals. Wires with different specification are contained between two terminals. Giving priority to the specification when performing wiring must cause repetitive operations.

2) *Frequent operation switch*: During wiring, process instruction reading, wire selecting, locating the two terminals on form board, wiring and determining the length, cutting, fixing, tagging and other operations will be performed item by item. Workers not only need to identify and deal with a variety of information, but also need to operate various manufacturing objects.

3) *Lack of measure standard*: Due to the information processing and operation contents both decided by workers, work efficiency largely depends on the technical level and experience of workers, combining with the problem of inconsistent tensity, resulting in the lack of measure standard for quality and cost.

Currently, development for equipment demonstrates diversified and short cycle characteristics due to the quick update of new technology, and along with it, harness

structure becomes more and more complex, which demands higher manufacturing efficiency objectively. In addition, technology of three-dimensional digital prototyping has been applied widely, thus, information about topology and harness length can be obtained by the digital prototype [2], which lay a foundation for improving manufacturing technology. To change the situation mentioned above, proposed a scheme to improve the wiring technology.

III. OPTIMIZATION FOR AIRCRAFT HARNESS WIRING TECHNOLOGY

For more reasonable operation method to improve production efficiency, ECRS Principle was used to seek the direction of process optimization, that is eliminate, combine, rearrange and simplify [3]. After a comprehensive analysis of wiring procedure's working flow and displaying ways of human-machine interface, put forward an optimization scheme by adjusting related working stations and importing equipment, under the background of equipment upgrading.

A. Optimization in working flow

On one hand, harness design involves large amount of data and the processing is complex. In general, a normal training plane usually contains more than 8000 wires, kinds of wires up to 20 and number of terminals more than hundreds. There is a large space for efficiency promoting. And on the other hand, wires between any two terminals have essentially the same routing path and fixation method though they are of different numbers, specifications and lengths, which provides a basis for operation combining and rearrangement [4].

1) *Combine*: Combine wiring operations in an action since the existent repetition. In other words, get the correct number and specification of single-core wires and multi-core wires between the certain two terminals ready respectively in advance, which will be assembled to a group ahead. When processing, workers can only perform the wiring at a time after locating the terminals of the group single-core wires or multi-core wires on form board. It can not only reduce the operating times and conversion time, but also avoid the resulting visual and muscle fatigue of workers.

2) *Rearrange*: To assemble the single-core wires into a group, there is a need that adjust current printing scheme. Firstly, add equipment for wire selection automatically, which will be connected with the MES system in production line and read the process instruction. Then, the equipment will income wires according to the requirements about number and specification of wires between terminals, performing printing mixed. Next, cut and wind these single-core wires on a special spool to form a group that will be sent to the next wiring working station. As for multi-core wires, we add a new working station specifically to group the proper multi-core wires manually, and wind them in the same way.

Through the above method of "Pre-grouping", basic time of wiring procedure is reduced. Meanwhile, printing order is adjusted and new working station is added to transfer part of the workload from wiring phase to other phases, aiming at operation balancing.

B. Optimization in human-machine interface

In the harness production line, human-machine interface is an important medium for getting information, on which instruction is displayed in an acceptable form for workers. During the production, both the computer screen and the form board can be seen as human-machine interface of information output type. Friendly interface helps users understand what it means more correctly and more quickly.

However, the existing human-machine interface is just a kind of static display, furnishing comprehensive information but not intuitive. Workers need to select the certain wire and locate the corresponding terminals performing wiring according to the information distributed in the complicated tables. In addition, locating a specific terminal on the static form board is quite difficult when the harness is large and includes lots of terminals. Such human-machine interface is inconsistent with the object oriented principle and can't respond to different processing objects properly. Along with the "Pre-grouping" mentioned above, adjust the human-machine interface as follow:

1) *Eliminate*: Eliminate the step that search the terminals by workers. On one hand, for wires have being grouped together, print two-dimensional code or RFID electronic tag on their spool, to record related information about the group of wires [5, 6]. On the other hand, replace the traditional form board with the type of electron induction. When wiring, related information will be firstly obtained by scanning the two-dimensional code, then the corresponding terminals, along with routing path, will be highlighted on form board under the control of PLC system. Thus, workers can master the wiring contents directly without checking the complicit tables on screen.

2) *Simplify*: Currently connects of human-machine interface is just a copy of process instruction. With "Pre-grouping" and wireless sensing devices, there is no need to process the information item by item. As information interactive interface, it displays process information as well as incepts the feedback from users [7]. Having simplified the processing tables which show part of the instruction only for multi-core wires, the emphasis is placed on the monitoring and feedback of the production line. On the software interface, workers confirm the complete with a mouse after each working procedure, so that the production system can update the state of harness manufacturing real-timely.

Thus, by redesigning the software and hardware, production line will form an information networking system, realizing instruction displaying by intelligent and dynamic human-machine interface.

IV. SIMULATION VERIFICATION

This paper based on analysis for wiring technology of an actual enterprise, proposed improved suggestions from process flow and assistant equipment. Whether the optimized processing technology is good enough to meet the production requirements and improve efficiency or not, remains to be further verified. Simulation technology can simulate state of production line and collect statistical data, previewing and evaluating for different schemes. In the study, we make use of simulation platform eM-Plant for discrete manufacturing system, modeling production line process to verify the feasibility and validity of the optimized scheme [8].

Before modeling, determine processing parameter of each working station with the second watch measuring time method [9]. The measure object is the basic and overhead time to process a single processing object, like a wire. The specific data are listed in Table II.

TABLE II
WORKING TIME OF VARIOUS STATIONS IN WIRING PROCEDURE

Working station	Operation content	Duration time (s)
SA1	Printing on single-core wires with laser printer.	4.4
SA2	Reading processing information about wiring on the screen.	12.4
SA3	Selecting a certain wire with corresponding wire number among lots of single-core or multi-core wires.	17.5
SA4	Locating corresponding terminals' positions of the wire on formboard.	8.3
SA5	Wiring and cutting.	28.6
SA6	Binding with binding materials.	55.4
SA7	Wearing marker sleeve for each wire.	13.6
SA8	Pre-grouping for multi-core wires	14.2

According to the working flow and parameters (processing time, number of processing objects at a time, etc.), build the process route model before and after optimization, as shown in Fig.3 and Fig.4. With the modeling elements in eM-Plant, such as "SingleProc" and "Assembly", which are used to represent the working steps operated by workers and machines separately, one or more elements form the corresponding working station, and the series parallel of working stations express the wiring procedure.

Assume that the harness to be processed contains 100 single-core wires, 100 multi-core wires and 11 terminals, run the simulation and open statistical function. After a time, simulation results of station utilization can be obtained in chart [10], as shown in Fig.5 and Fig.6. Due to the addition of new working station and the elimination of terminals locating manually, there are a few differences between the two figures on abscissa. By comparison, we found that operation balancing rate was increased though part of the station utilizations has a slight decline. And according to the statistical results, the total time to complete the procedure with new scheme is 1h20min while time of the original is 2h35min, which has reduced the working procedure time by 48%. And the larger the harness size is, the more obvious the efficiency improvement.

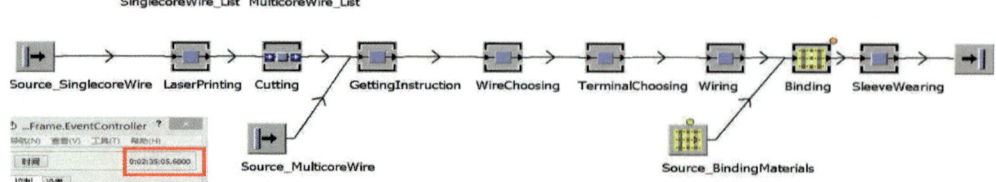

Fig.3. Simulation model of wiring procedure before optimization

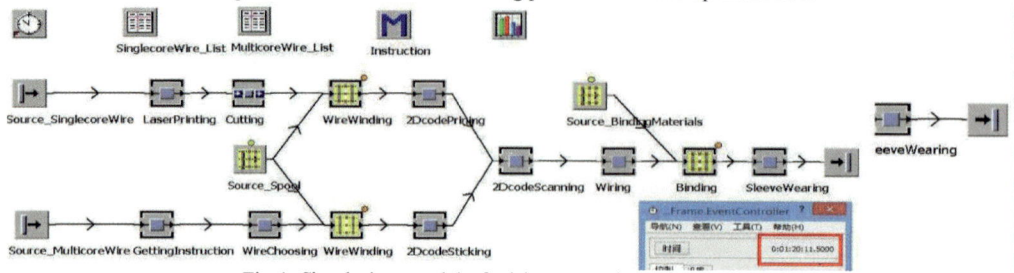

Fig.4. Simulation model of wiring procedure after optimization

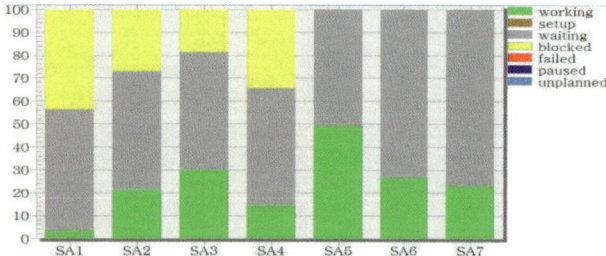

Fig.5. Simulation result of station utilization before optimization

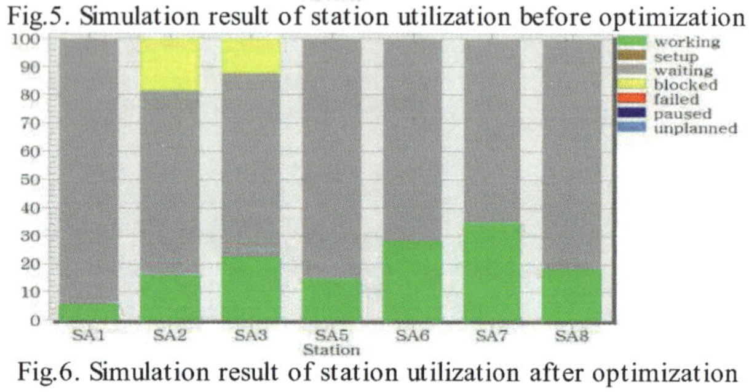

Fig.6. Simulation result of station utilization after optimization

V. CONCLUSION

This paper analyzed the wiring process of harness production with the method of ECRS Principle, proposed an optimization scheme and established simulation model in eM-Plant for verification. Statistical results have shown the validity of the proposed scheme. However, time parameters of each working station is calculated from measured value, which should be calculated in detail to obtain more accurate results, considering the instability of manual operation and the uncertainty of new added equipment parameters. And for "Pre-grouping" scheme, assembling mode for wires was simply discussed as winding on spool. If there are more economic and convenient methods need to be further designed. In addition, the paper put forward improving scheme mainly for the wiring procedure, while the process should be

optimized from global analysis, which certainly offer a direction for further research.

REFERENCE

[1] Zhang Xinfeng, and Yang Diange, "Invariant wiring harness for vehicle electrical device connection" (in China), *Journal of Tsinghua University (Science and Technology),* vol. 2, pp. 281-284, 2009.

[2] J. M. Ritchie, G. Robinson, P. N. Day, R. G. Dewar, R. C. Sung and J. E. Simmons, "Cable harness design, assembly and installation planning using immersive virtual reality," *Virtual Reality,* vol. 14, no. 4, pp. 261-273, 2007.

[3] Zhu Huabing, Wang Long, Tu Xueming, and Yu Feng, "Balance improvement of motor assembly line based on ECRS Principle and process reorganization" (in China), *Machinery Design & Manufacture,* no. 1, pp. 224-226+229, 2013.

[4] Guo Fu, Li Sen, and Dai Chunfeng, "Flow process analysis method used in printing and dyeing production line improvement application" (in China) *Industrial Engineering,* vol. 5, pp. 62-64, 2006.

[5] J. C. Hung, "Using active RFID to realize Ubi-media system," *Journal of Networks,* vol.6, no.5, pp.743-749, 2011.

[6] J. T. Lin, Hou Jiangliang, Chen Weiching, and Huang Chihao, "An RFID application model for the publication industry: a Taiwan perspective," *International Journal of Electronic Business Management,* vol.3, no.2, pp.129-139, 2005.

[7] Xiao Chao, and Chen Shuhong. "Transmission assembly planning study based on SLP and eM-Plant" (in China), *Journal of Engineering Design,* vol. 17, no. 6, pp. 430-434, 2010.

[8] Lu Yao, Xu Kelin, Gao Yang, and Liu Gaokun, "The research on the improvement of the production line balance of T factory" (in China), *Manufacturing Automation,* vol. 32, no. 2, pp. 111-114, 2009.

[9] Feng Xiaowen, Huang Anxiang, Gao Shenyu and Ye Peihua, "Rational design for human-computer interface of guide

control system" (in China), Journal of System Simulation, vol.23, no.S1, pp.102-105, 2011.

[10] Wang Junjun. and Wu Yongming, "Simulating production lines of recycling WEEE products in eM-Plant," *Advanced Materials Research*, vol. 97, pp. 2287-2290, 2010.

A Multi Objective Stochastic Optimization Case of Heavy Cargo Transport Scheme

Shi-qi Tong*, Jian-ru Zhang, Yue Lu

School of Transportation Management, Dalian Maritime University, Dalian China

(tsqxf@126.com)

Abstract - **To solve the difficulties resulted from uncertainty differences in multi objective optimization of the heavy cargo transport scheme; a multi objective stochastic optimization case of key equipment multimodal transport scheme for a chemical construction project was given. A multi objective stochastic optimization model with decision network planning and multi objective programming was established. A solution method, which is combined with utility function theory and upper bound minimal model, was put forward to solve the model. The result shows that the optimization model and the solution method can solve the multi objective optimization problem of heavy cargo transport scheme effectively even when uncertainties are different. And the case study is of highly practical value.**

Keywords - **Heavy cargo transport, multi objective, optimization, stochastic**

I. INTRODUCTION

Heavy cargo transport project is a particular transport activity for goods whose appearance and weight go beyond the general standard. When the heavy cargo transport scheme was designed, some processes can be developed several viable alternatives (such as different handing technologies or transport routes etc) [1]. The scheme selection was often treated as a multi objective problem. Besides that, every alternative has its unique uncertainty [2]. So how to consider the uncertainty differences among those alternatives should be studied when the scheme was optimized.

There were lots of fruits [3-6] in optimization study for heavy cargo transport scheme. But the studies about multi objective stochastic optimization were not found. In this paper, a multi objective stochastic optimization [7] case of key equipment multimodal transport scheme for chemical construction project was given. By using decision network planning [8] and multi objective programming, a multi objective stochastic model for the multimodal transport scheme optimization was presented. Multi attribute utility function [9] and upper bound minimal model [10] were applied to solve the model. The result shows that the model and the solution method, which were constructed for the case, can properly solve the heavy cargo transport problem.

II. OPTIMIZATION MODEL FOR MULTIMODAL TRANSPORT SCHEME

A. Project context

A petrochemical plant was constructed in city *A*. A large hydrogenation reactor for this plant was needed to transport from a manufactory in city *B* to the construction site. It was relatively difficult to design the transport scheme because the reactor is a heavy cargo instrument and the transport process is no standardized. So the petrochemical enterprise entrusted a project logistics company to design a transport scheme. After research, a multimodal transport plan (see Fig.1) from city *B* to city *A* was designed.

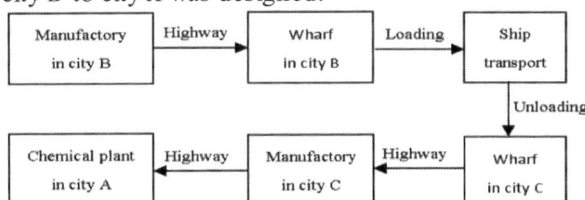

Fig.1. Multimodal transport Plan

The transport plan of the equipment was divided into 9 working processes (see Fig.2), which are as follows: ① Loading in manufactory in city *B*; ② Transporting to the wharf in city *B* by highway; ③ Ship loading; ④ Shipping by sea to the wharf in city *C*; ⑤ Ship unloading; ⑥ Transporting by highway to the manufactory in city *C*; ⑦ Loading again after the manufacturing accomplishment; ⑧ Transporting by highway to the construction site in city A; ⑨ Unloading in construction site.

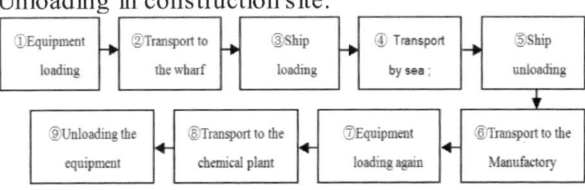

Fig.2. Multimodal transport processes

B. decision network planning model

Analysis of each process in Fig.2 found that process ②, ③, ⑤ and ⑧ can be developed viable alternatives. And the time, cost and security level of alternatives in each process were different. Besides that, every alternative has its own unique uncertainty in process ③, ⑤ and ⑧. Those uncertainties should be expressed in the model. So a decision network planning model, which can express different sample spaces, was used to build the model of multimodal transport scheme (see Fig.3).

In Fig.3, "○" indicates one process of the project. The three numbers from left to right on "○" stand for time (unit: hour), costs (unit: thousand RMB) and safety (the safety measure can be set the dimensionless number from 0 to 1). "△" indicates one decision point, in which one of the viable alternatives should be chosen. Each "○" after "△" stands for a viable alternative. "□" indicates uncertainty point of the viable alternative. "○" after "□" stands for a discrete sample of the viable alternative. All the "○" after "□" composed the random sample space of the viable alternative. The number on the arrow, which links "□" and "○", stands for the probability of "○". The actual meaning of each "○" were shown in TABLE I.

E. Qi et al. (eds.), *Proceedings of the 21st International Conference on Industrial Engineering and Engineering Management 2014*, Proceedings of the International Conference on Industrial Engineering and Engineering Management, DOI 10.2991/978-94-6239-102-4_73, © Atlantis Press and the authors 2015

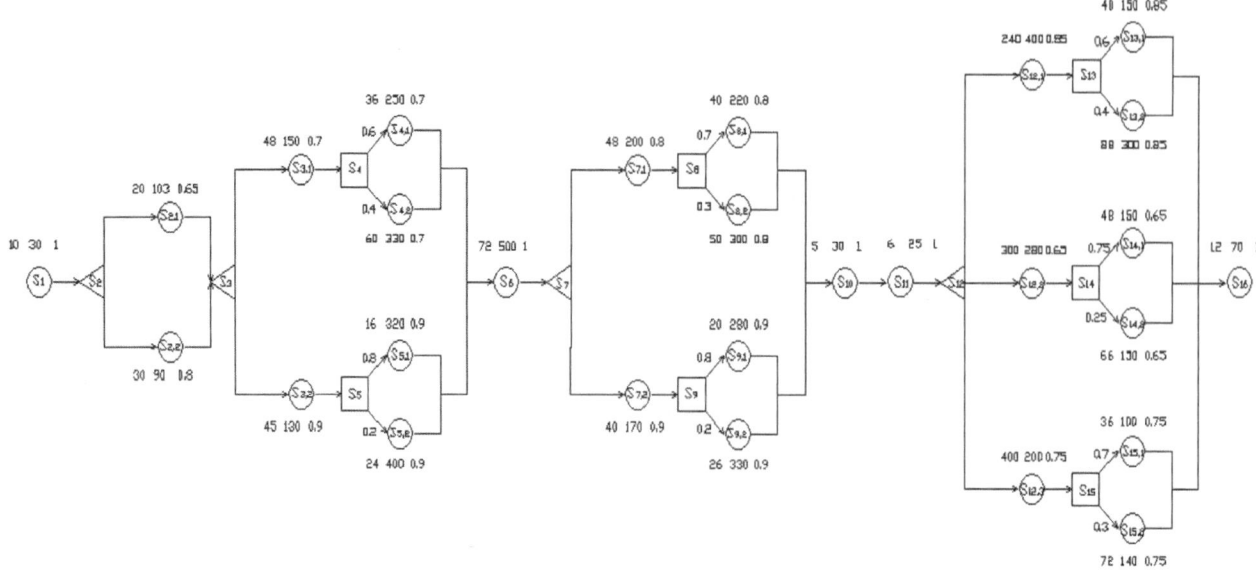

Fig.3. Decision network planning model of multimodal transport scheme

TABLE I
FEASIBLE ALTERNATIVE FOR TRANSPORT PROCESS

Decision points	Viable alternatives	Actual meaning
S_2(②)	$S_{2,1}$	Highway transportation line 1
	$S_{2,2}$	Highway transportation line 2
S_3(③)	$S_{3,1}$	Ro-ro scheme for loading [11]
	$S_{3,2}$	Hoisting scheme for loading [12]
S_7(⑤)	$S_{7,1}$	Ro-ro scheme for unloading
	$S_{7,2}$	hoisting scheme for unloading
S_{12}(⑧)	$S_{12,1}$	Highway transportation line 1
	$S_{12,2}$	Highway transportation line 2
	$S_{12,3}$	Inland water transport line 3

C. Optimization objective function of the model

(1) Time optimization goal. Because of the time restriction of petrochemical plant construction scheme, postponement was strictly prohibited. And the owner of the petrochemical plant wished that the transport project should be finished ahead of time as possible. So the time optimization goal of the decision network planning was formulated in the following mathematical model:

$$\text{Min } P_T = \begin{cases} 0 & T \leq T_E \\ T - T_E & T_E < T \leq T_L \\ \infty & T \geq T_L \end{cases} \quad (1)$$

where P_T is time optimization goal; T is the finish time of the scheme; T_E is the earliest finish time that the owner wished; T_L is the latest finish time limit.

(2) Cost optimization goal function. The lowest cost is another major goal of the project. According to the decision network planning model[13], the cost optimization goal was formulated in the following mathematical model:

$$\text{Min } P_C = \sum_{i=1}^{m} \sum_{j=1}^{k(i)} C_{i,j} d_{i,j} \quad (2)$$

where P_C is time optimization goal; $C_{i,j}$ is the cost of the $S_{i,j}$; $d_{i,j}$ is the decision variable of $S_{i,j}$; If $S_{i,j}$ was chosen then $d_{i,j}=1$,

otherwise $d_{i,j}=0$; m is the number of decision points; $k(i)$ is the viable alternative number in S_i.

(3) Safety optimization goal function. A safety measure $\mu_{i,j}$ ($\mu_{i,j} \in (0, 1)$) was set to estimate the safety level of a transport process. If $\mu_{i,j}$ is closer to 1 then safety level of a transport process is higher. So the safety optimization goal of the decision network planning was formulated in the following mathematical model:

$$\text{Max } P_\mu = \sum_{i=1}^{m} (\omega_i \cdot \sum_{j=1}^{k(i)} \mu_{i,j} d_{i,j}), \quad \sum_{i=1}^{m} \omega_i = 1 \quad (3)$$

where P_μ is safety optimization goal; ω_i is the weight of S_i to embody the importance of safety level in the transport scheme.

According to function (1), (2) and (3), the optimization goal of the decision network plan was formulated in the following mathematical model:

$$P = Sup(P_T, P_C, P_\mu) \quad (4)$$

III. THE SOLUTION METHOD

A. The multi attribute utility function for function (4)

According to function (4), the optimization of transport scheme could be treated as a multi objective problem. In this paper, multi attribute utility function theory was used to transform the function (4) into a single objective problem. P_T, P_C and P_μ were defined as the variables of multiple attribute function. The function $u = u(P_T, P_C, P_\mu)$ were defined as multi attribute utility function of multimodal transport scheme. The bigger the value of u is, the higher the degree of optimization is. According to decomposition theorem of multi attribute utility function theory, u was formulated in the following mathematical model:

$$u(P_T, P_C, P_\mu) = k_T u(P_T) + k_C u(P_C) + k_\mu u(P_\mu) \quad (5)$$

$$u(P_T) = \begin{cases} 1 & T \le T_E \\ \dfrac{T_L - T}{T_L - T_E} & T_E \le T \le T_L \end{cases} \quad (6)$$

$$u(P_C) = \frac{P_C^{\max} - P_C}{P_C^{\max} - P_C^{\min}} \quad (7)$$

$$u(P_\mu) = \frac{P_\mu - P_\mu^{\min}}{P_\mu^{\max} - P_\mu^{\min}} \quad (8)$$

$$k_T + k_C + k_\mu = 1 \quad (9)$$
$$k_T, k_C, k_\mu \ge 0; \quad (10)$$

where $u(P_T)$, $u(P_C)$ and $u(P_\mu)$ respectively stands for the single variable utility function of time, cost and safety; k_T, k_C, k_μ is the corresponding weight coefficient of $u(P_T)$, $u(P_C)$ and $u(P_\mu)$; $P_C^{\max}$ and $P_C^{\min}$ is the maximum and minimum value of function (2); $P_\mu^{\max}$ and $P_\mu^{\min}$ is the maximum and minimum value of function (3).

Hamming Distance was used to measure optimization degree of the transport scheme. The ideal solution and the negative ideal solution were set to be $u(T_E, P_C^{\min}, P_\mu^{\max})$ and $u(T_L, P_C^{\max}, P_\mu^{\min})$. Then the distance from $u(T, P_C, P_\mu)$ to $u(T_E, P_C^{\min}, P_\mu^{\max})$ is

$$D_j^+ = |u_j(T, P_C, P_\mu) - u(T_E, P_C^{\min}, P_\mu^{\max})| \quad (11)$$

The distance from $u(T, P_C, P_\mu)$ to $u(T_L, P_C^{\max}, P_\mu^{\min})$ is

$$D_j^- = |u_j(T, P_C, P_\mu) - u(T_L, P_C^{\max}, P_\mu^{\min})| \quad (12)$$

So the function (4) was converted into

$$Max\ R_j = D_j^- / (D_j^+ + D_j^-), \quad (13)$$
$$0 \le R_j \le 1, j = 1, 2, \ldots, n$$

where R_j stands for the relative adjacent degree between $u(T, P_C, P_\mu)$ and $u(T_E, P_C^{\min}, P_\mu^{\max})$.

B. The stochastic optimization function for uncertainty

In Fig.3, there are 24 viable alternatives for the transport scheme. And each alternative has 8 possible situations. So the optimization of the transport was still a stochastic programming problem[14]. To solve the problem, an upper bound minimal model, which is based on the stochastic programming theory, were put forward. The upper bound minimal model is a stochastic optimization model to solve the minimum (or maximum) value of the objective function when the probability of the owner's satisfaction β has been given[15].

In this transport project, the decision goal of the upper bound minimal model is: owner of the petrochemical plant required that R_j was get the maximum value when the joint probability distribution of time and cost $\beta_j \ge \beta$. According to the requirement of the owner, the value of $\beta = 0.9$. Then the optimization goal of the decision network plan was turned into the following mathematical model:

$$Max\ R_j = D_j^- / (D_j^+ + D_j^-) \quad (14)$$
$$St. \begin{cases} \beta_j^r(P_T^{(r)} \le P_T, P_C^{(r)} \le P_C) \ge 0.9 \\ 1 \le r \le 8, 1 \le j \le 24 \end{cases} \quad (15)$$

where r is the possible situation number of alternative; j is the number of alternative.

C. The calculation results

According to the construction time requirements of the petrochemical plant, the value of T_E and T_L is 600 hours and 800 hours. When $\beta = 0.9$, calculation results of 24 viable alternatives were shown in TABLE II (because of the length strict, some alternatives were not shown in TABLE II). It can be seen in TABLE II that $P_C^{\min} = 1998$, $P_C^{\max} = 2408$, $P_\mu^{\min} = 0.8292$ and $P_\mu^{\max} = 0.9125$. By the above data, function (5) were turned into

$$u(T, P_C, P_\mu) = k_T \bullet \frac{800 - T}{800 - 600} + k_C \bullet \frac{2408 - P_C}{2408 - 1998}$$
$$+ k_\mu \bullet \frac{P_\mu - 0.8292}{0.9125 - 0.8292} \quad (16)$$

where $k_T = 0.3$, $k_C = 0.3$ and $k_\mu = 0.4$ was gained by consulting the experts.

TABLE II
THE CALCULATION RESULT OF 24 VIABLE ALTERNATIVES

j	viable alternatives	T	P_C	P_μ
1	$S_{2,1}, S_{3,1}, S_{7,1}, S_{12,1}$	649	2358	0.8625
…	…	…	…	…
10	$S_{2,1}, S_{3,2}, S_{7,2}, S_{12,1}$	580	2408	0.9125
…	…	…	…	…
14	$S_{2,2}, S_{3,1}, S_{7,1}, S_{12,2}$	689	2138	0.8292
…	…	…	…	…
18	$S_{2,2}, S_{3,1}, S_{7,2}, S_{12,3}$	775	1998	0.8625
…	…	…	…	…
22	$S_{2,2}, S_{3,2}, S_{7,2}, S_{12,1}$	590	2378	0.9125
23	$S_{2,2}, S_{3,2}, S_{7,2}, S_{12,2}$	622	2138	0.8792
24	$S_{2,2}, S_{3,2}, S_{7,2}, S_{12,3}$	734	2018	0.8958

All the alternatives in TABLE II were calculated by using function (14) and (16). And the result shows that Scheme 22($S_{2,2}, S_{3,2}, S_{7,2}, S_{12,1}$) is the best solution. And the corresponding parameter calculation results are: $T = 590$, $P_C = 2378$, $P_\mu = 0.9125$ and $R_j = 0.72$.

IV. CONCLUSION

To solve the stochastic problem in multi objective optimization of heavy cargo transport scheme, a multi objective stochastic optimization case was given. The empirical results show that the idea of modeling and solving, which were put forward in this paper, can optimize the heavy cargo transport scheme in consideration of the uncertainty differences among the viable alternatives. In practice, uncertainty differences among the viable alternatives are common problems in decision making of heavy cargo transport scheme. So the model and solution in this paper has great application value.

ACKNOWLEDGMENT

This manuscript benefited from the critical comments and helpful suggestions of the anonymous referees. The work is funded by the Ministry of Education Doctoral Program of the New Teachers (No. 2012Z0289) and the Dr. Foundation Program in Liaoning province (No. 20121027).

REFERENCES

[1] W. Nuo, *Introduction to project logistics* (in Chinese), Beijing, China: Chemical Industry Press, 2007, pp.15-20.

[2] W. Nuo, T. Shiqi., X. Cunxiao, "Uncertainty of Large Scale Project Logistics Based on The Decision Network Planning Technique, (in English)," in *International Conference on Transportation Engineering*, Chengdu, China, pp. 3673-3678.

[3] L. Jian, "Study on the route selection scheme and model about heavy cargo highway transportation" (in Chinese), *Journal of Xihua University*, vol. 232, no. 4, pp.71-76, 2013.

[4] T. Shiqi, W. Nuo, X. Chunxiao, L. Hang, "Time-cost-security Tradeoff Optimization in Project Logistics Based on Decision Network Planning (in English), " at the *14th International Conference on Industrial Engineering and Engineering Management*, Tianjin, China, pp.206-210.

[5] T. Shiqi, W. Nuo, X. Chunxiao, "Multi-objective tradeoff optimization model of implementation scheme in project logistics" (in Chinese), *Computer Integrated Manufacturing Systems*, vol. 14, no. 11, pp. 2312-2316, 2008.

[6] Petraska, A., Palsaitis, R., Batarliene, N., Bazaras, D., "Evaluation Criteria System of the Routes for Super Heavy and Oversized Cargo (in English)," in *Proceedings of the 15th International Conference Transport Means 2011*, Kaunas, Lithuania, pp.236-239.

[7] Z. Weiyan, L. Chuan, Z. Jingling, L. You, W. Wanliang, "Novel algorithm for multi-objective vehicle routing problem with stochastic demand" (in Chinese), *computer Integrated Manufacturing Systems*, vol. 18, no. 3, pp. 523-530, 2012.

[8] T. Shiqi, W. Nuo, G. Lei, "Expandation both of decision unit structure and its optimization method in decision network planning" (in Chinese), *Journal of Dalian Maritime University*, vol. 33, no. 4, pp. 65-76, 2007.

[9] Y. Yaohong, W. Yingluo, W. Nengmin, "Fuzzy Tradeoff Optimization of Time, Cost and Quality in Construction Project" (in Chinese), *Systems Engineering Theory & Practice*, no. 7, pp. 112-117, 2006.

[10] R. Xiaoxia, "Model and Algorithm Study for Uncertain optimization problem (in Chinese)," Ph.D. dissertation, Shandong University, Jinan, China, 2005.

[11] Paulauskas, V., Paulauskas, D., "Heavy Cargo Transportation by Ro - Ro Ships (in English)," at the *16th International Conference - Transport Means 2012*, Kaunas, Lithuania, pp.25-26.

[12] W. Kong, H. Du, "Study of optimum scheme in hoisting large-scale reinforced concrete long post (in English)," in *2010 International Conference on Computer and Communication Technologies in Agriculture Engineering (CCTAE 2010)*, Chengdu, China, pp.361-364.

[13] W. Nuo, *Network Network planning technology and Its Expanded Study* (in Chinese), Beijing, China: China Communications Press, 1999, pp.56-70.

[14] Y., Yun, L. Jie, L. Zukui, T. Qiuhua, X. Xin, Floudas, C.A., "Robust optimization and stochastic programming approaches for medium-term production scheduling of a large-scale steelmaking continuous casting process under demand uncertainty (in English)," at the *23rd European Symposium on Computer Aided Process Engineering*, Lappeenranta, Finland, pp.165-185.

[15] L. ChiaYen, C. ChenFu, "Stochastic programming for vendor portfolio selection and order allocation under delivery uncertainty" (in English), *OR SPECTRUM*, vol. 36, no. 3, pp.761-797.

Study on Simulation-based Green Evaluation Methodology of Discrete Manufacturing System

Zheng-ping LUO, Qian-wang DENG*

State Key Laboratory of Advanced Design and Manufacturing for Vehicle Body, Hunan University, Changsha, China
(deng_arbeit@163.com)

Abstract - **Digital factory technology is a supporting technology of discrete manufacturing system modeling and simulation. However, it only analyzes the manufacturability and efficiency of discrete manufacturing processes, lacking of the evaluation of green performance of manufacturing system. By studying the resource consumption and environment emission characteristics in the process of discrete manufacturing enterprise workshop, the concept of simulation evaluation on green performance in the production process of workshop has been put forward based on the concept of digital factory. In the meantime, the framework and related information models of simulation-based green performance evaluation system have been established based on multi agent technology. The proposed system is aiming at the effective evaluation of the relevant green indicators in manufacturing system, thus realizing the integral optimization of the technical, economical and green performance of discrete manufacturing system.**

Keywords - **Agent based modeling, digital factory, discrete manufacturing system, evaluation of green performance**

I. INTRODUCTION

For a long time, discrete manufacturing enterprises promoted the process of rapid economic development by manufacturing various products. However, their inefficient mode of production have caused a huge amount of energy and material resources waste and produced CO_2 emissions and other hazardous emissions, e.g. SO_2 [1]. It is well acknowledged that these processes could bring tremendous damages to natural environment and public health. With sustainable growth of energy and raw material prices in the past few years other than gradually increased environmental awareness, highly attention is paid to the establishment of sustainable manufacturing systems [2]. As the key part of enterprise resources consumption and environmental impact, the decrease of manufacturing system environmental impact has become a hotspot in the field of green manufacturing.

Sustainable discrete manufacturing system design needs to meet the optimal objectives of high efficiency, low cost, high quality, the high utilization rate of resources and the minimum environmental impact [3]. The multiple optimization goals will naturally increase the complexity of manufacturing system design. Before the actual construction of manufacturing system, it is necessary to apply digital factory technology to analyze and optimize the performance of discrete manufacturing system. Traditional digital technology only treat some indicators related to the manufacturing system production

efficiency and product manufacturability as optimization targets, certain green indicators such as energy consumption, environmental emissions, material consumption are ignored.

The paper is organized as follows: firstly, input and output model of discrete manufacturing system is set up through the analysis of resources consumption and environment emissions characteristics in discrete manufacturing system based process flow; secondly, the concept of green simulation evaluation of discrete manufacturing process is put forward based on the concept of digital factory; at last, a green production simulation evaluation system model is built based on multi agent technology.

II. THE GREEN PERFORMANCE ANALYSIS OF DISCRETE MANUFACTURING SYSTEM

The production of mechanical products and parts highly depends on discrete manufacturing system. Basic components of typical discrete manufacturing system can be divided into production environment, production equipment, production materials and production operators [4]. Discrete manufacturing system is an input-output system [5], its input resources related to green performance consist of material, energy, etc; and its internal contains the process activities, production equipment and production process control, output resources are made up of products, waste and pollution and so forth as it shows in Fig.1.

Discrete manufacturing system mainly has the following three kinds of green characteristics:

i. High energy consumption and low utilization rate of equipment. Research shows that the energy consumption for actual processing accounts for only 14.8% of the total energy consumption in a large mechanical processing production line [6].

ii. Fast material flow with a serious material waste. Manufacturing systems consists of both productive materials and non-productive materials. Productive materials contain the direct components of products such as work blanks, parts and semi-finished products, while non-productive materials consist of cutting fluid, cutters and lubricants [7].

iii. High environmental emissions. The environmental emissions of production process include CO_2, hazardous gases, liquid pollution, solid waste emissions and physical emissions [8].

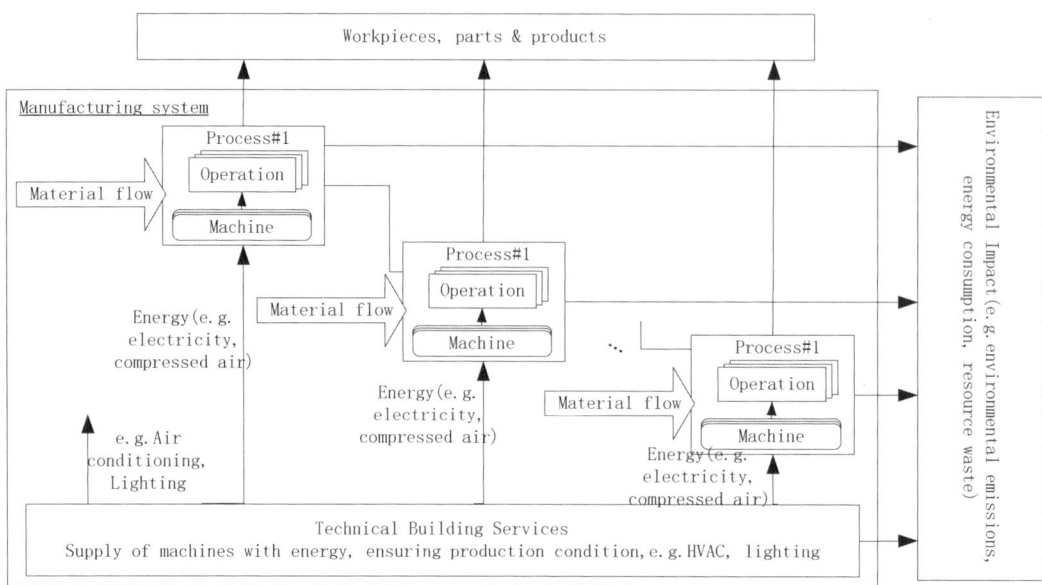

Fig.1. Input-output model of manufacturing system

III. GREEN EVALUATION BASED ON DIGITAL FACTORY TECHNOLOGY

A. Digital Factory

Digital factory [9] is referred to as the use of information technology and computer simulation to model real-world manufacturing possesses for the purpose of analyzing and understanding them. Digital factory does not result in material or products, but rather in information on products and technical parameters of the manufacture. Digital factory enables analysis of the manufacturing process and control system behavior within a virtual manufacturing system prior to the implementation of the manufacturing process and suitable control approach into a real-life manufacturing system. This helps to avoid direct implementation of the control system into the manufacturing system, which may cause failures in the manufacturing system due to potential disadvantages of the manufacturing system structure or lack of coordination between control parameters.

Digital factory enables feasibility evaluation of the production process plan in the manufacturing system, evaluation of manufacturing system design as well as optimization of the manufacturing process.

The application of digital factory technology such as virtual manufacturing, virtual assembly greatly facilitates the optimization of discrete manufacturing system and production process. Using digital factory technology can assist engineers in optimizing the defects of manufacturing system before the construction of system in time so as to attain a more reasonable production line layout and production process. It is very beneficial to reduce production cost and improve product manufacturability.

B. The concept of green evaluation through production process simulation

In view of the optimal goals of efficient and environmental manufacturing system, the green data and indicators can be integrated into the manufacturing system simulation. The concept of green evaluation through Digital Factory is shown in Fig. 2.

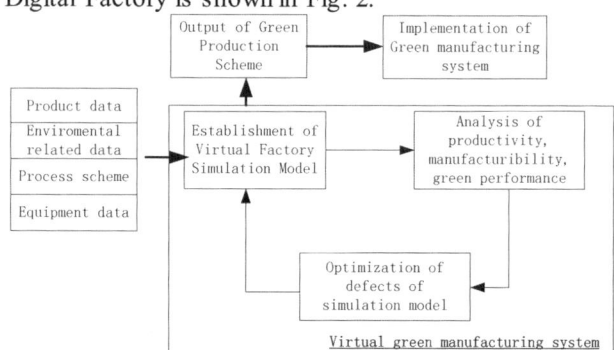

Fig.2. Concept of green production simulation

Based on the concept of digital factory, materials consumption, energy consumption and environmental emissions parameters related to environmental indexes are integrated into the production process simulation, then the environmental impact of the system operation process can be evaluated quantificationally by the method of relevant environmental impact assessment such as LCA [10]. At last, a manufacturing system is designed with the thought of high efficiency and environment protection.

C. Required functions of green production simulation

Green production simulation evaluation mainly aimed at the workshop production process, its main required functions include:

1) Precise 3D simulation of production scenario and each process;
2) Material resources consumption of production processes, e.g. raw materials consumption, auxiliary materials consumption;
3) Energy consumption of production processes, e.g. electricity, heat energy, fossil fuels;
4) Pollutants emissions of production processes, e.g. scraps emissions, liquid emissions and gas emissions.

IV. THE ARCHITECTURE OF GREEN PRODUCTION PROCESS SIMULATION EVALUATION SYSTEM

A. Multi agent Technology

Multi agent system is widely applied in the fields of computer science, network communications, artificial intelligence, engineering design and scientific computing nowadays [11]. There is no universal definition of an agent. However, an agent can be defined as follows: "An autonomous component that represents physical or logical object in the system, capable to act in order to achieve its goal, and being able to interact with other agents, when it does not possess knowledge and skills to reach its objective alone." Because of its autonomy, interactive, collaborative and initiative characteristics, it is introduced into the collaborative product design and simulation research as a kind of new system simulation tools and the basis of a new information system framework [12]. Multi agent technology mainly studies a set of autonomous agents to solve the problems of complex control in the distributed dynamic environment through interaction, cooperation, competition and negotiation behaviors [13].

Manufacturing system is a kind of complex dynamic system. Its internal contains a large number of frequent interactive processes, which is difficult to adopt analytic methods and numerical analysis methods to cope with. The system simulation method is the most effective way by far. Agent based modeling can describe the complexity of the production simulation evaluation system with green attributes.

B. The Architecture of Agent-based green evaluation system

The Architecture of Agent-based green evaluation system (AGES) is shown in Fig.3. The architecture of the system consists of three layers: agent control layer, virtual simulation layer, data support layer.

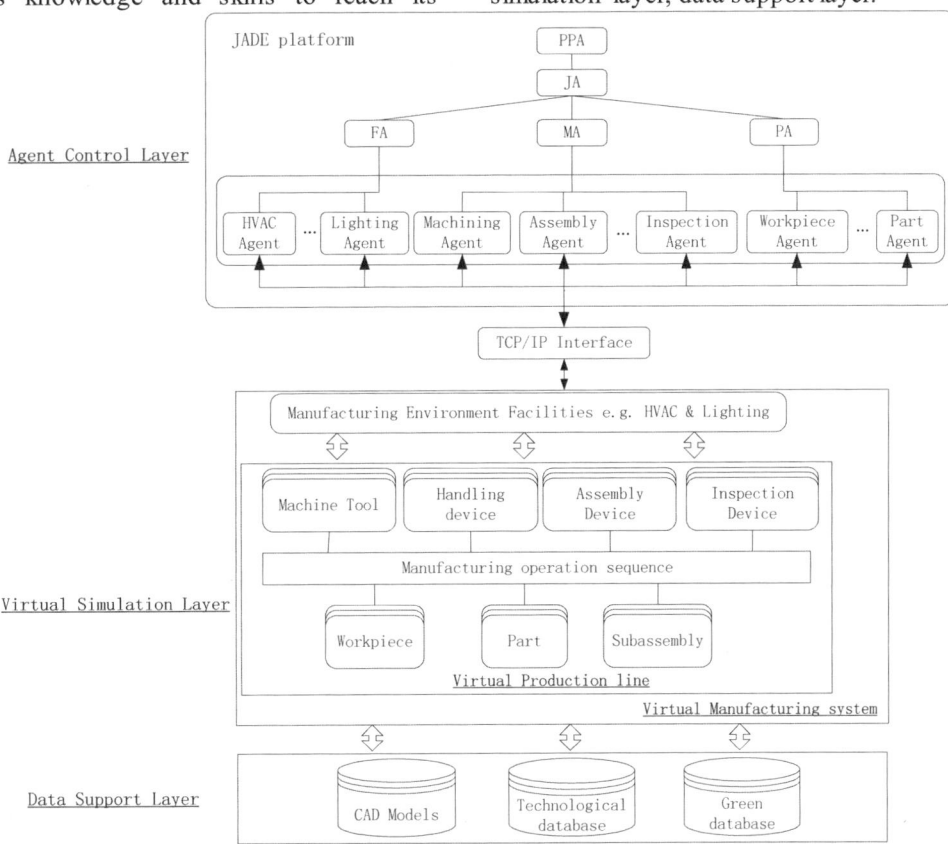

Fig.3. The structure of AGES

The functions of agent control layer mainly control the system simulation process design and operation, including manufacturing processes planning, operational motion definitions of machines, creation of components of product and machines, the feedback of manufacturing information related to simulation targets. The virtual simulation layer is used to model manufacturing system including manufacturing system layout, different levels 3D production scenario operation. Data support layer is made up of different kinds of databases and knowledge bases e.g. machine 3D model base, LCA database, energy calculation knowledge base.

C. Agent modeling of Green production simulation evaluation system

The key problem of multi agent system is to determine the granularity of Agents [14]. Its basic principle is to make the system simple, easy to control. According to manufacturing processes, the system includes five types of agents: Process Planning Agent, Job Agent, Machine Agent, Facility Agent and Product Agent.

1. Process Planning agent

Process Planning Agent (PPA) is the main control agent in the manufacturing system, which develops the sequence of manufacturing processes depending on the product characteristics including technical and technological characteristics of a product as well as the material of components forming a product, their input dimensions and mutual position.

2. Job Agent

Job Agent (JA) is also a kind of control agent which in cooperation with PPA and MA defines the sequence of operation phases (set-up, processing, idle, standby) of machine. In the process of defining the sequence of operation phases, JA performs the following tasks: determination of the sequence of operation phases, assignation of operation phases to the machines, description of operation phases and time value determination of operation phases.

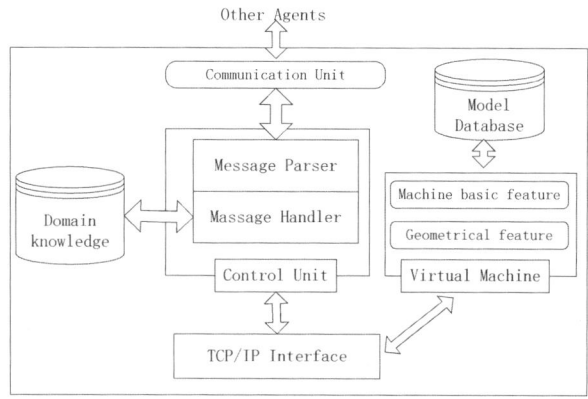

Fig. 4. Structure of Machine Agent

3. Machine Agent

Machine Agent (MA) is a representative of the manufacturing entity in a manufacturing system. Its structure is shown in Fig.4. The machines in manufacturing system mainly include processing machine tools, conveyors, industrial robots, weld guns, AGVs, inspection equipment. Each Agent stands for a machine. MA communicates with JA through its communication unit to decide how the machine accomplishes the operations. The MA mainly is composed by three parts: communication unit, control unit, virtual model. The

virtual model mainly contains two kinds of information, which is the geometrical information of a machine e.g. size, position and machine basis information e.g. ID, name.

4. Facility Agent

Facility Agent (FA) defines the technical facility which ensures the production environment e.g. HVAC facilities, lighting. Its structure is similar with MA and it also has its operation phases. The difference between FA with MA is that FA does not have specific 3D operational motions in the production scenario.

5. Product Agent

Product Agent (PA) is the representative of workpiece or part which composes product. In actual production process, product is formed through the workpiece processing and parts assembly. PA mainly contains information about its basic feature and structural feature. Basic feature includes ID, name, weight, cost, material; and structural feature contains geometrical shape, position and orientation of virtual CAD model in the global coordination.

D. Cooperation mechanism of agents

The model of agent cooperation within the coordination mechanism of control in manufacturing system operation is shown in Fig.5. Messages communicated among agents with coordination mechanisms of control are basic messages of JADE ACL (Agent Communication Language), which is in accordance with the FIFA standards [15].

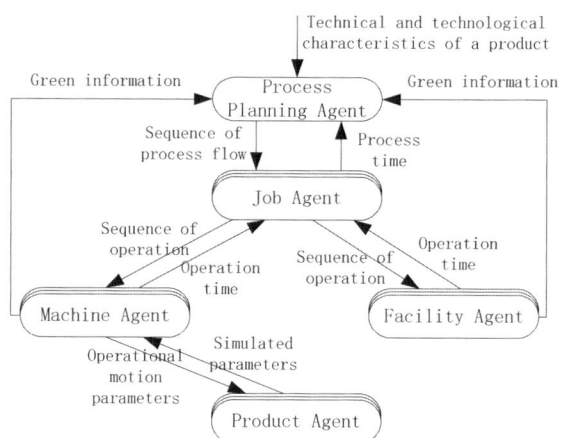

Fig.5. Coordination mechanism of control in manufacturing system operation

The operational procedure of green evaluation is described as follows:

(1) PPA obtains information on technical and technological characteristics of the manufactured product and then defines the sequence of process flows of manufacturing system and monitors the operation stages realizations.

(2) Each process will contain several job sequences which will be executed in different machines. Based on

the sequence of process flows, PPA will inform JAs and assign the execution of process contents to different JAs. JA defines its operation sequence and finds appropriate MA to perform the operation sequences.

(3) When MA receives the messages of operation sequences, it will define its parameters to execute every operational motion. Meanwhile, FA receives the messages of supply operation information. It will stimulate its operational procedure.

(4) When different levels operational motions are set, than execute the layout of MAs, PAs and FAs in the virtual simulation interface.

(5) When the virtual layout of manufacturing system is accomplished, PPA stimulates the manufacturing dynamic simulation. The green information is collected from different entity agents.

(6) Finally, according to the collection information, the green performance of manufacturing system is analyzed based on specific algorithms.

V. CONCLUSION

This paper puts forward the concept of green manufacturing simulation evaluation, and then a green production simulation evaluation system structure was designed based on multi agent technology. The system integrated green data about the components of manufacturing process into their models, then not only can plant layout the validation of product manufacturability and production efficiency be reacted but also green production status of the operation levels of manufacturing system can be evaluated and optimized.

This article only put forward the idea of green evaluation system based on Agent technology, the system development is not processed so far. The establishment of the database about green performance and manufacturing process, the establishment of a green evaluation knowledge base need further research.

ACKNOWLEDGMENT

The authors would like to thank the financial support offered by National High-tech Research and Development Program (863 Program) (No. 2013AA040206), and the National Natural Science Funds (No. 71473077).

REFERENCES

[1] J. Zhigang, Z. Hua, "The theory system and its implemented strategy for green reproducing (in Chinese)," *China Mechanical Engineering*, vol. 17, no.24, pp. 2573-2576, 2006.

[2] W. Xiaozhen, Z. Hua, "Green production process implemented architecture for manufacturing enterprises (in Chinese)," *Computer Integrated Manufacturing Systems*, vol. 16, no. 1, pp. 70-75, 2010.

[3] A.D. Jayal, F. Badurdeen, O.W. Dillon, I.S. Jawahir, "Sustainable manufacturing: Modeling and optimization challenges at the product, process and system levels," *CIRP Journal of Manufacturing Science and Technology*, vol. 2, pp. 144-152, 2010.

[4] W. Junfeng, L. Shiqi, L. Jihong, "A survey on energy efficient discrete manufacturing sytem (in Chinese)," *Journal of Mechanical Engineering*, vol. 49, no. 11, pp. 89-97, 2013.

[5] J. Heilala, S. Lind, et al, "Simulation-based sustainable manufacturing system design," *in Proc. of the 2008 Winter Simulation Conference*, Austin, pp. 1922-1930.

[6] T. Gutowski, M. Branham, J. Dahmus, "Thermodynamic analysis of resources used in manufacturing processes," *Environ. Science and Technology*, vol. 43, no. 5, pp. 1584-1590, 2009.

[7] H. Xiaohui, "Research on Characteristics and acquisition methods for running-status of job-shop manufacturing equipments (in Chinese)," Ph.D. dissertation, Chongqing University, China, 2012.

[8] K.A. Hossain, F.I. Khan, K. Hawboldt, "E-Green-A Robust Risk-Based Environmental Assessment Tool for Process," *Eng. Chem. Res.*, vol. 46, pp. 8787-8795, 2007.

[9] M. Gregor, S. Medvecky, J. Matuszek, "Digital factory,"*Journal of Automation, Mobile Robotics & Intelligent Systems*, vol. 3, no. 3, pp. 123-132, 2009.

[10] S. Thiede, Y. Seow, J. Andersson, B. Johansson, "Environmental aspects in manufacturing system modelling and simulation—State of the art and research perspectives," *CIRP Journal of Manufacturing Science and Technology*, vol. 6, pp. 78-87, 2013.

[11] Z. Jie, *Agent-based scheduling and control of manufacturing system*, Beijing, NJ: National Defense Industry Press, 2013, pp. 40-41.

[12] P. Leitão, "Agent-based distributed manufacturing control: a state-of-the-art survey," *Engineering Applications of Artificial Intelligence*, vol. 22, no. 9, pp. 79–91, 2009.

[13] Z. Jie, G. Liang, L. Peigen, Application of multi-agent technology in advanced manufacturing, Beijing, NJ: Science Press, 2004, pp. 125-336.

[14] W. Yankai, G. Baiwei, D. Jiuhui, "Research of MAS-based collaborative simulation (in Chinese)," *Journal of System Simulation*, vol. 23, no. 8, pp. 1610-1613, 2011.

[15] M. Jovanovic, S. Zupan, M. Starbek, I. Prebil, "Virtual approach to holonic control of the tyre-manufacturing system," *Journal of Manufacturing Systems*, vol. 33, pp. 116-128, 2013.

The Monitoring Path Forecasting Method in Digital Main Control Room of Nuclear Power Plant and its Verification

Hong Hu[1,2,*], Li Zhang[2], Kebing Liao[2], Cannan Yi[2]

[1]College of Nuclear Science and Technology, University of South China, Hengyang, China

[2]Ergonomics and Safety Management Institute, Hunan Institute of Technology, Hengyang, China

(fengzhisu16@163.com)

[1]Abstract - **It is difficult for operators in main control room (MCR) of nuclear power plant (NPP) to determine the next possible monitoring object when they are monitoring digital human-machine interface (DHMI) parameter information process, so monitoring delay or transfer error would occurs. According to this situation, a forecasting path programming method (FPPM) in monitoring process was proposed based on human factor reliability. Firstly, forecasting path model in monitoring process is proposed; then a FPPM which covers path tree algorithm building based on information filtration and searching path algorithm is presented; finally, the calculation method of transferring path successful probability was raised. For future verification for the proposed method, taking 3K00118YMA DHMI in SGTR (Stream Generator Tube Rupture) event process of NPP as monitoring task source of t time, the transferring path of next monitoring object was obtained successfully to minimize the risk of monitoring error and improve the efficiency of monitoring.**

Keywords – **Digital control room, forecasting path programming method, monitoring, nuclear power plant, SGTR**

I. FOREWORD

Today, many NPPs are using or plan to use digital control system (DCS) [1] in MCR, which is characterized by digital information display, highly integrated human-machine elements, multilevel data processing and automation operation from the view of human-machine interface. The operators sometimes cannot judge next possible monitoring object which would lead to monitoring delay or transfer error in the monitoring digital human-machine interface (DHMI) parameter information process in nuclear power plant (NPP) [2-3].

With regard to operators' monitoring process in MCR, this paper proposes a FPPM method to predict the next monitoring object according to the current system state so that operators are able to select and reach the next valid monitoring object more rapidly and exactly, it will provide much helping with reducing human errors, improving monitoring efficiency, and also contributing to analyze the driving mechanism of operators' monitoring activities and to optimize the DHMI.

II. FPPM MODEL IN MONITORING PROCESS

FPPM in monitoring process refers to how to rapidly and exactly obtain the next monitoring object each time after monitoring current state or parameter information for NPP system. Investigation and interviews with operators show that there is logical correlations existed between current monitoring object and the next one, namely, the selection of the next monitoring objects depends on the current system state or parameter information. That is, operators' monitoring transfer only relates to the current state or parameter information, so Markov process is available to simulate. As the whole monitoring task process is changeable, namely, the monitoring path vary with operators' monitoring transfer, this paper proposes path selection algorithm and the calculation method of Markov transfer path success probability. Forecasting path with high relevance will be selected out of all possible paths as the next monitoring object, a dynamic programming model (Fig.1) is applied to describe the process.

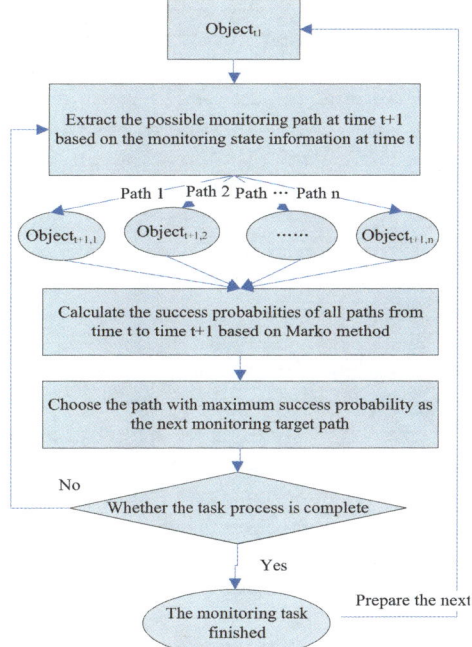

Notations:

Object$_t$: the state information of the monitoring object at time t

Object$_{t+1,n}$: the state information of the would-be monitoring object n at time t+1

Fig.1. Flowchart of FPPM model in monitoring DHMI of NPP

1 National nature Science Foundation (Nos. 70873040, 71071051, research projects of Ling Dong Nuclear Power Company Ltd. (KR70543), Innovation Ability Construction Projects based on the new Industry- Academy- Research Cooperation of Hunan Province (2012GK4101)

III. FPPM ALGORITHM IN THE MONITORING PROCESS ABOUT DHMI OF NPP

Fig.1 shows that FPPM algorithm searches effective programming through multi paths, obtains next possible monitoring object with maximum success probability and optimal way by dynamic programming method based on current monitoring object. The FPPM algorithm is specified as follows:

At t time, start from the initial object $Object_t$:

Ini_Filter(t,t+1,task_info): screen out information being irrelevant to the task information of current time t and next time t+1, namely, the preliminary selection;

IV. KEY ALGORITHM OF FPPM IN THE MONITORING PROCESS ABOUT DHMI OF NPP

Algorithm proposed in Chapter 3 shows that there are two key steps in the algorithm: Ini_Filter(t,t+1,task_info) and Next_Obj_Pro(t,t+1,n). We will interpret them in this part.

A. Ini_Filter(t,t+1,task_info) Algorithm

As operators have to face tremendous information in the monitoring process, it is essential to sieve out the irrelevant part to simplify search path and to reduce the exaggerated calculation time.

How to sieve out irrelevant part? As we know, tasks (information) with high similarity are relevant, while the others with low similarity are irrelevant. So the key step for operators' screening process is to evaluate tasks' similarity. Generally, operators judge the degree of similarity according to the index pertinence or the semantic similarity or the attribute resemblance of tasks. For example, Gao Ting discussed the grey correlation degree of customers' scarification under business model based on index system [4]; Zheng Yuhua adopted risk index to research the relevance of petroleum engineering projects [5]; Cui Qiwen studied similarity based on sememe [6]. Inaccuracy would appear when we calculate the tasks' similarity; it would become impossible to extract similarity factors, so the calculation of similarity or association is not dealt in this study to keep the theme. This paper applies multi-branched model to screen.

(1) Build_tree(n): Set up a tree structure with n nodes in database. It is a process that should be analyzed by experts with repeated training and studying. The steps of building tree are: First, mark every tree to differ tasks, cause each tree has its own tree structure. Every node should be marked with its number and task feature, the root node denotes the task source.

Second, start from the root node, if found that the current node information and some next task information is correlated, we choose the next correlated task node as the child node of current node. Third, take the child node founded in step forward as parent node and then try to find its child node, the search process would be circled in the same way introduced above until there is no relevant child node anymore. Thus a tree structure is formed, but it still need be repeated trained, studied and revised to form an expert system.

(2) Search(tree,relative_edge): Find out all of child nodes to the current task and mark them in terms of tasks' type and features. About the method of marking, two arrays are applied to respectively restore the information of current node and its child node sequences. The computer pseudo-code expressions of this process are specified as follows:

(3) build_calculation_path(parent_node,child_node): Get the information of the parent node and child nodes orderly from Crrent_array and Child_array to form the calculation path based the algorithm listed above, then we can achieve the success probability of every path.

B. Next_Obj_Pro(t,t+1,n) Algroithm

This part aims to calculate transfer path success probability from current node (parent node) to next possible object (child node). In monitoring DHMI, operators' monitoring transfer is influenced by relevant information of DHMI and monitoring manipulation, so it could be assumed that operator's monitoring is data-driven. If licensed operators have had enough knowledge to monitor, it can be concluded that the next monitoring state generally only relates to the current monitoring state, and that the monitoring transfers merely base on the current information, so operators' monitoring transferring has strong Markov property.

It is generally acknowledged that Markov consists of quadruples (X, A, P, R), where X stands for states set, A for paths set, P for transfer probability between two states, R for expectation value [7]. The mathematical formulation of Markov is [8,9]: as to any integer n and any extraneous variable xi, if:

$$P (X_1 = x_{i1}, X_2 = x_{i2}, \cdots, X_n = x_{im}) > 0 \qquad (1)$$

then,

$$P (X_{n+1} = x_{in+1} \mid X_1 = x_{i1}, X_2 = x_{i2}, \cdots, X_n = x_{in})$$
$$= P (X_{n+1} = x_{in+1} \mid X_n = x_{in}) \qquad (2)$$

As to the transfer way from current state to the next, there are only two factors considered in this paper: one is the task information state, while the other is the decision process of operators. Transfer successful or not is decided by these two factors. So it is consistent with Multiplication Principle of probability theory, the transfer path error probability can be defined as:

failure_Tranfer_path_probility(t,t+1)=
P(rask_{t+1}|rask_t,rask_{t-1}, \cdots,rask_1)*p(task_{t+1}|decision_t,
decision_{t-1}, \cdots,decision_1) \qquad (3)

Where $rask_t$ is the current task information state at time t, $rask_{t+1}$ is the information of possible task at time t+1, $decision_t$ is decision process of the operators' selecting the path in monitoring. Eq. (3) indicates that the transfer process is mainly affected by current task state and operators' decision upon selecting the path in monitoring. The current task state reveals the physical properties of the system at that moment, while decision of selecting path is about the operators' mental activities. According to Eq. (2), we find that the transfer-influenced factors at time t+1 are only concerned with factors at time t, so Eq. (3) can be simplified as:

$$failure\_Tranfer\_path\_probility(t,t+1)$$
$$=P(ras\,k_{t+1}|ras\,k_t)*p(tas\,k_{t+1}|decision_t) \quad (4)$$

thus the transfer path success probability is:

$$Succ\_Tranfer\_path\_probility(t,t+1)$$
$$=1- P(ras\,k_{t+1}|ras\,k_t)*p(tas\,k_{t+1}|decision_t) \quad (5)$$

How to calculate the success probability according to Eq.(5)? We can see from Eq.(5) that if we know the methods to calculate $P(ras\,k_{t+1}|ras\,k_t)$ and $p(tas\,k_{t+1}|decision_t)$, we can easily get the value. So we will discuss their calculations followed.

(1) $P(ras\,k_{t+1}|ras\,k_t)$ Algroithm

As mentioned in the previous analysis, $p(task_{t+1}|decision_t)$ refers to the error probability caused by the transfer process of task information state from time t to time t+1, Δt denotes the time space between two states, in terms of relevant study [2], the error probability of Δt is obtained (Table I):

TABLE I
CALCULATION FORMULAE OF TRANSFERRING FAILURE RATIO BETWEEN TWO CONSECUTIVE STATES

$P\{task_j(t)|task_j(t-1)\}$		$task_j(t-1)$	
		0	1
$task_j(t)$	0	$e^{-Fp(t)\Delta t}$	0
	1	$e^{Fp(t)\Delta t}-1$	1

Note: Where $task_i(t)=0$ denotes the current normal state, $task_i(t)=1$ the current abnormal state, while $Fp(t)$ the monitoring error probability of $task_i(t)$.

(2) $p(task_{t+1}|decision_t)$ Alogroithm

The term $p(task_{t+1}|decision_t)$ refers that the operators' monitoring transfer to next object depends on operators' current decision error, namely, the value is the error probability at time t, so, $p(task_{t+1}|decision_t)$ and $p(decision_t)$ are mathematically equivalent, that is:

$$p(task_{t+1} \mid decision_t) \Leftrightarrow p(decision_t) \quad (6)$$

The decision process is decisive to transfer in that decision error will consequently lead to transfer error and this process is mainly influenced by the physical properties of task information and operators' individual factors. This decision process influenced by multi-factors can be simplified as [7]:

P(decision_t)=P(decision_t|task character, human_factors)(7)

Actually, Eq. (7) is a calculation under multi-conditions. According to relevant studies [7], the conditional probability with many parent nodes can be solved by conditional probability with single parent node, and the expressions are listed below:

$$P(n = S_{N_i} \mid M_1 = S_{M_1,P_1}\, M_2 = S_{M_2,P_2} ......M_K = S_{M_K,P_K}) \quad (8)$$

$$= \lambda \sum (\prod_{j=1}^{j=k} P(N = S_{N_i} \mid M_j = S_{M_j,P_i}))$$

Where,
$$\lambda = \sum_{i=1}^{n} P(N = S_{N_i} \mid M_1 = S_{M_1,P_1},$$
$$M_2 = S_{M_2,P_2} ......M_K = S_{M_K,P_K}$$

How to judge whether a factor could be the decision influencing factor? Based on the author's study and

experience in NPPs, analysis of the interviews with the operators, other experts' studies and the characteristics of this research itself [10-13], we conclude 6 factors being taken into consideration (Table II).

Toward to the factors in Table II, combine Eqs. (7) and (8), then the further derivation expression of P(decision) is as following:

$$p(decision_t) = \sum_{i=1}^{i=6}[\sum(\prod_{j=1}^{k} p\,(decision_t \mid d_j))] \quad (9)$$

Where i refers to the factor count, j refers to the state of factors.

TABLE II
THE CORRELATION FACTORS THAT INFLUENCE OPERATORS' DECISION

affecting factors	Variables	affecting factors	Variables
Operators' knowledge and experience	d1	Operators' training level	d2
Task complexity	d3	Decision support system	d4
Stress level	d5	Time stress	d6

V. CASE STUDY

This paper take SGTR accident of NNP as an example and its 3K00118YMA DHMI as the information source node at time t in the monitoring process. The laboratory equipments used in the experiment are eye tracking system (Tobii), virtual workstation, and accident simulation screen (developed by soft Visual studio.net). In terms of the FPPM method constructed above, the specific process in this case are:

(1) Build_tree(n): set up tree structure of the monitoring path based on FPPM method;

(2) Take the DHMI of 3K00118YMA as the initial monitoring object at time t;

(3) According to Search(tree,relative_edge) algorithm, find out all of child nodes screen to the parent node screen(3K00118YMA) from the monitoring path tree of SGTR, then the parent node screen and the next possible child node screens, that are child node screen b (3K00121YMA) , child node screen c (3K00119YMA), child node screen d (3K00120YMA), child node screen e (3K00123YMA), and child node screen f (3K00122YMA).

(4) Get the tree structure of the parent node and the child node according to build calculation_path (parent_node, child_node) (Fig.2).

(5) Next_Obj_Pro(t,t+1,n) algorithm For ease of description, the transfer path success probability is separated into 2 parts: $P(ras\,k_{t+1}|ras\,k_t)$ and P(decision). The follows are the respective calculations of them: $P(ras\,k_{t+1}|ras\,k_t)$ algroithm

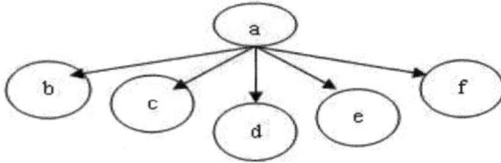

Fig.2. The father node and its child nodes selected in SGTR accident

When accident happens, the current task state and the next task state are abnormal, Eq. (10) can be obtained according to Table I:

$$p(task_j(t) \mid task_j(t-1)) = e^{F_p(t)\Delta t} - 1 \qquad (10)$$

Parameters of parent node and each child node are obtained by experimental analysis and listed in Table III:

We can obtain the P(raskt+1|raskt) value of node pair (the parent node with a child node) according to Eq. (10) and Table III (See Table IV).

TABLE III
PARAMETER VALUES OF PARENT NODE AND SOME CHILD NODES OBTAINED FROM EXPERIMENT

$a \rightarrow b$		$a \rightarrow c$		$a \rightarrow d$		$a \rightarrow e$		$a \rightarrow f$	
$F_p(t)$	Δt	$F_p(t)$	Δt	$F_p(t)$	Δt	$F_p(t)$	Δt	$F_p(t)$	Δt
0.0008	29	0.0001	20	0.005	47	0.003	39	0.001	35

TABLE IV
THE RESULTS OF THE P(rask$_{t+1}$|rask$_t$) VALUE OF NODE PAIR (THE PARENT NODE WITH A CHILD NODE)

Transferring path	$a \rightarrow b$	$a \rightarrow c$	$a \rightarrow d$	$a \rightarrow e$	$a \rightarrow f$
P(rask$_{t+1}$|rask$_t$)	0.023	0.002	0.265	0.124	0.035

1) P(decision$_t$) algroithm

Among the 6 decision affecting factors in Table II, d1,d2,d5 and d6 are unrelated to the execution of the task but to the operators' themselves, that is, though it is irrelevant to the path, there would be different states for them in monitoring. In terms of [9] and [14], we can achieve the values of d1, d2, d5 and d6 of different states, Table V can be obtained based on Eq. (7).

TABLE V
THE ERROR PROBABILITY OF FACTOR d1, d2, d5 AND d6, AND THEIR SUM OF PRODUCTS OF DIFFERENT STATE

d_1			j=1	j=2	j=3	d_2	j=1	j=2	j=3
			0.25	0.002	0.001		0.01	0.02	0.001
Decision	j=1	0.1	0.025	0.00004	0.00001		0.003	0.0004	0.00003
	j=2	0.02	\multicolumn{3}{c}{P(decision/d_1)= 0.02505}	\multicolumn{4}{c}{P(decision//d_2)= 0.00343}
	j=3	0.01							

d_5			j=1	j=2	j=3	d_6	j=1	j=2	j=3
			0.03	0.01	0.003		1.0	0.1	0.01
Decision	j=1	0.1	0.003	0.0002	0.00003		0.1	0.002	0.0001
	j=2	0.02	\multicolumn{3}{c}{P(decision//d_5)= 0.00323}	\multicolumn{4}{c}{P(decision//d_6)= 0.1021}
	j=3	0.01							

Note: j=1, 2, 3 refers to the state bejing high, middle and low.

According to the actual situation, factor d3 and d4 only have one sort of state, but their values vary with DHMI. Combined with expert judgment and actual experience in NPP, the values can be obtained in accordance with different transfer paths, the results are listed in Table VI as follows:

TABLE VI
THE ERROR PROBABILITY OF FACTOR d3 AND d4 IN DIFFERENT PATHS

Transferring path	$a \rightarrow b$	$a \rightarrow c$	$a \rightarrow d$	$a \rightarrow e$	$a \rightarrow f$
P(decision/d$_3$)	0.06	0.001	0.9	0.6	0.2
P(decision/d$_4$)	0.001	0.005	0.1	0.08	0.04

In accordance with Table V, Table VI and Eq. (9), the values of P(decision$_t$) of each transfer are listed in Table VII.

TABLE VII
THE ERROR PROBABILITY OF P(decision$_t$)

Transferring path	$a \rightarrow b$	$a \rightarrow c$	$a \rightarrow d$	$a \rightarrow e$	$a \rightarrow f$
P(decision$_t$)	0.1948	0.13981	1.13381	0.8138	0.3738

2) The results of Succ_Tranfer_path_probility(t,t+1)

Based on Table IV and VII, Eq. (5), the success probability of each path is listed in Table VIII:

TABLE VIII
THE SUCCESSFUL PROBABILITY OF EACH TRANSFERRING PATH

Transferring path	$a \rightarrow b$	$a \rightarrow c$	$a \rightarrow d$	$a \rightarrow e$	$a \rightarrow f$
Succ_Tranfer_path_probility(t,t+1)	0.996	0.998	0.7	0.899	0.987

(6) Max(t+1,n)

Select the maximum transfer probability from Table VIII, we can find that the maximum is 0.998.

(7) Transfer_route(t+1)

Based on step (6), we can know that the next transfer path would be(a)→(c), that is, the parent node screen (a) transfers to the child node screen (c).

(8) END().

According to the FPPM method for DHMI of NPP proposed in this paper, operator's transfer path was obtained successfully through monitoring 3K00118YMA in the SGTR accident. The forecasting result is in consistent with path transferring (frame (a) transfers to the frame (c) in Fig.2) of responding manipulation instruction given by operators in an actual SGTR accident of a NPP, as well as the transferring path achieved by video investigation and eye tracking system analysis when operators handle with SGTR accident in simulator. Therefore, the FPPM algorithm proposed in this paper is available.

VI. CONCLUSION

This paper proposed FPPM method for the monitoring transferring process of DHMI of NPPs, which consists of programming mode, execution flow and key algorithm and so on. Then, the method is applied to analyze SGTR accident of digital NPP and the predicted monitoring transferring path which is consistent with actual situation. Thus the method proposed herein does helpful to decrease operator's monitoring errors, which will also contributes to analyze the driving mechanism of operators' monitoring activities, to train simulated for monitoring behavior, to optimize the digital man-machine interface, to analyze the monitoring behavior on other fields: radar, intelligent robot, aerospace, and so on.

However, there exists some limitations as to this method, for example, we only select 6 factors that influences operators' decision, undoubtedly, there are more than 6 factors that would influence decision; also the value of the factors adopted from the existed data or experts judgments; meanwhile, the formation of the tree structure still need improving to reduce the complexity of inquiry.

REFERENCES

[1] Yu J B. "The characteristics of the digital control system in nuclear programming", *Modern Science*, 2008, vol.22, pp.48-49.

[2] Jiang J J, Zhang L, Wang Yq, Peng Y Y, Zhang K, He W. "Markov reliability model research of monitoring process in digital main control room of nuclear power programming", *Safety Science*, 2011, vol.49, pp.843-851.

[3] Jiang J J, Zhang L, Wang Y Q, Yang D X, Zhang K, He W. "Association rules analysis of human factor events based on statistics method in digital nuclear power programming", *Safety Science*, 2011,vol.49, pp.946-950.

[4] Gao T. "Gray correlation assessment of customer satisfaction on c2c e-business model", *Modern Business Trade Industry*, 2011, vol.24, pp.368-369.

[5] Zheng Y H, Luo D K. "Gray correlation assessment of the risk of petroleum engineering project based on utility theory", *Project Management Technology*, 2011, vol.9, pp.100-103.

[6] Cui Q W, Xie F. "An improved computational method for method for conceptual semantic similarity in domain onthlogy", *Computer Applications and Software*, 2012, vol.29, pp.173-175.

[7] Luiz Guilherme Nadal Nunes, Solon Venâncio de Carvalho,Rita de Ca´ssia Meneses Rodrigues. "Markov decision process applied to the control of hospital elective admissions", *Artificial Intelligence in Medicine*, 2009, vol. no.47, pp.159-171.

[8] Ran J J, Zhao Y J, Liang C. "The application of the prediction of precipitation state based on weighted Markov chain", *Yellow River*, 2006, vol.28, pp.32-34.

[9] Seunghwan Kim, Yochan Kim, Wondea Jung. "Operator's cognitive, communicative and operative activities based workload measurement of advanced main control room". *Annals of Nuclear Energy*, 2014, vol.72, pp.120-129.

[10] H.Putzer, R.Onken. "COSA-a generic cognitive system architecture based on a coginitive model of human behavior", *Cognition, Technology & Work*, 2003, vol.5, no.2, pp.140-151.

[11] Oliver Sträter, Heiner Bubb. "Assessment of human reliability based on evaluation of plant experience: requirements and implementation", *Reliability Engineering and System Safety*, 1998, vol.63, no.2, pp.199-219.

[12] Y.H.J. Chang, A. Mosleh. "Cognitive modeling and dynamic probabilistic simulation of operating crew response to complex system accidents. Part 2: IDAC performance influencing factors model", *Reliability Engineering and System Safety*, 2007, vol.92, pp.1014-1040.

[13] Donald Mender. "The implicit possibility of dualism in quantum probabilistic cognitive modeling". *Behavioral and Brain Sciences*, 2013, vol.36, no.3, pp.298-299

[14] Alyson G.Wilison, Aparna V. Huzurbazar. "Bayesian networks for multilevel system reliability", *Reliability Engineering and System Safety*, 2006, vol.92, no.10, pp.1413-1420.

Modeling Technology on Aircraft Cable Harness in Virtual Assembly Environment

Jing Yan [1,*], Wei Hong [1], Lu Ju [1], Bi Lan [2], Xu-hui Li [2]

[1] Department of Mechanical and Electrical Engineering, Nanjing University of Aeronautics and Astronautics, Nanjing, China

[2] National Engineering and Research Center for Commercial Aircraft Manufacturing, Commercial Aircraft Corporation of China, Ltd, Shanghai, China

(yanjing@nuaa.edu.cn)

Abstract - **Aircraft cable harness virtual assembly technology has an important influence on improving the efficiency of the aircraft assembly and performance of the machine. Modeling on aircraft cable harness is elementary and difficult. For aircraft cable harness elements, there are two kinds of the most common process constraints: suspension constraint and minimal bending constraint. The catenary theory has been proposed to model the elements under the suspension constraint accurately. For the elements under the minimal bending constraint the robot arm has been chosen to build the dynamic model to simulate the lying process. By controlling the parameters of the robot arm to meet the minimal bending radius requirement, the cables' geometric changes can be realized in time. The Aircraft Cable Harness Virtual Assembly System (ACHVAS) has been developed and the cable harness elements under the two constraints have been demonstrated in the platform to verify effectiveness of the two models.**

Keywords - **Aircraft cable harness, catenary theory, minimal bending radius constraint, robot model, suspension constraint, virtual assembly**

I. INTRODUCTION

Cable is an important component of the current signal transmission in mechanical and electrical products. In the class of complex electromechanical products in aircraft, cable exists in the way of harness. Statistics, assembly occupies half-time of the entire aircraft manufacturing cycle. Assembly process design is the foundation of modern aircraft industry, it is the bridge that connecting aircraft design and manufacture of aircraft. Harness assembly is an indispensable part, the process is not only time-consuming, costly, but also the wiring result is one of the important objects in aircraft troubleshooting and maintenance. Wiring quality directly affects the reliability, efficiency and economy of the aircraft during its service. For example, the U.S. General Electric Company analyzed and summarized the reason of air parking event during the use of ever developed engine. It found that 50% is due to external piping damage, cables damage and sensor damage[1]. According to a report of the quality problems of a batch of missiles in aerospace, cable fault accounted for about 20% of total failures [2].

Virtual reality technology is the information platform that can improve the efficiency of harness assembly and optimize wiring results. Using virtual reality technology to simulate the process of harness assembly, it has a positive and realistic significance on improving the overall quality of the aircraft. For harness assembly, study abroad has commenced. Literatures [3-5] revolve research work around COSTAR harness design immersive virtual reality system. The platform of system software uses WorldToolKit and the hardware devices include SGI Octane2, HMD, Pinch Gloves and Flock of Birds. On this platform, harness design is divided into three steps: cable path planning, cable bundled into a harness and cable modifications. Specifically, the cable is connected by the given key segments in space by a cable designer, each cable segment coupled to the cable path. Cables can be edited, including the key selection or increased, the cable bend around obstacles, increasing the plug, fasten; finally tied into bundles. Literature [6] provides an overview of the digital analog cable wiring installed as the core design techniques and tools for data management using three expressions of a harness tree, using discrete control points on the harness geometric modeling [2], which is the same with COSTAR. Literature [7] provides three types of spacecraft harness, including power lines, data lines and signal lines. In a very high speed impact of the vulnerability of the 10 experiments were analyzed to obtain that the main reason of harness fault is the damage of mechanical structure, the form of destruction includes material melts, the gasification and material bending, but the relationship between the severity fault of harness and mechanical damage is not clear. Literature [8] studied maximum force of security facilities - security cable in the balance recovery test and relations of restore the critical balance.

Aircraft harness has proprietary features [9, 10]. Currently, little research has dedicated to virtual assembly of aircraft harness. In this paper, two common aircraft wiring standard states are addressed: harness suspension constraints and minimum bending radius of the free end of the assembly process to achieve its reproduction. On the basis of the analysis of two kinds harness anchor points status, derivation of theoretical modeling were carried out, and it gives harness hierarchical data structure. On the design of Aircraft Cable Harness Virtual Assembly System

[1] Fund of National Engineering and Research Center for Commercial Aircraft Manufacturing (The project No. is SAMC13-JS-15-024).

(ACHVAS) virtual reality platform, programming to achieve aircraft harness assembly simulation process on the constraints of these two common state.

II. MODELING ON ELEMENTS UNDER THE SUSPENSION CONSTRAINT

When fixed, cable harness is usually required to ensure to be smooth and avoid being crossed and distorted. There is a kind of cable harness element that is the cable harness element under suspension constraint. For the type element, cables between two fixed points are not allowed to pull too tight and are required to have specific slack value. As is shown in Fig.1, r is deflection, x_2 is the center pitch of the clamp. In this case, the stress state of the harness hanging constraints at both ends, and exposed only to gravity. This paper proposes the use of catenary model constraints on the harness hanging modeling. As is shown in Fig.1, local coordinate system X direction is taken as the endpoint connection of harness fixing constraint; the constraint point refers to the positive direction of left to right. Y-axis is taken as the direction of gravity, which is positive on the orientation; according to the right-hand rule, you can determine the Z-axis and its positive direction, point P_1 (point O) is origin point.

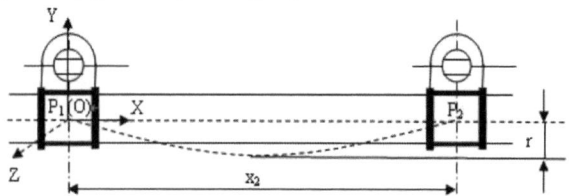

Fig.1. Elements under suspension constraint

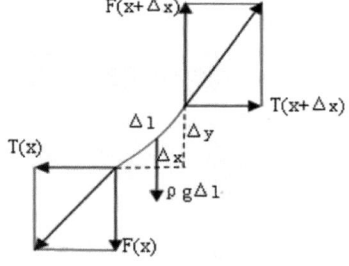

Fig.2. Forces that act on micro-segment of the elements

We can obtain from the force balance condition of the object,

$$\begin{cases} T(x) - T(x + \Delta x) = 0 \\ F(x + \Delta x) - F(x) - \rho g \Delta l = 0 \end{cases} \quad (1)$$

Formula (1) shows T (x) = T, it is a constant. Fig.2 shows,

$$\Delta l = \sqrt{(\Delta x)^2 + (\Delta y)^2} \quad (2)$$

Put formula (2) into (1), dividing Δx on both sides,

$$\frac{F(x + \Delta x) - F(x)}{\Delta x} = \frac{\rho g \sqrt{(\Delta x)^2 + (\Delta y)^2}}{\Delta x} \quad (3)$$

When $\Delta x \to 0$,

$$F'(x) = \rho g \sqrt{1 + [f'(x)]^2} \quad (4)$$

Fig.2 shows

$$F(x) = T \cdot f'(x) \quad (5)$$

Combine equation (4) and (5),

$$f(x) = \frac{T}{\rho G} \cdot \cosh\left(\frac{\rho g}{T} x + c_1\right) + c_2 = a \cdot \cosh\left(\frac{x}{a} + c_1\right) + c_2 \quad (6)$$

The parameters a, c1 and c2 are constants in the establishment of the local coordinate system in this paper. We can determine c_1, c_2 from the left end point coordinates (0, 0) to the right point coordinates (x_2, 0).When the deflection of the harness is known as r, assuming a uniform harness, the harness has a midpoint coordinates ($x_2/2$, -r) can be calculated as a.

Establishing harness equation y=f(x), using the formula (5), the value and direction of the two endpoints of the cable can be predicted.

As shown in the Fig.3, in the built virtual reality environment, the world coordinate system is $O_W X_W Y_W Z_W$ and the cable's local coordinate system is OXYZ. Corresponding to Fig.1, O point that P_1 point coordinates in the global coordinate system is (P_{1x}, P_{1y}, P_{1z}) and P_2 coordinates in the world coordinate system is (P_{2x}, P_{2y}, P_{2z}). The transformation process is, first, local coordinate system rotates 180° around the X_W relative to the world coordinate system. Then local coordinate system rotates θ around the Y-axis, according to "left to right" principle, when given at any point in the local coordinate system coordinates (x, y, 0), which coordinates (x_w, y_w, z_w) in the world coordinate system can be calculated using the homogeneous transformation,

$$\begin{pmatrix} x_w \\ y_w \\ z_w \\ 1 \end{pmatrix} = \begin{pmatrix} \cos\theta & 0 & \sin\theta & P_{1x} \\ 0 & -1 & 0 & P_{1y} \\ \sin\theta & 0 & -\cos\theta & P_{1z} \\ 0 & 0 & 0 & 1 \end{pmatrix} \begin{pmatrix} x \\ y \\ 0 \\ 1 \end{pmatrix} \quad (7)$$

When $x_2 = |P_1 P_2| \neq 0$, the angle θ in equation (7) is determined by equation (8),

$$\theta = arctg \frac{P_{2z} - P_{1z}}{P_{2x} - P_{1x}} \quad (8)$$

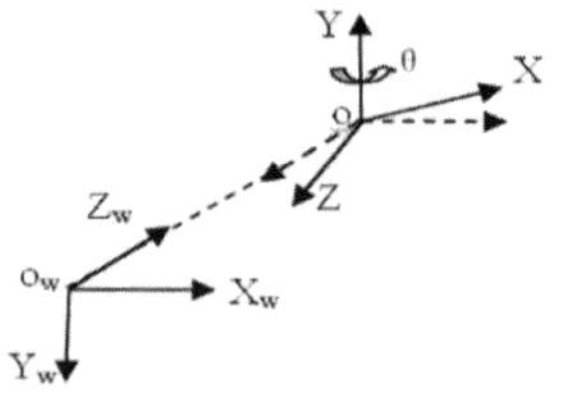

Fig.3. Schematic illustrating conversion between WCS and LCS about the elements

III. MODELING ON ELEMENTS UNDER THE MINIMUM BENDING RADIUS CONSTRAINT

It is often the case that harness and harness branch - cable wiring bending, closely associated with them is an important technical indicators of harness and cable: minimum bending radius. The meaning and purpose of the index is to control the cable harness and its branches due to a sharp bend or causing excessive stretching line failure, in order to improve the quality and manufacturability of harness. As is shown in Fig.4, the harness is bundled together of seven cable of the same diameter. One branch of cable bends at the free end which is not to be fixed, the minimum bending radius of the cable requires n times of the diameter d, typically n = 10.

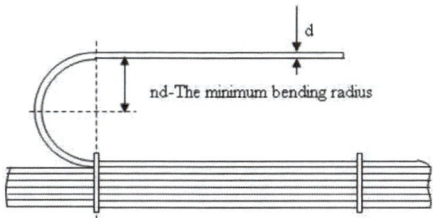

Fig.4. Elements under minimum bending radius constraint

Harness bending and cable bending assembly process are typical installation requirements, but no part of the contents discussed in detail in previous literature and discussion. This paper proposes a robot model to simulate minimum bending radius of harness and cable .On this basis, we further propose geometry changes of the harness assembly process.

PUMA560 robot [11-12] coordinates of each link is shown in Fig.5, the icon sizes such as d_2 are the D-H parameters. Using this model, the structure of the simulation is shown in Fig.4 and the branch bend is shown in Fig.6.

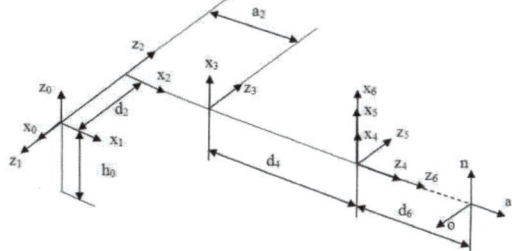

Fig.5. Robot arm model and its parameters

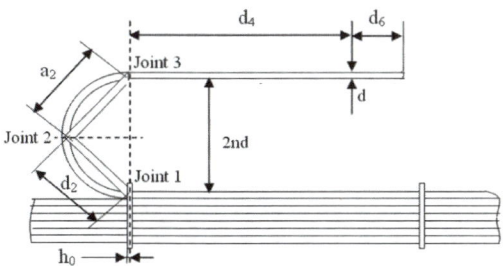

Fig.6. Relationship between minimum bending radius and parameters of the robot arm

As shown in the Fig.6, the robot model parameters from the calculation shows,

$$a_2 = d_2 \geq \sqrt{2}nd \qquad (9)$$

$$d_4 + d_6 \leq l - \pi nd - h_0 \qquad (10)$$

The parameter h_0 is the constraint length of the cable harness, d_4 is the length from robot arm elbow to wrist, d_6 is the length of the gripper in the end. Allocating in a certain proportion l is the effective

length of the crotch portion of the harness.

$$l = \pi nd + d_4 + d_6 + h_0 \qquad (11)$$

Fig.6 shows the structure of a robot model with PUMA 560, when a_2, d_2 values satisfy equation (9) to meet the requirements of the minimum bending radius of the harness branch. However, the model uses only lever 1 and lever 2 to simulate the bending part of harness, apparently too stiff. Using three bars with the same length and two joints 2', 3' are used to institute the original lever 1 and lever 2 to refine the bending part shown in Fig.7. Using the method, the length l is unchanged.

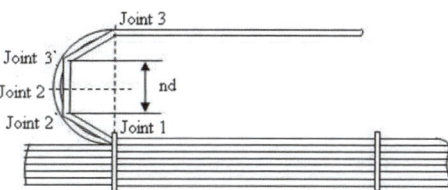

Fig.7. Refining the robot model to improve simulation precision

As shown in the Fig.7, the length of the three rods is $\geq$ nd. The mechanical structure of the robot model to ensure minimum bending radius conditions, in order to smooth the harness, each joint point in Fig.7 will be drawn with B-spline curves. In this virtual reality platform, the algorithm simulation results are shown in Fig.8.

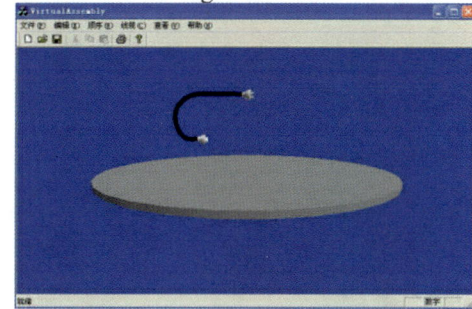

Fig.8. Simulation screenshot of the elements under minimum bending radius constraint

IV. MODELING ON CABLE HARNESS GENERAL SIMULATION PROCESS

Simulation of the harness assembly process characterized in that the geometry of the harness can make a quick response, and can avoid the obstacle. This process is similar to the robot arm operating modes, harness mobile endpoints, representing the end of the robot arm gripper. Harness no interference to the target position, which is the end of the robot gripper reaching the designated point of inverse kinematics solution, involving the existence of solutions, under conditions of multiple solutions to meet the requirements and avoid obstacles shortest travel requirements. Harness geometry, corresponding to the arm of the robot manipulator arm posture to reach the point of space. Fig.9 shows a comparable between the two. In view of the above analysis, we propose the use of a robot model in harness assembly process.

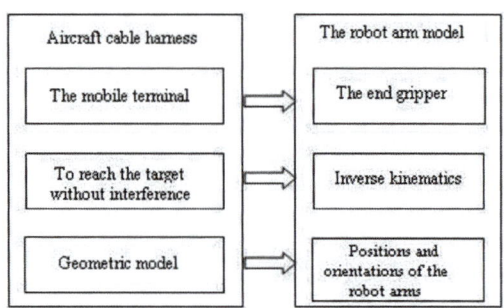

Fig.9. Similarity comparison between cable harness assembly process simulation and robot arm work mode

Harness assembly process model should meet three conditions. First, the rigid body and space are similar, should have six degrees of freedom. Second, harness assembly process remains the same length; the joints should be rotating joints, without moving the joints. Third, to meet the harness virtual assembly process requirements, the model should have closed inverse kinematics solution. Based on this, in order to maintain consistency with the model, we use PUMA560 model structure in the assembly process of harness. Model parameter assignment is the same with Part 3 of this article. Fig.10 and 11 are achieved the minimum bending radius of the harness assembly using the model shown in Fig.8.

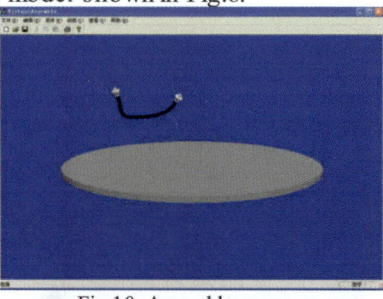

Fig.10. Assembly process

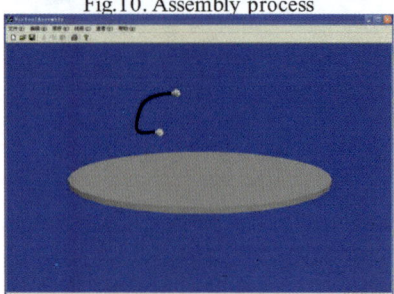

Fig.11. Assembly process

V. CABLE HARNESS DATA STRUCTURE AND MODELS SIMULATION EXAMPLES

From the main harness, aircraft harness can separate many different levels of sub-harness or cable and the structure is complex. In order to make clear recording level, data of harness and consider programming convenience. In this paper, a robot modeled on the basis of the connection structure [13], proposing harness data structure shown in Fig.12. In

Fig.12, each of which has only two branches, the left branch of its sub-branches, plus a series, right branch tied to its branch level. Specifically, looking down from the left branch of Fig.12, showing the harness segments expanding hierarchy, such as the "main harness", "2nd level harness A", "3rd level harness A1", "4th level harness A11".The right branch, the "main harness" does not have the same level. The "2nd level harness A" is the same level with "2nd level harness B" and "2nd level harness C". Fig.13 is a specific example of the data structure of Fig.12; Fig.14 is a data structure of the harness.

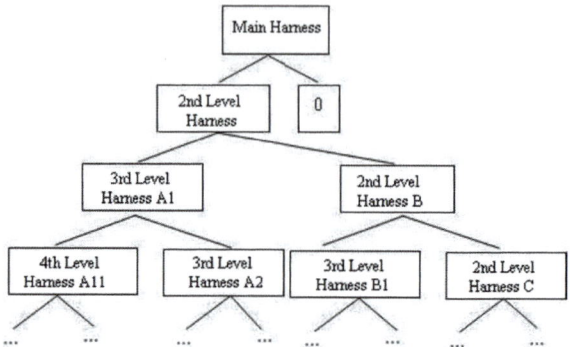

Fig.12. Multi-level data structure diagram of cable harness

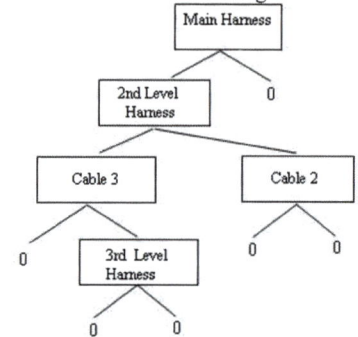

Fig.13. Multi-level data structure instance

In this paper, VC++ language and WTK developed a virtual assembly system based on MFC interface -Aircraft Cable Harness Virtual Assembly System (ACHVAS).On the platform, based on a common aircraft assembly structure shown in Fig.4.This article discusses the harness to build the model and conducts a harness assembly process simulation, as is shown in Fig.14 and 15.

In Fig.14, the harness is constituted by three cables, the diameter of main harness is 6mm and the diameter of 2 level harness is 4mm. The diameter of the 3 level harness, cable 2 and cable 3 are 2mm. Two spacing shown in the clamps from left to right are 35 mm, 300 mm, respectively, corresponding deflection is 2 mm, 10mm. The bending radius of cable2 and cable 3 are both 20mm. Fig.15 is a screenshot of the harness assembly, the harness hanging constricting portion has been modeled, the cable 2 and the assembly process of the cable 3 is formed in the minimum bending radius. Fig.16 is a diagram of the clamp of harness in Fig.14 which is fixed on the machine panel.

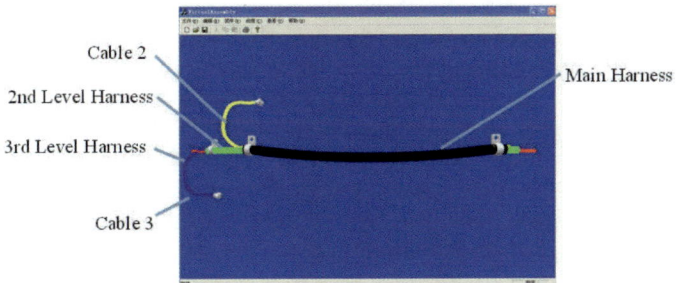

Fig.14.Modeling example of the two kinds of elements in VR

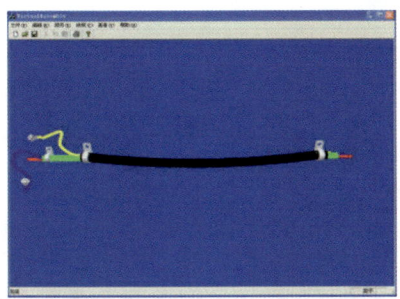

Fig.15. Example of cable harness assembly process simulation

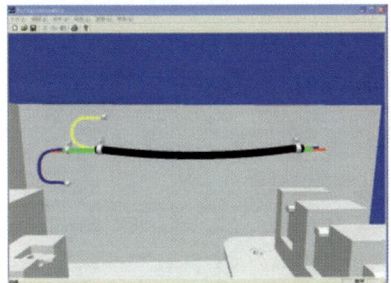

Fig.16. Example of cable harness installation

VI. CONCLUSIONS

1) Based on features of aircraft cable harness assembly process, two kinds of cable harness elements have been proposed: elements under the suspension constraint and elements under the minimum bending radius constraint.

2) For the cable harness elements under the suspension constraint, the catenary theory has been proposed to model.

3) For the cable harness elements under the minimum bending radius constraint, the robot arm model has been proposed. The relationship between minimum bending radius and parameters of the robot arm has been discussed. Using the model, the dynamic geometry of cables can be displayed in time.

4) In the developed platform ACHVAS, a cable harness assembly simulation instance has been demonstrated.

REFERENCES

[1] LIU Jianhua, ZHAO Tao.Motional Cable Harness Physical Characteristic Oriented Modeling and Kinetic Simulation Technology in Virtual Environment [J]. Mechanical Engineering, vol. 47, no. 9, pp. 117-124, 2011.

[2] LIU Jianhua, WAN Bile, NING Ruxin. Realization technology of cable harness process planning in virtual environment based on discrete control point modeling method [J]. Mechanical Engineering, vol. 42, no. 8, pp. 125-130, 2006.

[3] Raymond C.W. Sung, James M. Ritchie, Graham Robinson, Philip N. Day, J.R. Corney, Theodore Lim. Automated design process modelling and analysis using immersive virtual reality [J]. Computer-Aided Design, vol. 41, no. 12, pp. 1082-1094, 2009.

[4] F.M Ng, J.M Ritchie, J.E.L Simmons, R.G Dewar. Designing cable harness assemblies in virtual environments [J]. Journal of Materials Processing Technology, vol. 107, no. 1-3, pp. 37-43, 2000.

[5] G. Robinson, J.M. Ritchie, P.N. Day, R.G. Dewar. System design and user evaluation of Co-Star: An immersive stereoscopic system for cable harness design [J]. Computer- Aided Design, vol. 39, no. 4, pp. 245-257, 2007.

[6] Wei Shang, Jian-hua Liu, Ru-xin Ning, Jia-shun Liu.A Computational Framework for Cable Layout Design in Complex Products [J]. Physics Procedia, vol. 33, pp. 1879-1885, 2012.

[7] R. Putzar, F. Schaefer, M. Lambert. Vulnerability of spacecraft harnesses to hypervelocity impacts [J]. International Journal of Impact Engineering, vol. 35, no. 12, pp. 1728-1734, 2008.

[8] Marc-André Cyr, Cécile Smeesters. Maximum allowable force on a safety harness cable to discriminate a successful from a failed balance recovery [J]. Journal of Biomechanics, vol. 42, no. 10, pp. 1566-1569, 2009.

[9] Aircraft assembly process - cable laying [S]. HB/Z223.16-2002.

[10]A. Andreu, L. Gil, P. Roca .A new deformable catenary element for the analysis of cable net structures [J]. Computers and Structures, vol. 84, pp. 1882–1890, 2006.

[11]XIONG Youlun. Robot technology [M]. WU HAN, Huazhong University of Science and Technology Press, 2011.

[12]HAN Jianhai. Industrial robots [M]. WU HAN, Huazhong University of Science and Technology Press, 2009.

[13]WEI TIAN Xiusi. Humanoid robot [M]. BEI JING, Tsinghua University Press, 2007.

Implementing Low Carbon Improvement Practices in Supply Network

Yongjiang Shi, Jialun Hu*

Institute for Manufacturing, University of Cambridge, U.K.

(ys@eng.cam.ac.uk, *jh693@cam.ac.uk)

Abstract - **This paper aims to discuss the low carbon practices development in manufacturing companies. As the environmental-friendly pressure and internal energy-saving motivation are growing, firms start to implement de-carbonization projects across their supply chain. This study conducts case studies to these projects in 11 companies. A process framework of implementing low-carbon improvement practices is generated via the case analysis. Before building up the framework, the constituents within the framework are also discussed, including the three constructs which lead to classification of improvement projects, the process of project portfolio generation, internal supporting system and external supports. This novel study covers the absence of method to design de-carbonization action in operation management field. Also it expands the understanding to continuous improvement for environmental purpose. Practically the process framework can guide CSR or carbon managers to design corporate carbon strategy.**

Keywords - **Carbon reduction, sustainable practices, supply network**

I. INTRODUCTION

Climate change is emerging as a major challenge for modern society [1]. Government, business, and wider society all have a shared responsibility to tackle the issue. Under multiple pressures from government legislation and society, companies have to embrace low carbon management [2-4]. It has been shown that emerging economies suffer more conflicts between environmental protection and economic development due to their relatively larger population and lower technology industrial level. For the firms in emerging economics, copying the traditional industrial development 'route' is a relative simple way, but it is not sustainable under the climate change issue. And they face more difficulties to conduct the de-carbonization due to lack of supporting infrastructure [5], sufficient technology and knowledge [6], and last but not least, their regional legislation pressure [7]. All these issues make the engaging of firms in developing economies complicated.

Therefore this research focus on the excellent companies in developed and emerging countries—UK, Taiwan, and China which performed some initial practices in de-carbonization.

Until 2000s the focus of the global warming problem has driven the researchers' attention from the broad arena of the 'Green' issue into a more specific 'low carbon' stream, that is considering the issues of risks [8], measurement [9], mitigation [10] in the different industries [11-15]. From the supply chain/network perspective, there are researches exploring the de-carbonization in different stages of supply chain. Hua et al.

[16] investigated how firms manage carbon footprints in inventory management under the carbon emission trading mechanism. Holweg et al. [17] assessed the global sourcing risk considering the carbon offset costs. In supplier engagement, Jira and Toffel [18] discussed factors associated with the suppliers' willingness to disclose carbon information to buyers. Other issues have also been discussed, including aggregate loading in energy management in the production phase [19], waste management [20], etc. Smith and Ball [21] focused on the factory level, and developed an "MEW" (material, energy, waste) flow model to provide guidance on systematically analyzing manufacturing facilities, and to assist with the identification and selection of improvement opportunities. Choi [22, 23] discussed the impact of the carbon footprint tax on retailer's sourcing decisions. According to literature, the low carbon supply chain practices can be categorized into 4 general stages:

- Product: product design, packaging design, user phrase performance, waste Management, product recycle/reuse;
- Procurement: sourcing, supplier management, transportation from supplier, waste management;
- Production: internal generic manufacturing process, waste management;
- Logistics: transportation to clients, reverse logistics, transportation packaging design.

These four categories are consistent with one of the three dimensions that emerged from the case data. However there are few literature that drill into the practical needs of designing carbon emission reduction action plans, which are addressed in this paper.

The structure of the paper is planned as follow: the research design is firstly presented to explain the method adopted and the cases selected. Then the data in cases improvement practices are presented in Table I, including a summary of practices, idea generation processes and the supporting system to enable creating improvement recommendations. According to the cross case analysis, an initial framework to classify improvement practices is generated. Based on analysis shown in the data, a framework for the improvement process is then presented, which discusses the various approaches made to generate project ideas and also, which is just as important, the internal organizational supporting system to facilitate these projects. Finally discussion and conclusion are presented.

II. RESEARCH DESIGN

This research aims to answer the following research question:

E. Qi et al. (eds.), *Proceedings of the 21st International Conference on Industrial Engineering and Engineering Management 2014*, Proceedings of the International Conference on Industrial Engineering and Engineering Management, DOI 10.2991/978-94-6239-102-4_77, © Atlantis Press and the authors 2015

What can firms do to improve carbon emission performance in their supply network?

Semi-structure interviews were conducted in a number of companies. Table I shows the main cases and some of the case data. In this study, the case companies were identified based upon their perceived performance in low-carbon supply-network operations, and to provide a cross-section of industries. The selected industries can be divided into two types: assembly oriented, such as ICT, and process oriented, such as beverages, and steel manufacturing. Selecting industries that have higher carbon-emission impacts on the supply network, such as the ICT and chemicals industries, was the second-priority. The third criteria was that the case companies should focus on manufacturing rather than services, and this criteria meant that the majority of case companies are in the Great China Area, due to the current global manufacturing landscape.

Data was collected mainly through semi-structured, in-depth interviews with managers between June and August, 2013, with the aim of obtaining personal insights from managers, as well as through documents and archives to ensure construct validity. The interview list and questions were conducted using pre-designed guidelines, as shown in Table II. Most of interviewees are middle and senior-level managers in sustainability, supply chain and manufacturing departments. Due to the word limit of the conference requirement, the detail of each firm's practices is not presented. The Table III is an example of case Lenovo.

III. RESULTS

A. An Initial Framework to Classify Reduction Projects

Cross-case analysis is conducted in the reduction projects, three constructs of reduction practices emerge which is shown in the Fig.1.

The themes concluded from the detailed practices of each case are listed in under the '1st order Categories'. These themes are further classified into 2nd order categories according their characteristics. Sub categories are eventually grouped into three constructs: supply network stages, the factors that links to carbon emission, and the type of change.

Supply Network Stages
Carbon Emission Performance Improvement Choice
Types of Changes

So an initial framework to classified reduction practices can be generated according the above-mentioned three constructs: supply network stages, emission reduction factors, and types of changes. These three constructs are independent to each other (shown in Fig.2).

B. A Framework of Carbon Emission Improvement Process

The three-construct framework for reduction projects can be used by practitioners to generate a solution idea. However this systematic way to devise a solution is a 'Top-down' approach, at the same time the bottom-up

approach to generate reduction projects also emerges from the case data. Table I summarizes the portfolio generation methods, and internal/external supporting system in each cases.

1) Portfolio Generation

From the case data, both top-down and bottom-up approach to generate the emission reduction projects can be found.

Top-down approach refers to that the improvement knowledge comes from external resources. Therefore the solution project portfolio will be generated by the environmental-affair team and deploy top-down to each departments—R&D, production, logistics, etc. The sources of these external carbon reduction knowledge include mainly three types:

-International Standards and guidelines: GHG protocol, ISO 14064/14067, etc. Lenovo uses GHG protocol to guide its reduction projects ideas;

-Industry specific reduction knowledge: WSA (World Steel Association)'s carbon reduction guideline for steel manufacturers. China Steel Corporation gained many support from WSA.

-Practitioner tools and guidelines: cleaner production principles, PDCA continuous improvement.

These resources can be obtained from multiple external entities, including NGO, Business associations, consultants, industrial standard organization, Government bureaus, academic institutions, technology service provider, supply network partners, etc.

Bottom-up approach is when the improvement knowledge comes from internal resources—field engineering, staff at direct business line. So the solution project portfolio will be proposed by these front-line employees and then approved by the senior management group, who will then deploy the solution. Emerging from the case studies there are two types of methods gaining this kind of knowledge:

-Encouraging front-line engineers and staffs to propose new practices based on their operation experience. In most cases monetary reward are used, e.g. Tung ho Steel set up 'Staff Green Improvement Proposal System'.

-Investing R&D project for Energy efficiency, such as British Sugar invested average 1.8 million GBP per year on energy & material efficiency projects on production, suppliers, transportation & packaging.

In most the case companies, these two approaches are in fact applied simultaneously to generate the potential reduction project portfolio. The top-down approach can give focal firms a systematic view of the carbon reduction requirement and a possible solution, but the solutions are in general not feasible for the focal firm; On the other hand, the bottom-up approach provides practical, specific and an applicable solution. The bottom-up approach can be very innovative, such as Tsingtao Brewery who collects CO_2 generated in wort fermentation to substitute the gas which is used to seal beer bottles. However the bottom-up approach can be so limiting to a firm's current operation that other aspects of reduction projects are neglected. Therefore a special environmental –affair team is needed to perform as a hub absorbing external knowledge and internal proposals, and makes the decision afterwards.

2) Internal Management Structure

The above-mentioned special team plays a key role to enable the projects to be implemented, however, there are other components in the system to support, as shown in the case data. From the 'internal supporting system' column in Table I, a pattern in organizational structure emerges, as shown in Fig.3.

The support from the top-level management group is important for implementing the improvement project, which ensures sufficient resources can be allocated to these projects. A vice-president-level officer is normally assigned to manage the carbon management issue as whole. Many case companies have set up virtual carbon management committees, which consist of the CEO, vice president for the carbon issue, and senior managers in each of the functional units-R&D, production, procurement, etc. These senior managers report the overall progress of carbon projects, and they are assigned with tasks to cooperate with the carbon special task team. This carbon management committee have regular meetings, but are not actually involved in the project implementation. The operating team is the 'Carbon Special Task Team' which facilitates reduction projects from beginning to end.

A carbon special task team is set up for coordination. The team normally has a few members to cover the different projects or different plants (if the focal companies have these, such as Tsingtao Brewery). The team acts as the carbon expert in the focal firm who best know carbon management and the related standards. The carbon targets should be proposed by the team to the carbon committee. Absorbing knowledge from external sources and reviewing the solutions proposed by front-line staff in each functional units, the team has to make decisions to select reduction projects, and then choose tutors who support and monitor the project implementation in the function units or multiple plants. This team normally belongs to the EHS (Environmental, Health & Safety) department, which is separated from other departments. However, it can also belongs to production departments or product QA (Quality Assurance) departments, depending on the focus of firm. E.g., Tsingtao Brewery sets its team under production due to Tsingtao's focus on boosting energy efficiency in the production stage.

3) Improvement Process

A process model to implement improvement practices can be summarized (shown in Fig.4).

The measurement to corporate a supply chain level emission is the first step and it sets the baseline which is needed to create carbon reduction targets and plans. The plans and targets are normally proposed by the 'carbon special team' and determined by carbon management committee.

There are three steps to implement the reduction projects:

-Generate portfolio of projects

The carbon special task team collects and generates the project portfolio, using both the 'Top-down' approach and 'Bottom-up' approach.

-Assess projects

The project portfolio needs to be assessed and some of these projects are selected to implement it. The firm sets selection criteria according to its own context and targets. Feasibility, cost & benefit, payback period, and risks are four common aspects to consider. In the Tsingtao Brewery case, five criterion are used: the innovativeness of the project, feasibility & complexity, carbon reduction potential, economic return and cost.

-Execute projects

The selected projects are executed by different departments and supervised by the carbon special task team. In the AUO case, a tactic is applied--comparative experiment of reduction project was conducted in two plants to test its utility.

After implementing the projects, feedback and continuous improvement are in need. Case companies devise monthly, quarterly, or half-annual reviews and report to the senior management group in order to gain timely feedback. Both the functional departments and carbon special team are responsible for the project's success, and achieving the reduction target as well. Continuous improvement tools-e.g. P-D-C-A (Plan, Do, Check, Action) is applied, together with annual audits and "Green Award" to cultivate the environmental protection culture of the firm.

IV. CONCLUSION

Based on the case study, this paper discusses the improvement practices in case companies, idea generation processes and the supporting system to enable creating improvement recommendations. Classification of reduction projects are then proposed to serve as the basis for the improvement process framework. This classification can help managers to design potential carbon reduction projects. Also top-down and bottom-up approaches are two common methods to guide firms. Then this study investigates the organizational structure of implementing carbon reduction projects, from internal operating units to external knowledge support. Based on all these constituents, a process framework of implementing improvement practices is generated. This framework contributes to the links between low-carbon requirement and operations management field, from the practices perspective. The related small number of investigated cases limits the generosity of the study. The future research should conduct a larger scope of survey to further test the framework.

REFERENCE

[1] The Climate Change Challenge - Carbon Trust [WWW Document], n.d. URL

http://www.carbontrust.com/resources/reports/advice/the-climate-change-challenge (accessed 6.5.14).

[2] Weinhofer, G., Hoffmann, V.H., 2010. Mitigating Climate Change - How Do Corporate Strategies Differ? Bus. Strateg. Environ. 19, 77–89.

[3] Busch, T., Pinkse, J., 2012. Reconciling stakeholder requests and carbon dependency: What is the right climate strategy. A stakeholder approach to corporate social responsibility: Pressures, conflicts, reconciliation. Aldershot, UK: Gower.

[4] Okereke, C., Russel, D., 2010. Regulatory pressure and competitive dynamics: Carbon management strategies of UK energy-intensive companies. California Management Review 52, 100–124.

[5] Jeswani, H.K., Wehrmeyer, W., Mulugetta, Y., 2008. How warm is the corporate response to climate change? Evidence from Pakistan and the UK. Business Strategy and the Environment 17, 46–60.

[6] Binh, H.D., Khang, D.B., n.d. Business Responses to Climate Change in Developing Countries: A Conceptual Framework.

[7] Dahlmann, F., Brammer, S., International Association for Business and Society, 2013. Reducing Carbon Emissions Worldwide: MNCs and Global Environmental Performance. Proceedings of the International Association for Business and Society 24, 144–152.

[8] Austin, D., Rosinski, N., Sauer, A., Duc, C.L., 2004. Quantifying the financial risks and opportunities of climate change on the automotive industry. Corporate Environmental Strategy 11, 2–233–2–250.

[9] Petersen, A.K., Solberg, B., 2002. Greenhouse gas emissions, life-cycle inventory and cost-efficiency of using laminated wood instead of steel construction. Case: Beams at Gardermoen airport. Environmental Science and Policy 5, 169–182.

[10] Lema, A., Ruby, K., 2006. Towards a policy model for climate change mitigation: China's experience with wind power development and lessons for developing countries. Energy for Sustainable Development 10, 5–13.

[11] Subak, S., Craighill, A., 1999. The contribution of the paper cycle to global warming. Mitigation and Adaptation Strategies for Global Change 4, 113–135.

[12] Floros, N., Vlachou, A., 2005. Energy demand and energy-related CO2 emissions in Greek manufacturing: Assessing the impact of a carbon tax. Energy Economics 27, 387–413.

[13] Huntzinger, D.N., Eatmon, T.D., 2009. A life-cycle assessment of Portland cement manufacturing: comparing the traditional process with alternative technologies. Journal of Cleaner Production 17, 668–675.

[14] Chaudhary, H., Bhagat, S., Gulrajani, M.L., 2009. Carbon footprints of a garment manufacturing unit. Journal of the Textile Association 70, 175–182.

[15] Song, J.-S., Lee, K.-M., 2010. Development of a low-carbon product design system based on embedded GHG emissions. Resources, Conservation and Recycling 54, 547–556.

[16] Hua, G., Cheng, T.C.E., Wang, S., 2011. Managing carbon footprints in inventory management. Int. J. Prod. Econ. 132, 178–185.

[17] Holweg, M., Reichhart, A., Hong, E., 2011. On risk and cost in global sourcing. International Journal of Production Economics 131, 333–341.

[18] Jira, C. (Fern), Toffel, M.W., 2013. Engaging Supply Chains in Climate Change. M&SOM 15, 559–577.

[19] Ngai, E.W.T., Chau, D.C.K., Poon, J.K.L., To, C.K.M., 2013. Energy and utility management maturity model for sustainable manufacturing process. Int. J. Prod. Econ. 146, 453–464.

[20] Koh, S.C.L., Gunasekaran, A., Tseng, C.S., 2012. Cross-tier ripple and indirect effects of directives WEEE and RoHS on greening a supply chain. Int. J. Prod. Econ. 140, 305–317.

[21] Smith, L., Ball, P., 2012. Steps towards sustainable manufacturing through modelling material, energy and waste flows. Int. J. Prod. Econ. 140, 227-238.

[22] Choi, T.-M., 2013. Carbon footprint tax on fashion supply chain systems. Int. J. Adv. Manuf. Technol. 68, 835-847.

[23] Gibbert, M., Ruigrok, W., Wicki, B., 2008. What passes as a rigorous case study? Strategic Management Journal 29, 1465–1474.

TABLE I
CASE DATA AND ANALYSIS

	Project Portfolio Generation	Internal Supporting System	External Support
Lenovo	1. Top-down way—following the GHG protocol as reference to consider firm's practices that generate Scope 1, 2, 3 emissions. These guideline is passed to operations department by GEA (General Environmental Affairs Department) 2. Bottom-up way—project ideas generated by managers in departments with first-hand experiences 3. Priorities of projects: three levels, energy efficiency, renewable energy, and purchase renewable energy credits/carbon offsets.	1. CEO supports the climate change policy, one Vice president is assigned as Chief Sustainability Officer (CSO) 2. A platform—Global Environmental Affairs—across the whole business group to implement carbon reduction projects, under the management of (CSO) 3. In each domestic offices and plants, coordinators in separate EA department and representatives in product, production, and other departments to support GEA 4. Follow ISO14000 standard to build up internal supporting system	"Cooperation with NGO, Government Bureaus, International Organizations, and Academic Institutions to gain rich resources on technology and knowledge on carbon issue"

Tsingtao Brewery	CEO's low carbon awareness spreads to all levels of staff, stimulating the low-carbon idea generation process naturally and self-consciously. Bottom-up approach. Employee at every level, especially engineers and technicians, are encouraged with monetary reward to submit new proposals on boosting energy & resources efficiency, via running a proposal competition.	1. The CEO strongly supports the carbon issue and considers low carbon into main business strategy. 2. Low carbon issue is managed under manufacturing department, and a special team is set up for targets, plans and monitor. 3. 6 specialist formed the team in the headquarters, and one specialist in each brewery factory. The carbon specialists are also evaluated yearly. And these specialists are promised with faster-track on promotion. 4. General manager was the first person responsible for the factory's environmental-protection management; every levels of environmental officers have to communicate with the general manager on a regular basis. 5. The participation of senior management ensures sufficient resources allocated to carbon management projects—the carbon projects have priority on getting funding, human resources compared to other projects.	Carbon footprinting project is supervised by external consultant
ZTE	1. ZTE reduction projects identification comes from the principles of Cleaner Production, 2. The internal CSR team conducted the cleaner production assessment, which collects the operation data and analyse the un-balance between materials in input/output activities. Specialists can figure out potential cleaner production options from those unbalanced items 3. Eight potential aspects are analysed: Resources, Energy, Technology, Equipment, Process Control, Products, Waste, Management and Personnel	1. In 2007 a special CSR team is built up to implement the CSR architecture of ZTE 2. A Vice President was named specially as representative of ZTE CSR system Specially for carbon issue, ZTE sets up a committee for energy saving and carbon emission reduction. The task for the committee includes two streams: one for the corporate level improvement and the other for the product level. 3. Four specialists were allocated equally into the corporate-level team and product-level team.	1. ZTE has obtained the LCA carbon analysis software "EMMi" to support product carbon footprinting. 2. For reduction projects, due to business confidential issue, ZTE only relies on internal capacity for evaluating and implementing these projects.
Acer	1. Bottom-up approach. Acer does not rely on the sustainable office to generate carbon emission reduction options. 2. Each department is responsible to identify the reduction solutions by their internal capability. And these projects need to comply with the overall carbon reduction target. 3. Due to tough market competition, Acer has a corporate culture with highly sensitive to the trend of global consumer's attitude. This culture influence every staff to consider environmental friendly practices in their daily work, which is part of continuous improvement conduct.	1. Acer sets up an environmental management committee to manage environmental issue as a whole 2. A group of 7 people consist the Acer sustainable office, aiming for coordination between departments and special task groups on those environmental, health and social issues.	1. External sources include research institute such as Industrial Research Institute of Taiwan, industry association such as Electronic Industry Citizenship Coalition (EICC), consultants, third-party auditors such as SGS consultants for the on-site auditing in suppliers' plants, and government organizations such as the special program of Taiwan Industrial Developing Bureau.
TSMC	1. Top-down approach: After the committee determines the carbon emission reduction target, each business department gives out individual implementation plans according to the target. Before the plans are reviewed, the Industrial Safety and Environmental Protection Technical Board provides the guidelines about industry requirement in worldwide scope and government regulation; 2. Each year, the Industrial Safety & Protection Technical Board conducts internal audit to examine the results of these practices. 3. TSMC's internal rules on energy efficiency improvement and polluted gas control practices can be transferred. 4. Bottom-up approach: EHS units in each plants also proposed reduction ideas and a competition with monetary reward is given. Good practices will be copied in all plants. 5. PDCA principle is applied	1. TSMC forms a special committee to manage the environmental-related issues, and there are several teams to implement detailed practices. 2. TSMC's environmental management organization consists of: the central Environmental, Safety & Health Planning unit; 3. The Industrial Safety and Environmental Protection Technical Board; 4. In each manufacturing facility there is designated Industrial Safety and Environmental Protection unit to cooperate with the central ESH department 4. Each business department gives out individual reduction project plans according to TSMC overall corporate target. Then the central ESH Planning unit will decide and approve these plans.	1. External auditing service providers are employed, and they give out improvement advice to TSMC. 2. TSMC works with World Semiconductor Council, so industry-level best practices information are obtained to support improvement-options generation.

China Steel	1. Bottom-up approach. Groups that consists of lined engineers and senior engineers from different departments are formed to generate optimization ideas/options. This type of meeting with line engineers are organized 4-5 times every year in order to collect timely ideas of operation optimization. 2. After the meeting, these ideas/options are accessed by Office of Energy & Environmental Affairs, which is directly headed by CSC Group Chairman & Vice Chairman.	1. CSC set up the CSC Group Committee for Energy and Environmental Promotion in April of 2011, with the Chairman of CSC acting as the chairman of the Committee to implement carbon-related tasks with PDCA. The committee consists of several sub-committees to cover multiple environmental issues, which are under daily manage of Office of Energy and Environmental Affairs 2. Within production department, CSC set up the "Energy-Conservation Committee" headed by Vice President. Three specific teams are responsible for the implementation of energy conservation and carbon reducing affairs within all the plants.	1. China Steel works closely with Taiwan Industrial Technology Research Institute (ITRI) to gain carbon emission measurement knowledge. 2. CSC gained the most important information and knowledge from the involvement in World Steel Association, especially in its Climate Action programme. 3. CSC regards participation in International Co-operative R&D programs is a good way to enhance knowledge exchanges and prepare for new trends. At present, the World Steel Association promotes co-operative projects to reduce the CO2 emissions from iron making processes by 30~70%.
AUO	1. The senior management—Directors and Vice Presidents, who are the top managers for each region, organize seasonal meetings for idea exchanging between regions 2. The production managers for each production line, forms Cross Function Team (CFT) to enable experience sharing. There are different solutions generated in different plants, serving to resolve the same energy-saving problem. 3. AUO conducts comparative experiments on two plants based on CFTs to test potential reduction projects. But this type of innovation mainly focuses on factory infrastructure improvements.	All the carbon management issues are managed under the EHS (Environmental, Health & Security) department in AUO. The company does not specially set up a new committee to manage related practices.	1. AUO cooperates with Industrial Technology Research Institute of Taiwan (ITRI) for carbon management knowledge and information. 2. AUO participated in an environmental assessment of ICT Industry at the MIT. The "Product Attribute to Impact Algorithm" Project helped AUO stay up-to-date on developments in international carbon footprint calculation science, and eventually develop the fast-calculation tools for measuring emission of flat panel products.
BenQ	Top-down approach. The requirement of carbon reduction target is allocated to each product lines. Under each product line, each department identify the detailed implementation options.	1. The corporate sets up CSR team in headquarter to manage related issues. 2. The carbon-related issues are directly managed in Product Technology Centre. This product technology centre consists of procurement centre, supply chain centre, quality management centre, etc. 3. The carbon management target is set up by CSR team, and approved by General Manager.	The LCA software SIMPRO is used for assisting estimation of production carbon footprint under new potential design.
Tungho Steel	1. Bottom-up approach: The green innovation practices proposals come from the production line staff. A system called "Staff Green Improvement Proposal System" is set up. Every year 2-3 proposals in the system are put in practices, with a share of economic benefit to the staff proposed the solution.	1. "GHG emissions inventory Committee" was established, with president as head of committee and director of R&D technology department as vice head. 2. The technology department in factory is the key division to be charge.	1. In the PCF project, external consultants are involved to help Tung Ho to conduct the process, including the knowledge support from the IDB (Industrial Development Bureau). 2. Steel Association in Taiwan has gathered steel manufacturers to share GHG emission reduction experience. 3. There are also numbers of cleaner production knowledge sharing between TungHo and other business associations, such as Taiwan Green Productivity Foundation.
British Sugar	1. Waste oriented approach 2. British Sugar has generated highly innovative ways to fully utilize all the raw materials into sustainable products. 3. British Sugar holds the continuous improvement principle, and invested an average of 1.8 million per year into R&D projects onto production, suppliers (farmers), transportation and packaging, etc. 4. Bottom-up approach: Any individual can put forward energy reduction ideas for consideration by the senior Leadership Team.	1. An Energy Management Team is set up in British Sugar to cover the overall carbon management issue. 2. The Energy Management Team controls the investments into new techniques and daily energy conservation initiatives.	1. British Sugar works with the consultants in North Energy Ltd and participated in pilot project of Carbon Trust. But the core improvement projects are conducted solely by British Sugar internal engineering team

WWF LCMP	1. WWF's external partners were teamed up for 1 year to generate the best practices action list. These partners are capable on industry-related technical knowledge on carbon emission management. The best practice plan follows the principle of ISO14000. SMEs involved in the projects are guided with the plan and these specialist hired by WWF LCMP program. 2. SMEs gain experience from supply chain. Getting the technical "know how" is very important for manufacturers to improve their technical competence.	WWF relies on external technical expert for detailed engineering knowledge. It aims to play the roles of motivator, coordinator, and monitor.	WWF get the fund from Green Dragon Foundation to support the operation of LCMP. It has obtained lots of connection resources from Hong Kong Productivity Council. The Ecofys is the technical partners to WWF, serving as external consultants as well in implementation.

TABLE II
SEMI-STRUCTURE CASE INTERVIEW QUESTIONS

Construct	Questions
Improvement	According to carbon strategy, what projects are implemented to reduce carbon emissions? Explain the timeframe, scope, attendees, procedures, project management and results of these projects. The criteria to select decarbonization projects, taking consideration of cost, feasibility, etc. How does the result of measurement link to the improvement of project selection? What the carbon reduction projects have been conducted in different supply-chain stages (product design, product packaging, procurement, logistics, distribution, etc.), and in what projects has the firm cooperated with supply chain partners – suppliers, service providers, distributors, etc.? Does the firm consider carbon reduction as a factor in supply chain re-design? Does the firm implement low-carbon supply-chain management systems?

TABLE III
LENOVO CARBON EMISSION REDUCTION PROJECTS SUMMARY

Aspects	Reduction Projects Details	Themes
Product Design	• Reuse of material in end-of-life products (Lenovo avoided in FY 2012/13 more than 27,050 MT CO2e thanks to recycling end-of-life electronic products. (This result is calculated according to the US EPA Waste Reduction Model) • Packaging -Lighter and smaller products -minimize the use of packaging material consumption per box -more compact and reusable packaging materials • Energy Efficient/Low Carbon Products -product design to generate less carbon emission of product in the customer user phrase	-Reuse waste -Packaging Material usage -Energy efficiency in user phrase
Procurement	• Supply Chain/Supplier Management -Lenovo engage Top Tier 1 suppliers into Carbon/Water Reporting Tools Coalition (Electronics Industry Citizenship Coalition carbon/water reporting tool, online platform from 2013) -Lenovo plans to add an evaluation of potential supplier climate change performance and strategy, as a differentiator in the procurement process. -Lenovo meets annually with its primary suppliers to share low carbon views and requirements. (Lenovo held a "Lenovo Environmental Affairs and Specifications Communication" conference for over 500 suppliers in Beijing, Shanghai and Shenzhen, China in October 2012, to engage suppliers and share Lenovo's requests on green product design, product carbon footprinting)	-Change of managerial requirement in its procurement
Production	• Manufacturing/Production -Area optimization-integrating and modifying assembly lines, -Reducing PC on-line testing time -Consolidation of operations -Use local manufacturing facilities in the Americas, Europe and Asia	-Optimize production process in new management way
Logistics	• Inward plant logistics -reduce transportation miles incurred and improve reuse of packaging and shipping materials. -In 2012 Pallet Pooling project: recycle and consumption reduction of wooden pallet recycle -bulk shipping alternatives -Use of low carbon shipping methods via trucks, rails, or sea-flights, instead of air-flight (During 2009, Lenovo shifted 7 percent of notebooks from air transport to ocean transport) -Working closely with its shipping partners to implement fuel efficient shipping standards. -In 2012 Lenovo begins collecting and calculating, or estimating product transportation emissions data via using DHL carbon data dashboard, and other 3 key carries data on shipment, these 4 key carriers represent majority of Lenovo's worldwide global logistics spend. -Regional distribution facilities allow for lighter loads, load consolidation and full-truck-load shipments.	-Material usage reduce -Reuse and recycle -new managerial tactics without using new technology -In pursue of saving fuel to reduce emission
Other	• Plant Environment -Installation of low energy lighting and related electrical equipment, -Energy efficiency improvements to HVAC system and chillers, -Eliminating or improving usage of transformers and air compressors • Behaviour Change in Plants -Building management adjustments that turn lights/HVAC on later in morning and off earlier in the afternoon -Signs/training for turning lights, laptops off • Behaviour Change in Office -Reduction in the number of company operated vehicles -Summer Hours program -LEED Commercial Interiors Gold Certification for a new office in Milan, Italy -ENERGY STAR® certification for Morrisville, NC buildings	-Change equipment and lighting with new technology - Energy saving in plant building and office building -Individual employee behaviour change

	• Industry Specific (Data centres) -Improving data centres energy efficiency • Measurement to the Product Carbon Footprint		
Energy	• Renewable Energy---Scope 2 -Hot water solar system was implemented on some buildings in Chinese facilities -solar lamps were installed for parking lot lights in Beijing -Lenovo installed solar panel arrays at the manufacturing site in Shanghai in 2012. It could save around 10-15% of site's annual electricity consumption and reducing GHG emissions by more than 400 MT CO_2e yearly. • Purchase of renewable energy externally -Purchase of carbon credits, from Climate Action with a little over 5,450 carbon offsets from a renewable energy – biomass waste to energy	-Substitute energy source—renewable energy -Enabled by new technology -Applied to production site and buildings	
External Network	• External Engagement for the low carbon economy transition -Lenovo participated in the Catalyzing Corporate Supply Chain Carbon Footprint Reporting in China's Export Industries project and has been engaged with the World Bank in their project named 'Spontaneous Promoting Green Travel' promoting more environmentally friendly employees' commuting practices. --Involved in the CDP Investor project and supply chain project	-Building a general network to work on emission reduction -Indirect emission reduction practice	

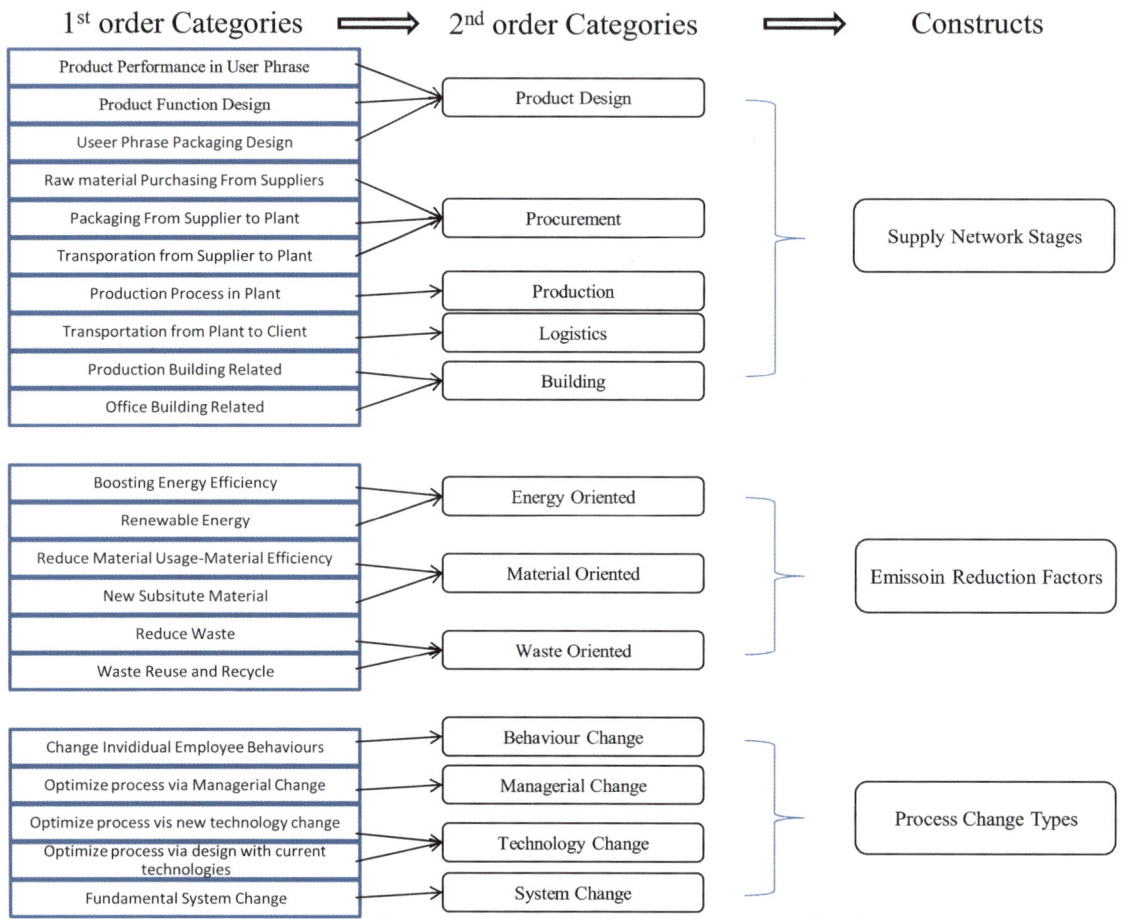

Fig.1. Three Constructs Coded from Data of Improvement Practices

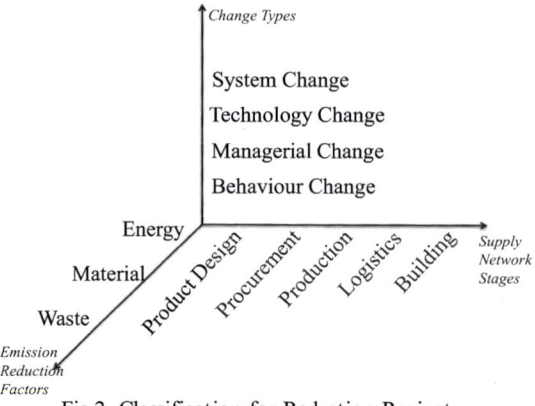

Fig.2. Classification for Reduction Projects

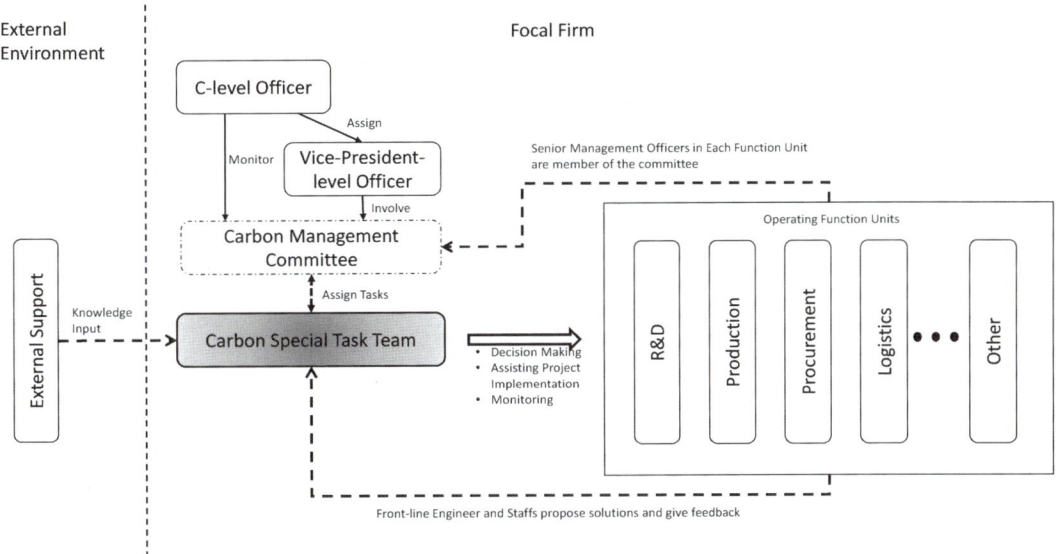

Fig.3. A General Model of Carbon Management Organization Structure

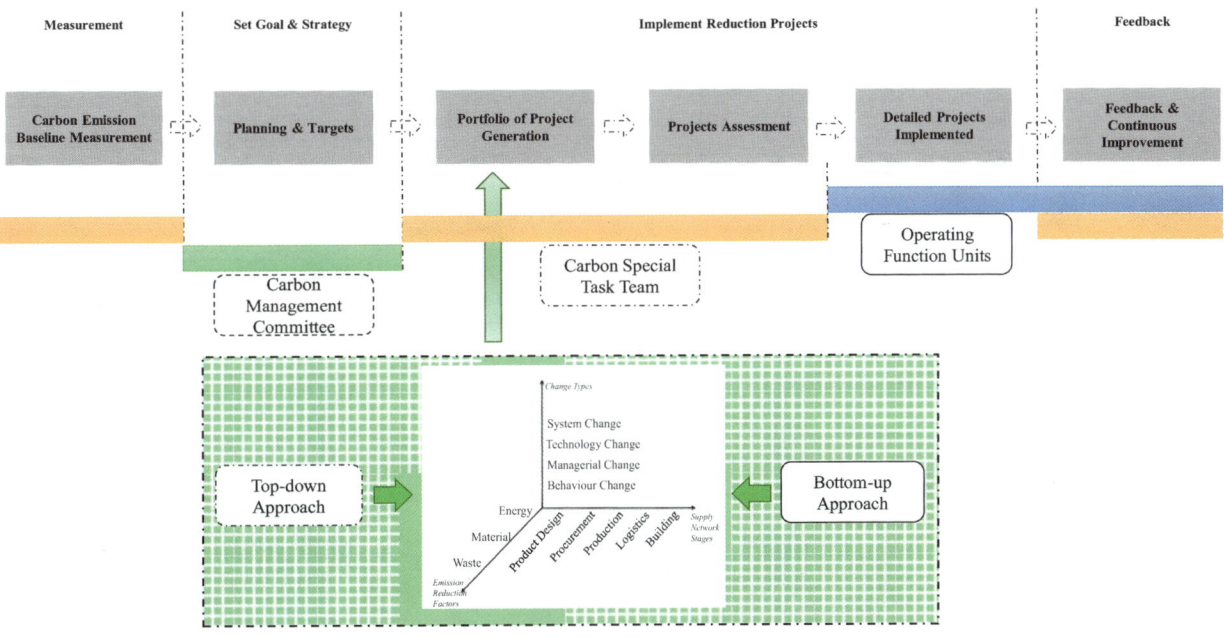

Fig.4. A Process Framework of Implementing Improvement Practices

The Application of Standard Hour in Commercial Aircraft Manufacturing Industry

Shunlong Jiang

Development & Planning Department, Shanghai Aircraft Manufacturing Co., Ltd. (SAMC) China   Industrial Engineering

(jiangshunlong@126.com)

Abstract - **The paper illuminates a signification part in the application of industry engineering that is using standard hour to guide and monitor the aircraft production through the examples of commercial aircraft enterprise. This method achieves notable results through long time practice and application. In order to popularize the methods and theory of standard hour application, the paper analyzes and compares the application status of standard hour in domestic and oversea enterprises, distinguishes the difference between the standard hour and man-hour, and points out the problems in the application of standard hour in domestic IE area. The paper aims at emphasizing the correct application skills and methods of standard hour, pushing the improvement in reducing the production cost and increasing production efficiency.**

Keywords - **Application of standard hour, commercial aircraft fabrication, IE methods**

I. INTRODUCTION

It is an indispensible method to determine the standard hour of fabrication in IE application, which has direct relationship with the production cost and efficiency of an enterprise, especially in the environment that market economy promotes the development. The accuracy of SH calculation and estimation also has direct relationship with the economic destiny of an enterprise, which may cause irreversible damage if something is wrong. Therefore, an IE, who participates in determining SH of aircraft fabrication/assembly in Commercial Aircraft Corporation of China Ltd ("COMAC" for short), shall pay high attention to the job, fully understand each steps in the fabrication/assembly, know all production sequences and operation processes, and finally calculate the working-hour data which will meet the need of production, and form an effective standardization of working-hour through the continuous observation, analysis and comparison.

II. STANDARD HOUR

Standard Hour, "SH" for short, is the fundamental data in the research and application of IE. It provides the reliable basis for the calculation and evaluation of production efficiency and the improvement of organization. It is the adopted production lead time which is common used method in the world. At the same time, the application of SH can also promote the optimization of production labor or lead time, and standardization of operation processes.

A. What is SH

A Standard Hour is the normal time for a normal man to do a normal job under normal conditions. The normal man refers to the representative operator, who shall have the certain knowledge, skill, physical strength and working experience to complete the work per given procedures. The normal condition means that work method and condition shall follow certain standards, such as cutting speed and depth, including equipment, tool, material supply, environment and etc. The normal speed refers to the operating speed under the normal conditions, for example, step length is 550cm, and step time is 0.6s.If all kinds of conditions mentioned above can be satisfied, it will achieve the requirement of standard time.

B. The Definition of SH

A Standard Hour is the normal time for a normal man to do a normal job under normal conditions. The standard is established by applying the basic elemental hours to each element of the job, as planned, considering material, dimensions, tolerance and tooling. Where the operation is peculiar and not covered by the Rate Guide, the estimated is made by empirical methods. To the foregoing standard, the following allowances are applied. When perishable tool are involved, apply a tool allowance to the machining elements (based on normal tool life and specific blueprint for machining).

Example:

1) Elemental Standard + Tool Allowance = Elemental Base Standard   (1)

$$\text{Tool Allowance} = 0.067 \times 3\% = 0.002$$

Elemental Base Standard = 0.002+0.067=0.069

2) To the elemental standard, adjusted for tool allowance, add standard allowance as follows

(1) fatigue   4%
(2) personal needs   2%
(3) rest time   4%
(4) time tickets   1%

Example:

① fatigue(10%)

$$0.069 \times 0.1 = 0.007 \qquad (2)$$

② personal needs(2%)

$$0.069 \times 0.02 = 0.001 \qquad (3)$$

③ rest time(4%)

$$0.069 \times 0.04 = 0.003 \qquad (4)$$

④ time tickets(1%)

$$0.069 \times 0.01 = 0.001 \qquad (5)$$

3) Standard time = element standard hour + fatigue allowance + personal allowance + rest

E. Qi et al. (eds.), *Proceedings of the 21st International Conference on Industrial Engineering and Engineering Management 2014*, Proceedings of the International Conference on Industrial Engineering and Engineering Management, DOI 10.2991/978-94-6239-102-4_78, © Atlantis Press and the authors 2015

allowance + time tickets allowance
= 0.069+0.007+0.001+0.003+0.001=0.081 (6)
Note: The same general procedure applies to assembly.

The summation of the foregoing represents the basis standard and is attainable at any unit, assuming all elements of the job have been included in the estimate [1].

C. The Main Features of SH

The features of SH include:

1) Objective and high accuracy with sufficient basis.

2) Stable, not affected by the change of operation. And only need to amend changed time if operation is changed;

3) Determine SH of a task in advance, but not evaluate the operation speed and individual effort extent in operation measurement;

4) Record operation methods in detail, and get the time for each element action. So, it can reasonable improve the operation.

5) Stopwatch is not necessary. Determine the SH time before the work starting and draw up a operation instruction;

6) The best way is to use PTS method to balance the production line;

The determination of SH should be a convincing process based on scientific analysis and judgment. It mainly consists of set-up time, operating time and allowance time.

1) Set-up time refers to the preparing time consumed prior to operation.
Set-up time = actual measured set-up time× evaluation coefficient

2) Operation time refers to the working time directly consumed in one task.
Operation time = actual measured operation time× evaluation coefficient

3) Allowance time can be divided to fatigue allowance and standard allowance:

(1) Fatigue allowance: 3~20%
in which, light 3~5% medium 6~10% heavy 11~15% very heavy 16~20%

(2) Standard allowance:
in which, rest time 4% physical needs 2% record work card 2% clean 2% rework 1% [2]

D. The Main Application of Standard Hour

1) Evaluation for the operation method

(1) It can be used to evaluate and improve the current operation method, and further inspire the conception for improve operation.

(2) It emphasizes on evaluation the operation method beforehand. It can evaluate any operation method prior to the production though predetermined action time standard.

(3)It can evaluate the recommended tooling, fixtures, tool designs and equipments, all of which are

the biggest reason for the change of operation method. Evaluate operation method is to evaluate these items.

(4) It can be the auxiliary reference for the product design.

(5) It can train the operators to operate per new method, and evaluate the capability of the operators.

2) It establishes the time standard

(1) It can determine the action time by predetermined action, and synthesize all of them will be operation time standard.

(2) Most time standard of actual operation action can be determined by elemental action time. Therefore, sort out all operation actions will quickly form a time standard (IE usually uses fabrication/assembly time standard evaluation chart).

(3) It can avoid the unnecessary loss due to the improper assessment conducted by non-proficient researcher.

(4) All predetermined time data is objective and easily and can be widely used.

E. The Estimate of Standard Hour

There are types of methods to determinate SH. In different manufacturing pattern we can adopt different methods to evaluation various SH. These methods have their own advantages and disadvantages, and have their special application. Therefore, to determination of SH shall consider following factors:

1) For SH which have high precious requirement, it is better to choose the continuous observation method (also called stopwatch method), PTS method, and historical data method to determine SH.

2) As SH need to keep consistency and stability requirement, it is better to use indirect method, because the direct method may affect the consistency of SH by the subject factor of observer.

3) If the task happens in different site, normally, it will choose historic data method and analogy method to determine SH.

4) For the long production cycle work, it is better to choose work sampling method and the continuous observation method (stopwatch method) combining with the work sampling method to determine SH.

5) For the short production cycle work, it is better to use PTS or video analysis method.

6) For the high repetitive production, it is better to use time standard or video analysis method.

7) For the low or seldom repetitive production, it is better to use continuous observation method (stopwatch method), work sampling method and historic data method.

8) For the team work or other complex work, normally, it is better to use video analysis method and work sampling method.

F. The Evaluation of Standard Hour

According to the characteristics of Shanghai Aircraft Manufacturing Co. Ltd, we use the SH pattern to control the labor hour cost in aircraft fabrication and assembly. The establishment of SH is one of essential data which must be adopted in the application of IE. It is the time required for completing a job by a qualified and trained operator per regulated operation criterion in normal working speed under a certain assembly condition.

Also, the main basic data for evaluating standard time is actual working-hour, efficiency coefficient or improvement rate.
The formula is
- SH= normal operation × evaluation coefficient × (1+allowance coefficient)   (7)
- First article working-hour= ST/1000 × target performance (%)   (8)

Proficient coefficient

$$= \text{double index} \times \sqrt{\frac{\text{working hour of one machine}}{\text{planned working - hour}}} \quad (9)$$

- Performance coefficient = SH / predetermined working-hour of each aircraft   (10)

At present, SH are mainly divided into two categories.

1)   SH of Fabrication Outline (FO)

The fabrication of an aircraft starts from the manufacture of a large number of parts and components. Each part shall have a FO, and IE must draw up the single working-hour and set-up time of each process in every FO. Then, the FO will be released to production control department. Under the normal circumstance, according to usual practice of foreign aviation industry, the SH of part fabrication will use the fabrication time of the 100th aircraft.

2)   SH of Assembly Outline (AO)

As well, the end of aircraft parts and components fabrication is means entering into the assembly stage. The primary task in assembly is to draw up an Assembly Outline (AO). After the AO drawn up by the planner, IE will determinate the SH of each AO, according to the working station and chart sequence. The SH of assembly will used the assembly time of the 1000th aircraft. There are two ways to monitor the SH of assembly, one is working station status chart (125A), and the other is working station chart change request (CCR).

In the fabrication process of advanced regional aircraft, the evaluation of compensation hour is very importation, because the aircraft includes not only many domestic manufactured products, but also many oversea purchased items, especial, the system parts are almost purchased from aboard. Therefore, the compensation for products is very important. If the products supplied by the supplier has quality problem or can't be delivered on time, it will claim for indemnity per provisions contract international convention. This work must be included in the procurement contract, and IE must following this kind of delivery status. Once it happens, IE shall evaluate and calculate the compensation hour from the aspects of working-hour postponed, and additional management hour, etc, and submit the data to the

related department for communication with the supplier, in order to achieve the target of reducing the procurement cost.

III. DIFFERENCES BETWEEN STANDARD HOUR AND MAN-HOUR

The domestic enterprises, especial the large scale state-owned hi-tech enterprise (not include cooperative business, joint-venture, and foreign sole proprietorship) have many different with the developed countries and Japanese enterprises in the classification standard of working-hour. Though China has already reformed and opened to the outside over 30 years, and transferred from the planned economy to the market economy, in fact, the residual ideology of planned economy still leave in many domestic enterprises and have influenced the development of market economy. Just likes an old dead tree which has deep root but cannot sprout, these enterprises have advanced equipments, but their technical management is still lag behind. The serious conflict in the enterprises has been the evaluation of between the working-hour and production efficiency, which causes the unbalance situation between the input and output of products. The market economy emphasizes on standardization, regulation and normalization. But domestic enterprises standard has the problem of disunity, and its application are lack of strong basis. From following chart we can see the comparison and different between China, western countries and Japan in working- hour (Fig.1).

The Application of Standard Hour and Standard Time
First of all, there are essential differences between standard hour and standard time in IE application. The premise to distinguish them is to clearly know the application concepts of standard time. But, the standard time method applied in many domestic enterprises is just to add a certain allowance coefficient to the actual operation time, which just forms a unify standardization extent. The application of SH is to adopt predetermined time (call PTS method). This is to resolve the production operation into element actions firstly, next to analyze and compare it with historic experience data collected before, and then to transfer the elemental action time (actual time) to predetermined working-hour, and finally use the performance coefficient to get the SH. The application of SH is suitable to large-scale, long lead time products, such as aerospace industry, shipbuilding industry, automobile industry and large-size equipment fabrication enterprises. The basic problem of SH application is to determine the performance coefficient. If there is only SH, and does not consider or neglect the performance coefficient, the SH will have no meanings in IE application. Therefore, performance coefficient is the ration between effective working-hour and actual working-hour, which is the necessary data in the application of IE in production and can be used to

analyze the issues how to enhance the production capability and keep balance of labor. As well known, standard hour is the working-hour standard which is determinate under "4 normal statues", per "2 operation requirement", and in "2 nothing more than". It is an ideal value which cannot be randomly changed. But in actual production, it is difficult not to change it, so that we use "performance coefficient" to evaluation the labor, production lead time and etc, in order to keep the SH reasonable. And, once the performance coefficient is including in the SH, it will allow IE to continuously obverse and record the operator who repeats the operation for a certain job. And then, the IE can determinate the performance coefficient of one job even a whole ship-set of aircraft, after analyzing and comparing the historic status. At the same time, the performance coefficient can be used to analyze the reason which may affect the increase of the production efficiency and help to find the restrict reason which may be technical issue or management problem. The

performance coefficient can provide persuaded and scientific basis for the effective improving actions, which is the unique point in the application of standard hour. In the manufacturing area of enterprises, the IE can combine the SH and learning curve and use the SH to control the whole production Takt-time. For aerospace industry, since 1990s, the IE has been implemented in the domestic production, and the application of SH has played an important role in the production cost reducing, efficient increasing and labor balancing in the batch fabrication of parts and assembly. And the SH can be used to predict the milestone process in the target working-hour curve from the first batch of product to a certain batch of product which will achieve the line of balance, and also to predict the starting and finishing date of each working-station, forming waterfall plan chart. Moreover, the standard hour can be used to make up many kinds of chart, which is key point that every enterprise will attach importance to. See Fig.2 and 3.

China	Man-hour						non-Man-hour		
	set-up finish time	operating time		set-up & rest time			non-production time	shutdown loss for enterprises	shutdown loss for operators
		basic time	aux. time	set-up time		rest and physical			
				organization	technique				
Western	Standard Hour								non-Standard Time
Japen	adjust time	operating time		allowance time					avoiable postpone
		machining	handle working	fatigue allowance		unavoible postpone	political allowance		
				stable	flexible				

Fig.1. Comparison between China, westerns countries and Japan in working-hour

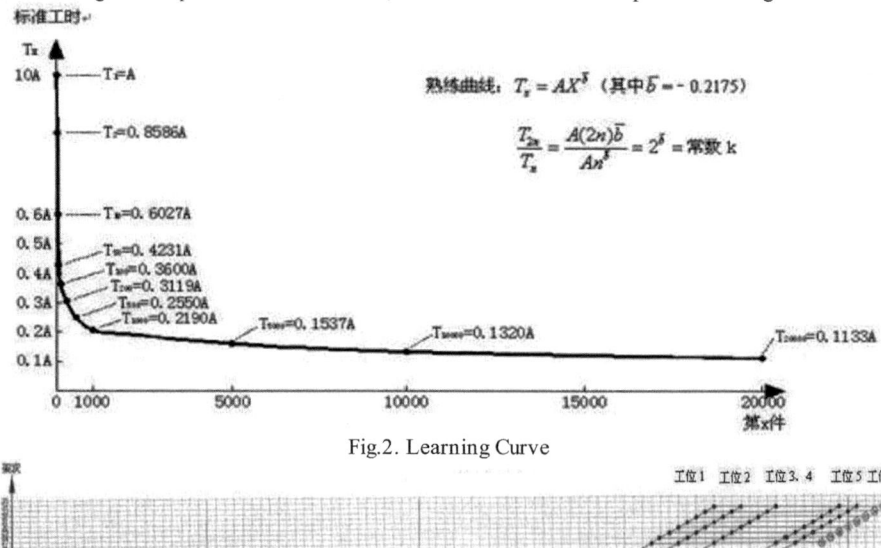

Fig.2. Learning Curve

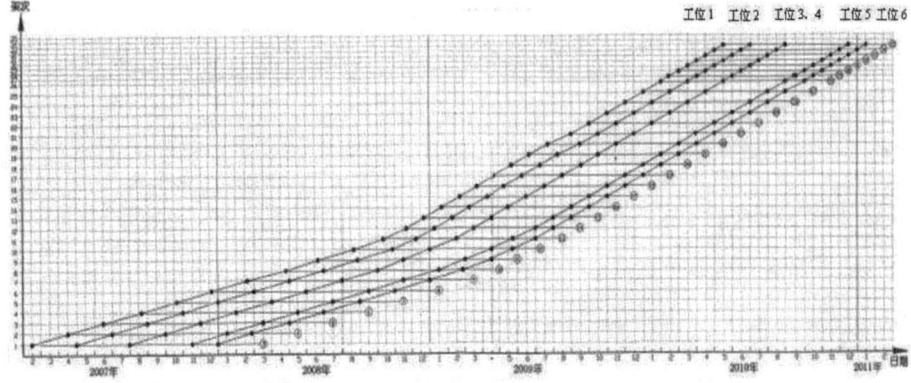

Fig.3. Product delivery waterfall plan

IV. THE BENEFIT OF SH APPLICATION

Since 1985, not only in 35 ship-set MD-82/90 cooperation project, but also at present research ARJ700 advanced regional jet, SAMC has learned and introduced advanced management concepts from foreign enterprises, and proved through many practices that the enhance of market economy must rely on the application of IE. In the aircraft fabrication enterprises, it is necessary to relay on the SH when monitoring and guiding for production efficiency and cost.

1) To implement SH management method will let working-hour quota management become the effective method for the scientific factory management. The existing domestic time standard is disunity (such as Chinese standard, industry standard, military standard, and enterprise standard). But, foreign large-scale hi-tech enterprises fully implement SH evaluated by IE in the work shop, and combine the application of SH, efficiency and improvement rate. The way using SH can become the main scientific reference for checking production status, drawing up operation schedule and labor schedule, avoiding the external interference at same time.

2) If SH will be applied in the whole process from parts fabrication to assembly, it is not necessary to amend the work norm every year. Because once SH is determined, it will not be changed, otherwise, there are significant modification in planning. If SH is used, the conflict between the enterprise and workers caused for the change of work norm can be avoided

3) IE can use learning curve method in schedule to supervise the whole production cycle. The method to estimate the performance coefficient per learning curve coefficient, and to evaluate the working-hour cost of products through SH will be an unify and scientific calculation formulation for an enterprise in new product development, contract negotiation, cost control and labor hour budget. Therefore, in the market economy, the standard hour is more suitable for the enterprise production.

V. PROSPECTS

As well known, in the aircraft manufacturing, the evaluation of all working-hours is basis on SH not only for a large number of parts fabrication, but also the complex and trivial components and final assembly, even the painting and fly test, except the fabrication of tooling. It is the job of the IE planner to determine the SH of each step in FO/AO. Through 20 years hard work, SAMC has went through the brilliant achievement in the cooperation with USA Douglas Aircraft Manufacturing Co in MD-82 and MD-90 projects. At present in ARJ-700 advanced regional jet and upcoming C919 truck liner program, IE is the supporting point for the whole aircraft manufacturing process, the consulter of working-hour evaluation and production plan supervision, and will be the guider of

smooth production. Facts have proved that the achievements of the application of SH by IE cannot be ignored by anybody, because it stood the test of long term of time in actual application. For example, during the researching phase (2006-2008) of ARJ21, its SH underwent many times serious audits by the working-hour auditing specialists group organized by the previous Aviation Industry Corporation of China, and the headquarter of Commercial Aircraft Corporation Ltd, and the final audit conclusion was the SH of each FO/AO in each working station of the whole aircraft was scientific and accurate. Looking to the future, the popularization and application of SH in domestic enterprises under current market society must bring a way for reducing production cost and increasing efficiency.

REFERENCES

[1] USA Douglas Aircraft Manufacturing Co, Industrial engineering and Standard Hour, 1969, A.1.0

[2] Shanghai Aircraft Manufacturing Co Ltd MD-82 Inspiration of Engineering Management [J] Beijing: Economy Management Publisher, 1990, 3(9): 227-228.

Comfort Analysis of Automobile Seats Based on 3D Human Models in SolidWorks

Ping ZHANG*, Ya WEN, Quan YUAN, Gen-wei ZHANG

College of Architecture and Art, Hefei University of Technology, Hefei, China
(Zhangp163@163.com)

Abstract - **This article describes a virtual human model designed by software SolidWorks to test the comfort and ergonomic characteristics of automobile seats. More importantly, this study may evaluate different designs of automobile seats with ergonomic parameters. In the process, the author established 3D digital human models based on ergonomic shape characteristics. Secondly, the author used the human models to obtain the ergonomic shape comfort parameters of automobile seats. Finally, matching the seat with the human model to get the recommended comfort values. So it is available and useful to establish contact surfaces and contact points , to analyze seat surface morphology as well as to give the virtual test and evaluation if the automobile seat is comfortable or not.**

Keywords - **Automobile seat, comfort analysis, ergonomic shape, human models**

I. INTRODUCTION

With the development of software technology, digital human models become common in product design. These models help designers assess the effectiveness, comfort and safety of various products through human-computer interaction (HCI). By using SolidWorks software, designers can use three-dimensional (3D) digital human models to adjust the structure and geometry of automobile seats. To obtain accurate seat size data, 3D body models must accurately mimic the human sitting posture on the automobile seat [1]. Therefore, the links, activities and restrictions of each module must accord strictly with parameters for ergonomics design. By matching models in SolidWorks, design data can be drawn intuitively, and designers can adjust form and contours. This saves time in terms of experimental procedures and simplifies the design process while retaining accuracy. Researchers can then visually evaluate seat design for comfort [2][3].

II. ERGONOMICS HUMAN MODEL RESEARCH BASED ON AUTOMOBILE SEAT

A. 3D digital human models based on ergonomic shape characteristics

In this study, we matched virtual human models to actual human forms for accuracy in simulation. Therefore, we set up multiple percentiles of the human model [4]. In addition, control dimensions of each part should be limited in order to facilitate changes. A 3D digital human model is mainly built with a loft entity, resulting in pyramidal limbs, spherical articular, and back torso modules following basic human body curve and scapula salient points [5][6].

Since SolidWorks is mainly used for mechanical design, we simplified the model to resemble a machine tool based on the skeletal form of the human body skeleton. We divided the model into 15 parts: head, neck, trunk, left upper arm, left former arm, left hand, right upper arm, right former arm, right hand, left thigh, left shank, left foot, right thigh, right shank and right foot.

In order to ergonomically design the automobile seat back, we added a torso module with the appropriate back curve, including thoracic curvature, lumbar curvature, cervical curvature and sacral vertebrae curvature. Therefore, the assembly process followed natural parameters for human physiological bending.

B. Parametric design of 3D human model dimensions

The first step in the design process was to create a model according to one percentile, and then modify the dimensions continually with the configuration function. The torso module was important to consider, since it plays a key role in the model, with other parts connected to it. Fig.1 shows the 3D digital human model [7][8].

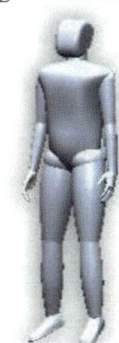

Fig.1. 3D digital human model

III. ERGONOMIC SHAPE OF AUTOMOBILE SEAT

Ergonomic design of automobile seats is dependent on the sitting position of the human body. When the seat shape curve fits the four physiological states of natural bending, the user feels the most comfortable. A comfortable automobile seat conform to the human body can prevent driver fatigue as well as spinal damage and disease from incorrect posture, along with improving comfort. Automobile seat ergonomics mainly refers to form and size related to the human body parameters, such as the seat surface profile, the basic point of angle, height and other data. The surface contour determines the ergonomic surface of the human body, and the size of the surface is an important index that affects comfort. In body pressure distribution and electrical experiments, changing surface morphology, yields important data [9][10].

E. Qi et al. (eds.), *Proceedings of the 21st International Conference on Industrial Engineering and Engineering Management 2014*, Proceedings of the International Conference on Industrial Engineering and Engineering Management, DOI 10.2991/978-94-6239-102-4_79, © Atlantis Press and the authors 2015

A. Module characteristics of automobile seat ergonomic shape

Automobile seat ergonomics related to shape mainly focus on the appearance of the car seat, such as the size and height of the headrest, the size of the backrest, the position of waist, the design of cushions and the whole angle adjustment [11].Therefore, according to ergonomic shape, seats can be divided into the headrest, backrest and cushion sections. The backrest includes the shoulder rest and the waist area. Automobile seat design involves the shape of ergonomics form, along with adjusting parameters based on the above five modules [12]. For example: for a percentile, first the fixed H-point is selected with the angle of the backrest for the torso and thigh at 95° to 115°; The width of the backrest should wrap in the shoulder on both sides and support shoulders. The location of the waist should be at the second and third lumbar vertebra. The depth of the cushion should satisfy the premise of not pressing the popliteal area. Therefore, automobile seat design aligns seat shape to human body size by optimizing the envelope and curved surface, improving the waist height adjustment, adjusting the size of shoulder rest, changing the cushion surface, etc. [13][14][15].

B. Ergonomic shape comfort parameters of automobile seats

Automobile seats have direct contact with the human body. Therefore, automobile seat comfort is considered as one of the most important parts of automobile design of the seat contributes not only to the aesthetics and ride of the vehicle, but is also related to comfort and safety. The most important role of automobile seat is to support drivers' and passengers' bodies ease and comfort in vehicle operation.

TABLE I
COMFORT PARAMETERS OF AUTOMOBILE SEAT
ERGONOMICS FORM

Comfort index of automobile seat Ergonomics form	Headrest	Contour morphological parameters
		Height
		Angle
	Backrest	Contour morphological parameters
		Angle
		Shoulder support
	Lumbar Support	Height
		Contour morphological parameters
	Cushion	Depth
		Contour morphological parameters

According to the above-mentioned ergonomics shape module, the parameters related to automobile seat comfort can be divided into the following aspects: contour morphological parameters, angle, height and depth. The headrest, backrest and cushion involve these parameters, are closely related to seat comfort and ergonomic shape. According to these parameters, the comfort parameters of automobile seats' ergonomic shape can be obtained, as shown in the Table I. Size and contour data on automobile seats that meet the targets can be obtained through experiments. This yields overall design size data of automobile seats and the extent to which they meet drivers' sitting comfort requirements [16].

IV. COMFORT ANALYZIS OF AUTOMOBILE SEATS BASED ON 3D DIGITAL HUMAN MODELS

A. Research on comfort with 3D digital human models matched with automobile seat: Using contour and seat backrest angle as an example

Morphology and geometry of car seats determine the user's sitting posture according to drivers' sitting comfort. Researchers analyze and evaluate seat comfort according to these factors. Fig.2 shows the recommended comfort values for humans in a seated position.

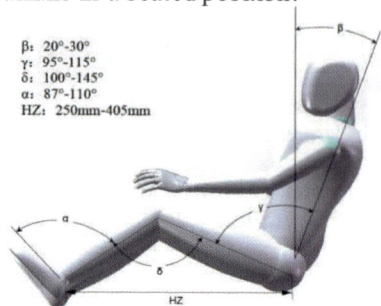

β: 20°-30°
γ: 95°-115°
δ: 100°-145°
α: 87°-110°
HZ: 250mm-405mm

Fig.2. Recommended comfort values of seated human body

SolidWorks software can be used to establish a 3D model of the human body and car seat. With the SolidWorks assembly environment, the designer can match the human model to the car seat to obtain the sitting posture and the angles of sitting comfort for analysis. On the basis of existing research on domestic vehicles, backrest lumbar support must coincide with human lumbar curvature and provide support. As shown in Fig.3, by matching the human model to the car seat, we can determine whether the lumbar support fits the human body. As shown in Fig.4, we can adjust the model view to evaluate consistency and shoulder support.

Fig.3. Lumbar support view

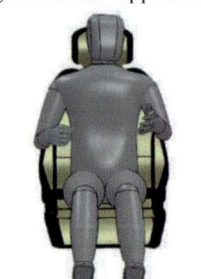

Fig.4. Shoulder support view

B. Exporting the size of the automobile seat model and the human model

Since each part of the 3D human model has its own data axis, when a human model is sitting, certain angles, heights and widths lie in the datum axis between the limbs. Therefore, in the SolidWorks assembly environment, we need to apply only the "Dimension" function to display the size comment. As shown in Fig.5, sequentially adding β, γ, δ, α and HZ parameters in the SolidWorks yields the joint angles of sitting comfort. Accordingly, SolidWorks can be used effectively in determining the export manufacturing dimensions of automobile seats [2].

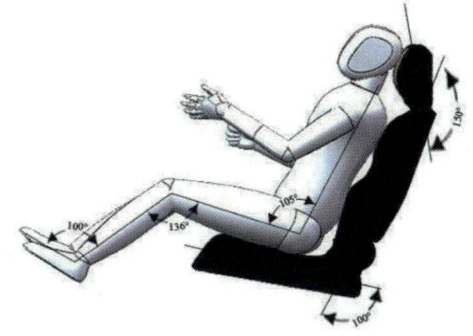

Fig.5.Size of automobile seat models and human models

V. CONCLUSIONS

This study explored the virtual dimension matching for virtual human models and 3D models of automobile seats is dong based on SolidWorks software. We verified several parameters of the automobile seat comfort index, and obtained a set of effective size recommendations for the ergonomic shape of automobile seats. This work provides an effective 3D model method for the research of automobile seat comfort. It also provides a reliable basis for comfort evaluation. 3D models for body posture simulation matching methods offer effective ways to design automobile seat and the perform comfort evaluations. Other parameters such as percentile body dimensions can be modeled for comfort analysis of automobile seats according to different scales of the human body. This is a time and cost-effective way to improve design, comfort and safety of automobile seats, meeting the comfort requirements according to various stakeholders.

ACKNOWLEDGMENT

Ping Zhang thanks Hefei University of Technology who provided the project of automobile seat ergonomic shape to us.

Ping Zhang also thanks Pei-Zhou Sun, Ding-Dan wen, Xing-Yu Chen, Quan Yuan, Wei-Guo Yang, Yu-Mei Zhao, Li-Chao Wang, Tian-Nan Chi and Yun-Xu Gu, for their helpful advice and useful suggestions on this thesis and the experiment.

REFERENCES

[1] HUANG Lu, YANG Yue, PENG Bo, "Three-dimensional Parametric Modeling of Human Body for Analysis of Seat Comfort". Journal of Engineering Graphics, vol.1, Pp.10-15, 2011. (In Chinese)

[2] TANG Xiao-hong, HUANG Lu, YANG Yue, "Three-dimensional human body modeling based on SolidWorks for analysis of seat comfort". Journal of Central South University of Forestry & Technology, vol.3, pp.133-137, 2010. (In Chinese)

[3] YANG Hai-Bo, JIANG Hong-Mei, WANG Xin. "Research of health office chair design based on Ergonomics". Design, vol.2, pp.24-25, 2012. (In Chinese)

[4] Ma Weiyin, Kruth J P. "Parameterization". Computer-Aided Design, 1995, (09):663-675.

[5] Y.L. Ding: Ergonomics (Beijing Institute of Technology Press, 2000: 20-25) (In Chinese)

[6] Alvin R.Tilley: The measure of man & woman: human factors in design (Tianjin University Press, 2008: 11-17)

[7] Y. Cao: Proficient posts of SolidWorks 2007 (Chemical Industry Press, 2008: 338-339) (In Chinese)

[8] L.J. Wang and X.G. Yuan: Computer Applications and Researches, 2005:194-195 (In Chinese)

[9] Y. Li. Computer Applications and Software, vol.25, 2008: 78-79 (In Chinese)

[10] M Ji and CH.Q. Zeng: Mechanical Engineer, 2009: 43-45 (In Chinese)

[11] S.F. Gao and CH.L. Zhang: Machinery Design & Manufacture, 2006:134-135 (In Chinese)

[12] J.D. Ren and Z.J. Fan: Automotive Engineering, vol.28, 2006:647-651 (In Chinese)

[13] J.CH. Wang: Ergonomics in products design (Chemical Industry Press, 2004:104-118) (In Chinese)

[14] CH.R. Liu: Applications of ergonomics (Shanghai: People's Art Press, 2004:115-125) (In Chinese)

[15] J.F. Huang, W. Tao, G. Zhao and P. Li: Standard for Applications of Virtual Reality in Man-machine Engineering (Chinese Journal of Ship Research, vol.3, no.6, Dec. 2008.) (In Chinese)

[16] S. Julier, J. Uhlmann and H. f. Durrant-Whyte: A new method for the nonlinear transformation of means and covariance in filters and estimators (IEEE Trans a C, 200, 45(3): 477-482)

A New Optimization Procedure to Design Campus Bus System

Guo-yi FU

College of Management and Economics, Tianjin University, Tianjin, China

(1280730306@qq.com)

Abstract - **A new optimization procedure is established to design the campus bus system. The bus stops, route of bus and bus frequency are determined at the same time. The problem is formulated as a multiple objective programming problem by considering the benefits of both the bus company and passenger. The variation of origin-destination (OD) is taken into account. In order to solve this novel optimization problem, a hybrid two-stage heuristic algorithm is designed which is based on particle swarm optimization and ant colony optimization. Numerical experiments are conducted. The results show the effectiveness of the algorithm and optimization procedure. It is reasonably believed that the model has potential applicability in real bus system design.**

Keywords - **Ant colony optimization, bus frequency optimization, bus stops selection, campus bus service design, particle swarm optimization**

I. INTRODUCTION

Campus bus is the bus that only moves in the campus to provide convenience for students and teachers. It has become one of the most important means of transportation in a school these years. The campus bus system is concerned with different interests including the bus company and passengers. There is limited research on campus bus. The optimization of bus system consists of several small sub-problems, which attract a great deal of attention in the past several decades.

According to the decomposition of Park et al. [1], the optimization of bus system includes the following main sub-problems: bus stop selection, bus route generation, route scheduling (bus frequency). Regarding the bus stop section, bus stops are determined by using a combination of a set covering algorithm and a traveling salesman problem algorithm [2]. An integer programming model was proposed to determine the set of bus stops to minimize the total distance travelled by all buses [3]. A bi-level optimization model for locating bus stops is designed to minimize the social cost of the overall transport system [4].

For the purpose of bus routing design, a multiple school buses routing problem is considered and heuristic approaches are proposed [5]. A simulation optimization method is given for Campus Bus Routing, which allows the vehicle divert from its current destination [6]. An optimization system that synthesizes aspects of previous approaches into a scalable, flexible, intelligent agent architecture is presented [7]. For the review of school bus routing problem, see the work of Park et al. [1].

To schedule a bus route, genetic algorithm is designed for bus frequency optimization [8, 9]. On account of route optimization of bus dispatching, genetic ant algorithm (GAA) is used for solution of bus scheduling [10]. In order to improve the service level of urban public transportation and the efficiency of the vehicles resources, a new multi-objective synthesis optimization scheduling model is set up [11]. Time-space network based approach is applied to solve regional bus scheduling planning problem [12, 13]. Multi vehicle types are taken into account in bus schedule in [14]. Works about school bus schedule problem can be referred to [15-17].

Most studies mentioned above focus on urban bus and school bus. There is limited research about optimization of campus bus system. The main differences between campus bus and others are as follows:

(1) Student trips are quite different among different dates. For example, OD trips from dormitory and teaching building reduces a lot in holidays. However, researches above are conducted in a deterministic environment with a set OD trips. The variation of OD trips is not considered. In campus, people with similar travel habits are naturally gathered, and the OD trips are simpler than in urban transportation system and can be obtained easily. It offers a convenience to change the bus schedule in various situations.

(2) OD trips can changes suddenly because of some events such as club activities. The sudden change is not considered in past studies. Since the number of bus lines in campus is small, it is possible to adjust the schedule.

This paper aims to fill these gaps. In addition, most studies regard the bus stop section, bus route and bus schedule as three independent problems. For instance, when solving the bus routing problem, the bus stops are usually given. Nonetheless, to attain the objective of minimum cost or maximum benefits, the solutions of three problems are associated with each other, and should not be considered separately. Our study attempts to propose a new procedure to optimize the campus bus system. Bus stops, bus routes and bus frequency are determined at the same time.

The rest of this paper is organized as follows. In section 2, a description of the problem is given and the problem is formulated as a multiple objective programming problem by considering the benefits of the bus company and passenger. The variation of origin-destination (OD) is taken into account. In section 3, a hybrid two-stage heuristic algorithm is designed which is based on particle swarm optimization and ant colony optimization. In section 4, case study is conducted in a simulated network to demonstrate the effectiveness of proposed model and algorithm. Section 5 concludes and summarizes the main outcomes in this paper.

E. Qi et al. (eds.), *Proceedings of the 21st International Conference on Industrial Engineering and Engineering Management 2014*, Proceedings of the International Conference on Industrial Engineering and Engineering Management, DOI 10.2991/978-94-6239-102-4_80, © Atlantis Press and the authors 2015

II. THE OPTIMIZATION OF CAMPUS BUS SYSTEM

A. Problem description

As mentioned in section 1, the optimization of bus system consists of bus stop selection, bus routing design and bus frequency design. In order to solve these problems at the same time, there are three kinds of decision variables. The first one is the location of bus stop, the second one is the route of bus, the third one is the bus frequency. Since the OD trips in campus are relatively simple, only one bus line is needed in most cases and the direction of bus is single. Therefore, the route of bus is determined when the bus stops are given. The problem is simplified to determine the bus stops and bus frequency. The bus system should make both bus company and passenger satisfactory. The objective is set to minimize the total cost.

We assume the time-sliced OD trips. For each time interval of a day and each link, OD demand is assumed to be stable from a long term perspective. To reflect the variation of OD trips on different dates, various OD situations are all considered in the objective function. In our study, arrival rate is used as the reflection of OD trips. The costs of bus stops on different nodes are same.

B. Mathematical formulation

The model is considered on a fully connected and directed graph denoted by $G(N; A)$ where N is the set of nodes corresponding to road intersections and A is the set of links corresponding to connections between pairs of intersections. The nodes in the network are the candidate bus stops. Whole time horizon was partitioned into T time intervals. $\rho_{k,j}$ is the arrival rate of on stop j at time interval k. The arrival rate is priori information. cb is the setup cost of bus stops. df is the departure fee of a bus. Q is the number of different OD situations. $\Delta t_{k,q}$ is the departure interval in time interval k on the q-th situation.

For the perspective of bus company, the objective is to minimize the cost which should consider the departure fees of bus and the setup cost of bus stops. The arrival rate is assumed to be subjected to uniform distribution. The objective function is shown in

$$\min \sum_{j \in J} cb + df \sum_{q=1}^{Q} \sum_{k=1}^{T} \frac{w_q T_{k,q}}{\Delta t_{k,q}} \quad (1)$$

where J is the set of bus stops, w_q is the weight of the OD situation q, $T_{k,q}$ is the length in time interval k on the q-th situation.

Regarding the benefits of passengers, the cost of waiting and the penalty of passengers missing a bus should be considered. As the arrival rate is subjected to uniform distribution, the arrival passengers in a departure interval is $\rho_{k,j}\Delta t_{k,q}$ and the average waiting time is

$\frac{\Delta t_{k,q}}{2}$. Therefore, the cost of waiting is $\frac{\rho_{k,j}\Delta t_{k,q}^2}{2}$. The objective function is shown in

$$\min \alpha_1 \sum_{q=1}^{Q} \sum_{k=1}^{T} \sum_{j \in J} \frac{w_q \rho_{k,j} * \Delta t_{k,q}^2}{2} + \alpha_2 pc \sum_{q=1}^{Q} \sum_{k=1}^{T} \sum_{l \in (N-J)} w_q \rho_{k,j} * T_{k,q} \quad (1)$$

where α_1 and α_2 are the weight parameter that control the effect of the two aspects, pc is the penalty cost parameter.

In terms of above description, the total cost is formulated based on aforementioned 2 aspects as follows:

$$\min \gamma(\sum_{j \in J} cb + df \sum_{q=1}^{Q} \sum_{k=1}^{T} \frac{w_q T_{k,q}}{\Delta t_{k,q}})$$

$$+ \mu(\alpha_1 \sum_{q=1}^{Q} \sum_{k=1}^{T} \sum_{j \in J} \frac{w_q \rho_{k,j} * \Delta t_{k,q}^2}{2} + \alpha_2 pc \sum_{q=1}^{Q} \sum_{k=1}^{T} \sum_{l \in (N-J)} w_q \rho_{k,j} * T_{k,q}) \quad (2)$$

where γ and μ are the weight parameter that control the effect of bus company and passenger.

The constraint is shown in

$$\Delta t_{min} \le \Delta t_{k,q} \le t_{max} \quad \forall k \le T, \forall q \le Q \quad (3)$$

III. HYBRID TWO-STAGE HEURISTIC ALGORITHM

The bus system optimization problem is NP hard problem. A hybrid two-stage heuristic algorithm is proposed to solve it. Proposed model asks for computation of both bus stop location and bus frequency. Ant colony optimization (ACO) is a problem solving technique inspired by the behavior of ants in finding paths from the nest to food first proposed by Dorigo et al. [18] in 1996. ACO performs very well at finding optimal or near-optimal locations. However ant colony algorithm is not suitable for solving continuous problem which refers to bus frequency decision making in our model. Particle swarm optimization originally introduced by Kennedy et al. [19], on the contrary, is a population-based stochastic approach that is suitable for solving continuous optimization problems. A hybrid algorithm that combined ant colony algorithm and particle swarm optimization is designed to solve our problem. The set of bus stops is determined by ant colony algorithm. Particle swarm optimization is applied to figure out optimal bus frequency on given bus stop location. Fitness function value is returned to ant colony algorithm for updating pheromone and next round iteration.

A. Generation of bus stop location

An individual ant simulates a solution of bus stop location, which is denoted by a vector. Each element in the vector is a node ID that means there is a bus stop on the node. The vector is constructed by incrementally selecting candidate node from the list of nodes that are allowed to choose until the candidate node only includes the destination depot. The first bus stop is set on the origin depot. In the constructing procedure of an ant, a node is deleted from the list once selected. The state transition rule in (5) is used to give the probability with which the ants decide to choose.

$$S = \begin{cases} \arg\max_{m \in allowedlist} \tau_{md}^{\alpha} \eta_{m}^{\beta} & h \leq h_0 \\ s & h > h_0 \end{cases} \quad (4)$$

where S is the next chosen node which is determined by right hand side of (4). *allowedlist* is the candidate node set that ants can choose for next bus stop location. d is the number of iteration. Two factors τ, η are the pheromone and visibility respectively. α and β are the relative influence of the pheromone and visibility. h is a random variable between 0 and 1. h_0 is a predetermined parameter($0 \leq q_0 \leq 1$). s is determined by a probability P which is further decided by following:

$$P = \begin{cases} \dfrac{\tau_{sd}^{\alpha} * \eta_{s}^{\beta}}{\sum_{m \in allowedlist} \tau_{md}^{\alpha} * \eta_{m}^{\beta}} & s \in allowedlist \\ 0 & otherwise \end{cases} \quad (5)$$

After all the solutions are generated, the solutions are transfer to PSO to calculate the optimal bus frequency of given bus stop location.

B. Generation of optimal bus frequency

The optimal bus frequency is determined by PSO. The solution is the set of $\Delta t_{k,q} (k \leq T, q \leq Q)$. Particle moves toward to optimum in terms of velocity and position. Bus frequency should not exceed a predefined maximum and minimum value as mentioned in (3). It is guaranteed by (8). At each iteration, particles' velocity and position are updated in terms of following equations:

$$v_{i,d} = Zv_{i,d-1} + C_1 rand() * (pbest_{i,d-1} - busstop_{i,d-1})$$
$$+ C_2 rand() * (lbest_{i,d-1} - busstop_{i,d-1}) \quad (6)$$

$$busstop_{i,d} = \begin{cases} busstop_{i,d-1} + v_{i,d}, \Delta t_{min} \leq busstop_{i,d-1} + v_{i,d} \leq \Delta t_{max} \\ \Delta t_{min}, busstop_{i,d-1} + v_{i,d} \leq \Delta t_{min} \\ \Delta t_{max}, busstop_{i,d-1} + v_{i,d} \geq \Delta t_{max} \end{cases} \quad (7)$$

where d represents the d-th generation for PSO algorithm. $pbest_{i,d-1}$ is the personal optimal solution found by i-th particle among its own historical solutions and $lbest_{i,d-1}$ is the local optimal solution. $v_{i,d}$ is the velocity of i-th particle of d-th generation. C_1 and C_2 are positive constants.

After the number of iterations is reached, the objective function value is calculated in terms of (2) used the bus stop location generated by ACO and the optimal bus frequency generated by PSO. The objective function

value is returned to ACO to update the pheromone and start the next iteration of ACO.

C. Update of pheromone

The procedure of update pheromone is as follows. First, pheromone updating is conducted by reducing the amount of pheromone on all nodes in order to simulate the natural evaporation of the pheromone and to avoid premature convergence. Pheromone is evaporated according to following rule:

$$\tau_m^{d+1} = \rho_1 * \tau_m^d \quad (8)$$

where ρ_1 is a parameter controlling pheromone evaporation.

Then, the best solution is employed to update pheromone as (10) on nodes in the solution vector.

$$\tau_m^{d+1} = \rho_2 * \tau_m^{d+1} + \delta * f_{best} \quad (9)$$

where ρ_2 and δ are predetermined parameters, f_{best} is the best solution value among the d-th generation solutions. After the update of pheromone is done, an iteration of ACO is completed. As shown in Fig.1, ant colony algorithm builds bus stop solutions in every iteration. PSO aims to generate the optimal bus frequency solution of the given bus stop solutions.

```
Set parameters for PSO and ACO respectively
while ACO termination condition not met do
    Construct bus stop locations
    Pass the constructed set of bus stop locations to PSO
    Initialize bus frequency solution particles for PSO
    while PSO termination condition not met do
        Evaluate all particles
        Update pbest and lbest
        Update velocity and position for each particle
    end while
    Return optimal bus frequency solution and fitness func-
    tion value to ACO
    Update pheromones
end while
```

Fig.1. The procedure of Hybrid two-stage heuristic algorithm based on PSO and ACO

IV. CASE STUDY

The bus system optimization problem is tested on the following simulated campus transportation network shown in Fig.2. There are 16 nodes which represent the candidate bus stops on this network. Numbers on the nodes are the nodes IDs. Node 1 is the origin depot and node 16 is the destination depot. The proposed hybrid two-stage heuristic algorithm is employed to solve this problem. Time horizon of each OD situation is partitioned into 12 time intervals. Duration of each time interval is 1 hour. Time horizon is from 7 am to 19 pm. Two kinds of OD situations are applied to reflect the variation of OD trips: one is weekday denoted by $q = 1$. The other is weekend denoted by $q = 2$. The weight of them is $w_1 = 5$ and $w_2 = 2$ respectively. Δt_{min} is 3 minutes, and

Δt_{max} is 60 minutes. The arrival rate is assumed to be known. In our implementation, each component (bus company and passenger) in the objective function is standardized. Therefore, the maximum value for each component is 1 and the total maximum value of objective function is 2.

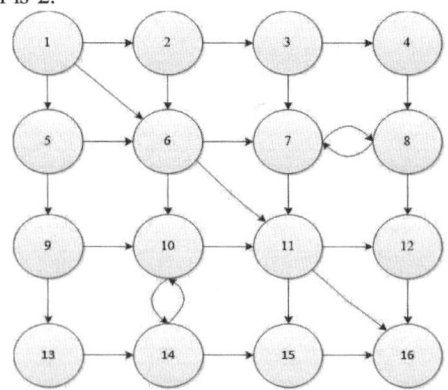

Fig.2. Simulated campus network

TABLE I
PARAMETER SETTING OF ACO

Parameter	Description	Value
D_{ant}	Number of ACO iterations	50
N_a	Number of ants	15
α	Influence strength of pheromone	1
β	Influence strength of visibility	1

TABLE II
PARAMETER SETTING OF PSO

Parameter	Description	Value
D_{pso}	Number of ACO iterations	1000
N_p	Number of particles	20
l	Number of neighborhood	3
C_1	Weight of pbest	1
C_2	Weight of lbest	1
v_0	The initial velocity	1.29169
Z	Inertia parameter	Ranging from 0.9 to 0.5

The parameters used in the algorithm are shown in

Table I. Parameters of C_1, C_2 and v_0 in Table II are

optimized by genetic algorithm. The number of iterations is set in terms of
Fig.3.

A. Solve the sub-problems at same time

In this section, experiments are conducted to demonstrate the efficiency of solving bus stops selection, bus routing design and bus frequency design at the same time. Two experiments are done in the simulated campus network.

First, the problem is solved by model proposed in this paper. Same bus frequency is used for weekdays and weekends. The solution of bus stops location is [1 5 6 14 15 16] and bus frequencies of all time intervals are shown in Table III. The objective function value is 0.885.

Then, we apply the traditional way to determine the bus frequency for comparison. The locations of bus stops are assumed to be known. Same number of bus stops with previous experiment is chosen to guarantee the cost of bus stops is the same. The bus stops are determined to set up on nodes [1 4 7 13 15 16] based on the value of arrival rate. Same objective function (2) is applied to optimize the bus frequency of the given bus stop location. The total cost is 1.045, which is much higher than which of the previous experiment. The necessity of solving bus stops selection, bus routing design and bus frequency design at the same time is demonstrated.

TABLE III
BUS FREQUENCY

Time Interval	Departure Interval (minute)
7-8	8.6
8-9	8
9-10	8.2
10-11	8.2
11-12	7.9
12-13	8.7
13-14	8.4
14-15	8.5
15-16	8.3
16-17	8.6
17-18	3
18-19	8.7

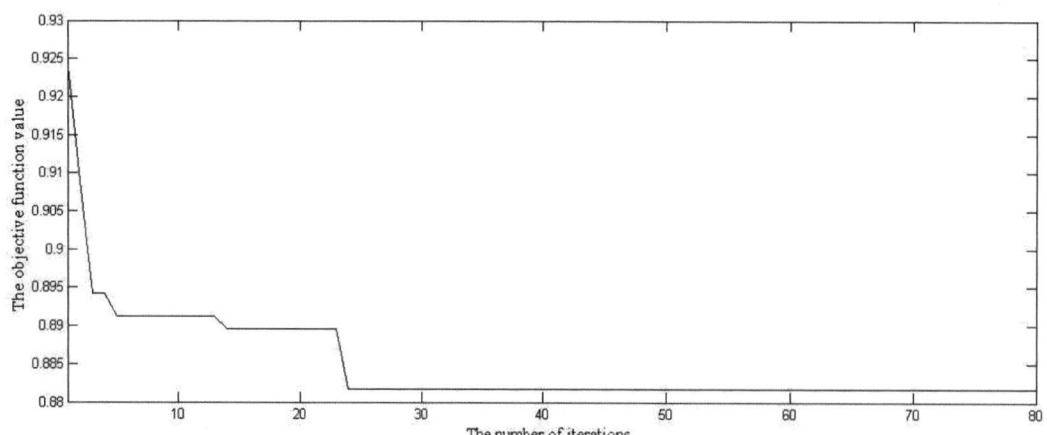

Fig.3. The ACO iterations

B. *Use different bus frequency*

In campus, OD trips are quite different on weekdays and weekends. In section 4.1, same bus frequency is used on the two different situations. As the number of bus line is small, it is totally possible to apply different bus frequency in different OD situations. In this section, two different bus frequencies are obtained by the model. The length of PSO solution vector is doubled. First 12 elements are the bus frequency for weekdays and the other elements are for weekends. The solution of bus stops location is [1 2 6 14 15 16] and bus frequencies of all time intervals are shown in Table IV

TABLE IV. The adjustment of bus frequency is in Table V. The bus stop on node 5 is replaced by the bus stop on node 2 in this case.

It is noted that the solution of bus stops is different from section 4. It further illustrates that bus stops, bus routes and bus frequency have impacts on each other. The bus frequency for weekends is different from bus frequency for weekdays. The total cost is 0.87, which is smaller than 0.885. Applying different bus frequencies for various OD situations results in the reduction of total cost.

TABLE IV
DIFFERNET BUS FREQUENCY

Time Interval	Departure Interval for Weekdays(minute)	Departure Interval for Weekends(minute)
7-8	8.1	7.3
8-9	8.1	3.7
9-10	8	7.9
10-11	7.7	8
11-12	7.4	11
12-13	6.6	7.5
13-14	8.1	10.8
14-15	9.6	8.1
15-16	7.4	7.4
16-17	8	5.7
17-18	8.7	7.3
18-19	8.4	7.3

TABLE V
ADJUSTMENT OF BUS FREQUENCY

Time Interval	Departure Interval (minute)
7-8	7.3
8-9	3.9
9-10	7.5
10-11	8.5
11-12	12
12-13	7.2
13-14	10.5
14-15	3
15-16	7.5
16-17	6
17-18	3
18-19	7

V.   CONCLUSION

The optimization of bus system will reduce the cost of the bus company and offer more convenience for passengers. This problem consists of three sub-problems: bus stop selection, bus route design and bus frequency design. Most studies focus on one of the sub-problem and ignore the connection among them. Campus bus system is

in a quite different situation from urban bus system. In campus, passengers' travel habit is similar to each other. Only one bus line is used in this paper. Therefore, bus stop section and bus route become the same problem. When the bus stops are set, the route of the bus is determined. The optimization problem asks for computing for both bus stops and bus frequency. Model is formulated with a multiple objective to satisfy both bus company and student passengers. Since the variation of OD in campus can be obtained, different OD situation is applied in the model.

Two decision variables are supposed to be determined at the same time. ACO performs very good at finding optimal or near-optimal locations. However ant colony algorithm is not suitable for solving continuous problem which refers to bus frequency decision making in our model. Particle swarm optimization is suitable for solving continuous optimization problems. A hybrid algorithm that combined ant colony algorithm and particle swarm optimization is designed to solve our problem. The set of bus stops is determined by ant colony algorithm. Particle swarm optimization is applied to figure out optimal bus frequency on given bus stop location. Fitness function value is returned to ant colony algorithm for updating pheromone and next round iteration. The algorithm performs good in case study.

In the numerical experiments, the proposed algorithm's advantage of solving the three sub-problems together is demonstrated. Two different OD situations-weekday and weekend are applied in the model. The usage of different bus frequency for different OD situations results in a reduction of total cost. In addition, a sudden OD increase is simulated on one node, the adjustment of bus frequency is demonstrated simple and effective.

It is reasonably believed that the model has potential applicability in real bus system design. In further research it is recommended to apply multiple bus lines in the model and apply the model in real network.

REFERENCES

[1] J. Park, B. Kim, "The school bus routing problem: A review", *European Journal of operational research,* Vol. 202, No. 2, 2010, pp. 311-319.

[2] R. Bowerman, B. Hall, P. Calamai, "A multi-objective optimization approach to urban school bus routing: Formulation and solution method", *Transportation Research Part A: Policy and Practice,* Vol. 29, No. 2, 1995, pp. 107-123.

[3] P. Schittekat, M. Sevaux, K. Sorensen, "A mathematical formulation for a school bus routing problem," in *Service Systems and Service Management, 2006 International Conference on,* 2006, pp. 1552-1557.

[4] Á. Ibeas, B. Alonso, O. Sainz, "Optimizing bus stop spacing in urban areas", *Transportation research part E: logistics and transportation review,* Vol. 46, No. 3, 2010, pp. 446-458.

[5] M. Spada, M. Bierlaire, T. M. Liebling, "Decision-aiding methodology for the school bus routing and scheduling problem", *Transportation Science,* Vol. 39, No. 4, 2005, pp.

477-490.

[6] W. Wang, X. Hu, L. Wu, Y. Fang, "A simulation optimization approach to campus bus routing with diversion," in *Innovative Computing, Information and Control (ICICIC), 2009 Fourth International Conference on*, 2009, pp. 713-716.

[7] J. J. Blum, T. V. Mathew, "Intelligent agent optimization of urban bus transit system design", *Journal of Computing in Civil Engineering,* Vol. 25, No. 5, 2010, pp. 357-369.

[8] B. Yu, Z. Yang, J. Yao, "Genetic algorithm for bus frequency optimization", *Journal of Transportation Engineering,* Vol. 136, No. 6, 2009, pp. 576-583.

[9] T. De-rong, W. Jing, L. Han-bo, W. Xing-wei, "The optimization of bus scheduling based on genetic algorithm," in *Transportation, Mechanical, and Electrical Engineering (TMEE), 2011 International Conference on*, 2011, pp. 1530-1533.

[10] M. Tang, E. Ren, C. Zhao, "Route Optimization for Bus Dispatching Based on Genetic Algorithm-Ant Colony Algorithm," in *Information Management, Innovation Management and Industrial Engineering, 2009 International Conference on*, 2009, pp. 18-21.

[11] X. Zhang, L. Zhou, "Study on synthesis optimization of vehicles scheduling for urban public transportation," in *Machine Learning and Cybernetics, 2009 International Conference on*, 2009, pp. 2692-2696.

[12] D. He, S. Guo, "Regional Bus Scheduling Planning Based on Time-Space Network," in *Third International Conference on Transportation Engineering (ICTE)*, 2011.

[13] J. Li, K. Larry Head, "Sustainability provisions in the bus-scheduling problem", *Transportation Research Part D: Transport and Environment,* Vol.14, No.1, 2009, pp. 50-60.

[14] A. A. Ceder, "Public-transport vehicle scheduling with multi vehicle type", *Transportation Research Part C: Emerging Technologies,* Vol.19, No.3, 2011, pp.485-497.

[15] A. Fügenschuh, "Solving a school bus scheduling problem with integer programming", *European Journal of Operational Research,* Vol.193, No.3, 2009, pp.867-884.

[16] A. Fügenschuh, "A set partitioning reformulation of a school bus scheduling problem", *Journal of Scheduling,* Vol. 14, No. 4, 2011, pp. 307-318.

[17] B. Kim, S. Kim, J. Park, "A school bus scheduling problem", *European Journal of Operational Research,* Vol. 218, No. 2, 2012, pp. 577-585.

[18] M. Dorigo, V. Maniezzo, A. Colorni, "Ant system: optimization by a colony of cooperating agents", *Systems, Man, and Cybernetics, Part B: Cybernetics, IEEE Transactions on,* Vol. 26, No. 1, 1996, pp. 29-41.

[19] J. Kennedy, R. Eberhart, "Particle swarm optimization," in *Neural Networks, 1995. Proceedings, IEEE International Conference on*, 1995, pp. 1942-1948.

EWMA Control Chart of NOx Atmospheric Environmental Monitoring System

Hu-sheng Lu[1], Jian-jing Zhen[2,*]

[1]Key Laboratory of Integrated Exploitation of Bayan Obo Multi-Metal Resources, Inner Mongolia University of Science and Technology, Baotou, China

[2]College of Economics and Management, Inner Mongolia University of Science and Technology, Baotou, China

(zhenjianjing2007@126.com)

[1]*Abstract* - **A new approach to detect out-of-control status of air quality effectively was introduced in this paper. A useful identification method of an exponentially weight moving average (EWMA) monitoring model was built, based on analyzing the statistical characterization of NOx of air quality parameters, with the relationship among the parameter of control limit, the smoothing coefficient, and the average run length of the EWMA control chart discussed. The NOx concentration data from January 1st to September 30th in 2012 were used and an illustration of environmental monitoring was given to demonstrate how to draw a chart. The results show that with the new approach the abnormal state of air quality could be detected earlier when the process mean had a smaller offset.**

Keywords - **Air quality, EWMA control chart, statistical process control**

I. INTRODUCTION

According to the characteristics of administrative divisions and functional area, combined with the accessibility of monitoring points, the corresponding air automatic monitoring points are set to monitor the parameters of air quality. Sulfur dioxide, nitrogen dioxide, carbon monoxide and inhalable particles, which are monitored every day throughout a year of 365 days or 366 days[1], are the targets in the monitoring program. The common methods for monitoring sulfur dioxide are: formaldehyde absorbing-pay rose aniline hydrochloride, spectrophotometry, pulse fluorescence; formaldehyde absorbing pararosaniline hydrochloride, pulse fluorescence spetrophotometry are for monitoring nitrogen dioxide; gas filter correlation method is for monitoring carbon monoxide; and weight method and β ray method are for monitoring inhalable particles. The routine indexes from the monitoring data with both the fixed threshold alarm valve and the changeable alarm valve are worked out to identify the abnormity of air quality and an alarm is made [2]. But the data from the equipment for air automatic monitoring are not plentiful so that the threshold setting is not accurate and the air quality forecast is not accurate, either, and the error alarm may occur. It is very significant to how to increase the prediction accuracy of air quality from the small sample data, similar to that from the large sample data.

SPC (statistical process control) is a process control tool by dint of a mathematical statistics method [3]. By analysis of the production process, signs of systematic factors are detected according to the feedback and some measures are taken to eliminate their influence to keep the process in control with nothing but random factors. A control chart, one of the most important tools in SPC, is used to detect whether the production process is in control. When the process variation is caused only by accident, the process is in statistical control. Otherwise, the process is out of control, then the possible reasons needs to be identified and be eliminated.

SPC has found good application in accurate control of the production process with the small sample data [4~5], especially in mechanical manufacturing process quality control. In addition, good results have been achieved in drug testing [6], the company's financial crisis [7], cosmetics [8] and food quality control [9]. Air monitoring and monitoring process of mechanical processing have generality, and the samples of air quality parameters present normal distribution, and air quality parameters meet the application conditions of EWMA (exponentially weighted moving average) control chart, so SPC is prospective in air quality monitoring. In this paper, some study is done with the NOx concentration data from January 1st to September 30th in 2012.

II. ANALYSIS OF PARAMETERS OF THE AIR QUALITY OF THE ENVIRONMENT

With the EWMA control chart to monitor the air quality parameters, the parameters must meet the modeling requirements of EWMA control chart. Taking the daily concentration data of NOx at a monitoring site in 2012 as an example, the characteristics of the sample parameters are analyzed to test whether SPC can be used for real-time monitoring. As the NOx histogram shown in Fig.1, the distribution is similar to a bell shaped curve with 0.045 as a symmetry center. In order to accurately determine whether the daily concentration data of NOx presents the normal distribution, the probability of NOx is mapped in Fig.2. The daily concentration data of NOx is almost a straight line, which proves that normal distribution and NOx concentration are fitting. So it is assumed that air quality parameters present normal distribution.

Further data fitting must be done about probability density function of normal distribution of NOx data from January 1st to September 30th, as shown in Fig.3. Fig.3 shows that the air quality parameters of NOx present normal distribution, so SPC can be used for monitoring modeling of air quality parameters.

[1] Fund Project: Project supported by the National Natural Science Foundation (71162027).

E. Qi et al. (eds.), *Proceedings of the 21st International Conference on Industrial Engineering and Engineering Management 2014*, Proceedings of the International Conference on Industrial Engineering and Engineering Management, DOI 10.2991/978-94-6239-102-4_81, © Atlantis Press and the authors 2015

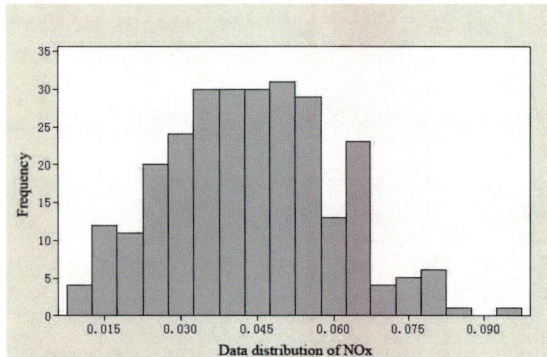

Fig.1. The concentration of NOx histogram

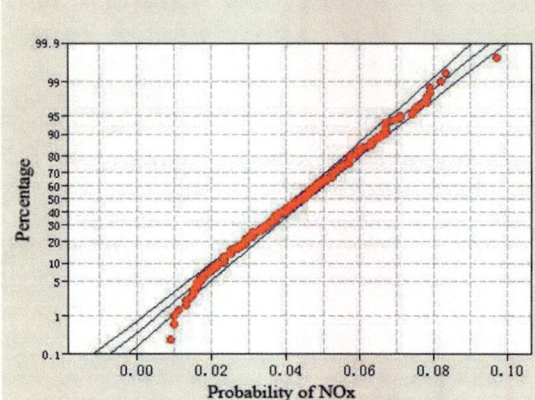

Fig.2. The concentration of NOx probability graph

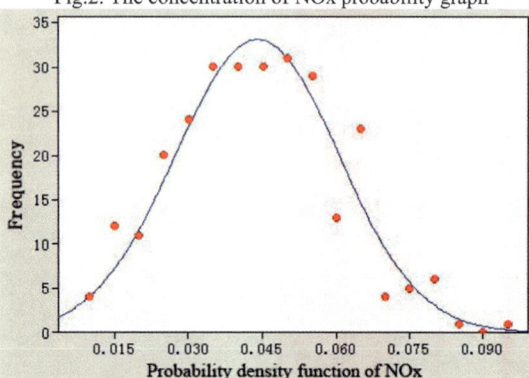

Fig.3. NOx probability density function

III. EWMA AIR QUALITY MODELING

A. Statistical process control

Generally statistical process control (SPC) functions by means of a control chart. The control limit of a control chart is used to distinguish the accidental fluctuation and abnormal fluctuation. Such control charts are common as Shewhart control chart, exponentially weighted moving average (EWMA) control chart, and cumulative sum (Cusum) control chart, etc. Shewhart control chart is efficient to distinguish the accidental factors and abnormal factors. It hits home that with the analysis of air quality parameters, parameters present normal distribution, with the small mean deviation of the process, while the EWMA control chart is sensitive to small shifts in the process [10]. So it is a good choice to apply the EWMA control chart to the concentration data of NOx for monitoring.

B. EWMA control chart

Let the sample sequence of NOx air quality parameters is $\{X_t\}$, and X_t is an independent random variable sequence, namely, $X_t \sim N\ (\mu_t, \sigma_N^2)$, μ_t represents that the desired value of sample sequence shifts along with time variation. Z_t is the EWMA statistic at time t , then

$$Z_t = \omega X_t + (1-\omega)Z_{t-1} \qquad (1)$$

Z_t is the state value at time t, and ω is the smooth coefficient or weight factor, and $0 < \omega < 1$, $Z_0 = E\ (X) = \mu_0$. Z_t is all the data in the form of weighted average, and replace Z_{t-1} on the right side in the equation (1) repeatedly, then

$$Z_t = \omega X_t + (1-\omega)Z_{t-1} = \omega X_t + \omega(1-\omega)X_{t-1} \\ + (1-\omega)^2 Z_{t-2} \qquad (2)$$

$$Z_t = \omega \sum_{i=0}^{t-1}(1-\omega)^i X_{t-i} + (1-\omega)^t Z_0 \qquad (3)$$

Clearly: $\omega \sum (1-\omega)^i + (1-\omega)^t = 1$

Through the equation (3) an EWMA statistic variance is got:

$$\sigma_R^2 = \sigma_x^2 (\frac{\omega}{2-\omega})\left[1-(1-\omega)^{2t}\right] \qquad (4)$$

When ω is close to 1, the weight of the current value in EWMA increases, the weight of former value in EWMA decreases. When $\omega = 1$, the EWMA control chart turns into individual $\overline{X}$ control chart. Thus the control limit of EWMA control charts is:

$$UCL_t = \mu + L\sigma_x \sqrt{(\frac{\omega}{2-\omega})\left[1-(1-\omega)^{2t}\right]} \qquad (5)$$

$$LCL_t = \mu - L\sigma_x \sqrt{(\frac{\omega}{2-\omega})\left[1-(1-\omega)^{2t}\right]} \qquad (6)$$

Among them, L is the control limit parameter, the control limit of EWMA control chart is wider with the increasing of t, and is close to a stable value. At this time, the control limits can be expressed as follows:

$$UCL_t = \mu + L\sigma_x \sqrt{\frac{\omega}{2-\omega}} \qquad (7)$$

$$LCL_t = \mu - L\sigma_x \sqrt{\frac{\omega}{2-\omega}} \qquad (8)$$

Noted for the smaller t, the equation (5) and (6) are used to determine the upper and lower control limits of the EWMA control chart.

C. Estimation of EWMA control chart parameters

Offset δ , control limit parameter L, and smoothing coefficient ω , are subjective to air parameter fluctuations. The operators or experts should know such factors as the accuracy of the air parameters fluctuation, monitoring capability, and weather conditions in the parameter collecting and then with the help of experts' knowledge, experience and judgment the evaluation can be done. The more accurate the values are the better to the modeling as well as monitoring of air quality [11~12]. Generally, the principle of selecting the smoothing parameters ω is to minimum errors between the predicted values and actual ones, namely, with a minimum square error.

In theory, simple selection can be done with heuristics. When the statistics fluctuate slightly, ω is from 0.1 to 0.3 to increase the original forecast weight; if statistics are volatile, ω is from 0.6 to 0.8 to increase the weight of new forecast. In order to have a good evaluation, several different values of ω can be used to compare [13]. Besides, simple selection can be done with the maximum ARL_0 and the minimum square error. Usually, when the parameters of the EWMA control chart are $\omega = 0.1$, $L = 2.7$, the EWMA control chart and the CUSUM control chart function similarly [14]. Or when $\omega = 0.2$, $L = 3$, it also functions well [15].

D. Performance evaluation of control chart

Usually ARL (ARL_0 and ARL_1) is an important index of evaluating the performance of control charts. It refers to the mathematical expectation of sampling times from being out of control to sending off out-of-control signals in the control chart [16]. For a statistical process control chart, when the process is in control (offset $\delta = 0$), the control chart has a bigger ARL_0; while the process is out of control (offset $\delta > 0$), the control chart has a smaller ARL_1.

IV. ANALYSIS OF NOx EWMA CONTROL CHART

In the air there are such nitrogen oxides as nitrous oxide, nitric oxide, nitrogen dioxide, nitrogen trioxide, the large percentage of which is nitric oxide and nitrogen oxide, with nitrogen oxide (NOx) denoted. The percentage of NOx in the air is one of the key parameters affecting air quality parameters, and also an important index of air quality. Meanwhile, the percentage of NOx in the air can show how the air pollution is. NOx also stimulate human respiratory organs, which is one cause leading to more and more respiratory diseases such as bronchial asthma. Nitrogen and oxygen compounds in the air can be induced to have photochemical reactions with organic compounds, producing the photochemical smog, and nitric acid and nitrate come into being, harming water and soil in the form of rainfall. Therefore, it is very significant to analyze and monitor air quality parameters of NOx.

In this paper a dynamic quality control method of EWMA control chart is used to monitor the NOx concentration data. The data brought at the same site form a sample, with their mean values of reflecting the change of NOx. Therefore, it is believed that there is no setting error in the data acquisition process, namely, the mean target value of 0.04392, and the standard deviation of

$\overline{\sigma} = 0.01653$. With experiences, let L=3, $\omega = 0.2$, then

$$Z_t = 0.2 X_t + (1 - \omega) Z_{t-1}$$

$$UCL_t = 0.04392 + 3 \times 0.01653 \times$$

$$\sqrt{(\frac{0.2}{2 - 0.2})\left[1 - (1 - 0.2)^{2t}\right]}$$

$$LCL_t = 0.04392 - 3 \times 0.01653 \times$$

$$\sqrt{(\frac{0.2}{2 - 0.2})\left[1 - (1 - 0.2)^{2t}\right]}$$

The dynamic quality monitoring on air quality parameters of NOx is shown in Fig.4, telling that the process is out of control. In the out-of-control chart, there are some data presenting continuous upward or downward trend before they are beyond the upper or the lower control limit, which proves the process is out of control, with a warn given, indicating that the air quality parameters are abnormal, so there should be more concerns.

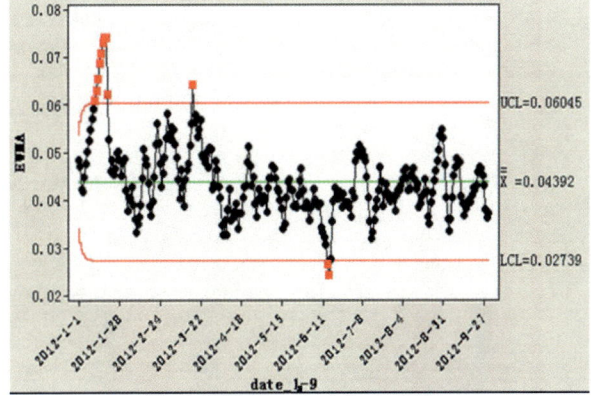

Fig.4. EWMA control chart of NOx

With further investigation and discussion with the environmental protection bureau, over NOx concentration is mainly caused by the followings: firstly, industrial production emissions; secondly, transportation, especially of which the vehicle exhaust emissions consist of such substances as carbon monoxide, sulfur monoxide, nitrogen oxide etc., harmful to human health, and causing serious air pollution; thirdly, cooking stoves and heating boilers. In winter, a large number of NOx produced by residential heating are emitted into the air, and in the meanwhile, the atmosphere has poor oxidation performance, so nitrogen oxide pollution is more and more serious.

V. SUMMARY

Human life and development is closely related to the air quality and ecological system. In order to protect the atmospheric environment and make sure of harmonious coexistence between human and nature, monitoring and forecasting the trend of air quality timely and accurately is an important project to be solved. A large number of data samples are required in monitoring air quality parameters. However, with poor monitoring instruments and economic conditions of monitoring, it is difficult to achieve the desired sample size. In this paper, with SPC of various and small-sized data, a dynamic statistical model of air quality parameters is built to monitor air quality parameters. The new approach proves feasible and prospective with practical cases.

REFERENCES

[1]   Baotou City Environmental Monitoring Station.Baotou City Environmental Quality Report (2012), 2012.

[2]   Xia Yue-qing, Wang Jian-hua. A Comparative Study and Strategy Discussion on Environmental Air Quality Monitoring of Pudong New Area [J]. Environmental Monitoring in China, 2012, 28(6):89-93.

[3]   Yi Ting-hua,Guo Qing, Li Hong-nan. The Research on Detection Methods of GPS Abnormal Monitoring Data Based on Control Chart [J]. Engineering Mechanics, 2013, 30(8): 133-141.

[4]   Niu Zhan-wen, Chen Tian-jun, Liu Xiao-nan. Statistical Process Control (SPC) for Quality Control in Multi-Product-and-Small-Batch Production [J]. Industrial Engineering Journal, 2010, 13(4): 100-103.

[5]   Li Xin-yun, Yuan Xiang-dong, Ren Wei-bin. Judge strategy of heat treatment process quality for multi-specification and small-batch production [J]. Heat Treatment of Metals, 2013, 38(9): 114-115.

[6]   Zhou Jian-min, Cao Feng-xi. Applications of Shewhart Control Charts in Drug Testing [J]. Chinese Journal of Modern Applied Pharmacy, 2014, 31(1): 116-119.

[7]   Chen Lei, Ren Ruo-en. Theory and Application of TSDA and EWMA in Financial Distress Prediction [J]. Journal of Systems & Management, 2009, 18(3): 241-248.

[8]   Zhang Tai-jun, Chen Chao-mao, Zhang Yi, Li Tao, Lin Tian-xian, Xie Fu-feng,Liu Di. Statistical quality control of cosmetics [J]. China Surfactant Detergent and Cosmetics, 2013, 43(5): 397-401.

[9]   Liu Rui,Wei Yi-min, Zhang Bo. Quality Control of Dried Noodle Processing Based on Statistical Process Control (SPC) [J]. Food Science, 2013, 34(8):43-47.

[10] Yu Lei, Liu Fei. Application of EWMA control chart in stability analysis of MSA [J]. Journal of Systems Engineering, 2008, 23(3): 381-384.

[11] Chen Yan, Yin Jian-jun, Xiang Zu-feng, et al. SPC monitoring method in environmental parameters of coastal water [J]. Environmental Science & Technology, 2012, 35(8): 124-128.

[12] Xue Li. Economic Design of Variable Sampling Intervals EWMA Control Chartsunder Geometric Distribution [J]. Operations Research and Management Science, 2013, 22(4): 126-132.

[13] Haq Abdul. A New Hybrid Exponentially Weighted Moving Average Control Chart for Monitoring Process Mean [J]. Quality and Reliability Engineering International, 2013, 29(7): 1015-1025.

[14] Hawkins Douglas M, Wu Qi-fan. The CUSUM and the EWMA Head-to-Head [J]. Quality Engineering, 2014, 26(2): 215-222.

[15] Chen Yong, Tang Ke-feng, Lin Fei-long, Liu Chun-yan. The Bottleneck Analysis and Balance Improvement of Flow Line in the Enterprise's CD Section [J]. Industrial Engineering and Management, 2008, 13(1): 112-115.

[16] Li Zhong-hua, Zou Chang-liang, Gong Zhen, et al. The computation of average run length and average time to signal: an overview [J]. Journal of Statistical Computation and Simulation, 2014, 84(8): 1779-1802.

Research of Electricity Enterprise Credit Assessment Combining Fuzzy Multiple Attribute Decision Making with Improved Analytic Hierarchy Process –Taking Baoding Electricity Supply Company as Example

Xin-Bo Dai[*], Yuan-sheng Huang

Department of Economics and Management, North China Electric Power University, Beijing, China
(lianlianfushi@126.com)

Abstract - **Under the competitive environment of electricity market, electricity customers will have more options and power grid companies will face more and more intense market competition. This paper studies the utility produced by customer value and customer credit to grid enterprises with the improved analytic hierarchy method and the combination of fuzzy multiple attribute decision making method. Power big customer credit evaluation model is established. The customer relationship from power supply enterprises in Baoding city is taken as an example on the empirical analysis, to verify the effectiveness of the proposed method.**

Keywords - **Customer credit, fuzzy multiple attribute decision making, improved analytic hierarchy process, power market**

I. INTRODUCTION

Customer relationship management is referred to CRM. At present there is a lot of study of power supply enterprise CRM. According to the characteristics of electric power industry, the literature [1-3] discusses the connotation of CRM and the existing problems, and puts forward the implementation strategy of the power supply enterprise CRM. The literature [4] is proposed to seek valuable customers, and provide personalized service. Power supply enterprises should focus on implementing the value management of big customer, and avoid the loss of big clients. The literature [5] points out that power supply enterprises implement customer value analysis and differentiation marketing. So they can improve enterprise competitiveness, and finally establish the evaluation model of the power customer. The literature [6] points out that under the environment of electricity market, the change which electric power enterprises face is setting up customer service center to implement CRM information management. The literature [8] builds customer credit management system based on the 5c theory. And group decision method and linear weighted method are applied to implement credit risk assessment. According to the actual situation of power Supply Company, the credit management system is designed. For electricity customers in arrears and the reasons, the literature [9, 10] designed electricity customer credit evaluation model. Most of the current domestic electric power enterprises did not establish a perfect customer relationship management (CRM) system. A part of power supply enterprises established customer information system, but basic data is not real and information is incomplete. The enterprise CRM system is difficult to play real role. In the future under the competition environment of electricity market,

electricity customers will have more options. But as the main body of market trade, power supply enterprises , how to avoid operation risk, gain maximum profit and guarantee to maintain the stability of customers, credit evaluation is required to electricity customers and customer relationship management (CRM) need be strengthened, in order to realize effective control of customer resources and prevention of risk. Based on utility theory, this paper studies the customer value and customer credit for power supply enterprises. Use the improved analytic hierarchy method and the fuzzy multiple attribute decision making method, customer relationship from power supply enterprises in Baoding city is taken as an example, and the two combination methods are used for the empirical analysis. Through the simulation, the correct conclusion is drawn.

II. METHODOLOGY

A. The improved AHP model

Using the concept of optimal transfer matrix, analytic hierarchy process is improved and satisfies the requirement of consistency naturally, in order to avoid blindness and reduce the adjusted times by estimation. The main steps of improvements are as follows:

The first step: To establish evaluation index system

The second step: To construct judgment matrix

Through pairwise comparison, determine the relative importance of each element compared to a particular element on one level. Two comparative judgment matrixes are constructed as:

$$A = (a_{ij})_{n \times n}, a_{ij} > 0, a_{ij} = 1/a_{ji} \quad (1)$$

In the formula, a_{ij} denotes the scale of importance of two indexes.

The third step: Based on the modified AHP method, calculation of all levels evaluation index weight.

After index weight is calculated based on the modified AHP method, consistency check need not be done. The calculation steps are as follows.

(1) Judgment matrix is fabricated to get quasi optimal matrix A^*, as shown in Fig.1.

(2) Root method to A^* be used to solve the feature vector of A^*.

First of all, the elements of matrix A^* multiply by lines. The following formula is concluded that:

$$M_i = \prod_{j=1}^{n} a_{ij} \qquad (2)$$

In the formula, $i = 1, 2, \cdots, n$.

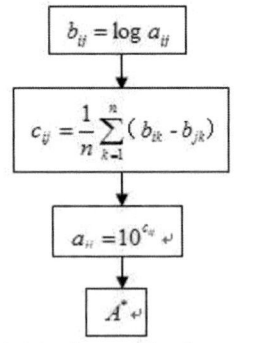

Fig.1. Matrix modification process

(3) The products are done n-th root respectively. It can be obtained:

$$W_i' = M_i^{1/n} \qquad (3)$$

Then the normalized processing is done to root vector $W' = (W_1', W_2', \cdots, W_n')$. The calculation formula of namely the sorting weight vector W is obtained as:

$$W_i = \frac{W_i'}{\sum_{j=1}^{n} W_j'} \qquad (4)$$

Index weight is obtained: $W = (W_1, W_2, \cdots, W_n)$.

The fourth step: Calculate the comprehensive weight of each evaluation index

The weight based on the second layer index is multiplied by the corresponding weight of the first layer and the comprehensive weight of each evaluation index is obtained.

B. Fuzzy Multi-Attribute Decision

Taking $X = \{x_1, x_2, \cdots, x_p\}$ as evaluation object collection and taking collection $C = \{c_1, c_2, \cdots c_n\}$ as evaluation indexes, evaluation object x_j is evaluated by index c_i. And the triangular fuzzy function a_{ij} of x_j about index c_i is obtained. All evaluation objects are evaluated by all the indexes and the fuzzy evaluation matrix $A = (a_{ij})_{n \times p}$ is formed.

To enable comparability between various indexes, the effects of different dimensions to evaluation results must be eliminated. So the evaluation matrix $A = (a_{ij})_{n \times p}$ need be transformed into standardized $R = (r_{ij})_{n \times p}$ n×p. Among them $r_{ij} = (r_{ij}^l, r_{ij}^m, r_{ij}^s)$.

The index values which are the bigger the better are processed as follows:

$$r_{ij}^l = a_{ij}^l / a_{i\max}^l, a_{i\max}^l = \max\{a_{i1}^l, a_{i2}^l, \cdots, a_{in}^l\}$$
$$r_{ij}^m = a_{ij}^m / a_{i\max}^m, a_{i\max}^m = \max\{a_{i1}^m, a_{i2}^m, \cdots, a_{in}^m\} \qquad (5)$$
$$r_{ij}^s = a_{ij}^s / a_{i\max}^s, a_{i\max}^s = \max\{a_{i1}^s, a_{i2}^s, \cdots, a_{in}^s\}$$

The index values which are the smaller the better are processed as follows:

$$r_{ij}^l = a_{i\max}^l / a_{ij}^l, a_{i\min}^l = \min\{a_{i1}^l, a_{i2}^l, \cdots, a_{in}^l\}$$
$$r_{ij}^m = a_{i\max}^m / a_{ij}^m, a_{i\min}^m = \min\{a_{i1}^m, a_{i2}^m, \cdots, a_{in}^m\} \qquad (6)$$
$$r_{ij}^s = a_{i\max}^s / a_{ij}^s, a_{i\min}^s = \min\{a_{i1}^s, a_{i2}^s, \cdots, a_{in}^s\}$$

The expectation value of the element r_{ij} in the matrix R is expressed with r_{ij}^a. Then:

$$r_{ij}^a = \frac{1}{2}[(1-a)r_{ij}^l + r_{ij}^m + ar_{ij}^s] \qquad (7)$$

In the formula, $0 \le a \le 1$. The choice of a value depends on the attitude of the evaluators. $a > 0.5$ means that evaluators are more optimistic. $a = 0.5$ means that evaluators are neither optimistic nor pessimistic. $a < 0.5$ means that evaluators are more pessimistic. Expected value evaluation matrix $R^a = (r_{ij}^a)_{n \times p}$ is obtained.

The expectations for comprehensive indicators are:

$$Z_j^a = \sum_{i=1}^{n} r_{ij}^a \times W_i, (j = 1, 2, \cdots, p) \qquad (8)$$

Among them, r_{ij} is the element of Qi king evaluation matrix. W_i is the weight calculated by the improved AHP method.

TABLE I
INDEX OF CREDIT RATING SYSTEM

Customer credit evaluation index system C			
Operating state C1	Commercial credit C2	Cooperative credit C3	Envoi-mental conditions C4
Sales profit rate C11 Asset-liability ratio C12 Rate of Return on net assets C13 Return on Total Assets Ratio C14	History arrears amount C21 Cumulative arrears number ratio C22 Repay electricity rates C23 Illegal conduct electricity C24 Corporate image C25	Scheduling cooperation records C31 Electricity fill out record of cooperation C32	Industry development C41 Market position C42

III. COMBINATION MODEL SIMULATION

A. Construction of Credit Evaluation Index

The company is evaluated by management capacity situation, business credit status, cooperative credit conditions and environment conditions, as shown in Table I.

In combination with the actual situation of domestic electric power industry, in this paper, the electric power client credit is mainly divided into four levels. They are A, B, C and D from high to low in turns. The lower the level is, the greater the credit risk which the power supply enterprise faces is, as shown in Table II.

TABLE II
CREDIT LEVEL STANDARD

Credit level	Composite indicator expectations	Credit status
A	0.8~1	Good credit rating. The enterprise have excellent credit record. Good operating performance. No stealing behavior. Pay bills on time. Arrears little risk. Good credibility.
B	0.6~0.8	General credit worthiness. There is a certain lack of risk.
C	0.4~0.6	Poor credit degree. Operating conditions are susceptible to the influence of uncertain factors. Don't pay the electricity bills on time, or there are stealing behavior. A lot of credit risk.
D	0~0.4	Very poor credit degree

TABLE III
CREDIT EVALUATION INDEX DATA

	Index	Company 1	Company 2	Company 3
Operating status	Sales profit ratio (%)	(7.22,7.22,7.22)	(14.55,14.55,14.55)	(10.06,10.06,10.06)
	Asset-liability ratio (%)	(36.80,36.80,36.80)	(46.16,46.16,46.16)	(48.24,48.24,48.24)
	ROE (%)	(48.24,48.24,48.24)	(10.35,10.35,10.35)	(8.76,8.76,8.76)
	Return on Total Assets Ratio (%)	(2.96,2.96,2.96)	(7.12,7.12,7.12)	(6.51,6.51,6.51)
Commercial credit	History arrears amount (Ten thousand yuan)	(15.67,15.67,15.67)	(20.54,20.54,20.54)	(16.56,16.56,16.56)
	Cumulative arrears number ratio (%)	(8.32,8.32,8.32)	(4.09,4.09,4.09)	(3.75,3.75,3.75)
	Repay electricity rates (%)	(50.87,50.87,50.87)	(92.02,92.02,92.02)	(87.93,87.93,87.93)
	Illegal conduct electricity	(0.26,0.32,0.38)	(0.16,0.22,0.28)	(0.12,0.19,0.25)
	corporate image	(0.46,0.55,0.63)	(0.72,0.85,0.91)	(0.64,0.88,0.94)
Co-operative credit	Scheduling cooperation records	(0.72,0.83,0.91)	(0.65,0.72,0.84)	(0.69,0.77,0.89)
	Electricity fill out record of cooperation	(0.56,0.59,0.62)	(0.83,0.90,0.96)	(0.78,0.83,0.95)
Environmental conditions	Industry development	(0.32,0.46,0.54)	(0.69,0.77,0.85)	(0.45,0.47,0.53)
	Market position	(0.20,0.27,0.35)	(0.60,0.65,0.76)	(0.36,0.39,0.44)

This article selected three companies of Baoding city which had greater consumption as the examples. The three companies were selected as the application of credit evaluation method. And there are five experts invited to participate in the evaluation. The five experts were assumed equally important. Though the evaluation of five experts and the survey data of the enterprises, the raw data of each index is gained. And the value of each qualitative index and the evaluation value of qualitative index were denoted with triangular fuzzy number. The credit evaluation index of each electricity customer is obtained, as shown in Table III.

B. Credit evaluation index weight calculation

The weights of secondary level index in the credit evaluation index system are taken for example below. The improved AHP method is used to calculate the index weight.

For the primary index of operating status, according to the experts, the importance of the secondary indicators in operating status indicators is shown in Table IV.

TABLE IV
OPERATING STATUS INDICATORS COMPARISON

	Sales profit rate	Asset-liability ratio	Rate of return on net assets	Return on Total Assets Ratio
Sales profit rate	1	3	1	1/2
Asset-liability ratio	1/3	1	1/3	1/4
Rate of return on net assets	1	3	1	1/2
Return on Total Assets Ratio	2	4	2	1

C. Establish fuzzy evaluation model

Based on Table IV, the fuzzy evaluation matrix is gained and the judgment matrix is constructed.

$$A = \begin{bmatrix} 1 & 3 & 1 & 1/2 \\ 1/3 & 1 & 1/3 & 1/4 \\ 1 & 3 & 1 & 1/2 \\ 2 & 4 & 2 & 1 \end{bmatrix}$$

The quasi optimal matrix A^* is gotten. By calculation, it is gained:

$$A^* = \begin{bmatrix} 1 & 2.72 & 1 & 0.56 \\ 0.37 & 1 & 0.37 & 0.29 \\ 1 & 2.72 & 1 & 0.56 \\ 1.8 & 3.47 & 1.8 & 1 \end{bmatrix}$$

The elements in A^* is multiplied by line and it is gained:

$$M = \begin{bmatrix} 1.52 \\ 0.04 \\ 1.52 \\ 11.24 \end{bmatrix}$$

Then the formulas are used to gain:

$$W' = \begin{bmatrix} 1.11 \\ 0.45 \\ 1.11 \\ 1.83 \end{bmatrix}$$

Finally, through the normalized processing, it is gained:

$$W = \begin{bmatrix} 0.25 \\ 0.10 \\ 0.25 \\ 0.40 \end{bmatrix}$$

For other secondary and primary indexes, the same method is used, and the weight of each index is gained.

$$A = \begin{bmatrix} (7.22,7.22,\ 7.22) & (14.55,14.5\ 5,14.55) & (10.06,10.0\ 6,10.06) \\ (36.80,36.8\ 0,36.80) & (46.16,46.1\ 6,46.16) & (40.24,40.2\ 4,40.24) \\ (4.32,4.32,\ 4.32) & (10.35,10.3\ 5,10.35) & (8.76,8.76,\ 8.76) \\ (2.96,2.96,\ 2.96) & (7.12,7.12,\ 7.12) & (6.51,6.51,\ 6.51) \\ (15.67,15.6\ 7,15.67) & (20.54,20.5\ 4,20.54) & (16.56,16.5\ 6,16.56) \\ (8.32,8.32,\ 8.32) & (4.09,4.09,\ 4.09) & (3.75,3.75,\ 3.75) \\ (50.87,50.8\ 7,50.87) & (92.02,92.0\ 2,92.02) & (87.93,87.9\ 3,87.93) \\ (0.26,0.32,\ 0.38) & (0.16,0.22,\ 0.28) & (0.12,0.19,\ 0.25) \\ (0.46,0.55,\ 0.63) & (0.72,0.85,\ 0.91) & (0.64,0.88,\ 0.94) \\ (0.72,0.83,\ 0.91) & (0.65,0.72,\ 0.84) & (0.69,0.77,\ 0.89) \\ (0.56,0.59,\ 0.62) & (0.83,0.90,\ 0.96) & (0.78,0.83,\ 0.95) \\ (0.32,0.46,\ 0.54) & (0.69,0.77,\ 0.85) & (0.45,0.47,\ 0.53) \\ (0.20,0.27,\ 0.35) & (0.60,0.65,\ 0.76) & (0.36,0.39,\ 0.44) \end{bmatrix}$$

Based on the analysis above, A is converted into standardized R matrix. Based on the formula above, the decision matrix Ra of R expectations is obtained as follows:

Commanding a=0.5, according to the formulas above, the comprehensive expectations of three companies can be obtained:

$Z_1^{0.5} = 0.045+0.04+0.038+0.062+0.1+0.023+0.072+0.04+0.019+0.06+0.072+0.02+0.012 = 0.635$;

$Z_2^{0.5} = 0.09+0.032+0.09+0.15+0.076+0.046+0.13+0.059+0.029+0.053+0.11+0.05+0.03 = 0.945$;

$Z_3^{0.5} = 0.062+0.031+0.076+0.137+0.095+0.05+0.125+0.07+0.028+0.057+0.104+0.031+0.018 = 0.884$;

$$R^{0.5} = \begin{bmatrix} 0.496 & 1 & 0.691 \\ 1 & 0.797 & 0.763 \\ 0.417 & 1 & 0.846 \\ 0.416 & 1 & 0.914 \\ 1 & 0.763 & 0.946 \\ 0.45 & 0.917 & 1 \\ 0.553 & 1 & 0.96 \\ 0.525+0.1a & 0.805+0.07a & 1 \\ 0.635+0.015a & 0.985-0.015a & 0.945+0.055a \\ 1 & 0.885+0.01a & 0.945+0.01a \\ 0.66-0.015a & 1 & 0.93+0.025a \\ 0.53+0.09a & 1 & 0.63-0.015a \\ 0.375+0.065a & 1 & 0.60-0.01a \end{bmatrix}$$

The credit levels of three enterprises are gained. The credit level of enterprise 1 is in general level and belongs to the grade B. Enterprise 2 and Enterprise 3 are in a higher credit level and belong to the grade A. But the credit level of enterprise 2 is higher than Enterprise 3.

IV. CONCLUSION

Finally, in order to reduce the credit risk of three electricity customers above, we should provide different services for the customers of different credit levels. For the customers of higher electricity credit, we should fully guarantee the power supply reliability, and give preferential policies, such as giving electricity consumption, providing free maintenance service, repairing priority and adopting other auxiliary policies at the same time. For Enterprise 1 whose credit level is lower, credit should be strengthened.

REFERENCES

[1] Li Chunying. The application of client relation management in the power supply enterprise [J]. Journal of Chifeng University, 2009, 25(2):103-104.

[2] Lv Jian. The research of applying client relation management system of power supply enterprise [J]. Power demand, 2004, 6(2):19-21.

[3] Feng Jia, Li Ning, Chen Yu. Power client credit management and system design. Deepen application column of Zhejiang power ERP, 2010, 8(9):76-79.

[4] Li Pin. Design and implementation of power customer credit evaluation system [J]. Measurement and control technology, 2010, 29(7): 98-102.

[5] Zhang Li. Design and implementation of power supplying enterprise client relationship management system [J]. Technology economy and management research, 2006, (2): 68-69.

[6] Tianjin city electric power company. Power marketing REVIEW [M]. Beijing: Electric power press of China, 2004.

[7] Zhang Wenquan. Discuss on the electricity market again. North China electric power university, 1998, (4).

[8] Li Liping. Discuss of successful customer relationship management applications. Enterprise research, 2011(4).

[9] Yin Hongwei. Struction and research of power supplying enterprise client relationship management system. Power information, 2003, 1(4).

[10] Xue xiang. Analysis of electric power information automation system based on CRM. Jiangsu motor engineering, 2003, 22(4)

SMT Optimization Based on the Cellular Genetic Algorithm

Xuan DU, Deng-qiao LI[*], Li-li SUN

Mechanical and Power Engineering, China Three Gorges University, Yichang 443002, China

(liqiao_jiao@163.com)

[1]Abstract - **This paper studied the surface mounting process optimization of the turret SMT machine. Firstly, it decomposed the issue into two sub-problems, which includes problems of component placement sequence and feeder arrangement. It built a surface mounting process optimization model based on the analysis of actual engineering application. The objective function of the model is PCB assembly in the shortest time. Afterwards, it solved the model using cellular genetic algorithm. The paper designs a two-dimensional segmented decimal encoding while neighbor structure using typical von Roy type. It used the improved ordered crossover and adaptive mutation in the process of genetic operations together with local search strategy in the algorithm based on the encoding. Finally, through the example, it can be shown that the algorithm in this paper are better in solving efficiency and results performance compared with the traditional genetic algorithm, which proved the effectiveness and superiority of the algorithm.**

Keywords - **Cellular genetic algorithm, component placement sequence, feed tank layout, interactive interface, surface mounting**

I. INTRODUCTION

The SMT (Surface Mount Technology) has been widely used in PCB assembly as the rapid development of electronics manufacturing; it is especially important considering how to improve production efficiency of current PCB assembly production line. Being the bottle device, the efficiency improvement of turret SMT machine can be quite significant.

Currently, for this optimization problem, it mainly adopts the intelligent optimization algorithm such as traditional GA[1], taboo search algorithm [2-3], ant colony algorithm [4], ant colony-shuffled frog leaping algorithm [5-6], distributed search algorithm [7], difference algorithm [8], which could obtain better optimization effect in a certain range and time. Additionally, there are other intelligent methods such as simulated annealing algorithm [9].

Presently, we mainly concentrate on one of the subproblem and neglect the whole correlation of multi objective or build optimization model of many subproblems separately, then change multi objective to single objective problems applying weighing method, which can't assure objectivity of weighing coefficient allocation. Cellular Genetic Algorithm (CGA) is a branch of GA [10], which is improved based on cellular automation and GA, it has simple model, easy realization and fast convergence, and present its great performance

in practical combination optimization problems [11-12]. To seek better optimization, this paper adopted CGA to solve problems. Built corresponding assembly process optimization mathematic model by analysis of placement workmanship process of turret SMT machine considering real situation of component size, multi layout of the same type of component in the feeder, CGA helped to solve component placement sequence and feeder layout optimization problems.

II. OPTIMIZATION MODEL BUILDIND AND DECRIPTION OF MOUNTING PROCESS

A. Description of mounting process

This essay researched mounting process optimization of turret SMT machine. This machine is composed of turret with placement head, feeder, PCB workbench, whose structure is simple, mounting speed is fast, easy fulfill of little size component. Its mounting process is: first, the workbench loading PCB moved to the mounting position of the first component, concurrently, feeder moved to where the feeder tank of the first component was, the placement head on the turret picked up the first component, then turret rotated one step to pick up the second component, repeatedly doing like this, if there are M placement heads, when picking up the $(M/2+1)$ component, placement head place the first one simultaneously; afterwards, PCB moves to where the second one is, meanwhile feeder tank moves to the $(M/2+2)$ component position, cycle like this till all components on PCB board are placed. The component on turret can identify, adjust orientation, straw switch automatically the component in the mounting process.

B. Definition of sub objective function

Picking up and mounting of component are done simultaneously in mounting process. The time of three parallel movement mechanisms include feeder's moving time, workbench moving time along x-y and turret rotation time. The longest time of the three determine the place time of a certain component, PCB assembly time is the sum of all components mounting time. This paper aimed at shortening PCB assembly time and improving efficiency.

C. Mounting process optimization model of turret SMT machine

One single component's assembly time in the working process of turret SMT machine is:

$$T(s) = \max\left\{t_t(s), t_f(s), t_r(s)\right\} + t_p(s)$$

---

[1] Funded projects: The master's degree paper of China Three Gorges University outstanding fund (Project No. 4014PY020)

In the equation, Moving time of PCB workbench:

$$t_t(s) = \max\left\{\frac{|x_{s+1}-x_s|}{V_x s_s}, \frac{|y_{s+1}-y_s|}{V_y s_s}\right\}$$

V_x, V_y represent workbench moving speed along x and y axis separately, s is component number, s_s represent the slowest speed factor of workbench, N is amount of component.

In the equation, Moving time of feeder shelf:

$$t_f(s) = \left(|F_{s+1}-F_s|\right)t_s, \quad s=1,2,\cdots,N$$

F_{s+1}, F_s represent feeder tank position of s+1 and s component separately, t_s represents time of feeder shelf moving one feeder tank.

Rotation time of turret and Component picking up time: $t_r(s), t_p(s)$.

Feeder layout and component placement sequence have interrelations, traditional solution by QAP and TSP is not fit for such kind of optimization, integrated optimization model aimed at component placement sequence and feeder layout has been established to achieve process optimization, the least assembly time is its objective function, this model is:

$$\min T = \sum_{s=1}^{N} t_p + \sum_{s=1}^{N} \max\left(\sum_{i=1}^{N}\sum_{j=1}^{N} t_t(s)x_{i(s-1)}x_{js}, \sum_{i=1}^{N}\sum_{j=1}^{N}\sum_{t=1}^{P}\sum_{h=1}^{L}\sum_{k=1}^{L} t_f(s)x_{i(s+M/2)}x_{j(s+M/2+1)}x_{tih}x_{tjk}, t_r(s)\right)$$

$$s.t. \begin{cases} \sum_{i=1}^{N} x_{is}=1, \sum_{s=1}^{N} x_{is}=1 \\ \sum_{h=1}^{L} x_{tih}\geq 1, \sum_{k=1}^{L} x_{tjk}\geq 1 \\ i\neq j, P\leq L \\ i=1,2,\cdots,N; j=1,2,\cdots,N; s=1,2,\cdots,N \\ t=1,2,\cdots,P; h=1,2,\cdots,L; k=1,2,\cdots,L \end{cases}$$

In the equation: N represent component quantity; M represent placement head quantity; P represent component type quantity; L represent feeder tank quantity; t represent component type; i, j represent component number; h, k represent feeder tank number; x_{is} is 0-1 variable, 1 means placement sequence time of component i is s, otherwise 0; $x_{tik}=1$ means component i belongs to type t, and it is in feeder tank k, otherwise 0 [6].

III. SOLUTION AND REALIZATION OF CGA

CGA is a combined intelligent evolutionary algorithm of GA and cellular automation. Basic principle: put initialized population of traditional GA onto circular connected mesh topological structure, each individual is interacted with its neighbor only, namely, the breeding cycle is limited to neighbors. One breeding cycle includes the following steps: choose two parents from neighbor of one individual according to definite rules, then, use parents to do genetic operator operations (mutation, crossover); lastly, replace current individuals with newly generated ones according certain principles.

A. Chromosome code and decoding

The assembly optimization process include placement sequence and feeder tank layout, in chromosome code process , considering their interaction and resolution efficiency it is required to reflect component placement sequence and feeder layout in one chromosome, the component placement sequence can be showed by component number [13-15], feeder layout can be determined by component type. For we considered component dimension and the same component being placed to many feeder tanks in actual assembly conditions, the feeder number could better show position of different type of component on feeder shelf. This paper applied two dimensions decimal segmented code method to describe relationship between every subproblem, chromosome structural diagram is showed as Fig.1.

	元件类型								元件类型					
1	6	4	1	2	3	1	5	1	1	2	4	5	1	3
1	9	8	4	7	2	5	3	6	6	2	3	5	1	4
	元件贴装顺序编码段								供料槽布置编码段					

Fig.1. Chromosome segmented code structural diagram

B. Genetic operation
1) Fitness calculation

This optimization model built here is to achieve placement sequence and feeder layout and the goal is to make PCB assembly time shortest .so it is necessary to consider PCB assembly time when calculating fitness function. The fitness evaluation function is:

$$T = \sum_{s=1}^{N}\left(\max\left(t_t(s), t_f(s), t_r(s)\right) + t_p(s)\right)$$

2) Strategy choice

Choice is made to identify recombination or mutation individuals and how many offspring would generate by chosen individuals. Tournament selection and optimal conservation strategy were used here. Banal tournament selection was adopted to choose mutation and crossover individuals from parents then used the optimal conservation strategy, individuals with greater fitness were chosen to generate next generation while maintaining the population size.

3) Crossover and mutation

The chosen individual is limited to its neighbor, which is the most critical operation in crossover process and the difference between CGA and traditional GA, here we adopted neighbor structure of Von Neumann model, each individual contains four neighbors, randomly choose one from neighbor as crossover parent in every crossover process to generate offspring by mutation. Improved order crossover was used directed at this code method so as to guarantee chromosome effectiveness, namely, guarantee placement sequence and feeder layout tank code without repetition. Divide component code and feeder tank into two segments to do crossover, generate two new offspring, they inherited part information of two parents chromosome and got recombined genes, made good crossover operation.

This paper applied two different mutation methods in the mutation process, exchange mutation and turnover mutation, various mutation methods could generate more new chromosome, to some degree, and it can avoid possibility of early getting into local optimal. With the increasing of genetic generation, the chromosome difference would become smaller, increasing the mutation possibility is good for getting rid of local optimal and obtaining global optimal, and so, self adoption was used here.

C. Algorithm achievement

The specific steps:

Step1: set CGA parameters mainly include population size, genetic count, crossover possibility and mutation possibility;

Step2: initiate the population. Generate size binary decimal segmented chromosome to initiate according to chromosome code rules and requirements.

Step3: define neighbor. Place the initiated population to two dimensional mesh topology structure, find corresponding neighbor separately for the convenient of subsequently genetic operation.

Step4: calculate fitness, evaluate function by fitness, and calculate population fitness value.

Step5: fitness comparison. Use binary tournament selection method, maintain individual with bigger fitness to do next crossover and mutation operation.

Step6: crossover operation. Use order crossover to generate new population.

Step7: mutation operation. Use exchange and turnover mutation together with self adoption mutation possibility.

Step8: conserve the optimal ones in every population after genetic operation.

Step9: repeat step4 to 8 till accomplish all set iteration; get the most optimal ones and the biggest fitness value.

IV. EXPERIMENT AND ALGORITHM ANALYSIS

A. The experimental data preparation

Use MATLAB to program; adopt GA and CGA, test 3 PCB boards' data. Test objective are 12 vertical turret SMT machine of placement head, the machine's basic performance parameters are: PCB movement speed along x, y is 60mm/s, feeder shelf movement speed is 60mm/s, a feeder tank's width is 15mm, turret rotation speed is 0.25s/30°.

To testify effectiveness and advantages, the basic parameters are set as: initiated population p size=100, genetic generation count=800, crossover possibility pc=0.6, mutation possibility pm=0.2.

The obtained results are showed in Table I after solving by GA and CGA, component quantity and type information on 3 PCB boards are recorded in Table I, run 30 times, get statistical average fitness and optimal values, and improved efficiency of CGA comparing with GA Fig.2 is result convergence got by two different algorithms, mainly include the most optimal fitness convergence and average fitness convergence.

TABLE I
ALGORITHM EFFECTIVENESS TESTIFIED RESULTS

PCB No	quantity	type	GA		CGA		Improved efficiency	
			average/s	optimal/s	average/s	optimal/s	average%	optimal%
PCB1	65	22	35.667	33.515	30.395	28.433	14.78	15.16
PCB2	71	31	38.975	37.000	32.135	29.500	17.55	20.27
PCB3	30	18	14.521	13.166	13.075	12.306	9.96	6.53

B. Results the model

It can be seen from Table I, the mounting process optimization efficiency of CGA is superior to that of GA, 3PCB boards containing different amount of component, use two algorithms, average an optimal time got by CGA are obviously less than that of GA; and with the expansion of problem scale, the resolution efficiency would increase more evidently, PCB2 has the most component quantity and type, the comparison result illustrate that the improved efficiency are 17.55%and 20.27% separately.

Fig.2 showed 3 different PCB optimization results,

CGA convergence speed is higher than that of GA, and the final results are better than GA, consequently, CGA could raise settle efficiency of mounting optimization to some extent, thereby directing practical production and increasing equipment production efficiency.

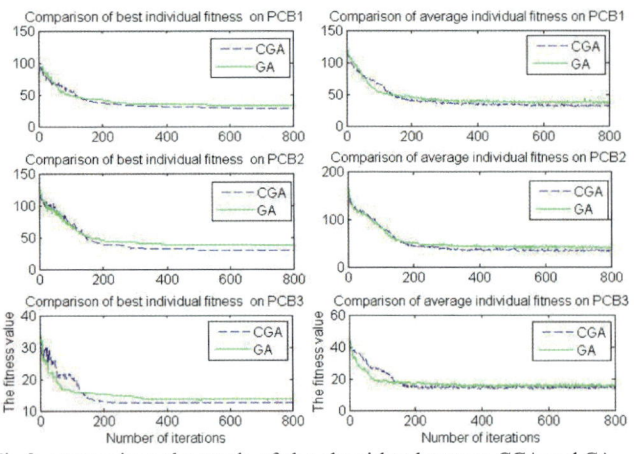

Fig.2. comparison the result of the algorithm between CGA and GA

C. Results analysis

By comparing the algorithm results from the Table I and Fig.2 it showed that at some extent the using of cellular genetic algorithm to solve this kind of problem will be better than traditional genetic algorithm. Especially in solving large-scale problems the method in this article will be more superior to other methods on performance. Why the method was superior to other methods. There were several reasons; firstly CGA is a combined intelligent evolutionary algorithm of GA and cellular automation, which inherit great performance as well as some characteristics of cellular automation [12]. It introduces "dynamic" factors to system, and has randomness, parallelism and global characteristics when handling problems especially has great robustness, convergence, high search efficiency aimed at dynamic problems. Secondly after considering assembly process fault, it has dynamics and randomness, so adopting CGA to solve problems would better present its advantages and better handle these problems.

V. CONCLUSION

This paper researched mounting process optimization of turret SMT machine, mainly included component placement sequence and feeder layout. Considering dynamic factors of equipment fault, the layout of the same type of component on many feeder tanks and influence on feeder tank layout by component, mathematic optimization model aimed at shortest time of mounting time was established based on the knowledge of mounting process. GA and CGA were used to optimize and solve, got final results. Through experiment analysis comparison, it proved CGA's advantages in dynamic optimization problems and can achieve better optimized results, further testifying its effectiveness and advantages.

REFERENCES

[1] DU Xuan, LI Zongbin, JIA Xiaochen. Component Placement Process Optimization for Dual-Gantry Placement Machine Based on Hybrid Genetic Algorithm [J]. Journal of Xi'an Jiaotong University, 2009, 43(5): 80-84.

[2] ZENU Youjiao, JIN Ye. Component placement sequence optimization for surface mounting machine based on genetical algorithm [J]. Computer Intergrated Manufacturing Systems, 2004, 10(2): 205-208.

[3] LUO Jaxiang, LUO Shuhao, WU Xinsheng. RLS based on tabu search algorithm for SMT placement sequence optimization [J]. Journal of South China University of Technology (Natural Science Edition), 2012, 40(3): 74-80.

[4] WANG Jun, LUO Jiaxiang, HU Yueming. Mount Process Optimization for Chip Mounter Based on Improved Ant Colony Algorithm [J]. Computer Engineering, 2011, 37(14): 256-258.

[5] CHEN Tiemei, LUO Jiaxiang, Hu Yueming. Mounting sequence optimization on surface mounting machine using ant-colony algorithm and shuffled frog-leaping algorithm [J]. Journal of Control Theory & Applications, 2011, 28(12): 1813-1820.

[6] DU Xuan, LI Zongbin, CAO Xinqin, et al. Component placement process optimization for chip shooter machine based on genetic algorithm [J]. Journal of Xi'an Jiaotong University, 2008, 42 (3): 295-299.

[7] WANG Jun, LUO Jiaxiang, HU Yueming. A hybrid algorithms for optimization of pick-and-place equipment [J]. Computer Measurement & Control, 2011, 19(3): 603-605 (in Chinese).

[8] LU Junying, ZHU Guangyu. Surface mounting multi-objective optimization based on grey entropy correlation analysis [J]. Computer Integrated Manufacturing Systems, 2013, 19(4): 766-772.

[10] ZHANG Yi, WAN Xinyu, ZHENG Xiaodong ZHAN Teng. Machine tool spindle design based on improved cellular multi-objective genetic algorithm [J]. Computer Engineering and Applications, 2013, 07, 31.

[11] ZHANG Yi, LU Chao, ZHANG Hu, FANG Zifan. Workshop layout optimization based on differential cellular multi-objective genetic algorithm [J]. Computer Integrated Manufacturing Systems, 2013, 19(4): 727-734.

[12] LI Junhua, LI Ming. Convergence Analysis of Cellular and Convergence Rate Estimate Genetic Algorithms [J]. PR & AI, 2012, 25(5): 874-878.

[13] CORNE D W. The Pareto envelope based selection algorithm for multi-objective optimization [J]. Lecture Notes in Computer Science, 2000, 1917: 839-848.

[14] PAVLOV V V. Polychromatic sets and graphs for CALS in machine building [M]. Moscow, Russia: Moscow Stankin Press, 2002.

[15] Liu Haimingl, Yuan Pengl, Luo Jiaxiang, Hu Yueming. Optimization, A lgorithm for Workload Balancing in Surface Mounting Lines [C]. Proceedings of the 32nd Chinese Control Conference. July 26-28, 2013, Xi'an, China.

Automatic Detection System of Surface Defects on Metal Film Resistors Based on Machine Vision

Jia-wei KE*, Yao-guang HU, Jing-qian WEN, Lin-wei MAO

School of Mechanical Engineering, Beijing Institute of Technology, Beijing, China

(kejiawei55@163.com)

Abstract - **An automatic detection system of surface defects based on machine vision was developed. Metal film resistors were chosen as model materials. A CCD camera captures three images while the metal film resistor rotate once to get its full surface image. Using histogram equalization and median filter, the image contrast was improved and contours were smoothed. Global threshold segmentation distinguished the band from the background. Then the anchor point was found in the upper-left corner of the resistor. Band areas were extracted according to the anchor point. In situations where band areas were found, proposed algorithm recognized defects in terms of the number of band contours and the times from black to white in each pixel-width line. Depending on the above results, PLC controls the motion of the relay, implementing the classification of qualified resistors and defective resistors. The system was tested to detect defects in different surroundings and showed a success rate of more than 95%. Future work is being done to design an automatic feeding mechanism to improve feeding speed.**

Keywords - **Image processing, machine vision, metal film resistors, surface defects**

I. INTRODUCTION

The inspection of bands on resistors is an important index in judging whether a resistor is qualified or not. The identification of bands is directly related to the inspection and replacement of resistors. The traditional visual check by human eyes was labor-intensive and inefficient. Besides, it was influenced by workers and some defects can't be identified correctly. Because of the small size of resistors, visual fatigue may easily occur after long working hours, which increases unreliable factors to inspection accuracy. Therefore, applying surface defect inspecting system to replace visual check can make up for the deficiency of the traditional inspection method.

Machine vision system (MVS) is an image processing system which transforms destination object into picture intelligence and conveys it to an appropriative image processing system [1]. MVS deals with the signals to get the target characteristic and then control the motion of custom equipment based on the judging results. Machine vision technique is widely used in industrial fields like automatic inspection, nondestructive examination and precision metrology [2]. MVS aims at using the machine in place of the human eye to make judgments. System structure of machine vision is shown in Fig.1 [3].

In foreign countries, machine vision is widely used on printed circuit board (PCB), the classification and identification of agricultural products and optical measuring of strip width and edge inspection in stainless steel. An annealing and pickling line has been developed

by Ricci, M et al, the system contextually exploits a magnetic imaging system designed and realized for the specific application [4].

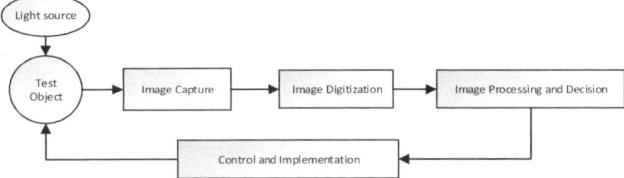

Fig.1. System structure of machine vision

The domestic study on machine vision has been paid great attention since 1990s. Compared to foreign countries, although there was some development in machine vision, domestic research is still in the stage of low end application. Chen Yong put forward and designed an automatic real-time inspection system which can identify the surface defect of metal work piece without surface damage [5].

Currently, MVS is mainly used to identify defects on the plane, while research on the surface of small cylindrical products is relatively little [6]. In this paper, maximum diameter of RJ resistors is only 2.5mm. Because it is a cylinder, a camera can't capture the full surface at one time. Therefore, the key technique is how to collect complete surface and accurate identification of surface defects.

II. MATERIALS AND METHODS

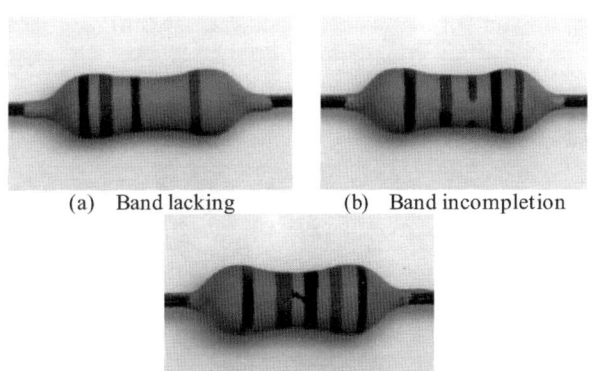

(a) Band lacking   (b) Band incompletion

(c) Band merging

Fig.2. Surface defects of the metal film resistor

The metal film resistors used in this paper are military products, which require being checked one by one rather than sampling inspection. Surface defects of metal film resistors focused on the band, including three types of defects: band lacking, band incompletion and band merging (Fig.2 (a)-(c) respectively).

The key techniques of the machine vision system are illumination system, image acquisition system, image preprocessing, image segmentation and image recognition.

E. Qi et al. (eds.), *Proceedings of the 21st International Conference on Industrial Engineering and Engineering Management 2014*, Proceedings of the International Conference on Industrial Engineering and Engineering Management, DOI 10.2991/978-94-6239-102-4_84, © Atlantis Press and the authors 2015

These key techniques used in this paper are described as follows.

A. Illumination System

The performance of the illumination system is essential to the machine vision system. It not only plays the role of illumination. what's more important is to produce a significant difference between the detection areas and background and avoid reflection [7, 8]. Thus, a respectable illumination system will reduce the difficulty of processing images and increases the precision of inspection. Ring-shaped LED light is used as light source in this illumination system. LED lights are usually adopted as the illumination source in MVS for its low power consumption, long life and excellent optical properties. Ring-shaped lights are conducive to the research object to get uniform distribution of light [9]. As the detected surface of the metal film resistor is irregular, reflection is inevitable. Therefore, how to eliminate reflection is the main point in this illumination system. The solution is that adding a polarizer on the lens and the light source respectively, original figure and the figure with polarizer are shown in Fig.3.

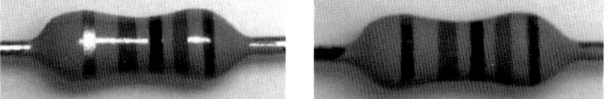

(a) Original figure   (b) Figure with polarizer
Fig.3. Original figure and the figure with polarizer

B. CCD Camera

The key parameters of the CCD camera are resolution, pixel depth and shutter speed. Resolution means the amount of picture elements contained within a unit length. Generally speaking, the higher resolution, the more picture elements contained, so the image will be clearer [10]. But it will take up more time for image processing in return. Pixel depth means bits used for saving each pixel, Open CV library function requests 8 bit image. And shutter speed determines how many pictures can be taken per second. It can be adjusted according to the moving speed of the target object [11].

The acA640-100gm digital camera from Basler was chosen as the image sensor in this system. Its resolution is 659-by-494 and the fastest frame rate reach 100 fps, which meets the requested speed of inspection. The digital camera can acquire the digital signals of images directly without image acquisition card. The image captured by the digital camera will be sent to an industrial computer by Gigabit Ethernet.

To capture high quality images, an industrial lens matched with the camera is also requested. Focus which determines the field of view is the key parameter of the lens. There exist two types of lens: the zoom lens and the prime lens. Compared with the zoom lens, the best advantage of the prime lens is the high speed of focusing and the stable quality of the images [12]. In addition, the distance between the lens and the metal film resistor is fixed. So the prime lens is a better choice for this system. Then the focus of the lens is given by Eq. (1) as follows:

$$f = wD/W \ or \ f = hD/H \qquad (1)$$

Where f is the focus, D is the distance between the focus and the metal film resistor, w is the width of the CCD, W is the width of the metal film resistor, h is the height of the CCD, H is the height of the metal film resistor.

According to the given condition, the size of the CCD is 1/4 inch (3.2*2.4mm). The distance between the lens and the metal film resistor is 250mm. The max length of the captured field is about 26mm. The focus of the lens calculated by Eq. (1), $f = wD/W$=3.2x250/26≈30.8mm. So a 25mm prime lens was chosen as the lens in the CCD camera.

C. Image Acquisition

As the metal film resistor is cylindrical, one camera can't capture the full surface at one time. To solve this problem, as shown in Fig.4. Plate on the chain pushes the metal film resistor moving on the track. The metal film resistor goes around the wire at this time [13]. When the metal film resistor comes to the slope, the metal film resistor goes around its body. And when the metal film resistor reach sensor under CCD, the CCD starts to capture three images continuously with the resistor one revolution.

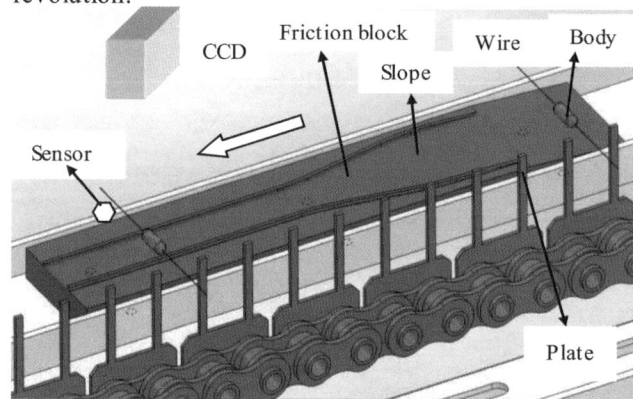

Fig.4. Image acquisition method

D. Image Processing and Image Recognition

The flow diagram of image processing and image recognition is shown in Fig.5 [14, 15]. Firstly, preprocessing aims at enhancing the image contrast and signal-noise ratio, histogram equalization and median filter were used in this study for image preprocessing, the results are shown in Fig.6 (b). Image segmentation is the technology that segments the image into several areas. As the image captured is the gray and white image and the key issue is to distinguish bands and the other areas, global threshold segmentation is suitable enough. And the effect image is shown in Fig.6 (c). This system was designed to detect the surface defect of bands of the metal film resistor. Image recognition aims at extracting the band area. The key issue is to find the upper-left corner (point C) of the band, and then intercept the band area with a fixed size rectangle according to the upper-left corner, as shown in Fig.6 (d).

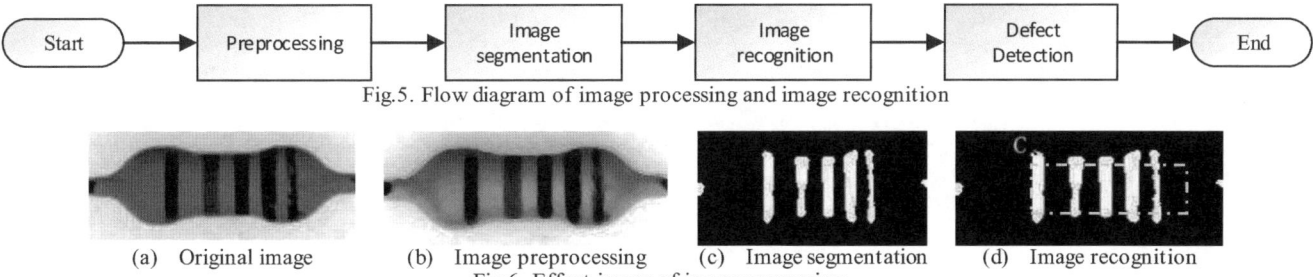

Fig.5. Flow diagram of image processing and image recognition

(a) Original image (b) Image preprocessing (c) Image segmentation (d) Image recognition

Fig.6. Effect image of image processing

Lastly and most importantly, defect detection determines the specific type of defects. As mentioned above, band lacking, band incompletion and band merging are three main defects. In this study, the number of band contour (white areas in Fig.7, expressed in N) and the times from black to white in each pixel-width line (expressed in T) are two important indicators. Fig.7 (a) shows that the qualified resistor has 5 band contours and 5 times from black to white in each pixel-width line. When N and T are both less than 5, it represents band lacking, as shown in Fig.7 (b). When N is greater than or equal 5 and T is less than 5, band incompletion occurred in Fig.7 (c). When N is less than 5 and T is greater than 5, band merging, as shown in Fig.7 (d).

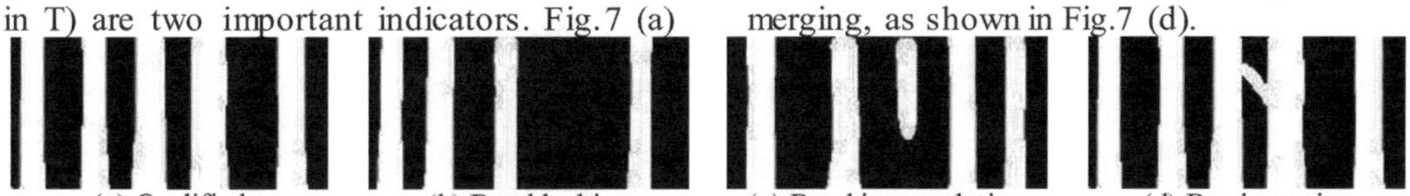

(a) Qualified (b) Band lacking (c) Band incompletion (d) Band merging

Fig.7. Chart of defect detection

III. RESULTS AND DISCUSSION

The linear speed of the chain is 22.2mm/s. As the maximum diameter of the metal film resistor is 2.5mm, it moves forwards at 7.85mm for the metal film resistor rotate once. In addition, the CCD camera captures 3 images while the metal film resistor rotate once. Thus, the shutter speed needed for this system is 118ms. The image captured by the CCD camera is sent to the computer for analysis. The experiment result showed that the detection speed is approximately one second and the accuracy of defect detection is 95.074%, which meets the requirements of a real-time detection system. Then judgment results are transmitted to the PLC control unit. The PLC controls the motion of the relay, open or close depended on judgment results. Thus, the resistors are separated into different containers, achieving detecting and sorting the metal film resistors automatically. System structure schematic is shown in Fig.8.

Detection of Multiple Resistors:

The algorithm above achieved the detection of one resistor. In order to improve detection speed, adding the number of resistors in a single acquisition is a good choice. Because of the limitation of the lens, the acquisition area is limited. With the 25mm prime lens, three resistors are the limit, as shown in Fig.9. Although smaller focus has a bigger scope, the image of resistor may distort. Compared with the above algorithm, the only difference is that the image including three resistors should be separated into three parts. There are three images for each resistor as the same as the previous method. Defects detection method is also the same as the algorithm for one resistor. Our algorithm is simple and practical, because parameter N and T are not difficult to get quickly. So our algorithm can detect surface defects with high speed and perfect accuracy.

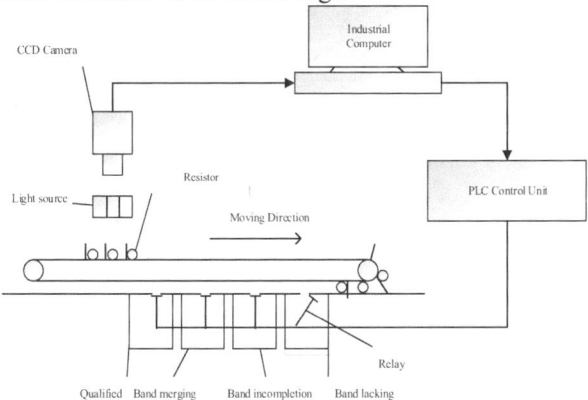

Fig.8. System structure schematic

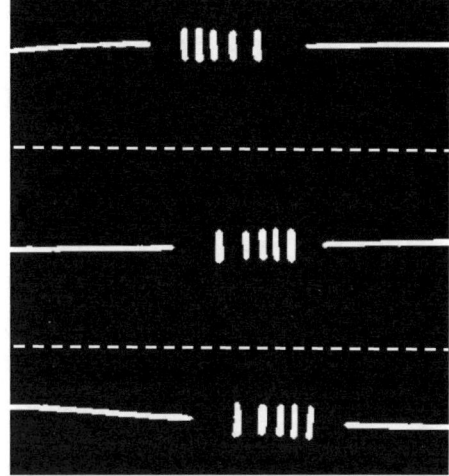

Fig.9. Multiple resistors acquisition

Our algorithm is suitable for surface defects detection of cylindrical objects such as RJK resistors. In addition, the threshold of the global threshold segmentation is capable of adjusting according to its surroundings.

IV. CONCLUSION

In this study, we implemented machine vision technique to detect the surface defects of the metal film resistor. The machine vision system can be implemented in the finished product surface inspection line of the metal film resistor. Precise positioning of the upper-left corner of the band area is the most important component. Experimental results showed that the proposed system can detect up to three resistors in a cycle, and the defect recognition is accurate. The system is fast enough to be implemented in real-time applications.

The system can be applied to cylindrical products and are efficient in capturing full surface image with one camera. This reduces cost and overcomes the drawback of previously proposed system. Our future research will focus on designing an automatic feeding mechanism to improve feeding speed.

ACKNOWLEDGMENT

This research is supported by the National 863 Program of China (Project No.2013AA040402). The authors would like to thank all members of the Industrial Engineering laboratory of Beijing Institute of Technology for discussion.

REFERENCES

[1] H. K. Mebatsion and J. Paliwal, "Machine vision based automatic separation of touching convex shaped objects," Computers in Industry, vol. 63, pp. 723-730, 2012.

[2] L. Zhou, V. Chalana, and Y. Kim, "PC-Based Machine Vision System for Real-Time Computer-Aided Potato Inspection," John Wiley & Sons, vol. 9, pp. 423-433, 1998.

[3] L. Wang and y. Shen, "Design of Machine vision applications in Detection of defects in high-speed bar copper," E-Product E-Service and E-Entertainment, vol. 10, pp. 362-365, 2010.

[4] M. Ricci, A. Ficola, M. L. Fravolini, L. Battaglini, A. Palazzi, P. Burrascano, et al., "Magnetic imaging and machine vision NDT for the on-line inspection of stainless steel strips," Measurement Science and Technology, vol. 24, p. 025401, 2013.

[5] Y. Chen, "The algorithm research and software design of the surface defect inspecting system based on machine vision," Master dissertation, Tianjin University, Tianjin, 2006.

[6] Z. Wei and Z. Li, "Surface Defects Detection System of Electronic Component Base on Image Sensor," (in Chinese), Instrument Technique and Sensor, vol. 6, pp. 53-57, 2011.

[7] B. Liu and S. Wu, "Automatic Detection Technology of Surface Defects on Plastic Products Based on Machine Vision," presented at the Mechanic Automation and Control Engineering (MACE2010), Wuhan, 2010.

[8] H. Kim, S. Kim, and J. Kim, "Mixed-color Illumination and Quick Optimum Search for Machine Vision," International Journal of Optomechatronics, vol. 7, pp. 207-221, 2013.

[9] A. Montenegro Rios, D. Sarocchi, A. L. Valdivieso, and Y. Nahmad-Molinari, "Machine Vision for Size Distribution Determination of Spherically Shaped Particles in Dense-Granular Beds, Oriented to Pelletizing Process Automation," Particulate Science and Technology, vol. 29, pp. 356-367, 2011.

[10] D. Branscomb and D. G. Beale, "Fault detection in braiding utilizing low-cost USB machine vision," Journal of the Textile Institute, vol. 102, pp. 568-581, 2011.

[11] D.-B. Perng, S.-M. Lee, and C.-C. Chou, "Automated bonding position inspection on multi-layered wire IC using machine vision," International Journal of Production Research, vol. 48, pp. 6977-7001, 2010.

[12] H. İ. Çelik, L. C. Dülger, and M. Topalbekiroğlu, "Development of a machine vision system: real-time fabric defect detection and classification with neural networks," Journal of the Textile Institute, vol. 105, pp. 575-585, 2014.

[13] Y. Sun and H.-r. Long, "Adaptive detection of weft-knitted fabric defects based on machine vision system," Journal of the Textile Institute, vol. 102, pp. 823-836, 2011.

[14] C.-C. Wang, B. C. Jiang, Y.-S. Chou, and C.-C. Chu, "Multivariate analysis-based image enhancement model for machine vision inspection," International Journal of Production Research, vol. 49, pp. 2999-3021, 2011.

[15] E. Scavino, D. A. Wahab, A. Hussain, H. Basri, and M. M. Mustafa, "Application of automated image analysis to the identification and extraction of recyclable plastic bottles," Journal of Zhejiang University SCIENCE A, vol. 10, pp. 794-799, 2009.

Ontology Model for Assembly Process Planning Knowledge

Zhicheng Huang[1], LihongQiao[1,*], NabilAnwer[2], Yihua Mo[3]

[1]Department ofIndustrial&Manufacturing SystemsEngineering, BeihangUniversity, Beijing, China
[2]LURPA, ÉNS de Cachan, 94235 France
[3]Beijing Institute of Aerospace Systems Engineering, Beijing, China
(hzc8366@126.com, *lhqiao@buaa.edu.cn,anwer@lurpa.ens-cachan.fr,pmyhp@163.com)

Abstract - **Assembly process planning is a highly knowledge-intensive work.As collaborative design and manufacturing is getting increasingly popularespecially for complex assembly products,assembly process planning knowledge model should be comprehensive, recognizable and reusable. Ontology meets therequirements as a semantic tool providing a source of sharedand precisely defined terms that can be utilized to describeboth knowledge and concepts.Many researchers have studied the ontology modeling for assembly process planningdomainand theymainly focus on the geometry information, tolerance type and manufacture environmentrespectively.This paper presents an assembly process design knowledge ontology consideringassembly requirement, spatial information, assembly operation and assembly resource. It has covered almost every important conceptrelated to assembly process planning knowledge.**

Keywords - **Assembly process planning,knowledge, ontology modeling**

I. INTRODUCTION

Assembly process planning (APP) is highly knowledge-intensive,involving assembly sequence planning, assembly path planning, resources and tools choosing, etc[1]. Modeling knowledge of assembly process planning is the foundation of decision-makingmechanism. Study onacquisition, management, retrieval, sharing and reusing of knowledgeisbecomingincreasingly important.As collaborative designand manufacturing is getting increasinglypopularespecially for complex assembly products. APP knowledge model should be comprehensive, recognized and reusable. This requests a knowledge modeling tool to be extendible and be able to make precisedefinition[2].

Ontology is a type of semantic tool providinga source of sharedand precisely defined terms that can be used to describeboth knowledge andconcepts[3]. Ontology can be expressed in standard formal languages like XML which would ensure the sharability and cooperation. These characteristics make ontology a notableknowledge modeling tool in many domains like medical science[4], digital library[5], manufacturing[6], etc. Ontology modeling for APP knowledge has always been the research focus.Fiorentini[7] et al. showed that the ontological assembly model can help in achieving various levels ofinteroperability as required to enable the full potential ofProductLifecycle Management(PLM).Kim[8] et aldesigned a collaborativeassembly design and information-sharing environmentscalled Assembly Design Browser based on assembly design ontology model.Designers are no longer merely exchanging specificgeometric data, but rather more knowledge about design andthe product development process[9]. Ontology fits all the requirements of APP knowledge modeling.

II. OVERVIEW OF RELATED RESEARCH

A.Ontology

An ontology is the representation of knowledge based on conceptualization in a formal and explicit manner[10], in another word, explicit, formal specifications of terms in thedomain and of the relations among them. The advantage of ontology is that it offers the concepts and their relations in a domain in a commonly agreed and formal expression that is machine-readable[11] and it has the reasoning capability that makes the implicit informationexplicit[12].

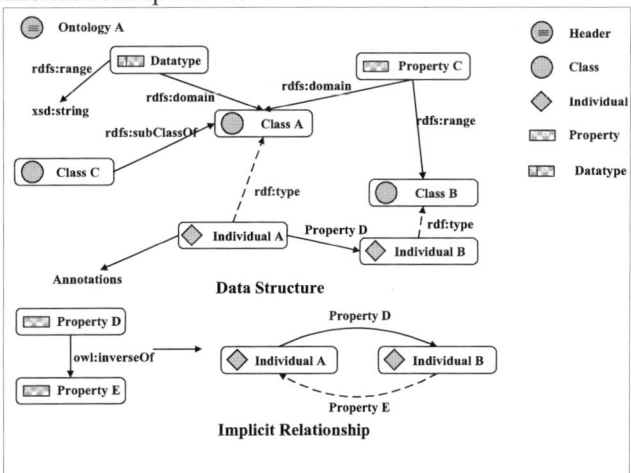

Fig.1.Sketch of ontology model

Ontology models are expressed as documents like OWL DL(WebOntology Language - Description Logic). Take the OWL document as anexample[13], each documentconsists of an ontology header, annotations, classes and propertydefinitions (more formally referred to as axioms), facts about individuals, anddatatype definitions, as Fig.1 shows.An ontology header is a resource that represents the ontology itself.Annotations are statements (triples) that have annotation properties as predicates.A class describes a set of resources that share common characteristics or are similar in some way which is used to define a concept. Individuals are instances of classes and are linked to classes via properties. A Property is a resource that is used as a predicate in statements thatdescribe individuals which canbe used to state relationships between individuals, or betweenindividuals and data values.Datatypes in OWL represent ranges of data values. OWL 2 allows you to define your own

E. Qi et al. (eds.), *Proceedings of the 21st International Conference on Industrial Engineering and Engineering Management 2014*, Proceedings of the International Conference on Industrial Engineering and Engineering Management, DOI 10.2991/978-94-6239-102-4_85, © Atlantis Press and the authors 2015

complexdatatypes that are explicitly enumerated or defined using facet restrictions(value range restrictions).

B.Ontology Modeling for Application Domain

There are some modeling methods for practical application domain based on ontology that have madesignificant progress like TOVE(Toronto Virtual Enterprise) method for engineering product and design[14], Skeletal Methodology proposed by Mike Ushold and MichealGruninger[15], KACTUS(Modeling Knowledge about Complex Technical Systems for Multiple Use) Project Method developed in Esprit Project[16], etc. These methods were summarized throughreverse engineeringin independent cases in the background of diverse domain and each of them had special emphasis. But a common methodology does not exist yetconsidering differences in the specific field and concrete engineering. There are five principles provided by Gruber [17] in 1995 generally believed to be very influential:

(1) Clarity: An ontology should effectively communicate the intended meaning of defined terms and definitions should be objective.

(2) Coherence: An ontology should be coherent: that is, it should sanction inferences that are consistent with the definitions.

(3) Extendibility: An ontology should be designed to anticipate the uses of the shared vocabulary. It should offer a conceptual foundation for a range of anticipated tasks, and the representation should be crafted so that one can extend and specialize the ontology monotonically.

(4) Minimal encoding bias: The conceptualization should be specified at the knowledge level without depending on a particular symbol-level encoding.

(5) Minimal ontological commitment: An ontology should require the minimal ontological commitment sufficient to support the intended knowledge sharing activities.

The principles for modeling a standard knowledge ontology based on those methodologies and principles mentioned could be summarized as:striving to cover all the content in the domain, describe all the concepts of data model and function model as well as the relationships, transformations and operations between the concepts in common definitions; Ensuring the correctness, normalization and simplicity of the knowledge model so that uniform knowledge base could be built which is the base of accessible and efficient data exchanging in later applications.

C. Related Research

Ontology has been used to model assembly knowledge domain knowledge in various researches already. Ontology-based researches take advantage of the capabilities to structure concepts and to connect them with part models, and make use of reasonersto set up inference rules to ensure the consistencyof the assembly description or extract informationthat is not readily available in the dataset describingan assembly, and some

researchers describe geometric features of entitiesunanimously utilizing standard specifications.

Kim[8]et al havedeveloped an assembly design (AsD) ontology to describe the specificationof assembly design. Investigated terms included Product, Assembly, AssemblyComponent, Part, Sub-assembly, Assembly Feature, FormFeature, Joint, Joint Feature, Mating Feature, etc. The definitions for assembly design termswere analyzed by Kim. For example, the definitionof an assembly feature in engineering design was "a group ofassembly information", which included form features, jointfeatures, mating relations, assembly/joining relations, spatialrelationships, material, engineering constraints, etc. The AsD ontology model was built based on these concepts. In themodel, six classes of assembly design concept were defined as shown in Fig.2: Material, Product, Feature, Spatial Relationship, Manufacturing and Degrees of Freedom. The concept of Product included Part and Assembly, the Feature concept included Feature for Part and Feature for Assembly, the Manufacturing was designed to include Manufacturing Process and Joining Process.The assembly design ontology model wasdesigned as shown in Fig.2 after further classification and subdivision.

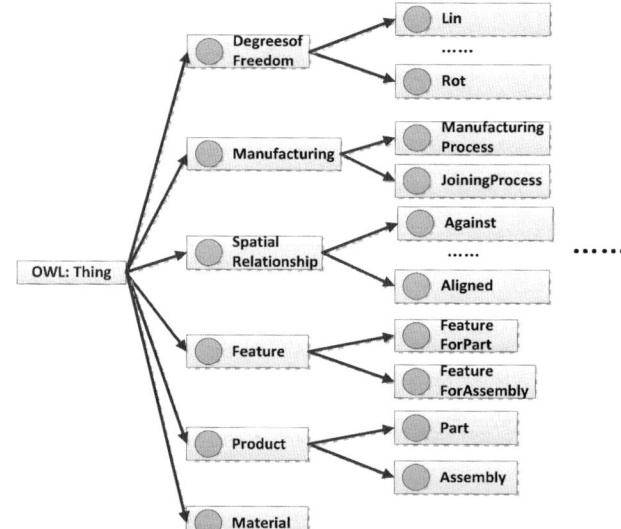

Fig.2.Part ofAsD ontology class hierarchy

The AsD model alsoincludedproperties that represent classcharacteristics, domains, and ranges that represent classrelationships as well asinference rules to query AsD information selectively like the assembly relation and spatial relation of chosen parts as well.This AsD formalism was also used in Assembly Design Browser-a collaborativeassembly information-sharing environments because of thesharing and reusability of ontology.

However, many important concepts were still left implicit or not defined in this assembly process ontology model, implied assembly constraints,and tolerance cannot be easily obtained and expressed. While Krima'swork[18, 19]focused on the geometry data informational in ontology model.

Krima and Barbau[18, 19] et al. proposed a way to enable the exchange of product data through a product lifecycle based on ontology betweendifferent designers and technologists. The ontology model is called OntoSTEP in their work using OWL-DL (WebOntology Language - Description Logic) to describe the Standard for Exchange of Product modeldata (STEP) (ISO 1030)[20].STEPmainly focuses on product management data andgeometry informationstill evolving to meetthe needs of modern Computer-Aided Design (CAD), Computer-Aided Engineering (CAE), and Product Data Management (PDM)systems.The STEP APs are defined using the EXPRESS (ISO 10303-11)[21] languagewhich is developed to enhanceproduct modeling and provide support to describe "the information required for designing, building,and maintaining products." The concept of entity in EXPRESS is similar to the concept of a class in object-orientedmodeling. After translating the main concepts in EXPRESS into ontology, three entities were described: product, productcategory and product related category. These entities and instances of entities mapped respectively to OWL classes andindividuals, detailed geometry data informational could then be described in the ontology model. In order to get better descriptive power, Krima built additional concepts in the model like Data Type, Aggregations, Select, Enumeration, Abstraction, Inheritance and Uniqueness Clauses to define the non-geometry information describe in EXPRESS.

In a given example by Barbau and Krima[18]et al., STEP AP203 was used to create a 3D CAD model of aproduct, while CPM(Core Product Model) and OAM(Open Assembly Model)were used to represent the functionaldecomposition of this product and the relationships between theparts. After geometric definition, integration of geometry and non-geometry information, the model designed could be stored in ontology and queriesabout retrieving the parts that areconnected to a particular part via a fixed connection could runsuccessfully.

Their workcovers a larger descriptionof a product with an ontological description incorporatingthe geometry levels, structure levels and even function levels in a unanimous approach by using international standard. But the authors showed that not all concepts of STEPcould be rigorously defined which would lead to limitations in detecting inconsistencies because of the compatibility.Thepracticalmanufacture environment information like assembly process information and resource information is not considered enough in their work as well as the tolerance.Related researchbyZhong[19] et al.focused on the assembly tolerance whichshowed that assembly tolerance types could be automatically generated based on ontology.

Zhong[22] et al.constructed an extended assembly tolerance representation model by introducing a spatial relation layer aiming at reducing the uncertainty and supporting thesemantic interoperabilityin assembly tolerance specification design. The assembly tolerance representational ontology model consisted of three layers: part layer, assembly feature surface layer and spatial relation layer.The classstructure in assembly tolerance representational ontology model is showninFig.3.

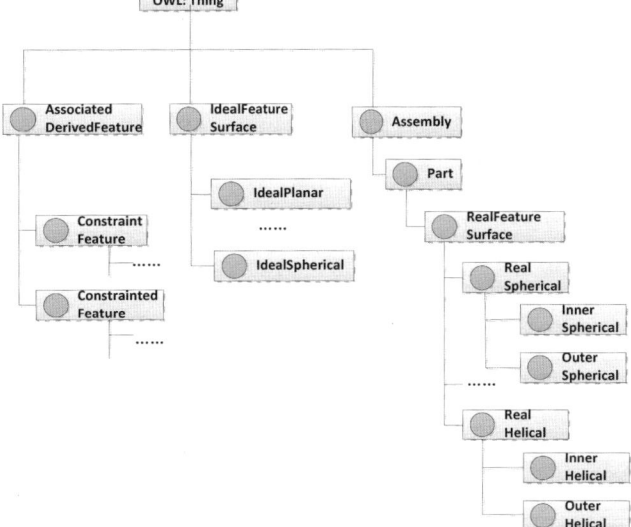

Fig.3.Classes in the ontology model for assembly tolerance representations

Assembly tolerances mainly included geometrical tolerance, angletolerance, and linear dimensional tolerance. Geometrical and dimensional tolerances of parts can be determinedby the assembly tolerances of the corresponding product. The authors classified the tolerance types according to the number of associated datums. Tolerances not associated with datums but instead associated with idealshapeswere called form tolerance, including straightness, flatness,roundness, cylindricity, profile any line, and profile any surfacetolerances.Tolerances needing one or more datumsto control the scope of theirchangeswere called positional tolerances, including circularrun-out, total run-out, parallelism, perpendicularity, angularity,position, concentricity (coaxiality), symmetry, angle, and linear dimensionaltolerances.Following consideration of the functional tolerances, the ontology model also coveredspatial relations.These were defined to be the Object propertiesused to connect certain geometry elements. Then this model could describe assembly tolerance explicitly. With this ontology model and SWRL rules defined, assembly tolerance types could be automatically deduced.

Lemaignanand Siadat[23] et al. presented an ontology model of manufacturing domain called MASON(MAnufacturing's Semantics ONtology) aimed to draft a common semantic net in manufacturing domain.MASON emphasizes data formalization and sharingparticularly in an open manufacturing environment. Three kinds of classes were defined in this work: (1) Entities, are the common helper concepts which provide the concepts to specify the products. (2)Operations, relate to process description which cover all processes linked to manufacturing in a wide acceptation. (3) Resources, stand for the whole set of manufacturing linked resource.They also defined properties to connect the conceptsconsidering the real assembly process planning works. It has already been used in Automatic Cost Estimation research as well as Multiagent module for manufacturing system. This work provided shared and precisely defined knowledge and concepts asshownin Fig.4.

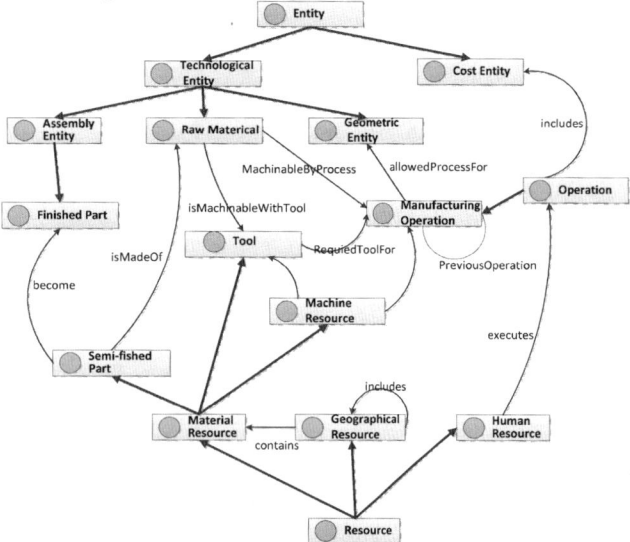

Fig.4.Overview of the ontology's main classes and object properties[23]

Lemaignan's workpresented a different view to define the domain of assembly which focuses on the practicalmanufacturingcondition.

III. AN ONTOLOGY MODEL FOR APP KNOWLEDGE REPRESENTATION

This paper aims at proposing an ontology model for APP knowledge in order to ensureconsistency of interface definitions among differentdesigners and technologists in a collaborative design and manufacturing environment[24]. Based on the principle of comprehensiveness and reusability, this paper subdivides the content and necessary information of APP knowledge into assembly requirement, spatial information, assembly operation and assembly resource, these concepts need to beaccuratelydefined in a common way.The APP knowledge ontology model is as shown in Fig.5.

The assembly requirement is the critical information designers attempt to present. It contains assembly structure, geometry entity, assembly constraint and tolerance. In the ontology model, the assembly structure is used to describe structural relationships within a product based on the concepts of part, component and product as well as the relationships between these concepts. Geometry entity requires to be defined unambiguously as the carrier of assembly constraint and tolerance, this work could refer to the Onto STEP[20] which could define geometrical elements in the use of international standard. Assembly constraint and tolerance can then be defined as distance and angle requirement between two certain geometrical elements. This would greatly enhance the ability to describe tolerance and other geometry information.

Spatial information presents the location, pose and movement of objects in 3D space which is closely connected to the assembly path planning and assembly sequence planning. The location and pose of objects could be completely defined by the absolute coordinate and the relative coordinate fixed on the part if deformation is not considered. Then the spatial motion could be resolved into the Translation of original point and the revolution around the axes of relative coordinate. These definitions could help Assembly Path Planning in the future.

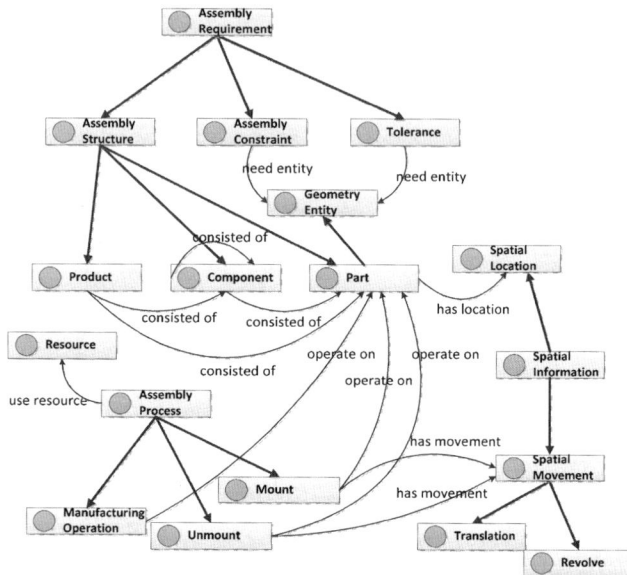

Fig.5.Main classes and object properties of APP knowledge model

Assembly operations are the basic elements of assembly process. These concepts defined with standard terms are linked to concepts of available resource aiming to describe the assembly capacity of a certain manufacture environment. This part should cover all the operations in assembly processes and the resources needed. Furthermore, the ability of certain operations and resources could be appended to this APP knowledge ontology model making it possible for automatic assembly process planning.

The ontologymodelhas covered important concepts of assembly process knowledge, and it could clearly express geometry and non-geometry information including tolerance, assembly structure, assembly process and resources that are not considered integrated before.A lot of work remain needed to finish this model and to realize its application. But it is worthy to be developed.

IV.CONCLUSION

Assembly process planning knowledge modeling is getting critically important.As this paper shows, ontology modeling for assembly domain has achieved many significant results. Most researchers focus ontolerance,practicalmanufacturing condition or spatial motion adequate consideration in their work. This paper presents an assembly process design knowledge ontology model covering all the important concepts related to assembly process planning likeassembly requirement, spatial information, assembly operation and assembly resource.A lot of work remain needed to be completed, but this model would greatly push the development of the assembly process planning automatically.

ACKNOWLEDGMENT

This work is supported by research project (A0320131002) and projecton digital assembly process planning. The authors would also like to thank Beijing Municipal Education Commission (Build a Project) for its support.

REFERENCES

[1] Y.X. Zhu, L.H.Qiao, N. Anwer. Geometrical simulation framework of assembly process based on ontology.*Computer Integrated Manufacturing Systems,* 2013. 19(5), 972-980.

[2] F.Boussuge, J. C. Léon,S.Hahmann, et al. An analysis of DMU transformation requirements for structural assembly simulations[C]//*Int. Conf. ECT.* 2012.

[3] R. Xu, X. Dai, et al. "Research on the construction method of emergency plan ontology based-on owl."*Proceedings of the 2009 International Symposium on Web Information Systems and Applications, Nanchang, China.* 2009.

[4] S. Liaw, et al. "Towards an ontology for data quality in integrated chronic disease management: a realist review of the literature."*International journal of medical informatics*82.1 (2013): 10-24.

[5] S.Dasgupta, A.Bagchi. Controlling Access to a Digital Library Ontology-A Graph Transformation Approach [J]. *International Journal of Next-Generation Computing*, 2014, 5(1).

[6] C.Li.Ontology-Driven Semantic Annotations for Multiple Engineering Viewpoints in Computer Aided Design. *Diss. University of Bath*, 2012.

[7] X.Fiorentini, I Gambino, V Liang, et al. *An ontology for assembly representation* [J]. 2007.

[8] K. Y.Kim, D. G. Manley., &H. Yang. Ontology-based assembly design and information sharing for collaborative product development.*Computer-Aided Design*, 2006. 38(12), 1233-1250

[9] L.Qiao, S. Kao,& Y. Zhang, Manufacturing process modelling using process specification language.*The International Journal of Advanced Manufacturing Technology*,2011. 55(5-8), 549-563.

[10] T.R.Gruber. A translation approach to portable ontology specifications [J]. *Knowledge Acquisition*, 1993, 5(2): 199–220

[11] D. Fensel, M. L. Brodie. Ontologies: A Silver Bullet for Knowledge Management and Electronic Commerce [M]. *2nd edition. Springer*, 2003

[12] Y. Kitamura, M.Kashiwase., M. Fuse, &R. Mizoguchi. Deployment of an ontological framework of functional design knowledge.*Advanced Engineering Informatics*, 2004. *18*(2), 115-127.

[13] J. Hebeler, M. Fisher, R.Blace, & A. Perez-Lopez. *Semantic web programming*. John Wiley & Sons. 2011. pp104-pp106.

[14] M. S. Fox, M.Gruninger. Enterprise modeling [J]. *AI magazine*, 1998, 19(3): 109.

[15] M.Uschold, & M. Gruninger. Ontologies: Principles, methods and applications.*The knowledge engineering review.* 1996. 11(02), 93-136.

[16] G. Schreiber, B. Wielinga, W. Jansweijer. The KACTUS view on the 'O'word[C]//IJCAI workshop on basic ontological issues in knowledge sharing. Springer, 1995: 159-168.

[17] T. R.Gruber. Toward principles for the design of Ontologies used for knowledge sharing [J], *International Journal of Human-Computer Studies* 1995, 43(5-6):907-928.

[18] R. Barbau, S. Krima, S. Rachuri, A. Narayanan, X.Fiorentini, S.Foufou, & R. D. Srira. OntoSTEP: Enriching product model data using ontologies.*Computer-Aided Design*, 2012. 44(6), 575-590.

[19] S. Krima, R. Barbau, X Fiorentini, et al. Ontostep: Owl-dl ontology for step[J]. *National Institute of Standards and Technology, NISTIR*, 2009, 7561.

[20] STEP, ISO. "10303."*Standard for the Exchange of Product Model Data.*

[21] ISO, TC. "184/SC 4, ISO 10303-11: 1994 Industrial automation systems and integration-Product data representation and exchange-Part 11: Description methods: The EXPRESS language reference manual."*International Organization for Standardization* (1994).

[22] Y. Zhong, Y. Qin, M. Huang, et al. Automatically generating assembly tolerance types with an ontology-based approach[J]. *Computer-Aided Design*, 2013, 45(11): 1253-1275.

[23] S. Lemaignan, A.Siadat, J. Y.Dantan, &A. Semenenko, MASON: A proposal for an ontology of manufacturing domain. *InDistributed Intelligent Systems: Collective Intelligence and Its Applications, 2006. DIS 2006. IEEE Workshop on* (pp. 195-200). IEEE.

[24] L. H. Qiao& Y.X. Zhu. A Multiple View Assembly Process Representation Based on Three Dimensional Models.*Advanced Materials Research*,2011. *189*, 1625-1630.

Energy Consumption Forecast Model of Urban Residential Buildings in Hainan

Li-ping Fu[1], Xiao-gang Li[2], Yuan Liu[1,*]

[1]Public Resources Management Research Center, Tianjin University, Tianjin, China
[2]HNSP Electric Power CO., LTD, Hainan, China
(kingtian12345@163.com)

Abstract - **To study energy consumption influencing factors of urban residential building in Hainan, on the basis of questionnaires which inquire the actual energy consumption of the target residents, and by means of correlation analysis and multiple regression analysis through SPSS software, an empirical study shows that the resident population, per capita building area, the sum number of air-conditioner and computer and the air conditioning cooling method have impacts on energy consumption of influencing factors the multiple linear regression model was drown. The fitting effect was tested with calendar year lystatistics of actual energy consumption; the results show that the model has high accuracy, and energy saving strategies of residential building are given out according to the forecast model.**

Keywords - **Building energy consumption, building energy saving, forecast model, residential building**

I. INTRODUCTION

Building energy consumption accounts for one-third of the societal energy consumption in China, and has a great energy-saving potential [1]. In recent years, with the rapid development of urbanization, economic development, people's incomes and living standards, Sino building energy consumption has increased dramatically. From 1996 to 2006 building energy consumption increased from 0.243 billion tce (ton of standard coal equivalent) to 0.563 billion tce, which represents a 1.3-foldincrease. In 2006, urban residential building energy consumption was 0.255 billion tce, or 45 percent of the total building energy consumption. That is to say, residential building energy consumption accounts for a large proportion of the total building energy consumption [1, 2]. While, residential building energy consumption is affected by lots of factors from home and abroad, which has some charicteristics, such as time-variance, complexity, randomness, regionalism. Due to the different environments and conditions, the residential building energy consumption is various from area to area.

Since China's residential building energy consumption research for the industrial city, no for the tourist city. Therefore, for the special environment of Hainan, to study of changes in overall building energy, however, is more important to decide the whole energy saving direction. In this paper, the resident population, per capita floor area, the sum number of air-conditioner and computer and the air conditioning cooling method, are discussed to analyze the factors and the extent to which they influence the growth of energy consumption by residential building in tourist city's rapid urbanization process.

Treating the residential building energy consumption as a relatively independent statistical object is good for systematically accumulating the underlying data for residential building energy consumption and understanding the basic conditions of residential building energy consumption. It can also provide forceful data supporting energy mix readjustment and energy policy formulation [3]. At the same time, civil building energy systems are closely related to national or regional energy and environmental policies, and an energy demand model is also the foundation of making strategy and plans for the entire industry development [4]. Therefore it is necessary for us to understand the characteristics of each area of residential building energy consumption to provide the correct guidelines for building energy efficiency by analysis and study of residential building energy demands.

Residential building energy consumption is affected by both domestic and international factors, and has households, building cases, ownership of appliances, thermal environment requirements [5]. According to a specific area expanded by sample surveys. However, the study sample of the residential building energy consumption in Hainan province is too small to involve annual energy consumption and establish a high energy reliability prediction model. Therefore, on the basis of expanding the sample of energy consumption, we research the annual influencing factors of Hainan residential building energy consumption and establish a residential building energy consumption prediction model with analysis [6-11]. This has important implications for Hainan province to develop energy planning and energy policy. Meanwhile, this paper provides a right direction for the research and practice of residential building energy-saving. At the same time, there are great practical significant for the global work of energy-saving and consumption-reducing.

II. QUESTIONNAIRE DESIGN AND DATA PROCESSING

The required objective data are taken from research data and the Statistical Yearbook; the required subjective data are taken from 2014, According to the characteristics and population distribution of Hainan residential buildings, select 1000 household surveys from Haikou, Sanya, Boao, Qionghai and Wenchang, get 729 valid questionnaires. Subjective investigation and objective consumption by one match. We use SPSS 16.0 software to analyze the correlation between annual residential building energy consumption and these variables.

E. Qi et al. (eds.), *Proceedings of the 21st International Conference on Industrial Engineering and Engineering Management 2014*, Proceedings of the International Conference on Industrial Engineering and Engineering Management, DOI 10.2991/978-94-6239-102-4_86, © Atlantis Press and the authors 2015

Subjective questionnaire design considerations of user information, the basic situation of the building, lifestyle, energy awareness, ownership of household appliances, therefore, it has a high degree of reliability and validity. User information including home resident population, age distribution and the total household income; the basic situation of buildings, including building area, age of the building, building orientation, building type, residential floors; lifestyle including the use of air conditioning in the meantime, air temperature setting, the indoor temperature and the overall feeling of comfort, winter heating mode; Awareness of energy including whether to use saving lamps; Ownership of household appliances including the number of computers, air-conditioner, refrigerators, heaters, electric water heaters, electric rice cookers.

III. EMPIRICAL ANALYSIS

Correlation analysis is to study whether there is some kind of interdependence between phenomena, and it specially investigates the phenomenon of dependency direction and its associated degree of correlation; also it develops a statistical method correlation between random variables. Simple correlation analysis between two variables is to study the degree of linear correlation between them and it is a common statistical method with a set of appropriate statistical indicators. Correlation coefficient is the representative indicators to measure the degree of linear correlation and dependency direction between variables. Its characteristics can be summarized as: first, the two variables in a correlation analysis are equal, regardless of the independent variable and dependent variable with only one correlation coefficient; second, there are both positive and negative correlation coefficients, the sign reflects the direction of correlation: positive sign reflects the positive correlation and the negative correlation with the minus; third, the two variables for calculating a correlation coefficient are random ones. The most commonly used Pearson correlation coefficient expression is as follows:

X_3 Where $\gamma_{xy} \in [-1,1]$; $\overline{X}$ and $\overline{Y}$ are respectively the mean value of X and Y; X_i and Y_i are respectively the ith observables of X and Y.

Partial correlation analysis refers to that it analyzes the degree of linear correlation of the two of several variables and calculates the partial correlation coefficients under the condition of controlling of the influence from other variables. Partial correlation analysis, also known as net correlation analysis, which analyzes the linear correlation between the two variables, parameters used for the net under the control of the linear correlation coefficient affect from other variables conditions.

Assuming X_1, X_2, X_3 three variables, eliminating the influence of variable X_3, with a partial correlation

coefficients between the variables X_1 and X_2, the partial correlation coefficient denoted $r_{12,3}$, variable 3 is called the control variables, Calculated as:

$$r_{12,3} = \frac{r_{12} - r_{13}r_{23}}{\sqrt{1-r_{13}^2}\sqrt{1-r_{23}^2}}$$

A. Correlation analysis of influencing factors

In residential buildings, building energy consumption mainly including air conditioning, lighting, appliances, cooking and hot water terminal energy consumption. Many factors affect building energy consumption, usually involving buildings situation, householder situation, the ownership of household equipment and lifestyle with energy saving concepts. The survey from five aspects, and accordingly design a survey indicators of annual household income, building area, building age, residential floors, building type, the resident population and so on. In order to analyze whether there is correlation between these variables and energy consumption and their related degree, for each variable a simple correlation analysis.

B. Simple correlation analysis

We use SPSS 16.0 software to analyze the correlation between annual residential building energy consumption and these variables, the result shown in Table I. As shown in Table I: although the separate two correlation coefficients of annual energy and the number of household equipment are different, all appliances usage and intensity of use are basically related to the annual energy; and our analysis results are more consistent with the actual results. Meanwhile according to the significance level p, such seven variables have a significant linear correlation with annual energy consumption: the resident population, the per capita building area, building type, energy-saving and insulation measures the building used, summer air conditioning cooling mode, the total number of and the daily use time of air conditioning and computer [12-13].

Simple correlation analysis can't control the influence of other variables, sometimes cannot accurately reflect the relationship between things, but can be partial correlation analysis. Therefore, it is necessary to determine by partial correlation analysis of the annual energy consumption of the simple-depth analysis of relevant variables. Do partial correlation analysis for the resident population, the per capita building area, building type, energy-saving and insulation measures the building used, summer air conditioning cooling mode, the total number of and the daily use time of air conditioning and computer, total energy consumption of each variable and the partial correlation analysis results was shown in Table II.

TABLE I
CORRELATION ANALYSIS ON ENERGY CONSUMPTION AND THE VARIABLES

Project	Variable	Correlation coefficient	Significant level P	Number of samples
Income Householder situation	Total household income	0.054	0.152	729
	Resident population	0.406**	0	729
Buildings situation	Building's era	-0.076	0.063	729
	Building area	0.678**	0	729
	Building orientation	0.035	0.448	729
	Building type	0.114**	0.003	729
	Residential floors	0.069	0.102	729
	Energy-saving insulation measures	0.446**	0	729
Lifestyle	Summer cooling mode	0.010	0.708	729
	Summer air conditioning use time	0.070	0.068	729
	Summer air conditioning cooling mode	0.318**	0.001	729
	Winner air conditioning use time	0.106	0.143	729
	Winner air conditioning heating mode	0.036	0.318	729
	The complete baths a day	-0.03	0.616	729
	Bathing way	0.042	0.412	729
The ownership of household equipment	Computer boot time a day	-0.004	0.928	729
	TV boot time a day	0.096*	0.009	729
	Number of refrigeration and air conditioning in summer	0.546**	0	729
		0.269**	0	729
	Air conditioning of heating in winter	0.180**	0	729
	Several other equipment of heating in winter	0.166**	0	729
	Number of cooker	0.299**	0	729
	Number of computers	0.135**	0	729
	Number of TV	0.178**	0	729
	Number of information equipment			
Energy saving concepts	The usage of energy saving lamps	0.046	0.242	729
	Electric appliances on standby condition	-0.035	0.373	729

Annotate: *represents 0.05 significant correlation; **represents 0.01 significant correlation.

TABLE II
TOTAL ENERGY CONSUMPTION OF EACH VARIABLE AND THE PARTIAL CORRELATION ANALYSIS RESULTS

Variable	Partial correlation	Significant level	Number of samples
The resident population	0.403**	0	729
The per capita building area	0.591**	0	729
Building type	-0.024	0.597	729
Energy-saving and insulation measures the building used	0.426**	0.004	729
Summer air conditioning cooling mode	0.396**	0.003	729
The total number of and the daily use time of air conditioning and computer	0.017	0.682	729

Annotate: * represents 0.05 significant correlation; ** represents 0.01 significant correlation.

TABLE III
THE INTRODUCTION OR ELIMINATE VARIABLES

Model number	Enter variables	Excluding variables	Method
1	X_3	X_1, X_2, X_4	Stepwise regression method
2	X_3, X_1	X_2, X_4	Stepwise regression method
3	X_3, X_1, X_2	X_4	Stepwise regression method
4	X_3, X_1, X_2, X_4		Stepwise regression method

TABLE IV
MODEL COEFFICIENT OF EACH TEST

Model number	Parameters	Non-standardized regression coefficients B	Standard error	Standardized coefficient	T-test	Significant level
1	Constant term	714.693	47.379		7.039	0.014
	The sum number of air-conditioner and computer	110.648	16.616	0.408	3.316	0.026
2	Constant term	480.420	79.819		3.412	0.028
	The sum number of air-conditioner and computer	94.322	16.842	0.442	3.210	0.004
	Resident population	83.499	20.683	0.156	2.127	0.016
3	Constant term	-178.824	103.494		-1.072	0.082
	The sum number of air-conditioner and computer	61.628	14.945	0.240	2.078	0.006
	Resident population	191.709	22.353	0.775	4.183	0.034
	The per capita building area	15.379	1.937	0.663	4.193	0.040
4	Constant term	-407.544	122.942		-1.752	0.011
	The sum number of air-conditioner and computer	41.736	16.336	0.186	1.408	0.034
	Resident population	191.472	22.038	0.483	4.639	0.004
	The per capita building area	15.399	1.741	0.624	4.431	0.015
	Summer air conditioning cooling mode	120.308	34.406	0.228	2.045	0.022

C. Energy consumption forecast model of residential buildings

According to the results of the partial correlation analysis, select four variables to influence to establish regression model. The particular form of multivariate linear regression model (Factor is the resident population X_1, the per capita building area X_2, the total number of and the daily use time of air conditioning and computer X_3, summer air conditioning cooling mode X_4):

$$Y = b_0 + b_1 X_1 + b_2 X_2 + b_3 X_3 + b_4 X_4 + \varepsilon$$

Select the independent variables by gradual regression fitting regression model, method as shown in Table III.

The Table IV shows the various models of non-standardized regression coefficients B, standard error, standardized coefficient, T-test and significant level. The four models of each variable significance level is less than 0.100, so it has statistical significance. The constant term of model 4 is -407.544, the non-standardized regression coefficients of sum number of air-conditioner and computer is 41.736, the standardized coefficient is 0.186, the non-standardized regression coefficients of resident population is 191.472, the standardized coefficient is 0.483, the non-standardized regression coefficients of the per capita building area is 15.399, the standardized coefficient is 0.624, the non-standardized regression coefficients of summer air conditioning cooling mode is 120.308, he standardized coefficient is 0.228.

In order to verify the fitting effect of the model. Based on the model 4 to back to the generation of test sample data, to obtain the annual energy consumption non-standardized predicted values of each sample. Use predictive value minus statistical quantity get the difference between the predicted and actual values. Absolute difference is less than 1004 (the regression equation to predict the value of standard deviation) to meet the requirements, otherwise not eligible, to get meet the requirements 460 samples, do not meet the requirements 166 samples, the historical coincidence rate was 73.5%. Based on the above analysis, to determine the final regression equation is:

$$Y = 407.544 + 41.736 X_3 + 191.472 X_1 + 15.399 X_2 + 120.308 X_4$$

As can be seen from the above model: annual energy consumption are positively correlated with the resident population, the per capita building area, summer air conditioning cooling mode and the total number of and the daily use time of air conditioning and computer, consistent with the partial correlation analysis results.

The resident population, the per capita building area, building type, energy-saving and insulation measures the building used, summer air conditioning cooling mode, the total number of and the daily use time of air conditioning and computer is the main factor affecting the energy consumption of residential buildings can be seen from the analysis results, this is similar to the results in the literature [1, 3, 6]. However, analysis shows that the income variable and annual energy consumption does not

have a significant linear correlation, this is inconsistent with the conclusions of the literature [2, 5]. Because many living facilities which can only be owned by high-income groups now can also be owned by low-income groups with the improvement of people's living conditions. In addition, Hainan residents income is closer, this has also been reflected in the frequency analysis of the survey results.

IV. CONCLUSION AND SUGGESTION

This paper proposed a predicted model for predicting residential building energy consumption in Hainan province that is in line with the actual data trends. It can be simultaneously used in: formulating Hainan province residential building energy-saving measures and standards; guiding and organizing the residential construction industry; helping to optimize the structure of residential buildings in Hainan. This study focused on both qualitative and quantitative analysis which improved the insufficient of previous research on Hainan building energy consumption that is limited to qualitative analysis, and this development can make the future predicting study for Hainan building energy consumption more scientific.

However, only 26 indicators which impact the residential buildings energy consumption are considered in this paper. Other indicators have not been taken into account, so in the future, we can add more other indicators into this model as a correction function.

REFERENCES

[1] CAE Consulting Project, Tsinghua University Building Energy Research Center. Annual Report on the Development on China Building Energy 2009; China Construction Industry Press: Beijing, China, 2009; pp. 28–36 (In Chinese).

[2] Jiang, Y.; Lin, B.R.; Zeng, J.L.; Zhu, Y.X. The Energy Saving of Residential Building; China Construction Industry Press: Beijing, China, 2006; pp.3–8 (In Chinese).

[3] Lei, Y.R. The Study on Chongqing Residential Building Energy Consumption Forecast Method; Master Thesis, Chongqing University, Chongqing, China, 2008; pp. 18–26 (In Chinese).

[4] Li, B.Z. Sustainable Response to the Urbanization in China. J. Cent. South Univ. Technol. 2007, 14, 1–7.

[5] LI Bai-zhan. Sustainable response to the urbanization in China 2007 (z3).

[6] Sari R; Soytas U The growth of income and energy consumption in six developing counties 2007 (35).

[7] Sinor R; Westphal F S; Lamberts R Regression analysis of electric energy consumption and architectural variables of conditioned commercial buildings in 14 Brazilian cities 2011.

[8] Swan, L.G.; Ugursa, V.I. Modeling of end-use energy consumption in the residential sector: A review of modeling techniques. Renew. Sustain. Energy Rev.2009, 13, 1819–1835.

[9]   Bentzen, J; Engsted, T. A revival of the autoregressive distributed lag model in estimating energy demand relationships. Energy 2001, 26, 45-55.

[10]   Hirst, E. A model of residential energy use. Simulation1978, 30, 69-74.

[11]   Crompton, P.; Wu, Y.R. Energy consumption in China: Past Trends and Future Directions. Energy Econ. 2005, 27, 195-208.

[12]   Chen, W.Y.; Wu, Z.X. Study on China's future sustainable energy development strategy using MARKAL model. J. Tsinghua Univ. Sci. Technol. 2001, 41, 103–106 (In Chinese).

[13]   Chen, X.G.; Pei, X.D. Artificial Neural Network Technology and Its Applications; Electric Power Press: Beijing, China, 2003, pp.168–181.

Research on CBR-RBR Fusion Reasoning Model and its Application in Medical Treatment

Jiang SHEN [1], Jin XING[1], Man XU[2,*]

[1]Department of Management and Economy, Tianjin University, Tianjin, China
[2]Department of Industrial Engineering, Nankai University, Tianjin 300457, China
(td_xuman@nankai.edu.cn)

Abstract - **The phenomenon exist in the reasoning process of medical decision-making system are insufficient or incomplete information, information singularity, imprecise measurement, low reliability, susceptible outside interference, which make the inference efficiency and accuracy lower in complex medical decision-making system, with the existence of risk influencing factor, such as low levels of knowledge sharing and human error. In this paper, we attempt to build CBR/RBR knowledge base and the process of acquainting and transferring knowledge, research the decision-making tools based on robustly inference fusion, in order to improve the quality and efficiency of health care decision-making system.**

Keywords - **CBR/RBR, complex systems, fusion mechanism, robust**

I. INTRODUCTION

In recent years, Department of Anesthesiology/Critical Care Medicine, Johns Hopkins University School of Medicine, Columbia University biomedical intelligence professional decision-making and Cognition Lab, Harvard University Medical School, etc., are carrying out extensive researches in the field of medical and health systems management on how to improve the quality of care and reduce the incidence of medical procedures to avoid adverse events, improving medical decision-making system's robustness, describing the status quo from the perspective of theory and practice and discussing appropriate solutions in deep.

The relative scarcity of medical resources and extremely uneven regional distribution has become the biggest problem restricting the level of medical treatment. In 2008 our clinic visits and hospitalization passengers reached 3.108 billion and 114.83 million, massive patient medical demand and the relative lack of medical resources made a huge conflict. Only real solutions to primary care staff's lack of expertise and experience on the treatment can really solve the current problems that the overall health system quality and efficiency is not high in our country, and can high-end medical resources play the biggest role in improving the health of the people.

II. PURPOSE

The project intends to conduct robust research on quality and efficiency of health care decision-making system, explore the decision-making mechanism of the complex health care system, and establish the fusion mechanism of combining static and dynamic, in order to improve the quality and efficiency of health care decision-making system from the perspective of system robustness.

III. LITERATURE REVIEW

A. Complex Systems of Information Fusion Model

The issue of vulnerability brought by the uncertainty of complex systems of knowledge will show on the conflict reasoning, while the use of information fusion method will resolve the conflict system. Hall et al. [1] (2001) using D-S fusion method successfully solved the problem of conflicts of heterogeneous systems, and achieved good application effect on decision-making of military-target's information fusion. Sun R et al. [2] (2008) using modified data fusion method based on D-S evidence theory resolved the data conflict in the system. Dasarathy BV [3] (2009) optimized the JDL and made Dasarathy functional model, which is capable of in-depth analyzing the law and mechanism of integration and good to reasoning optimization of complex systems. In China, HAN Chong et al. [4] (2006) studied the conceptual model of multi-source information fusion using the main information fusion model. Wang Xin et al. [5] (2008) applied the information fusion model for the study of complex systems dynamic performance. Pandey B [6], et al. (2009) uses intelligent algorithms to study the physical characteristics and behavior reasoning information fusion between entities.

B. Information and Knowledge Acquisition, Transfer Process Quality and Efficiency

Process-aware information systems, data mining techniques and rough set methods are currently used to obtain system information and knowledge. In order to effectively get medical decisions in the system of multi-source heterogeneous information, Cao et al. [7] (2004) used a human-computer interaction semi-automatic knowledge acquisition method based on the theory of formal ontology to obtain medical information. Uramoto N, et al. [8] (2010) using data mining approach to explore the clinical documentation related knowledge. Reilly et al. [9] (2007) improved the transmission channels of medical information by eliminating the clinical path of structural factors which are not reliable. Meloni, et al. [10] (2009) used mutual information technology to reduce the transmission of information nodes clinical databases, achieving a timely and consistent messaging.

C. CBR/RBR Fusion Reasoning

Taking vulnerability of the CBR-RBR fusion reasoning into account, robustness of the system can be characterized by the ability to maintain its system when

E. Qi et al. (eds.), *Proceedings of the 21st International Conference on Industrial Engineering and Engineering Management 2014*, Proceedings of the International Conference on Industrial Engineering and Engineering Management, DOI 10.2991/978-94-6239-102-4_87, © Atlantis Press and the authors 2015

facing internal structure or function of the external environment changes (Baker JW [11], 2008), thus the system robustness analysis methods will help solve the problem of vulnerability of the system. Golding, et al. [12] (1996) the first pioneered the use of case similarity threshold (RBR/CBR-Hybrid) to solve the problems of the rules of singular, reasoning for RBR and CBR separately and using decision comparison of competitive strategy for both, this is a non-fusion strategy. CBR/RBR fusion reasoning method using Harmonic module requires expert knowledge in special field Kumar K.A, et al. [13] (2009), while influenced by the human factor in practical applications. In the case of rule-based retrieval system (Tung YH [14], et al., 2010), the rules clustering method effectively reduces the retrieval space and improve the retrieval efficiency, but the result may be a non-global optimal solution. Luengo [15], et al. (2010) analyzed the relationship between the CBR rule-based reasoning system's behavior and knowledge uncertainty, combined with the relationship given to build fusion strategy, but did not solve the issue of vulnerability fusion system from the perspective of system robustness analysis. Patel VL [16] et al (2009) pointed out that RBR will continue to play an important role in the future of health care decision-making system.

IV. RESEARCH ON FUSION MECHANISM OF CBR / RBR AND KNOWLEDGE FUSION DECISION-MAKING TOOLS

A. C1: Decision-Making Processes in Complex Medical Decision-Making System CBR/RBR Fusion Mechanism

Due to the uncertainty, multi-criteria setting, complexity and other characteristics of decision-making, there is a need for the study of multi-source decision-making information collected through a variety of information channels of complex medical decision-making system, such as the transfer mechanism and their integration in a heterogeneous space reasoning mode (Fig.1). This study uses a broad spatial data structures, such as the integration space for knowledge representation. On the one hand study its static structure, on the other hand, use functional methods to study its dynamic performance. Comprehensively study the information field of energy spatial decision the harmonic maps and collaborative optimization pass laws medical decisions knowledge acquisition and knowledge reasoning, etc., as shown in Fig.2.

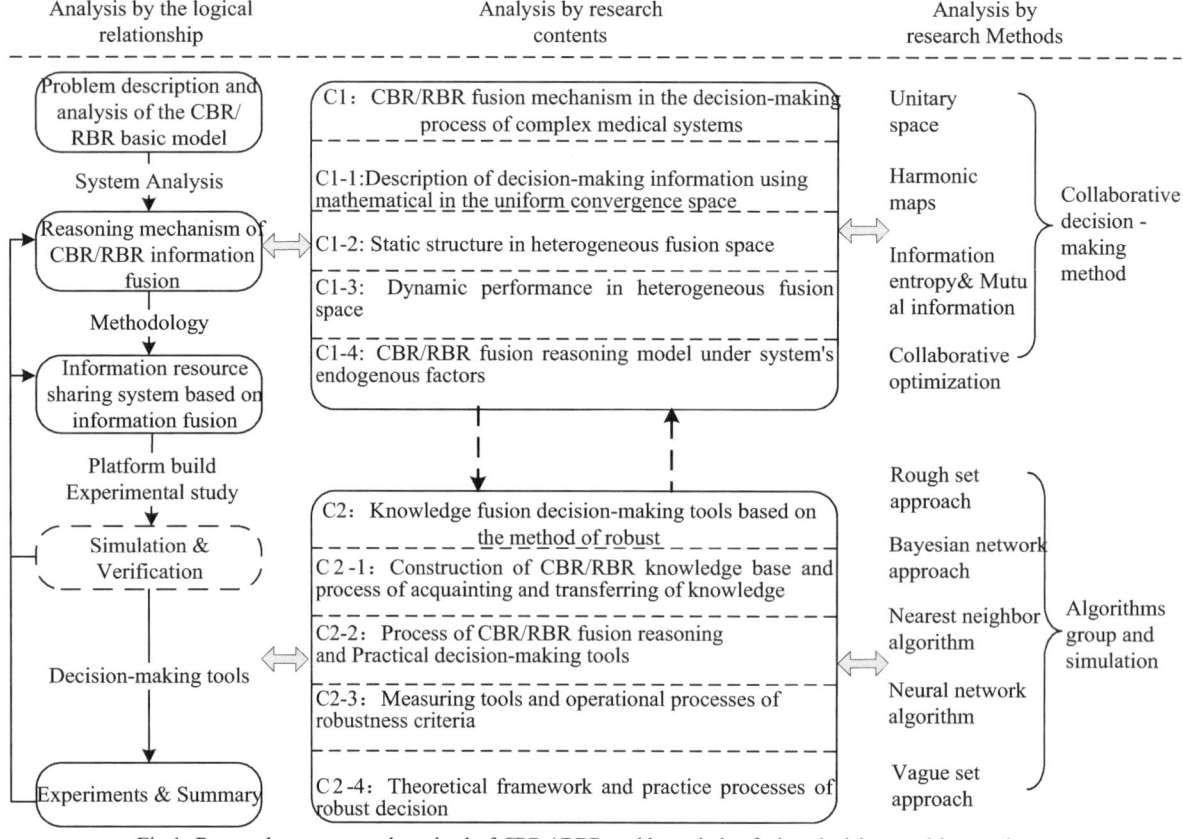

Fig.1. Research content and method of CBR / RBR and knowledge fusion decision-making tools

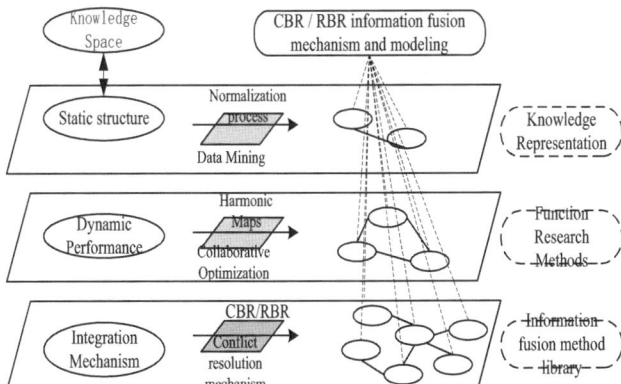

Fig.2. Complex medical decision-making process of the system CBR/RBR fusion mechanism

C1-1: study the decision-making information which is mathematically described in a unified fusion space, including uncertainty, uncontrollable and heterogeneous characteristics of the information element;

C1-2: study the static structure of heterogeneously integrated space, such as the correlation and mapping relationship between knowledge source, unstructured characterization of tacit knowledge, etc.;

C1-3: study dynamic performance of the heterogeneous integration of space, such as information channel, information path distance and the information transfer function when fusion space is seen as a field of information, as well as the transfer cost, quality and efficiency and other factors of decision-making knowledge in the information field.

C1-4: Study CBR/RBR fusion reasoning model influenced by endogenous factors in the system, segmentation studies include: identification and classification of system the endogenous factors, reasoning information acquisition and transfer law, pure, integration and fusion model of CBR and RBR hybrid reasoning mechanism, and the convergence space resource strategy in the separation and conflict resolution mechanism.

B. C2: Research and Development of Knowledge-Based Decision-Making Tool Integration Robust Methods

For China's national conditions aiming at the problem of relatively scarce medical resources and set up community health care policy, this study intends to establish a medical knowledge decision-making tool set support platform based on robust fusion, to further improve the quality and efficiency of medical decision-making research. This study will cooperate with hospital in international leading level, such as Tianjin Teda international cardiovascular hospital in medical knowledge sharing and management decision-making, the robustness of the empirical research system quality and efficiency of health care decisions. Development of this study was to establish the medical decision-making case base and rule base, knowledge acquisition and transfer modules, embedded integration reasoning module, vulnerability identification, interference transfer module and robustness criteria group decision-making modules, etc. The study can be decomposed into specific points:

C2-1: Build CBR/RBR Knowledge acquisition and transfer of knowledge and processes;

C2-2: Development CBR/RBR fusion reasoning processes and practical decision-making tools;

C2-3: Design and Development robustness criteria group measure tools and operational processes;

C2-4: forming a theoretical framework and practical decision-making process robustness.

C. Problem description and CBR/RBR basic pattern analysis

With the statistical tools and data mining methods, the data from the medical field and typical patient cases could be analyzed. The results of analysis is used to extract the key information in a medical emergency process, and also the results is helpful to understand the system complexity of the medical emergencies, the state space of information, the state awareness, heterogeneous spatial data structure, expression and synergistic integration of medical knowledge, detail description of system information. Medical decision-making system and vulnerability researches under the knowledge sharing environment with time constraints is based on this section. Finally, temporal information transfer efficiency, and the source and the interference frequency of multi-task information transfer could be presented in this analysis.

In the multi-source heterogeneous information integration and data associated research, C1-1 uses methods such as neural network data association (NNDA) and joint probabilistic association method (JPDA) to integrate and associate the information about diagnostic decisions, sparse target, high density medical decision making environment. C1-2 can be used to abstracting the classifying the multi-source heterogeneous information sources in the medical process, building the relationships among the information, separating the incomplete, disordered, and random information, intensively identifying effective information, dealing it with fuzzy optimization process, and finally to achieve comprehensive integration and correlation of multi-source heterogeneous information. C1-3 focuses on CBR/RBR basic model, such as a simple CBR or RBR reasoning model, CBR/RBR binding model, CBR/RBR mixed model and CBR/RBR integrated model, which can be used to analyze their reasoning applicable conditions in decision-making, advantages and disadvantages of reasoning results as well as the internal mechanism of the reasoning process.

V. CONCLUSION

CBR/RBR information fusion mechanism mainly focused on the research of intelligent integration model and reasoning algorithm based on CBR/RBR. Through building information fusion RBR rule repository based on Vague sets rules, repository-based CBR cases and medical experts repository of information and on-site control of the expected feedback information for comparison, adjustment and processing, integration of multi-parameter information and determine and solve system problems with real-time processing and uncertain factors.

CBR/RBR fusion model is used to build the reasoning mechanism of medical process through researches about intelligent integrated reasoning algorithm such as CBR/RBR, Bayesian networks, and dynamic fuzzy neural network. This section of research is an important reasoning model and method on medical decision-making system and vulnerability researches under the knowledge sharing environment with time constraints.

ACKNOWLEDGMENT

This research was supported by the National Natural Science Foundation of China (Grant No. 71171143), and the Fundamental Research Funds for the Central Universities (Grant No. NKZXB1458), China.

REFERENCES

[1] Hall L D, Llinas J, Handbook of Multisensor Data Fusion. Boca Raton, FL, USA: CRC Press, 2001

[2] Sun R, Huang HZ, Miao Q. Improved information fusion approach based on D-S evidence theory [J]. Journal of Mechanical Science and Technology, 2008, 22(12): 2417-2425

[3] Dasarathy BV. Multisensor, Multisource Information Fusion: Architectures, Algorithms, and Applications 2007 [C]// Society of Photo-Optical Instrumentation Engineers (SPIE) Conference Series. 2009, 7345.

[4] Chongzhao Han, Hongyan Zhu. Multi-source information fusion [M]. Tsinghua University Press, 2006.

[5] Xin Wang, Taifan Quan. Information fusion system improved BA model and the network dynamics [J]. Harbin Institute of Technology Journal, 2007, (5): 737-741

[6] Pandey B, Mishra R B. Knowledge and intelligent computing system in medicine [J]. Computers in Biology and Medicine, 2009, 39(3): 215-230

[7] Cao C, Wang H, Sui Y. Knowledge modeling and acquisition of traditional Chinese herbal drugs and formulae from text [J]. Artificial Intelligence in Medicine, 2004, 32(1): 3-13

[8] Uramoto N, Matsuzawa H, Nagano T, et al. A text-mining system for knowledge discovery from biomedical documents [J]. IBM Systems Journal, 2010, 43(3): 516-533.

[9] Reilly J., Newton R., Dowling R. Implementation of a first presentation psychosis clinical pathway in an area mental health service: the trials of a continuing quality improvement process [J]. Australas Psychiatry, 2007, 15(1):14-18

[10] Meloni A, Ripoli A, Positano V, et al. Mutual information preconditioning improves structure learning of bayesian networks from medical databases [J]. IEEE Transactions on Information Technology in Biomedicine, 2009, 13(6): 984-989.

[11] Baker J W, Schubert M, Faber M H.On theassessment of robustness [J]. Structural Safety, 2008, 30(3): 253-267

[12] Golding A, Rrosenbloom PS. Improving accuracy by combining rule-based and case-based reasoning [J]. Artificial Intelligence, 1996, 87(1-2): 215-254.

[13] Kumar K.A., Singh Y., Sanyal S. Hybrid approach using case-based reasoning and rule-based reasoning for domain independent clinical decision support in ICU [J]. Expert Systems with Applications, 2009, 36(1): 65-71

[14] Tung Y-H, Tseng S-S, Weng J-F, et al. A rule-based CBR approach for expert finding and problem diagnosis [J]. Expert Systems with Applcations, 2010, 37(3): 2427-2438

[15] Luengo J, Herrera F. Domains of competence of fuzzy rule based classification systems with data complexity measures: A case of study using a fuzzy hybrid genetic based machine learning method [J]. Fuzzy Sets and Systems, 2010, 161(1): 3-19

[16] Patel V L, Shortliffe E H, Stefanelli M, et al. The coming of age of artificial intelligence in medicine [J]. Artificial Intelligence in Medicine, 2009, 46(1): 5-17

A Selection Algorithm of Survey Lines and the Interactive Mode of the Sounding Curve

Lin-peng LYU[1], Juan LI[1], Jin-sheng ZHANG[1], Zhi-peng TIAN[1], Hong-tao BAI[2,*]

[1]College of Software Engineering, Jilin University, Changchun, China
[2]Center for Computer Fundamental Education, Jilin University, Changchun, China
(baiht@jlu.edu.cn)

Abstract - **To solve the data interaction fault of three-dimensional CSAMT model, a three-dimensional selection algorithm of survey lines is proposed. The survey lines are generated in a batch, based on the pre-given direction and distance error. By comparing the distances of survey points with the error, a series of survey lines can be formed by turns. The computing load is primarily concentrated on the sort of distances of survey points. In addition, we introduce an interactive mode of the sounding curve for single survey point displayed in a group of one survey line. The outliers of the observation data curve can be revised by intercepting, when editing the sounding curve of one specific point.**

Keywords - **CSAMT, interactive mode, survey line selection, sounding curve**

I. INTRODUCTION

The Controlled Source Audio-frequency MagnetoTellurics (CSAMT), which has high reliability and is less affected by environmental factors because of the powerful artificial source, plays an important role in ore exploration [1-2]. It has been successfully used in a series of exploration areas, such as metal mining, petroleum, geothermal resources and groundwater since its inception [3-7].

However, in actual measurement, observation data will still be affected by the some environmental factors, such as the location of survey points [8]. As a result, some outliers will be introduced to the data, which make forward and inversion and other related works cannot be carried out properly after data collection. Moreover, the data whose magnitude is pretty high make it necessary to use computer to identify and modify the outliers. At present, American SCSINV two-dimensional inversion tool and a piece of software for one-dimensional inversion developed by the Phoenix-Geophysics Company in Canada are the most popular ones used in the CSAMT [9-10]. To deal with the interaction problem involved, a user-friendly interactive mode is proposed when we edit the multi-parameter sounding curve of a single point.

Before our discussion of this interactive mode, we also present a selection algorithm of survey lines based on a rectangular point map. The multi-parameter sounding curves of a single point are often browsed as a group of survey line, so selecting survey line from a large number of survey points is a fundamental work. Therefore, an efficient selection algorithm for lines will be very necessary.

II. SELECTION ALGORITHM

A. Algorithm flow

This algorithm selects a cluster of points to form one survey line which is called based on a given slope in the coordinate system. It is a survey line that forms when a row of survey points just stand in line in the plane or they can be consider to be in line within the given error range. Forming strategic of survey line is introduced as follows.

Assume that there forms a dot matrix of $m \cdot n$. Take it for example when the slope of the survey line is greater than 0. We choose the top left (or bottom right) point as the initial survey point, through which the first survey line is formed at the given slope. Then, calculate all values of distance $d_i (i = 1, 2, ..., m \cdot n, d_i \geq 0)$ between the initial line and each point in dot matrix, and sort in ascending order according to the distance. Especially, the distance from the initial point to the first survey line is zero.

Under control of the particular error value, select those points whose distances are less than the error and add them to the initial survey line, that is to say those points is regarded as components of the initial line. After the selection, pick up the first point that has not been added to the initial line, whose distance d is beyond the specific error ε. Then take it as our start point to form next survey line. Same as mentioned, add the rest points to a second survey line whose distances are within the same error, so the next line completes. Repeat the method before, until each point in the dot matrix belongs to a line in turn.

If the slope is lower than zero, we take the bottom left (or top right) point in the dot matrix as the first survey point, then calculate the initial line through that point in the given slope. Especially, when the slope equals to zero (or doesn't exist), just put all those points with the same y-axis (or x-axis) to one survey line in turn. Algorithm flow is shown in Fig.1.

As is shown, in the process of algorithm a series of survey lines will form based on information of survey points and the specific slope. The position of a survey line is defined by the start point and other points whose distances are within the error are added to the line like a "Bucket" [11]. In order to improve the efficient of the algorithm, we sort the distances between the initial line and each point in dot matrix to filter points that meets our requirement. When a new survey line is complete, we don't update the distances of the rest points. We update the error by adding current error a value of distance between the last survey line and the current one instead.

E. Qi et al. (eds.), *Proceedings of the 21st International Conference on Industrial Engineering and Engineering Management 2014*, Proceedings of the International Conference on Industrial Engineering and Engineering Management, DOI 10.2991/978-94-6239-102-4_88, © Atlantis Press and the authors 2015

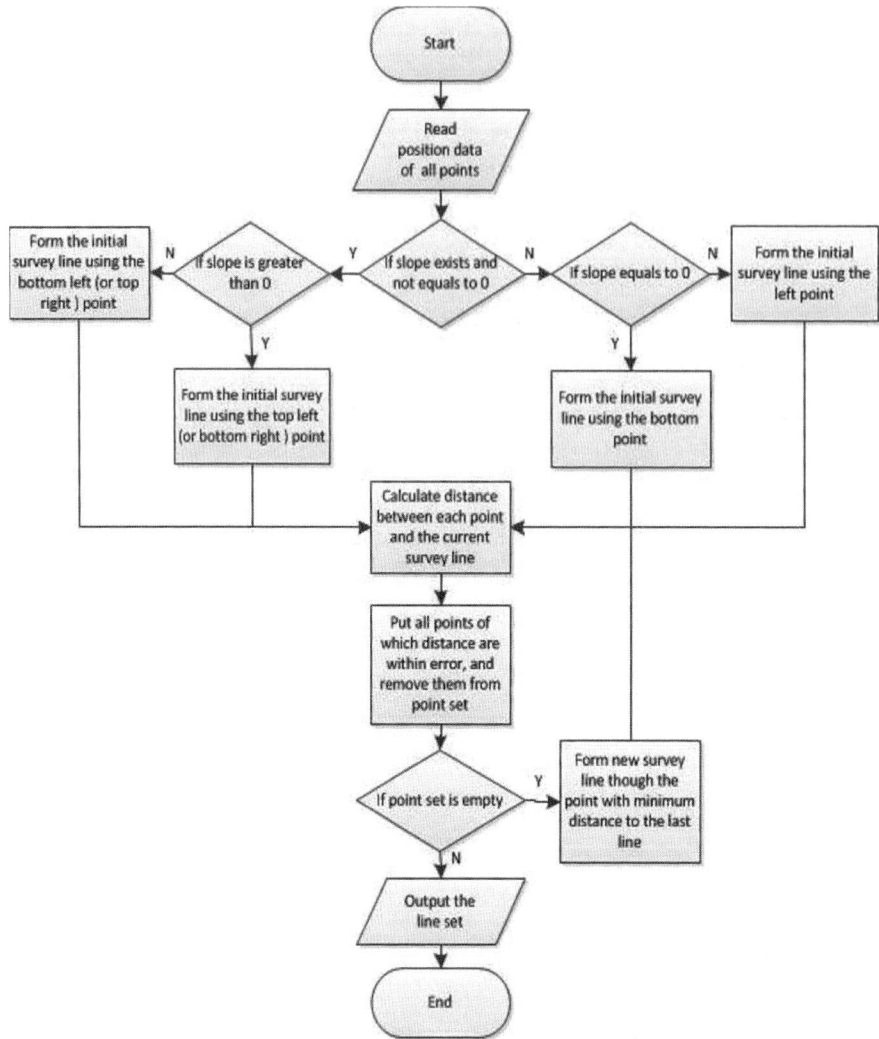

Fig.1. Algorithm flow chart

The algorithm itself just adapts to the selection of dot matrix or tidy shape. But in the real exploration, survey points are usually placed in a dot matrix shape [12].

B.   Time complexity analysis

In this section, we analyses the time complexity of the algorithm. *O* is a method of incremental representation of time complexity [13]. Assuming that the input size of algorithm is *n*, which means the amount of points is *n*. Then start the analysis.

Firstly, read the point set and select the top-left (or top right) point according to slope. This will take $O(n)$. Then the time complexity of calculating the distance of each point to the initial survey line is $O(n)$. Afterwards, sort the list of distances in ascending order using the most efficient sort algorithm, such as quick sort [14]. Time complexity of the sorting process is $O(n \cdot log_2 n)$. But as mentioned before, we don't need to recalculate the distances between rest points and the current survey line when new survey line is complete. On the contrary, just update the value of error   by adding a distance value of last two survey lines. The time complexity will be decreased from $O(n (n+1)/2)$ to $O(n)$. Therefore, including

the sorting process, the average of time complexity reduces from $O(n (n+1)/2)$ to $O(n \cdot log_2 n)$ with the best condition of $O(1)$. A more detailed time complexity analysis of each stage of the algorithm is showed in Table I.

TABLE I
THE TIME COMLEXITY

STEP	Condition		
	Best	Worst	Average
Select the initial point	$O(n)$	$O(n)$	$O(n)$
Calculate distances	$O(n)$	$O(n)$	$O(n)$
Sort distances	$O(n \cdot log_2 n)$	$O(n^2)$	$O(n \cdot log_2 n)$.
Form survey line	$O(n)$	$O(n)$	$O(n)$
Update error	$O(n^{1/2})$	$O(n)$	$O(n)$
Total	$O(n \cdot log_2 n)$	$O(n^2)$	$O(n \cdot log_2 n)$

III. THE INTERACTIVE MODE

A.   Introduction of multi-parameter sounding curve

Based on the data of one measuring point surveyed by CSAMT in different frequencies, we can get multi-parameter sounding curve of the corresponding point. The data here can be the electric field strength as well as the magnetic one. It reflects the electric distribution of underground layers, which contributes on subsequent multidimensional forward and inversion [15-16].

When modeling the curve in coordinate system, we let the data of one parameter, namely the electric field strength or magnetic one, to be on the x axis and the frequency data to be on the y axis. Because the whole curve has already been depicted on the screen, it will be possible to identify wrong data fast and accurately by considering the overall tendency of the curve. After the identification process, computer's efficient computation can help researchers to correct outliers effectively. Obviously, this curve model will support researcher to find out outliers easily and rectify them accurately and efficiently.

B. Curve characteristics

In CSAMT, the electric field strength (E) and the magnetic field strength in (H) of the field stimulated by galvanic couple in homogeneous half space are two important parameters. And in this condition, only E_x, which represents the electric field strength in x-direction and H_x which represents the magnetic field strength in z-direction can reflect the underground electrical structure to maximum extent [17]. So we just put those two parameters into consideration. According to Maxwell Equations and Wave Equation in the condition of homogeneous media, the electric formula of homogeneous half space is given below [17]:

$$E_x = \frac{I \cdot AB \cdot \rho}{2\pi r^3}\left[e^{ik_1 r}\left(1 - ik_1 r\right) + \left(3\cos^2\theta - 2\right)\right] \quad (1)$$

The magnetic equation of homogeneous half space is:

$$H_z = \frac{iI \cdot AB \cdot \rho}{2\pi\mu_0\omega r^4}\cdot\sin\theta\left[e^{ik_1 r}\left(-3 - 3ik_1 r - k_1^2 r^2\right) + 3\right] \quad (2)$$

In those two formulas, symbol I is the current intensity stimulated by galvanic couple and AB is the distance between two sources of a galvanic couple. Symbol ρ is the subsurface resistivity. Symbol μ_0 is the magnetic permeability of rocks. Symbol r denotes distance between each measuring point and the field source. Symbol ω represents the angular frequency of electromagnetic waves and its formula is:

$$\omega = \frac{2\pi}{T} \quad (3)$$

where T is the cycle of electromagnetic waves.

In addition, symbol θ is the angle between the galvanic couple and the axis of the earth. Symbol k_1 is a constant of the wave and the formula is:

$$k_1 = \sqrt{\pm i\mu_0\omega\sigma_1} \quad (4)$$

And also, the cycle T can be represents by the formula:

$$T = \frac{1}{f} \quad (5)$$

As a result of (3), (4) and (5), (1) and (2) can be written as follows:

$$E_x = \frac{I \cdot AB \cdot \rho}{2\pi r^3}\cdot \quad (6)$$
$$\left[e^{ir\sqrt{\pm 2\pi i\mu_0 f\sigma_1}}\left(1 - ir\sqrt{\pm 2\pi i\mu_0 f\sigma_1}\right) + \left(3\cos^2\theta - 2\right)\right]$$

$$H_z = \frac{iI \cdot AB \cdot \rho}{2\pi\mu_0 2\pi fr^4}\cdot\sin\theta\cdot \quad (7)$$
$$\left[e^{i\sqrt{\pm 2\pi i\mu_0 f\sigma_1}r}\left(-3 - 3ir\sqrt{\pm 2\pi i\mu_0 f\sigma_1} \pm 2\pi i\mu_0 f\sigma_1 r^2\right) + 3\right]$$

According to (6) and (7), we can concluded that the curves of E_x and H_x are smooth and continuous, and are derivable at any points where the value of frequency is larger than zero (f>0). Fig.2 shows two curves of E_x, which are apparently smooth and continuous.

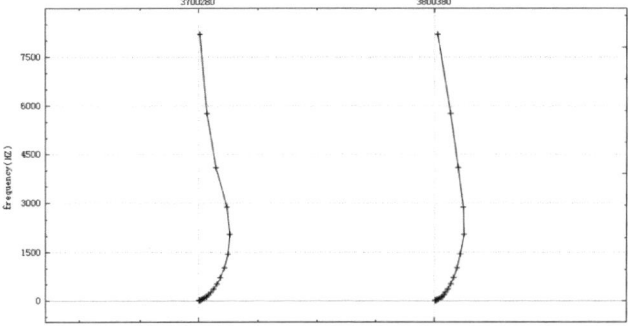

Fig.2. Smooth curves

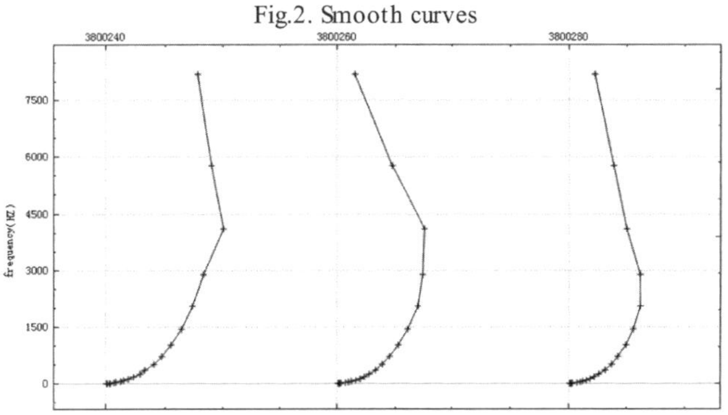

Fig.3. Abnormal curves

However, because of the outliers brought by CSAMT, the curve may not be completely smooth-the curve would not be derivable in some points. Fig.3 depicts the abnormal curve caused by the brought-in outliers.

C. Interactive mode

Base on the curve's characteristics introduced in last section, we've studied a correction mode of the curve that can accurately revise the outliers. We, now, will use the first curve showed in Fig.3 as an example to illustrate this interactive mode.

Firstly, select the curve that you want to edit. After that, a vertical line, which is used for later correction, will be showed in the left of this curve. Fig.4 demonstrates this situation. Then move this line to targeted location, where the points located in right side should be on this line after correction, using left and right keys of keyboards (showed in Fig.5). Finally, correct the selected curve. After the correction action, all points in the right of the interception line will be reset on the line. As a result, the observation data of points whose data values are large than the value indicated by this line will be assigned to the line's date value (showed in Fig.6).

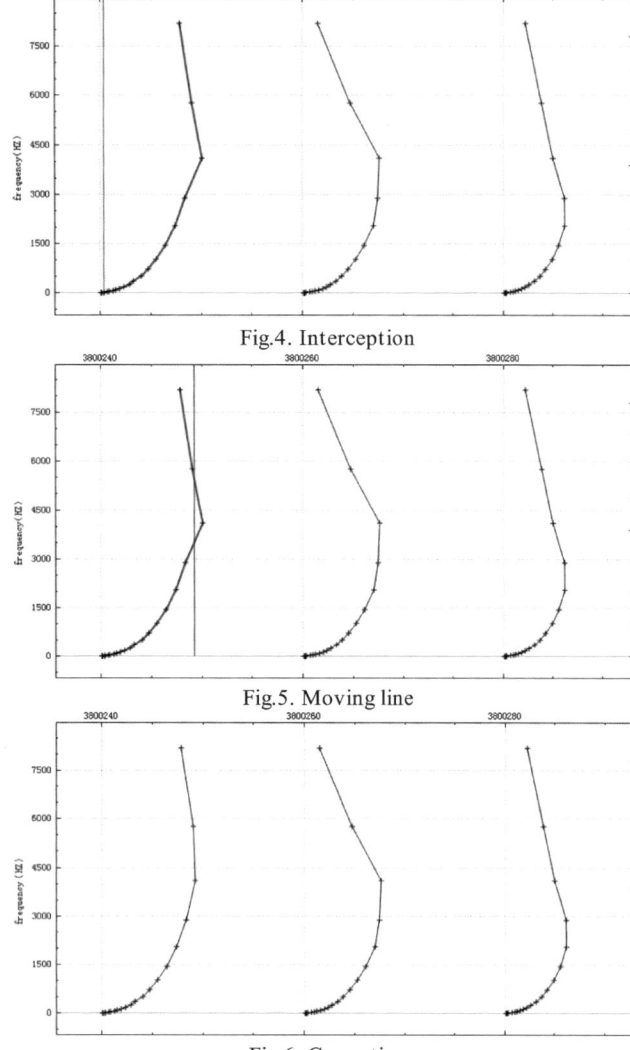

Fig.4. Interception

Fig.5. Moving line

Fig.6. Correction

However, when we edit such curve showed in Fig.7 (the first line) that the abnormal data value is less than the expected right one, these three steps mentioned before would not realize our correction work. Base on this problem, we also propose another operation conducted on the abnormal curve – reversion (showed in Fig.8).

After the reversion operation, the three steps mentioned before can achieve the correction work. The result is demonstrated in Fig.9. And finally, reverse the targeted curve again to the original place as showed in Fig.10.

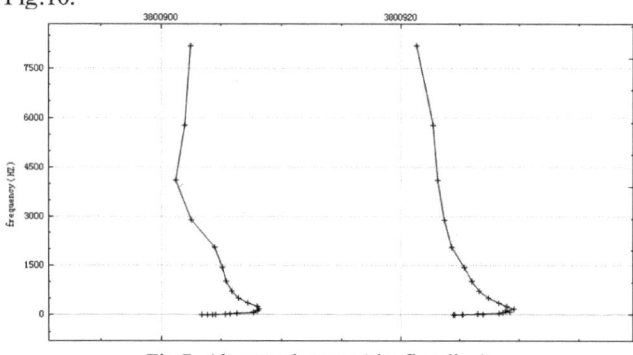

Fig.7. Abnormal curve (the first line)

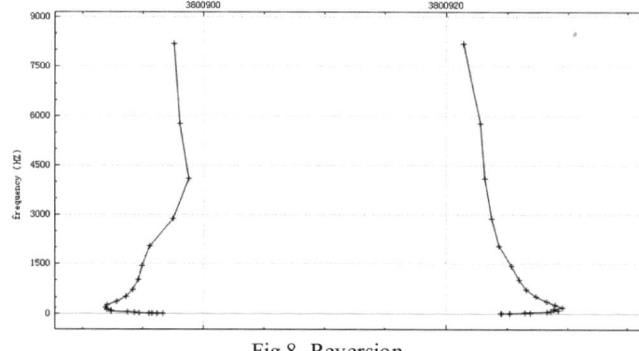

Fig.8. Reversion

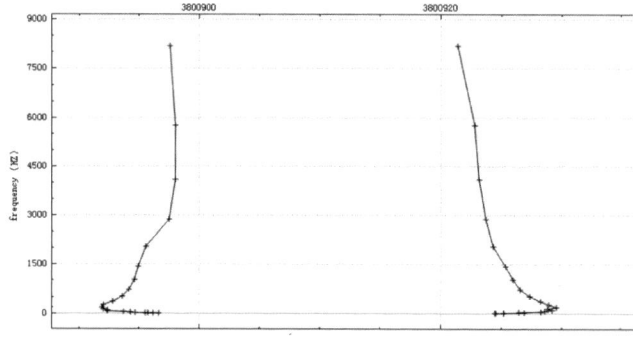

Fig.9. The result of correction before final reversion

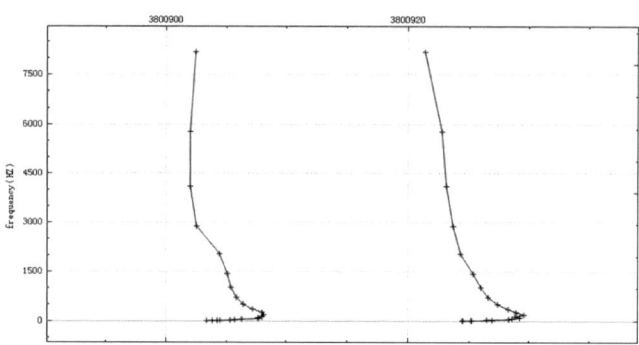

Fig.10. The result of correction after final reversion

IV. CONCLUSION

In this paper, to puzzle out the efficiency problem of selecting survey lines faced in the process of modeling CSAMT observation data on computers, we design an efficient selection algorithm. We clarify a reasonable survey-line-selection strategy and avoid constantly updating the value of distance from each survey point to a specific survey line by resetting the error instead. The time complexity of this algorithm is $O\ (n\cdot log_2n)$. Meanwhile, we also brought up a friendly interactive mode to solve the data-editing difficulty. We propose a strategy grounded on the interception to edit the curve. And this editing style holds the advantage of batch edition and simple procedure.

Our work suggests some fantastic directions for further study. For example, designing new survey line selection strategies and then, based on this scheme, realizing efficient selection algorithm of survey lines, can improve the efficiency of selection further. Additionally, devising more friendly interaction mode of other kinds of data curves in terms of their features would be a direction that is worth researching deeply.

ACKNOWLEDGMENT

This work was supported by "Fundamental Research Funds" of Jilin University (201103134) and National Science Foundation of China, NSFC (61272208).

REFERENCE

[1]   M. S. Zhdanoy, P. P. de Lugao and O. Portniaguine, "Fast and Stable Two-dimensional Inversion of Magneto telluric Data," *presented at the 4th International Congress of the Brazilian Geophysical Society.* 1995.

[2]   N.B. Boschetto, G.W. Hohmann, "Controlled-source audio frequency magneto telluric responses of three-dimensional bodies," *Geophysics*, vol. 56, no. 2, pp. 255-264, 1991.

[3]   M.J. Unsworth, X. Lu and M. D. Watts, "CSAMT exploration at Sellafield: Characterization of a potential radioactive waste disposal site," *Geophysics*, vol. 65, no. 2, pp. 1070-1079, 2000.

[4]   Wu Lu-ping, Shi Kun-fa, Li Yin-huai and Li Song-hao, "Application of CSAMT to the search for groundwater," *Chinese Journal of Geophysics*, vol. 39, no. 5, pp. 712-717, 1996.

[5]   Yu Chang-Ming, "The application of CSAMT method in looking for hidden gold mine," *Chinese Journal of Geophysics*, vol. 41, no. 1, pp. 133-138, 1998.

[6]   S. N. Sheard, T. J. Ritchie, K. R. Christopherson, and E. Brand, "Mining, environmental, petroleum, and engineering industry applications of electromagnetic techniques in geophysics," *Surveys in Geophysics*, vol. 26, no. 5, pp. 653-669, 2005.

[7]   S. K. Sandberg, G. W. Hohmann, "Controlled-source audiomagnetotellurics in geothermal exploration," *Geophysics*, vol. 47, no. 1, pp. 100-116, 1982.

[8]   Huang Zhao-hui, Di Qing-yun and Hou Sheng-li, "CSAMT static correction and its application," *Progress in Geophysics*, vol. 4, 2006.

[9]   N. R. Carlson, M. P. Phillip and A. U. Scott, "Applications of controlled source and natural source audio-frequency magnetotellurics to groundwater exploration," presented at the *18th EEGS Symposium on the Application of Geophysics to Engineering and Environmental Problems*, 2005.

[10] L. Sito, J. Farbisz, M. Stefaniuk and M. Wojdyla, "Application of CSAMT Method to Water Resources Recognition in High Urbanized Areas," presented at the 73rd EAGE Conference & Exhibition, 2011.

[11] N. L. Max, "Atmospheric illumination and shadows," *ACM SIGGRAPH Computer Graphics*. vol. 20, no. 4, pp. 117-124, 1986.

[12] Xue Guo-Qiang, Chen Wei-Ying, and Zhou Nan-Nan Li Hai, Zhong Hua-sen, "Understanding of Grounded-Wire TEM Sounding with Near-Source Configuration," *Journal of Geophysics & Remote Sensing*, vol. 2, no. 1, 2013.

[13] M. A. Weiss, "Algorithm analysis," in *Data structures and algorithm analysis in C*, Pearson Education Asia, 2002, ch. 3, pp. 30-46.

[14] C. A. Hoare, "Quicksort," *The Computer Journal*, vol. 5, no. 1, pp. 10-16, 1962.

[15] Shi Lin-hua, "Forward Calculation and analysis of electrical sounding curves of three inclined layers," *Geophsysical and Geochemical Exploration*, vol. 10, no. 2, pp. 104-114, 1986.

[16] Shi Lin-hua, Chen Bao-hua, "A method of inverse interpretation of electric sounding curve from geoelectrical observatory," *Northwestern Seismological Journal*, vol. 10, no. 1, pp. 38-43, 1988.

[17] Di Qing-yun, Wang Ruo, "*Controlled source audio frequency magneto telluric data inversion and Application,*" Science Press, 2008, pp. 13-14.

The Study of Long-term Electricity Load Forecasting Based on Improved Grey Prediction

Hong Wang [1], Kun Yang[2,*], Lin-yan Xue[2], Shuang Liu[2]

[1]Department of Electronic Information Engineering, Hebei University, Baoding 071002, China
[2]Department of Quality Technology Supervision, Hebei University, Baoding 071002, China
(lianlianfushi@126.com)

Abstract - **This paper first introduces modeling theory of GM (1,1) grey equidimensional filling vacancies, and establishes GM (1,1) grey forecasting model of equidimensional filling vacancies by using power demand load of Shan Dong Province from 2004 to 2011, then forecasts power load demand of Shan Dong Province in 2012 and 2013 for two years. When compared with the actual electricity load, that grey forecasting model of equidimensional filling vacancies has high accuracy in long-term power load forecasting.**

Keywords - **GM (1, 1), grey theory, grey forecasting model of equidimensional filling vacancies, power load forecasting**

I. INTRODUCTION

With the rapid development of China's electric power industry, power load forecasting techniques have increased more attention, and become an important field of power systems. Load forecasting is a scientific basis for decision-making. To some extent, long-term load forecasting results determine the future development of power system planning period. Currently, the electricity market is in the phase of reform, higher requirements will be put forwarded after reform of the electricity market for long-term load forecasting. Therefore, the precision of load forecasting for power system development is of great significance [1-4].

Currently the forecast for the load has got a more systematic study. Recently, there are several methods of studying electricity load forecasting technology: Trend extrapolation forecasting technique, Regression model to predict technology, Time series forecasting techniques, Grey Prediction and so on. Among them, because of its small sample size, the higher prediction accuracy, Grey prediction technology has been widely used in the short, medium and long-term power load forecasting. On this basis, there are many applications have been proposed to improve the grey prediction to predict the load. While, GM (1, 1) models has its limitations as other forecasting methods. When the data greater than the degree of dispersion, that is to say, the greater the grey scale data, the prediction accuracy worse; and it is not suitable for predicting long-term forecast for several years. There are some improvements in the GM (1, 1) model [5-7], for example, the use of genetic algorithm, Particle swarm optimization model for the grey model. This combination of GM (1, 1) model with the other-dimensional grey-scale to predict the power load is designed and it is fully demonstrated that the improved Grey GM (1, 1) model

can improve the long-term load forecasting accuracy than the traditional one [8-10].

II. LITERATURE REVIEW

In 1982, Professor Deng Julong, a famous scholar in China, proposed and developed Grey system. When modeling the grey model of load forecasting, we usually use historical data to establish differential equations, which is used as the power load forecasting models. As the power load, there are many uncertain factors affect the system load changes, which is called gray system, so the performance does not seem to load change law. With the help of Grey theory, these seemingly erratic historical data, after generated by the cumulative, showed a clear exponential growth law compared with the original value. At the same time, differential equations have the form of exponential form. Therefore, after the use of differential equations to fit exponential growth to generate regular data column, then conducting the load forecast, the final regressive reduction method to generate the actual load forecasts. GM (1, 1) model is one of the simplest models in gray system theory [11, 12].

Deng (1982) focuses on model uncertainty and information insufficiency in analyzing and understanding systems via conditional analysis, prediction and decision-making. The Grey method has numerous applications, as any issue of the Journal of Grey System will testify. Extensive research has been done to attempt to explain the phenomenon of geography, geology, agriculture and earthquakes. Meanwhile, other researches have studied social phenomenon including financial operating performance, stock markets, supply and demand for electronic power, and the market for air travel and management decisions. The GM (1, 1) model uses the most up-to-date data to predict future values, and poor forecasting may result when the data are random with central symmetry. In this paper, GM (1, 1) equidimensional filling vacancies model is applied to predict load, the effectiveness of which is verified by a real case of computation [13-16].

III. METHODOLOGY

A. GM (1, 1) Model

The GM (1, 1) is one of the most frequently used grey forecasting models, which was developed by Deng (1982).

Step 1: Assume the original series to be $X^{(0)}$

E. Qi et al. (eds.), *Proceedings of the 21st International Conference on Industrial Engineering and Engineering Management 2014*, Proceedings of the International Conference on Industrial Engineering and Engineering Management, DOI 10.2991/978-94-6239-102-4_89, © Atlantis Press and the authors 2015

$$X^{(0)} = \left(x_1^{(0)}, x_2^{(0)}, ..., x_n^{(0)}\right) \qquad (1)$$

Where n is the observed number.

Step 2: A new sequence $X^{(1)}$ is generated by the accumulated generating operation (AGO).

$$X^{(1)} = \left(x_1^{(1)}, x_2^{(1)}, ..., x_n^{(1)}\right) \qquad (2)$$

Where $x_t^{(1)} = \sum_{i=1}^{t} x_i^{(0)}, t = 1, 2, ..., n$

Step 3: Establish a first-order differential equation

$$\frac{dX^{(1)}}{dt} + aX^{(1)} = u \qquad (3)$$

Step 4: From step 3, we can have

$$\hat{x}^{(1)}(t+1) = (x^{(0)}(1) - \frac{u}{a})e^{-at} + \frac{u}{a} \qquad (4)$$

Where

$$B = \begin{bmatrix} -\frac{1}{2}(x_1^{(0)} + x_2^{(0)})1 \\ -\frac{1}{2}(x_2^{(0)} + x_3^{(0)})1 \\ \vdots \\ -\frac{1}{2}(x_{n-1}^{(0)} + x_n^{(0)})1 \end{bmatrix}, Y_n = \begin{bmatrix} x_2^{(0)} \\ x_3^{(0)} \\ x_4^{(0)} \\ \vdots \\ x_n^{(0)} \end{bmatrix},$$

Step 5: By applying the inverse AGO (IAGO)

$$\hat{x}^{(0)}(t+1) = \hat{x}^{(1)}(t+1) - \hat{x}^{(1)}(t) \quad \text{to} \quad (4), \quad \text{the}$$

predictions for original data series can be obtained:

$$\hat{x}^{(0)}(t+1) = (1 - e^a)(x_1^{(0)} - \frac{u}{a})e^{-ak}, \; t=0,1,2... \quad (5)$$

B. GM (1, 1) Equidimensional Filling Vacancies Model

Type primary headings in capital letters roman (Heading 1 tag) and secondary and tertiary headings in lower case italics (Headings 2 and 3 tags). Headings are set flush against the left margin. The tag will give two blank lines (26 pt) above and one (13 pt) beneath the primary headings, 1½ blank lines (20 pt) above and a ½ blank line (6 pt) beneath the secondary headings and one blank line (13 pt) above the tertiary headings. Headings are not indented and neither are the first lines of text following the heading indented. If a primary heading is directly followed by a secondary heading, only a ½ blank line should be set between the two headings. In the Word programmed this has to be done manually as follows: Place the cursor on the primary heading, select Paragraph in the Format menu, and change the setting for spacing after, from 13 pt to 0 pt. In the same way the setting in the secondary heading for spacing before should be changed from 20 pt to 7 pt.

Because of the first-order differential equation which is used in Gray GM (1, 1) model is exponential; GM (1, 1) model is applied to a strong exponential load forecasting. But it requires the data is equidistant, adjacent, no jumping, and requires the latest data as a reference point, the earliest data is dispensable, but the latest data to be added. For these reasons, the application of Gray GM (1,1) Model is limited. So in this paper, the author would work to improve this.

The essence of Grey forecasting model of equidimensional filling vacancies GM (1, 1) is to get each new forecast data into the original data, while removing one of the earliest data, thereby maintaining the same number of data. Then, use the sample sequence with the times to rebuild gray GM (1, 1) model to predict the next value. Repeat the above process, forecast one by one, elected one by one. When using it to predict, it can replenish the use of new information and increase degree of gray plane albino. Finally, load forecasting accuracy will be improved significantly. The modeling process is as follows:

First of all, handling the data respectively by 1-AGO and 1-IAGO as the traditional GM (1, 1) model, then the corresponding time series is obtained.

By 1-AGO and 1-IAGO as the traditional GM (1, 1) model, then the corresponding time series of GM (1, 1) model is as following.

$$x^{(0)}(k+1) + \frac{1}{2}a\left[x^{(1)}(k+1) + x^{(1)}(k)\right] = u$$

$$\hat{x}^{(0)}(k+1) = \hat{x}^{(1)}(k+1) - \hat{x}^{(1)}(k) = (1 - e^{\hat{a}})$$

$$\left(\hat{x}^{(0)}(1) - \frac{\hat{u}}{\hat{a}}\right)e^{-\hat{a}k} \quad (k = 0,1,2,...,n-1)$$

The original sequence and other by-dimensional grey-scale dynamic process:

Remove $x^{(0)}(1)$, add $\hat{x}^{(1)}(n+1)$. Thus, the original data sequence becomes:

$$x^{(0)} = \left\{x^{(0)}(2),...,x^{(0)}(n), \hat{x}^{(0)}(n+1)\right\}$$

Based on this adjusted data sequence, re-use the traditional GM (1, 1) model to predict the next value. Finally, repeat the above steps until the final demand forecast results is satisfied.

IV. THE CASE STUDY

A. GM (1, 1) Model

There are sufficient power resources in Shandong. By the end of 2004, the province's power generation installed capacity reached 32.92 million kilowatts, the total electricity consumption to complete 164 billion kWh. As of the end of 2006, the province's total installed generating capacity reached 50.05 million kilowatts. The province's electricity supply and demand relative tends to the overall balance of the surplus, showing the trend of oversupply. Shandong Power Grid is the only province in the six that has independent power grid. Therefore, we selected Shandong Province as objects for analysis. In the modeling process, we selected electricity consumption in Shandong Province from 2004 to 2011 as the raw data, and the 2012, 2013 electricity consumption data as two years of testing the merits of the standard model. The power consumption of Shandong province from 2004 to

2013 as Table I:

TABLE I
THE POWER CONSUMPTION OF SHANDONG PROVINCE FROM
2004 TO 2013 UNIT: KWH)

Year	2004	2005	2006	2007	2008
Throughput	1000.71	1104.53	1241.74	1395.72	1639.92
Year	2009	2010	2011	2012	2013
Throughput	1911.61	2272.07	2596.05	2726.97	2941.07

(Date from China Bureau of Statistics)

After using matlab to get the original series and after series of accumulated generating trends, the exponential growth trend can be found. That is to say, we can use GM (1, 1) to predict.

With the help of matlab, the results could be calculated as below (Fig.1):

$$A = \begin{bmatrix} \hat{a} \\ \hat{u} \end{bmatrix} = \begin{bmatrix} -0.14864 \\ 834.51820 \end{bmatrix}$$

Therefore:

$$\hat{x}^{(1)}(k+1) = 6615.07e^{0.14864k} - 5614.36$$

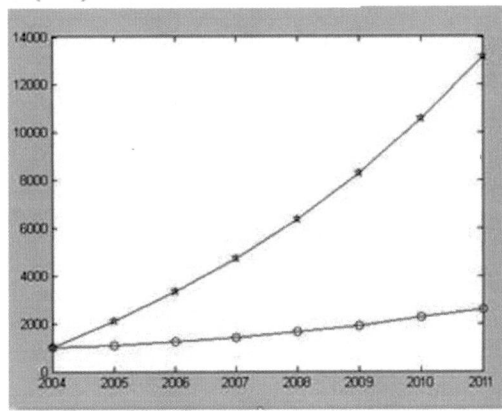

Fig.1. The fitting value of GM(1,1) model and the original data

From 1-IAGO to get the grey prediction model that is:

$$\hat{x}^{(0)}(k+1) = 6615.07(1 - e^{0.14864})e^{0.14864k} \quad (k = 0, 1, 2, \ldots)$$

The original data and the predictive data from 2004 to 2011 for Shandong as Table II:

TABLE II
GM (1, 1) PREDICTIVE VALUE, ACTUAL VALUE AND THE
RESIDUAL

Year	Original data	Forecast	Residual
2004	1000.71	913.68	-87.03
2005	1104.53	1060.10	-44.43
2006	1241.74	1229.98	-11.76
2007	1395.72	1427.09	31.37
2008	1639.92	1655.78	15.86
2009	1911.61	1921.12	9.51
2010	2272.07	2228.99	-43.08
2011	2596.05	2586.19	-9.86

Avoid combining SI and CGS units, such as current in amperes and magnetic field in worsteds. This often leads to confusion because equations do not balance dimensionally. If you must use mixed units, clearly state the units for each quantity that you use in an equation.

B. GM (1, 1) Equidimensional Filling Vacancies Model

Predict the 2012 and 2013 loads of Shandong Province with other by-dimension method improved gray GM (1, 1) model. The results are as Table III:

TABLE III
COMPARING THE PREDICTIONS BETWEEN TRADITIONAL GM
(1, 1) PREDICTIONS BETWEEN TRADITIONAL GM (1, 1) AND
IMPROVED MODEL

Traditional Grey Model			
Year	Real value	Forecast	Residuals	Residual rate
2012	2726.97	3000.63	273.66	10.0353%
2013	2941.07	3481.49	540.42	18.3749%

(CONTINUED)

Improved Grey Prediction Model		
Year	Forecast	Residuals	Residual rate
2012	3000.23	273.26	10.0206%
2013	3479.72	538.65	18.3148%

There are two indicators about difference test after test: poor ratio of posterior "c" and small error probability "p". "c" is smaller, the model is better; "p" is greater, the model is better.

TABLE IV
ACCURACY ASSESSMENT MODEL

Accuracy class	p	c
First grade: Good	>0.95	<0.35
second grade: Qualified	>0.80	<0.5
Third grade: Reluctantly	>0.70	<0.65
Fourth grade: Failure	<=0.70	>=0.65

Table V
THE POSTERIOR MARGIN OF THE TWO MODELS

	p	c
Traditional GM (1,1) model	3.62	0.06767
Improved GM (1,1) model	4.28	0.03999

From the Table IV and Table V, we found that the accuracy of the improved model was better than the original grey model. However, the accuracy of the improved model was better than the one with improvements. Improved gray model for 2012 and 2013 achieved significantly higher accurancy than the conventional gray prediction.

V. CONCLUSION

There are many factors affect power load which belongs to the gray areas of the system. In this paper, the author improved GM (1, 1) model and introduced grey forecasting model of equidimensional filling vacancies. At the end, the author predicted the Shandong Province in 2012 and 2013 the electricity load by using the improved GM (1, 1) forecasting model, and compared the results with the actual electricity load. However, the greater the need to predict the amount, the greater the computation it is. For this reason, the model still needs to be improved.

ACKNOWLEDGMENT

This paper is supported by Hebei Province Social Science Foundation the Grant No.HB12GL073, and Hebei Province Educational department Science Foundation the Grant No.GH121003, and National Natural Science Funds of China (Grant No. 11104058), and Hebei province natural science foundation of China (Grant No. A2011201155).

REFERENCES

[1] Deng J.L, Control problems of Grey System, Wuhan, Huazhong University of Science and Technology Press, Wuhan, 1990

[2] Chao, H. W, Predicting tourism demand using fuzzy time series and hybrid grey theory, Tourism Management, Vol.25, 2004, P367-374

[3] Sun Jihu, Forecasting model of coal requirement quantity based on grey system theory, Journal of China University of Mining & Technology, vol.11, No.2, 2001, P192–195

[4] Cuifeng Li, Study on Theory of the Grey Markov Chain Method and Its Application, 2006 IMACS Multiconference on "Computational Engineering in System Applications" (CESA) Conf, P1742-1746

[5] Sun Caizhi , Zhang Ge, and Lin Xueyu, Model of Markov Chain with weights and Its Application in Predicting the Precipitation State, Systems Engineering- theory & Practice, vol.23, Apr.2003, pp.100-105

[6] Jianli Ding, Guansheng Tong. Real-time Sub-time Early Warning of Airport Scheduled Flight Delay Base on Immune Algorithm[C]. IEEE Computer Society, 2008: 430-435.

[7] Zhou Zuliang, Yin Chunwu, Application of Gray Metabolic Forecast Model in the Prediction of the Cotton Output in China, Journal of Anhui Agricultural Sciences, 2011 .

[8] Yufeng Tu, Michael O.Ball,Wolfgang S. Estimating Flight Departure Delay Distributions A Statistical approach with Long-term trend and short term pattern. Journal of the Amerian Statistical Association. March2008, Vol.103, No.481.

[9] Y. Wang, Q.B. Song, MacDonell Stephen, Integrate the GM(1, 1) and Verhulst models to predict software stage-effort, IEEE Transactions on Systems, Man, and Cybernetics-Part C: Applications and Reviews, 39 (6) (2009)

[10] Shi Yumei, Chen Yongcheng, Ma Benxue, Li Zhiyong. Application of Metabolic GM (1, 1) Model in Corps' Agricultural Machinery Total Power Forecasting. Agricultural Equipment & Vehicle Engineering, 2006.

[11] Wang Zhaoyang, Xu Qiang, Fan Xuanmei, Zeng Jinhua. Application of renewal gray GM (1, 1) model to prediction of landslide deformation with two case studies. Hydrogeology & Engineering Geology 2009.

[12] Jing Guoxun, Heng Xianwei. Study on the Prediction of Coalmine Gas Emission Quantity Based on the Comparison between Gray System and One Element Linear Regression. Safety and Environmental Engineering, 2010.

[13] Mei Mudan, Yuan Shujie, Wei Wei. Prediction of Coal Mine Disastrous Accidents Based on Gray GM (1, 1) Model and its Algorithm. Coal Technology, 2010.

[14] Erdal Kayacan, Grey system theory-based models in time series prediction, Journal of expert systems with applications, Vol.37, March 2010, P1784-1789.

[15] Tzu-Li Tien, A new grey prediction model FGM (1, 1), Journal of Mathematical and Computer Modelling, Vol.49, April 2009, P1416-1426.

[16] Zhao hai-qing (2007) Research and Application of Equivalent Dimensions Additional Grey Correct Model, Operations Research and Management Science, 1(2), 97-99.

Research and Application of an Intelligent Decision Support System

Xiaoqing ZHOU[1], Jiaxiu SUN[2,*], Shubin WANG[1]

[1]Center of Computing, China West Normal University, Nanchong, China

[2]Colleges of Business, China West Normal University, Nanchong, China

(zhousun123@163.com, *369656820@qq.com)

Abstract - **This paper discusses a new decision-support system that integrates data warehouse, knowledge warehouse and model warehouse. Contrast to the fixed model of the old decision-support system and it's limited application, the new system can overcome the shortcoming of the old system efficiently, and also it can simplify model-obtaining and coding. So the new system strengthens the effectiveness, intelligence and efficiency of the decision.**

Keywords - **Decision-support system, data warehouse, data mining, knowledge warehouse, model warehouse**

I. INTRODUCTION

Although DSS (Decision-Support System) can supply timely, accurate and scientific information, the most advanced SDSS (Spatial Decision Support System) has defects [1]. SDSS integrates the traditional and the new DSS (including data warehouse, OLAP (On-Line Analysis Processing), data mining, data base, and ES), so it can solve many questions. But due to the fixed model of the model warehouse, which cannot adjust according to the change of the condition parameter, the application of the SDSS is limited [2, 3]. So the paper is tries to introduce a new decision-system system that is based on the data warehouse, knowledge warehouse and model warehouse. The new system can update the knowledge of the knowledge warehouse freely by using knowledge warehouse and date warehouse. And also the new system can strengthen the effectiveness, intelligence and efficiency of the decision by the management of the MWMS and the study of the system.

II. MAIN MODULES

The system is composed of model warehouse, knowledge warehouse, method base, data warehouse, OLAP and data mining modules.

A. Model Warehouse

Model warehouse has the following functions: management with classification, memory the necessary model (including using date-mining model) and comprehensive model parameter (in order to choose out the proper model) [4]. Machine-detecting technology integrated the artificial intelligence (AI) can accomplish model creating by computer by simulating the data of the date warehouse/ database. The Self-study algorithm

by the nerve network of the model-study can adjust the model fine and update the parameter to get the optimal practical model, so the model can keep in chorus with the fact. Flexible software development technology integrated Software Engineering supports model-coding. Model management system is to manage model of the model system and to call/operate model.

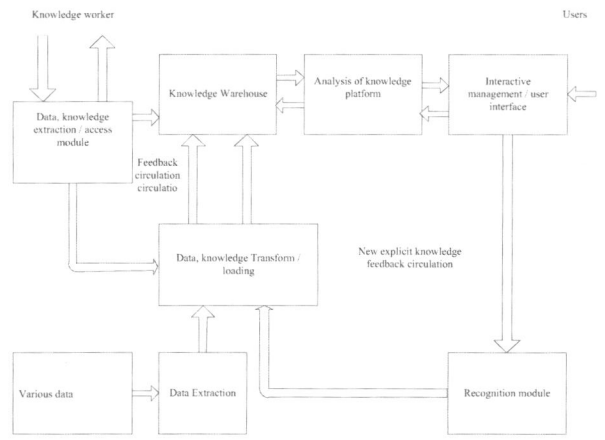

Fig.1. knowledge warehouse system

B. Knowledge Warehouse

Knowledge Warehouse has the function of obtaining, clearing/transforming/coding, organizing, memorizing, adjusting, and propagating knowledge [5]. KW can accomplish the function by expanding system structure of the date warehouse. KW is composed of six components. ① Knowledge/date-obtaining module. It is to switch recessive knowledge to dominant knowledge, which is to say to get recessive knowledge from decision-maker. ② Two feedback loops. One is between knowledge-obtaining module and knowledge-memorizing module. The other is between Extract-Transform-Load (ETL) module and mutual management module. And it is to memorizing the knowledge, which has been verified by the system and to update knowledge warehouse timely. ③ ETL module. It is similar with the corresponding module of data warehouse. ④ Knowledge warehouse module. One of the main components is Knowledge Base Management System, which accomplishes the analysis both by knowledge warehouse and model warehouse. ⑤ Analysis worktable. It is composed of task controlling, conclusion getting and technology-Management modules. ⑥ Interface module. It is to handle the interaction between KBMS and user's interface. The knowledge warehouse system is as following Fig.1.

C. Method base

Method should be based on the model and be adjusted according to the model in order to calculate. But one model can have several methods [6]. Method base is to supply method for DSS problem model to calculate. And method base management system is to add, delete, revise and search method and to give service for model solving.

D. Data warehouse and OLAP

OLAP is one kind of data warehouse application, and it is based on data warehouse [7]. So it can provide decision-makers with analysis results by analyzing and handling. Data warehouse organizes data according to function requirement, the use and granularity of DSS. The key point of OLAP is how to organize data to satisfy user's multi-dimension data analysis.

E. Data-mining

Data-mining module is to mine data to get the needed knowledge according to the model, method and knowledge provided by relevant warehouse. And the result of data mining can be used as new knowledge and model to solid knowledge warehouse and model warehouse.

F. Problem solving and interactive system

Problem solving module is to solve problem by using knowledge, model, method and knowledge of relevant warehouse. Non-structure problem, which cannot be structured, may be solved by deduction system.

III. THE FRAMEWORK AND STRUCTURE OF NEW DECISION-SUPPLY SYSTEM

Fig.2 is the structure of DSS, which integrated DW, KW, MW, MB, OLAP, Data-mining and Problem Solving system. Data mining, knowledge-deduction center, model-creating units of model warehouse are the intelligence center of DSS which strengthens intelligence property of DSS. And problem solving and interactive system are the function center.

DSS comprises three main parts. The first one is the integration of MW system, DS system and DW system. And it is the basis of decision-support system to provide assist decision-making information of Quantitative analysis (Model Calculation). The second

one includes DW and OLAP, which extract spatial data and information from DW. The third one is the integration of experts system and data mining system. Data mining mines knowledge from DB and DW and puts it into knowledge warehouse of experts system, then experts system analyzes. The three parts are integrated. Users can choose one part for decision, either two or three according to the fact. The traditional DSS chooses the first part, IDSS chooses the first part and the third part, and the new DSS chooses the second part and data mining of the third part. The new DSS integrates the three parts by using problem solving and interactive system can give better assist decision-making decision.

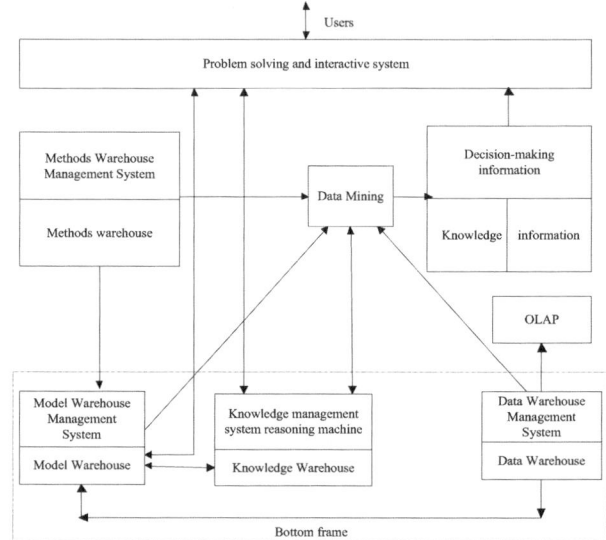

Fig.2. Integrated DW, KW, MW of DSS architecture diagram

Generally speaking, three integrated parts; three warehouses and the application of closed cycle free back and the introduction of MW system is the characteristic of the framework, which makes it more intelligent.

IV. KEY TECHNOLOGY TO ACCOMPLISH THE DECISION-SUPPORT SYSTEM

A. Data mining and text mining

Data mining is to find out unobvious pattern and to acquire needed knowledge in order to help enterprises make decision more scientific and more accurate by analyzing and handling large number of data. Text mining is to acquire valuable information from all kinds of text information. The text source can be Web, fax, E-mail, example and other kinds of text. The decision-makers can extract useful information according to rules and guides who have been defined advanced to make decision.

B. Modeling

Workers should define objective function, decision variables and its' weight. They also should make definite restrictive conditions and coefficients of variables according to the decision variables. So the model that comprises the elements (decision variables, coefficients, restrictive conditions and objective function) can reflect the invisible knowledge obviously.

C. Decision-support tools

It is the tool which uses existed knowledge to help make decision. It needs much technology and tools, including AI, expert system, software engineering, knowledge search tool, knowledge explaining tool and multi-dimensional tool and so on [8].

D. Intelligence-support technology

It includes: ① Model warehouse system should be designed to accomplish it's function. ② Interface: All parts are jointed by interface. Model, data and knowledge are separated parts that should be integrated. So the interface is very important. Interface should have the function of saving and extracting data, calling and operating the model, and knowledge reasoning. ③ System integrated: an integrated system should integrate all parts by words according to the fact.

V. EXAMPLES

Here is an example for a domestic large-scale machinery limited corporation. The old quality management decision-support system of the corporation cannot adjust effectively due to the fixed model coefficients and cause low efficiency [9, 10]. In order to increase production the decision must be changed. Generally, quality breakdown and quality cost will be the core after checking relevant files and investigating the decision-makers. And the analysis should be started from suppliers, manufactories, employees, products and time.

As for products, they should be analyzed by one product or product classification. But one unit can analyze manufactories, employees and suppliers. Time itself is a dimensional data. The quality analysis for every employee, product, supplier and manufactory can be made yearly, quarterly and monthly and the result (graph or table) can help make decision. But if the model coefficients can not adjusted finely, the above result cannot be obtained. The new decision-support system that integrated KW, MW, and DW can be developed and it is based on Windows2000, SQLServer2000, and Excel 2000. The new system can revise model coefficients automatically according to the change of time, employee, product and manufactory. It also can call method of

method base and handle by OLAP; the results are as Fig.3. So decision-makers will know the worse part and can make better decision by getting the right reason.

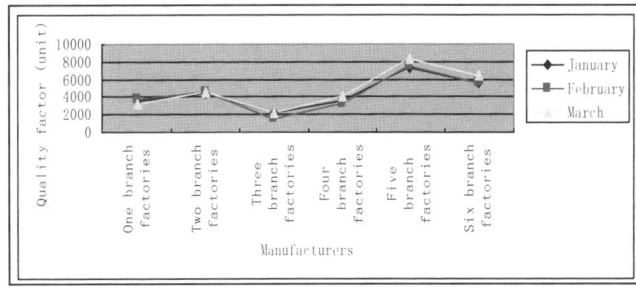

Fig.3. Product qualities pursues every branch factory comparatively

REFERENCES

[1] Chen Song-can, ZhuYu-lian, Zhang-Daoqiang, et al. Feature Extraction Approaches Based on Matrix PaRem: MatPCA and MatFLDA [J]. Pattern Recognition Legers, 2005, 26(8): 1157-1167.

[2] RAO Yi-ning, LIU Qiang, DU Xiao-li, YE Peng, Research and Design of Extensible Knowledge Database Model Applied to Intelligent Chinese Search Engine [J], Application Research of Computers 2006, 23(6): 223-226.

[3] ZHAO Han, DONG Xiao-hui, FENG Bao-lin, WU Zhao-yun, Modeling and Application on Decision Support System Based on Knowledge Warehouse [J], Journal of Systems & Management, 2008, 17(3): 327-331.

[4] Feng Qing, Yu Suihuai, Yang Yanpu, Product DSS Model Based on Cloud Service [J], China Mechanical Engineering, 2013, 24(15): 2013-20159.

[5] Yang Fenfen, Wang Ying, The Research and Design of the Decision Support System about Agricultural Machinery [J], Journal of Agricultural Mechanization Research, 2014, (3): 35-38.

[6] LIU Bo-yuan, FAN Wen-hui, XIAO Tian-yuan, Development of Decision Support System, Journal of System Simulation, 2011, 23(7): 241-244.

[7] Xu Wei, Based on Knowledge Discovery Mechanism of Enterprise Decision Support Systems Research [D], 2013, 12.

[8] LI Bing-nan, ZHAO Dong-zhi, JIANG Xue-zhon, Conceptual design of emergency decision support, Marine Environmental Science, 2014 33(3):418-423

[9] Afarwal S, Mozafari B, Panda A, Milner H, Madden S, Stoica I. Blink DB: queries with bounded errors and bounded response times on very large data [C]// ACM. Proceedings of the 8th ACM European Conference on Computer Systems.2013:29-42

[10] Zheng Y,Zhou X.Computing with spatial trajectories [M]. Springer-verlag New York Inc. 2011

Design and Development of Parametrical Modeling Module for Warheads

Zhi-fang Wei [1,*], Hui Zuo[2], Fang Wang[1], Xiao-guang Li[1]

[1]College of Mechatronic Engineering, North University of China, Taiyuan, China
[2]No 208 Research Institute of China Ordnance Industries, Beijing, China

(lilac1974@139.com)

Abstract - **The software platform for damage efficiency optimization of small arms provides convenient conditions for designers to complete the demonstrating, designing work and optimization of projectiles. Based on Solidworks 2007, by VC++, and using OLE technology, procedure module of parametric modeling of warhead is designed and developed in this paper, which is integrated into the software platform for damage efficiency optimization of small arms, consequently parametric modeling and automatic output of structural characteristic parameter for warhead are realized on the optimization platform.**

Keywords - **Integration, OLE technology, parametric modeling, small arms, software platform, solidworks, warhead**

I. INTRODUCTION

The software platform for damage efficiency optimization of small arms was designed and developed to provide a convenient, unified design platform for designers of small arms ammunition, on which such designing activities as structural design, calculation, analysis, and performance evaluation can be effectively integrated together, consequently the design efficiency and quality would be greatly improved, at the same time development period for small arms ammunition products would be immensely shortened [1-3]. Parametric structure design is an important functional module of the platform [4]. In this paper, a lot of geometrical structure models of typical bullets including pistol bullets and rifle bullets and minor-diameter bullets were created, and saved in the corresponding databases, at the same time the secondary development technology on Solidworks based on OLE technology was studied, and then the parametrical structure design module for warhead was designed and developed using the VC ++ Programming tools, and was integrated in the platform by studying the integration technology. So，on the platform, designers can quickly generate the three-dimensional structure models for warheads by selecting and parametrically modifying the appropriate warhead models in the database, and output the structural sizes and structural characteristic parameters of the bullet to the subsequent calculation program.

II. FRAME DESIGN

A. Requirements analysis

Using the warhead structural parametrical design module, designers can quickly complete the parametrical structure design for warheads of pistol bullet, rifle bullet, and other minor-diameter projectiles, they can also automatically calculate and output the geometry sizes and structural characteristic parameters of kill element model. The module was integrated into the software platform for damage efficiency optimization of small arms, so the correlative models and parameters in the database of the platform can be obtained and used by the designers.

B. System frame

The system frame of warhead parametrical design module was shown in Fig.1, including user layer, application layer and data layer.

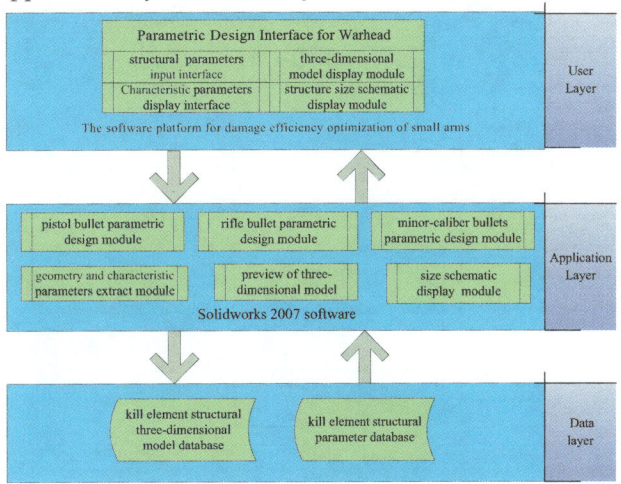

Fig.1. System frame of warhead parametric design module

User Layer: the design process would be guided in the form of design wizards, by designing and establishing a friendly man-computer interaction interface.

Application Layer: it is the core module, including pistol bullet parametrical design module, rifle bullet parametrical design module, minor-diameter bullets parametrical design module, the extracting and outputting module of structural characteristic parameters and outline sizes for pistol bullet, rifle bullet and minor-caliber bullets, the real-time preview module for three-dimensional models, and the display module of sketch map of structure sizes for warheads.

Data layer: the database was designed to save the geometrical structure models and correlative parameters of current domestic and foreign warheads and minor-caliber bullet.

III. CREATING OF MODELS DATABASE FOR KILL ELEMENT STRUCTURE

A. Principle of parametrical structure design

In this paper, parametric modeling for warhead structure was realized by modifying dimensions of the corresponding template model [5, 6]. Template models were created in advance and saved in the corresponding database. The parametric modeling principle was described as shown in Fig.2: the appropriate model template was selected from the models database, the modified dimension parameters were inputted by designers, then the first part of the template assembly was opened, the corresponding size was confirmed and modified by the size name previously defined in the template, and then the part was reconstructed and closed, other parts were also modified in turn in the same way, consequently the right structure assembly model was created. On the other hand, the structure characteristic parameters of the warhead including mass, center of mass, moment of inertia and so on were automatically calculated and outputted to subsequent calculating modules, which could be displayed in the user interface, or saved in the named document according to different needs. It was realized by programming using Solidworks API functions

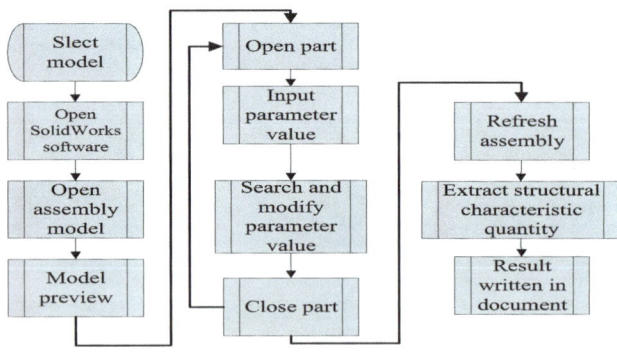

Fig.2. Parametric modeling process of warhead

B. Creation of models database for kill element structure

As the basis of parametric modeling, and used to parametrically construct a variety of 3-D models again, the template models would be current and accurate. This is to say, the template models would contain various warhead structure, and describe fine structural characteristics.

With Solidworks software, a structure characteristic can be created by a variety of modeling methods, so, the modeling regulation of template model must be previously made, as follows:

(1) There were many good tools calculating the center of mass and the moment of inertia for the part in Solidworks software, in order to extract correct and credible parameters and to accord with the engineering practical requirements, the center of the bottom of warhead was defined as the origin and the axis of bullet as Y axis.

(2) The template model was used to create the new model by parametrically modified the corresponding sizes, and their geometrical topology between characteristics must be unchangeable during the reconstruction of the model, so the appropriate dimensions and topology relation for the template model to parameter-drive would be defined by analyzing geometry and dimensioning information for the part.

(3) To meet many designers' requirement, some design regulations were made as follows: the shape of circular arcs of bullet tip was determined by arc radius and arc height and position of the arc vertex, and the structure of the tail cone was determined by cone length and cone angle, and the chamfer for bullet bottom was created using an independent chamfer feature, and so on.

(4) Considering the connection with the internal and external ballistics and other subsequent calculation program, some design regulations were made. For example, the warhead circular arc could be created by the ways of three points on circular arc, or two points on arc together with circular arc center, or two points on arc together with radius. Because arc radius and height values would be used for subsequent calculation procedures, the way of two points on arc together with radius would be used to create a circular arc.

(5) In order to simplify the structure, some design regulations were made. For example, the thickness of shells were different among the bottom part, the column part, and the tip part of the bullet, but the small differences were not enough to cause remarkable errors to the entire structure, so, after building the exterior of bullet shell, the inner surface of bullet shell was created in the form of overall offset of the exterior.

(6) Because the reconstruction of the model was achieved by parametrically modified the corresponding sizes of the template model, and some sizes would be automatically output to other subsequent calculation program according to the size name, the appropriate name should be previously determined for the corresponding sizes for the template model.

In this paper, a model database was created, containing template models of pistol bullet, rifle bullet and minor-caliber projectiles. The schematic diagram of the structure and size of a pistol bullet was as shown in Fig.3, and the three-dimensional model of the bullet was shown in Fig.4. At the same time, the sizes pictured in Fig.3 were named as shown in Table I.

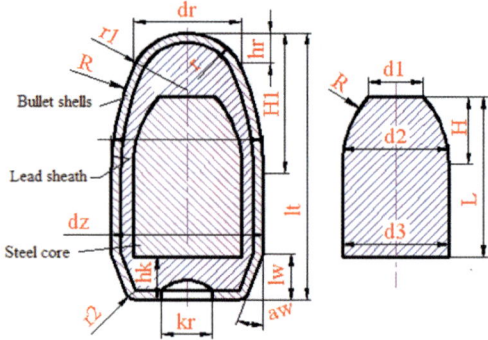

Fig.3. Schematic diagram of a pistol bullet size

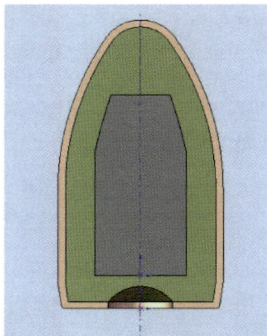

Fig.4. Three-dimensional model diagram of pistol bullet

TABLE I
DIMENSIONS VARIABLE OF PISTOL BULLET

Size Name	Dimensions variable (with steel core) in Solidworks
Warhead length (lt)	lt@ sketch_dk_py
Tail cone length (lw)	lw@Sketch_dk_pno
Tail cone angle (aw)	90-aw@ sketch_dk_py
Warhead arc height (H)	H@ sketch_dk_py
Warhead cylindrical diameter (dz)	dz/2@ sketch_dk_py
Arc rotating body big end diameter (dh)	dh/2@ sketch_dk_py
Bullet bottom diameter (kr)	kr@ sketch_dk_py
Warhead tip radius (r)	r@ sketch_dk_py
Curved portion radius (R)	hu_R@sketch_dk_py
Warhead shell thickness (tz)	tz1@sketch_dk_py
bullet bottom chamfer(r2)	D1@ bullet bottom chamfer
Steel core length (L)	L@sketch_gx_py
Steel core diameter (d3)	d3/2@sketch_gx_py
Steel core diameter (d2)	d2/2@sketch_gx_py
Steel core diameter (d1)	d1/2@sketch_gx_py
Steel core arc height (H)	H@sketch_gx_py
Steel core radius of arc (R)	R@sketch_gx_py
Distance from the center of mass to bullet bottom (hk)	D1@ bullet bottom distance

IV. DESIGN AND DEVELOPMENT OF THE PARAMETRIC DESIGN PROCEDURE

Both COM and OLE ways could be used in the secondary development for Solidworks software [7-9]. COM technology was more popular and widely used, and was named as internal development model, in which a dynamic link library (DLL) file was created, loaded and used by SolidWorks software in the form of plug-in files. On the other hand, OLE technology, or the object embedding and linking technology, was rarely used currently and named as external development mode, in which an executable file (.exe) was created, could run independent of SolidWorks platform and call functions of SolidWorks software. The latter is suitable for the model in which SOLIDWORKS should be integrated into and called by other programs as third-party software.

Because the warhead parametric design module would be integrated into and called by the software platform for damage efficiency optimization of small arms, the asynchronous development model of OLE technology was used in this paper. This was also the main innovation of the paper.

A. Import and export of models in Solidworks software in asynchronous mode

Before opening the model file, the connection between warhead parametric design software and Solidworks software should be created, then the specified assembly file could be opened by using Solidworks API function. The procedure code was shown as follows:

```
ISldWorks m_SldWorks ; //////SolidWorks object definition
    BOOL panduan;
    panduan=m_SldWorks.CreateDispatch("SldWorks.Application",NULL);
    /// Open SolidWorks software
    if(panduan==TRUE)
    {
    m_SldWorks.SetVisible(TRUE); // SolidWorks is defined visible after opening
    }
    else
    {
    AfxMessageBox("SolidWorks Can not run, please check whether SolidWorks is installed ");
    }
    m_SldWorks_refer=m_SldWorks;// M_SldWorks obtained is defined as global variables.
    // Open the assembly file
    long type=swDocASSEMBLY;// Definition of document type
    LPDISPATCH modDisp0;// Definition of intermediate variables
    file_path="D:\\Warhead model \\Warhead model library \\pistol bullet\\ pistol bullet_without steel core \\Assembly drawing_pistol bullet_with steel core.SLDASM ";
    //Close the assembly file
    pModelDoc=NULL;
    m_SldWorks_refer.CloseDoc(file_path);
    CDialog::OnCancel();//Close the assembly file
```

B. Achievement of dimension-driven function for parametric model

The function was to find and determine the corresponding size according to the size name and to modify to the specified value after opening a part. In following procedure code, m_pn_lt is the name of a variable.

```
    LPDISPATCH P1;
    IDimension h1;
    P1=m_iModelDoc1.Parameter("lt@sketch_dk_pno");// Define the name of the specified size got.
    h1.AttachDispatch(P1);
    h1.SetValue (m_pn_lt);// modify the size with a new value again
```

C. Calculation and output of structural characteristic for parts in Solidworks in asynchronous mode

The function was mainly to calculate and output the equatorial moment of inertia, polar moment of inertia, mass and the distance from center of mass to bullet bottom.

```
VARIANT Vmass;
VariantInit(&Vmass);
Vmass.vt=VT_ARRAY;
Vmass=pModelDoc.GetMassProperties();
SafeDoubleArray tezheng(&Vmass);
m_tz_a=(float)(tezheng[8]*1E9);//    Equatorial
moment of inertia
m_tz_c=(float)(tezheng[7]*1E9);// Polar moment
of inertia
m_tz_m=(float)(tezheng[5]*1E3);//Mass
m_tz_lcd=(float)(tezheng[1]*1E3);//  Distance from
center of mass to bullet bottom
```

V. INTEGRATION OF THE PARAMETRIC DESIGN MODULE INTO THE PLATFOEM [10-12]

A. Module Interface Design

By Visual C ++ 6.0, a dialog-based application was created using MFC AppWizard (exe), and Solidworks.tlb file in the SolidWorks installation directory was imported into the application program MFC Class Wizard, then the share of functions and data between application procedure and SolidWorks software was achieved, at the same time, SolidWorks library file was added to the .CPP and .h file in the dialog application procedure, so the asynchronous call of SolidWorks by the platform could be realized.

B. User Interface Design

User interface was the communication platform between procedures and users, through which users' input data could be got, output data of procedures could be displayed, sketch map of size could be shown, and the three-dimensional structure model could be previewed in time. For example, the model selection interface of the warhead parametric modeling software was designed as shown in Fig.5.

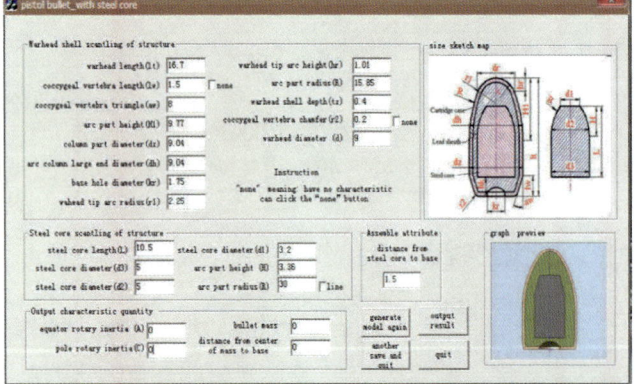

Fig.5. Parametric design interface of pistol bullet with steel core

C. Data Interface Design

To take full advantage of valid data of typical products, and to simplify user's input work, the parametric design module was designed based on the database of the software platform for damage efficiency optimization of small arms. Through the data interface designing, the appropriate data of typical products could be conveniently obtained by designers. The relationship between parametric design module and platform database was as shown in Fig.6.

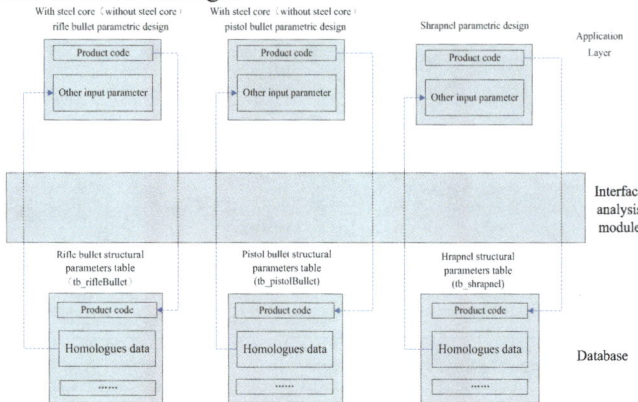

Fig.6. Relationship between parametric design module and platform database

VI. CONCLUSION

In this paper, the warhead structure parametric design module was studied and developed, and integrated into the software platform for damage efficiency optimization of small arms. Using the module, the warhead structure parametric modeling could be achieved in the platform system, and the characteristic parameters of killer element model could be calculated and outputted automatically and transferred to the correlative analysis program.

REFERENCE

[1] Hongyu Xie, Weihua Zhang, Web-based integrated design platform for solid rocket motor, Journal of Propulsion Technology, vol. 28, no. 1, pp. 108-112, 2007. (In Chinese)

[2] Wen Zhao, Haidong Chen, Collaborative digital missile design technology based on virtual prototype, Missiles and Space Vehicles, vol. 31, no. 4, pp. 23-28, 2005. (In Chinese)

[3] Hongwei Zhou, Quan Li, The design and implementation of integration framework in the weapon system concept design, Journal of national university of defense technology, vol. 24, no. 4, pp.91–95, 2002. (In Chinese)

[4] Haoxuan Jin, Study and realization of parametric design subsystem in CAD system, Mechanical & Electrical Engineering Magazine, no. 3, pp.1-3, 2002. (In Chinese)

[5] Xiuqin Huang, Development of three-dimensional standard parts library of rolling bearing based on solidworks, Journal of Changzhou Institute of Technology, vol. 22, no. 3, pp.6-9, 2009. (In Chinese)

[6] Kailing Li, Naikun Sun and Lianfu Zhu, The development of an injection mold CAD system based on solidworks, Journal of Shandong University of Technology, vol. 34, no. 2, pp.22-26, 2004. (In Chinese)

[7] Zhusheng Yan, The study of key technologies in pre-development of solidworks, China Science and Technology Information, pp.146-148, 2006. (In Chinese)

[8] Weiming Li, Shufen Liu, Pre-development technology based on Solidworks platform", *Machinery*, vol. 41, no.4, pp.24–26, 2003. (In Chinese)

[9] Qiang Zhu, The study of pre-development technology of part characteristic parameter extraction based on Solidworks, *Machinery*, vol. 42, no.479, pp.38–40, 2004. (In Chinese)

[10] Sun S.X., Zhao J.L, "Developing a Workflow Design Framework Based On Dataflow Analysis", IEEE Proceedings of the 11th International Conference on the 41st Annual Hawaii International Conference on System Sciences, pp.8-19, 2008.

[11] Jingzhi Guo, Zhuo Hu, Chi-KitChan, Yufeng Luo, Chun Chan, "Document-Oriented Heterogeneous Business Process Integration through Collaborative E-Marketplace", ACM Proceedings of Tenth International Conference on Electronic Commerce Austria, August 1 pp.9-22, 2008.

[12] Xiao Zhou, Hong Jia, Yanlin Lu, Weilong Ding, "Product Model Data Exchange Technology of Heterogeneous Systems in Collaborative Design Environmen", IEEE Proceedings of the International Conference on Artificial Reality and Telexistence Workshops, pp.145-148, 2006.

Study on Measure of Matching Degree between Management Platform and Information Platform in the Context of IS Success Based on Kullback-Leibler Divergence

Qian WANG [1], Jin SHI [2,*], Feng LIANG [1]

[1] Department of Industrial Engineering, Nankai University, Tianjin 300457, China
[2] Department of Logistic Management, Nankai University, Tianjin 300457, China
(shijin@nankai.edu.cn)

Abstract - **Aim to solve the lack of quantitative method on measure of information system success and IT project management, the article advance the principle and related technique on measure of matching degree between management platform(MP) and information platform(IP) based-on Kullback-Leibler divergence(KLD). Firstly, it discuss the essential of enterprise modeling technique based on information flow, and then introduce and analysis the "Critical Nodes" which will contribute to build the architecture of evaluation index system from the management reference model which derives from enterprise modeling process. After that, it investigates how to describe the probability, distribution and parameter of the uncertainty existing in the Critical Nodes. And then, it gives out the method and procedure on using KLD method to calculate the difference between MP and IP on certain Critical Node and then integrate the whole index value via Analytic Network Process (ANP) method. Finally, it gets the perspective on development and application of the quantitative method.**

Keywords - **Evaluation of IS success, IT project management, Kullback-Leibler Divergence, modeling based-on information flow, measure of matching degree**

I. INTRODUCTION

During the informatization process in China, it is hard to solve the "input-output" model and to assess the promotion strength of Information Systems (IS) on enterprise Management System (MS). The lack of accurate evaluation method on implemented IT projects also leads to blind decision making on setting up a new project and misleading on informatization strategy.

The very shortage of quantitative evaluation method on IS effectiveness is mainly involving feasibility analysis, assessment on milestone points, and evaluation at project acceptance stage.

We raise an evaluation technique based on Kullback-Leibler Divergence method, aimed to solve the conundrum of quantitative method in feasibility analysis stage of IT project management, to measure the correspondence between MS and IS, then give the answer to implementation strategy of software product's selection and Business Process Management (BPM).

In the past several decades, researchers carried out ample studies in effectiveness evaluation of IS, those considerable studies by using from qualitative methods to hybrid of qualitative and quantitative methods, but pure quantitative method is rare.

As the founder of the theory of Information, C. E. Shannon and W. Weaver [1] divided the measurement of information output into three layers: technical layer, semantic layer and effectiveness layer. The effectiveness layer shows the effect of the information on the receivers. Based on the model of Shannon, Richard Mason [2] separated the output measurement into technical layer, semantic layer, function layer and influence layer, the influence layer equals to the effectiveness layer in Shannon's model. Mason concentrated his analysis on influence layer, further divided this layer into multistage, from low to high: receive, accept, retain, integrate, evaluate, apply, change behavior, change in system performance. Mason is the first one to study the integration of information technology (IT) and MS, the achievement beyond the limitation of information theory and played vital effect for subsequent researchers.

Robert W. Zmud [3] provided a category method by using indices such as performance, use of Management Information Systems (MIS), and user's satisfaction, to evaluate the influences of human on IS. Peter Keen [4] discussed the long term influences of MIS on business model and social development and raise the necessary of measuring the effectiveness of IS, and then advanced the concept of independent variables and dependent variables used in informatization evaluation. The independent variables are individual factors in the processing of assessment, dependent variables are the evaluation indices. Different independent variables will produce different result on same domain. Blake Ives [5] contributed further research on user satisfaction.

William H. DeLone and Ephraim R. McLean [6] raised the famous D&M model, in which the taxonomy posits six major dimensions or categories of IS success: quality of IS, quality of information, use, user satisfaction, individual impact, organizational impact. Peter B. Seddon [7] revised D&M model and made it more reasonable to meet the up-to-date situation of Informatization. Arun Rai [8] investigated by using questionnaire to improve and perfect the research achievements of D&M model and Seddon's model. Rajiv Sabherwal et al. [9] extended D&M model and increase indices of user training and user experience and established a theory model.

Wen-Hsien Tsai et al. [10] developed a conceptual framework for investigating how Enterprise Resource planning (ERP) selection criteria are linked to IS success, and raised four influence factors: consultant's suggestion, a certified high-stability system, compatibility between the system and the business process, and the provision of best practices. Tsai pointed out that the knowledge ability and service ability of consultant company will play vital influence on IS success.

Qian Wang [11-12] advanced the theory of positive matching between management platform and IS platform based on enterprise business process analysis, and raised the principle and baseline of matching assessment. Wang also put forward ERP implementation capability evaluation model (ERP-ICEM) aimed at solving the measurement on the implementation ability in ERP Project Management (ERP-PM), ERP-ICEM can realize the quantitative measuring on the ability of coordinating among Enterprise and Consultant Company and ERP software suite. Wang [13] advanced the concept of "information particle" and, based on it, proposed the method of studying the energy consuming and the features of Information Flow (IF) in the context of informatization from micro perspective, and raised the fundamental conceiving of information effectiveness measurement based on IF.

Outside the academia, enterprises and governments also carried out lots of works on IS success measurement, the important ones are: The Oliver Wight ABCD Checklist for Operational Excellence [14-16], US standardization institute Benchmarking Partners proposed ERP project evaluation system, Korean proposed the evaluation architecture of informatization index. IDC's Information Society Index (ISI), Harvard University's Networked Readiness Index, (NRI), China's manufacturing industry index by NIEC and Tianjin University are the rest important achievement in this field [17].

The deficiency of the current researches are apparent and ubiquitous:

➤ Quantitative ability is not enough. The complex of MS heavily increase the difficult of quantitative analysis and evaluation, the exist methods and models are more qualitative and lack of deep analysis capability on complexity and non-linear characteristics of management system and IS.

➤ The objects and independent variables and factors to be assessed are a bit of macro, and cannot reflect the inner influence of IT/IS contributes to the management process.

➤ Direct evaluations of informatization effectiveness upon IS are not easy to be achieved. Operational indices, such as financial data, influence and user satisfaction, are indirectly and dependently.

II. KULLBACK-LEIBLER DIVERGENCE

Kullback-Leibler Divergence (KL-Divergence), also called information gain or information Divergence or Relative Entropy, is a non-symmetric manner to measure the difference between two probability distributions of P and Q, is proposed by Solomon Kullback and Richard

Leibler at 1951 [18-19]. KL-Divergence aims to measure the difference of two probability distribution within same event space. The significance of KL-Divergence in information field is to measure the expected number of extra bits required to code samples from P when using a code based on Q, rather than using a code based on P.

Concerning discrete random variables, probability distribution P and Q have their probability mass function (pmf) $P(i)$ and $Q(i)$ respectively, then the definition of KL-Divergence from Q to P is:

$$D(P \| Q) = \sum_i P(i) \log \frac{P(i)}{Q(i)} \quad (1)$$

For continuous random variables, $p(x)$ and $q(x)$ are the probability density function (pdf) of probability distributions of $P(x)$ and $Q(x)$ respectively. Then the definition of KL Divergence from Q to P is:

$$D(P \| Q) = \int p(x) \log \left(\frac{p(x)}{q(x)} \right) dx \quad (2)$$

In the above functions, P represents the true distribution of the data, Q represents the theory distribution of the data or approximate distribution of P.

KL-Divergence have some features as below:

➤ Non-negative, namely $D(P \| Q) \geq 0$, or $D(Q \| P) \geq 0$.

➤ Non-symmetric, namely $D(P \| Q) \neq D(Q \| P)$.

➤ Not satisfy the triangle inequality conditions, namely $D(A \| B) + D(B \| C) > D(A \| C)$ is not guaranteed.

KL-Divergence method is mainly used in Statistics and Information Science and play vital roles in the field of Information Searching and statistical natural language. We invite KL-Divergence into the field of evaluation of IS effectiveness and deploy it to measure the difference between MS and IS within the certain information flow space and achieve the quantitative measurement of the matching degree of MS and IS platforms.

III. MEASUREMENT OF PLATFORMS MATCHING BASED-ON INFORMATION FLOW

A. Platforms and Information Flow

Management Platform (MP) is the term that reflects and includes management method, manner, process, auditory control, feedback system, mechanism and cultural of management system of enterprise operation.

Information Platform (IP) is the term that reflects and includes all the contents, such as hardware and software and related environment, of Information Technology and Information Systems (IT/IS) which aim to improve efficiency and quality of operation management.

The matching degree between MP and IP is the vital determining factor that reflects how the IS runs smoothly and earns reasonable benefits to the revenue. The higher the degree of matching, the larger the IS effectiveness be improved and more significant the management level will be reached. IS harvests valuable achievement dependently and must through the promotion of efficiency of MS, so to measure the effectiveness of IS must through MS indirectly.

Information flow act as instruction through whole management process and undertake the task of forward planning and backward controlling message. The operational efficiency of MS relies on efficiency and quality of information flow. Information flow is the object and subject of information systems and the efficiency's carrier and core of content. Information flow have same meaning and expression in both MS and IS, also the communicational media among MS and IS.

Information flow has features as below:

➤ Having same meaning and content in integrality and semantic in dual platforms

➤ In the context of informatization, complete information flow process is divided by two platforms, the dividing manner is related to the architecture of IS, the interfaces between two platforms act as the communication channel for continuing the flow of information particle.

➤ Information flow is more objective rather than subjective and can be used for objective analysis and depicting.

➤ From information theory and Statistics, uncertainty and randomness can be described by using probability and its knowledge

➤ Spatial and temporal attributes are very important features contained in information flow, they are also the critical indices to measure the efficiency of management processes.

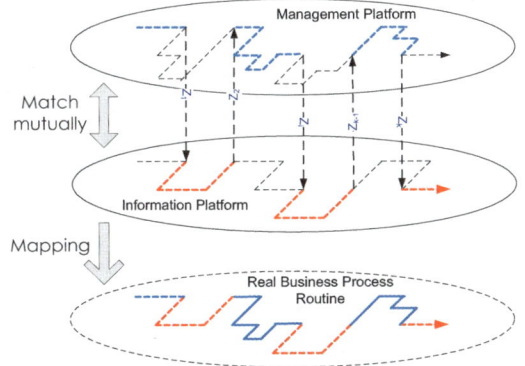

Fig.1. Existence and running manner of Information Flow among Management Platform and Information Platform

B. Critical Nodes in Architecture

Both MS and IS have their integrated architecture and running processes based-on information flow. In traditional mode, information flow which exist in management platform is deal by manual, after using IS, processing of information flow will be arranged by both management platform and information platform. Intuitively, information flow will be deal in sharply quick in information platform due to the advantages information technology. But unfortunately, because there are information islands and lack of artificial capability in information platform, so the information flow cannot be deal only by information platform but also need to be transfer to management platform and by manual help to finish it, especially in decision making procedure. As depicted in Fig.1, in the context of informatization, information flow distribute and run across the two platforms and a complete information flow procedure will jump from/into platforms, the outward manifestation of jump is manual processing or man-machine interaction, during the period, the format of information flow will be transformed to fit for the need of certain platform, this transformation will cost external management resource or energy and will offset or over the efficiency gained from using IT/IS, that will cause reducing of management efficiency and even chaos in business operation.

Fig.1 also mention that the flow path of information particle is not same in management platform and information platform respectively, some node in the whole flow define the content and function and features of a certain flow and never changed with the variety of management optimization and informatization. These nodes are the milestone to measuring the efficiency of the processes so that we call such nodes as "critical nodes in Information Flow" or just "critical nodes". Critical nodes express as the superficial characteristics of information flow, can be used to realize the approximate description upon critical process and core value of information flow.

Based on information model technology, we can get reference models from two separate platforms. Due to semantic homogeneity and integration of information flow, hence the difference between the two reference models is architectural, such that the measurement of platforms transform the question of measuring the architectural difference of reference models. To identify the difference of reference models can be realized by investigating critical nodes. Critical node must cover the main links or segments or contents inside information processes. Information flow can be treated as the motion of information units. Within the critical nodes, information units have several status: arriving, processing, outputting, waiting, each status have its statistical features and can be used for probability description upon its uncertainty, namely, reaching probability distribution and its parameters, then we can calculate the divergence by using KL-Divergence method, this is the quantitative foundation to achieve platform measurement and assessment.

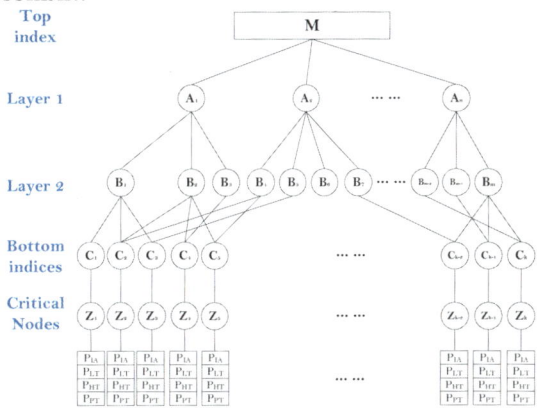

Fig.2. Architecture of evaluation system based on Information Flow

Critical nodes scatter within the functions of enterprise management systems, corresponding to certain business process, at these nodes the data can be collected and analyzed, to achieve the calculation of KL-Divergence, hence it can be used into evaluation index systems as bottom indices, so the index system can be model from bottom to top which is quite different from traditional manners. Fig.2 depict an example of 3-layer architecture, in which the critical nodes and bottom indices are one-to-one matched.

C. KL-Divergence at Critical Nodes

We define Σ as domain of management systems and Γ as domain of information systems, set up the models in Σ and Γ respectively, then we get critical nodes with same numbers in the two models, and marked as Z_i^Σ and Z_i^Γ, $i = 1,2,...,n$. Because Z_i^Σ and Z_i^Γ are corresponding appearance, so mark them as $Z_i = (Z_i^\Sigma, Z_i^\Gamma)$.

The vital factors, of selecting critical nodes Z_i, is based on its features of statistics and information flow, and also concerning the factors of information flow motion process, information density, waiting time and effectiveness of being processed, are showed as below: interarrival time P_{IA}, lead time (waiting time) P_{LT}, processing time P_{HT} and post processing time (waiting time) P_{PT}.

Concerning $Z_i = (Z_i^\Sigma, Z_i^\Gamma)$, the critical nodes of Management Platform Z_i^Σ has its corresponding factors P_{IA}^Σ, P_{LT}^Σ, P_{HT}^Σ, P_{PT}^Σ, and their relative probability density function (pdf) are $f_{IA,i}$, $f_{LT,i}$, $f_{HT,i}$ and $f_{PT,i}$; the critical nodes of Information Platform Z_i^Γ has its corresponding factors P_{IA}^Γ, P_{LT}^Γ, P_{HT}^Γ, P_{PT}^Γ, and their relative pdf are $\varphi_{IA,i}$, $\varphi_{LT,i}$, $\varphi_{HT,i}$, $\varphi_{PT,i}$. As to critical pair node $Z_i = (Z_i^\Sigma, Z_i^\Gamma)$, KL-Divergence can be reached at factor P_{IA} from IS to MS is:

$$D_{Z_i}^{IA} = D_{Z_i}(f_{IA,i} \| \varphi_{IA,i}) = \int f_{IA,i}(x)\log\left(\frac{f_{IA,i}(x)}{\varphi_{IA,i}(x)}\right)dx \quad (3)$$

Likewise, KL-Divergence upon factors P_{LT}, P_{HT} and P_{PT} can also be reached from IS to MS are:

$$D_{Z_i}^{LT} = D_{Z_i}(f_{LT,i} \| \varphi_{LT,i}) = \int f_{LT,i}(x)\log\left(\frac{f_{LT,i}(x)}{\varphi_{LT,i}(x)}\right)dx \quad (4)$$

$$D_{Z_i}^{HT} = D_{Z_i}(f_{HT,i} \| \varphi_{HT,i}) = \int f_{HT,i}(x)\log\left(\frac{f_{HT,i}(x)}{\varphi_{HT,i}(x)}\right)dx \quad (5)$$

$$D_{Z_i}^{PT} = D_{Z_i}(f_{PT,i} \| \varphi_{PT,i}) = \int f_{PT,i}(x)\log\left(\frac{f_{PT,i}(x)}{\varphi_{PT,i}(x)}\right)dx \quad (6)$$

At the critical pair node Z_i, the KL-Divergence from IS to MS can be calculate by below function

$$D_{Z_i} = \mu_{i1}D_{Z_i}^{IA} + \mu_{i2}D_{Z_i}^{LT} + \mu_{i3}D_{Z_i}^{HT} + \mu_{i4}D_{Z_i}^{PT} \quad (7)$$

u_{i1}, u_{i2}, u_{i3}, u_{i4} is the weight corresponding to P_{IA}, P_{LT}, P_{HT} and P_{PT} at critical pair node Z_i, and $\sum_{j=1}^{4}\mu_{ij} = 1$. Obviously, D_{Z_i} is the index value of C_i, the sub-index at bottom level, namely $M_{C_i}^s = D_{Z_i}$, s is the number of level of index system.

D_{Z_i} stands for the difference between business software suite and management level of certain enterprise, is an quantitative measurement at software selection stage.

D. Using ANP method to determine weight

Analytic Network Process (ANP), raised by T. L. Saaty [20] based on Analytic Hierarchy Process (AHP), is a more general form of the AHP used in multi-criteria decision analysis, is based on network feedback architecture and could be used in measuring complicate systems. ANP depict the relationship of the elements by using a depressed and network method and allow the existence of inter-dependent and feedback mechanism, hence it can be used to solve the problem in the fields of society, administration and decision making due to its reality (Fig.3).

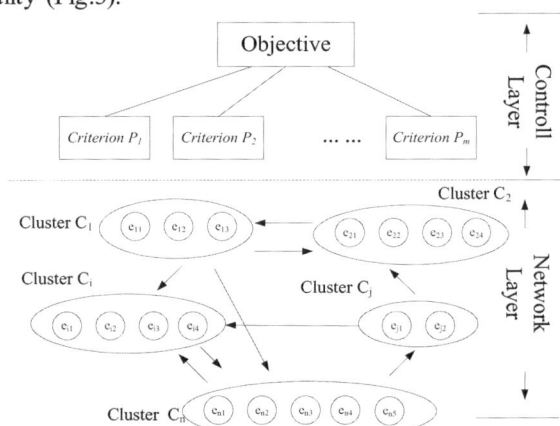

Fig.3. Network Architecture of ANP

ANP contains two part of system elements, control layer and network layer. The control layer includes objective and criterions. All criterions are independent and controlled by objective. Control elements might not consist decision criterion but have to consist at least one objective. The weight with control layer can be obtained by using AHP method. Network layer is consist of elements reflecting by control layer and forms the inter-action network architecture [21-23].

The procedure of ANP is:

1) Pairwise comparison matrices and priority vectors. This is done through pairwise comparisons by asking "How much importance/influence does a criterion have compared to another criterion with respect to our interests or preferences?" The relative importance value can be determined using a scale of 1 to 9 to represent equal importance to extreme importance.

2) Super matrix formation and determining limit super matrix. Obtain the weighted super matrix by multiplying the normalized matrix

$$W = \begin{array}{c} \\ \\ C_1 \\ \\ C_2 \\ \\ \vdots \\ \\ C_n \\ \\ \end{array} \begin{array}{c} \\ e_{11} \\ e_{12} \\ \vdots \\ e_{1m_1} \\ e_{21} \\ e_{22} \\ \vdots \\ e_{2m_2} \\ \vdots \\ e_{n1} \\ e_{n2} \\ \vdots \\ e_{nm_n} \end{array} \begin{bmatrix} W_{11} & W_{12} & \cdots & W_{1n} \\ W_{12} & W_{22} & \cdots & W_{2n} \\ \vdots & \vdots & \vdots\vdots\vdots & \vdots \\ W_{n1} & W_{n2} & \cdots & W_{nn} \end{bmatrix}$$

(8)

3) Calculating the results

Evaluation index architecture raised in this paper, based on information flow, is multi-dimension and cubic, each level indices are dependent and feedback, its architecture satisfies the prerequisite of implementation of ANP method. But due to its complex, the calculating work is pretty huge such that have to via computer tools, for example Super Decision maybe a better choice for that.

IV. IMPLEMENTATION

According to our assumption, to achieve the match measuring and analysis work for informatization system need under below processes:

1) Modeling the reference models of management system and information system respectively, which based on information flow
2) Applying the reference models to draw out critical nodes and depict the attribute of such nodes
3) Using critical nodes as bottom layer component, raise the index architecture down-to-up
4) Determining the probability distribution and corresponding parameters of critical nodes via historic data analyzing or sampling method
5) Calculating the value of critical nodes via KL-Divergence method
6) Obtaining the weight of each level indices via ANP method
7) Aggregating the entropy value down-to-up and using the sum number as the evidence and reference of analyzing.

In this paper, we mainly focus on raising a novel method by using KL-Divergence method to realize quantitative evaluation, so we do not talk about how to model the reference model and critical nodes, some of them are still under researching, some will be described in another paper.

Now we assume that the probability density function $f_{IA,i}$ at critical node Z_i^Σ upon the factor P_{IA}^Σ is exponential distribution, namely $f_{IA,i} = \lambda e^{-\lambda x}$, the probability density function $\varphi_{IA,i}$ at critical node Z_i^Γ upon the factor P_{IA}^Γ is also exponential distribution, namely $\varphi_{IA,i} = \mu e^{-\mu x}$, then we have

$$D_{Z_i}^{IA} = D_{Z_i}(f_{IA,i} \| \varphi_{IA,i}) = \int \lambda e^{-\lambda x} \log\left(\frac{\lambda e^{-\lambda x}}{\mu e^{-\mu x}}\right) dx = 1 - \frac{\mu}{\lambda} - \ln\frac{\lambda}{\mu}$$

If $\varphi_{IA,i}$ belongs to normal distribution with parameter (μ, σ), namely $\varphi_{IA,i} = \frac{1}{\sigma\sqrt{2\pi}} e^{-\frac{(x-\mu)^2}{2\sigma^2}}$, then we have

$$D_{Z_i}^{IA} = D_{Z_i}(f_{IA,i} \| \varphi_{IA,i}) = \int \lambda e^{-\lambda x} \log\left(\frac{\lambda e^{-\lambda x}}{\frac{1}{\sigma\sqrt{2\pi}} e^{-\frac{(x-\mu)^2}{2\sigma^2}}}\right) dx$$

$$= \log\lambda + \frac{1}{2}\log 2\pi + \frac{1}{2}\log\sigma^2 + \frac{1}{2\sigma^2}(\frac{2}{\lambda^2} - \frac{2\mu}{\lambda} + \mu^2)$$

Obviously, if the distribution and parameter are known, the calculation of KL-Divergence is very easy to finish, so by using the method we provided in this paper, the work of platform matching measurement and evaluation of Informatization can be carries out efficiency.

V. CONCLUSION

IS construction consists two stage: optimization of management process and implementation of IS. The former one is essential and last one is the guarantee for improving the management level of certain enterprise. The two stage functionally integrated in the integrity of management processes, and unified in the consistence of process of information flow formally.

In this paper, the $D(P \| Q)$ can be treat as the difference between enterprise management software and management practice, hence it can be the measuring evidence for software choice and selection. Then $D(Q \| P)$ can stand for the difference among the current level of management platform and management software, thus it can be the design evidence of BPM or software reprogramming or adjusting.

The research achievement of this paper did is still under studying, in future we will carry on applying research such that it can be used to solve the real problem to which enterprise facing. The difficult of further investigation exist in two fields, one is model method based on information flow and the abstraction of critical nodes, second is how to assign the weight via ANP method. The first problem will depend on hard work on integrating and innovating current management modeling or enterprise modeling method. The second problem will benefit from using computer programming or business mathematical software.

ACKNOWLEDGMENT

The proposed model and results obtained in this study are personal viewpoints of the authors and the participants.

This work was supported by Fundamental Research Fund of Education Department of China (NKZXB10105) and National Science Foundation of China (71271122).

REFERENCES

[1] C.E. Shannon, W. Weaver, *The Mathematical Theory of Communication*, Urbana, IL: University of Illinois Press, 1929

[2] R.O. Mason, "Measuring Information Output: A Communication System Approach", Information & Management,1978: 1(4): 219-234

[3] R.W. Zmud, "An Empirical Investigation of the Dimensionality of the Concept of Information", *Decision Sciences*, 1978, 9(2):187-195

[4] P.W. Keen, "Information Systems and Organizational Chang", *Communications of the ACM*,1981, 24(1):24-33

[5] B. Ives, M.H. Olson, J.J. Baroud. "The Measurement of User Information Satisfaction", *Communications of the ACM*, 1983, 26(10):785-793

[6] W.H. Delone, E.R. McLean. "Information Systems Success: The Quest for the Dependent Variable", *Information Systems Research*, 1992,3(1):60-95

[7] P.B. Seddon, "A Respecification and Extension of the DeLone and McLean Model of IS Success", *Information Systems Research*, 1997, 8(3):240-253

[8] A. Rai, S.S. Lang, R.B. Welker. "Assessing the Validity of IS Success Models: An Empirical Test and Theoretical Analysi", *Information Systems Research*, 2002, 13(1):50-69

[9] R. Sabherwal, A. Jeyaraj, C. Chowa. "Information System Success: Individual and Organizational Determinants", *Management Science*, 2006, 52(12):1849–1864

[10] Tsai Wen-Hsien, Lee Pei-Ling, Shen Yu-Shan, et al, "A comprehensive study of the relationship between enterprise resource planning selection criteria and enterprise resource planning system success", *Information & Management*, 49(2012): 36-46

[11] Wang Qian, *The Study of Matching Theory between Information Platform and Management Platform and the Measurement of the Capability in ERP Project*, Tianjin: Tianjin University, 2004.2

[12] Wang Qian, Qi Er-shi, Ding Jie, "Study of Information Dual-platform Matching Theory Based on Region Mapping Model", *Industrial Engineering*, 2008, 11(3):10-15

[13] Wang Qian, Huang Shuang-xi, Zheng Yi-song, "Microcosmic Value-Analysis of Information Flow faced Informatization Benefit-Evaluation", *Manufacturing Automation*, 2007, 29(12):1-6

[14] Zhou Yu-qing, Liu bo-ying, Yang Bao-gang, et al, *Principle and Application of ERP*, Beijing: China Machine Press, 2002:147-159

[15] Zhao Ling-jia, Tang Zi-xuan, *Bible of Enterprise Informatization: ERP/PDM/CAPP*, Beijing: Press of Qinghua University, 2002:60-117

[16] Oliver Wight International, Inc., *The Oliver Wight ABCD Checklist for Operational Excellence*, 5th Edition, October 2000

[17] Qiu Hui-Jun, Huang Peng, You Xian-ju, *Comparing among Six IS Evaluation Architecture*, Beijing: Electronic Technology Information Research Institute. MIIT. 2003.3

[18] S. Kullback, R.A. Leibler, "On Information and Sufficiency". *Annals of Mathematical Statistics*, 1951, 22 (1): 79–86

[19] S. KULLBACK, *Information Theory and Statistics*, New York: Dover Publications Inc., 1968

[20] T.L. Saaty, *Decision Making with Dependence and Feedback*. Pittsburgh:RWS Publication, 1996

[21] GONG Jun-tao, LIU Bo, SUN Lin-yan, et al, "Analytic Network Process and Application for Supplier Selection", *Industrial Engineering Journal*, 2007, 10(2):77-80,92

[22] Liu Rui, Yu Jian-xing, Sun Hong-cai et al, "Introduction to the ANP Super Decisions Software and Its Application", *Systems Engineering-Theory & Practice*, 2003, 23(8):141-143

[23] Wang Lian-fen, "The Theory and Algorithm of Analytic Network Process", *Systems Engineering-Theory & Practice*, 2001, 21(3):44-50

Part III
Engineering Management

A New Algorithm for the Risk of Project Time Based on Monte Carlo Simulation

Xing Bi*, Zhiyuan Chen, Lei Li

School of Management, Tianjin University, Tianjin, China

(bistar@126.com)

Abstract - **The risk of project time was analyzed and a new algorithm was proposed based on the previous outcomes and Monte Carlo Simulation. The computing process under the new algorithm was demonstrated combing with an example, which suggested that the new algorithm could reduce the calculated amount effectively and thus provide a reference for the risk management of project completion time.**

Keywords - **Completion time, crystal ball, Monte Carlo Simulation, minimum crucial path, risk management**

I. INTRODUCTION

Construction project is the most common and typical project type, quality objectives, duration and cost targets are three main objectives for project management. In the risk management, risk management for the duration can't be ignored. For a construction project, the completion is restricted by many factors, on the management of completion, we generally use traditional method, which assumes the completion of each process is identified, and then find the critical path by drawing a network, and finally calculate the completion time. However, there are a lot factors affecting the duration in practice, dominant and implicit, the actual results are often highly different from the theoretical results, which could lead to mistakes in management decisions.

There has been a lot of outcomes for the risk management of project, Gao Ying used Monte Carlo simulation to analyze the relevant theory of PERT and MC, made simulation risk assessment for the duration under uncertain conditions, and identified the uncertain risk factors which played a key role [1], Meng Wenqing, Zhang lining proposed a project risk assessment method based on fuzzy network, this method can better reflect the actual situation of the project [2]; Gao Feng, Chen Yingwu established the risk management decision model. Some essential risk factors of the project duration such as activity overlapping, activity iterations, uncertain activity durations and non- consumable resource amount were integrated into the decision model. A simulation optimization method based on adaptive genetic algorithm was proposed considering the characteristics of the model [3]; Fan Chengyu introduced fuzzy triangular numbers into the network schedule, established a fuzzy network project schedule [4], Han Shangyu, Li Hong studied the duration risk management in road construction phase, combined with risk analysis theory, and established a new road project duration risk management calculation method, improved the accuracy of the analysis during the construction period [5]; Liu Xiaoju, Wang Yue presented a lead-time risk transfer algorithm for the calculation of the

total lead-time risk of the system integration project, provided sufficient decision support for system integration project [6]; Zhou Fangming, Zhang Mingyuan combined BP neural networks, genetic algorithms, principal component analysis to propose a model based on PCA-GA-BP to predict the project time risk [7]; Ding Zengxin, Sun Donggen, considered the fuzziness and randomness of the construction project, present a fuzzy and probability analysis method for risk evaluation of the project, and achieved good results [8]. The above study are based on existing mathematical methods, including fuzzy decision, principal component analysis, BP neural network, to build a model to evaluate the risk of completion of the project, although it's reasonable, but the algorithm is designed very complex, large computation, The actual operation is not easy, exists some limitations.

Based on the above findings, we propose a new method for predicting risk, and prove the rationality of the algorithm using the Monte Carlo simulation method, combined with an example. Articles distributed as follows: In part II, We propose the algorithm and the basic principles of Monte Carlo simulation; Part III is a numerical example; Part IV is Conclusion and instruction.

II. ALGORITHM INTRODUCTION AND THE BASIC PRINCIPLES OF MONTE CARLO SIMULATION

A project contains a number of processes, sometimes the completion of these processes is uncertain, which may result in the fact that the completion of project becomes very short or very long, then we need to do risk analysis on the completion of the project [9]. First, we introduce a concept:

The minimum critical path: when the time of completion of the step floats by a certain probability distribution in a range, we calculate the critical path by the lower bound of the time required by each process step. For the project which the duration is certain, the minimum critical path is the critical path [10]. First we introduce a conclusion:

Lemma 1: Let the set $A = \{a_1, a_2, \ldots, a_n\}$, $B = \{b_1, b_2, \ldots, b_n\}$ In the set, $a_i \leq b_i (i = 1,2,\ldots,n)$, $V_{i=1}^n a_i \leq V_{i=1}^n b_i$

$$(1)$$

Proof: Let $a_{k_1} = V_{i=1}^n a_i$, $b_{k_2} = V_{i=1}^n b_i$ $\qquad (2)$

So

$$b_{k_2} = V_{i=1}^n b_i \geq b_{k_1} \geq a_{k_1} = V_{i=1}^n a_i \qquad (3)$$

Then lemma is proved.

If set A in the above lemma seen as a set of the completion time of the minimum critical path, then we can know the minimum critical path is a lower bound for the project completion time. So for the project of which the completion time is uncertain, you can use the Monte Carlo method to simulate the distribution of the completion time. Since the minimum time for each path is less than or equal to the time of the minimum critical path, and will not decrease the number of days resulting from change in the critical path, the change is often caused by the time change of some paths, so first calculate the minimum critical path, and then use the maximum time of each step to calculate the time of all paths (denoted as S), when the time some paths in S take is longer than the minimum critical path, they are likely to be the critical path, denote these paths as set D, so that in a simulation, the maximum of the time every path in set D takes is the time needed by the critical path. The following briefly describes the basic principles of Monte Carlo simulation.

Monte Carlo simulation is built on the basis of probability theory. For a real problem, the factors it contains may be uncertain, and obey a certain probability distribution, under the combined effect of these factors, the result may be uncertain, due to the complexity of the problem, the probability distribution which the results obey often can't be determined by theoretical calculations, so the result need to be determined by the method of random sampling, each sample corresponds to a result, according to the law of large numbers, if the number of samples is large enough, it can approximate the true result by these samples [11].

For a random variable, it is critical to get the sample, inaccurate samples may lead to wrong conclusions. The method of sampling usually is : first create a model, for example, to construct a space geometry, and each sample can be seen as a point of the geometry, then obtain the sample points through computer programming. But sometimes the modeling process can be complicated, then we need additional methods, that is to get sample of corresponding random variable by random number. Random number is also a sample, they obey [0,1] uniform distribution. Why can it generate any sample of random variables through random number? Its principle is based on the following important conclusion:

Lemma 2: Let X be a continuous random variable with distribution function $F(x)$, and there is an inverse function, then F(x) obey [0, 1] uniform distribution.

Proof: Let $Y = F(x)$, obviously, $0 \leq F(x) \leq 1$, and monotone increasing.

When $y \geq 1, P\{Y \leq y\} = P\{F(x) \leq y\} = 1$ (4)

When $y < 0, P\{Y \leq y\} = P\{F(x) \leq y\} = 0$ (5)

When $0 \leq y < 1$,

$P\{Y \leq y\} = P\{F(x) \leq y\} = P\{X \leq F^{-1}(y)\} = F[F^{-1}(y)] = y$ (6)

That is

$$F(y) = \begin{cases} 1, & y \geq 1 \\ y, & 0 \leq y < 1 \\ 0, & y < 0 \end{cases}$$ (7)

Probability density is

$$f(y) = \begin{cases} 1, & 0 \leq y < 1 \\ 0, & else \end{cases}$$ (8)

Therefore, Y obeys [0, 1] uniform distribution. For example, let $X \sim N(\mu, \sigma^2)$, so $y = \frac{x-\mu}{\sigma}$, then $Y \sim N(0,1)$, as to the random number r, look-up table can get $\Phi(y_0) = r$, let $\frac{x-\mu}{\sigma} = y_0$, namely $x = \mu + \sigma y_0$ is the corresponding sample of random number r. In practical problem, random variables are mainly continuous, of course, we can't rule out the discrete case. In the discrete case, the distribution function is piecewise, randomly falls within a range, stipulate the left or right end of the range as the sample value [12]. Based on the above analysis, an example is given below.

III. EXAMPLE

Under normal circumstance, the calculation of using Monte Carlo methods to solve the problem is great, here we can use the Crystal ball software to solve the problem [13].

Certain engineering processes information in following Table I:

TABLE I
CERTAIN ENGINEERING PROCESSES INFORMATION

Process name	Precedence Activity	Shortest time	Longest time	Distribution
A	—	2	4	triangular distribution (2.5)
B	A	6	8	Truncated Normal Distribution(7,0.01)
C	A	3	3	—
D	B	3	3	—
E	B,C	3	5	Uniform Distribution
F	E	1	2	Uniform Distribution
G	A	2	4	triangular distribution (3.5)
H	G	1	3	Truncated Normal Distribution(1.5,0.04)

Note: In the table above, the value in brackets in the distribution row indicates the most likely value, the mean and variance.

The correlation coefficient of Process B and E is 0.7, the correlation coefficient of Process F and A is -0.3. Project network is as following Fig.1:

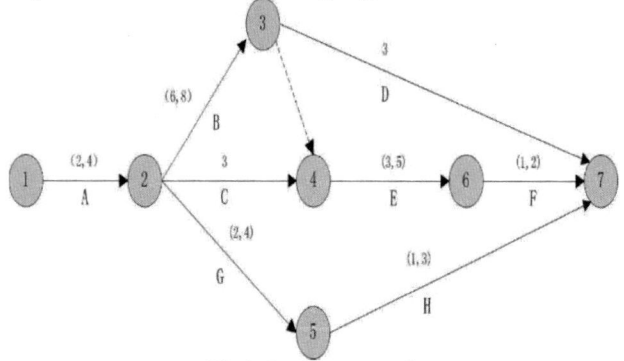

Fig.1. Project network

The smallest critical path: A → B → E → F, the duration is 12. Use the maximum duration of processes, there are three paths of which the duration is bigger than the minimum duration of the critical path, namely: A → B → D, the duration is 15 A → C → E → F, the duration is 14 A → B → E → F, the duration is 18, 1000 times simulation results are as following Fig.2: (existing an outlier, 999 showed)

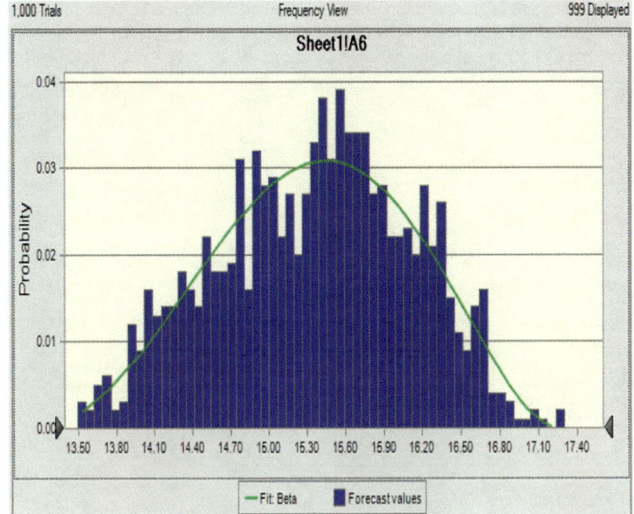

Fig.2. Simulation results.

From the figure, we can see that the project is most likely to be completed in 15-16 days, and its distribution is closed to the Beta distribution as following Fig.3.

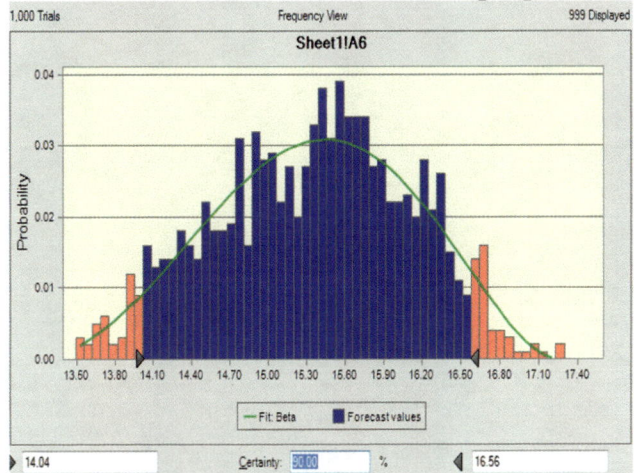

Fig.3.The distribution

And there is 90% probability that the project could be completed in 14 to 16.5 days. Following are the statistics and standard values of the distribution fitting as Fig.4:

Statistic	Fit: Beta	A6
Trials	---	1,000
Base Case	---	12.00
Mean	15.36	15.36
Median	15.38	15.42
Mode	15.45	---
Standard Deviation	0.76	0.76
Variance	0.58	0.58
Skewness	-0.1264	-0.1262
Kurtosis	2.39	2.39
Coeff. of Variation	0.0497	0.0497
Minimum	13.04	13.50
Maximum	17.26	17.62
Mean Std. Error	---	0.02

Fig.4. Statistics and standard values of the distribution fitting

In the figure above, the middle column is the standard data, the right column is the fitting data, we can see the fitting data of indexes such as mean, variance, median, kurtosis and others fit the standard data very well, and the average standard deviation is 0.02, so we can

approximately think that the project completion obeys the Beta distribution [14].

The risk that affects the completion of the project is analyzed in the following Fig.5:

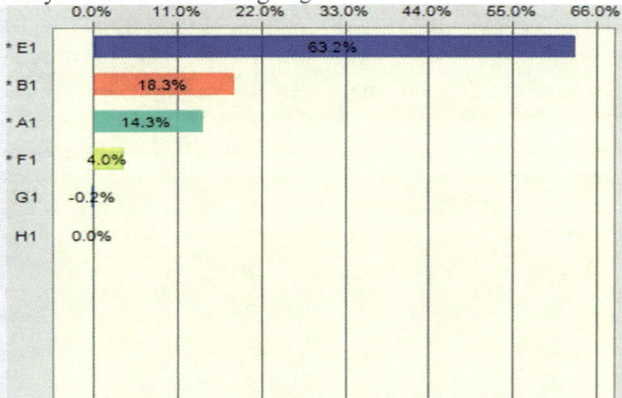

Fig.5.The variance of the completion of project

The variance of the completion of project can be seen as the size of the risk, the risk of a large variance project correspondingly becomes larger. The figure above is the contribution to the variance of completion of each process, it can be seen, the cumulative contribution of process E, B, A is more than 95%, we may say that the risk of completion of the project is mainly generated by these processes. So, in actual production, we should strengthen management on these three steps, to prevent large deviations.

Then analyze the sensitivity, as is shown in the following Fig.6:

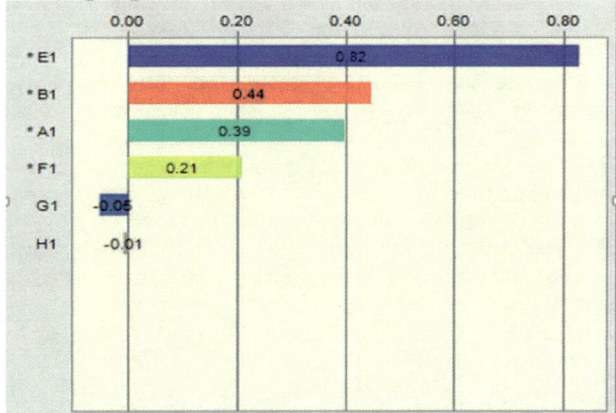

Fig.6.The sensitivity

In the figure above, the correlation coefficient of process E and the completion of project is 0.82, which means that the completion time for process E increases each step by 1, the total duration increases 0.82, equally, process B and A also have some impacts on the completion time. Take all the analysis together, process E is essential for the completion of the project [15], the construction process should strengthen the management on process E.

IV. CONCLUSION

This paper compares the results of previous studies and presents a new algorithm for the risk management of the project completion from a different perspective. There are many algorithms based on the Monte Carlo Simulation of the project completion. We can also simulate every path, then take the longest path as the critical path. But for complex projects, this method will be of great amount of calculation. The proposed method is firstly weeding out the path which may not be the critical path by prior estimation, and then simulate the remaining path, so that we can effectively reduce the amount of calculation to obtain reasonable results. In addition, this paper presents a numerical example. This example is not complicated, but enough to explain the problem. As to the engineering which possesses more processes, the proposed algorithm has its advantage.

REFERENCES

[1] Gao Ying. Management of the risk of construction project time [D].Zhe Jiang University. 2006:96-103

[2] Meng Wenqing, Zhang Lining, Li Wanqing. Real estate portfolio mode based on information entropy [J]. Journal of Hebei Institute of Architectural Science and Technology. 2005, 02: 73-75+79

[3] Gao Feng, Chen Yingwu. The risk management decision model of the model project duration and adaptive genetic algorithm [J]. Journal of national university of defense technology. 2005, 06:106-112

[4] Fan Chengyu. Analysis of the management of project time risk [J]. Engineering and construction. 2006, 03: 259-261.

[5] Han Shangyu, Li Hong, Hong, Baoning. Risk analysis method for project duration of highways during construction period and its application [J]. Chinese Journal of Geotechnical Engineering. 2013, S1: 264-268.

[6] Liu Xiaoju, Wang Yue. Evaluation and control of the lead - time risk of the system integration project with risk transfer algorithm [J]. Operations research and management science, 2004, 01:38-43.

[7] Zhou Fangming,Zhang Mingyuan,Yuan Yongbo. Risk of project time based on PCA-GA-BP [J]. Journal of Engineering Management Oct. 2011, 05: 534-538.

[8] Din Zhenxin. Sun Donggen, He Guang. Hu Jinsong. Fuzzy and probabilistic method for project risk evaluation [J].Henan Science. 1998, 01: 23-28.

[9] Riffis F H. (Bud), John V. Farr: Construction planning for engineers.120-123

[10] Daud Nasir, Brenda McCabe, Loesie Hartono, Evaluating risk in construction-schedule model (ERIC-S): construction schedule risk model [J]. Journal of Construction Engineering and Management, September/October, 2003. 78-82

[11] Wang Renchao, Ou Yangbin, Monte carlo simulation of project network planning and schedule risk analysis [J] Computer Simulation. 2004.4:143-147

[12] Abbasi G Y, Mukattash A M. Crashing PERT networks using mathematical programming [J]. International Journal of Project Management, 2001(19): 181- 188.

[13] Wang Zhongwei. Realizing project's Monte carlo simulation by EXCEL [J] Journal of Guang Dong Communications Polytechnic. 2005.3: 101-103

[14] Ma Xiaomeng, Ma Bin, ZHU Jiwei, Simulation and evaluation of the risk of international EPC project based on MCM crystal ball model [J]. Yellow River. 2013.11:121-123

[15] He Ying, Discussion on management of construction investment project duration risk [J]. China Urban Economy, 2011. 11:240-241

Investment Risk Analysis of Gas and Pipeline Construction Project Based on the Gray Analytic Hierarchy Process

Lin-lin Mu*, Xing Bi

Department of Manage and Economics, Tianjin University, Tianjin, China

(mulinlin119@163.com)

Abstract - **The oil and gas pipeline projects are often faced with risks of many aspects in the process of development and construction. The article describes the process of the gray analytic hierarchy process for risk analysis, and calculates the comprehensive evaluation index of the project via a real case, namely the overall risk of investment projects, so that the investors make right investment decisions according to the results of the risk analysis to make sure the safe state of the pipeline project construction maintain an acceptable risk level, which helps to form security for the safety after putting into production.**

Keywords - **Investment risk, oil and gas pipeline projects, risk analysis, the gray analytic hierarchy process**

I. INTRODUCTION

In recent years, China has developed rapidly in the oil and gas pipeline construction. The national oil and gas pipeline network has formed initially. The gas supplying pattern of West-East natural gas transmission, sea-land gas transmission and supplying from nearby has been formed, as well as a relative perfect regional natural gas pipeline network [1].

The oil and gas pipeline projects are in the high-speed development period. However, there are risks from many aspects in the oil and gas pipeline projects. Economic losses and social impacts are very large once a pipeline accident takes place. The oil and gas pipeline projects generally need a long time to conduct planning, feasibility studies, designing construction and the operation. The situation and goals of risk management are not the same in the different stages of this process, as well as the risk factors we face. There are many uncertainties in the investment phase of the project. The probability of the risk occurring is great, but the cost of risk treatment at this time is relatively low [2] [3]. That is to say, we can get the greatest opportunity to avoid damage by low-cost risk analysis. The cost of handling risk factors or risk events increases sharply due to lots of projects being invested with the conducting of the project, despite the decreasing of uncertainty [4].

Thus, in the investment phase of the project, we make a comprehensive risk analysis of the project, so that investors can make investment decisions that whether we apply limited funds to pipeline construction, and they can deal with potential problems in time, in order to keep the safety state at an acceptable risk level, and avoid the serious losses caused by the risks in the follow-up work.

II. METHODOLOGY

Currently, there are some achievements in the risk analysis of oil and gas pipeline projects, but the studies on the risk analysis of pre-investment stage of oil and gas pipeline projects are absent, and there is a lack of specific methods of risk analysis. Therefore, in this article, based on Gray analytic hierarchy process [5], we obtain the risk factors in the oil and gas pipeline project investment stage by calculating, and sort on risk factors, finally calculate the gray comprehensive evaluation value of the risk of the project investment phases, namely the overall investment risk of the project [6], so as to provide basis for the investment comprehensive risk evaluation of the project. This helps the implementation of decisions in the investment stage, and helps to take appropriate risk management measures.

A. Risk Identification

First of all, the classification and identification of risk in the project investment stage are needed before the project investment risk evaluation. When identifying oil and gas pipeline project investment risk, we give an open-ended questionnaire [7] about listing investment risks of oil and gas pipelines to the experts, and then consult the experts about the results, give them questionnaire about the results with the options of strongly agree, agree, neutral, do not agree, strongly disagree, finally obtain consistent conclusion as the identified risk by the summary of the experts' opinions.

B. Sort of the Risk Factors

We use the identified risks to establish the risk evaluation index system, represent target layer corresponding investment decision project risk by U, guidelines layer elements by Ui, and index layer elements namely various risk factors by Uij. The judgment matrix represents the fact of relative importance among the relevant elements in this layer for some elements in last layer. We assume that element A_k in layer A and elements B_1, B_2, …, B_n in next layer are linked, and make judgment matrix in the TABLE I.

In Table I, b_{ij} represents the numerical expressions of B_i's relative importance for B_i comparing to A_k, which is called scale value. The value of bij is usually among 1, 3, 5, 7, 9 and their reciprocals. The meanings of the scale values are list in the TABLE II.

TABLE I
JUDGMENT MATRIX TABLE OF HIERARCHY ANALYSIS

A_k	B_1	B_2	…	B_n
B_1	b_{11}	b_{12}	…	b_{1n}
B_2	b_{21}	b_{22}	…	b_{2n}
…	…	…	…	…
B_n	b_{n1}	b_{n2}	…	b_{nn}

E. Qi et al. (eds.), *Proceedings of the 21st International Conference on Industrial Engineering and Engineering Management 2014*, Proceedings of the International Conference on Industrial Engineering and Engineering Management, DOI 10.2991/978-94-6239-102-4_94, © Atlantis Press and the authors 2015

TABLE II
TABLE OF SCALE VALUE

Scale value	Meaning
1	The two elements have equal importance
3	The former element is slightly more important
5	The former element is obviously more important
7	The former element is strongly more important
9	The former element is extremely more important
2, 4, 6, 8	intermediate value of the judgment above
reciprocals	The ratio of importance of element j and I is $b_{ji=1}/b_{ij}$ if the ratio of importance of element i and j is b_{ij}

Then we calculate the weights W according to the scoring results given by the experts [8]. We collect the experts' questionnaires, and fill the scale values into the judgment matrix table according to the scale of the table. And then we obtain the characteristic root and eigenvector for matrix B which meet the condition $BA=K_{max}A$. Finally we obtain eigenvector $A= (A_1, A_2, A_3)$ by normalization, name the weight value of each factor.

Consistency verification of the judgment matrix is needed to see if the eigenvector above is reasonable weight distribution. The verification formula is (1).

$$CR=CI/RI \qquad (1)$$

In the formula, CR represents the random consistency ratio of the judgment matrix, and CI is the general consistency index of the judgment matrix. The formula of CI is (2).

$$CI=(\lambda_{max}-n)/(n-1) \qquad (2)$$

In the formula, λ_{max} represents the maximum characteristic root of the matrix, and RI is the average random consistency index of the judgment matrix. The RI values of 1-9 orders judgment matrix are visible in the TABLE III.

TABLE III
VALUE OF THE AVERAGE RANDOM CONSISTENCY INDEX

n	1	2	3	4	5	6	7	8	9
RI	0	0	0.58	0.90	1.12	1.24	1.32	1.41	1.45

Judgment matrix P has satisfactory consistency if its $CR < 0.1$, or $\lambda_{max} = n$, $CI= 0$, otherwise we should adjust the element in P to make sure it has satisfactory consistency [9]. In this way we get the arrangement of the risk factors.

C. Assessment of Comprehensive Risk

We mark for the risk indicators according to the occurring possibility of risk. According to the grey system theory [10], we set 5 grey category grades respectively representing "low risk", "relatively low risk", "ordinary risk", "relatively high risk" and "high risk". The grades are represented by $e=1,2,3,4,5$,which correspond marking range 0-2, 2-4, 4-6, 6-8, 8-10, and they form vector $Z=[1,3,5,7,9]$.

Grey category $e=1$ (low risk), whitened weight function f_1 [11] is formula (3).

$$f_1(d)=\begin{cases} 1 & d = 0 \\ 2-d & d \in [0,2] \\ 0 & d \notin [0,2] \end{cases} \qquad (3)$$

Grey category $e=2$(relatively low risk), whitened weight function f_2 is formula (4).

$$f_2(d)=\begin{cases} d/2 & d \in [0,2] \\ 2-d/2 & d \in [2,4] \\ 0 & d \notin [0,4] \end{cases} \qquad (4)$$

Grey category $e=3$(ordinary risk), whitened weight function f_3 is formula (5).

$$f_3(d)=\begin{cases} d/4 & d \in [0,4] \\ 2-d/4 & d \in [4,8] \\ 0 & d \notin [0,8] \end{cases} \qquad (5)$$

Grey category $e=4$(relatively high risk), whitened weight function f_4 is formula (6).

$$f_4(d)=\begin{cases} d/6 & d \in [0,6] \\ 2-d/6 & d \in [6,12] \\ 0 & d \notin [0,12] \end{cases} \qquad (6)$$

Grey category $e=5$(high risk), whitened weight function f_5 is formula (7).

$$f_5(d)=\begin{cases} d/8 & d \in [0,8] \\ 1 & d \in [8,+\infty) \\ 0 & d \notin [0,+\infty) \end{cases} \qquad (7)$$

So we can obtain the gray evaluation coefficient x_{ij} belonging to the evaluation grey category e as follows.

$$x_{ij}=\sum f_e(d_{ij}) \qquad (8)$$

The overall gray evaluation value is denoted by X.

$$x=\sum x^e_{ij} \qquad (9)$$

r^e_{ij} denotes the gray evaluation weight of index factors' gray category e , then:

$$r^e_{ij}= x^e_{ij}/x \qquad (10)$$

The gray evaluation matrix of the index factor is:

$$R_i=\begin{bmatrix} r_{i1} \\ r_{i2} \\ \vdots \\ r_{ij} \end{bmatrix} = \begin{bmatrix} r_{i1}^{1} & \cdots & r_{i1}^{f} \\ \vdots & \ddots & \vdots \\ r_{ij}^{1} & \cdots & r_{ij}^{f} \end{bmatrix}$$

We make a synthesized evaluation of the vector consisting of the index factors, then gray evaluation weight matrix B of various risks is the product of the weight vector and R_i. And the gray evaluation weight matrix of each evaluation gray category is as follows.

$$R=\begin{bmatrix} B_1 \\ B_2 \\ \vdots \\ B_m \end{bmatrix} = \begin{bmatrix} b_{11} & \cdots & b_{1f} \\ \vdots & \ddots & \vdots \\ b_{m2} & \cdots & b_{mf} \end{bmatrix}$$

Synthesized evaluation weight vector of layer U for the pipeline evaluation is denoted by B, then:

$$B=A\times R \qquad (11)$$

Accordingly, we can obtain the synthesized evaluation value [12] of the air and oil pipeline project investment risk as follows.

$$W=B\times Z^T \qquad (12)$$

III. RESULTS

A. An Example of Air and Oil Pipeline Project Investment Risk

A pipeline project to be constructed is the connecting pipelines between air pipeline A and B. The length of the whole pipeline is estimated to be 93.0km, and the pipeline. L485 helix (or straight) submerged arc welded steel pipe is applied to constructing the pipeline. The design pressure is 10.0MPa. 7 valve chambers need to be constructed, 2 of which are monitoring and controlling valve chambers, and the others 5 are monitoring valve chambers. The two stations above should be reformed and no new station is built.

Polyethylene composite structure coating layer and impressed current cathodic protection are applied to the whole pipeline, with two cathodic protection stations build along the pipeline. There are two large crossings of rivers along the expected pipeline route, as well as 12 medium crossing projects, 1 crossing of high speed rail being under viaduct, 4 crossings of freeway, 1 crossing of state road and 6 crossings of provincial highways. The main landscape along the pipeline is paddy fields.

B. Results of Risk Identification

The experts obtain the results according to the practical project situation, questionnaires and related literature [13] [14]. The investment risk of this air and oil pipeline project can be divided into five categories according to the character of the risk factors.

(1) resource risk

Resource risk mainly consists of the reserve and production of oil and gas, funds and capital structure.

(2) management risk

Management risk mainly consists of irrational management system, construction program plans, irrational resource allocation and so on.

(3) security risk

Security risk mainly consists of design flaws, pipeline corrosion, the third party damage and so on.

(4) economic risk

Economic risk mainly consists of annual transportation and market supply and demand.

(5) environmental risk

Environmental risk mainly consists of natural disasters, policy and economic situation.

Establish risk assessment indicator system [15] as Fig.1 according to the identified factors. Corresponding investment decision risk in target layer is represented by U. Factors in guideline layers are resource risk, management risk, security risk, economic risk, environmental risk, which are represented by U_i (i=1,2,3,4,5), and factors in indicator layer are represented by U_{ij} (i=1,2,3,4,5; j=1,2...9)

C. Results of Sorting Risk Factors

We obtain the weights of the risk factors by experts marking method. For example, the result of mark and calculation for expert 1 is in the TABLE IV-IX.

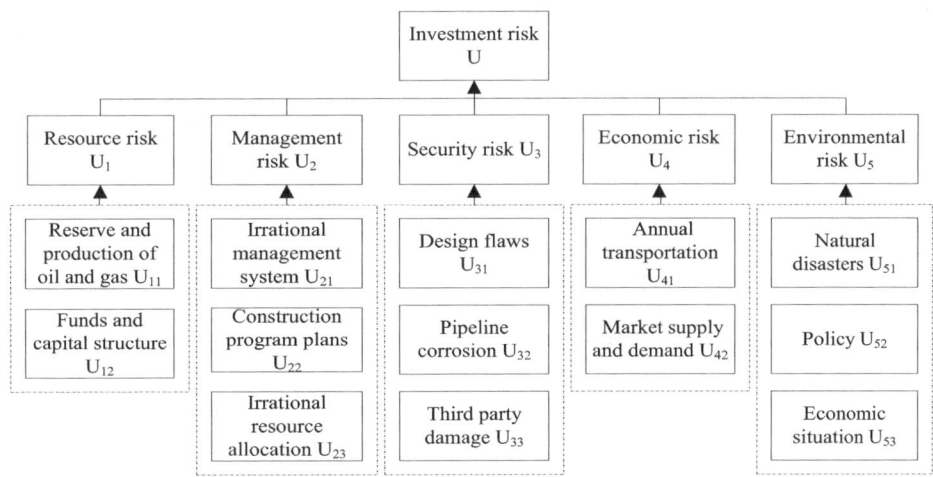

Fig.1. Investment risk indicator system

TABLE IV
JUDGMENT MATRIX AND WEIGHT CALCULATION OF LEVEL TWO INDICATOR

U	U_1	U_2	U_3	U_4	U_5	Weight W
U_1	1	1/2	1/7	1/3	1/5	0.0497
U_2	2	1	1/5	1/2	1/3	0.0842
U_3	7	5	1	5	3	0.5063
U_4	3	2	1/5	1	1/2	0.1306
U_5	5	3	1/3	2	1	0.2293
CR=0.0214<0.1 λ_{max}=5.096						

TABLE V
MATRIX AND WEIGHT CALCULATION OF RESOURCE RISK

U_1	U_{11}	U_{12}	Weight W
U_{11}	1	1/2	0.3333
U_{12}	2	1	0.6667
CR=0<0.1 λ_{max}=2			

TABLE VI
JUDGMENT MATRIX AND WEIGHT CALCULATION OF MANAGEMENT RISK

U_2	U_{21}	U_{22}	U_{23}	Weight W
U_{21}	1	1/5	1/3	0.1047
U_{22}	5	1	3	0.6370
U_{23}	3	1/3	1	0.2583
CR=0.037<0.1 λ_{max}=3.0385				

TABLE VII
JUDGMENT MATRIX AND WEIGHT CALCULATION OF SECURITY RISK

U_3	U_{21}	U_{22}	U_{23}	Weight W
U_{21}	1	3	2	0.5396
U_{22}	1/3	1	1/2	0.1634
U_{23}	1/2	2	1	0.2970
CR=0.0088<0.1 λ_{max}=3.0092				

TABLE VIII
JUDGMENT MATRIX AND WEIGHT CALCULATION OF ECONOMIC RISK

U_4	U_{11}	U_{12}	Weight W
U_{11}	1	2	0.6667
U_{12}	1/2	1	0.3333
CR=0<0.1 λ_{max}=2			

TABLE IX
JUDGMENT MATRIX AND WEIGHT CALCULATION OF ENVIRONMENT RISK

U_5	U_{21}	U_{22}	U_{23}	Weight W
U_{21}	1	1/5	1/4	0.0974
U_{22}	5	1	2	0.5695
U_{23}	4	1/2	1	0.3331
CR=0.0236<0.1 λ_{max}=3.0246				

We can obtain the calculation result of weight vectors in guideline layer and target layer from the 5 experts by the method above, and fill the table below with the result. The table is weight values summary table of each indicator. Then we calculate the arithmetic average value the weight vectors from the 5 experts. The final weights of each assessment indicator are list in the TABLE X. Thus we sort the risk weights in descending order. The result is listed in TABLE XI.

TABLE X
WEIGHT VALUES

weight	Experts number					W
	1	2	3	4	5	
U_1	0.0497	0.1525	0.0523	0.0546	0.3976	0.1413
U_2	0.0842	0.0523	0.0887	0.4360	0.0767	0.1476
U_3	0.5063	0.2621	0.1525	0.1543	0.1617	0.2474
U_4	0.1306	0.0887	0.2621	0.0898	0.0417	0.1226
U_5	0.2293	0.4443	0.4443	0.2653	0.3223	0.3411
U_{11}	0.3333	0.1667	0.2500	0.7500	0.3333	0.3667
U_{12}	0.6667	0.8333	0.7500	0.2500	0.6667	0.6333
U_{21}	0.1047	0.2297	0.1958	0.1220	0.1571	0.1619
U_{22}	0.6370	0.1220	0.4934	0.6483	0.5936	0.4988
U_{23}	0.2583	0.6483	0.3108	0.2297	0.2493	0.3393
U_{31}	0.5396	0.5954	0.5396	0.6483	0.5917	0.5829
U_{32}	0.1634	0.1283	0.2970	0.2297	0.0751	0.1787
U_{33}	0.2970	0.2764	0.1634	0.1220	0.3332	0.2384
U_{41}	0.6667	0.2500	0.6667	0.1667	0.3333	0.4167
U_{42}	0.3333	0.7500	0.3333	0.8333	0.6667	0.5833
U_{51}	0.0974	0.0936	0.1396	0.1958	0.1634	0.1380
U_{52}	0.5695	0.6267	0.5278	0.4934	0.2970	0.5029
U_{53}	0.3331	0.2797	0.3325	0.3108	0.5396	0.3591

TABLE XI
INVESTMENT RISK FACTORS SORT TABLE

Risk Sub-factor	Risk weight
Policy U_{52}	0.1715
Design flaws U_{31}	0.1442
Economic Situation U_{53}	0.1225
Capital and capital structure U_{12}	0.0895
Construction program plan U_{22}	0.0736
Market supply and demand U_{42}	0.0715
Third party damage U_{33}	0.0590
Reserve and Production of oil and gas U_{11}	0.0518
Annual delivery amount U_{41}	0.0511
irrational resource allocation U_{23}	0.0502
Natural Disasters U_{51}	0.0471
Pipeline corrosion U_{32}	0.0442
Management system U_{21}	0.0239

TABLE XII
INVESTMENT RISK INDICATORS MARKING TABLE

score	Expert number				
	1	2	3	4	5
U_{11}	4	2	1	2	2
U_{12}	4	4	1	6	2
U_{21}	2	4	0	4	2
U_{22}	6	8	5	6	6
U_{23}	4	6	4	6	6
U_{31}	8	6	6	4	4
U_{32}	5	4	4	4	6
U_{33}	6	5	5	6	6
U_{41}	1	2	1	0	4
U_{42}	4	6	4	6	6
U_{51}	6	4	8	4	2
U_{52}	8	10	8	8	9
U_{53}	6	8	8	6	7

TABLE XIII
OIL AND GAS RESERVE AND PRODUCTION RISK GRAY EVALUATION WEIGHT VECTOR TABLE

Score gray category	4	2	1	2	2	Σ	Evaluation weight
1	$f_1(4)$	$f_1(2)$	$f_1(1)$	$f_1(2)$	$f_1(2)$	1	0.0956
2	$f_2(4)$	$f_2(2)$	$f_2(1)$	$f_2(2)$	$f_2(2)$	3.5	0.3347
3	$f_3(4)$	$f_3(2)$	$f_3(1)$	$f_3(2)$	$f_3(2)$	2.7500	0.2629
4	$f_4(4)$	$f_4(2)$	$f_4(1)$	$f_4(2)$	$f_4(2)$	1.8334	0.1753
5	$f_5(4)$	$f_5(2)$	$f_5(1)$	$f_5(2)$	$f_5(2)$	1.375	0.1315
				Σ		10.4584	

D. The Gray Comprehensive Evaluation Value

The experts mark for the measurement of the risk value according the gray theory. Marking survey table is as the TABLE XII.

We calculate the gray evaluation weight vector of the risk evaluation indicators by the table dispatching method .The calculation result of oil and gas reserve and production risk gray evaluation weight vector is as the TABLE XIII.

The gray evaluation weight vector of U_{11} obtained from the TABLE XIII is:

$$r_{11}=(0.0956,0.3347,0.2629,0.1753,0.1315)$$

We can obtain the other gray evaluation weight vectors in the table in the same way. Thus we obtain the gray evaluation weight matrixes of U_1, U_2, U_3, U_4, U_5.

$$R_I=\begin{bmatrix} 0.0956 & 0.3347 & 0.2629 & 0.1753 & 0.1315 \\ 0.0934 & 0.1401 & 0.3035 & 0.2646 & 0.1984 \end{bmatrix}$$

$$R_2 = \begin{bmatrix} 0.1053 & 0.2105 & 0.3158 & 0.2105 & 0.1579 \\ 0 & 0 & 0.2118 & 0.4235 & 0.3647 \\ 0 & 0 & 0.3158 & 0.3910 & 0.2932 \end{bmatrix}$$

$$R_3 = \begin{bmatrix} 0 & 0 & 0.2857 & 0.3810 & 0.3333 \\ 0 & 0 & 0.3644 & 0.3563 & 0.2794 \\ 0 & 0 & 0.2769 & 0.4000 & 0.3231 \end{bmatrix}$$

$$R_4 = \begin{bmatrix} 0.3214 & 0.2143 & 0.2143 & 0.1429 & 0.1071 \\ 0 & 0 & 0.3158 & 0.3910 & 0.2932 \end{bmatrix}$$

$$R_5 = \begin{bmatrix} 0 & 0.0968 & 0.2903 & 0.3226 & 0.2903 \\ 0 & 0 & 0 & 0.3617 & 0.6383 \\ 0 & 0 & 0 & 0.4878 & 0.5122 \end{bmatrix}$$

Thus we obtain the gray evaluation weight matrixes of each risk according to the product of investment risk weights and R_i, and then obtain the gray evaluation weight matrix of each evaluation gray categories.

$$R = \begin{bmatrix} 0.0942 & 0.2115 & 0.2886 & 0.2318 & 0.1739 \\ 0.0170 & 0.0341 & 0.2639 & 0.3780 & 0.2070 \\ 0 & 0 & 0.2977 & 0.3811 & 0.3212 \\ 0.1339 & 0.0893 & 0.2735 & 0.2876 & 0.2157 \\ 0 & 0.0133 & 0.0401 & 0.4016 & 0.5450 \end{bmatrix}$$

According to the weight vector A, then we can calculate the investment risk comprehensive evaluation weight vector of this oil and gas pipeline project as follows.

$$B = \begin{bmatrix} 0.0322 & 0.0504 & 0.2006 & 0.3551 & 0.3617 \end{bmatrix}$$

Finally we calculate the gray comprehensive evaluation value W of the oil and gas pipeline project investment risk.

$$W = \begin{bmatrix} 0.0322 & 0.0504 & 0.2006 & 0.3551 & 0.3617 \end{bmatrix} \times$$

$$\begin{bmatrix} 1 & 3 & 5 & 7 & 9 \end{bmatrix}^T = 6.93$$

IV. DISCUSSION

The comprehensive evaluation value of the project risk is 3.28 based on the gray analytic hierarchy process. It shows that the overall risk of this project is relatively high. It can provide decision basis for the project investors. Besides, managers should make prevention and control of the risks beforehand according to the sort of the risk grades in the project reality, in order to realize risk minimization and benefits maximization.

V. CONCLUSION

The construction of oil and gas pipeline in China is in the period of prosperity. However, oil and gas pipeline projects are often with high risk and large uncertainty, and are faced with interferences and threats from resource, economy, environment and so on. These risks often result in huge economic losses and social impact.

While analyzing and summarizing oil and gas pipeline project investment risk, the paper uses the risk management theory to establish risk analysis of ideas. In the investment stage of the project, we establish risk evaluation index system, and assess the overall risk of the project. The results can help to guarantee the safety in the operation of the project and take appropriate risk management measures.

REFERENCES

[1] Ying Li, Guoyi Li, "Status and development of oil and gas pipeline construction" (in Chinese), *West China*, vol. 8, no. 14, pp.6-8, 2009.

[2] Di Zhao, "Investment risk assessment of early-stage venture capital" (in Chinese), master dissertation, Jilin University, Jilin, China, 2005.

[3] Sichuan Petroleum Administration, "Pipeline Risk Management" (in Chinese), Monograph, Petroleum Industry Press, 1995.

[4] Hongli Pan, "Research on pipeline risk management approach" (in Chinese), *Natural Gas and Petroleum*, vol. 24, no. 1, pp.9-14, 2006.

[5] Wenqing Zhao, Zhouxin Qian, "Marketing risk evaluation based on GAHP" (in Chinese), *Anhui University of Technology*, vol. 23, no. 1, pp. 101-105, 2006.

[6] Jiu Song, Dawei Dong, Guoan Gao, "Decision Models' Research based on GAHP and gray relational degree" (in Chinese), *Journal of Southwest Jiaotong University*, no. 4, pp. 463-466, 2002.

[7] Jie Xiao, Lin Bu, Tingmei Sun, "Investigation and plan of open questionnaires" (in Chinese), *Neijiang Technology*, no. 1, pp. 50-51, 2010.

[8] Yi Dai, Xuezhi Yang, Hongping Luo, Liwen Yan, "Priori information fusion method for the numerical control system reliability evaluation - expert scoring method" (in Chinese), *Journey of Tianjin Vocational Technical Normal University*, vol. 22, no. 3, pp.6-8, 2012.

[9] Jiane Chang, "Studies to determine the weights by AHP" (in Chinese), *Wuhan University of Technology*, vol. 29, no. 1, pp.153-155, 2007.

[10] Xiaofen Zhao. "Overview of gray system theory" (in Chinese), *Journey of Jilin Education Institute*, vol. 27, no. 3, pp.152-154, 2011.

[11] Julong Deng, "Grey Prediction and Decision" (in Chinese), Monograph, Huazhong University Press, 1998.

[12] Yunfei Feng, Fanbo Meng, Fubin Zhu, Yang Chen, "Pipeline Risk Based on GAHP" (in Chinese), *Oil and Gas Storage and Transportation*, vol. 32, no. 12, pp.1289-1294, 2013.

[13] Na Tian, Baodong Chen, Fan Yang, Qisheng Chen, "Grey Correlation Analysis in Oil and Gas Pipeline Risk Evaluation" (in Chinese), *Petroleum Engineering Construction*, vol. 32, no. 1, pp.17-21, 2006.

[14] Xilong Zhu, Yuhui Li. "Analysis and evaluation of oil and gas pipeline projects' investment decision risk" (in Chinese), *Petroleum Machinery*, vol. 34, no. 2, pp.74-76, 2006.

[15] Denghua Zhong , Jianshe Zhang, Guangjing Cao, "Risk analysis methods for Projects based on AHP" (in Chinese), *Journey of Tianjin University*, vol. 35, no. 2, pp.162-166, 2002.

Risk Analysis for Chinese CDM Projects Seller Enterprises Based on Multi-step Fuzzy Comprehensive Evaluation

Xing BI[*], Chi WANG

Collage of Management and Economics, Tianjin University, Tianjin, China
(bistar@126.com)

Abstract - **The risk analysis of the enterprise toward Chinese clean development mechanism (CDM) projects is proposed in this paper, which has important theoretical and practical significance to the relevant enterprises and sustainable development of China. Based on the analysis of risk factors, which extracting indexes from three aspects of the national level, multilateral trade level and the enterprise level, the risk index system is established. Then based on the fuzzy comprehensive evaluation method and analytic hierarchy process (AHP), the risk analysis model is established and applied to analyze. The research in this paper will help Chinese CDM projects seller enterprises to strengthen risk prevention and management in CDM projects, and promote the health development of CDM projects in China.**

Keywords – **CDM, Fuzzy Comprehensive Evaluation, project management, risk analysis**

I. INTRODUCTION

With the Kyoto Protocol come into force, the development and implementation of the Clean Development Mechanism (CDM) projects brings additional capital and advanced technology to Chinese enterprises. The CDM projects is benefit to both the global greenhouse gas emissions and China's sustainable development [1]. However, due to the obstacles of philosophy, law and some other aspects, domestic enterprises, especially small and medium-sized ones, do not have deep understanding of the CDM, and strange to the risks specific to the CDM, which bring a lot of potential risks to enterprises participated in this new state trading [2].

China has become a major strategic country of developed countries to carry out CDM projects [3]. Data form shows that, as of April 2014, a total of 5048 CDM projects has been approved by the national development and reform commission, including 3793 projects in the CDM executive board (EB) successful registration. And 1389 projects get certified emission reductions (CERs) [4]. As a country with world's largest amount of CDM registered projects and largest expected annual emission reductions, however, China is far behind the western countries in the studies of method, the carbon finance market development, designated operational entity (DOE) registration and other aspects, and the imbalance between buyers and sellers brings a number of risks to Chinese enterprises [3,5].

II. RISK ANALYSIS OF CHINESE CDM PROJECTS SELLER ENTERPRISES

Sun Zeng-qin thought that domestic enterprises in clean development mechanism projects are still faced with the basic project risks, political risk and legal changes as well as its own unique risk like regulatory approval, CREs delivery, etc., and analyzed the risk control path to the risks above in qualitative point [6]. Wang Dan and Wang Ning divided CDM project risk assessment index into pre-investment period, investment and construction period, operation period and other policies, then quantified CDM projects risk with the method of fuzzy cluster analysis and triangular fuzzy number analytic hierarchy process [7]. Ma Jian-ping and Zhuang Gui-yang thought that there are five major risk event: approval defeat, validation returned, registration fail, reporting bias and breach of agreement. In addition, the owners also face financial crisis, uncertain post-Kyoto international emissions reduction mechanisms and other systemic risk [8]. Matsuhashi R analyzes the various risks CDM projects exist, including price risk, certification risk, baseline risk and country risk. Then, the financing of securitization to diversify risk, issue project bonds and diversified investment are put forward as effective ways to control risk [9].

The fuzzy comprehensive evaluation was proposed by Professor L. A. Zadeh in 1965 [10]. It is used to repent the uncertainty of things based a comprehensive evaluation of mathematical methods. It has many advantages in the use of multi-level fuzzy comprehensive evaluation method to analyze the risk of CDM projects seller enterprises. Firstly, it is suitable for the study of the fuzzy nature. Secondly, multi-level is beneficial to determine the importance and membership of objects to be evaluated. Finally, it can evaluate well for both subject and object factors.

III RISK ASSESSMENT MODEL BASED ON MULTI-LEVEL FUZZY COMPREHENSIVE EVALUATION

A. Construction of Risk Evaluation Index System

Based on the risk factors analysis of Chinese CDM project seller enterprise, build risk analysis index system with fuzzy comprehensive evaluation and AHP from three levels: the national level, the multilateral trade level and the corporate level, and apply analysis with this model.

We set the risk of Chinese CDM project seller enterprise as the goal layer, which is represented by A. Then divide it into three intermediate goal, the national level (B1), the multilateral trade level (B2) and the corporate level (B3). The factors in national level are embodied in economic conditions (C1) and policy conditions (C2); the factors in multilateral trade level are embodied in international trade (C3) and international policy (C4); the factors in corporate level are embodied in registration and approval (C5) and project implementation (C6). Finally, select fifteen indicators as the basic factors, which are classified and combined to their upper level. These indicators are represented by Di (i=1, 2... 15). The index system is shown in Fig.1, 2, 3.

B. Risk Evaluation Model

There are 6 basic steps in fuzzy comprehensive evaluation [11].

(1) Determine the fuzzy evaluation index set U

$$U = (U_1, U_2, U_3) = (U_{11} \sim U_{15}, U_{21} \sim U_{25}, U_{31} \sim U_{35})$$

(2) Determine evaluation set V

$$V = (V_1, V_2, V_3, V_4, V_5) = (\text{very good, good, fair, poor, very poor})$$

Get value of set V:

$$V = (V_1, V_2, V_3, V_4, V_5) = (10, 8, 6, 4, 2)$$

It means that very good get the point of 10.

(3) Determine the single factor evaluation matrix (membership matrix) R

The matrix element r_{ij} in R means the membership of something to the v_j grade fuzzy subset from the view

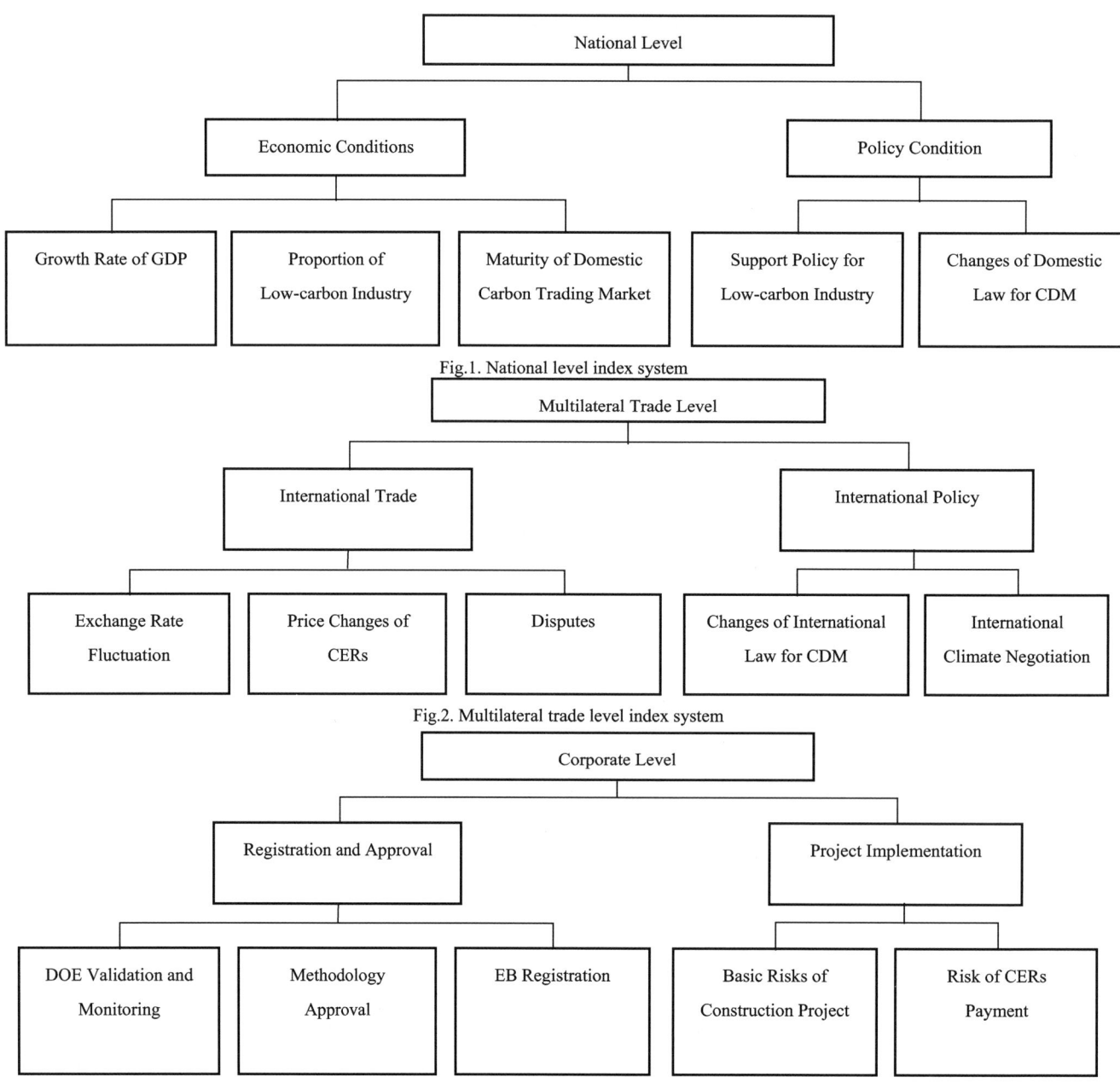

Fig.1. National level index system

Fig.2. Multilateral trade level index system

Fig.3. Corporate level index system

of u_i. And the value range of membership is [0, 1].

(4) Determine the weight of each index set A

According to the degree of importance of each factor, give appropriate weight on each index indicators, to compose a weight set of evaluation factors.

$$A_k(a_{k1} \ a_{k2} \ ...a_{k3}), \text{ and } \sum_{j=1}^{n} a_{kj} = 1 \qquad (1)$$

Determine the weight is key of comprehensive evaluation. Analytic Hierarchy Process (AHP) is an effective method to determine the weight, especially for the complex problems which are difficult to analyze by quantitative indicators. So we choice AHP in this study to determine the weight of indicators in each level.

The specific steps to determine the weight use of AHP:

1) Establish the hierarchy model.

2) Establish pairwise comparison judgment matrix.

Use the 1-9 scale method on the weight assignment, as shown in Table I.

TABLE I
1-9 SCALE METHOD

Comparative scale	Meaning
1	Two elements compared, the same important
3	Two elements compared, one is slightly more important than the other
5	Two elements compared, one is more important than the other
7	Two elements compared, one is much more important than the other
9	Two elements compared, one is extremely more important than the other
2,4,6,8	Intermediate value between the two adjacent analyzing

It will get the relative importance of each element.

3) Consistency check

Calculate eigenvalue of each comparison judgment matrix λ_{max} to check consistency.

$$C.R. = \frac{C.I.}{R.I.} \qquad (2)$$

$C.I.$ is the consistency index;

$$C.I. = \frac{\lambda_{max} - n}{n - 1} \qquad (3)$$

The value of $R.I.$ is showed in Table II.

TABLE II
AVERAGE RANDOM CONSISTENT INDICES

order	1	2	3	4	5	6	7	8
R.I.	0	0	0.58	0.89	1.12	1.26	1.36	1.41

When $C.R. < 0.10$, comparison judgment matrix has satisfactory consistency. When $C.R. > 0.10$, comparison judgment matrix has to be weighted again.

4) Calculate the weight combination of each layers

According to the upper weight, calculate the combination weight of each elements to the top layer. Then calculate the consistency for the whole hierarchy

model.

(5) Calculate fuzzy comprehensive membership value set B [12]

The single factor evaluation matrix in some level is R_k, According the weight set A_k, the fuzzy comprehensive membership value set of each level is

$$B_k = A_k R_k \qquad (4)$$

(6) Model [13]

After normalization, we get $[B_1; B_2; B_3]$.

According to the weight set for the top level A, we can get the comprehensive evaluation index for the top goal. (After normalization).

$$B = AR \qquad (5)$$

The final risk score S is

$$S = VB \qquad (6)$$

The comprehensive membership S is the evaluation score for object U.

IV. APPLICATION

Taking a municipal solid waste incineration power generation CDM project as an example [14-15]. Use the risk analysis model for application.

Get the integrated weight of all levels by AHP. As shown in Table III.

TABLE III
INTEGRATED WEIGHT OF ALL LEVELS

Level	Weight	Indicator	Weight
National Level	0.105	Growth Rate of GDP	0.062
		Proportion of Low-carbon Industry	0.079
		Maturity of Domestic Carbon Trading Market	0.243
		Support Policy for Low-carbon Industry	0.320
		Changes of Domestic Law for CDM	0.296
Multilateral Trade Level	0.637	Exchange Rate Fluctuation	0.126
		Price Changes of CERs	0.119
		Disputes	0.239
		Changes of International Law for CDM	0.145
		International Climate Negotiation	0.370
Corporate Level	0.258	DOE Validation and Monitoring	0.306
		Methodology Approval	0.200
		EB Registration	0.138
		Basic Risks of Construction Project	0.052
		Risk of CERs Payment	0.304

According to weight of indicators, combined with evaluation of fuzzy numbers, we can get the evaluation vector of all aspects after normalization $[B_1; B_2; B_3]$.

$$\begin{bmatrix} 0.0678 & 0.2581 & 0.3460 & 0.2584 & 0.0697 \\ 0.0000 & 0.0629 & 0.1559 & 0.3975 & 0.3837 \\ 0.0990 & 0.1686 & 0.2852 & 0.3156 & 0.1316 \end{bmatrix}$$

So the comprehensive evaluation index after normalization B.

$$[0.0327 \quad 0.1106 \quad 0.2092 \quad 0.3618 \quad 0.2857]$$

After calculation, the final score is of this project is

4.5. It means that, the project is in the general level of risk, and this enterprise have some larger potential risks.

V. CONCLUSION

This paper analyses the influence factors to Chinese CDM project seller enterprises from the level of national, multilateral trade and corporate, and constructs a risk analysis model with fuzzy comprehensive evaluation and AHP. Combined with a project case, an empirical research based on risk analysis model is carried out. After calculation, the final risk score is 4.5. It can be concluded that the project is in the general level of risk, and this enterprise have some larger potential risks. This model has a certain operability, but the system should continue to improve in practice for more accurate risk prediction.

REFERENCES

[1]   Zhou Sheng, Dong Qing, "The uncertainty risk of CDM project development", *Ecological Economy*, 2009, no. 9, pp. 26-29.

[2]   Shen Zeng-qin, "Discussion of seller specific risk control in Chinese CDM project", *Practice in Foreign Economic Relations and Trade*, 2013, no. 6, pp. 27-30.

[3]   Zhuang Gui-yang, "Low carbon economy is guiding the direction of the world economic development", *World Environment*, 2008, no. 2, pp. 34-36.

[4]   Clean Development Mechanism in China, http://cdm.ccchina.gov.cn/.

[5]   He Hai-xia, "Research on China carbon finance market in the post Kyoto era", *Contemporary Economy Management*, 2013, vol. 35, no. 2, pp. 93-97.

[6]   Sun Zeng-qin. "Research on domestic enterprise's risk identification and risk control of clean development mechanism projects in China", *Ecological Economy*, 2013, no. 10, pp. 144-149.

[7]   Wang Dan, Wang Ning, "Simulation of CDM project risk assessment based on fuzzy analysis", *Resources & Industries*, 2011, no.4, pp.50-54.

[8]   Ma Jian-ping, Zhuang Gui-yang, "Risk factor identification and avoidance measures in CDM project development", *Journal of Huazhong University of Science and Technology (Social Science Edition)*, 2011, no. 2, pp. 93-98.

[9]   Matsuhashi R., Fujisawa S., Mitamura W., Momobayashi Y., Yoshida Y., "Clean development mechanism projects and portfolio risks." *Energy*, 2004, vol.29, no.9, pp. 1579-1588.

[10] Li Lie-nan, Li Jing-Lei, "Application of multi_step fuzzy comprehensive evaluation", *Journal of Harbin Engineering University*, 2002, no. 3, pp. 135-138.

[11] Meng Guang-wu, Zhang Xing-fang, "The operational rule of fuzzy numbers", *Fuzzy Systems and Mathematics*, 2001, vol. 5, no. 3, pp. 25-29.

[12] Tang Bing-yong, Wang Wen-jie, Zheng Fei, "Fuzzy mathematics method and its application", Beijing: Coal Industy Press, 1992, pp. 33-35.

[13] Lin Qing-quan, "Modern fuzzy management mathematical methods", Shanghai: China Textile University Press, 1999, pp. 57-61.

[14] Hu Xiu-lian, Jiang Ke-juan, Cui Cheng, "Urban domestic waste incineration for electricity generation-A CDM case study", *Energy of China*, 2002, no,7, pp. 22-28.

[15] Yang Jing, Ma Xiao-xi, "Guangzhou municipal solid waste incineration for electricity generation-a CDM case study", *Renewable Energy Resources*, 2006, no. 1, pp. 62-65.

Research on the Risk Identification for the International Engineering Contracting EPC Projects

Tang Tang

Collage of Management and Economics, Tianjin University, Tianjin, China

(450835875@qq.com)

Abstract - **The International Engineering Contracting EPC projects have the characteristics of high risk, intense complexity and numbers of uncertainties. Thus, as the basis of the risk assessment, risk response and risk control, risk identification can achieve the goal of systematically, dynamically identify, classify, and analyze the risk, thereby contributing to risk aversion and reducing the risk of loss. In this paper, through the literature review and survey analysis, filter out the risks which has the larger-probability of occurrence, the higher-degree of harm and the weaker-management-capacity, establishing the list of risk factors of international project contracting EPC project. Subsequently, this paper suggests the use of interpretative structural modeling (ISM) to prepare a hierarchical structure as well as the interrelationships of these risks, exploring the mechanism of interaction between the risks with the steps as followed, establishing the adjacency matrix and reachability matrix according to the bivariate correlation between the risk factors to, and dividing into elements of level, so as to provide theoretical and practical instruction as well as to offer the EPC contractor additional evidence on the decision-making.**

Keywords - **Engineering procurement construction (EPC), interpretative structural modeling (ISM), risk identification**

I. INTRODUCTION

During the year of 2013, China's foreign contracted projects turnover has come to $ 137.14 billion, with an increase of 17.6%; At the same time, the new contract amount of the foreign contracted projects has come to $ 171.63 billion, with an increase of 9.6%, including the number of projects in billions of dollars is 392, which represents an increase of 63 over the same period of last year. Taking the market distribution situation into consideration, the traditional markets of Asia, Africa remain a major area of foreign project contracting, and the situation in those countries and regions of concentrated market continue unrest in recent years. In addition, the trade protectionism forces of the international market rises, the capacity of corporation finance requirements generally increases in the post-crisis era. Also, global climate change has made the health, safety and environmental issues received unprecedented attention. These problems are making the risk of international projects greatly improve. With an increasingly large scale of projects, technology increases complex, the project integration (survey, design, procurement, construction, operation) trend is increasingly evident, and the international engineering EPC model is becoming main stream as well.

Our companies' business towards the international engineering contracting is expanding rapidly, nevertheless the profitability has stayed at the low condition, and even some of the projects remains are far below the international average level. The complex project environment, the lack of experience in the implementation of major projects and ability in risk management have resulted, directly or indirectly, various types of problems occurrence during the project implementation process, such as the quality of the project cannot be guaranteed, the project schedule delays, and low profits remains. Compared with general engineering projects, international EPC projects are much more susceptible to the influence of many environmental factors change. Since there are many uncertainties during the implementation process, therefore, the risk analysis towards the nature and the impact of the risk factors which may have effect on the projects and the effective identification for the interaction mechanism seem particularly important.

Chihuri, Pretorius (2010) [1], Fang D.P, Wang S.Q. (2010) [2] took the method of brainstorming and expert interviews to identify the risk factors related to security from the projects in South Africa and the Beijing Olympic venues respectively. Mañelele, Muya (2008) [3] got the risk factors identified in the construction project in Zambia through brainstorming. Sun Y. (2008) [4] focused on identifying and assessing the inherent security risks of Beijing Olympic venue construction by brainstorming, interviews and questionnaires with 27 experts from universities and administrative, construction sector. Ke Y.G., Wang S.Q. (2011) [5] figured out 20 major risk factors for Chinese PPP projects through two rounds of the Delphi method. Li Y.B., Tang H. (2011) [6] set sea-sand projects of the Guangzhou Asian Games as an example, building large clusters project risk analysis matrix with WBS-RBS model so as to identify and analysis. The above risk identification take the use of qualitative methods, however, for more complex systems of large projects, entirely qualitative identification will generate a very large solution space expansion, while the explanation towards the behavior of the system will also become extremely difficult. Therefore, the method of quantitative identification can effectively reduce parts of computing ambiguity. Jaruskova. D. (2009) [7] applied FTA and RTA in the risk quantification process of tunnel excavation implementation, which is prone to omissions and errors when applied to complex systems. Fang X.J. (2010) [8] established a quantitative analysis model with the method of the fuzzy graph theory, accomplishing the security risks identification of highway tunnel project. Ohtaka H., Fukazawa Y. (2010) [9] used the cyclic causal model identify the main risk of the major issued projects under the system integration environment. Lin Y., Zhou

E. Qi et al. (eds.), *Proceedings of the 21st International Conference on Industrial Engineering and Engineering Management 2014*, Proceedings of the International Conference on Industrial Engineering and Engineering Management, DOI 10.2991/978-94-6239-102-4_96, © Atlantis Press and the authors 2015

Z. (2011) [10] focused on the impact of design changes on supply chain risk, assessing of the Chinese SPV (Special Purpose Vehicle, SPV / company) supply chain risk. Currently, the project risk analysis application has become a combination of quantitative and qualitative methods. Cheng Y., Liu Z.B. (2011) [11] figured out the large cross Shallow highway tunnel construction risk, combining the expert investigation method with analytic hierarchy process.

In addition, facing the complex causal linkage between risk factors, some scholars have studied the relationship between risk factors through mathematical methods and models. Eunchang Lee (2009) [12] took the ship-building projects for example, and established a Bayesian network risk assessment process applying the Bayesian network to the risk assessment. Zhou G.H. and Peng B. (2009) [13] took use of Bayesian network to analyze the quality-related risk factors of Beijing-Shanghai high-speed railway construction project. Recently, many basic probability of occurrence of events in research are based on determined value representing. In fact, the external environment where each projects constructs is different from each other, and constantly changes, why it is difficult to determine the numerical expression of the probability of an event occurrence. Lu Y. (2010) [14] pointed out that we can take advantage of expert knowledge and expert judgment to the semantic variable so as to change the vague description-probability of an event into the triangular fuzzy number or trapezoidal fuzzy number, and then accomplish risk probability forecast through Defuzzication and Bayesian network inference technology. Lyer and Mohammed (2010) [15] identified the relationship between the life-cycled risk factors of Indian highway PPP project by using ISM model and MICMAC analysis, which found out that delayed payments, cost overruns and schedule overruns has the strongest correlation with the other 14 risk factors. In this paper, we establish a list of risk factors of the international EPC contracting through literature review and survey methods, apply ISM model to study the interaction mechanism of risk factors, setting a delivered structural diagram of risk factors in the international EPC project for intuitive interpretation and analysis.

II. LIST OF RISK FACTORS OF THE INTERNATIONAL EPC CONTRACTING

The social environment of Africa, Southeast Asia and the Middle East is quite volatile, and the natural environment there is much severe, where the project's general contractor is faced with more complex risk environment. According to the occurrence stage of risk factors of international EPC contracting project, their frequency of occurrence, the degree of harm and corporation's risk management capabilities are investigated and analyzed. Through access to a large number of references and background information, we initially divide the risk factors for international EPC projects into 3 levels in accordance with project

environmental risk, project participants risk and project management risk consisting of 19 categories and 37 risk factors. The main decomposition structure is as follows.

1) Project environmental risk

Including the possible risk factors from the international EPC projects in the external environment, The main categories are as follows: (1) political risk (risk of political instability, risk of poor geopolitical relationship); (2) socio-cultural risk (risk of social security chaos, religious and cultural differences); (3) natural environmental risks (force majeure risk, risk of adverse environmental construction); (4) marketing risk (exchange rate risk, inflation risk, interest rate fluctuations risk); (5) legal environment risk (risk of the host country imperfect legal system, risk of the host country law changing).

2) Project participant risk

Including the risk factors which project participants may bring in the international EPC projects' internal and external interactive environment. The major categories are as follows: (1) subcontract management risk (risk of poor coordination of subcontractors, risk of subcontractor undue performance); (2) contract management risk (risk of unstandardized contract management, risk of contract changing); (3) organizations and people risk (leadership / management / technical staff risk, risk of labor disputes).

3) Project management risk

The risk factors include these international EPC project management contractor may face in the internal environment. The major categories are as follows: (1) the tender decision risk (risk of mistakes for project selection, risk of inadequate project investigation); (2) capital risk (project financing / guarantee availability of risk, risk of insufficient expected capital on the early stage of the project); (3) design risk (risk of design errors, risk of inconsistent design, risk of defective design, risk of design delay); (4) procurement risk (risk of procurement plans management failures, materials/ equipment procurement risk, risk of inadequate clearance of imports, risk of unreasonable procurement and construction convergence, the owners' undue interference towards procurement); (5) technical risk (risk of insufficient attention to their environment, construction technology risk, technology transfer risk); (6) insurance/guarantee/credit risk (inadequate insurance policy risk, risk of inadequate guarantee security measures); (7) risk of completion and acceptance (commissioning risk, transfer risk).

III. IDENTIFICATION TOWARD THE INTERNATIONAL EPC PROJECT RISK

Even Risk factors of international EPC projects has been initially screened, because it has certain subjectivity, so it ultimately affects the results of analysis and evaluation. Thus, according to the selected risk factors above, the risk factor questionnaire of international EPC project is designed with the purpose of deeper scientific screening. The questionnaire consists of following three parts:

(1) Expert background information. It is collected for statistics of the resource of the information, understanding the implementation of the risk identification in industry, enterprise as well as leadership management, technology level, and timely feedback the research information to respondents.

(2) Instructions for filling out the questionnaire. The definition of risk factors and rule for risk assessment involved in questionnaire should be explained exactly, ensuring that the investigators effectively fill in the questionnaire, which is based on the scene set properly in advance.

(3) Parts of risk assessment. According to the second part of the assessment rules, investigators give every risk factor a score for the occurrence phase of international EPC risk factors, the occurrence frequency, and degree of harm and enterprise's risk management capability. The scoring method is as shown in Table I-III.

TABLE I
GRADING RULES FOR OCCURRENCE FREQUENCY OF RISK

Score	Frequency	Definition
7	Frequently	Once a week
6	Very likely	Once a month
5	Likely	Once a fiscal year
4	Sometimes	Once 3 fiscal years(enterprise),once a year(industry)
3	Seldom	Once 10 fiscal years(enterprise),once 5 years(industry)
2	Rarely	Once over 10 fiscal years(enterprise),once over 5 years(industry)
1	Never	Never happen

TABLE II
GRADING RULES FOR DEGREE OF HARM OF RISK

Score	Degree of harm
4	Extremely seriously
3	Seriously
2	Generally seriously
1	Slightly

TABLE III
GRADING RULES FOR ENTERPRISE'S RISK MANAGEMENT CAPABILITY

Score	Capability	Definition
5	Strongly capable	Staff have a strong sense of risk management; enterprises has established a proper risk management system and early warning and rapid response mechanisms; the risk control measures can be adjusted according to changes in real-time situations
4	Highly capable	Staff have a high sense of risk management; enterprises has established a risk management system; the risk control measures can adapt to the impact of the risk factors
3	Generally capable	Staff have a basic sense of risk management; the risk management system can analyze and assess the risk factors; the risk control measures can solve some of the problems the risk factors cause
2	Poorly capable	Personnel poor awareness of risk management; the risk management system cannot correctly analyze and assess risk factors; risk control measures cannot be taken to respond effectively to the consequences of the risk factors
1	Incapable	No awareness of risk management; No risk management system established

The survey is taken in the form of an online form, which will be sent to each of the respondents by e-mail. The respondents of the survey were selected to be the staff of international engineering contracting business.

The questionnaires of the survey were 100 copies total, recovered 71 copies withdrawn, 67 copies of valid questionnaires. The effective rate of the questionnaires was 94.37%.

The terms of respondents' business are much balanced to meet the designed requirements of the questionnaire, which is required to cover the relevant departments involved in the project implementation process so as to more scientifically reflect risk factors. In addition, respondents' work experience in the field of international projects for more than seven years accounts for nearly 30%, while for 4-7 years accounts for more than half of the total copies. Therefore, these project management staff has extensive experience, then the information provided can be of high reliability and accuracy.

A. The occurrence phase of international EPC risk factors

The occurrence phase of each international EPC project risk factor differs from each other, corresponding to its whole staged life cycle of the project. Most of the project's environmental risks may have the possibility of occurring in the four stages of the project implementation process, while marketing environment risks were more likely to occur at the project implementation phase. It is whether the external environment risk of the project (including natural and social aspect) is stable that should be fully investigated during the decision-making phase of the project to decide bid or not. On the contrary, the marketing environment risks are associated with the construction, procurement, personnel organization and other aspects of the project management and can have much more influence on the implementation process of the project.

Then, project participant risks and project management risk have a more obvious nature of stages compared to the project environmental risk. The tender decision risk and capital risk mostly occur in the pre-project, such as evaluation and programming phase, at the same times, the rest of the risks occur quite often in the implementation process and the final acceptance phase.

B. The occurrence frequency and degree of harm of international EPC risk factors

The occurrence frequency of international EPC risk factors is closely related to the environment where the project exists, project participants' condition and the enterprise's risk management capability as well. According to survey statistics, the results have to be examined with the consistency test (T-test). After that we calculate the mean and variance of the probability of occurrence of each risk factors, screening these risk factors of which the mean is less than the median average of the mean ($M_1 = 4.20$) and of which the variance is greater than the median average of the variance ($M_2 = 1.20$) That is to say, we screen these risk factors which may occur at the smaller probability and about which the investigators has greater disagreement with each other. The total comes to eight, including: risk of poor geopolitical relationship, risk of social security chaos, risk of adverse environmental construction, mistakes for project selection risk, risk of inadequate project investigation, technology transfer risk, risk of poor coordination of subcontractors, risk of labor disputes.

The degree of harm risk factors cause is estimated, based on the experience of each of the respondents, such as the project situation involved and their sensitivity towards the risk factors. Similarly, the results has to be conformed with the consistency test (T-test).Then we calculate the mean and variance of the degree of harm risk factors cause, screening these risk factors of which the mean is less than the median average of the mean ($M_3 = 2.04$) and of which the variance is greater than the median average of the variance ($M_4 = 0.56$). Namely, we screen these risk factors which may cause a little problem and about which the investigators has greater disagreement with each other. The total comes to eight, including: risk of poor geopolitical relationship, risk of the host country law changing, risk of inadequate project investigation, risk of defective design, the owners' undue interference towards procurement, technology transfer risk, inadequate insurance policy risk, commissioning risk.

C. Enterprise's risk management capability

Enterprise's risk management capability is mostly judged by the level of proper risk management system and early warning and rapid response mechanisms. Accordingly, this article also calculate each of the various risk factors for enterprise's risk management capability, screening these risk factors of which the mean is less than the median average of the mean ($M_5 = 3.80$) and of which the variance is greater than the median average of the variance ($M_6 = 0.56$), which means we screen the risk factors which enterprise remains proper risk management system and early warning and rapid response mechanisms for and about which the investigators has greater disagreement with each other. That accounts seven factors, including: risk of poor geopolitical relationship, interest rate fluctuations risk, risk of insufficient expected capital on the early stage of the project, risk of defective design, risk of inadequate clearance of imports, the owners' undue interference towards procurement, risk of contract changing.

D. Establishing a list of risk factors of the international EPC project

We can get further screening with the comprehensive statistical data above, consisting of the occurrence frequency, the degree of harm, and enterprise's risk management capability of each risk factors. This article finally obtains 20 remaining risk factors, constituting the international EPC contracting projects risk factors systems, with 17 factors screened.

1) Project environmental risk

(1) political risk (risk of political instability); (2) socio-cultural risk (religious and cultural differences); (3) natural environmental risks (force majeure risk); (4) marketing risk (exchange rate risk, inflation risk); (5) legal environment risk (risk of the host country imperfect legal system).

2) Project participant risk

(1) Subcontract management risk (risk of subcontractor undue performance); (2) contract management risk (risk of unstandardized contract management); (3) organizations and people risk (leadership / management / technical staff risk).

3) Project management risk

(1) Capital risk (project financing / guarantee availability of risk); (2) design risk (risk of design errors, risk of inconsistent design, risk of design delay); (3) procurement risk (risk of procurement plans management failures, materials/ equipment procurement risk, risk of unreasonable procurement and construction convergence); (4) technical risk (risk of insufficient attention to their environment, construction technology risk); (5) insurance/guarantee/credit risk (risk of inadequate guarantee security measures); (6) risk of completion and acceptance (transfer risk).

IV. FACTOR ANALYSIS TOWARD THE INTERNATIONAL EPC PROJECT RISK

Based on the international engineering contracting EPC project risk list established above, the data of 20 risk factors to the respondents on the list is collected again to analyze and research. In this paper, the explanation structure model (ISM) and principal component analysis method is adopted to explain and analyze the mechanism of action between risk factors and the importance index.

Assume that the international engineering contracting EPC project risk system for R, Risk factors for R_i, so $R = (R_1, R_2, \cdots, R_n)$. In which, R_1 refers to the political risk, R_2 refers to the socio-cultural risk, R_3 refers to the natural environment risk, R_4 refers to the marketing risk, R_5 refers to the legal environment risk, R_6 refers to the capital risk, R_7 refers to the design risk, R_8 refers to the procurement risk, R_9 refers to the technical risk, R_{10} refers to the credit risk, R_{11} refers to the subcontract management risk, R_{12} refers to the organizations and people risk, R_{13} refers to the insurance/guarantee/credit risk, R_{14} refers to the risk of completion and acceptance. According to the identified risk factors set $R = (R_1, R_2, ..., R_n)$ to determine the relationship between the risk factors, where if there is a direct binary relation, the corresponding values for 1; Otherwise, the corresponding values for 0. According to the results of actual investigation and study to determine the logical relationship between risk factors, adjacency matrix A is established.

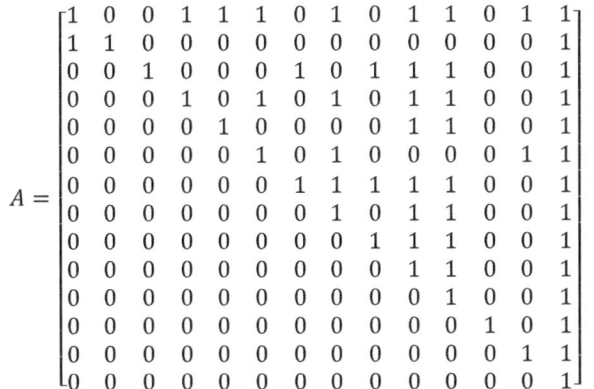

$$A = \begin{bmatrix} 1 & 0 & 0 & 1 & 1 & 1 & 0 & 1 & 0 & 1 & 1 & 0 & 1 & 1 \\ 1 & 1 & 0 & 0 & 0 & 0 & 0 & 0 & 0 & 0 & 0 & 0 & 0 & 1 \\ 0 & 0 & 1 & 0 & 0 & 0 & 1 & 0 & 1 & 1 & 1 & 0 & 0 & 1 \\ 0 & 0 & 0 & 1 & 0 & 1 & 0 & 1 & 0 & 1 & 1 & 0 & 0 & 1 \\ 0 & 0 & 0 & 0 & 1 & 0 & 0 & 0 & 0 & 1 & 1 & 0 & 0 & 1 \\ 0 & 0 & 0 & 0 & 0 & 1 & 0 & 1 & 0 & 0 & 0 & 0 & 1 & 1 \\ 0 & 0 & 0 & 0 & 0 & 0 & 1 & 1 & 1 & 1 & 1 & 0 & 0 & 1 \\ 0 & 0 & 0 & 0 & 0 & 0 & 0 & 1 & 0 & 1 & 1 & 0 & 0 & 1 \\ 0 & 0 & 0 & 0 & 0 & 0 & 0 & 0 & 1 & 1 & 1 & 0 & 0 & 1 \\ 0 & 0 & 0 & 0 & 0 & 0 & 0 & 0 & 0 & 1 & 1 & 0 & 0 & 1 \\ 0 & 0 & 0 & 0 & 0 & 0 & 0 & 0 & 0 & 0 & 1 & 0 & 0 & 1 \\ 0 & 0 & 0 & 0 & 0 & 0 & 0 & 0 & 0 & 0 & 0 & 1 & 0 & 1 \\ 0 & 0 & 0 & 0 & 0 & 0 & 0 & 0 & 0 & 0 & 0 & 0 & 1 & 1 \\ 0 & 0 & 0 & 0 & 0 & 0 & 0 & 0 & 0 & 0 & 0 & 0 & 0 & 1 \end{bmatrix}$$

With the aid of MALTAB software, the reachable matrix M is obtained as follows:

$$M = \begin{bmatrix} 1 & 0 & 0 & 1 & 1 & 1 & 1 & 0 & 1 & 1 & 0 & 0 & 1 & 1 \\ 1 & 1 & 0 & 1 & 1 & 1 & 1 & 0 & 1 & 1 & 0 & 0 & 1 & 1 \\ 0 & 0 & 1 & 0 & 0 & 0 & 1 & 1 & 0 & 1 & 0 & 0 & 0 & 1 \\ 0 & 0 & 0 & 1 & 1 & 0 & 1 & 1 & 0 & 1 & 0 & 0 & 1 & 1 \\ 0 & 0 & 0 & 0 & 1 & 0 & 1 & 1 & 0 & 1 & 0 & 0 & 1 & 1 \\ 0 & 0 & 0 & 0 & 0 & 1 & 1 & 1 & 0 & 1 & 0 & 0 & 1 & 1 \\ 0 & 0 & 0 & 0 & 0 & 0 & 1 & 0 & 1 & 1 & 0 & 0 & 1 & 1 \\ 0 & 0 & 0 & 0 & 0 & 0 & 0 & 1 & 0 & 0 & 1 & 0 & 1 & 1 \\ 0 & 0 & 0 & 0 & 0 & 1 & 0 & 1 & 0 & 0 & 0 & 0 & 1 \\ 0 & 0 & 0 & 0 & 0 & 0 & 0 & 0 & 1 & 0 & 0 & 0 & 1 \\ 0 & 0 & 0 & 0 & 0 & 0 & 0 & 0 & 0 & 1 & 0 & 0 & 1 \\ 0 & 0 & 0 & 0 & 0 & 1 & 1 & 1 & 0 & 0 & 1 & 0 & 1 \\ 0 & 0 & 0 & 0 & 0 & 0 & 0 & 0 & 0 & 0 & 1 & 1 \\ 0 & 0 & 0 & 0 & 0 & 0 & 0 & 0 & 0 & 0 & 0 & 0 & 1 \end{bmatrix}$$

According to the reachable matrix M, the following sets of R_i are obtained: $P(R_i)=\{R_j|\ m_{ij}=1\}$; $Q(R_i) = \{R_j|\ m_{ij}=1\}$; $P(R_i) \cap Q(R_i) = T(R_i)$. In which, $P(R_i)$ for reachable set, $Q(R_i)$ for leading set, $T(R_i)$ for common set. Through calculation, when $P(R_i)$ and $T(R_i)$ contain the equal number of factors at the same time, the most superior unit is obtained. Then, delete the rows and columns which contain the factors in $T(R_i)$ set from the original reachable matrix M. By the same token, the second level unit is obtained. In turn, the factors can be divided for ladder structure.

According to the above process, the factor set in this system is described as a hierarchical model with 7 ladders structure, Level 1: L1= {R14}; Level 2: L2= { R10, R11, R13}; Level3 L3= { R7, R8, R9 }; Level4: L4= {R3, R5, R6, R12}; Level5: L5= {R4}; Level6: L6= {R1}; Level7: L7= {R2}.

According to the order R14, R10, R11, R13, R7, R8, R9, R3, R5, R6, R12, R4, R1, R2 which obtained through the displacement subsystem of decomposition results, the adjacency matrix is rearranged and the structure matrix S is established.

$$S = \begin{bmatrix} 1 & 0 & 0 & 0 & 0 & 0 & 0 & 0 & 0 & 0 & 0 & 0 & 0 & 0 \\ 1 & 1 & 0 & 0 & 0 & 0 & 0 & 0 & 0 & 0 & 0 & 0 & 0 & 0 \\ 1 & 0 & 1 & 0 & 0 & 0 & 0 & 0 & 0 & 0 & 0 & 0 & 0 & 0 \\ 1 & 0 & 0 & 1 & 0 & 0 & 0 & 0 & 0 & 0 & 0 & 0 & 0 & 0 \\ 1 & 1 & 1 & 0 & 1 & 1 & 1 & 0 & 0 & 0 & 0 & 0 & 0 & 0 \\ 1 & 1 & 1 & 0 & 0 & 1 & 1 & 0 & 0 & 0 & 0 & 0 & 0 & 0 \\ 1 & 1 & 1 & 0 & 0 & 0 & 1 & 0 & 0 & 0 & 0 & 0 & 0 & 0 \\ 1 & 1 & 1 & 0 & 1 & 0 & 1 & 1 & 0 & 0 & 0 & 0 & 0 & 0 \\ 1 & 1 & 1 & 0 & 0 & 0 & 0 & 0 & 1 & 0 & 0 & 0 & 0 & 0 \\ 1 & 1 & 1 & 0 & 0 & 9 & 0 & 0 & 0 & 1 & 0 & 0 & 0 & 0 \\ 1 & 1 & 1 & 0 & 1 & 1 & 1 & 0 & 0 & 0 & 1 & 0 & 0 & 0 \\ 1 & 1 & 1 & 1 & 0 & 1 & 0 & 0 & 0 & 1 & 0 & 1 & 0 & 0 \\ 1 & 1 & 1 & 0 & 1 & 0 & 0 & 0 & 1 & 1 & 0 & 1 & 1 & 0 \\ 1 & 1 & 1 & 0 & 0 & 0 & 0 & 0 & 1 & 0 & 0 & 1 & 1 & 1 \end{bmatrix}$$

After the division of the above, the hierarchical structure diagram of international engineering contracting EPC project risk system is obtained, as shown in Fig.1.

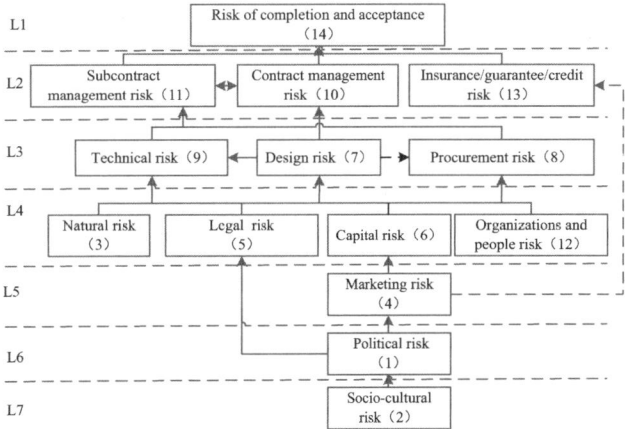

Fig.1. Schematic diagram of the recursive structure of risk factor for the international EPC project

As shown above, socio-cultural risk in systematic risk includes the phenomenon of social unrest and the differences between religious and cultural are both likely to lead to political risk. Moreover, the poor political instability and geopolitical relations caused by political risk also makes the host country market environment changes, the exchange rate, interest rate wave, intensification of inflation, and caused the capital risk of project. The systemic risk represented by natural environment risk, legal environment risk, capital risk and organizations and people risk in fourth level risk factors directly influences the design management, procurement management and technology management of the project, which increase the probability of risk on the stage. In the meantime, the contract management risk and subcontract management risk brought by the project design, procurement and construction error or unreasonable will lead to the final completion of settlement risk with the insurance/guarantee/credit risk together: delay period, increase the cost, reduce the income and even the handover difficulties which is caused by the external environment changes referring insufficient demand and market price fluctuations, and by the internal management problems referring the lack of experience and technical complex.

V. CONCLUSION

Taking the great risk factors, complexity and many of the uncertainty of the international EPC project into consideration, this article compares the merits and scope of the identification and analysis of risk factors according to the relevant research review, then determines the risk factor identification and analysis methods used in this study combined with characteristics and the risk-share principle of international EPC projects. Through literature review, we divide risk factors of international EPC contracting engineering projects into project environmental risk, project participants risk, project management risk, preliminary figuring out 37 risk factors. Further, we screen 20 risk factors out of the 37 factors, and obtain these corresponding to the higher probability of occurrence, higher degree of harm and weaker enterprises' risk management through questionnaire, so as to build a list of international EPC contracting engineering projects risk factors. Then this paper makes the method of ISM model, to build up the Adjacency Matrix and Reachability Matrix, which is on the basis of the correlation of risk factors, and to divide levels of factors. In the last, this paper prepares a delivered structural diagram of risk factors in the international EPC project for visual interpretation and analysis.

REFERENCE

[1] Chihuri S., Pretoriu.L. Managing Risk for Success in a South-African Engineering and Construction Project Environments [J]. South African Journal and Industrial Engineering, 2010, 21(2):63-77.

[2] Zhu D.F., Fang D.P., Wang S.Q. Development of Risk Measurements for 2008 Beijing Olympic Venue Construction [J]. Journal of Engineering Management, 2010, 21(2): 23-28.

[3] Mañelele I., Muya. M. Risk Identification on Community-based Construction Projects in Zambia [J]. Journal of Engineering, Design and Technology, 2008, 6(2): 145-161.

[4] Sun Y., Safety Risk Identification and Assessment for Beijing Olympic Venues Construction [J]. Journal of

Management in Engineering, 2008, 24 (1): 40-47.

[5] Ke Y.J., Wang S.Q. Understanding the risks in China's PPP projects: ranking of their probability and consequence [J]. Engineering, Construction and Architectural Management, 2011, 18(5):481-496.

[6] Li Y.H., Tang H. The WBS-RBS Risk Identification Method and Its Application to Large Cluster Projects [J]. Construction Economy, 2011, 8:31-34.

[7] Sejnoha. J., Jaruskova. D. Risk Quantification for Tunnel Excavation Process [J]. Proceedings of World Academy of Science, Engineering and Technology, 2009, 58:101-109.

[8] Zhou Z., Fang X.J. Application Study on Fuzzy Influence Diagram in Highway Tunnel Construction Safety Risk Assessment [A]. International Conference on Management Science & Engineering, Melbourne, 2010, 272-279.

[9] Ohtaka H., Fukazawa Y. Managing Risk Symptom: A Method to Identify Major Risks of Serious Problem Projects in SI Environment Using Cyclic Causal Model [J]. Project Management Journal, 2010, 41(1): 51-60.

[10] Lin Y., Zhou Z. The Impacts of Product Design Changes on Supply Chain Risk: A Case Study [J]. International Journal of Physical Distribution & Logistics Management, 2011, 41(2): 162-186.

[11] Cheng Y., Liu Z.B. Risk Recognition of Construction of Large-span and Shallow Buried-highway Tunnels based on Analytic Hierarchy Process [J]. Chinese Journal of Geotechnical Engineering, 2011, 33(1): 191-195.

[12] Lee E, Park Y, Shin J G. Large Engineering Project Risk Management Using a Bayesian Belief Network [J]. Expert Systems with Applications, 2009, 36(3): 5880-5887.

[13] Zhou G.H., Peng B. Analysis of Quality Management Risk in Large Construction Projects based on Bayesian Belief Network: A Case Study of Beijing-Shanghai High-speed Railway Project [J]. China Soft Science, 2009, 9:99-106.

[14] Lu Y., Li Q.M. Safety Risk Prediction of Subway Operation based on Fuzzy Bayesian Network [J]. Journal of Southeast University (Natural Science Edition), 2010, 40(5): 1110-1114.

[15] Lyer K.C. Hierarchical Structuring of PPP Risks Using Interpretative Structural Modeling [J]. Journal of Construction Engineering and Management, 2009, 136(2): 151-159.

The Relationship between Top Management Team and Firm Performance - Data from China GEM Listing Corporation

Peng Yang[1, 2], Su-ying Gao[1, *], Long Xu[1], Chun-ying Song[1]

[1]School of Economics Management, Hebei University of Technology, Tianjin, China

[2]College of Mathematics and Information Technology, Xingtai University, Hebei, China

(sue2007@hebut.edu.cn)

Abstract - **Top management team plays an important role in the survival and the development of the firms. The authors study the 249 China GEM Listing Corporation, analyze top management team's age, education, salary and other aspects by means of empirical test and investigate the effect to firm performance. The results show that the average age of top management team is significantly negative related to firm performance. There is significantly positive relationship between the profits of top management team and firm performance. In the corporate governance the authors select two important indexes to do T test, find that the independent directors proportion has significantly improving effect to firm performance, but the two power setting of the chairman and the general manager has not obvious influence.**

Keywords - **Corporate governance, firm performance, op management team**

I. INTRODUCTION

Decision and action of people play a key role in enterprise development. As a senior human capital top management team is the source of sustainable competitive advantage [1]. Through the strategic decision-making process they have a significant impact on the development of firm performance in a long term. Top management team is more important than the grass-roots management. According to the Resource-Based View (RBV,Barney, 1991), the senior management with the characteristics of VRIO is the key to keep the core competitiveness of the firms.

II. LITERATURE REVIEW AND HYPOTHESIS

Top management team is a core group which decides development of the firms [2]. Generally top management team includes the members of the board of directors, the general manager, deputy general manager, president, vice president, chief financial officer, chief engineer, chief economist, the secretary of the board of directors [3].

The average age of top management team is one of the important human capital indexes. Young top management team is not only willing to take risks, but also willing to try to change [4]. GEM Listing Corporations are mostly high-tech and young enterprises, need to break through and innovate. Therefore put forward the

Fund Project: National Natural Science Fund Project (71172153); Hebei Natural Science Fund Project (G2014202233);Hebei Social Science Fund Project (HB10XGL203)

hypothesis 1: for GEM Listing Corporations, the average age of top management team has a negative impact on firm performance.

The long term executive team can often exchange, members can share knowledge and information. It is good to enhance team cohesion and improve the profitability of enterprises [5]. So the long average tenure of top management team can reduce the internal conflicts and strengthen communication. Therefore put forward the hypothesis 2: the average tenure of top management team has a positive impact on firm performance.

Higher education level means more effective information which top management team possesses, and it is good for the development of company. Chuntao Li, Xiaowei Kong deemed that there was a significant positive correlation between management degree and firm performance [6]. Hui Zhang, Tongliang An thought that the correlation was weakly positive between management educational background and firm performance [7]. Hypothesis 3: the average education level of top management team has a positive impact on firm performance.

If top management team has abundant resources to solve problems, they are likely to make high quality decisions [8]. The large-scale top management team with a lot of information and resources are better than the small-scale top management team. Therefore put forward the hypothesis 4: the size of top management team has a positive effect on firm performance.

The age heterogeneity of top management team can complement each other. It effectively enhances the firm performance [9]. But the education level heterogeneity office produces communication barriers and conflict each other. This is difficult to reach a consensus, eventually led to the decision to delay [10]. Therefore put forward the hypothesis 5: the heterogeneity of top management team has a significant effect on firm performance.

If top management team process more stock, their personal interests will more closely contact with firm performance and they will more actively work hard [11]. Cong Wang found that top management team shareholding significantly influenced the performance of enterprises and the enterprise benefits mainly depended on the efforts of top management team [12]. Therefore put forward the hypothesis 6: the shareholding ratio is positively related to firm performance.

Some American scholars analyzed hundreds of business companies in recent 15 years, researched that executive compensation and corporate performance had strong correlation [13]. Another scholar thought that ROE and executive compensation has a significant linear relationship. Therefore put forward the hypothesis 7: the average salary of top management team has a positive effect on firm performance.

The independent directors proportion represents the level of company governance, can bring equal benefits for all the shareholders [14]. To some extent the implementation of the independent director system affects firm performance. Tongying Liang selected the private listing corporations of Shanghai in 1998-2003, found that the independent directors proportion is positively correlated with private listing corporation performance. Therefore put forward the hypothesis 8: the proportion of independent directors has a significant impact on firm performance.

Two rights separation of the chairman of the board and general manager is easy to form the democratic atmosphere. The two rights separation is effective to resist the management risk [15]. But there are also the different opinions of some scholars. Donaidosn considered two rights merging can improve the efficiency of decision-making. Based on the two aspects, we present the following hypothesis 9: the two rights separation of the chairman and the general manager has a significant impact on enterprise performance.

III. METHODOLOGY

A. Sample Selection

Study selects Chinese GEM Listing Corporation as the samples. Based on the CSMAR database, we collect the data and screen the original indexes in order to reduce error. We remove the samples in which Return On Equity(ROE) is negative, finally get the 249 samples of GEM Listing Corporation in 2012. In this paper, we do the descriptive statistics, correlation analysis, multiple regression analysis and T test by means of SPSS17.0.

B. Variable Setting

Return On Equity (ROE) is a chief indicator of firm performance. So we select Return On Equity as dependent variable with the symbol "Y".

Independent variables about top management team include average age (X_1), average tenure (X_2), average level of education (X_3), team size (X_4), age heterogeneity (X_5), education heterogeneity (X_6), shareholding ratio (X_7) and average salary (X_8).

Control variables are scale of the firms (D_1) and asset-liability ratio (D_2).

Corporate governance variables are the independent director proportion and the two power settings of the chairman and the general manager.

C. Model Specification

Based on the analysis of previous research, we propose a model

$$Y = \alpha + \beta_1 X_1 + \beta_2 X_2 + \beta_3 X_3 + \beta_4 X_4 + \beta_5 X_5 + \beta_6 X_6 + \beta_7 X_7 + \beta_8 X_8 + \beta_9 D_1 + \beta_{10} D_2 + \varepsilon$$

Y is Return On Equity (ROE) of Chinese GEM Listing

Corporation. α is the constant term, $\beta_1 - \beta_{10}$ is the regression coefficient, ε is residual value.

IV. RESULTS

A. Descriptive Statistical Analysis

TABLE I
DESCRIPTIVE STATISTICS

	N	Min	Max	mean	Std.Deviation
return on equity	249	0.03	0.42	0.1738	0.07293
average age	249	37.42	52.57	45.567	2.96377
average tenure	249	2.18	7.09	3.0545	0.59307
average level of education	249	2	4.38	3.2945	0.41236
team size	249	2	13	6.18	1.865
age heterogeneity	249	0.09	0.31	0.187	0.04647
education heterogeneity	249	0.23	0.87	0.659	0.09859
shareholding ratio	249	11.87	16.24	14.709	0.57127
average salary	249	0.0	0.9	0.3967	0.22204
scale of the firms	249	18.81	22.5	20.675	0.59878
asset-liability ratio	249	0.02	0.75	0.1836	0.13486

From Table I, the minimum rate of ROE is 3%, the maximum is 42%. The difference is very big. We can see that GEM enterprises are in the fierce market competition with income instability and large risk. We find that the average age is 45 years old, the standard deviation is 2.96 and the discrete degree is very small. This shows that the most people of top management team in GEM Listing Corporation are middle-aged. The average tenure is 3.0545 and this shows that top management team is instable. The average educational level of top management team is 3.2945, between undergraduate and master's degree, it shows that most members of top management team have received a good education. Age heterogeneity is 0.187, in the low level, indicating that top management team members' age difference is very small. Descriptive statistics shows that top management team has high education levels, young age structure. Top management team members are energetic and the firms have the potential. But the average tenure is not long, indicating that top management team members transform job frequently. This may be related to the company startup.

B. Correlation Analysis

TABLE II
CORRELATION ANALYSIS

	Y	X_1	X_2	X_3	X_4	X_5	X_6	X_7	X_8
Y	1								
X_1	-0.188**	1							
X_2	-0.100*	0.070	1						
X_3	-0.030	-0.154**	0.085	1					
X_4	-0.014	-0.005	-0.010	0.050	1				
X_5	0.047	-0.008	-0.054	-0.225**	-0.056	1			
X_6	0.025	0.056	-0.219**	-0.130*	-0.096	0.032	1		
X_7	0.027	0.030	-0.012	0.260**	0.401**	-0.069	-0.017	1	
X_8	0.190**	-0.178**	-0.003	-0.141*	0.044	0.023	-0.019	-0.165**	1

Note: ** means significant at 1% level,* means significant at 5% level.

From Table II, ROE has a certain correlation with each variable. The relation between ROE and average age is significant in 5% level, so is ROE and average salary. Furthermore, ROE is significantly related with the average tenure in 1% level. In addition, ROE only has the weak correlation with other variables.

At the same time, the variables also exists the significant relationship each other, such as the average age (X_1) and the average level of education (X_3). The average tenure (X_2) and education heterogeneity (X_6) is significantly negative correlation. The average level of education (X_3) and shareholding ratio (X_7) have significant positive correlation, indicating that education is better, the income of members is higher.

C. Regression Analysis

TABLE III
REGRESSION ANALYSIS

		coefficient	t	Sig.
	constant term	0.651	3.43	0.001
independent variables	average age	-0.003	-1.868	0.05
	average tenure	-0.011	-1.493	0.137
	average level of education	-0.005	-0.386	0.7
	team size	-0.001	-0.512	0.609
	age heterogeneity	0.112	1.138	0.256
	education heterogeneity	0.004	0.087	0.931
	shareholding ratio	0.025	2.607	0.01
	average salary	0.045	2.154	0.032
control variables	scale of the firms	-0.034	-3.948	0.00
	asset-liability ratio	0.032	0.94	0.348

From Table III, hypothesis 1 is supported. The average age of top management team is significant at the 5% level. Its coefficient is -0.003, indicating that the average age of top management team has a negative impact on firm performance. Young members of top management team are willing to accept new challenges, new knowledge and new management concepts. This will promote the firm performance in Chinese GEM Listing Corporation. But hypothesis 2 and 3 is not supported.

Hypothesis 4 is not supported. Results display team size is not the main factor affecting performance. The team quality and efficiency should be the key of firm performance improvement.

Hypothesis 5 is not supported. The heterogeneity of top management team does not pass the test, indicating that its impact is limited. With the rapid development of economy, enterprises often organize top management team members to study technical, financial knowledge in order to cope with the market change from minute to minute. Thus it weakens the differences between them.

Hypothesis 6 and 7 is supported. Shareholding ratio of top management team is significantly positive at 1% level. The average salary of top management team is significantly positive at the 5% level. Good salary and high proportion of shares will enhance working enthusiasm of senior managers. The result also shows that the incentive policy has a great impact on firm performance.

D. Corporate Governance

(1) The important indicator of corporate governance is independent director proportion. We sort all cases according to the value of independent director proportion from big to small. We select the 30 enterprises in front as good governance enterprises, meantime choose the last 30 enterprises as poor governance enterprises. We do T test of this 60 enterprises and compare whether there are the significant differences on the average variables.

TABLE IV
T TEST OF INDEPENDENT DIRECTOR PROPORTION RANKING TOP 30 FIRMS AND LAST 30 FIRMS

		F	Sig.	t	Sig.
return on equity	equal variance	6.274	0.015	-1.998	0.050
	unequal variance			-1.998	0.050
average age	equal variance	4.211	0.045	1.749	0.086
	unequal variance			1.749	0.086
average tenure	equal variance	0.391	0.534	-0.521	0.604
	unequal variance			-0.521	0.604
average level of education	equal variance	2.461	0.122	0.167	0.868
	unequal variance			0.167	0.868
team size	equal variance	4.593	0.036	-16.23	0.000
	unequal variance			-16.23	0.000
age heterogeneity	equal variance	0.380	0.540	0.967	0.338
	unequal variance			0.967	0.338
education heterogeneity	equal variance	0.174	0.678	1.478	0.145
	unequal variance			1.478	0.145
shareholding ratio	equal variance	0.465	0.498	-3.526	0.001
	unequal variance			-3.526	0.001
average salary	equal variance	0.007	0.935	-2.072	0.043
	unequal variance			-2.072	0.043

From Table IV, hypothesis 8 is supported. There is the significant difference of firm performance between good governance enterprises and poor governance enterprises, indicating that the independent director proportion significantly influences firm performance. At the same time, two kinds of enterprises also have the obvious difference in the team size, the shareholding ratio and the average salary of top management team.

(2) We set the two power settings of the chairman and the general manager. Symbol "1" expresses that the chairman and general manager is one person, symbol "2" expresses that the chairman and general manager is not performed by the same individual. This is an important index to reflect the corporate governance. We do T test to compare the mean value of firm performance.

TABLE V
T TEST OF THE TWO POWER SETTINGS

		F	Sig.	t	Sig.
return on equity	equal variance	2.644	0.105	1.126	0.261
	unequal variance			1.127	0.261

From the data in Table V we can see that Sig value is not significant at the 5% level. The two the right setting of the chairman and the general manager has not influence on firm performance. Hypothesis 9 is not supported.

V. CONCLUSION

The authors examine the impact of top management team in China on firm performance with the empirical method, and get the following conclusions. First, the average age of top management team has significantly negative impact on firm performance. At present, GEM Listing Corporations are mainly high-tech enterprises and need innovation actively. The young members dare to take a risk and possess competitive advantage. Second, there is the significantly positive correlation between the benefit of top management team and firm performance. The benefit of top management team stimulates the enthusiasm and the initiative of the members. Third, we do not find that education has an obvious impact on firm performance. It not only because that the gap of samples is small, but also because the current education system has some deficiencies. It may also reflect some blindness at the time of introduction of a high degree of managers. Fourth, in terms of corporate governance, we find that the proportion of independent directors has a significant impact on firm performance, indicating that the independent director system plays a positive role on corporate governance. But the two rights setting about the chairman and the general manager has no significant effect, indicating that the restriction mechanism of GEM Listing Corporation is not good. This may have a certain impact on enterprise development.

REFERENCES

[1]   Dejun Cheng, Shuming Zhao. "High involvement work system and firm performance: The effect of human capital specificity and dynamic environment" (in China), Management world, 2006(3), pp.86-93.

[2]   Gang Wei,Naige Yang. "Senior management incentive and corporate performance" (in China), Securities market Herald, 2000.3, pp.19-29.

[3]   Tihanyi L, EllstrandAE, DailyCM, "Composition of the top Management Team and firm International diversification". Journal of Management, 2000(35), pp.503-538.

[4]   Smith,K.G., Smith,K.A. & Olian.J.D. "Top management team demography and process：The role of social integration and communication". Administrative Science Quarterly, 1994(39), pp.412-438.

[5]   C.M.Daily, J.L.Johnson. "Sources of CEO Power and Firm Financial Performance: A Longitudinal Assessment". Journal of Management, 1997, 23(2).

[6]   Chuntao Li, Xiaowei Kong. "Empirical research on managers educational level and operating performance of the listing Corporation" (in China), Nankai Economic Studies, 2005(1).

[7]   Hui Zhang, Tongliang An. "Empirical analysis on Chinese listing Corporation board education level distribution and company performance" (in China). Economic Science, 2005, (5).

[8]   Ying Wang. "Enterprise manager human capital structure and corporate performance" (in China). Statistics and decision making. 2004, (12).

[9]   KilduffM, AngelmarR, MehraA. "Top management team diversity and firm performance: Examining the role of cognitions". Organization Science. 2000, 11(1). pp.21-34.

[10] Hamid, Mehran. "Executive Compensation Structure, Ownership, and Firm Performance". Journal of Financial Economics. 1995(38), pp.163-184.

[11] Hall.Brian.J and Jeffrey.B.Liebman. "Are CEOs Really Paid Like Bureaucrats?". Quarterly Journal of Economics. 1998(34). pp.131-146.

[12] Cong Wang, Qi Li. "Discussion on executive compensation design in listing Corporation" (in China), Communication of Finance and accounting. 2011(3). pp.8-19.

[13] SimonsTL, PelledLH, SmithKA. "Making use of difference: diversity, debate, and decision comprehensiveness in top management teams". Academy of Management journal. 1999, 42(6). pp.662-673.

[14] Ying Wang, Jiancheng Guan. "Study on the relationship between the top management of our enterprises, innovation strategy and business performance" (in China). Journal of management Engineering. 2003.

[15] Gang Wei. "Senior management incentive and Listing Corporation operating performance" (in China). Economic Research. 2000, (3).

The Correlation of Urban Cluster and Cultural Industry Cluster

Guo-jun Chai[1,2]

[1]Economic Management and Economic Department of Technology, Tianjin University, Tianjin, China
[2]Science Department, Inner Mongolia University of Finance and Economics, Hohhot, China
(nmstar@vip.sina.com)

Abstract - **The development of urban cluster and cultural industry cluster complement each other. Exploring the inner correlation between the development of the urban cluster and cultural industry cluster from the quantitative perspective and disclosing their interior rules of development, may have great significance for the urban designing, urban orientation, urban developing direction, urban formats distribution and urban industrial structure adjustment. Firstly, in this paper, for the collaborative linkage between city agglomeration economy development and the development of cultural industry are analyzed theoretically. Then, in the empirical study, the study objects of this empirical research are the urban cluster of HBE (Hohhot, Baotou and Erdos). Through co-integration test and calculation of granger causality test, the research demonstrates the correlation between the development of urban cluster and cultural industry cluster. The results show that the development of cultural industries must be based on the cluster development as the focal point. Only through cultivating clusters and amplifying cluster effects, can speed up their development.**

Keywords - **Cultural industry cluster, comovement relation, urban cluster of HBE, urban cluster**

I. INTRODUCTION

Currently, there are no specific policies on urban functional orientation, urban features designing and industry structure adjustment among urban clusters during the development and planning of urban clusters and culture industries [1]. Urban clusters lack of efficient coordination and interaction [2-3]. The linkage function between the development of the urban cluster and cultural industry cluster has not been well played. As a result, many isolated plans and strategies are formulated, which lead to the similarities of city orientation and the simple repetition or imitation of the culture industries' projects. Many researchers discuss this phenomenon in different areas [4-6]. By taking the urban clusters of HBE (Hohhot, Baotou and Erdos) as examples, this paper focuses on the correlation between the development of the urban cluster and cultural industry cluster.

II. THE ORETICAL RESEARCH ON THE RELATIONSHIP BETWEEN THE URBAN CLUSTER (UC) AND CULTURAL INDUSTRY CLUSTER (CLC) DEVELOPMENT

A. The economic development of urban clusters is the base and carrier of cultural industry clusters development

When modern industry develops to a certain stage, the mass development of the cultural industry will be an inevitable result of full development of market economy.

From the initial production to the final consumption, the cultural product and service always need the support of a strong economic foundation, and the economic development of urban clusters is the base and carrier of cultural industry cluster development.

(1) The economic development of urban clusters provides a strong support to the cultural industry cluster development. By surveying and analyzing the basic situation of urban residents in Inner Mongolia in 2012, it is shown that the per capita income in the region of HBE urban cluster reached 37,526 Yuan in 2012, and 34,984 of which was disposable income. Meanwhile, the Engel's coefficient of resident families in HBE urban cluster region had decreased to about 30%, which shows that the proportion of resident consumption for the basic necessities of survival has decreased while the need for culture consumption in spiritual levels is rising, so the economic development of the urban cluster provides strong support for the cultural industry cluster development.

(2) The economic development of the urban cluster promotes the infrastructure construction of the cultural industry. Investment in the cultural infrastructure construction of HBE urban cluster has been gradually increased. In the first Plenary Session of the 12th Inner Mongolia People's Congress in 2013, the municipality government put forward that special financial support to cultural industry from the level of municipality government will be doubled. From the analyses, it is concluded that with the rapid development of economy, the infrastructure for cultural industry development of HBE urban cluster has been gradually improved.

(3) Changes in the consumption level and structure of the urban cluster stimulate the increase in the demand of cultural consumption. The increase in residents' consumption level will promote the increase in the demand for cultural products and services, and then push the development of cultural industry. As above mentioned, the residents' consumption demands in HBE urban cluster has basically changed from survival requirements to spiritual requirements. The economic development of HBE urban cluster promotes the changes of residents' consumption level and structure, which stimulates residents' cultural demands and enhances the potential development of cultural industry on the one hand, and opens up more space for the development of cultural industry on the other hand.

E. Qi et al. (eds.), *Proceedings of the 21st International Conference on Industrial Engineering and Engineering Management 2014*, Proceedings of the International Conference on Industrial Engineering and Engineering Management, DOI 10.2991/978-94-6239-102-4_98, © Atlantis Press and the authors 2015

B. The Cluster Development of the Cultural Industry Plays a Significant Role in Promoting the Economic Development of the Urban Cluster

The economic development history in developed countries shows that at the initial period of urban cluster economy development, agricultural economy plays a leading a role, then the industrial economy, and then the knowledge-based economy, namely the cultural economy. The above empirical analysis in the above subsection has proven that the cultural industry development obviously promotes the HBE unban cluster. Detailed analysis is given as follows.

(1) New economic growth points are formed with the rapid development of cultural industry. The GDP per capita of HBE city cluster is now more than 10,000 USD. With the promotion of economic development, from 2008 to 2012 (see Table I in section III), the cultural industry output had reached an average growth rate has reached 17.5%, the cultural industry had contributed the economic development in the rate of 4%, and the cultural industry has become a new economic growth point.

(2) It improves the proportion of tertiary industry, increases employment opportunities, and optimizes the economic structure. The aggregation and development of the cultural industry in HBE urban cluster has stimulated the rapid development of employment. The cultural industry of HBE urban cluster has many categories, of which culture and entertainment, cultural tourism, network services, radio and television, and animation industry have rapid employment growth. The rapid cluster development of the cultural industry increases the employment opportunities, enhances the tertiary industry proportion in the whole economy, and then optimizes the economic structure of urban cluster.

(3) It expands domestic consumption and promotes the transformation of economic growth mode. The consumption of cultural products and services is higher level consumption to meet the spiritual needs of the consumers, and will greatly promote the transformation of economic growth mode. According to the development law of international cultural industries, when the GDP per capita reaches 5,000 USD, the consumption of cultural products and services will be continually doubled. The GDP per capita of HBE urban cluster is now more than 10,000 USD. Prosperity and development of the cultural industry is developing this part of the potential purchasing power; through the consumption of cultural products and services, the transformation of economic growth mode will be quickly realized.

III. EMPIRICAL ANALYSIS ON THE CORRELATION BETWEEN THE DEVELOPMENT OF THE URBAN CLUSTER OF HBE (HOHHOT, BAOTOU AND ERDOS) AND CULTURAL INDUSTRY CLUSTER

In order to demonstrate the beneficial theory viewpoint further, we selected HBE as the center of the city group as the research object for empirical analysis. In March 2013, the government of Inner Mongolia Autonomous Region officially approved the execution of the Planning of Hohhot, Baotou and Erdos Urban Clusters (2010-2020) [7]. The HBE (Hohhot, Baotou and Erdos) urban cluster in this planning contains the major cities of Hohhot, Baotou, Erdos ,Ulanqab, Bayannur, Wuhai and the major league of Alashan, totally including 48 leagues, counties, cities and districts. The cities of HBE are the center and engine of the cluster, which drives the development of the other cities of Ulanqab, Bayannur, Wuhai and the league of Alashan.

A. Indicators and Data Selection

For exploring the correlation of urban cluster and cultural industry cluster of HBE, this paper chooses two indicators named by GDP of and CIV (cultural industry value). Considering the effectiveness and collect ability, the sample data for the economy development of HBEUC (Hohhot, Baotou and Erdos urban cluster) are chosen from GDP data of the relevant cities. The sample interval of the econometric analysis is from 1992-2012, totally 21 sample values. In order to testify the co-integration relationship between the two variables, this paper employs the logarithmic linear model to obtain the natural logarithm of the HBEUC GDP and HBEUC CIV, the data is given in Table I as follows.

TABLE I
HBEUC GDP AND CIV (UNIT: ONE HUNDRED MILLION YUAN)

Year	HBEUC GDP[a]	HBEUC CIV[a]
2012	11118.292	195.905
2011	9985.861	176.748
2010	8116.709	145.477
2009	6773.370	122.566
2008	5908.257	108.167
2007	4466.679	81.506
2006	3438.231	63.208
2005	2715.558	49.336
2004	2114.760	39.642
2003	1660.879	31.563
2002	1349.730	25.025
2001	1191.783	24.213
2000	1070.304	21.051
1999	959.172	20.073
1998	877.970	15.627
1997	802.151	17.278
1996	711.457	15.663
1995	596.000	12.608
1994	483.345	9.603
1993	373.993	7.657
1992	293.236	6.719

a. The original data of the research is from Inner Mongolia Statistics Yearbook (2002-2013, China Statistics Publication), Inner Mongolia Economics and Social Investigation Yearbook (2002-2013, China Statistics Publication) and 2013 China Culture Industry Annual Development Report.

B. Empirical Analysis

Firstly, we give the unit root test by use ADF method [8] for the time series of the GDP and CIV, the results are showed in Table II and III.

Table II shows GDP has unit root when the significance level is below 5%, which is non-stationary series [9]. The first order difference sequence of GDP (1) series has no root unit when the significance level is below 5%, which implies the series is stationary. Table III shows CIV has unit root when the significance level is below 5%, which is the nonstationary series. But its first order difference sequence of CIV (1) series has no root unit when the significant level is below 5%, which is the stationary series. So, we can use the GDP (1) and CIV (1) to explore the correlation of urban cluster and cultural industry cluster of HBE.

TABLE II
UNIT ROOT TEST RESULTS OF GDP

Time series types	ADF test value	Significance level	Critical value	Test results
GDP	1.420998	1%	-3.920350	Not stationary
		5%	-3.065585	Not stationary
		10%	-2.673459	Not stationary
GDP(1) (first order difference)	-3.509601	1%	-3.959148	Not stationary
		5%	-3.081002	stationary
		10%	-2.681330	stationary

TABLE III
UNIT ROOT TEST RESULTS OF CIV

Time series types	ADF test value	Significance level	Critical value	Test results
CIV	0.329412	1%	-3.808546	Not stationary
		5%	-3.020686	Not stationary
		10%	-2.650413	Not stationary
CIV (1) (first order difference)	-3.824648	1%	-3.831511	Not stationary
		5%	-3.029970	stationary
		10%	-2.655194	stationary

Taking GDP (1) as explanatory variable, and CIV (1) as explained variable, we derive the model heteroscedasticity and correlation by OLS method as follows [10-13]:

GDP(1) =3.73367905117+1.06008924721*CIV(1)   (1)

Now we will consider the causal relationship among economic variables by GDP Granger causality test [14-15], GDP and CIV. Using software Eviews6.0, we derive the followings results:

TABLE IV
GRANGER CAUSALITY TESTS OF AND CIV

Null Hypothesis	F-Statistic	Prob.
GDP does not Granger Cause CIV	5.03541	0.0384
CIV does not Granger Cause GDP	0.16925	0.6859

Table IV shows that when the lag is 1, GDP Granger causes CIV, which means the economic development of HBE cities Granger causes the CIV development of HBE; when lag is 1, CIV does not causes GDP, which means the CIV development of HBE does not causes the economic development of HBE cities (GDP).

The above analyses of co-integration and Granger test proves that the economic development of urban clusters promotes the cultural industry clusters development. There is no doubt that the economic development of HBE cities Ganger causes the cultural industry cluster development of HBE. The result of the Granger causality test shows that the cultural industry cluster development is not Granger causes of the economic development of urban clusters, but it doesn't mean that the cultural industry cluster development has no effect on the economic development of HBE cities. This paper needs to give some special explanation that the economic development of HBE cities depends on the rise and development of the region's pillar industries, so the development of HBE urban cluster fundamentally relies on the energy and chemistry industries, and the cultural industry is not the pillar industry of the HBE urban cluster. Therefore, it cannot be concluded that the CIV development causes the HBE cities development. From this perspective, the reliability and validity of this empirical test is quite satisfactory. We cannot say that the CIV development causes the economic development of HBE cities, but it is undeniable that the cultural industry cluster development promotes or stimulates the economic development of HBE cities. Based on the above co-integration test model, it can be seen that the cultural industry cluster development plays a positive role in promoting the economic development. So, the economic development of HBE cities has a joint relationship with the cultural industry cluster development. In this paper, the relationship is more formally defined: as a basic motive force for cultural industry development, the economic development of the HBE cities promotes the cultural industry cluster development, while the cultural industry cluster development plays a good role in stimulating the economic development of HBE cities, so they have cooperative and joint relation.

From the above theoretical and empirical analysis, the aggregation and development of the cultural industry effectively promotes the economic development of the urban cluster, while the latter further enhances the quality of the cultural industry cluster and also makes the cultural industry cluster development more solid and effective. However, there appear many problems in the process of rapid development; for instance, the development model of cultural industry cluster region still remains and follows the mode of the traditional industrial parks or technology parks, and the cultural industry cluster region is real estate oriented and operated with simple mode, lacking development of clustering effects, so there needs further research.

IV.   CONCLUSION

The paper considers the inner correlation between the development of the urban cluster and cultural industry cluster from the quantitative perspective, and finds their interior rules of development for the urban designing, urban orientation, urban developing direction, urban formats distribution and urban industrial structure adjustment. We first analyze the collaborative linkage between city agglomeration economy development and the development of cultural industry, and then, we give the empirical analysis on the urban cluster of HBE (Hohhot, Baotou and Erdos). The research derives the conclusions that the development of cultural industries must be based on the cluster development.

REFERENCES

[1] Allen J. Scott, "The cultural economy of cities: theory, culture and society", 2002.

[2] Huilin Hu, "Current Characteristics and Trends in the Development of China's Cultural Industry", Research on Development, vol. 25, 2006, pp. 129–134.

[3] Shimei Yao, Mingying Zhu, and Zhenguang Chen, "City clusters of China", Press of University of Science and Technology of China, 2001.

[4] Weide Ren, "Research on HBE city cluster integration", Inner Mongolia University, 2012.

[5] Xiaoping Dong, "Strategic thinking on promoting the cultural industry to be a Pillar industry of migrant workers, Theoretical Research, 2011.

[6] Lang Ye, "China's cultural industry development annual report", Pecking University Press, 2012.

[7] Wu Lan, "Actively Explore the Way to Promote Cultural Industry Promotion in Minority Regions", Inner Mongolia Daily, 2010.

[8] S. E. Said, D. A. Dickey, "Testing for Unit Roots in Autoregressive-Moving Average Models of Unknown Order", Biometrika, vol. 71, no. 3: 599–607, 1984.

[9] Hegwood Natalie, Papell, H. David, "Are Real GDP Levels Trend, Difference, or Regime-Wise Trend Stationary? Evidence from Panel Data Tests Incorporating Structural Change", Southern Economic Journal vol. 74, no. 1, pp. 104–113, 2007.

[10] D. N. Gujarati, D. C. Porter, Basic Econometrics, Boston: McGraw-Hill Irwin, 2009, pp.400.

[11] Furno, Marilena, "The Glejser Test and the Median Regression". Sankhya – the Indian Journal of Statistics, Special Issue on Quantile Regression and Related Methods, vol. 67, no. 2, pp. 335–358, 2005.

[12] H. E. T. Holgersson, G. Shukur, "Testing for multivariate heteroscedasticity". Journal of Statistical Computation and Simulation, vol. 74, no. 12, pp. 879, 2004.

[13] C. Tofallis, "Least Squares Percentage Regression," Journal of Modern Applied Statistical Methods, vol. 50, no. 7: pp. 526–534, 2010.

[14] Gujarati, N. Damodar, Porter, C. Dawn, Causality in Economics: The Granger Causality Test. New York: McGraw-Hill, 2009, pp.652–658

[15] Bressler, L. Steven; Seth, K. Anil, "Wiener–Granger Causality: A well established methodology," NeuroImage, vol. 58, no. 2, pp. 323–329, 2010.

Effect of New CEO Power on Managerial Compensation Gap: A Conceptual Model

Shuo Wang, Chang-zheng Zhang*

School of Economics and Management, Xi'an University of Technology, Xi'an, China, 710054
(zcz7901@163.com)

Abstract- **A new CEO has very different motivations and faces more strict monitoring strength compared to a senior CEO. Therefore, logically new CEOs should have different manipulation choices in managerial compensation setting process. Therefore, a conceptual model focusing on manipulation effects of new CEO's power on managerial compensation gap (MCG) is proposed and analyzed by adopting both inductive and deductive Methods. The results show that under each specified condition of the four kinds of power structures for a new CEO, there are actually different manipulation effects of CEO power on MCG. The conclusions are of great meanings in understanding the differences between the power application patterns of new CEO and senior CEO. However, the further empirical study needs to be accomplished in the future.**

Keywords - **CEO power, conceptual model, manipulation compensation gap, new CEO**

I. INTRODUCTION

CEO change has always been a hot topic both in practice and theory. In facts, since the year of 2000, CEO change has been getting more and more popular. In 2012, there are more than 10% firms which changed their CEOs. Such facts show that CEO change has been a common corporate governance issue that cannot be ignored just as before. With the proposal and execution of a series government policies emphasizing and enhancing the free competition among different firms, no matter state-owned or privately-owned, bigger or smaller, labor-intensive or technology-intensive, it is can be concluded that the market competition will be necessarily getting more heavier and new knowledge innovation will be more frequent. Faced with such a dynamic and complex business environment, top management teams (TMT) members are needed to make more effort to run professional business knowledge. That means knowledge coordination needs (KCN) within TMT gets higher and higher with time goes by, and consequently firms expect more capability and competence from their CEO. Since most of the existing CEOs, as the pivotal role of firms' economic success and strategic change, have not reached the expectation, in the near future, many of the firms will pay more attention to changing a new CEO hoping to seeking for firms' competitive edge by improving the knowledge coordination (KC) quality.

At the beginning of CEO change, the balance of responsibility, power and benefits during the governance practice within firms is up to be redefined. However, due to the relative absence of good theoretical research, among the new CEO, the boards, the shareholders, the debtors, or the public, each group cannot confirm how to reallocate the CEO power and complete good decision-making behavior on the issue of executive compensation. Consequently, the change of CEO rarely reaches the expected purpose of value-creating. Therefore, the study on the manipulation effect of CEO power on managerial compensation gap (MCG), a critical dimension of managerial compensation that has not been explained enough as far [1], is of good help to deal with such a practical issue.

II. LITERATURE REVIEW

Study on the relationship between CEO tenure and firm performance and research on the effect of CEO power on MCG may be the two groups of literature that are most related to this topic.

A. Relationship between CEO tenure and firm performance

Talking of the relationship between CEO tenure and firm performance, there are two schools of views, respectively direct relationship and indirect relationship.

The former group regards that there exists a direct relationship between CEO tenure and firm performance, linear or nonlinear. The linear relationship between the two variables may show as a negative one [2], or a positive one [3]. The nonlinear relationship can be viewed as inversed-U-shaped curve model which has been accepted by many of the scholars as the foundation of further exploration [4].

According to Hambrick et al, CEO's attention focus on CEO behavior mode and eventually firm performance which will represent different features with the tenure goes on. They propose that CEO tenure can be divided into five phases, respectively nomination period, exploration period, mode-choosing period, mode-focusing period and non-functional period. In their opinions, each phase has different effect on the relationship between CEO tenure and firm performance. Specifically speaking, a CEO who can experience all the five periods will reach the highest performance during the mode-focusing period, while get relatively lower performance both during the nomination period and non-functional period. That is to say, a new CEO will do worse than an older CEO. The reason may be that on their early age of CEO position, CEOs' competence has not reached the highest level and

This research was supported by the Scientific Research Foundation of Ministry of Education of the PRC under Grant 14YJA630089 and the National Natural Science Foundation of China under Grant "71272118".

they have to learn and accumulate knowledge about CEO tasks [5]. In addition, CEO has not enough information on the running of firm business and the ability of the executives. That is to say, a new CEO has poor capability on improving firm performance. On the contrary, in their late period during the tenure, CEOs always insist too much in using the out-of-date management mode depending on the narrow and unsmooth information channels, and simultaneously the interest on job tasks begins to decrease heavily. As a result, firm performance will get poorer with the tenure goes too much. The second group holds that there exists a deeper relationship between CEO tenure and firm performance. Under this opinion, the mediating role of CEO attitude on change and the risk preference of TMT between CEO tenure and firm performance has been discussed empirically.

According the related research, the role of CEO attitude on change has been discussed. For example, Meyer concludes that with the tenure gets longer, the influence of CEO on firm running will get heavier, and the CEO will intend to follow the beaten track which will depress organizational change and lead to the depravation of the match between firm and environment [6]. In the whole, during the former period of CEO tenure, new CEO can change the organization in each field according to the demands of external environment. With the tenure growing, CEO always tends to be passive and wants to keep the status quo no matter how the environment changes. The explanation of the scholars is that, the long tenure provides CEO more power and opportunity to promote, nominate and select the executives who have the similar mental modes with him or her, and exclude the ones who hold different opinions, and eventually the homogeneity of TMT gets higher and higher [7]. In the same time, in longer tenure of CEO, the information selecting mechanism obeying the favor of CEO will be established and enhanced which will provide CEO with much information far away from the actual environment change. Such information increases the power of CEO in controlling firms and maintaining the status quo, and simultaneously lowers the adaptability of changing with the environment with the tenure goes too far. Consequently, the relationship between CEO tenure and firm performance appears as a non-linear relationship.

Just shown as above, with the longer tenure of CEO, the attitude on change will direct the adjustment of firm strategy and the relevant organizational change which has obvious relationship with firm performance. Actually the direct study on the attitude on change of CEO will benefit to the understanding of the effect mechanism of CEO tenure on firm performance. For example, Miller takes the Hollywood film firms as the examples and proves such a relationship among CEO tenure, CEO attitude on change and firm performance [8]. In the relationship, CEO attitude on change is actually the mediating variable.

Drawing on the views of upper echelons theory, the characteristics of TMT can explain the change of firm performance to a better degree [9]. TMT risk propensity can directly influence the strategic decision-making and

firm performance. Simsek provides an empirical model which actually explains the non-linear relationship between CEO tenure and firm performance more exactly [10]. In the model, TMT risk propensity and internal entrepreneurship behavior are introduced as the mediating variables. The empirical data analysis shows that CEO tenure has negative effect on TMT risk propensity, and TMT risk propensity has positive effect on internal entrepreneurship behavior, while the later has non-linear effect on firm performance. Most of the all, all the above relations are significant statistically. Therefore, the indirect relationship between CEO tenure and firm performance has been explained in detail.

B. Effect of CEO power on MCG

CEO power, which can be named as managerial discretion according to many other scholars, mainly refers to the influence ability of CEO on corporate governance system, governance process and governance performance representing as the actual influencing degree of CEO on each strategic decision-making items [11]. The study on managerial compensation from the perspective CEO power theory begins at 1980s, develops at 2000s, and gets mature in recent three years. At present, CEO power theory has been one of the main-stream research views on managerial compensation field, which nearly exceeds the value of the traditional optimal contract theory in this field. Specifically speaking, the literature on the relationship between CEO power and MCG mainly comes to three competitive conclusions.

Firstly, the positive view of CEO power on MCG regards that to enhance MCG is helpful to emphasize the authority and status of CEO within TMT, and what is more, it is good to lower monitoring cost; The last but not the least, to improve MCG can improve executives' expected benefits when they outperform in the competition, which actually can increase good competition among different executives. Therefore, CEO with higher power always likes to improve MCG [12]. Secondly, the inversed U-shaped effect of CEO power on MCG demonstrates that excessive MCG may affect the relation quality among executives and morale, further affect good competition, though appropriate improvement of MCG can enhance internal competition among executives. At this time, CEO's good choice is to constrain the enlargement of MCG. Therefore, there is an inversed U-shaped relationship between CEO power and MCG [13]. Third, there is no clear effect of CEO power on MCG. Since the theory circles have not reached a consensus on this issue, many CEOs have no clear guidance when they try to manipulate MGC, so practically they probably give up such manipulation behavior. Especially under the condition of financial distress, CEOs will not try to take high risks of adjusting MCG too much [14].

The existing literature makes great contributions on the issue of MCG. However, there is still a great shortage on the understanding of the relationship between CEO power and MCG. In the literature at present, the difference in features and motives of CEO power between new CEO and old CEO has not been reached, and the effect of new CEO power on MCG has not been explored in detail. The existing studies on CEO power mainly take the view of "selfish CEO" as the basic assumption, which may be appropriate for senior CEOs but not good enough for new CEOs. Specifically, from two perspectives, the "selfish CEO" assumption is not appropriate for new CEO. Firstly, from the subjective environment, new CEOs will be monitored more strictly and they will have relatively narrower action room than senior CEOs; Secondly, from the objective will of new CEOs, they have higher competence as a CEO and care more about their long-termed career benefits than the short-termed material interests than the senior CEOs [15], and consequently they have stronger will of making the firms better. Therefore, the motive of new CEO power running may be "altruism CEO", which means that new CEOs will do things according to the maximization of firms value. The conceptual model will be constructed under such a radically new assumption.

III. CONCEPTUAL MODEL ON THE RELATIONSHIP BETWEEN NEW CEO POWER AND MCG

New CEOs will have different capability and motives of utilizing their power relative to the senior CEOs. Based on the new radical assumption on the manipulation activities of new CEO power on MCG, i.e., new CEOs have motives of improving firm performance rather than merely caring about their own benefits, the conceptual model on the relationship between new CEO power and MCG is constructed as Fig.1.

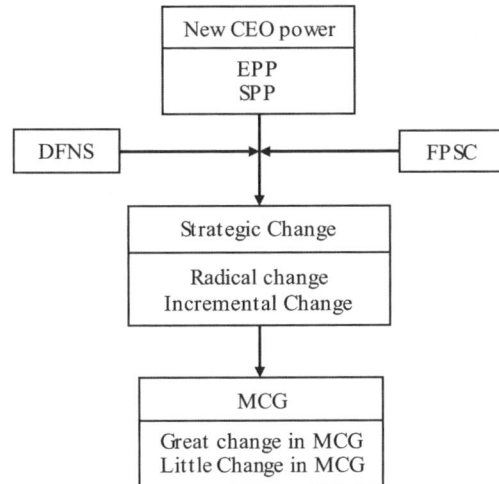

Fig.1. Conceptual model on relationship between new CEO power and MCG: FPSC and DFNS as the moderating variables

According to Fig.1, different new CEO power allocation will lead to different strategic changes, so that the strategic change of the whole firm will lead to the change of MCG arrangement. New CEO power is the independent variable, strategic change is the mediating variable, and MCG change is the dependent variable.

What is more, the different features between new CEO power and senior CEO power (DFNS) and firm performance of the senior CEO (FPSC) are the two moderating variables. New CEO power can be discussed separately in two relatively independent sub-variables, respectively, structural and positional power (SPP) and expert and prestige power (EPP). A new CEO will has four kinds of power allocation, respectively new CEO with higher SPP and higher EPP, new CEO with higher SPP and lower EPP, new CEO with lower SPP and higher EPP, and new CEO with lower SPP and lower EPP.

A. Condition 1: new CEO with higher SPP and higher EPP

Under this condition, new CEO not only has enough expert capability to initiate a strategic change, but also has absolute administrative power to hold a strategic change with clear confidence in pursuing excellent performance. Consequently, the probability of initiating a radical strategic change confirming to new CEO's personal values and opinions is nearly to 1. According to the fundamental rule of HRM practices, to carry out a strategic change, the managerial compensation strategy has to be modified in order to answer the demands of strategic change. That is to say, MCG, as one critical dimension of managerial compensation strategy, has to be changed due to the strategic change. Specifically speaking, there are two main changes in MCG:

One is that the original order of the compensation level within TMT will be destroyed and re-ranked. Since new CEO has good capability in recognizing executives' competence and performance, he ore she will pay more attention to the actual behavior and attitudes of the executives and try to give fair evaluation. According to the evaluation results, each executive will get the right rewards. The dispersion of executive compensation will naturally change a lot compared with the MCG in former CEO tenure. Higher EPP provides new CEO with such capability and confidence in ranking the contribution of each executive. Therefore, the rank of the compensation level within TMT is not regular but may be random, since it is up to the performance of each executive.

The other change is that the highest compensation level of the executives (other than CEO) is very near to the CEO compensation level, while is much higher than the second highest compensation level of the other executives. Two reasons lead to such fact, respectively, (1) New CEO wants to execute his or her new strategy, and such execution needs the complete support from executives who are in charge of the functional departments mostly confirming to the strategic change or being important for the success of the strategic change. Consequently, new CEO has to provide enough incentive to the right executive of "core strategic meanings" by setting up both obviously higher absolute compensation level and clearly larger relative compensation gap between the other executives; (2) New CEO has to release some signals that the executives who are not like the new strategy or nor willing to accept the lead of new CEO will be punished. Hence any executives who are viewed as target will be firstly punished by an obviously lower compensation package which tells him or she that "you are not welcomed: leave or change, it is up to you".

B. Condition 2: new CEO with higher SPP and lower EPP

Under this condition, new CEO objectively can manipulate enough resources to carry out a strategic change, but subjectively he or she has no enough confidence and capability in initiating strategic change. In our opinions, whether or not to initiate a strategic change when a new CEO with higher SPP and lower EPP comes, it is up to the status of FPSC. When FPSC is good, new CEO will not change the existing strategy. In fact, the purpose of the board in selecting such a new CEO with relatively lower EPP mostly may be maintaining the extant strategy and performance steadily. In other words, the board does not expect the new CEO to change too much. In confirming to such a choice, MCG will not be changed significantly. When FPSC is poor, new CEO has to change the existing strategy in order to show his or her value to firm performance improvement, otherwise, new CEO will be changed again. In executing such strategic change, since new CEO has no enough capability and confidence, he or she has to motivate the executives to provide new directions and good wisdoms. Hence, new CEO will try to identify two groups of executive within the TMT, respectively good executives and bad executives. For good executives, new CEO wants to absorb them into the "core team" and thus can make full use of the group wisdoms in order to offset his or her own capability shortage. Therefore, new CEO will provide the "good" executives a relatively higher level compensation than before and the gap among them is very small; For bad executives, new CEO with lower EPP will give them relatively lower compensation than before, and the gap between the two groups will be rather big. If the bad ones do bad things to the new strategy implementation, new CEO with higher SPP can dismiss them; instead, if the bad ones do good things or make good contributions to new strategy, new CEO with higher SPP can reevaluate them and then improve the compensation level.

C. Condition 3: new CEO with lower SPP and higher EPP

Under this condition, new CEO will be very unsatisfied with his arrangement, since the match between his ore her capability (& prestige) and the delegated power is unbalanced. For new CEO, whether or not to initiate a strategic change, it is up to the status of DFNS. When DFNS is insignificant, that is to say, the former CEO with higher EPP is allocated with lower SPP too, new CEO will intend to accept this status with lower dissatisfaction since the former CEO similar with him or her has been treated just like this. Hence, new CEO will try to make strategic changes in order to prove his or her higher capability. Since new CEO has no enough formal and positional power in carrying out strategic changes, for

example, dismissing executives, or allocating resources, new CEO will have to mostly depend on informal influences derived from CEO characteristics and prestige by adopting good communication and other excellent management methods. Consequently, new CEO with lower SPP and higher EPP will keep a relatively similar compensation package among executives in order to seek for harmonious atmosphere, and simultaneously adopt the other leadership methods which need higher capability and techniques of new CEO, something like effective monitoring, recognizing and rewarding other than merely relying on monetary countermeasures, trying to execute the strategic effectively. When DFNS is significant, i.e., the former CEO is of lower EPP, new CEO will feel very sad and angry. At this time, new CEO will not pay attention to firm running or strategic change, but focus on the self-interest satisfying in order to offset the dissatisfaction. Since new CEO has specialized expert and skill, i.e., tacit knowledge, in firm running, new CEO has some methods of improving his or her own compensation level without receiving any punishment or blame. Under such self-interest motivation, the final result is that new CEO with lower SPP and higher EPP will improve the whole compensation level of TMT, and enlarge compensation gap between CEO and other executives.

D. Condition 4: new CEO with lower SPP and lower EPP

Under this condition, new CEO has received the balanced treatment from the board. In general, new CEO with lower EPP and lower SPP not only has no confidence and motivation to initiate a strategic change, but has no positional and operational power to implement a strategic change. However, whether such new CEO chooses to make a strategic change or not is up to the status of FPSC. When FPSC is relatively lower, new CEO has to initiate a strategic change just because the board wants him or her to do that. Each group of a firm wants to get a good performance, especially the board, since the shareholders will give the board heavy pressure about the performance. When a firm faces poor performance and then chooses to change a new CEO, the purpose is actually very clear that such change of CEO should lead to a big improvement of firm performance. If a new CEO can not make strategic change, the board will naturally regard the new CEO as incompetent to the CEO position. Therefore, no matter new CEO has the capability of holding a strategic change or not, he or she must initiate a change in strategy. Since new CEO has relatively lower EPP, he or she does not know how to implement the strategic change, instead, the executives must be considered to function well in helping new CEO carry out strategic change. Consequently the MCG will be changed, and such a change always shows as a general reduction in the whole compensation package of the TMT, while the compensation of some core executives will maintain the original level, and some other executives will lower their compensation to a degree. In the whole, MCG is weakened, especially the compensation gap between CEO and the core executives. When FPSC is relatively higher, new CEO has no any motivation to change the existing strategy. The optimal choice is to maintain the status quo. Therefore, MCG will not change too much. However, new CEO must do something new in executive compensation policy, otherwise the board or the executives will think that new CEO has no contribution to the new performance. The reasonable choice for a new CEO with lower EPP and lower SPP is to initiate spiritual motivation programs, such as EAP, Recognition & Rewards Programs, etc., or to turn to the so-called total compensation package (TCP) in order to motivate the executives and show his or her leadership competence.

IV. CONCLUSIONS

New CEOs have different motivations and favors compared to senior CEOs. A new CEO faces deeper pressure from stakeholders, and each group will expect certain behavior mode for new CEO, therefore, new CEO has different manipulation choices when he or she make use of the managerial power delegated from the board. The paper proposes a conceptual model which focusing on manipulation effect of new CEO's power on MCG. The result shows that under different conditions, there are different manipulation effects. Specifically speaking, (1) Under the condition of CEO with higher EPP and higher SPP, One change in MCG is that the original order of the compensation level within TMT will be destroyed and re-ranked, and the other change is the highest compensation level of the executives (other than CEO) is very near to the CEO compensation level, while is much higher than the second highest compensation level of the other executives; (2) Under the condition of CEO with higher SPP and lower EPP, when FPSC is good, MCG will not be changed significantly; When FPSC is poor, MCG between "good" executives and "bad" executives will be enlarged than before; (3) Under the condition of CEO with lower SPP and higher EPP, when DFNS is insignificant, new CEO with lower SPP and higher EPP will keep a relatively similar compensation package among executives in order to seek for harmonious atmosphere, when DFNS is significant, new CEO will be more apt to enlarge the compensation gap between CEO and the other executives; (4) Under the condition of CEO with lower SPP and lower EPP, when FPSC is relatively lower, the MCG will be changed, and such a change always shows as a general reduction in the whole compensation package of the TMT, and MCG is weakened, especially the compensation gap between CEO and the core executives; when FPSC is relatively higher, MCG will not change too much. However, new CEO will do something new in executive compensation policy, such as to emphasize spiritual motivation than former CEO. The paper only provides the logic reference based on conceptual model, while further empirical study needs to be accomplished in the near future.

ACKNOWLEDGMENT

Authors thank the support by the Scientific Research Foundation of Ministry of Education of the PRC under Grant 14YJA630089 and the National Natural Science Foundation of China under Grant "71272118".

REFERENCE

[1] Garner, J. L., Harrison, T.D., "Boards, Executive Excess Compensation, and Shared power: Evidence from Noprofit Firms," *Finiancial Review*, vol.48, no.4, pp. 617-643, 2013.

[2] Hambrick, D.C., Geletkanycz, M A, and Fredrickson, J W., "Top executive commitment to the status quo," *Strategic Management Journal*, vol.14, no.6, pp. 401-418, 1993.

[3] Bergh, Donald D., "Executive retention and acquisition outcomes," *Journal of Management*, vol.27, no.5, pp. 603-622, 2001.

[4] Finkelstein, S., and Hambrick, D C., *Strategic leadership: Top executives and their effects on organizations*, Minneapolis, St Paul: West Publishing, 1996, pp. 121-135.

[5] Niamh, B., "Boards of directors and firm performance: Is there an expectation gap," *Corporate Governance*, vol.14, no.6, pp. 577- 593, 2006.

[6] Meyer, Marshall W., "Leadership and organizational structure", *The American Journal of Sociology*, vol.81, no.3, pp.514-542, 1975.

[7] Mintzberg, R., *Power in and around organizations*, Englewood Cliffs: Prentice Hall, 1983, pp.11-28.

[8] Miller, D, and Shamsie, J., "Learning across the life cycle," *Strategic Management Journal*, vol.22, no.8, pp.725-745, 2001.

[9] Hambrick, D.C., and Mason, P., "Upper echelons: The organization as a reflection of its top managers," *Academy of Management Review*, vol.9, no.2, pp. 193-206, 1984.

[10] Simsek, Zeki, "CEO tenure and organizational performance: Testing a nonlinear intervening model," *Academy of Management Best Conference Paper*, BPS, 2004.

[11] Bebchuk, Lucian Arye, Fried, Jesse M, and Walker, David I, "Managerial Power and Rent Extraction in the Design of Executive Compensation," *University of Chicago Law Review*, vol.69, pp.751-846, 2002.

[12] Patrick McClelland, and Tor Brodtkorb, "Who Gets the Lion's Share? Top Management Group Pay Disparities and Powerful CEOs," *American University of Sharjah, SBMWPS: 04-04/2013*, unpublished.

[13] Ricardo Correa, and Ugur Lel, "Say on Pay Laws, Executive Compensation, CEO Pay Slice, and Firm Value around the World," *International Finance Discussion Papers, Number 1084*, July 2013, unpublished.

[14] Lu Hai-fan, "CEO power, top executives' pay gap and company performance in financial crisis," *Finance and Trade Research*, vol.3, pp. 116-124, 2012(in Chinese)

[15] Craig Crossland, and Guoli Chen, "Executive accountability around the world: Sources of cross-national variation in firm performance-CEO dismissal sensitivity," *Strategic Organization*, vol.11, no.1, pp. 78-109, 2013.

Social Capital of Private Businessman Class in China

Jia-di Fu*, Guan-jian Qiu

School of Marxism, Wuhan University of Technology, Wuhan, China, 430070

(whutjd@163.com)

Abstract – **This article aims to discuss the operational logic of the merging class of private businessman, with an emphasis on the important role of social capital. It begins with a review of the development of social capital and of the previous research on social capital of private businessman in China. Based on this, the concept of social capital is clearly defined, which lays a foundation for the following analysis. In order to analyze the emergence of private businessman class, Robert Vegeta's argumentation method is adopted. Three theoretical hypotheses and nine inferences are put forward to demonstrate the inner logic of social capital of private businessman in China.**

Keywords – **Private businessman class, social capital, social network**

I. INTRODUCTION

How can private businessmen become a new class? We need to explore theoretical sources from the origin to seek its legality. The concept of legality is firstly proposed by Max Weber [1]. He points out that there must be an authority in an organization, community, association or tribe. He contends that authority gains its legality and acceptability by three ways: 1. personal charisma attracting people to follow; 2. tradition, such as power succession; 3. law and reason, such as election.

The emergence of a new social stratum should be attributed to specific historical conditions. Many economic, social and cultural factors may influence its formation [2]. In different disciplines, there are corresponding theories to explain its emergence logic. With limited ability, the author only focuses on the role and operational logic of social capital in the emergence of private businessman class.

II. METHODOLOGY AND DISCUSSION

Chinese society is a society of relation and network. Operating in the network, private businessman class is also endowed with its features. They establish a network of relations with a focus on themselves according to different pattern. In a society influenced by Confucius ethics, they exploit necessary social capital to strengthen the core competitiveness of the company and to acquire more resources. In order to solve these problems, the operational logic of social capital of the class should be closely tested. Here we adopt the reasoning inference used by Robert Alan Dahl in his book, *Introduction to Democratic Theory*.

A. The emergence of private businessman class is under the dual influence of institutional and technology environment

The theory of technology and institutional environment is put forward by John Meyer [3], a Neo-institutionalism scholar. He made a research about the educational system in the states in America. The result unveiled that the educational system in America was featured by the separation of power. However, the educational systems in different states were very similar, with a tendency of institutional isomorphism. He put forward a question that why different organizations had similar inner system and institution. He tried to solve the problem by the relationship between organization and environment. He contended that environments facing organization can be divided into technology environment and institutional environment. Organization is a product of both technology requirement and institutional environment.

1) Institutional environment has an embedded relationship with the emergence of private businessman class

Institutional Environment refers to the taken-for-granted social facts, such as legal system, cultural expectation, social constitutions and mind surrounding an organization. Institutional environment entails organization to comply with "legality" mechanism and to adopt the widely-accepted organization form and measures, regardless of its efficiency. For example, a company takes part in public welfare activities to improve its social status and recognition.

Institutional environment is embedded in all the processes of private businessman class and is exogenous. It is not only the driving force for the emergence of the class, but also the limiting factor impeding its development. As a result, how to accept and to deal with external environment is essential to the development of the class.

2) Technology environment is endogenous with the emergence of private businessman class

The technology environment of a company entails efficient organization. That is to organize production according to the principle of maximization. If a bank (environment) gives loans to a company, it entails the company to have the ability of highly efficient production and operation and of repayment. Customers (environment) want to buy cheap products of high quality, which also entails efficient production.

For a private company to develop, the production mode should be changed to establish a modern enterprise system, de-familization and improvement of technological ration. Since the industrial revolution, the dramatic growth of production efficiency has perfectly embodied the production rationalization. Production rationalization is reflected not only by the invention and application of machines, but also by the scientific and rational production process. The rationality of human beings is reflected not only in production, but also in organization to manage production, namely the bureaucracy of modern society. By analyzing the common features of modern large-scale administrative management system, Marx Weber explored the origin of its efficiency and put forward the "ideal type" organization from: power hierarchy, regulations, impersonalization and professionalization. As a result, technology environment is endogenous with the emergence of private businessman class.

B. Social capital of private businessman class is the bridge coordinating institutional environment and technology environment

As has been mentioned, social capital is a productive resource to promote the cooperation between social actors. It helps individual to make profits from social network or other connections in social structure. Network, trust and regulation is the key elements of social capital. According to system theory, here comes the second hypothesis.

1) Company is a stable connection between social groups and organizations and their members. Individuals can get scarce resources from social groups and organizations by this stable connection.

Political scientists pay special attention to the social capital formed by "social member relations". A leading figure is Robert Putnam from Harvard University. His study on the America after WWII unveiled that as the number of members of voluntary group and organization decreased over the years, social capital in the country also declined. (Putnam, 1995)[4]. The research on non-government organization by Chen Jianmin and Qiu Haixiong revealed that group members can acquire more social capital because of membership and so get more resources. (Chen Jiamin and Qiu Haixiong, 1999) [5].

2) The interpersonal social network of private businessman class is developed from the interaction, communication, contact and exchange between people.

In two important articles, Mark Granovetter discussed how the information in labor market disseminate through social network and how interpersonal trust is established, strengthened and developed by social network (Granovetter 1973 [6], 1985 [7]).

Lin Nan emphasized that social resources, such as power, wealth and reputation are embedded in social network. People in lack of these resources can acquire (borrow) them from the social network. The usage of social resources is an effective way of achieving instrumental target. (Lin 1982 [8], 1990 [9]).

3) Social capital of private businessman is connected with institutional and technology environment by vertical, horizontal and social linkage.

Vertical linkage of a company refers to its connection with higher authority, local government departments and subordinate companies and departments. This vertical linkage tends to be upward, which aims to acquire scarce resources from the "upward". Vertical linage is not a unique characteristic of socialist planned economy. For non-public companies newly-emerged in a period of economic transformation, some of them are affiliated with certain government department. Some are influenced by the vertical structure of government because of their cooperation with state-owned and collective companies. Others are under direct supervision of a government department, such as some township enterprises.

Horizontal linkage of a company refers to its connection with other companies. The linkage can be various, such as business relation, cooperation relation, debtor-creditor relation and holding relation. There was horizontal linkage between companies even in the time of planned economy. Its function was not only information communication, but also the final guarantee to deal with resource shortage and emergency. During the time of economic transformation, as independent financial units, companies are expanding their horizontal linkage. If horizontal linkage increases in scale and number, the company will gain more effective information and have more choices, which promotes its development. If horizontal linkage decreases in scale and number, the company will be secluded with few opportunities. It can develop only in a limited space.

Although companies operate within economic filed, companies and their managers live in a broader social environment. Social communication and connection of company owners are not the property of a company, but they are the necessary wealth. It is because non-economic social communication and connection of company managers are the channels for the company to communicate information and establish trust with the exterior. They are also informal mechanisms for companies to acquire scarce resources and to compete for business projects.

C. Party organization in non-public companies is an important social capital for private businessman class.

In his study on the institutional school of organizational sociology, Zhou Xueguang [10] points out that the school uses the concept of legality to emphasize an authority relation established on social recognition. The mechanism of legality is institutional power to lead or force organizations to adopt the legal organizational structure or measures in external environment, which can also be called "the logic of social recognition" or "the logic of accountability". In the book, *How Institution Thinks*, Mary Douglas contends that institution limits people's thinking mode and behavior pattern and shapes their thinking. Organizations have to gain legality by the ways recognized by external environment.

1) In order to comply with the institutional environment, private businessman should take the initiative to set up party organization.

In terms of party construction in private companies, there is no legal footing, but there are institutional regulation by the party and state. Along with the development of non-public economy, the state's attitude toward party membership of private businessmen changes in logic: acquiescence – opposition – allowance – encourage. Party-joining of private businessman and the establishment of party organizations in private companies are the requirements of institutional environment. Party branches are established in non-public companies nationwide. Until March 8th, 2012, with the issue of Opinions about Strengthening and Improving Party Construction in Non-public Companies (trial) by General Office of the CPC Central Committee, the party has realized organization coverage in large-scale non-public companies.

2) Organization convergence is a necessary logic of party construction in different places.

Dimaggio and Powell contend that reasonable choice in history and efficiency mechanism play an important role in organization convergence. However, in modern society, organization convergence originates in the institutional environment of organizations and is realized through three mechanisms: force mechanism, imitation mechanism and social regulation mechanism.

3) Party construction in private companies connects social capital of private businessmen.

Firstly, as an important carrier of company culture and political training, party organization connects and promotes interpersonal relationship in the company. Hylton Mayo, Rothlisberger and Dixon, forerunners of Interpersonal Relationship School from America, conducted the famous Hawthorne test in the Hawthorne factory of American Western Electric Company from 1927 to 1932. After a series of observation and experiments, they realized that the mental condition of organization members is influenced by interpersonal relationship and mental condition influences people's

behavior and productivity. They also found that in formal organizations, there are informal relations based on emotion and practical functions. Chester I. Barnard, the founder of Social System School of western modern management theories put forward "organization balance" in the book, *the Function of Manager*. He contended that the existence of organization is determined by the balance between contribution and satisfaction of organization members. Workers should not only be given material reward, but also mental satisfaction. Party organizations connect formal and informal relations and give workers humanistic care to strengthen cohesion and centripetal force.

In addition, party organization is an importation carrier of the symbiosis of institutional environment and technology environment. Party organization is a symbol of legality of party in non-public companies. It reflects an inoculation and control of ideology. It influences the culture of a company by organization convergence and interference, in order to control non-public companies. By using this institutional force, private businessman class can expand its vertical linkage with the party and government, in order to acquire more scarce resources to support the development of the company.

III. CONCLUSION

The article discusses the operational logic behind the emergence of private businessman class with an emphasis on the important role of social capital in its emergence. It begins with a review of the development of social capital and of the previous research on social capital of private businessman in China. Based on this, the concept of social capital is clearly defined, which lays a foundation for the following analysis. In order to analyze the emergence of private businessman class, Robert Vegeta's way of argumentation is adopted. Three theoretical hypotheses and nine inferences are put forward in order to demonstrate the inner logic of social capital of private businessman in China.

ACKNOWLEDGMENT

Financially supported by National Social Science Fund (NO.10BDJ029).

REFERENCES

[1] Anonymous. Max Weber and the Idea of Economic Sociology [J]. The American Journal of Economics and Sociology, 1999 (3): 549-550.

[2] Compiled by the Central Literature Press. Selective Collection of Important Literature After 15th National Congress (the second volume) [M]. Beijing: The Central Literature Press, 2011: 163.

[3] Meyer John W, Brian Rowen. Institutionalied Organizations: Formal Structure as Myth and Ceremony [J]. American Journal of Sociology, 1977(83): 340-363.

[4] Robert, Putnam. Bowling Alone America's Declining Social Capital [J]. Journal of Democracy, 1995(6): 65-78.

[5] Jianmin Chen, Haixiong Qiu. Social Organization, Social Capital and Social and Economic Development [J]. Sociological Study, 1999(4): 65-74.

[6] Granovetter, Mark. The Strength of Weak Ties [J]. American Journal of Sociology, 1973(78): 1360-80.

[7] Granovetter, Mark. Economic Action and Social Structure: The Problem of Embeddedness [J]. American Journal of Sociology, 1985(91): 481-510.

[8] Lin, Nan. Social Resources and Instrumental Action [M]. Beverly Hills, CA: Sage Publications, 1982: 131-47.

[9] Lin, Nan. Social Resources and Social Mobility: A Structural Theory of Status Attainment [M]. New York: Cambridge University Press, 1990: 247-71.

[10] Xueguang Zhou. Handout of Sociology in Tsinghua University: Ten Lectures on Organizational Sciology [M]. Social Sciences Academic Press, 2003: 6-11.

Risk Management of International DCS Projects Based on Multilevel Fuzzy Comprehensive Evaluation Model

Jia-sheng Xiao, Rui Miao*, Yu-dong Si

School of Mechanical Engineering, Shanghai Jiao Tong University, 800 Dongchuan Road, Shanghai, China, 200240

(xiaojs@xinhuagroup.com, *miaorui@sjtu.edu.cn, 654786911@qq.com)

Abstract - According to the characteristics of international DCS project that benefit and risk coexist, the multilevel fuzzy comprehensive evaluation model covering Delphi, AHP (Analytic Hierarchy Process) and Fuzzy Comprehensive evaluation Model will be introduced into the international DCS project risk analysis. Taking Salah Al-Din Fuel Gas Power Plant DCS project In Iraq as an example, we successfully carries on the comprehensive evaluation of risk model. The evaluation results shows strong systemic and practical. It also provides strong support for project risk management and decision-making.

Keywords - AHP, Delphi, international project, risk

I. INTRODUCTION

With the rapid economic development of developing countries, Chinese DCS (Distributed Control System) project exported to overseas countries is increasing year by year. At the same time, we have to see that DCS projects not only help the companies to gain great economic benefits but also brings high risks to them due to the complexity and uncertainty of the project. The existing international DCS project risk research [1-3] is mainly based on the identification of risk factors and countermeasures research. Based on the research of international project risk management [4-6], the paper establishes the multilevel fuzzy comprehensive evaluation model covering Delphi, AHP and FCM. We take the 2*630MW Fuel Gas Power Plant DCS project in Salah Al-Din, Iraq as an example, using multilevel fuzzy comprehensive evaluation model for the purpose of the comprehensive risk degree of the effective evaluation.

II. ESTABLISH THE EVALUATION INDEX SYSTEM OF INTERNATIONAL DCS PROJECT RISK

This paper refers to many data management project of the international DCS, collects DCS senior experts experience extensively. Considering the factors of international DCS projects covering environment, construction characteristics, status of project management, we determined the suitable evaluation index of international DCS project risk management, and set up the international DCS risk evaluation index system as shown in Fig.1.The risk assessment index system have three levels including the target layer, criterion layer and index layer. The target layer is the project risk management (A); the criterion layer include project environmental risk (A_1), project participants risk (A_2), technology risk (A_3) and project management risk (A_4); the index layer contains sixteen risk items including political risk (C_{11}), economic risk (C_{12}) etc.

III. ESTABLISH THE MULTILEVEL FUZZY COMPREHENSIVE EVALUATION MODEL

A. Establish the evaluation index system of project risk management A , evaluation set V and the index weight vector W, expressed as : $A = \{A_1, A_2, A_3, A_n\}$; $V = \{v_1, v_2, \cdots, v_m\}$; $\{W = W_1, W_2, \cdots, W_n\}$, where n, m, respectively index grading number and evaluation rank number. The vector W meet $\sum_{i=1}^{n} w_i = 1$ [7] .

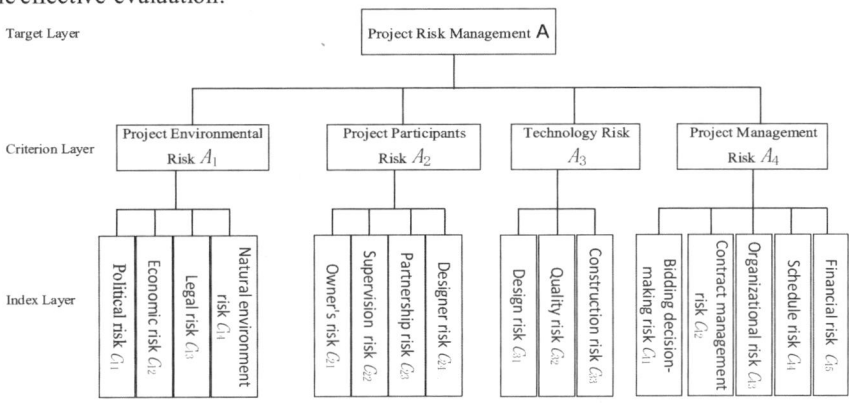

Fig.1. International DCS Risk Evaluation Index System

B. Using AHP to determine the weight of evaluation index set of each risk index in A, $w_i(i=1, 2, \cdots ,n)$, Specific as:

(1) Using Delphi constructs pairwise comparative judgment matrix $A = \left(a_{ij}\right)_{n \times n}$, where using 1~9 scale method for assignment a_{ij} [8].

(2) Using the root method to calculate the weight of lower index on the upper index. First, calculate the product of the elements in each line of the judgment matrix $M_i = \prod_{j=1}^{n} a_{ij}$; Then calculate the geometric mean $\overline{W_i} = \sqrt[n]{M_i}$; Normalized the judgment matrix A to get the weight coefficient $W_i = \frac{\overline{W_i}}{\sum_{i=1}^{n} \overline{W_i}}$ of the item i.

And according to the $AW = \lambda_{max}W$ calculation to determine the maximum eigenvalue and eigenvector of

E. Qi et al. (eds.), *Proceedings of the 21st International Conference on Industrial Engineering and Engineering Management 2014*, Proceedings of the International Conference on Industrial Engineering and Engineering Management, DOI 10.2991/978-94-6239-102-4_101, © Atlantis Press and the authors 2015

the judgment matrix. Finally verify the numerical validation after normalization processing whether meet the consistency requirements, if it meets the criterion, then use the numerical value as the weights of indicators of the layer to upper layer. Among them, the consistency index of $CI=\frac{\lambda_{max}-n}{n-1}$, the consistency ratio $CR=CI/RI$, the RI as the mean random consistency index, RI value are shown in Table I. If random consistency ratio CR<0.10, it means that the consistency of judgment matrix is satisfied, that is to say the weight meets the requirements [9].

(3) Establish the fuzzy relationship matrix $R|A=$

$(r_{ij})_{n \times m}$, using Delphi for membership assignment. Using fuzzy comprehensive evaluation principles, carries on the comprehensive operation step by step. The next level fuzzy comprehensive evaluation comes the first, and conclude the next level of fuzzy comprehensive evaluation result $B_i = W_i \times R_i$; Then, using the higher level evaluation matrix $B = (B_1, B_2, \cdots, B_n)^T$ to obtain the total evaluation vector $D=W*B$ using fuzzy comprehensive evaluation.

(4) Based on the comprehensive score values $P = D \times V^T$ to rate the risk level.

TABLE I
THE MEAN RANDOM CONSISTENCY INDEX AND THE MODULE RELATION

Module	1	2	3	4	5	6	7	8	9
RI	0	0	0.58	0.90	1.12	1.24	1.32	1.41	1.45

IV. RISK COMPREHENSIVE EVALUATION OF SALAH AL-DIN FUEL GAS POWER PLANT DCS PROJECT

A. Constructs the multilevel fuzzy comprehensive evaluation factor set

According to the DCS project risk evaluation index system, setting up the project multilevel fuzzy risk evaluation index system which is consisted of 4 primary indicators and 16 secondary indicators: The primary evaluation index set of project risk management $A=\{A_1,A_2,A_3,A_4\}=\{$Project environment risk, Project participants risk, Technology risk, Project risk management$\}$; The secondary evaluation index set $A_i = \{C_{i1}, C_{i2}, ..., C_{is}\}, (i = 1,2,3,4)$, where s denotes the number of evaluation index in the sub-index set.

B. Determine the evaluation set for degree of risk

According to the severity of the DCS project risk,

the evaluation set V can be divided into five dimensions: $V=\{0.9,0.7,0.5,0.3,0.1\}=\{$ Very high, High, Medium, Low ,Very low$\}$. Quantitative evaluation of standard design is shown in Table II [10].

C. Calculate the weight of each risk factor

With the abundant expert resource, consistency judgment matrix is finally constructed by several rounds of Delphi. For example $A - Ai$ code for the target layer of the judgment matrix:

$$A = \begin{bmatrix} 1 & 3 & 4 & 2 \\ 1/3 & 1 & 2 & 1/2 \\ 1/4 & 1/2 & 1 & 1/3 \\ 1/2 & 2 & 3 & 1 \end{bmatrix}$$

Risk weighting data is calculated by EXCEL software programming. It is fast, accurate and reliable. Specific calculation data can refer to the Table III. In other levels of risk factors of weight vector is calculated with the weight, as listed in Table IV.

TABLE II
GRADING STANDARD OF RISK ASSESSMENT

Evaluation value	$P \geq 0.8$	$0.6 \leq P < 0.8$	$0.4 \leq P < 0.6$	$0.2 \leq P < 0.4$	$P \leq 0.2$
Risk grading	Very high (VH)	High (H)	Moderate (M)	Low (L)	Very low(VL)

TABLE III
JUDGMENT MATRIX CALCULATION TABLE OF A-Ai

Project risk	A_1	A_2	A_3	A_4	$\prod_{j=1}^{n} a_{ij}$	$\sqrt[n]{M_i}$	W_i	Aw_i	Aw_i/W_i	CI=(I=n)/(n-1)	CR=CI/RI
Environmental risk	1	3	4	2	24.0000	2.2134	**0.4668**	1.8840	4.0356		
Participants risk	1/3	1	2	1/2	0.3333	0.7598	**0.1603**	0.6453	4.0262		
Technical risk	1/4	1/2	1	1/3	0.0417	0.4518	**0.0953**	0.3847	4.0366		
Management risk	1/2	2	3	1	3.0000	1.3161	**0.2776**	1.1174	4.0255	0.0103	0.0116
						4.7411			4.0310		

TABLE IV
FUZZY COMPREHENSIVE EVALUATION OF AND THE SALAH AL-DIN DCS PROJECT IRAQ

Total evaluation index	Criterion layer index	Second layer index	Weight	Membership degree evaluation				
				VH 0.9	H 0.7	M 0.5	L 0.3	VL 0.1
Project risk management A 0.5970	Project environment risk A₁ 0.6843	C_{11}	0.5731	4	4	2	0	0
		C_{12}	0.2532	3	4	2	1	0
		C_{13}	0.1082	2	3	4	1	0
		C_{14}	0.0655	0	1	3	2	4
	Project participants risk A₂ 0.5077	C_{21}	0.4747	1	3	4	2	0
		C_{22}	0.2551	0	3	5	2	0
		C_{23}	0.1072	0	1	4	3	2
		C_{24}	0.1630	0	1	5	3	1
	Technical risk A₃ 0.4414	C_{31}	0.5584	0	2	5	2	1
		C_{32}	0.3196	0	2	4	3	1
		C_{33}	0.1220	0	1	3	4	2
	Project management risk A₄ 0.5551	C_{41}	0.0655	0	2	4	3	1
		C_{42}	0.2017	0	3	4	3	0
		C_{43}	0.1164	0	2	5	2	1
		C_{44}	0.3587	2	3	4	1	0
		C_{45}	0.2577	1	4	3	2	0

D. Established each level's fuzzy relationship matrix

Using Delphi established each level's fuzzy relation matrix. When it comes to the actual assignment, we can invite 10 experts to evaluate the degree of membership according to the DCS enterprise management and technical resources. Evaluation of the results in Table IV.

According to Table III, we can know the 4 two level index fuzzy relation matrix $R_i (i = 1,2,3,4)$, Take the evaluation of project environment risk as an example a, according to the number of experts on every risk factors after the judgment is very high, high, medium, low, very low calculated membership each comment, the corresponding the fuzzy relation matrix can be expressed as:

$$R_1 = \begin{bmatrix} R_1| & c_{11} \\ R_1| & c_{12} \\ R_1| & c_{13} \\ R_1| & c_{14} \end{bmatrix} = \begin{bmatrix} 0.4 & 0.4 & 0.2 & 0 & 0 \\ 0.3 & 0.4 & 0.2 & 0.1 & 0 \\ 0.2 & 0.3 & 0.4 & 0.1 & 0 \\ 0 & 0.1 & 0.3 & 0.2 & 0.4 \end{bmatrix}$$

E. Comprehensive evaluation of DCS project risk factors risk step by step

Comprehensive evaluation of the risk of operational level two risk factors:

$B_1 = W_1 \times R_1$
$= (0.3269,0.3695,0.2282,0.0492,0.0262)$

Similarly, we can know the evaluation results of the other risk assessment factors $B_2 \sim B_4$. Primary risk factors in comprehensive evaluation of operation risk. The comprehensive evaluation of the results obtained by the above evaluation index subset $B_1, B_2, \cdots, B_n$, constitute a level fuzzy evaluation matrix $B = (B_1, B_2, \cdots, B_n)^T$.

$D = W_A \times B$
$= (0.1873,0.3152,0.3267,0.1368,0.0340)$

F. Rating according to comprehensive score value of Salah Al-Din Fuel Gas Power Plant DCS project risk

Calculate respectively with each index and overall project risk fuzzy comprehensive evaluation results and we can get the following the risk value. Project environment risk comprehensive evaluation value:
$P_{A_1} = B_1 \times V^T =$
$(0.3269,0.3695,0.2282,0.0492,0.0262) \times$
$(0.9,0.7,0.5,0.3,0.1)^T = 0.6843$

Project participants risk comprehensive evaluation value: $P_{A_2} = B_2 \times V^T = 0.5077$. Technical risk comprehensive evaluation value: $P_{A_3} = B_3 \times V^T = 0.4414$. Project management risk comprehensive evaluation value: $P_{A_4} = B_4 \times V^T = 0.5551$

According to the calculation results, we can know that the results of the risk assessment of Salah Al-Din Fuel Gas Power Plant DCS project "project environment risk" as "high" belongs to the H class; the other three risk evaluation results as "moderate" belongs to the class M. Project overall risk comprehensive evaluation value: $P_D = D \times V^T = 0.5970$. The overall risk level as "moderate", belongs to the M class risk. Based on the overall evaluation of risk, through risk avoidance, risk mitigation, risk transfer and risk retention, risk prevention and risk reserves, such as the comprehensive utilization of risk coping strategies, targeted to develop the Salah Al-Din Fuel Gas Power Plant DCS project various risk measures, to provide a powerful guarantee for the smooth implementation of the project.

V. CONCLUSION

In summary, the multilevel fuzzy comprehensive evaluation model simplifies the complicated calculation, and gives accurate values of risk evaluation index quickly. The model evaluation results is easy to calculate and master. It has strong systemic and practical to provide strong support for project risk management and decision-making. It can be applied in not only risk management of normal DCS projects in overseas thermal power plants, but also in green field including nuclear power, wind power and distributed energy stations. It will help with avoiding the project risk and raising the level of project process control.

ACKNOWLEDGMENT

The authors gratefully acknowledge the financial support of the eighth innovation practice program of Shanghai Jiao Tong University (IPP8035).

REFERENCES

[1] L.F.Alarcon, D.B.Ashey, A.S.Hanily, K.R.Molenaar, R.Ungo, "Risk Planning and Management for the Panama Canal Expansion Program", *Journal of Construction Engineering and Management,* ASCE, vol. 137, pp. 762-771, October 2011.

[2] Bao-chen Yang, Yue Chen. "Integrated Risk management for EPC Projects." *Industrial Engineering Journal.* vol. 14, no. 5, pp.52-57, 2011.

[3] Jun-yi Shao, Kun-tao Dong, Han Guo, Hui Zhao. "Research on Risk Evaluation for International Construction Projects." *Journal of Engineering Management,* vol. 25, no. 2, pp.187-190, 2011.

[4] Dong-xiao Niu, Quan-yong Bai, "Nuclear Investment Risk Assessment Research Based on Dynamic Fuzzy AHP." *East China Electric Power,* vol.40, no.12, pp.2113-2116, 2012.

[5] Ning Lu, Min Zhao. "Research on International Engineering Project Investment Risk Based on the fuzzy Comprehensive Evaluation," *Value Engineering,* vol.30, no.33, pp.67-68, 2011.

[6] Yi-fan Xi, Xing-xin Nie, Chao Wang, "Research on Risk Evaluation Model of Construction Project," *Journal of Natural Disasters,* pp. 89-92, 2009

[7] Shu-bin Gao, Zi-xian Liu, "Evaluation Method of the Government Service Process Based on Fuzzy and Grey Correlation Analysis Model," *Journal of Tianjin University,* pp. 79-82, 2012.

[8] Ling Wang, Jian-lin Liu, Ji-wei Zhu. "Risk Evaluation of International EPC Hydropower Project Based on AHP-MF Model," *Journal of Engineering Management,* vol.4, pp. 82-86. 2012

[9] Gao Yunli, Li Hongnan, and Zhang Guojun, "Dynamic Fuzzy Evaluation of Construction Project Risk Based on Cooperation," *Journal of Dalian University of Technology,* pp. 405-408, 2010.

[10] S.Ebrahimnejad, S.M.Mousavi, H.Seyrafianpour, "Risk identification and assessment for build-operate-transfer projects: A fuzzy multi attribute decision making model", *Expert Systems with Applications,* vol.37, pp. 575-586, 2010.

Claim Practice for the Drainage of Foundation Pit of Huangjinping Hydroelectric Station

Li TU, Li-min JIA *

Department of Economics and Management, Three Gorges University, Yichang, China
(384144118@qq.com)

Abstract - On the one hand, change claims of hydroelectric engineering are the effective means for the contractor to reduce losses and improve profit. On the other hand for the owner, if the engineering change claim work is managed well, it will reduce project costs and improve project management efficiency. The article will introduce the case of the frequent drainage items of dam foundation pit change claims in the Dadu River Huangjinping hydroelectric station. And it will analyze how to do the engineering change claims from the contractor perspective.

Keyword - Cost claim, engineering change claims, Huangjinping hydroelectric station, the frequent drainage items of dam foundation pit

I. INTRODUCTION

Engineering change mainly refers to the measures that change the original contract documents to ensure the smooth implementation of the project which is due to the change of contract status. And it always comes with the measure to adjust the contract price and duration. In the process of the project, due to the complexity of the construction environment and the design drawings which is not deep enough, the difference comes between the original design and actual construction [1]. The differences will increase the duration, added fee, and make the contractor suffering losses. Then the contractor could make a claim to the owner to make up for the loss [2]. The article will combine the construction condition change that the water seepage increases in the dam foundation pit, which causing the change claim case. It will analyze the concrete process of the claim and provide a reference for the actual claims.

II. CLAIMS CASE STUDY

A. Basic Situation

Huangjinping Hydroelectric Station is located in the upper reaches of the Dadu River. There is a planning of the main stream in the Dadu River - "three base and twenty-two grade". Huangjinping Hydroelectric Station is the eleventh grade power station of the hydroelectric planning. It is developed by the mix way of the reservoir dam and "one station and two factories". The hydroelectric station plants total installed capacity of 850MW (800MW of the large plant and 50MW of the small plant). The frequent drainage items of dam foundation pit belonging to the construction of diversion and water control. Although bid contracts stipulate explicitly about the frequent drainage project of dam foundation pit, but in the actual construction of one year, the actual drainage volume is much greater than the design drainage quantity. The contractor has invested a lot of manpower, material resources and money to accomplish the task, and it makes a huge lose to the contractor. So the contractor proposes the claim for the losses. If none of the contractor's ability results the project costs raise, and it enable the contractor to increase additional costs or suffer economic losses, then the contractor can propose an expenses claim according to the provisions of the contract.

In the case, the frequent drainage project of dam foundation pit officially started from May 6, 2012. The contractor set 1# pumping station in the right of downstream foundation pit according to the construction organization design requirements. But in the later process, the 1# pumping station couldn't meet the needs. Then they added the 2# pumping station in the left of upstream foundation pit due to the construction of cutoff wall for foundation pit in 2012 December. From May, 2012 to May, 2013, the actual drainage volume was much greater than the design drainage capacity, which was 3.9 million m^3. The original design documents had provisions on this item, it is showed in the TABLE I.

TABLE I
PROVISIONS IN THE ORIGINAL DESIGN DOCUMENTS

Daily average drainage volume of frequent drainage in the foundation pit	Pay way
<=4000m³/h	Pay for the total contract price
>=4000m³/h	Pay for the station price

B. Preparation for Claim

After the contractor found the situation in the actual construction process, they timely communicated with supervision. The employer, supervision and the design units were also very concerned about the work. As of August, 2013, they had organized conferences seven times, issued fifteen copies of special documents and instructions, required the contractor to do the frequent drainage items of dam foundation pit well in the coming flood season. The contractor site quantity visa work actively, and collected the bills of quantities,

E. Qi et al. (eds.), *Proceedings of the 21st International Conference on Industrial Engineering and Engineering Management 2014*, Proceedings of the International Conference on Industrial Engineering and Engineering Management, DOI 10.2991/978-94-6239-102-4_102, © Atlantis Press and the authors 2015

meeting records and construction instruction.

C. Claim Values

According to the general conditions of contract clause 39.1 (1) - (1) During performance of the contract, the supervisor can indicate the contractor for the following types of change according to the needs of the project. ① Increase or decrease the content in the contact; ② adding any extra work required to complete the project. If there is none of the supervisor's instruction, the contractor shall not alter the project. The equipment increase and volume increase of the drainage project, which are in response to the requirements of the construction environment change and approved by ministry of supervision. They belong to engineering change.

The contractor firstly calculated the added fee of drainage equipment, according to the actual drainage equipment which was put as the actual construction documents. Because the part that the volume was ≤ 4000m3/h, which was far more than the plan volume of 3.9 million in the technology design documents, so the new part according to the actual drainage work hour which was visa by the supervision, multiplied by the corresponding price, should be given compensation costs.

The contractor then reported well written claim report to the owners, and added the original data, the complete proof of claim, claim items and claim cost, and timely communicated with the owners to enter the stage of the claim negotiation.

III. THE STRATEGIES AND SKILLS OF CLAIMS

In order to get the construction contract of one engineering project in the fierce market competition, the contractors will not only make full use of the advantages in management level, construction techniques and mechanical equipment to win the bidding with low interest policy and dig the potential of the enterprise as far as possible, but also find all claim opportunities that might appear in the process of construction through field investigation, analysis of the contract conditions in the tender documents and the design drawing, to get extra profits by claiming for compensation during the construction [3]. This strategy is legal and reasonable as well. The particularity of hydropower engineering determines the complexity and particularity of the engineering project contract management, that even the most detailed contract documents being supposed to may have omissions places due to the complicated engineering construction, and the unpredicted accident risk that might happen gives a chance to the contractors of the claim opportunity [4]. With this strategy, the contractors can turn the disadvantageous factors of low bid price into advantages, and regain the initiative in construction.

A. Establish a Claim Leader Team

The work of claims is very complex and detailed with wide covers and long cycle, and need cooperation among different staffs and departments. Establishing a specialized claim team for the great compensation claiming event plays an important role in this work, and the team's work determines success or failure. Usually, the claim team is leaded by the Project Manager, including experienced experts and contract management staffs concretely [5]. During the construction of the Huangjinping Hydroelectric Station, the claim team established by the contractor (Jiangnan Water Conservancy and Hydroelectric Company), is leaded by the Project Manager, and making up by experienced section chief and members in Contract Planning Department, with the cooperation of other technical section, finance section and on-site construction personnel, the team starts collecting the claim evidence from the on-site interference incidents analysis, and communicate with supervisor and first party, to carry out a comprehensive claim work [6].

B. Collect the Enough Claim Evidence

Evidence is the premise of effective claim, the claim evidence usually comprising ① the contract documents, bidding documents, construction scheme, construction drawings; ② correspondence such as change of instruction, notice; ③ the various meeting records; ④ the construction schedule and the actual construction schedule record; ⑤ the construction site of the project file, the construction records of site; ⑥ engineering visa; ⑦ engineering photos and video. The claim must be reasonable, but achieving the point we must understand the original data, ensure the accuracy and integrity of original data, in order to make the claim more reliable, success rate higher.

C. Using Right Methods

Striving for individual claims, avoiding comprehensive claims. Individual claim refers to the claim that appears to interfere with the original contract in engineering construction, contractor for this event claim. Comprehensive claim refers to claims that had been proposed in the construction process but unresolved, and contractor would put forward a total claim to reports to the owner's claim. So many disturbance events would appear when disposing comprehensive claims, and the difficulties are complex, some of the evidence lose efficacy, and that may result in difficulties in analysis and calculation which make claims processing and negotiations are difficult. And the amount of comprehensive claim is larger, sometimes may require the contractor to make greater concessions to solve. Due to contract events is relatively simple,

Individual claim is simple, and responsibility analysis and the calculation is not too complex, the amount is smaller than comprehensive claim, both sides tend to reach an agreement, the claim may successful easily. Therefore, the contractor must master the favorable opportunity in construction claims, and strive to individual claim, to solve every item in the process of construction alone [7, 8].

D. Make Good Communication with the Owners and Supervision

The claim should be settled amicably, prevent antagonism, and pay attention to communication with supervising engineer. Claims disputes are inevitable .The dispute cannot sensibly solved may result in some unsolved problems pending. Especially the contractor should particularly cool, prevent antagonism, and do their best to amicable settle claims. Supervision engineer is the witness of the construction process, striving for the engineer's understanding is the key settle claims. Through supervision engineers to establish a good communication bridge with owners, owners' final goal is to complete the quality and quantity of the project during the construction period, so they willing to have a friendly communication with conductors, do not cause unnecessary delays in engineering, with the reasonable claim, the owner is would acceptable [9-11].

IV. CONCLUSION

The claim is a legitimate measure. In practice, the construction enterprise should pay special attention to the claim, achieve special claim responsibility system, collect perfect evidence compensation claims, use the correct methods, and communicate with the owners, supervision well. If the contractor tightly around the four points to carry out the work, the success rate of claim will be greater.

REFRENCE

[1]   Yongqiang Chen, Shuibo Zhang. "International engineering claim". Beijing: China architecture & building press, pp. 56-57, 2008.

[2]   Wenjie Liu. "Construction project cost composition of the construction claims and the calculation method". Engineering and construction, Vol 5, No 20, pp.542-544, 2006.

[3]   Jie Shen. "Engineering evaluation". Nanjing: Southeast University Press, pp.329-332, 2005.

[4]   Yongqiang Chen, Shuibo Zhang. "International engineering contract management". Beijing: China architecture &building press, pp. 308-309, 2011.

[5]   Fugang He. "The research &system development of hydroelectric engineering construction claims decision analysis". Chengdu: Sichuan University, 2006.

[6]   Xiaoyu Si, Gang Guo, "Analysis of compensation-claiming in supervision cost", Journal of Hebei institute of architecture and civil engineering, Vol 23, No 10, pp.121-123.

[7]   Jian Liang. "International engineering construction claims", Beijing: China building industry press, 2002, pp.25-27.

[8]   Yaohuang Guo, Yaping Wang. "Engineering and claims management", Beijing: China railway publishing house, 1999, pp.75-78.

[9]   Xingyu Zhu, Xiuqin Wang. "International project management fee claim", International economic cooperation, No.3, pp.66-70, 2009.

[10]   Xiaoyong Yuan. "Investigate of contractor for claims of the profits and headquarters management fee in international project contracting", Academic forum, No 8, pp. 99-104, 2008.

[11]   Hu Cheng. "Construction contract management and claims". Nanjing: Southeast University Press, 2009, pp.110-112.

Investigation on Work Stress and Effective Coping Style among Post-80s Generation

Ya-rong WANG[1], Jing-ya AN[1,*], Zi YANG[2]

[1]School of Economics and Management, Inner Mongolia University of Science and Technology, Baotou, China
[2]Sycp Beijing Consultation Service Company, Beijing, China
(anjingyazhi@163.com)

Abstract - **With the increasingly growing social competition, people's pace of life and work accelerates greatly, which causes rising social concern about the work stress of the post-80s generation, especially those with higher education background. They gradually start their careers, and act as the backbone or reserve force in the course of enterprises' growth. Therefore social attention should be paid to their work stress and problems related to stress. This paper has acquired the basic data about the specific post-80s generation, including the data of their work stress, stress coping styles and psychological health status. Based on the data processing results of SPSS 18.0, series of conclusions have been drawn ultimately.**

Keywords - **Coping style, post-80s, psychological health, work stress**

I. INTRODUCTION

As a result of growing intensified social competition, the pace of work and life accelerates increasingly, which leads to much attention on work stress. A survey conducted by the United Nations International Labor Organization shows that psychological stress will become one of the most serious health problems of the 21st century. Based on a research, conducted by the national trade union in 2010, 70% of 1698 workers born in 1980s complain about relatively huge work stress, 20% of whom thinking they are under extremely considerable work stress. In addition, studies have shown that excessive work stress will not only harm individuals' physical and psychological health, but also affect their work status and performance. The post-80s generation refers to the group of people born between 1980 and 1989, about 204 million people, accounting for 1/7 of China's total population. Meanwhile, they gradually start their careers, acting as the backbone or reserve force during the growth period of enterprises. Therefore, we need to pay special attention to their work stress, coping styles and psychological health status, especially those with higher education background.

II. LITERATURE REVIEW

Originally, stress is a concept of physics. Since it was introduced to the social sphere, it has attracted the attention of medicine, psychology, sociology and other disciplines, from studies of stress events to related stress source analysis, then to stress coping styles and finally to stress effect on health. Consequently, prominent and influential achievements have been produced. Lazarus (1966) believes that stress is a special relationship between people and environment and that there exists a cognitive evaluation process between stress and people's reaction to stress, during which people will firstly evaluate the outside environment to see whether it's challenging or threatening, and then evaluate their own response to it. When individuals think they will fail to deal with the outside challenging or threatening events, stress is produced [1]. In the 1960s, the focus on stress events shifted from work environment to work role. For a long period, people have been exploring new ways to cope with work stress on organizations and individuals, work and family, salary and occupational health [2].Gradually, multi-angle stress scales aimed at measuring and analyzing work stress emerged. Among the foregoing scales, the scale of occupational stress index (OSI) developed by Cooper and his colleagues is dominant and widely used [3].

Stress is objective and it is incapable to determine its production force. As a mediator to adjust and cushion stress, stress coping style can affect the force of stress felt by individuals. Stress coping styles can be divided into four kinds, namely positive action, avoidance, attack and laissez-faire (Lazarus, 1966) [4].There are six kinds of stress coping styles, namely self-blame, fantasy, retreat, rationalization help-seeking and problem-solving type (Ji-hua Xiao, 1996) [5]. Stress coping types can also be classified in to positive and negative styles (Omit, 2006) [6]. However, Ya-ning Xie (1996) developed a simple stress coping scale on the basis of simplified and modified foreign stress coping scales because above-mentioned scales are not suited for the Chinese people. In this scale, the Cronbach's Alpha is 0.9. Positive and negative coping styles are the two factors affecting the validity test, and the variance of the two factors is 48.9% [7]. Domestic researchers adopt the simple stress coping scale developed by Ya-ning Xie to do researches about stress coping styles of different groups.

Of course, work stress can influence people's physical, psychological conditions as well as behaviors and moderate pressure can impel workers to do what they can to deal with problems, in order to keep good balance between work stress and psychological status. Meanwhile, excessive work stress will trouble body functions, thus having negative influence on people's psychological condition. Due (1994) suggests that huge work stress has significant negative correlation with bad mood, poor health and strong dissatisfaction towards work and stress of some work causes not only such physical problems as infection, headache, and respiratory diseases but also bad psychological condition, such as depression ,anxiety, and much poorer health [8]. Chang Hsiu- hua (2007) finds that work stress and fatigue for long time can cause health problems, including heart diseases and strokes [9]. Consequently, the force of work stress can be measured according to the level of psychological health status.

Derogatis (1965) designed the Discomfort Scale, which became the Brief Symptom Inventory through simplification, also called Symptom Check list 90 (SCL-90) or Hopkin's Check List (HSCL) [10]. This scale, one of the world's most famous psychological health scales, is widely used both at home and abroad. Domestic researchers adopt this scale to study work stress and related health problems on nurses, IT knowledge workers, teachers and other groups. The results of above studies have offered a paradigm scale for this paper's research about work stress, stress coping style and psychological health status of post-80s.

III. SURVEY PROCEDURES

A. Sampling and Instruments

All participants are post-80s, working in different industries of domestic companies all over China. Under socio–demographic characteristics, post-80s' age, gender, education and work experience are considered. Totally, 400 questionnaires are distributed by e-mail in this sample survey. Of all these 400 participants, 275(68.75%) have provided effective questionnaires, whose basic information is as follows (see TABLE I).

TABLE I
SOCIO–DEMOGRAPHIC CHARACTERISTICS OF POST-80S

	Classification	Number	Percentage (%)
Gender	male	136	49.5
	female	139	50.5
Education	college	35	12.7
	undergraduate	222	80.7
	Postgraduate and above	18	6.5
Work experience	less than one year	51	18.5
	one to three years	205	74.5
	more than three years	19	6.9

Work stress source scale uses occupational stress index (OSI) designed by Cooper and his colleagues. The initial index scale consists of 27 items, among which the items, whose correlation coefficients are less than 0.30, have been deleted after analyzing the correlation between the theme and items. In the process of reliability analysis, items with lower internal consistency have been deleted. Then, we use principal component analysis to extract the common factors and delete items whose load values are less than 0.5 through exploratory factor analysis. Eventually, the index scale contains 23 items and each item uses 5-point Likert scale: "strongly disagree" (coded1), "inclined to disagree" (coded2), "no idea" (coded3), "inclined to agree" (coded4) and "strongly agree" (coded4). The higher the scores are, the greater, the perceived work stress is. All participants are required to objectively fill in the scales according to true facts. This study shows the value of Cronbach's Alpha is 0.840 and split-half reliability coefficient is 0.708, and this indicates a good internal consistency of the scale.

Coping Scale refers to simple coping style developed by Ya-ning Xie including 20 items as regards positive and negative coping styles, adopts 4-point scale scoring method : 'never' (coded0), 'occasionally' (coded1), 'regular' (coded2) and 'sometimes' (coded3). Participants are required to reflect their attitudes and measures faced with work stress and frustration. The value of Cronbach's Alpha is 0.875, which indicates a good internal consistency of the scale.

Psychological health scale adopts the Symptom Check list 90 (SCL-90) compiled and revised by Derogatis by using 5-point scale scoring method: 'none' (coded1), 'light' (coded2), 'medium' (coded3), 'a little bad' (coded4) and 'serious' (coded5). The value of Cronbach's Alpha is 0.971, which indicates a good internal consistency of the scale.

B. Primary analysis

Work stress source data is processed by SPSS 18.0. And both the Bartlett test of sphercity (1771.416 at p=0.000) and the Kaiser-Meyer-Olkin measure of sampling adequacy (KMO=0.806) indicate that there are sufficient inter-item correlations within the data for factor analysis. Factor analysis extracts 7 factors, whose eigenvalues are above 1 and loadings above 0.5, and these seven factors accounts for 62.265% of variance. These seven factors are economic reward stress, occupation development stress, work stress, life role stress, organizational atmosphere stress, interpersonal stress and work-family conflict stress. By summing up products of each factor coefficient value and the average value of each item, we get the average value of each factor, and find economic rewards stress (3.823), career development stress (3.373), the stress of work itself (3.353) and role stress(3.013) are fields putting heavier stress on post-80s.

The one – way ANOVA analysis about post-80s, by keeping gender, education and work experience as the control variable respectively, showed that gender is a significant factor affecting organizational atmosphere, economic reward and work-family conflict stress. Compared with males of post-80s, females have lower stress in economic reward as well as organizational atmosphere, but higher stress in work-family conflict (see TABLE II). Although education level has close relationship with the stress form interpersonal relationship, organizational atmosphere and work itself, and so does work experience with the stress of work itself, life role and interpersonal relationship, conclusions remain to be further verified because the sample population of post-80s with college and postgraduate education background is relatively small, and so is the sample population of post-80s with over 3 years' work experience.

TABLE II
SIGNIFICANT DIFFERENCE BETWEEN GENDER AND WORK STRESS ($\bar{x} \pm s$)

Stress	Gender		F
	Male	Female	
Organizational atmosphere	2.76±0.62	2.57±0.62	6.51*
Economic reward	3.91±0.54	3.73±0.57	7.17**
Work-Family conflict	2.38±0.51	2.43±0.48	4.86*

According to the descriptive statistical analysis of the way of stress coping styles, the mean and standard deviation of positive coping styles are respectively 1.978 and 0.708, which means that post-80s sometimes take positive stress coping style; the mean and standard deviation of negative coping style are respectively 1.089 and 0.718, which means that post-80s occasionally take negative stress coping style. When gender is the only variable in the research, regardless of education background and work experience, we can see males of post-80s are more inclined to adopt negative coping style with comparison to females (see TABLE III).

TABLE III
GENDER EFFECT ON COPING STYLES ($\bar{x} \pm s$)

	Gender		
	Male	Female	F
Positive coping style	1.98±0.32	1.97±0.32	0.09
Negative coping style	1.07±0.33	1.10±0.32	0.35*

According to the result of descriptive statistical analysis of psychological health, the average values of depression, anxiety, interpersonal sensitivity are respectively 3.100, 2.877 and 2.712, showing post-80s are in the medium-to-quite-slight degree in such three aspects. The average values of hostility and obsessive-compulsive disorder arc respectively 2.527 and 2.476, showing post-80s are in the quite slight degree in such two aspects. The average values of paranoia, somatization, psychological illness and terror are respectively 1.638, 1.417, 1.347 and 1.294, showing post-80s are in the very-slight-to-none degree in the four aspects. Gender factor shows that there are no significant differences in psychological health between males and females of post-80s. However, education level and work experience, when used as the control variables, will bring about huge differences in paranoia, somatization, anxiety, interpersonal sensitivity and terror. But this result needs further verification by enlarging the sample size of post-80s with college and postgraduate education background and with more than three years' work experience.

C. Correlation analysis

Results of correlation analysis on work stress and psychological health show no significant correlation between the stress of work itself and terror, the role of stress and hostility, interpersonal stress and terror, but significant positive correlation between other work stress factors and psychological health indicates that the heavier the work stress on post-80s are, the severer their psychological health problems are. Based on the correlation analysis, multiple linear regression analysis is gradually conducted, with the seven factors of work stress source as independent variables and psychological health as dependent variable (see TABLE IV). As can be seen in TABLE IV, these seven work stress factors can explain the total 13.2% variation of psychological health, among which the stress of role (β=0.198), the stress of work itself (β=0.240), career development (β=0.248) and work-family conflict (β=0.250) are elected into the regression

equation, emerging as prominent predictors of psychological health.

TABLE IV
MULTIPLE REGRESSION ANALYSIS ON PSYCHOLOGICAL HEALTH AND WORK STRESS

	Std β	T	Significance
Stress of role	0.198	3.165	0.002
Work itself	0.240	2.152	0.003
Career development	0.248	3.235	0.000
Work - family conflict	0.250	4.000	0.000

R^2=0.132; Adjusted R^2=0.110; F=5.819; Sig.=0.000

Standard regression equation: Psychological reaction = 0.198* Stress of role +0.240 * Work itself + 0.248* Career Development +0.250 * Work-family conflict

Result of correlation analysis on work stress and coping style indicates that there is significant negative correlation between interpersonal relation, organizational atmosphere, work-family conflict and positive coping style, namely, the higher stress 80s feel in interpersonal relation, organizational atmosphere and work-family conflict, the weaker tendency they will choose positive coping style; there is significant positive correlation between stress of role, career development, work itself and negative coping style, namely, the bigger stress in life role, career development and work stress itself, the stronger tendency they will choose negative coping style, which predicts negative coping well. Based on correlation analysis, multiple regression analysis is conducted, with the seven factors of work stress source as independent variables and two coping styles as dependent variable (see TABLE V). It is observed from TABLE V that the seven work stress factors can explain the total 9.9% variation in negative coping style, among which the stress of life role (β=0.284), the stress of work itself (β= 0.252), and career development (β=0.325) are elected into the regression equation, emerging as outstanding predictors of negative stress coping style.

TABLE V
MULTIPLE REGRESSION ANALYSIS ON NEGATIVE COPING STYLE AND WORK STRESS

	Std β	T	Significance
Stress of role	0.284	4.460	0.000
Work itself	0.252	3.320	0.001
Career development	0.325	2.355	0.000

R^2=0.099; Adjusted R^2=0.076; F=4.201; Sig.=0.000

Standard regression equation: Negative coping = 0.284* Stress of role +0.252 * Work itself + 0.325* Career Development

Result of correlation analysis on psychological health and coping style indicates that there is significant negative correlation between somatization obsessive-compulsive disorder and positive coping style, namely, post-80s will have lighter somatization and obsessive-compulsive disorder with preference to positive coping style; there is significant positive correlation between interpersonal sensitivity, depression, anxiety, hostile, psychological illness, horror and negative coping style, namely, post-80s will have more serious interpersonal sensitivity, depression, anxiety, hostile, psychological illness and horror when they adopt the negative coping style. Based on the correlation analysis, multiple regression analysis is conducted, with coping style as independent variables and psychological health as dependent variable (see TABLE VI). The result of TABLE VI shows that the two coping style factors can explain the total 8.2% variation in psychological health. Besides, only negative coping style, with positive and predictive effect on psychological health (β=0.283), is taken into the regression equation.

Correlation analysis on three types of scales indicates that coping style is a proper mediating variable. TABLE VII shows the change of equation for introducing coping style: stress of role has reduced to 0.139 from 0.198, work itself has reduced to 0.232 from 0.240, career development has reduced to 0.126 from 0.248, work-family conflict has reduced to 0.240 from 0.250. These figures demonstrate

that negative coping style of post-80s acts as a mediator between the stress of role, career development, work-family conflict stress and psychological health. Some scholars have also confirmed that coping style is an important mediator between stress sources and responses, which is of great value for the study of individual's physical and psychological health.

TABLE VI
MULTIPLE REGRESSION ANALYSIS ON PSYCHOLOGICAL HEALTH AND COPING STYLE

	Std β	T	Significance
Negative coping style	0.283	4.866	0.000
R^2=0.082; Adjusted R^2=0.075; F=12.100; Sig.=0.000			
Standard regression equation: Psychological reaction = 0.283* Negative coping style			

TABLE VII
MULTIPLE REGRESSION ANALYSIS ON PSYCHOLOGICAL HEALTH AND WORK STRESS

	Std β	T	Significance	Std β	T	Significance
	Step1			Step2		
Stress of role	0.198	3.165	0.002	0.139	2.171	0.031
Work itself	0.240	2.152	0.003	0.232	2.272	0.000
Career development	0.248	3.235	0.000	0.126	3.245	0.000
work – family conflict	0.250	4.000	0.000	0.240	3.874	0.000
Negative coping style				0.209	3.541	0.000
R^2=0.132; Adjusted R^2=0.110; F=5.819; Sig.=0.000				R^2=0.172; Adjusted R^2=0.144; F=6.103; Sig.=0.000		

Step1 Regression equation: Psychological reaction = 0.198* Stress of role +0.240 * Work itself +0.248*Career Development+0.250 * Work-family conflict

Step2 Regression equation: Psychological reaction = 0.139* Stress of role +0.232 * Work itself +0.126*CareerDevelopment+0.240* Work-family conflict+ 0.209*Negative coping style

Correlation analysis on three types of scales indicates that stress coping style is also a buffer variable in affecting the strength of the relationship between work stress and psychological health. When setting work stress factors as independent variable, psychological health as dependent variable, and coping style as moderating variable, we carry out hierarchical regression analysis with product terms owing to the continuity of both independent and dependent variables. TABLE VIII shows that the value of R^2 of second regression analysis is higher than that of the first regression analysis and significant interaction between coping style and work itself is produced also with a high interaction coefficient (above 0.05). The result indicates that coping style takes moderating effect on work itself and psychological health. The same method of data processing also show coping style moderates, to some extent, stress of role, career

TABLE VIII
COPING STYLE'S MODERATING ON THE STRESS OF WORK ITSELF AND PSYCHOLOGICAL HEALTH

	Std β	T	Significance	Std β	T	Significance
	Step1			Step2		
Coping style	0.047	1.785	0.043	0.276	2.185	0.030
Work itself	0.194	3.261	0.001	0.596	2.913	0.004
Coping style* Work itself				0.480	2.054	0.041
R^2	0.040			0.054		
R^2 change				0.014		

development, interpersonal relationship, organizational atmosphere, economic reward and psychological health (omit TABLE).

IV. CONCLUSION

According to sample survey data analysis of post-80s'work stress, stress coping styles and psychological health and their relationship, we get the following conclusions:

Work stress is classified into seven groups, namely economic reward stress, occupational development stress, stress from work itself, life role stress, organizational atmosphere stress, interpersonal stress and work-family conflict stress, among which the first four are relatively higher by comparison of average values. Compared with males of post-80s, females feel smaller stress in economic reward as well as organizational atmosphere, but larger stress in work-family conflict.

According to coping style analysis, post-80s sometimes take active coping style and occasionally take positive coping style.

According to psychological health data analysis, post-80s show medium-to-slight degree in depression, anxiety and interpersonal sensitivity; quite slight degree in hostility and obsessive-compulsive disorder; slight-to-none degree in paranoia, somatization, psychological illness and horror.

According to correlation analysis between work stress and related psychological health, the bigger stress post-80s feel, the more serious psychological health problems they have, among which life role stress, occupation development stress, the stress of work itself and work-family conflict predict psychological health well.

According to correlation analysis between work stress and coping style, the larger stress post-80s feel in interpersonal relation, organizational atmosphere and work-family conflict, the weaker tendency they will choose positive coping style; the larger stress in life role, career development and work stress itself, the stronger tendency they will choose negative coping style, which predict negative coping style well.

According to correlation analysis between coping style and related psychological health, post-80s will have lighter somatization and obsessive-compulsive disorder with preference to positive coping style and more serious interpersonal sensitivity, depression, anxiety, hostile, psychological illness and horror with more towards negative coping style. Negative coping style has a positive predictive effect on psychological health.

Results of analysis on coping style, work stress and psychological health shows that negative coping style can not only mediate the relation between stress of role, work itself, career development, work-family conflict and psychological health, but also can adjust stress of economic reward, career development, work itself, life role, organizational climate and interpersonal relationship.

Research needs to be further expanded in the college education, graduate education, work experience of less than one year and more than three years, aiming at verifying the work stress of post-80s' with different education background and work experience on the basis of unbalanced data. Furthermore, the above conclusion need to be verified by exploring their work stress causes in combination with post-80s growing and living background via in-depth interviews.

REFERENCES

[1] R. S. Lazarus, Psychological stress and the coping process, New York, 1966, Mcgraw-Hill.

[2] Ari Väänänen, Erkko Anttila, Jussi Turtiainen, and Pekka Varje, "Formulation of work stress in 1960-2000: Analysis of scientific works from the perspective of historical sociology", *Kaohsiung Journal of Medical Sciences*, no. 27, pp.144-149, 2011.

[3] C. L. Cooper, S. J. Sloan, and S. Williams, Occupational stress indicator: Management guide, Windsor, 1988, NFER-Nelson.

[4] Yi-tuo Ye & Yan-e Shen, "General overview of coping and coping style (in Chinese)", *Psychological Science*, vol. 25, no. 6, pp.755-756, 2002.

[5] Ji-hua Xiao & Xiu-feng Xu, "Study on validity and reliability of Coping Style Questionnaire(in Chinese)", *Chinese Psychological Health Journal*, vol. 10, no. 4, pp. 164-168, 1996.

[6] C. E. Amiot, D. J. Terry, N. L. Jimmieson, V. J. Callan, "A Longitudinal Investigation of Coping Processes During a Merger: Implications for Job Satisfaction and Organizational Identification", *Journal of Management*, vol. 32, no. 4, pp. 552-574, 2006.

[7] Ya-ning Xie, "Preliminary study on reliability and validity of simple coping style scale(in Chinese)", *Chinese Journal of Clinical Psychology*, vol. 6, no. 2, pp. 114-115, 1998.

[8] J. K. Dua, "Job Stressors and their effects on physical health, emotional health, and job satisfaction in a university", *Journal of Educational Administration*, vol. 2, no. 1, pp. 59-78, 1994.

[9] Chang Hsiu-hua, "Work stress and death from overwork (in Chinese)", *SCIENCE & TECHNOLOGY INFORMATION*, no. 21, pp.12-14, 2007.

[10] A. Hessel, M. Geyer, and E. Brähler, "Psychiatric problems in the elderly--standardization of the Symptom Check List SCL-90-R in patients over 60 years of age", *Zeitschrift fuer Gerontologie und Geriatrie*, vol. 34, no. 6, pp.498-508, 2002.

Board independence and Performance of Listed SMEs in China: based on the Perspective of Concentrated Ownership

Qing-mei TAN[1,*], Pan LIU[1], Meng-ying JING[1], Xiao-fang DONG[2]

[1]College of Management and Economics, Tianjin University, Tianjin, China

[2]Tianjin Intelligent Technology Engineering Center of Finance and Taxation, Tianjin, China

(tanqm@tju.edu.cn)

Abstract - **Based on the sample of listed firms on SMEs board in China during the period 2004-2010, this paper investigates the impact of board independence on firm performance in the presence of significant ownership concentration. This paper does not find the evidence that the board independence of listed SMEs is endogenous. The results show that there is no significant relationship between performance of listed SMEs and board independence, which means more independent directors cannot improve the performance of listed SMEs in China. There is significantly positive relationship between ownership concentration and board independence, while there is no significant relationship between ownership concentration and performance of listed SMEs in China.**

Keywords - **Board independence, ownership concentration, return on assets, small and medium-sized enterprises, Tobin's Q**

I. INTRODUCTION

The role of independent directors in the protection of shareholders has long been a subject of much debat. There are many empirical studies on the relationship between board independence and firm performance. However, most of previous studies are based on the data of large firms, and the empirical results are mixed. Especailly in China, most studies on board independence and firm performance are based on the data of firms listed on the main board, namely large listed firms.

As an important part of Chinese capital market, the SMEs board plays important role in finance, corporate governance and industrial development for small and medium-sized enterprises(short for SMEs). Compared with firms listed on the main board in China, the board structure and ownership structure have the following obvious characteristics: firstly, the proportion of independent directors of listed SMEs is lower, and most of the independent directors of listed SMEs have professional expertise in accounting or finance, as a consequence, the independent directors provide more advisroy than monitoring function in listed SMEs in China(Jiang, 2010)[1]. Secondly, there is high ownership concentration in listed SMEs, the most pervasive agency conflict in listed SMEs is between controlling shareholders and minority shareholders. Controlling shareholders may use their controlling right to expropriate wealth from minority shareholders. What's more, boards of firms with high ownership concentration may tend to be mostly comprised of directors who represent the large owner's or manager's intrests. Since the controlling shareholders have important impact on the board composition, the ownership concentration has direct impact on the independence of independent directors. Thus, ownership concentration has indirect impact on firm performance. The independent directors can play effective supervision function, and reduce the agency

problem between minority shareholders and managers in firms with low ownership concentration. However, the cotrolling shareholders can reduce the independence of the board using their voting right in frims with high ownership concentration. Thirdly, most of the listed SMEs are private firms, there is no obvious multilevel pyramid ownership structure of listed SMEs in China. Consequently, compared with firms listed on the main board, the relationship between board independence and performance may be different in listed SMEs.

This paper aims at investigating the impact of the board independence on performance of listed SMEs under high ownership concentration based on the data of firms listed on SMEs board in China during the period 2004-2010. Our paper sheds light on the understanding of the relationship between board independence, ownership concentration and performance of firms listed on SMEs board in emerging markets.

II. LITERATURE REVIEW

There are many studies on the impact of board independence on firm performance, but the results are mixed. Hermalin and Weisbach(1991), Mehran(1995), and Klein(1998) all report insignificant relationship between frim performance and board independence, using various measures of firm performance[2-4]. Erickson et al.(2005) also find that greater board independence does not have significant impact on firm value, and that poorly performing firms will increase the proportion of outside directors in the subsequent periods[5]. However, Agrawal and Knoeber (1996), Bhagat and Black(2001) both find that the board independence has significantly negative impact on firm value meausred by Tobin's Q[6, 7]. Further, Dehaene, Vuyst and Ooghe(2001) find a significantly positive relationship between return on equity and the number of external directors, which means that board independence has positive impact of firm's accounting performance[8]. Krivogorsky(2006) find that the board independence can increase firm performance measured by return on assets, Tobin's Q and market to book ratio, using data from 87 European firms[9]. Lefort and Urzua(2007) find that the increase in the proprtion of outside directors can increase firm value, using a four-year, 160-company pannel data[10].

In the recent year, some Chinese scholars studied the relationship between board independence and firm performance based on the data of listed Chinese firms on the Main board. Wang, Zhao and Wei(2006) find a significantly positive relationship between independent directors and firm performance[11]. However, Li and Sun(2007) do not find that there is significant relationship between the board independence and firm performance[12]. Hao and Zhou(2010) get the different result, they find that the increase in the proportion of independent directors does not necessarily lead to the improvement of firm performance, but the performance improvement helps to the decline in the independence of the board[13].

III. RESEARCH DESIGN

A. Data and Sample

The impact of the board independence on performance of listed SMEs is tested using financial data from a sample of listed SMEs in China which are selected by several criteria: (1) Since accounting earnings are more probably manipulated in ST firms and *ST firms, ST firms and *ST firms are both excluded; (2)Since there is difference between financial statements and capital structure of financial firms and non-financial firms, financial firms are excluded, also excluded are public services firms and firms with missing financial data or abnormal financial data. The requirement yields a sample of firms consisting of 1223 firm-year. All the financial data are obtained from CSMAR database.

B. Variables

1) *Firm performance*: Tobin's Q and *ROA* are used to measure the market performance and accounting

performance of listed SMEs, respectively. Tobin's Q is the ratio of the sum of the market value of equity and the book value of total debt divided by the book value of total assets, *ROA* is return on assets.

2) *Board independence*: board independence (*BI*) is measured by the fraction of independent directors on the board. The higher the fraction means the stronger board independence.

3) *Ownership concentration*: we use the fraction of voting rights held by the top five shareholders to measure ownership concentration of listed SMEs (*OC*).

4) *Control variables*: we consider the following control variables to filter out their influences on performance of listed SMEs in China: financial leverage (*LEV*) is measured by the ratio of total debt to total assets; assets structure (*AS*) is the ratio of fixed assets to total assets; firm size (*FS*) is measured by the natural logarithm of total assets; firm growth (*GR*) is measured by the growth rate of operating revenue; R&D intensity (*R&D*) is the ratio of R&D expenditure to total assets. What's more, industry dummy variable (*I*) and year dummy variables (*Y*) are both controled.

The definitions of all the variables can be found shown in Table I.

TABLE I
DEFINITIONS OF VARIABLES

Variables	Definitions
Tobin's Q	The ratio of market the sum of the market value of equity and the book value of total debt divided by the book value of total assets
ROA	Return on assets
Board independence	The fraction of independent directors on the board
Ownership concentration	The fraction of voting rights held by the top five shareholders
Financial leverage	Total debt-to-assets ratio
Assets structure	The ratio of fixed assets to assets
firm size	The natural logarithm of assets
firm growth	The growth rate of operating revenue
R&D intensity	The ratio of R&D expenditure to assets

C. Regression Model

There is high ownership concentration in listed SMEs in China. Since the controlling shareholders have an important impact on the board composition, ownership concentration has direct impact on the independence of independent directors. Thus, ownership concentration has indirect impact on firm performance. To examine the endogeneity of board independence in listed SMEs, we establish a linear regression equation like (1).

$$BI = \beta_0 + \beta_1 PER + + \beta_2 CR_5 + \sum_{i=3}^{8} \beta_i CV + \varepsilon \quad (1)$$

If there is no evidence that the board independence and firm performance are endogenous. We will investigate how board independence impacts on performance of listed SMEs based on the perspective of exogeneity. Then we establish a linear regression equation like (2).

$$PER = \beta_0 + \beta_1 BI + \beta_2 CR_5 + \beta_3 BI \times CR_5 + \sum_{i=4}^{10} \beta_i CV + \varepsilon \quad (2)$$

Where *PER* is firm performance, which can be meassured by Tobin's Q and *ROA*, respectively; *BI* is board independence, which refers to the proportion of independent directors account for board directors; CR_5 is ownership concentration measured by the fraction of voting rights held by the top five shareholders; *CV* represents control variables, including finnacial leverage (*LEV*), asset structure (*AS*), firm size (*FS*), firm growth (*GR*), R&D intensity (*R&D*), respectively; *I* and *Y* represent industry dummy variables and year dummy variables, respectively; β is regression coefficient and ε is error term.

If there is a significant evidence that board independence and firm performance are endogenous. We will establish simultaneous equations based on (1) and (2), and investigate the relationship between board independence and firm performance using three-stage least square method.

IV. EMPIRICAL RESULTS

A. Descriptive Statistics

Table II presents the descriptive statistics of the variables. The mean and standard deviation are 2.089 and 1.331, respectively. Such a result confirms that there is a great difference between market performance of different listed SMEs, and the market performance of listed SMEs are overvalued to some extend. The mean and standard deviation of ROA are 0.061 and 0.064, respectively. It

indicates that there is a great difference between accounting performance of different listed SMEst. The mean of board independence comes up to 0.363, indicating that the proportion of independent directors is low and reach to the 1/3 base line required by CSRC. The maximum and mean of the proportion of the top five shareholders both reach as high as 0.819 and 0.507, however, the minimum and standard deviation are only 0.171 and 0.128, which shows that there is high ownership concentration in listed SMEs in China, and the most pervasive agency conflict in listed SMEs is between controlling shareholders and minority shareholders.

TABLE II
DESCRIPTIVE STATISTICS

Variables	N	Min.	Max.	Mean	S.D.
Q	1223	0.815	10.853	2.089	1.331
ROA	1223	-1.189	0.389	0.061	0.064
BI	1223	0.143	0.600	0.363	0.048
CR_5	1223	0.171	0.819	0.507	0.128
LEV	1223	0.013	0.898	0.372	0.184
AS	1223	0.002	0.920	0.247	0.150
FS	1223	19.204	24.505	20.916	0.787
GR	1223	-0.941	58.803	0.342	2.125
R&D	1223	0.000	0.110	0.001	0.007

Where N is the number of observations; S.D. is sample standard deviation of variables. Q and ROA are Tobin's Q and return on assets; BI is board independence; CR_5 is ownership concentration; LEV represents finnacial leverage; AS is asset structure; FS is firm size, GR is firm growth; R&D is R&D intensity.

Total debt-to-assets ratio of the mean of 0.372, illustrating the low financial leverage of listed SMEs in China. The maximum and minimum value of fixed assets-to-assets ratio are 0.920 and 0.002, respectively, showing that there are significant diffrence of the assets structures in listed SMEs. The maximum and minimum value of firm size reaches 24.505 and 19.204, while the standard deviation is only 0.787. The result indicates that the sizes of listed SMEs are more or less similar. The

mean and standard deviation of growth rate of operating revenue are respectively 0.342 and 2.125, implying that there are great differences among the growth of listed SMEs in China. The maximum of R&D expenditure-to-assets ratio is only 0.110, what's more, some SMEs don't have R&D expenditure in the period of 2004 -2010 at all. It shows that R&D expenditure of listed SMEs is relatively less.

TABLE III
PEARSON CORRELATION COEFFICIENTS

	Q	ROA	BI	CR_5	LEV	AS	FS	GR	R&D
Q	1	0.310**	0.010	-0.198**	-0.250**	-0.061*	-0.116**	-0.024	0.036
ROA		1	-0.017	0.102**	-0.375**	-0.216**	0.016	-0.035	0.018
BI			1	0.098**	-0.013	-0.024	0.017	0.053	0.041
CR_5				1	-0.078**	-0.170**	0.117**	0.020*	0.023
LEV					1	0.212**	0.440**	0.009	-0.020
AS						1	-0.020	-0.011	-0.051
FS							1	-0.006	0.028
GR								1	-0.009
R&D									1

Coefficients marked with *, ** are significant at 5% and 1%, respectively. Q and ROA are Tobin's Q and return on assets; BI is board independence; CR_5 is ownership concentration; LEV represents finnacial leverage; AS is asset structure; FS is firm size, GR is firm growth; R&D is R&D intensity.

B. Correlation Analysis

Table III shows the Pearson correlation coefficients among variables employed in our study. There is not siginificant relationship not only between board independence and Tobin's Q, but also between board independence and ROA. It means there may not be correlation between board independence and firm performance. Ownership concentration is positively related to Tobin's Q and negatively related to ROA, which provides preliminary evidence for the correlation between ownership concentration and firm performance of listed SMEs. Ownership concentration is positively related to board independence, which means there may be correlation between ownership concentration and board independence. In addition, there are significant re-

lationships between most of control variables and Tobin's Q and ROA.

C. Endogeneity Test of Board Independence

In order to investigate the endogeneity of board independence in listed SMEs, based on the panel data of listed SMEs in the period 2004-2010, regression analysis of (1) is carried on as follows. The regression results are represented in Table IV. There is significantly positive relationship between board independence and ownership concentration, which implies that firms with higher ownership concentration have stronger board independence. Tobin's Q and ROA both have no significant impact on board independence. In addition, there are no significant relationships between all control variables and board independence. The results indicate that the board independence and firm performance are not endogenous. As a consequence, this paper will investigate that how board independence impacts on performance of listed SMEs based on the perspective of exogeneity.

D. The Influence of Board Independence on Performance of Listed SMEs

To examine the influence of board independence on performance of listed SMEs in China, using Tobin's Q and ROA as market performance and accounting performance, based on the panel data of listed SMEs in the period 2004-2010, regression analysis of (2) is carried on as follows. The results are shown Table IV. There are not significant relatioshipis between board Tobin's Q and board independence, which indicates that the increase of proportion of independent directors cannot improve the market performance of listed SMEs in China. What's more, the board independence has no significant impact on ROA, which menas that the increase of proportion of independent directors cannot improve the accounting performance of listed SMEs in China. There is also no significant relationship between the percentage of shares held by the top five shareholders and Tobin's Q, ROA, which implies that high ownership concentration of listed SMEs has no significant influence on firm performance. The regression coefficients of the cross-term of board independence and ownership concentration are both not significant, which further verifies that board independence and ownership concentration both have no significant influence not only on market performance but also on accounting performance of listed SMEs in China.

Financial leverage has significantly negative influnce onTobin's Q and ROA, which means the decrease of debt ratio will improve both the market performance and accouting performance of listed SMEs in China. There is no statistically significant relationship between assets structure and Tobin's Q, while there is significantly negative relationship between ROA and assets structure, which means that higher fixed assets-to-assets ratio brings worse accounting performance. There is significantly negative relationship between firm size and Tobin's Q, while there is significantly positive relationship between firm size and ROA, which means larger firm size will improve the accounting performance but destroy the market performance of listed SMEs in China. Firm growth and R&D intensity both have no significant influence on market performamnce and accounting performance of listed SMEs in China.

TABLE IV
REGRESSION RESULTS

	BI			Tobin's Q			ROA		
(constants)	0.341*** (0.000)	0.338*** (0.000)	0.355*** (0.000)	6.525*** (0.000)	6.621*** (0.000)	6.512*** (0.000)	-0.237*** (0.000)	-0.236*** (0.000)	-0.234*** (0.000)
BI				1.968 (0.591)	1.951 (0.556)		-0.178 (0.543)		-0.180 (0.301)
Tobin's Q	0.003 (0.265)		0.001 (0.423)						
ROA		-0.025 (0.451)	-0.028 (0.259)						
CR$_5$	0.031*** (0.000)	0.036*** (0.000)	0.038*** (0.001)		-1.377 (0.511)	-1.365 (0.631)		-0.074 (0.457)	-0.077 (0.231)
BI×CR$_5$					-2.605 (0.649)	-2.623 (0.667)		0.237 (0.385)	0.242 (0.562)
LEV	-0.001 (0.528)	-0.002 (0.453)	-0.002 (0.799)	-1.463*** (0.003)	-1.453*** (0.000)	-1.468*** (0.000)	-0.165*** (0.000)	-0.169*** (0.000)	-0.165*** (0.000)
AS	-0.001 (0.714)	-0.001 (0.864)	-0.001 (0.919)	-0.167 (0.491)	-0.160 (0.475)	-0.156 (0.490)	-0.052*** (0.000)	-0.057*** (0.000)	-0.050*** (0.000)
FS	0.002 (0.309)	0.001 (0.435)	0.000 (0.946)	-0.188*** (0.001)	-0.198*** (0.000)	-0.185*** (0.000)	0.022*** (0.000)	0.025*** (0.000)	0.021*** (0.000)
GR	0.002** (0.049)	0.001* (0.058)	0.001* (0.094)	0.013 (0.491)	0.016 (0.632)	0.011 (0.474)	-0.001 (0.127)	-0.001 (0.179)	-0.001 (0.364)
R&D	0.231 (0.423)	0.236 (0.325)	0.248 (0.212)	3.010 (0.552)	3.016 (0.543)	3.123 (0.576)	0.046 (0.849)	0.044 (0.653)	0.046 (0.253)
I	NO	NO	NO	YES	YES	YES	NO	NO	NO
Y	NO	NO	NO	YES	YES	YES	YES	YES	YES
F	45.150	49.150	42.150	48.572	48.865	48.572	45.079	46.521	45.364
Adjusted R^2	0.315	0.319	0.303	0.384	0.389	0.384	0.388	0.390	0.389

Coefficients marked with *, **, ***are significant at 10%, 5% and 1%, respectively. Q and ROA are Tobin's Q and return on assets; BI is board independence; CR$_5$ is ownership concentration; LEV represents finnacial leverage; AS is asset structure; FS is firm size, GR is firm growth; R&D is R&D intensity. I and Y are industry dummy variable and year dummy variable respectively.

V. CONCLUSIONS

High ownership concentration is a noticeable trait in listed SMEs in China. Ownership concentration can impact the independence of independent directors, and then can impact the firm performance indirectly. This paper investigates how the board independence impact on firm performance using a sample of listed firms on SMEs board in China during the period 2004-2010. Endogeneity test indicates that Tobin's Q and ROA both have no significant relationship with board independence, which means that board independence and firm performance of listed SMEs are not endogenous. The regression results based on the perspective of exogeneity indicate that board independence has no significant influnece on both Tobin's Q and ROA, which implies that board independence has no significant influence on firm performance. There is significantly positive relationship between ownership concentration and board independence, while there is no significant relationship betweeen ownership concentration and performance of listed SMEs in China. The results means that the ownership concentration has no impact on performance of Chinese listed SMEs.

ACKNOWLEDGMENT

We acknowledge financial support from National Natural Science Foundation of China(No. 71002104) and Tianjin University Innovation Funds.

REFERENCES

[1] X. Y. Jiang, "The governance of listed private companies on SMEs board in China", *Working Paper of Shenzhen Stock Exchange*, 2010-7-7

[2] B. E. Hermalin, M. S. Weisbach, "The effects of boards composition and direct incentives on firm performance", *Financial Management*, vol. 20, no.2, pp. 101-112, 1991.

[3] H. Mehran, "Executive compensation structure, ownership, and firm performance", *Journal of Financial Economics*, vol. 38, no. 2, pp. 163-184, 1995.

[4] A. Klein. "Firm performance and board committee structure", *Journal of Law and Economics,* vol. 41, pp.275-299, 1998.

[5] J. Erickson, Y. W. Park, J. Reising, H. H. Shin, "Board composition and firm value under concentrated ownership: the Canadian evidence", *Pacific-Basin Finance Journal*, vol. 13, no.4, pp. 387-410, 2005.

[6] S. Bhagat, B. Black, "The non-correlation between board independence and long-term firm performance", *Journal of Corporation Law*, vol. 27, pp. 231-274, 2001.

[7] A. Agrawal, C. R. Knoeber, "Firm performance and mechanisums to control agency problems between managers and shareholders", *Journal of Financial and Quantitaive Analysis,* vol. 31, pp. 377-397, 1996.

[8] A. Dehaene, V. D. Vuyst, H. Ooghe, "Corporate performance and board structure in Belgian companies", *Long Range Planning ,*vol.34, pp. 383-398, 2001.

[9] V. Krivogorsky. "Ownership, board structure, and performance in continental Europe", *The International Journal of Accounting,* vol. 41, pp. 176-197, 2006.

[10] F. Lefort, F. Urzua, "Board independence, firm performance and ownership concentraton: evidence from Chile", *Journal of Business Research*, vol.61, no.6, pp.615-622, 2008.

[11] Y. T. Wang, Z. Y. Zhao, X. Y. Wei, "Does independence of the board affect firm performance?", *Economic Research*, no. 5, pp. 62-72, 2006.

[12] W. A. Li, W. Sun, "An empirical study on accumulation effect of board govenance upon corporate performance: the evidence from China listed companies", *China Industrial Economy*, no. 12, pp.77-84, 2007.

[13] Y. H. Hao, Y. X. Zhou, "Board structure, corporate governance and performance: the empirical evidence based on dynamic endogeneity", *China Industrial Economy*, no. 5, pp.110-120, 2010.

Study on Prediction of Urbanization Level Based on GA-BP Neural Network
——Taking Tianjin of China as the Case

Gang Hao [1, 2]

[1] Department of Management Science, Tianjin University of Finance and Economics, Tianjin, China
[2] Department of Finance, Tianjin University, Tianjin, China
(hotertony@126.com)

Abstract - **In this paper, we establish evaluation index system of urbanization level, The three main factors and comprehensive factor score are obtained by factor analysis, we impot the BP neural network improved by the genetic algorithm. Urbanization level three variables prediction model and single variable prediction model are established respectively, Through the comparative analysis of different forecast models found that modified BP neural network three variables prediction model by genetic algorithm is superior to other prediction model in the aspect of nonlinear fitting capability and prediction precision, finally, we utilize the model to make a short-term prediction for the urbanization level of Tianjin.**

Keywords - **BP neural network, factor analysis, genetic algorithm, urbanization level**

I. INTRODUCTION

As one of the core content of modern and marks, urbanization is an important part of the process of human social development [1], it is the objective requirement of the development of the modern market economy, and recognized as one of the most significant social and economic phenomenon in the 20th century widely. Research shows that urbanization is helpful to boost domestic demand and promote economic structural adjustment, improve the economic benefit, rational utilization of resources, significantly reduce the income gap between urban and rural areas [2] and stimulate economic growth [3]. But urbanization also brings a series of social problems and ecological environment problems [2, 4], Such as the loss of the farmers on the land and the increase of carbon emissions [5] and air pollution is aggravated and so on [6]. Therefore, Scientific forecasting the future development of urbanization level has a strong practical significance to develop the urban development planning and urban construction planning [7].

Some scholars establish different kind of urbanization level forecast models on the basis of econometric theory, are mostly linear model [8, 9]. However, in the process of the urbanization level forecast, nonlinear, time-varying and uncertainty widely exist [10], it is difficult to set up linear, determined functional relation, The linear model is difficult to grasp of nonlinear phenomena the urbanization level system, inevitably leads to the error of the prediction. Although some scholars did a lot of improvement in linear models, such as building phase linear model and parameter time-varying linear model, but the result is not ideal. So, the researchers begin to seek some nonlinear tools to establish the prediction model. The BP neural network possesses the advantages of parallel computing, distributed information storage, fault tolerant ability, adaptive learning function, etc [11]. It has superiority in dealing with nonlinear problem than econometrics, Compared with traditional methods, it describes the function relationship is more close to actual [12, 13], the prediction accuracy is higher, the reliability is stronger [14]. So some scholars put forward on the basis of BP neural network to establish urbanization level prediction model [10, 15]. But as a result of the BP neural network training algorithm is the gradient descent method in essence, it is more complex to minimize the objective function, this leads to its two defects [16]: the learning process of slow convergence speed; easy to fall into local minimum point. In order to overcome the above defects, some scholars proposed the improved optimization algorithm, improving training function [17] and simulated annealing method [18], and so on. Although the above methods also has certain feasibility, but always can't completely solve these defects. Therefore, this study attempts to adopt GA to optimize the BP neural network. Genetic algorithm has strong adaptability and robustness, especially suitable for high dimension, multipolar point not differentiable, continuous or discrete space to global search optimization solution [19]. But the shortcoming of this algorithm is unable to accurately determine the position of the optimal solution nearby the optimal solution, the ability of its local search space of fine-tune is poorer, and the BP neural network for local search more effective. We can first use genetic algorithm to optimize, when the search gets narrow, we can use the BP neural network make an accurate solution. So, using genetic algorithm to optimize the BP neural network is the perfect tool to escape from local minimal value, and to accelerate the network convergence speed.

Therefore, in order to accurately predict the level of urbanization, this paper constructs the evaluation index system of urbanization level, The three main factors and comprehensive factor score are obtained by factor analysis, we impot the BP neural network improved by the genetic algorithm. Urbanization level three variables prediction model and single variable prediction model are established, respectively, through the comparative analysis of different forecast models to determine the optimal prediction model, and we make a short-term prediction for the urbanization level of Tianjin.

E. Qi et al. (eds.), *Proceedings of the 21st International Conference on Industrial Engineering and Engineering Management 2014*, Proceedings of the International Conference on Industrial Engineering and Engineering Management, DOI 10.2991/978-94-6239-102-4_105, © Atlantis Press and the authors 2015

II. GA-BP NEURAL NETWORK MODEL

BP neural network, at the beginning of the training will link weights and threshold and be initialized a random number between [0, 1], this without optimization random initialization can result in slow convergence speed, the search process is easy to fall into local optimal solution. Therefore, introducing GA to optimize the initial weights and threshold of neural network can make the model has good astringency and exactness. Specific steps for:

Step1: Given population size as P, $W = (W_1, W_2, \ldots W_p)^T$ is the initial population of randomly generated individuals, due to the determination of initial population has a great influence on the global optimization of GA, so using linear interpolation function to generate a real vector $w_1, w_2 \ldots w_p$ of individuals W_i as a chromosome. It's length is

$$S = RS_1 + S_1 S_2 + S_1 + S_2 \qquad (1)$$

where R is the number of input layer neurons, S_1 is the number of hidden layer neurons, S_2 is the number of output layer neurons, every individual of population $W_i = (w_1, w_2, \ldots w_s)$ $i = 1, \ldots P$ represents the BP neural network weights and thresholds.

Step2: Determine the evaluation function of the individual.

The first step's chromosomes do an assign for the weights and threshold of BP neural network, input samples input the neural network, then, get a network output value $\hat{y}_i$, the response value $fitness$ and the average fitness value $\hat{f}$ are defined as follows:

$$fitness_i = \sum_{j=1}^{M-1} (\hat{y}_j - y_j)^2, (i = 1, 2, \cdots P) \qquad (2)$$

$$\bar{f} = \frac{\sum_{i=1}^{P} fitness_i}{P} \qquad (3)$$

Where $\hat{y}_i$ is the network output value, y_i is the desired output, P is the population size.

Step3: Selecting operation; Calculate the fitness value of each individual,

Adaptation degree proportion method was applied to selection of chromosomes in each generation of population, the individuals of bigger fitness have higher probability inherit to the next generation.

Select the probability of each chromosome p_i,

$$p_i = \frac{f_i}{\sum_{i=1}^{P} f_i}, (i = 1, 2, \cdots P) \qquad (4)$$

Where $f_i = 1/fitness_i$, P is the population size.

Step4: Interlace operation. Because individual adopts real coding, the Cross artificial methods use real number crossing method. The Crossover operation of the K gene W_k and the L gene W_l at j as follows:

$$\begin{cases} w_{kj} = w_{kj}(1-b) + w_{lj}b \\ w_{lj} = w_{lj}(1-b) + w_{kj}b \end{cases} \qquad (5)$$

Where b is the random number on the interval $[0,1]$.

Step5: Mutation operation. Select the i individual j gene to make mutation operation,

$$w_{ij} = \begin{cases} w_{ij} + (w_{ij} - w_{max})f(g), r \geq 0.5 \\ w_{ij} + (w_{min} - w_{ij})f(g), r < 0.5 \end{cases} \qquad (6)$$

$$f(g) = r_2(1 - \frac{g}{G_{max}}) \qquad (7)$$

where W_{max}, W_{min} are the upper and lower bound of W_{ij}, r is the random number on the interval $[0,1]$, r_2 is a random number, g is the current number of iterations, G_{max} is the biggest evolution algebra.

Step6: Insert the new individual in the population P, Calculate the new fitness function value of the individual.

Step7: If you don't find the optimal individual, you should return to Step3, if you find the optimal individual, it should be decoded as the links weights and threshold of BP neural network, using the training data to train BP neural network.

III. THE CONSTRUCTION OF EVALUATION INDEX SYSTEM OF URBANIZATION LEVEL

Reference to previous research results [20, 21, 22], following the pertinence, gradation and comprehensive, scalability, dynamic, the principle of comparability, availability, etc. On the basis of several rounds of screening for a large number of statistical index, we put forward the framework of evaluation index system of urbanization level, which is comprised of the economic level of urbanization, population urbanization level of urbanization and regional landscape, lifestyle urbanization level, the state of the environment index of urbanization level 5 class, 15 secondary indexes and 17 tertiary indicators. As shown in Table I:

TABLE I
THE EVALUATION INDEX SYSTEM OF URBANIZATION LEVEL

Objective level	Criterion layer	Indicator layer
Economic urbanization level	Economic strength	Per capita gross domestic product (yuan)
	Industrial structure	The proportion of GDP the third industry（%）
	Economic outgoing	The total freight（tons）
Population urbanization level	Population structure	The third industry accounted for the proportion of total employment（%）
	Population size	Density of population（People/km2）
		Natural population growth rate（%）
The urbanization level of regional landscape	Urban ecology	Green coverage ratio（%）
	Traffic condition	Per capita public transportation vehicles（vehicle）
Lifestyle urbanization level	Electric power supply	Annual per capita electricity consumption（kw/h）
	Gas Supply	Annual per capita living with natural gas and fuel gas volume（m³）
	Municipal water	Water consumption per capita（t）
	Cultural development	Per capita library collection（volume）
	Living condition	Per capita housing area（m²）
The state of the environment urbanization level	Environmental quality	Ambient air quality rate（%）
	Environment control	Treatment rate of domestic sewage（%）
		Industrial solid waste comprehensive utilization rate（%）
	Environmental build	Town life sewage treatment rate（%）

IV. EMPIRICAL ANALYSIS

(1) The original data collection

When the sample is selected to build the prediction model, first, sample data to be in a relatively fixed environment; followed by, sample selection should cover the information of the system characteristics, and provide more comprehensive boundaries information to give the neural network. In this paper, taking Tianjin city as the case, we select its data of the level of urbanization evaluation from 1995 to 2011, these data are from the Statistical Yearbook of Tianjin, China City Statistical Yearbook.

(2) Data analysis and variable selection

When we use the three main factors to measure the level of urbanization in Tianjin, it is difficult to intuitive comparison, so it need a level of urbanization to reflect a comprehensive index [23]. Therefore, this paper presents a comprehensive factor scores (see Table II in the fifth column), which can be used to measure the level of urbanization in a given year. Its computation formula is:

$$F = \frac{\lambda_1}{\lambda_1 + \lambda_2 + \lambda_3} F_1 + \frac{\lambda_2}{\lambda_1 + \lambda_2 + \lambda_3} F_2 + \frac{\lambda_3}{\lambda_1 + \lambda_2 + \lambda_3} F_3 \quad (8)$$

Where λ_i, $i = 1, 2, 3$ is the main factor variance contribution rate.

From this, we can use GA-BP Neural network prediction model of three-variable, based on the primary factor of urbanization level in previous years data forecasts for future years the level of urbanization of factor scores.

2009~2011 Years of data on the three main factors enter into factory trained GA-BP Neural networks can be obtained 2013 Years of data on the three principal components of the urbanization level of predictive values for, And then formula (8), You can get 2013 Annual comprehensive factor score of urbanization level of Tianjin forecast data 1.5328.

TABLE II
THE MAIN FACTOR SCORE AND COMPREHENSIVE FACTOR SCORE

year	F_1	F_2	F_3	F
1995	-1.2172	-1.6745	-1.0169	-1.2897
1996	-0.9790	-1.5977	-0.7224	-1.0787
1997	-0.8624	-1.0363	-0.4814	-0.8568
1998	-0.6115	-0.9410	1.4106	-0.459
1999	-0.4414	-0.6645	1.6713	-0.2571
2000	-0.8589	1.0663	0.4503	-0.3189
2001	-0.6495	0.9982	1.6069	-0.0635
2002	-0.7327	1.4636	0.5046	-0.1446
2003	-0.6100	1.1557	-0.1366	-0.194
2004	-0.1976	0.7801	-1.0941	-0.0936
2005	0.0134	0.8939	-1.1679	0.0663
2006	0.3428	0.6778	-0.6855	0.2998
2007	0.7323	0.3163	-1.3734	0.4169
2008	1.1451	-0.0279	-0.6656	0.7056
2009	1.4325	-0.4644	0.5755	0.9477
2010	1.6503	-0.4927	0.7156	1.1062
2011	1.8437	-0.4528	0.4090	1.2134

V. CONCLUSIONS

First, compared to the classic BP neural network forecast model, based on genetic algorithm optimized BP neural network forecasting model of urbanization level of prediction, to accelerate model convergence, improve forecast accuracy.

Second, compared with single neural network prediction model of neural network prediction model of three variables although there are some disadvantages in convergence, its nonlinear fitting more powerful. Forecasts are more accurate.

Third, genetic algorithm to improve BP neural network prediction model of three variables can converge at a fast pace, and its most powerful nonlinear fitting, precision high nonlinear relationship in the process of urbanization can be predicted accurately describe them.

REFERENCES

[1] Chaolin Gu, Liya Wu. Progress in research on Chinese urbanization [J]. Frontiers of Architectural Research, 2012, 1(2): 101-149.

[2] David E. Bloom, David Canning, Günther Fink. Urbanization and the Wealth of Nations [J]. Science, 2008, 319(8): 772-775.

[3] Yan Han, Hualin Nie. China's urbanization level and the regional economic growth difference empirical research [J]. Urban problem, 2012(04):22-26.

[4] Yangfan Li, Yi Li, Yan Zhou. Investigation of a coupling model of coordination between urbanization and the environment [J]. Journal of Environmental Management, 2012, 98(5): 127-133.

[5] Madlener R., Sunak Y. Impacts of urbanization on urban structures and energy demand: What can we learn for urban energy planning and urbanization management? [J]. Sustainable Cities and Society, 2011, 1(1): 45–53.

[6] Qingsong Wang, Xueliang Yuan, Jian Zhang. Key evaluation framework for the impacts of urbanization on air environment – A case study [J]. Ecological Indicators, 2013, 24: 266-272.

[7] Jun Xiong.On some issues of the international comparison of China's urbanization - China's urbanization lag [J]. Population Science of China, 2009, (06): 32-40+111.

[8] Gang Ding, Pingping Zhao. The level of urbanization based on PDL model prediction [J]. Statistical research, 2005, (03):45-48.

[9] Jianming Ren,Hui Sun. Provincial level of urbanization prediction methods comparison - Take as in Beijing [J]. Urban city, 2006, (03): 15-19+38.

[10] Hui Sun, Yuan Wang, Dongming Zhang. Urbanization in Tianjin based on neural network analysis and prediction [J]. Tianjin University (Social Science Edition), 2006, (01): 68-71.

[11] Wei Wang, Jingtong He, Jianxun Zhang. Artificial neural networks in nonlinear economic forecasting [J]. Journal of Systems Engineering, 2000, (02): 202-207.

[12] Song Li, Lijun Li, Yongle Xie. Genetic algorithm to optimize BP neural network to predict short-term traffic flow Chaos [J]. Control and Decision, 2011, (10): 1581-1585.

[13] Xinghui Zhang, Shengzhi Du, Zengqiang Chen. Principal Component Analysis in Neural Networks economic prediction [J]. Quantitative & technical economics, 2002, (04): 122-125.

[14] Rasit K. Design and performance of an intelligent predictive controller for a six-degree-of-freedom robot using the Elman network [J]. Information Science, 2006, 176(12): 1781-1799.

[15] Zhiyi Guo, Gang Ding. Urbanization Prediction Methods - to apply the BP neural network model as an example [J]. Population and economy, 2006, (06): 3-8.

[16] Melanie T. Young, Susan M. Blanchard, Mark W. White. Using an artificial neural network to detect activations during ventricular fibrillation [J]. Computers and Biomedical Research, 2000, 33(1): 43-58.

[17] Vogl T. P., Magis J. K., Zigler A. K., Accelerating the convergence of the back-propagation method [J]. Biological Cybernetics, 1988, 59(4): 246-264.

[18] Shuangyin Liu. Economic forecasting models of immune neural network artificial fish [J]. Computer Engineering and Applications, 2009, (29): 226-229.

[19] Zhaoyang Chen, Lequn Hu, Hequn Wan. Establishing of economic forecasting model based on the neural network genetic algorithm [J]. Forecast, 1997, (01): 69-71.

[20] Min Wei, Guoping Li. Based on the comprehensive level of urbanization TAL systems Measurement Model [J]. Systems engineering, 2004, (07): 50-55.

[21] Sujie Meng, Xu Huang. Evaluation index system of rural urbanization in Beijing [J]. Urban city, 2004, (04): 40-44. (Chinese)

[22] Puming Liang. Research and Measurement of the particularity of Chinese urbanization [J]. Statistic research, 2003, (04): 9-15.

[23] Gengtian Zhang. Discussion on the establishment of the index system of urbanization [J]. Urban problems, 1998, (01): 6-9.

Research on the Effect of the Grain Financial Direct Subsidies Based on Grey Incidence Analysis

Xiao-zhuo Wei [1,2,*], Li-fu Jin [1], Jun-min Wu [2]

[1] School of Management, Jiangsu University, Zhenjiang 212013, China

[2] School of Economics and Management, Jiangsu University of Science and Technology, Zhenjiang 212003, China

(wwwxxx0915@sina.com)

Abstract - **This paper examined the closeness of the incidence between the grain output and the four grain financial direct subsidies as well as between farmers' income and the four subsidies through the grey incidence analysis method. The empirical results showed that the grain direct subsidies, the subsidies for growing superior grain cultivators and the general subsidies for agricultural production supplies among the four subsidies had strong incidence with the total grain output and the per capita agricultural income, while the subsidies for purchasing agricultural machinery and tools had weak incidence. We also analyzed possible reasons for the weak incidence of the subsidies for purchasing agricultural machinery and tools and provided suggestions to increase its incidence.**

Keywords - **Degree of grey incidence, decision analysis, effect of the subsidies, grain financial direct subsidies**

I. INTRODUCTION

The grain financial direct subsidies have been the focus of the Chinese agricultural subsidies policy, consisting of four subsidies: (1) the grain direct subsidies, (2) the subsidies for growing superior grain cultivators, (3) the subsidies for purchasing agricultural machinery and tools and (4) the general subsidies for agricultural production supplies for the purpose to guarantee the grain security and increase the farmers' income. It is important to evaluate how well the subsidies have worked and whether it's necessary to adjust the structure ever since the grain financial direct subsidies were granted in 2004. The existing research has been focused mainly on the grain output (the production increment) [1]-[2], the farmers' income (the income increase) [3]-[4] or the combination of both (the production increment and the income increase) [5]-[6] which led to many valuable suggestions and comments on the policy structure and the system construction [7]-[9]. However, few quantitative studies have been conducted in China on the degree of incidence of the grain output and the farmers' income with the grain financial direct subsidies. In fact, a search of the China National Knowledge Infrastructure (CNKI) showed that only two relevant studies [10]-[11] have been published till the end of June, 2014. In this paper, based on the statistic data from the "China Rural Statistical Yearbook-2013" [12] and other yearbooks, we applied the grey incidence analysis method to investigate the closeness of the incidence between the grain output and the four grain financial direct subsidies as well as between farmers' income and the four subsidies. The purpose is to identify weaknesses for further improvement so as to optimize the structure of the grain financial direct subsidies and realize the policy goals.

II. THE GREY INCIDENCE ANALYSIS MODELLING

The grey incidence analysis method in the grey system theory focuses on small samples and poor information. It allows valuable information to be extracted from the mining of some known information. Thus, it is suitable for policy evaluation, particularly for the early stage of policy implementations. This method will cater for the situation where the grain financial direct subsidies in China were only implemented for a short time and large scale panel data have not been accumulated. Next we briefly describe the steps of grey incidence analysis [13]:

The first step is to determine the parent sequence $X_0(k)$ and the subsequence $X_i(k)$, and compute initial images of all sequences, i.e. have dimensionless method on the primary data. Let $X'_i=X_i / x_i(1)=(x'_i(1), x'_i(2), ..., x'_i(n))$, $i=0, 1, 2, ..., m$.

The second step is to compute difference sequences. Set $\Delta_i(k)=| x'_0(k)-x'_i(k) |$, $\Delta_i=(\Delta_i(1), \Delta_i(2), ..., \Delta_i(n))$, $i=1, 2, ..., m$.

The third step is to compute the difference between the two extremes. Set $M = \max_i \max_k \Delta_i(k)$, $m = \min_i \min_k \Delta_i(k)$.

The fourth step is to compute the incidence coefficients. $\gamma_{0i}(k) = \dfrac{m + \xi M}{\Delta_i(k) + \xi M}$, $\xi \in (0, 1), k=1, 2, ..., n$; $i=1, 2, ..., m$.

The fifth step is to compute the degree of grey incidence. $\gamma_{0i} = \dfrac{1}{n} \sum_{k=1}^{n} \gamma_{0i}(k)$, $i=1, 2, ..., m$.

Generally, the degree of grey incidence can be grouped into three categories [14]-[15]: 0~0.35 means weak incidence, 0.35~0.7 means moderate incidence, while 0.7~1 means strong incidence.

III. THE DATA SOURCE OF THE TOTAL GRAIN OUTPUT, THE PER CAPITA AGRICULTURAL INCOME AND THE GRAIN FINANCIAL DIRECT SUBSIDIES

Because the 2014 versions of some yearbooks (such as the "China Rural Statistical Yearbook-2014" and the "Finance Yearbook of China (2014))" haven't been published as of the end of June, 2014, all of the yearbooks used in this paper are actually the 2013 versions with all of the data involved subject to the information as of the end of 2012. The relevant data of the total grain output (TGO (ten thousand tons)), the per capita agricultural income (PCAI (yuan)), the grain direct subsidies (GDS (billion yuan)), the subsidies for growing superior grain cultivators (SGSGC (billion yuan)), the subsidies for purchasing agricultural machinery and tools (SPAMT (billion yuan)), the general subsidies for agricultural production supplies (GSAPS (billion yuan)) and the total amount of the four subsidies (TAFS (billion yuan)) covered in the grain financial direct subsidies during 2004~2012 are shown in Table I.

TABLE I
THE RELEVANT DATA OF THE TOTAL GRAIN OUTPUT, THE PER CAPITA AGRICULTURAL INCOME AND THE GRAIN FINANCIAL DIRECT SUBSIDIES

Year	TGO	PCAI	GDS	SGSGC	SPAMT	GSAPS	TAFS
2004	46947	1056.5	116	28.5	0.7	null	145.2
2005	48402	1097.7	132	38.7	3	null	173.7
2006	49804	1159.6	142	41.5	6	120	309.5
2007	50160	1303.8	151	66.6	20	276	513.6
2008	52871	1427.0	151	123.4	40	716	1030.4
2009	53082	1497.9	151	198.5	130	795	1274.5
2010	54648	1723.5	151	204	154.9	716	1225.9
2011	57121	1896.7	151	220	175	860	1406
2012	58958	2106.8	151	199	215	1078	1643

Data source: data of the total grain output (TGO) has been obtained through the processing and calculation of the relevant data in the "China Rural Statistical Yearbook" (2013) with the data for the per capita agricultural income (PCAI) obtained from the processing and calculation of the relevant data in the "China Rural Statistical Yearbook" (2005~2013). As to the total amount of the four subsidies (TAFS) during 2004~2006, it has been processed and calculated according to the relevant data in the "China Grain Development Report" (2005~2007) with the total amount of the four subsidies (TAFS) during 2007~2012 processed and calculated according to the relevant data in the "China Rural Statistical Yearbook" (2013). Moreover, the annual categorical data of the four subsidies (GDS, SGSGC, SPAMT, GSAPS) have been obtained through the selection, processing and calculation of the relevant data from the "China Grain Development Report", the "China Rural Statistical Yearbook", the "Finance Yearbook of China" and the public news over the years.

IV. THE DEGREE OF GREY INCIDENCE ANALYSIS FOR THE TOTAL GRAIN OUTPUT & THE PER CAPITA AGRICULTURAL INCOME AND THE GRAIN FINANCIAL DIRECT SUBSIDIES

Taking the total grain output as an instance, we consider the total grain output (X_1) as the parent sequence (the reference sequence) for the grey incidence analysis with the four subsidies covered in the grain financial direct subsidies acting as the subsequence (the comparative sequence), which includes the grain direct subsidies (X_2), the subsidies for growing superior grain cultivators (X_3), the subsidies for purchasing agricultural machinery and tools (X_4) and the general subsidies for agricultural production supplies (X_5) for the grey incidence analysis.

Since the general subsidies for agricultural supplies were not granted until 2006, there was no data related to this index in Table I from 2004 to 2005 to make the acquired incidence data featured with no comparability with the other data. Hence, our analysis was based on the relevant data from 2006 to 2012. According to Table I, we could find that

The total grain output: $X_1=(x_1(1), x_1(2), x_1(3), x_1(4), x_1(5), x_1(6), x_1(7))=(49804, 50160, 52871, 53082, 54648, 57121, 58958)$;

The grain direct subsidies: $X_2=(x_2(1), x_2(2), x_2(3), x_2(4), x_2(5), x_2(6), x_2(7))=(142, 151, 151, 151, 151, 151, 151)$;

The subsidies for growing superior grain cultivators: $X_3=(x_3(1), x_3(2), x_3(3), x_3(4), x_3(5), x_3(6), x_3(7))=(41.5, 66.6, 123.4, 198.5, 204, 220, 199)$;

The subsidies for purchasing agricultural machinery and tools: $X_4=(x_4(1), x_4(2), x_4(3), x_4(4), x_4(5), x_4(6), x_4(7))=(6, 20, 40, 130, 154.9, 175, 215)$;

The general subsidies for agricultural production supplies: $X_5=(x_5(1), x_5(2), x_5(3), x_5(4), x_5(5), x_5(6), x_5(7))=(120, 276, 716, 795, 716, 860, 1078)$.

A. Compute initial images

By $X'_i = X_i / x_i(1)=(x'_i(1), x'_i(2), x'_i(3), x'_i(4), x'_i(5), x'_i(6), x'_i(7))$, $i=1, 2, 3, 4, 5$, we can get
$X'_1=(1, 1.0071, 1.0616, 1.0658, 1.0973, 1.1469, 1.1838)$; $X'_2=(1, 1.0634, 1.0634, 1.0634, 1.0634, 1.0634, 1.0634)$; $X'_3=(1, 1.6048, 2.9735, 4.7831, 4.9157, 5.3012, 4.7952)$; $X'_4=(1, 3.3333, 6.6667, 21.6667, 25.8167, 29.1667, 35.8333)$; $X'_5=(1, 2.3000, 5.9667, 6.6250, 5.9667, 7.1667, 8.9833)$.

B. Compute difference sequences

By $\Delta_i(k)=|x'_1(k)-x'_i(k)|$, $i=1, 2, 3, 4, 5$, we can get
$\Delta_2=(0, 0.0562, 0.0018, 0.0024, 0.0339, 0.0835, 0.1204)$; $\Delta_3=(0, 0.5977, 1.9119, 3.7173, 3.8184, 4.1543, 3.6114)$; $\Delta_4=(0, 2.3262, 5.6051, 20.6008, 24.7194, 28.0198, 34.6495)$; $\Delta_5=(0, 1.2929, 4.9051, 5.5592, 4.8694, 6.0198, 7.7995)$.

C. Compute the difference between the two extremes
$M = \max_i \max_k \Delta_i(k)=34.6495$; $m = \min_i \min_k \Delta_i(k)=0$.

D. Compute the incidence coefficients
If we set $\xi=0.5$, we can get
$$\gamma_{1i}(k) = \frac{m + \xi M}{\Delta_i(k) + \xi M} = \frac{17.32475}{\Delta_i(k) + 17.32475}, i=2, 3, 4, 5,$$
$k=1, 2, 3, 4, 5, 6, 7$, and then
$r_{12}(1)=1$, $r_{12}(2)=0.9968$, $r_{12}(3)=0.9999$, $r_{12}(4)=0.9999$, $r_{12}(5)=0.9980$, $r_{12}(6)=0.9952$, $r_{12}(7)=0.9931$;
$r_{13}(1)=1$, $r_{13}(2)=0.9667$, $r_{13}(3)=0.9006$, $r_{13}(4)=0.8233$, $r_{13}(5)=0.8194$, $r_{13}(6)=0.8066$, $r_{13}(7)=0.8275$;
$r_{14}(1)=1$, $r_{14}(2)=0.8816$, $r_{14}(3)=0.7556$, $r_{14}(4)=0.4568$, $r_{14}(5)=0.4121$, $r_{14}(6)=0.3821$, $r_{14}(7)=0.3333$;

$r_{15}(1)=1$, $r_{15}(2)=0.9306$, $r_{15}(3)=0.7793$, $r_{15}(4)=0.7571$, $r_{15}(5)=0.7806$, $r_{15}(6)=0.7421$, $r_{15}(7)=0.6896$.

E. Compute the degrees of grey incidence

$$\gamma_{12}=\frac{1}{7}\sum_{k=1}^{7}\gamma_{12}(k)=0.9976;\quad \gamma_{13}=\frac{1}{7}\sum_{k=1}^{7}\gamma_{13}(k)=0.8777;$$

$$\gamma_{14}=\frac{1}{7}\sum_{k=1}^{7}\gamma_{14}(k)=0.6031;\quad \gamma_{15}=\frac{1}{7}\sum_{k=1}^{7}\gamma_{15}(k)=0.8113.$$

F. Results Analysis

The grain direct subsidies have the greatest incidence with the total grain output, which has been less related with the subsidies for growing superior grain cultivators and the general subsidies for agricultural production supplies with the subsidies for purchasing agricultural machinery and tools having the least incidence.

Similarly, take the per capita agricultural income as the parent sequence with the grain direct subsidies, the subsidies for growing superior grain cultivators, the subsidies for purchasing agricultural machinery and tools and the general subsidies for agricultural production supplies acting as the subsequence. After making the calculation according to the above-mentioned steps, we shall obtain the degrees of grey incidence, which are 0.9820, 0.8879, 0.6035, and 0.8184, respectively. In this way, we could find out that the grain direct subsidies have the highest incidence with the per capita agricultural income, which has been less related to the subsidies for growing superior grain cultivators and the general subsidies for agricultural production supplies with the subsidies for purchasing agricultural machinery and tools having the weakest incidence.

According to the empirical results, among the four subsidies covered in the grain financial direct subsidies, three of them (i.e., the grain direct subsidies, the subsidies for growing superior grain cultivators and the general subsidies for agricultural production supplies) are strongly correlated to the total grain output and the per capita agricultural income, which at the same time are moderately correlated to one of the four subsidies (i.e., the subsidies for purchasing agricultural machinery and tools). These findings illustrate that in general, positive effects have been achieved through the grain financial direct subsidies from the perspective of grey incidence analysis.

In China, the grain direct subsidies guarantee the grain farmers' income from the perspective of the prices of the output products, and the general subsidies for agricultural production supplies guarantee it from the angle of the prices of the input products, which both belong to the comprehensive income subsidies; and the subsidies for growing superior grain cultivators improve the grain competitiveness from the perspective of the quality of the grain, and the subsidies for purchasing agricultural machinery and tools improve it from the angle of the grain production efficiency, which both belong to the production special subsidies. The empirical results reveal that among the comprehensive income subsidies, both the grain direct subsidies and the general subsidies for agricultural production supplies have contributed not only to the realization of the direct target (the increase in the farmer's income), but also to the achievement of the indirect goals (the guarantee of the grain security). Among the production special subsidies, both the direct targets (the guarantee of the grain security) and the indirect goals (the increase in the farmer's income) have been realized through the subsidies for growing superior grain cultivators. However, it doesn't work very well in the realization of the goals through the subsidies for purchasing agricultural machinery and tools.

V. POLICY SUGGESTIONS

The reason that there's weak incidence separately between the subsidies for purchasing agricultural machinery and tools and the total grain output & the per capita agricultural income is mainly because that:

The farmers haven't had a good cognition on this subsidy policy, which is not a general preferential policy with a long time lag. Meanwhile the subsidies haven't been featured with large scale and there's no guarantee on the fund for the management of the subsidies. Also the personnel in the farm machinery management department haven't been characterized with good service awareness and some of the subsidized agricultural machinery and tools haven't performed very well with weak adaptability to make it hard to balance the fairness and efficiency of the subsidy policy. In addition to these, together-conspired bidding is an outstanding problem during the public ceremony for wagging numbers and lotting, making it hard to supervise the prices of the subsidized agricultural machinery and tools. Meanwhile it's very hard to conduct the acceptance on such subsidized agricultural machinery and tools with unsatisfactory technical services and poor after-sales service having been provided.

In order to increase the degree of incidence of the subsidies for purchasing agricultural machinery and tools with the total grain output and the per capita agricultural income, we would like to suggest the following measures to be taken: (1) From the perspective of the policy environment. Strengthen the propaganda about the subsidy policies to guarantee the farmers' rights to know, to select and to supervise. Meanwhile improve the relevant policies and regulations to maintain the continuity of the policies in addition to the exploration on the diversified investment mechanism such as the introduction of private capital to work out and improve the credit policies and insurance policies for the purchase of agricultural machinery and tools so as to develop a subsidy method for the simultaneous development of revenue and finance. (2) From the perspective of the policy implementation. Increase the financial subsidies by working out different subsidy standards according to the specific conditions to increase the variety of the subsidized agricultural machinery and tools. Then during the budget presentation, make sure to allocate appropriate funds for the management work done by the administrative staff. The government should organize public biddings, publishing the bid-winning enterprises and the prices of the agricultural machinery and tools in addition to the severe punishment of those enterprises that have violated the regulations. For those farmers who have applied for the subsidies for the acquisition of agricultural machinery and tools, they should purchase the agricultural machinery and tools first and then apply for the subsidies to solve the problem of bidding collusion. Also it's necessary to improve the mechanism to handle the complaints on the biddings. (3) From the perspective of the agricultural mechanization development. Improve the planning for the mechanization development of the agriculture and strengthen the R&D of the new agricultural machinery and tools and the new technologies to optimize further the equipment structure of the agricultural machinery. Meanwhile it's necessary to demonstrate and popularize the new agricultural machinery and tools to guarantee the safety, applicability and reliability in addition to the improvement of quality of the after-sales service and the establishment of the scrapping and recalling system for the agricultural machinery and tools. Also make sure to increase the skill training for the operators of the new agricultural machinery and tools to enhance the production efficiency of the agricultural machinery and tools in addition to the promotion of the moderate-scale management on the lands to develop agricultural machinery cooperation for the purpose to bring the agricultural mechanization into full play.

ACKNOWLEDGMENT

This research is supported by: the National Natural Science Foundation of China (71303096), the National Social Science Foundation of China (11BSH067), the Fund for Humanities and Social Science Research Provided by the Ministry of Education (12YJA790060); Jiangsu University Innovation Project for Doctoral Students (CX10B_017X); the Key Subject of the Quality Project for the Research of Social and Scientific Application in Jiangsu Province (14SWA-002); the Key Pre-research Fund for the Research of Humanities and Social Science by Jiangsu University of Science and Technology (2010JG107J); the Project for the Research of the Social and Scientific Application in Zhenjiang City (ZSKZ [2014] NO. 16).

REFERENCES

[1] Y. F. Chen, Z. G. Wu, T. H. Zhu, L. Yang, G. Y. Ma, and H. P. Chien, "Agricultural policy, climate factors and grain output: evidence from household survey data in rural China (Periodical style)," *Journal of Integrative Agriculture*, 2013 (1): 169–183.

[2] W. R. Zang, "The impact evaluation study of Chinese food financial direct subsidy policies on food quantitative security (Thesis or Dissertation style)" (in Chinese), Ph.D. dissertation, Ya' an: Sichuan Agricultural University, 2012.

[3] H. J. Gu, *Peasants' role differentiation and the income distribution effect of agricultural policy* (Book style) (in Chinese), Beijing: China Social Sciences Press, 2013.

[4] J. Han, "Research for farmers' income and subsidy policy in major grain-producing areas in China (Thesis or Dissertation style)" (in Chinese), Ph.D. dissertation, Beijing: Chinese Academy of Agricultural Sciences, 2010.

[5] S. P. Zhang, "Institutional research of the coordination between grain growth and increasing farmers' income in China (Thesis or Dissertation style)" (in Chinese), Ph.D. dissertation, Beijing: Party School of the Central Committee of C.P.C, 2011.

[6] S. Zheng, D. Lambert, S. S. Wang, and Z. G. Wang, "Effects of agricultural subsidy policies on comparative advantage and production protection in China (Periodical style)," *Chinese Economy*, 2013 (1): 20–37.

[7] M. D. Zhu, G. Q. Cheng, "China's agricultural policies its level of support, its effect of subsidy and charactaristics of its structure (Periodical style)" (in Chinese), *Management World*, 2011 (7): 52–60.

[8] J. G. Zhan, "The performance evaluation and system frame of grain subsidy policies in our country (Thesis or Dissertation style)" (in Chinese), Ph.D. dissertation, Changsha: Hunan Agricultural University, 2012.

[9] J. K. Huang, X. B. Wang, R. Scott, "The subsidization of farming households in China's agriculture (Periodical style)," *Food Policy*, 2013 (8): 124–132.

[10] S. R. Zhang, H. Y. Li, "Grey incidence analysis of agricultural subsidy structure based on the grain output and grain income (Periodical style)" (in Chinese), *Development & Research*, 2011 (1): 86–89.

[11] X. Zhou, J. Huang, "Gray incidence analysis of production subsidy and income subsidies impact on food security (Periodical style)" (in Chinese), *Hunan Agricultural Machinery*, 2012 (5): 111–113.

[12] Rural Social and Economy Investigation Division of National Bureau of Statistics, *China Rural Statistical Yearbook—2013* (Book style) (in Chinese), Beijing: China Statistics Press, 2013.

[13] S. F. Liu, N. M. Xie, *Grey System Theory and Its Application (Version VI)* (Book style) (in Chinese), Beijing: Science Press, 2013.

[14] Z. H. Cao, J. M. Hao, L.T. Liang, "Gray incidence analysis on major grain output and input elements of Huang-Huai-Hai Plain (Periodical style)" (in Chinese), *Research of Agricultural Modernization*, 2008 (3): 310–313.

[15] J. G. Fan, "Gray incidence analysis on grain output and its main input elements in Shaanxi province from 1983 to 2004 (Periodical style)" (in Chinese), *Agricultural Research in the Arid Areas*, 2007 (3): 209–212.

The Role of External Knowledge Search in Firms' Innovation Performance: Evidence from China

Hang Wu

Business School, East China University of Political Science and Law, Shanghai, China
(wuhang0503@163.com)

Abstract - **Open innovation has emerged as the main innovation mode in the era of knowledge-driven economy. However, existing research pays little attention to the precise classification of international and local knowledge search, and ignores their joint impact on innovation performance. This study examines how international knowledge search and local knowledge search independently and jointly help enhance innovation performance in China's emerging economy. A survey of 219 firms indicates that international knowledge search and local knowledge search are positively related to innovation performance. Specifically, local and international knowledge search interact to positively predict innovation performance. Theoretical and managerial implications are discussed.**

Keywords - **External knowledge search, international knowledge search, innovation performance, local knowledge search**

I. INTRODUCTION

With the increase of technology change speed and competitive intensity, innovation has evolved as a vital development strategy for firms to survive in the world market. Through introducing new products, firms can meet customer's need and sustain market shares [1]. However, innovation is a kind of costly and risky activity, which challenges the firm's resource stock and external knowledge search capability [1][2][3].

Existing academic studies have found that external knowledge, knowledge that spans a firm's boundaries, is vital for success in innovation [4][5]. Chesbrough [6] puts forward the concept of open innovation to describe the importance of knowledge search outside the firm boundary, such as introducing advanced technology, and recruiting employees in the world market. The central idea is that open search for new knowledge helps a firm access external sources of knowledge, overcome the risk of blind spots and avoid unexpected changes in the market and technology [6] [7] . Using a large-scale sample of industrial firms, Laursen and Salter [2] find that searching widely and deeply is curvilinearly (taking an inverted U-shape) related to innovation performance. Based on Chinese sample, Chen et al. [8] discover the same result. In spite of the large body of evidence documenting the influence of external knowledge search on innovation outcomes, most of previous studies have focused their attention just outside the firm boundary, without a precise classification on the outside sources of knowledge. In fact, external knowledge search can be classified in terms of scope: international knowledge search and local knowledge search [9]. Both search strategy sometimes interact with each other. Therefore, existing research does not uncover the outside knowledge sources and explore the independent and joint effect of international knowledge search and local knowledge search on firms' innovation.

To address these research gaps, this article examines how international knowledge search and local knowledge search independently and jointly are related to innovation performance of Chinese manufacturing firms. We argue that entering into international market can help firms obtain novel resources and learning opportunity which finally will promote innovation, searching locally provide firms with cost and speed advantage. Most importantly, both knowledge searching source complement each other in predicting better product innovation. This research contributes to existing open innovation literature by classifying the knowledge source and examining the interacting effect of both external knowledge source. The research result can help guild the external knowledge search and improve innovation performance.

II. THEORY AND HYPOTHESES

Although researchers in open innovation field do not make a precise classification on external knowledge source, research in knowledge search gives some insight. Scholars have identified two main external search strategies: local search and boundary-spanning search. Local search refers to immediate refinement and proximate alternatives to the firm's R&D activity, searching for solutions in a firm's current geographic and technological vicinity [10]. Boundary-spanning search involves searching for solutions beyond the neighbourhood, which allows firms to access more technological opportunities and acquire new technologies that are not available through local search, emphasizing experimental and distant innovation [11] [12]. In this study, we refer local knowledge search to search for new knowledge within a firm's national boundaries, refer international knowledge search to search internationally for new knowledge.

A. *International Knowledge Search and Innovation Performance*

Research in international business has found that entering into international markets can provide firms in emerging countries with a lot of valuable innovation resources [13][14][15]. Especially with the rise of faster and cheaper communication technologies and integration of global markets, innovative knowledge is becoming more globally dispersed and firms are increasingly willing to operate globally. Through foreign network partners, firms in emerging countries take advantage of key sources of innovation knowledge from around the world to launch their products more quickly [16]. Kafouros et al. [17] argue that firms can establish cooperation ties with foreign suppliers, competitors, customers and intermediaries to enter into international knowledge source. Zahra et al. [14] empirically discover that foreign cooperation ties enhance the depth, scope and speed of technology learning. Kafouros and Forsans [18] argue that knowledge sourced internationally is novel and sophisticated, and can greatly improve innovation when integrated into firm's existing knowledge pool. Accordingly, we propose:

Hypothesis 1: International knowledge search has a positive relationship with innovation performance.

B. Local Knowledge Search and Innovation Performance

Although international knowledge search has been emphasized as an important way to enhance innovation performance, there is an alternative, less-explored perspective that highlights the role of local network partners. In fact, firms tend to search locally, because of the greater ease and lower cost. In their study of small Argentine firms, Mesquita and Lazzarini [19] find that collaboration with local suppliers, competitors, and customers creates collective efficiencies that help firms overcome internal infrastructure limitations, obtain complementary resources, and cospecialize their resources and competencies to create cost-based competitive advantages and faster product innovation. Zhang and Li [20] demonstrate that ties with local service intermediaries broaden the scope of firm's external innovation search and reducing their search cost, which finally enhance innovation performance. Patel et al. [21] argue that local knowledge search create a kind of geographic proximity based advantage, enhancing innovation performance resulting from the increased access to knowledge via spillovers. As innovation is an interactive process and the exchange of tacit information is favored by face-to-face contact, local network collaboration can be vital for technology-based firms [22]. Accordingly, we propose:

Hypothesis 2: Local knowledge search has a positive relationship with innovation performance.

C. The Joint Impact of International and Local Knowledge Search on Innovation Performance

International knowledge search and local knowledge search complement each other in contributing to better innovation performance. On the one hand, international knowledge search can help firms avoid the competence trap resulting from local knowledge search. Although local knowledge search can provide firms with cost and speed advantage in product innovation, such knowledge search is restricted to the local sources [12]. Excessive emphasis on local knowledge search can result in competence trap and core rigidities. International knowledge search can provide firms with diverse and novel knowledge, unavailable in domestic market, which helps a firm recognize the importance of new foreign technology and consider how can be combined to improve a firm's innovation performance [12]. On the other hand, local knowledge search decreases the uncertainty of international knowledge search. Compared to local knowledge search, firms searching internationally face a lot of risk and uncertainty because of the different institutional contexts, unknown knowledge source and unexpected outcome [23]. These problems raise serious challenges for firms relying on international search. Local search can alleviate these problems by encouraging firms to rapidly internalize the knowledge it gains from international sources and adapt it to local customer need. Accordingly, we propose:

Hypothesis 3: International and local knowledge search interact to promote innovation performance.

III. METHODOLOGY

A. Sample and Data Collection

Our data was obtained from questionnaires sent to manufacturing firms mainly in Shanghai, Beijing, Zhejiang, Hubei, Jiangsu province, China. We chose China as a research laboratory because it is the world's largest emerging economy and possesses an unprecedented development pace. In order to survive and succeed in this rapidly changing market, Chinese manufacturing firms are under great pressure to search out locally and internationally to acquire advanced technologies and new ideas [8] [24]. Therefore, intense competition, open innovation culture and international orientation all make Chinese manufacturing firms particularly suitable for studying the relationship between local and international knowledge search and innovation. We selected the top managers from each of the sampled firms as the key informant because they can provide more reliable and valid data. Following various efforts, including both formal and informal contacts with the selected firms, we received 219 valid questionnaires, with a response rate of 41.71%. Such a high response rate reflects our extensive use of local social networks. Among the responding firms, 29.68% were in the electronic information, 20.55% in special equipment manufacturing, 12.33% in transportation equipment manufacturing, 11.42 percent in ordinary machinery manufacturing, 9.13% in metal product, and 16.89% classified as others (e.g., textile and clothing).

Following the approach recommended by Podsakoff et al. [25] and Chang et al. [26], we use Harman's one-factor test to check for the presence of common method variance. We subjected all the key measures to a factor analysis and then determined the number of factors accounting for the variance in the measures. The results of the tests indicate no single factor accounted for a majority of the variance, which showed no sign of a common method bias.

B. Measures

1) Dependent variable

Innovation performance (IP) is the dependent variable. We referred to the understanding and measurement items of Chen et al. [8] and Zhang and Li [20] and used six items to measure innovation performance. We asked them to rate the extent to which their firms were successful relative to their major competitors in terms of: number of new products; the ratio of new products sales to total sales; the speed of new product development; the success ratio; the number of patent applications; and the novelty of new product.

2) Independent variables

International knowledge search (IKS) and local knowledge search (LKS) are the independent variables. On the basis of work by Laursen and Salter [2] and Chen et al. [8] to measure international and local knowledge search, we asked the respondents to indicate the extent to which their firms had close relationships with international or local (1) customers, (2) competitors, (3) suppliers, (4) universities, and (5) innovation intermediary, such as talent search firms, law firms, technology service firms, and so on. International knowledge search was quantified by combining data on five international sources of knowledge or information used in product innovation; local knowledge search was quantified by combining data on five local sources of knowledge or information used in product innovation.

3) Controls

We controlled for the following variables. First, we use the natural log of the number of full-time employees to measure firm size (FS). Second, we use the number of years since the firm was founded to measure firm age (FA). Third, a dummy variable was created for state

ownership (SO), taking the value 1 if the firm was controlled by state capital, and zero otherwise. Fourth, R&D intensity (R&D) is measured as R&D expenditures as a percentage of sales (R&D/sales). Finally, we introduce five industry dummy to control the industry effect.

IV. EMPIRICAL RESULTS

A. Assessment of the Measures

Table I presents the means, standard deviations, and correlations among the variables examined in the study. International knowledge search ($r=0.436$, $p<0.01$) and local knowledge search ($r=0.377$, $p<0.01$) are positively correlated with innovation performance. International knowledge search is positively correlated with local knowledge search ($r=0.163$, $p<0.05$).

TABLE I
DESCRIPTIVE STATISTICS AND PEARSON CORRELATIONS

	Mean	SD	1	2	3	4	5	6	7
1.FS	7.56	1.40	1						
2.FA	15.54	9.16	.405**	1					
3.R&D	0.05	0.02	.003	.030	1				
4.SO	0.36	0.48	-.018	-.046	-.026	1			
5.IKS	3.99	1.37	.151*	.026	.250**	-.064	1		
6.LKS	4.60	1.06	.147*	-.060	.236**	.045	.163*	1	
7.IP	3.98	1.66	.211**	.015	.365**	-.027	.436**	.377**	1

**$p<0.01$; *$p<0.05$ (two-tailed test).

B. Reliability and Validity of the Constructs

We assessed the reliability of the multi-item constructs with Cronbach's alpha. The alpha values for international knowledge search is 0.964, and for local knowledge search is 0.921, and for innovation performance is 0.985, which indicate a good reliability (see Table II).

TABLE II
MEASUREMENT SCALES AND PROPERTIES

Constructs/Measurement items	Standardized loadings	AVE	Cronbach's alpha
International knowledge search			
1.Establishing friendly cooperation relationship with foreign customers	0.898		
2.Monitoring daily operation and innovative behavior of foreign competitors actively	0.893		
3.Establishing healthy cooperation ties with foreign suppliers or vendors	0.914	0.846	0.964
4. Establishing tight cooperation ties with foreign university and research institutions	0.905		
5. Establishing tight cooperation ties with foreign innovation intermediary	0.985		
Local knowledge search			
1.Establishing friendly cooperation relationship with local customers	0.848		
2.Monitoring daily operation and innovative behavior of local competitors actively	0.828		
3.Establishing healthy cooperation ties with local suppliers or vendors	0.850	0.700	0.921
4. Establishing tight cooperation ties with local university and research institutions	0.818		
5. Establishing tight cooperation ties with local innovation intermediary	0.840		
Innovation performance			
1.The number of new products	0.954		
2.The ratio of new products sales to total sales	0.961		
3.The speed of new product development	0.949	0.916	0.985
4.The success ratio	0.955		
5.The number of patent applications	0.968		
6.The novelty of new product	0.956		

All standardized coefficient loadings are significant at $p<0.001$.
AVE = Average variance extracted for each multi-item construct in the research model.

We conducted a confirmatory factor analysis to assess the convergent and discriminant validity of the multi-item constructs. Our three-factor confirmatory factor analysis model fits the data well, with all indices meeting the respective criteria ($\chi^2 = 193.299$; $\chi^2/df =$

1.914, NNFI = 0.973, comparative fit index (CFI) = 0.980, incremental fit index (IFI) = 0.980, root mean square error of approximation (RMSEA) = 0.065). All items loaded significantly on their corresponding latent construct, thereby providing evidence of good convergent validity, as shown in Table II. The results also support the discriminant validity of international knowledge search, local knowledge search and innovation performance, as the AVE of each construct was far greater than the corresponding inter-construct squared correlations.

C. Multiple Regression Analysis

Table III presents the results of the hierarchical multiple regression models. Model 1 only included controls. Model 2 added the main effects of international knowledge search. Model 3 added the main effects of local knowledge search. Model 4 added both main effects of international knowledge search and local knowledge search. Model 5 added the interaction term of international knowledge search and local knowledge search in model 4. To reduce the potential problem of multicolinearity, international knowledge search and local knowledge search were mean-centered prior to the creation of interaction terms [27]. All the Max VIF (variance inflation factor) is far less than 10, which indicate multicollinearity is not a problem in our study.

TABLE III
RESULTS OF STANDARDIZED REGRESSION ANALYSIS

	Model 1	Model 2	Model 3	Model 4	Model 5
Ind 1	0.036	0.019	0.009	-0.004	-0.004
Ind 2	0.083	0.048	0.027	-0.001	-0.036
Ind 3	-0.055	-0.046	-0.095	-0.083	-0.090
Ind 4	0.083	0.061	0.029	0.014	-0.021
Ind 5	0.068	0.057	0.020	0.014	-0.006
FS	0.253***	0.196**	0.192**	0.144*	0.114¥
FA	-0.106	-0.082	-0.074	-0.055	-0.081
R&D	0.382***	0.297***	0.316***	0.242***	0.150*
SO	-0.015	0.004	-0.030	-0.010	0.021
IKS		0.329***		0.310***	0.292***
LKS			0.273***	0.249***	0.180**
IKS×LKS					0.278***
R²	0.202	0.299	0.266	0.352	0.405
F	5.892***	8.883***	7.538***	10.218***	11.673***
Max VIF	2.250	2.253	2.261	2.263	2.263

*** p<0.001; ** p<0.01; * p<0.05; ♀ p<0.1 (two-tailed test).

Hypothesis 1 proposes that international knowledge search is positively related to innovation performance. The results in Model 2 suggest that international knowledge search is positive and significant (b=0.329, p<0.001). Hypothesis 1 is supported. Hypothesis 2 proposes that local knowledge search is positively related to innovation performance. The results in Model 3 suggest that local knowledge search is positive and significant (b=0.273, p<0.001). Hypothesis 2 is supported. The results in model 4 also proves that international knowledge search (b=0.310, p<0.001) and local knowledge search (b=0.249, p<0.001) are positively related to innovation performance. Hypothesis 1 and hypothesis 2 are supported again.

Hypothesis 3 proposes that international knowledge search and local knowledge search interact to promote innovation performance. The results in Model 5 suggest that the interaction term is positive and significant (b= 0.278, p<0.001). Hypothesis 3 is supported.

V. DISCUSSION AND CONCLUSION

A. Main Findings

The importance of external knowledge search has been demonstrated in a lot of theoretical and empirical studies on open innovation [2] [6]. However, existing research doesn't make a precise classification on external knowledge search and ignores to examine the independent and joint effect of each external knowledge search. In view of this, this study examined the relationships between international knowledge search, local knowledge search and their innovation performance.

With data on a sample of manufacturing firms in China, we found that international knowledge search and local knowledge search have significant positive relationships with firms' innovation performance. This finding is consistent with open innovation studies of Laursen and Salter [2] and Chen et al. [8]. However, this study deepens existing studies and finds that firms can not only search locally, but also search in the international market.

We also found that international and local knowledge search interact to promote innovation performance. This result suggests that international knowledge search and local knowledge search can strengthen each other. This finding enriches the open innovation theory and emphasizes the need to search both locally and internationally.

B. Theoretical and Practical Implications

The main contribution of this study is to further uncover the mechanism of open innovation. Existing research just proves that firms need to open, and search outside the firm boundary, but ignores to answer the question that whether search at local or international market [2] [8]. This study fills the research gap and discovers that international knowledge search and local knowledge search interact to promote innovation performance.

The findings of this study offer some important implications for managers and policy makers. Managers need to carefully consider local and international knowledge search sources so that they can maximize the innovation benefits while minimizing the search costs and risks. For policy makers, they should enact some encouraging policies to push local firms to search in the international market.

C. Limitations and Future Research Directions

This study has some limitations that need to be addressed in future studies. On the one hand, we just obtain one-year data, and can not prove the casual relationships between local and international knowledge search and innovation performance. Future research can use panel data to justify the relationship. On the other hand, we just consider the interactive item of both knowledge search, ignore to examine the contextual conditions. Future research can deepen in this aspect.

ACKNOWLEDGMENT

This paper is supported by soft science research project in zhejiang province (2014C35012).

REFERENCES

[1] J. Wu, "The effects of external knowledge search and CEO tenure on product innovation: evidence from Chinese firms", *Industrial & Corporate Change*, vol.23, no.1, pp. 65-89, 2014.

[2] K. Laursen, and A. Salter, "Open for innovation: The role of openness in explaining innovation performance among UK manufacturing firms", *Strategic Management Journal*, vol.27, no.2, pp.131-150, 2006.

[3] H. Garriga, G. Von Krogh, and S. Spaeth, "How constraints and knowledge impact open innovation", *Strategic Management Journal*, vol.34, no.9, pp. 1134-1144, 2013.

[4] L. Rosenkopf, and A. Nerkar, "Beyond local search: Boundary-spanning, exploration, and impact in the optical disk industry", *Strategic Management Journal*, vol.22, no.4, pp. 287-306, 2001.

[5] R. Katila, and G. Ahuja, "Something old, something new: a longitudinal study of search behavior and new product introduction", *Academy of Management Journal*, vol.45, no.6, pp. 1183-1194, 2002.

[6] H. W. Chesbrough, *Open innovation: The new imperative for creating and profiting from technology*, Cambridge, MA: Harvard Business Press, 2003.

[7] T. H. Clausen, T. Korneliussen, and E. L. Madsen, "Modes of innovation, resources and their influence on product innovation: Empirical evidence from R&D active firms in Norway", *Technovation*, vol.33, no.6/7, pp. 225-233, 2013.

[8] J. Chen, Y. Chen, and W. Vanhaverbeke, "The influence of scope, depth, and orientation of external technology sources on the innovative performance of Chinese firms", *Technovation*, vol.31, no.8, pp. 362-373, 2011.

[9] S. K. Kim, J. D. Arthurs, A. Sahaym, and J. B. Cullen, "Search behavior of the diversified firm: The impact of fit on innovation", *Strategic Management Journal*, vol.34, no.8, pp. 999-1009, 2013.

[10] R. R. Nelson, and S. G. Winter, *An evolutionary theory of economic change*, Cambridge, MA: The Belknap Press, 1982.

[11] L. Rosenkopf, and A. Nerkar, "On the complexity of technological evolution: Exploring coevolution within and across hierarchical levels in optical disc technology", In J. Baum & W. McKelvey (Eds.), Variations in organization science: In honor of D.T. Campbell. (pp. 169–183). Thousand Oaks, CA: Sage, 1999.

[12] M. A. Hitt, R. E. Hoskisson, and H. Kim, "International diversification: Effects on innovation and firm performance in product-diversified firms", *Academy of Management Journal*, vol.40, no.4, pp. 767-798, 1997.

[13] S. A. Zahra, R. D. Ireland, and M. A. Hitt, "International expansion by new venture firms: International diversity, mode of market entry, technological learning and performance", *Academy of Management Journal*, vol.43, no.5, pp. 925-950, 2000.

[14] O. R. Mihalache, J. J. J. P. Jansen, F. A. J. Van Den Bosch, and H. W. Volberda, "Offshoring and firm innovation: The moderating role of top management team attributes", *Strategic Management Journal*, vol.33, no.13, pp. 1480-1498, 2012.

[15] M. Subramaniam, and N. Venkatraman, "Determinants of transnational new product development capability: testing the influence of transferring and deploying tacit overseas knowledge", *Strategic Management Journal*, vol.22, no.4, pp. 359-378, 2001.

[16] M. I. Kafouros, P. J. Buckley, J. A. Sharp, and C. Wang, "The role of internationalization in explaining innovation performance", *Technovation*, vol.28, no.1/2, pp. 63-74, 2008.

[17] M. I. Kafouros, and N. Forsans, "The role of open innovation in emerging economies: Do companies profit from the scientific knowledge of others?", *Journal of World Business*, vol.47, no.3, pp. 362-370, 2012.

[18] L. F. Mesquita, and S. G. Lazzarini, "Horizontal and vertical relationships in developing economies: Implications for SMEs' access to global markets", *Academy of Management Journal*, vol.51, no.2, pp. 359-380, 2008.

[19] Y. Zhang, and H.Y. Li, "Innovation search of new ventures in a technology cluster: The role of ties with service intermediaries", *Strategic Management Journal*, vol.31, no.1, pp. 88-109, 2010.

[20] P. C. Patel, S. A. Fernhaber, P. P. Mcdougall-Covin, and R. P. V. D. Have, "Beating competitors to international markets: The value of geographically balanced networks for innovation", *Strategic Management Journal*, vol.35, no.5, pp. 691–711, 2014.

[21] R. J. Funk, "Making the most of where you are: Geography, networks, and innovation in organizations", *Academy of Management Journal*, vol.57, no.1, pp. 193–222, 2014.

[22] J. Wu, and Z. Wu, "Local and international knowledge search and product innovation: The moderating role of technology boundary spanning", *International Business Review*, vol.23, no.3, pp.542-551, 2014.

[23] C. Lin, P. Lin, F. M. Song, and C. Li, "Managerial incentives, CEO characteristics and corporate innovation in China's private sector", *Journal of Comparative Economics*, vol.39, no.2, pp. 176–190, 2011.

[24] Y. Lu, L. Zhou, G. Bruton and W. Li, "Capabilities as a mediator linking resources and the international performance of entrepreneurial firms in an emerging economy", *Journal of International Business Studies*, vol.41, no.3, pp.419-436, 2010.

[25] P. M. Podsakoff, S. B. MacKenzie, J. Y. Lee, and N. P. Podsakoff, "Common method biases in behavioral research: a critical review of the literature and recommended remedies", *Journal of applied psychology*, vol.88, no.5, pp. 879-903, 2003.

[26] S.-J. Chang, A. van Witteloostuijn, and L. Eden, "From the Editors: Common method variance in international business research", *Journal of International Business Studies*, vol.41, no.2, pp. 178-184, 2010.

[27] L.S. Aiken, and S. G. West, *Multiple Regression: Testing and Interpreting Interactions*, Sage: Newbury Park, CA. 1991.

Analysis of High-Tech Industry's Development in Shandong Province Based on PLS

Xu Yang[1],[*], Xing-yuan Wang[2]

[1]School of Science, Shandong Jianzhu University, Jinan 250101, P. R. China
[2]School of Management, Shandong University, Jinan 250100, P. R. China
(Yangxu2011@sdjzu.edu.cn)

[1] Abstract - **From the perspective of high-tech industry's development in Shandong province, this paper firstly identifies the influencing factors of high-tech industry's development in Shandong province. Then those factors and the process of development in 16 years (1997-2012) are analyzed by clustering analysis using the data from the statistical yearbook. Moreover, this paper creates a model to predict the trend of high-tech industry's development in Shandong province by partial least squares (PLS) regression. After the analysis of regression coefficient by clustering analysis, importance of each factor is gained, and based on the result we can forecast the tendency of high-tech industry's development in Shandong province in the future. Ultimately, the research results can be the decision basis to promote the development of high-tech industry in Shandong province rapidly.**

Keywords - **High-tech industry, influencing factors, partial least squares regression, Shandong province**

I. INTRODUCTION

Shandong province is located in the eastern coast of China and on the lower Yellow River with rich resources and large population. It is one of the richest economic provinces in China. Recently, the economic growth of Shandong province is rapid and stable. In 2013, above scale high-tech industry output value was ￥3,958,274,000,000 which 14.72% more than that in 2012, and made up 30.23% of above scale industrial output value. There are 3,067 enterprises in Shandong province which 1.7 times more than that in 2012. In the same year, Shandong province was rated as one of the three most comprehensive competitive provinces in China.

Compared with the traditional manufacturing industry, high-tech industry has many features, such as high R&D input, high return, high risk, high permeability, high acceleration and low energy consumption, less pollution and other distinctive features. The strong driving power of high-tech industry has become an important force in regional economic growth and social development. The development of high-tech industry does not comply with the mode of the traditional manufacturing's development, and it can boost the industry competitiveness significantly. At present high-tech industry is still growing up rapidly, and now it has become the focus of worldwide competitive competence. The development of high-tech industry in Shandong province with prominent features mainly

reflected in the pharmaceutical manufacturing industry, electronic and communication equipment manufacturing industry. For the past few years, domestic and foreign scholars on the distribution of high-tech industry, growth, innovation ability, contact of traditional industries, human resources, influencing factors and evaluation methods carried out extensive research. Those researches obtained many encouraging results [1-7]. However, most researches focus on high-tech industry in a nation but a province. Now the lack of comprehensively quantitative study on large data for influencing factors of high-tech industry's development in Shandong province will be paid more attention.

High-tech industry's development is a systematic engineering. In recent 10 years, such as Guangdong, Jiangsu, Shanghai and other domestic developed provinces' high-tech industry developed rapidly. Especially, high-tech industry's scale in the Pearl River Delta, the Yangtze River Delta and the Beijing-Tianjin region are already very large. Therefore, the significant advantage of competitiveness of high-tech industry in those areas brings great pressure to high-tech industry's development in Shandong province. The study [8] claimed that the scale of high-tech industry in China is already the third in the world, but the profitability of high-tech industry declined, so did high-tech industry in Shandong province. So, those problems are great challenges to high-tech industry in Shandong province. Then here comes some questions: How to develop high-tech industry in Shandong province significantly? What are the influencing factors of high-tech industry's development in Shandong province? How important is the contribution of each influencing factor to high-tech industry of Shandong province? Analyzing those problems needs a systematic quantitative analysis, so that we can clarify the causal relationship between them and know their importance. To research those problems may help promote high-tech industry, even economic, development in Shandong province significantly.

Based on the above analysis, this paper adopts the partial least square regression to create a model predicting the tendency of high-tech industry's development in Shandong province. According to the data for 16 years (2002-2013) from statistical yearbook of high-tech industry, we work out the importance of each factor, and find out some main factors affecting most. The research may provide the reference for the formulation of high-tech industry's development strategy in Shandong Province.

---
[1] The paper supported by NSFC: 71272121

II.   ANALYSIS OF INFLUENCING FACTORS OF HIGH-TECH INDUSTRY'S DEVELOPMENT IN SHANDONG PROVINCE

Regional economy, science and technology funds, fixed assets, human resources, enterprise system, enterprise scale, the scale of higher education, export ability and so on may possibly affect the development of high-tech industry. Outstanding regional economy, sufficient R&D funds and professional employees, lots of enterprises and reasonable regulations may promote the healthy development of high-tech industry. For instance, the eastern coastal area of Shandong province is a developed region of economy with abundant resources. On the contrary, economy of the central and western area is backward and weak. In 2012, the total output value of high-tech industry in Qingdao is ￥3,977,000,000 and the expenditure of basic research cost ￥952,780,000. However, the total output value of Heze which located in the western of Shandong province is only ￥2,863,000,000 and the expenditure of basic research cost ￥4,590,000. The development of regional economy needs scientific research coinciding with development of high-tech industry. Consequently, with a high level of scientific researches the development of high-tech industry will feel just like a fish in water.

Based on the information of management, R&D, related activities and fixed asset investment in China's high-tech industry statistical yearbook. This paper selects eight indexes to be the influencing factors of high-tech industry's development in Shandong province or quantitative analysis. The factors which need qualitative analysis, such as policy and regime, won't be considered in this paper.

III.   THE DEVELOPMENT TREND OF HIGH-TECH INDUSTRY IN SHANDONG PROVINCE PREDICTION MODEL

A. The Variable Selection and Index Explanation

According to the analysis of high-tech industry's development in Shandong province, the following indexes are eventually chose to be the variables of empirical analysis in the development trend of high-tech industry in Shandong province prediction models.

Y_1 –revenue; Y_2 –main business income; X_1 –the quantity of Enterprises; X_2 –internal R&D expenditure; X_3 –R&D full-time; X_4 –The quantity of patent; X_5 –The new fixed assets; X_6 –The quantity of college students; X_7 –Terminal energy consumption; X_8 –The average quantity of annual employee

B. Create the Partial Least Squares Regression Prediction Model [9]

The partial least squares regression (PLS) combines the multiple linear regressions (MLR), the principal component analysis (PCA) and the canonical correlation analysis comprehensively so that it may solve the data distortion problem caused by variable multiple correlation. Also, it can determine the importance of each influence factor accurately. For example, if the quantity of dependent variable is 1, we would create the model by extracting the principle component t and add the restriction that the principle component of X should relate to Y.

C. The Standardization of Data

We define F_0 as the standardize variable of the dependent variable:

$$F_0 = Y^* = \frac{Y - \bar{y}}{s_Y} \tag{1}$$

$\bar{y}$ is the average value of Y. s_Y is Y's standard deviation. And we define E_0 as the standard matrix of the independent variable set:

$$E_{ij} = X^* = \frac{X_{ij} - \bar{X_j}}{s_j}, i=1,2,\cdots,n; j=1,2,\cdots,m \tag{2}$$

$\bar{X_j}$ is the average value of the number J independent variable, and s_j is the standard deviation of the number J independent variable.

D. The Main Effective Components Extraction

We extract the principal component t_1 from E_0. $t_1 = E_0 w_1$, $||w_1|| = 1$. We define w_1 as the first principal axis of E_0. Then we work out the regression equation of F_0 and t_1:

$$F_0 = t_1 r_1^T + F_1 \tag{3}$$

The Regression coefficient vector is

$$r_1 = \frac{F_0^T t_1}{\|t_1\|^2} \tag{4}$$

F_1 is the residual matrix of the regression equation. Next we test the effectiveness of the cross. If the result of the testament is effective ($Q_1^2 \geq 0.0975$), the calculation will be continued. If the result is non-effective, we will extract one component t_1. In the same way, we will gain h principal components by h iterative computations, and the regression equation is

$$F_0 = t_1 r_1 + t_2 r_2 + \cdots + t_h r_h + F_h \tag{5}$$

The equation (5) is the model which we create by the partial least squares regression. Finally, we gain the regression equation by replacing F_0 and E_{0j} with Y^* and X^*:

$$Y^* = \alpha_1 X_1^* + \alpha_2 X_2^* + \cdots + \alpha_m X_m^* \tag{6}$$

IV.   ANALYSIS OF THE TENDENCY OF HIGH-TECH INDUSTRY'S DEVELOPMENT IN SHANDONG PROVINCE

This paper selects 16 years statistical data from China's statistical yearbook of high-tech industry and Shandong's statistical yearbook (2002-2013) [10,11] for research and adopts the software called DPS [12] to analyze.

A. Correlation Analysis

First of all, standardize the raw data for simple correlation analysis. Then gain the correlation coefficient matrix between the variables (TABLE I).

TABLE I
CORRELATION COEFFICIENT MATRIX BETWEEN THE VARIABLES

	X_1	X_2	X_3	X_4	X_5	X_6	X_7	X_8	Y_1	Y_2
X_1	1	0.8481	0.7586	0.8070	0.7758	0.9647	0.9561	0.9659	0.8236	0.8880
X_2	0.8481	1	0.9827	0.9892	0.9803	0.8601	0.9313	0.9547	0.9890	0.9906
X_3	0.7586	0.9827	1	0.9866	0.9840	0.7717	0.8601	0.8960	0.9829	0.9667
X_4	0.8070	0.9892	0.9866	1	0.9803	0.8227	0.9037	0.9280	0.9934	0.9846
X_5	0.7758	0.9803	0.9840	0.9803	1	0.7837	0.8704	0.9070	0.9825	0.9633
X_6	0.9647	0.8601	0.7717	0.8227	0.7837	1	0.9662	0.9538	0.8184	0.8832
X_7	0.9561	0.9313	0.8601	0.9037	0.8704	0.9662	1	0.9853	0.9077	0.9522
X_8	0.9659	0.9547	0.8960	0.9280	0.9070	0.9538	0.9853	1	0.9370	0.9732
Y_1	0.8236	0.9890	0.9829	0.9934	0.9825	0.8184	0.9077	0.9370	1	0.9897
Y_2	0.8880	0.9906	0.9667	0.9846	0.9633	0.8832	0.9522	0.9732	0.9897	1

From TABLE I, we can recognize that the correlation coefficient between Y_1 revenue and Y_2 main business income is 0.9897 so that they will play the same role in the analysis. Therefore, this paper chooses Y_1 revenue to be the dependent variable Y. The correlations between variables are very high. Every independent variable may explain the dependent variable efficiently.

B. Clustering Analysis

This paper let "year" be the ordinate and "influencing factors" be the abscissa for clustering analysis (Fig.1). If we divide 16 years into 5 parts, we would find that (1) the years in 1997-2001 Just during the "ninth five-year plan" period in China is the first part; (2) the years in 2002-2004 during the "tenth five-year plan" period is the second part; (3) 2005-2007 years during the former part of the "eleventh five-year plan" period is the fourth part; (4) 2008-2010 years during the latter part of the "eleventh five-year plan" period is the fourth part; (5) the last part is the years between 2011 and 2012 just during the former part of the "twelfth five-year plan" period. This classification reflects that high-tech industry's development in Shandong province is growing frequently and stably. What's more, the rank of developing speed is $(5) \rightarrow (4) \rightarrow (3) \rightarrow (2) \rightarrow (1)$.

If switch the coordinate, the result showed on Fig.2 is gained. Then we divide the influencing factors into 5 parts: (1) the quantity of enterprise, college students, patents, new assets; (2) the quantity of R&D employees; (3) the cost of energy; (4) R&D spending; (5) the annual average quantity of employees.

C. Regression Analysis

After standardizing the original data, we get the result of regression. The clustering analyzed result of regression coefficient is showed on the Fig.3. All kinds of model's error sum of squares and Press statistic when we extract different principle components are showed in TABLE II. Through the data analysis, R^2 (coefficient of determination) from different models may show that when extract the 5 principle components, the regression model's fitting degree are high and the error sum of squares is on the decline.

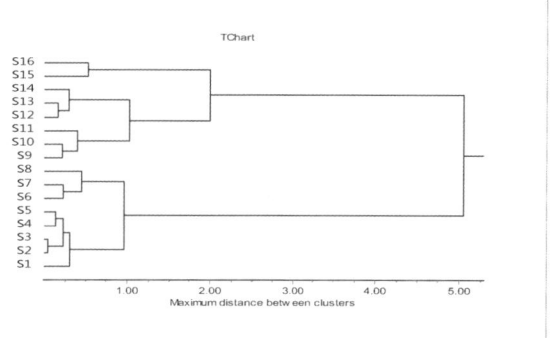

Fig.1. Annual systematic Clustering figure

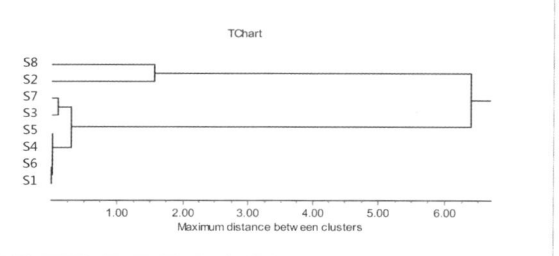

Fig.2. The systematic clustering figure of influencing factors

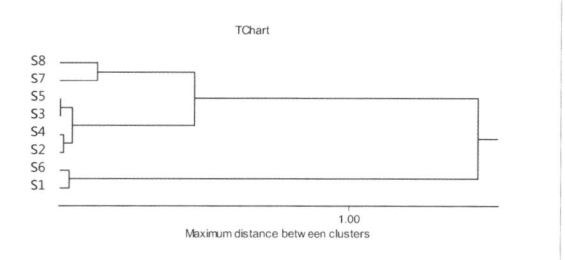

Fig.3. The systematic clustering figure of regression coefficient

TABLE II
THE RESULT OF THE STANDARD REGRESSION COEFFICIENT

Principal components	1	2	3	4	5
Error sum of squares	0.7696	0.1712	0.0947	0.0767	0.0726
Press statistic	1.1145	0.2139	0.1747	0.2778	0.3362
Determination coefficient R^2	0.9487	0.9886	0.9937	0.9949	0.9952

However, the independent variable X_6's (college students) weight is negative so that it cannot explain the dependent variable. Therefore this paper determine one principle component (correlation coefficient is positive, and the cumulative contribution rate is 91%).

Ultimately, the standardized regression equation of high-tech industry's development in Shandong province and the regression equation (7) and (8) of ordinary variables are created:

$$Y^* = 0.1124X_1^* + 0.135X_2^* + 0.1342X_3^* + 0.1356X_4^* + 0.1341 X_5^* + 0.135X_6^* + 0.1239X_7^* + 0.1279X_8^* \quad (7)$$

$$Y = -141.64 + 0.0506X_1 + 0.0001X_2 + 0.0044X_3 + 0.0222X_4 + 0.2206X_5 + 0.0506X_6 + 0.004X_7 + 0.0002X_8 \quad (8)$$

V.   CONCLUSIONS AND RECOMMENDATIONS

A.   Conclusions

As shown by TABLE II, the minimum R^2 (coefficient of determination) value of PLS model is 0.9487. It means that the fitting degrees of regression models are high. So results of analysis are proved to be credible. The rank of the regression coefficient values are $X_4 > X_2 > X_3 > X_5 > X_8 > X_7 > X_1 > X_6$. According to the result of clustering analysis, we obtain 5 parts: part 1 (X_4 and X_2), part 2 (X_3 and X_5), part 3 (X_8), part 4 (X_7), part 5 (X_1 and X_6). Therefore, this paper offers four important influencing factors for the development of high-tech industry in Shandong province: (1) The most important factors are the quantity of patents and internal R&D spending; (2) The secondary important factors are R&D full-time equivalent of personnel and new fixed assets; (3) The tertiary important factors are terminal energy consumption and the annual average quantity of employees; (4) The last important factors are the quantity of enterprises and college students.

Consequently, the trend of high-tech industry's development in Shandong province is healthy, stable and rapid.

B.   Some Recommendations

The analysis suggests that the government should give full play to its functions and enterprises should play a great role in innovations. Strengthen the university-industry cooperation. Perfect the reward system of patent application and encourage the whole society to participate in the invention; Attract the capital, technological talents and increase the investment; Adjust the university personnel training mode and appropriately control the scale of enrollment; Don't pursuit the quantity of high-tech enterprise; Continue to develop the energy-saving and low consumption industries. Based on those suggestions, eventually implement the rapid, stable and sustainable development of high-tech industry in Shandong province.

REFERENCE

[1]   Han-hui Hu, Lang-feng Wang. Ecology investigation on China's high-tech industries and manufacturing industries evolution law [J]. Studies in Science of Science, 2009, 27(10): 1523-1527. (Chinese)

[2]   Shi Dan, Xiao-bin Li. Factors analysis and data test on high-tech industry development [J]. China Industrial Economics, 2004, (12):32-39. (Chinese)

[3]   Yu-lin Zhao, Cui-hong Ye. Empirical analysis on the growth stage and its transition of high-tech industries [J]. Science of Science and Management of S.&T., 2011, 32(5): 92-101. (Chinese)

[4]   Hai-yan Shi, Gen-nian Tang. The agglomeration of high-tech industry and its effectiveness in the eastern coastal area [J]. R&D Management, 2013, 25(2): 22-28. (Chinese)

[5]   Choi B. High technology development in regional economic growth [M]. England: Ashgate Publihing Company, 2003.

[6]   Xiao-li Li, Lian-cai Hao, Hong Rong. Study on human resources of high-tech industry in Shandong Province [J]. Science & Technology Information, 2012(35): 96-97. (Chinese)

[7]   Gu Jing, Wei-xian Xue. Study on collaborative innovation of the high-tech industry [J]. Science & Technology Progress and Policy, 2012, 29(22):84-89. (Chinese)

[8]   Fu-qian Fang, Zhang Ping. Analyzing input-output efficiency of the high-tech industries based on DEA [J]. China Soft Science, 2009(7): 48-56. (Chinese)

[9]   Hui-wen Wang. Partial least-squares regression-method [M]. Beijing, National Defense Industry Press, 1999. (Chinese)

[10] National Bureau of Statistics. China statistics yearbook on high-tech industry 2002-2012 [M]. Beijing: China Statistics Press, 2002-2012. (Chinese)

[11] Shandong Provincial Bureau of Statistics. Shandong statistics yearbook 2002-2012 [M]. Beijing: China Statistics Press, 2002-2012. (Chinese)

[12] Qi-yi Tang, Ming-guang Feng. DPS data processing system [M]. Beijing: Science Press, 2007. (Chinese)

Game Analysis of Engineering Quality Control Base on Owner and Superintendent

Hui GAO[1,*], Zhao-yi YE[2], Hai-rong FENG[1]

[1] Department of Industrial Engineering of Business School, Sichuan Normal University, Chengdu Sichuan 610101, China
[2] China National Chemical Engineering No.7 Engineering Company Limited, Chengdu Sichuan 610101, China
(gooher@163.com)

Abstract - **Engineering projects' quality control has always been of great concern. How to get better effect of engineering quality is worth to be discussed. Firstly, the rational behavior of the owner and the supervisor is analyzed. Then based on the two parties of owner and supervisor, the complete information static game model is build up in engineering quality supervision. At last by solving the model we have the mixed strategy Nash equilibrium point. Based on the mixed strategy Nash equilibrium, we analyze the rational behavior of the owner and the supervisor, and provide some suggestions of quality control to the owner.**

Keywords - **Engineering supervision, game theory, Nash equilibrium, quality management**

I. INTRODUCTION

Articles which use game theory to analyze the effectiveness of the quality management are not few. In 2000, Fei-lian ZHANG, Meng-jun WANG, Ji-zu ZHOU, Hao-jun YU [1] established a game model which is about project quality control between the supervisor and the builder, and proposed penalty coefficient concept and provided some suggestions. Yi CHEN [2] analyzed the behavior of prosecutors and builders by using the game approach to obtain the thresholds about inspection probability and illegal probability. Later Fei-lian ZHANG et al[3], Gao-ming ZHU et al[4], Peng CHEN et al [5], Ping-hui TU [6], Zhi-fang HUANG et al [7] have discussed the effective supervision and control mechanism with the game theory between the supervisor and the builder. Kong-ling LIU [8], Huang XI et al[9] used game theory to analyze the decision-making behavior of the owner, the supervisor and the contractor. Zhong-ke ZHOU [10] used the game theory to analyze the safety countermeasures tripartite behavior of China's coal mine.

In the framework of the game analysis, these essays under different assumptions, constructed different revenue function, and explained the actual status of quality control by using the results of the analysis, and made useful suggestions about improving the quality management. But the analysis background information is too much to ignore the practical problems. There are some flaws that the results of the analysis have owned.

Disagreement of quality objectives between the owner and supervisor revolves in interest [11]. Owners concern about engineering's quality and products, and do not care about the supervisor's cost and profits. The supervision unit concern about cost and profit. The self-profit maximization is both parties' fundamental objective.

Owners want to ensure the quality of the project. This is not only depended on the correct contractor, but also depended on the owner's agent--the superintendent. Therefore, this paper will do some game analysis about the interactive decision-making behavior between the owner and the supervisor, and hope to provide some suggestion to improve the actual quality control effect.

Based on the existing research literature, we will analyze the game behavior between the supervisor and the owner. With the actual situation of the management, we will build new revenue function of both sides using new function variables.

II. GAME ANALYSIS

Game theory is the mathematical study of strategic interactions among the participants [12]. Here we use the static game model of complete information to analyze interactive decision between owner and supervisor.

A. Construction of the game model

Some explanations:

A static game (G) has n participants, each participant's strategic space, such as $S_1, S_2, \cdots S_n$, respectively, each player's gain is $u_1, u_2, \cdots, u_n$, the game's strategic formula is denoted by [13].

$$G = \{S_1, S_2, \cdots S_n; u_1, u_2, \cdots, u_n\}.$$

1) Participants, $n = 2$: in our game model, there are two participants: one is the owner, and the other is supervision unit.

2) The owner's strategic space:
$$S_1 = \{\text{random inspection, no inspection}\}$$

The supervisor's strategic space:
$$S_2 = \{\text{trying to supervision, no trying to supervision}\}$$

In order to facilitate analysis in mind, we marked:
$X_1 = $ random inspection, $X_2 = $ no inspection,
$Y_1 = $ trying to supervision, $Y_2 = $ no trying to supervion,

3) Assuming that the parties did not know each other's actions before making decisions, both sides' behavior can be considered simultaneously, so our model is complete information static game.

4) Two sides revenue function

To analyze the revenue function of both sides, here we have some of the following assumptions:

1) When supervisor do not fulfill their duty, he still need a minimum cost, the cost amount showed with C_1; When supervisor do their duty, they pay even more, the

E. Qi et al. (eds.), *Proceedings of the 21st International Conference on Industrial Engineering and Engineering Management 2014*, Proceedings of the International Conference on Industrial Engineering and Engineering Management, DOI 10.2991/978-94-6239-102-4_109, © Atlantis Press and the authors 2015

cost is showed as LC_1, here $L \geq 1$, it indicates the efforts degree of supervisor.

2) When owner have no inspection, he hasn't any pay, so his cost is considered 0; while owner do some checks, his cost is C_2.

3) If supervision unit do their duty, owner will have more income by reducing the life cycle cost of the project etc., and this amount represented as R. In order to praise the supervisor's endeavor, the owner will give a bonus to supervisor, the reward amount expressed as k_1R, $0 < k_1 < 1$.

4) If supervision unit don't do his duty, a lot of quality problems will occur. In this case we assume that if the owner does check, he will surely find out the problem. The defect will bring some losses to the owner in later, the losses marked as F. Owner will punish supervision, the fine amount is denoted: k_2F, and $k_2 > 1$, k_2 is called punishment factor. Here, $k_2F > R > F + C_2$, otherwise the owners' check is meaningless.

B. establish payoff matrix

Through above analysis, we can draw the owner and supervisor's payoff matrix, as shown in Table I.

TABLE I
PAY OFF MATRIX OF OWNER AND SUPERVISOR

Supervisor	Owner	
	Random check	No random check
Do duty	$(k_1R - LC_1, \quad R - k_1R - C_2)$	$(-LC_1, \quad R)$
Not do duty	$(-k_2F - C_1, \quad k_2F - F - C_2)$	$(-C_1, \quad -F)$

C. solving the game model

The core of the model is the Nash equilibrium problem.

1) Pure strategy equilibrium

Because there are only two parties in this model, and the less number of strategic space, so we can use a simple dash method for solving this model.

As a rational "economic man", when the owner choose random inspection, supervisor will preferred the strategy which is fulfilling his duty, because only by doing so he will have more earnings.

Similarly, owners get their available choices. All those choices are underlined appear in Table II.

TABLE II
PAY OFF DASH RESULTS MATRIX OF OWNER AND SUPERVISOR

Supervisor	Owner	
	random check	No random check
On duty	$(\underline{k_1R - LC_1}, \quad R - k_1R - C_2)$	$(-LC_1, \quad \underline{R})$
Not on duty	$(-k_2F - C_1, \quad \underline{k_2F - F - C_2})$	$(-C_1, \quad -F)$

From the table, we do not have a pure strategy equilibrium of the model. But Nash (1950) proved the mixed strategy Nash equilibrium's existence of any finite strategic game [14]. Here we go solving mixed strategy equilibrium of the game model.

2) The mixed strategy Nash equilibrium

The owners and supervisor's mixed strategy marked separately as $P_1 = (x, 1 - x)$, $P_2 = (y, 1 - y)$, that means the probability of the owner's random inspection S_{11} is marked as x, and the probability of no random inspection S_{12} is $(1 - x)$. Similarly, we set the parameter about supervisor. Now we have the result about a two-dimensional discrete joint distribution as follows:

TABLE III
JOINT PROBABILITY DISTRIBUTIONS

S_{1I}	S_{2J}	
	S_{21}	S_{22}
S_{11}	$P_{11} \times P_{21} = X \times Y$	$P_{11} \times P_{22} = X \times (1 - Y)$
S_{12}	$P_{12} \times P_{21} = (1 - X) \times Y$	$P_{12} \times P_{22} = (1 - X) \times (1 - Y)$

We use Table III and Table I to obtain the owners' expected revenue function:

$$V_1(P_1, P_2) = \sum_{i=1}^{2}\sum_{j=1}^{2} p_{1i} \times p_{2j} \times u_1(s_{1i}, s_{2j})$$
$$= xy(R - k_1R - C_2) + x(1 - y)(k_2F - F - C_2)$$
$$+ (1 - x)y\,R + (1 - x)(1 - y)(-F) \tag{1}$$

Similarly, we have the supervisor revenue's expectation:

$$V_2(P_1, P_2) = \sum_{i=1}^{2}\sum_{j=1}^{2} p_{1i} \times p_{2j} \times u_2(s_{1i}, s_{2j})$$
$$= xy(k_1R - LC_1) + x(1 - y)(-k_2F - C_1)$$
$$+ (1 - x)y(-LC_1) + (1 - x)(1 - y)(-C_1) \tag{2}$$

Because we assume that all participants are rational "economic man", then the problem is to solve the following simultaneous optimization problem:

$$\begin{cases} \max_{0 \leq x \leq 1} V_1(x, y) = \sum_{i=1}^{2}\sum_{j=1}^{2} p_{1i} \times p_{2j} \times u_1(s_{1i}, s_{2j}) \\ \max_{0 \leq y \leq 1} V_2(x, y) = \sum_{i=1}^{2}\sum_{j=1}^{2} p_{1i} \times p_{2j} \times u_2(s_{1i}, s_{2j}) \end{cases} \tag{3}$$

By solving the simultaneous equations, we get the owner's mixed strategy equilibrium:

$$P_1^* = \left(\frac{LC_1 - C_1}{k_1R + k_2F}, 1 - \frac{LC_1 - C_1}{k_1R + k_2F} \right) \tag{4}$$

and we have supervision's mixed strategy equilibrium

$$P_2^* = \left(\frac{k_2F - C_2}{k_1R + k_2F}, 1 - \frac{k_2F - C_2}{k_1R + k_2F} \right) \tag{5}$$

III. MEANINGS of the MODEL RESULTS

A. From owner's mixed strategy equilibrium, we can have some results:

1) When $x < x*$, $\dfrac{\partial V_1(x, y)}{\partial x} < 0$, in order to get the revenue maximization, the owner will not have checks;

If the owner do not check, supervision's optimal choice is not fulfilling their duty.

2) When $x > x*$, $\dfrac{\partial V_1(x, y)}{\partial x} > 0$, in order to gain the revenue maximization, the owner will take random strategy; then the supervisor will fulfill their duty.

3) When $x = x*$, $\dfrac{\partial V_1(x, y)}{\partial x} = 0$, the owners will have same earning whether he takes checking strategy or not, so the owner will have a random action.

B. Supervisor's mixed strategy equilibrium solution shows:

1) When $y < y*$, $\dfrac{\partial V_2(x, y)}{\partial y} < 0$, in order to have maximum benefits, supervisor does not fulfill his duty, and then the optimal behavior of the owner is random check.

2) When $y > y*$, $\dfrac{\partial V_2(x, y)}{\partial y} > 0$, in order to have maximum benefits, supervisor will do his duty, then the optimal behavior of the owners is not random check.

3) When $y = y*$, $\dfrac{\partial V_2(x, y)}{\partial y} = 0$, whether supervisor is dutiful or not, there is no impact on his earnings, so he will have a random action.

Through the above analysis we can see that when party A tend to an action for increasing revenue, then party B will take self-serving actions, thereby reducing party A's revenue, and therefore the best choice is the mixed strategy equilibrium. Rational people would choose an action to pay a minimum cost while the income keeps constant. So we think the owner will choose not inspection action with the probability ($x*$), and the supervision will choose not fulfilling his duty with probability $y*$.

IV. SUGGESTIONS

1) From the owner's equilibrium we find that the probability of owner's not check has a positive correlation with the supervisor's endeavor, which is consistent with common sense; and the probability of owner's not check has a negative correlation with the supervisor's fines. This indicates that the owner may curb supervisor's adverse behavior through positive incentives and negative punishment.

2) From the equilibrium we see the probability of supervisor's not fulfilling duty is negatively correlated with the amount of the reward. So owners may guide the supervisor to fulfilling duty with positive incentive mechanism, and then reducing the probability of adverse action.

More and more attention has been paid to the project management [15] [16], we hope our work will be useful to the practical project quality management.

ACKNOWLEDGMENT

This work was supported by MOE (Ministry of Education in China) Project of Humanities and Social Sciences (Project No.13YJC630202), the Scientific Research Foundation of the Education Department of Sichuan Province, China (No.14ZA0026 and No.14ZB0027).

Furthermore, we are thankful to Ke-liang WU who is a student of Minnesota University for his help about this paper's translation.

REFERENCES

[1] Fei-lian ZHANG, Meng-jun WANG, Ji-zu ZHOU, Hou-jun YU. Game Theory in quality control of engineering project [J]. Journal of Changsha Railway University, No2, Jun. 2000: 27-30 (CHINESE)

[2] Yi CHEN. Application of Game Theory in quality inspection of construction engineering [J]. Journal of Chongqing Jianzhu University, Vol.24, No.6: 77-79 (CHINESE)

[3] Fei-lian ZHANG, Li LIU, Wu-zhou DONG, Wei ZHANG. The application of Game Theory in project safety management [J]. Systems Engineering Nov, 2002.11, Vol.20, No.6: 33-37 (CHINESE)

[4] Gao-ming ZHU, Xi-jun WANG, Meng-jun WANG. Application of Game Theory in project management. Journal of Changsha Railway University, No1 Mar. 2002: 60-63 (CHINESE)

[5] Peng CHEN, Yu-hua WU, Yu-cheng JIN. Application of Game Theory in quality management of engineering project [J]. Journal of East China Jiaotong University. Aug., 2004, Vol.21, No.4 (CHINESE)

[6] Ping-hui TU. A Game analysis on safety management of engineering project [J]. Journal of Wuhan University of Science and Engineering. Dec. 2004 NO.8: 90-91 (CHINESE)

[7] Zhi-fang HUANG. Guo-hua WEN, Yue-sheng HUANG. A Game analysis on quality management of engineering project [J]. Journal of Hunan City University (Natural Science) Dec. 2005 Vol. 14 No.4:12-14 (CHINESE)

[8] Kong-ling LIU, Game Theory analysis of quality control based on the owners of the project perspective [J]. Science & Technology Progress and Policy. 2010.10: 40-43(CHINESE)

[9] Xi HUANG, Chun-feng WANG. Game analysis and countermeasure study of engineering supervision [J]. China Population. Resources and Environment Vol.16, No.4, 2006: 90-93 (CHINESE)

[10] Zhong-ke ZHOU, Liang XU. Analysis of three-parties Game Theory in mining safety supervision in China [J]. Journal of Safety Science and Technology, Vol.2, No.4, Aug. 2006: 96-100 (CHINESE)

[11] Shou-kui HE, Hong-yuan Fu. An economic explanation of the engineering quality risk and risk prevention [J]. Journal of Chongqing Jianzhu University Vol.28, No.6, Dec. 2006: 106-110 (CHINESE)

[12] Wei-ying ZHANG. Game Theory and Information Economics [M]. Shang Hai: Shang Hai People's Publishing House, 1996, P4 (CHINESE)

[13] Guang-jiu LI. A Primer in Game Theory [M]. Jiang Su: Jiang Su university press, 2008. P.33-34 (CHINESE)

[14] Ze-Ke WANG, Fei GE. Nash equilibrium [M]. Shang Hai: Shang Hai Science and Technology University Press. 2009. P.3 (CHINESE)

[15] Avraham Shtub, Jonathan F. Bard, Shlomo Globerson. Project management: processes, methodologies, and economics, Second Edition. 2002.

[16] Project Management Institute. A Guide to the Project Management Body of Knowledge.

Study on Marketing Strategy for Aviation Manufacturing Enterprise Based on Domestic Military Market

Li Zhang[1,2], Fajie Wei[1], Shan Lu[1,*]

[1]School of Economics and Management, Beihang University, Beijing 100191
[2]Avic Aerospace Life-Support Industries, Ltd, Hubei 413002
(shinelu914@sina.com)

Abstract - **Recent years, government is actively promoting innovation in technology defense industry by breaking the closed system and introducing the market competition mechanism, hoping to make full use of the whole social resources and improve weaponry and equipment production capacity. To seek for survival and development, aviation manufacturing enterprises need to adjust to the new environment. In-depth study of the characteristics of the military market and the establishment of an effective marketing system are required to vigorously develop the military market and improve the efficiency of enterprise itself.**

Keywords - **Aviation manufacturing enterprises, military market, marketing strategy**

I. PREFACE

Since the reform, with the deepening of the market economy as well as changes in the new era of military strategy, China's national defense equipment construction has undergone profound changes [1-3]. The military market overall situation changes dramatically: mandatory planning is no longer the procurement while contract is; private enterprises actively participate in the market competition while state-owned monopoly disappeared. Corresponding with the above changes are implement procedures in terms of competition, evaluation, supervision, incentive and restraint mechanisms have become more standardized and rational [4].

With the National Development No. 37 promulgating, military aviation manufacturing enterprises are facing survival problems in the more intense competition market [5, 6]. To win the market share and grasp the opportunity of rapid development, aviation manufacturing enterprises must change the mechanism, the ideas, the methods and strategies of military market as soon as possible to adapt to the rapidly changing domestic military market.

II. CHARACTERISTICS OF THE DOMESTIC MILITARY MARKET AND ITS IMPACT ON MARKETING

A. Characteristics of the domestic military market

The military is a special commodity, with attributes of public goods, which invisibly affects the entire public utility, regardless of whether individuals willing to spend [7]. In China, military procurement is implemented by army which is commissioned by the government. Military market follows not only laws of commodity markets but also the regular patterns in non-perfectly competitive market.

Military is a special products in non-competitive market [8-10]. Take the military aircraft as an example. First, military identifies its user —army —and clearly defines whether the user is the Air Force, Navy or Ground Force. Second, the seller is also identified. In current layout of the aviation industry, the phenomenon of airlines fixed production and monopoly is obvious in military market. Third, the market is in the lack of entry mechanism because of the planned economy background for many years, especially in the military R&D and manufacturing market. In addition, since weapons and equipment require full life-cycle services, military enterprises exit barriers are high.

B. Impact on the marketing activities
1) Government is in a dominant position

As the buyer is the government, national will and interests guide the military markets. According to the National macroeconomic conditions, the macro-environment, financial ability, international economic and political environment, military strategy and economic development strategies, the government makes the plan of weapons and equipment development or recent demand as the guidance of the military market [11]. Generally, unlike a perfectly competitive market participants, aviation manufacturing military enterprises are in the state of "passive" marketing. They rarely guide consumer demand or develop new markets.

2) Trading with Long-term stability

In the military market, the number of sellers is also extremely limited and relatively fixed. This makes the traders trade in a closed circle. Long-term stability of the transaction has two sides. On one hand, it helps sellers to maintain relationships with military customers. On the other hand, it is not conducive to technological innovation and cost savings.

3) Military customers with specific consumption characteristics

Because the military is a special products in non-competitive market, marketing tools such as flexible market pricing, advertising, promotions and etc. as well as the classic 4P (product, price, place, promotion) marketing mixing strategy do not apply to the military market. In contrast, military sales is influenced by product quality, reliability, delivery and support service. In this case, the military's marketing chain is shorter than general commodity which has basically neither intermediate links nor initiative on product pricing strategy.

III. PROBLEMS IN MILITARY AVIATION MANUFACTURING ENTERPRISE MARKETING MANAGEMENT

With the military market becoming increasingly competitive, problems of aviation military enterprises marketing strategy under the existing planning system have been expose to the air, which restricting the development of enterprises. The main problems are reflected in:

A. Lack of modern marketing philosophy

Under the planned economy, most of the enterprises did specific work according to the government plan. The trading activity cannot be counted as marketing in the modern sense. With having stepped in the market economy era, weapons and equipment procurement becomes increasingly market-oriented. However, many military aviation military enterprises still stuck in the idea of "Marketing is equal to sell". The ideas of daily operations should meet the army's needs and the consciousness of user-oriented marketing are missing. Instead, they pay lots of attention on "delivery" regardless of the cost. Besides, they hold the opinion of "marketing is the mission of marketing department", which results in each department simply views marketing issues from their own perspective.

B. Marketing management system is imperfect

Currently, the widespread problem of aviation military enterprises is that the marketing management system is not perfect. First, there is no full-time military marketing department. A number of military companies anchored marketing department in other functional department. What is worse, many companies still do not have a dedicated marketing department. Second, the marketing function is not perfect: no market research, planning, developing capabilities. Relatively simple way of selling, inadequate marketing resources investment, weak integration of sales network, inefficient incentive mechanism and etc. make it difficult to improve marketing

C. Military marketing strategic and planning is not clear

Due to the influence of the planned economy way of thinking, some aviation military enterprises never analyze the market in professional methods or propose a specific marketing strategy. Besides, there is no complete annual marketing plan to guide sales and marketing promotion. The only marketing activity is to call or visit the army relative office, asking their annul plan and then set out a few marketing goals in a small-grouped discussion. But actually, the annual marketing plan should include the construction and

maintenance of military marketing, the military customer relationship management, marketing planning activities, market demand demonstration, new product development and etc. There's a lot to do.

D. Military marketing institutional mechanisms is unsound

First, there are too many levels from business leaders to marketing manager. Second, both the business processes and the departmental responsibilities are vague. Thirdly, evaluation and incentive mechanisms are imperfect. Finally and the most importantly, integration and process flow of market information is not well identified. Not only the internal collaboration is not emphasized as it should be, but also the point of how important the mutual information sharing is not fully understand. Basically, all departments operate independently. Therefore, the market sector misses a lot of information. Even the market information within the department is often twisted or distorted.

E. The sense of initiative competition in procurement is weak

At first, implementation of the procurement of weapons and equipment is under the control of mandatory procurement plan, which any competition between businesses and enterprises does not exist. After the reforms, mandatory plans supply relationship had gradually changed to contract ordering relationship. Open tender and inviting bidding have been introduced to military procurement. However, this competitive procurement methods are mostly used in general military supplies, not in weapons and equipment purchases due to the impact of the defense industry fixed layout, which resulting in military enterprises ignorance of studies on aviation military market demand to create market opportunities. Recently, the competition among aviation military enterprises rarely reflected in the aviation models or products, but more reflected in the competition for limited military spending. In other words, companies, which have a thorough study on military needs and take the initiative to guide and to meet military needs at the same time, will get greater benefits on military assignment.

IV. SUGGESTION ON MARKETING STRATEGY FOR AVIATION MILITARY ENTERPRISES

National Development No. 37 has been promulgated and implemented, which has a significant impact on all military enterprises, including aviation military manufacturing enterprises. On the basis, this part of the paper is mainly talking about what kind of military aviation business marketing strategy should be developed and how to develop it. We will talk about the issue from two different aspects.

A. Overall marketing Capacity

To meet the requirements of market, enterprises should focus on the "big marketing" concept to make up for the lack of capacity in some areas in the past so as to enhance their overall marketing capabilities. The following aspects should be taken into consideration.

1) Strengthen the capacity of research and demonstration on military demands

To develop military industries requires the ability to find what the customer needs. To meet the needs of army, we must strengthen the military market research functions first as well as improve the military market research system to realize the goal of grasping market opportunities, guiding the market demands, promoting the formation of military products. Second, outsourcing or uniting the related research institutions (or universities) could be taken into the pattern to carry out feasibility studies. For instant, enterprises may strengthen the cooperation with relevant institutions or universities, taking use of their influence to communication with the army in a better way. Finally, on the basis of the full study on military demands, military products marketing strategies should be developed and the market public relations measures should be carried out, especially with government and army. The ultimate goal is to strive to make the enterprises military development plan taken into the army's long-term development plan, to ensure the sustainable development of military operations.

2) Broaden the financing channels and enhance the military's cost competitiveness

Aviation military enterprises can take advantage of the nation promoting the reform of military investment environment, setting up new company with external specialized companies, social capital to undertake non-core business. What's more, enterprises should strengthen the impact and control on the upstream and downstream industry chain through the establishment of supplier entry standards system in order to improve engineering outsourcing ability and to attract a wide range of social resources integrated into its industrial system. By attracting external resources, aviation military enterprises could reduce their investments due to the increasing of unit production costs rising and enhance the products competitiveness.

3) Enhance the capabilities of military repairing

Most aviation military enterprises have a complete set of system from designing, manufacturing, materials purchasing to quality assurance and support service, companied by a fixed maturity raw material procurement channels, relatively stable supplier relations and technical information and experience on military modification and maintenance. All of the above are surely the advantages of military repairs. So why not integrate internal resources and combine with external resources to form a complete military aviation maintenance entity? The entity will successfully participate in the military reform and restructuring.

B. Overall marketing system

1) Construction of a modern marketing system

Facing the increasingly fierce competition, aviation military enterprises must lay the foundation of marketing work. The establishment of a modern marketing system and improving demand management, product management, R & D management, marketing promotion, customer service processes are required. In addition, the marking system, the R&D system, sales system and supply chain system should be combined together as a whole marketing system so that all departments and units could play a role in the marketing system.

2) Improve the marketing management

From R&D to sales, the military enterprises and customer are inseparable. Therefore, military enterprises should strengthen the whole process of marketing management. In addition to an annual contract, enterprises should spare no effort on pre-research and model development tasks, which is the key of follow-up product ordering and opening up new markets.

3) Strengthen marketing strategy planning

Marketing departments should carry out military operations positioning analysis, competitor analysis, internal and external environmental analysis, marketing strategy objectives (including sales revenue and growth, market share growth and other strategic objectives) and project planning (including liability leadership, target market, marketing team, competitors, competitive strategy, work plans, etc.).

4) Implementation of information technology on dynamic customer relationship management

Under the current circumstances, aviation military enterprises must be well aware of the customers and follow-up their needs in time. Even the product updates should regularly be introduced to the customer to make them repeat customers. Without complete customer information management systems, this is impossible. As a whole, enterprises should focus on the product's entire value chain and improve customer perceived value. Good customer relationship management using the management information system is the first step.

5) Establish incentives of marketing

In order to scientifically evaluate the performance of military sales staff who working in the marketing process and fully mobilize the enthusiasm, the evaluation and incentives measures should be proposed. Military project can be divided in to work units. Each unit of work can be divided into several independent expression, measurements and evaluation units in basic. All projects must have clear deadline. The actual completion time would be compared with the deadline as the basis for the assessment so that sales staff get their rewards or punishment.

REFERENCES

[1] Li Ming, Mao Jingli. Equipment procurement theory and practice [M]. National Defense Industry Press, Beijing: 2003.

[2] Liu Lin. AVIC market management system construction and practice of [Z], 2012.

[3] Hu Jinlin, Liu Yongqing. Reform of market sale management of the military enterprises [J]. Military Economic Research, 1998 (11).

[4] Guo Hong. Marketing elements in military market [J]. Modern economy, 2008 (8).

[5] Philip Kotler. Marketing Management [M]. Shanghai People's Publishing House, 2003.

[6] Li BenHui, Deng Desheng. Corporate marketing planning practice [M]. Chinese Economic Publishing House, 2008.

[7] Wang Xiyuan. Study on market mechanism to promote and assure the innovation and development of war industry [J]. Technology and innovation management, 2008(3).

[8] Zhang Yuexian, Ma Qinhai and Liu Ruping. A literature review on the relationship among expectancy disconfirmation customer enotion and customer satisfaction [J]. Management Review, 2004(4).

[9] The State Council, the Central Military Commission. Advices on The establishment and improvement of military and civilian combination of weaponry research and production systems (National Development [2010] No. 37) [Z], 2010.

[10] Lin Zuoming. Grasp the pulse of the era adhere to the user-oriented marketing and make efforts to open up a new situation in the company [Z], 2011

[11] Ma Zeping, Zhai Zhigang. To achieve synergistic integration of the business group's customer resource intensive management [Z]. National compilation of enterprise management innovation, enterprise management Press, 2011.

Analysis on Key Influence Factors of Chinese Special Equipment Safety Supervision Capacity

Zhu Zeng[1,*], Yun Luo[1], Jie Feng[2], Yu-qian Sun[1], Yan-peng Yang[1]

[1]School of Engineering and Technology, China University of Geosciences (Beijing), Beijing, China
[2]Information & Communication Research Department, China Electric Power Research Institute, Beijing, China
(zengzhu2466@163.com)

Abstract - **Improving the capacity of special equipment safety supervision is significant to ensure the safety of special equipment. Special equipment safety supervision capacity was defined based on organizational capability theory. 20 factors that influence the capacity of special equipment safety supervision were identified by the method of literatures metrology and expert investigation. By calculating the Spearman correlation coefficient between the influence factors and safety performance index, the number of accidents per ten thousand sets of equipment, 10 key influence factors that correlate significantly with special equipment safety supervision capacity were finally identified. Calculation results showed that, on the overall trend, the influence factors of special equipment safety supervision capacity presented negative correlation with safety performance index. The capacity of special equipment safety supervision can be improved and the accident probability can be reduced through increasing the input of supervision resources, improving the quality of supervisors and performing strict supervision.**

Keywords - **Correlation analysis, influence factors, special equipment safety supervision, safety supervision capacity, safety performance**

I. INTRODUCTION

Special equipment in China includes boilers, pressure vessels, elevators, lifting appliances, passenger ropeways, large amusement devices and automobiles which is related to safety and has high risks [1]. Special equipment is closely related to national economic construction and people's life [2]. Also, the safety of special equipment is directly related to the safety of people's lives and properties, as well as the stability of economic operation [2]. As one of the important means to ensure the safety of special equipment, special equipment safety supervision is attracting more and more attention from the whole society [3].

According to organizational capability theory, influence factors of organizational capability include: organization resources and functions to perform [4-6]. In China, the organizations responsible for special equipment safety supervision include: special equipment safety supervisory institutions and the inspection agencies [3]. The capacity of safety supervision is directly influenced by the resource allocation and the function implementation status of these two kinds of organizations. Therefore, based on the organization feature of special equipment safety supervision and organizational capability theory, special equipment safety supervision capacity was defined as: the ability that special equipment safety

supervision organization equips with necessary basic resources, performs relevant necessary supervision and inspection to ensure the safety of special equipment. Aimed to study special equipment safety supervision capacity systematically, the initial 20 influence factors were determined by the method of literature metrology and expert investigation [7-9], see Table I.

The main purpose of special equipment safety supervision is to prevent and reduce the special equipment safety accidents and improve special equipment safety performance. The number of accidents per ten thousand sets of equipment is always used as the basic index to measure special equipment safety performance in China. Studying the correlation between the influence factors of special equipment safety supervision capacity and the number of accidents per ten thousand sets of equipment through correlation analysis can provide a scientific basis for improving special equipment safety supervision capacity and safety performance.

II. METHODS

A. Data collection

Statistical data related to special equipment safety status from 2006 to 2010 in China were selected as sample [10-14], see Table I.

According to the same sources in Table I, the data of safety performance index of special equipment was summarized, see Table II.

B. Analytical method

Correlation analysis [15] is a method that analyzes the correlation degree between two variables and describes the degree of relationship between the two variables. It reflects the magnitude of changes of one value when the other value is controlled. Correlation coefficient is the index that used to measure the degree of correlation between variables. Suppose there are two sequences: $a = (a_1, a_2, \wedge, a_n)$, $b = (b_1, b_2, \wedge, b_n)$. According to the principle of statistics, the correlation coefficient between the two sequences is:

$$r(a,b) = \frac{\mathrm{cov}(a,b)}{\sqrt{D(a)D(b)}} = \frac{\sum_{i=1}^{n}(a_i - \overline{a})(b_i - \overline{b})}{\sqrt{\sum_{i=1}^{n}(a_i - \overline{a})^2 (b_i - \overline{b})^2}} \quad (1)$$

Among: $\overline{a}$, $\overline{b}$ are the average of the two

E. Qi et al. (eds.), *Proceedings of the 21st International Conference on Industrial Engineering and Engineering Management 2014*, Proceedings of the International Conference on Industrial Engineering and Engineering Management, DOI 10.2991/978-94-6239-102-4_111, © Atlantis Press and the authors 2015

sequences respectively, n is the sample number of the two sequences.

The value range of correlation coefficient r is

$-1 \leq r \leq 1$ and the relationship between correlation coefficient and correlation is shown in Table III.

TABLE I
THE ANNUAL STATISTICS OF THE INFLUENCE FACTORS OF SPECIAL EQUIPMENT SAFETY SUPERVISION CAPACITY IN CHINA

Three-level index of special equipment safety supervision capacity (number)	2006	2007	2008	2009	2010
Number of supervisory organ (D_1)	2984	2994	3012	3049	3092
Number of supervisory personnel (D_2)	8907	9119	9363	9624	9874
Number of certified supervisory personnel (D_3)	7208	7068	6821	8007	8194
Proportion of certified supervisory personnel (D_4)	80.93	77.51	72.85	83.20	82.99
Number of supervisory personnel per ten thousand sets of equipment (D_5)	22.06	20.57	17.97	16.52	15.25
Number of supervision and inspection (D_6)	2114600	2707000	2380000	776000	716000
Number of times for supervision and inspection of ten thousand sets of equipment (D_7)	5237.66	6106.20	4567.17	1332.05	1105.54
Number of rectification for hidden dangers (D_8)	303000	363000	418000	237000	198000
Number of rectification for hidden dangers of ten thousand sets of equipment (D_9)	750.50	818.82	802.13	406.83	305.72
Number of inspection agency (D_{10})	2903	2889	2802	2485	2503
Number of staff of inspection agency (D_{11})	23969	25056	53797	56372	57533
Number of certified inspection personnel (D_{12})	19209	19651	28737	51823	52573
Proportion of certified inspection personnel (D_{13})	80.14	78.43	53.42	91.93	91.48
Number of staff with college degree or above (D_{14})	18232	19669	30355	33588	35897
Number of engineering technicians (D_{15})	19834	20746	34704	35463	36028
Number of intermediate and above engineers (D_{16})	10564	10998	16619	16929	17021
Number of supervision and inspection for product safety performance (D_{17})	1090000	1310000	1320000	1120000	3490000
Number of supervision and inspection for installing safety performance (D_{18})	306500	428200	540700	569600	604900
Number of periodical inspection (D_{19})	2204500	2241800	2506900	2696400	3036600
Proportion of periodical inspection (D_{20})	84.5	95.6	89	95.2	96.1

TABLE II
THE ANNUAL STATISTICS OF SPECIAL EQUIPMENT SAFETY PERFORMANCE INDEX IN CHINA

special equipment safety performance index	2006	2007	2008	2009	2010
Number of accidents per ten thousand sets of equipment /number $\cdot$ Million^{-1}	0.74	0.58	0.59	0.65	0.46

TABLE III
THE RELATIONSHIP BETWEEN CORRELATION COEFFICIENT AND CORRELATION

Correlation	Negative correlation	Positive correlation
Low correlation or uncorrelated	$-0.3\sim0.0$	$0.0\sim0.03$
Medium correlation	$-0.5\sim-0.3$	$0.3\sim0.5$
Significant correlation	$-1.0\sim-0.5$	$0.5\sim1.0$

TABLE IV
CORRELATION COEFFICIENT BETWEEN SAFETY SUPERVISION CAPACITY INFLUENCE FACTORS AND SAFETY PERFORMANCE INDEX

Three-level index number	D_1	D_2	D_3	D_4	D_5	D_6	D_7	D_8	D_9	D_{10}
Correlation coefficient	-0.200	-0.600	-0.600	0.100	0.600	0.100	0.300	0.200	0.477	0.300
Significance level (two-tailed)	0.774	0.285	0.285	0.873	0.285	0.873	0.624	0.747	0.450	0.624
Three-level index number	D_{11}	D_{12}	D_{13}	D_{14}	D_{15}	D_{16}	D_{17}	D_{18}	D_{19}	D_{20}
Correlation coefficient	-0.600	-0.600	0.100	-0.900	-0.900	-0.600	-0.900	-0.600	-0.600	-0.100
Significance level (two-tailed)	0.285	0.285	0.873	0.37	0.37	0.285	0.37	0.285	0.285	0.873

Common correlation coefficients include: Pearson product-moment correlation coefficient which requires random variables that need verified to conform normal distribution; Spearman correlation coefficient is a nonparametric rank statistical parameter (has nothing to do with the distribution) and is used widely in practice; Kendall rank correlation coefficient is the index which reflects classified variable's correlation. It can be used when the two classified variables are orderly classified.

Nonparametric correlation analysis was carried out on related ordinal variables, the values ranged from -1 to 1 and the analysis method was suitable for square table. Therefore Spearman rank correlation analysis was selected to confirm the correlation between the influence factors of special equipment

safety supervision capacity and safety performance index.

C. Correlation analysis

The data of the influence factors of special equipment safety supervision capacity in Table I and the data of safety performance index, the number of accidents per ten thousand sets of equipment in Table II were selected to conduct Spearman nonparametric rank correlation analysis. Take D_1 for example, calculation process is as follows:

According to formula (1), put the value of D_1 into variable a, put the value of P_1 into variable b, then the correlation coefficient between D_1 and P_1 is calculated, r=-0.600. The negative correlation coefficient indicates that D_1 and P_1 are negative correlated, and the lower significance level (two-tailed) indicates that the analysis result is reliable.

III. RESULTS

As the quantity of samples was significant, SPSS18.0 was used to conduct the analysis, the calculation results were shown in Table IV.

IV. DISCUSSION AND RECOMMENDATIONS

The correlation coefficients in Table IV show that the number of accidents per ten thousand sets of equipment and D_2, D_3, D_{11}, D_{12}, D_{14}, D_{15}, D_{16}, D_{17}, D_{18}, D_{19} have significant negative correlation. It indicates that the relative quantity of accidents reduce with the increase of supervision resource investment and inspection human resource, the improvement of supervisors' quality and the amount and quality of inspection. These ten indexes and special equipment safety supervision capacity are positive correlated. The capacity of special equipment safety supervision increases with the growth of these ten influence factors. On the other side, the calculating results show that D_4, D_5, D_6, D_7, D_8, D_9, D_{10}, D_{13} and the number of accidents per ten thousand sets of equipment have positive correlation. Furthermore, it appears that D_1, D_{20} and the number of accidents per ten thousand sets of equipment have negative correlation, but the significance level of correlation result is too high to confirm the correlation. So D_1, D_4, D_5, D_6, D_7, D_8, D_9, D_{10}, D_{13} and D_{20} cannot be selected as the key influence factors of special equipment safety supervision capacity.

V. CONCLUSION

According to organizational capability theory and the structure of special equipment safety supervision organization in China, special equipment safety supervision capacity has been defined. 10 key influence factors of special equipment safety supervision capacity which appear significant negative correlation with the number of accidents per ten thousand sets of equipment were confirmed through correlation analysis. Special equipment safety supervision capacity can be strengthened and special equipment accidents can be prevented through increasing the input of supervision and inspection resources, improving the quality of supervisors, strengthening the rectification of equipment hidden dangers, intensifying the level of inspection and performing stricter inspection.

ACKNOWLEDGEMENT

This study was funded by a grant of key projects in the National Science & Technology Pillar Program during the Twelfth Five-year Plan Period (2011BAK06B06).

REFERENCES

[1] State Council of the People's Republic of China, Special equipment safety supervision regulations (in Chinese), Beijing: China Legal Publishing House, 2009.
[2] JIANG Shu-jun, DING Ri-jia, HAO Su-li, Research on risk evaluation index system construction of special equipment using unit (in Chinese), Industrial Safety and Environmental Protection, vol.38, no.7, pp.42-44, 2012.
[3] WANG Qin-ping, The research of special equipment safety monitors (in Chinese), Beijing: China Metrology Press, pp.121-158, 2006.
[4] Grant R M, The resources-based theory of competitive advantage: implications for strategy formulation, California Management Review, vol.33, no.3, pp.114-135, 1991.
[5] Amit R, Schoemaker P J H. Strategic assets and organizational rent, Strategic Management Journal, vol.14, no.1, pp.33-46, 1993.
[6] Helfat C E, Peteraf M A, The dynamic resource-based view: capability life cycles, Strategic Management Journal, vol.24, no.10, pp.997-1010, 2003.
[7] FENG Jie, LUO Yun, ZENG Zhu, CUI Gang, HUANG Qiang-hua, LENG Hao, et al, Correlation analysis between special equipment safety performance and safety supervision (in Chinese), China Safety Science Journal, vol.22, no.2, pp.170-176, 2012.
[8] LIANG Jun, CHEN Guo-hua, Discussion on establishment of special equipment risk management system and its key issues (in Chinese), China Safety Science Journal, vol.20, no.9, pp.132-138, 2010.
[9] General Administration of Quality Supervision, Inspection and Quarantine of the People's Republic of China. Special equipment safety development strategies outline (in Chinese), 2010.
[10] General Administration of Quality Supervision, Inspection and Quarantine of the People's Republic of China, Special equipment accident report 2006 (in Chinese), 2007.
[11] General Administration of Quality Supervision, Inspection and Quarantine of the People's Republic of China, Special equipment accident report 2007 (in Chinese), 2008.
[12] General Administration of Quality Supervision, Inspection and Quarantine of the People's Republic of China, Special equipment accident report 2008 (in Chinese), 2009.
[13] General Administration of Quality Supervision, Inspection and Quarantine of the People's Republic of China, Special equipment accident report 2009 (in Chinese), 2010.
[14] General Administration of Quality Supervision, Inspection and Quarantine of the People's Republic of China, Special equipment accident report 2010 (in Chinese), 2011.
[15] YANG Jin-xiu, HU Wang-lian, Principles of statistics (in Chinese), Changsha: Central South University Press, pp.258-265, 2007.

An Empirical Study of Environmental Kuznets Curve in China

YANG Zhoumu[1,2,*], WANG Wenping[1], YANG Yibo[2], FANG Fen[3]

[1]School of Economics and Management, Southeast University, P. R. China, 211189

[2]School of Mathematics and Statistics, Nanjing University of Information Science and Technology, P. R. China, 210044

[3]Department of Foundation, Jinling Institute of Technology, P. R. China, 211169

(Yangzhoumu1978@163.com)

Abstract - **Environment Kuznets Curve (EKC) hypothesis is a classic expression of describing the relationship between economy and environment. Based on the theory of EKC, this paper established an econometric model between economic development and environmental pollution according to the data of sulfur dioxide emission and per capita GDP of China over the period of 1998 to 2012. Regression analysis shows that relationship between the sulfur dioxide emissions and per capita GDP displays a typical N shape, and the present situation of China is at the rising stage after two tunning points. Economic development of China is still at the expense of sacrificing environment to some extend. Pollution has been rising along with the development of economy, and this trend will not change in the short term. Transforming and upgrading from factor-driven and capital-driven to innovation-driven is particularly important for economic transformation and upgrading in China.**

Keywords - **EKC, economic development, environment pollution, per capita GDP, sulfur dioxide emissions**

I. INTRODUCTION

With the collapse of resources and deterioration of environment, research on the relationship between the economy and environment has been widespread concerned among scholars of different fields as an important social science problem. EKC theory was first mentioned in the research on relationship between environmental change and economic growth. In 1991, Grossman and Kruger [1] analyzed the data of urban air quality estimated by the Global Environmental Monitoring System (GEMS), finding that the inversed-U shaped relationship existed with respect to sulfur dioxide and soot emission. Then, as leaders, Grossman and Kruger [2] empirically studied the relationship between economic development and environment: in the early stage of development, environmental quality is easily deteriorated with the process of economic development; When the economy develops to a certain level, deterioration of the environment will reach its peak; After that, with the increase of per capital income and the development of economy, environmental pollution will get ease owing to structure and technology effect [3], and then environment quality will improve gradually. After Grossman and Kruger, many empirical studies showed that the inverted-U shaped relationship between most environmental quality indicators and per capita income does exist. Shafik and Bandyopadhyay [4] sthdied the relationship between ten environmental factors and per capita income in 149 countries by a logistic form and found that inverted-U shape existed in the relations between the density of sulfur dioxide and suspended particles and per capita income. Selden and Song [5] selected a cross-national panel of data on emissions of four air pollutants: suspended particulate matter, carbon monoxide, sulfur dioxide and nitrogen oxides, discovering that per capita emissions of these four pollutants exhibited inverted-U relationships with per capita GDP and forecasting that emissions in global world would continue growing rapidly over the next several decades. Of course, various doubts have been exercised on the theory of Environmental Kuznets Curve (EKC) [6-11]. Alexi [12] suggested that research of EKC could be used to control the population. He chose two models with different population quantity and found that these two models displayed two different turning points. Besides, he suggested the future research of EKC can turn to study how population influence the EKC model. Wu Haiying and Zhang Shenglin [13] modified the mathematical model of EKC-theory with a cubic equation of higher goodness-of-fit, finding that "U" type curve existed between the waste water, solid waste and waste gas emissions and per capital GDP. Liu Genyao [14] constructed respectively the linear, quadratic and cubic model to analyze and finally found that the cubic model has a best fitting effect.

It is well known that the sulfur dioxide is one of the atmospheric pollutants and is also the main component of acid rain. Even though many countries haven't take strict policy regulations on its emissions, lots of scholars have suggested reducing its emissions to protecting environment. Among literatures studying the relationship between economic development and environment, number of scholars favor in sulfur dioxide [15, 16]. This paper will choose the sulfur dioxide emissions alone as the environmental pollution indicator to explore the relation between economy and environment by analyzing the evolution rule of sulfur dioxide emissions and per capita GDP.

II. SULFUR DIOXIDE EMISSION AND PER CAPITAL GDP IN CHINA

A. Sources of data

The data of per capita GDP and sulfur dioxide emission required in this paper are taken from *China statistic yearbook*. Data of sulfur dioxide emission during the period of 1998 and 2012 are based on industrial and life sulfur dioxide emission, summaried by author.

B. Descriptive statistics of data

Fig.1 presents the trends of SO_2 emissions and per capita GDP during the period of 1998 and 2012.

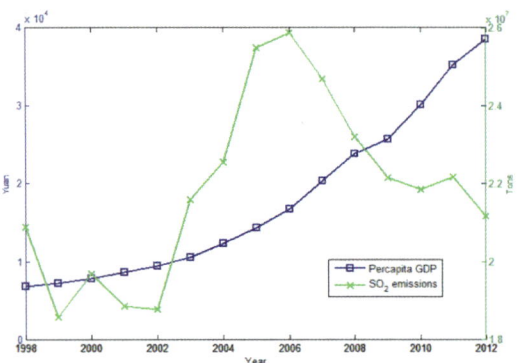

Fig.1. SO_2 emissions and per capita GDP

III. THEORETICAL MODEL AND THE RESULT OF DATA FITTING

A. Model

Referring to literature [3], this paper explores a cubic regression model:

$$y_{it} = \alpha_i + \beta_1 x_{it} + \beta_2 x_{it}^2 + \beta_3 x_{it}^3 + \beta_4 z_{it} + \mu_{it}, i = 1,2,\cdots,N.t = 1,2,\cdots,T.$$

In Eq. above, y_{it} is the depended variable of environmental degradation, x_{it} is the independent variable of income, z_{it} reflects other variables that may affect y_{it}, α_i is the constant term, and β_i are the estimated coefficients of the explanatory variables. The μ_{it} represents the error term.

B. Results of fitting

As show in Table I, we can see the cubic regression equation is

$$y_{it} = 6.136 \times 10^6 + 2.424 \times 10^3 x_{it} - 0.101 x_{it}^2 + 1.248 \times 10^{-6} x_{it}^3, i = 1,2,\cdots,N.t = 1,2,\cdots,T.$$

The t-test sig=0.003.

The relationship between SO_2 emissions and per capita GDP discussed above is represented in Fig.2.

TABLE I
MODEL SUMMARY AND PARAMETER ESTIMATES

Dependent Variable:SO2emissions

Equation	Model Summary					Parameter Estimates			
	R Square	F	df1	df2	Sig.	Constant	b1	b2	b3
Cubic	.707	8.852	3	11	.003	6.136E6	2.424E3	-.101	1.248E-6

The independent variable is Percapita GDP

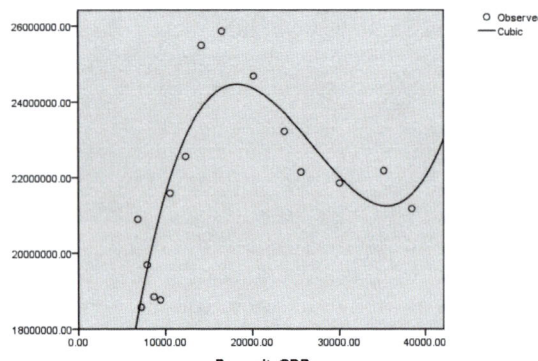

Fig.2. SO_2 emissions and per capita GDP

IV. CONCLUSIONS

According to the analysis of regression, the coefficient of quadratic term of the model which explores the relationship between sulfur dioxide emission and per capital GDP is less than zero, and the coefficient of cubic term is greater than zero. Therefore, the curve fitted by this model is N shape. Calculating the first order

derivative, we can find that the first turning point is located at per capital GDP ￥17937 during 2006 and 2007, i.e, after the first turning point appearing between 2006 and 2007, sulfur dioxide emission showed a trend of decline. The scend turning point appeared during the year of 2010 and 2011,which is located at per capital GDP ￥33870,and then the trend of sulfur dioxide emission turned to rise again.

The above results show that measures taken for environmental protection are effective and we should continue adhering to the governance and improvement of environment, but there still exists a certain repeatability for sulfur dioxide emission. Economic growth in China is still at the expense of the environment to some extent. Pollution has been rising along with the development of economy, and this trend will not change in the short term. Transforming and upgrading from factor-driven and capital-driven to innovation-driven is particularly important for economic transformation and upgrading in China.

ACKNOWLEDGMENT

This research was financially supported by Major Bidding Project of the National Social Science

Foundation of China (Grant No. 12&ZD207), the National Natural Science Foundation of China (Grant No. 70973017, 71172044, 71273047), Research Fund for the Doctoral Program of Higher Education of China (Grant No. 20120092110039), Major Project of Philosophy and Social Science Research of the Higher Education Institutions of Jiangsu Province (Grant No. 2014ZDAXM002).

REFERENCES

[1] Grossman, G.M, Krueger, A. B. Environmental impacts of a North American Free Trade Agreement [A]. National bureau of economic research working paper 3914, NBER [C], Cambridge MA., 1991.

[2] Grossman, G.M., Krueger, A.B. Economic growth and the environment [J]. The Quarterly Journal of Economics, 1995, 110(2), 353-377.

[3] Dimitra Kaika, Efthimios Zervas. The Environmental Kuznets Curve (EKC) theory - Part A: Concept, causes and the CO2 emissions case [J]. Energy Policy, 2013, 62: 1392-1402.

[4] Shafik, N, & Bandyopadhyay, S. Economic growth and environmental quality: time series and cross country evidence. Background Paper for the World Development Report 1992 [M], The World Bank, Washington, DC. , 1992.

[5] Selden T. M., Song D. Environmental quality and development: is there a Kuznets curve for air pollution emissions? [J]. Journal of Environmental Economics and management, 1994, 27(2): 147-162.

[6] Focacci A. Empirical analysis of the environmental and energy policies in some developing countries using widely employed macroeconomic indicators: the cases of Brazil, China and India [J]. Energy Policy, 2005, 33(4): 543-554.

[7] Vehmas J, Luukkanen J, Kaivo-Oja J. Linking analyses and environmental Kuznets curves for aggregated material flows in the EU [J]. Journal of Cleaner Production, 2007, 15(17): 1662-1673.

[8] Luzzati T, Orsini M. Investigating the energy-environmental Kuznets curve [J]. Energy, 2009, 34(3): 291-300.

[9] Kijima M, Nishide K, Ohyama A. EKC-type transitions and environmental policy under pollutant uncertainty and cost irreversibility [J]. Journal of Economic Dynamics and Control, 2011, 35(5): 746-763.

[10] Henriques C O, Antunes C H. Interactions of economic growth, energy consumption and the environment in the context of the crisis–a study with uncertain data [J]. Energy, 2012, 48(1): 415-422.

[11] Bimonte S. Public goods, environmental quality and the EKC – the 'unsaid' of the intensity of use indices [J]. International Journal of Sustainable Economy, 2012, 4(2): 167-180.

[12] Alexi T. Accounting for population in an EKC for water pollution [J]. Journal of Environmental Protection, 2013.

[13] Wu Haiyin, Zhang Shenglin. Empirical study on the relationship between economic development in the west area and environmental equality [J]. Social Sciences in Ningxia, 2005, (5): 29-33.

[14] Liu Genyao, Sheng Long. Study on the relationship between economic growth and environment pollution based on the hypothesis of environmental Kuznets curve-take Zhejiang Province as an example [J]. Journal of Industrial Technological Economics, 2012(4): 28-35.

[15] FAN Jin, Hu Hanhui. Studies and applications of Environmental Kuznets Curve (EKC) [J]. Mathematics in Practice and Theory, 2002, 32(6): 944-951.

[16] Bao limin. On Some modern interpretations of epicures' philosophy [J]. Fudan Journal (Social Sciences), 2004, 2: 87-94.

Based on Credibility Measure The Force of Mortality in Life Insurance Actuarial

Min-ying YUAN, Wen-wen HAN, Da-jun SUN[*]

Department of Mathematics and Information Sciences, Langfang Teachers College, Langfang, China

(ymy8219@sina.com)

Abstract - **In order to establish continuous life insurance actuarial model theoretical system based on credibility measure, this paper defined the force of mortality based on credibility measure, and deduced the expressions of future lifetime distribution function, survival function, density function and actuarial notation by the force of mortality. Accordingly, it established models of continuous life functions based on credibility measure.**

Keywords - **Credibility measure, force of mortality, residual life distribution function, survival function**

I. INTRODUCTION

Life insurance actuarial uses the method of modern mathematics and mathematical statistics to quantitatively analyze and research the premiums, reserves, and cash value in life insurance. The two basic types involved in were the discrete and the continuous. The essential difference is the discreteness and continuity of the uncertainty of life in life insurance. In the continuous life insurance actuarial models, one of the basic problem is to define the force of mortality and to express the life functions by force of mortality. The force of mortality is the vital evidence to price the premium, and the foundation to construct theoretical system of continuous life insurance actuarial. The force of mortality shows the possibility that one man dies at a certain time, and used to be defined by probability measure, as traditional life insurance actuarial models are built in probability space [1-5]. Probability measure is the one that fulfills countable additive, yet countable additive is a very harsh condition, and hardly does the uncertainty in life insurance which one man dies at a certain time satisfy this condition. Thus, we need to find a non-additive measure, more widely-used and flexible than probability, to measure the uncertainty of man's survival or death. To deal with the complexity of the uncertainty of man's survival or death, The reference [6] substitutes non-additive quasi probability measure for probability measure, and deduces the basic formula of life insurance actuarial models on proposed probability space, further to make the application of the theoretical model of life insurance actuarial broader and more efficient. Credibility measure is an uncertain measure, more widely-used than probability measure. It possesses the advantages like self-duality and subadditivity, which make it a new method to describe fuzziness. The reference [7-8] points out that the uncertainty of human lives in life insurance perform more like fuzziness, and believes that human lives are fuzzy variables on credibility space. Using credibility measure, it expands life insurance actuarial theory onto credibility space. On the theories of references [7-10], this paper discusses the force of mortality in life insurance actuarial on the basis of the credibility measure, defines the force of mortality also on the basis of credibility, and gives the expressions of life distribution function, survival function, density function and actuarial symbol based on the force of mortality. Thus, it lay the foundations for the establishment of continuous life insurance actuarial models based on credibility measure.

II. PRELIMINARY

Definition 1 [11] Let Θ be a nonempty set, and let $\mathrm{P}(\Theta)$ be the power set of Θ. Each element $A \in \mathrm{P}(\Theta)$ is called an event. A set function $Cr: \Theta \to [0,1]$ is called a credibility measure if $(i) Cr\{\Theta\} = 1$; (ii) $Cr\{A\} \leq Cr\{B\}$ whenever $A \subset B$; (iii) $Cr\{A\} + Cr\{A^c\} = 1$ for any event A; (iv) $Cr\left\{\bigcup_i A_i\right\} = \sup_i Cr\{A_i\}$ for any events $\{A_i\}$ with $\sup_i Cr\{A_i\} < 0.5$.

The triplet $(\Theta, \mathrm{P}(\Theta), Cr)$ is called a credibility space [12].

Definition 2[13] A fuzzy variable is a function from credibility space $(\Theta, \mathrm{P}(\Theta), Cr)$ to the set of real numbers.

Definition 3[14] The credibility distribution $\Phi: \Re \to [0,1]$ of a fuzzy variable ξ is defined by

$$\Phi(x) = Cr\{\xi \leq x\}$$

Definition 4[7] Let $T(0)$ be anyone's life span. The life distribution function is defined by

$$\Phi_0(x) = Cr\{T(0) \leq x\} \ (x > 0).$$

$\Phi_0(x)$ represents the credibility that a neonatal (i.e., 0 years old) died in the next x years.

Definition 5 [7] Let $T(x)$ be x years old one's residual life span. The residual life distribution function is defined by

$$\Phi_x(t) = Cr\{T(x) \leq t\} \ (t > 0).$$

$\Phi_x(t)$ represents the credibility that a x years old one died in the next t years. In fact, it reflects the credibility that a 0 years old one died before $x + t$ years old in condition that who lives x years old.

Definition 6 [7] Let $T(0)$ be anyone's life span. The Infant survival function is defined by

$$S_0(x) = 1 - \Phi_0(x) = Cr\{T(0) > x\}(x \geq 0)$$

$S_0(x)$ represents the credibility that at the age of 0, people in the x years old still survive.

Definition 7 [7] Let $T(x)$ be x years old one's residual life span. The survival function of a person age x years old is defined by

$$S_x(t) = Cr\{T(x) > t\}(t > 0)$$

$S_x(t)$ represents the credibility that x years old one in the $x+t$ years old still survive, $S_x(t) = 1 - \Phi_x(t)$.

Definition 8 [15] The credibility density function $\phi: \mathfrak{R} \to [0,+\infty)$ of a fuzzy variable ξ is a function such that

$$\int_{-\infty}^{+\infty} \phi(y)\,dy = 1, \Phi(x) = \int_{-\infty}^{x} \phi(y)\,dy, \forall x \in \mathfrak{R}$$

where Φ is the credibility distribution of the fuzzy variable ξ.

Theorem 1 [7]

$$Cr\{x < T(0) \le x+t \mid T(0) > x\}$$

$$= \begin{cases} \dfrac{\Phi_0(x+t)}{1-\Phi_0(x)}, & \text{if } \Phi_0(x{+}t) < 0.5 - 0.5\Phi_0(x) \\[2mm] 1 - \dfrac{1-\Phi_0(x+t)}{1-\Phi_0(x)}, & \text{if } \Phi_0(x+t) > 0.5 + 0.5\Phi_0(x) \\[2mm] 0.5, & \text{otherwise} \end{cases}$$

The definition of actuarial symbol on the credibility space [7]:

$_t q_x$ —The credibility which a person ages x won't be alive after $x+t$ years old, that is $_t q_x = \Phi_x(t)$.

$_t p_x$ —The credibility which a person ages x is still alive after $x+t$ years old, that is $_t p_x = 1 - _t q_x$.

$_{t|u} q_x$ — The credibility which a person ages x dies between $x+t$ and $x+t+u$ years old, that is

$$_{t|u} q_x = Cr\{t < T(x) \le t+u\}$$

Theorem 2 [8] Let $(\Theta, P(\Theta), Cr)$ be a credibility space, $_{t|u} q_x$ represents the credibility which a person ages x dies between $x+t$ and $x+t+u$ years old. Then

$$_{t|u} q_x = [1 - _t q_x] \wedge _{t+u} q_x.$$

III. FORCE OF MORTALITY μ_x

This section mainly discusses the expression of life distribution function, survival function, density function and actuarial symbol based on the force of mortality. Firstly this paper gives the definition of the force of mortality on credibility space.

A. The definition of Force of Mortality Based on Credibility Measure

Definition 9 Let $(\Theta, P(\Theta), Cr)$ be a credibility space. Then the force of mortality of x years old man is defined by

$$\mu_x = \frac{Cr\{T(x) \le t\}}{t} \quad (t > 0).$$

That is

$$\mu_x = \frac{Cr\{T(x) \le t\}}{t} = \frac{Cr\{x < T(0) \le x+t \mid T(0) > x\}}{t}.$$

Form theorem 1, we obtain

$$\mu_x = \begin{cases} \dfrac{\Phi_0(x+t)}{[1-\Phi_0(x)]t}, & \text{if } \Phi_0(x+t) < 0.5 - 0.5\Phi_0(x) \\[3mm] \dfrac{1 - \dfrac{1-\Phi_0(x+t)}{1-\Phi_0(x)}}{t}, & \text{if } \Phi_0(x+t) > 0.5 + 0.5\Phi_0(x) \\[3mm] \dfrac{0.5}{t}, & \text{otherwise} \end{cases}$$

$$\dots\dots\dots\dots(*)$$

Especially when $x = 0$, since $\Phi_0(0) = 0$, we obtain

$$\mu_0 = \begin{cases} \dfrac{\Phi_0(t)}{[1-\Phi_0(0)]t}, & \text{if } \Phi_0(t) < 0.5 - 0.5\Phi_0(0) \\[3mm] \dfrac{\Phi_0(t) - \Phi_0(0)}{[1-\Phi_0(0)]t}, & \text{if } \Phi_0(t) > 0.5 + 0.5\Phi_0(0) \\[3mm] \dfrac{0.5}{t}, & \text{otherwise} \end{cases}$$

$$= \begin{cases} \dfrac{\Phi_0(t)}{t}, & \text{if } \Phi_0(t) < 0.5 \\[2mm] \dfrac{\Phi_0(t)}{t}, & \text{if } \Phi_0(t) > 0.5 \\[2mm] \dfrac{\Phi_0(t)}{t}, & \text{if } \Phi_0(t) = 0.5 \end{cases}$$

We have $\mu_0 = \dfrac{\Phi_0(t)}{t}$, namely $\Phi_0(t) = \mu_0 t$.

B. The Expression of Life Functions Based on μ_x

We will deduce some expressions of life distribution function, survival function, density function and based on the force of mortality. The following deduces related expressions by $(*)$.

(1) If $\Phi_0(x+t) < 0.5 - 0.5\Phi_0(x)$, then by using

$$\mu_x = \frac{\Phi_0(x+t)}{[1-\Phi_0(x)]t}, \text{ we obtain}$$

$$1 - \Phi_0(x) = \frac{\Phi_0(x+t)}{\mu_x \cdot t}.$$

By the Theorem1, we have

$$\Phi_x(t) = \frac{\Phi_0(x+t)}{1 - \Phi_0(x)} = \frac{\Phi_0(x+t)}{\Phi_0(x+t)} \cdot \mu_x \cdot t = \mu_x \cdot t,$$

$$S_x(t) = 1 - \frac{\Phi_0(x+t)}{1 - \Phi_0(x)} = 1 - \frac{\Phi_0(x+t)}{\Phi_0(x+t)} \cdot \mu_x \cdot t = 1 - \mu_x \cdot t,$$

$$\phi_x(t) = \mu_x.$$

Especially when $x = 0$, we have

$$\Phi_0(t) = \mu_0 \cdot t, \; S_0(t) = 1 - \mu_0 \cdot t, \; \phi(t) = \mu_0.$$

(2) If $\Phi_0(x+t) > 0.5 + 0.5\Phi_0(x)$, then by using

$$\mu_x = \frac{1}{t}\left[1 - \frac{1 - \Phi_0(x+t)}{1 - \Phi_0(x)}\right], \text{ we obtain}$$

$$1 - \frac{1 - \Phi_0(x+t)}{1 - \Phi_0(x)} = \mu_x \cdot t.$$

Further, we obtain

$$\frac{1 - \Phi_0(x+t)}{1 - \Phi_0(x)} = 1 - \mu_x \cdot t.$$

By the theorem1, we have

$$\Phi_x(t) = 1 - \frac{1 - \Phi_0(x+t)}{1 - \Phi_0(x)} = \mu_x \cdot t,$$

$$S_x(t) = \frac{1 - \Phi_0(x+t)}{1 - \Phi_0(x)} = 1 - \mu_x \cdot t,$$

$$\phi_x(t) = \mu_x.$$

Especially when $x = 0$, we have

$$\Phi_0(t) = \mu_0 \cdot t, \; S_0(t) = 1 - \mu_0 \cdot t, \; \phi(t) = \mu_0.$$

(3) Otherwise, by using $\mu_x = \dfrac{0.5}{t}$. By the Theorem1, we have

$$\Phi_x(t) = 0.5 = \mu_x \cdot t,$$

$$S_x(t) = 0.5 = 1 - 0.5 = 1 - \mu_x \cdot t,$$

$$\phi_x(t) = \mu_x.$$

Especially when $x = 0$, we have

$$\Phi_0(t) = \mu_0 \cdot t, \; S_0(t) = 1 - \mu_0 \cdot t, \; \phi(t) = \mu_0.$$

Based on the analysis above, we have

$$\Phi_x(t) = \mu_x \cdot t, \; S_x(t) = 1 - \mu_x \cdot t, \; \phi_x(t) = \mu_x.$$

$$\Phi_0(x) = \mu_0 \cdot x, \; S_0(x) = 1 - \mu_0 \cdot x, \; \phi(x) = \mu_0.$$

C. The Expression of Actuarial Symbol Based on μ_x

By the theorem 2, we have the expression of actuarial symbol that

$$_t q_x = \mu_x \cdot t,$$

$$_t p_x = 1 - \mu_x \cdot t.$$

$$_{t|u} q_x = \left[1 - {}_t q_x\right] \wedge {}_{t+u} q_x = (1 - \mu_x \cdot t) \wedge \left[\mu_x \cdot (t+u)\right]$$

$$= \begin{cases} 1 - \mu_x \cdot t, & \text{if } 1 - \mu_x \cdot t < \mu_x \cdot (t+u) \\ \mu_x \cdot (t+u), & \text{if } 1 - \mu_x \cdot t > \mu_x \cdot (t+u) \end{cases}$$

$$= \begin{cases} 1 - \mu_x \cdot t, & \text{if } 1 < \mu_x \cdot (2t+u) \\ \mu_x \cdot (t+u), & \text{if } 1 > \mu_x \cdot (2t+u) \end{cases}$$

Especially when $t = 1$, we have

$$q_x = \mu_x, \; p_x = 1 - \mu_x.$$

When $u = 1$, we have

$$_{t|u} q_x = {}_{t|} q_x = \begin{cases} 1 - \mu_x \cdot t, & \text{if } 1 < \mu_x \cdot (2t+1) \\ \mu_x \cdot (t+1), & \text{if } 1 > \mu_x \cdot (2t+1) \end{cases}.$$

IV. CONCLUSIONS

This paper defines the force of mortality based on credibility measure, and obtains the regular and simple expressions of the life distribution function, survival function, density function, actuarial notation, etc, based on the force of mortality. Thus, we are able to transform the calculation of survival model from a complicated operation using an expression of life distribution function to an simple arithmetic that uses the force of mortality. Meanwhile, it lays a foundation for deeper discussion on the continuous life insurance actuarial on account of credibility measure.

ACKNOWLEDGMENT

This work is supported by the Soft Science Research Project of Science and Technology Department of Hebei Province of China (No.10457292), the Natural Science key project of Langfang Teachers' College of China (No.LSZZ201306).

REFERENCES

[1] N.L. Bowers, H.U. Gerber, J.C. Hickman, D.A. Jones and C.J. Nesbitt , *Actuarial mathematics, The Society of Actuaries*. Itasca, IL, 1986.

[2] Lei YU, *Life Insurance Actuarial Science*. Beijing: Beijing University Press, 1998. (in Chinese)

[3] M.Y. Dorfman, S.W. Adelman, *Life Insurance*, 2nd ed. Dearbom Financial Publishing, Inc, 1992.

[4] Fu-sheng XONG, Zhi-zhong SHEN, *Life Insurance Actuarial Science*. Wuhan: Wuhan University Press, 2006. (in Chinese)

[5] Yan WANG, *Life Insurance Actuarial Science*. Beijing: Renmin University of China Press, 2008. (in Chinese)

[6] Hai-jun LI, Ming-hu HA, "The basic formula and model of life insurance actuarial on the quasi probability space." (in Chinese), *Journal of Hainan Normal University*, vol. 21, no. 4, pp. 357–361, Dec. 2008.

[7] Min-ying YUAN, Da-jun SUN, "Distribution of curate future lifetime and model of life actuarial on credibility space," presented at the *2011 IEEE 18th International Conference on Industrial Engineering and Engineering Management*, Changchun, China, 2011, 9.

[8]   Min-ying YUAN, Da-jun SUN, "The further decomposition of actuarial notation's expression on credibility space," presented at the *2011 3rd International Asia Conference on Informatics in Control, Automation and Robotics*, Shenzhen, China, CA, 2011, 12.

[9]   Min-ying YUAN, "*Model of life actuarial on credibility space* (in Chinese)," M.A. dissertation, Management Science and Engineering, Hebei University, Baoding, China, 2011.

[10] Min-ying YUAN, Da-jun SUN, "Life distribution function expression of actuarial notation on credibility space," presented at the *The 2nd International Conference on Management Science and Engineering*, Engineering Technology Press, Hong Kong, CA, 2011, 10.

[11] Bao-ding LIU, Yan-kui LIU, "Expected value of fuzzy variable and fuzzy expected value models, " *IEEE Transactions on Fuzzy Systems*, vol. 10, no. 4, pp. 445-450, 2002.

[12] Bao-ding LIU, "A survey of credibility theory," *Fuzzy optimization and decision making*, vol.5, no.4, pp. 387-408, 2006.

[13] Bao-ding LIU, *Uncertainty Theory*, 2nd ed. Berlin: Springer-Verlag, 2007. pp. 91.

[14] Bao-ding LIU, *Theory and Practice of Uncertain Programming*, Heidelberg: Physica-Verlag, 2002.

[15] Bao-ding LIU, Rui-qing ZhAO, Gang WANG, *Uncertion Programming with Applications*, Beijing: Tsinghua University Press, 2008, pp. 148.

The Strategic Management of Creative Products based on Bertrand Competition Model

Jian-rong HOU[1,*], Xiao-feng ZHAO[2]

[1]Antai College of Economics and Management, Shanghai Jiaotong University, China
[2]College of Business, University of Mary Washington, Virginia, USA
([*]jrhou@sjtu.edu.cn, xzhao@umw.edu)

Abstract - **Based on the assumption that two types of heterogeneous consumer groups exist in the market (one only having the creative product demand, another not caring about creative characteristics of products), we present a Bertrand competition model between creative products and ordinary products. The results show that firms with creative products in the existence of common market have a greater profit than in the non-existent of ordinary product under certain conditions. It suggests that creative product firms and ordinary product firms can work together in a symbiotic marketing system.**

Keywords - **Bertrand model, creative products, game theory, symbiotic marketing strategy**

I. INTRODUCTION

The creativity not only improves product differentiation, but also increases the added-value of products and improves the market competitiveness of firms because it has become a part of product value and production process. At present, China's creative industry market is still not mature and has not formed a complete and efficient industry chain. "Creativity products exist, but there is no creative industry" [1-3]. Creative products have their own value only in the consuming process. Creative product demand refers to the number of consumers who are willing and able to buy such products at a variety of possible prices in a certain period of time. The consumption of the creative products is highly elastic. Consumer economy levels and personal preferences will make the creative product demand instable. The unpredictability of the demand will make the industry participants and observers confused [4]. However, research on the marketing of creative products is rare. Current researches emphasize only on the creative side of creative products and symbiotic marketing strategy is usually excluded, which leads to the low efficiency of marketing activities.

Based on the assumption that two types of heterogeneous consumer groups exist in the market (one only has the creative product demand and another has ordinary product demand), we present a Bertrand competition model between the creative products and ordinary products. The results show that under certain conditions creative product firms in the existence of the market of ordinary products have a greater profit than in the nonexistence of ordinary products. Symbiotic marketing is a marketing strategy objective by which two or more than two firms can improve their efficiency and enhance market competitiveness through the sharing of marketing resources. The results of this study indicate that the existing creative products firms and other ordinary product providers can work together in a symbiotic marketing system.

II. RELATED WORK

Research on high-end consumer promotions show that the collaboration of one high-end brand and other high-quality brands in a product line will be beneficial to the firms in order to protect the interests of the market, which improves the value of other products with high-quality brand [5]. An empirical study from Soberman and Parker (2004) showed the existence of heterogeneous consumer groups, some consumers are willing to pay high prices for those advertised (brand) products, while others believe that the values of own brand products and well-known brand products are the same [6].

Pauwels and Srinivasan showed that the intrusion of own brand products would improve high-quality brand's profits because consumers could think the quality of famous brand products will be far better than its own brand [7]. Although the basic situation of different industries is not always the same, the conclusion is used as reference in this study. Coughlan and Soberman discussed whether to establish the direct stores which are independent of major retail outlets [8]. According to their conclusions, the direct stores would be beneficial to independent retailers.

Chen and Riordan (2007) developed the model of monopolistic competition in the horizontal differentiation level [9]. In some cases, a new entrant will increase the profits of current existing firms.

Different from the above researches, this research intends to explore the relationship between the profits of firms and types of creative products and ordinary products, as well as the competitiveness in a vertical horizon.

III. THE BASIC MODEL

We assume that there are two different types of products (h and l), h and l represent the creative products and ordinary products respectively. Supposed that there is a straight line with distance of [0, 1] unit. Firm 1 is located in position 0, and firm 2 is located position 1. The two firms produce product h at zero marginal cost. The product' difference is divided into the different points on the straight line. Two types of consumer groups (creative product market H and common market L) lie in

E. Qi et al. (eds.), *Proceedings of the 21st International Conference on Industrial Engineering and Engineering Management 2014*, Proceedings of the International Conference on Industrial Engineering and Engineering Management, DOI 10.2991/978-94-6239-102-4_114, © Atlantis Press and the authors 2015

uniform distribution in [0, 1] interval. There are two kinds of difference between the two market segments. First, the transportation cost product h is different. H market consumers usually pay higher transportation cost(t_H), while the L market consumers (usually) pay low transportation cost(t_L), $t_H > t_L$. L-type consumers are more sensitive to price and have higher price elasticity than H consumers. For convenience, let t_H value is 1, then $t_L < 1$. Second, for two different consumer groups, the willingness to pay is different. H-type consumers only need product h and L consumers are indifferent to the difference of product h and l. If the total consumer market is 1, the number of H is λ and the number of L consumers is $1-\lambda$. Each consumer' demand is at most one unit. If a consumer located in x would buy the product h, his utility function is as follows:

$$u_j = \begin{cases} s_j - t_j x^2 - p_1 & \text{if purchasing from firm 1} \\ s_j - t_j (1-x)^2 - p_2 & \text{if purchasing from firm 2} \end{cases} \quad (1)$$

s_j represents the value of ideal product of consumer j, $t_j (\cdot)^2$ represents that consumers in $j(H,L)$ need to spend the cost of transportation, p_j is product' prices of firm i (i=1,2). Assume that s_j ($j = H, L$) is high enough. From formula (1), the consumers located in $x_j(p_1, p_2)$ to buy products h from two enterprises are neutral, i.e.:

$$x_j(p_1, p_2) = \frac{p_2 - p_1 + t_j}{2t_j} \quad (2)$$

A. Market without Ordinary Products

In this case, the game is a simple Hotelling duopoly model. Each consumer in L and H would buy one unit of product from firm 1 or firm 2. From the formula (2), firm 1's demand function is D1 and firm 2's demand function is D2.

$$D_1(p_1, p_2) = \begin{cases} 1 & p_2 - p_1 \in [1, +\infty] \\ \lambda x_H(p_1, p_2) + (1-\lambda) & p_2 - p_1 \in (t_1, 1) \\ \lambda x_H(p_1, p_2) + (1-\lambda)x_L & p_2 - p_1 \in [-t_L, t_L] \\ \lambda x_H(p_1, p_2) & p_2 - p_1 \in (-1, -t_L) \\ 0 & p_2 - p_1 \in (-\infty, -1) \end{cases} \quad (3)$$

$$D_2(p_1, p_2) = 1 - D_1(p_1, p_2)$$

Consider whether there exists a Nash equilibrium. In this situation, the profit function of enterprises is as follows:

$$\pi_1(p_1, p_2) = \frac{p_1(t_L + (1 - \lambda + \lambda t_L)(p_2 - p_1)}{2t_L}$$

$$\pi_2(p_1, p_2) = \frac{p_2(t_L + (1 - \lambda + \lambda t_L)(p_1 - p_2)}{2t_L}$$

First order derivative is as follows:

$$p_1' = p_2' = \frac{t_L}{1 - \lambda + \lambda t_L}$$

$$\pi_1(p_1', p_2') = \pi_2(p_1', p_2') = \frac{t_L}{2(1 - \lambda + \lambda t_L)} \quad (4)$$

Formula (4) shows that firm 2 has two choice when $p_i' = t_L / (1 - \lambda + \lambda t_L)$:

1) Set p_2, $p_2 - p_1' \in (t_L, 1)$, which means that firm 2 would abandon the sale of products in the market in L, and focus on the market H;

2) Set p_2, $p_2 - p_1' \in (-1, -t_L)$, which means that firm 2 would fully take the market L (if $p_2 = p_1' - 1$, it fully take the two consumers L and H in the market).

To check whether firm 2 will take the two options, we consider the following maximization problem:

(i) $\max\limits_{p_2} p_2 \lambda(1 - x_H)$, $s.t.$ $x_L \geq 1$

(ii) $\max\limits_{p_2} p_2(\lambda(1 - x_H) + (1 - \lambda))$, $s.t.$ $x_L \leq 0$

The first equation' solution exists if and only if:

$$(1 - \lambda)(1 - 3t_L) - 2\lambda t_L^2 > 0 \quad (5)$$

If the inequality does not hold, the solution in formula (4) is more favorable to firm 2. When the inequality holds, the price and profit will be:

$$p_2'' = \frac{1 - \lambda + (1 + \lambda)t_L}{2(1 - \lambda + \lambda t_L)} \quad (6)$$

$$\pi_2(p_1 p_2'') = \frac{\lambda(1 - \lambda + (1 + \lambda)t_L)^2}{8(1 - \lambda + \lambda t_L)} \quad (7)$$

The pi'' is a derivative of the solution. From the formula, it is easy to see that $p_2'' - p_1'$ is less than 1. If $\pi_1(p_1', p_2') < \pi_2(p_1', p_2'')$, the equilibrium price from formula (4) exists. This can also be expressed as:

$$(1 - \lambda)(\lambda - 2(2 + \lambda)t_L) > \lambda(3 + \lambda)t_L^2 \quad (8)$$

This means firm 2 will never take this option.

If either formula (5) or formula (8)does not hold, (p_1', p_2') would not be the equilibrium price. If the condition of price equilibrium in formula (8) holds, then it will satisfy the following inequality:

$$(1 - \lambda)(\lambda - 2(2 + \lambda)t_L \leq \lambda(3 + \lambda)t_L^2$$

$$\Rightarrow t_L \geq \frac{-(2 + \lambda)(1 - \lambda) + 2\sqrt{1 - \lambda}}{\lambda(\lambda + 3)} \quad (9)$$

From formula (9), we can see, for a given λ, if t_L is sufficiently close to 0, (p_1', p_2')would not be the equilibrium price. When t_L is close to 0, p_i' (i=1, 2) be close to 0. A firm would set higher price than pi' under the condition, so the firm only provides the creative product market with products.

The value range of t_L in formula (9) depends on the parameter λ. The right side of the inequality is a concave function of variable λ. If formula (9) does not hold, there will be an upward bias. The deviation includes the following three aspects: ① the loss of ordinary product consumers; ② the reduction of supply for creative product consumers; ③ the increase of consumers' income of the creative product market. The former two kinds of consequences have negative effects, but the last one is a positive utility. If λ (the proportion of consumers from creative products market) is very small, the utility of ① is greater than ② or ③, the minimum value of t_L will become very low. When λ value increases, the effect of type③ becomes more important.

B. Market with Many Ordinary Product Firms

We assume that there are l-type production firms at the end of each line. Since the price p_1 decreases to be zero, the two firms do not want to sell products to the market of L. In this case, the game becomes a duopoly. Demand function can be depicted as follows:

$$D_1 = \lambda x_H (p_1, p_2) \quad D_2 = \lambda (1 - x_H (p_1, p_2))$$

The profit function of enterprises can be described below.

$$\pi_1(p_1, p_2) = \frac{\lambda p_1 (p_2 - p_1 + 1)}{2}$$

$$\pi_{21}(p_1, p_2) = \frac{\lambda p_2 (p_1 - p_2 + 1)}{2}$$

After the calculation, we can obtain the optimal solution.

$$p_1^* = p_2^* = 1$$

$$\pi_1(p_1^*, p_2^*) = \pi_2(p_1^*, p_2^*) = \frac{\lambda}{2} \quad (10)$$

Where pi* is the equilibrium price of firm i.

C. Comparative Analysis

From the above results, option 2 is more favorable than option 1. According to equation (4) and (10) we can obtain the inequality:

$$\frac{\lambda}{2} > \frac{t_L}{2(1 - \lambda + \lambda t_L)} \Leftrightarrow t_L < \frac{\lambda}{1 + \lambda} \quad (11)$$

If and only if both formulas (9) and (11) hold, the two creative firms favor adequate supply of ordinary products.

In the above model, the heterogeneous consumer groups (H and L) are the driving factors. If no ordinary product firms exist, creative product firms are unable to

keep higher prices. Because all firm want to obtain a large number of consumers from the L (price sensitive consumers), a slight reduction in price will attract many price sensitive consumers. Once the price decreases, two firm's prices will drop significantly. As a result, the loss of price decrease is greater than the gain of the sale increase. If ordinary product firm exist and p1is sufficiently low, the entry of the ordinary product firms into creative product market is beneficial to creative firms because it makes the ordinary product market become unprofitable to creative product firms and creative products can avoid price decrease at the beginning.

IV. CONCLUSION

Based on Bertrand competition model, this research demonstrates the existence of low price ordinary product may be beneficial to existing high price creative product firms, which makes the creative firms only sell their products to the creative products consumers. The increase of price makes up for the loss of the creative product sales. When the ordinary product market doesn't exist, the equilibrium price of creative products market will be higher, otherwise the ordinary product market will make the equilibrium price and profit decline. Our conclusion indicates that the existing creative product firms and ordinary product firms can work together in a symbiotic marketing system.

REFERENCES

[1] Zhang Xiaoming, Chinese report on the development of culture industry, Social Sciences Academic Press; first edition 2014
[2] Wang Yumei. The profit model of creative industry based on value chain Management, 2006.5.
[3] Wang Chuanlei, Tan Xing, Xie al. Study on the problems and Countermeasures of creative industry [J]. China productivity research, 2007, (6).
[4] Hotho, S., & Champion, K. (2011). Small businesses in the new creative industries: innovation as a people management challenge. Management Decision, 49(1), 29-54
[5] Randall, T., K. Ulrich, D. Reibstein. 1998. Brand equity and vertical product line extent. Marketing Science. 17, 356-379.
[6] Soberman, D. A., P. M. Parker. 2004 Private labels: Psychological versioning of typical consumer products. International Journal of Industrial Organization, 22, 849-861
[7] Pauwels, K., S. Srinivasan. 2004. Who benefits from store brand entry? Marketing Science. 23, 364-390.
[8] Coughlan, A. T., D. A. Soberman. 2005. Strategic segmentation using outlet malls. International Journal of Research in Marketing, 22, 61-86.
[9] Chen, Y., M. H. Riordan. 2007. Price and variety in the spokes model. Economy [J]. 117, 897-921.

The Collaborative Management Analysis of Organization Coordination Network in Major Scientific and Technological Engineering

Xin-wen HE[1,*], Yan WANG[1], Ji-xun XIN[1], Guang-ming HOU[2]

[1] School of Management, Minzu University of China, Beijing, China

[2] School of Management and Economy, Beijing Institute of Technology, Beijing, China

(10808040@bit.edu.cn)

Abstract - **Major scientific and technological engineering need many organizations to implement. In order to deeply analyze the problem of organization coordination of major scientific and technological engineering, we use organization theory, coordination theory and network organization theory to solve it, which includes the implementation and control process such as an assessment of gaps in the generation process of collaborative management and identification for coordination opportunities, pre-evaluation of elements, communication, integration of elements, the selection and management of order parameter and some comparison and feedback of the result.**

Keywords - **Collaborative management, major scientific and technological engineering, organization coordination network**

I. INTRODUCTION

In order to deeply analyze the problem of organization coordination of major scientific and technological engineering [1-3], we need the integrated use of organization theory, coordination theory and network organization theory to analyze the process of collaborative management [4, 5]. Collaboration management is the full use of information and knowledge to overcome barriers of communication to generate multiplier effect with individual functions. It is very necessary to add the process analysis of collaborative management for us to fully understand and grasp the motivation, operation conditions and management modes. Because it has an effect on organization coordination network and affect its survival or change at a certain state.

II. THE GENERATION OF COLLABORATIVE MANAGEMENT

A. To determine the relationship between organization coordination's management objectives and network objectives in major scientific and technological engineering

The object of the collaborative management implementation is organization coordination network of major scientific and technological engineering, whose goals are consistent with the objectives of collaborative management. Confirming the relationship between them is the assumptions to analysis the process. In fact, no mater how much difference between the co-management goals and the objectives of the network, they point to the content or nature of a deep identity, all in pursuit of the overall functional effects and the maximize value. Therefore, we believe that they are consistent and this is

also the same in the daily activities [6]. For example, every organization coordination network has its own goal, which is achieved by imposing a certain degree of methods about organization management but also for achieving goals of organization coordination network of major scientific and technological engineering.

B. Check the operating conditions of organization coordination network in major scientific and technological engineering

Although the co-management objectives is consistent with the goal of organization coordination network which is to pursue multiplier effect, it needs to contrast the operating conditions, such as we can recognize its real level of development to find the gap between the ideal level of development and make the resource into full effect.

C. Assess the gap between real level development and ideal level development of organization coordination network in major scientific and technological engineering

The purpose of contrasting its operational status is to understand the gap between the ideal and reality level of development in time or a period about the organization coordination network of major scientific and technological engineering. How to determine the gap between them? The analysis of the target is not specific but general organization coordination network of major scientific and technological engineering. We always establish the coordinates to determine two curves named ideal and reality level of development to assess the gap. The result shows that if the two curves are very close, we can indicate that the network itself has good state of coordination organization and there is no need to manage. On the contrary, it indicates that it is necessary to collaborative management to achieve its goals and a multiplier effect [7, 8].

D. Use collaborative management to solve or shorten the gap

From the above we can determined the gap between the reality development and the ideal level of development of the organization coordination network in major scientific and technological engineering, besides the size of gap can reflect its operation states. The relationship between them is that the greater the gap, the worse of its running condition and the network is unstable. And vice versa counter is. The purpose of the implementation of collaborative management is to narrow the gap between the ideal and reality level of development, which is decided by its nature and characteristics of collaborative management [9].

III. THE IMPLEMENTATION OF COLLABORATIVE MANAGEMENT

The above four aspects are the generate motivation of collaborative management, that is why we need to joint management. The generation process of collaborative management is only coordination in ideology and it shows the necessity and feasibility of collaborative management. We must analyze how to implement the collaborative management to change the ideological coordination to actual collaborate behavior. The process of collaborative management includes:

A. The identification of collaborative opportunities

The identification of collaborative opportunities mainly aims to seek opportunities for synergies in the implementation of collaborative management. The identification is the breakthrough of the collaborative management and it can help to achieve the desired effect with many methods and tools. Meanwhile, the identification is the foundation of the follow-up actions. The implementation of collaborative management is based on the identification of the collaborative opportunities.

B. Pre-value the collaborative value of elements

Pre-valuing the collaborative value of elements is making the evaluation for value or contribution of the elements coordination in the process, basing on the identification of collaborative chances. Its effect shows in two ways: the first is it can compare the costs and value of collaboration in the process though the evaluation of the collaboration value, thus we can know the importance of elements collaboration in the whole collaboration process. The second is that it can determine the value of collaborative elements in the collaboration process in advance, which is benefit for us to distribute the benefits and to ensure the follow-up actions to run smoothly [10-13].

C. Communication

Communication is the foundation of the successful implementation of collaborative management. No communication, no coordination, no chances to achieve the objectives in the organization coordination network of major scientific and technological engineering. Communication plays a bridge or link role in unifying the behaviors of the network organization. , which is the basis of any problems about organization management. Collaborative opportunities identification and the value evaluation can achieve its proper value and ensure the implementation of collaborative management run smoothly only though the deep and effective communication. All the above can make the organization understand, recognize and reception clearly to translate into conscious act of organization.

D. Integration of elements

The integration of elements is an orderly process of organization coordination network of major scientific and technological engineering and it is also a process to balance choice and coordinate elements to achieve the collaborative management objectives, which is basing on collaborative chances identification, the pre-valuation of collaborative value and communication. The integration of elements aims to excavate the strength of each subsystem or elements and to make up the shortage in the organization coordination network of major scientific and technological engineering [14]. Its effect is to improve or break the restricted link to make the collaborative elements develop the best functions, thus we can achieve the overall goals of system. The mode, principle and implementation of the integration are the contents that must be studied, for they are related to the realization of collaborative management effects and whether we can create value in the process.

E. Order parameter selection and management

In coordination theory, order parameter is a measure of the degree of macro-order system, which dominates the system from disorder to order [15]. As long as we can determine the order parameter of the organization coordination network in major scientific and technological engineering, we can grasp its development direction by a series of methods and means. The aim of elements integration is to produce the desired order parameter and make it play dominant role. Ultimately we can double the overall function of the system, namely the creation of synergies. We need to create a favorable environment for its effects to play. In the government-led major scientific and technological engineering management process, the government's macro-control is the order parameter for the major scientific and technological engineering. Therefore we can control the direction of its development though the setting of certain powers or the binding of interests legally.

F. The comparison and feedback of the results

In the dominant of order parameter, the organization coordination network will change from the disordered state of instability toward a new steady state ordered, thus it can have new time, space and functional structural to achieve overall functional effects which is the ideal result of the collaborative management. However, we need to according to the feedback to determine whether it is the effect we pursue. If it is, then we achieve the objectives of collaborative, otherwise, we need to return to the beginning of collaborative management to reconsider.

IV. COLLABORATIVE MANAGEMENT CONTROL

As an indivisible part of collaborative management process, control plays an important role in the successful implementation of collaborative management. Control is the rule of collaborative management process, without it, the process will not achieve the objectives of collaborative management .The control in the process of collaborative management have two types: the control in generation process and implementation process of collaborative management [16].

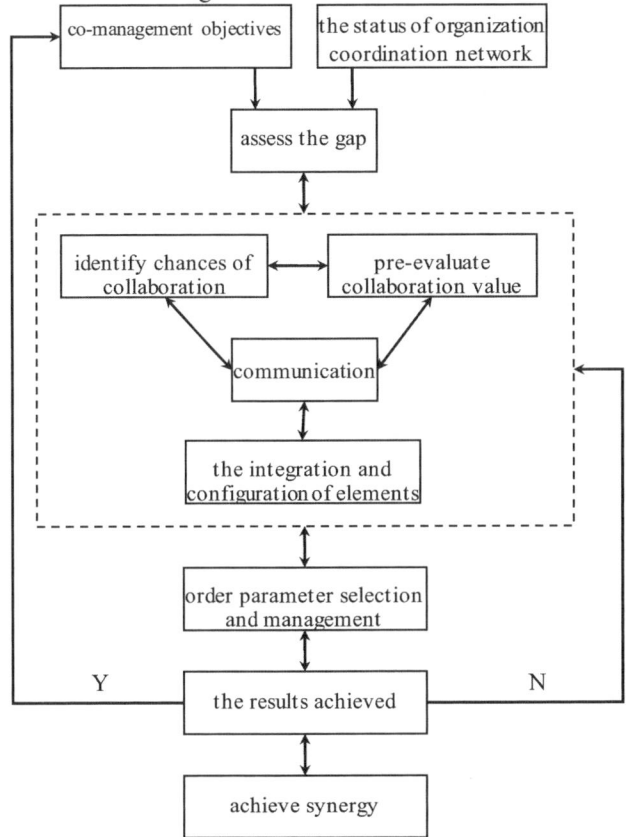

Fig.1. The collaborative management process of organization coordination network in major scientific and technological engineering

According to the analysis of ideas and contents in the collaborative management process, we design a framework for collaborative management process, as is shown in Fig.1. The three processes of collaborative management are interrelated and interactive ,as they play their respective functions and effects together, leading to achieve the goal of collaborative management finally which is the realization of synergies. Although the generation process is a sense of collaboration in collaborative management process, it is the starting point of the whole process. Therefore, the generation process of collaborative management is a prerequisite for the implementation process to run. However, the implementation process of collaborative management can produce the expected order parameter across a series of specific operations such as identification for collaborative chances, the pre-evaluation of elements' collaborative value, communication, the integration and configuration of elements and information feedback of the results, which bases on the generation process of collaborative management. Thus the order parameter can lead the whole system to develop orderly and stably. The implementation process of collaborative management turns the coordination into reality in the generation process finally. As an important guarantee to achieve the effects of collaborative management, the control process of

collaborative management throughout the whole process whether in the generation process or in the implementation. The control process plays an important part to ensure synergies to carry out smoothly.

V. CONCLUSION

Every organization coordination network of the major scientific and technological engineering needs to achieve the goal, which is also the ultimate pursuit of the co-management implementation. However, the realization of objectives is closely related to the operation status for the organization coordination network of major scientific and technological engineering. Besides, the operation status reflects the level of development of the organization coordination network of the major scientific and technological engineering. In this case, the organization coordination network of the major scientific and technological engineering may reach a higher level of development, thus the primary problem that must be solved is to search effective management methods and tools to shorten the gap. The collaborative management objective is an effective way to shorten gap for it is not only consistent with the goal of organization coordination network for the major scientific and technological engineering, but it also has an advantage over the traditional management. However, it needs a series of processes to shorten the gap and control in the generation and implementation process of collaborative management. The processes includes identification of collaborative opportunities, the pre-evaluation of elements' collaborative value, communication, elements to achieve integration, order parameter selection and management, the comparison and feedback of results and so on.

ACKNOWLEDGMENT

This research work relates to collaborative innovation model and mechanism for major science and technology engineering, and was supported by "the Fundamental Research Funds for the Central Universities" (2014MDGLXYQN10), "the National soft science research program of China" and "the National Natural Science Foundation of China" (71173016).

REFERENCES

[1] Xin Wen He, Yan Wang, Guang Ming Hou, "Construction of Organization Coordination Network of Major Scientific and Technological Projects," in Proc. 3rd International Asia Conference on Industrial Engineering and Management Innovation, Beijing, 2012.

[2] Xin Wen He, Yan Wang, "The relationship measurement model analysis of Organization Coordination Network of Major Scientific and Technological Project based on authority links," Advances in Information Sciences and Service Sciences, vol. 4, no. 23, pp. 325-333, 2012.

[3] Xin Wen He, Yan Wang, Guang Ming Hou, "The Measurement of Complexity of Organization Coordination Network of Major Scientific and Technological Projects,"

in *Proc. Logistics, Informatics and Services Sciences*, Beijing, 2012.

[4]   Guang Ming Hou, *the Introduction to Organization System of Science*. Beijing: Beijing science press, 2006.

[5]   Wei An Li, *Nerwork Organizations—the New Trends of Network Development*. Beijing: Beijing science press, 2003.

[6]   Ha ken, *Introduction to Coordination*. Translation by Zhangjiyue and Guozhian. Northwest university research Center, 1981.

[7]   Xin Wen He, Yan Wang, Guang Ming Hou, "The Operation Research on Authority-Linked Management Model of Organization Coordination Network of Major Scientific and Technological Project," in *Proc. Logistics, Informatics and Services Sciences*, 2012.

[8]   Xin Wen He, Yan Wang, "Mechanism analysis of division and coordination of Organization Coordination Network of Major Scientific and Technological Project based on contract links," *Advances in Information Sciences and Service Sciences*, vol. 4, no. 23, pp. 318-324, 2012.

[9]   Yan Wang, Xiao Hong Chen, Xin Wen He, "The Measurement of Complexity of Organization Coordination Network of MSTP," *Advances in Information Sciences and Service Sciences*, vol. 4, no. 23, pp. 334-342, 2012.

[10] Hong Jun, Ke Tao, "The study of complex adaptive of the network organization," *China management science,* no. 12, pp. 123–126, 2004.

[11] Wu de sheng, "Network organization management: perspectives based on relationship contact," *Tianjin social science,* no. 5, pp. 68-71, 2005.

[12] Hao sheng, "The confidence, contact and network organization governance mechanisms," *Tianjin social science,* no. 5, pp. 64-67, 2005.

[13] Shi guang hua, "The network relationship contact and virtual organization," *Logistics technology,* no. 9, pp. 32-37, 2002.

[14] AMIR R, "Modelling Imperfectly Appropriable R&D via Spillovers," *International Journal of Industrial Organization,* no. 18, pp. 1013-1032, 2000.

[15] Zeng De Ming, Ren Lei, "The Collaborative Research and Government Policy of High-tech Business," *System Project,* no. 9, pp. 59-63, 2000.

[16] Zhou Qing, "The Analysis of Business Collaborative R&D Network," *Business Research,* no. 1, pp. 25-27, 2006.

Research on the Relationship between Reduction of State-owned Shares and Enterprise Value
——The Empirical Data from A-share Listed Companies of China

Yang Xiang, Rong Fu[*]

Department of Accounting, Sichuan Normal University, Chengdu, China

(371090945@qq.com, *80693091@qq.com)

[1]*Abstract* - **Reduction of state-owned shares is in an important measure in the process of deepening the reform of enterprises. Whether the reduction is realized enterprise value appreciation, remains to be more evidence. So far, views of the relationships between the proportion of state-owned shares and enterprise value are actually inconsistent. It's still hard to give the final conclusion of the impacts on the enterprise value, which brought by the reduction of state-owned shares. This article selects all the A shares of Shanghai & Shenzhen stock exchange, which proportion of state-owned shares has decreased greater than or equal to 5% one-time since 2008, and after reduction the proportion is still greater or equal to 50% of the listed companies as samples. Using standard event study method to analysis cumulative abnormal returns of the 71 sample enterprises. But the results did not support the positive effect of reduction of state-owned shares.**

Keywords - **Cumulative abnormal return, enterprise value, event study method, reduction of state-owned shares**

I. THE BACKGROUND AND PROCESS OF REDUCTION OF STATE-OWNED SHARES IN CHINA

A. The background of reduction of state-owned shares

1) The domestic background of the reduction of state-owned shares in China

State-owned share is composed of state shares and the national legal person share. In the early stages of the securities market of our country, state-owned enterprises to raise funds and increase energy through issuing shares in the capital market, but also formed the state-owned shares of major, not the special equity structure of listed circulation. With the market economy gradually establish and the continuous development of the securities market, the state-owned shares in the process of "a dominant" more and more outstanding, not only seriously restrict the development of our economy and the establishment of modern enterprise system, brought many disadvantages to the corporate governance structure, but also infringes the interests of small and medium-sized investors. So then the government has adopted a series of measures to reduce the proportion of state-owned shares and realize its circulation. Our country many economists believe that reduction of state-owned shares is the measure to improve the business performance of enterprises, improve the corporate governance mechanism, but also to raise a lot of social security funds, thus reduction of state-owned shares is the inevitable choice of deepening reform on the road in our country [1].

2) The international background of the reduction of state-owned shares in China

After many countries have been carried out a reduction of state-owned shares, the purpose is basically the same. Usually in order to improve the operational efficiency of enterprises, improve the ownership structure of listed companies, improve the country's fiscal situation, to reduce the degree of government intervention to expand the scope of the market mechanism, and other political and economic goals [2].

In order to cut the deficit and solve financial problems, the British government began to implement privatization in competitive industries, establish innovation system such as undersell to ensure that privatization of powerful push. There were 55 enterprises privatized in British during 1977-1996, through calculating the Abnormal returns of these companies, found that in the first years after the privatization of the enterprise average cumulative Abnormal return of 21%, the second year of 30%, in the fifth year is up 57%, explain the value of the state-owned enterprise privatization of enterprise brings positive effect.

In facing serious losses of state-owned enterprises and the slow development in economic, Argentina's government began to implement the privatization in 1989. Argentina's government also issued relevant laws and regulations, starting from the public utilities privatization, then push to the industrial sector, including utilities partial privatization play an important role. The privatization practice in Argentina for more than ten years, greatly promoted the development of the economy.

France, as one of the larger proportion of the national state-owned economy in the developed countries, in order to achieve the separating government from enterprises, set up the national bureau of participation to manage 71 large state-owned enterprises. National bureau of participation instead of the government to exercise shareholder rights, not directly involved in the daily management of the enterprise, but assigned national representatives stationed in the board of state-owned enterprises. In order to encourage the management of state-owned enterprises more efficient management of the enterprise, has taken equity incentive and performance bonuses and other incentive system. After the privatization, the government has been gradually withdrawn from competitive industries, currently distributed in the public service sector, and now the performance of state-owned enterprises in France are not worse than private enterprises.

From the practice of the above countries in terms of SOE reform, although with different characteristics, as the

[1]11SA047

core of reduction of state-owned shares are undoubtedly effective initiatives. Because of this, the reform of SOE in China also opens a feast of reduction of state-owned shares.

B. The process of reduction of state-owned shares in China

Reduction of state-owned shares is in order to achieve the separating from government and enterprises, and make the enterprise become economic subject in market economy, laying the foundation for a fundamental transformation of the economic system and modern enterprise system, is an important move in deepening the reform of the enterprise. The reduction of state-owned shares by shrinking stock circulation, auction, state-owned shares allotment, equity transfer of creditor's rights, stock repurchase and so on [3]. It has experienced four stages roughly:

The first stage is the exploration of the pilot of the reduction of state-owned shares. The first transfer of state-owned shares and the first state-owned shares repurchase in our country are both occurred in 1994. In September 1999 through the "decision on the SOE reform and development issues" by a clearly stated, "without prejudice to the state-controlled premise, appropriate to reduce some state-owned shares." In December, state-owned shares placing pilot started [4].

The second stage is the formal implementation of the reduction of state-owned shares. On June 12, 2001 sets of the "reduce state-owned shares to raise social security funds management interim measures", corresponding to this is the stock market reaction is strong, and sharply lower. Until June 23, 2002 reduction of state-owned shares is the pilot was forced to suspend.

The third stage is the equity division reform of the reduction of state-owned shares. The reduction did not cease with the temporarily stop of the pilot. On April 29, 2005, the SFC issued on the pilot reform of non-tradable shares of listed companies related issues notice, officially launched pilot marks the equity division reform [5]. So far the reform of non-tradable shares of listed companies in our country has been basically completed, solve the problem of the full circulation.

The fourth stage is pushed forward the reduction of state-owned shares. On April 20, 2008 the SFC issued "listed company terminate the restricted stock share transfer guidance". On June 19, 2009, the state council decided to implement the transfer of state-owned shares in domestic securities market. In 2013 the third plenary session of the eighteenth emphasized on the road for the reform of SOE, to improve the system of state-owned assets investors, optimize the evaluation system of state-owned assets, to speed up the state-owned capital withdraw never has the advantages of industry and make more into the private economy.

SOE reform in China has adopted more than 30 years, has obtained certain achievement. Decrease the proportion of state-owned economy, but the total assets growth significantly, control and influence also significantly increased. At the same time, improve the structure of the state-owned economy; improve the market competitiveness of state-owned enterprises. But there are still some disadvantages, such as the relationship between the government and the enterprise responsibility has not completely separate, SOE are not fully in accordance with the corporations to run, SOE is lack of innovative leading to the lack of market competitiveness, etc.

II. THE INFLUENCE OF REDUCTION OF STATE-OWNED SHARES TO THE ENTERPRISE VALUE

A. The measure of enterprise value and its related research

Enterprise value refers to the business forecast free cash flow to the weighted average cost of capital is the present value of the discount rate discount [6]. Commonly used to measure the enterprise value of the method is mainly the book value and market value, among them, the book value refers to the enterprises listed in the balance sheet shows the value of the assets, can get directly from the enterprise statement. But due to different enterprise, or the same enterprise in different accounting period, the accounting policy adopted by the may also be different, be enterprise managers more easily tamper with the report data, etc., so that greatly reduced the book value of practicality.

And in the practice of enterprise value assessment and theory study, people often choose to measure the value of the enterprise market value. Market value is to point to in the market to sell the enterprise can get the price. If companies sell in the market, based on the rational economic man hypothesis, enterprise's sale price is the market value of the enterprise. In an efficient securities market, the stock price of the enterprise is the estimate of the market for corporate equity value, namely the price of stock market can accurately and completely reflect all information related to the price, including the enterprise the past performance and future development potential, etc. So, use market value can more directly and objectively reflect the recognition of the value of the company for people, but this kind of method shall not apply to the enterprise which equity isn't circulated in the securities and exchange market.

In theory study, for an enterprise to measure the change of market value, most scholars abroad use cumulative abnormal return (CAR) to examine the enterprises' stock price. There are also many domestic scholars use cumulative abnormal returns to evaluate the enterprises value.

B. The influence of reduction of state-owned shares to the enterprise value
1) Related research of domestic scholars

Despite the reduction of state-owned shares as a national policy implemented in our country, but domestic scholars views on the relationship between the proportion of state-owned shares and corporate performance are very inconsistent [7], mainly divided into two opposing camps.

Side of the research results to support the reduction of state-owned shares, they think that state-owned shares is negatively related to the performance of listed companies, the relationship between the reduction of state-owned shares is bring positive effect to the enterprise value. But on the other side of the research conclusion is state-owned shares is positively correlated with the performance of listed companies [8], the relationship between the reduction of state-owned shares is value for the enterprise to bring negative effects. Among them, Li-hui Tian (2005) considered the relationship between the proportion of state-owned shares and corporate performance showing a U-shaped curve [9].

2) Related research of foreign scholars

Foreign scholars for the study of the relationship between state-owned shares and corporate performance have not reached a consensus. The first is that there is no correlation between state-owned shares and corporate performance. The second is considered a positive correlation between the state-owned shares and corporate performance. The third is considered a negative correlation between the state-owned shares and corporate performance. So far, views by domestic and foreign scholars about what is the relationship between the proportion of state-owned shares and enterprise value is not consistent, about the effects of reduction of state-owned shares is the enterprise value how is hard to say.

According to the existing literature collection, although scholars study for the relationship between the reduction of state-owned shares and corporate performance are very rich, at the same time, adopt the method of cumulative abnormal returns for calculation analysis of the performance of mergers and acquisitions, more for the relationship between the reduction of state-owned shares and the enterprise value of direct research literature, especially the empirical research literature is less. This article try to calculate and analysis the CAR before and after of the reduction of SOE, to further explore the impact of reduction of state-owned shares is to the enterprise. Measure the value of SOE after reduction the main content of the evaluation, as well as whether or not to continue for the future implementation of reduction of state-owned shares is the policy to provide the reference.

III. EMPIRICAL TEST ON THE REDUCTION OF STATE-OWNED SHARES OF LISTED COMPANIES AND ENTERPRISE VALUE IN OUR COUNTRY

A. Research methods

In this paper, using standard event study to test the effect for the reduction of state-owned shares, taking the market model to measure the normal returns of enterprise stock. Set the reduction day as the event date, namely t0. Set [-20, 20] for the event window, set [-100, -21] for estimate period.

B. Sample selection

In order to explore the influence of reduction of state-owned shares to the enterprise value, referring to Henk Berkman in "Improving corporate governance where the state is the controlling block holder: Evidence from China" the article adopt the method of choosing sample [10], choose the listed on the Shanghai stock exchange and Shenzhen stock exchange among all the A shares, since 08, proportion of state-owned shares one-time reduced is greater than or equal to 5%, and after reduction, proportion of state-owned shares accounted still greater than or equal to 50% of the listed companies as the sample.

For the dispersed equity enterprise, even if the major shareholders hold the proportion of shares less than 50%, but may still be in control. But for the different enterprises, don't lose control may have differences. So this article uses the simple principle, selection of state-owned shares of 50% or more as a symbol of SOE.

There are total 80 enterprises meet the sample selection criteria by screening CSMAR database, of which 9 companies due to the incomplete data could not be empirical research then have been dropped, after excluding these enterprises, a total of 71 valid samples.

C. The empirical process
1) Build the model:

Calculate the stock real returns:

$$Rs_t = \frac{Pi_t - Pi_{(t-1)}}{Pi_{(t-1)}} \qquad (1)$$

Calculate the market real returns:

$$Rm_t = \frac{Index_t - Index_{(t-1)}}{Index_{(t-1)}} \qquad (2)$$

This paper used in the process of empirical Shanghai a-share index and Shenzhen composite a-share index returns as A market returns, because the selection of sample companies from a-share companies.

By using the market model, calculate the normal returns of stocks:

$$\widetilde{Rs}_t = a + b * Rm_t \qquad (3)$$

The stock window abnormal returns:

$$AR_t = Rs_t - \widetilde{Rs}_t \qquad (4)$$

Average abnormal returns:

$$AAR = \frac{1}{n} \sum_{i=1}^{n} ARi \qquad (5)$$

The CAR of the stock in the window period [t1, t2]:

$$CAR = \sum_{t=t_1}^{t_2} AR_t \qquad (6)$$

All the sample average cumulative abnormal returns (ACAR):

$$ACAR = \frac{1}{n} \sum_{i=i}^{n} CARi \qquad (7)$$

Involved in this paper, all the data are derived from CSMAR database, [-100, 20] research data according to the manual interception, using Eviews6 measurement software for statistical analysis, at the same time with Microsoft Excel as auxiliary.

2) The empirical process

First phase is calculated each sample's stock real returns and market real returns based on the 80 days of data, then carries on the least squares estimation of each sample enterprise parameter to establish the model, and then substitute the window data into the model to calculate the sample companies normal returns during the event window period, abnormal returns are according to real returns minus normal returns. Finally, summarized the sample data, got all the sample AAR and ACAR.

In order to describe in detail for empirical process, with 000011 deep properties company A as an example to illustrate this. On January 24, 2013, the company occurred the reduction of state-owned shares, namely 2013-01-24 as the event date. According to the method described choose 80 session's data to estimate the sample period from CSMAR database, namely on Aug. 24, 2012 to Dec. 21, 2012; Then capture data from 41 days as the event window, on Dec. 24, 2012 to Feb. 28, 2013.

Firstly using ADF unit root test to examine (3) $\widetilde{R}s_t = a + b * Rm_t$ model of the independent variable of Rmt level sequence, to optimize the fitting degree and improve the accuracy of the results of the model, the results shown in Fig.1 as follows:

		t-Statistic	Prob.*
Augmented Dickey-Fuller test statistic		-8.229972	0.0000
Test critical values:	1% level	-4.078420	
	5% level	-3.467703	
	10% level	-3.160627	

Fig.1. 00001 Rmt sequence ADF stationary test

Can be seen from the above, ADF unit root test statistics for 8.229972, less than 1% under the level of 4.07842, so it can significantly reject the null hypothesis, that there is no unit root.

Secondly using Eviews6 software measurement least-square method to estimate the parameters of the model, the result is shown in Fig.2:

Sample: 1 80
Included observations: 80

	Coefficient	Std. Error	t-Statistic	Prob.
C	0.001082	0.001961	0.551786	0.5827
RMT	1.182644	0.132263	8.941607	0.0000

Fig.2. Estimated regression model

Set up the sample estimates of the regression model:

$$\widetilde{R}s_t = 0.001082 + 1.182644 * Rm_t$$

Thirdly carrying on the residual error test. White heteroscedastic inspection at first:

Heteroskedasticity Test: White

F-statistic	0.355210	Prob. F(2,77)	0.7022
Obs*R-squared	0.731351	Prob. Chi-Square(2)	0.6937
Scaled explained SS	1.684732	Prob. Chi-Square(2)	0.4307

Fig.3. White heteroscedastic test of regression model

Shown in Fig.3, according to the test of white statistical value of 0.7022 is greater than the significance level of alpha value, therefore refused to exist heteroscedastic null hypothesis, that there is no residual difference variance.

Then B-G residual autocorrelation test:

Breusch-Godfrey Serial Correlation LM Test:

F-statistic	1.334464	Prob. F(2,76)	0.2694
Obs*R-squared	2.714086	Prob. Chi-Square(2)	0.2574

Fig.4. B-G residual autocorrelation test.

According to the Fig.4 B-G residual autocorrelation test statistic value p=0.2694 is greater than the significance level, α=0.2574 can refuse to return to the original assumption, namely there is no regression. And because the regression statistics of 2.079422 D-W, can think that there is no residual autocorrelation.

According to the above steps after the inspection, considered that 000011 estimated regression model has good prediction effect. According to the established model to calculate abnormal returns and CAR, the results are as Fig.5:

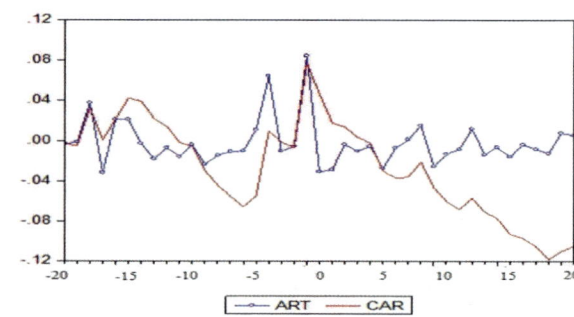

Fig.5. 00001's AR & CAR.

By the above it can be seen during [-5, 0], the company's abnormal returns of stock price fluctuation is bigger, reduce the previous peak of 0.0844, underweight fell sharply to 0.0308 on that day, five days after the holdings of shares has negative abnormal returns, then has hovered near zero; CAR in volatile during [-10, 0], reached the peak 0.0772 on the day before the reduction. Underweight showed a trend of accelerated decline, after 18 days fell to the lowest 0.1178.

3) Build the model:

According to the above method one by one to calculate the samples of abnormal returns and the CAR, getting the sequence of AR and CAR, then according to the formula (5), (7) stated to calculate the all samples AAR and ACAR, the results are as Fig.6:

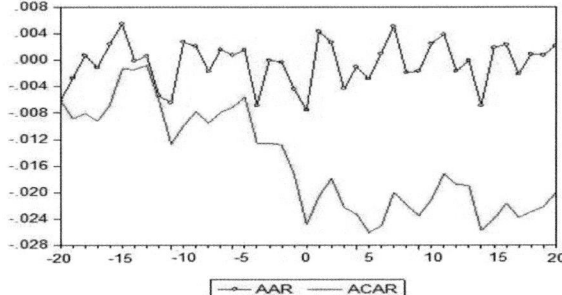

Fig.6. All samples' AAR & ACAR during [-20, 20]

From the above it can be seen that the AAR has been zero value fluctuates up and down, wave 0.0055 a 15 days before reducing, the lowest 0.0076 occurred at the reducing date, AAR has no obvious trend; ACAR during the window period has been less than zero, especially [5, 0], a sharp decline, fell to 0.0249 when t0. In general, the reduction of state-owned shares 71 before the event did not bring our expectations are ACAR, AAR in both before and after the reduction is roughly in the normal range fluctuations, ACAR has obvious decline.

IV. RESEARCH CONCLUSIONS AND DEFICIENCIES

This article system elaborated the current popular standard event study at home and abroad, the application of the reduction of state-owned shares is to calculate the happen abnormal returns and CAR in the enterprises are studied the impact of reduction of SOE. This article selects the listed on the Shanghai stock exchange and Shenzhen stock exchange among all a-share companies, since 08, proportion of state-owned shares reduced one-time is greater than or equal to 5%, and after reducing the proportion of state-owned shares accounted still greater than or equal to 50% of the enterprise's share, using the standard event study method, to calculate its estimated period [-100, -21], window [- 20, 20] the AAR and ACAR, after the comparative analysis it is concluded that reduction of state-owned shares didn't bring positive effect to the enterprise value conclusion. Get this conclusion of the empirical testing; I think the possible reasons are the following:

Firstly, our capital market lack of effectiveness. This paper selects the market model to calculate the normal returns, but our securities market is not perfect enough, market model is not necessarily the most appropriate, thus in the process of calculating beta coefficient may produce larger error, will cause the results of a certain deviation.

Second, the system construction of reduction of state-owned shares in China is still a gap. Since proposed the reduction of state-owned shares, government issued the corresponding system is relatively small. However, the market situation of our country needs the appropriate system to be constraints and specifications.

Third, the study sample may have limitations. Due to the ability and the time limit, this article does not research on other listed companies except a-share listed companies. And changes in the book value of the sample companies are not being studied, ignoring the relationship between state-owned shares and enterprise book value.

REFERENCE

[1] F. Hu, Y. Wang, "The empirical study on the relationship between the reduction of state-owned shares and listed company management performance" (in Chinese), *Nankai Management Review*, 2004(01): 64-68.

[2] L. Huang, Y. Teng, "Experience analyses with draw lessons from reduction of state-owned shares abroad" (in Chinese), *Fujian BBS*, 2004(5): 38-41.

[3] Y. Shangguan, "Reduction of state-owned shares is the inevitable choice of the state-owned enterprise capital structure management" (in Chinese), *Mechanical Management Development*, 2007(10): 177-178.

[4] J.Chen, "Research on Chinese equity division reform" (in Chinese), *China's Water Transport,* 2006 (1): 113-116.

[5] M. Zhang, "Analysis of consideration issues in equity division reform" (in Chinese), *Modern Enterprise Culture*, 2008(5): 9-10.

[6] Y. Liu, L. Zhu, "The empirical study of corporate governance and enterprise value" (in Chinese), *Management Review,* vol. 23, no. 02, pp. 45–52, 2011.

[7] Y. Ding, W. Gu, "The empirical research on the relationship between the reduction of state-owned shares and business performance of listed companies, based on the research of listed companies in Anhui province" (in Chinese), *Science and Technology Economic Market*, 2008 (4): 42-43.

[8] Y. Liu, Z. Huang, E. Tse, X. He, "The empirical research on China's ownership structure and corporate performance of listed companies" (in Chinese), *Economic and Management Research*, 2011(2): 24-32.

[9] L. Tian, "The u-shaped curve on the influence of state-owned equity of listed companies performance and theory of government shareholders' dual tactics" (in Chinese), *Economic Research*, 2005(10): 48-58.

[10] H. Berkman, R. Cole, J. Fu, "Improving Corporate Governance Where the State is the Controlling Block Holder: Evidence from China", *European Journal of Finance*, Forthcoming V, 2012, (Jan).

Research on Health Service Cost Accounting Based on the Application of TDABC

Li Luo, Feng-Jiao Wang*, Wei Cheng, Fang Qing, Ting Zhu

Business School, Sichuan University, Chengdu, China

(wfj735823255@163.com)

Abstract – **China's health cost is increasing rapidly, the component of cost is complex, and the method of cost accounting is confusing. In order to better manage and control healthcare costs, we explore the use of Time-Driven Activity-Based Costing (TDABC) in this study, and describe the particular procedure of application in healthcare, track expenses incurred in the hospital, and then simply apply this method to a kind of surgery in West China Hospital as a case. In the end, we attempt to point out the advantages and adaptability of this method for cost accounting in hospital.**

Keywords - **Cost accounting, cost-driver rate, health service, TDABC**

I. INTRODUCTION

Along with the development of healthcare industry, an aging of population and the changes of disease spectrum, healthcare costs have been rising at alarming rates for decades, bring about a heavy social and global economic burden on communities and countries. How to lower cost and meet citizens' needs for health service is a common challenge when global countries facing health care reform. The costs of healthcare in the United States are the highest in the world, in 2012 U.S. the total amount of healthcare costs was nearly 18% of GDP and continue to rise [1]. In China, the total cost of healthcare is 502.5 billion in 2001, it reached 2.78 trillion in 2012, and the cost of healthcare are growing faster than the growth of GDP, by 18.8% and 7.7% [2].

In the hospital systems, cost construction of health service is complex, including direct costs, indirect costs, administration costs, etc. Department shall be the unit for most of the current hospital cost accounting system, healthcare managers often allocate their costs to procedures, departments, and services not based on the actual resources used to deliver health services but on how much they charged. Patient pathway are complex, often involving in multiple departments, consequently, it is difficult to estimate the range of cost clearly for outpatient visit or disease cost. Under the condition of the lack of accurate cost accounting system, health services charge deviation cost, medical insurance claims deviate costs, and the difficulty and expense of medical services have long plagued our society, they are the two major problems in China's health system.

Under the background of the rapid growth and complex management in healthcare costs, this paper devoted to manage and control the cost better, mainly

research on how to use the method of cost accounting Time-Driven Activity-Based Costing (TDABC).

II. METHODOLOGY

In order to overcome the difficulties inherent in traditional Activity-Based Costing (ABC), professor Kaplan at Harvard University (2004) [3] put forward a new cost accounting method TDABC based on the theory of ABC, which uses time equations to resolve the assignment of resources to activities. At present, there are more than one hundred companies in foreign countries who implemented the method TDABC successfully, and have achieved good results. Brugemann (2005) [4] and Everaet (2008) [5] applied the theory and method of TDABC to logistics, compared the costs allocation accuracy with traditional ABC, and proved that TDABC can improve the profitability of company-understanding of customer's profitability. Pernot (2007) [6] used TDABC in the library, which proved that this kind of cost accounting method can clearly calculate and reduce the service cost of cooperation. Demeere (2009) [7] tested the method of TDABC in healthcare contexts, and they implemented an interesting case study in outpatient clinic, in which they pointed out that the method can improve medical service of hospital and provide better supply chain. Szychta (2010) [8] suggested the method of TDABC is a powerful tool for profitability analysis which is applied mainly in the service industry and is suitable for time equations. French (2013) [9] described the use of TDABC in a clinic setting to quantify the value of process improvements in terms of cost, time and personal resources. The research showed TDABC allows for quantification and evaluation of value in process improvements. Chan (2005) [10] compared the two kinds of approaches, traditional ABC and TDABC, in which they proved TDABC can provide more accurate, prompt and decision-related cost information. Min (2007) [11] implicated TDABC to calculate the cost in logistic companies, proved that it is suitable for the complex environments and easy to update. Huang Cheng li (2009) [12] have taken patients' nursing cost accounting inside hospital departments as an example, mainly focusing on the advantages and application in healthcare industry of TDABC by comparing with traditional ones. Zhao (2011) [13] made a literature review from ABC to TDABC, and then pointed out the direction of research in the future. Wang (2013) [14] built a total cost allocation model based on TDABC to explore how to use it to improve the hospital cost accounting system, which concluded the method plays positive role in raising the level of hospital cost management and provides useful references and

---

[1] This study is sponsored by the National Nature Science Foundation of China (71131006, 71172197), and funded by Sichuan University (skqy201411).

scientific basis for the formulation and revision of medical service fee.

According the findings of those studies, TDABC assigns the overhead costs only into the one time equations, comfort and deal with complicated operation processes. Furthermore, the model is easy to update, discovers the possibility of unused capacity and provide more accurate cost information, which is suitable for application in the medical services. The famous professor at Harvard business school, Michael E Porter (2011) [15] pointed out that the poor costing system and insurance reimbursement are the root causes, when considering America's problem about the rapid growth of healthcare costs, and he proposed to apply TDABC for balancing the cost and outcomes. He explained the enlightenment of how to establish a new costing measure system with consideration to value maximization.

Although the scholars have explored the applications of TDABC in some aspects, including the applications in the healthcare industry. But scholars has not paid enough attention to the relevant research about the application process and the practical application in the healthcare industry and published only a few papers on it. Therefore, in order to better control the rapid increasing of medical cost, this paper attempts to research on the principle of TDABC, and describes the particular procedures of application in healthcare contexts, and then uses it in a certain surgery as case study. In the end, we try to point out the advantages and adaptability of this method for cost accounting in hospital.

III. APPLICATION AND ANALYSIS

A. Example

Compared to the method ABC, the main improvement of TDABC is that introducing the time equation to the method ABC, regarding time as the basis of allocating resource costs, and conducting reliable estimation of the effective operation time and the unit activity time-consuming to calculate the cost share of the unit activity. The advantage and breakthrough of TDABC method depend on the estimation of time; moreover, TDABC's application needs to estimate the amount of complete time of the unit activity rather than the ratio that the complete time of the unit activity divided by the whole work time. TDABC model mainly calculates cost driver rate, regarding the time as calculating basis. Cost

driver rate equals that the capacity cost rate * times the unit activity cost of activity capacity (time); in the equation, capacity rate equals that the whole resource costs are divided by expected available capacity (time). So the whole costs can be calculated by the following equation (1).

$$C(t) = \sum_{i=1}^{n} r_i t_i \qquad (1)$$

The other advantage of TDABC is that the time differences required for different activity types are put into the time equation. Increasing the number of items of the model can calculate the time required for different services. Operation (2) time-consuming denoted by T.

$$T = a_0 + a_1 + a_2 + \cdots + a_i \qquad (2)$$

a_1 is the standard time of based activity, a_i is the extra activity time. For example, ordinary patient registration time is 2 minutes, however the initial registration patient needs 2 + 1 minutes.

We take calculating the total costs of certain types of patients A seeing the doctors as a simple example applying the TDABC method. Assuming the processes and paths a patient experienced only need to use three kinds of resources, respectively administrative staff, nurses and doctors, as shown in Fig.1.

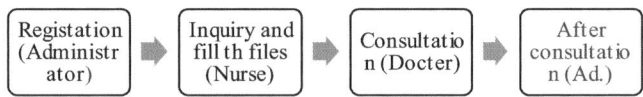

Fig.1. The main activities to see a doctor

We estimate the parameters required in the calculation, the time of providing services to patients needs 0.3 hours for administrative staff, the time of providing check-ups spends 0.4 hours by nurse, the time of providing advice and diagnosis needs 0.25 hours by doctor. Then we can calculate the capacity cost rate. For nurse in this example, we calculate the total resource costs including not only the salary and welfare but also the indirect costs generating in the service process such as management, space, equipment, communications and other costs. By calculating the total cost of the nurses is 18,000 Yuan per month, the work days per month removing the number of days without work is 20 days all the same, the work time per day removing the rest, meeting and teaching is 6 hours.

TABLE I
THE CALCULATION OF TDABC

Personnel	Capacity cost rate	The quantity of time	Capacity cost	The quantity of time	Capacity cost
Administrator	80	0.3h	24	0.3h+0.05h	28
Nurse	150	0.4h	60	0.4h+0.1h	75
Doctor	340	0.25h	85	0.25h+0.2h	153
Total cost		169		256	

Thus, estimated available capacity adds up to 120 hours. Nurse's capacity cost rate is 150 Yuan per hour, in the same way, we can calculate the capacity cost rate of administrative staff is 80 Yuan per hour, the doctor is 340

Yuan per hour. The total cost of patient A in the hospital = 0.3 * 80 + 0.4 * 150 + 0.25 * 340 = 169 Yuan. As it is shown in the Table I, if there is a special case of the patient (such as the existence of complex ill conditions of patients), the variation of time is added into time equation to calculate the total cost, the total cost is 256 Yuan. Using the time equation method TDABC automatically allocates resource cost to each operating. TDABC calculates the complex activity costs with a simple method, and can be flexibly updated all the same, and can provide accurate cost information for managers to meet management needs.

B. Analysis

TDABC method is applied to medical field, in which expected available capacity is denote as the amount of actual work time removing the amount of idle time for activity object. In order to account the total cost generated by providing services to patients with a certain type of disease. Exact cost system must take into account all the resources that will be used as providing health services to patients, which usually involves a variety of different resources: personnel, equipment, space, medicine and supplies, etc. Applying the method TDABC, by estimating the capacity cost for each resource and the time-consuming activity, we can exactly measure the total cost generating in the medical system. Applying TDABC to re-measure cost in the medical environment, we design the initial steps which are shown in TABLE II:

TABLE II
THE PROCEDURE OF TDABC IN HEALTHCARE

STEPS	COST ACCOUNTING MODEL
1	Selected disease.
2	Determine the process map and the activities contained in the process.
3	Obtain the time estimates of each activity in the process.
4	Determine the cost of providing health care resources.
5	Estimate actual capacity of each resource and calculate the capacity cost rate.
6	Calculate the total cost.

1) Select disease.

Firstly, what disease and what kind of disease condition would be chosen clearly to calculate the cost, in the beginning stage, we can choose a kind of simple disease without the complications which has relatively mature and stable treatment techniques. After maturely accounting the simple disease, we further take into account a kind of complex disease which has the complications and comorbidities. At this point, on the basis of a single disease, we conduct a more detailed cost accounting for complex disease with all kinds of complications.

2) Determine the process map and the activities contained in the process.

Process map covers the major pathways of patient treatment process, the changes of pathway which will also be reflected in the process map. Second, make sure the required resources for each activity in the process map, which includes direct and indirect resources, such as personnel, equipment, drug and supply, as shown in Fig.2 (does not show all resource information).

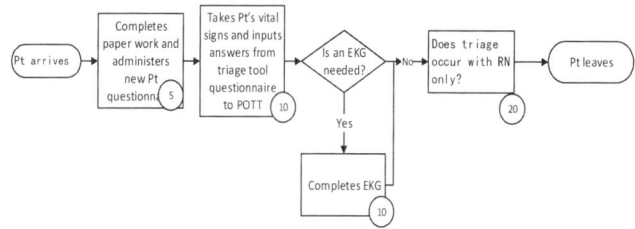

Fig.2. The Process of Anesthesia Assessment Center (From Katy. 2013[9])

3) Obtain the time estimates of each activity in the process.

Based on the process map, we determine the time that patient experienced in every step as shown in the process or estimate the time of every resource used. If the process time is short and has small variance, standard time can be adopted. Considering the time-consuming and unpredictable flow, we can adopt the exact register time. The time estimates can be figured out by the professional medical team, with the further improvement of RFID and data information systems, we can also achieve the time through more scientific methods such as statistical regression. Fig.1 shows the required activity time in the process.

4) Determine the cost of providing health care resources.

Estimate the direct costs of a variety of related resources in patient treatment, as well as indirect costs and ancillary costs that must be considered. Including human resource costs (described in the above example), as well as equipment costs and depreciation expenses, medical supplies, operation and management costs, etc.

5) Estimate actual capacity of each resource and calculate the capacity cost rate.

For example, the actual capacity of the staff is equal to working hours per day minus the rest, teaching, meetings and other time that has nothing to do with the patient, etc. similarly needing to estimate exact capacity of the resources as well as equipment. It is generally believed that the actual capacity account for 80% -85% of theory capacity, it also can use more scientific methods such as data analysis and other statistical regression to determine the actual capacity. In order to calculate capacity cost rate, the total cost of resource consumption in the previous step divided by the actual capacity, we can get a ratio that is the capacity cost rate.

6) Calculate the total cost.

We get the cost driver rate of each operation through the capacity cost of individual resources that each patient used in the process (including related auxiliary costs) multiply by the time of the resources patient used. To plus the patient cost of various activities in the process is equal to the total healthcare cost of patients.

C. Case study

This paper applied the method simply considering a kind of otolaryngology surgery as a case in West China hospital. Through the data analysis of 147 cases within a year of the surgery, with the total expense 625,255 thousand Yuan. There are four time points in the cases of surgical procedures, which can be divided into three processes, including preparation time, surgery time and clean-up time. We have obtained the figure of these three time periods based on data analysis, showed in TABLE III.

TABLE III
THE TIME OF EACH PROCESS

ITEM	TIME (Ave.)
Preparation time	45min
Surgery time	125min
Clean-up time	17min
Total time	187min

The procedure of preparation can be completed by two nurses, surgery performed by two nurses and a doctor, cleaning performed by two nurses, and then by calculating the capacity cost rate of them, we can get the capacity cost rate of nurse is 5.21YUAN/min and doctor is 11.45YUAN/min. Other calculation about cost is shown in the TABLE IV below.

TABLE IV
THE CALCULATION OF COST

ACTIVITY	Capacity cost rate① (YUAN/min)	Time② (min)	Total time ③=②*147 (min)	Total cost ④=①*③ (YUAN)
Preparation (2N)	10.42	45	6615	68928.3
Surgery (2N+1D)	21.87	125	18375	401861.25
Clean-up (2N)	10.42	17	2499	26039.58

Comparing the calculated cost and actual cost information, there is a quite big difference, accounting for a 20.54% difference of the calculated cost actual total cost. As the following TABLE V.

TABLE V
THE DIFFERENCE BETWEEN THE CALCULATED COST AND ACTUAL COST

ITEM	TOTAL COST(Yuan)
Calculated cost	496829.13
Actual cost	625255.67
Gap between them	128426.54

In addition to the existence of calculation errors, we concluded that through the calculation there is the presence of underutilized capacity in the surgery, which provides a basis for the subsequent process optimization and cost control. According to the results, we could concentrate on process improving and cost controlling in the future research.

IV. DISCUSSION

The advantages of using the method TDABC in healthcare contexts are as following:

(1) Using TDABC for healthcare cost accounting is easy to operate. Although there are many processes and activities in the hospital, healthcare belongs to service sector which the work time can be considered as the main cost driver. We can not only observe or analysis the service time of personnel and equipment, but also can obtain time estimates for each process. TDABC is able to encompass special aspects of particular activity into the

one time equation so the model can be quickly updated. And the principle of TDABC is easy to understand, not complex, so the method TDABC is applicable in a hospital.

(2) TDABC is helpful to enhance the management and efficient to improve the capacity utilization. The complexity and uncertainty of health services have greatly increased the difficulty of cost management in hospitals, the method TDABC can estimates and calculate the time of complex service times accurately based on the rich data resources of ERP and HIS system, which makes it easy to calculate the cost of each unit activity. Applying the method can create the necessary conditions for health-service-price reform and DRGs. Meanwhile, the use of TDABC can objectively reflect the consumption of resources; simultaneously, it also discovers the possibility of unused capacity, improves operational process, respects interaction between time drivers, and detects the process without value in the way of trace of costs and changes in every activity.

(3) TDABC can assess the performance of hospital personnel while providing a scientific basis for improving process efficiency and saving cost. The application of TDABC can calculate the capacity cost of each physician and staff, and combining with the assessment of patient outcomes etc., it can be used as medical staff performance evaluation index. And then, after healthcare managers understand the cost clearly, they can recognize that they should allocate the personnel of high capacity cost to focus on high-tech level of work skills.

TDABC not only helpful of the design of new accounting and information system, but also brings the opportunities to improve the value in healthcare. However, there is a big question that whether the method can be used in healthcare industry comprehensively. For example, how to take all costs incurred fully into account with this method, how to divide the unclear activities, and whether this is appropriate to use the standard time. These problems need further research and exploration.

V. CONCLUSIONS

For a long time, the high cost of health services, the rapidly rising costs bring enormous pressure to government and patient, TDABC is used to calculate the actual cost of hospital and to adjust the charge and insurance payments, as well as plays a positive role in health reform. In many hospitals abroad, it have begun in the trial run and expanded the range of application, but whether the method TDABC is suitable for the special complex situation in China, and how to become suitable for China's healthcare cost accounting model also requires more exploration. Finally there are some potential areas for further research.

(1) Combining DRGs together to establish a new cost -payment model. DRGs is known as diagnosis-related

group, which is a kind of medical resource consumption intensity as the axis of grouping a case classification, and also is recognized as one of the more advanced payment in the world currently. Combining the DRGs with TDABC costing mechanism to establish a new accounting system according to the disease, forming a medical insurance method based on disease, and then building bundle payments mechanism for health service in above foundation are absolutely important for health care service systems.

(2) Combined with the clinical path. Clinical pathways, also known as care pathways, critical pathways, integrated care pathways, are one of the main tools used to manage the quality in healthcare concerning the standardization of care processes. It has been shown that their implementation reduces the variability in clinical practice and improves outcomes. In recent years, clinical pathways were vigorously promoted in hospitals; the ministry of health in December 2009 formulated and issued the "clinical pathway management guidelines (trial)"and "clinical path management pilot scheme". The implementation and promotion of Clinical Pathway provide more realistic foundation for application of TDABC.

REFERENCES

[1] Fuchs VR. The gross domestic product and health care spending [J]. N Eng. J Med. Vol.369, pp.107-109, 2013.

[2] China statistical yearbook 2013.China Statistics Press [M].

[3] Kaplan, R.S. Anderson, S.R. Time-driven activity-based costing [J]. Harvard Business Review. Vol.82, pp.131–138, 2004.

[4] Bruggeman W, Moreel K. Activity-Based Costing in Complex and Dynamic Environments: The Emergence of Time-Driven ABC [J]. Controlling, Vol.16, No.11, pp.597-602, 2004.

[5] Everaert P, Bruggeman W, Sarens G, et al. Cost modeling in logistics using time-driven ABC: Experiences from a wholesaler [J]. International Journal of Physical Distribution & Logistics Management. Vol.38, No.5, pp. 172-191, 2008.

[6] Pernot E, Roodhooft F, Van Den Abbeele A. Time-driven activity-based costing for inter-library services: a case study in a university [J]. The Journal of Academic Librarianship. Vol.33, No.5, pp.551-560, 2007.

[7] Demeere N, Stouthuysen K, Roodhooft F. Time—driven activity—based costing in an outpatient clinic environment: Development, relevance and managerial impact [J]. Health Policy. Vol.92, pp.296-304, 2009.

[8] Szychta A. Time-Driven Activity-Based Costing in Service Industries [J]. SOCIAL SCIENCES. Vol.67, No.1, pp.49-60, 2010.

[9] Katy, E. French. Heidi W, Albright. Measuring the value of process improvement initiatives in a preoperative assessment center using time-driven activity-based costing [J]. Healthcare. Vol.1, pp.136–142, 2013.

[10] CHEN Yu-Qing, YAN Lin. Analysis and comparison of two types of Time-driven Activity-based Costing. Journal of Northeastern University (Social Science) [J]. Vol.7, No.5, pp342-345.Sep.2005.

[11] MIN Heng-feng. The Logistics Cost Accounting Based on Time-Driven Activity-based Cost. Logistics Sci-Tech [J].Vol.30, No.6, pp.93-95.2007.

[12] HUANG Cheng-li. ZHU Wei-wei. Application of time—based ABC approach in patients' nursing cost accounting. Chinese Hospital Management [J]. Vol.29, No.2, pp.60-62. 2009.

[13] ZHAO XI, LI YAGUANG1, QI JIANMIN. A Review of Time-driven Activity-Based Costing: Technique, Implementation, and Revelation. Journal of Xidian University (Social Science Edition) [J]. Vol.22, No.3, pp.32-39. May, 2012.

[14] WANG Jie, GUO Yu-Hai, DAI Zhi-min. The Application of Time Driven Activity-Based Costing Approach in the Mode of Hospital Total Cost Accounting. Chinese Health Economics [J]. Vol.32, No.5, pp90-92. 2013

[15] Kaplan, R.S. Porter, M.E. The big idea: How to solve the cost crisis in healthcare [J]. Harvard Business Review. Vol.9, pp.46-64, 2011.

Analysis of Operation Mode of Automobile Implantation Marketing

Shao-ling WU*, Jiang-hua WANG, Jing-wen ZHOU

School of Economics and Management, Hubei University of Automotive Technology, Shiyan, China

(wuming1122@163.com)

Abstract - **As the world car industry grows, car market competition intensified. How to find new marketing mode has become the car companies actively seeking a major breakthrough. The desalination advertisement publicity of dominant trait, will own brand and product clever film and television media, network game implanted sports events, education and training, and some popular form of mass, often can have the ideal brand spread and product promotion effect. This topic is trying to analyze the characteristics of implantable marketing, approach, the principle, construct a car implantable marketing mode of operation analysis its limitations and put forward some countermeasures.**

Keywords - **Automobile implantation marketing, operation mode, operation strategy**

I. INTRODUCTION

With the continuous development of world automobile industry, competition in the automobile markets is more and more severe [1, 2]. Good marketing strategy becomes an important index in measuring the development prospect for an automobile company. It seems that traditional marketing modes cannot satisfy the severely competitive automobile markets [3-5]. And automobile companies are compelled to explore new transmission and marketing modes. Implant marketing is developing rapidly abroad [6]. And with its heat, Chinese automobile companies are starting to explore new types of marketing modes [7]. Automobile implant marketing, refers to the marketing means to strategically integrate automobile products, brands, representative visuals, sounds, symbols, even service contents into media, activities, and events, via reproduction of the scene, unknowingly leave good impression of products and brands on people's minds, and reach brand promotion and product marketing [8, 9].

II. PRINCIPLE OF AUTOMOBILE IMPLANT MARKETING

A. Implant carrier should be consistent with brand positioning

During the operation process of automobile implant marketing, it is necessary to select implant matrix compatible to brand features and market positioning. If implant marketing promotion is conducted randomly, expected effects cannot be reached. And it may also make the brand image looks undetermined, even damage the brand image. Mazda Atenza in movie "Go Lala Go", as another example, is positioned at fashionable

white-collar. Mazda Atenza naturally integrated into the scenes of the movie, displayed the white-collars preference towards the brand and model, and led market consumption. As senior model, Mazda Atenza speedster integrates the advantages of sports car and limousine, and reflects the perfect combination of personalized appearance, superb sports and high level equipment [10]. It is consistent with the qualities of the main characters in the representative story. And it demonstrates the taste of elites for vitality, and enjoyment of driving.

B. Implant brand should possess certain reputation

In utilizing automobile implant marketing, it should be taken into consideration that the implant brand should have certain reputation. Such is human nature that when you watch a movie, television series, entertainment program, sports game, or plan online games, main attention is paid to the characters and story line in the movie or television series, interesting things and passion in sports games, and scenes and experiences in games. Brands and symbols hidden are not proactively paid attention to. Yet, if the implant brand is familiar to the audience, the impression will be profound. Take the logo of "three pointed star" of Mercedes-Benz in tennis tournaments for example, it is because the brand of Mercedes-Benz itself possesses certain popularity, the fans are more easily to remember and pay attention to the logo [11].

C. Implant brand or product should have individualized differentiation

The fact that implant automobile products or brands possess personalized differentiation is a supporting condition for the success of implant marketing. Automobile implant marketing is a kind of comparatively hidden advertising promotion. It is not as evident and exposed as traditional advertisements. How can the audiences be enabled to notice it when watching the implanted matrix, and stimulate their sub-consciousness to achieve the purpose of promotion?

This requires the implant products or brands to possess unique personalized features, so that it can be distinguished from other products of the same type. Hence, requirements for implant are high. Products or brands must be highly compatible with implant matrix, harmonious with story line, and able to present its sown style and personality. For example, in television series "Love Stories at the Countryside 3", Chevrolet Sail sedan is the scooter for Xiao-meng WANG and

E. Qi et al. (eds.), *Proceedings of the 21st International Conference on Industrial Engineering and Engineering Management 2014*, Proceedings of the International Conference on Industrial Engineering and Engineering Management, DOI 10.2991/978-94-6239-102-4_118, © Atlantis Press and the authors 2015

Yong-qiang XIE. Via tenaciously struggling and pursuit of the characters in the story, the marketing target of Sail brand is stressed at Chinese families in growth [12, 13].

D. Implant marketing activities should be conducted consistently

One of the biggest features of automobile implant marketing is the hidden of information. Consistently keep implant activity going on is a feasible and effective way to ensure the continuous demonstration of products and brands. For example, the world known "three pointed star" logo of Benz, had kept 12 years of brand naming and sponsorship for ATP Tennis Masters tournament ever since 1996. Hyundai-KIA has been associated with the World Cup ever since 1999. Education and training for Chinese technical workers are conducted by Toyota one step after another. These successful cases tell us, if automobile companies wish to obtain the richest returns from implant marketing, the implant activity must be carried on. Consequently, implants in movies and television series by automobile companies follow this rule, and possibly visualized implant products repeatedly to reach expected effects without arousing dislike of the audiences [14, 15].

III. AUTOMOBILE IMPLANT MARKETING CHANNELS

A. Movie and TV Media Implant

Automobile and television programs, the two greatest inventions in the late 19th century are connected seamlessly after their births. The manufacturing of a good television program demands a lot of money. And the transmission and popularity of a good television program is borderless. Automobile companies cannot neglect this superb promotional platform. Integrating automobile products and brands into movies and television works, can effectively bring out brand image, and display the functions of the products. The unique left-right parallel four-wheels drive system and horizontal motor of Subaru grant drivers with free driving experience. Via implanting in movie "You Are the One", Subaru perfectly combines the essence of automobile products with modern emotional life, makes the automobile live, and in the view of many of the audiences.

B. Internet Game Implant

Since the entry of the internet in China market back in the 1990s, network media is occupying a more and more important position among public media in China. According to relevant data, the number of internet users is increasing rapidly in China. The total number of internet users of China ranks number one in the world. Speedy development of network media, offers more business opportunity for automobile companies, and

creates more sufficient channels and space for promotion and marketing. Delivering product information to target consumer groups via network media is a feasible and effective way.

For automobile industry and network media, the generation with the development of internet will become the mainstay of the society. They gradually grow to become potential consumer group that cannot be neglected for automobile consumption. Network games, as the medium that best affect game players, unknowingly affect and lead the potential consumption group. In 2009, FAW Volkswagen and network game operator Shanghai Joyzone Network Co., Ltd. formed in cooperation, and implanted high performance "Sagitar" newly launched in the market into large-scale network racing game "Racing" run by Joyzone. The communication platform of network game with extensive pool of audiences was fully utilized. Brand new medium and presentation methods were adopted, to display the quality performance of "Sagitar" to potential consumer group among Chinese game players.

C. Sports Events Implant

Basketball, football and other sports have won the hearts of people. Large scale sports events such as the World Cup, the Olympics, and the Asian Games have grabbed the eyeballs of audiences around the world. Implanting company image, products and brands via naming and sponsorship in such major events, not only establishes company image rapidly, but also improves brand reputation and company influence.

According to estimates, under similar investment spending, the returns of implanting in sports events to automobile companies are of multiples to traditional advertising modes. Such feasible and effective promotional mode is more and more favored by automobile companies.

Shanghai Volkswagen was one of the excellent partners for the 2008 Beijing Olympic Games. It successfully implanted company brand in Olympic events, and obtained fruitful results. The company culture philosophy "Pursuit of Excellent, Fight to become the First" of Shanghai Volkswagen, is consistent with the spiritual level of Olympics "Higher, Faster, Stronger". The theme dynamic of "Share Olympics, Heart for Excellence" integrated company philosophy, consumer vision, and Olympic spirit, and established the company image of "offering quality service for the public". After the Olympics ended, based on authoritative network reputation statistics, the reputation weighted index of Volkswagen was 73.99, ranking the 6th in many of the partners. From this we can tell, that implant in Olympics won good reputation for Shanghai Volkswagen, and set up solid company image.

D. Education Training Implant

Automobile is an expensive product with high cost in purchase and use. In selecting the purchase of automobiles, people have to weigh against its after-sales services. Company image can be established and brand promotion can be conducted through nurturing huge batch of excellent technical workers, improving after-sales services quality, and obtaining trust and reliance of customer group. Placing company brand and product implant into the cradle of technical workers, namely, middle and high level vocational education, will surely and effectively accelerate the development of the company.

Toyota Motor Corp took the lead. As early as 1994, it introduced T-TEP (Toyota Technical Training Program) in China. As an advanced after-sales services technical staff training system, this program has development win-win in over 400 schools and over 50 countries around the world, and the professional and rich technical training experiences of Toyota Motor Corp is shared. For 18 years in China, particularly after the founding of FAW Toyota, Toyota Motor Corp has consistently paid special attention on T-TEP. FAW Toyota provide teaching facilities and equipment for schools in cooperation of T-TEP, periodically conduct lecture training, and make effort to transmit update automobile technical information to schools.

IV. AUTOMOBILE IMPLANT MARKETING OPERATION MODE

Message is different from message in traditional advertisement. Message in implant advertising is composed by the integration of implant advertisement and implant matrix. They are related, and inseparable. Implant matrix is the medium, as well as the message. The better integration of product advertisement with the implant matrix, the better the transmission is, and the better the promotional effects are. Audiences tend to place trust of the medium and favorability in the implant product itself.

The difference between implant marketing mode and traditional advertising mode is that effective delivery of message to the audience requires a process for being activated, the association with traditional advertisement. What implant marketing affects is the sub-consciousness of the audience. How to activate such sub-consciousness so as to compel the consumers to conduct purchase demands association with traditional advertisements. If it lacks association, message is not effectively delivered to the audience. And the implant marketing is not bringing its expected function.

During the transmission process, there are noises generated, which disturb the transmission to consumers, and result in distortion and compromise of message obtained by audience.

Feedback mechanism is comparatively weak. From receiving implant advertisement message to activating sub-consciousness, to forming consumption attitude, and to the final decision and behavior of consumption, is a long process. The audiences are not necessarily consumers. After the above process, some transform to consumers, some transform to potential consumers, and the rest are non-consumers. Therefore, only those who become final consumer are possible to provide feedback on products or advertisements. Then, how can advertisers collect such feedback information is difficult. Given that this is a long and interactive promotional process, it cannot be determined via data to prove that the final consumers are transformed from implant advertisement audiences. Thus, implant advertisement lacks a feasible and effective evaluation mechanism. We can only refer the method of observation from the study of transmission to conduct indirect observation. The number of people who watched the advertisement can be preliminarily estimated through television ratings. And for sure, it is not accurate. It is an issue worth consideration for advertisers upon how to suggest the establishment of a perfect implant advertisement evaluation mechanism and feedback mechanism.

V. AUTOMOBILE IMPLANT MARKETING STRATEGY

A. Innovative Advertising Form, In-depth Implant

Many implant advertisements only place the products in the matrix, in simple form, and stiff way. When implanting, it is suggested that advertisers and producers of implant matrix conduct in-depth communication, innovate on advertisement for, and find ways that are creative and easy to be recognized by audience. In this way, advertisements are done "invisibly". It is also recommended that automobile advertisers need to find compatible implant matrix with the image and brand of the automobile products, and utilize all possible things to highlight the products and brands, so that they can be remembered by the audience.

B. Insist on Integration of Product with Characters in the Story/Play

When selecting implant matrix, the integration of product and characters in the story/play should be insisted. Attention should also be paid to rationality and authenticity. Under the premises that the quality of matrix is ensured, implant point for automobile products can be found, so that the demonstration of every implant of product is consistent with the story, and its own features can be embodied. For example, the implant of GEELY series car models in "The Drive of Life" is successful. Before the shooting of the TV series, GEELY Automobile Company participated in early stage of writing the scripts. GEELY Automobile is a national brand. It embodies the efforts of cohesive, self-reliant, and hard-working Chinese people. The main characters in the TV series represented these excellent qualities of the Chinese nation. This TV series skillfully integrated the brand features of GEELY with the story line, won the appreciation of the audience, and achieved desired effect.

C. Pursue Compatibility of Target Customer and Implant Matrix

In the selection of implant matrix, automobile companies should first conduct in-depth comprehensive analysis on its contents and target customers. With the utilization of various channels to understand the age, gender, financial status, taste, and habits of target customers, and via comparison with brand appeal, it can be decided whether to have in the implant from compatible level. Only when the compatible level of target customer group and implant matrix audience is high, can the effectiveness of advertisement be improved.

REFERENCES

[1] Yu-feng YUAN, "Automotive Marketing" [M], Beijing: *Machinery Industry Press*, 2005, ch. 9, pp. 195-216.

[2] Wen-lin GAN. "Development of automobile marketing direction by the implantable marketing" [J]. *Popular Science and Technology*, 2011, (06).

[3] Wen-lin GAN. "Look at the development of automobile marketing direction by the implantable marketing" [J]. *The popular science,* 2011, (06).

[4] Hang Xi, Liu Jie. The antagonistic effect of implantable marketing [J]. *The vitality of enterprises*, 2010, (09).

[5] Hu Po. Implantable marketing, a surprise move to create a myth [J]. *Commercial culture*, 2011, (02).

[6] Liu Fan. Clumsy shouts and crazy exposure "fate of the call transfer" in the Product placement of [J]. *Journal of Beijing Film Academy*, 2008, (04).

[7] Liu Guoyan,Yi Shizhi. Research on Product placement operation strategy [J]. *China Market,* 2010, (36).

[8] Peng Hui. The false and true: symbolic language movie Product placement. [J]. *Film literature*, 2011, (12).

[9] Jing Lin, Wu Sizong. Research into the marketing based on consumer psychology [J]. *Market modernization*, 2010, (27).

[10] Zhao-xin Li. Research on advertising communication mode of implantable [D]. *Shandong University*, 2010.

[11] Chen Anyi. TV drama Product placement of [D]. *Nanjing Arts Institute,* 2011.

[12] Yuan Zhang. Product placement and avoid weaknesses [J]. *The News Sentinel,* 2010, (08).

[13] Tong Xiao. Forms of Product placement in the reality television show. [D]. *Renmin University of China*, 2008.

[14] Liang-ji Liu. Research on brand equity influence movie ads [D]. *Jinan University*, 2011.

[15] Ning-ning Xu. The audience psychology study film implanted in advertising [D]. *Hebei University*, 2010.

Pricing of Chinese Defaultable Bonds Based on Stochastic Recovery Rate

Ping Li, Jing Song*, Bojie Chen
School of Economics and Management, Beihang University, Beijing, China
(songjing.heng@163.com)

Abstract - In order to price corporate bonds more precisely, a stochastic recovery rate (RR) has been used in this paper instead of a fixed RR used by most researchers. The main contribution is to consider the negative correlation between the recovery rate and default probability by using an exponential function. Based on this function and reduced-form model, a pricing formula is obtained and solved by a characteristic equation. At last, this paper selects seven Chinese corporate bonds to do empirical research. The estimated parameters showed that a fixed RR will lead to great errors and thus stochastic RR is necessary. The results also indicate that the hazard rate has positive correlation with the interest rate, which means that the level of interest rate is still the main reason influencing a firm's credit risk. And the theoretical price obtained from the model fits the market price very well, which demonstrates the validity of the model.

Keywords - Defaultable bond, reduced-form model, stochastic recovery rate

I. INTRODUCTION

Since August 2007, corporate bonds have been widely traded in China, and its advantages have been shown in the capital market. The relatively high credit risk makes the pricing of corporate bond hard. Recently, many researches about corporate bond pricing have been done. Generally, there are two approaches to model the credit risk. The first approach is the structural model initiated by Merton (1974) [1], who use Black-Sholes model to price corporate bonds by assuming default is based on the evolution of a firm's assets and liabilities. Following Merton, Geske (1977) [2] and Geske and Jonson (1984) [3] used structural model to price normal bonds rather than zero-coupon bonds. Merton and his followers assumed that default events can only happen at maturity, which in fact is not the case. Black and Cox (1976) [4] improved this and allowed defaults to occur at any time. All of the above assumed that the holders of the bonds can get a fix sum of payoff when default occurred, Longstaff and Schwartz (1995) [5] extended this form and supposed the payoff was a proportion of the bond value. The second model is the reduced-form model, which takes default and recovery rate as exogenous variables. Reduced-form model was first introduced by Jarrow and Turnbull (1992) [6]. Thereafter, in the framework of reduced-form model, Jarrow, Lando and Turnbull (1995, 1997) [7-8] created a Credit Risk Transfer Markov Model, Duffie and Singleton (1999) [9] applied Duffie's Affine model to value contingent claims subject to default risk.

In practice, reduced-form model is more widely used in pricing credit risk because of its simplicity. However, it assumes that recovery rate is a constant which actually is not the case. Furthermore, it ignores the negative correlation between the probability of default (PD) and recovery rate. Therefore, the pricing results deviate from the market price. Bakshi, Madan and Zhang (2006) [10] demonstrated that the default probability and recovery rate have a negative correlation and the recovery rate is stochastic. Krekel (2008) [11] assumed recovery rate to be in four discrete states: 60%, 40%, 20% and 0%, which were too subjective and may differed with the reality.

Our paper focuses on corporate bonds' pricing assuming that the recovery rate is stochastic in the framework of reduced-form model, and provides a theoretical reference for investors. The paper is organized as follows. Section II and III give the preliminary models and our model, respectively; giving the solution of our model by using a characteristic equation. An empirical study for seven active Chinese corporate bonds is presented in Section IV. Section V concludes the paper.

II. PRELIMINARIES PRICING MODEL FOR DEFAULTABLE BONDS

Currently, there are many types of reduced-form model, among which affine model has more advantage for its efficiency to get PD. Affine model was derived from Duffie & Kan's (1996) [12] affine process and firstly applied to price default bond by Duffie and Singleton (1999) [9]. It gives the pricing model by solving characteristic equation's ODE (Ordinary Differential Equation). In this paper, we use affine model as the basic model.

To make it simple, we use RFV (Recovery Rate of Face Value) [13] to couple the recovery rate function, i.e., recovery rate is corresponding to a percentage of the bond value.

Default density $h(t)$ (also called risk ratio) denotes the conditional per year, so PD (probability of default) is

$$PD = 1 - \exp(-\int_0^t h(s)ds) \qquad (1)$$

Assume that $h(s)$ is a constant λ, then we can write the default intensity as

$$\lambda = -1/t \cdot \ln(1 - PD) \qquad (2)$$

Based on this, we can estimate the risk intensity λ_i and get a function of recovery rate and λ_i. The empirical study shows that PD and recovery rate have a negative correlation. Logarithmic model and exponential model can finely depict this relation. Consider that PD is the integral of risk ratio, so

[1] This work was supported by the National Natural Science Foundation of China (No. 71271015, 70971006).

we assume that the function of recovery rate and risk ratio is as follows:

$$R(t) = \beta_0 + \beta_1 e^{-h(t)} \qquad (3)$$

Where, when $h \to 0, R \to \beta_0 + \beta_1$; when $h \to \infty, R \to \beta_0$, so we must make constrains $0 \le \beta_0 + \beta_1 \le 1$ and $\beta_0 > 0, \beta_1 > 0$.

III. OUR PRICING MODEL

According to the previous discussion, we will price the corporate bonds by using stochastic recovery rate function on the consumption of RFV and reduced-form model.

Typically, corporate bonds have a face value F, maturity time T, and coupon payment. We use non-decreasing function $C(t)$ to denote accumulated coupon payment until time t. Default happens at time τ, by which time investors can get a recovery $y(\tau)$. Define a Heaviside function $\chi(t)$ as follows:

$$\chi(t) = \begin{cases} 1 & t \ge \tau \\ 0 & \text{otherwise} \end{cases} \qquad (4)$$

Defaultable corporate bond can be defined as $(F, T, C(t), y(t))$. Assume that it belongs to a filtration produced by a Brownian motion $(G_t, 0 \le t \le T)$. The probability space of $(\Omega, \Im, P)$ satisfies the normal condition, and $(\Im_t, 0 \le t \le T)$ denotes the arrival of information. $\Im_t$ is generated by default time process.

Assume that $b(t) \equiv \exp(\int_0^t r(s)ds)$ is a cumulative function of the money market account, where the spot rate is $r(t)$. According to Duffie (1996) [8], no arbitrage principle demands a probability measure Q, which is equivalent to probability measure P in the risk neutral world. Thus, all assets' discounted processes are martingales, and thus defaultable corporate bond with maturity time τ has a price P as follows:

$$P(t,T)_{\tau > t} = E^Q \{ \int_t^T \frac{b(t)}{b(u)} (1 - \chi(u)) dC(u)$$
$$+ \frac{b(t)}{b(t+\tau)} (1 - \chi(t+\tau)) F \qquad (5)$$
$$+ \int_t^T \frac{b(t)}{b(u)} (1 - \chi(u)) y(u) d\phi(u) \mid G_t \}$$

Where, E^Q is the expectation under the probability measure Q. This price can be observed when we are aware of the default which hasn't happened, so we simplify $P(t,T)_{\tau > t}$ as $P(t,T)$. The first integral of Formula (5) illustrates that as long as the default didn't happen, investors can receive a coupon payment; if default occurs, payment will stop. The second part explains that the investors can get a payment of the face value when there is no default. The last integral states a receiving when default happens.

Formula (5) is based on an assumption that there is a positive value process $h(t)$, i.e., risk ratio process, which make sure $\chi(t) - \int_0^t (1 - \chi(u)) h(u) du$ is a martingale. In this case, we refer $h(t)$ as a risk neutral process, i.e., $h(t)dt$ is a default probability on the interval of $(t, t+dt)$. Stochastic process is adapted to $\{G_t\}$, so we get:

$$E^Q[(1 - \chi(u)) \mid G_u] = \exp(-\int_0^u h(s)ds) \qquad (6)$$

We apply iterative calculation to the expectation of equation (6), calculating the expectation of G_u and the expectation of G_u. Then we can get $P(t,T)$ as follows:

$$P(t,T) = E^Q \{ \int_t^T \frac{b(t)}{b(u)} \exp(-\int_t^u h(s)ds) dC(u)$$
$$+ \frac{b(t)}{b(t+\tau)} \exp(-\int_t^T h(s)ds) F \qquad (7)$$
$$+ \int_t^T \frac{b(t)}{b(u)} \exp(-\int_t^u h(s)ds) y(u) h(u) du \mid G_t \}$$

Different from Formula (5), Formula (7) eliminates discrete process $\phi(t)$, and simplifies the pricing of defaultable corporate bonds by using other discounting rate and cash flows. With regard to continuous coupon payment $\{C(u) : u > t\}$, the pricing function is as follows:

$$P(t,\tau) = E^Q \{ \int_t^T \exp(-\int_t^u [r(s) + h(s)]ds) \cdot c(u)du \}$$
$$+ E^Q \{ \exp(-\int_t^T [r(s) + h(s)]ds) \} \cdot F \qquad (8)$$
$$+ E^Q \{ \int_t^T \exp(-\int_t^u [r(s) + h(s)]ds) \cdot y(u) \cdot h(u)du \}$$

We define the recovery $y(t)$ as a proportion of the face value of the defaultable corporate bond as follows:

$$y(t) = \omega(t) * F \qquad (9)$$

According to previous discussion, we assume RR as a function of the risk ratio:

$$R(t) = \beta_0 + \beta_1 \exp(-\beta_2 h(t)) \qquad (10)$$

Where, $\beta_0, \beta_1, \beta_2 \ge 0$ and $0 \le \beta_0 + \beta_1 \le 1$. If $\beta_1 = 0$, then RR is a constant.

Assume that the risk-free interest rate process $r(t)$ conforms to the CIR model:

$$dr(t) = k(\theta - r(t))dt + \eta \sqrt{r(t)} dW(t) \qquad (11)$$

Where, $k, \theta, \eta > 0$, and k is the *mean reversion rate*; parameter θ is the *long-term mean*.

According to Duffie and Singleton (1997, 1999) [14], Duffie (1999) and Longstaff, Mithal and Neis (2005) [15], we assume the risk ratio process (under the risk neutral measure) to be $h(t) = \lambda_0 + \lambda_1 r(t)$, where, $\lambda_0 > 0$ and λ_1 denotes the correlation coefficient between risk ratio and interest rate.

Since RR and risk ratio are both the function of short-term interest rate process, we built a characteristic equation as follows:

$$J(t,u;\phi,v) \equiv E^{Q}\left\{\exp\left(-\phi\int_{t}^{u}r(s)ds + vr(u)\right)\right\} \quad (12)$$

The closed-form solution of characteristic equation is:

$$J(t,u;\phi,v) = \exp[\alpha(t,u;\phi,v) - \beta(t,u;\phi,v)r(t)] \quad (13)$$

$$\gamma = \sqrt{k^2 + 2\phi\eta^2}$$

$$\alpha(t,u;\phi,v) = \frac{2k\theta}{\eta^2}\left[\begin{array}{l}\ln\gamma + k(u-t)/2 \\ -\ln\left(\begin{array}{l}\gamma\cosh\left(\gamma(u-t)/2\right) \\ +(k-v\eta^2)\sinh\left(\gamma(u-t)/2\right)\end{array}\right)\end{array}\right]$$

$$\beta(t,u;\phi,v) = \frac{v\gamma\coth\left(\dfrac{\gamma(u-t)}{2}\right) - vk + 2\phi}{\gamma\coth\left(\dfrac{\gamma(u-t)}{2}\right) + k - v\eta^2}$$

Use this characteristic equation to get $P(t,T)$, the result is:

$$\begin{aligned}P(t,T) &= \int_{t}^{T}e^{-\lambda_0(u-t)} \times J(t,u;1+\lambda_1,0) \times c(u)du \\ &+ F \times e^{-\lambda_0(T-t)} \times J(t,T;1+\lambda_1,0) \\ &+ F\beta_0\lambda_0\int_{t}^{T}e^{-\lambda_0(u-t)} \times J(t,u;1+\lambda_1,0)du \\ &+ F\beta_0\lambda_1\int_{t}^{T}e^{-\lambda_0(u-t)} \times J_v(t,u;1+\lambda_1,0)du \\ &+ F\beta_1\lambda_0 e^{-\beta_2\lambda_0}\int_{t}^{T}e^{-\lambda_0(u-t)} \times J(t,u;1+\lambda_1,-\beta_2\lambda_1)du \\ &+ F\beta_1\lambda_1 e^{-\beta_2\lambda_0}\int_{t}^{T}e^{-\lambda_0(u-t)} \times J_v(t,u;1+\lambda_1,-\beta_2\lambda_1)du\end{aligned} \quad (14)$$

Where, to differentiate $J(t,u;\phi,v)$:

$$\begin{aligned}J_v &= \exp\left[\alpha(t,u;\phi,v) - \beta(t,u;\phi,v)r(t)\right] \\ &\cdot\left[\alpha_v(t,u;\phi,v) - \beta_v(t,u;\phi,v)r(t)\right]\end{aligned} \quad (15)$$

Where,

$$\alpha_v(t,u;\phi,v) = \frac{2k\theta\sinh\left(\dfrac{\gamma(u-t)}{2}\right)}{\gamma\cosh\left(\dfrac{\gamma(u-t)}{2}\right) + (k-v\eta^2)\sinh\left(\dfrac{\gamma(u-t)}{2}\right)},$$

and $\beta(t,u;\phi,v) = \dfrac{v\gamma\coth\left(\dfrac{\gamma(u-t)}{2}\right) - vk + 2\phi}{\gamma\coth\left(\dfrac{\gamma(u-t)}{2}\right) + k - v\eta^2}$.

IV. EMPIRICAL RESULT

A. Data

To estimate the parameters of the model illustrated above, we choose seven Chinese defaultable corporate bonds from industries of finance and real estate, material chemistry, energy, transportation, paper industry, manufacture and retail whose stocks are traded on Shanghai Stock Exchange. The data is from Wind; sample period is from January 4, 2012 to December 31, 2012. The sample information is shown in TABLE I. For convenience, we use overnight interest rates as instantaneous interest rates.

B. Parameter Estimation

We need to estimate four parameters: risk ratio parameters λ_0, λ_1 and recovery rate parameters β_0, β_1. We denote $P_{model}(t_n,T)$ the theoretical price of corporate bonds and $P_{market}(t_n,T)$ the market price. Then we can get the parameters by minimizing the following formula:

$$\min_{\lambda_0,\lambda_1,\beta_0,\beta_1}\left(\frac{1}{N}\sum_{n=1}^{N}\left(\frac{P_{market}(t_n,T) - P_{model}(t_n,T)}{P_{market}(t_n,T)}\right)\right)^2 \quad (16)$$

We conduct the minimization through the method of genetic algorithm and use monthly data as a sample, thus we can get 12 sets of parameters for each corporate bond, i.e., $\{\lambda_0(t),\lambda_1(t),\beta_0(t),\beta_1(t)\}$. We calculate maximum, minimum, mean and standard deviation (SD) for each parameter, which are summarized in TABLE II.

From TABLE II we can see that, except for **11DQ01**, all else bonds have a small λ_0, which is corresponding to China's high credit rating for corporate bonds.

In the case of mean, means of λ_1 have both positive and negative values. $\lambda_1 > 0$ demonstrates that risk ratio and interest rate have a positive correlation; $\lambda_1 < 0$ indicates that profitability is relatively higher when the interest rate is bigger.

Meanwhile, all β_1 are relatively high, which shows that the recovery rate of bond will go down fast when the default probability arises. In this case, fixed recovery rate will lead to a large error.

C. Pricing Result and Comparison

In this part, we apply the parameters to get theoretical prices of **YG bond** at some certain time. Thereafter, we make a comparison with the market price, which is shown in Fig.1.

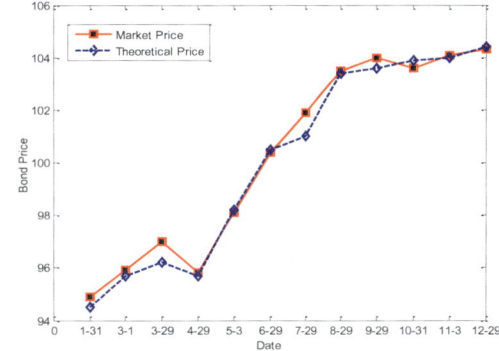

Fig.1. Comparison of theoretical price and market price of **YG bond**

As showed in Fig.1, the theoretical price fits market price very well and the fitting has no significant error. On the whole, the average theoretical price is a little higher than the average market price. However, the trends of the two are consistent, i.e., the theoretical price obtained from our model can precisely reflect the trend of market price, which ensures that our pricing model can be used in the practical investment decision.

V. CONCLUSION

In this paper, the pricing of defaultable corporate bonds is studied using reduced-from model with stochastic recovery rates. To explain the stochastic recovery rate (RR), a function between RR and interest rate are constructed.

Then we obtain the pricing formula for defaultable corporate bonds. In order to solve the formula, we build a characteristic equation and get a closed-form solution. Finally, an empirical study was conducted. Results of high β_1 show that fixed recovery rate will lead to a large error and stochastic recovery rate is necessary. Finally, the theoretical price and market price are compared to demonstrate that our model can precisely reflect the trend of market price and works well in practical investment decision.

ACKNOWLEDGMENT

This work was supported by the National Natural Science Foundation of China (No. 71271015, 70971006).

TABLE I
SAMPLE INFORMATION OF CORPORATE BONDS

Name	Code	Issue date	Maturity date	Period	Coupon rate	Pattern of payment	Total amount (billion)
08JF bond	122011	2008-7-24	2013-7-24	5	8.20%	once a year	10
08BC bond	122013	2008-7-18	2013-7-18	5	8.20%	once a year	17
09GH bond	122021	2009-8-26	2016-8-26	7	6.95%	once a year	10
10YG bond	122060	2010-12-22	2017-12-22	7	7.09%	once a year	8
11DQ01	122092	2011-8-18	2013-8-18	2	5.48%	once a year	40
11ZY bond	122088	2011-8-31	2016-8-31	5	7.50%	once a year	7
11PD02	122126	2012-3-1	2017-3-1	5	8.50%	once a year	22

TABLE II
PARAMETER ESTIMATION FOR EACH BOND

bonds industries		08 JF bond material chemistry	08 BC bond Finance and real estate	09 GH bond energy	10 YG bond paper industry	11 DQ01 transportation	11 ZY bond industry	11 PD02 retail
λ_0	Min	0.0156	0.0260	0.0002	0.0173	0.0247	0.0237	0.0112
	Max	0.1846	0.1955	0.0872	0.0581	0.4845	0.0796	0.0378
	Mean	0.0884	0.0883	0.0343	0.0443	0.1260	0.0539	0.0294
	SD	0.0487	0.0522	0.0270	0.0144	0.1278	0.0155	0.0080
λ_1	Min	-0.3558	-0.3353	-0.6668	-0.5679	-0.8301	-0.5759	-0.2333
	Max	0.3376	0.9037	0.1984	0.1147	0.1621	0.2609	0.2646
	Mean	0.0777	0.1762	-0.3079	-0.2969	-0.5161	-0.4219	-0.0987
	SD	0.2359	0.3380	0.3135	0.2118	0.2687	0.2592	0.1737
β_0	Min	0.0028	0.0508	0.1299	0.1094	0.0938	0.1167	0.0975
	Max	0.4685	0.4584	0.9597	0.7275	0.3069	0.2734	0.2628
	Mean	0.2218	0.2872	0.4032	0.3722	0.1969	0.2255	0.2098
	SD	0.1196	0.1251	0.2727	0.2604	0.0720	0.0481	0.0549
β_1	Min	0.0029	0.1945	0.0183	0.1770	0.3322	0.6975	0.1618
	Max	0.7375	0.7199	0.7630	0.6542	0.7859	0.8681	0.4314
	Mean	0.4917	0.3910	0.2811	0.3571	0.5710	0.7463	0.2666
	SD	0.2462	0.2036	0.2369	0.1466	0.1833	0.0539	0.0834

REFERENCES

[1] R. C. Merton, "On the pricing of corporate debt: The risk structure of interest rate," *Journal of Finance*, vol. 29, no. 2, pp. 449-470, May. 1974.

[2] R. Geske, "The valuation of corporate liabilities as compound options," *Journal of Financial and Quantitative Analysis*, vol. 12, no. 4, pp. 541-552, Nov. 1977.

[3] R. Geske and H.E. Johnson, "The Valuation of Corporate Liabilities as Compound Options: a Correction," *Journal of Financial and Quantitative Analysis*, vol. 19, no. 2, pp. 231-232, Apr. 1984

[4] F. Black and J. Cox, "Valuing corporate securities: Some effects of bond indenture provision," *Journal of Finance*, vol. 31, no. 2, pp. 351-367, May. 1976.

[5] F.A. Longstaff and E.S. Schwartz, "A simple approach to valuing risky fixed and floating rate debt," *Journal of Finance*, vol. 50, no. 3, pp. 789-819, Jul. 1995

[6] R. Jarrow and S. Turnbull, "Credit risk: Drawing the analogy," *Risk Magazine*, vol. 5, no. 9, pp. 63-70, 1992.

[7] R. Jarrow and S. Turnbull, "Pricing derivatives on financial securities subject to credit risk," *Journal of Finance*, vol. 50, no. 1, pp. 53-58, Mar. 1995.

[8] R. Jarrow, D. Lando and S. Turnbull, "A Markov Model for the term structure of credit risk spread," *Review of Financial Studies*, vol. 10, no. 2, pp. 481-523, Feb. 1997.

[9] D. Duffie and K. Singleton, "Modeling term structures of default risky bonds," *Review of Financial Studies*, vol. 12, no. 4, pp. 687–720, Apr. 1999.

[10] D. Madan, G. Bakshi and F. X. Zhang "Understanding the role of recovery rate in default risk models: Empirical comparisons and implied recovery rate rates," unpublished.

[11] M. Krekel, "Pricing distressed CDOs with base correlation and stochastic recovery," Working paper, May. 2008.

[12] D. Duffie and R. Kan, "A yield - factor model of interest rates," *Mathematical Finance*, vol. 6, no. 4, pp. 379-406, Oct. 1996.

[13] D. Lando, "On Cox processes and credit risky securities," *Review of Derivatives Research*, vol. 2, no. 2-3, pp. 99-120, 1998.

[14] D. Duffie and K. J. Singleton, "An econometric model of the term structure of interest - rate swap yields," *The Journal of Finance*, vol. 52, no. 4, pp. 1287-1321, 1997.

[15] F. A. Longstaff, S. Mithal and E. Neis, "Corporate yield spreads: Default risk or liquidity? New evidence from the credit default swap market," *The Journal of Finance*, vol. 60, no. 5, pp. 2213-2253, Oct. 2005.

A Lightweight Workflow Model for Performance Evaluation System of Customer Managers in the Bank

Ji-cheng Liu

Institute of Information Science and Engineer, Henan University of Technology, Zhengzhou, China

(ljcyu@163.com)

Abstract - **In the performance evaluation system, it is necessary to transfer data files between some roles for supplying missing data and auditing these data. Based on the analysis of business process, the paper designs a lightweight workflow model. The establishment process of the model is discussed in detail. The paper presents how to define the workflow process and properties related to the workflow. The simulations to verify the model are performed, and the result shows the model is valid. This workflow mode has been applied in this system.**

Keywords - **Lightweight workflow model, properties, simulation, workflow process**

I. INTRODUCTION

The purpose of the workflow is breaking a big business activity in an enterprise down into small tasks, which will be finished by different roles according to predefined business rules and order, achieving automatic assignment of the business activity. The workflow can improve the productivity of the enterprise and reduce costs. A list of tasks is assigned to roles rather than specific users, a number of roles are supplied for this purpose.

The Workflow Management Coalition (WfMC) proposes the reference workflow model in [1], including organization model, information model, and process model. The organization model describes the relationships between individuals and departments in an enterprise; it is used to define the roles in a workflow. The information model is related to the data used in a workflow, such as business process, business rules, the status of the workflow and so on. The process model manages a workflow, controls the execution of activities in predefined order [2].

There are many workflow models. Reference [3] designs a relationship-based lightweight workflow engine, the model, the organization model and the information model are expressed through relation structure, the process model is implemented by executing store procedures, triggers provided by RDBMS (Relational Database Management System). In [4], the paper analyzes the financial workflow features, and proposes the model of financial workflow based on the mapping relationship between financial workflow diagram and Petri network. A workflow model of product design process is introduced in [5], the model is generated dynamically according to the components instance of Product Structure Tree and its membership relation at run time. Reference [6] proposes a kind of workflow model based on service orientation, through calling the WEB services, enables the workflow engine to finish the workflow instantiation defined by enterprise customers.

The system is to assess the performance of the customer managers in the bank in one score period, such as the performance of loans, savings and so on. The assessment needs a large amount of data; the data are input or imported from external files. But some data are not complete, it is necessary for users to supply missing data. Supplementary data can't be directly saved to the database before being checked by users owning audit right. Reviewing different data needs different rights. Only the data approved can be passed into the system, the data refused will be return. In order to supply missing data, the system uses the workflow method to transfer data files between roles.

There are some workflow engines such as Shark [7-9], OSWorkflow [10-12], jBPM [13-16] and so on, but they are too heavy for this system. A flexible, easy to use, lightweight workflow model is developed to meet the needs of the system. The paper describes how to establish the model. The rest of this paper is structured as follows. In Section II, the requirement of the system for the workflow is introduced. How to establish the lightweight workflow model is described in Section III, including how to define the workflow process and other properties related to the workflow. Finally, Section IV presents preliminary conclusion.

II. THE WORKFLOW IN PERFORMANCE EVALUATION SYSTEM

In this performance evaluation system, when users import data to assess the performance, some data are missing. Using the loan contract as an example, a contract is often finished by more than one team; one team consists of two customer managers. When importing the loan contract from external source, there is only one manager, this manager is responsible for supply other mangers. There are others miss data, such as the lending rate, the floating rate, how to allocate performance among customer managers and so on. Some roles must supply missing data. After passing the review, the data can be saved to the database.

There are six different kinds of data file needed to be transferred between users to supply data. A workflow needs many roles to participate in. Sometimes, the same role may be assigned different activities in different position in the same workflow. The workflow model should consider the situation. Moreover, when some data input by a role are not approved by another role that audits these data, the data file cannot continue to be

E. Qi et al. (eds.), *Proceedings of the 21st International Conference on Industrial Engineering and Engineering Management 2014*, Proceedings of the International Conference on Industrial Engineering and Engineering Management, DOI 10.2991/978-94-6239-102-4_120, © Atlantis Press and the authors 2015

forwarded to the next stage, it should be sent back to the role making mistake. After correcting errors, the data file continues to move through from this role.

There may be many instances of a kind of workflow running at the same time, we called an instance as a To-do, a To-do may be running or completed. For example, when the loan contract has only one manager, the system initiates one To-do for this manager.

The system may find the next role to process a To-do based on business rule, and then assign activities to one of users owning that role, it also can manage running To-dos and ended To-dos.

III. THE WORKFLOW MODEL

Consider the system has such roles, r_i ($i \in [1,9]$), w_j ($j \in [1,6]$) denotes six kinds of workflow. A workflow moves through a number of roles, there is role-specific data for each of these roles, the set $s(r_i, w_j)$ ($i \in [1,9]$, $j \in [1,6]$, $r_i \in role$, $role$ is a collection of all roles taking part in w_j) is the data that r_i should supply or audit. Every workflow w_j has many To-dos, $t_k w_j$ ($k \in Z$, Z is the set of natural number) is the kth To-do of w_j.

Role r_1, r_2 and r_3 take part in w_1 in turn. For w_2, the data file moves through r_1, r_2, r_1 and r_3, r_3 audits the data supplied by r_1 and r_2.

A. Defining the workflow process

The workflow process is the order in which activities of the workflow must be performed. When roles participating in a workflow are different, we can define the process by joining role names together in the business order using one comma as separator. So, the process of w_1 is " r_1, r_2, r_3 ", the data set that r_i should supply is $S(r_i, w_1)$ ($r_i \in \{r_1, r_2, r_3\}$). For r_1, the data is $S(r_1, w_1)$. From the process definition, the system can find the data that should be supplied by the roles that login users own, the reason is that the workflow moves through each role only once, tasks assigned to each are unique and unambiguous. The system also can determine how to forward the data file based on the process: who is the next actor?

But for w_2, this method is not suitable. r_1, r_2, r_1 and r_3 are assigned activities in turn, r_1 participate in this workflow twice. If using the same method to represent the process, i.e., " r_1, r_2, r_1, r_3 " denotes the workflow process of w_2. The data set that r_1 should supply is $S(r_1, w_2)$, for r_2 the data is $S(r_2, w_2)$. In fact, when users owning r_1 login, it is impossible for the system to determine the data set r_1 should supply, because r_1 supply data before r_2, also after r_2. Activities before r_2 's tasks or these after r_2 's, which should be assigned to r_1? According to the

business rule, the order is r_1, r_2, then r_1, the system can't assign tasks before and after r_2 as a whole to r_1.

In order to solve the problem, when a workflow moves through a role twice or more, we can declare a *step* property for all To-dos to avoid ambiguity. *step* denotes the index the next role in the workflow process, starting from zero. Now the role-specific data is $S(r_i, w_j, step)$. When the system initiates one new To-do, $step = 0$. According to the index, *step*, the system can infer the next role from the workflow process. After a role finishes his tasks, the next role is after him, he increases *step* by 1. Doing so, each role has a different *step*. Even for the same role, $S(r_i, w_j, step)$ is different because of the difference of *step*.

The simulation to validate the model is presented as follows. For $t_1 w_2$, one of w_2 's To-dos, the workflow process is " r_1, r_2, r_1, r_3 ", in initial state, $step = 0$. The execution process for $t_1 w_2$ is shown as follows:

- In the initial state, $step = 0$.
- r_1 login, input data in $S(r_1, w_2, 0)$, $step+ = 1$.
- r_2 login, input data in $S(r_2, w_2, 1)$, $step+ = 1$.
- r_1 login, input data in $S(r_1, w_2, 2)$, $step+ = 1$.
- r_3 login, audit data in $S(r_3, w_2, 3)$, $step = -20$.

In the above process, r_1 input data in $S(r_1, w_2, 0)$ and $S(r_1, w_2, 2)$, they are different. So, introducing *step* property eliminates the ambiguity when the flow moves through a role many times. In the last step, $step = -20$, -20 is used as the flag that the workflow has finished. According to the value of *step*, we can know the status of one To-do: running or completed.

To make it easy for find To-dos for a role, *role* property is introduced to the model, it denotes the next role, which can be inferred from *step* and the workflow process.

B. The workflow rollback

The data supplying needs to be reviewed, when they are not approved, the workflow moves backward to a specified activity in the workflow and tasks that have already executed are reset. To support the rollback, *isReturn* property is used as an indicator of the workflow direction: forward or backward. When *isReturn<0*, a To-do is being sent back. The reason of rejection is given in *returnReason* property. A role can sent the flow back to its previous role or any of roles that has executed tasks. To support jumping back to already passed roles, the step jumping backward needs to be saved. When a role decides to jump back to previous executed steps, he selects one of previous passed roles as the target, and then the system counts how many steps from this role to the selected role. We negative the value of steps, assign it to *isReturn* property, and then set

$$step+ = isReturn$$

At the same time, the system assigns a new role for $role$ property according to the new value of $step$. Let us take an example, for $t_1 w_2$, the process is "r_1, r_2, r_1, r_3", when r_3 audit data and decide to sent this To-do back, it chooses first r_1 as target, there are three steps between first r_1 and itself, so

$$isReturn = -3.$$
$$stetp+ = is\,Re\,turn.$$
$$role = "r_1"$$

So far, every To-do has $step$, $role$, $isReturn$ and $returnReason$ property.

When the workflow is reverted back, the system processes a To-do as follows.

r_1 login, the system searches all To-dos for this role according to role property of a To-do.

while(having more To-dos for r_1){

get one To-do.
if(isReturn < 0){
display the message in returnReason, then returnReason ="".
isReturn = 0 .
}
supply or review data.
if(rollback this To-do){
choose one of already passed roles to jump back, count how many steps from the role to oneself in the workflow process.
let isReturn property equal to the negative value of steps.
set returnReason property
step+ = isReturn, compute and set role property.
continue
}else{
if(To-do ends) {
step = -20, role ="".
continue
}
step+ = 1, compute and set role property.
}
}

C. Simulation

The simulation to verify the model when the workflow rollback is presented as follows. When $t_1 w_2$ is reverted back by r_3, the workflow process is:

- In the initial state, $step = 0$, $role = "r_1"$, $isReturn = 0$, $returnReason = ""$.
- r_1 login, input data in $S(r_1, w_2, 0)$, $step+ = 1$, $step = 1$, $role = "r_2"$, $isReturn = 0$, $returnReason = ""$.

- r_2 login, input data in $S(r_2, w_2, 1)$, $step+ = 1$, $step = 2$, $role = "r_1"$ $isReturn = 0$, $returnReason = ""$.
- r_1 login, input data in $S(r_1, w_2, 2)$, $step+ = 1$, $step = 3$, $role = "r_3"$ $isReturn = 0$, $returnReason = ""$.
- r_3 login, audit data in $S(r_3, w_2, 3)$, refuse and revert back to one of r_1, r_2, r_3, supposing the first r_1 , $isReturn = -3$, $returnReason = "some\,error"$, $step+ = isReturn$, $step = 0$, $role = "r_1"$.
- r_1 login, the system displays the message in $returnReason$, $isReturn = 0$, $returnReason = ""$, input data in $S(r_1, w_2, 0)$, $step+ = 1$, $step = 1$, $role = "r_2"$.
- Other steps are the same with these without being reverted back.

IV. CONCLUSION

After analyzing the workflow requirements in the performance evaluation system of customer managers in the bank, the paper designs a lightweight workflow model. Firstly, the paper defines the workflow process, joining the names of all roles taking part in a workflow in business order. When a workflow moves through a role twice or more, the system can't determine tasks that should be assigned to a role only according to the workflow process. $step$ property is added to the model, its value is the index that a role is in the workflow process, the index is unique. When a workflow moves through a role twice, this role has different position, so two indexes are different. The system can determine role-specific tasks. It is possible that a To-do is reverted back because of some mistakes, so, the property, $isReturn$, is used to an indicator whether a To-do is being rollbacked. Simulations to verify the model are performed; the result shows that model is valid. The model is used in this system; the system can effectively process To-dos, assign tasks for different roles.

ACKNOWLEDGMENT

This work is supported by the Doctor Science and Research Foundation of HAUT under Grant No.2006bs006.

REFERENCES

[1] W. Hollings, "Workflow Management Coalition: The Workflow Reference Model. Document Number WFMC - TC00-1003," Brussels, 1994.
[2] P. Barna, F. Frasincar, and G. Houben, "A workflow-driven design of web information systems," in *Proc. 6th International Conference on Web Engineering, ICWE'06,* Palo Alto, CA, United states, 2006, pp. 321-328.

[3] Q. He, G. Li, and L. Liu, "Relation-based lightweight workflow engine" (in Chinese), *Journal of Computer Research and Development*, vol. 38, no. 2, pp. 129-137, 2001.

[4] J. Chen, L. Han, D. Xiong, and J. Luo, "Design and implementation of financial workflow model based on the petri net," in *Proc. Future Computer and Control Systems, FCCS 2012,* Changsha, China, 2012, pp. 495-500.

[5] S. Li, X. Shao, and J. Chang, "Dynamic workflow modeling oriented to product design process" (in Chinese), *Computer Integrated Manufacturing Systems*. vol. 18, no. 6, pp. 1136-1144, 2012.

[6] A. Li, "Research on the workflow model designing of enterprise management informationization based on service-oriented," in *Proc. 2010 International Conference on e-Education, e-Business, e-Management and e-Learning, IC4E 2010,* Sanya, China, 2010, pp. 641-644.

[7] C. Shuang, L. L. Liu, and H. F. Zhang, "Research and Implementation of Multi-step Process Rollback Method"(in Chinese), *Computer Engineering*, vol. 36. no. 21, pp. 67-70, 2010.

[8] Z. B. Ruan, "Research workflow system based on workflow engine Shark"(in Chinese), *Computer Knowledge and Technology*, vol. 5, no. 21, pp. 5716-5717, 2009.

[9] W. J. He, G. Q. Li, J. Zhang, W. Y. Yu, and J. B. Xie, "Study on WPS implement framework with Shark workflow," in *Proc. 9th International Conference on Grid and Cloud Computing, GCC 2010,* Nanjing, Jiangsu, China, 2010, pp. 511-516.

[10] J. L. Zhang and K. Chen, "Research and Application of OSWorkflow in Office Automation System" (in Chinese), *Software Guide.* vol. 10, no. 3, pp. 94-95, 2011.

[11] S. Y. Wang and H. L. Liu, "Equipment Management System based on OSWorkflow" (in Chinese), *Journal of Changchun University of Technology (Natural Science Edition)*. vol. 32, no. 5, pp. 443-448, 2011.

[12] Y. Y. Zhao, H, Liu, and Q. Pan, "Research on OSWorkflow-based workflow management system integration strategy," in *Proc. 2nd International Conference on Information Management and Engineering, ICIME 2010,* Chengdu, China, 2010, pp. 431-434.

[13] A. J. Xu, "Research and implementation of JBPM workflow management system"(in Chinese), *Computer Technology and Development*, vol. 23, no. 12, pp. 100-104, 108, 2013.

[14] W. X. Gu, Q. Wang, and T. R. Xu, "Study and design of jbpm based workflow management system"(in Chinese), *Computer Applications and Software*, vol. 26, no. 5, pp. 104-106, 2009.

[15] Y. F. Sun, "Design and implementation of equipment management system based on the JBMP and lightweight J2EE," in *Proc. 2013 International Conference on Precision Mechanical Instruments and Measurement Technology, ICPMIMT 2013,* Shenyang, Liaoning, China, 2013, pp. 3283-3286.

[16] B. Han and D. M. Xia, "Research and design of document flow model based on JBPM workflow engine," in *Proc. 2009 International Forum on Computer Science-Technology and Applications, IFCSTA 2009,* Chongqing, China, 2012, pp. 336-339.

On the Influencing Factors of Equity Incentives of Listed Logistics Companies in China

Sha LV*, Qin-xian DONG

School of Business, Sichuan Normal University, Chengdu, China
(shasa27@hotmail.com)

Abstract - The research on the influencing factors of equity incentives is a hot topic for the research of equity incentives in Chinese listed companies. This paper uses the equity incentive level as the explained variable, while uses company size, ownership concentration, equity balance degree, company growth and free cash flow as the explanatory variables to study the influencing factors of equity incentives in listed logistics companies. The research results indicate that company size has significant positive correlation with equity incentives, while ownership concentration has no significant negative correlation with equity incentives, but company growth and equity balance degree have significant negative correlation with equity incentives.

Keywords - Equity incentives, logistics industry, multiple regression analysis, ownership concentration

I. INTRODUCTION

Managers' equity incentives has been widely used in management practice in every country. Equity incentives is an effective way for an enterprise to achieve the long-term incentive to the management. It makes the management and the enterprise share the consistent interest and solves the agency problem in the listed companies effectively. The principal-agent theory holds that the interest relationship between the managers and shareholders can be integrated through the equity incentives and the contradiction of the separation of ownership and management right can be solved to a certain extent, and helping to achieve the goal of maximizing value of the enterprise [1].

Currently, the logistics industry has been recognized as the artery of the national economic development and one of the basic industries in China. Logistics plays an increasingly important role in operation of the social economy and becomes the third pillar and the third profit source of the national economy development, following manufacturing and commerce. Therefore, this paper selects 20 listed companies in the logistic industry as the sample, using the multiple regression analysis to analyze the determinants of equity incentives.

II. LITERATURE REVIEW

The foreign and domestic researches mainly concentrate on two aspects: one is on how the equity incentives influence the operating results; and the other is on factors affecting the equity incentives implementation.

Foreign scholars began researching on equity incentives in 1976. Subsequently, western scholars studied this topic systematically. Jensen and Murphy (2003) studied the relationship between the equity incentives and future earning and found equity incentives had no linear relationship with future earning [2]. Studies by Tzioumis (2008) argued that the equity incentives had negative correlation with property owned by the chief executive officer (CEO) and the age of CEO[3]. Other scholars also analyzed the influencing factors of equity incentives from different angles, such as company size (Konari, 2006) [4], corporate governance characteristics (Rosenberg, 2003) [5], or personalities of CEO (Attaway, 2000; Phillip and Cyril, 2004) [6][7].

Chinese scholars also discussed the influencing factors of equity incentives from different aspects. Li and Liu (2010) conducted the study on 92 listed companies in the period 2006-2009, empirical analysis showed that the lower ownership concentration, the more selective tendency; the more cash rewards for executives, the more selective tendency [8]. Li and Hui (2013) took A-share listed companies as samples and found that the bigger the size, the more selective tendency [9]. According to He and Wang (2011), the company's risk was significantly positive related to equity incentive level, while equity balance degree and free cash flow had significantly negative correlation with equity incentives level [10].

III. HYPOTHESES DEVELOPMENT

This paper discusses the influencing factors of equity incentives from the following five aspects: the company size, ownership concentration, equity balance degree, the company growth and free cash flow.

For the company with large size, corporate management is more complex than comparing with small size. This leads to the lack of supervision, which would result in serious agency problem. The company would have stronger motivation of using equity incentives to reduce agency costs. Thus, the first hypothesis is:
H_1: Company size is positively related to equity incentive level;

If listed company's equity are mostly concentrated in the top ten shareholders or even more concentrated in the top one shareholder, they can supervise managers more effectively. Meanwhile, they do not want to spread their control of the equity through equity incentives. In the following, second hypothesis is:
H_2: Ownership concentration is negatively related to equity incentive level;

The equity balance degree would have significant effect on the behavior of the company's largest shareholder. The higher the shareholding by the other shareholders, the more it can reduce the probability which the top one shareholder would infringe the interests of other shareholders. To maximize the value of stakeholders, the company would reduce the long term incentives for management. Hence, the third hypothesis is:

H₃: Equity balance degree is negatively related to equity incentive level;

High growth companies are generally in the high speed development stage, in order to achieve higher future development goals; the company needs the implementation of equity incentives program to increase managers' enthusiasm for work. Hence, the fourth hypothesis is:

H₄: Company growth is positively related to equity incentive level;

The last key factor that we test is free cash flow. Yermack (1995) used the U.S. data to verify that the companies which were lack of cash flows were more likely to use stock options [11]. The company executes incentive stock options, not only reduces the salary paid in cash to the managers, but also gets cash flow when managers enforce their stock rights. Therefore, the fifth hypothesis is:

H₅: Free cash flow is negatively related to equity incentive level.

IV. METHODOLOGY AND DATA

A. Sample and Data

This paper takes 20 listed companies in the logistic industry which carried equity incentives in the period 2011 -2013 as sample to do the empirical study of equity incentives. The sample data is collected from Wind, which is a premier financial database on Chinese companies.

B. Selection of variables

Based on the relevant literature, this paper uses the equity incentive level as the explained variable, while use company size, ownership concentration, equity balance degree, company growth and free cash flow as the explanatory variables to study their impact on the equity incentives (As shown in Table I).

TABLE I
DEFINITION OF VARIABLES

Variable	Names	Symbolic	Computation methods
Explained variable	Incentive levels	Y	Number of managers' equity incentives ÷number of total equity
Explanatory variables	Company size	X_1	Natural log of total assets at the balance sheet date
	Ownership concentration	X_2	Shareholding by the top ten shareholders
	Equity balance degree	X_3	Shareholding by the top nine shareholders ÷shareholding by the top one shareholder
	Company growth	X_4	Weighted average rate of return on equity
	Free cash flow	X_5	Cash and cash equivalents

C. Multiple Linear Regression Model

We use regression model to test our hypotheses. Then this paper sets explained variable Y as dependent variable, and the explanatory variables X_1, X_2, X_3, X_4, X_5 as independent variables, then the multiple linear regression model can be represented as shown:

$$Y = \beta_0 + \beta_1 X_1 + \beta_2 X_2 + \beta_3 X_3 + \beta_4 X_4 + \beta_5 X_5 + \varepsilon \quad (1)$$

Where β_0 is the constant term; β_1, β_2, β_3, β_4, β_5 are regression coefficients of each variable; ε is the error term.

V. EMPIRIAL FINDINS

A. Descriptive Statistics

Table II reports descriptive statistics for the explained and explanatory variables in our multiple linear regression models.

TABLE II
DESCRIPTIVE STATISTICS

	Number	Mean	Median	Std. deviation	Variance	Min	Max
Y	60	0.137728	0.07505	0.16163	1.1735475	0	0.6397
X_1	60	21.38278	21.24576	1.07116	0.0050095	19.80503	24.81405
X_2	60	0.622763	0.61525	0.12694	0.2038410	0.4134	0.9039
X_3	60	0.467335	0.4686	0.12515	0.26779832	0.0946	0.6289
X_4	60	0.136718	0.1287	0.07530	0.55081278	0.0163	0.3401
X_5	60	9.84E+08	3.09E+08	2.7E+09	2.75682595	2 .4473744	1.33E+10

In Table II, we can see that the Mean and the Median for most of the explanatory variables tend to be equal and be suitable for regression analysis. But the Mean for free cash flow is 9.84E + 08; the Median is 3.09E + 08. The difference between the two figures is substantial and may not be suitable for regression analysis. Furthermore, its variance is 2.75682595, which is the largest one among the variables. So, free cash flow is not suitable for further explanation regression analysis and is ignored.

B. Correlation Analysis

This paper uses financial data of sample companies as the observation, in the multiple regression analysis, there may be problems of multicollinearity among the variables, so the correlation analysis is conducted to check whether the presence of multicollinearity among the variables, and whether they are suitable for multiple regression analysis. Correlation analysis results are shown in Table III.

TABLE III
CORRELATION ANALYSIS OF MAIN VARIABLES

	Y	X_1	X_2	X_3	X_4
Y	1				
X_1	-0.11633	1			
X_2	-0.20371	0.075726	1		
X_3	0.413178	-0.4589	-0.19421519	1	
X_4	-0.20203	0.32699	0.276947333	0.07090411	1

As can be seen from Table III, the correlation coefficient between variables are below 0.5. This indicates that the possibility of the existence of multicollinearity is little and it is appropriate for a multivariate linear regression analysis.

C. Multiple Linear Regression Analysis

The regression results of explained variable Y with explanatory variables X_1, X_2, X_3, X_4, X_5 is output by SPSS17.0 as shown in Table IV, Table V, Table VI.

TABLE IV
MODEL SUMMARY

Multiple R	R Square	Adjusted R Square	Std. Error	N
0.512551	0.262708	0.209087	0.143743	60

TABLE V
ANOVA

	df	Sum of Squares	Mean Square	F	Significance F
Regression	4	0.404923	0.101231	4.899337	0.001884
Residual	55	1.136418	0.020662		
Total	59	1.541342			

TABLE VI
COEFFICIENTS

	Coefficients	Std. Error	t Stat	P-value	Lower 95%	Upper 95%
Intercept	-0.81826	0.525124	-1.55822	0.124916	-1.87063	0.234111
X Variable 1	0.035015	0.021779	1.607762	0.113614	-0.00863	0.07866
X Variable 2	-0.0412	0.159329	-0.2586	0.796911	-0.3605	0.2781
X Variable 3	-0.691125	0.181142	-3.815372	0.000347	-0.328108	1.054142
X Variable 4	-0.65868	0.288947	-2.2796	0.026533	-1.23774	-0.07962

According to the regression coefficients in Table VI, we can draw the multiple linear regression model.

$$Y = -0.81826 + 0.035015X_1 - 0.0412X_2 + 0.691125X_3 - 0.65868X_4 \qquad (2)$$

R = 0.512551, R^2 = 0.262708, Adjusted R^2 = 0.209087, the regression equation has better degree of fitting. There is a strong linear relationship between Y and X_1, X_2, X_3, X_4.

F-test: F=4.899337>F0.05(5,52)≈2.422, and Significance F=0.001884<α=0.05. These show between Y and X1, X2, X3, X4 have significant linear relationship.

β_1=0.035015>0, β_2=-0.0412<0, β_3=-0.691125<0, β_4=-0.65868<0, These show X_1 has linear positive correlation with Y, while X_2, X_3, X_4 have linear negative correlation with Y.

t-test: $t_{0.025}$≈2.0141, the $|t_1|$=1.607762<$t_{0.025}$, $|t_2|$=0.2586<$t_{0.025}$, $|t_3|$=3.815372>$t_{0.025}$, $|t_4|$=2.2796>$t_{0.025}$, These show β_3 and β_4 pass the t-test, β_1 and β_2 do not pass t-test. It indicates that X_1 has no significant positive correlation with Y; X_2 has no significant negative correlation with Y; X_3 and X_4 have significant negative correlation with Y.

VI.CONCLUSION

Through the multiple regression analysis of relevant data for sample companies, we can draw the following conclusions:

1) Company size has no significant positive correlation with equity incentive level, hypothesis 1 holds. With the increasing size of the company, it is more difficult for shareholders to monitor the managers and agency cost will become larger. Under this situation, the company is willing to execute equity incentives. Since the equity incentives in listed logistics companies is still in the exploratory stage, company size and incentive level have certain positive correlation, but not significant.

2) Ownership concentration has no significant negative correlation with equity incentive level, hypothesis 2 holds. Lower the degree of ownership concentration, stronger the shareholder's willingness to monitor the managers through the equity incentives. No matter how imperfect the system of professional managers is in China, it may affect the enthusiasm which listed logistics companies execute the equity incentives to some extent, thus the effect of ownership concentration to equity incentive level is limited.

3) Equity balance degree has significant negative correlation with equity incentive level, hypothesis 3 holds. When equity balance degree is low, major shareholders can totally control the company, while the agency cost is low, since the implementation of higher equity incentives might dilute their earnings; they are less likely to implement equity incentives.

4) Company growth has significant negative

correlation with equity incentive level, hypothesis 4 does not hold. The main reason is like this: the sample data selected only three years, while the equity incentives pays attention to the long-term incentive effect, therefore it may be the long-term interests of the company let the management to give up some short-term behavior, making equity incentive level and the company's growth to present a negative correlation.

REFERENCE

[1]   M. Jensen and W. Meckling, Theory of the firm: managerial behaviour, agency costs and ownership structure, Journal of Financial Economics, vol. 3, no.4, pp. 305-360, 1976.

[2]   M. C. Jensen, and K. J. Murphy, Performance pay and top management incentives, Journal of Political Economy, vol. 2, pp.225-264, 2003.

[3]   K. Tzioumis, Why do firms adopt CEO stock options? evidence from the United States , Journal of Economic Behavior and Organization, vol. 68, no. 1, pp. 100–111, 2008.

[4]   Konari Uchida., Determinants of stock option use by Japanese companies, Review of Financial Economics, vol.15, pp.251-269, 2006.

[5]   M. Rosenberg, Stock option compensation in Finland: an analysis of economic determinants, contracting frequency, and design. European Financial Management Association 2003 Annual Meeting, 2003.

[6]   M. C. Attaway, A study of the relationship between company performance and CEO compensation, American Business Review, vol. 18 , pp.77-86, 2000.

[7]   M. Phillip, and T. Cyril, The implications of firm and individual characteristics on CEO pay, European Management Journal, vol. 22, pp.27-41, 2004.

[8]   Yue-Mei LI, Tao LIU, Study of effect factors of equity incentives. Journal of Shaanxi University of Science & Technology (Natural Science Edition) (Chinese), vol. 1, pp.153-158, 2010.

[9]   Bing-xiang LI , Xiang HUI. Empirical research on influencing factors of equity incentive of listed companies in China. Finance and Accounting Monthly (Chinese), vol.4, pp.33-36, 2013.

[10] Wei HE, Meng-yi WANG, Research on decisive factors of stock option incentive plan for management in Listed companies, Accounting and Finance (Chinese), vol. 2, pp.46-49, 2011.

[11] D. Yermack, Do corporations award CEO stock options effectively? Journal of Financial Economic, vol. 39, pp.237-269, 1995.

Origins and Trends of the Multinational's R&D Internationalisation Research- a Systematic Literature Review

Xingkun Liang[1,*], Zhutian Xing[2], Yongjiang Shi[1]

[1]Institute for Manufacturing, Department of Engineering, University of Cambridge, Cambridge, the UK
[2]Department of Information Management, Peking University, Beijing, P.R. China
(*xl345@cam.ac.uk, xzt@pku.edu.cn, ys@cam.ac.uk)

Abstract - **This paper adopts bibliometric analysis on 621 literatures on multinational corporations' (MNC) R&D internationalisation from various fields to identify key patterns of this research topic. Bibliographic data of these literatures are collected from ISI Web of Knowledge with specific search criteria (e.g. from 1965 to 2013). A generic approach of bibliographic analysis is implemented to present visualised and organised results, which enable a systematic review on previous studies. We found that (1) research on internationalisation of R&D originated from three different fields, *i.e.*, strategic management, innovation management and international business, and (2) research frontiers are emerged in most recent studies that can be categorised into four key areas. Bibliometric analysis provides a possibility to identify and understand the intellectual structure of research on R&D internationalisation in latest 35 years. Based on these findings, we analytically proposed an agenda for future studies on R&D internationalisation.**

Keywords – **Bibliometrics, global innovation strategy, international R&D network, R&D internationalisation, systematic literature review**

I. INTRODUCTION

The growing trend of global R&D activities in MNCs triggers research on internationalisation of R&D in management and international business areas. This topic has been well studied in recent decades, as many papers published to explore related issues. The increasing number of literature on R&D internationalisation enables bibliometric analysis for identifying key patterns for future research. Bibliometric analysis is usually used for capturing the big questions and identifying frontier topics from published literatures in a visualised and quantitative way [1]. This paper will present a systematic review on internationalisation of R&D with bibliometric methods.

Bibliometric methods have been commonly used to identify intellectual structures of literature, such as strategic management [e.g. 2], innovation research [e.g. 3], management information system [e.g. 4] and, operations management [e.g. 5]. It is a widely adopted and valid method of summarising and reviewing relevant literature on various topics in one discipline or across subjects [6]. Based on the large volumes of publications on the topic of internationalisation of R&D, it became necessary and useful to have a bibliometric analysis on the topic. This paper is aimed to achieve the following objectives: (1) cocitation analysis to understand theoretical foundations of R&D internationalisation research, and (2) bibliographic coupling analysis to identity frontier in R&D internationalisation areas in recent ten years. With bibliographic analysis, this paper systematically summarises current research as well as maps research gaps and future directions that are potential to become critical in future R&D internationalisation research.

II. METHODOLOGY

A. Data Collection

Bibliographic data are collected from the ISI Web of Knowledge (WoK) [7]. The criteria for searching in ISI WoK are as follows: (1) title including internationali* R&D or internationali* innovat* (2) title including globali* "research and development" (3) title including "international R&D" or "international Research and Development" (4) title including "global R&D" or "global Research and Development" (5) title including globali* R&D or globali* innovat* (6) title including internationali* "research and development" (7) keywords including "R&D network" or "research and development network".

This search includes a comprehensive set of literature on R&D internationalisation. In terms of spelling differences, for instance, 'globalisation' and 'globalization', the search criteria are set with an asterisk (*). This symbol is attached to certain terms to address such spelling issue. For example, 'globali*' means words starting with 'globali' will be matched in the search, including 'globalised' and 'globalized', 'globalising', 'globalizing', 'globalisation', 'globalization', *etc*.

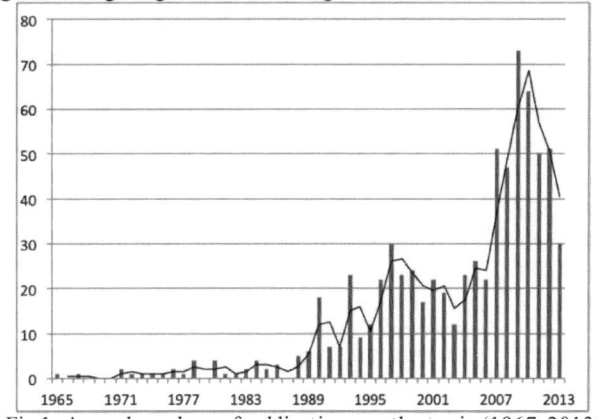

Fig.1. Annual numbers of publications on the topic (1967-2013)

The search with the above criteria was made in January 2014. In total, 726 papers were found on WoK. Bibliometric analysis methods is implemented on these 726 papers with their 15915 references. The first paper explicitly on this topic was published in 1967. By the word 'explicitly', we mean the paper that matched the search criteria. Other papers might also discuss R&D

E. Qi et al. (eds.), *Proceedings of the 21st International Conference on Industrial Engineering and Engineering Management 2014*, Proceedings of the International Conference on Industrial Engineering and Engineering Management, DOI 10.2991/978-94-6239-102-4_122, © Atlantis Press and the authors 2015

internationalisation but did not label titles with internationalisation or globalisation or other related terms in the search criteria. The paper within the criteria that get published annually is shown Fig.1 below. There is an increasing trend for papers being published on this topic.

The number of publications peaked at around 1999, when internationalisation of R&D became more common in various industries. The other summit was in 2009. The distribution of the number of papers published indicates it may be proper to set ten-year for time scaling.

B. Bibliometric Methodology

Bibliometrics is a method associated statistical analysis on counts and citing links between different articles in a body of literature [8]. Different techniques of bibliometrics are suitable for studies with different objectives, such as cocitation analysis for mapping the critical questions [9], co-author for identifying knowledge exchange patterns [10], and bibliographic coupling for revealing frontier research streams [11].

Two techniques are used in the research. First, cocitation analysis is applied to find out common references that are cited by two different papers. The more similar two papers' references are, the more likely both papers are discussing very relevant topics. Cocitation analysis is adopted to seek for big questions on R&D internationalisation when it emerged. Bibliographic coupling is also used to map recent frontiers in R&D internationalisation research [12]. Bibliographic coupling is to identify relationships between different papers according to their shared references. These similar references are the intellectual basis for the citing papers.

Pieces of software are used for data processing, analyzing and visualizing, including the Science of Science (Sci2) Tool, Matlab 2013a and SPSS 20.0.

The study follows generic bibliometric procedures [13]. Data from WoK are first extracted into a paper citation network. Papers among top 100 in cocitation or bibliographic coupling are included in analysis.

Cocitation analysis follows with guidance from [14] and focuses on earlier decades. Cocitation matrices are calculated and transformed into proximity matrices using Pearson's R. Bibliographic coupling follows similar approaches but examines the latest ten years. Multivariate analysis techniques, i.e., factor analysis, cluster analysis and multidimensional scaling (MDS) are applied to the proximity matrices in both analyses.

Initially, factor analyses of both periods are implemented with no specification, which explores the natural number of factors. Parallel analysis is used to identify the optimal number of factors in each factor analysis. Then, factor analysis is performed again with the optimal numbers of factors.

Cluster analysis is performed to further understand the intellectual structure of R&D internationalisation research in both periods [15]. Dendrogram using Ward linkage in SPSS is used for clustering [16]. The cutting-off boundary is set with a distance of $y=5$. Results of

factor analysis and cluster analysis are used to identify key clusters of research domains for triangulation.

To visualize the results, MDS maps (ALSCAL in SPSS) are generated from the proximity matrices of both periods, presenting the knowledge base of these publications in two dimensions [17].

There are certain specifications on bibliographic coupling. First, papers used for analysis are only those with at least ten links to other papers [18]. The linkage of papers with citing/cited relationship is weighted by betweenness centrality [19].

III. RESULTS

A. Cocitation Analysis

There are only 98 papers on R&D internationalisation from 1965 to 1993. Three of them are eliminated in analysis because of no cocitation with others.

Initial factor analysis on the proximity matrix derived 8 factors explaining 96.89% of variances. As showed in Table I, there are 5 factors in the cocitation proximity matrix that have eigenvalues exceeding corresponding ones in the parallel analysis. Hence, it is more appropriate to limit the number of factors to five. Factor analysis with fixed five factors is performed so that papers are grouped according to the largest loading, as showed in Table II.

The factor analysis of fixed factors on publications from 1965 to 1993 is effective. Over 90% of the total variance can be explained by these five factors and over 78% items have loadings over 0.7 or below -0.7.

TABLE I
FACTOR LOADING AFTER ROTATION

Factor	Rotation Sums of Squared Loadings					
	1965-1994		2004-2013		Parallel Baseline	
	T	% Va	T	% V	T	% V
1	40.781	42.48	39.466	39.864	3.926	40.82
2	20.34	63.668	34.13	74.339	3.607	61.115
3	13.806	78.049	6.473	80.877	3.514	74.885
4	6.697	85.025	5.302	86.233	3.191	81.247
5	5.52	90.775	2.455	88.713	3.068	86.899
6	2.627	93.512	1.853	90.584	2.961	89.473
7	1.716	95.3	1.558	92.157	2.883	91.235
8	1.531	96.894				

a T: Total variance explained, sic passim;
b % V: Cumulative percentage of variance, sic passim;

Clustering analysis shows a consistent result with factor analysis. The first cluster covers all papers of Factor 2 and Factor 3. The third cluster includes most papers of Factor 1 while Cluster 2 contains all papers of Factor 4 and 5, as well as a small portion of Factor 1.

The MDS map (Fig.2) with R^2 = 95.215% shows clearly that these factors are nearly isolated and located in different parts of the map. These results indicate there are possibly three antecedents that are the origins where studies on R&D internationalisation emerge. Within the three main streams, several small directions emerge.

TABLE II
PRINCIPLE COMPONENT FOR THE PERIOD 1965-1993
(PUBLICATIONS ARE PRESENTED BY THEIR IDENTICAL NUMBERS FOR SHORT)

F1					F2		F3	F4	F5
1	17	33	55	75	10	68	2	9	66
3	18	37	56	76	23	72	7	11	53
4	19	39	57	77	29	83	24	32	
5	20	40	59	80	38	86	42	34	
6	21	43	60	84	41	87	78	35	
8	22	47	61	85	44	88	79	36	
12	25	48	67	95	45	89	81	66	
13	26	49	69		46	90	82		
14	27	50	70		58	94	91		
15	28	51	71		62	68			
16	30	52	73		63				
17	31	54	74		64				
T	42.643				20.591		15.344	6.182	3.307
%V	44.420				21.449		15.983	6.440	3.445

KMO=0.248, Bartlett's P<0.001; Total Variance Explained=91.738%
All publications are with loadings greater than 0.4 or less than -0.4.

TABLE III
PRINCIPLE COMPONENT FOR THE PERIOD 2004-2013
(PUBLICATIONS ARE PRESENTED BY THEIR IDENTICAL NUMBERS FOR SHORT)

F1				F2				F3	F4
4	26	49	79	2	32	62	89	43	1
6	27	50	83	3	33	63	91	46	7
8	28	51	85	5	35	64	92	58	17
10	31	53	88	9	36	69	93	61	68
14	34	55	90	11	41	70	94	67	73
15	37	57	96	12	45	72		84	86
16	38	65	97	13	48	74		95	
19	39	66	98	18	52	75		99	
20	40	71		21	54	80		100	
22	42	76		23	56	81			
24	44	77		29	59	82			
25	47	78		30	60	87			
T	37.924			36.334				6.931	5.680
%V	43.656			41.826				7.979	6.539

KMO=0.424, Bartlett's P<0.001; Total Variance Explained=86.869%
All publications are with loadings greater than 0.4 or less than -0.4.

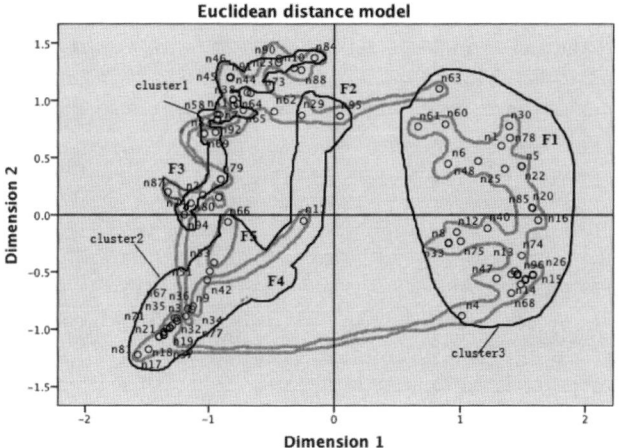

Fig.2. MDS map of cocitation correlation matrix (1965–1993)
(Stress=0.12414)

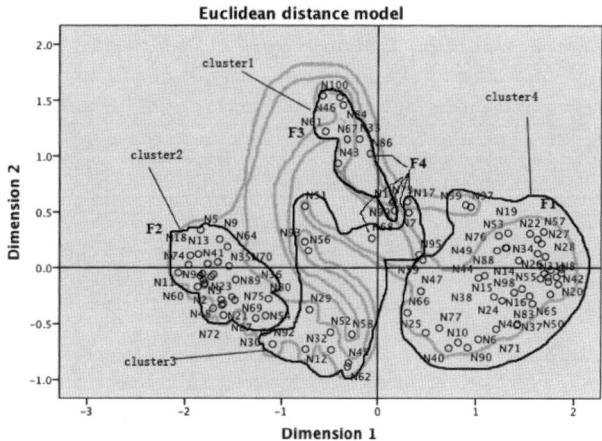

Fig.3. MDS map of bibliographic coupling matrix (2004–2013)
(Stress=0.07419)

Cluster analysis reveals four clusters from all publications in the period, consistent with factor analysis.

B. Bibliographic Coupling Analysis

Bibliographic coupling is used to analyze 437 publications publications together with their 11857 references from 2004 to 2013. Similar strategy for factor analysis is implemented. There are 7 factors in the initial factor analysis while 4 factors in comparison with parallel analysis. Factor analysis of fixed four factors is performed, as reported in Table III. This factor analysis is also effective. The total variance explained is over 85% and 82% loadings are below -0.7 or over 0.7.

The first cluster covers all publications of Factor 3 whereas the fourth includes all publications of Factor 1. The second cluster consists of most papers of Factor 2. The third cluster contains all publications of Factor 4 and part of Factor 2.

The MDS map (Fig.3) with $R^2 = 99.082\%$ of this period is also provided. Four clusters of literature constitute the main frontiers of the up-to-data research in R&D internationalisation that evolves from the original research questions. All these results will be used as guidance for the systematic review of literature.

IV. DISCUSSION

A. Big Questions in R&D Internationalisation Research

Three antecedents of R&D internationalisation research are identified, that is, the three clusters in cocitation analysis.

International Business. Cluster three (Factor 1) includes all publications that have strong linkages to international business (IB) theory, including the classic and generic IB works [e.g. 20-23]. These works served as a foundation for other research topics of the cluster. Later researchers focused specifically on internationalisation of R&D but following the paradigm of these classic works. In this cluster, R&D internationalisation is a branch of international business. R&D internationalisation can be explained as a result of R&D FDI [24]. The main aims of R&D internationalisation research are (1) tracing and characterizing R&D FDI [e.g. 25-27] and (2) how headquarters to designate the best roles and responsibilities for overseas R&D centres [28] and (3) how to collaborate with overseas R&D centres [29-30]. The first aim embeds with traditional IB research while the next two aims are focusing on how can MNC's R&D internationalisation be explained by Dunning's paradigm as well as Bartlett & Ghoshal's transnational framework.

Innovation Research. Publications in cluster one explained R&D internationalisation from an innovation management (IM) perspective. R&D internationalisation is a new phenomenon for innovators to use global resources to make new breakthroughs. Overseas R&D activities are to innovate in a global context. This point of view is operationally oriented to understand R&D internationalisation and strongly associated with traditional theories of IM and R&D management. In particular, studies in this cluster focus on ways to organise global innovation activities. The main research topics include allocation of overseas R&D resources, for example [31], coordinating and integrating R&D activities of both home and overseas [e.g. 32-33] and creation of value from global knowledge assets [34].

Strategic Management. Cluster two mainly explained the internationalisation of R&D, as a strategic management (SM) decision, part of a firm's global strategy. Main research question accordingly is how firms can obtain competitive advantages through R&D internationalisation. Porter's famous works [35-36] on strategy are both shown in the graph. Important topics examined from the SM angle include global R&D strategy [37], strategic decision on R&D internationalisation, in particular the location strategy [38-39]. Decisions on R&D internationalisation cannot be simplified as location or FDI decisions, because R&D internationalisation can be a strategic issue of scanning information on advanced technology and rivals [40]. Thus, this angle explained R&D internationalisation complementary to the IB perspective.

These three perspectives are critical to provide a holistic picture of R&D internationalisation research in early ages. IM and SM angles offers a comprehensive picture of management views on internationalisation of R&D from both operational and strategic levels whereas IB cluster helps understand R&D internationalisation from an economic angle, which emphasizes economic motivations and patterns of R&D FDI and is complementary to the management views.

B. Frontiers in R&D Internationalisation Research

According to the bibliographic analysis on recent publications, four groups of literature are identified. These four groups are frontiers of recent research.

Managing International R&D Networks. Publications on Cluster Four (mainly Factor 1) focus on unfolding management issues on inter-firm international R&D networks at strategic, operational and dynamic level. Strategically, managing international R&D networks requires researchers to understand capabilities [41] and identify new strategies [e.g. 42-43] of these networks. These strategies that aim to help firms manage and design R&D networks, advocate for a play-structure dual relationship. Empirical research [44] has been conducted to understand impacts of dual relationships (source networks and recipient networks) on firms' innovation performances. Some other research tries to understand the impact of positions [e.g. 45], structural strength and density [e.g. 46], as well as typological structures [e.g. 47] on firms' innovation capabilities. Hub firms linked with many innovators are of advantage in innovation. Such linkages can be formal and/or informal tiers.

From a longitudinal perspective, structural analysis of R&D networks has been focused on evolution and dynamics issues [e.g. 48-49]. Evolutions of R&D networks unfold processes of embedding R&D networks into foreign markets. It also informs stages to build up and operate R&D networks.

The last subgroup focuses on managing international R&D network at an operational level. These publications emphasize governance options in terms of scopes and modes of international R&D networks [50]. These options, varying for many reasons, have impacts on firms' innovation performances [51]. Firms need to be able to embed into local environment in operating R&D networks for superior innovation performances [52]. The process of local embeddedness requires interacting with local intellectual property protection policies [e.g. 53] and regional social capital [52].

R&D Internationalisation- New Practices. The second cluster mainly covers papers of Factor 2 that focus on understanding up-to-date new models of global innovation in R&D internationalisation. Many new forms of R&D internationalisation have been observed and studied, such as, international R&D service [54], appropriability of R&D [55], global R&D alliances in emerging countries [e.g. 56-58] and international R&D teams [59]. The new practices can be categorized to three types, first new R&D internationalisation practices from emerging countries, then, new global innovation models (most internally designed by firms) and a new approach to interact with local contexts for global innovation.

These studies enrich the landscape of R&D internationalisation by linking R&D internationalisation research with other fields such as business model and project management. Some works [e.g. 60] try to provide a more integrated view to understand practices of R&D internationalisation, for instance, how to choose proper forms of R&D internationalisation [61]. They also provide an operational perspective to understand the location, management techniques and performance of establishing R&D labs overseas.

Spillover of International R&D Networks. The first cluster (Factor 3) is mainly on spillover effects of the international R&D networks. The spillover effects of R&D FDI was illustrated by Japan's FDI to Korea [62]. However, the case found insignificant spillover effects. A possible answer to this question may be the differentiation of the direct (e.g. patents) and indirect (e.g. intermediate goods) spillover, where the direct spillover effect is significant [63]. Further, effects of international R&D spillover have been examined through a long run and a short run [64]. Spillover effects are significant in many cases. However, technological gaps of countries can be enlarged due to different level of R&D input in host and home countries [e.g. 65-66]. These studies consider the interplay of R&D network and local R&D recourses, emphasize that such interplay is key channels for spillover effect [67].

Small and Medium Sized Enterprises (SMEs) and R&D Networks. The last is Cluster Three that consists of papers of Factor 4 and some of Factor 2. This cluster focuses on explaining how SMEs can make use of global R&D networks at an inter-firm level to exploit their own entrepreneurial talents [68]. SMEs are usually specialised in a very specific sector. They usually can obtain competitiveness in innovation and performance on the specialisation. Hence, SMEs are encouraged to participate in international inter-firm R&D networks [69] to win out in global markets. Key moderators to facilitate the participation of SMEs can be the use of social knowledge and national innovation system [70-71]. These clusters comprehend previous studies because it helps to provide specific views on SME while previous studies are mainly focused on large MNCs.

To conclude, most frontier studies on international R&D networks are targeted at three subjects, *i.e.*, MNCs, SMEs and policy-makers. MNCs are leaders of international R&D networks and have to develop more complex strategies based on structures of R&D networks. SMEs are major participants in R&D networks and have to make use of local knowledge and social capitals. From government perspectives, they need consider spillover effects of R&D internationalisation to facilitate local innovation capability development and economic growth.

V. CONCLUSION

Bibliographic analysis outlines the intellectual structure of R&D internationalisation. Some remarkable characteristics of R&D internationalisation research have been summarised. Here, some new research directions are provided based on previous discussions.

First of all, most recent research focuses on inter-firm R&D networks [e.g. 41-53]. While one frontier in R&D internationalisation research is to summarise new practices of firms from emerging economies [e.g. 56], it may indicate a revisit to examine intra-firm R&D networks. The intra-firm network of R&D department within one MNC had been carefully ten years ago [72]. However, most situations have been changed substantially. For instance, MNCs are expanding R&D facilities to emerging countries. They need take advantages of all the internal R&D units worldwide to leverage external R&D resources globally. Thus, there is a need to integrate internal R&D networks (intra-firm level) and external ones (inter-firm level). This is because the former one unfolds R&D operating mechanisms within a firm and the later one indicates collaboration approaches with other firms. Integration of internal and external networks can clarify the interplay of internal and external R&D networks, which can improve innovation performance and effectively links R&D, global markets and firm's strategy. To integrate intra-firm and inter-firm R&D networks can be an essential future direction for research.

Secondly, current research also focuses on capturing internal attributes of R&D networks [e.g. 50-53], for instance, R&D network structures, dynamics of R&D network and configurations of R&D networks. The dual relationship [44] indicates the necessity to push the research boundary to external attributes of R&D networks. Two external attributes are in particular need of research. On one hand, there neglects investigations on the interaction of focal firm's R&D management and R&D network governance, both at inter-firm and intra-firm levels. R&D networks are consisting of R&D units that are managed by R&D departments of MNCs. In this sense, R&D network is implicitly influenced by their headquarters' characteristics in R&D management and strategies. On the other, R&D networks are in certain contexts, including business and industrial environment, technological development and other infrastructure conditions (e.g. information systems). More research can be conducted to understand influences of contextual issues on R&D networks. Based on relevant contextual studies, it can be possible to find out some key contingence shaping structures and strategies of R&D networks. This is a complementary angle to the current configuration approach of understanding R&D networks.

Many research [e.g. 54-61] summarised various new forms of R&D internationalisation across industries and countries. These studies provide fragmented views on R&D network, which offers an opportunity to integrate them to a holistic view. The integration can reveal whether there exists general approach for managing R&D internationalisation. Influential theories on R&D internationalisation may be generated from this integration that can provide a rudimentary understanding on R&D internationalisation.

This research has its limitations as well. The intermediate periods (from 1994 to 2003) of the publications have been left out for analysis. Meanwhile, more studies can be conducted to understand this period, which enables us to find out an evolutionary process of research on R&D internationalisation.

ACKNOWLEDGEMENT

We thank for the research funds supported by China Scholarship Council, Research and Development Management Association and Cambridge University Engineering Department. We are also grateful to David Probert, who offers very insightful comments on the earlier draft. All errors remain ours.

REFERENCE

[1]   McCain, K. W. 1990. Mapping authors in intellectual space: A technical overview. Journal of the American Society for Information Science, 41(6): 433-443.

[2]   Ramos-Rodriguez A.R. and Ruiz-Navarro J, 2004. Changes in the Intellectual Structure of Strategic Management Research: A Bibliometric Study of the Strategic Management Journal, 1980–2000; Strategic Management Journal, 25: 981-1004.

[3]   Shafique, M. 2013. Thinking inside the box? Intellectual structure of the knowledge base of innovation research (1988-2008). Strategic Management Journal, 34(1): 62-93.

[4]   Culnan MJ, 1986. The Intellectual Development of Management Information Systems, 1972-1982: A Co-Citation Analysis, Management Science, 32, (2), 156-172.

[5]   Pilkington A., Meredith J, 2009, The evolution of the intellectual structure of operations management—1980–2006: A citation/co-citation analysis, Journal of Operations Management, 27, 185-202.

[6]   Tsai Wenpin, Wu Chia-hung, 2010. Knowledge Combination: A Cocitation Analysis, Academy of Management Journal, 53(3), 441-450.

[7]   Cobo M.J., López-Herrera A.G., Herrera-Viedma E. and Herrera F, 2011. Science Mapping Software Tools: Review, Analysis, and Cooperative Study Among Tools, Journal of the American Society for Information Science and Technology, 62(7):1382-1402.

[8]   Culnan MJ, 1986. The Intellectual Development of Management Information Systems, 1972-1982: A Co-Citation Analysis, Management Science, 32, (2), 156-172.

[9]   Small, H., 1973. Co-citation in the scientific literature: a new measure of the relationship between two documents. Journal of the American Society for Information Science, 24, 265-269.

[10]   White, H., Griffith, B.C., 1981. Author co-citation: a literature measure of intellectual structure. Journal of the American Society for Information Science, 32, 163-171.

[11]   White, H.D., McCain, K., 1998. Visualizing a discipline: an author cocitation analysis of information science, 1972–1995. Journal of the American Society for Information Science, 49, 327-355.

[12]   Persson, O. 1994. The intellectual base and research front of JASIS 1986–1990. Journal of the American Society for Information Science, 45(1), 31-38.

[13]   Börner, K., Chen, C., & Boyack, K. (2003). Visualizing knowledge domains. Annual Review of Information Science and Technology, 37, 179–255.

[14]   McCain, K. W. 1990. Mapping authors in intellectual space: A technical overview. Journal of the American Society for Information Science, 41(6): 433-443.

[15]   De Moya Anegón, F., Jiménez Contreras, E., & De La Moneda Corrochano, M. 1998. Research fronts in library and information science in Spain (1985–1994). Scientometrics, 42(2), 229–246.

[16]   McCain, K. W. 1990. Mapping authors in intellectual space: A technical overview. Journal of the American Society for Information Science, 41(6): 433-443.

[17]   Shafique, M. 2013. Thinking inside the box? Intellectual structure of the knowledge base of innovation research (1988-2008). Strategic Management Journal, 34(1): 62-93.

[18]   Glänzel W, & Czerwon H.J. 1996. A new methodological approach to bibliographic coupling and its application to the national, regional and institutional level. Scientometrics, 37(2). 195–221.

[19]   Chen, C. 2003. Mapping Scientic Frontiers, the Quest for Knowledge Visualization. London: Springer-Verlag

[20]   Dunning, J. H. 1992. Multinational Enterprises and the Globalization of Innovatory Capacity. In L. H. O. Granstrand, S. Sjölander (Ed.), Technology Management and International Business. Internationalization of R&D and Technology: 19–51. , Chichester John Wiley and Sons.

[21]   Casson, M. C., Pearce, R.D., Singh S.. 1992. Business Culture and International Technology: Research Managers' Perceptions of Recent Changes in Corporate R&D. In H. L. Granstrand O., Sjölander S., (Ed.), Technology Management and International Business. Internalization of R&D and Technology: 117–135. Chichester: John Wiley and Sons.

[22]   Bartlett, C. A., & Ghoshal, S. 1989. Managing Across Borders: The Transnational Solution: Boston, Harvard Business School Press.

[23]   Prahalad, C. K., & Doz, Y. L. 1987. The multinational mission: Balancing local demands and global vision: Free Press (New York and London).

[24]   Mansfield, E., Teece, D., & Romeo, A. 1979. Overseas Research and Development by US-Based Firms. Economica, 46(182): 187-196.

[25]   Allen, T. J. 1977. Managing the flow of technology: Technology transfer and the dissemination of technological information within the R & D organization (Book). Research supported by the National Science Foundation. Cambridge, Mass., MIT Press, 1977. 329 p.

[26]   Hewitt, G. 1980. Research and Development Performed Abroad by United-States Manufacturing Multinationals. Kyklos, 33(2): de308-327.

[27]   Hakanson, L. 1990. International decentralization of R&D-The organizational challenges. In C. A. Bartlett, Y. L. Doz, & G. Hedlund (Eds.), Managing the Global Firm. London; New York: Routledge.

[28]   Ghoshal, S., & Bartlett, C. A. 1988. Creation, Adoption, and Diffusion of Innovations by Subsidiaries of Multinational Corporations. Journal of International Business Studies, 19(3): 365-388.

[29]   Kanno, M. 1968. Effect on Communication between Labs and Plants of the. Transfer of R&D Personnel. Unpublished Master Thesis, MIT.

[30]   Perrino, A. C., & Tipping, J. W. 1989. Global Management of Technology. Research-Technology Management, 32(3): 12-20.

[31] Lall, S. 1979. The International Allocation of Research Activity by US Multinationals. Oxford Bulletin of Economics and Statistics, 41(4): 313-331.

[32] Hakanson, L. 1981. Organization and Evolution of Foreign R-and-D in Swedish Multinationals. Geografiska Annaler Series B-Human Geography, 63(1): 47-56.

[33] Granstrand, O., Fernlund, I. 1978. Coordination of Multinational R and D - Swedish Case-Study. R & D Management, 9(1): 1-7.

[34] Teece, D. J. 1986. Profiting from Technological Innovation - Implications for Integration, Collaboration, Licensing and Public-Policy. Research Policy, 15(6): 285-305.

[35] Porter, M. E. 1986. Competition in global industries: Harvard Business Press.

[36] Porter, M. E. 1990. The Competitive Advantage of Nations: Free Press, New York.

[37] Casson, M. 1991. Global research strategy and international competitiveness: B. Blackwell.

[38] Howells, J. 1990. The Location and Organization of Research-and-Development - New Horizons. Research Policy, 19(2): 133-146.

[39] Malecki, E. J. 1980. Corporate Organization of R and D and the Location of Technological Activities. Regional Studies, 14(3): 219-234.

[40] Herbert, E. 1989. Japanese R and D in the United-States. Research-Technology Management, 32(6): 11-21.

[41] Hagedoorn, J., Roijakkers, N., & Van Kranenburg, H. 2006. Inter-firm R&D networks: The importance of strategic network capabilities for high-tech partnership formation. British Journal of Management, 17(1): 39-53.

[42] Freel, M., & de Jong, J. P. J. 2009. Market novelty, competence-seeking and innovation networking. Technovation, 29(12): 873-884.

[43] Dhanaraj, C., & Parkhe, A. 2006. Orchestrating innovation networks. Academy of Management Review, 31(3): 659-669.

[44] Zhao, Z., Anand, J., & Mitchell, W. 2005. A dual networks perspective on inter-organizational transfer of R&D capabilities: International joint ventures in the Chinese automotive industry. Journal of Management Studies, 42(1): 127-160.

[45] Aalbers, R., Dolfsma, W., & Koppius, O. 2013. Individual connectedness in innovation networks: On the role of individual motivation. Research Policy, 42(3): 624-634.

[46] Bertrand-Cloodt, D., Hagedoorn, J., & Van Kranenburg, H. 2011. The strength of R&D network ties in high-tech sectors - a multi-dimensional analysis of the effects of tie strength on innovation performance. Technology Analysis & Strategic Management, 23(10): 1015-1030.

[47] Salavisa, I., Sousa, C., & Fontes, M. 2012. Topologies of innovation networks in knowledge-intensive sectors: Sectoral differences in the access to knowledge and complementary assets through formal and informal ties. Technovation, 32(6): 380-399.

[48] Balland, P. A. 2012. Proximity and the Evolution of Collaboration Networks: Evidence from Research and Development Projects within the Global Navigation Satellite System (GNSS) Industry. Regional Studies, 46(6): 741-756.

[49] Karna, A., Taube, F., & Sonderegger, P. 2013. Evolution of Innovation Networks across Geographical and Organizational Boundaries: A Study of R&D Subsidiaries in the Bangalore IT Cluster. European Management Review, 10(4): 211-226.

[50] Oxley, J. E., & Sampson, R. C. 2004. The scope and governance of international R&D alliances. Strategic Management Journal, 25(8-9): 723-749.

[51] Fang, G., & Pigneur, Y. 2007. The integrative model of international innovation network and performance. Cambridge: Univ Cambridge, Inst Manufacturing.

[52] Song, J., Asakawa, K., & Chu, Y. 2011. What determines knowledge sourcing from host locations of overseas R&D operations? A study of global R&D activities of Japanese multinationals. Research Policy, 40(3): 380-390.

[53] Rutten, R., & Boekema, F. 2007. Regional social capital: Embeddedness, innovation networks and regional economic development. Technological Forecasting and Social Change, 74(9): 1834-1846.

[54] Martinez-Noya, A., Garcia-Canal, E., & Guillen, M. F. 2012. International R&D service outsourcing by technology-intensive firms: Whether and where? Journal of International Management, 18(1): 18-37.

[55] Di Minin, A., & Bianchi, M. 2011. Safe nests in global nets: Internationalization and appropriability of R&D in wireless telecom. Journal of International Business Studies, 42(7): 910-934.

[56] Asakawa, K., & Som, A. 2008. Internationalization of R&D in China and India: Conventional wisdom versus reality. Asia Pacific Journal of Management, 25(3): 375-394.

[57] Li, J. T., & Kozhikode, R. K. 2009. Developing new innovation models: Shifts in the innovation landscapes in emerging economies and implications for global R&D management. Journal of International Management, 15(3): 328-339.

[58] Li, J. T. 2010. Global R&D Alliances in China: Collaborations With Universities and Research Institutes. Ieee Transactions on Engineering Management, 57(1): 78-87.

[59] Ambos, B., & Ambos, T. C. 2009. Location choice, management and performance of international R&D investments in peripheral economies. International Journal of Technology Management, 48(1): 24-41.

[60] Corsaro, D., Ramos, C., Henneberg, S. C., & Naude, P. 2012. The impact of network configurations on value constellations in business markets - The case of an innovation network. Industrial Marketing Management, 41(1): 54-67.

[61] Ito, B., & Wakasugi, R. 2007. What factors determine the mode of overseas R&D by multinationals? Empirical evidence. Research Policy, 36(8): 1275-1287.

[62] Kwon, H. U. 2005. International R&D spillovers from Japanese to Korean manufacturing industry. Hitotsubashi Journal of Economics, 46(2): 135-147.

[63] Lee, G. 2005. International R&D spillovers revisited. Open Economies Review, 16(3): 249-262.

[64] Bottazzi, L., & Peri, G. 2007. The international dynamics of R&D and innovation in the long run and in the short run. Economic Journal, 117(518): 486-511.

[65] Wang, M., & Wong, M. C. S. 2012. International R&D Transfer and Technical Efficiency: Evidence from Panel Study Using Stochastic Frontier Analysis. World Development, 40(10): 1982-1998.

[66] Cermeno, R., & Vazquez, S. 2009. Technological Backwardness in Agriculture: Is it Due to Lack of R&D, Human Capital, and Openness to International Trade? Review of Development Economics, 13(4): 673-686.

[67] Cermeno, R., & Vazquez, S. 2009. Technological Backwardness in Agriculture: Is it Due to Lack of R&D,

Human Capital, and Openness to International Trade? Review of Development Economics, 13(4): 673-686.

[68] Cook, G. A. S., & Pandit, N. R. 2005. Clustered High-Technology Small Firms and Innovation Networks: The Case of Post-Production in London. Oxford: Elsevier Science Ltd.

[69] Cappellin, R., & Wink, R. 2009. International Knowledge and Innovation Networks: Knowledge Creation and Innovation in Medium-Technology Clusters. Cheltenham: Edward Elgar Publishing Ltd.

[70] Dodgson, M., Mathews, J., Kastelle, T., & Hu, M. C. 2008. The evolving nature of Taiwan's national innovation system: The case of biotechnology innovation networks. Research Policy, 37(3): 430-445.

[71] Fiedler, M., & Welpe, I. M. 2011. Commercialisation of technology innovations: an empirical study on the influence of clusters and innovation networks. International Journal of Technology Management, 54(4): 410-437.

[72] Von Zedtwitz, M., & Gassmann, O. 2002. Market versus technology drive in R&D internationalization: four different patterns of managing research and development. Research Policy, 31(4): 569-588.

APPENDIX

List of 95 Papers Used in Cocitation Analysis (1965-1993)

1 Ronstadt R., 1977, Res Dev Abroad Us Mu
2 Imai K, 1989, 135 Hit U I Bus Res
3 Markusen Jr, 1984, J Int Econ, V16, P205, Doi 10.1016/s0022-1996(84)80001-x
4 Mansfield E, 1980, Q J Econ, V95, P737, Doi 10.2307/1885489
5 Hirschey Rc, 1981, Oxford B Econ Stat, V43, P115
6 Vernon R, 1966, Q J Econ, V80, P190, Doi 10.2307/1880689
7 Perrino Ac, 1989, Res Technol Manage, V32, P12
8 Casson M, 1992, Technology Managemen, P117
9 Griliches Z, 1979, Bell J Econ, V10, P92, Doi 10.2307/3003321
10 Franko Lg, 1989, Strategic Manage J, V10, P449, Doi 10.1002/smj.4250100505
11 Schwartz Nancy L., 1982, Market Structure Inn
12 Bartlett C. A., 1989, Managing Borders
13 Hakanson L, 1983, Govt Multinationals
14 Rugman Am, 1981, Can Public Pol, V7, P604, Doi 10.2307/3549490
15 Ondrack Da, 1983, Govt Multinationals
16 Hewitt G, 1980, Kyklos, V33, P308, Doi 10.1111/j.1467-6435.1980.tb02637.x
17 Mansfield E, 1979, Rev Econ Stat, V61, P49, Doi 10.2307/1924830
18 Spencer Bj, 1983, Rev Econ Stud, V50, P707, Doi 10.2307/2297771
19 Gruber W, 1967, J Polit Econ, V75, P20, Doi 10.1086/259235
20 Hakanson L, 1981, Geogr Ann B, V63, P47, Doi 10.2307/490997
21 Posner M.v., 1961, Oxford Econ Pap, V13, P323
22 Lall S, 1979, Oxford B Econ Stat, V41, P313
23 Prahalad C.k., 1987, Multinational Missio
24 Tushman Ml, 1977, Admin Sci Quart, V22, P587, Doi 10.2307/2392402
25 Hakanson Lars, 1986, Managing Int Res Dev
26 Zejan Mc, 1988, Studies Behavior Swe
27 Ronstadt Rc, 1976, J Int Bus Stud, V9, P7
28 Cordell Aj, 1971, Multinational Firm F
29 Aharoni Y., 1966, Foreign Investment D
30 Pearce R., 1989, Int Res Dev Multinat
31 Linder S. B., 1961, Essay Trade Transfor
32 Scherer F., 1980, Ind Market Structure
33 Dunning Jh, 1992, Technology Managemen, P19
34 Johnston J., 1972, Econometric Methods
35 Grabowski Hg, 1968, J Polit Econ, V76, P292, Doi 10.1086/259401

36 Brander Ja, 1983, Bell J Econ, V14, P225, Doi 10.2307/3003549
37 Arrow K., 1962, Rate Direction Inven, P609
38 Mensch G., 1979, Stalemate Technology
39 Hughes Ks, 1986, Eur Econ Rev, V30, P383, Doi 10.1016/0014-2921(86)90050-4
40 Hakanson L, 1980, Multinationella Fore, P4
41 Sciberras E, 1987, R&d Manage, V17, P15, Doi 10.1111/j.1467-9310.1987.tb00044.x
42 Dunning Jh, 1979, Oxford B Econ Stat, V41, P269
43 Wortmann M, 1990, Res Policy, V19, P175, Doi 10.1016/0048-7333(90)90047-a
44 Teece Dj, 1986, Res Policy, V15, P285, Doi 10.1016/0048-7333(86)90027-2
45 Mansfield E, 1984, Ieee T Eng Manage, V31, P122
46 Vernon R, 1979, Oxford B Econ Stat, V41, P255
47 Rugman Am, 1982, Columbia J World Bus, V17, P58
48 Perlmutter Hv, 1969, Columbia J World Bus, V4, P9
49 Dixit A, 1988, Int Competitiveness, P149
50 Hedlund G, 1974, Managing Relationshi
51 Gruber W, 1967, J Political Econ Feb
52 Hakanson L., 1989, Managing Global Firm
53 Servan-schreiber J.-j., 1968, Am Challenge
54 Fischer Wa, 1979, J Int Bus Stud, V20, P453
55 Hewitt Gk, 1983, Multinationals Techn
56 Cordell Aj, 1973, Long Range Plann, V6, P22, Doi 10.1016/0024-6301(83)90181-4
57 Swedenborg B, 1988, Svenska Ind Utlandsi
58 Dunning J.h., 1958, Am Investment Brit M
59 Bergholm F, 1985, Multinationella Fore
60 Granstrand O., 1978, R & D Management, V9, Doi 10.1111/j.1467-9310.1978.tb00124.x
61 Porter M E, 1990, Competitive Advantag
62 Hood N., 1982, Multinational Busine, V2, P10
63 Ronstadt Rc, 1978, J Int Bus Stud, V9, P7, Doi 10.1057/palgrave.jibs.8490647
64 Us Tariff Commission, 1973, Impl Mult Firms Worl
65 Behrman Jn, 1970, National Interests M
66 Kamien Mi, 1975, J Econ Lit, V13, P1
67 Howe Jd, 1976, Can J Econ, V9, P57, Doi 10.2307/134415
68 Cordell Aj, 1971, Special Study Sci Co, V22
69 Hakanson L, 1980, Multinationella Fore
70 Loury Gc, 1979, Q J Econ, V93, P395, Doi 10.2307/1883165
71 Feenstra Rc, 1982, J Polit Econ, V90, P1142, Doi 10.1086/261115
72 Caves Re, 1971, Economica, V38, P1, Doi 10.2307/2551748
73 Steele Lw, 1975, Innovation Big Busin
74 Hedlund G, 1986, Hum Resource Manage, V25, P9, Doi 10.1002/hrm.3930250103
75 Mansfield E, 1977, Production Applicati
76 Katrak H, 1973, Oxford Econ Pap, V25, P337
77 Hakanson L, 1988, R&d Manage, V18, P217, Doi 10.1111/j.1467-9310.1988.tb00588.x
78 Behrman Jn, 1980, Overseas Activities
79 Edstrom A, 1977, Admin Sci Quart, V22, P248, Doi 10.2307/2391959
80 Hufbauer Gc, 1966, Synthetic Materials
81 Demeyer A, 1989, R&d Manage, V19, P135
82 Allen T., 1977, Managing Flow Techno
83 Behrman J.n., 1980, Overseas R D Activit
84 Mansfield E, 1979, Economica, V46, P187, Doi 10.2307/2553190
85 Keesing Db, 1967, J Polit Econ, V75, P38, Doi 10.1086/259236
86 Caves Richard E., 1982, Multinational Enterp
87 Buckley P.j., 1976, Future Multinational
88 Porter M. E., 1986, Competition Global I, P15
89 Creamer D. B., 1976, Overseas Res Dev Us
90 Howells J, 1990, Res Policy, V19, P133, Doi 10.1016/0048-7333(90)90043-6
91 Demeyer A, 1989, Insead8962 Work Pap
92 Ghoshal S, 1988, J Int Bus Stud, V19, P365, Doi 10.1057/palgrave.jibs.8490388
93 Kanno M, 1968, Thesis Mit Cambridge
94 Robinson R, 1988, Int Transfer Technol
95 Behrman Jn, 1980, Columbia J World Bus, V15, P55

List of Papers Used in Bibliographic Coupling (2004-2013)

1 Cappellin R, 2009, NEW HORIZ REG SCI, P1
2 Asakawa K, 2008, ASIA PAC J MANAG, V25, P375, DOI 10.1007/s10490-007-9082-z
3 Ambos B, 2004, J WORLD BUS, V39, P37, DOI 10.1016/j.jwb.2003.08.004
4 Sammarra A, 2008, J MANAGE STUD, V45, P200
5 Moncada-Paterno-Castello P, 2011, IND CORP CHANGE, V20, P585, DOI 10.1093/icc/dtr005
6 Aalbers R, 2013, RES POLICY, V42, P624, DOI 10.1016/j.respol.2012.10.007
7 Fiedler M, 2011, INT J TECHNOL MANAGE, V54, P410, DOI 10.1504/IJTM.2011.041582
8 Lhuillery S, 2011, IND INNOV, V18, P105, DOI 10.1080/13662716.2010.528936
9 Belderbos R, 2008, J JPN INT ECON, V22, P310, DOI 10.1016/j.jjie.2008.01.001
10 Oxley JE, 2004, STRATEGIC MANAGE J, V25, P723, DOI 10.1002/smj.391
11 Di Minin A, 2012, EUR MANAG J, V30, P189, DOI 10.1016/j.emj.2012.03.004
12 Erkelens R, 2010, LECT NOTES BUS INF, V55, P82
13 Loof H, 2009, J EVOL ECON, V19, P41, DOI 10.1007/s00191-008-0103-y
14 Boschma R, 2010, P120
15 Rutten R, 2007, TECHNOL FORECAST SOC, V74, P1834, DOI 10.1016/j.techfore.2007.05.012
16 Fritsch M, 2010, ANN REGIONAL SCI, V44, P21, DOI 10.1007/s00168-008-0245-8
17 Balas N, 2012, ROUT STUD INNOV ORG, V19, P301
18 Sanna-Randaccio F, 2007, J INT BUS STUD, V38, P47, DOI 10.1057/palgrave.jibs.8400249
19 Ryan MP, 2010, WORLD DEV, V38, P1082, DOI 10.1016/j.worlddev.2009.12.013
20 Dhanaraj C, 2006, ACAD MANAGE REV, V31, P659
21 Belderbos R, 2013, J INT BUS STUD, V44, P765, DOI 10.1057/jibs.2013.33
22 Rampersad G, 2010, IND MARKET MANAG, V39, P793, DOI 10.1016/j.indmarman.2009.07.002
23 Chuang WB, 2010, ASIAN ECON J, V24, P305, DOI 10.1111/j.1467-8381.2010.02041.x
24 Lowe MS, 2012, J ECON GEOGR, V12, P1113, DOI 10.1093/jeg/lbs021
25 Ensign PC, 2009, P1, DOI 10.1057/9780230617131
26 Cantner U, 2011, P366
27 Li HW, 2010, P76
28 Hanaki N, 2010, RES POLICY, V39, P386, DOI 10.1016/j.respol.2010.01.001
29 Mudambi R, 2010, ADV INTL MANAGEMENT, V23, P461, DOI 10.1108/S1571-5027(2010)0000023025
30 Song J, 2011, RES POLICY, V40, P380, DOI 10.1016/j.respol.2011.01.002
31 Bojanowski M, 2012, J TECHNOL TRANSFER, V37, P967, DOI 10.1007/s10961-011-9234-7
32 Penner-Hahn J, 2005, STRATEGIC MANAGE J, V26, P121, DOI 10.1002/smj.436
33 Carayannis EG, 2009, INT J TECHNOL MANAGE, V46, P195
34 Li YP, 2009, SER OPER SUPP CH MAN, V3, P823
35 van Beers C, 2008, RES POLICY, V37, P294, DOI 10.1016/j.respol.2007.10.007
36 Castellani D, 2013, J INT BUS STUD, V44, P649, DOI 10.1057/jibs.2013.30
37 Hagedoorn J, 2006, BRIT J MANAGE, V17, P39, DOI 10.1111/j.1467-8551.2005.00474.x
38 Colombo MG, 2009, STRATEG ENTREP J, V3, P346, DOI 10.1002/sej.78
39 Zeng DM, 2009, P519, DOI 10.1109/BIFE.2009.123
40 Karna A, 2013, EUR MANAG REV, V10, P211, DOI 10.1111/emre.12017
41 Ito B, 2007, RES POLICY, V36, P1275, DOI 10.1016/j.respol.2007.04.011
42 Gilsing V, 2005, EUR MANAG REV, V2, P179, DOI 10.1057/palgrave.emr.1500041
43 Kwon HU, 2005, HITOTSUB J ECON, V46, P135
44 Macpherson A, 2005, INT J TECHNOL MANAGE, V30, P49, DOI 10.1504/IJTM.2005.006345
45 Frost TS, 2005, J INT BUS STUD, V36, P676, DOI 10.1057/palgrave.jibs.8400168
46 Coe DT, 2009, EUR ECON REV, V53, P723, DOI 10.1016/j.euroecorev.2009.02.005
47 Hagedoorn J, 2005, J INT BUS STUD, V36, P175, DOI 10.1057/palgrave.jibs.8400122
48 Veliyath R, 2011, MANAGE INT REV, V51, P407, DOI 10.1007/s11575-011-0079-y
49 Balland PA, 2012, REG STUD, V46, P741, DOI 10.1080/00343404.2010.529121
50 Fang G, 2007, P167
51 Love JH, 2009, IND INNOV, V16, P273, DOI 10.1080/13662710902923776
52 Laperche B, 2008, P2281
53 Konig MD, 2011, J ECON BEHAV ORGAN, V79, P145, DOI 10.1016/j.jebo.2011.01.007
54 Li JT, 2010, IEEE T ENG MANAGE, V57, P78, DOI 10.1109/TEM.2009.2028324
55 Buchmann T, 2012, ELGAR ORIG REF, P466
56 Iwata S, 2006, IEEE T ENG MANAGE, V53, P361, DOI 10.1109/TEM.2006.877448
57 Dilk C, 2008, MANAGE DECIS, V46, P691, DOI 10.1108/00251740810873455
58 Helble Y, 2004, R&D MANAGE, V34, P605, DOI 10.1111/j.1467-9310.2004.00366.x
59 Corsaro D, 2012, IND MARKET MANAG, V41, P54, DOI 10.1016/j.indmarman.2011.11.017
60 Lehrer M, 2011, J INT MANAG, V17, P42, DOI 10.1016/j.intman.2010.08.001
61 Bottazzi L, 2007, ECON J, V117, P486, DOI 10.1111/j.1468-0297.2007.02027.x
62 Branzei O, 2011, ASIAN BUS MANAG, V10, P9, DOI 10.1057/abm.2010.34
63 Li JT, 2005, TECHNOL ANAL STRATEG, V17, P317, DOI 10.1080/09537320500211367
64 Pavlinek P, 2012, ECON GEOGR, V88, P279, DOI 10.1111/j.1944-8287.2012.01155.x
65 Bertrand-Cloodt D, 2011, TECHNOL ANAL STRATEG, V23, P1015, DOI 10.1080/09537325.2011.621294
66 Edwards-Schachter M, 2013, INT J TECHNOL MANAGE, V62, P128, DOI 10.1504/IJTM.2013.055162
67 Bosetti V, 2008, ENERG ECON, V30, P2912, DOI 10.1016/j.eneco.2008.04.008
68 Dodgson M, 2008, RES POLICY, V37, P430, DOI 10.1016/j.respol.2007.12.005
69 Li JT, 2009, J INT MANAG, V15, P328, DOI 10.1016/j.intman.2008.12.005
70 Haakonsson SJ, 2013, J ECON GEOGR, V13, P677, DOI 10.1093/jeg/lbs018
71 Hurmelinna-Laukkanen P, 2012, EUR MANAG J, V30, P552, DOI 10.1016/j.emj.2012.03.002
72 Di Minin A, 2011, J INT BUS STUD, V42, P910, DOI 10.1057/jibs.2011.16
73 Cook GAS, 2005, P165
74 Liu JJ, 2010, INT J TECHNOL MANAGE, V51, P409
75 Bergek A, 2010, RES POLICY, V39, P1321, DOI 10.1016/j.respol.2010.08.002
76 Konig MD, 2012, GAME ECON BEHAV, V75, P694, DOI 10.1016/j.geb.2011.12.007
77 Gkypali A, 2012, IND CORP CHANGE, V21, P731, DOI 10.1093/icc/dtr057
78 van der Valk T, 2011, TECHNOL FORECAST SOC, V78, P25, DOI 10.1016/j.techfore.2010.07.001
79 Salavisa I, 2012, TECHNOVATION, V32, P380, DOI 10.1016/j.technovation.2012.02.003
80 Ernst D, 2009, POL STUD, V54, P1
81 Ambos B, 2009, INT J TECHNOL MANAGE, V48, P24
82 Wang J, 2012, INNOV-MANAG POLICY P, V14, P192
83 Freel M, 2009, TECHNOVATION, V29, P873, DOI 10.1016/j.technovation.2009.07.005

84   Cermeno R, 2009, REV DEV ECON, V13, P673, DOI 10.1111/j.1467-9361.2009.00520.x

85   Baum JAC, 2010, MANAGE SCI, V56, P2094, DOI 10.1287/mnsc.1100.1229

86   Barros PP, 2008, INT J HEALTH CARE FI, V8, P301, DOI 10.1007/s10754-008-9042-2

87   Martinez-Noya A, 2012, J INT MANAG, V18, P18, DOI 10.1016/j.intman.2011.06.004

88   Corsaro D, 2012, IND MARKET MANAG, V41, P780, DOI 10.1016/j.indmarman.2012.06.005

89   Higon DA, 2012, RES POLICY, V41, P592, DOI 10.1016/j.respol.2011.12.007

90   Zhao Z, 2005, J MANAGE STUD, V42, P127, DOI 10.1111/j.1467-6486.2005.00491.x

91   Liu MC, 2012, RES POLICY, V41, P1107, DOI 10.1016/j.respol.2012.03.016

92   Li XY, 2013, INT BUS REV, V22, P639, DOI 10.1016/j.ibusrev.2012.12.002

93   Kurokawa S, 2007, RES POLICY, V36, P3, DOI 10.1016/j.respol.2006.07.001

94   Cantwell J, 2005, REG STUD, V39, P1, DOI 10.1080/00343400520003 20824

95   Sala A, 2011, INT J TECHNOL MANAGE, V53, P19, DOI 10.1504/IJTM.2011.037236

96   Cowan R, 2009, ACAD MANAGE REV, V34, P320

97   Liou DY, 2007, IN C IND ENG ENG MAN, P1950, DOI 10.1109/IEEM.2007.4419532

98   Smart P, 2007, INT J OPER PROD MAN, V27, P1069, DOI 10.1108/01443570710820639

99   Kurumoto JS, 2011, IFIP ADV INF COMM TE, V362, P207

100 Wang M, 2012, WORLD DEV, V40, P1982, DOI 10.1016/j.worlddev.2012.05.001

Research on Investment Capacity Model for Power Grid Enterprises

Lie-xiang HU[1], Qian XU[2], Xiao-fen LU[2], Qiong WANG[3,*], Ming FU[3], Hong-hao QIN[3]

[1]State Grid Zhejiang Electric Power Company, Hangzhou, China
[2]State Grid Zhejiang Electric Power Company Economic Research Institute, Hangzhou, China
[3]North China Electric Power University, Beijing, China
(553877009@qq.com)

Abstract - **The power grid investment has the characteristics of huge amount of investment, long payback period, and it is a very big spending of the power grid enterprises. Therefore, it has the very strong practical significance for the power grid enterprises to quantitatively study and accurately grasp their investment capacity, and then formulate a reasonable investment plan, use the financial resources as a whole, make financial benefit play the leading role. In this paper, combined with the financial management theory, a quantitative model is conducted which uses two constraints of target profits and assets liability rate to study the power grid investment capability. The calculation results show that the proposed model can accurately reflect the investment capacity of power grid enterprise, and it has strong applicability. What's more, it provides the scientific basis for power grid enterprises to invest reasonably and increase the rate of return effectively.**

Keywords - **Calculation model, investment capacity, investment demand, power grid enterprises**

I. INTRODUCTION

Grid is one of the vital basis of national economic construction, it has very important significance in the support of industrial production, maintaining social stability and safeguarding residents' development. With the rapid growth of demand for electricity, the investment scale of power grid construction appears sharply rising trend year by year, but investment spending soared. So as the debt-to-assets ratio of power grid enterprises often climb, it not only makes enterprise financing scale be restricted, also affects the enterprise's financial soundness. In order to promote the rationality of the power grid enterprise investment planning, ensure an orderly investment, and promote the healthy development of the enterprise, the construction of a scientific investment capacity calculation model, and the accurate measurement of investment capacity of power grid enterprises has an strong practical significance.

At present, the research on grid investment capacity is less. Reference [1] introduces concepts of return on investment, operating coefficient, etc. to build a quantitative model for power grid enterprise investment ability; Reference [2] using theories like grey correlation analysis, GM 1, 1) model for grid investment capacity has carried on the forecast analysis, and obtained the high accuracy; Reference [3] from two aspects of investment benefit analysis and investment estimates to measure investment ability with strong innovation. All these research results above calculate the investment capacity of power grid in their respective, but not on the composition

of the investment capacity of the mining and analysis, losing scientific nature.

As a result, this paper, on the basis of previous studies, combining with the theory of financial management and considering the target profits and asset-liability ratio constraints, make a quantitative research for the composition of the investment capacity of power grid enterprises, and to build a model which can intuitively reflect the enterprise investment ability. It can make the power grid enterprise clear their own investment ability, the overall usage of financial resources, and scientific investment decisions.

II. INVESTMENT CAPACITY CALCULATION MODEL

A. Calculation principle

The investment capacity of power grid enterprises is effected by two constraints of target profit and asset liability ratio. The investment capacity is respectively calculated under the above two constraints, the smaller is served as the annual investment capacity, namely the investment scale within the limit of financial ability.

B. Calculation model [4-7]

1) Calculation model with the constraint of target profit

The investment capital which the company can put to use is mainly composed of two parts, the remaining amount of profit after deducting the target profit and depreciation. Combined with the requirements of minimum capital rate, the actual investment capacity can be calculated by the investment capital. The calculation formula is as follows:

$$\text{The actual investment capacity} = \frac{\text{The investment capital}}{\text{minimum capital rate}} \quad (1)$$

$$\text{The investment capital} = \text{total profit} - \text{the target profit} + \text{depreciation} \quad (2)$$

2) Calculation model with the constraint of asset liability ratio

The repayment of the loan amount is set according to the actual company repayment plan, the constraint of asset liability ratio is determined by rising one percentage point than asset liability ratio at the beginning of the year, and the maximum is not more than 75%. The asset liability ratio can be expressed as the following formula:

$$\frac{I_{NTOD} + M_L - R_{EPD}}{I_{NTOA} + N_P + M_L} = C_{DAR} \qquad (3)$$

Where, I_{NTOD} is total liabilities at the beginning of the year, I_{NTOA} is total assets at the beginning of the year, M_L is the maximum loan amount, R_{EPD} is repayment loans, N_P is net profit, C_{DAR} is the constraint of asset liability ratio.

Transform the above formula into an expression of X, showed as follows:

$$M_L = \frac{C_{DAR}(I_{NTOA} + N_P) + R_{EPD} - I_{NTOD}}{1 - C_{DAR}} \qquad (4)$$

Here, the actual investment capacity mainly includes the total profit, depreciation and bank loans three parts, the formula is as follows:

$$\text{The actual investment capacity} = \text{total profit} \\ + \text{depreciation} + M_L \qquad (5)$$

III. EMPIRICAL ANALYSIS

This paper uses a power company as an example. According to the company in 2006-2013 financial data, we calculate the corresponding data considering the requirements of investment capacity calculation model. Then we calculate the maximum investment ability respectively under the two constraints of the target profit and the asset-liability ratio, and take the smaller values of both as the final investment ability.

According to the provisions of the latest power grid enterprise investment policy [8], the enterprise has its own investment funds accounted for 25% of the total investment amount, therefore, the minimum capital ratio is 25%. Combining the specific requirements for the power company, profit requirement is set to 2 billion yuan, the asset-liability ratio is set to one percent higher than that of initial asset-liability ratio, the income tax rate is 25%. The investment capacity of two models of the calculation results are shown in Table I.

TABLE I
THE INVESTMENT CAPACITY OF THE MODEL CALCULATION RESULTS ONE HUNDRED MILLION YUAN

model	2006	2007	2008	2009	2010	2011
Ensure the target profit	174.4	157.6	157.3	74.4	285.5	307.2
Ensure that target asset-liability ratio	87.6	92.8	99.3	112.5	139.4	155.5

For intuitively describing the relationship between investment capacity and the actual investment, we compare the company's actual investment ability with the actual investment of data analysis in 2006-2013, as shown in Fig.1.

From Fig.1, we can see that the power company's investment ability present a trend of increasing year by year, while the actual investment has slight trend of fluctuations because of the power grid construction and technical transformation. From 2006 to 2013, the electric power company's actual investment is in the scope of the investment ability. This indicates that the power of the

company's operating performance can basically meet the annual capital construction investment.

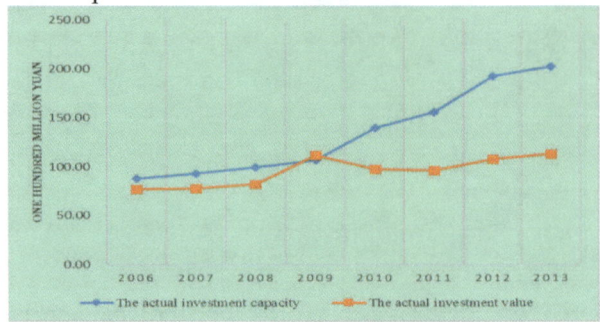

Fig.1. Comparison between the actual investment capacity and the actual investment value in 2006-2013

But in 2009, the actual investment exceeds the largest investment capacity. The investment capacity can't meet the annual demand for investment, and the actual investment is in the jump. If its long-term investment capacity can't meet the actual investment, it will inevitably bring power grid investment insufficiency and overweight the grid company financial pressure. As a result, it will led to affect Power grid companies as well as the regional electric power development seriously. Therefore, this paper deals with the special phenomenon in 2009 with detailed observation and analysis, then consider that there is three following reasons.

A. International economic situation [9-11]

In 2008, the global financial crisis broke out, then the domestic and international economic situation was undergoing profound changes. The province's economy, export, processing, and small and medium-sized enterprises as the main characteristics, was influenced a lot. Electric power demand has close relation with the development of national economy, which is positive relationship. When economic development rapidly, the electricity demand is large, and the grid enterprise's revenue is rising; on the other hand, if the economic downturn, social power consumption reduced and electricity grid enterprise's revenue will also be cut accordingly. Electricity demand changing leaves behind the change of the economic situation. In 2009, the province's electricity demand fell, then the growth of electricity sales amount is clearly slowing with an increased risk of arrearage, main business revenue growth is lesser, so as to make the annual investment ability increase slower than that of 2008, which brought the huge challenge for the operation and development of power grid enterprises.

B. Cost control losing

Before 2009 the company's line loss rate remain at 3% around which is a high line loss, making the line loss cost highly, until 2010 the line loss management level improved, the line loss began to decline, and reduce the cost. Before 2010 the company's management cost grew rapidly year by year, and by 2010, it reduce from 800 million yuan to 350 million yuan, which greatly improved the company profit space, so as to make the investment ability increased. Thus the company's cost management level, including line loss management, has great influence on investment capacity. The company should control the cost in a reasonable manner, guarantee the profits, and ensure continued investment ability.

C. Uncontrollable natural disasters [12-15]

Through the research, the province was in the southern areas of ice disaster, continuous low temperature sleet freezing weather caused varying degrees of damage to all levels of power grid, such as transmission lines appearing fold tower, tower accidents, substation tripping shutdown, etc., which led to funds of repairing and reconstruction, technical transformation have risen sharply. For ice and snow disasters causing such a serious impact to all levels of line, we investigate its reason, mainly had the following three aspects: one is the actual thickness of ice surpass circuit designed ice thickness; the second is the original procedures of circuit design technology require related load, especially the value of longitudinal uneven tension, was low; the third is that line was not to avoid adverse terrain, thus the location of weather was cold. After the ice disaster, the company continuously pushed for deepening the work of melting ice in order to promote transmission lines ice reconstruction project. Considering lines importance and power outages, the investment amount was huge for the corresponding circuit implementation differentiation of melting ice, reinforcement and reconstruction. This is also one of the important reasons in the province power grid investment increasing in 2009.

D. Strategic decisions of Expansion of domestic demand and growth of promotion

Although the financial crisis has effect on the real economy of the province and the consumption growth of whole society is slowing, this province is still in the stage of industrialization and urbanization. In 2009, the province determined the government-led investment 183.9 billion which will spur social investment of more than 500 billion. Stimulating domestic demands generate heavy effect on economic growth.

In order to satisfy the demand of social economic development, this provincial company seriously implements the state and the province strategy on expanding domestic demand and promotes economic growth. This province accelerates the annual power grid construction, increases the investment dynamics, integrates resources, focuses on promoting the construction of key projects and improves the main network structure so as to ensure the enough power which supplies for the province's economic and social development.

Among them, 500 kV power transmission and transformation investment project is the biggest one. In 2009, this provincial electric power company has built six 500 kV power transmission and transformation projects, expanded the four projects, and started seven projects. Then the production of the substation capacity and the

length of the line to achieve the goal "ShuangQian", power grid construction scale hit a record high, the investment amounted to a 65.04% increase over 2008.

In total, the electric power company's actual investment increased rapidly in a short time, the actual investment growth is slowing in 2009. The combination of these two aspects that make the company's investment ability can't meet the annual investment demand.

IV. CONCLUSION

This paper uses the theory of financial management, considers the constraints of the target profit and asset liability ratio, and builds investment capacity calculation model for power grid enterprises, which displays the quantitative approaches of the investment capability more intuitively. The case study shows that the model is applicable, the calculation is simple, and it can objectively reflect the power grid company's investment structure and investment ability. The proposed model in this paper provides the decision basis for the grid enterprises to make scientific plans, and it has huge value of application and popularization.

REFERENCE

[1] Zhao Huiru, Fu Liwen. Quantitative Analysis of Investment [J]. Water Resources and Power, 2012, 04:191-194.

[2] Wei Zijie. The calculation method research and software implementation of the power grid investment based on Grey Theory [D]. University of Electronic Science and Technology of China, 2013.

[3] Dai Wenbo. Design and Implementation of the Investment Ability Analysis System for Power Grid Enterprises [D]. North China Electric Power University, 2012.

[4] Zhang Xueqiang. The application of power grid investment capability measurement model in the county power supply enterprises [J]. China Business, 2013, 19: 31.

[5] Wang Qiang. Analysis on the investment capacity of media listing corporation [J]. Money China, 2014, 03: 104-105.

[6] Cao Yang. Study on the economic evaluation of electric grid invests [D]. Zhejiang University, 2005

[7] Luo Guo-liang, Zhang Xin-ying. Research and Application on the Evaluation of the Investment Benefit of Grid Enterprise [J]. East China Electric Power, 2010, 38(11): 1655-1658

[8] State Grid Corporation of China. Urban power grid construction standards and Policy Research Report material assembly [R]. Beijing: State Grid Corporation of China, 2006.

[9] Liu Chang, Gao Tie-mei. Characteristics of Electricity Industry Cycle Fluctuation and Influential Factors of Electricity Demand Based on Business Analysis and the Error Correction Model [J]. Resources Science, 2011, 01: 169-177.

[10] Xia Hua-li, Ye Jin-shu. Grid Investment Fluctuation Factors and Growth Effect [C]. Excellent papers of power project cost management. 2011:9.

[11] Li Jun. Analysis of the factors of power investment risk based on the interpretative structural model [J]. China Market, 2011, 09: 31-32+34.

[12] Bian Rong. Reason Analysis and Countermeasures of Ice Disaster in Zhejiang Power Grid [D]. North China Electric Power University, 2010.

[13] Zhu Tian-hao, Gu Jun-qiang, Bian Rong, etc. Analysis and report of Zhejiang power grid ice graph partition [R]. Hangzhou: State Grid Zhejiang Electric Power Company, 2008

[14] Pan Bin, Zhang Yong-jun, Huang Hui. Modeling for Multi-Factor Risk Assessment on Ice Disaster of Power System [J]. Power System Technology, 2012, 05: 102-106.

[15] Chai Xu-jing, Liang Xi-dong, Zeng Rong. An Improved Calculation Method for Power-Transmitting Capability Curve of AC Transmission Line [J]. Power System Technology, 2005, 24: 20-24.

Research on Correlation Measurement of Combined Inventory Financing Price Risk

Yun-ke Sun[1,2,*], Da-cheng Guo[3]

1School of Management and Economics, Beijing Institute of Technology, Beijing, China
2Development and Planning Office, Beijing Municipal Commission of Education, Beijing, China
3Institute of Education, Beijing Institute of Technology, Beijing, China

(sunyk_qd@163.com)

Abstracts - **In order to measure the time-variant correlation of the combined inventory financing price risks, the improved time-varying Copula model is introduced, especially discussing the dynamic evolution process of the time-varying correlation Copula model parameters. The empirical analysis results show that the effectiveness of time-varying SJC Copula model is better than the time-varying normal Copula function when describing the asymmetrical correlation and tail correlation.**

Keywords - **Combined Inventory Financing, Correlation Measurement, Time-varying SJC Copula model**

I. INTRODUCTION

For enterprise that owns several kinds of inventory, combined inventory financing means to use one or several kinds of inventory to finance. By means of combined inventory financing, banks are able to unified manage several kinds of inventory of the enterprise, disperse risk by analyzing collateral prices changes, and provide more reasonable pledge rate to the financing enterprises [1].

As to the pledge price, the actual research is more concerned about price change or yield of the pledge [2]. There are several reasons. Firstly, price is an absolute value, it can't provide the practical concern of assets essence -- the value and investment opportunity information. Secondly, from a statistical point of view, due to the instability of the price, the variance of the price will increase with time, which means the variance is infinite. Some properties of the price series, such as non stationary, make the statistical modeling more complex, difficult. But the yield series are stationary random sequence with mean reversion, they have unconditional constant mean and limited variance, therefore they have some statistical properties, such as stationary, ergodicity and easy to model.

The logarithm yield rate is usually used to indicate yield rate. Yield rate $r_t = \ln(1 + R_t) = \ln\left(\frac{P_t}{P_{t-1}}\right) = \ln P_t - \ln P_{t-1}$, in which P_t is the price at time t. Logarithmic rate of return has good statistical properties, such as the calculation of multi period yield just need to add the yield rates of each period .The theoretical scope of logarithmic yield rate is $(-\infty, +\infty)$.

Price risk of combined inventory financing is much more complex than single inventory financing;

correlation analysis is the core of price risk measurement of combined inventory financing. The common correlation analysis methods are: Granger causality analysis method, the linear correlation coefficient analysis [3-4]. But they have some limitations [5]. The Copula function is a kind of multivariate and nonlinear correlation statistical analysis tool, which can be used to analyze the nonlinear and non symmetrical relationship between variables, commonly used in correlation analysis of asset combination.

II. DEFINITION AND PROPERTY OF COPULA FUNCTION

As correlation analysis and multivariate statistical analysis tools, Copula function is proposed by Sklar in 1959 [6], he pointed out that a multivariate joint distribution can be decomposed into several one variable marginal distribution and a Copula function [7-8] which shows the correlation structure between variables. Because Copula function can connect marginal distribution and joint distribution of the variables together, it's also called "contiguous function" or "dependent function". In 1990s, Copula function is used in the economic field, and has got lots of attention and promotion [9]. The definition and properties of the Copula function are introduced as behind.

$C(u_1, u_2, \cdots, u_n)$ is defined as a n-variable joint distribution function, the marginal distribution function u_i of $C(u_1, u_2, \cdots, u_n)$ is a uniform distribution function in the interval $[0,1]$, then $C(u_1, u_2, \cdots, u_n)$ is named as n-variables Copula function, it can be indicated as:

$$C(u_1, u_2, \cdots, u_n) = P(x_1 \le u_1, x_2 \le u_2, \cdots, x_n \le u_n) \quad (1)$$

According to the definition, the basic properties of n-variable Copula function are as followed [10]:

(1) $0 \le C(u_1, u_2, \cdots, u_n) \le 1$

(2) n-variable Copula function is bounded in n-dimensional space, which means for any variable u_i, on the condition of other variables unchanged, Copula function value will increase or constant along with the variable u_i increasing;

(3) $\forall u_i \in [0,1]$, the marginal distribution

E. Qi et al. (eds.), *Proceedings of the 21st International Conference on Industrial Engineering and Engineering Management 2014*, Proceedings of the International Conference on Industrial Engineering and Engineering Management, DOI 10.2991/978-94-6239-102-4_124, © Atlantis Press and the authors 2015

function u_i of $C(u_1, u_2, \cdots, u_n)$ meet the condition of $C(1, 1, \cdots, u_n, 1, \cdots, 1) = u_n$, $i = 1, 2, \cdots, n$.

In addition, Copula function has many useful properties, such as uniform continuity, boundedness. The boundary of Copula function is also known as the Frechet Hoeffding boundary, the expression is:

$$\max\left(\sum_{i=1}^{n} u_i - d + 1, 0\right) \le C(u_1, u_2, \cdots, u_n) \le \min(u_1, u_2, \cdots, u_n) \quad (2)$$

According to Sklar theorem, Copula function has the character of existence and uniqueness [6].

If the marginal distribution $F_1(x_1), F_2(x_2), \cdots, F_n(x_n)$ is continuous, then the form of the copula function is unique. Sklar theorem proves that, for a joint distribution function $F(x_1, x_2, \cdots, x_n)$, its marginal distribution function and correlation structure can be considered separately. The marginal distribution functions $F_1(x_1), F_2(x_2), \cdots, F_n(x_n)$ contain the information of each variable's marginal distribution, while $C(u_1, u_2, \cdots, u_n)$ contains the correlation structure of the random variables.

III. CONSTRUCTION OF THE TIME-VARYING COPULA FUNCTION

The correlation between combined stock returns will change according to the internal and external market factors changes, namely the degree of correlation between yields or correlation structures is dynamic, the assumption of Copula function parameter constant is not consistent with the actual. Therefore consider the introduction of dynamic Copula model. Time-varying correlation or correlation pattern, reflected in the research of dynamic Copula, is mainly the following two categories: time varying Copula model and variable structure Copula model, this article focuses on the characteristics of time-varying parameters, study the dynamic changes of yields correlation of combined inventory by the study on time varying Copula model. As the pattern of this model won't change, only the parameters of Copula function is time-varying, we needn't consider the variable edge distribution modeling, just focus on the study of Copula function parameter's time-varying characteristics.

A. The evolution equation for time-varying parameters

When the correlation parameter of Copula function is constant, this kind of Copula function is called static or normal Copula, when the correlation parameter is time-varying; it's called time-varying Copula function. Therefore the dynamic evolution of the correlation parameter of Copula function is the key to construct the time-varying Copula model. The method to establish correlation parameter model that changes with time is similar to ARCH model.

Suppose the correlation parameter is ρ, the evolution equation of the parameters is:

$$\rho_t = f(\omega, t) \quad (3)$$

Formula 3 describes the process of correlation parameters changing with time. ω is a set of the correlation parameter fluctuation factor, and t represents time. The basic idea of the parameter evolution equation is estimating parameter value in next period according to the known information.

From the above analysis, the dynamical evolution of equation parameters can also be established according to the one to one corresponding relationship between the Copula function parameters and the tail dependence coefficient or correlation measurement.

The first person studying on the time-varying Copula model is Patton [11], he uses an equation similar to ARMA (1, 10) to describe the parameters of two variables normal Copula function, namely the dynamic evolution process of the Pearson correlation coefficient. Based on Patton's research, Goorbergh, Genest and Werker [12] put forward that the parameters evolution process of the Copula function depends on by the evolution process of the rank correlation coefficient τ:

$$\tau = \gamma_0 + \gamma_1 \ln \max(h_{1t}, h_{2t}) \quad (4)$$

In which h_{1t} and h_{2t} are used to indicate the yield rate fluctuation of two kinds of asset at the moment t, the evolution equation describes the asset correlation changes at different levels. $\gamma_1 > 0$ reflects that any market large fluctuation will promote the correlation of the assets.

From above, the evolution equation of Copula parameters can be determined according to the correlation measure and the one to one correspondence relationship between the correlation measure and the Copula function parameters.

B. Time varying normal Copula function

When studying the general characteristic of time series, it usually assumes that correlation coefficient is constant for describing sequences correlation. According to the definition of linear correlation coefficient and the character of normal Copula function, the static correlation coefficient ρ is actually a Pearson linear correlation coefficient of variables $\Phi^{-1}(u)$ and $\Phi^{-1}(v)$. Suppose ρ is variable correlation coefficient. Patton proposed a process similar to ARMA (1, 10) to describe the dynamic evolution process of normal Copula function correlation coefficient ρ; this paper carries on the expansion [13]:

$$\rho_t = \bar{\Lambda}\left(\omega_\rho + \beta_\rho \rho_{t-1} + \alpha_\rho \times \frac{1}{q}\sum_{i=1}^{q}\Phi^{-1}(u_{t-i})\Phi^{-1}(v_{t-i})\right) \quad (5)$$

In which $\Phi^{-1}(\cdot)$ is the inverse function of standard normal distribution function, function

$\bar{\Lambda}(x) = \frac{1-e^{-x}}{1+e^{-x}}$ guarantee ρ_t within the interval $[-1,1]$, u_t and v_t are sequence obtained by the probability integral transform of the original time series. The lag correlation coefficient ρ_{t-1} in Formula 5 can be used to describe the correlation coefficient changes,

The mean value changes of the product of $\Phi^{-1}(u_{t-i})$ and $\Phi^{-1}(v_{t-i})$ with lagging order q can reflect the correlation coefficient: when the product of $\Phi^{-1}(u_{t-i})$ and $\Phi^{-1}(v_{t-i})$ are positive, it can illustrate that $\Phi^{-1}(u_{t-i})$ has positive correlation with $\Phi^{-1}(v_{t-i})$. When the product of $\Phi^{-1}(u_{t-i})$ and $\Phi^{-1}(v_{t-i})$ are negative, it can illustrate that $\Phi^{-1}(u_{t-i})$ has negative correlation with $\Phi^{-1}(v_{t-i})$.

C. the time-varying SJC Copula

Normal Copula function can't describe the characteristics of variables' tail correlation and symmetry correlation, and the Joe-Clayton Copula function is a good solution to this problem, the distribution function is [14]:

$$C^{JC}(u,v;\kappa,\gamma) = 1 - \left(\left\{ \left[1-(1-u)^\kappa\right]^{-\gamma} + \left[1-(1-v)^\kappa\right]^{-\gamma} - 1 \right\}^{-1/\gamma} \right)^{1/\kappa} \quad (6)$$

In which, the correlation coefficient $\gamma > 0$, coefficient $\kappa \geq 1$.

The parameters and tail correlation coefficient of Joe-Clayton Copula exist one to one correspondence relation:

$$\tau^U = 2 - 2^{1/\kappa} \quad (7)$$
$$\tau^L = 2^{-1/\gamma} \quad (8)$$

In which, τ^U is upper tail correlation coefficient, τ^L is lower tail correlation coefficient. From formula 7 and 8, it shows the upper tail correlation coefficient is completely determined by the parameter κ, and the lower tail coefficient is completely determined by the parameters γ.

A major drawback of Joe-Clayton Copula function is that when the two tail correlation metric is equal, the simplified function form set of the model cause that Joe – Clayton Copula function still exists a certain degree of asymmetry [15]. A more reasonable model is to determine the non symmetry by measure the tail correlation. In order to solve the problem that Joe-Clayton Copula functions measuring the tail dependence, Patton modified the Joe - Clayton Copula model, and put forward the SJC (Symmetrized Joe-Clayton, SJC) Copula model, its form is as following:

$$C^{SJC}(u_t,v_t|\tau^U,\tau^L) = 0.5(C^{JC}(u_t,v_t|\tau^U,\tau^L) + C^{JC}(1-u_t,1-v_t|\tau^U,\tau^L) + u_t + v_t - 1) \quad (9)$$

In which u_t and v_t are sequence obtained by the probability integral transform of the original sequence, τ^U is upper tail correlation coefficient, τ^L is lower tail correlation coefficient. The parameters and the tail correlation coefficient of the time-varying SJC Copula function have one to one responding relation, and can describe the asymmetric correlations, the tail correlation of the variables and the dependence at the time of dramatic market volatility.

The parameters and the tail correlation coefficient of the time-varying SJC Copula function have one to one responding relation:

$$\tau^U = 2 - 2^{1/\kappa} \quad (10)$$
$$\tau^L = 2^{-1/\gamma} \quad (11)$$

In which, τ^U is upper tail correlation coefficient, τ^L is lower tail correlation coefficient. From formula 10 and 11, it shows the upper tail correlation coefficient is completely determined by the parameter κ, and the lower tail coefficient is completely determined by the parameters γ.

Based on the one to one correspondence relation between the dependence parameter and the tail dependence parameter, as long as the evolutionary process of the tail dependence coefficient is determined, the dynamic evolution process of the SJC Copula function's dependence parameters can be determined. According to the time-varying characteristic of tail dependence coefficient, Patton defined the time-varying equation of SJC Copula function's tail dependence coefficient.

The dynamic evolution process of the SJC Copula function's tail dependence coefficient is:

$$\tau_t^U = \Lambda\left(\omega_U + \beta_U \tau_{t-1}^U + \alpha_U \times \frac{1}{q}\sum_{i=1}^{q}|u_{t-i} - v_{t-i}|\right) \quad (12)$$

$$\tau_t^L = \Lambda\left(\omega_L + \beta_L \tau_{t-1}^L + \alpha_L \times \frac{1}{q}\sum_{i=1}^{q}|u_{t-i} - v_{t-i}|\right) \quad (13)$$

In which, u_t and v_t are sequence obtained by the probability integral transform of the original sequence. Function $\Lambda(\cdot) = \frac{1}{1+e^{-x}}$ is logistic transform function; its function is to guarantee that the tail dependence coefficient is within interval $[0,1]$ at anytime. Formula 12 and 13 simulate a ARMA $(1,q)$ model, the right of the model contains a exogenous variables and autoregressive terms $\beta_U \tau_{t-1}^U$ or $\beta_L \tau_{t-1}^L$.

As The tail correlation coefficient of SJC Copula function corresponds to two parameters of the Copula function one-to-one, the parameter of Copula function can be calculated by calculating the tail dependence coefficient:

$$\kappa_t = \kappa(\tau_t^U) = \left[\log_2(2-\tau_t^U)\right]^{-1} \quad (14)$$

$$\gamma_t = \kappa(\tau_t^L) = \left[\log_2(\tau_t^L)\right]^{-1} \qquad (15)$$

Therefore, as long as the evolution process of the tail dependence coefficient is determined, the dynamic evolution process of Copula parameters is determined as well.

IV. CONCLUSION

This paper mainly studies the price risk measurement of correlation combination inventory financing; the correlation analysis is one of the core problems in in measuring price risk of the combined inventory financing. Copula function is a kind of multivariate statistical analysis and nonlinear correlation analysis tool; it can be used to analyze the nonlinear and non symmetrical relationship of variables, commonly in the correlation analysis of combination asset. This paper introduced time-varying Copula model to describe time-varying characteristic of inventory portfolio yield dependence, focusing on the analysis of the time varying SJC Copula models and the time varying normal Copula model, and discusses the dynamic evolution process of time-varying Copula model's dependence parameters. Case study shows that the time-varying correlation Copula plays better performance in characterizing the yield sequences correlation relationship than static Copula model.

REFERENCE

[1] Burman R W, Practical Aspects of Inventory and Receivables Financing (Periodical style). Law and Contemporary Problems, 1948, 13 (4): 555-565.

[2] Biederman D, Logistics financiers (Periodical style). The Journal of Commerce, 2004, 4: 40-42.

[3] Zhang Xiaoting, Copula technology and financial risk analysis (Periodical style). Statistical study, 2002, 4: 48-51.

[4] Wang Beibei, Research on Inventory Financing Price Risk Measurement and Control (Dissertation style). Beijing Institute of Technology, 2012

[5] Li Jing, Copula Function and Its Application in the Dependence Analysis between Stock Markets (Dissertation style). HUST, 2007.

[6] Sklar A, Fonctions de répartition à n dimensions et leurs marges (Book style). Publications de l'Institut de Statistique de L'Université de Paris, 1959.

[7] Roger B Nelsen, An introduction to copulas (Book style). New York: Springer, 2006.

[8] Corsten D, Lenz M, Klose M, Logistics services providers and information-based logistics services: an explanatory study (Periodical style). Logistic Management, 2002, 4(1): 45-50.

[9] Li Jing, Copula theory and its application in finance market (Dissertation style). Huazhong University of Science and Technology (Dissertation style), 2007

[10] Wei Yanhua, Zhang Shiying, Copula Theory and Its application in Financial Analysis (Book style). Beijing; Tsinghua University Press, 2008.

[11] Andrew J Patton. Modelling Time-Varying Exchange Rate Dependence using the Conditional Copula (Thesis or Dissertation Style). University of California at San Diego, Economics Working Paper Series with number 542493, 2001.

[12] Goorbergh R W J, Genest C, Werker B J M, Multivariate Option Pricing Using Dynamic Copula Models (Thesis or Dissertation style). Tilburg University, Center for Economic Research Discussion Paper, No. 2003-122, 2003.

[13] Ralph dos Santos Silva, Hedibert Freitas Lopes, Copula, marginal distributions and model selection: a Bayesian note (Periodical style). Statistics and Computing. 2008, (3):50-54.

[14] Harry Joe, Multivariate models and dependence concepts (Book style). London: Chapman& Hall, 1997.

[15] Frey R, MeNeil A J, Copula and credit models (Book style). The Risk Metrics. 2001.

Research on FDI and China's Investment Promotion Polices in the Post Financial Crisis Era

Yun-ke Sun[1,2,*], Da-cheng Guo[3]

[1]School of Management and Economics, Beijing Institute of Technology, Beijing, China
[2]Development and Planning Office, Beijing Municipal Commission of Education, Beijing, China
[3]Institute of Education, Beijing Institute of Technology, Beijing, China
(sunyk_qd@163.com)

Abstract - **The outbreak and spread of International financial crisis has caused great effect on the global foreign direct investment. As an important foreign direct investment destination, China faces severe challenges, how to introduce and use foreign capital, changing China's economic development mode is an important subject for the government. From the perspective of investment fields, investment location selection, competitive strategy and knowledge spillover, this paper analyses the changes and influence of foreign direct investment, combining the analysis, discusses our investment promotion polices after the financial crisis.**

Keywords - **FDI, investment promotion polices, international financial crisis**

I. INTRODUCTION

Foreign direct investment (FDI) has made an important contribution on economic and social development since China's reform and opening-up. By the end of 2008, China's actual use of FDI reached $852.61billion, the eastern coastal areas become the focus of FDI, many regional headquarters or R&D center of Multi-National Corporation appeared in Beijing, Shanghai, Jiangsu, Guangdong and other area. Since the American subprime mortgage crisis, the global economy is facing severe challenges. The economic crisis has spread from the financial sector to the real economy. Multi-National Corporation's global business caused great difficulties. Grasp the new trend of foreign direct investment, formulate reasonable investment policies, changes the present situation of foreign focus only on low-end processing trade in Chinese. Attract more Multi-National Corporation to set up R&D institutions in China. It plays an important role for China to improve the quality of foreign capital utilization, promote industrial transformation and upgrading. China government proposed, utilization of foreign capital to optimize structure, broaden the channels, improve quality, perfect soft environment for investment, protecting the legitimate rights and interests of investors. Strengthening the work of introduce intelligence, talent and technology, encourage foreign enterprises to set up R&D center in China, promote the innovation of science and technology by drawing lessons from international advanced idea, system and experience. In the context of main countries in the world gradually out of the economic crisis, to explore the changes of FDI and change the concept of investment, has vital significance for China's development in economic and society.

II. INTRODUCTION OF RELATED RESEARCH

Sollow (1957) in the study of creative economic growth theory, an important part of economic growth "input" was decomposed of technology, capital, labor, FDI etc [1]. Romer (1990) think, FDI accelerate economic growth by strengthening of the role of human capital [2]. Grossman and Helpman (1991) stressed that FDI enhanced the competition and promoting innovation, which leads to the improvement of scientific and technological, and improve productivity, then achieve long-term economic growth target [3]. Borensztien (1998) studied the role of FDI drive the economic growth based on increasing product categorics, and believes that the foreign capital was more efficient than domestic capital [4]. Reis (2001) proposed foreign direct investment model of knowledge spillover, the foreign R&D activities can produce effect in two aspects of positive and negative. If the world interest rates higher than the domestic interest rate, foreign R&D activities will have a negative impact on the economy, and if the world interest rates lower than domestic interest rates, foreign R&D activities will have a positive impact on economic growth [5].

Specific to the impact of foreign direct investment on China, scholars study the role of China's economic growth and technology progress played by FID from the region, industry, enterprises and other aspects. He Jie (2000), Pan Wenqing (2003) and Tu Taotao (2008) think that spillover effects of FDI on China's industrial sector should be effect by the threshold level of local economic. Simply raise economic openness of an area is meaningless, even with a negative value. The positive spillover effect must be based on the level of economic development, infrastructure, improving their technical level and expanding market scale [6-8]. Jiang Dian-chun (2004), Zhang Yu Zhang Cheng (2011) built two stages game model to analyze the influence on the R&D capability of China enterprise from Multi-National Corporation, think that if domestic enterprises and Multi-National Corporation can be closely linked, China can realize technological progress through the acquisition of knowledge spillovers [9-10]. Zhong Chang-biao (2010), Guo Feng, Hu Jun (2013) did empirical research by using the data of China provinces, arrival at a conclusion that FDI not only regional spillover, but also indirectly led to the improvement of productivity in other areas [11-12]. Sheng Lei (2010), Li Wuwei, Cao Yong (2012), Zhang Zhengang, Hu Qiling (2012) studied foreign R&D activities had influence on independent innovation ability of Chinese domestic enterprises. The study show, foreign R&D activities enhance the capability of independent innovation of enterprises China. But foreign capital in different industry is in different degree, the spillover effects of foreign R&D have great difference [13-15].

From the above research, the influence of FDI on the host country mainly in promoting economic growth, improving the technology level and promoting expand market scale. But there are also many negative effects, for example, compel host country enterprises out of the market by technical monopoly and squeezing market. There is a big risk of excessive dependence on foreign direct investment on the host country's economy. But in the economic fluctuation period, great changes have taken place in the Multi-National Corporation management behavior, may have a negative impact on the host country. How to grasp the direction of foreign investment after the crisis should be paid attention. Therefore, it's needed to strengthen the research on FDI under the influence of economic crisis, and have important practical significance for the host country to avoid market risks and improve technology level.

III. IMPACT OF THE INTERNATIONAL FINANCIAL CRISIS ON FDI IN CHINA

Under the influence of international financial crisis, in order to reduce the cost and increase new product development efforts, many Multi-National Corporation adjust development strategies, to adjust and optimize the business developing country. Specific to China, the foreign direct investment amounted to China little decrease. But there are great changes in some areas, such as the FDI industry and location selection, knowledge spillover effect of FDI, the competition strategy of Multi-National Corporation and domestic enterprises.

A. The areas of foreign investment is growing, emerging industries and high-end service industry has become a new investment hot spot.

In the international financial crisis, China's FDI industry is changing. Especially a lot of restricted areas will be open to foreign capital, promote the expansion of foreign direct investment scope. At present, foreign direct investment in China generally focused on electronic information, communication, medical, automotive and other industries, emphasis on manufacturing industry. Under the influence of the financial crisis, China is faced with the challenges of industrial structure upgrading, some of the emerging industry has been rapid development in the country's strong support, such as environmental protection, new

energy, new material, bio medicine and other fields will become the new hot spot of FDI. On the other hand, in order to reduce the unemployment rate, the western countries put forward regression and manufacturing reengineering idea, transform and upgrade the traditional manufacturing industry. It provides opportunities for China to carry out R&D, marketing, logistics outsourcing in service industry. Many Multi-National Corporation and financial institutions want to get rid of the financial crisis as soon as possible, will accelerate the reorganization and integration, improve the core competitiveness, the more business includes part of the core business outsourcing to other countries and regions. This has provided an important opportunity for China to use foreign capital to develop modern service industry.

B. The unbalanced of regional eased, FDI transfer to the central and western trend obviously.

For a long time, China eastern coastal region has a large number of FDI, while the western region to attract foreign direct investment less, FDI was in very unbalanced state. With the coastal area prices of labor cost and resource rising, the investment situation at present mainly focused on foreign direct investment of East will change, to invest in the central and Western regions. The Multi-National Corporation in the western region of Chongqing, Chengdu, and Xi'an set up many branches. The regional layout optimization and balanced development of foreign investment in China is conducive to China's regional coordinated development.

C. Foreign enterprises and domestic funded enterprises cooperation is showing a variety of forms and increasing R&D cooperation behavior

Under the influence of the financial crisis, foreign investment funds seriously. As far as possible, use of resources, foreign enterprises and domestic enterprises in the same industry will expand cooperation. Foreign enterprises will put some monopoly technology sharing and domestic enterprises, promote the upgrading and transformation of their own. Foreign companies carry out comprehensive cooperation with domestic enterprises from product design, manufacturing, marketing and logistics, in order to reduce operating costs and enhance market competition ability. In order to maintain the foreign enterprises in the key of technology monopoly, foreign firms will tend to develop vertical cooperation with enterprises in the industry chain, to enhance the whole value chain through sharing the cost of research and development. Cooperative R & D will be a new competition strategy taken by foreign enterprises in this situation, provides an important opportunity for domestic enterprises to acquire advanced technology through cooperative R & D.

D. Foreign direct investment makes knowledge spillover effects in China increased

Foreign direct investment is the main channel of knowledge spillovers. As FDI adjust the scale of investment in China, the conduction effect will be further enhanced. Under the influence of international financial crisis, foreign enterprises will use more advanced technology to enhance their competition ability of the production, and get more market share. So, more advanced products will be produced in China and knowledge spillover effect will be more obvious. And this process will be accompanied by knowledge spillover effect. In addition, to enhance the innovation capability of Chinese enterprises, to resist the impact of the economic crisis, the Chinese government has increased the R&D investment and subsidies, especially the construction of public innovation platform to enhance the absorptive capacity of enterprises in our country, narrowing the enterprise technical gap between China and developed countries, which will further enhance FDI spillover effect on China's enterprises.

IV. POLICY RECOMMENDATIONS

With the increasing pressure in adjustment of industry structure in China, China's investment policy will change from focusing on quantity to quality. Foreign preferential policies also are adjusted, and the unified tax rate for domestic and foreign enterprises is the obvious sign. We need to change the way of working, changing from giving foreign preferential measures to giving foreign better service environment. Give FDI positive guide, and guiding FDI play a greater role in China's economic restructuring and independent innovation.

A. Actively guide the Multi-National Corporation set up R&D centers in China, and encourage foreign investment in emerging industries.

For a long time, we focus on the growth amount of attracting foreign investment, while ignoring technology requirements of FDI. In the context of building innovation oriented country, it's needed to actively guide foreign investors to set up R&D institutions in china. The foreign research and development institutions incorporate regional innovation system in our country, and give appropriate subsidies and preferential. The state should attract foreign investment in R & D in the emerging field of new energy, new materials, bio pharmaceutical etc.. The government should guide foreign businesses to play a greater role in the emerging field, and promote the demonstration effect and the competition effect for the development and innovation of enterprises.

B. To create a good policy environment, promote balanced development of foreign direct investment in China's Regional.

Since the reform and opening, with many preferential policies and regional advantages, in the eastern coastal area is larger proportion to attract foreign direct investment. The eastern and western regions of the FDI showed uneven distribution situation. With the deepening of the western development, the western region will get more state support. Guide foreign direct investment in China's western region, to promote development of the West and plays an important role in the balanced development between the East and west. The western region should exert the advantage of resource and policy, do the preliminary research and industrial development planning, to guide more FDI to undertake the transfer of industries, to develop new energy, agricultural products processing, equipment manufacturing and other industries. Continuously improve the level of economic development of the western region.

C. Establish encouragement mechanism for innovations, promote the development of cooperation between domestic and foreign enterprises.

Under the influence of the financial crisis, foreign enterprises and domestic enterprises competition strategy are changing. Cooperation will replace the competition become an important relationship between enterprises. Government should give domestic enterprises appropriate subsidies, strengthen enforcement of intellectual property rights, encourage domestic enterprises and Multi-National Corporation to carry out cooperation in research and development, strengthen the construction of innovation platform, enhance the ability of independent innovation of domestic enterprises, narrow the technology gap with the advanced enterprises, enhance the absorption capacity of the domestic enterprises. Support the domestic and foreign enterprises in the field of key technology research and development cooperation, to promote the knowledge spillover between domestic enterprises and Multi-National Corporation.

D. Innovation investment, the foreign investment target and direction should be guided by market

From essentially said that investment is a commercial act, which belongs to the field of competition. The government should gradually fade out from investment, to the enterprise or other market main body. In accordance with the business plan and the laws of the market, the government should clear the role position in the investment process. With the diversification of foreign investment, we should adopt various promotion means, pay attention to merger investment, private equity fund equity investment, and overseas listed public investment in new investment ideas. Encourage enterprises to become the main body of investment.

E. Perfect the mechanism of foreign investment withdrawal and exit, reasonable promoting investment promotion work.

Since the outbreak of the financial crisis, the eastern region of China has a series of foreign retreating tide. Governments at all levels should strengthen warning and supervision of foreign investment, and actively help enterprises solve the difficulties and problems. For foreign-invested enterprises, give the necessary credit support with loan principle. Strengthening the legal construction of foreign economy, improve the utilization of foreign investment laws and regulations system. To guarantee the foreign investment enterprise management autonomy, safeguard the legitimate rights and interests of investors are not violated, protect the legitimate rights and interests of workers. Establish FDI exit mechanism, guide FDI improvement from quantity to quality.

F. Learn the advanced experience of developed countries, and encourage domestic enterprises to carry out overseas investment business.

Government should encourage domestic enterprises and foreign capital enterprises to strengthen cooperation actively. Play more attention to foreign strategic of assets and emerging industry. Learning the experience of foreign business in China, and encourage excellent domestic enterprises participation in foreign enterprise restructuring by purchasing, merger, shares and other forms. Seek the foreign advanced technology by reverse technology spillover effect. Absorption of foreign advanced technology, to meet the needs of China's current of lacking innovative ability, continuously improve the opening up level of China's economy and dealing with global challenges.

V. CONCLUSION

In the post crisis era, the investment is still playing an important role for China's regional economic growth. The government should change the ideas, to create a good policy environments, guide the direction of foreign investment to emerging industries with has broad prospects for development. The government should try to improve the technology level of domestic enterprises, encourage cooperation between domestic enterprises and foreign enterprises. Enhanced absorption of foreign advanced technology and management experience, and improve the quality and level of utilizing foreign capital in China. Through these efforts, FDI will play a greater role for the steady and rapid growth of China's economy and the adjustment of industrial structure.

REFERENCE

[1]   Solow R M, "Technical change and the Aggregate Production Function [J]", The Reviews of Economics and Statistics, Vol. 39, No. 3, pp. 312-320, Aug, 1957

[2]   Romer. Paul M. Increasing, Returns and Long-run Growth [J]. Journal of Politic Economy, Vol. 94, No. 5, pp. 1002-1037, Oct., 1986,

[3]   Grossman, Gene M and Elhanan Helpman, "Innovation and Growth in the Global Economy [M]," Cambridge: MIT Press, 1991.

[4]   Borensztien, E.,de Gregorio, J. and Lee, J.W. "How does foreign direct investment affect economic growth [J]," Journal of International Economics, Vol. 45, No. 1, pp. 115-135, Oct., 1998

[5]   Ana Balcao Reis, "On the welfare effects of foreign investment [J]," Journal of International Economics, Vol. 54, No. 2, pp. 411-427, Aug., 2001

[6]   He Jie, "Further quantify the spillover effect of foreign direct investment on the industrial sector Chinese [J]," The Journal of World Economy, Vol. 23, No. 12, pp. 29-36, Aug., 2000

[7]   P an Wen-qing, "The Spill-over Effects of FDI on China's Industrial Sectors: A Panel Data Analysis [J]" The Journal of World Economy, No. 6, pp. 3-7, Jun., 2003

[8]   Tu Tao-tao, "Spillover Effect of FDI in China Industrial Sector: An Analysis based on the quantile regression approach [J]," World Economy Study, No. 12, pp. 56-60, Dec., 2008

[9]   Jiang Dian-chun, "Influence of Multi-National Corporation on R&D ability of enterprises of our country: a model analysis [J]," Nankai Economic Studies, No. 4, pp. 56-60, Apr., 2004

[10] Zhang Yu, "Influence of transnational enterprise R&D activities on business innovation in China -An Empirical Study Based on manufacturing industry in China [J]," Journal of Financial Research, Vol. 377, No. 11, pp. 139-152, Aug., 2011

[11] Zhong Chang-biao, "Empirical Evidence on the Regional Spillover Effects of FDI in China [J]," Economic Research Journal, No. 1, pp. 80-89, Jan., 2001

[12] Guo Feng, Hu Jun, Hong Zhan-qing, "Study on spatial spillover effects of FDI and trade import [J]," Journal of International Trade, No. 11, pp. 125-135, Nov., 2001

[13] Sheng Lei, Does foreign R&D facilitate or dampen indigenous innovation? [J], Studies In Science Of Science, Vol. 28, No. 10, pp. 1571-1581, Oct., 2010

[14] LI Wu-wei, Cao Yong, "Dynamic Effects and Regional Difference of Foreign R&D on the Product Innovation Performance of China's Indigenous Enterprises [J]," Science Of Science And Management Of S.&T., Vol. 33, No. 9, pp. 21-27, Sep., 2012

[15] Zhang Zhen-gang, Hu Qi-ling, "Relationship between Foreign R & D and Innovation Capability of Local High-tech Enterprise: From Perspective of Technology Spillover [J]," Technology Economics, Vol. 31, No. 8, pp. 26-32, Aug., 2012

Study on Enterprise Comprehensive Budget Management System Dynamic Framework Design Based on TRIZ and BSC

Haicao Song[1,2,*], Shuping Yi[1]

[1]College of Mechanical Engineering, Chongqing University, Chongqing, China
[2]College of Mechanical and Electrical Engineering, Shihezi University, Shihezi, China
(songhaicao@sina.com)

Abstract - **Comprehensive budget management framework design is the foundation of enterprise management system construction. In order to make the comprehensive budget management be effectively coupled with business strategies, the contradictions of the comprehensive budget management system planning processes must be identified and resolved based on TRIZ theory. The company strategic targets are decomposed to CSF (critical success factors) and KPI (key performance indicators) targets based on BSC methods. So in this paper, the problem of enterprise comprehensive budget management system dynamic framework design is solved based on TRIZ, BSC, strategic principle of consistency, a case study on S company is researched.**

Keywords - **BSC, comprehensive budget management, strategic targets, TRIZ**

I. INTRODUCTION

Comprehensive budget management is a set of scientific indicators management control system, which is based on the company's business objectives, strategic targets, developing targets, layers of decomposition, budget control, coordination, and assessment [1]. The theorists and industries have increasingly and extensively pay attention to the comprehensive budget management. Scholars have conducted pioneering research for different forms of ownership enterprise budget management operating mode and selection, budget management system framework [2].

Enterprise budget management has gone through different stages of development, Su and Gao summarize three stages of foreign budget management, which are infancy period, growing period, depth differentiation stage in their books [3][4].

(1) Infancy period. This is mainly demand-driven. Major budget models are included: A. Imposed Budget; B. Single Budge. Major budget models are included: 1) Operating Budget. 2) Financial Budge. 3) Capital Budget.

(2) Growing period. This is mainly based on demand-and- management driven theory. The main modes of the budget are included: A. Comprehensive Budget: This mode analyzes Comprehensive Budget controlling by systems theory. B. Participatory budget: It can fully mobilize the enthusiasm of all employees, playing their ability to self-control [5] [6]. C. Zero Based Budgeting.

(3) Depth differentiation stage. This is mainly based on management-driven theory and method. The major budget models include: A. Activity Based Budgeting (ABB). Cooper, Kaplan (1991, 1998) raised Activity Based Costing (ABC) and the new budget method ABB [7] [8]. B. Kaizen Budgeting. Tanaka (1993) proposed Target Costing and Kaizen Budgeting [9]; C. Strategic budgeting.

(4) The foreign budget management new development is Beyond Budget. This stressed company objectives of objectivity and relevance, and proposed reasonable performance appraisal, and closely integrated the dynamic process of budget management and business operations [10].

Domestic budget management researches include: Su put forward that enterprise budget management has four main functions: planning functions, coordination functions, control functions and performance appraisal functions. Miao argued that people should re-recognize effect of the enterprise budget management on a strategic management. Pan et al. held that comprehensive budget management is an important component of enterprise structure, and it is the third level of legal instruments after "Company Law", "Articles of Association". Gao viewed that comprehensive budget management is strategic security system in consistency with the enterprise development strategy. And it is an operation and management system fully integrated enterprise operation flow, capital flow, information flow and human resources flow, which lie in the core position of the internal control system. Yu et al. considered that budget is an effective method to integrate enterprise resource, and not just as a means of control costs [11]. In accordance with the product life cycle theory Wang [12] summarized four modes of budget management which are capital budget, sales budget, cost budget and cash budget.

According to the above literatures, the author considers that a variety of theories, methods and techniques spur budget management mode changes. But there still exist some questions. Budget management and evolutionary degree of business process technology systems are in want of assessment index system. Enterprise's strategic objectives and budget targets information are not completely symmetrical, performance targets and budget management module lack close contact. These lead to poor control and strong subjectivity. And there is no integral and systematic management in budget management research. So a dynamic framework of comprehensive budget management should be established systematically in company.

II. METHODOLOGY

A. TRIZ (the theory of inventive problem solving)

TRIZ has been promoted by several enthusiasts as a systematic methodology or toolkit which provides a logical approach to solve the problem of creativity for innovation and inventiveness. For this purpose, TRIZ offers a comprehensive set of tools to analyse and solve problems in different perspectives. The main contents include technology evolution theory, analysis, and conflict resolution principle. Its core is a technical evolution principle; the principle believes that the technical system has been an evolution, and conflict resolution is the driving force of evolution [13].

Technology system evolution process is divided into four stages by TRIZ: infancy, growth, maturity, recession (As in Fig.1) [14], enterprise's comprehensive budget management system is evolving technology systems TRIZ provides specific and scientific guidance to enterprise's comprehensive budget management. It includes system evolution mode, the law of evolution, evolutionary lines and application models. Comprehensive budget management process is evolution from simple to complex, from lower to higher, from a single system to a multi-system[15][16].

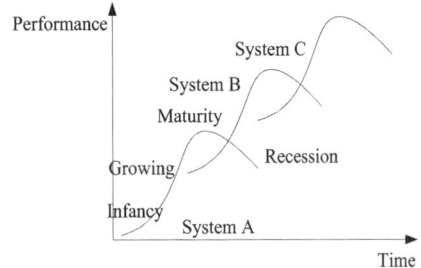

Fig.1. Evolutionary curve of Technology systems about comprehensive budget management

From Fig.1, if the system B is the next generation new system of the system A, system C is the next generation new system of the system B, in order to ensure that the sustainability of the system development, the new system should be put into use before recession of the previous one system at the time [17]. Enterprise comprehensive budget management is a dynamic system, and external and internal environment may impact on it constantly, so real-time monitoring, feedback and dynamic adjustment should be carried out in the implementation process.

B. BSC (Balanced Scorecard)

BSC is a new performance management system which implements organization's strategic as operational indicators and target value. BSC is designed to create "strategic guidance" performance management system, in order to ensure the effective execution of corporate strategy. BSC provides a framework for selecting multiple performance indicators related to strategic goals, Decomposed integrating traditional financial measures and non-financial measures four dimensions: customer, internal process, and learning and growth [18]. As in Fig.2.

In order to make the comprehensive budget management align effectively with business strategies, the contradictions of the comprehensive budget management system planning processes must be identified and resolved based on TRIZ theory. The company strategic targets were decomposed by CSF and KPI targets based on BSC methods. This study on enterprise comprehensive budget management system dynamic framework design is based

on TRIZ and BSC and strategic principle of consistency, a case study combined with S company is researched.

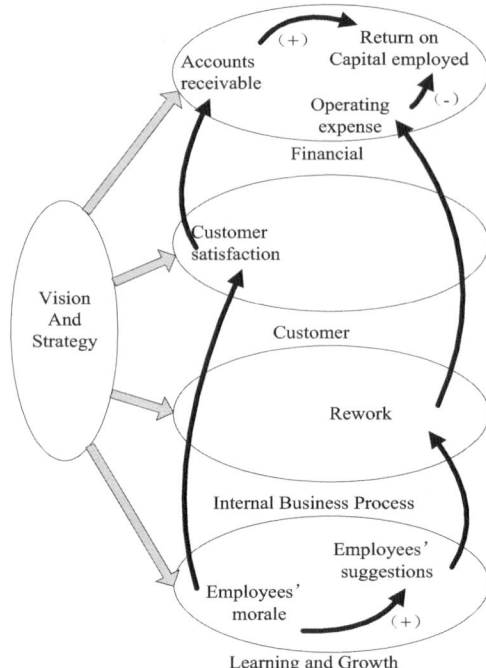

Fig.2. Balanced Scorecard: Four Perspectives

III. A CASE STUDY

S company is a modern aviation group enterprise involving design, development and mass customization production. S company has a marketing subsystem, a research and design institute, and three production bases. Its strategic objectives are "To build China's first and world-class standard fasteners Industry Company based on Research & Design, Production and Service." The company puts forward strong operation control mode "six uniforms, three chain loop", and adhere to professional production, industrial management, capital operation, international route "four business philosophy", and strongly implement market expansion, transformation extends, winning innovation, lean production, talent thriving enterprise, supply chain " six strategic".

A. Decomposition of strategic objectives by BSC

Under the SWOT strategic analysis tools and a face to face questionnaire, the development vista of S company is "to build China's first, world-class set of research & design, production and service as one of the high-end standard fasteners industry leader. S company's objectives: 1) the annual output value will reach more than 10 billion in the next five years. 2) The next decade (2011-2020) Objectives: Regular products rate will be 100%; Customer Satisfaction will be 95%; the rejection rate will be less than 3% (Based on the number of inputs and outputs meter).

1) BSC strategy map

The paper draws enterprise strategic map by analyzing the relationship of BSC from four dimensions. The enterprise will lay stress on using intangible assets such as human capital, information capital and organizational capital (Learning and growth), and efficiencies (Internal processes),innovating and establishing strategic advantages and benefiting the market (Customers), so as to achieve enhancement of shareholder value (Financial).

2) CFS and KPI identification

Decomposition of the strategic objectives is crucial for implementation of comprehensive budget management. And the assessment mechanism should be established so that the CFS and KPI can promote the formation of a full range of vision, objectives and plans, its KPI used to assess the target reached quantitative indicators and answer inspected basis of strategic implementation plan "how successful". Flow chart as in Fig.3 [19].

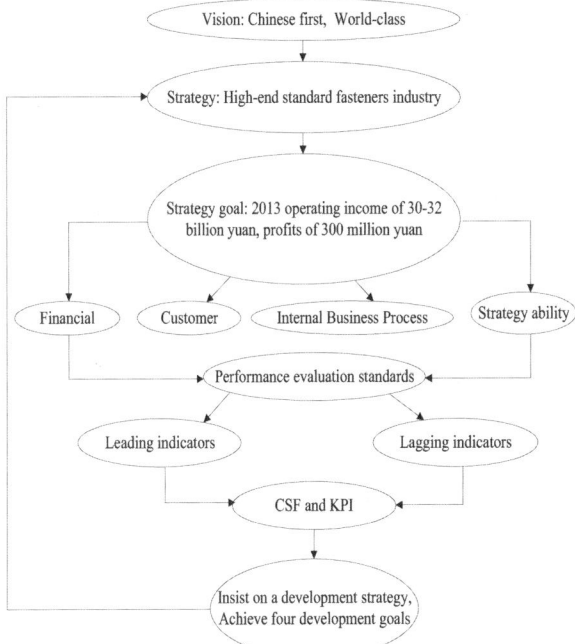

Fig.3. A company the BSC strategy of decomposition process

B. Problems of S company

The production of S company is characterized by a small batch, multi-class, and multi-model, multi-regional. The production bases are different from each other in products and life cycle of the products. The main problems of the company is that product delivery rate is low; Rejection rate is high; multi-regional collaborative management efficiency is low; the determination of the budget target is multiplied by the coefficient of the budget posts based on last year's business data, it did not take full account of the market changes. And different subsystems have different budget template, and they did not consider that calculation of the cost of materials is not uniform for the number of different types of products, and equipment depreciation, rejection, different regions of human cost differences accounting, ignoring the operating time of different products. They use a one-size-fits-all method to deal with different subsystems' budgets, and do not consider cost control methods for difficult to predict the costs, and do not clear low-carbon budget assessment index.

C. S Company dynamic process framework of comprehensive budget management

An overall budget management system is consisted of eight modules, including budgetary planning, budgeting implementation, measurement results of actual execution, audit metering data, variance analysis, feedback report, budget adjustments, reward and punishment. Each module is composed by technology, organization, behavior and environment four levels [19]. Which can be seen in Fig.4 [1].

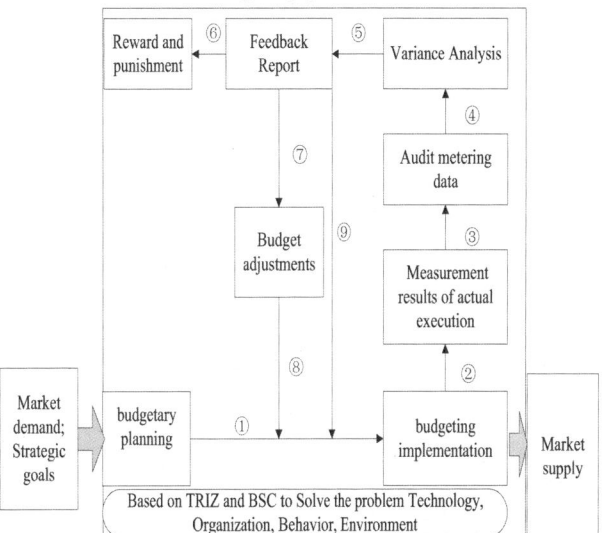

Fig.4. Dynamic framework of comprehensive budget Management system based on the of TRIZ and BSC methods

IV. CONCLUSION

In this paper a dynamic framework of comprehensive budget management system is established based upon theory of TRIZ and BSC, This paper identifies and resolves the contradictions of comprehensive budget management system planning processes and optimize comprehensive budget management system technical Route, decompose strategic targets of the company by the theory. Comprehensive budget management is a system. The company should pay attention to two key factors. One is to create an environment for the staff to participate in the company's development. Budget management involving all levels of business Organization requires agreement in thinking and cooperation of the employers and the employees. Another is to do well in the routine tasks. And strictly implement the budget requirements. At the same time the company must strictly implement budget management in accordance with the concrete requirements, and do well in tracking and controlling the products and information feedback and supervision.

ACKNOWLEDGMENT

The authors would like to thank all anonymous reviewers for their helpful comments and suggestions. This research is supported by Chongqing city science and technology research project (No: cstc2012gg-yyjs70009)

REFERENCES

[1] Wen Long Hou, Yan Hou, Ying He. Modern Master Budgeting Management [M], Beijing: Economics and Management Publishers, 2005, pp.1-38. (Chinese)

[2] Ai xiang Pan, Dong Li Jing. How to interpret comprehensive budget management [J], Finance and Accounting, 2002. 08, pp.30-32. (Chinese)

[3] Shou Tang Su. Target profit oriented Enterprise budget management [M]. Beijing: Economic Science Publishers, 2001, pp.24–29. (Chinese)

[4] Chen Gao. Enterprise budget management: Strategy Oriented [M]. Beijing: China Financial and Economic Publishing House, 2004: 2. (Chinese)

[5] Horngren C T, Foster G, Datar S M, et al. Cost accounting: a managerial emphasis [J]. Issues in Accounting Education, 2010, 25(4), pp.789-790.

[6] Garrison R H, Noreen E W, Brewer P C. Managerial accounting [M]. New York: McGraw-Hill/Irwin, 2003.

[7] Cooper R, Kaplan R S. Profit priorities from activity-based costing [J]. Harvard Business Review, 1991, 69(3), pp. 130-135.

[8] Cooper R, Kaplan R S. The promise-and peril-of integrated cost systems [J]. Harvard Business Review, 1998, 76(4), pp. 109.

[9] Tanaka T. Kaizen budgeting: Toyota's cost-control system under TQC [J]. Journal of Cost Management, 1994, 8, pp. 56-62.

[10] Hope J, Fraser R. Beyond budgeting: how managers can break free from the annual performance trap [M]. Harvard Business School Press, 2003.

[11] Shao Lu Zou. "Based on strategy-oriented Comprehensive budget management," Ph.D. dissertation, China. Management Science and Engineering. Zhong Nan University, 2004. (Chinese)

[12] Bin Wang. Enterprise budget management and mode research [J]. Accounting Research, 1999, 3(11), pp.20-24. (Chinese)

[13] Moehrle M G. What is TRIZ? From conceptual basics to a framework for research [J]. Creativity and innovation management, 2005, 14(1), pp. 3-13.

[14] Altshuller G S. Creativity as an exact science: The theory of the solution of inventive problems [M]. Gordon and Breach Science Publishers, 1984.

[15] Mann D, Dewulf S. Evolving the world's systematic creativity methods [J]. TRIZ Journal, 2002, 4, pp. 1-10.

[16] Jones E, Mann D, Harrison D D. An Eco-innovation Case Study of Domestic Dishwashing through the Application of TRIZ Tools [J]. Creativity and Innovation Management, 2001, 10(1), pp. 3-14.

[17] Yi Hong Luo, Yun Fei Shao. Study of enterprise information system framework design based on TRIZ and BSC [J]. JOURNAL OF MANAGEMENT SCIENCES IN CHINA, 2012, 15 (9), pp. 20-34. (Chinese)

[18] Kaplan R S, Norton D P. Using the balanced scorecard as a strategic management system [J]. Harvard business review, 1996, 74(1), pp. 75-85.

[19] Zeng Biao Yu. Study on Group Company budget management system framework [J]. Accounting Research, 2004, 8, pp. 22-29. (Chinese)

Research on the Model of People Being Led Satisfaction Based on Cadre Assessment

Quan-qing Li*, Xiao-shu Li

Zhengzhou Institute of Aeronautical Industry Management, Zhengzhou, P. R. China, 450015

(*liqqing@139.com, lxszh1015@139.com)

Abstract - **At present the common satisfaction models are all to evaluate product or organization, in this paper, it puts forward the satisfaction model which evaluates the individual. Based on the cadre assessment problems in the interior of enterprises and institutions, it proposes the model of people being led satisfaction, and points out the unique difficulties of evaluating the individual. It constructs the two levels index system which includes the whole work of working unit, the work in charge by leaders, leadership qualities ability and employees benefits, etc. And according to the role of leaders in leadership activities, working relations with appraisers, it puts forward the distribution principle of index weight.**

Keywords - **Index system, model, people being led, satisfaction**

I. INTRODUCTION

For any enterprises and institutions, the importance of the product or service quality is self-evident. With the development of science and technology and economy, people's demand for quality enhances unceasingly, the concept of quality has been continually changing [1]. The quality concepts with representative are: (1) The conformance quality. This quality concept emphasizes the conformity of product performance and technical standards, in line with the technology is in order to ensure the realization of the basic functions of products, this is the technical natures of quality, it is the management mode which takes production as the center. The quality control in this level for supplier reflects that the judgment of consumer on the quality is based on the material object form of products, its criteria is the objective technology, product is the carrier of quality. (2) Applicability quality. This quality concept is to suit for the need degree of customers as a measure basis, that is from the view of use to define the quality, it considers that the product quality is the degree that the product can successfully meet customer's need when it is used

The development of "applicability quality" concept, illustrates the people are on the understanding of the quality concept, the customer demand is gradually put in the first place. But the products quality to meet customer using needs is not necessarily make the customer satisfaction, so the quality concept has developed to "customer satisfaction quality", formed the new concept of quality [2].

In 1989, Sweden Statistics Bureau first applied the model and calculation method of C. Fornell doctor of America University of Michigan, designed "Swedes Customer Satisfaction Barometer", for short SCSB. The index covered more than 100 companies of 31 industries in Sweden. It is the first nationwide index model of the customer satisfaction [3]. Since then, American, Germany, New Zealand, Canada and Taiwan province of China, Europe and other countries and regions, has established the customer satisfaction index of national, regional or industry.

Obviously, in the interior of enterprises and institutions, employee satisfaction is very important for the enterprises and institutions to improve customer satisfaction, therefore, along with the rise of the research on customer satisfaction, employee satisfaction has attracted widespread interest in the world, it is one of the new hot spot in the satisfaction research. In 1935, Hoppock. R. proposed the concept of Job Satisfaction [4], it laid the foundation for later booming Employee Satisfactory. In our country, there are many experts and scholars and enterprises to carry on the all-round research on employee satisfaction [5, 6].

Initially, the model of customer satisfaction and employee satisfaction is only applied in the economic sector and enterprises, with the application range of the customer satisfaction index model is gradually widening, in some countries, including China, the public sectors have also begun to apply the customer satisfaction index model to evaluate public satisfaction. At present, in addition to the enterprise, satisfaction models have begun to study and application in the culture and health [7], the administrative departments [8], public service [3], and other departments and industries. Predictably, the application range of satisfaction model will be further expanded, the application departments will further increase.

II. PUTTING FORWARD THE MODEL OF PEOPLE BEING LED SATISFACTION

Satisfaction model is a calculation method and mode of simple, intuitive and easy operation. Satisfaction is a relative concept: beyond expectations is satisfaction, achieving expectations is basic satisfaction, lower than expectations is not satisfaction [9]. Customer Satisfaction is the feel of the degree that the customer for itself demands has been met [10]. It depends on the comparison of a product performance (or normalization of service process) to be understood by the customer with their expectations, if lower than the expectations of customer, customer will not be satisfied; If the performance is in accord with the expectations, the buyer will be satisfied. Anyhow, customer satisfaction is used to measure the indexes of customer satisfaction degree. Employees of enterprises and institutions are as same as customers, also have their psychological and physiological needs, employee satisfaction (is called ES for short) is the satisfaction degree of this need [5].

Satisfaction model is used to evaluate the performance of an object, at present, the evaluating objects of research and application are organizations, such as enterprise, administrative departments and government. Customer satisfaction is to evaluate the achievements of organization from the organization exterior, and the employee satisfaction is to evaluate the organization's performance from the organization interior. So far, it has not found satisfaction evaluation model for the individual.

In the modern society, to evaluate the individual is very ordinary things. For example, the election, recommendation, assessment leaders, is essentially the individual evaluation for leaders. Only those leaders who obtain good evaluation, will win the votes in the election, in the opinion polls wins recommended votes, in the annual assessment as well as various examination gets high marks. Therefore, using satisfaction model to evaluate the individual of leaders, is very necessary.

III. SATISFACTION MODEL OF THE PEOPLE BEING LED

Satisfaction model uses the simple data - satisfaction index to quantitatively express the good or bad of an object to be evaluated, there have been many related research and achievements. The satisfaction model of the people being led is constructed on the basis of these achievements.

A. Description of satisfaction model

In the existing satisfaction model, its mathematical description can be simply expressed as [10]:

Suppose the evaluation project set of customer satisfaction is:

$$F = (f_1, f_2, f_3, ..., f_n)$$

Therein f1, f2, f3,..., fn are the factors that influence satisfaction, that is the index system of satisfaction model.

Comprehensive evaluation vector as follows:

$$S = (s_1, s_2, s_3, ..., s_n)$$

Therein s_1, s_2, s_3, ..., s_n respectively express the satisfaction of project f_1, f_2, f_3, ... , f_n, these data are given by the personnel who participate in satisfaction evaluation, and the data after the necessary mathematical processing can be used effectively..

Weight of evaluation project as follows:

$$W = (w_1, w_2, w_3, ..., w_n)$$

Therein w_1, w_2, w_3, ..., w_n respectively express the weight of project f_1, f_2, f_3, ... , f_n.

So the satisfaction is:

$$CS = S \cdot W^T = (s_1, s_2, s_3,..., s_n) \cdot W^T$$
$$= s_1 \cdot w_1 + s_2 \cdot w_2 + s_3 \cdot w_3 + ... + s_n \cdot w_n \quad (1)$$

B. Concept of the people being led Satisfaction model

The people being led satisfaction that proposes in this paper refers to the satisfaction degree of the people being led to the leader in a enterprises and institutions, is when the people being led accept the leader's leadership, they have experienced the degree that the actual feelings compare with their expectations. Satisfaction index of the people being led is the evaluation that the people being led give leaders. When the people being led felt the leaders better than expected, they will feel very satisfied; if they felt the leaders do not reach their expectations, ,they felt dissatisfied.

In order to distinct and separate from general satisfaction model, set up the people being led satisfaction as PLS

C. Model features of the people being led satisfaction

Satisfaction is a kind of psychological feeling, like other satisfaction model, the people being led satisfaction has the common features of satisfaction:

1) Individuation: Satisfaction is related to expectation and perception, and the expectation and perception are closely related to the individual economic status, cultural background, evaluation motivation, personal character, psychological mood and other factors, therefore, different the people being led to the same leader will have different satisfactory evaluation.

2) Generalization: All the people being led have experience perception for the leaders work, all of them would give the evaluation whether they feel satisfied or not. So the satisfaction survey cannot spread to the individual user.

3) Integration: The satisfaction of the people being led is not only about leader individual, but also relates to the working unit and the leading body, including the nature of the unit, image, achievement, prospects, fighting capacity of the leading body, relationship between the team members.

4) Relativization: The satisfaction of the people being led varies with the objective conditions and subjective requirements, has stage characteristics.

In addition, as for the individual satisfaction evaluation, the satisfaction model of the people being led has its unique features:

1) Complexity: For evaluation work, evaluating the individual is the most difficult. This is because the human being itself is the most complex object, there is no way to find a suitable and general index system to evaluate people. The satisfaction model of people being led which is proposed in this paper can only evaluate the individual under the simplified environment.

2) Differentiation: Even using the same index system to evaluate the individual, people to participate in the evaluation may also give extreme incompatible evaluation, evaluation opinions would appear polarization. This is because evaluating the individual, especially evaluating leaders, inevitably admix into personal feelings, even the ideology factor.

D. Index system of the people being led satisfaction model

To build evaluation project set F, that is to build the index system, it is the first problem of satisfaction model.

The index system of the people being led satisfaction is composed of 4 first level indexes and 34 secondary indexes, as shown in TABLE I:

TABLE I
INDEX SYSTEM OF PEOPLE BEING LED SATISFACTION

Objective layer	The first level indexes	The second level indexes
People being led satisfaction	whole work of working unit	Values of working unit f_1
		Images of working unit f_2
		Strategy and prospects of working unit f_3
		Fair distribution of reward f_4
		Organization procedure fairness f_5
		Interaction fairness f_6
	In charge of the work	Work thinking f_7
		Work achievements f_8
	Qualities ability	Ideology and morality f_9
		Social morality f_{10}
		Professional ethics f_{11}
		Legal consciousness f_{12}
		Breadth of mind quality f_{13}
		Management knowledge f_{14}
		Professional knowledge f_{15}
		Decision-making ability f_{16}
		Coordination ability f_{17}
		Dealing with emergency ability f_{18}
		Communication ability f_{19}
		Assessment and prediction ability f_{20}
		Innovation ability f_{21}
		Using person ability f_{22}
	Employees benefits	Work environment f_{23}
		Work means f_{24}
		Working time f_{25}
		Job interest f_{26}
		Itself actualization f_{27}
		Salary distribution f_{28}
		Promotion f_{29}
		Welfare benefit f_{30}
		Opinions communication f_{31}
		Informal organizing activities f_{32}
		Conflict coordination f_{33}
		Conflict coordination f_{34}

There are 4 first-level indexes. The first index, the whole work of working unit points at the policy, system and work of involving the overall situation of the working unit, all of these are determined by the leading body. The second index--in charge of the work, it indicates that the work is determined by the leader individual, involving a range of work in the working unit, generally refers the work that the leaders to be evaluated are in charge of. The third index, qualities ability refers to the quality and ability that the leader individual has demonstrated in leadership activities. Here, the leadership quality does not refer to ability, morality, knowledge of the common people, but the radical mark that distinguish the leaders from the non-leaders, specifically refers to the necessary basic conditions of the leaders in leadership activities, also can be said it is the whole sum of knowledge, experience and behavior ability that the leader has demonstrated in the leadership activities, these qualities often play a key role in leadership activities. The leaders qualities will be change when the profession characteristics, job function, time background, work nature has changed, therefore, it is a dynamic concept in a constantly changing. The fourth index, is employees benefits. This is employee satisfaction of a narrow sense, it is a kind of psychological and physiological feelings and opinions which the employees go in for the work itself and working environment.

E. Correction of index weight of the people being led satisfaction

In the research field of satisfaction evaluation models, mostly according to the established index system, using methods of analytic hierarchy process or fuzzy evaluation to determine the index weight, then using a mathematical method to carry out evaluation [7]. The establishment of the index weight and the model solution, have had many applications [11], there is no longer discussion. However, we must pay attention to: the model of the people being led is aim at leader individual to evaluate, while the leader's role in the leadership activities is very different, for the leadership's performance assumes different responsibilities. For example, for the whole work of working unit, apparently, chief leaders have more responsibility than deputy leaders; again for instance, for a specific work, the leader who is in charge of the work, has more responsibility than other leaders who are not in charge of the work. Besides, for the employees benefits, the estimator's leaders who are in charge of manage them have more responsibility than the other leaders. In order to avoid to construct the respective index system for different roles, the satisfaction model of the people being led introduces role weight variable x, its essence is for the weight set W to carry on correction, when evaluate the leaders of having greater responsibility, the corresponding indexes are highlighted.

Considering the role in leadership activities and working relationship with evaluation people, the role weight can be divided into 4 kinds of circumstances: common weight X_1, chief leader weight X_2, director leader weight X_3, straight management leader weight X_4. Among them, straight management leader refers to the leader of the estimator oneself working department, director leader refers to the leader who is not in charge of the estimator oneself work. For example, for the estimators engaged in production management, the leader who is in charge of production is the straight management leader, to evaluate him applies to X_4, and the leader who is in charge of technology is director leader, to evaluate him applies to the X_3. There are:

$$x = (X_1, X_2, X_3, X_4)$$

To determine each value of x, should use the method to determine the W value. Obviously, there should be $X_1=1$, and $X_i > 1$ (i=2, 3, 4).

So, the index of the people being led satisfaction is:

$$PLSI = x \cdot CS = x \cdot S \cdot W^T \qquad (2)$$

Thereinto: the role variable x:

$$x = \begin{cases} X_2, & \text{chief leader} \ \& 1 \le i \le 6 \\ X_3, & \text{director leader} \ \& 7 \le i \le 8 \\ X_4, & \text{straight management leader} \ \& \ 23 \le i \le 34 \\ X_1, & \text{other one} \end{cases} \quad (3)$$

i is corresponding to f_i, w_i in formula (2) (see Table I).

In order to evaluate the leaders easily, the satisfaction PLS can be converted into relative value, such as the percentile scores.

IV. CONCLUSION

In the modern society, to evaluate public figures, assess and appoint leading cadres is a very common things, therefore, research satisfaction model to evaluate the individual is very necessary. This paper is aim at the evaluation problem of leading cadres in the enterprises and institutions, proposes the model of the people being led satisfaction, analyses the index system and the principle of weight distribution, provides the reference for the establishment of the individual satisfaction model.

REFERENCES

[1] Wang Li-zhi, "Evolution Paths of Quality Management", *Contemporary Economy & Management*, vol. 28, no. 4, pp. 5-7, 2006.
[2] Li Zheng-Quan, "Customer Satisfaction - A New Quality View", *World Standardization & Quality Management,* no.5, pp. 8-9, 2003.
[3] Liu Wu, *Research on the Public Services Customer Satisfaction Index Models*, Northeastern University Press, 2009, pp. 2-4, 6.
[4] ZHANG Chen-lu, "A Survey of the Research in the Degree of Satisfaction of Staff's Work", *Journal of Anhui University of Technology Social Sciences,* vol. 29, no. 3, pp. 62-64, 70, 2012.
[5] Yang Nai-ding, "The Study on the Model and Management of Employee Satisfactory Degree", *Chinese Journal of Management Science,* vol. 8, no. 1, pp. 61-65, 2008.
[6] Wang Wen-hui, Mei Qiang, "An Evaluation Model and Strategic Research about Employee Satisfaction", *Science & Technology Progress and Policy,* vol. 19, no. 11, pp. 131-133, 2002.
[7] Wang Jin-feng, Gao Xiao-ning, Feng Li-jie, "A Research on Satisfaction Evaluation Model for Service Quality of Higher Education", *Higher Education of Sciences*, no. 4, pp. 38-43, 112, 2013.
[8] He Liang, Li Jun, "The survey research of administrative service satisfaction", *Times Finance,* no.7, pp.55-58, 2010.
[9] Liu Yao, Zhang Dao-wei, "Employee satisfaction improvement based on 4P concept", *Enterprise Economy,* no.9, pp. 61-65, 2008.
[10] Xue Wei, Zhou Hong-ming, ZHeng Bei-rong, Li Feng-ping, "Optimization Model and Empirical Research of Customer Satisfaction Based on Quality Cost", *Industrial Engineering and Management*, vol.9, no.5, pp.64-67, 2004.
[11] Fan Ping, Xu Jie-yi, "Research on modeling method of customer satisfaction index", *Modern Business*, no.3, pp.17-20, 2013.

Correlation Analysis and Dynamic Evaluation on City Aquatorium Ecological Service Value
——Taking Xianning City of Hubei Province as an Example

Ping-fan LIAO, Xiao-yang YANG, Zhi CHEN*, Jin ZHANG

School of Resources Environmental Science and Engineering, Hubei University of Science & Technology, Xianning, P.R.China

(chzh1967@163.com)

Abstract - **City aquatorium is an all-important subsystem in the city complex ecosystem, and it has vital function on maintaining the ecological balance and developing economy in the city. Taking Xianning City as an example, and quoted from the water resources and social economic statistical data in Xianning City from 2000 to 2008, this paper is intended to evaluate the aquatorium ecological service value in Xianning City respectively using the function analysis method and equivalent factor method. After obtaining the test results of the two methods, we calculate the comprehensive index using the weighted average, and analyze its variation trend. We do grey correlation analysis through some of the selected economic indicators and aquatorium ecological service value, so as to reveal the correlation between aquatorium ecological service value and economic development. The results show that, (1) The aquatorium ecological service value in Xianning City from 2000 to 2008 shows a rising trend; (2) Aquatorium ecological service value is the most closely related to GDP; (3) Aquatorium ecological service value in Xianning City is large and diverse. Water conservation and daily water supply are the most particularly important aspects, having significantly influence on agricultural industry, industrial production and people's daily life. Through the dynamic evaluation and correlation analysis on aquatorium ecological service value in Xianning City, we expect to provide theoretical reference to related departments when they are doing regional ecological planning and ecological construction.**

Keywords - **City aquatorium, ecological service value, equivalent factor method, function analysis method, grey correlation analysis, Xianning City**

I. INTRODUCTION

Being an all-important subsystem in the city complex ecosystem, City aquatorium has vital function on maintaining the ecological balance and developing economy in the city, reflecting in ways of water supply, products production, leisure and recreation, air cleaning, climate adjusting and biodiversity maintaining, etc. With the rapid development of social economy in recent years, the negative influence on city aquatorium has become increasingly obvious. The contradiction between urban development and city aquatorium has now becoming increasingly highlighted. As the city aquatorium ecological service value is non-market value, it's hard to estimate and is not so closely related to external city development, thus, it's difficult to draw the public's attention. Serving as an important part of WuHan "1+8" city circle, Xianning City takes a role as "the back garden", thus, it is particularly important to strengthen ecological environment protection and ecological construction. How to protect the city aquatorium? How to make the best use of city aquatorium? How to coordinate city aquatorium and city development? All these questions have aroused great attention of scholars.

In the early 1970s, the Scientific concept about ecosystem services was proposed [1], it Has been highly concerned by the scholars at home and abroad, the research on ecosystem service function has also made a series of progress. Daily (in 1997) [2] systematically introduced the ecosystem service function. Costanza et al (in 1997) [3] divided and assessed the global ecosystem service function, it was divided into 17 types, and was estimated in monetary form according to the 10 biome. Ouyang zhiyun et al (in 1999) [4] investigated the service function of terrestrial ecosystems in China. According to the Costanza et al research results, Chen zhongxin et al (in 2000) [5] estimated China's ecosystem service function economic value in accordance with the area proportion. According to the ecosystem characteristics and the actual situation in the study area, many scholars improved the assessment methods and parameters in China, and evaluated different types and different scales of ecosystem service value [6-8].

From the domestic and foreign research, study of natural ecosystem service function and its economic value assessment has been launched in the global scope, but most research of ecosystem service function and its value is mostly in natural ecosystem such as the forest, grassland, wetland and water et al [9-16], the thematic evaluation of city aquatorium is less, and the static evaluation is most.

Taking Xianning City as an example, and quoted from the water resources and social economic statistical data in Xianning City from 2000 to 2008, this paper is intended to evaluate the aquatorium ecological service value in Xianning City respectively using the function analysis method and equivalent factor method. After obtaining the test results of the two methods, we calculate the comprehensive index using the weighted average, and analyze its variation trend. We do grey correlation analysis through some of the selected economic indicators and aquatorium ecological service value, so as to reveal the correlation between aquatorium ecological service value and economic development. Through the dynamic evaluation and correlation analysis on aquatorium ecological service value in Xianning City, we expect to provide theoretical reference to related departments when they are doing regional ecological planning and ecological construction.

E. Qi et al. (eds.), *Proceedings of the 21st International Conference on Industrial Engineering and Engineering Management 2014*, Proceedings of the International Conference on Industrial Engineering and Engineering Management, DOI 10.2991/978-94-6239-102-4_128, © Atlantis Press and the authors 2015

II. AQUATORIUM ECOLOGICAL SERVICE VALUE EVALUATIONS IN XIANNING CITY

A. Function Analysis and Estimation Results

Function analysis divides the service functions of city aquatorium ecosystem service functions into production function, regulating function, cultural function and support function 4 types of functions and 8 evaluation indexes. And these practices are mainly according to aquatorium ecological service functions and the mechanism in Xianning City, types and utility of ecosystem service that provided by the aquatorium ecological system. Respectively estimate on each index on the basis of clear indicators, we summarize and get the aquatorium ecological service value of each year in Xianning City.

Different type of city aquatorium ecological service function needs different evaluation methods. The economic value of aquatorium ecological service function in Xianning City can be divided into direct value and indirect value according to its benefit evaluation value. Direct value and indirect value adopts appropriate methods to evaluate respective service functions, in accordance with the theories and methods value of resource economics and ecological economics. For different service functions, there are different corresponding evaluation methods, but various methods are using the universal model: $M = f(E, D, P, \text{and } Q)$ [12-15]. In the formula, M is the economic value for all functions, E is price coefficient, D is action contribution amount, and P and Q are parameters. The calculation method of each service function value is as follows:

(1) Daily water supply: $M_1 = E_1 \times D_1$, in the formula, M_1 is water supply value (10^8 Yuan/a), E_1 is the total water supply (10^8 m^3), D_1 is average water in price Xianning City (Yuan/m^3).

(2) Organic matter production: $M_2 = E_2 \times D_2 \times P_2$. In the formula, M_2 is the total economic value of organic production (10^8 Yuan/a). As the aquatorium in Xianning City is mainly used for aquatic products production, so E_2 should be the aquatic product price factor (Yuan/kg), D_2 is aquaculture area, P_2 is aquatic product yield per unit.

(3) Flood control: $M_3 = E_3 \times D_3$. In the formula, M_3 is function value of flood controlling (10^8 Yuan/a), E_3 is the cost coefficient of building per m^3 reservoir capacity (0.67 Yuan), D_3 is the maximum storage capacity (10^8 m^3).

(4) Water conservation: $M_4 = E_4 \times D_4$. In the formula, M_4 is the functional value of water conservation (10^8 Yuan/a), E_4 is the cost coefficient of building per m^3

reservoir capacity (0.67 Yuan). D_4 is the city's total amount of fresh water resource, and we used the shadow engineering method for the calculation.

(5) Soil conservation: $M_5 = E_5 \times D_5 \times P_5 \times Q_5$. In the formula, M_5 is the functional value of soil conservation (10^8 Yuan/a), E_5 is the income of land acreage unit, D_5 is the total sediment accumulation of the lakes, P_5 is the average thickness of soil overburden (0.5 m), Q_5 is the average unit weight of soil (1.28 t/m^3). The materials show that the income of each land area unit in Xianning City is 25500 Yuan/hm^2 (using the annual agricultural output/the total cultivated area), the total amount of sediment deposition is 470.55×10^4 t (total lake and reservoir area×average annual sediment deposition of eastern plain lakes and reservoirs 1.16×10^4 t), using the opportunity cost approach to calculate.

(6) Purification: $M_6 = E_6 \times D_6 \times P_6$. In the formula, M_6 is the functional value of purification (10^8 Yuan/a), E_6 is the processing cost of N, P (N is 1.5 Yuan/kg, P is 2.5 Yuan/kg), D_6 is lake area, P_6 is the average N and P removal rate of lake unit area (N is 3.98 t/$km^2 \cdot$a, P is 1.68 t/$km^2 \cdot$a). The calculation is done by ecological value method.

(7) Entertainment: $M_7 = E_7 \times D_7$. In the formula, M_7 is the functional value of leisure and entertainment value (10^8 Yuan/a), E_7 is the cost coefficient of leisure and entertainment, here we mainly adopts the entertainment value of global lake ecological system researched by Costanza, and D_7 is lake area, it is calculated using the travel cost method.

(8) Living environment: $M_8 = E_8 \times D_8$. In the formula, M_8 is the functional value of providing living environment (10^8 Yuan/a), E_8 is the price factor of biological habitat. As all kinds of water on the earth are important habitats or shelter for wild lives, so according to Costanza's research results, we take the biological habitat value of wetlands. And D_8 is water area (river water area + lake area + reservoir area) in Xianning City, and we used the ecological value method to calculate.

Divide Xianning City's water area into three categories, the ecological service value into eight categories, we calculated the total aquatorium ecological service value in Xianning City in 2000 was 8.607 billion Yuan. From the above calculation model we can get (ignore the dollar currency rate interannual variability and RMB discount rate), the aquatorium ecological service value in Xianning City from 2000 to 2008 (Such as Table I).

TABLE I
THE AQUATORIUM ECOLOGICAL SERVICE VALUE BASED ON FUNCTION ANALYSIS METHOD (2000-2008)

Year	Daily water supply	Organic matter production	Flood control	Water conservation	Soil conservation	Purification	Entertainment	Living environment	Total value (10^8 Yuan)
2000	28.18	6.52	10.85	38.75	0.19	0.05	0.33	1.20	86.07
2001	30.61	7.56	8.61	37.33	0.20	0.05	0.34	1.20	85.90
2002	29.21	8.63	16.08	79.86	0.21	0.05	0.34	1.20	135.58
2003	26.37	9.98	11.50	68.88	0.24	0.05	0.34	1.20	118.56
2004	27.76	11.39	9.83	47.52	0.30	0.05	0.34	1.20	98.39
2005	26.63	13.37	12.10	44.11	0.33	0.05	0.34	1.20	98.13
2006	26.41	16.47	9.26	39.03	0.35	0.05	0.34	1.20	93.11
2007	27.80	19.59	8.21	30.00	0.37	0.05	0.34	1.20	87.56
2008	30.09	22.78	10.93	44.31	0.40	0.05	0.34	1.20	110.10

From Table I we can see that, among all the aquatorium ecological service value evaluation indexes in Xianning City, in addition to water conservation, there is small interannual variability. Furthermore, purification and living environment have no changes. In 2002, 2003 and 2008, precipitation in this city has great fluctuation. Surface runoff increased, lake water yield increased, fresh water resources skyrocketed which led to the rapidly increase of water conservation, so the daily water supply in the three years rose. As there was no big changes on urban water area, purification functions and living environments haven't changed, so we can just ignore them. The value of organic matter production and soil conservation grew steadily. Due to the variability of annual precipitation, the value of flood control was fluctuating. The variation trend of total aquatorium ecological service value is similar to that of water conservation in Xianning City, indicating that the main function aquatorium is water conservation in Xianning City. Therefore, it helps to the water conservation, humid air and climate regulating in Xianning City.

B. Equivalent Factor Method and Its Estimating Results

The evaluation indexes of equivalent factor method are mainly based on the potential contributions of ecological services generated by ecological system. According to the "equivalent factor table of ecosystem service value in China" summed up by Xie Gaodi and combined with the statistical data [6, 7, 16], we estimate an economic value of equivalent factor, to conclude the ecological service value of each unit area in Xianning City. The aquatorium ecological service value in Xianning City is product of the value of each unit area and the total water area.

Based on the ecosystem service value theory proposed by Costanza, and combined with the ecological service value coefficient calculation method that put forward by Xie Gaodi, we are able to estimate the aquatorium ecological service value change in Xianning City. The estimation formula is: $ESV=\sum W \times V$. In the formula, ESV is the ecosystem service value (Yuan); W is the total water area (hm^2) in Xianning City, V is the ecosystem service value (Yuan/hm^2) per unit area in Xianning City.

Xie Gaodi and his group divided service function of ecology into 9 categories, and they did a questionnaire survey to 200 ecologists in our country to summarize the "equivalent factor table of ecosystem service value in China". Ecological service value equivalent factor refers to the potential ecological service contribution ability produced by ecological system. The economic value of one equivalent factor equals to 1/7 market value of the average production of 1 hm^2 in that specific area.

According to statistics, the average food production in Xianning City from 2000 to 2008 is 4806.56 kg/hm^2, adopting the average food market price in 2008 ($3.4/kg), we can determine that the total economic value of an ecological service value equivalent factor is 2334.61 Yuan. In addition, according to the water equivalent standard in the "equivalent factor table of ecosystem service value in China", we can know that, service value of each water unit area is 107322.02yuan/hm^2.

According to the above conclusions and existing researches, and the specific circumstances in Xianning City, we can reach the equivalent factor table of aquatorium ecological service value in Xianning City (Such as Table II).

TABLE II
THE AQUATORIUM ECOLOGICAL SERVICE VALUE BASED ON
EQUIVALENT FACTOR METHOD (2000-2008)

Year	The aquatorium area ($10^4 hm^2$)				Total value (10^8 Yuan)
	Lake	Reservoir	River	Total	
2000	31.956	28.892	25.212	86.059	923.61
2001	32.050	28.891	25.208	86.150	924.58
2002	32.042	28.889	25.208	86.139	924.46
2003	32.036	28.879	25.208	86.124	924.30
2004	32.047	28.878	25.208	86.133	924.40
2005	32.120	28.879	25.208	86.207	925.19
2006	32.152	28.878	25.206	86.236	925.50
2007	32.168	28.879	25.206	86.254	925.70
2008	32.231	28.883	25.206	86.320	926.41

From Table II we can see that, the overall aquatorium ecological service value in Xianning City is growing steadily upward. The dropping interval from 2001 to 2003 is mainly because of the engineering destruction caused by flood disasters, and that led to a decrease of water area. Subsequently, the water area grew steadily, so that the service value of aquatorium ecological increased in Xianning City.

C. Weighting Calculation Results and Analysis

There is a big difference in the calculation results of the two above methods. The function analysis method calculates respectively according to the various functions, if there is any lack of data, it may lead to incomprehensive calculation or missing item. For example, the organic production in Xianning City only calculated the fishery production, not including other aquatic plants. Integrating all functions, equivalent factor method's calculation is general and lack of pertinence. Measured by average grain yield per unit area, it might lose accuracy in specific area.

Combining the above two methods the frequency of use in the field study, taking each year data to the weighted average. The weight of function analysis method is 70%, the weight of equivalent factor method is 30%, thus calculated the weighted average value of the aquatorium ecological service value in Xianning City from 2000 to 2008(Such as Table III).

TABLE III
THE AQUATORIUM ECOLOGICAL SERVICE VALUE BASED ON
WEIGHED AVERAGE METHOD (2000-2008)

Year	Function analysis method (Weight: 70%)	Equivalent factor method (Weight: 30%)	Weighed average value (10^8 Yuan)
2000	86.07	923.61	337.33
2001	85.90	924.58	337.50
2002	135.58	924.46	372.24
2003	118.56	924.30	360.28
2004	98.39	924.40	346.19
2005	98.13	925.19	346.25
2006	93.11	925.50	342.83
2007	87.56	925.70	339.00
2008	110.10	926.41	354.99

From Table III we can see that, the aquatorium ecological service value rise slowly in fluctuation, specifically in Xianning City from 2000 to 2008. Specific view, it had a smooth curve from 2000 to 2001; the ecological service value suddenly rose, ascensional range was large from 2001 to 2002; the overall trend of the curve declined from 2002 to 2007, but recovered to steady after 2004, and finally went back to the level of 2000; the ecological service value presented a fast growth from 2007 to 2008.

The aquatorium ecological service value fluctuated obviously in Xianning City from 2001 to 2002. The increase was because of the greatly enhanced overland runoff and water storage capacity caused by the rich precipitation. After 2002, because of the flood disaster, the water storage project was destructed to some extent, so the ecological service value dropped; After the government repairing on the water conservancy project from 2004 to 2007, theoretically the ecological service value shall recover steadily, but there were less rain in these years, so it declined slowly; due to the rich precipitation from 2007 to 2008, the service value of ecological show a sharp rise trend.

III. THE GREY RELATIONAL ANALYSIS OF AQUATORIUM ECOLOGICAL SERVICE VALUE AND ECONOMIC DEVELOPMENT

Grey correlation model is an important method of grey system analysis, and is a method of quantitative description and comparison for the system development situation. The meaning of grey relational analysis is to point out that in the system development process, if the change trend of the two factors is consistent, namely we say they have high synchronous change degree, and we say they have great association; on the contrary, the correlation is small.

For further reveal the correlation of the aquatorium ecological service value and economic development, in this paper we do gray correlation analysis through the selected economic indicators and water ecological service value. Considering the weighted average value of aquatorium ecological service value in Xianning City from 2000 to 2008, as the reference sequence, then select 3 economic indicators closely related to urban development: GDP (10^8 Yuan), the proportion of non-agricultural population (%), the total social retail sales of consumer goods per capita (Yuan/per person). Adopt data of the corresponding period as the comparative sequence.

The calculation was done according to the principles and methods of grey correlation model, and we can reach that three economic indicators of ecological service value and city development in Xianning City were 0.730477, 0.700679 and 0.531017. Namely, ecological service value has maximum correlation to GDP and the minimum correlation to the degree of total retail sales of consumer goods per capita.

Multifunctional water bodies in Xianning City are closely related to the urban industrial development. (1) The main contributions for the first industry are production and water supply. By the end of 2008, the stocking area in Xianning City are 49820hm2, a year-on-year increase of 14527hm2, having an amplification of

41%; among them, The stocking area for famous, special and superior animals are 39866.67hm2, accounted for 80% of the total area; the output of aquatic products is 176000t, representing an increase of 39000t, having an amplification of 28%; comprehensive output of fishery reached 2.03 billion Yuan. At the same time, water of that area also has played a significant role for farmland irrigation. In a broad sense, it is promoting agricultural development, stabilized agricultural output. On the other hand, the developing of agriculture, especially aquatic agricultural, also helped the full play of climbing ecological service value. (2) The main contribution for the second industry is mainly for industrial production. There are numerous food processing enterprises, textile enterprises and small manufacturing enterprises in Xianning City, so there is large water demand. In 2008, the total water supply in Xianning City was 1.512 billion m3; it provided enough water for multitudinous industrial production enterprises. At the same time, the industrial production pollution discharge has great polluted the water to some extent. The main water quality problem in Xianning City is mild eutrophication, and that would influence ecological service value into full play such as habitat providing, entertainment and leisure, etc; (3) Water function to the third industry is mainly for the entertainment and providing habitat. The surrounding water such as the axes Lake and LiangZi Lake are good. They are suitable for leisure, entertainments, and for creatures. We had countless ecological tourism projects in Xianning City, such as hot spring resources, lake scenic spots and so on, are creating value for the third industry. In 2008, the output value of tertiary industry reached 12.363 billion Yuan, accounted for 34% of the total output value. With the rapid development of tourism, the regional water environment pressure increases. The government is stressing on water environment regulation; adhere to the principle of "protection and development", so the water ecological service functions in Xianning City are expected to play sustainably.

IV. CONCLUSION AND DISCUSSION

(1) The aquatorium ecological service value in Xianning City from 2000 to 2008 shows a rising trend. The maximum value is 37.224 billion Yuan/a, and the minimum value is 33.733 billion Yuan/a, the average value is 34.851 billion Yuan/a. These data show that its value got a good play.

(2) Aquatorium ecological service value is the most closely related to GDP, the degree of association is 0.730477. It means that city water ecological service value has the largest contribution to GDP, at the same time, it is also most affected by GDP.

(3) Aquatorium ecological service value in Xianning City is large and diverse. Water conservation and daily water supply are the most particularly important aspects, having significantly influence on agricultural industry, industrial production and people's daily life. Therefore, the management of urban waters must be based on protection, so it can be fully used and sustainably

developed. In the direct value aspect, on the one hand, we must protect the water supply channels, and ensure the capacity for production and living and water quality. On the other hand, we should give full play to its function of breeding. We shall adopts the mixed model of multiple fish breeding, aquatic plants and fish intercropping, so that there would be effective circulation while the organic matter production and water system; In the indirect value aspect, to the exclusion of water conservation and flood control, we shall help to increase the entertainment value. Xianning City has wide waters, and surrounding environment is also very favorable. When strengthening the surrounding infrastructure construction and tourist attractions, we are able to create more pleasant leisure environment and higher economic value, so as to promote the rapid development of the tertiary industry.

ACKNOWLEDGMENT

Foundation: The National Natural Science Foundation of China, No.40961009, No.41071069; The Key Science Study Program of Hubei Provincial Department of Education, No.D20112802; The Key Humanities and Social Science Study Program of Hubei Provincial Department of Education, No.2012D124.

REFERENCES

[1] Holder J, Ehrlich P R. Human population and global environment [J]. American Scientist, 1974, (62): 282-297.

[2] Daily G C. Nature's service: Societal dependence on natural ecosystems [M]. Washington D C: Island Press, 1997, 12-13.

[3] Costanza R, Arge R D, Groot R D, Farber S, Grasso M, Hannon B, et al. The value of the world's ecosystem services and natural capital [J]. Nature, 1997, 387: 253-260.

[4] OUYANG Zhiyun, WANG Xiaoke, MIAO Hong. A primary study on Chinese terrestrial ecosystem services and their ecological economic values [J]. Acta Ecologica Sinica, 1999, 19(5): 607-613. (in Chinese)

[5] CHEN Zhongxin, ZHANG Xinshi. Value of ecosystem benefit in China [J]. Chinese Science Bulletin, 2000, 45(1): 17-22. (in Chinese)

[6] XIE Gaodi, ZHANG Yili, LU Chunxia, ZHENG Du, CHENG Shengkui. Study on valuation of rangeland ecosystem services of China [J]. Journal of Natural Resources, 2001, 16(1): 47-53. (in Chinese)

[7] XIE Gaodi, LU Chunxia, LENG Yunfa, ZHENG Du, LI Shuangcheng. Ecological assets valuation of the Tibetan Plateau [J]. Journal of Natural Resources, 2003, 18(2): 189-196. (in Chinese)

[8] OUYANG Zhiyun, ZHAO Tongqian, ZHAO Jingzhu, XIAO Han, WANG Xiaoke. Ecological regulation services of Hainan Island ecosystem and their valuation [J]. Chinese Journal of Applied Ecology, 2004, 15(8): 1395-1402. (in Chinese)

[9] ZHAO Tongqian, OUYANG Zhiyun, ZHENG Hua, WANG Xiaoke, MIAO Hong. Forest ecosystem services and their valuation in China [J]. Journal of Natural Resources, 2004, 19(4): 480-491. (in Chinese)

[10] JIN Fang, LU Shaowei, YU Xinxiao, RAO Liangyi, NIU Jianzhi, XIE Yuanyuan, ZHANG Zhenming. Forest ecosystem service and its evaluation in China [J]. Chinese Journal of Applied Ecology, 2005, 16(8): 1531-1536. (in Chinese)

[11] XUE Dayuan, BAO Haosheng, LI Wenhua. A valuation study on the indirect values of forest ecosystem in Changbaishan Mountain Biosphere Reserve of China [J]. China Environmental Science, 1999, 19(3): 247-252. (in Chinese)

[12] MAO Dehua, WU Feng, LI Jingbao, Pi Hongli. Evaluation on ecosystem services value of Dongting Lake wetland and ecological restoration countermeasures [J]. Wetland Science, 2007, 5(1): 39-44. (in Chinese)

[13] CHEN Peng. Evaluation on services value of wetland ecosystem in Xiamen City [J]. Wetland Science, 2006, 4(2): 101-107. (in Chinese)

[14] CAO Zhihong, XU Xinwang, WANG Yanlin. Evaluation of wetlands ecosystem services value of the Yangtse River region in Anhui Province [J]. Chinese Agricultural Science Bulletin, 2008, 24(8): 413-419. (in Chinese)

[15] OUYANG Zhiyun, ZHAO Tongqian, WANG Xiaoke, MIAO Hong. Ecosystem services analyses and valuation of China terrestrial surface water system [J]. Acta Ecologica Sinica, 2004, 24(10): 2091-2099. (in Chinese)

[16] YUE Shuping, ZHANG Shuwen, YAN Yechao. Impacts of land use change on ecosystem services value in the Northeast China Transect (NECT) [J]. Acta Geographic Sinica, 2007, 62(8): 879-886. (in Chinese)

Macro Management Innovation, Institution Environment and Enterprise's Internal Capacity
——A Case Study on Cultural Enterprises in Tianjin

Li-ping Fu[1], Jun-sheng Song[2], Tian-pin Shao[2,*]

[1]Public Resources Management Research Center, Tianjin University, Tianjin, China
[2] College of Management and Economics, Tianjin University, Tianjin, China
(tianpinshao@163.com)

Abstract - **Based on structural equation model (SEM), this article conducts an empirical study on the influence factors of enterprise's internal capacity using the survey data of 213 cultural enterprises in Tianjin. The results show that: macro management innovation (which consists of two parts: industrial policy and financial policy) as well as institution environment are proved to be the two main influence factors. Internal capacity relates significantly and positively with institution environment and industrial policy, while the active effects of industrial policy in Tianjin cannot reach the expected level. And the importance of financial policy hasn't been reflected. On this basis, this article puts forward policy suggestions for the development of cultural enterprises in Tianjin. For example, powerful legal system and policy system support should be set up especially financial policies and investment policies to create a sound micro environment for sustainable development.**

Keywords - **Cultural enterprises, factor analysis, structural equation model (SEM)**

I. INTRODUCTION

Different from general enterprises, cultural enterprises have a broader development perspective as a sunrise and green organization in our country.-Nowadays, the revitalization of cultural industry has been raised to national strategy [1]. According to the overall situation, cultural industry in Tianjin is lagged behind compared with which in developed cities in China. As a main market subject, the growth and expansion of cultural enterprises is the only way to development cultural industry. By researching the successful cases and methods of the culture-powerful states such as Japan, South Korea and the United States, it can be known that local government intervention plays a significant part in the development of cultural enterprises [2]. So exploring the influence factors of enterprise's internal capacity from the angle of public management and innovation, as well as the relationship between them will be important for perfecting relevant policies and regulations to create a favorable environment.

Recently, the relevant studies about cultural enterprises mainly concentrate on the influence factors (Zhao and Fan 2012; Ma and Bai, 2012; Meng 2013) as well as the problems and countermeasures (Zhang and Li 2012; Zhang 2013; Wang and Cheng 2013; Wang 2013; Jiang 2014). Most researches take qualitative analysis as the core, while there are less quantitative researches [3]. Li and Fan (2014) use the method of PCA to extract the principle components of various factor of Gansu province [4]. Using grey relation model, Wan and Zhang (2013),

Hao and Tang (2013) analyze the correlation degree between the influence factors of the cultural enterprises.

There are observed variables and latent variables among influence factors. And questionnaire always has measurement errors in the independent and dependent variables [3]. So, traditional regression methods cannot be used well to deal with this kind of problems. Therefore this article conducts an empirical study on the influence factors based on structural equation model (SEM). SEM, as a multiple statistical analysis method, which integrates factor analysis, path analysis and the multiple linear regression analysis, makes a research on the relation between independent and dependent variables [5]. Finally it completes the unity of the explanation research and the descriptive research using a combination of structural model and measurement model. Therefore, errors intrinsic to the testing systems are well compensated for by SEM. In this article, structural equation model is built to explore the critical influences of the internal capacity of cultural enterprises in Tianjin through questionnaire.

There are two possible innovations: firstly, this article explores the three influence factors of enterprise's internal capacity from the angle of public management and innovation. They are industrial policy and financial policy as well as the institution environment. Secondly, as an innovative method, SEM could be used in the research to explore the structural relationship between the influence factors.

II. THEORETICAL MODEL

Structural Equation Model (SEM) can be divided into two parts: structural model and measurement model. The matrix equations are shown below. Fig.1 shows the relationship between variables of SEM.

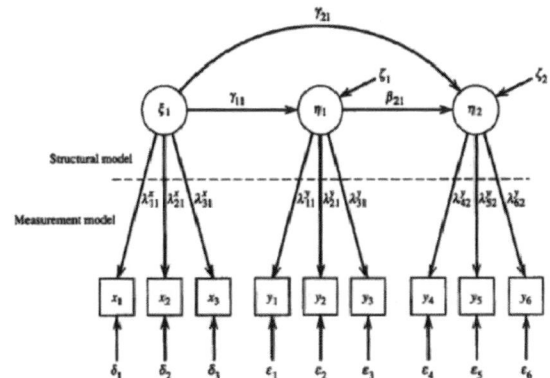

Fig.1. The Relationship between Variables of SEM

E. Qi et al. (eds.), *Proceedings of the 21st International Conference on Industrial Engineering and Engineering Management 2014*, Proceedings of the International Conference on Industrial Engineering and Engineering Management, DOI 10.2991/978-94-6239-102-4_129, © Atlantis Press and the authors 2015

$$X = \lambda_x\xi + \delta \qquad (1)$$
$$Y = \lambda_y\eta + \varepsilon \qquad (2)$$
$$\eta = B\eta + \Gamma\xi + \varsigma \qquad (3)$$

"(1)" and "(2)" are measurement models, "(3)" is structural model. X is for exogenous observed variables, ξ is for exogenous latent variables, λ_x is the factor loading matrix, δ is the residual of exogenous observed variable. Y is for the endogenous observed variables, η is for endogenous latent variables, λ_y is the factor loading matrix, ε is the residual of the endogenous observed variables. B and Γ represent the path coefficient. ς stands for the residual of SEM [6].

III. QUESTIONNAIRE DESIGN

In this part, scale design, data collection and sample statistics are finished. First of all, a pretest and in-depth interviews are conducted in 10 cultural enterprises to pick up the specific items. After the expert reviews, a small-scales test is carry out among 30 enterprises in Tianjin for verification. Then the final scale of influence factors comes into being. It has 5 subscales including 40 items. They are economic environment scale (6 items included), social environment scale (5 items included), technological environment scale (3 items included), policy environment scale (20 items grouped into financial policy, educational policy and industrial policy) as well as internal capacity scale (6 items included). Likert scale is used to quantify the results. We totally send out the questionnaire 250 cents in this survey and finally 213 are valid, the 213 respondents are sufficient. The effective returns-ratio is 85.2% and the sample statistics are shown in Table I.

TABLE I
SAMPLE STATISTICS

Classification of enterprises		Number of enterprises	Percentage
Group by income level	Annual revenue> 5 million yuan	174	0.817
	Annual revenue< 5 million yuan	39	0.183
	Aggregate	213	1
Group by ownership	State-owned enterprises	30	0.141
	Non-state-owned enterprises	183	0.859
	Aggregate	213	1

IV. EMPIRICAL ANALYSIS

A. Exploratory Factor Analysis (EFA)

First of all, before the questionnaire analysis, the reliabilities and validities of the scale is examined with SPSS16.0. Results show that the value of Cronbach's α is 0.943 for the entire scale and the Cronbach's α for 5 latent variables is 0.812 (for economic environment), 0.759 (for social environment), 0.768 (for technological environment), 0.947 (for policy environment) and 0.823 (for internal capacity). That means the reliabilities and

validities of this test meet the standards of psychological measurement.

Then exploratory factor analysis (EFA) should be used to conduct a research based on 40 items in order to find out the common factors as well as the correlation between influences and observations. Final result shows the scale structure is effective and suitable for factor analysis (KMO=0.921>0.5, sig<0.05).

Next, Maximum Likelihood (ML) is used in sampling, the factors are extracted by principal component analysis and rotated by Varimax. It turned out 8 factors' eigenvalues exceed 1. While the 3 items (the proportion of the added value of the industry in GDP, the market competition and the eco-environment of cultural enterprises) don't belong to any dimension for they have no convergence. At last, 6 factors' eigenvalues are found greater than 1 after cutting off the 3 items mentioned above. The final factor matrix shows that 13 items are taken out for their factor loads are all less than 0.5. The 24 remaining items converge to 5 factors at last. The result is shown in Table II.

The cumulative contribution of the 5 factors is 78.24%. That is to say the 40 items could be represented by the 5 common factors: enterprise's soft power (softpow), enterprise's hard power (hardpow), institution environment (instenv), industrial policy (indpol) and financial policy (finpol). Otherwise, the enterprise's soft power and hard power are merged into one factor: enterprise's internal capacity (incap). This article explores the relationship among enterprise's internal capacity and the other 3 common factors.

B. Research Hypothesizes

H1: There will be a significant positive relationship between enterprise's internal capacity and institution environment

As the micro carriers, enterprises play an important part of modern society coordinating with the institution environment such as social, economic and technological environment which bring new opportunities for them [7]. Facing with the opportunities and challenges, enterprises have been improving self-regulation capacities to adapt to the new circumstances. In addition, favorable technical environment could help enterprises to renew their concept and management to innovate production techniques by digesting, absorbing and recreating [8].

H2: Cultural industrial policy will relate significantly and positively with enterprise's internal capacity

Reasonable industry policies (such as talent policies, reward policies and land policies) occupy a core position in cultural enterprises, which can help distribute social resources through changing industrial structure and the layout to improve their competitiveness [9]. In addition, industrial policies can help to standardize market order and crack down the unjust competition to create a convenient and safe environment for consumers.

H3: Financial policy will relate positively and significantly with enterprise's internal capacity

As fund is the blood of an enterprise's development, financial policies play an important role through culturally-relevant tax policies, venture capital funds and financial allocation and so on. On one hand, it can guide the social capital flows in order to optimize the economic structure and stimulate innovation[10]. On the other hand, financial policies help enterprises to solve fund difficulties and improve the capabilities of withstanding risks. What's more, support of the interest rate and the exchange rate will help cultural enterprises to exports around the world with high technological products.

TABLE II
EXPLORATORY FACTOR ANALYSIS ON REGIONAL CULTURAL INDUSTRY

Items	Factors				
	softpow	hardpow	instenv	indpol	finpol
External environment (softpow1)	0.713				
Innovation consciousness and behavior (softpow2)	0.634				
Innovative talents availability (softpow3)	0.658				
Motivation mechanism for innovation (softpow4)	0.597				
Marketing of the products and service (softpow5)	0.636				
Technology level (hardpow1)		0.759			
Enterprise's financial competence (hardpow2)		0.635			
Regional agglomeration of cultural industry (instenv1)			0.687		
Market requirements of products and service (instenv2)			0.642		
Supplements of cultural products and service (instenv3)			0.697		
Competition and cooperation (instenv4)			0.621		
Maturity of the cultural medium (instenv5)			0.701		
Reward policies for of cultural products (indpol1)				0.609	
Preferential land policies for cultural industry (indpol2)				0.581	
Export subsidies for cultural products (indpol3)				0.612	
Policies of the talent introduction and training (indpol4)				0.628	
Good public service platform (indpol5)				0.615	
Cultural system reform (indpol6)				0.676	
Policies for intellectual property rights protection (indpol7)				0.713	
Culturally-relevant tax policies (finpol1)					0.568
Venture capital funds (finpol2)					0.573
Special fund policies (finpol3)					0.602
Discount loan of financial institution (finpol4)					0.794
Government procurement (finpol5)					0.556

C. Establishment and Verification of SEM

In this part, Generalized Least Squares method (GLS) is used in establishing SEM and the statistics tool AMOS17.0 is used to put out the route picture of structure equation.(Here the model path graph of structure equation is omitted.) The result of goodness-of-fit tests about the model is shown in Table III.

TABLE III
RESULT OF MODEL VALIDATION AND THE
GOODNESS-OF-FIT TESTS

Model	CMIN	DF	P	CMIN/DF	CFI	NFI	IFI	RM- SEA
Default Model	324.2	131	0	2.172	0.79	0.832	0.86	0.071

From Table III, it can be concluded that the result is not sufficiently good and the model should be modified (CFI=0.79<0.9, NFI=0.832<0.9). The correct model path graph is shown in Fig.2. And the result of goodness-of-fit tests is shown in Table IV.

TABLE IV
MODIFIED RESULT OF MODEL VALIDATION AND THE
GOODNESS-OF-FIT TESTS

Model	CMIN	DF	P	CMIN/DF	CFI	NFI	IFI	RM-SEA
Default Model	126.6	121	0.38	1.32	0.93	0.917	0.91	0.038

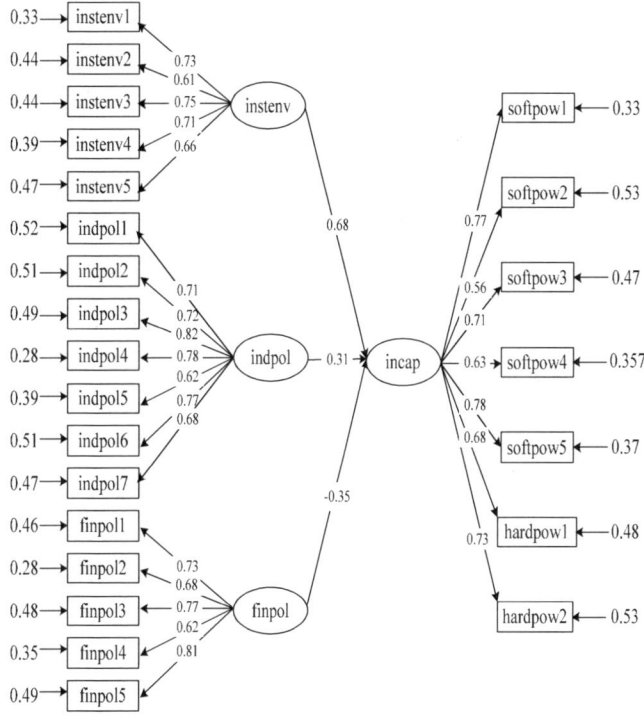

Fig.2. Model Path Graph of the Influence Factors upon Regional Cultural Industry

Table IV shows that the result is sufficiently good enough (P=0.38, CFI=0.93>0.9, NFI=0.917>0.9, RE–SEA=0.038<0.05). The estimate of statistical parameters is shown in Table V.

TABLE V
ESTIMATE OF STATISTICAL PARAMETERS

Path relationship	Standardized path coefficients	T	Significance level
Instenv → Incap	0.68	5.39	Significant under the 1% level
Finpol → Incap	-0.35	-1.8	non-significant under the 1% level
Inpol → Incap	0.31	2.86	Significant under the 1% level

V. RESULTS

Table V hows enterprise's internal capacity relates significantly and positively with institution environment (SPC=0.68), as well as the industrial policy (SPC=0.31). And it negatively correlates with the financial policy (SPC=-0.35), but the effects are insignificant (T=-1.8). That is to say institution environment plays a significant role in enterprise's internal capacity while the active effects of industrial policies in Tianjin cannot reach the expected level (SPC=0.31<0.68). The financial policy isn't significantly correlated with internal capacity, which can be explained that the importance of financial policy hasn't been reflected.

The industrial policy and financial policy belong to the macro policy innovation of the enterprise's management, so the 2 factors above can be described as the macro management innovation.

VI. CONCLUSION AND SUGGESTION

The political direction of the country plays a decisive role in the formation and development of cultural enterprises, for instance the macro management innovation and the institution environment. From the survey, the cultural industrial policies couldn't achieve the desired effect. That could be due to the fuzzy definition and the poor execution. For another, the lack of relevant government departments and powerful policy system support is not conducive. Above all, it can be concluded that the legal system and government policy especially the financial policy should be strengthened and optimized to create a good external environment for the development of cultural enterprises in Tianjin. Fund must come first, which is the biggest obstacles. Local government help to offer special fund and positive investment policies is of great importance to innovate service and management and to improve the investment environment in order to broaden its appeal constantly.

ACKNOWLEDGMENT

The authors wish to thank Dr. Li-ping Fu for her continuing care and guidance in the process of choosing topic and research, and Dr. Lan-ping He for her helpful discussion. Also the authors like to thank their families and schoolmates for their support during the work. What's more, the authors wish to thank the reviewers of this paper for their helpful and valuable suggestions. Give the sincerity appreciation and great respect to them.

REFERENCES

[1] Du Zhimin, Lei Xiaokang, "On the difficulties of the development of china's cultural industries and its countermeasures" (in Chinese), Chinese Public Administration, vol. 06, pp. 019, 2010.

[2] Xie Xuefang, Zang Zhipeng, "Foreign cultural industry development strategy based on government-led perspective [J/OL] (in Chinese)", Journal of Beijing Institute of Technology (Social Sciences Edition), vol.04, 2012, http://www.cnki.net/kcms/detail/11.4083.C.20120726.110 5.201204.23_004.html.

[3] Huang De-sen, Yang Chao-feng, "Research on the influence factors upon animation industry based on structural equation model" (in Chinese), China Soft Science, vol. 05, pp. 148-153, 2011.

[4] Li Xing-jiang, Fan Fan, "An empirical study of the factors influencing the development of cultural industry in Gansu province" (in Chinese), Journal of Lanzhou Commercial College, vol.01, pp. 67-72, 2014.

[5] Götz O, Liehr-Gobbers K, Krafft M. "Evaluation of structural equation models using the partial least squares (PLS) approach", Handbook of partial least squares, Springer Berlin Heidelberg, Press, 2010,pp. 691-711.

[6] Ong T F, Musa G. Examining the influences of experience, personality and attitude on scuba divers' underwater behavior: a structural equation model [J]. Tourism Management, 2014, 30: 1-14.

[7] Pan K, Ding T, "Political status of the private enterprise actual controller, the transition of institution environment, and loan contract", Journal of Shanxi Finance and Economics University, vol.07, pp: 008, 2012.

[8] Aghion P., Acemoglu D., Bursztyn L., & Hemous D, "The environment and directed technical change", NBER Working Paper, vol.21, 2011.

[9] Jiang Ling, "The basic policy types of China's cultural industry in recent ten years"(in Chinese), Journal of Jiangnan University (Humanities & Social Sciences Edition), vol.11, no. 01, pp. 90-97, 2012.

[10] Wei Peng-ju, "The desirability and fiscal policy of public founding for cultural industries" (in Chinese), Chinese Public Administration, vol.05, pp. 45-47, 2009.

The Demonstration on Influence Factors of Engineering College Students' Innovation Ability

Xiang-sheng MENG*, Zi-biao LI, Cun SONG

School of Economics and Management, Hebei University of Technology, Tianjin, China

(xshm@hebut.edu.cn)

Abstract - **Engineering Students' innovation ability is the source of national science and technology, and it has important theoretical and practical significance for us to investigate factors of college students' innovation ability. So, in this paper, the definition of engineering college students' innovation ability is given firstly. Then, some influence factors of engineering college students' innovation ability: business management ability, engineering application ability, innovation personality and cultural feature are summarized. After the test on reliability and validity, regression analysis is made for the students' innovation ability influence factors. Finally, some suggestions are given on cultivation of college students.**

Keywords - **Demonstration, engineering college students, influence factor, innovation ability**

I. INTRODUCTION

Innovation has become the main power of economic development and social progress [1]-[3]. Innovation ability of college students represents for a country or a region's innovation education and potential of future development [4]-[6]. The engineering college students' innovation ability mainly refers to the ability of realizing technical innovation in the industrial field. Engineering college students have professional advantages in the process of innovation. But the study on knowledge structure, adaptive ability, psychology characteristic and some other aspects are sometimes neglected, which in some extent will affect their innovation ability. In this paper, by carrying out students' reality investigation, we find the main factors influencing engineering college students' innovation ability, and give some advice on engineering college students' innovative training.

II. THE INFLUENCE FACTORS OF INNOVATION

Innovation is a complex system process with multi-subjects participation and many links fitted together. There are many influencing factors on the cultivation of college students' innovation ability [7]. The first of them is the cultural factor. The cultural influence on innovation consciousness and innovation efficiency is continuous. The second influencing factor is college students' personality [8]. In general, innovation personality is the innovative talents representing in thinking activity such as original thinking, thinking acuity, innovation perseverance and determination, practice ability, predictability and other personal qualities. It means that the innovative talents based on their experience to give new judgment, new ideas and new conclusions for the

new situation, new problems and new things. The third is the practical operation ability [9]. Practical operation ability mainly refers to the engineering college students' ability that can independently accomplish the corresponding operation, familiar with related fields, standard and complete the technical operation or achieve innovative ideas. Practical operation ability of college students directly determines their innovation output, thus affects their innovation ability. Finally, management ability is an important factor supporting the college students' innovation [10]. In developed country, mature entrepreneurship education institution establishes related management courses. About 86% entrepreneurs who make innovation entrepreneurship successfully have system education in the field of management. This shows the important position of management ability in innovation and enterprise education. The main related management skills of engineering college students are: technology commercialization ability, ability of sharing knowledge with team members, communication ability, etc.

III. THEORETICAL MODELS

According to the previous discussion, we get the causality model of the relation between innovation ability and the business management ability, engineering application ability, innovation personality, cultural feature (See Fig. 1).

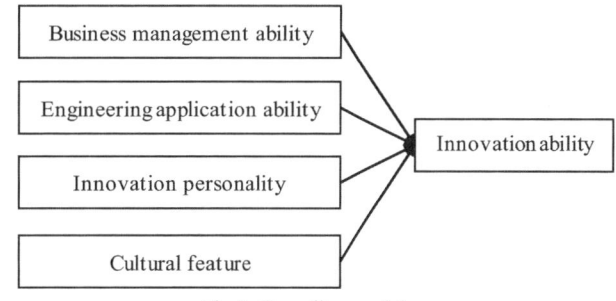

Fig.1. Causality model

Here we put forward the corresponding relationship assumptions between the four abilities and innovation ability:

H1: Business management ability has a significant influence on the innovation ability.

H2: Application ability of engineering operation has a significant influence on the innovation ability.

H3: Innovation personality has a significant influence on the innovation ability.

H4: Cultural feature has a significant influence on the innovation ability.

A. The selection of business management ability index

In this paper, we propose the following index (Table I).

TABLE I
BUSINESS MANAGEMENT ABILITY INDEX

	Item number	Index
Business management ability	Management 01	Features known the problem of intellectual property rights and contract issues
	Management 02	Understanding the customer and design requirements
	Management 03	The ability to make team members to have a consistent goal
	Management 04	Sharing knowledge in the team
	Management 05	The ability to know about team members' psychological needs

B. The selection of engineering application ability index

The engineering application ability is nothing more than an ability, which is based on engineering environment and supported by engineering to express engineering technology and engineering knowledge through engineering equipment. In this paper, we give the following index (Table II):

TABLE II
ENGINEERING APPLICATION ABILITY INDEX

	Item number	Index
Engineering application ability index	Operation 01	Thinking ability of the utilization of technology literature and using other information resources
	Operation 02	The ability of clearly stating your design and explaining the key factors and limitations
	Operation 03	The ability of using modern engineering tools and equipment
	Operation 04	The applying ability of specific materials, equipment, process and product characteristics and other knowledge
	Operation 05	The ability of elaborating engineering problems
	Operation 06	The ability of knowing quality problems
	Operation 07	The ability of skilled operating experiment or process

C. The selection of innovation personality index

There are many factors influencing college students' innovation ability, such as ability of innovation learning, innovation thinking ability, innovation capability and so on. They interact each other; they connect to each other and restrain each other. None is dispensable. The whole innovation ability quality is guaranteed only if the quality of each element's quality is ensured.

With the above discussion, we select the measurements as Table III:

TABLE III
INNOVATION PERSONALITY INDEX

	Item number	Index
Innovation personality	Personality 01	With a strong desire to accept new knowledge and new skills
	Personality 02	Like to use new method to solve problems
	Personality 03	Like multiangle thinking personality
	Personality 04	Learn a lot through mistakes and failures
	Personality 05	Independently assume the risk of trying new things
	Personality 06	Be adept at flexibly applying knowledge to the work

D. The selection of cultural feature index

We give the index as Table IV:

TABLE IV
CULTURE FEATURE INDEX

	Item number	Index
Culture feature	Culture 01	Official rank standard is not popular in the local culture.
	Culture 02	There are values of pursuing utility and advocating competition in the local culture.
	Culture 03	In local culture, there is a mentality of trying on new things.
	Culture 04	In the local market, compared with cheap consumer goods, people are willing to pay relatively higher price for innovative products.
	Culture 05	Has good tolerance with foreign culture.
	Culture 06	In the local culture, there is a high tolerance for failure.
	Culture 07	In the local culture, there is a good cultural atmosphere for cooperation and open policy.

E. The selection of innovation ability index

In this paper, we propose the following index (Table V).

TABLE V
INNOVATION ABILITY INDEX

	Item number	Index
Innovation ability	Innovation 01	Have the ability of quickly translating technology into products
	Innovation 02	The ability of integrating the existing technology and improving product design
	Innovation 03	Have development and design which are difficult to be imitated
	Innovation 04	The ability of proposing new patented invention
	Innovation 05	Own the ability of getting products into the market quickly

IV. STATISTICAL DESCRIPTIONS

A. Sample data distribution

There are 138 valid samples in this study, in which 10 people are engaged in electronic communication industry, 24 people are engaged in chemical material industry, 43 people are engaged in electrical and mechanical major, 13 people are engaged in software and IT industry, 7 people are engaged in biological pharmaceutical industry, the remaining 41 people are engaged in energy industry, real estate, consulting, food processing industry and other science technology majors. Among the 138 valid samples, 17 of them get the major college degree, 85 of them obtain the corresponding bachelor's degree, 28 of them obtain master degree in which receive doctor's degree. These coincide with the characteristics of engineering practice guidance.

B. The self-assessment on innovation ability

In statistical variables of the questionnaire, subjects are allowed to give self evaluation of their own innovation ability. Among all the 138 respondents, 17 of them think that their innovation ability is relatively poor, 59 of them think that their innovation ability is general, 31 of them think that they have strong innovation ability, the last 31 think that their innovation ability is very strong. About half of the respondents think that their innovation ability is general or weak. Overall, subjects have general evaluation of their own innovation ability.

C. Business management ability

In the statistical description section of the questionnaire, subjects are also asked to give self evaluation about the effect of their business management ability in innovation activities. Among all the 138 respondents, only 11 of them think that the effect of business management ability in the technical innovation is not obvious, and innovation ability is not important. 47 of them think that business management ability plays important role in innovative activities. 80 of them think that business management ability plays very important role in innovative activities. In general, engineering college students have higher consistent evaluation of business management ability, which to say that business management ability plays an important role in the innovation activities, especially in the case where technical talents change to management positions.

D. Methods to acquire new knowledge

In the statistical description section of the questionnaire, the ways of how to acquire new knowledge are also investigated. In most of the respondents, they obtain the corresponding new knowledge by at least two kinds of methods. Engineers are full of enthusiasm for new knowledge, which meets the time requirement of the engineer needing a lifelong learning ability. Among the respondents, 53% of them attend systematic training; 39% of them attend academic lectures to satisfy their desire for new knowledge; 34% of them often concentrate on the front in their spare time or independently study related knowledge; 29% of them pay attention to found problems and looking for possible breakthrough.

E. Statistical description of innovation ability

In the Table VI, we give a summary of statistical descriptions of the tested engineering college students' innovation ability. Each problem item's maximum, minimum, mean, median, median number and standard deviation are listed respectively. The mean of item 2 (Innovation 02), item 3 (Innovation 03) and item 5 (Innovation 05) is more than 3, moreover the mean of item 5 reaches 3.733, which shows that engineering students have a better ability to getting product into market. The lowest mean is 2.7333 reached by item 01 (Innovation 01), which means that the ability of translating technology into products is weak. The mean of item 4 (Innovation 04) is 3.000, which shows that students' ability of proposing new original innovation has average performance. But in general, engineering college students have higher innovation ability.

TABLE VI
DESCRIPTIVE STATISTICS

		Innovation 01	Innovation 02	Innovation 03	Innovation 04	Innovation 05
N	effective	138	138	138	138	138
	absence	0	0	0	0	0
Mean		2.7333	3.3333	3.2000	3.0000	3.7333
Median		3.0000	3.0000	3.0000	3.0000	3.0000
Mode		2.00ᵃ	3.00	3.00	3.00	3.00
Standard deviation		1.09978	.81650	.77460	.75593	1.03280
Minimum value		1.00	2.00	2.00	2.00	2.00
Maximum value		4.00	5.00	5.00	4.00	5.00
Percentile 25		2.0000	3.0000	3.0000	2.0000	3.0000
	50	3.0000	3.0000	3.0000	3.0000	3.0000
	75	4.0000	4.0000	4.0000	4.0000	5.0000

Note: There are many modes displaying minimum value.

V. MULTIVARIATE REGRESSION ANALYSIS

We put forward 4 assumptions of innovation process (assumption H1-H4). In this paper, innovation ability is the explanatory variable; business management ability, engineering application ability, innovation personality and cultural feature are independent variables. The regression model is established. And we give accurate judgment of the relationship between the dependent variable and independent variables. Analysis results of regression model are shown in Table VII.

In the regression model, the first one, which enters the regression model, is innovation personality. Its standardized regression coefficient is 0.450, and the independent variable's significance level is 0.000. This result shows that innovation personality has a significant positive influence on the engineering college students' innovation ability. That is, the stronger the level of innovative personality, the greater contribution of innovative personality to engineering college students' innovative ability. As a result, by the regression test results, H3 pass the hypothesis test, that is, innovation personality has a positive promotion to the engineering college students' innovation ability.

TABLE VII
REGRESSION ANALYSIS

Model	Standardized regression coefficient	T value	Significance level	Collinearity inspection	
				Admissible width	Expansion factor
Innovative personality	0.450	5.875	0.000	1	1
Innovative personality	0.450	6.458	0.000	1	1
Management ability	0.377	5.416	0.000	1	1
Innovative personality	0.450	6.989	0.000	1	1
Management ability	0.377	5.861	0.000	1	1
Engineering application ability	0.316	4.910	0.000	1	1
Innovative personality	0.450	7.474	0.000	1	1
Management ability	0.377	6.268	0.000	1	1
Engineering application ability	0.316	5.250	0.000	1	1
Cultural feature	0.271	4.501	0.000	1	1

In the regression model, the second one, which enters the regression model, is business management ability. Its standardized regression coefficient is 0.377, and the independent variable's significance level is 0.000. The above result shows that business management ability has a significant positive influence on the engineering college students' innovation ability. That is, the stronger business management ability they have the more outstanding engineering college students' innovative ability they possess. By the regression test results, H1 pass the hypothesis test, that is, business management ability has a positive promotion to the engineering college students' innovation ability.

In the regression model, the third one, which enters the regression model, is engineering application ability. Its standardized regression coefficient is 0.316, and the independent variable's significance level is 0.000. This result shows that engineering application ability has a significant positive influence on the engineering college students' innovation ability. That is, the stronger engineering application ability, the higher engineering college students' innovative ability. Thus assumption H2 is supported by our sample data, H2 pass the hypothesis test.

In the regression model, the fourth one, which enters the regression model, is cultural feature. Its standardized

regression coefficient is 0.271, and the independent variable's significance level is 0.000. This result shows that cultural feature has a significant positive influence on the engineering college students' innovation ability. That is, the more outstanding engineering application ability, the stronger engineering college students' innovative ability. Thus assumption H4 is supported by our sample data, H4 pass the hypothesis test.

Simultaneously, as shown in the table this model is significant in statistical significance (F = 35.744, P < 0.001), which indicates that the regression equation has a good interpretation effect. In a certain extent, the F value is increased as the independent variable increase. That is to say that the four independent variables should be included in the regression equation, and they can well explain the variance of the dependent variable's variance. Combining with the above regression analysis, we can see that the assumptions of the four independent variables all pass the hypothesis test. The study is consistent with expectations.

VI. THE RESULTS AND DISCUSSION

A. Analysis the results

This study investigate the students' innovation ability, we verify the model of the influence of business management ability, engineering application ability, innovation personality and cultural feature to innovation ability. The verification results show that the original hypothesis has been confirmed. The four stages have positive influence on engineering college students' innovation ability.

B. Training strategy

1) Management ability training

Business management ability can close the connection between technology and economy and society. Related management ability training courses, therefore, should be appropriately joined in the university education stage. It can make students to have technical innovation by using knowledge of relevant laws and regulations, information system, team management, innovative methods. This curriculum module is often opened with business college cooperation, focusing on new financing and management problems arising on new business start-up period.

2) Engineering application ability

Practice is the only way from ideal to reality. Therefore the students' ability training should lay particular emphasis on the ability of solving practical problems. It can be carried out by adding engineering training and setting up school-run factory.

3) Cultivate college students' innovation consciousness

Innovation consciousness belongs to the category of personality structure's attitude toward reality. Only one has a strong sense of innovation consciousness, can he dares to think things which predecessors never thought, and create a new career which predecessors never created.

ACKNOWLEDGMENT

This research is supported by ZH2011206 of Hebei Province.

REFERENCE

[1] OECD, "The knowledge-based economy", *OECD, Pari.* 1996.

[2] J. Fagerberg, D. C. Mowery, R. R. Nelson, "The Oxford handbook of innovation", *Oxford University Press*, 2004.

[3] L. Q. Jia, X. Liu, Y. L. Wang, "Research progress on new Schumpeterian school's technology innovation theory", *Forum on Science and Technology in China*, no. 5, pp. 39–41, 1995 (Chinese).

[4] C. Y. Jiang, "Innovation pole: the mission and pursue of research university", *China higher education*, no. 1, pp. 40-42, 2006 (Chinese).

[5] B. Martin, "The relationship between publicly funded basic research and economic performance: a SPRU review", *Report for HM Treasury*, SPRU, University of Sussex, 2011.

[6] N. Rosenberg, R. R. Nelson, "American universities and technical advance in industry", *Research Policy,* no. 23, pp. 323-348, 1994.

[7] H. Guo, Y. Mao, X. Bai, Z. Zeng, "A study of the impact of the enterprise on entrepreneurial intention of college student education", *Soft Science*, no .9, pp. 69-74, 2009 (Chinese).

[8] Q. Y. Wu, S. H. Ding, W. H. Hou. "A study on the impact of college students' personal characteristics on entrepreneurship tendency", *Pioneering with Science & Technology Monthly*, no. 6, pp. 30-31, 2008 (Chinese).

[9] H. Wan, H. H. Zhang. "The connotation and structure of college students' innovation ability--case and empirical study", *Journal of National Academy of Education Administration*, no. 2, pp. 81-86, 2012 (Chinese).

[10] Y. Xu, "Formation of an entrepreneurial competency scale", M. D. dissertation, *East China normal university, Shanghai, China*, 2011 (Chinese).

The Operational Risk Management Information System in Commercial Banks of China: Issues and Analysis

Shaoqiang Qu[1],[*], Yongping Ma[2]

[1] School of Accounting, Shandong Institute of Business and Technology, Yantai, China
[2] Yantai Rural Commercial Bank Company ltd., Yantai, China
(13688680961@163.com)

Abstract – **Operational risk is the second major risk faced by the commercial banks of China. So how to manage it effectively has always been a fundamental task for us. A consensus had now been reached that the operational risk management of commercial banks in China is a systematic engineering. The operational risk management strategy and operational risk management process, as two important factors in this system engineering, cannot do well without effective underpinning coming from other factors, for example, the operational risk management information system (ORMIS) and so on. In this paper, the operational risk management information system, which has strong supporting to the management of the operational risk of commercial banks in China, has been analyzed in its main functions, its information transmission routes and security. At last, some suggestions have been put forward in order to provide useful help to the operational risk management in commercial banks of China.**

Keywords – **Function and information transmission route, operational risk, security, the operational risk management information system (ORMIS)**

I. INTRODUCTION

The operational risk management in the commercial banks of China is a systematic engineering and how to run it effectively are the issues that we have to face. According to the system safety theory [1], everything can be seen as a system and is made up of smaller related systems. When a factor relating to human or material stops working or losses its function, risk event will happen. This theory tells us that the risk management is a systematic engineering. It relates to different factors including people, process, machine and environment, and so on. An organic whole is constituted by these factors. Once a part of the system has problems, there will be a paralysis of the entire system. Therefore for us, the risk management is not the responsibility of one or two individual sectors in the system, but a full range of work. As a systematic engineering for the operational risk management of the commercial banks in China, effective operational risk management strategy and management processes cannot do well without coordination of relevant factors, without mutual supporting of related factors. For the operational risk management information system (hereafter referred to as ORMIS) in the commercial banks of China, it is one of the most important factors that have a strong supporting to the effective operational risk management for the commercial banks of China.

The human society has entered the information society in today. The information has become an indispensable part of our daily management activities. The information theory, which is a part of the system theory, considered that any practical activities, putting aside the specific movement patterns relating to the matter and energy, can be simplified as multi streams, namely the stream of people, the stream of logistics, and the stream of financing, the stream of energy and the stream of information. The information stream, which can also be called information flow, plays a dominant role in all of these streams. It is the reasonable flowing of information in the system that makes the system to maintain its normal purposeful movement. In the internal of system itself, the information stream is also adjusting the amount, direction, speed and target of other streams, and controlling regular exercises of people and material. It is always an important factor of supporting to the efficient operating of the system. Hence, as the management systematic engineering for the operational risk management in commercial banks of China, the information flow also plays an important supporting role in the effectiveness of operational risk management.

In this paper, the operational risk management information system has at first been analyzed about its main functions, its information transmission routes and security by utilizing qualitative analysis. At last, some suggestions have been put forward in order to provide useful help to the sound operational risk management in commercial banks of China.

II. ISSUES IN THE MANAGEMENT INFORMATION SYSTEM IN THE COMMERCIAL BANKS OF CHINA

In China, the management information system of commercial banks is an application system based on bank information system and underpinned by network. It is a comprehensive management information system based on Intranet Network and so far had perfect office LANs among the head office, the first level branches and the second level branches of the bank. "Between the head office and the first level branches, the implementation of LANs interconnection is realized through the first level data communication network of the bank. And, between the first level branch and the second level branches, the realization of LAN interconnection is achieved through the second level data communication network of the bank." [2] In general, the management information system of commercial banks, which is combined organically with comprehensive banking business information system, has powerful security mechanisms, and rigorous authorization management, and meticulous operation rules. It has high operation stability coefficient.

E. Qi et al. (eds.), *Proceedings of the 21st International Conference on Industrial Engineering and Engineering Management 2014*, Proceedings of the International Conference on Industrial Engineering and Engineering Management, DOI 10.2991/978-94-6239-102-4_131, © Atlantis Press and the authors 2015

Currently, some achievements have been made in the construction of management information system of banks in China, especially in the aspect of hardware. But there are still insufficient issues due to various reasons, mainly in the following:

A. The Weakness in the Fundamental Work

For most commercial banks of China, the overall planning was not fully in place and the fundamental work was weak at the beginning of the construction of the information management system due to historical reasons. These resulted in the fact that the development of the information management system was begun basically in the different time and different management departments according to their respective needs. This would be bound to cause inconsistent standard of the information collection and exchange, which is faced with how to carry out the transmission of information between different banks, different management departments.

B. The Lagging of the Risk Monitoring

For commercial banks of China, the risk monitoring ways were behind times due to coming late of the ideas of operational risk management. This resulted in different degrees of attention to the operational risk management for different commercial banks. The result was that some banks were having better control to the risk information and some banks were being at preliminary stage. However, it was an indisputable fact that the risk monitoring way had been behind times.

C. The Lacking of the Risk Management Data

The data is the main load of the information flow, which shows that the risk information system is strongly dependent on the risk data. But it was a universal fact that the risk data management had not been in place for institutions at all levels of China's commercial bank. This is bound to affect effectively and efficiently the running of the operational risk management information system.

In order to provide stronger support to the operational risk management, the ORMIS (the operational risk management information system) which should be to run smoothly must be established without delay.

III. THE MAIN FUNCTIONS OF THE ORMIS IN COMMERCIAL BANKS OF CHINA

At present, operational risk is one of the main risks faced by the commercial banks of China. The effective and efficient management of it must rely heavily on the strong support of bank management information system. "Accurate, timely and comprehensive data along with robust, integrated information systems are an integral part of an effective risk management program. The firm's risk management systems must have the ability to capture and measure key risks in a globally integrated manner."[3] The

ORMIS is the subfield of the bank management information system in the aspects of operational risk management. The ORMIS is essential in the operational risk management process. It is the basis and the propellant of operational risk management. The process of scientific collection, induction and deduction about loss data are dependent greatly on the successful ORMIS for commercial bank, otherwise, the carrying forward of the operational risk management will be restricted greatly. At present, the ORMIS of international active banks are still in the process of exploration and construction. The degree of perfection and maturity of them cannot be mentioned in the same level with the market risk and credit risk management information system. Even if for HSBC, as a famous international active bank, the final statistical summary report is still based on manual for the operational risk management information which has been collected although through information system[4].

For China's commercial banks, the qualified ORMIS should include the information on operational risk management widely. Not only can it efficient collect and accumulate information relating to the management of operational risk from front desk operators of the banks, but also can provide information to the background and the management departments of the bank. It can be linked smoothly with other internal systems, and when necessary, connected with other external information systems unhinderedly, implementing relating data downloading and operating and other functions.

The main supporting functions provided by the ORMIS should include the technology adoptions and their feedbacks relating to the operational risk control self assessment, and the tracking of loss event, and the key risk indicators, and the qualitative or quantitative analysis and capital allocations.

A. The Control Self Assessment of the Operational Risk

For the control self assessment of operational risk, the ORMIS can provide powerful support for data archiving and assessing of operational risk, which includes business process, operational risk events and the corresponding control measures to the operational risk management. The ORMIS can also support the qualitative and quantitative analysis of the operational risk, also can provide support to the making of the plan of action and revising of it. For the outputting of operational risk management data, and the testing, reviewing and evaluating of self assessment process, the ORMIS can provide powerful support, as the same as to the bank's administrative management and audit trail.

B. The Tracking of Loss Events

It is very important to record timely loss events in the operational risk management process. The tracking on the existing and possible loss events can be carried out through the ORMIS. Relying on the commercial bank's ORMIS, the occurring time of loss events, the amount of losses and the mutual relationship of them can be treated with timely and effectively. In order to obtain perfect recording of loss events, the ORMIS can assist the correction of tracking if the previous records were not complete

C. The Key Risk Indicator

In the processes of operational risk management for commercial banks of China, some key statistical indicators and financial indicators are usually chosen so as to reflect the operational risk profile of the bank, to monitor the changes of the operational risk. The ORMIS can provide the tracking of some specific values relating to the management of operational risk. The obtaining of data from the data source can be accomplished with manual and automatic ways based on the ORMIS. Also, the functions of early warning indicators, the establishing of the benign interactive relationship of operational risk and its control can be carried out through the commercial bank's ORMIS.

D. The Qualitative or Quantitative Analysis

In the processes of operational risk management, the application of qualitative analysis and quantitative analysis is essential. But it should be verified whether the approaches of qualitative analysis and/or quantitative analysis are fully consistent with the needs of the bank. Therefore, the real-time tracking and the feedback of information on the approaches are necessary so as to obtain required information for further correction action. The ORMIS can provide this support strongly.

E. The Capital Allocation

Capital allocating to the operational risks faced by commercial bank is the imperative requirement of the Basel banking regulatory authorities. The commercial banks of China must attach great importance to it. The ORMIS can provide added risk exposure, can provide the data for the capital model used or some system functions, can promote continuous optimization of the calculation of capital allocation.

IV. THE INFORMATION TRANSMISSION ROUTES OF THE ORMIS

One of the characteristics of commercial banks in China is that multi-level organizations have been set up by much of the banks and these organizations are over a broad in geography. In such case, how to ensure the high speed, accurate and reliable transmission of information related to the operational risk management, namely the design of information transmission routes becomes an important issue that can not be ignored in the course of the ORMIS construction.

Based on the System Theory and authors' practical experiences many years in a commercial bank in China,

we think the integral information transmission routes of the ORMIS should be divided into two lines in the opposite direction respectively. One is the route of from bottom to top, which can be used to transfer risk information. Another is the up-to-down route, which can be used to transfer decision making related to the operational risk management. The two routes should be indispensable and depend mutually on and constitute an organic whole.

A. The Bottom to Top Transmission Route

The managers coming from the business sector and the first-line of the banks and being responsible for operational risk management is the starting point of operational risk information transmission route. They collect operational risk information relating to the business operation, including the occurring time and location of risk events, amount of loss, the relevant responsible person, the types of risk, and having been taken or will being taken remedial measures, and whether a large operational risk loss events and so on. These are basic and original information needed for the operational risk management. The requirements demanded on them should be comprehensive, accurate and timely. The managers, described above, should at once transfer such risk information collected to a higher level above them or the head office of the risk management department after they have collected them at any time.

After the operational risk information has been received through the ORMIS of the bank by the functional departments in charge of the operational risk management, the information needed to upload should be sorted out, together with further collection in these functional departments, including periodical verification and gathering analysis of such information coming from the managers on the business departments and front-line operational risk management. The analysis tools can include qualitative analysis methods and/or quantitative analysis tools. Among them, operational risk VaR model and stress testing model may be used to conduct the risk information depth processing, together with the testing of policies of some relevant functional departments and the testing of regulation rules and supervisory policies of the supervisory authorities, so as to analyze and evaluate the original and summary risk information integrally. Based on the ORMIS, the risk information and the relevant policy information and the evaluation information of the functional departments should be packaged in accordance with the requirements of relevant decision-making institutions relating to the operational risk management and then reported to the risk management department at higher level.

The terminal point of the bottom to top information transmission route is the decision making sectors relating to the operational risk management in the highest level, such as the bank's board of directors, board of supervisors, the risk management committee, and the functional department of the operational risk management in the head office of the bank. These decision making sectors should make further analysis and further evaluation of the information uploaded entirely and comprehensively, including the comparisons with the strategy of operational risk management of bank as a whole, and whether the collected information consistent with the provisions of the head office, and whether the management process and analysis tools appropriate, and whether preventive or remedial measures reasonable and whether the internal and external environment being changed, etc.. According to the above analysis and evaluation and the up-to-date requirements of the bank, these risk management decision making sectors make reasonable adjustments to the operational risk management strategy, tolerance of risk, process, infrastructure and environment, and at last, the final new operational risk management decision coming into being.

B. The Up-to-Down Transmission Route

The decisions relating to the operational risk management made by higher level decision making sectors of the bank should be passed to the appropriate risk management departments and business departments through the ORMIS in the form of operational risk management strategy, and operational risk management policies, operational risk management index, proposals approved, suggestions of management, and so on. Based on the risk management decision information received, along with the specific operational risk information obtained in specific business, the specific functional departments can convert such described information into specific operational risk management information. That is to convert generalized decision information into specific and quantitative decision information which has been provided with the practicability and maneuverability to the specific business departments, and to communicate it through various channels so as to become specific basis for the operational risk management.

The operational risk information transmission route and the transmission route for the management decision information is an organic whole, is a dynamic circulated process. Each circulation has the new information added, and some information feedbacks may occur in different transfer stage. In the changing external environment, all of these can prompt commercial bank' operational risk management effectively.

V. THE CONSIDERATIONS OF THE ORMIS SECURITY

For the commercial banks of China, the ORMIS is one of the most valuable assets. So its security is an issue that can not be ignored. In the process of operating of the ORMIS, we should not only ensure the ORMIS to provide safe supporting to the bank's effective management of operational risk, but also to ensure the safety of the ORMIS itself so as to prevent new operational risk coming from the ORMIS itself. Therefore, the particular attention should be paid to the following issues.

A. The Consideration to the Safety Culture

The commercial banks of China must attach enough importance to the development of safety culture imbedded in the ORMIS. This involves mainly the establishment of mutual restraint system, the power separation system and the lowest possible privileges system, and so on.

The ORMIS is a complex system. Sensitivity and key features are the characteristics of some functions and programs of the ORMIS, for example, the system initialization, and the network security construction, and the modification of operating system parameters and business continuity plan, and the installation of firewall, the setting up of master password, the carrying out of the emergency procedures, the accessing to back-up recovery resources and the establishment of security key. Their safe operation is vital to the whole ORMIS. Every one position of these must be operated by more than one person, cannot by only one person. Therefore, it is a matter of utmost urgency for the establishment mutual restraint system. The powers separation system is an important part of safety culture. In it different person should be in charge of different functions and operations, for such as system design and development, operation and maintenance procedures, computer operation, database management, security management, data security control, database and backup data. The powers separation also requires the job rotation for security management function and cross functional or trans-department training. In order to guard against fraud, we must design appropriate transaction process to ensure that no one can generate at the same time the input, the approval and the implementation.

In the minimum possible privileges system, the right of accessing and the system privileges must be authorized to the employees according to their functional needs and requirement of to complete their work. Anyone can not contact information and confidential system resources without limit because of his position. Only authorized person can be permitted to access to confidential information and to use legitimately the system resources. For the outsourcer of the key network and computer resources, if not properly managed, the network system would be faced by operational risk. Therefore, stricter control, supervision and entry restrictions must be made to the external person.

Senior management of the commercial banks of China must attach great importance to the setting up the bank safety culture. "It is the responsibility of management to create a security culture that is equipped to handle the pace of change, and remain motivated to further improve systems." [5]

B. The Consideration to the Maintaining Secrecy

In order to ensure the security of ORMIS, necessary secrecy for something is essential. Information stored in medium such as a card, hard disk, disk, tape, CD, film or written records and so on, are to be protected with secrecy. We must consider whether to encrypt the password when it is stored in the electronic, magnetic or visual media. For the information which was limited and confidential, we should consider whether to delete it permanently when it is beyond the period of validity. We should tell the employees often to change his password of entering network and email. In a word, we should always have sense of confidentiality and ensure the security of the ORMIS.

C. The Consideration to Firewall

A major feature of modern information management system is the widespread application of computer network. It is also for the ORMIS. Firewall is one of the key components of network security. It can isolate the connection between the commercial banks' internal network and the Internet. In the process of running of management information system, the internal network of the bank and the Internet must be isolated logically by the firewall structure. Usually, the internal network and external network must be isolated physically and logically. If possible, all of the inflow and outflow of data should be filtered and checked. In order to ensure the validity, all firewall must be qualified in the round under the attacked. In the active use of firewall to protect information system, we should at the same time also be aware soberly that, with the development of technology, the use of firewall protection measures and their future changing forms cannot make radical guarantee for the security of bank's management information system. Various infiltrators may have a way to break through the firewall limit and bring a variety of hidden trouble to the security of information system.

We should always bear in mind: being on guard against the ORMIS can greatly reduce the invading coming from the infiltrator and can greatly reduce the possible loss.

D. The Consideration to Emergency

One of the most important steps to ensure the security of information system is to keep a copy of the important information. Of course, in the long run, it is the best to have a clear emergency structure, which is to establish a comprehensive and clear and efficient emergency plan for the commercial banks of China. In this way, the sound and stable operation of the ORMIS in emergent situations can be ensured by the commercial banks of China.

IV. CONCLUSION

Based on above analysis, we can see that for the operational risk of the commercial banks of China, as a kind of comprehensive risk, the construction of the ORMIS, especially in the effective information

transmission routes, security in information, has a significant impact on the operational risk management. We must pay special attention to it so as to the sound and healthy development of the commercial banks of China.

ACKNOWLEDGMENT

Authors of this paper would like to thank all reference authors. Thanks to the supporting coming from the Shandong Institute of Business and Technology. Thanks to Yantai Rural Commercial Bank Company ltd.. Authors would like to thank the editors who work for this paper.

REFERENCES

[1] Zhi Zhuo, "Discourse upon the Theory of Risk Management", China Financial Publishing House, first edition, 2006, P. 20-38 (Chinese)
[2] Qinghong Shuai, Kuanhai Zhang "A Introduction to Financial Electronization", Southwestern University of Finance and Economics Press, first edition, January, 2005, P.85(Chinese)
[3] Marc Lore and Lev Borodovsky, "The Professional's Handbook of Financial Risk Management", Butterworth-Heinemann, A division of Reed Educational and Professional Publishing Ltd 2000, P.606
[4] Siqing Chen, "The General Theory of the Risk Management in Commercial Banks", China Financial Publishing House, August 2006, first edition, P.105
[5] Hans-Ulrich Doerig, Vice Chairman, "Operational Risks in Financial Services: An Old Challenge in a New Environment", January 2001, Partly Adjusted 2003, P.118.

The Complementary Effect of Knowledge Management Strategies on Firm Performance

Yi Li[1,*], Xue Zhou[1], Nan Zhou[2], Jun You[1]

[1]School of Economics and Management, Chongqing University of Posts and Telecommunications, Chongqing, China
[2]School of Foreign Language and Literature, Chongqing Normal University, Chongqing, China
([*]yili.cqupt@gmail.com, Zhoux.cqupt@gmail.com, Zhounan007@163.com, youjun@cqupt.edu.cn)

Abstract - **This study aims to test the relationship between knowledge management (KM) strategy and firm performance. An inductive approach is employed to refine six KM strategy orientations drawing on previous literatures. Based on survey data from 345 Chinese firms, structural equation modeling is applied to test the relationships between KM strategy orientations and firm performance. The research results demonstrate that (1) KM strategy orientations complement each other, and their complementarity is an essential intermediate link of KM strategy affecting performance, and that (2) the direct effect of each KM strategy orientation on performance is not significant.**

Keywords - **Complimentarity, firm performance, knowledge management strategy (KM strategy), structural equation model**

I. INTRODUCTION

Knowledge management (KM) strategies influence firm performance [1-3]. But "how performance effects of KM strategies form" is not well understood [2]. It is believed that KM strategy usually consists of many strategy orientations [4]. The relationship between these orientations and firm performance is controversial in large amount of researches [2]. Some of those researches argue that complementarities exist among KM strategy orientations, vital to the improvement of firm performance [5-6]; while others insist that no complementarities exist there, and suggest that KM strategy orientations separately and independently influence performance [7-8].

Based on survey data of 345 Chinese firms, we investigate the relationship between KM strategy and firm performance. The study finds that complementarity among KM strategy orientations influences firm performance significantly, and that performance effect of single orientation is not significant.

II. THEORETICAL FOUNDATIONS

A. The Concept of KM Strategy

We define KM strategy as "a set of firms' decisions on basic principles and characteristics of KM activities" by referring to relevant concepts of articles [4-5]. And we also refine and deduce six major KM strategy orientations building on two complementary constructs from [7] [9]. Each pair of these six orientations separately belongs to one of following three groups: knowledge acquisition, dissemination and application [10].

Knowledge acquisition. External orientation emphasizes acquiring external knowledge in public places, and the Internet etc. [11]. Internal orientation emphasizes acquiring internal knowledge embedding in organizational behaviors and processes.

Knowledge dissemination. Explicit orientation emphasizes knowledge coding and diverse forms of documents in helping employees acquire and share knowledge. Tacit orientation emphasizes interpersonal communication, boosting knowledge dissemination through communications and social networks.

Knowledge application. Exploratory orientation emphasizes exploring new knowledge, encouraging innovation and heavy investment of substantial resources on R&D. Exploitative orientation emphasizes the consolidation, integration and improvement of existing knowledge.

Current literatures about six KM strategy orientations are exhibited in Table I.

TABLE I
LITERATURES ABOUT KM STRATEGY ORIENTATIONS

| Literatures | Relevant KM strategy orientations | | | | | |
| | Knowledge acquisition | | Knowledge dissemination | | Knowledge application | |
	External	Internal	Explicit	Tacit	Exploratory	Exploitative
Dibella, Nevis and Gould [10]	●	●	●	●	●	●
Bierly and Chakrbarti [5]	●	●			●	●
Jordan and Jones [12]	●	●	●	●	●	●
Hansen, Nohria and Tierney [7]			●	●		
Zack [9]	●	●			●	●
Schulz and Jobe [13]			●	●		
Bierly and Daly[14]	●	●			●	●
Choi and Lee [6]			●	●		
Pai [8]	●	●			●	●
Choi, Poon and Davis [2]	●	●	●	●		
Kumar and Ganesh [15]			●	●		
Liu, Chai and James [16]			●	●		

E. Qi et al. (eds.), *Proceedings of the 21st International Conference on Industrial Engineering and Engineering Management 2014*, Proceedings of the International Conference on Industrial Engineering and Engineering Management, DOI 10.2991/978-94-6239-102-4_132, © Atlantis Press and the authors 2015

The extant researches on1007

The extant researches on1007

The extant researches on1007

The extant researches on

The extant researches on

1007

The extant researches on

1007

The extant researches on

The extant researches on

1007

The extant researches on

1007

The extant researches on

1007

The extant researches on

1007

The extant researches on

1007

The extant researches on

1007

The extant researches

B. Performance Effects of KM Strategy Orientations

The extant researches on "How firms could efficiently implement KM strategy orientations for higher performance" have provided two dominating viewpoints. Some of researches argue that firms should focus on single one orientation [7]. Others suggest that simultaneously developing multiple orientations would bring higher performance [6] [8-9].

Different perceptions of relatedness among KM strategy orientations give rise to the controversy among current researches [2]. It was also concluded that strategy orientations affect performance mainly in two ways: complementarity and non-complementarity [2]. Researches supporting "complementarity" [6] [9] argue that complementary relationship exists among KM strategy orientations, having synergistic effect in attaining organizational performance. The management implication of "complementarity" of KM strategy orientations is that balance among specific orientations is essential for profitability and performance. Researches supporting "non-complementarity" [7] suggest that there appear to be non-complementarity among KM strategy orientations when affecting firm performance. The management implication of "non-complementarity" with the perspective of KM strategy orientations is that pursuing individual orientations leads to better performance, instead of implementing a complementary set of orientations.

Thus, clarifying the relationships between KM strategy orientations and firm performance remains a valuable research issue.

Drawing on literatures of organizational economics, complementarity is used to describe a situation that one activity would increase pay-offs to the rest of activities when they are complementary [17]. There appear to be numerous empirical studies on complementarity of corporate management activities [18-21]. Adopting analysis logic of complementarity [20-21], we infer that complimentarity of KM strategy orientations comes from two parts: the sub-additive cost and super-additive value.

The use of versatile resources simultaneously in multiple KM strategy orientations creates sub-additive cost. For example, established computer network supports the external orientation (collecting information from the Internet and email customers) and the internal (internal forum, email system, etc.); culture learning promotes both knowledge exploration and exploitation, as well as the dissemination and sharing of explicit and tacit knowledge; and introducing qualified employees will strengthen all KM strategy orientations. Strengthening one orientation by investing with information technology or human resources, the rest of orientations will be strengthened at the same time, which means that investment in one orientation leads to reinforcement of all orientations, resulting in sub-addictive cost.

Synergistic effects of applying various orientations produce a "super-additive value". For example, knowledge about market environment and customer preference (external orientation) and ideas of product

designs (exploratory orientation) are mutually complementary, together ensuring the success of product development; dissemination of employees' tacit knowledge (tacit orientation), exploration and exploitation of employees' ideas (internal orientation) interact with and strengthen each other, jointly attributing to expanding the influence of knowledge.

Sub-addictive cost and super-addictive value are regarded as symbols of high level of performance. And they are brought by complimentarity of KM strategy orientations. Though the theoretical analysis suggests performance effect of complimentarity, it is also crucial to test the independent performance effects of individual orientations, by following the method proposed by [20-21]. Thus, we propose two competitive hypotheses:

H1a. Complementarity of six KM strategy orientations has a positive effect on firm performance.

H1b. KM strategy orientations separately and independently affect firm performance positively.

III. METHODOLOGY

A. Research Model

Based on the research hypotheses displayed above, our research model is set as Fig.1 including Fig.1 (a) and (b).

B. Measurement

Twenty four items of KM strategy come from literatures in Table I. The firm performance scale contains four items simplified from [22], which are applicable to describe the firm performance in Chinese context. And we introduce two control variables: firm size (staff population) and firm age.

The above items have been discussed repeatedly by two Management professors and three Management PhDs, and constituted a draft questionnaire.

Nineteen middle and senior managers from different industries voluntarily participated in pretest study in Chongqing. And according to their suggestions, we revised previous questionnaire. Then we invited serving MBA students to participant in a pilot test and sent out 200 questionnaires. We refined items again based on data from 113 valid questionnaires completed by informants, obtaining 24 items for KM strategy, 4 items for performance.

The KM strategy scale items describe current situations of their firms. The scale ranges from 1 (strongly disagree) to 5 (strongly agree). Firm performance scale items describe the performance compared to that of main competitors. The scale ranges from 1 (much lower) to 5 (much higher).

C. Sample

We sent 800 questionnaires to the companies in China's east, middle and west (including Shanghai, Beijing, Jiangsu, Fujian, Guangdong, Anhui and Chongqing, etc.). We recalled 436 questionnaires, and the response rate is 55%. The number of valid questionnaires is 345, and the valid rate is 71%. Eighty-two percent of respondents held management titles (from first line to executive level).

The proportions of questionnaires returned from China's east, middle and west are respectively 29%, 33%, 38%. Forty one percent of the responding firms operate in

manufacturing industries, while 59% of them operate in service industries. The proportions of state-owned, private, foreign-invested and other firms are respectively 50%, 20%, 26%, 4%. The average number of employees (average firm size) is 2037, and the average firm age is 12 years.

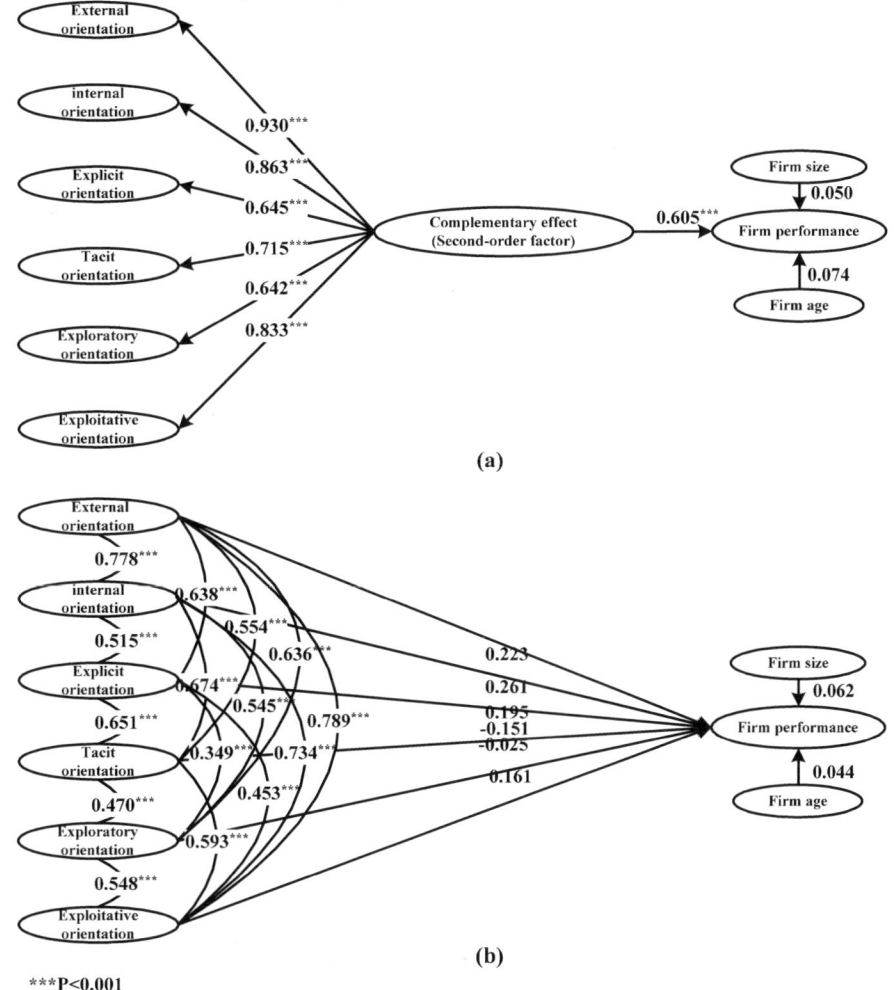

(a)

(b)

***P<0.001

Fig.1. Research model

IV. RESULTS

A. Reliability and Validity

We have done confirmatory factor analysis (CFA) for the 24 items of KM strategy. A six-factor CFA model shows acceptable Goodness-of-fit statistics ($\Delta\chi^2$ (237) = 538.606, RMSEA = 0.061, CFI = 0.967, NNFI = 0.962, IFI = 0.968). Standard factor loadings are between 0.477 and 0.849, which achieve comparatively high significance level (p<0.001) and manifest comparatively good convergent validity. The 95 percent confidence interval of the pairwise-related coefficients between factors does not contain 1 or -1; therefore it has a good discriminate validity. Cronbach's Alpha values for six orientations (0.692, 0.694, 0.772, 0.723, 0.810, 0.871) offer evidence of reliability.

Then, we measure a second-order factor model ($\Delta\chi^2$ (246) = 578.925, RMSEA = 0.063, CFI = 0.964,

NNFI = 0.959, IFI = 0.964), by specifying six first-order factors to a second-order factor. Standardized factor loadings of the second-order factor are all highly significant (p<0.001), varying from 0.639 to 0.927. Hence, the second-order factor model is acceptable.

Four items of firm performance are contained in an one-factor CFA model. Goodness-of-fit statistics are acceptable (χ^2 (2) = 2.083, RMSEA = 0.011, CFI = 1.000, NNFI = 1.000, IFI = 1.000). Standardized factor loadings of measurement items vary from 0.574 to 0.835, and t-values are all highly significant (p<0.001), demonstrating justification for convergent validity. Cronbach's Alpha value is 0.777, showing a preferable reliability.

We compare a seven-factor model (six KM strategy orientations and one firm performance dimension) with a Harman one-factor model to test the common method variance. The $\Delta\chi^2$ is significant (p<0.001). Thus, common method variance bias is acceptable.

B. Hypotheses Testing

Following the methodology of previous literatures [20-21], this study uses structural equation models in Fig.1 (a) and (b) to test H1a and H1b. Fig.1 (a) models complementarity as a second-order construct. The first-order factors respectively capture six orientations, while the second-order factor captures complementarity of all first-order factors, indicating that orientations affect firm performance via the second-order factor. Fig.1 (b) models the direct effects of six orientations separately on firm performance, which allows the pair-wise correlations of six orientations. There are two control variables "firm size (staff population)" and "firm age" in both models.

Goodness-of-fit statistics of the two models are acceptable. The second-order factor (complementarity) model (χ^2 (398) = 1043.16, RMSEA = 0.069, CFI = 0.95, NNFI = 0.94, IFI = 0.95) identically fits the first-order factor model (χ^2 (384) = 997.51, RMSEA = 0.068, CFI = 0.95, NNFI = 0.95, IFI = 0.95). The second-order factor model should be accepted because of its conciseness. As shown in Fig.1 (a), the complementarity (second-order factor) has a significant impact on firm performance (standard path coefficient = 0.605, p<0.001), and the finding supports H1a. As shown in Fig.1 (b), the direct effect of each KM strategy orientation on performance is not significant, and the findings do not support H1b. Collectively, H1 is accepted.

V. DISCUSSION

A. Contributions to Research and Practice

Researches on "How KM strategy influence firm performance" have long been involved in the controversies about its "black-box". Previous researches have theoretically stated and practically tested the crucial role of complementarity [2] [9]. But current empirical studies on this issue provide inconclusive results. The study has explained the reasons why KM strategy orientations complement with each other, providing empirical evidence for complementarity affecting performance and finding that complementarity is an essential intermediate link for performance effects of KM strategies. Also, the study offers a corresponding answer to the controversy of current researches.

The study implies that balancing resource allocation to KM orientations is conducive to increase firm outcome and long-term development.

B. Limitations and Future Work

Cross-sectional design is one of the study's limitations, because KM strategy needs a while to take effect. Longitudinal design is more conducive to in-depth analysis and exploration of the "black-box" of KM strategy affecting firm performance. Other mediators and moderators of performance effect of KM strategy also warrant more inclusive researches.

VI. CONCLUSION

Based on survey data from 345 Chinese firms, this study empirically tests the direct and the indirect relationships between KM strategy orientations and firm performance, and uncovers that (1) complementary effect exists among six KM strategy orientations, having a significant positive effect on firm performance, and that (2) KM strategy orientations individually have no significant effect on firm performance.

ACKNOWLEDGMENT

This work was supported by Grants from The National Social Science Fund of China (no. 12CGL049).

REFERENCES

[1]   P. F. Drucker, *Post Capitalist Society*, Harper Collins Publisher, New York, 1993, pp. 8-47.

[2]   B. Choi, S. K. Poon, and J. G. Davis, "Effects of knowledge management strategy on organizational performance: a complementarity theory-based approach", *Omega the International Journal of Management Science*, vol. 36, no. 2, pp. 235-51, 2008.

[3]   A. K. Singh, M. D. Singh, and B. P. Sharma, "Modeling of knowledge management technologies: an ISM approach", *IUP Journal of Knowledge management*, vol. 28, no. 2, pp. 203-11, 2013.

[4]   M. H. Zack, "Epilogue: developing a knowledge strategy", in *The Strategic Management of Intellectual Capital and Organizational Knowledge*, C.W. Choo and N. Bontis, Ed. Oxford University Press, New York, 2002, pp. 268-76.

[5]   P. Bierly, and A. Chakrabarti, "Generic knowledge strategies in the US pharmaceutical industry", *Strategic Management Journal*, vol. 17 (WINTER), pp. 123-35, 1996.

[6]   B. Choi, and H. Lee, "An empirical investigation of KM styles and their effect on corporate performance", *Information and Management*, vol. 40 no. 5, pp. 403-17, 2003.

[7]   M. T. Hansen, N. Nohria, and T. Tierney, "What's your strategy for managing knowledge", *Harvard Business Review*, vol. 77, no. 2, pp.106-17, 1999.

[8]   D. C. Pai, "Knowledge strategies in Taiwan's IC design firms", *Journal of American Academy of Business*, vol. 7, no. 2, pp. 73-7, 2005.

[9]   M. H. Zack, "Developing a knowledge strategy", *California Management Review*, vol. 41, no. 3, pp. 125-43, 1999.

[10]   A. J. Dibella, E. C. Nevis, and J. M. Gould, "Understanding organizational learning capability", *Journal of Management Studies*, vol. 33, no. 3, pp. 361-79, 1996.

[11]   C. Ramanigopal, "Knowledge management strategies for successful in Aerospace Industry", *Advance in Management*, vol. 5, no. 12, pp. 17-21, 2012.

[12]   J. Jordan, and P. Jones, "Assessing your company's Knowledge management style", *Lang Range Planning*, vol. 30, no. 3, pp. 392-98, 1997.

[13]   M. Schulz, and L. A. Jobe, "Codification and tacitness as knowledge management strategies: an empirical

exploration", *Journal of High Technology Management Research*, vol. 12, no.1, pp. 139-65, 2001.

[14] P. E. Bierly, and P. Daly, "Aligning human resource management practices and knowledge strategies", in *The Strategic Management of Intellectual Capital and Organizational Knowledge,* C.W. Choo and N. Bontis, Ed. Oxford University Press, New York, 2002, pp. 277-95.

[15] J. A. Kumar, and L. S. Ganesh, "Balancing knowledge strategy: codification and personalization during product development," *Journal of Knowledge Management*, vol. 15, no.1, pp. 118-35, 2011.

[16] H. Liu, K. H. Chai, and F. N. James, "Balancing codification and personalization for knowledge reuse: a Markov decision process approach," *Journal of Knowledge Management*, vol. 17, no. 5, pp. 8, 2013.

[17] P. Milgrom, and J. Roberts, "The economics of modern manufacturing: technology, strategy, and organization", *American Economic Review*, vol.80, no.3, pp.511-28, 1990.

[18] S. E. Black, and L. M. Lynch, "How to compete: the impact of workplace practices and information technology on productivity", *Review of Economics and Statistics*, vol. 83, no. 3, pp. 434-45, 2001.

[19] T. F. Bresnahan, E. Brynjolfsson, and L. M. Hitt "Information technology, workplace organization, and the demand for skilled labor: firm-level evidence", *Quarterly Journal of Economics*, vol. 117, no. 1, pp. 339-76, 2002.

[20] H. Tanriverdi. and N. Venkatraman, "Knowledge relatedness and the performance of multibusiness firms", *Strategic Management Journal*, vol. 26, no.2, pp. 97-119, 2005.

[21] H. Tanriverdi, "Performance effects of information technology synergies in multi-business firms", *MIS Quarterly*, vol. 30, no. 1, pp. 57-77, 2006.

[22] D. Wang, A. S. Tsui, Y. Zhang, and L. Ma, "Employment relationship and firm performance: evidence from the People's Republic of China", *Journal of Organizational Behavior*, vol. 24, no.5, pp. 511-35, 2003.

Analysis on Methods to Applying TRIZ to Solve Management Innovation Problems

Ya-qiang ZHANG*, Hong-mei LI

Department of Business Administration, Xingtai University, Xingtai Hebei, China
(tjhbzyq@126.com)

Abstract - **Being an effective way to technological innovations, basic ideas and methods of TRIZ can be applied to the methods of management innovations. There is difference between technical system and management system, but at the same time management innovations provide more space for the application of TRIZ, management problems having more solving paths and possible solutions. The key point of applying TRIZ in the method of management innovations successfully is to innovate TRIZ methods according to the characteristics of management system. This paper discusses the methodology advantages of applying TR1Z in management innovation, put forward the basic idea of the management specialization of TRIZ to management innovation, as well as the basic work of the management specialization of TRIZ.**

Keywords - **Conflict, management innovations, management specialization, TRIZ, TRIZ methodology**

I. INTRODUCTION

TRIZ is Theory of Inventive Problem Solving, first started by Soviet inventor, the chairman of Inventors Association, Genrich S. Altshuller, from 1946. It puts forward a complete system of theories and methods for inventive problem solving on the basis of studying two million high level patents around the world. It mainly aims at studying the principles and laws that human beings follow during innovation, invention and technical problem solving.

TRIZ provides a set of standardized tools to guide the direction and process of innovation [1, 2], such as Contradiction Matrix, invention Principles, technical evolution rules, Substance-Field analysis, the Innovation Algorithm (ARIZ) [3], and so on. Meanwhile, TRIZ provides a number of methods to overcome thinking inertia to train professionals' innovative thinking [3, 4], such as miniature person model, STC(Size-Time-Cost) operator etc. TRIZ allows humans' innovation activities have the potential to become a systematic process to be learned, trained and operated, and it is now recognized as the most systematic and mature innovation methods theory.

TRIZ have begun to receive academic attention in China, and get a better application in the field of engineering and technology, but less so in the field of management innovation. This paper attempts to explore the idea of using TRIZ for solving management problems.

II. APPLICABILITY OF TRIZ FOR MANAGEMENT INNOVATION

TRIZ regards the core of innovation is to solve conflict. Management innovation problems may be thought of as the elimination of management conflict. According to Ma Qingguo's definition, management is the arrangement made about a system and its consisting components in order to achieve a certain goal [5], before achieving the goal. This definition is quite suitable for the studies of management innovation methods: The "arrangement" is a state, the process of transition from one state to another state is management innovation [6]. Transformation of arrangement may result in conflicts between management elements, people can use TRIZ to define management problems and conflicts, to confirm management innovation direction until using the invention principles and tools to eliminate management conflicts.

However, TRIZ is the summary of the law of inventions in the field of engineering and technology. After all, technology systems and management system is different, the former mainly is about object system, while the latter is more of a system comprised of human and objects, which is becoming more complex because of human involvement. "Human" factor make management study face two difficulties, which are manager's image thinking and management situations [7]. Apart from under specific work environments, the various management elements are also under the influence of mode of thinking of the management body. The key to successfully apply TRIZ to management innovation is to carry out innovation to TRIZ method, combining the features of management object system.

What needs to be pointed out is that "human" factor provides more space for applying TRIZ into management innovation.

Firstly, as a kind of soft technology, management is still a technical system. The main characteristic of soft technology differs from hard technology is that its action target is the operator [8], the action mode of soft technology is limiting the actions and methods between operators and between operator and tools by rules systems, adjusting the way, method or process of operator by regulations and standards. Therefore, the management system is still a technology system, as a technological invention theory, TRIZ's majority outcomes still apply to management innovation.

Moreover, it is more flexible in the aspect of adjusting and regulating human, providing more opportunities for management innovation. From an

¹ This paper is subsidized by the Natural Sciences Fund in Hebei Province (№. G2013202184) and the Science and Technology Research Projects for Colleges and Universities in Hebei Province (№. QN2014322).

analysis of a large number of conflict-solving strategies, Extension theory finds out that objects can be analyzed from different angles [9], in which the concept of object is a much broader. In the aspect of materiality, objects have material part (real part) and non-material part (imaginary part); in the aspect of systematicness, objects have consisting parts and relationships (hard part and soft part); in the aspect of dynamic, objects have revealing part and hidden part; in the aspect of opposition, object have positive part and negative part about certain characteristic. Many innovations take advantage of objects' imaginary part, soft part, hidden part and negative part, which makes some conflict problems that hard to solve to transform. In management system, "human" factor is often reflected as imaginary part, soft part and hidden part of system, the adjustment of human contains more opportunities for innovation, this adjustment is more flexible, and often is no-cost or low-cost, which actually does benefit to system's evolution to IFR(the Ideal Final Result). For example, JIT innovation puts more emphasis on the continuous optimization of human and organizational aspects, such as Kanban, multifunctional workers, visualization, etc.

III. THINKING OF USING TRIZ FOR SOLVING MANAGEMENT PROBLEM

TRIZ is a kind of heuristic methodology, does not aims at each specific problem of users, generalizes specific problems as general problem and provides general solutions through the abstracted standard models. TRIZ offers a range of standard analysis models without technical jargon, such as engineering Parameters, Contradiction Matrix, Substance-Field analysis, ARIZ etc. Then TRIZ compares the general solution with specific problem and transforms to the solution of specific problem, realizes it in practical design and finally getting the practical solution of specific problem. The methodology of TRIZ for problem solving is shown in Fig.1.

A. Advantage of TRIZ Methodology

TRIZ is different from the conventional science and engineering methods, is very practical. Its methodology strengths can also be applied to management systems, is beneficial to solve management problems.

1) TRIZ is the Embodiment of Philosophy Principles

Opposition and unification law, from quantitative change to qualitative change law and negative of negative law, these three laws in Dialectics of Nature as the most common applied law are too macro and uneasy to operate. TRIZ's technological evolution modes and routes are a kind of reveal and summary of these objective laws which is easy to operate, and are also reflected in the other tools of TRIZ system. Actual designers do not need to focus on the internal relations between all evolution principals, simply need select and materialize it. In this way, TRIZ embodies general technology philosophy, also has become a practical design and invention tool.

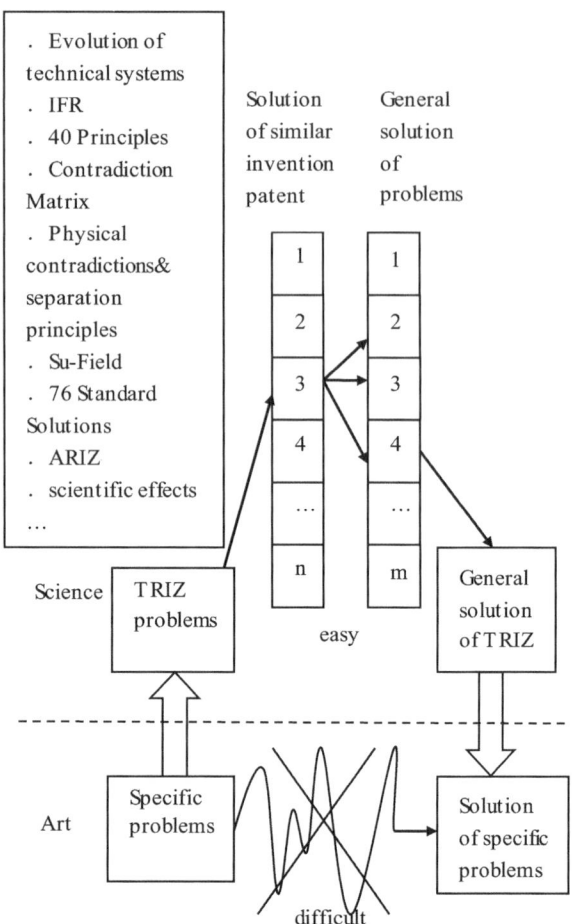

Fig.1. TRIZ methodology

2) TRIZ is a Methodology of Knowledge-based, People-oriented Innovation Problem Solving

The knowledge adopted by TRIZ are abstracted from the world-wide patents, TRIZ's design allows and supports many rounds of repeated use of these abstract knowledge. Based on abstract knowledge, TRIZ provides a set of standardized analysis models, which transform fuzzy original invention problems into simple problem models, and try to avoid using technical terms, so may more widely uses knowledge outside of company, industry, subject for problems solving.

For complex design problems, an inventive principle is not enough, in the selection process of principles, in-depth consideration of problems and experience are needed. In the transformation process from general solution to specific solution, in-depth thinking, ingenuity and experience about problems are also indispensable. Thus, TRIZ is based on tacit knowledge and experience, so is based on human, and TRIZ's application is linked with designers' experience and mature process. As pointed out by Qian Xuesen, quantitative methodology for dealing with complex systems is a combination of scientific theory, experience and expert judgment. This kind of methodology is a semi-empirical and semi-theoretical [10].

3) TRIZ is a Reverse Search Mechanism of "Target-Approach"

Different from the positive "condition-result" mode of current science and technology system, at the beginning of problem-solving, TRIZ confirms the direction and position (technological evolution, the Ideal Final Result) to avoid repeating exploration work in other methods, and then explores specific solutions with appropriate principles or laws, improving the efficiency of innovation.

TRIZ has many advantages, which makes the introduction of TRIZ into management system can be directly based on TRIZ's ideas, methods and frameworks. In particular, the stylized analysis tools about innovation problems, such as Substance-Field analysis, ARIZ algorithm, can be directly introduced in the analyzing and solving of management problems, because a thorough analysis of problem often means to solve the problem.

B. Management Specialization of TRIZ

Introducing TRIZ into management systems needs to make adjustments about TRIZ, according to the characteristics of management innovation. There are two possible directions, one is adding, amending the components and tools of TRIZ in view of management system, the other is carrying out necessary simplification to adapt practice in management field.

1) Revise of TRIZ for Management Specialization

TRIZ mainly is the summary of invention knowledge in engineering and technology field, though technical jargons are avoided as much as possible, there is still deep specialized "mark" of engineering and technology. A considerable part of inventive principles and analytical tools are applicable to management problems, but parts of evolution rules and innovation principles need to revise.

Evolution law of technical system is an important basis of TRIZ. Rule of "evolution to micro-level and application of field" need to adapt to the management system, summarizing different energy fields of management. S-curve evolution rule is changing due to human factor in management system. Specific amendments of evolution laws will lead to adjustment of other innovative principle and methods structure.

39 engineering Parameters and 40 Principle should make professional adjustments according to management problems, particularly designing the standard parameters about management system. Reference [11] has made relevant attempts.

Segmentation principle is also proved to have good usability in solving management contradictions. However, what is difficult for management arrangement is the segmentation of system factors, managers stress more on some sort of "art" of "balancing and focusing". This requires the introduction of relevant principles of balance and symmetry, guiding to solve management innovation problems.

2) Simplification of TRIZ for Management Specialization

Relevant surveys show that the use of management tools in enterprise is not optimistic [12]. The complex and complicated process, high demanding of resources, the limitation of environment and culture conditions is unfavorable to the use and success of traditional management tools. This also reflects the current management theory has wide applicability and constrains of implementation conditions, management disciplines is more like a mysterious jungle, rather than a familiar path. Simple, practical management innovation logic is more practical significant, is conductive to change the "management jungle" situations.

Given the using experience of management tools and techniques, TRIZ should constantly enrich its knowledge base and improve its tool system, at the same time, should carry out necessary simplification about management problems to facilitate the application of management practices. For example, after professional adjustments are made about general engineering Parameters and 40 Principles targeting at management problems, it is about to determine the Parameters and Principles that are more often used, confirm the priority and reorder, form a new simplified management Contradiction Matrix. Administrators can even skip the analysis and applications processes of Contradiction Matrix when solving certain problems, directly consider high priority invention principles, which can increase the efficiency of innovation.

3) Basic Work of Management Specialization of TRIZ

TRIZ in nature is a knowledge base about innovation rules and offers a range of principles and tools which can be easily retrieve by users. This knowledge base is constantly evolving and improving, new knowledge and systems are being added to TRIZ, such as, new symbol system of Substance-Field model, further development of conflict and solving technologies, improvement of ARIZ algorithm, integration of TRIZ, Robust Design and QFD(Quality Function Deployment) [13], simplified TRIZ model [14, 15], and so on. The management specialization of TRIZ is based on TRIZ framework, should also be a process having constant additions and amendments with regard to management innovation rules. It is hard to complete the theory work once for all, not a simply mechanical application of TRIZ in management system. Applying TRIZ to solve management innovation problem needs summarize management theories and management practice experience, refine general management innovation rules, especially summarize new evolution rules, innovation principles, standard solutions or effects about management problems formed by the influence of human factor, as pointed out above, the adjustment about human is more flexible, which provides more solving path and feasible solutions for management innovation.

IV. CONCLUSION

TRIZ, as a summary of invention & innovation rules in engineering and technology field, its fundamental ideas, methodologies and frameworks can be used by management innovation. TRIZ in essence is a knowledge base of rules related to invention and innovation, is also a methodology of solving innovation problems base on knowledge and human; its application in management innovation problems has more space, is more conductive for managers to use management knowledge and management experience, providing more solving path and feasible solutions for management innovation.

The application of TRIZ to solve management problems needs to target at the characteristics of management system, make professional adjustments to TRIZ, for example, selecting new management Parameters, setting up a management Contradiction Matrix, on this basis, carrying out necessary simplification about Contradiction Matrix and innovative Principles in order to facilitate the use of managers. In addition, TRIZ system about management innovation should also be a acknowledge base with continuous accumulation and enriching.

REFERENCES

[1] Yang Qingliang, *This Invention was Born - full Contacting with TRIZ* (in Chinese). Beijing: China Machine Press, 2006, pp. 11-24.

[2] Altshuller G. S., *Creativity as an exact science*. New York: Gordon & Breach Science Publishers, 1984, pp.16-25.

[3] Altshuller G S., *The Innovation Algorithm*. Worcester: Technical Innovation Center, INC., 1999.

[4] Altshuller G S., *And Suddenly the Inventor Appeared*. Worcester: Technical Innovation Center, INC., 1996.

[5] Ma Qingguo, "Several key problems of management science research in China" (in Chinese), *Management World*, no. 8, pp. 105-110, 2002.

[6] Zhang Dongsheng, Xu Man, and Yuan Yuan, "Study on the method of TRIZ-based management innovations"(in Chinese). *Studies in Science of Science*, vol. 23, no. Suppl., pp. 264-269, 2005.

[7] Li Huaizu, *Methodology of Management Research (2nd version)* (in Chinese). Xi'an: Xi'an Jiaotong University Press, 2004, pp. 18-30.

[8] Ma Qingguo, Hu Longji, and Yan Liang, "Redefining the Concept of Soft-technology" (in Chinese), *Science Research Management*, vol. 26, no. 6, pp. 99-105, 2005.

[9] Chou Cheng, Feng Junwen, and Guo Chunming, "Comparison of TRIZ and Extenics" (in Chinese), *Industrial Technology & Economy*, vol. 26, no.10, pp.105-107, 2005.

[10] Qian Xuesen, *On Systems Engineering (revised edition)* (in Chinese). Changsha: Hunan Science & Technology Press, 1988, pp. 1-7.

[11] Tian Xin, Ren Gongchang, "Research about the Conflict of Management System Based on TRIZ" (in Chinese), *Machinery Design & Manufacture*, no. 11, pp. 160-162, 2006.

[12] Andrew Cox, Chris Lonsdale, Joe Sanderson, and Glyn Watson, *The Right Tools for the Job*. Palgrave Macmillan, 2004.

[13] Tan Runhua, Wang Qingyu, "TRIZ-TRIZ Engineering, Tools and Development Trend" (in Chinese), *Journal of Machine Design*, no. 7, pp. 7-11, 2001.

[14] Kalevi Rantanen, and Ellen Domb, *Simplified TRIZ (2nd edition)*. Taylor & Francis Group, 2008.

[15] Zhou Jiehan, Xiong Guangleng, and Fan Wenhui, "Research and Progress on TRIZ: A theory of Inventive Problem of Solving" (in Chinese), *Manufacturing Automation*, vol. 24, no. 8, pp. 24-27, 2002.

Analysis of Hot-Topics and Prospect for Study on Theory of Financial Management in China

Yu-di Zhou, Rong Fu*

Department of Commerce, Sichuan Normal University, Chengdu, China

(80693091@qq.com)

Abstract - **Although the start of China's financial management is late, but after the development of about half a century, relevant theoretical research has matured, and research focus is changing. Based on present situation, this paper tries to count and analyze a lot of documents, which are published in Chinese academic journals from 2010 to 2012. We plan to research from two angles of literature, research content and method of our country in recent years, and summarizes the research achievements of this stage, forecasts the development trend of financial management theory, and provide a reference for research in the future.**

Keywords - **Financial management theory research, Hot-topic of Chinese financial management, statistics, trend analysis**

I. CHINA'S FINANCIAL MANAGEMENT THEORY DEVELOPMENT

Study on financial management theory in China began in twentieth Century 60 time, basically stagnated before the reform, but after developing for the past three and four years of rapid, it had fruitful results, and at different stages, research focus is also advancing with the times.

In the first decade of twenty-first Century, the theory of financial management in China entered into the accelerated development period. Research content and visual angle are diversified. Wang Hua-cheng (2010) [1] pointed out that the research of Chinese financial management theory has combined with the actual and features of China. Furthermore, many different research methods and interdisciplinary research perspectives coexist.

In second beginning of ten years of twenty-first Century, content of Chinese financial management is more rich, more widely tentacles, and at the same time, facing the vast literature, how to manage clear thinking, grasp the direction of the mainstream of the current and future is a problem. This paper attempts to analyze the 2010 to 2012 published in the core academic journals literature sort out the present research trend, to provide reference and inspiration to other scholars.

II. LITERATURE SOURCES AND SAMPLE SELECTION

Although there are many outstanding publications in the field, due to limited energy, this paper only selects the period from 2010 to 2012 published in the "accounting research", "management of the world", "economic research", "financial research" to the field of financial management literature as the research sample. According to the correlation literature and the theme, we selected 550 articles as the research sample.

III. ANALYSIS OF THE RESEARCH HOT-SPOTS OF FINANCIAL MANAGEMENT THEORY IN OUR COUNTRY

This paper research the hot-spots of Chinese financial management theories from the two aspects, the contents and methods of the research.

A. From Chinese financial management theory content

We refer to Wang Hua-cheng's (2010) classification method of financial management theory. Theories can be divided into four categories, and adjust the detail appropriately. In carefully read 550 papers, we found a part of the papers involve several research areas. Therefore, to ensure the rationality of the statistical results, we according to the classification of the study content, repeat count the papers which involves many research areas. All together, we have 681 papers for statistics, as Table I:

TABLE I
SAMPLE ACCORDING TO THE CONTENT OF THE FREQUENCY DISTRIBUTION

Category	theory of foundations							The general theory of business				Special business theory					other theories			
Details	basic category	Modern financial management method	value management	agency theory	corporate governance	market efficiency	information disclosure	Financing Managment	investment management	operation management	assignment management	International Finance	group financial management	strategic management	merger	The enterprise bankruptcy liquidation	GREEN	social responsibility	theoretical innovation	other
number	13	99	13	9	114	4	40	94	62	24	45	11	26	21	20	3	13	22	30	18
sum	292							225				81					83			
Percentage	42.88 %							33.04%				11.89%					12.19%			

Note: the basic category of financial management concepts, financial management objectives and functions; modern financial management method includes internal control, risk management, enterprise performance and performance evaluation

B. The basic theory of financial management

The research on the basic theory of China has broken through the traditional research scope, research content innovation. Research direction is mainly transferred to modern financial management, corporate governance, involving the two aspects of the article to the class of 73%.

First of all, we analyze the research on corporate governance.

The author believes that the scope of corporate governance should be attributed to the financial management for the reasons as follows: firstly, the objective of corporate governance is to reduce the agency cost, improving the efficiency of corporate governance, to maximize the value of enterprises, the financial management objectives are consistent; secondly, one of the main content of corporate governance is to deal with the relationship between the various interests related enterprise the financial management, and financial management also cover these content. Study on the categories as shown in Table II.

TABLE II
ON THE CORPORATE GOVERNANCE LITERATURE RESEARCH CONTENT IN THE SAMPLE

Content	Excitation	Ownership structure	Internal governance mechanism	External governance mechanism	Other	sum
2010	19	6	8	3	7	43
2011	7	8	13	1	3	32
2012	14	7	11	1	6	39
sum	40	21	32	5	16	114

From Table II, we can see that the literatures which research on corporate governance, incentive problems are Adequate. Most Chinese enterprises still prefer to use monetary incentive, but the theoretical research of equity incentive has been. Related samples 15 articles on enterprise equity incentive, they are distributed in three years were 5, 3, 7, of high heat. At the same time, because of the implementation of equity incentive in China since December 31, 2005, the CSRC "management measures" incentive listing Corporation ownership before officially started, so in practice there are many problems to be resolved. For example, our managers manipulate earnings have occurred, destroyed the positive effect of equity incentive; again, the equity incentive can effectively solve the discussion of shareholders and management problems of information asymmetry in the enterprises of our country has not concluded. This correlation is a hot issue for experts and scholars to study and further research direction.

Secondly, research on the methods of modern financial management.

Modern methods of financial management is an important part of the basic theory, the specific research contents as shown in Table III.

TABLE III
RESEARCH LITERATURE ON MODERN FINANCIAL MANAGEMENT METHOD IN A SAMPLE

content	Internal control	performance and performance evaluation	Risk management	other	sum
2010	11	13	3	2	29
2011	24	13	5	2	44
2012	13	10	2	1	26
sum	48	36	10	5	99

It can be seen from Table III the internal control in the category accounted for the largest proportion, and in this three years have maintained the high heat. A series of recent economic problems exposed the attention and consideration of the internal control of the enterprise, the relevant state departments have issued internal control guidelines for enterprise construction policies specification. Based on this, the study of internal control has not limited to stand in the "separation of incompatible positions" point of view, but innovation into enterprise culture, the philosophy of China traditional culture perspective, maintains that the internal control and "people-oriented" concept combination, make the internal control and the formation of a "people" interactive relations, institutional arrangements to become a maintenance of the legitimate rights and interests of the various stakeholders (Wang Hai-bing, 2011) [2], and the internal control into the enterprise culture, to be spontaneous in enterprise management needs.

Based on the theory of financial management, in addition to the relevant documents of corporate governance and modern financial management method, samples are part of this paper relates to the financial management of basic category, value management, agency theory, market efficiency, information disclosure and so on, but the research methods are usually based on the original quantitative analysis further or lateral ties for example, to discuss the information disclosure, the separate parsing is not traditional, but the green financial management, corporate social responsibility and other issues together.

TABLE IV
THE RESEARCH CONTENT OF COMMON BUSINESS LITERATURE

content	Financing management	Investment management	Operation management	Distribution management	sum
2010	29	15	5	17	66
2011	32	25	11	16	84
2012	33	22	8	12	75
sum	94	62	24	45	225
percentage	42%	28%	10%	20%	100%

C. The common business theory of literature

The common business theory has experienced in long history, it not be limited in purely academic discussion, rather than on the present situation, combining with the reality of China's economic development, the specific research contents as shown in Table IV.

Firstly, the research on financing management, investment management business are still the main.

Secondly, research on the universal service theory more practical.

For a long period of time, research on the financial management theory of the Western emphasis on the analysis of how to look for financial activities, with a strong tendency of pragmatism. With the thought of Western studies of communication impact, research in view of our country enterprise characteristic of financial management activities are more and more. For example, the financing problem of small and medium-sized enterprises to China and Chinese style folk lending relationship research, to explore our country enterprise income distribution in the balance of efficiency and fairness and the distribution structure of the sustainable development of China's national economy influence, has the strong Chinese characteristics.

Thirdly, to shorten the gap between research in china and related theory of other country.

Chinese scholars make full use of "advantage of backwardness", with "fast run with small pace" way to go steady in the frontier of research of foreign scholars approacheds. For example, the "pecking order theory", "the free cash flow theory", "contract theory", in our country on the market effect, the game between Chinese enterprise stakeholders, research on corporate earnings management issues.

Finally, Study on the content of "common business theory grafting" phenomenon more common.

The phenomenon of "cross research" between the specific content of universal service theory or general business theory and financial management theory often appear in. For example, the debt maturity about beam impact on executive compensation incentive intensity of [3], the enterprise again qualification of financing and corporate dividend policy relations and other issues of the [4].

D. Special business theory

Special business theory is aimed at specific times, specific conditions, specific economic activities of the main financial management theory, this part includes the research content as shown in Table V.

TABLE V
RESEARCH CONTENTS OF RELEVANT DOCUMENTS ABOUT THE SPECIAL SERVICE THEORY IN THE SAMPLE

content	International financial management	The group's financial management	Strategic management of enterprises	Enterprise merger	bankruptcy liquidation	sum
2010	2	7	8	6	1	24
2011	6	9	7	9	1	32
2012	3	10	6	5	1	25
sum	11	26	21	20	3	81

Research on the financial management of enterprise group can be divided into two types of samples, one is the study on the subject of corporate governance, investment management, financing management problems enterprises, relevant content and is similar with the individual enterprises; another is focused on Internal Capital Market Research company. Samples with 8 articles relating to the enterprise group's internal capital market and external financing, financing cost pressure, internal resource allocation efficiency, enterprise value promotion etc.. Conclusion the study shows that the current scholars of China's internal capital market mixed attitude. Therefore how to improve the internal capital market regulatory system, how to improve the internal free cash flow efficiency, how to prevent and eliminate the hidden crisis group issues need in future studies to obtain the reasonable solution.

E. Other theories

The research content of other theories more fragmented, sample articles related to this category is relatively small, but it does not exclude such contains a number of current financial management in our country in the field research, such as the green financial management.

The "green financial management" is refers to the enterprise financial management related to environmental protection and sustainable development content. The study of this issue 13 articles samples, three years were 5, 2, 6 sample articles, visible in recent years has maintained certain research heat. The green financial information disclosure is another important content in this category, and often associated with social responsibility together. The performance of green financial management information embodies the social responsibility of enterprises, there are certain impact on corporate performance, reputation, so it is common concern of all stakeholders. The past research on the green financial management in our country is mainly referring to foreign theories, but China has its own specific national conditions, and the sense of social responsibility of enterprises generally needs to be strengthened, the implementation of green financial management policy environment, material conditions have to be concerned about, how to encourage enterprises to disclosure of high quality green financial management information, to prevent disclosure formal, need further study.

In addition, the "other" theory also studied the content of corporate social responsibility, enterprise innovation, financial fraud, usually studies these problems is presented as a composite type, cross type mode.

F. From our country financial management theory research methods

The general research method in academic circles mainly include normative study and Empirical Study of two categories. From the angle of research methods, this paper a total of 550 literatures which sample, using empirical research methods of the total of 431, accounting for about 78% of the total number of samples, the possession of absolute advantage in number. And in these three years, the number of samples in each year using the empirical research method in the total number of samples in the proportion of the corresponding year were 73%, 79%, 83%, visible, an empirical wind gradually popular. Investigate its reason, first, with China's economic development, the capital market gradually improve, enterprise management more standardized. Based on this, the collection of relevant data is more easily and accurately, which provides conditions for the empirical research; second, a large number of problems in the financial management for quantitative analysis, suitable for the use of empirical research methods; third, the specific empirical research methods varied, can meet the needs of different research. The sample involved in the empirical research methods are mathematical and empirical research method, case analysis method, questionnaire survey method, descriptive statistics, as shown in Table VI. Among them, the mathematical and empirical research method with its low cost, operational characteristics, empirical research methods is the most commonly used sample. In addition, the case analysis method, especially multi case analysis method to comprehensive analysis, the analysis results more convincing advantage by some scholars favor. In recent years, often will study various empirical combined use, such as Liu Hao (2012) [5], the case analysis method with mathematical and empirical research method of combining study.

TABLE VI
STATISTICS THE NUMBER OF SAMPLES BY DIFFERENT METHODS OF EMPIRICAL RESEARCH ARTICLES INVOLVED

Empirical method	empirical study of math	Case analysis	Questionnaire investigation	Descriptive statistics	"combination"	other	sum
2010	108	12	4	5	2	0	131
2011	126	11	10	6	3	2	158
2012	114	15	1	4	7	1	142
sum	348	38	15	15	12	3	431

Note: "Combination" refers to combine different the empirical methods of this field.

IV. FUTURE DEVELOPMENT TREND OF THE RESEARCH ON THE THEORY OF FINANCIAL MANAGEMENT IN CHINA

Summary on the basis of previous studies, the author believes that the development trend of the theory of financial management of our country in the future will as the following [6-10]:

First, research on the theory of financial management in our country will relate to the actual situation of our country more closely.

Theoretical research can not be divorced from practice, or it will lose the meaning of the research. With the development of Chinese enterprises, we will have Chinese special theory of innovation more, but only to combine the theory study and actual situation of our country, in order to practical problems of financial management in China to provide more effective guidance [11].

Second, research that in line with the other country will become the focus of research in china, which is needs of China's economic development. Although we are at the special situation, but it can not be denied the leading role of foreign advanced management experience to us. On the other hand, because of the economic environment in China is different, the related theory is introduced to solve the specific problem of our country also need to prove, we research process should also pay attention to the actual situation of our country as a guide, not a reference to foreign research results.

Third, empirical research methods will continue to maintain the dominant position, at the same time, various research methods will coexist.

The empirical research method has an absolute advantage in the study of financial management theory in China, this advantage will continue to maintain. At the same time, in the empirical research in the future should be encouraged to study different methods of selection in different circumstances, and try to use different empirical studies to create favorable conditions. Construction quantity model study of mathematical demonstration should be combined with China's specific conditions for innovation, not blindly use foreign already constructed model [12].

Fourth, cross discipline of the phenomenon will become increasingly common

As China's economy continued to improve, the economy becomes more and more complex, simply using the knowledge of a certain subject to solve the problems in this area has become very thin. Cross disciplines tend to produce new research perspective, this phenomenon will be more prevalent in the future. The most common field is the integration of psychology and organizational behavior, the. There are a great span of disciplines, such as the problem of investment and geography cross the field, the internal control and the philosophy of crossing. At the same time, the cross of research contents in this field will be more common among.

In summary, this paper analyzes the hotspot in recent three years in the field of financial management, the trend and direction of the future research expectation in the characteristic foundation of summarizing the results of the present study is to provide reference for the follow-up study.

REFERENCE

[1] Wang Hua-cheng, Li Zhi-hua, Qing small right, to Yue, Zhang Wei-hua, Huang Xin-ran, China's financial management theory and the history of the future - "accounting research", published thirty years financial theory literature review [J], "accounting research", 2010 twelfth, pp. 17-23.

[2] Wang Hai-bing, Wu Zhong-xin, Li Wen-jun, Tian Guan-jun, the enterprise internal control framework for the interpretation and reconstruction [J], "accounting research", 2011 seventh, pp. 59-65.

[3] Chen Jun, Xu Yu-de, will focus on the interests of the creditors executive compensation? - the term listing Corporation debt of our country from the perspective of empirical evidence based on constrained [J], "accounting research", 2012 ninth, pp. 73-81.

[4] Wang Zhi-qiang, Zhang Wei-ting, listing Corporation's financial flexibility, re financing option with dividend to cater to the strategy of [J], "management world", 2012 seventh, pp. 151-163.

[5] Liu Hao, Tang Song, Lou Jun, the independent director supervision or consultation? -- banks against the background of independent directors on corporate credit financing effect of [J], "management world", 2012 first, pp. 142-156.

[6] Wang Hua-cheng, the financial management theory research: Retrospect and Prospect - 20 years after twentieth Century China a review of the theory of financial [J], "accounting research", 2001 twelfth, pp. 37-45.

[7] Wei Ming-hai, a new perspective of company financial theory study [J], "accounting research", 2003 third, pp. 53-57.

[8] Du Ying-fen, research frontiers in financial management theory: extending to the special field and other disciplines fusion [J], "Chinese society science news", 2009 November

[9] Nanjing University Research Institute of accounting and finance research group, explore the philosophical basis of the internal control system [J], "accounting research", 2012 eleventh, pp. 57-63.

[10] He Ping-lin, Shi Ya-dong, Li Tao. Data envelopment environmental performance analysis method based on - a case study of thermal power plants in China [J], "accounting research", 2012 second, pp. 11-17.

[11] Guo Fu-chu, China's financial reform practice and theory of the development of [J], "accounting communications", 2000 May, pp. 3-7.

[12] Liu Ping, "enterprise financial management development history of new China on" [J], "research" in Contemporary Chinese history, 1996 November, pp. 85-87.

Challenges Associated With Automation of NPO: Requirement Engineering Phase

Misha P Mathai*, Sunny Raikwar, Rahul Sagore, Shaligram Prajapat

International Institute of Professional Studies, Devi Ahilya University, Indore, INDIA

(*mishamathai91@gmail.com, raikwar.sunny23@gmail.com, rahul.sagore@gmail.com, shaligram.prajapat@gmail.com)

Abstract - **Social development, is an indicator of real growth, is the backbone of a nation. NPOs and NGOs working along with government play a pivotal role in social development. In the era of e-governance, handling resource mismanagement, capacity building and Founder's syndrome are challenging issues for a Non Profit Organization. A NPO working in various domains across different parts of country finds it difficult to handle complex hierarchical information management system, which deals with men, machine, service and money. After studying metrics like cost, time and effort on an NPO according to client requirements many problems came into picture. Thus, this paper addresses, sheds its light upon the hidden factors and challenges associated with automation of an NPO and reveals how to deal with dynamically changing requirements. This paper attempts to create guidelines and provides a future direction to automate any NPO in near future.**

Keywords – **Automation, capacity building, founder's syndrome, NPO, NGO, resource mismanagement**

I. INTRODUCTION

In this era of globalization, a nonprofit organization deals with various operations. Sunrise is one of the major nonprofit organizations of India, which handles over 10 million people associated directly or indirectly with organization. The organization maintains over 15,000 team members, handling day to day operations of organization. Organization handles over 45 projects located at different parts of the India. A bureaucratic NPO generally consists of 3 different levels: Top-management (Founder/Visionary Person); Middle level management (Trustees, Office administration, Management committee) and Operational Level (onsite employees, office employees, volunteers).Funding is most important aspect for any organization. Funding for organization is received from: Donations (Cash and kinds-Grains, materials) from people associated with organization and revenue generated from printing and publication of spiritual books.

The NPO follows a manual system for recording diurnal operations (like Inventory management, Employee work allocation, Leave management, Budget management, Public Relations, Printing Publication, Donation Estimates). It also manually maintains details of all people associated with organization. Records are maintained in isolated non-standardized local files (created as when required) in the form of excel sheets, word files and Google-form that do not generate reports. As a result, system has data redundant files which lead to inconsistent and ambiguous interpretation of information. In such a complex manual system, it gets challenging for developers to elicit useful information required to develop an automated system for an NPO.

Major problems of NPO are: complex system structure with diverse functionalities; large scale operations, different employee structure of an NPO: It manages both paid employees and unpaid employees (Individual Volunteer). Unpaid employees work in organization as per their wish (Fig.1).

Objective of the paper is to address issues associated with software development for a NPO, and propose an area of research where open source products for NPO are not freely available. Although, some features like employee management, inventory management are available in open source libraries (like Open Petra), they do not satisfy the requirements of a specific NPO. For example: existing open source Finance management system cannot make estimate of Donations obtained in Kind form(a sack of wheat, a kilogram of sugar, a blanket, few plates of homemade food offered etc.). We need an open source product made solely for NPO's which is flexible according to NPO requirements and easy to maintain.

Two major problems this paper addresses are:
- Rift between of organization and development team
- Problems development team face in finding a open source product for NPO

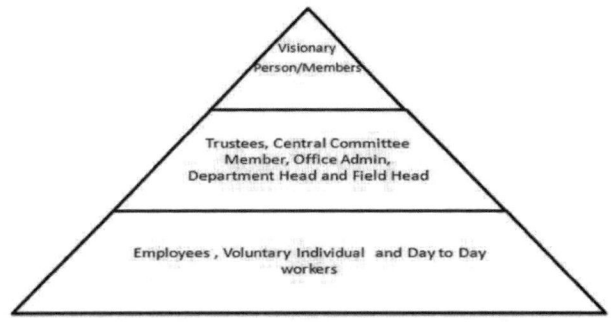

Fig.1. Organizational Structure

II. LITERATURE SURVEY

Project initiation is the first point to begin with before starting any project work together with identification of various stakeholders that directly or indirectly affects the project. Since, there are many factors associated with project development. Managing all those factors in an NPO is a cumbersome task therefore it is essential to sight a number of works done in this direction.

In [1], it exposes project development status in African countries. During their study, key observations identified were: (the top five factors of project success) client/customer satisfaction, by project team skill level, senior management buy-in, communication or project reporting, problem scope management and project delivery on time .

Project complexity arises as lack of comprehensive list makes it difficult not only for project managers but also for researchers to evaluate projects based on these factors [2]. Different projects have different set of requirements for example a coaching company and an NPO will have disparate requirements. So a careful study of such factors is required.

Adnane Belout reveals that considerably less attention is given to the human resource factor associated behavioral system of project management [3]. This study indicates how personnel factor plays a vital role in project success. Lack of communication between client and project team increases project complexities to a great extent. This results in changing requirement of the customer. In turn, project team is demotivated and it leads to delay in overall project development.

Camci, A. and Kotnour, T. identify the complexity of paradigm of systems. In a NPO with varied numbers of operations and excessive departmentalization, complexity of system is difficult to access. In such systems measures for assessing project technology complexity is required [4].

Martin, N.L.; Pearson, J.M.; Furumo, K.A. Their findings suggest that project size influences budget and project quality.While project complexity influences the use of specific project management practices [5]. So, we can stipulate that big projects involve lot of complexities which involve budget, quality, project management style. In a NPO with bulk transactions and proliferating member base system automation is challenging.

In data analysis by Yugue, R.T.; Maximiano, A.C.A. on 313 project managers indicated that the complexity of projects managed by the participants is generated by the criticality of the goals and can influence the frequency of use of processes and techniques related to the project planning and people management [6]. If goal is clear among the team .It would not be difficult to figure out how project planning and management is to be done. When we have a complex system with inter-related short term goals, it becomes challenging to manage such kind of system.

In case study of UNESCO, it was revealed that non-profit organizations find it difficult to manage resource efficiently. Reason being specific goals and organization environment, non-profit organizations require appropriate tools to manage their projects' risk [7]. This implies there is a paucity of resource management tools available for non Profit organizations.

1) Kong, E. in "Perceptions of Information Flow and Sharing in Non-profit Organizations" reveals that perspectives of people internal to organization make an impact on development of a balanced relational capital [8]. It can be inferred that development of a automated system requires active support of internal members of NPO. In our study, it was identified that NPO members do not corroborate with project team or developers to facilitate them with documents necessary (for example: donation list and different entities) for automated system development.

2) Organizational learning and the effective management of complexity reveals how organizational learning will improve its management of complexity [9]. Although, the paper refers to nonprofit private university, similar techniques are applicable to NPO as well.

It can be inferred from Lynda Rogerson; Derek Phair **paper that crowd outsourcing of a project can prove helpful and cost saving in handling complex systems.** The results of the study show that while there were inherent challenges engaging the crowd, there were also tangible benefits to the organization. These benefits included measurable cost savings and the ability to increase the number of projects worked [10].

Open source technology has progressed a lot since 1998.It has been adopted in disparate ways still existing research does not describe the context of organization studied, it fails to get benefit from related fields [11]. This indicates there are several disciplines unexplored.

"The promise and perils of a participatory approach to developing an open source community learning network" discusses about development ideals that bring notable benefits as well as significant challenges for the parties involved with project development. It also insinuates how mismatched expectations, budget squeezes, and slipped schedules have been attributed to the development approach being too participatory and too open [12].

Similarly, a paper reveals knowledge needs of a NPO indicates they work with limited number of resources. It also addresses their need for development of better knowledge management solutions [13]. It gives us an innuendo that volunteer motivation is main factor for recruitment and retentions. It implies that employee work as per their wish (unpaid/free employees).

TABLE I
BARRIER

Domain	Issues	Barriers
Organizational	Inter-organizational Coordination	Divergent goals Conflicting interests Lack of resources, Ineffective utilization of resources Ineffective joint assessment and planning
Information management	Information availability and Accessibility. Information quality Information Sharing Information distortion	Lack of sharing spirit Timeliness, Validation of information Relevancy of information Mismatch in time Combining information

There are several factors associated with technology and its use in non governmental institutions those factors include: financial barriers, technical barriers, communicational barriers, non conventional data, and different interpretations [14] (Table I).

Coordination barriers play a major role in an organization. Since an organization manages several disciplines these factors are crucial and require diligent attention. Coordination barriers [15] addressed by Carleen

Maitland, Louis Marie and Andrea H. Tapia can be presented as:

Non-profits organizations have certain impediments that are arduous to overcome – part time and volunteer workers, narrow specialization, little to no experience with project teams, and political problems [16].

From literature review it is obvious that there is heterogeneity and gap between expectations of NGO laws. The identified challenging issues are:
1. There is a dearth of open source software for NPO.
2. There are several financial, communicational barriers in a NPO.
3. Organization learning is related to project management complexity.
4. Lack of system understanding is making it difficult for developers to manage projects.
5. Existing open source projects do not describe the context in which organization is studied.
6. Project size influences project budget and complexity.
7. Lack of communication between client and project team increases project complexities.
8. Different projects have different requirements.
9. Customization of open source tool is difficult.

III. PRESENT SYSTEM

Since NPOs follow non-profit strategy therefore these organization are different from the common profit oriented organizations. In current system, following are the stakeholders of a nonprofit organization (Table II):

TABLE II
STAKE HOLDERS OF NPO

Stake Holders	Description
Visionary/Founder Members	A person or group of people who initially started the organization
Trustees	Include group of people are influential to work and keep the organization running. They help in generating funds for organization.
Organizing Committee	It has a team of experts who are involved in higher management of Organization
Office Admin and Department Heads	These are the middle level management employees who deal with day to day operations of organization.
Paid Employees	These consist of groups of employees, who are paid for their services in organization.
Individual Volunteer (Unpaid Employees)	These are those people who voluntarily work for the organization and they are not paid for it
Government	Since NPOs rely on the government funding and approvals for their projects so government is an important stakeholder for NPO/NGO
Associated Members	These are those people who are associated with Organization. They consist of Donor and beneficiary of the organization.

Working flow of NPO can be described through following Fig.2.

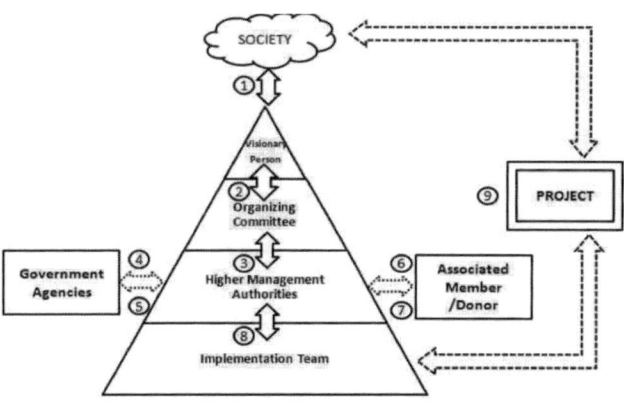

Non Profit Organization Model

Fig.2. The working model of a non profit organization

According to the Fig.2 the working model of a non profit organization can be divided into 9 steps. And these steps are as follows:
1. Problem is derived from society.
2. Visionary person finds the solution.
3. Organizing committee approves the solution.
4. Higher Management authorities (HMA) formulate the solution into project proposal and present it to government agencies for approval or funding.
5. Government Agency approves/funds the project proposal.
6. Higher Management Authorities ask associated member for donation for project.
7. Donor donates for project.
8. HMA provides a implementation plan to working team.
9. Working team completes the project and society is benefited.

Since feedback is involved at each and every level therefore bidirectional arrows are used.

Till now, almost all the day to day operations such as maintaining records for inventory, allocation of work to employees etc are handled manually or on isolated i.e. non-standardized local files (created as when required) in the form of excel sheets, word files, Registers and Google forms. As a result, there is lot of redundancy, inconsistencies and ambiguities of information. Getting useful information requires a lot of effort and sometimes it also requires instant creation of new files.

IV. PROPOSED SYSTEM

In order to integrate and fully automate working of such organization, well structured system is to be developed. An online, transparent automated system that can handle future demands is required. Desirables from An NPO-automated system are described in Table III:

TABLE III
PURPOSED MODULES

Required Module	Description
Public relation master Module	To maintain the information about all organization associated people so that they can have a central hub of information with them.
Human Resource module	To maintain team member's personal information, attendance, work allocated to the employees, the status of the work etc.
Calendar and Schedule module	To create and maintain the schedule of events and corresponding reminders and checks
Inventory Module	To manage the information about incoming and outgoing items (machine/equipments, material or items may be any glossary item, stationary or any other item)
Liaison Module	To manage all liaisoning issues such as permission from government for any event or project
Project Module	To maintain the all information about ongoing projects.
Account module	To maintain various accounts, sale and purchase, donations etc.
Report Module	To generate customized (dynamic) reports for the all above modules.

V. ISSUES AND CHALLENGES

There are 4 major factors which play a vital role in the development of an automated system. These factors are man, money, material and machine.

A. Problem related to Man

Human resource of any system can be broadly classified into two subcategories. One is development team of the system and another is the employees of the organization who are the user of the system. The development of any system depends upon the information provided to by the employees of the organization to the development team. The following are the problems which may occur in development of the system because of human resource (HR) employed in development.

1) Problem related to the employee of the organization
1. Resistance- towards automation. As employees are not tech savvy, they don't willingly help in system automation. They have a phobia that they will lose their importance/power in the organization.
2. Confusing information- Since employees are unwilling for automation they do not provide full and accurate information about the organization.
3. Out of box thinking-Employee in such organization ask for the functionalities beyond the technical feasibilities.
4. Demand of game-like tools: Since the employee are new to automation therefore the system should be made as intuitive as possible
5. HCI with Regional language- Since these kind of organizations have employees from various parts of country and the employees are not much educated therefore the system must be developed in different regional languages.
6. Quicker Deployment-Employee of organization ask for unrealistic deadlines for the development of the system.
7. Communication Gap-Proper communication between the employee and the development is also an issue.

2) Problem related to the Development team
1. Designing cannot be started before full and proper need identification/ preliminary survey is over.
2. The organization is scattered at various geographical locations, and understanding only one situations/location we cannot guarantee that we understood requirements comprehensively at other locations including field works. Hence, visit at all these places is recommended.
3. The system is very huge and complex (due to different types of projects, new field visits/events for example: dynamic speech for empowerment of society). Therefore, visionary/ expertise in developing technologies are required.
4. System study is a time consuming activity and to avoid repetition of any activity of system, proper study and time is necessary. Rapid development of the system is difficult since proper planning, executing and testing the system are required .Considerable time is required for all specified activities.(Client needs all things to be done in one week)
5. Massiveness of the system requires skilled and trained developers who are well equipped for mobile application development, application programmers and lot of experience. Apart from appropriate motivated development team, who are managing their academics, conveyance and stationary are also required (along with communication cost- for proper communication among the team members).

B. Problem related to Money

For any organization, money is backbone of system, its information is generally revealed only to trusted parties. Stipulating information to student-cum-developer may not be a policy of organization. After the discussion with admin we got to know that the actual flow of money in NGO/NPO from charity is the main source of income. Therefore, this resource must be used in such a way that maximum effectiveness and efficiency can be achieved. How these funds are coming? Where it is distributed? And how things are actually being managed? Answers are not clear. Similarly, turnover, input and output of publication department from money point of view is not clearly drawn, so in future some key decision oriented strategies may not be finalized due to shortage of information. Clear exposure of these kinds of details will be also beneficial.

From development side: To reduce the set up cost, maintenance cost and to make robust software, we have to focus on the open source technologies for the development of the desired system. So that money can be utilized in the hardware setup rather then fulfilling the software requirements. Developing and using easy software (making from scratch) may lead to a problem that increases complexity in the development of the system.

C. Problem related to Machine

Server, computers, networks, Biometric devices, CCTVs, VSATs, Cloud services Modem/Routers -All hardware and software issues lie under this category. Following are the challenges for handling machines of NPO projects. Finance for the devices, access rights, control and maintenance of the devices, driver and updating from software are major issues. Hardware setup has to be done in such way that minor change in the setup is required and future trends can be satisfied. Deciding and selecting appropriate software and hardware (within allowed financial position) have a huge impact on choosing technology for development of the system.

VI. CONCLUSION

This paper documents all the necessary steps carried out in the preliminary survey and requirements phase of automation of NGO. The complexity and diversity in the nature of NGO is the biggest challenge for the NGO automation. In India, general automated product for NPO/NGO is rarely available. So this paper would provides a guideline for development of such projects even for open source platform that can be customized for effective and efficient utilization with minimum available resource can be done.

VII. FUTURE WORK

After investigating all facts and identification of all stakeholders and charting out all activities to be carried out, one can finalize blueprint of tool. Further, agile based project development scheme can be used for staged implementation. With user intervention, modification in the development methodology and their possible corrections (dynamic requirements) can be handled. With continuous review process and feedbacks the system can be used for the betterment/benefit of society.

ACKNOWLEDGMENT

The authors would like to thanks to Sunrise Trust and team of Mr. Praveen Dhuri for his time and support. Authors would also like to thank Development Center of DAVV for providing all the infrastructure support for requirement engineering phase of this work.

REFERENCES

[1] Leon Uys, Pretoria, "Current status of project management in South Africa", *Management of Engineering & Technology, 2008. PICMET 2008.* Portland International Conference: 27-31 July 2008, Pp: 1285 - 1294

[2] Belassi, Tukel, "A new framework for determining critical success/failure factors in projects", *International Journal of Project Management 1996.* Volume 14, Issue 3, June 1996, Pp 141–151

[3] Adnane Belout, "Effects of human resource management on project effectiveness and success: Toward a new conceptual framework', *International Journal of Project Management 1998,* Volume 16, Issue 1, February 1998, Pp 21–26

[4] Camci, Kotnour, "Technology Complexity in Projects: Does Classical Project Management Work?" *Technology Management for the Global Future, 2006. PICMET 2006.* Conference: 8-13 July 2006 Pp: 2181-2186

[5] Martin, N.L. Pearson, J.M. Furumo, K.A, "IS Project Management: Size, Complexity, Practices and the Project Management Office", System Sciences, 2005. HICSS '05. Proceedings of the 38th *Annual Hawaii International Conference.* Date of Conference: 03-06 Jan. 2005 Pp: 234b

[6] Yugue, R.T., Maximiano, A.C.A., "Contribution to the research of project complexity and management processes" *Management of Innovation and Technology (ICMIT), 2012. IEEE International Conference.* Date of Conference: 11-13 June 2012 Pp: 668 - 673

[7] Cavalcanti, F.M., Santiago, L., "Risk Management and Expert Opinion Assessment at Non-Profit Organizations: the case of UNESCO" *Engineering Management Conference, 2006 IEEE Trans.* 17-20 Sept. 2006 Pp: 356 - 360

[8] Kong, E., "Perceptions of Information Flow and Sharing in Non-profit Organisations: A Relational Capital Perspective", Cooperation and Promotion of Information Resources in *Science and Technology, 2009. COINFO '09.* Fourth International Conference Date of Conference: 21-23 Nov. 2009 Pp: 162 - 167

[9] Alfonso, R.A., "Organizational learning and the effective management of complexity" Grey Systems and Intelligent Services (GSIS), 2011 IEEE International Conference Date of Conference: 15-18 Sept. 2011 Pp: 914 - 918

[10] Lynda Rogerson, Derek Phair,"Open crowdsourcing: leveraging community software developers for IT projects" Publisher: *Colorado Technical University* ©2012 Pp: 101

[11] Øyvind Hauge, Claudia Ayala, Reidar Conradi, "Adoption of open source software in software-intensive organizations - A systematic literature review" *Information and Software Technology* Vol. 52 Issue 11, November, 2010 Pp: 1133-1154

[12] Luke et al, "The Promise and Perils for the Pariticpatory approach to developing an open source community learning network" PDC 04 Proceedings of the eighth conference on *Participatory design: Artful integration: interweaving media, materials and practices* – Vol. 1 Pp 11 – 19

[13] John Huck, Rodney Al, Dinesh Rathi, "Finding KM solutions for a volunteer-based non-profit organization" *VINE*, Vol. 41 Iss: 1, pp.26 - 40 ISSN: 0305-5728

[14] William J. Craig, Trevor M. Harris, Daniel Weiner, "Community Participation and Geographical Information Systems" *CRC* In Press, 04-Apr-2002

[15] Maitland et al, "Information Management and Technology Issues Addressed by *Humanitarian Relief Coordination Bodies"*

[16] Arline Conan Sutherland, Jeff Sutherland, Christine Hegarty, "Scrum in Church: Saving the World One Team at a Time" AGILE '09 Proceedings of the 2009 *Agile Conference* Pp 329-332.

A Reinforcement Learning Method of Obstacle Avoidance for Industrial Mobile Vehicles in Unknown Environments Using Neural Network

Chen XIA[*], A. EL KAMEL

LAGIS (UMR CNRS 8219), Ecole Centrale de Lille, Villeneuve d'Ascq, 59650, France
(donaldgreg.x@gmail.com)

Abstract - **This paper presents a reinforcement learning method for a mobile vehicle to navigate autonomously in an unknown environment. Q-learning algorithm is a model-free reinforcement learning technique and is applied to realize the robot self-learning ability. The state-action Q-values are traditionally stored in a Q table, which will decrease the learning speed when large storage memory is needed. The neural network has a strong ability to deal with large-scale state spaces. Therefore, the neural network is introduced to work with Q-learning to ensure the self-learning efficiency of avoiding obstacles for industrial vehicles in unpredictable environments. Experiment results show that an autonomous mobile vehicle using the proposed method can successfully navigate to the target place without colliding with obstacles, and hence prove the self-learning ability of navigation in an unknown environment.**

Keywords - **Neural network, obstacle avoidance, Q-learning, reinforcement learning, unpredicted environments**

I. INTRODUCTION

Traditional fixed industrial robots always have a place in many various industries, while mobile robots have the capability to move around in their working environment. The mobility allows more flexibility in a great number of industrial applications, such as warehouse transportation and distribution. Therefore, increased interest in mobile robotics is spread across all industries, especially in hazardous applications.

Industrial mobile robots should have the autonomous navigation capacity in its working environment, which aims to find a collision free path from a starting point to a target point [1, 2], such as transporting materials from one position to another in a warehouse. Global path planning techniques can be applied when a complete knowledge of the environment is acquired. While robots are working in unknown environments, local path planning techniques, which rely on sensory information of the mobile robots, has successfully proven adequate in achieving the task of obstacle avoidance, such as fuzzy logic control [3], potential field method [4], genetic algorithms [5].

However, classic path planning methods have a common constraint that control strategy needs to be well designed by programmers. When the environment changes or the robot encounters situations that are not considered beforehand by designers, they may make no reaction but execute the predefined coping strategy. This may lead to dangerous or even fatal consequences, such as collision, crash. In this way, such robot is still a kind of machine that perfectly executes what it is taught without good adaptability. What we expect is a truly intelligent

autonomous robotic system. Therefore, mobile robots with self-learning ability become a hot research topic.

Robot learning is normally realized by the interaction between the robot and the surrounding environment. Reinforcement learning is a machine learning technique based on trial-and-error mechanisms, improving the performance by getting feedback from the environment [6, 7]. This paper enhances the learning ability of industrial mobile vehicles based on accumulated interacting experiences. Reinforcement learning provides so many learning methods, and Q-learning is one of them [7-9]. However, Q-learning is usually applied to discrete sets of states and actions. In real applications, the state spaces are continuous and large-scale spaces will bring the problems of the generation and the curse of dimensionality. Since neural network (NN) has a good generalization performance and can approximate any functions in any accuracy, it is natural that the implantation of neural network is one of the effective approach for solving the problem of discrete Q-learning generation [8, 10, 11].

Some similar researches use exactly the same environment both to train the robot's learning ability and to test the navigation skills [12, 13, 14]. However, what we really expect is that the robot learns the obstacle avoidance behaviors in some environments and can navigate independently in a completely new unknown environment, which will prove the feasibility and stability of the proposed navigation strategy.

This paper presents an approach of obstacle avoidance learning for mobile vehicles in an unknown environment by developing a neural network based Q-learning architecture (NNQL).

The rest of the paper is organized as follows. Q-learning design is described in Section II. Section III introduces the neural network and presents the architecture of the algorithm of neural network based Q-learning for mobile robot navigation. In Section IV, the experiment is conducted to show the simulation results of the proposed method. Section V draws the final conclusions.

II. Q-LEARNING DESIGN

Reinforcement learning is aimed at learning a mapping from the states of the environment to the robot behaviors. The robot doesn't need previous knowledge about the surrounding environment. It learns about the environment via interacting with it [9]. The structure of reinforcement learning in a mobile robot navigation problem is shown as Fig.1. The learning system perceives the environment and receives a state that describes it.

E. Qi et al. (eds.), *Proceedings of the 21st International Conference on Industrial Engineering and Engineering Management 2014*, Proceedings of the International Conference on Industrial Engineering and Engineering Management, DOI 10.2991/978-94-6239-102-4_136, © Atlantis Press and the authors 2015

Then the system chooses an action from the action space to execute and then moves to a new state. Then, the system receives an immediate evaluation reward as well as perceives the new state. The purpose of the learning system is to find a control policy that maximizes the expected amount of reward during the learning period [7].

Fig.1. The structure of reinforcement learning

A. Models for mobile vehicles and the environment

The mobile robot is assumed that has three wheels, one in front and two in back. The vehicle is equipped with eight sensors around to perceive the environment, and each sensor is responsible for one region of a range of 45°, as shown in Fig. 2. The total sensor detection range of θ is the interval $[0, 2\pi]$.

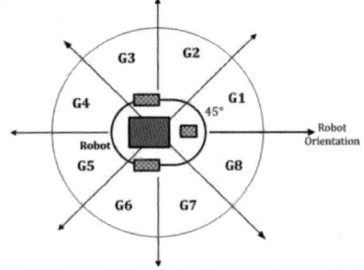

Fig.2. The vehicle and the detection regions of eight sensors

The working environment of the vehicle consists of its target and the obstacles, as shown in Fig.3. The initial vehicle location and the goal are predefined, where the robot will try to reach the goal with free collision path in spite of the presence of obstacles in the environment.

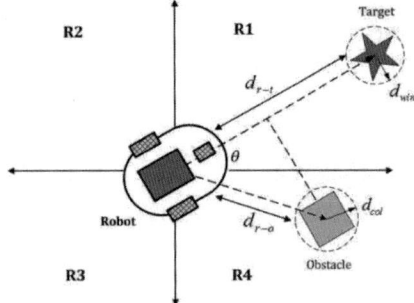

Fig.3. The working environment and the important distances

d_{r-t} is the distance between the robot and the target and d_{r-o} is the distance between the robot and obstacles.

Several basic assumptions concerning the vehicle and the navigation environment are made here [9]:

- The position p, the velocity v, of the robot are known at each time instant.
- The shapes of the obstacles in the robot's detection range are known for the robot at each time instant.
- The position of the target p_{tar} and the region of the target R_g are known at each time instant.

B. States and actions spaces of Q-learning

Q-learning is aimed at learning a mapping from the state input to the action output. In a navigation problem, the robot perceives the state from the environment by means of its sensors, and this state of environment is used by a reasoning process to determine the action to execute in the given state.

The state spaces can be completely defined by the robot sensor information which detects the relative or approximate distances and directions between him and the surrounding obstacles or the target. Hence, a state of environment can be expressed in a vector with eight components:

$$s_t = [d_1, d_2, d_3, d_4, d_5, d_6, d_7, d_8] \tag{1}$$

s_t is the state at instant t. $d_i, i = 1, 2, \ldots, 8$ are the reading distances of each sensor. If no obstacle is found, then $d_i = 0$.

The action spaces are defined by five moving actions of the robot: move forward, turn left at 30°, turn left at 60°, turn right at 30° and turn right at 60°. The five actions are based on the vehicle orientation.

C. The reward function

The reward function measures an immediate feedback for the action taken at a given state. It evaluates how good or how bad the taken action is at a specific situation. Before giving the reward function, one environment state is classified into four different properties, called the state property:

- Safe State (SS): a state where the robot has a low or no possibility of collision with surrounding obstacles.
- Non-Safe State (NS): a state where the robot has a high possibility of collision with some obstacles in the environment.
- Winning State (WS): one of the terminate states when the robot reaches its goal.
- Failure State (FS): one of the terminate states when the robot collides with obstacles.

The reward is given to the robot instantly when it moves from one state to a new state after executing one action. The reward function r is defined as follows:

- Moving from a Non-Safe State to a Safe State: $r = 0.3$.
- Moving from a Safe State to a Non-Safe State: $r = -0.2$.
- Moving from a Non-Safe State to a Non-Safe State but getting closer to the obstacles: $r = -0.4$.
- Moving from a Non-Safe State to a Non-Safe State and getting away from the obstacles: $r = 0.4$.
- Moving to a Winning State: $r = 1$.
- Moving to a Failure State: $r = -0.6$.

D. The Q-value function

The Q-value function expresses the mapping policy from the perceived environment state to the executing action. One state-action Q-value $Q(s_t, a_t)$ corresponds with one specific state and one action in this state. All the Q-values should be initially set to zeros. Then the

Q-values will be updated while training the robot and this can be interpreted as the robot is learning.

At each time instant t, the robot observes its current state of environment s_t, then selects an action a_t. A state-action Q-value $Q(s_t, a_t)$ is now produced. After performing the action a_t, the robot observes the subsequent state s_{t+1}, and also receives an immediate reward r_t. Last, the current $Q(s_t, a_t)$ is updated to its optimal Q-value $Q^*(s_t, a_t)$ according to the following Q-value function [7, 15]:

$$Q^*(s_t, a_t) = Q(s_t, a_t) + \alpha \left[r_t + \gamma \max_{a_i \in A} Q(s_{t+1}, a_i) - Q(s_t, a_t) \right] \quad (2)$$

α is the learning rate, set between 0 and 1. Setting it to 0 means that the Q-values are never updated, hence nothing is learned. Setting a high value means that learning can occur quickly.

γ is the discount factor with a range of 0 and 1. If γ is close to 0, the robot will tend to consider immediate reward. On the contrary, if γ approaches 1, the robot will take more future reward into account.

E. Action selection strategy

After obtaining the environment information, the vehicle selects an appropriate action to execute according to the action selection mechanism. During learning, it is important to try different actions as much as possible, but the robot has also to solve the dilemma between "exploration" and "exploitation". A good way is to select actions applying the Boltzmann probability distribution.

$$P(a|s) = \frac{e^{Q(s,a)/T}}{\sum_{b \in A} e^{Q(s,b)/T}} \quad (3)$$

T is a temperature parameter between 0 and 1 that determines the stochastic probability of action selection.

Then, during the robot navigation process, in order to exploit the most the policy, the robot takes greedy action selection according to the following equation:

$$a^*(s) = \arg \max_{b \in A} Q(s, b) \quad (4)$$

III. THE ALGORITHM OF NEURAL NETWORK Q-LEARNING

Traditional Q-learning is designed for the discrete set of states and actions. However, in the mobile robot navigation tasks where the state spaces are continuous due to the sensory inputs, a large-scale memory space is needed to store all the state-action pairs and the learning speed will decrease. In order to solve this dimensionality problem, the neural network is introduced, since it provides a good generalization performance as a universal function approximator.

A. Architecture of neural network Q-learning

In the proposed NNQL, a three-layer neural network replaces the traditional Q-table and approximates the Q-value function, as shown in Fig.4.

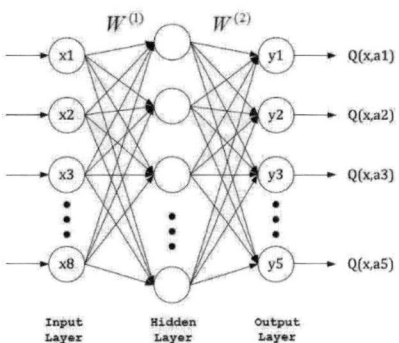

Fig.4. Three-layer neural network architecture

The inputs are the eight reading distances from the robot sensors, representing the perception of environment. The outputs correspond with the five action Q-values. The weight W_1 is used to connect the input layer and the hidden layer, and similarly, the weight W_2 links the hidden layer and the output layer. For the hidden units, the sigmoid function is used.

The feedforward neural network (FFNN) takes the responsibility to calculate the activations of the output units from the input units. Each output unit's activation represents the Q-value for the corresponding action in the given state. In FFNN, the weights W_1 and W_2 always keep unchanged.

B. Backpropagation neural network

The backpropagation neural network (BPNN) is designed to train the neural network by updating the weights W_1 and W_2. Only a neural network with fully trained weights can be used in robot navigation tasks. The weight changes are based on the network's error, the difference between its output for a given input and a target value, what the network is expected to output. The optimal Q-value in Eq. (2) is treated as the target for the output unit corresponding to the selected action. Gradient descent is used to optimize the network error [15].

C. The training process of neural network Q-learning

The neural network based Q-learning can be divided into two processes. The first one is the training process to train the robot learning ability, and the second one is the navigation process to use the trained policy to finish a navigation task.

Training the mobile robot is done by exposing it to different episodes of environments. The greater the number of episodes used to train the robot, the better will be the performance of the robot navigation. Each episode starts by perceiving the current state of the environment. The surrounding obstacle locations are supplied to the robot through its sensors. Once the current state is checked, if it is a Safe State the robot changes its orientation toward the target location, and moves one step forward trying to reach the target in the shortest path. If the current state is Non-Safe State, the robot inputs the current state into the FFNN and outputs all the possible Q-values. According to the Boltzmann action selection mechanism, the robot takes an action and moves to a new state. Then, the robot checks the resulting new state, gets the immediate reward and updates the Q-values accordingly. Then the updated Q-values are sent back to the neural network and the weights of the neural network get updated by using backpropagation algorithm.

Each episode has limited steps. The robot needs to reach the target within the steps. If the robot runs out of the steps and does not reach the target, or if the robot collides with an obstacle or reach the target, the episode is terminated and a new episode is started.

D. The navigation process using the learned algorithm

After training the robot, the resulting policy can be used by the robot for future navigation tasks in various environments.

The robot starts its navigation by finding its current environment state. If it is a Safe State the robot changes its orientation towards the target and moves one step forward. It continues moving until entering a Non-Safe region where adopts the trained policy. The robot uses the FFNN to generate all possible Q-values. The robot takes the action that has the biggest Q-value. After that, the robot finds its new current state and repeats the process until the robot reaches its goal or collide an obstacle.

IV. SIMULATION AND RESULTS

In order to evaluate the proposed method, the simulation experiments are carried out in MATLAB. The mobile vehicle is represented by a rectangle-shaped robot. It is equipped with 8 sensors to observe the environment, as shown in Fig. 1. The environment map has a size of 100 m × 100 m. The obstacles are randomly scattered in the environment and the robot has no prior knowledge of their numbers, sizes and positions. The initial position of the robot is placed at (20, 20) and the target position, a red circle in the map, is found at (90, 90). The velocity of the robot is fixed at 2m/s. The mission of the robot is to start from the initial position and to find an optimal path to arrive at the target position without any collision with any obstacles. If no obstacles are detected, the robot is designed to move directly to the target.

The neural network has three layers: 8 in the input layer, 6 in the hidden layer and 5 in the output. The input is eight reading distances from the robot sensors. The range of sensor detection is 10 m. The output is five action Q-values. The experiment parameters are selected as follows:

- Learning rate: $\alpha = 0.1$,
- Discount factor: $\gamma = 0.15$,
- Maximum temperature: $T_{max} = 0.8$,
- Minimum temperature: $T_{min} = 0.01$.

A. The training process simulation

1000 episodes are set in the learning process and each episode has a maximum of 500 moving steps. All episodes have different configurations of random obstacle positions. A new episode will be started in the following three situations:

- The robot finds a collision free path to the target;

- The robot collides with an obstacle or the map borders;
- The robot runs out the moving steps.

The training process will be terminated when all the learning episodes are finished. Some training episodes are shown in Fig.5.

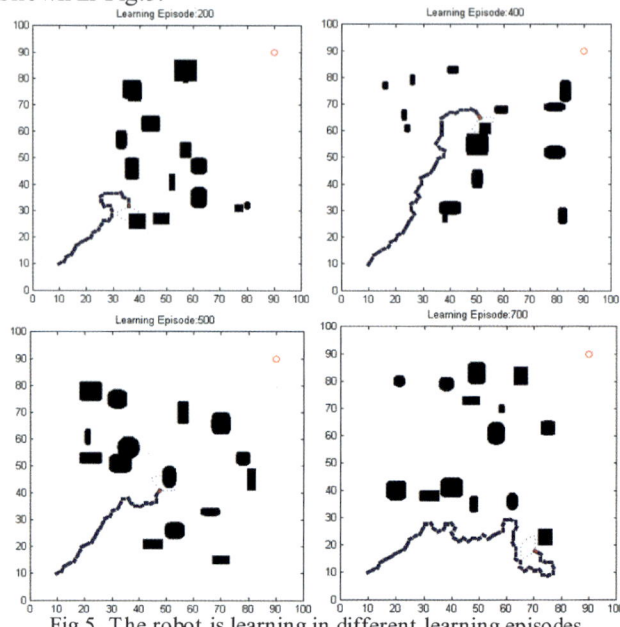

Fig.5. The robot is learning in different learning episodes

Since the proposed algorithm is a trial-and-error method, it is reasonable that a great number of episodes failed due to a collision. In these episodes, the robot learned to navigate in different environments and we can see that the robot tried different actions and the rewards evaluated this action decision, which can help the robot to correct the action selection in the future. During all episodes, the proposed algorithm adopted BPNN to train the weights W_1 and W_2 . When the learning process is completed, the weights have been well trained and can be used directly in the following process of robot navigation.

B. Robot navigation process simulation

In the navigation process, the weights W_1 and W_2 have converged to their optimal values. The robot has learned how to behave in front of obstacles. The FFNN is now only adopted, and the robot chooses the best fit action. Then it is the time for the mobile robot to demonstrate its intelligence of independent navigation in an unpredictable environment.

Fig.6 shows that the robot was exposed to a new unknown environment and it succeeded in getting to the target position in a collision free path.

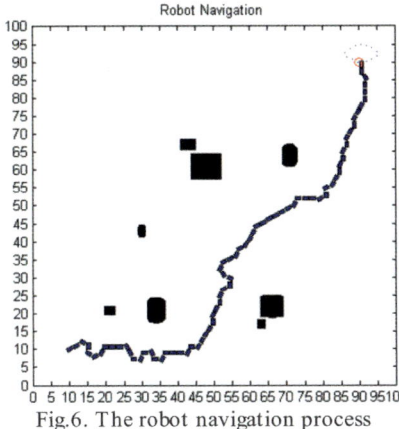

Fig.6. The robot navigation process

Then, in order to test the stability of the proposed method, the robot executed other navigation missions using the same weights and the policy, as shown in Fig.7.

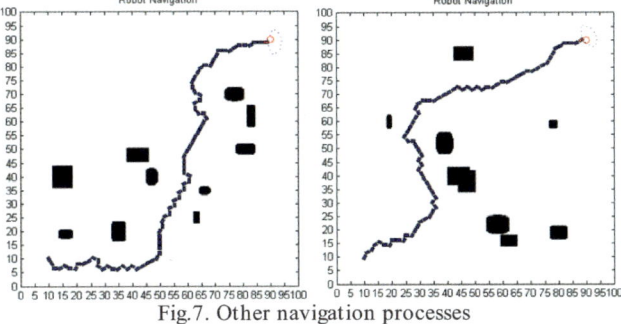

Fig.7. Other navigation processes

The robot navigation environments changed in the above two missions, and the robot completed both tasks successfully. The paths that the robot chose may not be the optimal ones due to a lack of complete knowledge of environment map, but they are still acceptable and perfect enough to meet our expectation. Therefore, the above experiments have proven the feasibility and the stability of the proposed NNQL algorithm.

V.   CONCLUSION

This paper explores the mobile vehicles navigation problem in the context of industrial applications by combining reinforcement learning and neural network. Q-learning is applied to enhance the self-learning ability of a mobile robot through trial-and-error interactions with an unknown environment. The neural network is integrated to store and train the large-scale Q-values. Therefore, an intelligent control strategy using neural network based Q-learning is implemented in the paper. The experiment results show the feasibility and the stability of the proposed method. The mobile vehicle can complete navigation tasks safely in an unpredictable environment and becomes a truly intelligent system with strong self-learning and adaptive abilities.

REFERENCES

[1]   D. Filliat and J.-A. Meyer, "Map-based navigation in mobile robots: I. A review of localization strategies," *Cognitive Systems Research*, vol. 4, no. 4, pp. 243–282,

Dec. 2003.

[2]   J.-A. Meyer and D. Filliat, "Map-based navigation in mobile robots: II. A review of map-learning and path-planning strategies," *Cognitive Systems Research*, vol. 4, no. 4, pp. 283–317, Dec. 2003.

[3]   S. M. Raguraman, D. Tamilselvi, and N. Shivakumar, "Mobile robot navigation using Fuzzy logic controller," in *2009 International Conference on Control, Automation, Communication and Energy Conservation, 2009. INCACEC 2009*, 2009, pp. 1–5.

[4]   S. S. Ge and Y. J. Cui, "Dynamic Motion Planning for Mobile Robots Using Potential Field Method," *Autonomous Robots*, vol. 13, pp. 207–222, 2002.

[5]   Y. Hu and S. X. Yang, "A knowledge based genetic algorithm for path planning of a mobile robot," in *2004 IEEE International Conference on Robotics and Automation, 2004. Proceedings. ICRA '04*, 2004, vol. 5, pp. 4350–4355 Vol.5.

[6]   S. J. Russell and P. Norvig, *Artificial Intelligence: A Modern Approach*, 3rd edition. Prentice Hall, 2010.

[7]   M. E. Harmon and S. S. Harmon, *Reinforcement Learning: A Tutorial*. 1996.

[8]   K. Macek, I. Petrovic, and N. Peric, "A reinforcement learning approach to obstacle avoidance of mobile robots," in *7th International Workshop on Advanced Motion Control, 2002*, 2002, pp. 462–466.

[9]   M. A. Kareem Jaradat, M. Al-Rousan, and L. Quadan, "Reinforcement based mobile robot navigation in dynamic environment," *Robotics and Computer-Integrated Manufacturing*, vol. 27, no. 1, pp. 135–149, Feb. 2011.

[10] B.-Q. Huang, G.-Y. Cao, and M. Guo, "Reinforcement Learning Neural Network to the Problem of Autonomous Mobile Robot Obstacle Avoidance," in *Proceedings of 2005 International Conference on Machine Learning and Cybernetics, 2005*, 2005, vol. 1, pp. 85–89.

[11] G.-S. Yang, E.-K. Chen, and C.-W. An, "Mobile robot navigation using neural Q-learning," in *Proceedings of 2004 International Conference on Machine Learning and Cybernetics, 2004*, 2004, vol. 1, pp. 48–52 vol.1.

[12] C. Li, J. Zhang, and Y. Li, "Application of Artificial Neural Network Based on Q-learning for Mobile Robot Path Planning," in *2006 IEEE International Conference on Information Acquisition*, 2006, pp. 978–982.

[13] J. Qiao, Z. Hou, and X. Ruan, "Q-learning Based on Neural Network in Learning Action Selection of Mobile Robot," in *2007 IEEE International Conference on Automation and Logistics*, 2007, pp. 263–267.

[14] J. Qiao, Z. Hou, and X. Ruan, "Application of reinforcement learning based on neural network to dynamic obstacle avoidance," in *International Conference on Information and Automation, 2008. ICIA 2008*, 2008, pp. 784–788.

[15] R. Rojas, The Backpropagation Algorithm in Neural Networks: A Systematic Introduction. Springer, 1996, pp. 151–182.

Printed by Printforce, the Netherlands